The Oosterschelde Estuary (The Netherlands): a Case-Study of a Changing Ecosystem

Developments in Hydrobiology 97

Series editor
H. J. Dumont

The Oosterschelde Estuary (The Netherlands): a Case-Study of a Changing Ecosystem

Edited by

P. H. Nienhuis & A. C. Smaal

Reprinted from Hydrobiologia, vols 282/283 (1994)

Kluwer Academic Publishers

Dordrecht / Boston / London

Library of Congress Cataloging-in-Publication Data

The Oosterschelde estuary : a case study of a changing ecosystem /
    edited by P.H. Nienhuis & A.C. Smaal.
        p.   cm. -- (Developments in hydrobiology ; 97)
    "Reprinted from Hydrobiologia, vols. 282/283."
    Includes bibliographical references and index.
    ISBN 0-7923-2817-5 (acid-free)
    1. Estuarine ecology--Netherlands--Oosterschelde. 2. Coastal
engineering--Environmental aspects--Netherlands--Oosterschelde.
3. Deltaplan (Netherlands)   I. Nienhuis, P. H. II. Smaal, A. C.
III. Series.
QH159.O58 1994
574.5'26365'0949242--dc20   94-14882

ISBN 0-7923-2817-5

Published by Kluwer Academic Publishers,
P.O. Box 17, 3300 AA Dordrecht, The Netherlands.

Kluwer Academic Publishers incorporates
the publishing programmes of
D. Reidel, Martinus Nijhoff, Dr W. Junk and MTP Press.

Sold and distributed in the U.S.A. and Canada
by Kluwer Academic Publishers,
101 Philip Drive, Norwell, MA 02061, U.S.A.

In all other countries, sold and distributed
by Kluwer Academic Publishers Group,
P.O. Box 322, 3300 AH Dordrecht, The Netherlands.

Front cover photographs: The storm-surge barrier, a compromise solution for the Oosterschelde estuary: Safety against storm-floods as well as nature conservation.
Photographs RIKZ, Middelburg and NIOO-CEMO, Yerseke.

Printed on acid-free paper

Printed in Belgium

Contents

THEME III: THE STRUCTURE OF THE BENTHIC SYSTEM

THEME IV: ECOLOGY OF THE SALT MARSHES

THEME V:
THE RESPONSE OF BENTHIC SUSPENSION FEEDERS TO ENVIRONMENTAL CHANGES

THEME VI: THE ANALYSIS OF A DYNAMIC SIMULATION MODEL FOR THE OOSTERSCHELDE ECOSYSTEM

THEME VII: THE HIGHER TROPHIC LEVELS

THEME VIII: THE OOSTERSCHELDE ESTUARY: AN EVALUATION OF CHANGES 561

Hydrobiologia **282/283**, 1994.
P.H. Nienhuis & A.C. Smaal (eds), The Oosterschelde Estuary (The Netherlands).

Preface

The Oosterschelde project is technically and scientifically unique. The building of the storm-surge barrier, together with the two secondary dams, represents an important breakthrough in marine civil engineering. The project offered ample opportunities for integrated physical, chemical, geological and biological research. Integration of the knowledge gained raised the entire project to the level of a case-study of a changing estuarine ecosystem, and demonstrated the effects of human interference in a non-polluted estuary. The main argument to build the storm-surge barrier was storm-flood protection, combined with conservation of the outstanding values of landscape and nature, fisheries included. The estuary is recognized as an international wetland, including sand- and mud-flats, feeding large numbers of migratory waterfowl, and further comprising deep tidal gullies, artificial rocky shores, salt marshes and inland brackish localities with specific aquatic plants and animals. The interest in storm-flood protection has recently gained momentum, owing to the wide international discussion on the impact of sea-level rise on society. Estuarine areas, the Oosterschelde included, are vulnerable to effects of sea-level rise.

In 1980 research groups of the Environmental Section of the Delta Department (former Tidal Waters Division, now National Institute for Coastal and Marine Management/RIKZ) of the Ministry of Transport, Public Works and Water Management and the Delta Institute for Hydrobiological Research (now Netherlands Institute of Ecology – Centre for Estuarine and Coastal Ecology) decided to start an ecosystem study, aiming at an analysis of the structure and functioning of the changing Oosterschelde estuary, in perspective of the future management of the water body (Project BALANS, from 1980 to 1988; Project EOS, from 1988 to 1991). The present volume of *Hydrobiologia* comprises 35 research papers, written by 56 authors, devoted to Oosterschelde estuary. The papers are mainly based on investigations carried out during the period 1980–1991, sustained by older investigations and long-term reviews. The core of the volume is extracted from the project EOS (1988–1991). The project meant a close cooperation between 40 to 50 scientists, technical assistants and water managers of several State Services, Universities and Private Agencies: National Institute for Coastal and Marine Management/RIKZ (former Tidal Waters Division) and Directorate Zeeland, Ministry of Transport, Public Works and Water Management; Netherlands Institute of Ecology – Centre for Estuarine and Coastal Ecology (former Delta Institute), Ministry of Education and Science; Netherlands Fisheries Research Institute and Institute of Forestry and Nature Research, Ministry of Agriculture, Nature Conservation and Fisheries; State University of Utrecht; State University of Ghent (Belgium); Free University of Brussels (Belgium); Institute of Nature Conservation, Belgium; Environmental Consultancy Agencies.

The initiative to publish a scientific justification of the Oosterschelde project originated in 1988. Kluwer Academic Publishers adopted the idea with enthusiasm, and the subsequent cooperation was very good. All papers were judged by international referees. Many people behind the scenes added substantially to the project. Only A. Bolsius (Yerseke) will be mentioned here. He carried out the classical art-work which enhances the uniformity in presentation of the book. Considering the size of the project and the employment of a considerable number of cooperating scientists, an impressive document came into existence. The papers contain useful information for both scientists and managers involved in comparable projects, dealing with major environmental changes in estuaries.

PIET H. NIENHUIS
AAD C. SMAAL

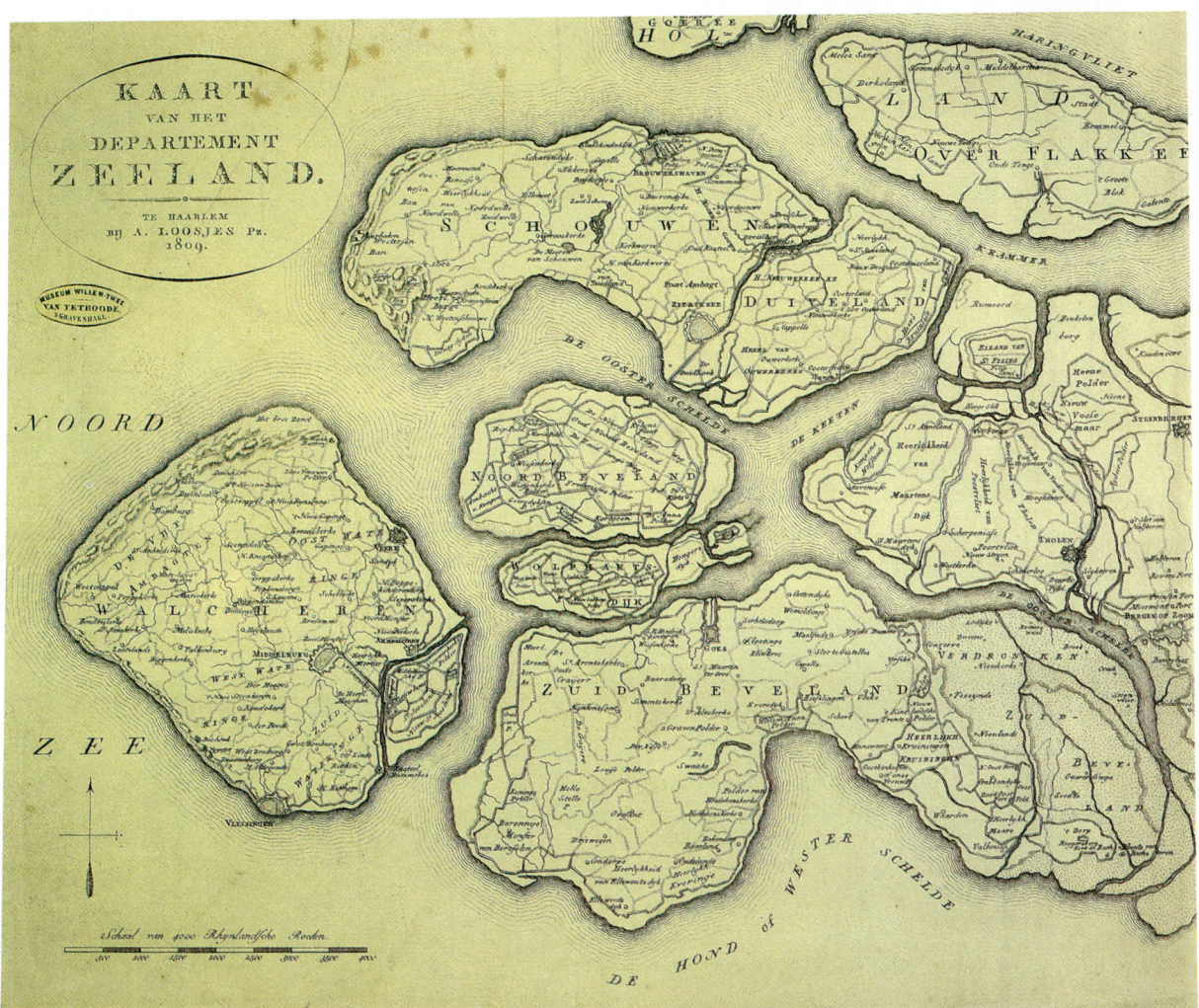

Fig. 1. Oosterschelde estuary is a dynamic system. Its configuration changed over historic times and it was only in the second half of the 19th and in the 20th century that the outline of the estuary was fixed by rigid seawalls and dikes. This map was drawn in 1809.

Fig. 2. 'De Ramp', the storm-flood disaster of February 1, 1953. The storm-surge barrier in the Oosterschelde estuary was the final part of the Delta Project, conceived as an answer to the continuous risk of flooding.

Fig. 3. A Landsat satellite image of Oosterschelde estuary, made during the pre-barrier period. North of Oosterschelde estuary is situated Grevelingen lagoon, a non-tidal, enclosed former estuary, containing extremely clear water (dark colour).

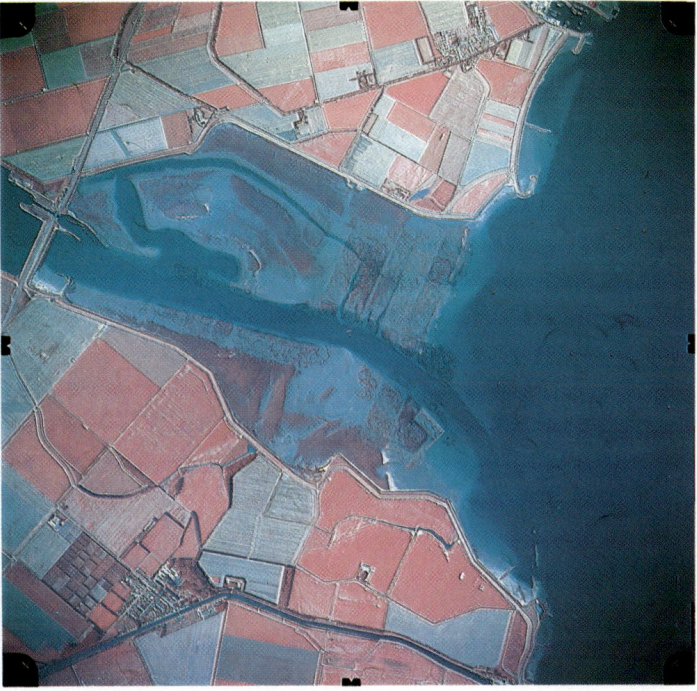

Fig. 4. False-colour aerial photograph of Zandkreek (2 × 2 km), a sheltered part of Oosterschelde estuary. The intertidal mudflats of Zandkreek, overgrown with macrophytes (red colour) contrast with the deep water of Oosterschelde proper. Photograph Ministry of Transport, Public Works and Water Management, 1984.

Fig. 5. Aerial picture of inland brackish localities (Dutch: inlagen) bordering the Oosterschelde estuary. These inland waters are an integrated part of the internationally recognized wetland. A 5-km wide bridge spanning the estuary is to be seen in the background. Photograph NIOO-CEMO, Yerseke.

Fig. 6. Prefabricated, concrete piers (pylons) form the framework of the storm-surge barrier. The piers have base dimensions of 25 × 50 m and have a maximum height of 43 m. Roughly 80% of the mass of the piers is now permanently under water. Photograph RIKZ, Middelburg.

Fig. 7. Gigantic vessels were used to transport the prefabricated piers, each roughly 40 m in height, from the coffer-dams to their final position, the barrier alignment across the estuary. Photograph NIOO-CEMO, Yerseke.

Fig. 8. The storm-surge barrier in open position. The steel gates in between the piers can be raised and lowered, but for most of the time they will be suspended above the waves. Photograph NIOO-CEMO, Yerseke.

Fig. 9. Aerial picture of the storm-surge barrier in the mouth of Oosterschelde estuary, at incoming tide. The North Sea is at the left-hand side. Photograph National Aerospace Laboratory, Amsterdam, 1989.

Fig. 10. The area of salt marshes suffered most from the construction of the civil-engineering works in and around the Ooster-schelde. Photograph NIOO-CEMO, Yerseke.

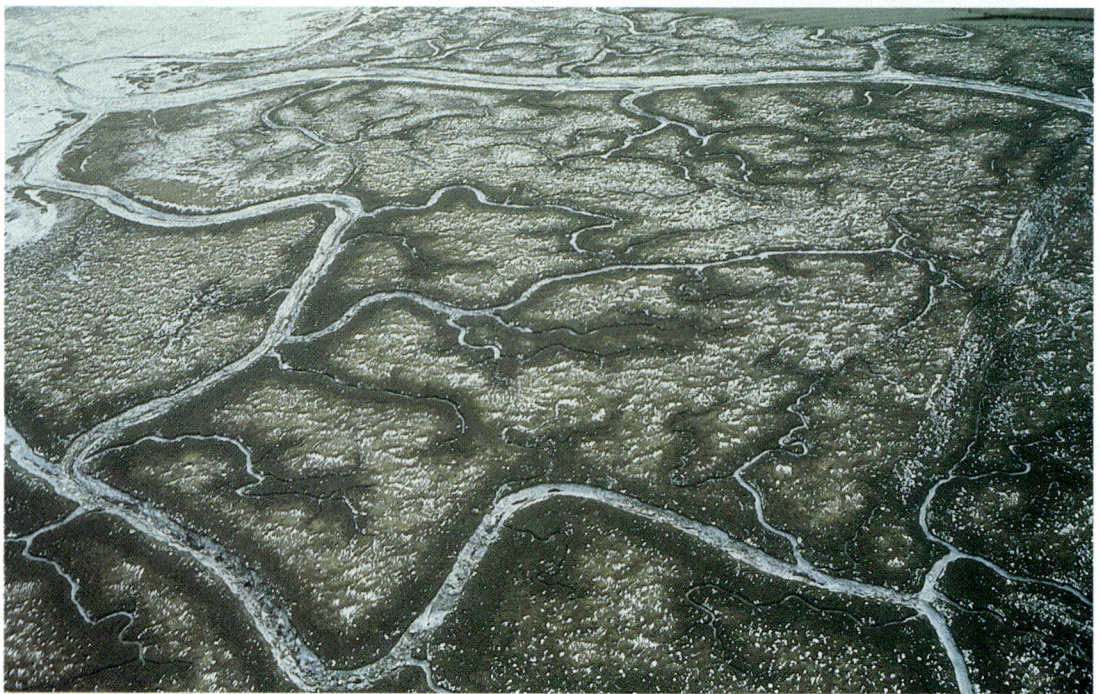

Fig. 11. The meandering pattern of salt-marsh creeks during winter. Photograph RIKZ, Middelburg.

Fig. 12. Vast intertidal mudflats are characteristic for Oosterschelde estuary. This mudflat is dominated by *Arenicola marina*. Photograph NIOO-CEMO, Yerseke.

Fig. 13. Intertidal sediments in Oosterschelde estuary range from coarse-grained shell debris to fine-grained silt. This picture shows seagrasses growing in a sheltered bay on soft, silty sediments. Photograph NIOO-CEMO, Yerseke.

Fig. 14. The sublittoral hard-substrate fauna of Oosterschelde estuary shows a remarkable diversity, owing to the good water quality. Photograph NIOO-CEMO, Yerseke.

Fig. 15. Man-made seawalls, artificial rocky shores, border the entire Oosterschelde estuary; they harbour a diversified flora and fauna. This picture is taken in the pre-barrier period, before the reinforcement of the dikes. Photograph NIOO-CEMO, Yerseke.

Fig. 16. SCUBA-divers offer direct observations of community structure and biomass distribution of sublittoral benthic assemblages. Photograph RIKZ, Middelburg.

Fig. 17. The research vessel 'Luctor' played an important role in many pelagic and benthic investigations in Oosterschelde estuary. Photograph NIOO-CEMO, Yerseke.

Fig. 18. Sampling with hand-operated corers is the classical approach in studies of intertidal zoobenthos of sand- and mudflats. Photograph RIKZ, Middelburg.

Fig. 19. The geomorphology and hydrodynamics of Oosterschelde estuary changed considerably as a result of the construction of the storm-surge barrier. One of the variables measured is the current velocity of the water mass. Photograph RIKZ, Middelburg.

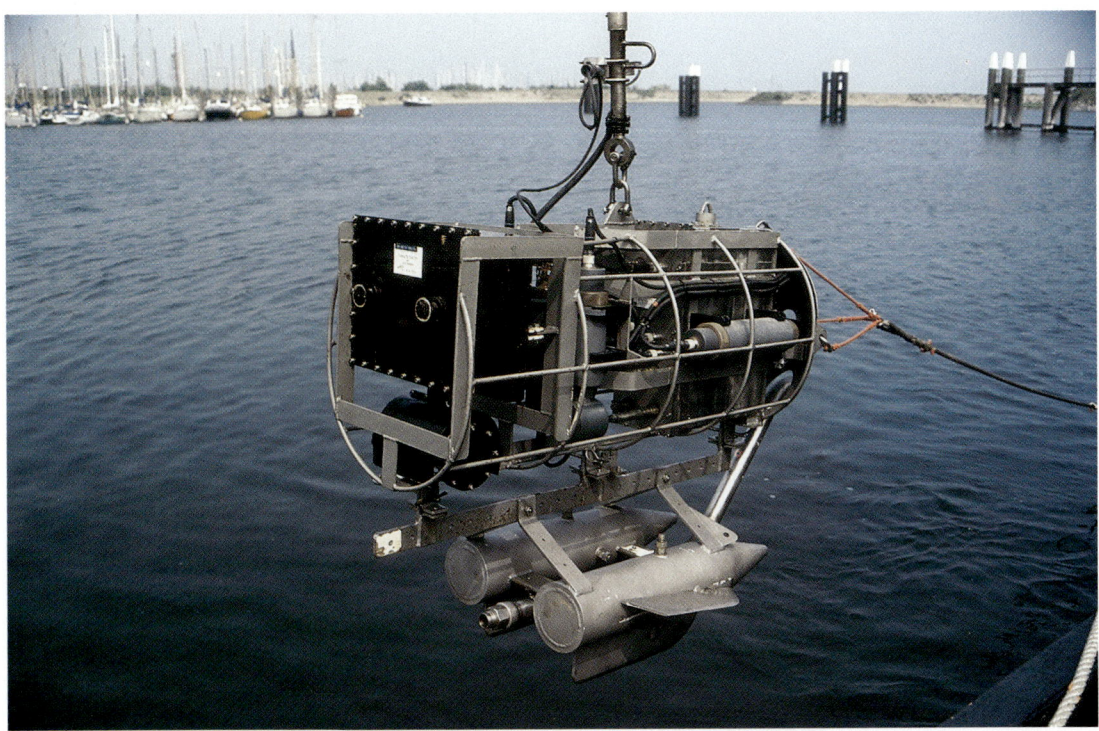

Fig. 20. Sophisticated instruments are lowered in the water mass to measure a number of water quality variables. Photograph RIKZ, Middelburg.

Fig. 21. Phytoplankton under the microscope. Phytoplankton is the most important primary producer in the estuary, providing energy to benthic filter feeders and, transformed in the foodchain, to numerous carnivores. Photograph NIOO-CEMO, Yerseke.

Fig. 22. Mussels, *Mytilus edulis*, the dominant filter-feeding bivalves in Oosterschelde estuary, are dependent on the supply of high-quality food, the phytoplankton. Photograph **RIKZ**, Middelburg.

Fig. 23. The mussel population in Oosterschelde estuary is controlled by mussel farmers. Photograph **NIOO-CEMO**, Yerseke.

Fig. 24. The Oosterschelde estuary is the only place in The Netherlands where oysters are cultivated and finally brought to the market. The oysters are temporarily stored in outdoor basins containing running seawater, the 'wet warehouses' at Yerseke. Photograph NIOO-CEMO, Yerseke.

Fig. 25. Hundreds of thousands of migratory waders winter in Oosterschelde estuary, feeding upon the rich macrozoobenthos biomass living in the intertidal flats. Photograph NIOO-CEMO, Yerseke.

Fig. 26. At present less than 20 harbour seals are living in and around Oosterschelde estuary. Providing optimal management, there is a potential for hundreds of seals in the estuary. Photograph RIKZ, Middelburg.

Figs 27–28. The interaction between a musselbed and the water column, the benthic-pelagic coupling, was studied with the Benthic Ecosystem Tunnel, a 12-m long plexiglass tunnel, placed over an intertidal bivalve bed. Samples are taken from the inflowing and outflowing water in order to measure the fluxes of dissolved and particulate matter. Lower picture (p. xxii) shows the tunnel at low water; upper picture (p. xxiii) shows two tunnels in submerged position. Photographs RIKZ, Middelburg.

Fig. 28. See p. xxii.

Fig. 29. The Reineck box-corer is used for quantitative sampling of permanently submerged macrozoobenthos communities. Photograph RIKZ, Middelburg.

Fig. 30. Recreational activities are increasing around Oosterschelde estuary. These sports-fishermen have chosen their beat at the seawall close to the storm-surge barrier. In the background the brick tower of a mediaeval village, Koudekerke, can be seen. This village drowned centuries ago during a severe storm flood, and only the church tower remained. Photograph NIOO-CEMO, Yerseke.

Fig. 31. Unpredictable severe winters have a dominant impact on the functioning of Oosterschelde ecosystem. Photograph NIOO-CEMO, Yerseke.

Hydrobiologia **282/283**: 1–14, 1994.
P. H. Nienhuis & A. C. Smaal (eds), The Oosterschelde Estuary.
© 1994 *Kluwer Academic Publishers. Printed in Belgium.*

1

The Oosterschelde estuary, a case-study of a changing ecosystem: an introduction

P. H. Nienhuis [1] & A. C. Smaal [2]
[1] *Netherlands Institute of Ecology, Vierstraat 28, 4401 EA Yerseke, The Netherlands;* [2] *National Institute for Coastal and Marine Management/RIKZ, P.O. Box 8039, 4330 EA Middelburg, The Netherlands*

Key words: Oosterschelde estuary, civil engineering works, storm-surge barrier, hydrographical changes, long-term ecological changes, community and ecosystem responses

Abstract

During the period 1980–1990 long-term physical, chemical and ecological studies were carried out, to study the changes induced by the building of a storm-surge barrier in the mouth of the Oosterschelde estuary and two large auxiliary compartment dams in the rear ends of the estuary. The storm-surge barrier was constructed in the mouth of Oosterschelde estuary (SW Netherlands) during the period 1979–1986. The barrier allows the tides to enter the estuary freely, and, on the other hand, the barrier guarantees safety for the human population and their properties when a stormflood threatens the area.

Oosterschelde estuary is isolated from the river input, the rear ends of the ecosystem were separated from the estuary by sea-walls and the strongly decreased tidal exchange with the North Sea induced sheltered circumstances. The Oosterschelde changed from a turbid estuary into a tidal bay, and yet primary production responses appear to be robust and resilient, and the biological communities showed only quantitative shifts from the dominance of specific species assemblages to other assemblages. In many cases predicted changes in the structure of the biological communities could not be verified owing to the large natural variability mainly caused by physical factors (e.g. temperature).

The Delta Project – Storm flood protection in the Oosterschelde

The Netherlands Delta Project covers the so-called Delta area, created by the rivers Rhine, Meuse and Scheldt. Like most delta-estuarine environments, this area represents, in its natural state, complicated ecosystems consisting of a complex hydrodynamic regime, with fast-flowing water masses in tidal channels, changing estuarine configurations, inhomogeneous tidal and subtidal sediments, and salt-marsh areas subject to periodic flooding. The history of the SW Netherlands is marked by a continuous struggle between man and the sea. Since the year 1000 man reclaimed salt-marsh areas and transformed those into agricultural land. But irregularly occurring stormfloods broke the man-built seawalls and recaptured parts of the gained land. On February 1, 1953, a northwesterly storm induced tides to 3 m above normal levels, breached approximately 180 km of coastal-defense dikes and flooded 160 000 hectares of polderland. 1835 people lost their lives in this large storm flood, more than 46 000 farms and buildings were destroyed or damaged, and approximately 200 000 farm animals were lost.

The Delta Project, formalized in 1957 by an act

of the Dutch parliament, was conceived as an answer to the continuous risk of flooding, which threatens lives and property in this low-lying region. Because of the low mean elevation and premium on space in The Netherlands, the Dutch have a long tradition of coastal-defense construction and land reclamation. The need for continuous coastal construction has intensified over the years as a result of population growth, land subsidence and rising sea levels. The core of the Delta Project called for the closure of the main tidal estuaries and inlets in the SW Netherlands, except for the Westerschelde where the existing dikes have been raised, for reasons of continued international shipping access to Antwerp (Huis in 't Veld *et al.*, 1984).

A prerequisite for the construction of the primary sea-walls in the mouths of the estuaries was the need to reduce tidal-current velocities in the estuaries, before the construction of the primary barriers could be undertaken. Tidal velocities were lowered by constructing secondary compartmentalisation barriers (Zankreekdam, Grevelingendam and Volkerakdam; Fig. 1) to reduce the extent of the Delta area subject to tidal influence. This resulted, in turn, in a reduced tidal volume and, therefore, lower current velocities through the main estuaries. The former estuaries Veersche Gat and Grevelingen were closed off from the North Sea by high sea-walls in 1961 and 1971, respectively, and turned into non-tidal lakes or lagoons filled with brackish or saline water, whereas the Haringvliet was closed in 1970 by the construction of large sluices, meant to function as an outlet for the rivers Rhine and Meuse (Fig. 1) (Knoester, 1984).

The original plan for the Oosterschelde estuary called for a dam across the mouth of the estuary, a distance of 9 km, to be finished in 1978. The tidal basin would than have been changed into a stagnant lake filled with – polluted – water from the river Rhine. But the final form of the present barrier differs drastically from the simple dam that has been envisaged originally. Through the 1960's and early 1970's, conservationists provoked an awareness in many people of the need to protect the area's outstanding natural resources and its unique tidal habitat, including an extensive shellfish industry, the only one in The Netherlands (Smies & Huiskes, 1981).

The Dutch government decided to change the design of the dam in 1974. After several years of desk studies the Dutch parliament accepted in 1976 a compromise solution: a storm-surge barrier. The barrier allows the tides to enter the estuary freely, thus safeguarding the tidal ecosystem, including the plant and animal communities. On the other hand the barrier guarantees safety for the human population and for the properties of the inhabitants when storm floods threaten the area. This design meant a turning point in the Dutch political decision-making process with regard to the natural environment (Knoester *et al.*, 1984). The storm-surge barrier was constructed between 1979 and 1986 in the western mouth of the estuary (Fig. 2). A series of 65 prefabricated concrete piers (pylons) form the framework of the barrier. The piers support 62 steel gates which can be dropped like portcullises to close off the estuary when danger threatens. The piers and the beams which tie them together have been built inside coffer-dams, and were transported by a special vessel, a pier transporter, to the barrier alignment across the estuary. The piers have base dimensions of 25×50 m^2 and a maximum height of 43 m; they weigh 18 000 tonnes each. The piers are located on the bottom of the sea on thick foundation mattresses, to prevent erosion of the bottom sediments. The largest part of the entire construction is permanently under water.

To provide for the long-term stability of the installed piers, the base of each pier has been covered with rubble and large stone blocks, and the piers were joined together with beams to make a single construction. The final superstructure of the storm-surge barrier carries a road across the estuary mouth (Fig. 3). The steel gates in between the piers can be raised and lowered, but for most of the time they will be suspended above the waves. But when extremely high tides – more than 3 m above Mean Sea Level – are predicted they will be lowered into the water, once or twice a year, to block off the estuary from the North Sea

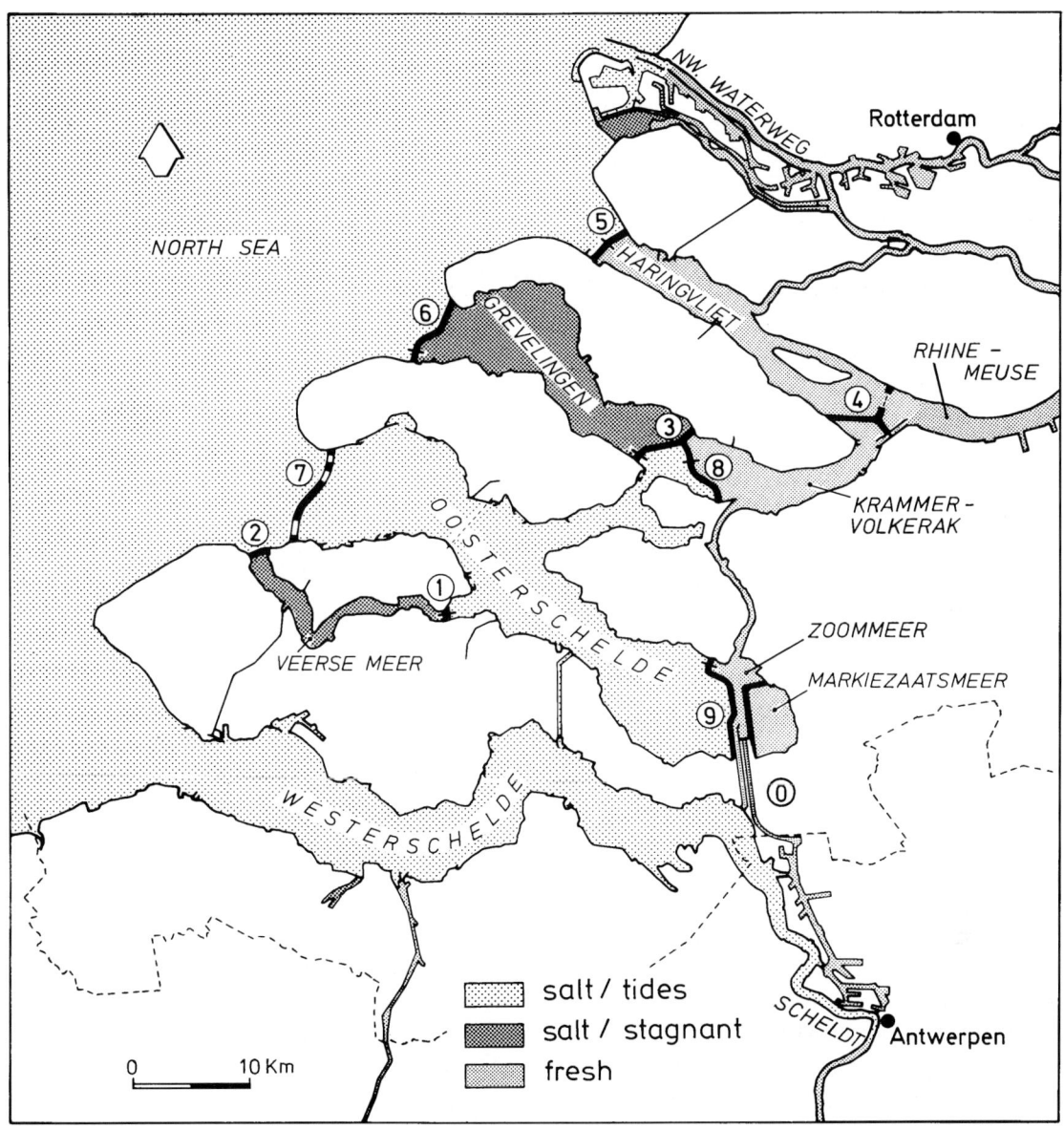

Fig. 1. Map of the Delta area of the rivers Rhine, Meuse and Scheldt in the SW Netherlands, with various waterbodies as resulting from the Delta project engineering scheme. 0 = Kreekrakdam, 1867; 1 = Zandkreekdam, 1960; 2 = Veersegatdam, 1961; 3 = Grevelingendam, 1964; 4 = Volkerakdam, 1969; 5 = Haringvlietdam, 1970; 6 = Brouwersdam, 1971; 7 = Oosterschelde storm-surge barrier, 1986; 8 = Philipsdam, 1987; 9 = Oesterdam, 1986. Markiezaatsmeer has been closed off from Zoommeer by Markiezaatsdam in 1983.

(Watson & Finkl, 1990; Van Westen & Colijn, 1994).

Although topographically separated from the main sea-wall, two auxiliary compartment dams, built between 1977 and 1987, form an indissoluble entity with the storm-surge barrier. The Oesterdam is an 11 km long dam in the rear end of Oosterschelde estuary, separating the saline sea-arm from the eastern freshwater compartment, the Markiezaatsmeer. The Zoommeer in between

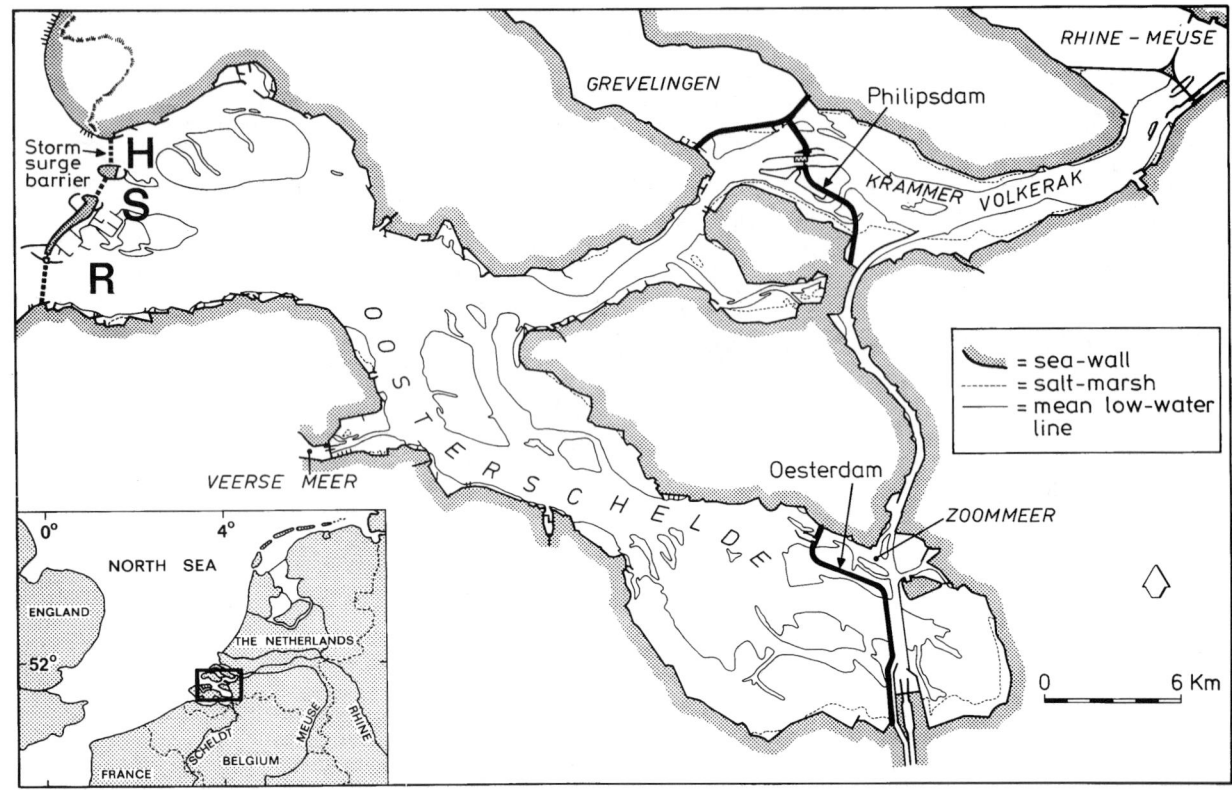

Fig. 2. Oosterschelde estuary in the SW Netherlands. R = Roompot; S = Schaar; H = Hammen. Krammer-Volkerak contains river water, Oosterschelde contains sea water. Locks in Philipsdam and Oesterdam indicated.

functions as an important shipping route between Rotterdam and Antwerp. The Philipsdam is 6 km long and separates the northern branch of Oosterschelde estuary from the newly created freshwater lake Krammer-Volkerak. The dam contains two large shipping sluices of 280×24 m and a smaller lock (Fig. 2).

The construction of the storm-surge barrier together with the compartment dams took place between 1977 and 1987. These 10 years can arbitrarily be divided in 3 periods, after 1984 roughly indicated by changes in the tidal amplitude (Fig. 5): (1) The pre-barrier period (1977–1984); (2) the construction period in which the prefabricated elements were positioned in the mouth of the estuary (1985–April 1987), and (3) the post-barrier period (April 1987 and later).

The Delta Works, and especially the storm-surge barrier, represent the state of the art of storm-surge protection. Newly developed designs and techniques, particularly with regard to foundation preparation and the use of foundation protection mattresses, reflected both innovative thought and effective communication between scientists specialized in hydrodynamics and geomorphology and civil engineers. The use of prefabricated components and methods to install heavy constructions in open water, subject to the effects of waves and currents, represent an important breakthrough in marine civil engineering (Watson & Finkl, 1990).

A case study of a changing ecosystem

The execution of the large hydrotechnical works in and around the Oosterschelde estuary offered ample opportunities to study the changes in the

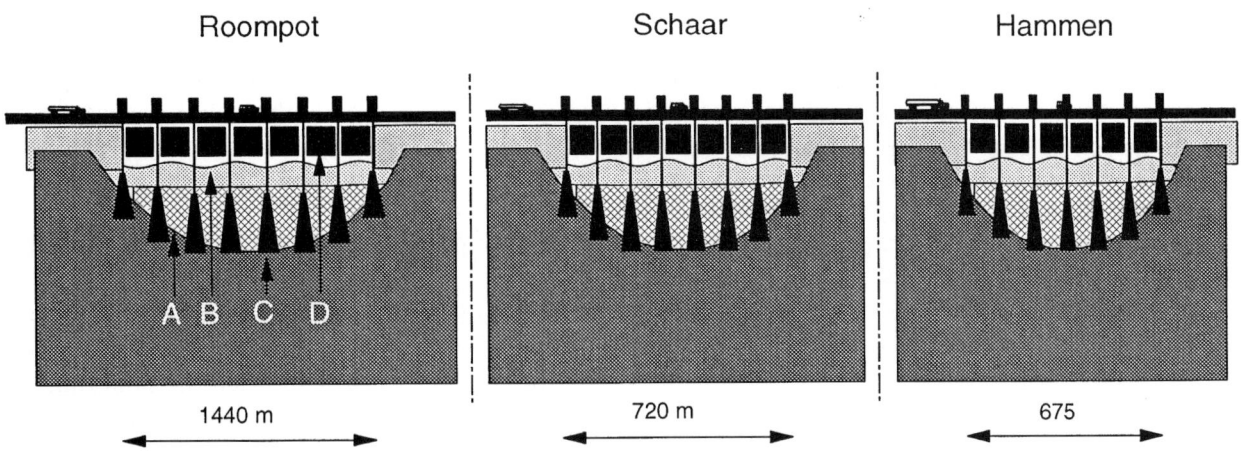

| Roompot | Schaar | Hammen |

A = sill (sediment)
B = tidal water
C = pier
D = steel gate

Fig. 3. Scheme of the storm-surge barrier in the seaward mouth of Oosterschelde estuary (see Fig. 2: main tidal branches Roompot, Schaar and Hammen).

aquatic ecosystem. Early biological inventories by collaborators of the Delta Institute for Hydrobiological Research (DIHO), and some other institutions, revealed the outstanding ecological qualities of the estuary (Elgershuizen *et al.*, 1979; Duursma *et al.*, 1982; Saeijs, 1982). In 1980 research groups of the Environmental Section of the Delta Department (now Tidal Waters Division) of the Ministry of Transport and Public Works and the DIHO (now NIOO) decided to start an ecosystem study, aiming at an analysis of the structure and functioning of the changing Oosterschelde estuary, related to the future management of the water body (project BALANS, from 1980 to 1988; project EOS, from 1988 to 1991).

This project is unique. It offered ample opportunities to carry out research in various spheres of interest: hydrography, hydrochemistry, autecology, community ecology, ecosystem ecology and fisheries. Integration of the knowledge gained, raises the project to a case study of a changing estuarine ecosystem. It demonstrated the effects of human interference in a non-polluted estuary,

compared with other marine ecosystems (Nienhuis *et al.*, 1994). Long-term responses of animal and plant communities and populations to environmental stress showed that discrimination between responses to human-induced stress and responses to natural stress is difficult to establish quantitatively. The majority of the cause-effect relations described come from correlative, descriptive work, which is not intended to offer causal explanations. A considerable number of documented fluctuations in population dynamics of dominant plants and animals cannot be directly attributed to the construction of the civil engineering works (Nienhuis & Smaal, 1993).

The present volume of Hydrobiologia comprises a number of research papers devoted to Oosterschelde estuary. The papers are mainly based on investigations carried out during the period 1980–1991, sustained by older investigations and long-term reviews. The core of the volume is distracted from the project EOS (1988–1991). This project meant a close cooperation between 25 to 30 scientists and water managers of several State Services, Universities and Private Agencies

(Tidal Waters Division, Ministry of Transport and Public Works; Delta Institute for Hydrobiological Research (now Netherlands Institute of Ecology), Ministry of Science and Education; Netherlands Fisheries Research Institute, Ministry of Agriculture and Fisheries, State University of Utrecht; State University of Ghent (Belgium); Environmental Consultancy Agencies).

Hydrography and hydrochemistry

The history of the Oosterschelde estuary is characterized by an increasing isolation from the river influences. The estuary changed in the course of time from a coastal plain estuary into a tidal bay. Already in 1867 the freshwater load from the Schelde river was diverted to the Westerschelde, by the closing of the Kreekrakdam. In 1969 the Volkerakdam was closed (Fig. 1), depriving the Oosterschelde from extensive amounts of fresh Rhine water: during winters with extremely high fresh-water discharges, salinities near Yerseke dropped to 10 promille chloride (Bakker, 1967).

The water balance over the period 1980–1989 (Fig. 4) shows that before 1987 the main fresh-water load (50 $m^3 s^{-1}$) entered the Oosterschelde via Krammer-Volkerak, through the sluices in the Volkerakdam, mainly derived from the river Rhine, and to a far less extent from the river Meuse. When the Philipsdam was finished (April 1987) the fresh-water load was further reduced to 10 $m^3 s^{-1}$ via the Krammer sluices. The brackish-water load from the adjacent Grevelingen lagoon (Grevelingenmeer) was less than 5 $m^3 s^{-1}$, except in the period 1985–1986, when the tidal range in the Oosterschelde estuary has been reduced drastically (Fig. 5) owing to manipulations with the flood gates in the storm-surge barrier. During that period the load from Grevelingen increased to 20 $m^3 s^{-1}$ via the siphon in the Grevelingendam. Diffuse loadings of slightly brackish water from agricultural run off – polder discharges – around the Oosterschelde played a minor role (5–8 $m^3 s^{-1}$). Precipitation and evaporation compensated eachother roughly. The water balance is closed by a net transport of water to the North Sea, not shown in Fig. 4. This transport amounted to

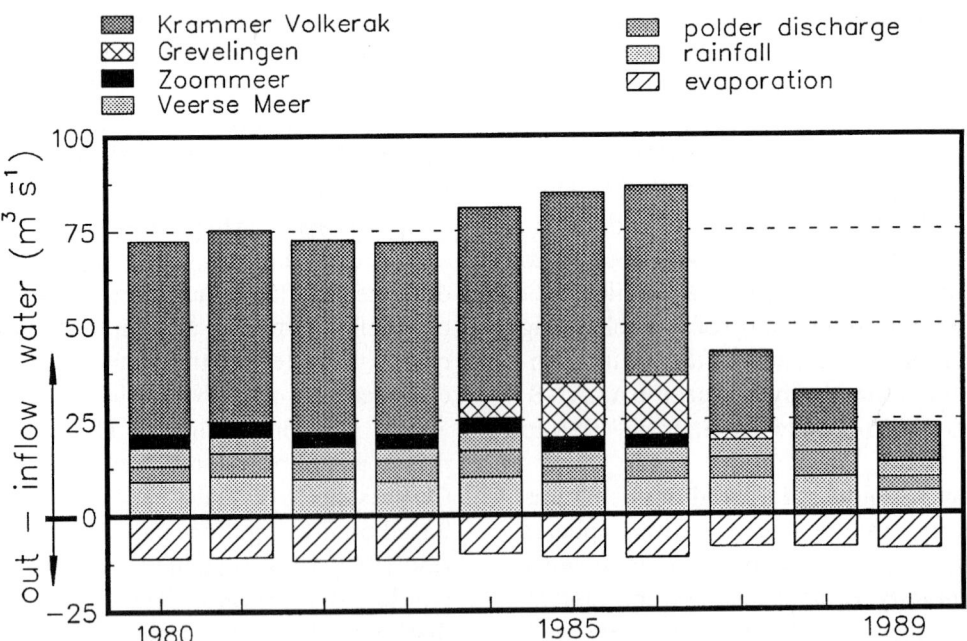

Fig. 4. Water budget of Oosterschelde estuary. The gains and losses of water (inflow and outflow) are indicated. The dominant factor, the tidal exchange (1230×10^6 m^3 tidal volume pre-barrier and 880×10^6 m^3 post-barrier; Table 1) is omitted from the figure (data Tidal Waters Division, Middelburg).

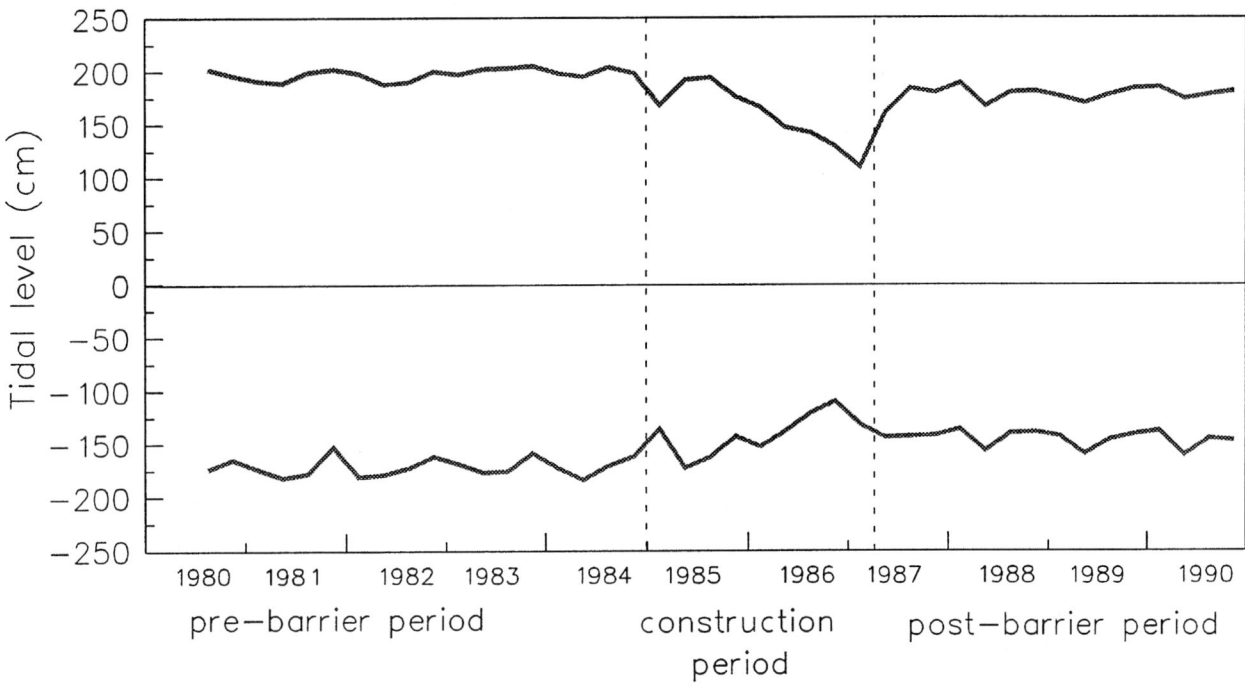

Fig. 5. Running average of mean high-water (MHW) and mean low-water (MLW) data, measured daily at Yerseke (central Oosterschelde). Data from Tidal Waters Division, Middelburg. The pre-barrier, the barrier construction and the post-barrier period are indicated.

50 (summer)–100 (winter) m^3 s^{-1}, and dropped sharply to 0–50 m^3 s^{-1} after April 1987.

Table 1 summarizes data on the changes in the hydrography of Oosterschelde estuary after the completion of the storm-surge barrier and the two auxiliary compartment dams. The total surface area of the saline tidal ecosystem decreased by 22%, mainly owing to building of the auxiliary dams. The surface area of the intertidal flats decreased by 36% and the area of salt marshes, mainly situated in the rear end of the estuary, by 63%. A new equilibrium between the strongly decreased tidal volume of water (minus 28%) and the cross section of the estuary (owing to the building of the barrier minus 78%) is developing. The former tidal gullies are too deep and the slopes of the intertidal flats are too steep now. A continuous erosion of the intertidal area occurs, and excess sand and silt are deposited in the tidal gullies. This process will continue for many years, leading to a predicted loss of 15% of the present surface area of the intertidal flats and to 15% of

the present salt marshes over a period of thirty years (Oenema, 1988; Smaal *et al.*, 1991; Smaal & Nienhuis, 1992; Vroon, 1994; Mulder & Louters, 1994; Ten Brinke *et al.*, 1994).

Table 1 also shows the decrease of the current

Table 1. Main hydrodynamic characteristics of the Oosterschelde Estuary before and after the completion of the coastal engineering works (data Tidal Waters Division, Middelburg).

	Pre-barrier	Post-barrier
Total surface, km^2	452	351
Water surface, (MWL) km^2	362	304
Tidal flats, km^2	183	118
Salt marshes, km^2	17.2	6.4
Cross section, barrier in open position, m^2	80000	17900
Mean tidal range, Yerseke, m	3.70	3.25
Max. flow velocity, m s^{-1}	1.5	1.0
Residence time, d	5–50	10–150
Mean tidal volume $m^3 \times 10^6$	1230	880
Total volume, $m^3 \times 10^6$	3050	2750
Mean freshwater load, m^3 s^{-1}	70	25

8

velocities after the building of the barrier, and the strong increase of the residence time of the water; both parameters express themselves most prominently in the remote rear ends of the estuary. From place to place the tidal amplitude shows minor differences. At Yerseke (Fig. 5) the tidal range was on average 3.7 m before 1985, and contracted to an average of 3.25 m after April 1987. Mean low water is approximately 20 cm higher now, than in the pre-barrier period, and mean high water is approximately 20 cm lower. From summer 1985 to April 1987 the tidal range has been manipulated by flood-gate control to permit the execution of hydraulic works in the back end of the estuary. During that period the tidal amplitude was artificially contracted to a minimum of 2.25 m (Fig. 5).

Table 2 summarizes data on the hydrochemistry of Oosterschelde estuary in the pre- and post-barrier period. The general tendency reveals an overall decrease of concentrations of inorganic nutrients, seston, particulate organic carbon, chlorophyll *a* and heavy metals as a consequence of the increased isolation from the influence of the main rivers. Table 2 shows average values, but in reality a wide variation over time is characteristic for most parameters. POC concentrations (Fig. 6) are given as an example: data ranging

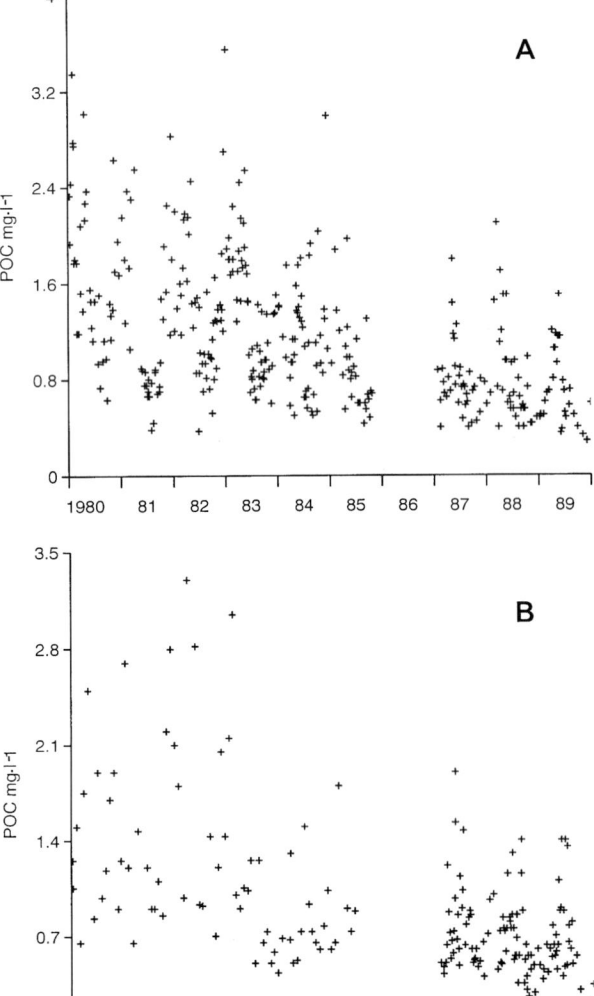

Fig. 6. Measured POC concentrations (mg l^{-1}) in the Oosterschelde western compartment (A) and eastern compartment (B) (data Tidal Waters Division, Middelburg).

between 0.2 and 3.5 mg l^{-1} showed both a decrease in average values as well as in the overall range after mid 1987. Secchi disc visibility as an expression of the penetration of sunlight into the water column, varied between 0.5 and 3.5 m in the western compartment, and between 0.5 and 3 m in the eastern compartment in the pre-barrier period, and increased to 1 to 4.5 m, only in the sheltered eastern compartment (Fig. 7). On average the water is clearer now than before the build-

Table 2. Average annual concentrations of some water-quality parameters in western and eastern part of the Oosterschelde Estuary in 1980–1984 and 1988 (until October). All units in g m^{-3} except for chloride (kg m^{-3}) and chlorophyll-*a* (mg m^{-3}); heavy metals consider the dissolved fraction only (data Tidal Waters Division, Middelburg).

	West			East		
	80/84	1988	Δ %	80/84	1988	Δ %
Chloride	16.9	17.1	+ 1	15.4	16.7	+ 8
Seston	26.8	12.9	− 50	22.0	5.9	− 75
POC	1.5	0.8	− 45	1.3	0.7	− 50
Chl-*a*	6.95	3.82	− 45	6.86	4.84	− 29
Silicate	0.39	0.32	− 20	0.55	0.29	− 45
Nitrate, NO$_2$ + NO$_3$	0.52	0.42	− 20	0.82	0.33	− 60
Phosphate	0.06	0.05	− 15	0.08	0.06	− 25
Cadmium	0.08	<0.01	− 88	0.19	0.02	− 89
Mercury	<0.01	<0.01	0	<0.02	<0.01	− 50
Lead	<0.4	<0.1	− 75	<0.6	<0.1	− 80
Copper	1.5	1.2	− 20	2.4	1.6	− 33

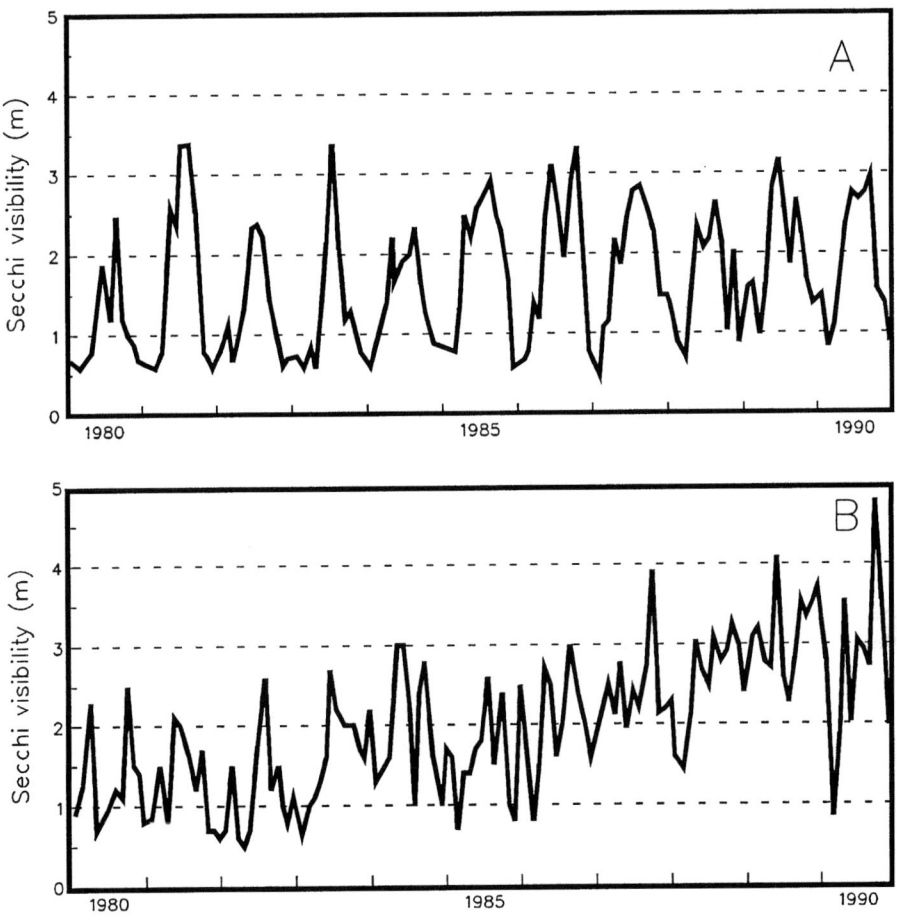

Fig. 7. Averaged Secchi disc visibility in the Oosterschelde western compartment (A) and eastern compartment (B) (data Tidal Waters Division, Middelburg).

ing of the storm-surge barrier. Already in 1984, before the final construction of the barrier, but after the fixation of the underwater sill in the mouth of the estuary, a decrease in the range of the extinction coefficient could be detected (Bakker & Vink, 1994).

Responses at the ecosystem level

Four main habitats have been distinguished in Oosterschelde estuary (Smaal & Nienhuis, 1992):

I. The open water mass. Following the changes in hydrodynamics, a shift in the phytoplankton assemblage has been observed from the heavily silicified diatoms to smaller diatoms and flagellates. The phytoplankton community changed from a typical turbid, estuarine community into a tidal bay or lagoonal community (Bakker *et al.*, 1990; Bakker *et al.*, 1994).

II. The intertidal sand- and mud-flats (Fig. 2). The within-habitat diversity of the intertidal sand- and mudflats, determined by the dominant macrobenthic communities, shows large year-to-year changes, depending a.o. on the availability of food, the success of the spatfall, the presence of predators and the occurrence of severe winters. The changes in hydrography and hydrochemistry, superimposed on the natural variability, resulting in a substantial decrease of the total surface area of the intertidal flats and a shift in sediment composition, did not alter the benthic communities qualitatively (Smaal & Nienhuis, 1992).

III. The supratidal salt marshes (Fig. 2). The strong decrease of the tidal range during prolonged periods in 1986 and early 1987 (Fig. 5) has irreversibly changed the supratidal salt marshes. Desiccation and aerobic mineralization of organic matter occurred, and a gradual shift in the zonation pattern of the halophyte communities was observed (De Leeuw *et al.*, 1994; De Jong *et al.*, 1994; De Jong & van der Pluijm, 1994; Vranken *et al.*, 1990). The changed tidal range, which introduced the process of erosion of the salt marsh cliffs, will continue in the near future, giving rise to the loss of 4 ha y^{-1} of salt marsh.

IV. The hard substrates of seawalls and dikes (Fig. 2). The restricted area of artificial stone substrates, covering the seawalls surrounding the estuary carries a diversified flora and fauna. The fixed zonation pattern changed its position, owing to the contracted tidal amplitude. The increased overall shelter of the estuary allows sessile animals and plants to grow in spots previously inaccessible (De Kluijver & Leewis, 1994; Meier & Waardenburg, 1994).

The dominant food chain in Oosterschelde estuary consists of phytoplankton (80% of the total primary production) on the primary level, benthic filterfeeding molluscs on the secondary level and carnivorous waterbirds (mainly waders) and man (mussel- and cocklefisheries and cultivation) on the tertiary level. A substantial part of the Hydrobiologia volume is devoted to primary production processes (Wetsteyn & Kromkamp, 1994) and to the structure of the pelagic community (Bakker *et al.*, 1994; Bakker & Vink, 1994; Bakker, 1994; Bakker & van Rijswijk, 1994; Tackx *et al.*, 1994). Changes in the structure of the phytoplankton assemblage reflect the concomitant changes in the environment more adequately than phytoplankton primary production. When averaged over the entire Oosterschelde estuary, from the pre-barrier to the post-barrier period, phytoplankton primary production ap-

peared to be a robust, integrated process, without notable changes (Wetsteyn *et al.*, 1990; Herman & Scholten, 1990; Smaal & Nienhuis, 1992). It is hypothesized that the decreased loading with nutrients after April 1987, has been compensated by the increased light transmittance through the water column, resulting in approximately the same level of annual primary production before and after the construction of the storm-surge barrier (Wetsteyn & Kromkamp, 1994).

The relation between the phytoplankton standing stock and the benthic filter feeders is of prime importance in the Oosterschelde estuary. For that reason much research has been devoted to the structure of the soft bottom macrozoobenthos communities (Meire *et al.*, 1994; Seys *et al.*, 1994), to the fluctuations in standing stock, growth rates and mortality rates of blue mussels (*Mytilus edulis*) and cockles (*Cerastoderma edule*) (Coosen *et al.*, 1994; Van Stralen & Dijkema, 1994), and to the nutrient regeneration capacity of filter feeders (Prins & Smaal, 1994). Filter feeders act as a driving force in the turnover of phytoplankton biomass and nutrients in the estuary. Filter feeders strongly reduce phytoplankton biomass. On the other hand filter feeders contribute significantly to the mineralization of deposited organic matter. Roughly 20 to 75% of the total nitrogen mineralization in soft sediments of Oosterschelde occurs in musselbeds, although these communities cover only 10% of the bottom of the estuary (Prins & Smaal, 1990). Hypothetically benthic filter feeders have the potential to enhance primary production in the overlying watercolumn (Prins & Smaal, 1990; Asmus & Asmus, 1991).

The standing stocks of the two main filter feeders, blue mussel and cockle, are largely manipulated by man. Each year thousands of tons of small mussels are harvested in the Wadden Sea and brought into the Oosterschelde, and stocked on specific mussel cultivation plots. A large percentage of these mussels dies before they reach their second or third year. A smaller number survives, and is harvested by mussel farmers (Dijkema, 1988). Roughly 50% of the mussel standing stock is annually harvested by man, and roughly 10% is consumed by water birds, mainly waders (Schekkerman *et al.*, 1994; Meire *et al.*, 1994).

Natural cockle beds in Oosterschelde estuary are increasingly exploited by man: from roughly 100 tons ash-free dry weight in 1982 to roughly 1000 tons in 1989 (Smaal & Nienhuis, 1992). Waders consume roughly 15% of the cockle standing stock on an annual basis. When cockle standing stocks are low, as is the case after a series of mild winters, competition may arise between man and waders for the same food source. The carrying capacity of the feeding grounds in Oosterschelde estuary for waders has decreased strongly, owing to the reduction of the intertidal area, and the reduced foraging period per tidal cycle. Increased competition for shellfish between waders and man is a further limit to the carrying capacity for waders (Smaal & Nienhuis, 1992).

Besides the main foodchain (phytoplankton – filter feeders – waders), several other aspects have been studied in the estuary, although not as intensively as the main compartments: microphytobenthos (De Jong *et al.*, 1994), zooplankton (Bakker, 1994; Bakker & van Rijswijk, 1994), meiozoobenthos (Smol *et al.*, 1994), hard substrate algae and macrofauna (De Kluijver & Leewis, 1994; Meier & Waardenburg, 1994), epibenthic fauna (Hostens & Hamerlynck, 1994), demersal fish (Hamerlynck & Hostens, 1994) and seals (Mees & Reijnders, 1994).

A substantial amount of work has been put in the construction of a mathematical simulation model of the main compartments of the estuary (model SMOES; Klepper, 1989; Klepper *et al.*, 1994; Scholten & van der Tol, 1994). Both averaged annual carbon models (Fig. 8) as well as sophisticated dynamic models have been presented (Scholten *et al.*, 1990), in order to predict ecological changes induced by the execution of the storm-surge barrier and the auxiliary dams. Figure 8 presents the average annual carbon budget of the main compartments of Oosterschelde estuary, in the pre-barrier and in the post-barrier period. The first impression of the model is a robust, stable carbon budget in which only minor changes have taken place (size of biomass boxes). Two significant changes in the carbon flows have

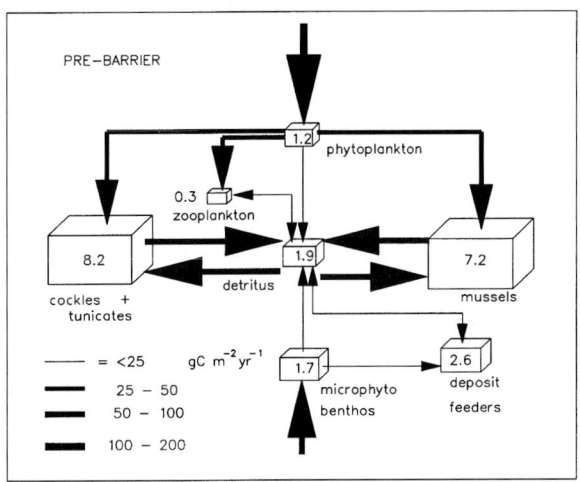

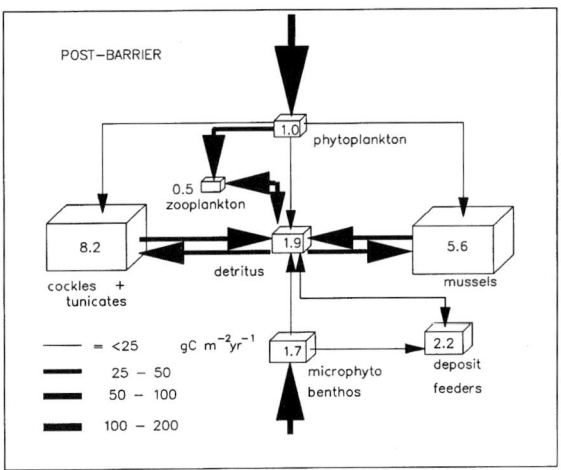

Fig. 8. Annual carbon budget of the Oosterschelde estuary in the prebarrier period and the post-barrier period. Boxes = biomass g C m^{-2}; Arrows = fluxes g C m^{-2} y^{-1} (Smaal & Nienhuis, 1992).

to be mentioned. (1) The carbon flow between herbivorous zooplankton and phytoplankton increased, due to the improved quality of the food, containing less silt in the post-barrier period than in the previous period. (2) The role of detritus as food for filter feeders decreased slightly. In the pre-barrier period turbulence kept much detritus suspended in the water column, and consequently much detritus was consumed by the filter feeders, but this was also disposed again as pseudofaeces and faeces.

The Oosterschelde project is one of the largest aquatic civil-engineering schemes of recent times.

The research investments in this project offer a unique series of documents from which an integration of physical, chemical and biological aspects at the level of the entire ecosystem can be put together. The picture arising, however, is far from complete. Oosterschelde estuary is isolated from the river input, the rear ends of the ecosystem were separated from the estuary by seawalls and the decreased tidal exchange with the North Sea induced sheltered circumstances. The Oosterschelde changed from a turbid estuary into a tidal bay, and yet primary production responses appear to be robust and resilient, and the biological communities showed only quantitative shifts from the dominance of specific species assemblages to other assemblages. In many cases predicted changes in the structure of the biological communities could not be verified owing to the large natural variability mainly caused by physical factors (e.g. temperature, insolation) (Nienhuis & Smaal, 1993; Nienhuis *et al.*, 1994).

Acknowledgement

This paper is communication No. **688** of the NIOO-CEMO, Yerseke, The Netherlands.

References

Asmus, R. M. & H. Asmus, 1991. Musselbeds: limiting or promoting phytoplankton? J. exp. mar. Biol. Ecol. 148: 215–232.

Bakker, C., 1967. Veranderingen in milieu en plankton van het Oosterscheldegebied. Vakbl. Biol. 47: 181–192. (In Dutch).

Bakker, C., 1994. Zooplankton species composition in the Oosterschelde (SW Netherlands) before, during and after the construction of a storm-surge barrier. Hydrobiologia 282/283: 117–126.

Bakker, C., P. M. J. Herman & M. Vink, 1990. Changes in seasonal succession of phytoplankton induced by the storm-surge barrier in the Oosterschelde (SW Netherlands). J. Plankton Res. 12: 947–972.

Bakker C., P. M. J. Herman & M. Vink, 1994. A new trend in the development of the phytoplankton in the Oosterschelde (SW Netherlands) during and after the construction of the storm-surge barrier. Hydrobiologia 282/283: 79–100

Bakker C. & P. van Rijswijk, 1994. Zooplankton biomass in the Oosterschelde (SW Netherlands) before, during and after the construction of a storm-surge barrier. Hydrobiologia 282/283: 127–143.

Bakker C. & M. Vink, 1994. Nutrient concentrations and planktonic diatom-flagellate relations in the Oosterschelde (SW Netherlands) during and after the construction of a storm-surge barrier. Hydrobiologia 282/283: 101–116.

Coosen, J., J. Seys, P. M. Meire & J. A. M. Craeymeersch, 1994. Effect of sedimentological and hydrodynamical changes in the intertidal areas of the Oosterschelde estuary (SW Netherlands) on distribution, density and biomass of five common macrobenthic species: *Spio martinensis* (Mesnil), *Hydrobia ulvae* (Pennant), *Arenicola marina* (L.), *Scoloplos armiger* (Muller) and *Bathyporeia* sp. Hydrobiologia 282/283: 235–249.

Coosen J., F. Twisk, M. W. M. van der Tol, R. H. D. Lambeck, M. R. van Stralen & P. M. Meire, 1994. Variability in stock assessment of cockles (*Cerastoderma edule* L.) in the Oosterschelde (in 1980–1990), in relation to environmental factors. Hydrobiologia 282/283: 381–395.

De Jong, D. J., Z. de Jong & J. P. M. Mulder, 1994. Changes in area, geomorphology and sediment nature of salt marshes in the Oosterschelde estuary (SW Netherlands) due to tidal changes. Hydrobiologia 282/283: 303–316.

De Jong, D. J., P. H. Nienhuis & B. J. Kater, 1994. Microphytobenthos in the Oosterschelde estuary (The Netherlands), 1981–1990; consequences of a changed tidal regime. Hydrobiologia 282/283: 183–195.

De Jong, D. J. & A. M. van der Pluijm, 1994. Consequences of a tidal reduction for the salt-marsh vegetation in the Oosterschelde estuary (The Netherlands). Hydrobiologia 282/283: 317–333.

De Kluijver, M. J. & R. J. Leewis, 1994. Changes in the sublittoral hard substrate communities in the Oosterschelde estuary (SW Netherlands), caused by changes in the environmental parameters. Hydrobiologia 282/283: 265–280.

De Leeuw, J., L. P. Apon, P. M. J. Herman, W. de Munck & W. G. Beeftink, 1994. The response of salt marsh vegetation to tidal reduction caused by the Oosterschelde storm-surge barrier. Hydrobiologia 282/283: 335–353.

Dijkema, R., 1988. Shellfish cultivation and fishery before and after a major flood barrier construction project in the Southwestern Netherlands. J. Shellfish Res. 7: 241–252.

Duursma, E. K., H. Engel & Th. J. M. Martens, 1982. De Nederlandse Delta. Natuur en Techniek, Maastricht, 600 pp. (In Dutch).

Elgershuizen, J. H. B. W., C. Bakker & P. H. Nienhuis, 1979. Inventarisatie van aquatische planten en dieren in de Oosterschelde, DIHO rapporten en verslagen 1979–3: 1–105. (In Dutch).

Hamerlynck, O. & K. Hostens, 1994. Changes in the fish fauna of the Oosterschelde estuary – a ten-year time series of fyke catches. Hydrobiologia 282/283: 497–507.

Herman, P. M. J. & H. Scholten, 1990. Can suspension-feeders stabilise estuarine ecosystems? In M. Barnes & R. N. Gibson (eds), Trophic relations in the marine environment. Proceed. 24th Europ. Mar. Biol. Symp., Aberdeen Univ. Press: 104–116.

Hostens, K. & O. Hamerlynck, 1994. The mobile epifauna of the soft bottoms in the subtidal Oosterschelde estuary: structure, function and impact of the storm-surge barrier. Hydrobiologia 282/283: 479–496.

Huis in 't Veld, J. C., J. Stuip, A. W. Walther & J. M. van Westen (eds), 1984. The closure of tidal basins. Delft University Press, 450 pp.

Klepper, O., 1989. A model of carbon flows in relation to macrobenthic food supply in the Oosterschelde estuary (SW Netherlands). Ph.D. Thesis University Wageningen, 270 pp.

Klepper, O., M. W. M. van der Tol, H. Scholten & P. M. J. Herman, 1994. SMOES: a simulation model for the Oosterschelde ecosystem. Part I: description and uncertainty analysis. Hydrobiologia 282/283: 437–451.

Knoester, M., 1984. Introduction to the Delta case studies. In B. A. Bannink, W. P. A. Broeders, M. Knoester, L. Lijklema & P. H. Nienhuis (eds), Integration of ecological aspects in coastal engineering projects. Wat. Sci. Techn. 16: 1–9.

Knoester, M., J. Visser, B. A. Bannink, C. J. Colijn & W. P. A. Broeders, 1984. The Eastern Scheldt Project. In B. A. Bannink, W. P. A. Broeders, M. Knoester, L. Lijklema & P. H. Nienhuis (eds), Integration of ecological aspects in coastal engineering projects. Wat. Sci. Techn. 16: 51–77.

Leewis, R. J., H. W. Waardenburg & M. W. M. van der Tol, 1994. Biomass and standing stock on sublitoral hard substrates in the Oosterschelde estuary (SW Netherlands). Hydrobiologia 282/283: 397–412.

Mees, J. & P. J. H. Reijnders, 1994. The harbour seal, *Phoca vitulina*, in the Oosterschelde: decline and possibilities for recovery. Hydrobiologia 282/283: 547–555.

Meijer, A. J. M. & H. W. Waardenburg, 1994. Tidal reduction and its effects on intertidal hard-substrate communities in the Oosterschelde estuary. Hydrobiologia 282/283: 281–298.

Meire, P. M., H. Schekkerman & P. L. Meininger, 1994. Consumption of benthic invertebrates by waterbirds in the Oosterschelde estuary, SW Netherlands. Hydrobiologia 282/283: 525–546.

Meire, P. M., J. Seys, J. Buijs & J. Coosen, 1994. Spatial and temporal patterns of intertidal macrobenthic populations in the Oosterschelde: are they influenced by the construction of the storm-surge barrier? Hydrobiologia 282/283: 157–182.

Mulder, J. P. M. & T. Louters, 1994. Changes in basin geomorphology after implementation of the Oosterschelde estuary project. Hydrobiologia 282/283: 29–39.

Nienhuis, P. H. & A. C. Smaal, 1993. The Oosterschelde (The Netherlands), an estuarine ecosystem under stress: discrimination between the effects of human-induced and natural stress. ECSA-ERF Proceedings, Olsen & Olsen (in press).

Nienhuis, P. H., A. C. Smaal & M. Knoester, 1994. The Oosterschelde estuary: an evaluation of changes at the ecosystem level induced by civil engineering works. Hydrobiologia 282/283: 575–592.

Oenema, O., 1988. Early diagenesis in recent fine-grained sediments in the Eastern Scheldt. Ph.D. Thesis University of Utrecht, Utrecht, 223 pp.

Prins, T. C. & A. C. Smaal, 1990. Benthic-pelagic coupling: the release of inorganic nutrients by an intertidal bed of *Mytilus edulis*. In M. Barnes & R. N. Gibson (eds), Proceed. 24th Europ. Mar. Biol. Symp., Aberdeen Univ. Press: 89–103.

Prins, T. C. & A. C. Smaal, 1994. The role of the blue mussel *Mytilus edulis* in the cycling of nutrients in the Oosterschelde estuary (The Netherlands). Hydrobiologia 282/283: 413–429.

Saeijs, H. L. F., 1982. Changing estuaries. Ph.D. Thesis Univ. Leiden: 1–413.

Schekkerman, H., P. L. Meininger & P. M. Meire, 1994. Changes in the waterbird populations of the Oosterschelde, (SW Netherlands) as a result of large scale coastal engineering works. Hydrobiologia 282/283: 509–524.

Scholten, H., O. Klepper, P. H. Nienhuis & M. Knoester, 1990. Oosterschelde estuary (SW Netherlands): a self-sustaining estuary? Hydrobiologia 195: 201–215.

Scholten, H. & M. W. M. van der Tol, 1994. SMOES: a simulation model for the Oosterschelde ecosystem. Part II: calibration and validation. Hydrobiologia 282/283: 453–474.

Seys, J. J., P. M. Meire, J. Coosen & J. A. M. Craeymeersch, 1994. Long-term changes (1979–89) in the intertidal macrozoobenthos of the Oosterschelde estuary: are patterns in total density, biomass and diversity induced by the construction of the storm-surge barrier? Hydrobiologia 282/283: 251–264.

Smaal, A. C., M. Knoester, P. H. Nienhuis & P. M. Meire, 1991. Changes in the Oosterschelde ecosystem induced by the Delta works. In M. Elliot & J. P. Ducrotoy (eds), Estuaries and coasts: spatial and temporal intercomparisons. ECSA 19 Symposium, Olsen & Olsen, Fredenborg: 375–384.

Smaal, A. C. & P. H. Nienhuis, 1992. The Eastern Scheldt (The Netherlands), from an estuary to a tidal bay: a review of responses at the ecosystem level. Neth. J. Sea Res. 30: 161–173.

Smies, M. & A. H. L. Huiskes, 1981. Holland's Eastern Scheldt estuary barrier scheme: some ecological considerations. Ambio 10: 158–165.

Smol, N., K. A. Willems, J. C. R. Govaere & A. J. J. Sandee, 1994. Composition, distribution and biomass of meiobenthos in the Oosterschelde estuary (SW Netherlands). Hydrobiologia 282/283: 197–217.

Tackx, M. L. M., P. M. J. Herman, P. van Rijswijk, M. Vink & C. Bakker, 1994. Plankton size distributions and trophic relations before and after the construction of the storm-surge barrier in the Oosterschelde estuary. Hydrobiologia 282/283: 145–152.

14

Ten Brinke, W. B. M., J. Dronkers & J. P. M. Mulder, 1994. Fine sediments in the Oosterschelde tidal basin before and after partial closure. Hydrobiologia 282/283: 41–56.

Van Stralen, M. R. & R. D. Dijkema, 1994. Mussel culture in a changing environment: the effects of a coastal engineering project on mussel culture in the Oosterschelde estuary (SW Netherlands). Hydrobiologia 282/283: 359–379.

Van Westen, C. J. & C. J. Colijn, 1994. Policy planning in the Oosterschelde estuary. Hydrobiologia 282/283: 563–574.

Vranken, M., O. Oenema & J. Mulder, 1990. Effects of tide range alterations on salt marsh sediments in the Eastern Scheldt, SW Netherlands. Hydrobiologia 195: 13–20.

Vroon, J., 1994. Hydrodynamic characteristics of the Oosterschelde in recent decades. Hydrobiologia 282/283: 17–27.

Watson, I. & C. W. Finkl, 1990. State of the art in storm-surge protection: The Netherlands Delta Project. J. coast. Res. 6: 739–764.

Wetsteyn, L. P. M. J., J. C. H. Peeters, R. N. M. Duin, F. Vegter & R. P. M. de Visscher, 1990. Phytoplankton primary production and nutrients in the Oosterschelde (The Netherlands) during the pre-barrier period 1980–1984. Hydrobiologia 195: 163–177.

Wetsteyn, L. P. M. J. & J. C. Kromkamp, 1994. Turbidity, nutrients and phytoplankton primary production in the Oosterschelde (The Netherlands) before, during and after a large-scale coastal engineering project (1980–1990). Hydrobiologia 282/283: 61–78.

Hydrobiologia **282/283**: 15–16, 1994.
P. H. Nienhuis & A. C. Smaal (eds), The Oosterschelde Estuary.
© 1994 *Kluwer Academic Publishers. Printed in Belgium.*

Theme I:
Hydrodynamic and geomorphological changes in the tidal system

Wilfried B. M. ten Brinke
Department of Physical Geography, Utrecht University, P.O. Box 80115, 3508TC Utrecht, The Netherlands

The Oosterschelde has been an eroding basin for centuries. Large sand losses along the basin due to disastrous storm surges in the 15th and 16th century have increased the Oosterschelde's tidal volume by at least 50%. More recently further changes in the hydrodynamics and the morphology of the Oosterschelde took place. These changes were due to the implementation of the Delta Project civil engineering works. The closure of the Grevelingen dam in 1964 and the Volkerak dam in 1969 increased the tidal volume in the Oosterschelde with 6%, which caused a considerable erosion and widening of the channels in the western part and the north-eastern branch of the Oosterschelde. Parallel to the basin erosion a distinct sedimentation at the Oosterschelde's outer delta took place. The morphological adaptions were still in progress during the construction of the Oosterschelde works. The erosion inside the basin and the expansion of the outer delta stopped when the storm-surge barrier and two auxiliary dams significantly reduced the tidal volume in the basin.

During the construction of the storm-surge barrier the cross-sectional area of the Ooster-schelde inlet gradually reduced from 80,000 m^2 to 17,900 m^2. When a cross-sectional area of 35,000 m^2 was reached in the middle of 1985, the tidal volumes in the channels and the tidal range started to decrease. At the end of 1986 a maximum reduction of the tidal volumes was reached. After the closure of the Philipsdam in the north-eastern branch in April 1987 (the Oesterdam was closed in October 1986), the tidal volumes and the tidal range increased again up to their final level.

The engineering works have reduced the tidal volumes in the channels by about 30% in the central and western part of the basin and by up to 80% in the north-eastern branch. The mean tidal volume of the basin decreased to 880 million m^3. Maximum depth-averaged current velocities in the channels decreased from 1.2 m s^{-1} to 0.8 m s^{-1} approximately. The tidal range reduced by about 12% from 3.70 m to 3.25 m at Yerseke. Fresh-water input into the system, already low before the construction of the works (70 m^3 s^{-1}), is now insignificant (25 m^3 s^{-1}). Salinity is fairly constant and high throughout the Oosterschelde (>30‰). The total basin area lowered from 452 to 351 km^2, and the intertidal area from 183 to 118 km^2. The salt marshes have suffered in an important way from the changes in hydrodynamical conditions: of the former 17 km^2, 11 km^2 is no longer subject to tidal influence.

It has been shown for tidal basins throughout the world that, in the case of a morphodynamic equilibrium, the tidal discharge relates to the cross-sectional area of the channels in a linear way. The morphology will adapt to the new conditions when the equilibrium between the tidal discharge and the morphology is disturbed. The relationship between the tidal discharge and the cross-sectional area can be used to quantify the amount of sediment needed to establish an equilibrium between the morphology of the Oosterschelde basin and the reduced tidal volumes in the channels. The sediment deficit in the present Oosterschelde is 400–600 million m^3. This equals about 2–3 times the sediment volume of the tidal flats inside the basin. Therefore, the adaption of the morphology of the basin depends on the sediment import from the coastal zone (North Sea).

The construction of the civil engineering works has strongly reduced the sand transport inside the basin. The sand exchange between the Oosterschelde and the near-coastal zone of the North Sea has dropped to almost zero. The channels of the outer delta are also out of equilibrium and therefore little sand is transported through the barrier into the basin. Any significant sand import into the basin is unlikely until this outer delta has reached a new morphodynamic equilibrium. On the other hand, fine-grained sediment is imported from the North Sea and settles in the channels throughout the basin. Thus, the Oosterschelde has changed from a (mainly) sand exporting into a fine-grained sediment importing basin.

The adaptation of the estuarine channel cross-sections to the reduced tidal volumes implies a decrease of the geomorphological gradient inside the basin. Both the height of the intertidal flats and the depth of the channels will reduce over the next decades to centuries. A lowering of the surface levels of the intertidal flats with 0.1 to 0.2 m has indeed been observed. The area of intertidal flats will reduce by approximately 10 to 15% in the next 30 years. Along with this general erosion the fine-grained sediment content of the

intertidal flats has also reduced. The fine-grained sediment has settled in the deepest parts of the channels, on some musselplots and in dead channel-ends. Unconsolidated fine-grained deposits in the channels throughout the basin, largely imported from the North Sea, illustrate the filling up of the channels in the absence of a significant sand import from the outer delta.

The changes in the hydrodynamic conditions, sediment transport and morphology affect the Oosterschelde ecosystem in several ways. The erosion of the intertidal flats results in an increase of the area of lower flats and a decrease of the area of higher flats. The shift, combined with an increase of the inundation frequency of the lower parts, has resulted in local reductions in availability of feeding grounds for birds of 10 to 30%. Muddy deposits on dike slopes have reduced the biomass of hard bottom macrozoobenthos. The reduction of the sand content in recent deposits on musselplots is reflected in a reduction of the firmness of the bottom which makes it difficult for mussels to stay on top of their own deposits. As a consequence of the reduced current velocities some parts of the Oosterschelde have become less suitable for the cultivation of mussels. On the other hand, mussels may now be cultivated in other parts of the basin where currents were too strong before the construction of the works. The reduction of the turbidity of the water has a positive effect on the primary production since it counterbalances the reduction of the nutrient input due to a reduced fresh-water input. A preliminary conclusion is that primary production is not reduced by dam construction.

Hydrobiologia **282/283**: 17–27, 1994.
P. H. Nienhuis & A. C. Smaal (eds), The Oosterschelde Estuary.
© 1994 *Kluwer Academic Publishers. Printed in Belgium.*

Hydrodynamic characteristics of the Oosterschelde in recent decades

Jacques Vroon
*National Institute for Coastal and Marine Management/RIKZ, P.O. Box 8039, 4330 EA Middelburg,
The Netherlands*

Key words: Oosterschelde estuary, hydrodynamic changes, storm-surge barrier

Abstract

Hydrodynamic conditions in the Oosterschelde have changed as a result of the Delta Project. Engineering works were carried out between 1960 and 1970 in the northern branch of the Oosterschelde. The Grevelingen dam was built in 1964 and the Volkerak dam was completed in the spring of 1969. With the completion of these dams, the northern branch of the Oosterschelde was cut off from other estuaries, and river flow was then regulated by a sluice complex in the Volkerak dam. Between 1980 and 1987 a storm-surge barrier was constructed across the mouth of the Oosterschelde, and two compartmentalisation dams were built in the eastern part of the basin. The storm surge barrier decreased the effective cross-sectional area at the mouth from 80 000 m^2 to 17 900 m^2. The compartmentalisation dams reduce the surface area of the Oosterschelde basin from 452 km^2 to 351 km^2. The hydraulic impact of the works is demonstrated with data from field measurements and data from model calculations.

Introduction

The Delta Project has changed the hydrodynamic regime of the Oosterschelde (Nienhuis & Smaal, 1994). Four phases can be distinguished:

1. The period 1958–1964: in 1958 work started on the Delta Project and by 1964 the Grevelingen dam had been completed, cutting of the northern reach of the Oosterschelde from the Grevelingen estuary (Fig. 1).
2. The period 1964–1969: in the spring of 1969 the Volkerak dam was completed and incorporated sluices to regulate river flow.
3. The period 1969 to mid-1985: preparations for the Oosterschelde project began, including the construction of work islands in the mouth and eastern part of the basin. The effect on the tide was mainly local.
4. The period mid-1985 to April 1987: the Oosterschelde project was completed. The final stage of the Delta Project consisted of construction of a storm surge barrier in the mouth of the Oosterschelde and two compartmentalisation dams in the eastern part of the basin. This period saw the most dramatic changes in hydrodynamics.

The impact of the Delta Project on hydrodynamic conditions is demonstrated by data from field measurements. Several discharge measurement surveys have been carried out: in 1959; in 1968 following completion of the Grevelingen dam; in 1972 after completion of the Volkerak dam (see Fig. 1); between 1980 and 1984, preceding the final civil works on the Oosterschelde project; and between 1987 and 1990 under the new regime. In addition to field data, data from model calculations are used to illustrate the changes in the period mid-1985 to April 1987.

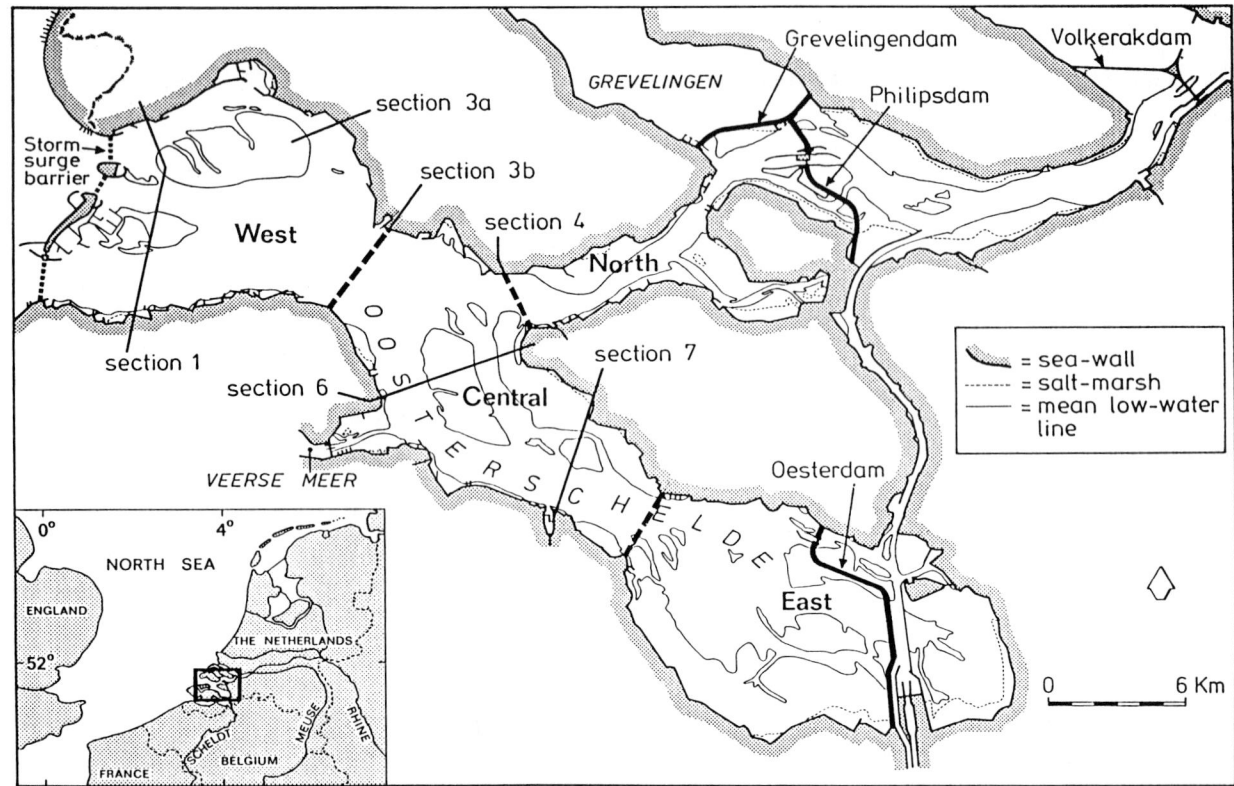

Fig. 1. Measured cross-sections during the discharge measurement surveys. Discharge measurements were carried out near the mouth (section 1), in the northern tidal channel in the western part, the Hammen channel (section 3a), at the end of the western part where the Oosterschelde narrows (section 3b), in the central part (section 6) and at the entrances of the northern and eastern parts (sections 4 and 7). Discharge measurements were carried out in the years 1959, 1968, 1972, in the periods 1980–1984 and 1987–1990.

Another factor responsible for hydraulic change is the general rise in sea level. Records for the period 1900–1980 indicate a relative rise in sea level of about 0.25 m and a natural increase in tidal amplitude of 3% to 4% (tidal records for Hook of Holland and Vlissingen, De Ronde, 1983). Sea level rise over the relatively short history of the Delta Project is therefore assumed to be negligible and secondary to the effects of the 18.6 year cyclic tidal variation and the engineering works.

With the closure of the Oester dam (November 1986) and the Philips dam (April 1987), compartmentalisation became a reality and the present hydrodynamic conditions have pertained ever since.

Mean high water outside the barrier is currently NAP (Dutch Ordnance Datum) + 1.54 m, and mean low water NAP −1.30 m. The storm surge barrier creates a discontinuity; the tidal range is greater on the seaward than the landward side, because of the resistance of the barrier. Tidal amplitude increases eastwards from the barrier, from a mean of 2.47 m near the barrier to 2.98 m in the northern branch and 3.39 m at the southeast end near the Oester dam.

Total mean tidal volume is about 880 million cubic metres. Maximum velocities on ebb and flood vary on average tides from 0.35 m s^{-1} in the northern branch, 0.75 m s^{-1} in the eastern part, and up to 1.0 m s^{-1} in the western part.

Freshwater discharges from the Oosterschelde are regulated; there is no natural river regime. The average freshwater load from sluices and from polder runoff is about 25 m^3 s^{-1} (Smaal & Boeije, 1991).

The residence time of the water is about 20 tidal periods near the mouth, 100 in the central part and roughly 200 near the Oesterdam and Philipsdam (Smaal & Boeije, 1991).

Material and methods

The hydraulic impact of the Delta works is demonstrated using field data of water levels and discharges. Data from model calculations are also used to show the influence of the most recent and most significant civil works in the Delta scheme, the Oosterschelde project. Both methods of data collection are described in this section.

Field data

Measurement of stage
Before 1980, there were only a few autographic water level recorders. Data were collected by observers and processing involved many steps. The statistics were presented as tables and graphs in an annual report.

Oosterschelde water levels have been monitored automatically since 1980, sometimes together with data on wind, temperature, salinity, flow velocities and waves. Telemetry equipment records the water stage at various locations every ten minutes and transmits the data to a central database. Here the records are validated, corrected if necessary, and stored. Application programmes enable the data to be presented in a variety of formats, and basic statistical analyses to be carried out.

The number of stations in the hydrometric network and the parameters they measured were constantly adjusted during the Oosterschelde project to correspond to the stage of the works.

Discharge measurements
Discharge measurements are made at strategic cross-sections, as shown in Fig. 1: near the mouth (section 1); in the central part (section 6); at the end of the western part where the Oosterschelde narrows (section 3b); and at the entrances to the northern and eastern parts (sections 4 and 7).

More detailed measurements were also made in the northernmost tidal channel in the western part, the Hammen channel (section 3a).

Variations in flow velocity and flow direction are measured in space and time across a section. Discharge during the flood and ebb period can then be calculated by integrating over the cross-section and with respect to time.

Most flow measurements are made with Ott-type propeller current meters operated from stationary vessels. Flow velocity is measured simultaneously at different depths at fixed positions along the section during a tidal cycle. The number of boats required depends on the width of the section and the bottom geometry. If the geometry is irregular, sharp velocity gradients occur and the boats have to be closer together. In the last discharge measurement survey carried out between 1987 and 1990, the number of boats varied from 5 in section 4 to 16 in section 1.

With so many boats required, measurements have to be made on each cross-section in turn. This means that the tidal conditions differ. To overcome this, two sets of gaugings are made under different tidal conditions (neap and spring tides). A linear relationship is then established between discharge and tidal range.

This rating curve is good enough to convert the measurements at different cross-sections to the same tidal conditions. The reliability of the rating curve correlating tidal range with tidal discharge was further improved during the Oosterschelde project by measuring a complete neap-spring tidal cycle. Automatic current meters (improved Flachsee) capable of storing data for long periods were used. The instruments are set at a single fixed depth across the section, which reduces the accuracy of the discharge measurement. The results are then corrected by correlation with a standard discharge measurement conducted from boats during the same period. Figure 2 shows the results before and after completion of the storm surge barrier.

Model calculations
Model calculations make it possible to generate missing data in space and time. Natural bound-

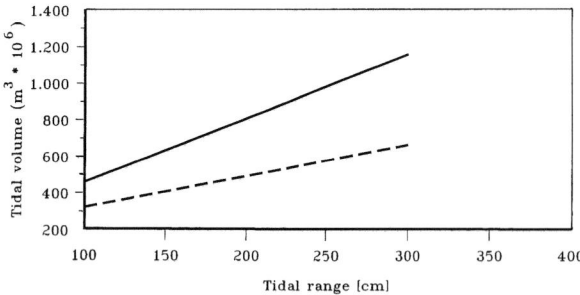

Fig. 2. An example of a result of a discharge measurement with automatic current meters (improved Flachsee). Correlation for the ebb period between tidal volume in section 3b and the tidal range on the North Sea. The solid line shows the relationship for the period 1980–1984 (corr. coeff. = 0.95) and the broken line the relationship for the period 1987–1990 (corr. coeff. = 0.90).

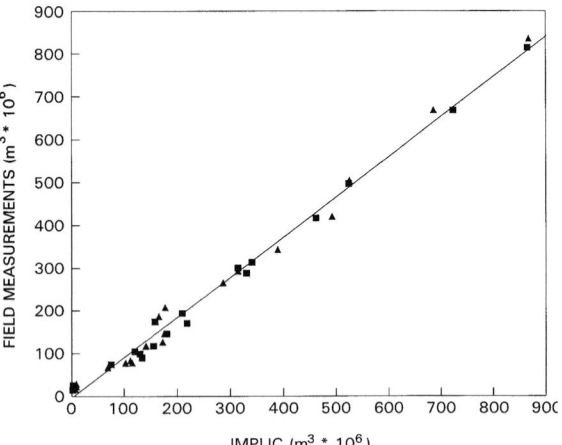

Fig. 3. Results of calculations with a one-dimensional tidal model (IMPLIC) model versus field measurements. Squares = ebb tidal volume; triangles = flood tidal volume).

ary conditions can be kept constant for all simulated project stages, avoiding differences in tidal conditions and meteorological effects. Field data were used to calibrate one-dimensional and a two-dimensional numerical tidal models (IMPLIC and WAQUA).

IMPLIC is relatively simple to handle. The time-dependent setting of the 62 gates in the storm surge barrier and the dimensions of the closure gaps in the compartmentalisation dams can be modelled simply and quickly. IMPLIC gives reliable simulations of tidal discharge and tidal range during work on the Oosterschelde project. Figure 3 shows IMPLIC results compared with field data.

WAQUA, on the other hand, provides a more detailed and accurate simulation of flow velocities

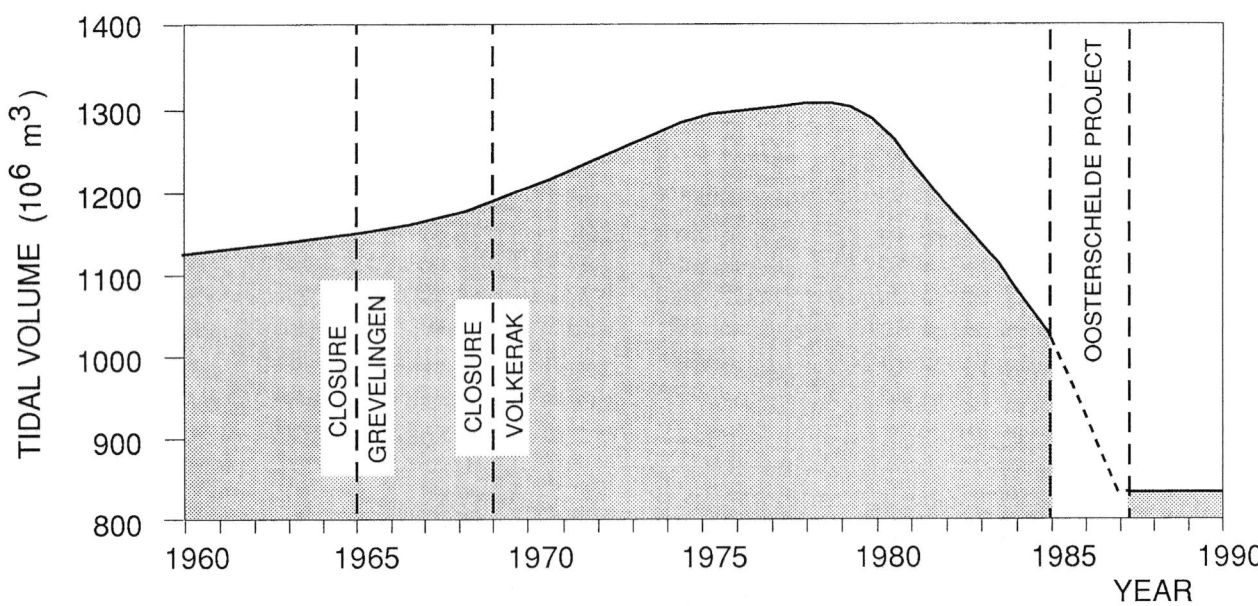

Fig. 4. Change in tidal volume at the entrance of the Oosterschelde in recent decades (Louters *et al.*, 1993).

in channels. The model is more difficult to handle and modelling takes longer.

Bottom geometry is kept constant for all simulations. From a morphological point of view, the project period is short and it can be assumed that no significant changes have taken place in bottom geometry.

Results

The change in tidal volume in recent decades at the entrance to the Oosterschelde is a good indicator of the hydraulic effect of the Delta works. Figure 4 (Louters *et al.*, 1993) shows the change with time. There was an increase in tidal volume after the completion of the Grevelingen and Volkerak dams, and a relatively large decrease once the Oosterschelde project was completed.

The Grevelingen and Volkerak dams (1964 and 1969)

In 1964, the Grevelingen dam separated the Grevelingen from the Oosterschelde. As a result flood peaks no longer reached the northern branch. The data in Fig. 5 (Louters *et al.*, 1993; Van den Berg, 1986) show that after the construction of the Grevelingen dam the flood volume in the mouth of the Oosterschelde and in the entrance to the northern branch increased from respectively 1130 and 195 million cubic metres in 1959 to 1180 and 250 million cubic meters in 1968. This increase could not have been due to the 18.6 year tide cycle which produces a variation in amplitude of about 3% of the tidal range all along the Dutch coast (De Ronde, 1983), since 1968 was just one year before a minimum value in this cycle. The increase can be attributed instead to the construction of the Grevelingen dam, which effectively added part of the tidal prism of the Grevelingen to the Oosterschelde system.

Figure 5 also shows that after the closure of the Volkerak dam the tidal wave in the northern branch amplified. This was due to resonance. The result was a further increase in the tidal prism.

The Oosterschelde project (completed 1987)

The final phase of the Delta Project was the construction of a storm surge barrier across the mouth of the Oosterschelde, and two compartmentalisation dams in the eastern part of the basin.

The storm surge barrier effectively decreased the cross-sectional area at the mouth from $80\,000$ m^2 to $17\,900$ m^2. Initially this was compensated for by an increase in flow velocity, and tidal discharge decreased very little. However, at about $35\,000$ m^2 a critical cross-sectional area was reached and decreases in both tidal discharge and tidal range became noticeable. The final reduction to $17\,900$ m^2 cross-sectional area represented a decrease in total tidal volume and tidal range (at Yerseke) of about 25%.

Such a large reduction in tidal range would be unacceptable from an ecological point of view and for the shellfish culture. For this reasons and to create a stagnant shipping route from Rotterdam to Antwerp, the surface area of the basin behind the barrier was reduced from 452 km^2 to 351 km^2 by building two compartmentalisation dams. These further decreased the tidal volume, but tidal range increased. After completion of the works, the total reduction in tidal volume was 30% at the mouth (25% due to the barrier and 5% to compartmentalisation), but the reduction in tidal range was only 13% (at Yerseke).

The hydraulic impact of the Oosterschelde project is not uniform. Figure 6 shows spatial variations in the reduction of tidal range and in the increase of the residence time.

Significant changes in geomorphology would not be expected over the relative short period of the Oosterschelde Project. Current velocities have therefore decreased in proportion to the decrease in discharge. This is shown in Fig. 7, which also shows a difference in the reductions in flow velocities between the northern and southern part of the basin, and between flood and ebb. Maximum reductions occur in the northern part (the Hammen channel and the northern branch) and during the ebb state of the tide. In Fig. 9 this is shown for the Hammen area, which is an important culture site of mussels.

22

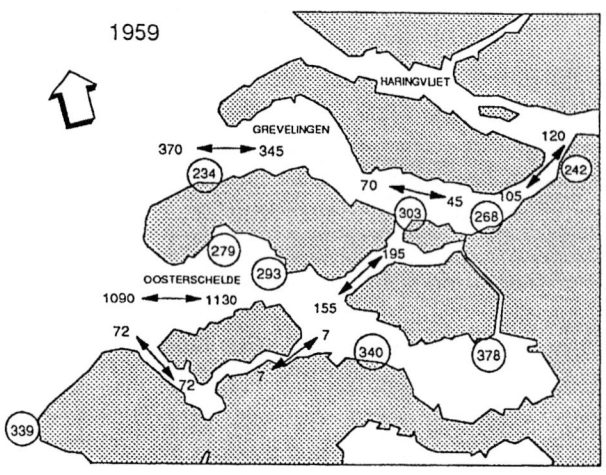

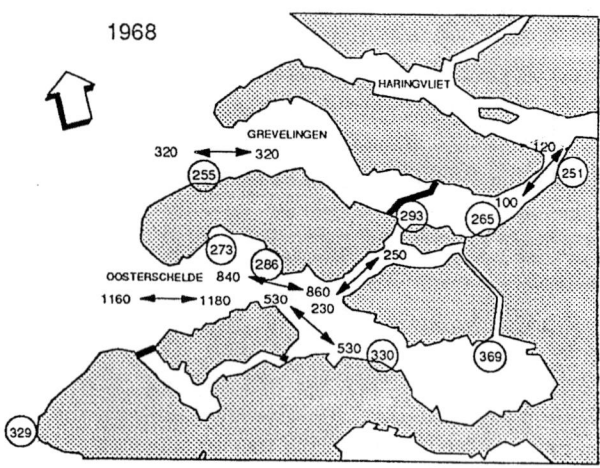

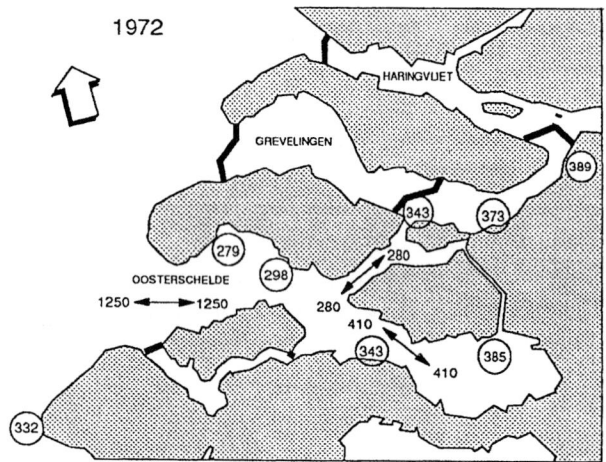

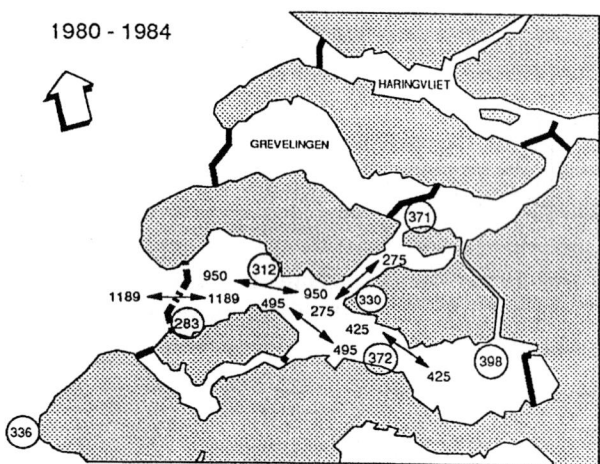

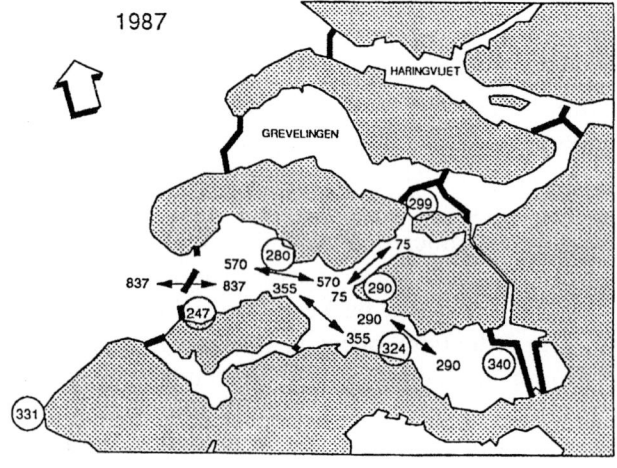

(280) TIDAL RANGE (cm)

837 EBB OR FLOOD
DISCHARGE ($\times 10^6$ m^3)

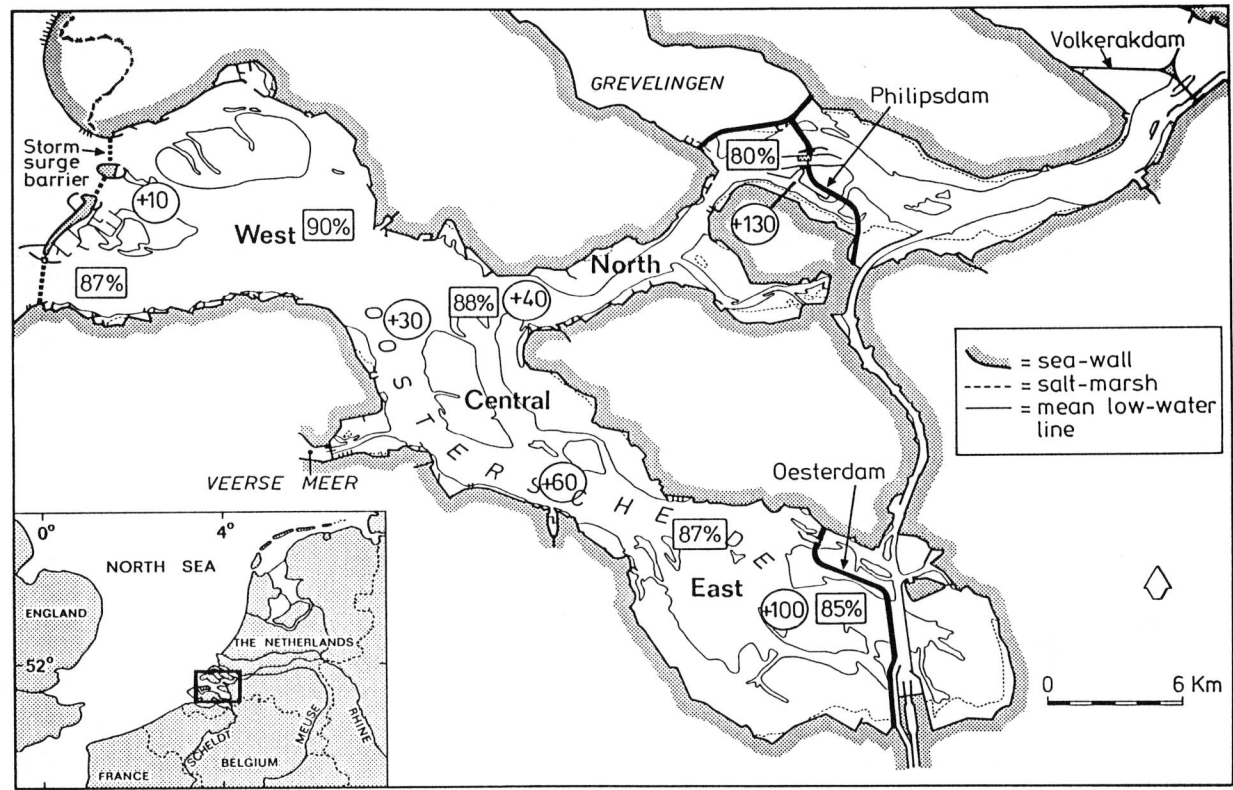

Fig. 6. The tidal reduction expressed as % and the increase in residence time expressed in tidal periods as a result of the Oosterschelde Project. Number in square = reduction in tidal range [%]; number in circle = increase in residence time [tidal periods].

The compartmentalisation dams were originally planned for completion simultaneously with the storm surge barrier. In the event, this changed for reasons of cost. The barrier was built first, so that it could be used to reduce flow velocities in the closure gaps of the compartmentalisation dams in the final stages of closure. This meant that the compartmentalisation dams could be sand rather than more expensive rockfill (Vroon *et al.*, 1988).

This phased construction produced several hydraulic sub-stages between mid-1985 and April 1987.

Mid-1985 to mid-1986

In this period there was a marked reduction in tidal range. By mid-1985, the cross-sectional area of the mouth of the Oosterschelde had reached the critical 35 000 m². From then on the reduction in tidal range accelerated. The reduction in tidal range ceased in mid-1986 when the cross-sectional area reached 17 900 m².

Mid-1986 to October 1986

This was a period of maximum reduction of tidal range. The storm surge barrier was in place, but the capacity of the basin behind had not yet been

---

Fig. 5. Results of discharge measurement surveys for 1959, 1968, 1972, 1980–1984 and 1987. 1959: initial situation; 1968: situation after the construction of the Grevelingen Dam (1964); 1972: situation after the construction of the Volkerak Dam (1969); 1980–1984: situation before the completion of the Oosterschelde Project; 1987: situation after the completion of the Oosterschelde Project (Louters *et al.*, 1993). Number in circle = tidal range [cm]; number with arrow = ebb or flood discharge [10⁶ m³].

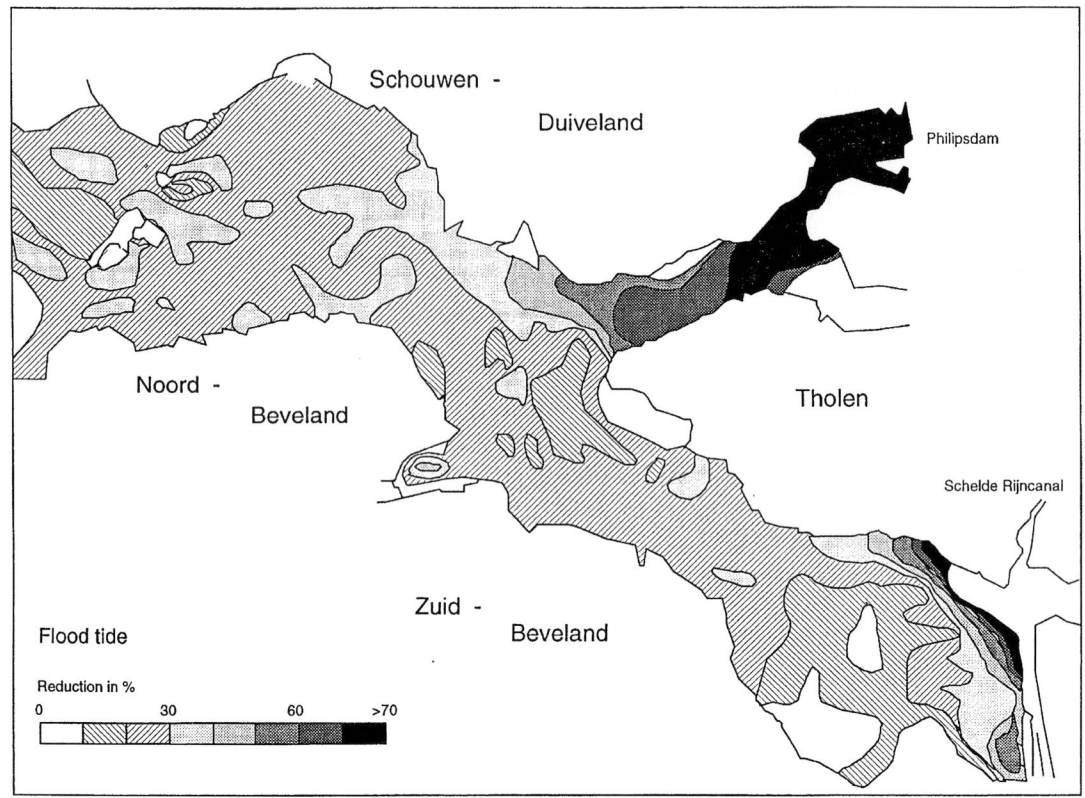

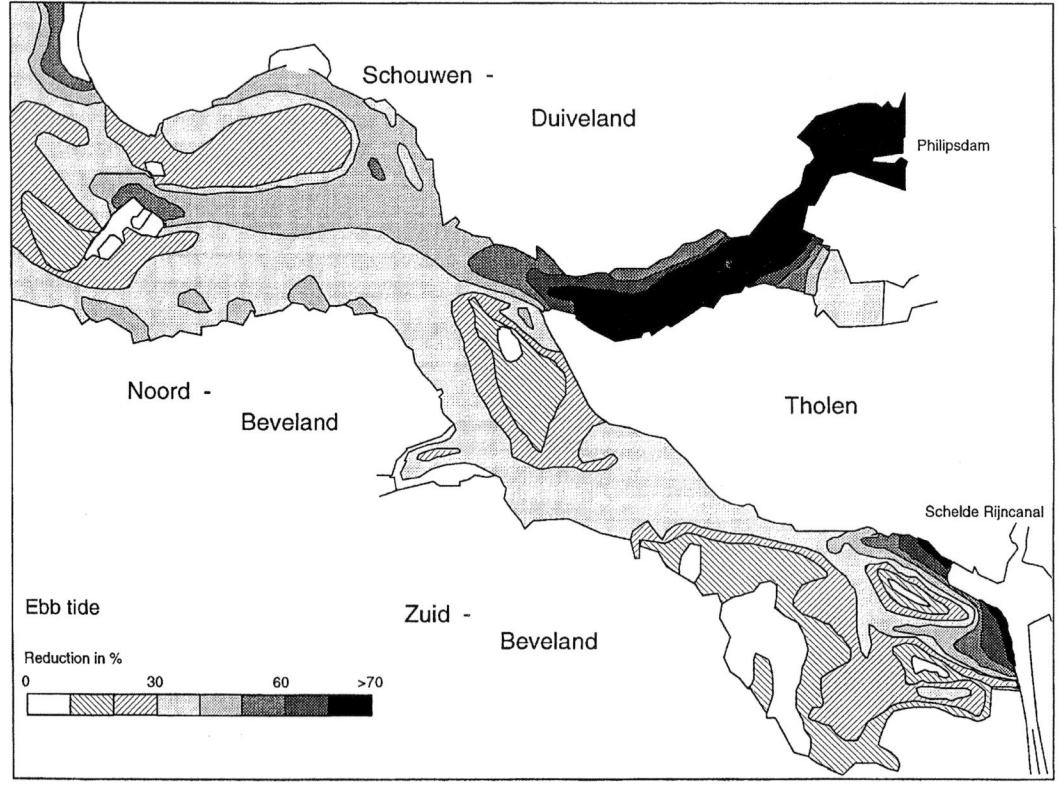

Fig. 7. The reduction in maximum flow velocities expressed as % as a result of the Oosterschelde Project.

decreased. In October 1986, the first of the two compartmentalisation dams – Oester dam – was finished.

October 1986 to April 1987

Although compartmentalisation was not yet complete and the basin surface area remained at 452 km², a slight increase in tidal range was predicted for this period. With the Oester dam finished, impounding of the Oosterschelde behind the compartmentalisation dams could only take place through a relatively small gap in the Philips dam. This means that, although the basin surface area remained the same, the basin capacity decreased.

In practice, however, there was no increase in tidal range because the barrier gates had to be partially lowered. This was necessary to reduce unexpectedly high flow velocities in the channel behind the compartmentalisation dams which had originally connected the northern and eastern parts of the basin. The Oester dam severed the connection and impounding could only take place from one direction. The large increase in flow velocity had not been predicted.

Table 1 shows changes in tidal volume reconstructed using a numerical tidal model (IMPLIC). Figure 8 shows the stages in the civil works and the hydraulic consequences. Hydraulic changes are represented by actual tidal range and theoretical tidal range. The difference is the reduction

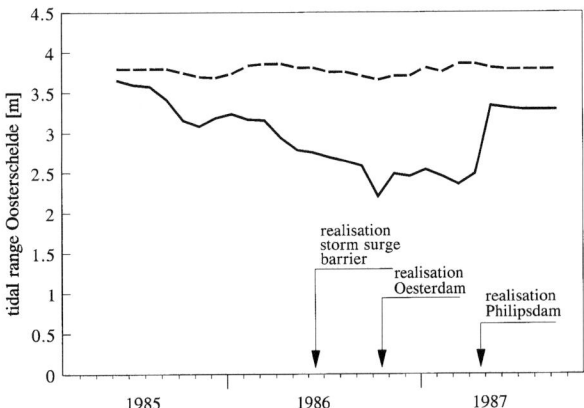

Fig. 8. The hydraulic consequences of a phased planning schedule. Solid line = actual tidal range Oosterschelde. Broken line = theoretical tidal range Oosterschelde (without works).

resulting from the engineering works. The theoretical tidal range is based on a correlation between a North Sea location unaffected by the works and an Oosterschelde location, during a period when the effect of the works was still negligible.

Discussion

The phased construction of the storm surge barrier and compartmentalisation dams of the Oosterschelde project meant relatively marked reductions in tidal range between mid-1985 and April 1987. One consequence of reduced tidal ranges is a lower inundation frequency in the higher zones of the estuary. The phasing of the project therefore had – and still could have – ecological consequences (De Jong *et al.*, 1994, De Jong & Van der Pluijm, 1994, de Leeuw *et al.*, 1994).

Once the Oosterschelde project was completed, maximum reductions in current velocity were observed in the northern part of the basin, the Hammen channel and the northern branch, and during the ebb state.

The large reduction in velocities in the northern branch can be explained by the tidal volume reduction from 280 to 80 million m³, due to the construction of the Philips dam.

Table 1. A reconstruction with a one-dimensional tidal model (IMPLIC) of the changes in tidal volume in the different cross-sections during the progress of the Oosterschelde project. (0): situation in 1980 before the start of the Oosterschelde project; (1): situation in mid-1985; (2): situation in mid-1986 (completion of Storm-Surge Barrier); (3): situation in October 1986 (completion of Oester dam); (4): situation in April 1987 (completion of Philips dam, present situation). Sections: see Fig. 1.

	Section 1	Section 3b	Section 4	Section 6	Section 7
(0)	100%	100%	100%	100%	100%
(1)	99%	99%	101%	97%	95%
(2)	75%	76%	78%	73%	70%
(3)	73%	72%	80%	67%	61%
(4)	69%	64%	30%	75%	68%

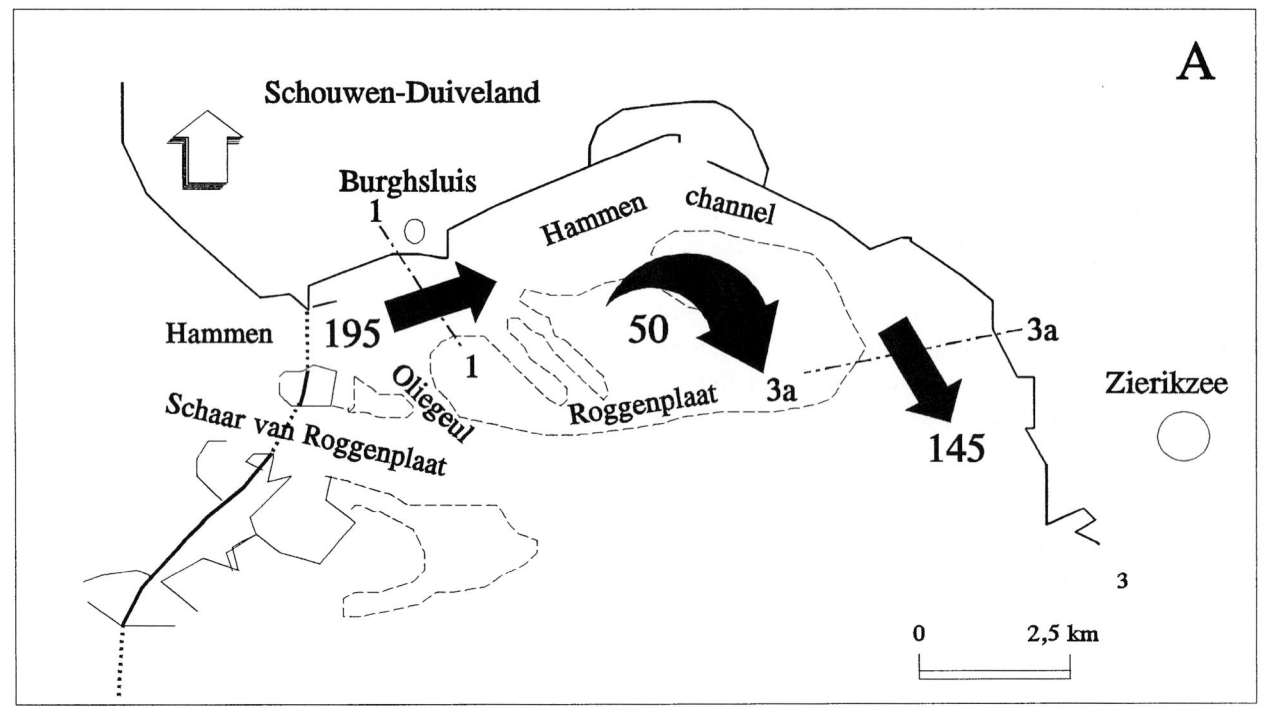

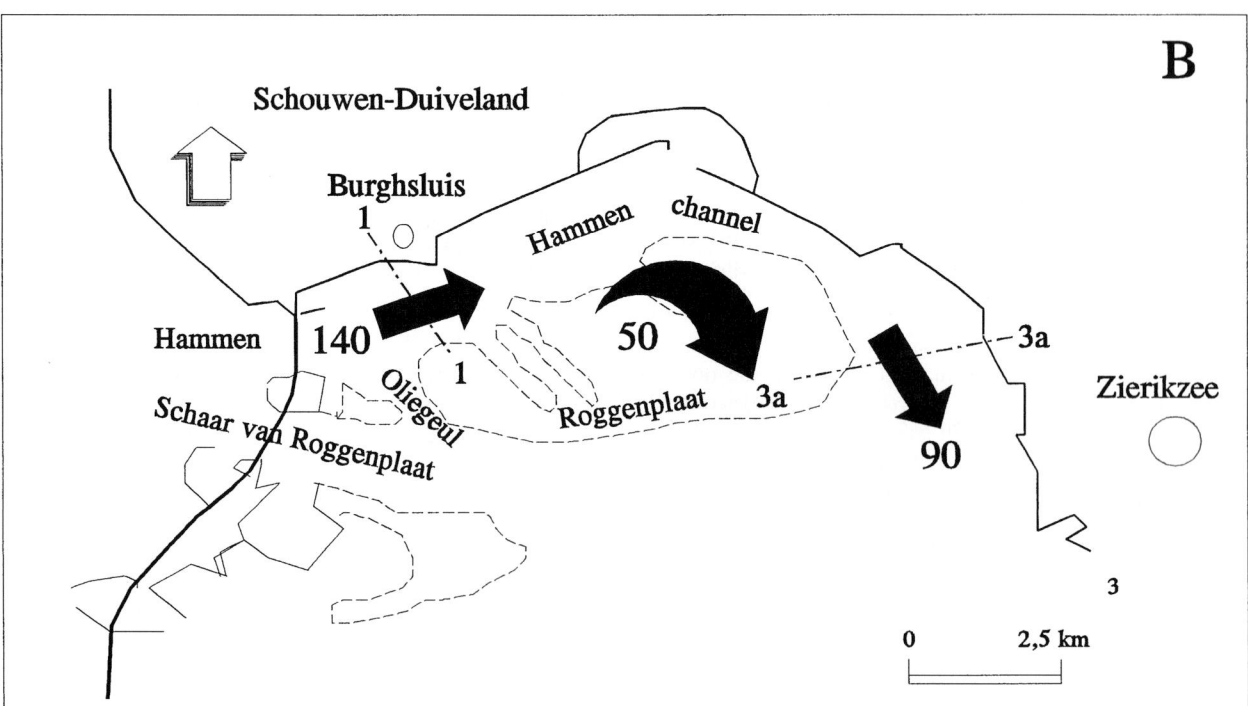

Fig. 9. The amount of water withdrawn from the Hammen channel by the Roggenplaat is relatively high in the new situation. For this reason there is a strong reduction in tidal volume, and therefore in flow velocity, in an easterly direction. A: situation before the Oosterschelde Project; B: situation after the Oosterschelde Project; $140 =$ tidal volume $[10^6 \, \text{m}^3]$; arrow = direction of flow during flood.

The marked velocity reduction in the Hammen channel can be explained by the difference in reduction between tidal range and volume. Tidal volume at the western entrance to the channel is about 30% lower. On the Roggeplaat, however, the reduction in storage is much smaller, because the reduction of high water level is only about 10% in this part of the tidal basin. The effect is an relative increase in the amount of water withdrawn from the Hammen channel by the Roggeplaat, as shown in Fig. 9. The result is a large reduction in tidal volume, and consequently in flow velocity, in an easterly direction.

One effect of the reduced tidal range is an increase in the period that the tidal zone is exposed to wave energy. This could lead to erosion of salt marshes. The effect might be exacerbated when the barrier is closed during a storm surge and the water level is unchanged for several hours.

A reduction in tidal volume also affects the residence time of the water. As expected, model calculations show an increase in this parameter. There is now less flushing with North Sea water. In the eastern part, for instance, residence time has doubled from 1.5 months to 3 months, and is longer than 3 months in the northern part.

Acknowledgements

The author is grateful to T. Louters, J. H. van den Berg and J. M. P. Mulder for permission to use their data.

References

De Ronde, J. G., 1983. Changes of relative mean sea level and of mean tidal amplitude along the Dutch Coast. In: Ritsema A. R. & A. Gurpinar (eds), Seismicity and seismic risk in the off-shore North Sea area, pp. 131–142.

De Jong, D. J., Z. de Jong & J. P. M. Mulder, 1994. Changes in area, geomorphology and sediment nature of salt marshes in the Oosterschelde estuary (SW Netherlands) due to tidal changes. Hydrobiologia 282/283: 303–316.

De Jong, D. J. & A. M. van der Pluijm, 1994. Consequences of a tidal reduction for the salt marsh vegetation in the Oosterschelde estuary (The Netherlands). Hydrobiologia 282/283: 317–333.

De Leeuw, J., L. P. Apon, P. M. J. Herman, W. De Munck & W. G. Beeftink, 1994. The response of salt-marsh vegetation to tidal reduction caused by the Oosterschelde storm-surge barrier. Hydrobiologia 282/283: 335–353.

Louters, T., J. H. van den Berg, J. M. P. Mulder, 1993. Large scale morphodynamic response of the Oosterschelde to implementation of the Delta Project. Journal of Coastal Research (in press).

Nienhuis, P. H. & A. C. Smaal, 1994. The Oosterschelde estuary, a case study of a changing ecoystem: an introduction. Hydrobiologia 282/283: 1–14.

Smaal A. C. & R. C. Boeije, 1991. Veilig getij, de effekten van de waterbouwkundige werken op het getijmilieu van de Oosterschelde. Nota GWWS 91.088. DGW/Directie Zeeland, Middelburg, 132 pp (in Dutch).

Van den Berg, J. H., 1986. Aspects of sediment and morphodynamics of subtidal deposits of the Oosterschelde (The Netherlands). Rijkswaterstaat, Communications no. 43/1986.

Vroon, J., A. van Berk, R. E. A. M. Boeters, R. 't Hart & J. J. P. Lodder, 1988. Logistic support for the sandfill operations in the Eastern Scheldt tidal basin. Oceanology '88: 261–272.

Hydrobiologia **282/283**: 29–39, 1994.
P. H. Nienhuis & A. C. Smaal (eds), The Oosterschelde Estuary.
© 1994 *Kluwer Academic Publishers. Printed in Belgium.*

Changes in basin geomorphology after implementation of the Oosterschelde estuary project

Jan P. M. Mulder & Teunis Louters
National Institute for Coastal and Marine Management/RIKZ, P.O. Box 20907, 2500 EX The Hague, The Netherlands

Key words: coastal engineering, tidal basin, estuaries, geomorphology, ecology, Oosterschelde estuary

Abstract

The completion in 1986/87 of an open storm-surge barrier in the inlet and of secondary dams in the landward parts of the Oosterschelde tidal basin (SW Netherlands) has had and will continue to have a significant impact on geomorphological developments. An analysis of historic data, and of recent detailed bathymetric and morphodynamic process data, indicates that former trends have reversed. At present the Oosterschelde is a sedimentation basin with a degrading intertidal area and silting up of channels. The continuing reduction in intertidal area, the decreasing geomorphological gradients, the increasing fine sediment content of channel deposits, combined with a general reduction in hydrodynamics, imply significant ecological effects.

Introduction

The Oosterschelde Project, completed in 1987, has been the last of a series of large coastal engineering projects aimed at the protection against flooding of the Delta area of the SW Netherlands. In the framework of this project the Oosterschelde basin has been partially closed off from the North Sea by a storm-surge barrier, and the total basin area has been reduced by the construction of dams in the landward parts of the south- and north eastern branches (Fig. 1). The barrier, completed in 1986, will be closed only during severe storm conditions and offers security against flooding. The landward dams, *i.e.* the Oesterdam and Philipsdam completed in 1986 and 1987 respectively, guarantee a tidal range which is meant to be sufficiently large to maintain natural values and fisheries in the basin (Knoester *et al.*, 1984). The Oosterschelde Project has induced abrupt

and large scale changes in hydraulic characteristics of the basin (Vroon, 1994). Significant changes in geomorphology are to be expected. The change in abiotic boundary conditions will affect the total ecosystem (Smaal *et al.*, 1991).

This paper concentrates on the geomorphological development in the basin. Special attention has been paid to the intertidal shoals and flats, representing the most characteristic habitats of the Oosterschelde ecosystem (Saeijs, 1982). The geomorphological development of salt marshes has been described by De Jong *et al.* (1994). An analysis of historic data has been used to illustrate the large temporal and spatial scale of geomorphological processes, and to derive an indicative prognosis of future developments. Detailed bathymetric observations and morphodynamic process data have been analysed to investigate changes at a smaller scale, including effects of implementation of the Oosterschelde Project dur-

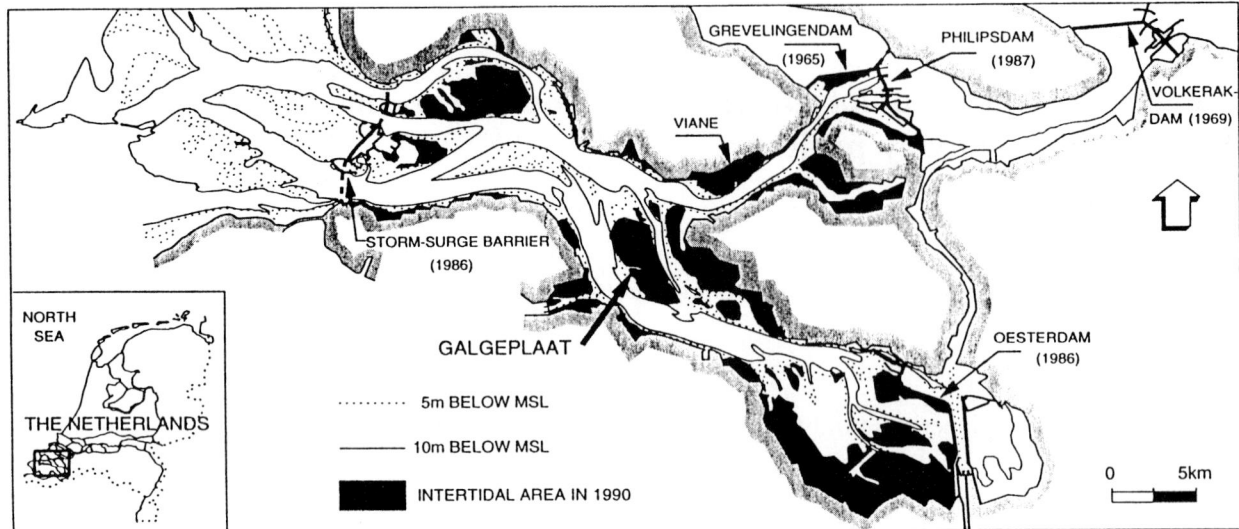

Fig. 1. The Oosterschelde tidal basin in the SW Netherlands; location of the major engineering works, main geomorphological characteristics and location of tidal shoal Galgeplaat and mudflats Slikken van Viane.

ing the last decade. Generally, modifications in geomorphology have been related to hydraulic changes.

Effects of coastal engineering works on estuarine geomorphology have been reported by other authors (e.g. Renger & Partenscky, 1981; Simmons & Hermann, 1972; Kjerfve, 1976; Eijsink, 1990). However, the scale of human interferences mostly is incomparable to the Oosterschelde case; moreover many cases concentrate on changes in basin sedimentation and channel development. Studies into effects on the intertidal area are rare.

Material and methods

Characteristics of the study area

The Oosterschelde tidal basin is part of the Delta area of the SW Netherlands. Implementation of the Oosterschelde Project has given the basin its present boundaries. Since 1987 the Oosterschelde is 351 km^2 large (Fig. 1). The geomorphology of the basin is characterised by tidal channels and intertidal sandy shoals, mud flats and salt marshes, representing a wide range of habitats to the ecosystem (Saeijs, 1982). The characteristics of channels and the intertidal area show a typical spatial variation. In landward direction mean basin depth decreases from *ca.* − 12 mNAP to −4 mNAP; mean channel depths from *ca.* − 30 mNAP to − 10 mNAP (NAP is the Dutch Ordnance Datum corresponding with mean sea level MSL). Maximum heights of the intertidal area amount *ca.* + 1 mNAP. The total intertidal area amounts 109 km^2, including 6.4 km^2 of salt marshes. Grainsizes in the basin range from fine silt (< 53 micron) in recent channel deposits (ten Brinke *et al.*, 1993) to medium coarse sands (250 micron) in more exposed areas. On shoals and tidal flats median grainsizes vary between *ca.* 125 and 210 micron. The hydraulic setting of the Oosterschelde since 1987 is characterised by the following spatial variation. In a landward direction Mean Low Water (MLW) decreases from −1.2 mNAP to −1.5 mNAP; mean tidal range increases landward from 2.5 m to 3.4 m. Maximum current velocities vary between 1-1.5 m s^{-1} in the channels, to *ca.* 0.2–0.4 m s^{-1} in shallow water conditions on flats and shoals. The mean tidal volume of the basin is *ca.* 0.9 million m^3 (Vroon, 1994). Dominant winds, generally occurring during autumn and winter, are from SW to

NW. Wave records over the last decade indicate that for these wind conditions, mean significant wave heights decrease in a landward direction from 0.4 to 0.1 m. The majority of short waves is locally induced by wind. The major wave energy flux is from SW (Louters *et al.*, 1993).

Research methods

Investigations into large scale geomorphological developments (*i.e.* at a basin scale and a time scale of several decades) have been based on an analysis of bathymetric data, combined with hydraulic information (see Louters *et al.*, 1993). Echo sounding maps of the study area have been recorded at yearly intervals since the early 60's. The accuracy in depth data of the sounding maps is estimated to be *ca.* 0.1–0.2 m, depending on waterdepth, bottom slope and mud content of bottom sediments (Louters *et al.*, 1993). From the higher parts of the intertidal area, where no sounding can be managed, levelling data have been collected with an estimated accuracy in heights of *ca.* 0.05 m. These depth data have been used to study cross-sectional developments and changes in sediment volume at 5 -yearly intervals during the period 1960–1990. The latter has been based on calculations from interpolated depth values using 50×50 m and 200×200 m grid squares, derived from the sounding maps. Tidal- and wave characteristics of the basin have been studied from data collected over the last decade at a number of fixed stations along the basin. Discharge measurements in monitoring transects across the basin over the period 1959–1990, have provided data on the spatial distribution of discharges in the Oosterschelde system (*cf.* Vroon, 1994).

A large temporal and spatial scale is inappropriate to study any effects of the, relative to this scale, recent implementation of the Oosterschelde Project. These effects have been investigated at smaller scales, based on an analysis of bathymetric and morphodynamic process data with a high resolution (see Kohsiek *et al.*, 1988; Louters *et al.*, 1993). The bathymetry of the Galgeplaat – a sandy shoal in the central part of the Oosterschelde (Fig. 1) – is being monitored at 43 observation plots, at fortnightly intervals since 1983. The location of the plots has been systematically distributed along 8 transects over the shoal (Fig. 8). The accuracy of sedimentation – erosion measurements at these plots has been estimated to be *ca.* 0.01 m. Morphodynamic parameters (current velocity, sediment concentration, wave height) of the intertidal area have been registered continuously at two stations situated at the Galgeplaat during the period 1983–1988. The sediment exchange between the tidal basin and ebb tidal delta has been estimated from sediment transport measurements in the vicinity of the Storm-Surge Barrier, in 1983 and 1987/1988. From different vessels in a transect, sediment transport has been registered simultaneously using acoustic measuring devices (ASTM). After completion of the Oosterschelde project, continuous sediment transport measurements (especially concerning fine grained sediment < 53 micron) have been carried out at a permanent station situated in the Storm-Surge Barrier itself (Ten Brinke *et al.*, 1994).

Results

Basin development during the last century (1880–1990)

The Oosterschelde's tidal volume has increased considerably at several occasions during the last centuries. An increase of at least 50% has been attributed to large land losses along the basin due to disastrous storm-surges in the 15th and 16th century. A further increase of tidal volume has occurred during the late 19th and early 20th century as an effect of civil engineering works in the hinterland of the basin, each favouring a more landward propagation of tides. Construction of the Grevelingendam (1965) and Volkerakdam (1969) – closing off the Oosterschelde from adjacent basins in the Delta–have resulted in an increase of tidal volume at the inlet of the Oosterschelde of *ca.* 6% (Van den Berg, 1986; Mulder,

1989). Implementation of the Oosterschelde Project has reduced the tidal volume of the basin by 30% (Vroon, 1994).

The Oosterschelde has been an eroding basin for centuries. The total basin erosion during the period 1872–1952 has been calculated to amount *ca.* 350 million m³ (Morra *et al.*, 1961). Parallel to the basin erosion in this period. a distinct sedimentation surplus has been observed at its outer delta, and an increase of geomorphological gradients inside the basin: a gradual deepening of channels and relative expansion of intertidal area (Fig. 2; Van den Berg, 1986; Mulder, 1989). The period 1960–1987 indicates a total sediment loss from the basin of *ca.* 120 million m³ (80 million m³ due to dredging, and 40 million m³ due to 'natural' processes), at the same time the sediment budget of its outer delta shows a surplus of *ca.* 34 million m³ (Louters *et al.*, 1993). Analysis of the grainsize distribution indicates that the major part of the sedimentation surplus in the outer delta is sand. The intertidal area during this period generally has increased in height (Fig. 3; Louters *et al.*, 1993), the tidal channels have further deepened and widened (Van den Berg, 1986). On a basin scale, over the period 1983–1989 no significant erosion is observed from hydrographic maps. In accordance with this observation, trans-

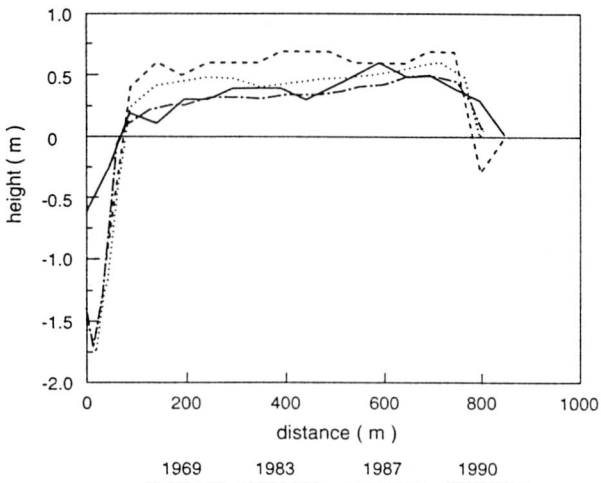

Fig. 3. Changes in a cross section over the shoal Slikken van Viane (*cf.* Fig. 1) during the period 1969–1990, generally indicating relative to Mean Sea Level (height = 0): a net accretion between 1969 and 1983, and a net erosion since 1983.

port measurements indicate that sand exchange between basin and outer delta has dropped dramatically in this period (Fig. 4). Observations on transport of fine sediments (particle sizes < 53 micron) show that a former export trend, after completion of the barrier has reversed into a small but significant, net yearly import of 0.5–1.0 million m³ (Ten Brinke, 1991; Ten Brinke *et al.*, 1994).

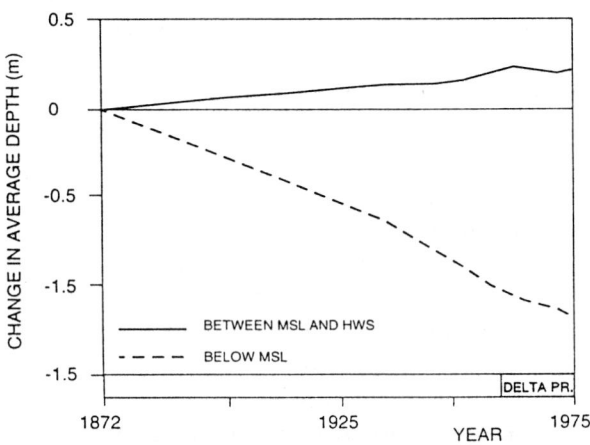

Fig. 2. Change in average depth of channels (below Mean Sea Level, MSL) and of the higher intertidal area (between MSL and High Water Slack, HWS = High Water Neap tides) in the Oosterschelde over the period 1872–1975. The period of the Delta Project is indicated. (modified after Mulder, 1989).

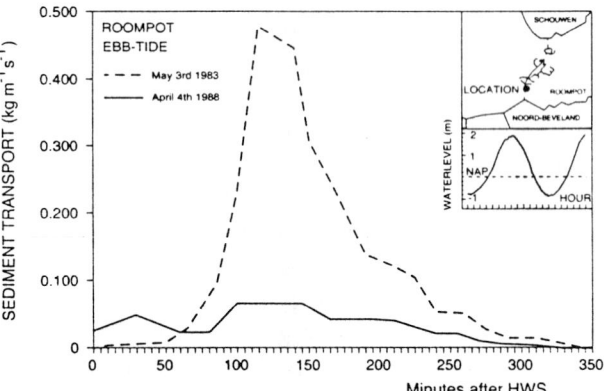

Fig. 4. Net sediment transports observed in 1983 and 1988, at a location near the storm surge barrier. HWS = High Water Slack. (from Louters *et al.*, 1993).

Channels, sandy shoals and mud flats during the last decade (1980–1990)

Construction activities of the storm-surge barrier and the landward dams between 1979 and 1987, have had immediate effects on hydraulic characteristics of the basin (Fig. 5; Vroon, 1994). During 1985 and 1986 progress in the barrier construction process gradually had diminished the cross sectional area of channels in the inlet of the basin. As a result also the tidal volume, tidal current velocities and the tidal range inside the basin, gradually decreased. Around the middle of 1986 these parameters had reduced to 75% of their original values, characteristic of the so called transition period (1985–1987). Completion of the Philipsdam in April 1987 has changed the hydraulic parameters to their present values: tidal volume and tidal current velocities presently amount to *ca.* 70% of their original values, the tidal range to *ca.* 87%. The reduction in tidal range has lead to a change in inundation frequency of the intertidal area. In the higher parts (above 0.5 mNAP) the inundation frequency has decreased by 2–5%, in the lower parts increased by 2–5%. Wave characteristics in the channels have not changed significantly due to the engineering works. However, the reduction of tidal range has resulted in a concentration of wave energy dissipation in a smaller vertical zone of the intertidal area.

Tidal channels at present, generally show a sedimentation trend as has been concluded from

widespread unconsolidated fine sediment layers (Ten Brinke, 1991; Ten Brinke *et al.*, 1994). Channel erosion is observed only very locally; most pronounced in this respect is the development of scour pits at both sides of the storm surge barrier since 1983/1984. This effect is due to the locally very large increase of velocity and turbulence of the tidal currents. The depths of the scour pits have increased in the order of 10 m between 1983 and 1991; no equilibrium has been established yet. The total amount of sediment eroded from the scour pits inside the Oosterschelde basin during this period, has been estimated to be *ca.* 3–5 million m^3. Most of this material has been deposited in oblong wedges in the channels at the landward side of the pits.

Sandy shoals and mud flats throughout the basin generally show a distinct erosion trend over the period 1983–1990. Levelling data indicate a lowering of surface levels with magnitudes between 0.1 and 0.2 m (*cf.* Fig. 3). Erosion appears to be dominant at each level in the intertidal zone (Fig. 6). Relatively, the highest erosion figures are shown by the higher shoal parts (above MSL), where over 20% of the original sand volume has been eroded since 1983. In an absolute sense, the largest amounts of sand have been lost from depth zones between MLW and MSL (= NAP). The shoal erosion is inducing a shift in available surface area per depth zone: above *ca.* − 0.5 mNAP surface areas have decreased since 1983, whereas below this level surface areas per depth zone have increased (Fig. 6). This development illustrates the general decreasing trend in depth gradients. The change in surface areas per depth zone, combined with the change in inundation frequencies induced by the reduction in tidal range, has resulted in local reductions in availability of feeding grounds for birds of 10 to 30% (Schekkerman *et al.*, 1994).

Local variations on a sandy shoal during respective years in the period 1980–1990

In shallow water conditions at sandy shoals and mud flats, the reduction of maximum current ve-

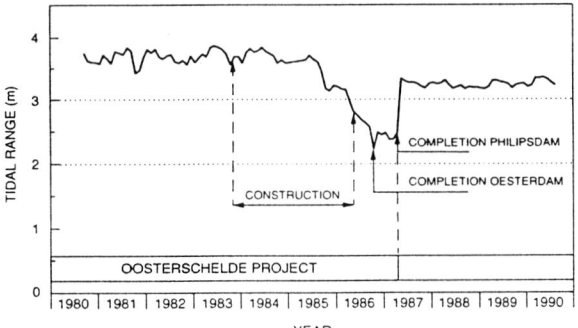

Fig. 5. Trend in mean tidal range in the central part of the Oosterschelde during the past decade, reflecting the direct effects of the engineering works.

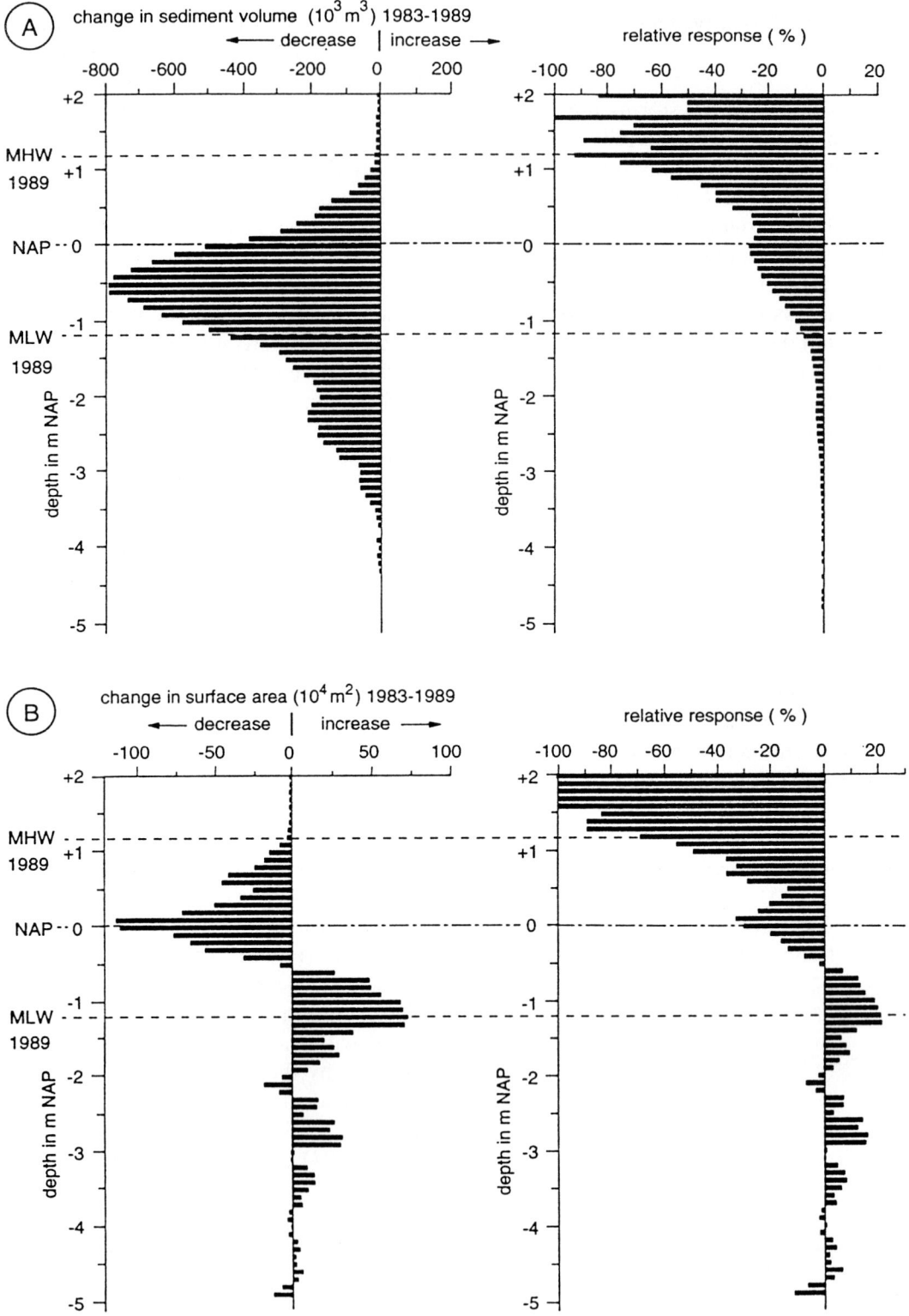

Fig. 6. Changes in sediment volume (A) and in surface area (B) of the Oosterschelde basin between 1983–1989, per depth interval of 0.1 m.

locities – common during flood – has been relatively larger than in channels. Flood velocities in channels generally have reduced to 70%, on shoals, from 70% down to 40% of the original values (De Vriend *et al.*, 1989; Vroon, 1994). Ebb velocities on shoals – dominated by water level gradients and phase differences between shoal and channel – have not changed significantly after completion of the engineering works (Fig. 7). Related to the velocity reductions during flood, the sediment transport capacity has dropped exponentially between 1984 and 1987 (Fig. 7; Louters *et al.*, 1993).

Sedimentation and erosion at an individual sandy shoal – the Galgeplaat – show characteristic spatial variations between locations at the upper and lower parts of the shoal, and between more or less wind – and wave exposed edges (Kohsiek *et al.*, 1988; De Vriend *et al.*, 1989). Until the middle of 1985, before implementation of the Oosterschelde Project had had any significant effect on hydraulics, this variation is illustrated by the typical distribution of sedimentation – and erosion dominated plots (Fig. 8). In this period the overall annual mean of all 43 observation plots on the shoal has shown a slight sedimentation surplus. Since 1985 the sedimentation

– erosion pattern is showing far less variation: erosion has become dominant at most locations and the overall sedimentation surplus has reversed into a net erosion of the shoal (Fig. 8).

Discussion

Long term basin development

The large-scale channel erosion of the Oosterschelde basin during the last centuries, in response to a successively increasing tidal volume, is illustrating an important morphodynamic mechanism. The tidal volume of a tidal inlet under conditions of geomorphological equilibrium, generally shows a linear relationship with the cross sectional area of its entrance (Fig. 9; O'Brien, 1969; Van de Kreeke & Haring, 1979). Similarly such relationship holds for individual channels within the basin (Gerritsen *et al.*, 1990; Van den Berg, 1986). The main implication of this relationship is that, in order to reach a new morphodynamic equilibrium, any significant shift in hydraulic conditions has to result in a proportional adaptation in channel cross section: erosion after a tidal volume increase, sedimentation after a decrease. Between

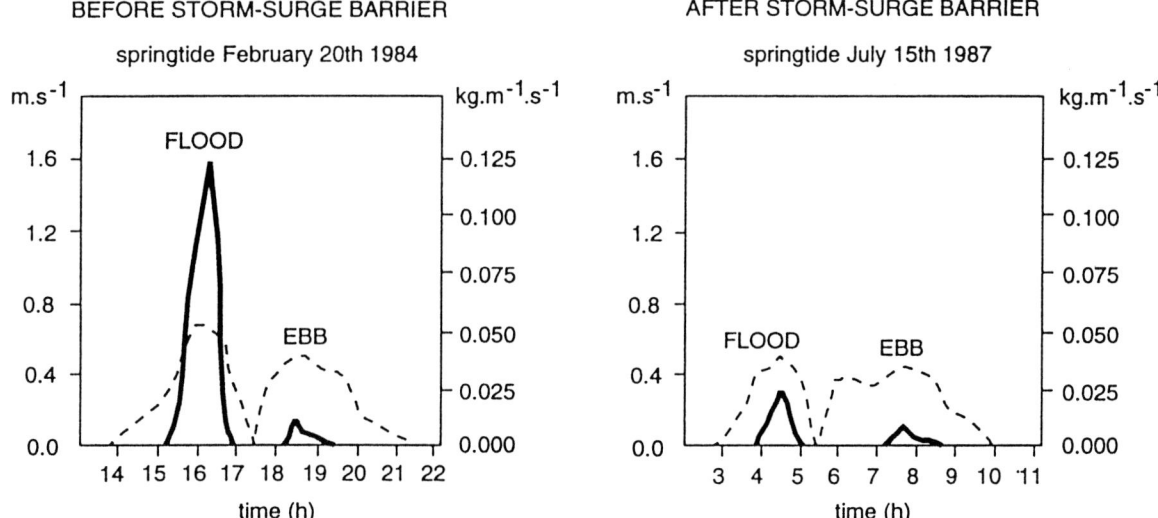

Fig. 7. Current velocity (left vertical axis; interrupted line) and sand transport (right vertical axis; solid line) in 1984 and 1987, as observed at a station near the NW edge of the shoal Galgeplaat (*cf.* Fig. 1) (from Louters *et al.*, 1993).

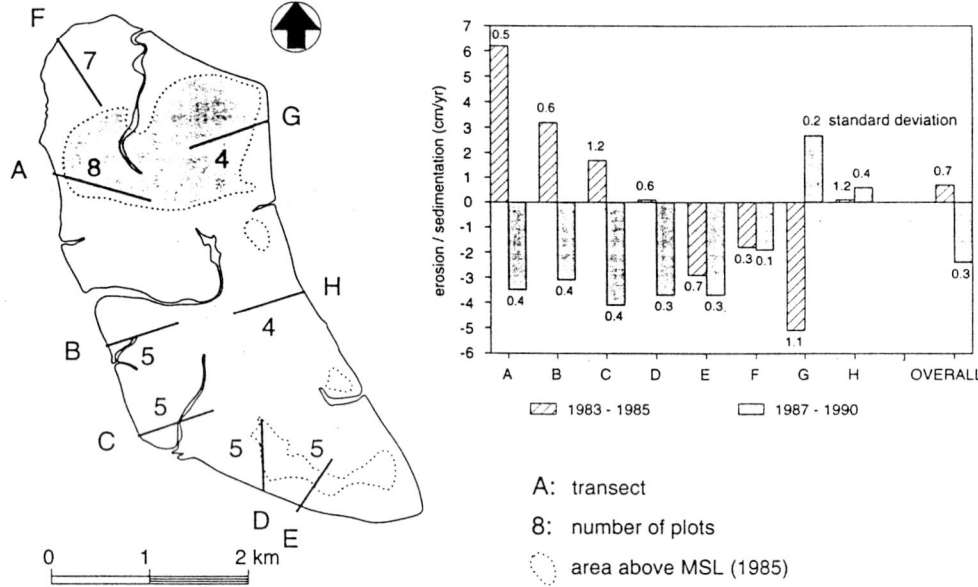

Fig. 8. Galgeplaat with location of 8 transects with 43 plots for detailed sedimentation – erosion observations. Annual mean and standard deviation of change in depth at the 8 Galgeplaat transects including an overall mean value, during the period before (1983–1985) and after (1987–1990) completion of the Oosterschelde Project.

1960 and 1970 the construction of the Greve-lingen- and Volkerak dams have induced a tidal volume increase in the Oosterschelde inlet. The system, close to the morphodynamic equilibrium in 1960, responded to this by channel erosion resulting in an increase of cross sectional areas between 1970 and 1980 (Fig. 9). The equilibrium relation shows that the cross sectional area of the Storm-Surge barrier itself is far too small to allow an equilibrium with the tidal volume. This illus-

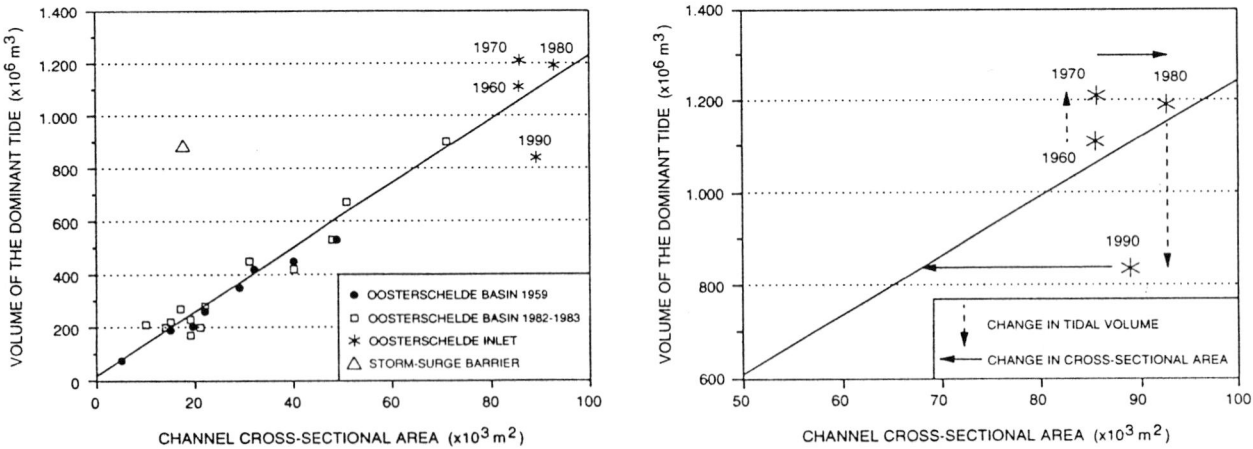

Fig. 9. Morphodynamic equilibrium relationship between tidal volume and cross sectional area for different tidal inlets (modified after O'Brien, 1969 and van den Berg, 1986), indicating the effects of changes in tidal volume and cross sections. The tidal volume increase in the Oosterschelde between 1965 and 1970, has induced a disturbance of the equilibrium leading to channel erosion between 1970 and 1980. Since 1985, the decrease in cross section at the Storm-Surge Barrier induces the development of scour pits, while the decrease of tidal volume induces sedimentation in the Oosterschelde channels.

trates the need for the extensive bottom protection works over 600 m at both sides of the barrier, and it explains the scour pits in the adjacent unprotected areas. Between 1985 and 1987 the tidal volume of the Oosterschelde has decreased by 30%. The given relationship indicates that the system will tend to adapt to the new situation by sedimentation. Based on the same relationship, the total amount of sediment needed to reach a new morphodynamic equilibrium in the Oosterschelde channels has been calculated to be *ca.* 400–600 million m³.

A sediment amount with a similar order of magnitude has been eroded from the Oosterschelde basin as an effect of tidal volume increase, over the period of one century (1872–1987). From this, an indication may be derived of the time required to establish a new morphodynamic equilibrium in the Oosterschelde after the recent tidal volume decrease. First, it has to be realized that the rate of geomorphological changes in a tidal basin with non-cohesive sediments, is related to the magnitude of sediment transport capacity by tidal currents; transport capacity is a power function of current velocity, with a power of magnitude 2–5 (e.g. Van Rijn, 1987; Baillard, 1981). Thus, after an increase of tidal current velocities resulting from an increase in tidal volume, the rate of geomorphological changes increases exponentially. The opposite occurs after a decrease. Then, if the transport of *ca.* 400 million m³ sediment in the former situation has taken one century, it follows that a transport of similar magnitude, in the latter (present) situation will take several centuries.

An illustration of the exponential reduction in transport capacity after 1987, is represented by the dramatic decrease in sand transport as has been observed both inside the basin (Fig. 7), and at the basin entrance (Fig. 4). The latter may seem contradictory as at this location current velocities have increased considerably due to barrier construction. However, this increase is mainly effective downstream of the barrier – both during flood and ebb –, thus creating flood dominated scour pits landward of the barrier and ebb dominated scour pits seaward. Upstream of the barrier, the

flow reduction over these pits might prevent any substantial sand transport through the barrier. Recent observations however, cannot proof that this mechanism up till now has been effective yet (Louters *et al.*, 1993). Both upstream and downstream of the pits, negligible sand transports have been registered. Obviously, transport capacities in the channels at the ebb tidal delta seaward of the storm-surge barrier, presently are too low, to allow a substantial sand transport in the direction of the basin. The ebb delta channels show, comparable to channels inside the basin, a large need of sediment. Until a new morphodynamic equilibrium has been established on the ebb tidal delta, any significant sand import into the basin is unlikely. This implies that external sand sources to meet the sediment needs of the Oosterschelde channels in the present situation, are negligible; for this purpose only internal sources remain *i.e.* sands from the intertidal areas. The general erosion of sandy shoals and mud flats, observed since 1987, illustrates that internal sources actually are supplying sediment to the channels.

Assuming that sediment import to the basin over the next 30 years will be of minor importance, and taking into account that the rate of geomorphological processes will decrease slightly in time (*cf.* Renger & Partenscky, 1981), the total erosive effect on intertidal areas has been estimated to amount *ca.* 1500 hectares (10–15% of the present total area) over the next 30 years (De Vriend *et al.*, 1989; Mulder, 1989).

Development of channels and intertidal area

The distribution of unconsolidated fine sediment layers in the deeper parts of the channels, indicates that fine sediments contribute significantly to meet the sediment need of the channels since 1987. This statement is supported by quantitative indications. The total contribution of sands to recent channel sedimentation has been estimated – from figures on the total erosion of intertidal areas since 1987 – to be similar to the magnitude of total fine sediments import since 1987. Import of fine sediments is supposed to be a main, but not the only source of recent fine channel deposits. Other sources are erosion of silt from intertidal areas, and primary production (Ten Brinke *et al.*, 1994).

The recent erosive trend of sandy shoals and mud flats resulting in a general decrease of geomorphological gradients, may be attributed to the major changes in hydrodynamic conditions during and after implementation of the Oosterschelde Project. This is illustrated by the discontinuities in the overall trends of the depth data at the observation plots of the Galgeplaat (Fig. 9), and of levelling data from other shoals in the basin (Fig. 3). These discontinuities are parallel to changes in hydraulics since 1983. A dominant factor in the observed geomorphodynamic process, is the dramatic drop in transport capacities of the tidal currents (Fig. 7). Formerly, flood currents especially during spring tide and moderate wind conditions, have been responsible for an overall net accretion of the shoal (Kohsiek *et al.*,1988; De Vriend *et al.*, 1989; Louters *et al.*, 1993). In the present situation due to the drop in transport capacities, these currents hardly carry any sediment and consequently sedimentation on the shoal has minimized. At the same time the average intensity of erosive forces on the shoals has changed far less dramatically: the stirring effect of waves, the transport capacity of wind driven currents over the shoal and gravitational forces at the sloping shoal edges are responsible for a net sediment transport from the shoal to the channel. These mechanisms explain the general reduction of geomorphological gradients with a dominating erosion of the intertidal area and with sedimentation in channels.

Due to the reduction in tidal range, more wave energy is dissipated in a narrower zone along the edges of the shoals. The stronger stirring effect at wind exposed edges, potentially might result in an increased erosion (De Vriend *et al.*, 1989). However, the observed overall rate of shoal erosion typical during storm conditions, has reduced by a factor 10 after completion of the barrier (Louters *et al.*, 1993). Obviously, the drop in transport capacities of the tidal currents effectively hampers erosion. In the present situation both average sedimentation- and erosion rates have reduced, which illustrates the general decrease in rate of geomorphological changes after completion of the Oosterschelde Project.

Conclusions

Implementation of the Oosterschelde Project has had and will continue to have a significant effect on the geomorphology of the basin. The formerly eroding basin enclosing expanding channels and aggrandizing sandy shoals, has changed into the opposite; since 1987 the Oosterschelde is a sedimentation basin with channels in demand of sediment, and with degrading shoals. Establishing a new morphodynamic equilibrium will take centuries. The present processes will dominate future geomorphological developments. Sedimentation in channels, will remain predominant. Any import of sand from the North Sea will remain insignificant until a new morphodynamic equilibrium has been established at the outer delta. Erosion of shoals and mud flats in the basin will continue at decreasing rates. The continuing reduction in intertidal area, the decreasing geomorphological gradients, the increasing fine sediment content of channel deposits, combined with a general reduction in hydrodynamics, imply significant ecological effects.

Acknowledgements

By order of the Ministry of Transport and Public Works, Directorate General Rijkswaterstaat, the present investigations have been carried out within the framework of a multidisciplinary research project. This project EOS, aimed at an evaluation of Oosterschelde developments, has been supervised by the Tidal Waters Division and the Directorate Zeeland of Rijkswaterstaat. Part of the geomorphological research presented in this paper, has been commissioned to the Rijksuniversiteit Utrecht, Department of Physical Geography. In this respect the authors are much indebted to M. E. Philippart and W. Jonkers.

A. W. Walburg and G. L. G. Stoové have assisted in the data analysis and preparation of the figures.

References

Baillard, J. A., 1981. An energetics total load sediment transport model for a plane sloping beach. J. of Geoph. Res., vol. 86, no. C11: 10938–10954.

De Jong, D. J., Z. de Jong & J. P. M. Mulder, 1994. Changes in area, geomorphology and sediment nature of salt marshes in the Oosterschelde estuary (SW Netherlands) due to tidal changes. Hydrobiologia 282/283: 303–316.

De Vriend, H. J., T. Louters, F. Berben & R. C. Steijn, 1989. Hybrid prediction of a sandy shoal evolution in a mesotidal estuary. In R. A. Falconer, P. Goodwin & R. G. S. Matthew (eds), Hydraulic and Environmental Modelling of Coastal, Estuarine and River Waters, Int. Conf. Bradford, England: 14: 145–156.

Eijsink, W. D., 1990. Morphological response of tidal basins to changes. Proc. 22nd Coast. Eng. Conf., ASCE, Delft, Vol. 2: 1948–1961.

Gerritsen, F., H. de Jong & A. Langerak, 1990. Cross sectional stability of estuary channels in the Netherlands. Proc. 22nd Int. Conf. Coast. Eng., ASCE, Vol. 3: 2922–2935.

Kjerfve, B., 1976. The Santee-Cooper: a study of estuarine manipulations. In M. L. Wiley (ed.), Estuarine Processes, Vol. 1, Academic Press, New York: 44–56.

Knoester, M., J. Visser, B. A. Bannink, C. J. Colijn & W. P. A. Broeders, 1984. The Eastern Scheldt Project. Wat. Sci. Tech., 16: 51–77.

Kohsiek, L. H. M., H. J. Buist, P. Bloks, R. Misdorp, J. H. van den Berg & J. Visser, 1988. Sedimentary processes on a sandy shoal in a mesotidal estuary (Eastern Scheldt). In P. L. de Boer et al. (eds), Tide-influenced sedimentary environments and facies. Reidel Publishing Company: 201–214.

Louters, T., J. P. M. Mulder, R. Postma & F. P. Hallie, 1991. Changes in coastal morphological processes due to the closure of tidal inlets in the SW Netherlands. Journal of Coastal Research Vol. 7, No. 3, Summer 1991: 635–652.

Louters, T., J. P. M. Mulder & J. H. van den Berg, 1993. Large scale morphodynamic response of the Oosterschelde system to implementation of the Delta Project. (submitted to Journ. of Coastal Research).

Morra, R. H. J., H. M. Oudshoorn, J. N. Svasek & F. J. de Vos, 1961. The movement of sand in the tidal region of the southwestern part of the Netherlands. Report Delta Committee, Vol. V: 328–380.

Mulder, J. P. M., 1989. The changing tidal landscape in the delta area of the South west Netherlands. In Hydro-ecological relations in the Delta Waters of the South-West Netherlands, TNO Committee on Hydrological Research, Proceedings and information No. 41: 71–88.

O'Brien, M. P., 1969. Equilibrium flow areas of inlets on sandy coasts. ASCE, J. Waterw. and Harbors Div., 95(WW1): 43–51.

Renger, E. & H. W. Partenscky, 1981. Sedimentation processes in tidal channels and tidal basins caused by artificial constructions. Proc. Coast. Eng., 1980, Vol III: 2481–2494.

Saeijs, H. L. F., 1982. Changing estuaries: A review and new strategy for management and design in coastal engineering. Rijkswaterstaat, Communications no. 32/1982.

Schekkerman, H., P. L. M. Meininger & P. M. Meire, 1994. Changes in the waterbird populations of the Oosterschelde (SW Netherlands), as a result of large-scale coastal engineering works. Hydrobiologia 282/283: 509–524.

Simmons, H. B. & F. A. Hermann, 1972. Effects of man made works on the hydraulic, salinity and shoaling regimens of estuaries. In B. W. Nelson (ed.), Environmental framework of coastal plain estuaries, The Geological Society of America, Memoir 133: 555–570.

Smaal, A. C., M. Knoester, P. H. Nienhuis & P. M. Meire, 1991. Changes in the Oosterschelde ecosystem induced by the Delta works. In N. Elliott & J. P. Ducotroy (eds), Estuaries and Coasts, spatial and temporal comparisons. Olsen and Olsen, Fredensborg, Denmark: 375–384.

Ten Brinke, W. B. M., 1991. Quantifying mud exchange between the Eastern Scheldt tidal basin and the North Sea. N. C. Kraus et al. (eds) Coastal Sediments '91, Vol. 1, ASCE: 760–774.

Ten Brinke, W. B. M., J. Dronkers & J. P. M. Mulder, 1994. Fine sediments in the Oosterschelde tidal basin before and after partial closure. Hydrobiologia 282/283: 41–56.

Van den Berg, J. H., 1986. Aspects of sediment- and morphodynamics of subtidal deposits of the Oosterschelde (the Netherlands). Rijkswaterstaat, Communications no. 43/1986.

Van de Kreeke, J. & J. Haring, 1979. Equilibrium flow areas in the Rhine Meuse delta. Proc. 16th Int. Conf. Coast. Eng., Vol. 3: 97–111.

Van Rijn, L. C., 1987. Mathematical modelling of morphological processes in the case of suspended sediments transport. Delft Hydraulics Communication no. 382.

Vroon, J., 1994. Hydrodynamic characteristics of the Oosterschelde in recent decades. Hydrobiologia 282/283: 17–27.

Wetsteijn, L. P. J. M. & J. Kromkamp, 1994. Turbidity, nutrients and phytoplankton primary production in the Oosterschelde (The Netherlands) before, during and after a large-scale coastal engineering project (1980–1990). Hydrobiologia 282/283: 61–78.

Hydrobiologia **282/283**: 41–56, 1994.
P. H. Nienhuis & A. C. Smaal (eds), The Oosterschelde Estuary.
© 1994 *Kluwer Academic Publishers. Printed in Belgium.*

Fine sediments in the Oosterschelde tidal basin before and after partial closure

Wilfried B. M. ten Brinke [1], Job Dronkers [2] & Jan P. M. Mulder [2]
[1] *Department of Physical Geography, Utrecht University, P.O. Box 80115, 3508 TC, Utrecht, The Netherlands;* [2] *National Institute for Coastal and Marine Management/RIKZ, P.O. Box 20907, 2500 EX, The Hague, The Netherlands*

Key words: Oosterschelde, Eastern Scheldt, fine sediments, estuary, budget, changes

Abstract

Changes in the budget of fine sediments in the Oosterschelde have been measured. These are related to the partial closure of the tidal basin. Before the engineering works, soil texture of most of the basin was sandy. After the works, unconsolidated fine sediments occurred at several locations throughout the Oosterschelde, mainly in the deeper parts of channels and on musselbeds. Fine sediments accumulate due to the reduction in current velocities. Most of the fine sediment comes from the North Sea; internal sources of fine sediments (primary production and erosion of intertidal flats) are of minor importance. Due to the works, the direction of net transport of fine sediments has changed from an export (before the works) into an import. A qualitative discussion of the underlying processes is presented. The changes in the budget of fine sediments have both positive and negative ecological consequences. Muddy deposits on dike slopes have reduced hard bottom macrozoobenthos. The reduced nutrients input due to reduced fresh water input has not resulted in a reduced primary production because it is counterbalanced by a decrease in turbidity.

Introduction

The physical and ecological evolution of the Oosterschelde, a tidal basin in the southwestern part of the Netherlands (Fig. 1), is to a large extent controlled by the behaviour of fine suspended sediments, here defined as all matter $< 50\ \mu$m in diameter. Particulate organic matter is the food for pelagic and benthic animals. Fine suspended matter influences the turbidity of the water and thereby the primary production. Fine sediments also constitute an important source of material for infilling of channels. In the Oosterschelde most of the sediment suspended in the water column is fine and cohesive. Adsorption of pollutants plays a minor role as there is little chemical pollution in the Oosterschelde.

The construction of a storm-surge barrier and two compartment dams (Fig. 1) has substantially changed hydrodynamics and thus the behaviour of fine sediments in the Oosterschelde. The change in hydrodynamics is attributed to a decrease in tidal volume by about 30%. The effects of man-made alterations on fine sediments transport in estuarine systems have been studied previously (Simmons & Herrmann, 1972; Kjerfve, 1976; Cronin *et al.*, 1977; Zhaosen & Peiyu, 1982; Kjerfve & Magill, 1989). However, these interventions are not comparable with the changes in the Oosterschelde. As far as we know, the effect

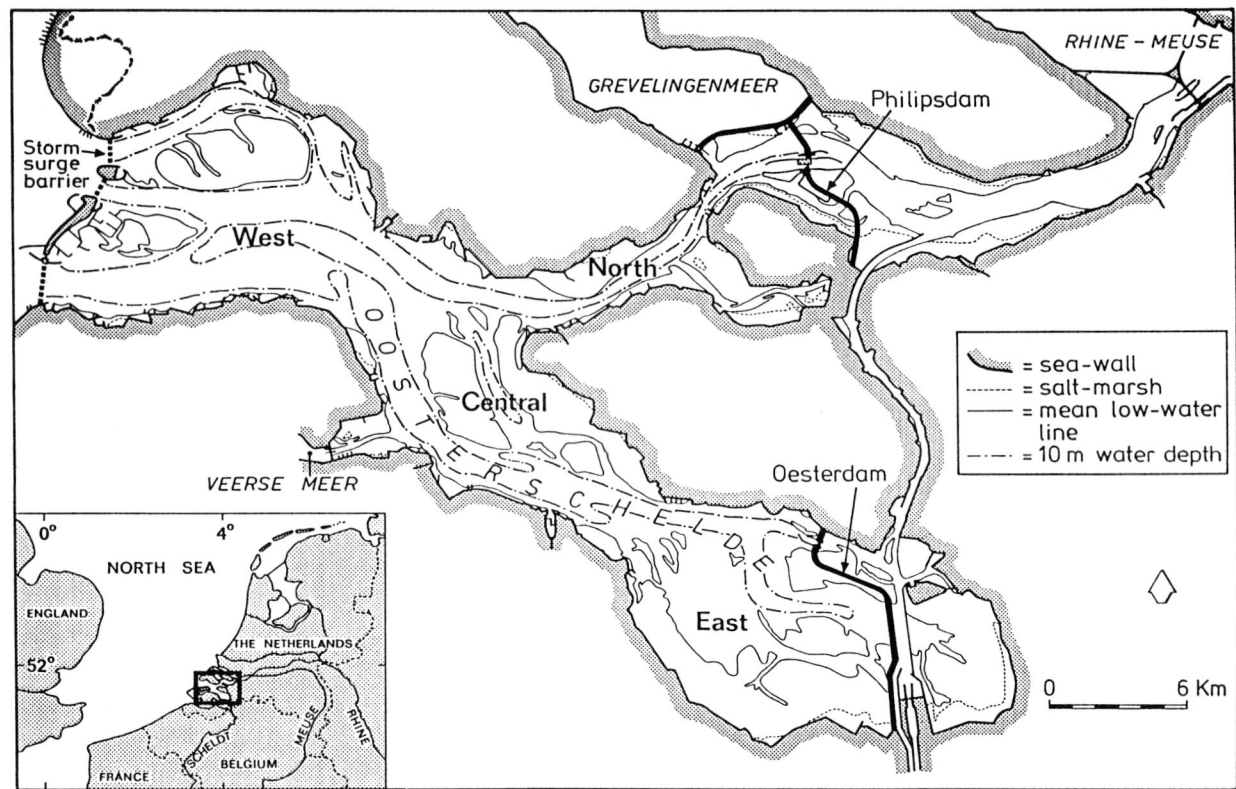

Fig. 1. The Oosterschelde tidal basin.

of tidal volume reduction on fine sediment transport in tidal basins has not been reported.

The aim of this paper is to describe how the change in hydrodynamics affects the large scale behaviour of fine sediments inside the tidal basin. Changes in the transport patterns of fine sediment in the Oosterschelde were studied in two ways. Firstly, the net suspended load transported through the inlet was measured both before and after the works. Secondly, changes in fine sediment distribution inside the basin were explored in channels, intertidal areas and salt marshes.

Area of research

As a result of major engineering modification (hereafter referred to as the works), current velocities in the Oosterschelde have been reduced by about 30% (Table 1), but locally much higher reductions (up to 80%) occur. The tidal range has been reduced by about 12%. As a result of this tidal reduction, wave energy dissipation is con-

Table 1. Morphological and hydrodynamic characteristics of the Oosterschelde.

	Pre-barrier	Post-barrier	% change
Total area (km^2)	452	351	-22
Intertidal area (km^2)	183	118	-36
Tidal volume (10^6 m^3)	1283	915	-29
Average current velocity (cm s^{-1})	120	80	-33
Residence time water (days)	50	100	$+100$
Fresh water input (m^3 s^{-1})	70	25	-63
Salinity (‰)	>25	>30	$+15$
Average depth (m)	8	8	0
Maximum depth (m)	55	55	0
Average tidal range (m) (at Yerseke)	3.70	3.25	-12
Average concentration Suspended matter (mg l^{-1})	25	15	-40

centrated on a smaller part of intertidal flats and salt marshes. Fresh-water input into the system, already low before the construction of the works (70 m³ s⁻¹), is now insignificant (25 m³ s⁻¹). Salinity is fairly constant and high throughout the Oosterschelde (>30‰). Fine suspended sedi-

ment concentration and thus turbidity is reduced strongly (Fig. 2) coinciding with the reduction of current velocities (Fig. 3A).

The civil engineering works were completed in April 1987. In this paper, the period after the works refers to the period after April 1987. Hy-

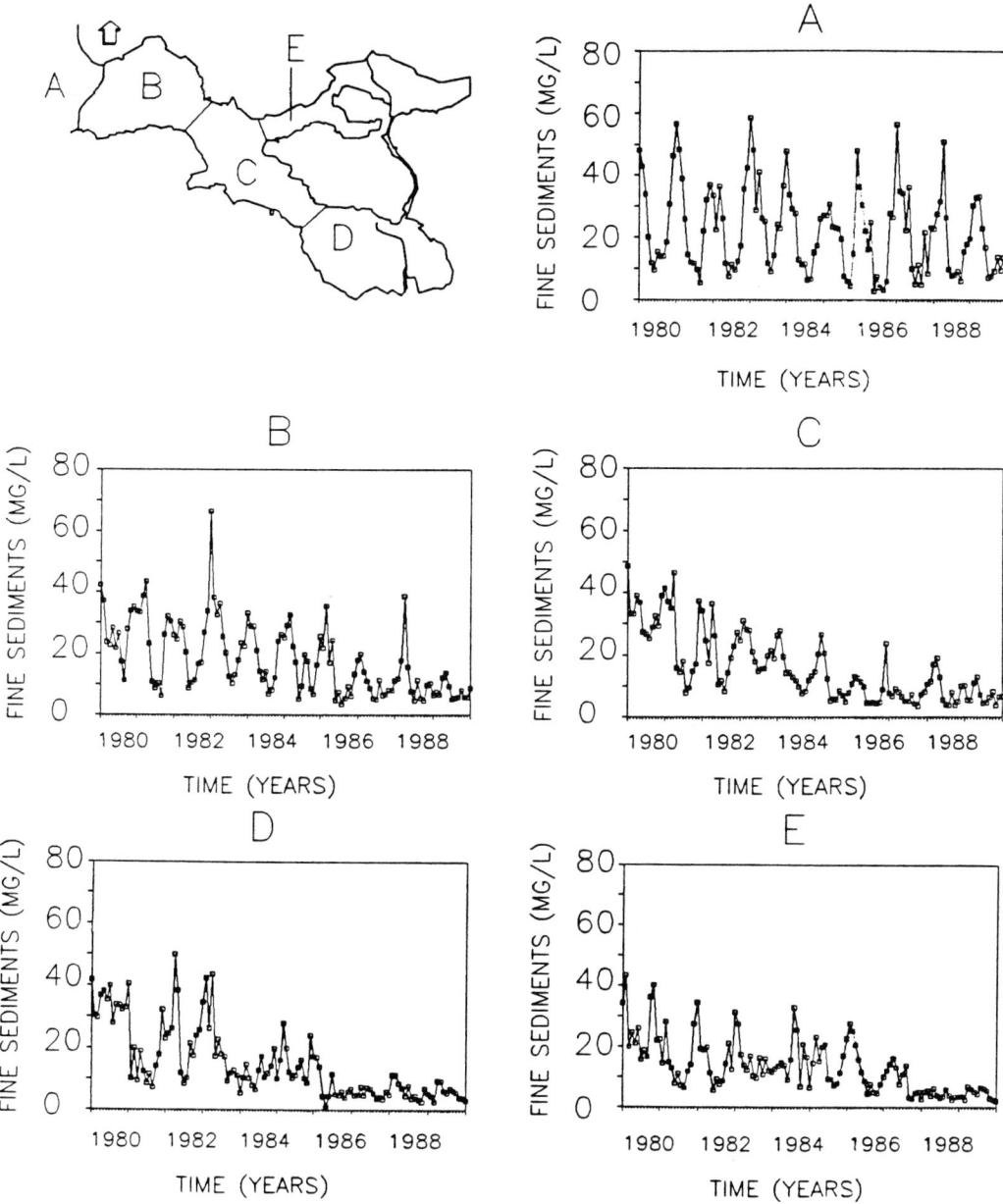

Fig. 2. Fine suspended sediment concentrations < 53 μm in 4 compartments of the Oosterschelde (west (B), central (C), east (D), and north (E)) and in the North Sea (A) during 1980–1989 at 1 metre below water surface.

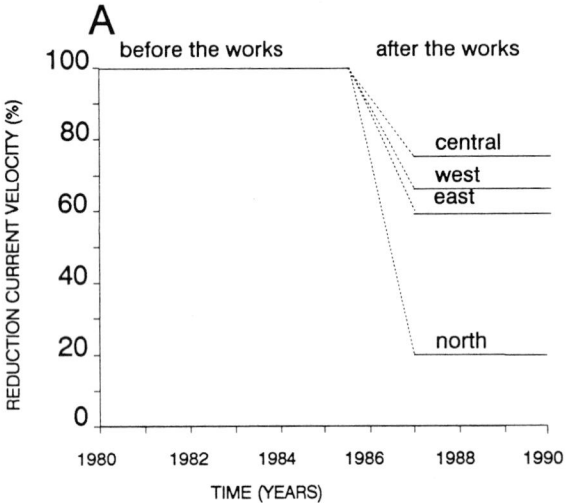

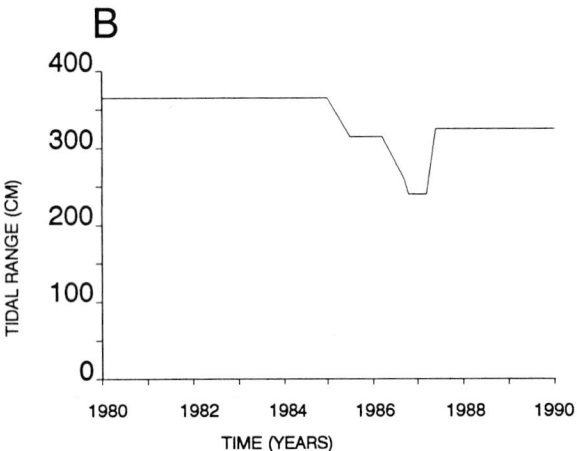

Fig. 3. The reduction in current velocity (A) and tidal range (B) due to the construction of a storm-surge barrier and 2 compartment dams.

drodynamics started to change in autumn 1985 due to the construction of the storm-surge barrier. The period before autumn 1985 is defined as the period before the works. The period autumn 1985–April 1987 is characterised by strongly reduced tidal ranges (Fig. 3B) and current velocities resulting from manipulations with the storm-surge barrier.

Because of the small fresh water input, estuarine, density induced circulation is absent or only weakly developed in the Oosterschelde (Dronkers & Zimmerman, 1982). Fine sediment concentra-

tion is fairly constant throughout the basin (Fig. 2). There is no clear turbidity maximum. Fine sediment transport studies in environments similar to the Oosterschelde are reported for American coastal embayments as well as for the Dutch Wadden Sea (Table 2). These are also mesotidal (tidal range = 2–4 m) environments characterised by a low fresh water input and relatively low suspended sediment concentrations. The suspended sediment is dominated by particles in the clay-silt size range. The behaviour of this sediment in the environments listed in Table 2 is mainly controlled by tidal motion. In addition, strong winds and biological processes (biodeposition, bioturbation etc.) influence fine sediments transport in varying degrees, causing seasonal differences in fine sediments behaviour.

Materials and methods

Net fine sediment exchange Oosterschelde–North Sea

In the period before the works, net fine sediment transport through the inlet of the Oosterschelde had been measured from vessels. A total number of 16 measurements were carried out during 1979–1982, spread throughout the year but always under rather calm weather conditions. All measurements were carried out during a complete tidal period (ebb + flood) at different phases within the neap-spring tidal cycle. Net fine sediment transport during a tidal period was calculated as a difference between ebb and flood transport. Corrections have been made to eliminate the influence of ebb/flood surplusses of water discharge (Dronkers, 1985). Different types of measurements were performed:

1) extensive transport measurements with a large number of vessels at fixed positions in the cross-section measuring current velocity every 15 minutes at 7–10, and sediment concentration every half an hour at 5 water depths.
2) extensive transport measurements with a large number of vessels at fixed positions in the

Table 2. References of fine sediment transport studies in similar environments as the Oosterschelde.

Reference	Area of research	Average concentration (mg l^{-1})	Average depth (m)	Salinity (‰)	Tidal range (m)	Maximum current velocity (m sec^{-1})
Oviatt and Nixon (1975)	Narragansett Bay	4	9	22–32	1.1–1.4	0.7 (near bottom)
Conomos & Peterson (1976) + Cloern & Nichols (eds) (1985)	San Fransisco Bay (southern reach)	25	6	15–30	1.7–2.7	1
Ashley & Grizzle (1988)	Great Sound, New Jersey	10–50	0.6	23.5–33.5	1.0–1.5	0.4
Wells (1988)	Cape Lookout Bight, North Carolina	20	7.5	>30	1.1–1.3	0.5 (near bottom)
Van Straaten & Kuenen (1957), Postma (1961, 1967), Dronkers (1984)	Wadden Sea, The Netherlands	30–75	3	>30	1.5	1
This article	Oosterschelde, The Netherlands	10–50	8	>30	3.3	1.5

cross-section measuring current velocity every 15 minutes at 7–10 water depths and sediment concentration every half an hour at 60% of the water depth from the surface.
3) reduced transport measurements with two vessels, each representing a part of the cross-section, measuring current velocity every 15 minutes at 7–10 and sediment concentration every half an hour at 5 water depths.
4) velocity measurements in two verticals every 15 minutes at 7–10 water depths and turbidity measurements in the transect between these two vessels by a sailing third vessel at a fixed water depth of 5 m below water level.

Fine sediment concentration was measured with optical turbidity meters (type HACH 2100 A or Monitek 160/131) which were calibrated with 1 litre water samples taken at several times (and different concentrations) during the measurements. From these water samples fine sediment concentration was determined by filtering through a 53 μm filter. OTT current velocity meters and Elmar current direction meters were used.

For the measurements described under 1), trend surfaces across the channels were calculated for current velocity and concentration, which were then multiplied (Elgershuizen & Stortelder, 1981). For the measurements described under 2), depth averaged current velocities were calculated and then multiplied by the concentration at 60% of the water depth. No spatial variation in sediment transport across the two parts of the cross-section was assumed for the measurements under 3). Corrections for the spatial variation in sediment concentration were made for the measurements under 4).

The net transports have been grouped into 4 seasons and then averaged to obtain estimates of net transport during the seasons. The incertainties in these transports have been estimated from the incertainties in the concentrations and current velocities across the channels (12 and 10% respectively; unpublished results). The incertainty in the ebb and flood transports therefore is 16%. When equal ebb and flood transports of $Q*C$ are assumed (Q = tidal volume (1283×10^6 m^3), C = concentration in winter (37 mg l^{-1}), spring (9 mg l^{-1}), summer (9 mg l^{-1}), and autumn (30 mg l^{-1}) respectively), which is nearly the case, the estimated incertainty in the net transport is $\sqrt{(2*(0.16*Q*C)^2)}$ per tide.

In the period after the works, net fine sediment transport was measured by vessels and a permanent station constructed at the storm-surge barrier. This station was employed for nearly two years during which an almost continuous registration of fine sediment concentrations of the water going in and out of the Oosterschelde was obtained. The period of registration contains several storms of different strengths and from different directions. Water was pumped from three depths through a Monitek 500 turbidity meter. This instrument was calibrated every two weeks with 1 litre water samples taken at regular time intervals. Correlations between turbidity and sediment concentration ($< 53 \ \mu$m) were generally good ($r > 0.8$).

Transport fluxes at the station were calculated by multiplication of depth averaged sediment concentration and ebb and flood volumes. These volumes were calculated by means of a numerical one-dimensional tidal model (IMPLIC). From the ebb and flood fluxes at the station, ebb and flood transports through the entire storm-surge barrier were calculated. Therefore the station had to be related to the cross-sectionally averaged concentration. This relation was obtained from measurements with vessels in the three channels in the mouth of the basin along transects parallel to the storm-surge barrier (Ten Brinke, 1991). The three channels were investigated simultaneously with measurements at the station. A total of 6 measurements during ebb and 6 measurements during flood has been carried out over 1989 under calm weather conditions. More details about the calculation procedure and a discussion of incertainties of the calculated transports can be found in Ten Brinke (1991).

Fine sediments in channels, intertidal areas and salt marshes

Channels
Fine sediments in the channels of the Oosterschelde (Fig. 1) have been investigated in a different way before and after the works. Before the works, the channels of the Oosterschelde have been sampled extensively by means of a Van Veen grab sampler (unpublished results; Oenema, 1988). These samples were taken by several investigators over the period 1961–1986 and refer to the upper 15 cm of the sediment. The samples were taken along transects across the channels at irregular intervals. For the entire Oosterschelde, some 1000 samples have been collected. From these samples the fine sediment content ($< 50 \ \mu$m) was determined partly by Sedigraph (Oenema, 1988) and partly by filtering through a 50 μm filter (unpublished results). Fine sediment content is expressed as percentage by weight.

The distance between the transects which were sampled, is large. A quite reliable interpolation is obtained by using information from side-looking sonar scans. Side-looking sonar shows the morphology of the sediment surface. A rippled surface indicates a sandy texture whereas a flat surface might be sandy or muddy.

After the works a completely different, more powerful technique was used to obtain information about fine sediments in the channels: double frequency echosounding (33 and 210 kHz, system ATLAS-2000). This method is based on a low frequency signal of 33 kHz penetrating a soft bottom and reflecting at a relatively hard layer somewhere in the bottom profile, and a high frequency signal of 210 kHz which is reflected at the sediment-water interface. With this instrument, soft bottoms rich in fine sediments can be traced as long as the upper layer consists of at least 10–15 cm of fine material.

In the Oosterschelde, several areas were pre-selected for investigation with double frequency echosounding. Van Veen grab samples were taken during these surveys in order to check the results. These samples showed that bottom sediment was seen as fine material when the fraction of fine sediment was at least 30% of all the sediment by weight. Also, this sediment had to be soft, containing at least 40% water by weight. Only recently deposited fine sediments can be discriminated using this type of echosounding.

Echosounding does not give any information about the accumulation rates. This information is needed for a budget of fine sediments in the

Oosterschelde. Accumulation rates have been obtained for 2 locations in the Oosterschelde: in a (subtidal) mussel cultivation area of the basin in the western part (8 m below MSL) (Fig. 8) and in the north-eastern branch near the Philipsdam (25 m below MSL) (Fig. 9). In September 1989, sediment cores of the muddy sediment were taken by SCUBA divers by twisting a PVC-tube into the sediment. From these cores samples were taken which were analysed for fine sediment content ($< 53 \mu m$) and organic matter in percentage by weight. The fine sediment fraction was determined by filtering through a 53 μm filter after pretreatment with H_2O_2 to remove organic matter. Organic matter was analyzed by oxidation with $K_2Cr_2O_7$. For the location near the Philipsdam, the initiation of mud deposition was known to be at dam closure in April 1987. The deposition history, and therefore the accumulation rate, of this sediment accumulated since April 1987 can be derived from the variation in organic matter content of the fine sediment with depth. Sediment that has settled during spring and summer contains more organic matter than sediment settled during autumn and winter. This seasonal variation in fraction of organic matter with depth can be used as an indication of continuity of sediment accumulation. For the location in the mussel cultivation area in the western part of the basin, net fine sediment accumulation was estimated by studying echosoundings over the period 1960–1989. Acoustic echosounding has been used in this area since 1960. Also Cs-137 and Cs-134 isotope analyses have been carried out to estimate the net sediment accumulation rate since 1986 (Tsjernobyl). Cesium-activity was measured by low-level gamma counting of dried sediment samples during 24 hours.

Intertidal flats
Fine sediments on intertidal flats have been investigated in the same way before and after the works. In the summer of 1985 (before the works) over 300 locations on 3 intertidal flats (one in the western, one in the central and one in the eastern part of the basin) were sampled in duplicate by twisting a tube with a diameter of 4.35 cm into the

soil and collecting a sediment mixture out of the upper 10 cm of the soil. The co-ordinates of these sampling locations were determined by laser position-finding (system Tachymeter with T-2000 distance meter). In the summer of 1989 (after the works), these locations were revisited and the same sampling procedure was performed. All the samples, both in 1985 and 1989, were analysed the same way. The sediments were treated with HCl and H_2O_2 to remove carbonates and organic matter, dried and sieved (nested sieves, mesh size = 53 μm).

Salt marshes
Fine sediments in salt marshes before the works have been studied by Oenema (1988). He used Cesium isotope analyses and artificial tracer layers to determine accretion rates in salt marshes. For more information about methods and analyses of this research we refer to Oenema (1988). The results of Oenema are used in this paper to estimate the fine sediments budget of the salt marshes before the works. The fine sediments budget of the salt marshes after the works has been studied by De Jong (see elsewhere in this volume).

Results

Net fine sediment exchange Oosterschelde–North Sea

Net transport of fine sediment between the Oosterschelde and the North Sea before the works is shown in Fig. 4. The scatter in the results of Fig. 4 is large. However, it is clear that on a yearly basis the Oosterschelde exported fine sediment to the North Sea. This calculated average export is 1.1×10^9 kg per year.

The net transport through the storm-surge barrier after the completion of the works is shown in Fig. 5. The results indicate that the Oosterschelde in its new shape imports fine sediment from the North Sea during all seasons. However, again the incertainty in these results is large. In fact, the standard error is so large that an export of fine

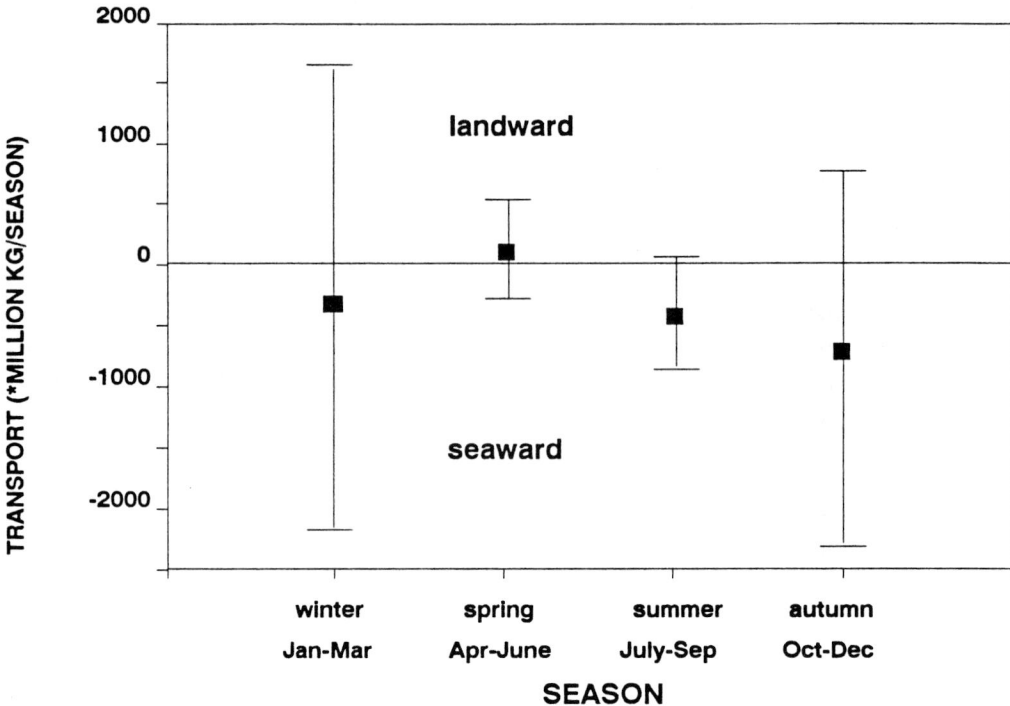

Fig. 4. Net transport (+ error bars) of fine sediment < 53 μm in the Oosterschelde inlet before the construction of the storm-surge barrier and compartment dams.

sediment cannot be excluded. The calculated average yearly transport is an import of 1.2×10^{9} kg per year.

Fine sediments in channels, intertidal areas and salt marshes

Channels

Before the construction of the civil engineering works, sediments in the channels of the Oosterschelde were mainly sandy (Fig. 6). Muddy deposits occurred near the fringes of the basin and in channels which were closed in earlier decades (Fig. 1). After the completion of the works, muddy deposits were observed at locations all over the Oosterschelde which were sandy in the past (Fig. 7). These locations are generally the deeper parts of channels (> 30 m depth) and areas with musselbeds.

The echosoundings show a net accumulation rate of the musselbeds in the western part of the basin of 11 cm y^{-1} (Fig. 8A). The Cs-137 and Cs-134 isotope analyses show that net sediment accumulation on the musselbeds since 1986 is 5.7 cm y^{-1} (Fig. 8B). The relatively low accumulation rate according to the Cesium isotopes does not necessarily point at reduced accumulation rates since 1986 but is probably caused by sampling inaccuracies. The echosoundings point at a continuous sediment accumulation during 1960–1989 (Fig. 8A). Besides, the variation of organic carbon and texture with depth (Fig. 8C) shows 4 peaks in the upper 30 cm, probably referring to spring-summer of the 4 years between 1986 and the moment of sampling (September 1989). The change in texture at 35 cm depth (Fig. 8C) refers to 1985/1986, when hydrodynamics in the Oosterschelde started to change (Fig. 3). Fine sediment content of the sediment has increased from 40–60% in the deposits before 1986 up to 90% after 1986.

Sediment accumulation near the Philipsdam (Fig. 9) is comparable to the western part (Fig. 8):

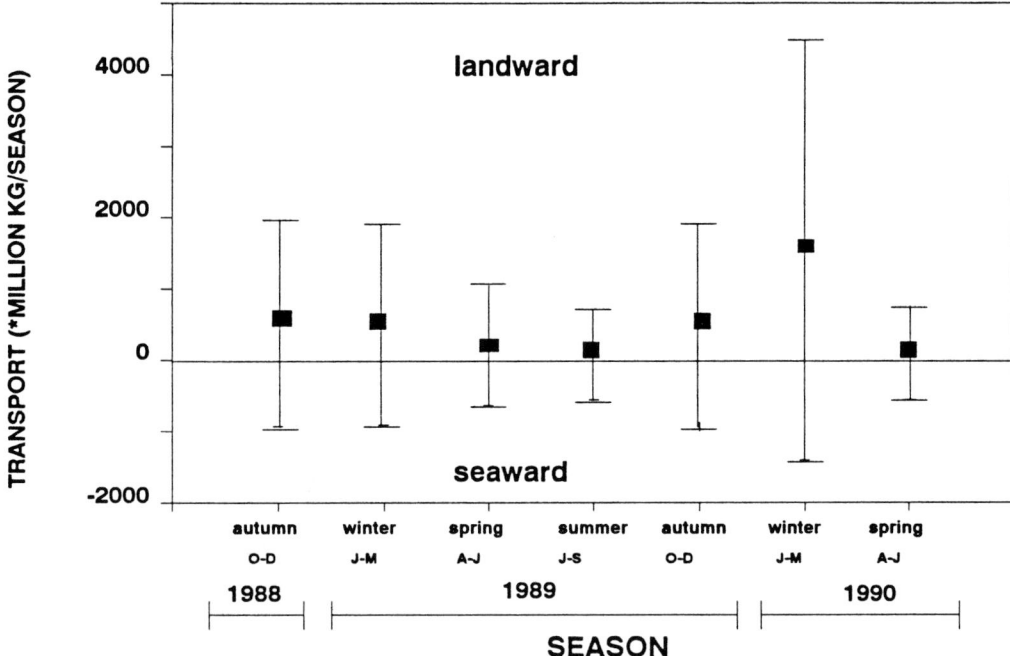

Fig. 5. Net transport (+ error bars) of fine sediment < 53 μm in the Oosterschelde inlet after the completion of the storm-surge barrier and compartment dams.

6–7 cm y^{-1}. The variation in organic matter clearly shows a continuous sedimentation since 1987.

The results from the echosoundings and the core-analyses can be combined to obtain estimates of the fine sediment accumulation in the channels. The area with muddy deposits is 1.5×10^7 m^2 and the accumulation rate in these areas is 5–10 cm y^{-1}. Dry soil bulk density is only 300–350 kg m^{-3} due to the high water content of the unconsolidated sediment. Fine sediment fraction is 90%. Thus, a net fine sediment accumulation rate of $(20–47) \times 10^7$ kg y^{-1} is calculated.

Intertidal flats
The changes in fine sediment content of intertidal sediments are only shown for one of the three intertidal flats which were investigated (Fig. 10). The other two areas (in the western and central part of the basin) showed little difference in sediment texture from 1985 to 1989. Sediments were very sandy with fine sediments content generally

less than 5% by weight. The intertidal flat in the eastern part of the basin (Fig. 10) however shows a remarkably different sediment texture in 1989 relative to 1985. Fine sediment content decreased an average of 1.6%. This decrease is statistically significant according to Wilcoxon Signed Ranks Test ($p < 0.05$). The amount of fine sediment that is eroded can be calculated assuming that (a) the most landward intertidal flat is representative of all the intertidal flats in landward branches of the Oosterschelde (5650×10^4 m^2), (b) sediment texture has changed in the upper 10 cm only and (c) dry sediment bulk density is 1500 kg m^{-3}. The computed amount of eroded fine sediment is 14×10^7 kg.

Salt marshes
Oenema (1988) found an average accretion rate of 1 cm per year for the Oosterschelde salt marshes before partial closure. Other estimates relevant for the sediment balance of the salt marshes are: the area of the marshes (1670×10^4 m^2), the dry soil bulkdensity (500 kg m^{-3}:

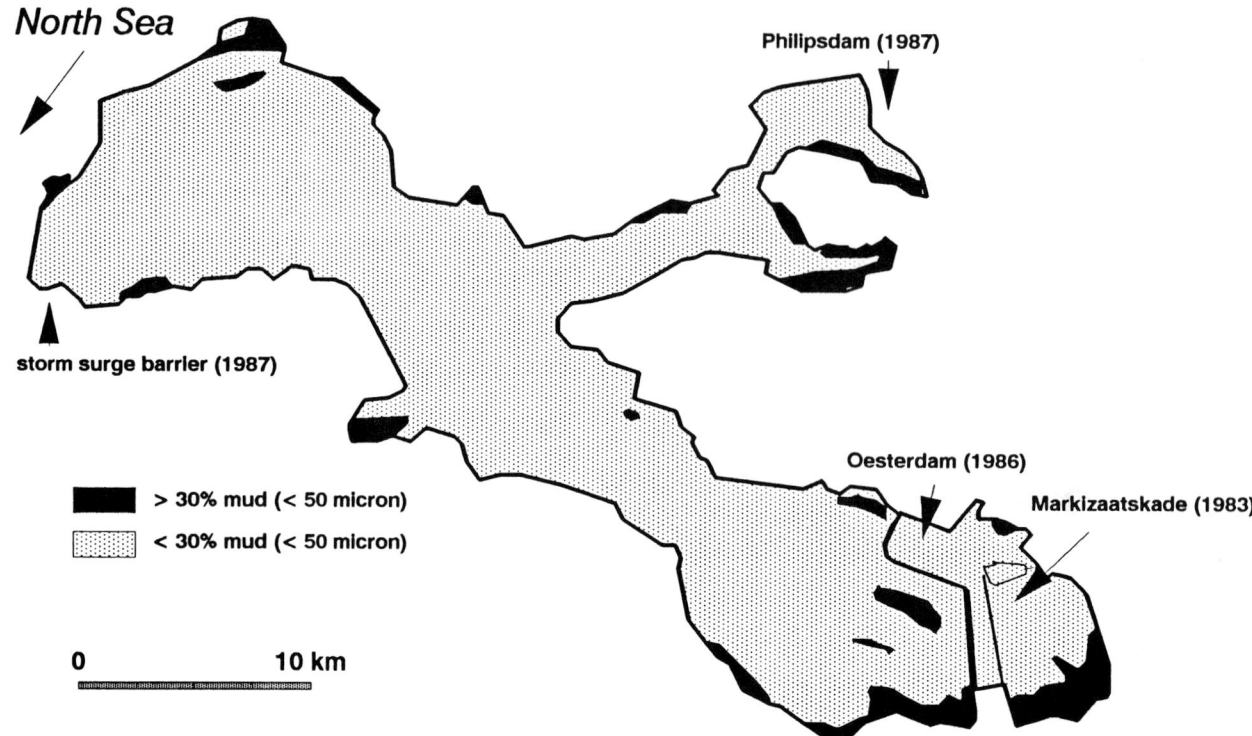

Fig. 6. Fine sediment content $< 53 \, \mu m$ of the sub- and intertidal bottom of the Oosterschelde before the construction of the storm-surge barrier and compartment dams (Van Veen grab sampler data from Elgershuizen and Oenema).

Oenema, 1988), the fine sediment fraction (70%: Oenema, 1988), the net cliff erosion (9000 $m^2 \, y^{-1}$: De Jong *et al.*, 1994) and the height of the cliffs (0.4 m). The net accumulation of fine sediments on salt marshes before the works is then found to be $5.7 \times 10^7 \, kg \, y^{-1}$.

Since the works the accretion rate has not changed much (De Jong, this volume) but the area of the salt marshes has been reduced about 60% to $640 \times 10^4 \, m^2$. The net erosion of salt marshes has increased to 45 000 $m^2 \, y^{-1}$ (De Jong *et al.*, 1994) due to increased wave attack. Net fine sediment accumulation on the salt marshes after the works is $1.2 \times 10^7 \, kg \, y^{-1}$.

Discussion

Changes in fine sediments budget

Sediment texture of the Oosterschelde used to be sandy except for fringes and dead channel-ends.

The works in the Oosterschelde have induced changes in the fine sediment distribution which were clearly observed within 3–4 years after completion of the works. In the major parts of the Oosterschelde the soil still is sandy. Muddy deposits are present at several locations throughout the Oosterschelde where soil texture was sandy in the past. Analyses of sediment cores show that the muddy deposits are made up mainly of very fine sediment (over 80% $< 53 \, \mu m$). Due to the reduction of current velocities sand transport has fallen to nearly zero.

Changes in large scale transport processes

The Oosterschelde was an eroding basin in the past (Van den Berg, 1986). The volume of water which it carried had increased since the Middle Ages mainly as an indirect result of human intervention. The increased tidal flow caused consid-

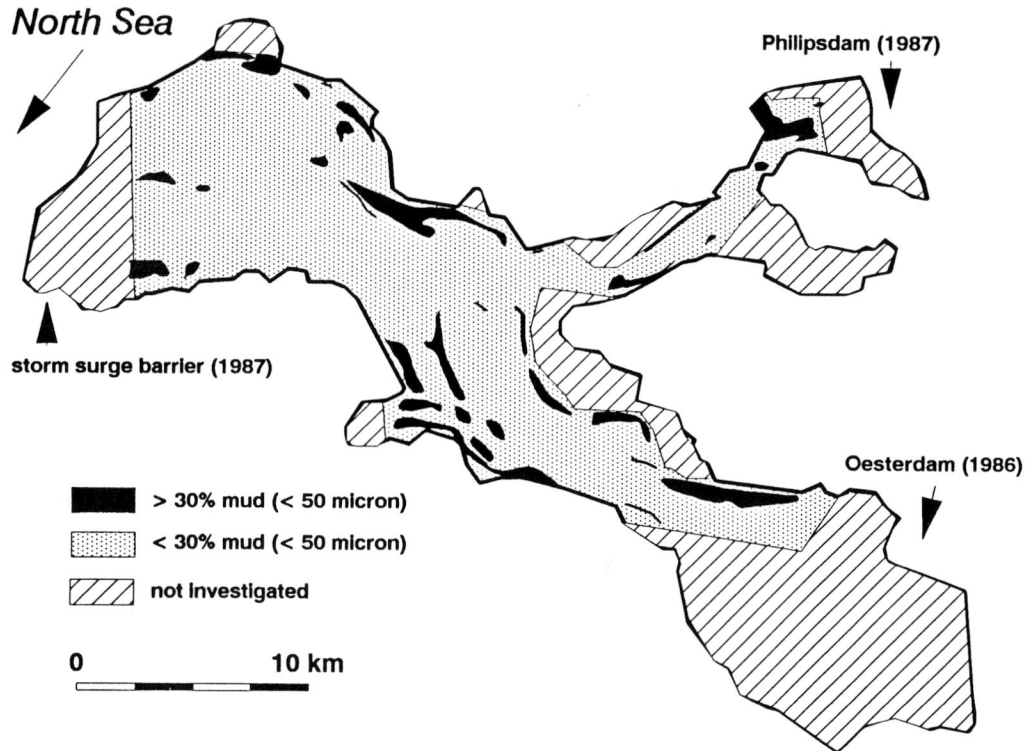

Fig. 7. Fine sediment content $< 53\ \mu$m of the subtidal bottom of the Oosterschelde after the completion of the storm-surge barrier and compartment dams (data from echosoundings with 2 frequencies verified by bottom samples).

erable deepening and widening of the estuary. Part of the eroded sediment consisted of clayey deposits. Over the last 20 years the average sediment export out of the Oosterschelde was 4.5×10^9 kg y^{-1}, consisting of 20% of fine (cohesive) particles (unpublished results). The seaward transport of fine sediment can be explained by the asymmetry of the tidal curve (Dronkers, 1985).

Since completion of the works channels have been filling-up with mainly fine sediments. Current velocities in the channels throughout the Oosterschelde are reduced. Apparently, bottom shear stress during ebb at several locations is not sufficient to resuspend sediment that has settled at water slack. Part of the sediment that comes into the basin during flood is trapped inside the channels.

Sources and Sinks

The echosoundings and core analyses showed that the net fine sediment accumulation in channels is $(20–47) \times 10^7$ kg y^{-1}. This is an estimate considering only muddy deposits. The sandy sediments throughout the Oosterschelde are probably also enriched in the amount of fine sediment but this enrichment cannot be quantified. The total accumulation of fine sediment in the Oosterschelde thus may considerably exceed $(20–47) \times 10^7$ kg y^{-1}.

Not only the channels (including the musselplots) are a sink for fine sediments. The salt marshes also trap fine sediments from the system. This deposition on salt marshes was relatively small in the past (5.7×10^7 kg y^{-1}) and has been reduced even further after completion of the works (1.2×10^7 kg y^{-1}) due to a reduction in the area of salt marshes and increased cliff erosion.

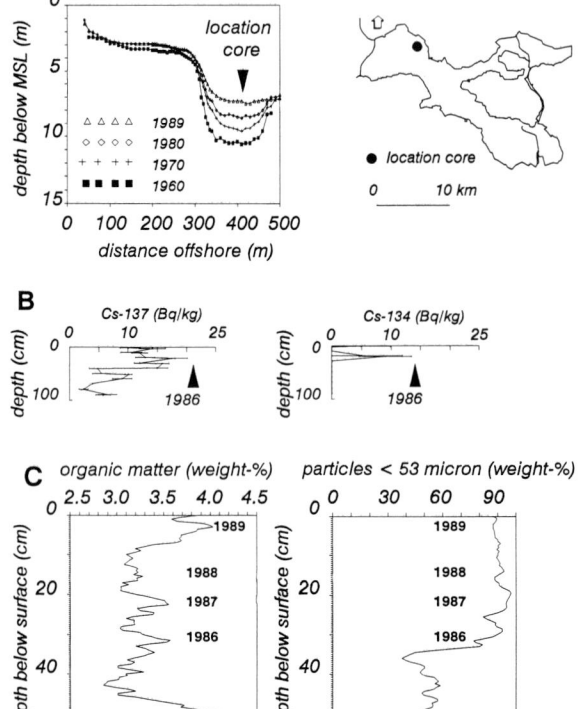

A

depth below MSL (m)

location core

△ △ △ △ 1989
◇ ◇ ◇ ◇ 1980
+ + + + 1970
■ ■ ■ ■ 1960

● location core

0   10 km

distance offshore (m)

B

Cs-137 (Bq/kg)

1986

Cs-134 (Bq/kg)

1986

C

organic matter (weight-%)   particles < 53 micron (weight-%)

1989

1988

1987

1986

1989

1988

1987

1986

Fig. 8. Echosoundings in a subtidal mussel cultivation area in the western part of the basin over the period 1960–1989 (A), and the activity of Cs-137 and Cs-134 (B) and the percentage of fine sediment < 53 μm and organic matter (C) in these muddy deposits.

As far as the origin of the fine sediment in the channels is concerned, two sources inside the Oosterschelde can be found. One of these sources is the intertidal flats. In the foregoing it was shown that during and after completion of the works 14×10^7 kg of fine sediments was eroded due to increased wave attack. The other source is primary production. During 1989 the total primary production was about 7.5×10^7 kg organic carbon (Wetsteyn, pers. comm.). This value represents some 15×10^7 kg algae since algae are approximately 50% organic carbon (Goldman, 1980). A large part of the primary production is mineralized (Wetsteyn, pers. comm.) or eaten in the water column. The sedimentation of fine particles derived from the primary production therefore is much less than 15×10^7 kg y^{-1}.

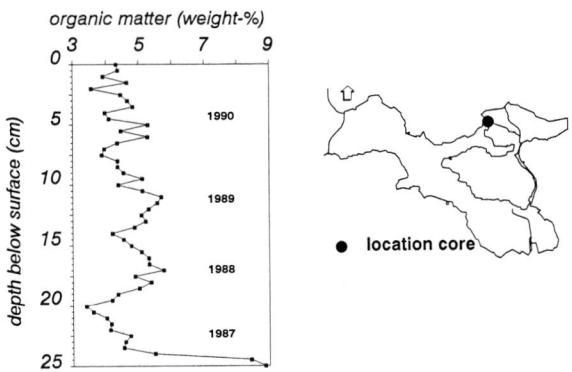

organic matter (weight-%)

depth below surface (cm)

1990

1989

1988

1987

Fig. 9. The percentage of organic matter in muddy deposits in a channel in the north-eastern branch of the basin, showing the seasonal cycle established after closure.

A comparison between the sources of fine sediment and the yearly deposition clearly shows that sources inside the Oosterschelde can by no means account for the net fine sediment accumulation in the basin. Thus, fine sediment must be imported from the North Sea. This import amounts to some $(20–50) \times 10^7$ kg y^{-1} or more. In fact, the import of 1.2×10^9 kg y^{-1} which has been computed from flux measurements near the storm-surge barrier seems realistic as an order of magnitude. The reversal of net fine sediment transport in the Oosterschelde inlet after the works agrees with the changes in fine sediment patterns inside the basin.

Settling from flocculation and biodeposition

Fine suspended sediment concentrations in the Oosterschelde after the works were reduced due to decreased current velocities. The reduction is largest in the landward branches. Despite these extremely low concentrations, a considerable accumulation of fine sediment occurs, for instance in the north-eastern branch. Suspended sediment in the Oosterschelde forms large aggregates which settle very fast at water slack (Fig. 11). Assuming that there is no resuspension of settled aggregates, the accumulation of fine sediment can be estimated. The settling velocity of aggregates is 1×10^{-3} m s^{-1}. The amount of sediment that settles at slack water is 2×10^{-3} kg m^{-3}. The

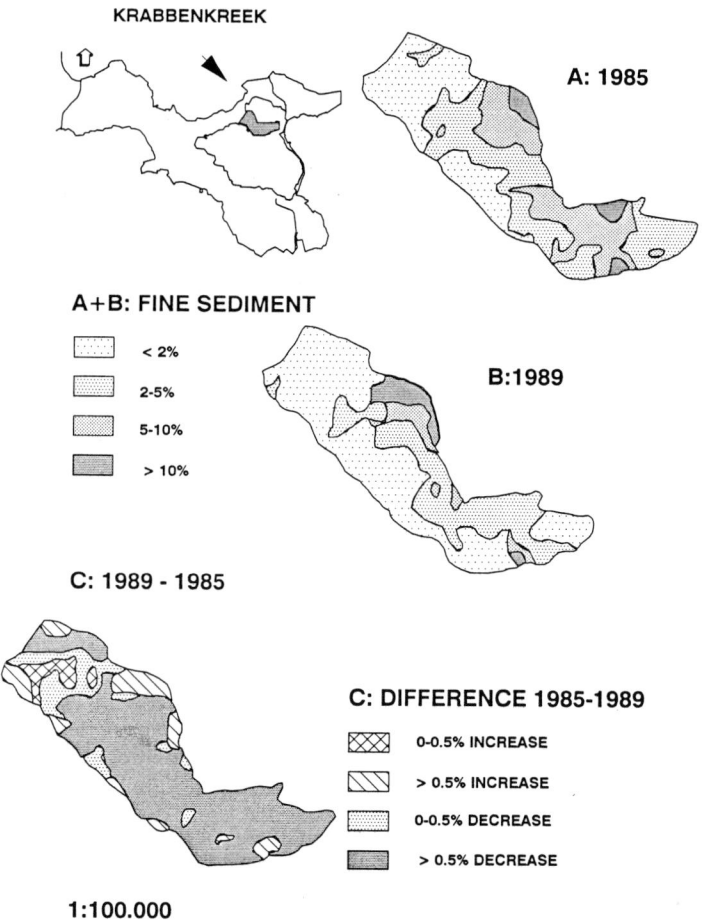

Fig. 10. Changes in fine sediment content $< 53\,\mu$m of the upper 10 cm of an intertidal flat in the eastern (landward) part of the basin from 1985 to 1989.

time during a year with favourable conditions for settling is $7 \times 365 \times 3600 = 9.2 \times 10^6$ s. Using these data and a bulk density of 350 kg m^{-3}, a sediment accumulation of 5 cm y^{-1} is calculated. This agrees very well with the results in Fig. 9. The assumption of no resuspension seems valid for the location in Fig. 9.

A part of the muddy deposits, which are mapped by means of echosounding, refers to biodeposition on musselbeds (Haven & Morales-Alamo, 1972; Rhoads, 1974). Mussels (and other filter feeders) filter the water and deposit the sediment as either faeces or pseusofaeces. The pseusofaeces are easily resuspended and thus hardly contribute to the accumulation of fine sediment (Risk & Moffat, 1977). The faeces, however, con-

sist of pellets which are more resistant to erosion than mud flocs (Rhoads, 1974). This faeces production can be an important mechanism for the deposition of fine sediment, especially in the Oosterschelde with its extensive mussel farming. The biodeposition of the mussels (*Mytilus edulis* L.) can be estimated from the filtration rate of mussels (2.5 l ind^{-1} h^{-1}: Prins, pers. comm.), the mussel density (330 ind m^{-2}), the mussel cultivation area (25×10^6 m^2) and the sediment concentration that is turned into faeces (5 mg l^{-1}: Widdows *et al.*, 1979). The biodeposition is 45×10^7 kg during spring and summer. Biodeposition during autumn and winter is hard to estimate. Some authors conclude that biodeposition at low water temperature is of minor importance

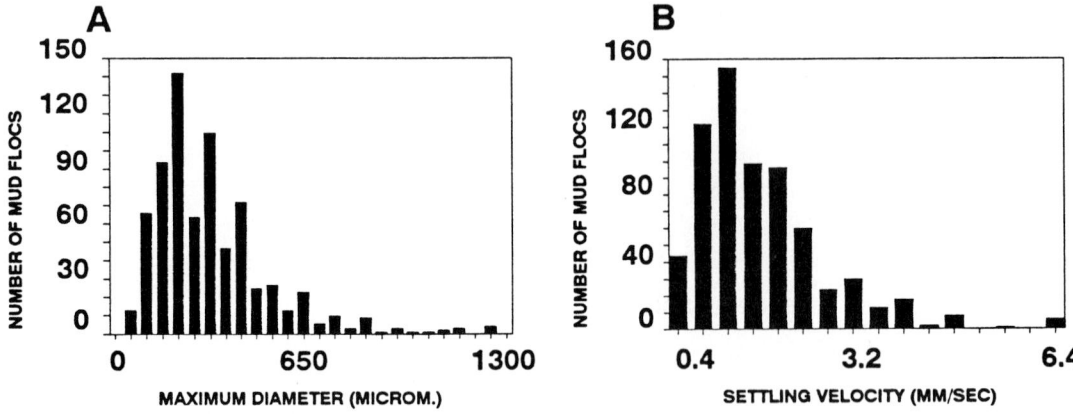

Fig. 11. Size(A) and settling velocity (B) of mud flocs in the Oosterschelde determined from registrations of a submersible video system developed by Van Leussen and Cornelisse (submitted).

due to low metabolic activity of benthic filterfeeders at low water temperature (Haven & Morales-Alamo, 1972; Risk & Moffat, 1977; Newell, 1979; Kautsky & Evans, 1987). However, a non-dependency of clearance rate on temperature has also been reported (Widdows *et al.*, 1979; Smaal *et al.*, 1986). The results of both Smaal *et al.* (1986) and Widdows *et al.* (1979) refer to a limited period of time (August–October and January–July respectively) and thus to a limited temperature range (10–18 °C and 8–16 °C respectively). Water temperature in the Oosterschelde, however, ranges from 0 to 20 °C. The results of Kautsky & Evans (1987) are probably most realistic, being the only experiments carried out under natural conditions during an entire year. According to Kautsky & Evans (1987) biodeposition in autumn and winter is low compared with spring and summer. The increase in biodeposition they found during stormy weather probably refers to pseudofaeces. Considering a threshold concentration of 5 mg l^{-1} that is turned into faecal pellets (Widdows *et al.*, 1979) and low filtration rates at low water temperature, storms will not result in extraordinary high faeces production. Nearly all the autumn biodeposits collected by Kautsky & Evans will be pseudofaeces which are not resuspended because they are captured in 50 cm deep sediment traps.

The biodeposition of 45×10^7 kg is of the same order of magnitude as the average fine sediment import from the North Sea during spring and summer (Fig. 5; 23×10^7 kg) and the results of the echosoundings and the core-analyses ((20–60) $\times 10^7$ kg y^{-1}). However, the biodeposition may be regarded as somewhat high in that only mussels are considered, whereas in the Oosterschelde other filterfeeders (especially cockles) are also present. In addition, sediment should settle quickly from flocculation in the water column. The estimates are thus affected by uncertainties in the following way:

1) the echosoundings were not carried out in some of the abandoned channels where part of the mussels is cultivated.
2) the echosoundings were carried out in the period April–June. Some of the biodeposits of the previous summer may have eroded and settled elsewhere (deep channels e.g.). The period April–June is probably too early to detect biodeposits of the current summer.
3) it was assumed that none of the faeces was resuspended.

According to the filtration rate, mussel density, and a dry bulk density of 350 kg m^{-3}, the biodeposition is 100 g m^{-2} day^{-1} during spring and summer (≈ 5 cm). This figure is comparable to the estimated settling from flocculation. It is hard to say how important the processes of biodeposition and settling from flocculation are in the fine sediment budget of the Oosterschelde.

Fine sediments on the relatively shallow mussel-plots are probably mainly (broken-down) faecal pellets. Settled aggregates will be resuspended for the greater part due to waves and fishing activities. On the other hand, a very large area of unconsolidated muddy deposits is in the deeper parts of channels where biodeposition does not occur and resuspension of settled aggregates is probably low. In view of the limited area of musselbeds and the fact that biodeposition in autumn and winter is probably of minor importance, it is unlikely that biodeposition by mussels dominates the fine sediment budget of the Oosterschelde.

Consequences of changes in fine sediments budget

Due to the construction of the works, the ebb and flood volumes in the channels of the Oosterschelde are no longer in equilibrium with the channel cross-section. Thus, the channels will fill-up with sediments until a new equilibrium is reached (Mulder & Louters, 1994). This filling-up will take place with the sediments that are available and which can be moved into the system. Apart from some erosion of intertidal flats (Mulder & Louters, loc. cit.) no sand is available. Sand import from the North Sea is very small (Mulder & Louters, loc. cit.) and will not increase until the channels seaward of the storm-surge barrier have adapted to the reduced ebb and flood volumes. This will probably take at least some 50 years. Thus, for the next decades most of the sediments accumulating in the channels will be fine-grained. This does not mean that all of the Oosterschelde will become muddy. In large parts of the basin the reduction in current velocities is relatively small. Currents in these areas are still strong enough to resuspend unconsolidated fine sediments.In the deeper parts of the channels and in the landward branches however, fine sediments will accumulate during the next decades.

Musselbeds are generally muddy due to the filterfeeding of the mussels. Nevertheless, the sand content of the sediments on musselplots was quite high before the works, creating a firm substrate for the mussels. After the works, the accumulat-ing sediments became very soft due to an extremely low sand content. It may be expected that in areas with very little resuspension during spring and summer some mussels will have difficulties in staying on top of their own deposits and perhaps even will suffocate.

The changes in fine sediments distribution have both positive and negative consequences for the ecology of the Oosterschelde. A positive consequence is that a decrease in nutrient input due to a reduced fresh water input, which normally would have resulted in a reduced primary production, is counterbalanced by a decrease in turbidity, so that primary production is not reduced by dam construction (Wetsteyn & Bakker, 1991). Negative consequences relate *i.a.* to the muddy deposits on dike slopes causing a reduction in hard bottom macrozoobenthos (De Kluijver *et al.*, 1994). An effect of the fine sediment reduction of the (landward) intertidal flats on macrobenthos species cannot be concluded at present (Coosen *et al.*, 1994).

Conclusions

The Oosterschelde has changed from a fine sediment exporting into an importing system due to civil-engineering works causing a reduction of the tidal prism. From transport measurements an average import of 1.2×10^9 kg y^{-1} is calculated which is the same order of magnitude as yearly fine sediment accumulation measured from bottom surveys inside the basin since the completion of the works. A small part of recent muddy deposits is derived from the primary production and erosion of intertidal flats. Fine sediment settles in the deeper parts of channels and on musselplots throughout the tidal basin. Biodeposition is an important mechanism for mud deposition but probably does not dominate fine sediments budget of the Oosterschelde.

Acknowledgements

This research was financially supported by Rijkswaterstaat. The manuscript benefited from critical reviews by K. R. Dyer & J. T. Wells.

56

References

Ashley, G. M. & R. E. Grizzle, 1988. Interactions between hydrodynamics, benthos and sedimentation in a tide-dominated coastal lagoon. Mar. Geol. 82: 61–81.

Cloern, J. E. & F. H. Nichols (eds), 1985. Temporal dynamics of an estuary: San Francisco Bay. Dr W. Junk Publishers, Dordrecht, 237 pp.

Conomos, T. J. & D. H. Peterson, 1977. Suspended-particle transport and circulation in San Francisco Bay: an overview. In M. Wiley, Estuarine Processes 2, Circulation, Sediments, and Transfer of Material in the Estuary, Academic Press, London: 82–97.

Coosen, J., J. Seys, P. M. Meire & J. Craeymeersch, 1994. Effect of sedimentological and hydrodynamical changes in the intertidal areas of the Oosterschelde estuary (SW Netherlands) on distribution, density and biomass of some common macrobenthic species. Hydrobiologia 282/283: 235–249.

Cronin, L. E., D. W. Pritchard, T. S. Y. Koo & V. Lotrich, 1977. Effects of the enlargement of the Chesapeake and Delaware Canal. In M. Wiley, Estuarine Processes 2, Circulation, Sediments, and Transfer of Material in the Estuary, Academic Press, London: 18–32.

De Jong, D. J., Z. De Jong & J. P. M. Mulder, 1994. Changes in area, geomorphology and sediment nature of salt marshes in the Oosterschelde estuary (SW Netherlands) due to tidal changes. Hydrobiologia 282/283: 303–316.

De Kluyver, M. J. & R. J. Leewis, 1994. Changes in the sublittoral hard substrate communities in the Oosterschelde estuary (SW Netherlands) caused by changes in the environmental parameters. Hydrobiologia 282/283: 265–280.

Dronkers, J. & J. T. F. Zimmerman, 1982. Some principles of mixing in tidal lagoons with examples of tidal basins in the Oosterschelde. In P. Laserre & H. Postma (eds), Coastal Lagoons, Proc. Int. Symp. Coast. Lagoons, Gauthier-Villars: 460–474.

Dronkers, J., 1984. Import of fine marine sediment in tidal basins. Neth. Inst. for Sea Res. Publ. Ser. 10: 83–105.

Dronkers, J., 1985. Tide-induced residual transport of fine sediment. In J. van de Kreeke (ed.), Physics of shallow estuaries and bays. Lecture Notes on Coastal and Estuarine Studies 16, Springer: 228–244.

Elgershuizen, J. H. B. W. & P. B. M. Stortelder, 1981. A direct measurement of the transport of organic matter in the Eastern Scheldt (S.W.-Netherlands) – application of a trend surface method. Proc. Dynamics of Turbid Coastal Environments, Halifax.

Goldman, J. C., 1980. Physiological aspects in algal mass cultures. In G. Shelef & C. J. Soeder (eds.), Algae Biomass, Elsevier/North-Holland Biomedical Press: 343–359.

Haven, D. S. & R. Morales-Alamo, 1972. Biodeposition as a factor in sedimentation of fine suspended solids in estuaries. In B. W. Nelson (ed.), Environmental framework of coastal plain estuaries, The Geological Society of America, Memoir 133: 121–130.

Kautsky, N. & S. Evans, 1987. Role of biodeposition Mytilus edulis in the circulation of matter and nutrients in a Baltic coastal ecosystem. Mar. Ecol. Prog. Ser. 38: 201–212.

Kjerfve, B., 1976. The Santee-Cooper: a study of estuarine manipulations. In M. L. Wiley (ed.), Estuarine Processes, Vol. 1, Academic Press, New York: 44–56.

Kjerfve, B. & K. E. Magill, 1989. Hydrobiological changes in Charleston Harbor, South Carolina. Proc. Sixth Symp. Coast. Ocean Man., Charleston: 2640–2649.

Mulder, J. P. M. & T. Louters. Changes in basin geomorphology after implementation of the Oosterschelde project. This volume.

Oenema, O., 1988. Early diagenesis in recent fine-grained sediments in the Eastern Scheldt. Ph.D. thesis, University of Utrecht, The Netherlands: 222 pp.

Oviatt, C. A. & S. W. Nixon, 1975. Sediment resuspension and deposition in Narragansett Bay. Estuar coast. mar Sci. 3: 201–217.

Rhoads, D. C., 1974. Organisms-sediment relations on the muddy sea floor. Oceanogr. Mar. Biol. annu. Rev. 12: 263–300.

Risk, M. J. & J. S. Moffat, 1977. Sedimentological significance of fecal pellets of Macoma balthica in the Minas Basin, Bay of Fundy. J. Sed. Petrol. 47: 1425–1436.

Simmons, H. B. & F. A. Herrmann, 1972. Effects of man-made works on the hydraulic, salinity, and shoaling regimens of estuaries. In B. W. Nelson (ed.), Environmental framework of coastal plain estuaries, The Geological Society of America, Memoir 133: 555–570.

Smaal, A. C., J. H. G. Verhagen, J. Coosen & H. A. Haas, 1986. Interaction between seston quantity and quality and benthic suspension feeders in the Oosterschelde, The Netherlands. Ophelia 26: 385–199.

Ten Brinke, W. B. M., 1991. Quantifying mud exchange between the Eastern Scheldt tidal basin (The Netherlands) and the North Sea. In N. C. Kraus et al. (eds), Coastal Sediments '91, ASCE, Seattle: 760–774.

Van den Berg, J. H., 1986. Aspects of sediment- and morphodynamics of subtidal deposits of the Oosterschelde. Rijkswaterstaat Communications, The Hague, 127 pp.

Van Leussen, W. & J. M. Cornelisse. The determination of the sizes and settling velocities of estuarine flocs by an under water video system. Neth. J. Sea Res. (submitted).

Van Straaten, L. M. J. U. & P. H. Kuenen, 1957. Accumulation of fine-grained sediments in the Dutch Wadden Sea. Geologie en Mijnbouw 19: 329–354.

Vroon, J. 1994. Hydrodynamic characteristics of the Oosterschelde in recent decades. Hydrobiologia 282/283: 17–27.

Wells, J. T., 1988. Accumulation of fine-grained sediments in a periodically energetic clastic environment, Cape Lookout Bight, North Carolina. J. Sed. Petrol. 58: 596–606.

Wetsteyn, L. P. M. J. & C. Bakker, 1991. Abiotic characteristics and phytoplankton primary production in relation to a large-scale coastal engineering project in the Oosterschelde (The Netherlands): a preliminary evaluation. In M. Elliot & J. P. Ducrotoy (eds), Estuaries and Coasts: Spatial and Temporal Intercomparisons, Olsen and Olsen: 365–373.

Widdows, J., P. Fieth & C. M. Worrall, 1979. Relationships between seston, available food and feeding activity in the common mussel Mytilus edulis. Mar. Biol. 50: 195–207.

Zhaosen, L. & G. Peiyu, 1982. Sedimentation associated with tidal barriers in China's estuaries and measures for its reduction. In V. S. Kennedy (ed.), Estuarine Comparisons, Academic Press, New York: 611–622.

Hydrobiologia **282/283**: 57–60, 1994.
P. H. Nienhuis & A. C. Smaal (eds), The Oosterschelde Estuary.
© 1994 *Kluwer Academic Publishers. Printed in Belgium.*

Theme II:
Structure and functioning of the pelagic system

Cees Bakker
Netherlands Institute of Ecology, CEMO, Vierstraat 28, 4401EA Yerseke, The Netherlands

Introduction

In the former (pre-barrier) Oosterschelde estuary, a coastal embayment of the Southern North Sea, a clear seasonality was observed of several environmental and biotic factors. Phytoplankton biomass and daily primary production reached a distinct summer peak under conditions of maximum irradiance; also zooplankton biomass was maximal during that time.

When a storm-surge barrier (in the mouth) and two auxiliary dams (in the Northern and Eastern compartments) were constructed, the physical and chemical environment changed drastically. In this chapter the responses of the planktonic biota in the present-day (post-barrier) Oosterschelde are recorded and analyzed. All data have been incorporated in a simulation model that is discussed elsewhere (Klepper *et al.*, 1994).

The pelagic system of the Oosterschelde has been studied (1) hydrographically: current velocities (Vroon, 1994), turbidity, transparency, salinity, nutrient concentrations; and (2) biologically: (a) on cell and organismic level: identification of phyto- and zooplankton species; (b) on population level: enumeration of cells and individuals, calculation of species population carbon contents; (c) on community level: species composition, species shifts, species succession, total phytoplankton carbon contents, chlorophyll concentration, primary production, total zooplankton biomass; and (d) on system level: basin averages, subdivided in values for the different compartments.

Structural aspects

This includes a detailed study of the seasonal change in species composition, an aspect that has been investigated thoroughly, notably in temperate coastal sea water. We are aware that the term 'species succession' has two meanings (Sommer, 1989). Firstly, the loose concept, defining succession as 'an allogenic shift arising from more permanent variation in physical structure' (Reynolds, 1980). Secondly, the restrictive concept defining succession as 'a true autogenic succession under relatively constant physical conditions, permitting structure segregation' (Reynolds, 1980). An example of the first was a species shift in the Oosterschelde phytoplankton that occurred as a consequence of increased water transparencies caused by stronger sedimentation of suspended material during and after barrier-construction (Bakker *et al.*, 1994). An example of the latter, implying effects of biotic interactions, is the increased abundance of flagellate species in the phytoplankton during summer in the Eastern compartment (Bakker & Vink, 1994). This could be ascribed partially by an increased copepod (*Temora*) biomass in this area (Bakker & Van Rijswijk, 1994), exerting size-selective grazing upon the phytoplankton assemblage. However, this may only be part of the explanation of the observed change in diatom-flagellate relations. Nutrient limitation in summer is supposed to have a stronger effect on diatom- than on flagellate growth and thus may play the first role. Species succession altered not only on a seasonal base, but notably on a long-term basis.

The strongly decreased current velocities and increased residence times during the barrier-construction period as well as the reduced tidal dynamics after the construction of the barrier (Vroon, 1994), especially occurred in the Eastern compartment and gave this part of the sea-arm a lagoonal character. Besides several phytoplankton species, also representatives of the zooplankton, especially rotifers of the genus *Synchaeta*, made profit from this condition (Bakker, 1994).

A Canonical Correspondence Analysis greatly helped to reveal that the phytoplankton assemblage more and more obtained a summer character, when the growth season was extended earlier and later in the year. The factor light did not only explain the seasonal pattern but also the long-term trend from pre-barrier to barrier-construction period (Bakker *et al.*, 1990). During the post-barrier years (i.e. from April 1987 onwards) not only the improved under-water light climate was maintained, but also the nutrient-salinity regime was changed: salinity remained constantly higher and nutrients, particularly nitrogen, were depleted. This resulted in a sea-arm with a tendency to oligotrophication and the phytoplankton assemblage obtained on a-seasonal trend (Bakker *et al.*, 1994).

An account is given of the species composition of the zooplankton (Bakker, 1994). An interesting feature is the changed succession of some *Acartia* species during summer in the Eastern compartment. *A. tonsa* was dominating during the estuarine pre-barrier years while *A. clausi* became abundant in the more saline post-barrier period.

An important event for suspension-feeding zooplankton in the Eastern compartment in post-barrier times is the higher phytoplankton *versus* decreased total particulate organic carbon concentration, causing a higher living to dead carbon ratio. This, and the larger residence times, may be held responsible for the increased zooplankton biomass in this area (Bakker & Van Rijswijk, 1994).

Special attention was paid to distinguish long-term changes caused by the barrier works, from year-to-year variation or interannual variability caused by heavy short-term effects like e.g.

storms, changed intensity of water exchange and severe winters. It has to be considered that the years of the barrier-construction period coincided with 3 subsequent severe winters. Consequently, some early spring diatoms, like *Biddulphia aurita*, were stimulated strongly in their development, a phenomenon that was experimentally confirmed earlier by Baars (1986).

Functional aspects

Data about nutrient concentrations and -ratios are reported both by Bakker & Vink (1994) and by Wetsteyn & Kromkamp (1994). Bakker & Vink (1994) deal with nutrient data (Si and N) from selected years (1983, 1986, 1988/89) and give information within a detailed time scale, while Wetsteyn & Kromkamp (1994) include also P-data, refer to all years of investigation and, therefore, present a complete account of all measurements. It is emphasized that Si could become limiting during the estuarine pre-barrier years and especially nitrate-N was depleted during the post-barrier period, both in summer and particularly in the Eastern compartment.

The use of annually averaged values enables the investigator to make interannual comparisons as well as comparisons between different water systems. However, a variable like annual primary production may vary very strongly within a system as Wetsteyn & Kromkamp (1994) demonstrated, making this factor of only limited value for comparison between systems. Rather it is the seasonal course and variation of a large number of variables, when followed regularly and frequently, that supplies the most valuable information about the system. In this case-study of the Oosterschelde it could be shown, e.g., that a characteristic feature of the basin, from the barrier-construction period onwards, was the earlier start of the spring bloom as a consequence of the strongly decreased tidal turbidity after the completion of the works. This did not lead to a higher total annual primary production, however, as production in summer did rather decrease as a consequence of nutrient limitation starting to play a role after the discharge of river water into the Northern branch diminished with 64% (Wetsteyn & Bakker, 1991).

Primary production in the central and eastern parts of the Oosterschelde did not change from pre- to post-barrier times and varied between *ca.* 180 and *ca.* 400 gC·m^{-2} y^{-1}. The first primary production measurements in the Oosterschelde, Eastern compartment, were done by Vegter & de Visscher (1987), reporting the annual values for pre- and barrier construction periods. Wetsteyn & Kromkamp (1994) extended the measurements to the other compartments and continued the work in post-barrier years; finally they combined and discussed all gathered data. Changes in photosynthetic physiological parameters suggested shade adaptation of the phytoplankton, but this does not agree with the improved light conditions and the reduced nutrient availability. This apparent inconsistency may be explained by the species shift that occurred. The conclusion is that annual primary production showed a large degree of homeostasis. In the Oosterschelde simulation model, also other patterns of energy flow, such as scope for growth of suspension-feeding zoobenthos and detritus production, did hardly change (Herman & Scholten, 1990). This insight is not new: in their comparative approach of the plankton ecology of subarctic Atlantic and Pacific oceans, Parsons & Lalli (1988) revealed large differences in ecological structure between the two areas but, in spite of this, similar annual productions of phytoplankton and macrozooplankton were demonstrated.

Nutrient limitation, combined with the longer residence times, caused the phytoplankton to become strongly dependent on nutrient regeneration processes and (increased) zooplankton grazing. In the Western compartment nutrients were not limiting, due to the exchange with the North Sea and the important benthic mineralization, mainly of mussels (Prins & Smaal, 1994).

Tackx *et al.* (1994) calculated phyto- and zooplankton biomass distributions for selected years throughout the period of investigation, in the Eastern compartment. Distribution patterns of these main plankton groups were irregular. The

data were used to test Sheldon's *et al.* (1977) model, predicting biomasses at higher from those at lower levels. Data and model predictions agreed in pre-barrier and barrier-construction years, but did not in post-barrier years, evidently because of the important changes in the zooplankton species abundance in this period.

Concluding remarks

The field data from the Oosterschelde phytoplankton studies led to a new understanding in *structural* ecosystem responses to the substantial interference in the environment during and after the construction of the storm-surge barrier.

The *functional* response of the pelagic ecosystem, however, did not essentially change. The Oosterschelde was characterized by functional stability or homeostasis.

Klepper *et al.* (1994) emphasized the stabilizing function of the suspension-feeding zoobenthos, mainly mussels and cockles. Interesting phenomena such as the changed diatom-flagellate relationships in the summer phytoplankton of the Oosterschelde, were already earlier demonstrated, cf. the marine mesocosm studies reported by Grice & Reeve (1982.).

References

Baars, J. W. M., 1986. Autecological investigations on marine diatoms. 4: *Biddulphia aurita* (Lyngb.). Brébisson et Godey – A succession of spring diatoms. Hydrobiol. Bull. 19: 109–116.

Bakker, C., P. M. J. Herman & M. Vink, 1990. Changes in seasonal succession of phytoplankton induced by the storm-surge barrier in the Oosterschelde (S.W. Netherlands). J. Plankton Res. 12: 947–972.

Bakker, C., 1994. Zooplankton species composition in the Oosterschelde (SW Netherlands) before, during and after the construction of a storm-surge barrier. Hydrobiologia 282/283 (Dev. Hydrobiol. 97): 117–126.

Bakker, C., P. M. J. Herman & M. Vink, 1994. A new trend in the development of the phytoplankton in the Oosterschelde (SW Netherlands) during and after the construction of a storm-surge barrier. Hydrobiologia 282/283 (Dev. Hydrobiol. 97): 79–100.

Bakker, C. & P. van Rijswijk, 1994. Zooplankton biomass in the Oosterschelde (SW Netherlands) before, during and after the construction of a storm-surge barrier. Hydrobiologia 282/283 (Dev. Hydrobiol. 97): 127–143.

Bakker, C. & M. Vink, 1994. Nutrient concentrations and planktonic diatom-flagellate relations in the Oosterschelde (SW Netherlands) during and after the construction of a storm-surge barrier. Hydrobiologia 282/283 (Dev. Hydrobiol. 97): 101–116.

Grice, G. D. & M. R. Reeve (eds), 1982. Marine Mesocosms, Biological and Chemical Research in Experimental Ecosystems. Springer-Verlag, 430 pp.

Herman, P. M. J. & H. Scholten, 1990. Can suspension-feeders stabilise estuarine ecosystems? In: M. Barnes & R. N. Gibson (eds), Trophic Relationships in the Marine Environment, Proc. 24th Europ. Mar. Biol. Symp., Aberdeen Univ. Press: 104–116.

Klepper, O., M. W. M. van der Tol, H. Scholten & P. M. J. Herman, 1994. SMOES: a simulation model for the Oosterschelde ecosystem. *Part I:* Description and uncertainty analysis. Hydrobiologia 282/283 (Dev. Hydrobiol. 97): 437–451.

Parsons, T. R. & C. M. Lalli, 1988. Comparative oceanic ecology of the plankton communities of the subarctic Atlantic and Pacific oceans. Oceanogr. Mar. Biol. Ann. Rev. 26: 317–359.

Prins, T. C. & A. C. Smaal, 1994. The role of the blue mussel *Mytilus edulis* in the cycling of nutrients in the Oosterschelde estuary (The Netherlands). Hydrobiologia 282/283 (Dev. Hydrobiol. 97): 413–429.

Reynolds, C. S., 1980. Phytoplankton assemblages and their periodicity in stratifying lake systems. Holarctic Ecology 3: 141–159.

Sheldon, R. W., W. H. Sutcliffe & M. A. Paranjape, 1977. Structure of pelagic food chain and relationship between plankton and fish production. J. Fish. Res. Board Can. 34: 2344–2353.

Sommer, U., 1989. Toward a Darwinian ecology of plankton. In: Sommer, U. (ed.), Plankton Ecology, Succession in Plankton Communities. Springer-Verlag: 1–8.

Vegter, F. & P. R. M. de Visscher, 1987. Nutrients and phytoplankton primary production in the marine tidal Oosterschelde estuary (The Netherlands). Hydrobiol. Bull. 21: 149–158.

Vroon, J., 1994. Hydrodynamic characteristics of the Oosterschelde in recent decades. Hydrobiologia 282/283 (Dev. Hydrobiol. 97): 17–27.

Wetsteyn, L. P. M. J. & C. Bakker, 1991. Abiotic characteristics and phytoplankton primary production in relation to a large-scale coastal engineering project in the Oosterschelde (The Netherlands): a primary evaluation. In: M. Elliott & J. P. Ducrotoy (eds), Estuaries and Coasts: Spatial and Temporal Intercomparisons. Proceedings of the ECSA 19 Symposium, Caen 1989. Olsen & Olsen, Fredensborg, Denmark: 365–373.

Wetsteyn, L. P. M. J. & J. C. Kromkamp, 1994. Turbidity, nutrients and phytoplankton primary production in the Oosterschelde (The Netherlands) before, during and after a large-scale coastal engineering project (1980–1990). Hydrobiologia 282/283 (Dev. Hydrobiol. 97): 61–78.

Hydrobiologia **282/283**: 61–78, 1994.
P. H. Nienhuis & A. C. Smaal (eds), The Oosterschelde Estuary.
© 1994 *Kluwer Academic Publishers. Printed in Belgium.*

Turbidity, nutrients and phytoplankton primary production in the Oosterschelde (The Netherlands) before, during and after a large-scale coastal engineering project (1980–1990)

L. P. M. J. Wetsteyn[1] & J. C. Kromkamp[2]
[1] *National Institute for Coastal and Marine Management/RIKZ, P.O. Box 8039, 4330 EA Middelburg, The Netherlands;* [2] *Netherlands Institute of Ecology, Centre for Estuarine and Coastal Ecology, Vierstraat 28, 4401 EA Yerseke, The Netherlands*

Key words: Oosterschelde, turbidity, nutrients, phytoplankton primary production

Abstract

Turbidity, nutrient concentrations and phytoplankton primary production were monitored in the Oosterschelde before, during and after the construction of a storm-surge barrier and two compartment dams.

Flow velocities and suspended matter concentrations decreased severely, causing an increased transparency of the watercolumn. In the eastern and northern compartments, the previously pronounced seasonal variation disappeared.

Reduction of the freshwater load and decreasing nutrient concentrations in the adjacent North Sea coastal waters resulted in lower nitrite + nitrate and silicate concentrations. Autumn phosphate concentrations remained at the same level as before the nutrient reduction. Silicate was a limiting nutrient during the pre-barrier period and nitrogen and silicate were limiting during the post-barrier period.

Annual patterns in chlorophyll-*a* concentrations in the western and central compartments showed no obvious trend; in the eastern and northern compartments higher values were measured from 1985 onwards.

Primary production during the period 1980–1990 varied between 176 and 550 g C m^{-2} yr^{-1}. The annual primary production in the western compartment had decreased, while in the central and eastern compartments annual primary production did not change: the formerly existing gradient disappeared. In the northern compartment higher chlorophyll-*a* concentrations and high annual production suggest that the phytoplankton could benefit from the increased transparency while nutrient concentrations were still high enough to support phytoplankton growth.

Changes in photosynthetic physiological parameters were observed which suggested shade adaptation. This is in contrast to improved light conditions and reduced nutrient availibility. The apparent incoherence with light-shade adaptation theory may be explained by the species shift that occurred.

As a result of the opposite effects of a more favourable light climate and a reduced nutrient availability, together with the resulting species shift, the annual primary production showed a large degree of homeostasis.

Introduction

The storm flood disaster in 1953 resulted in a large-scale coastal engineering project in the Dutch Delta area (Knoester *et al.*, 1984). In the Oosterschelde, the project involved the construction of a storm-surge barrier at its mouth (1982–October 1986) and two compartment dams, the

62

Oesterdam (1982–October 1986) and the Philips-dam (1985–April 1987) (Nienhuis & Smaal, 1994; Fig. 1).

During 1980–1984, a pre-barrier ecosystem study assessed the main carbon flows and integrated them into an ecosystem model (Klepper, 1989). Many data were also gathered during the period 1985–1986. In a recent ecosystem study from 1987–1990, the new, post-barrier situation was assessed. A preliminary evaluation of the impact of the civil-engineering works on the whole ecosystem is given by Smaal et al. (1991).

The storm-surge barrier decreased the cross-sectional area in the mouth from 80 000 to 17 900 m^2 and the mean tidal volume from 1240 × 10^6 to 880 × 10^6 m^3 (Vroon, 1994). After the completion of the barrier and the compartment dams a location-dependant reduction of the maximum flow velocities during ebb and flood was realized, varying from 25–86% (Wetsteyn &

Bakker, 1991: Table 3); residence times in the western compartment (Fig. 1) increased from 5 to 10 days, in the central compartment from 25 to 40 days, in the eastern compartment from 50 to 100 days and in the northern compartment from 35 to more than 100 days (Smaal & Boeije, 1991).

The annual mean freshwater load to the Oosterschelde from 1980 to April 1987 was ca 70 m^3 s^{-1} and after April 1987 ca 25 m^3 s^{-1}, which means a reduction of 64%, resulting in a nutrient load reduction, especially in the northern compartment (Wetsteyn & Bakker, 1991; Smaal & Boeije, 1991: Fig. 3.8).

The decreased flow velocities and reduced freshwater loads were expected to influence turbidity and nutrient concentrations and therefore phytoplankton growth and production.

In this paper data on turbidity, nutrients and phytoplankton primary production, covering the period 1980–1990, will be presented in order to

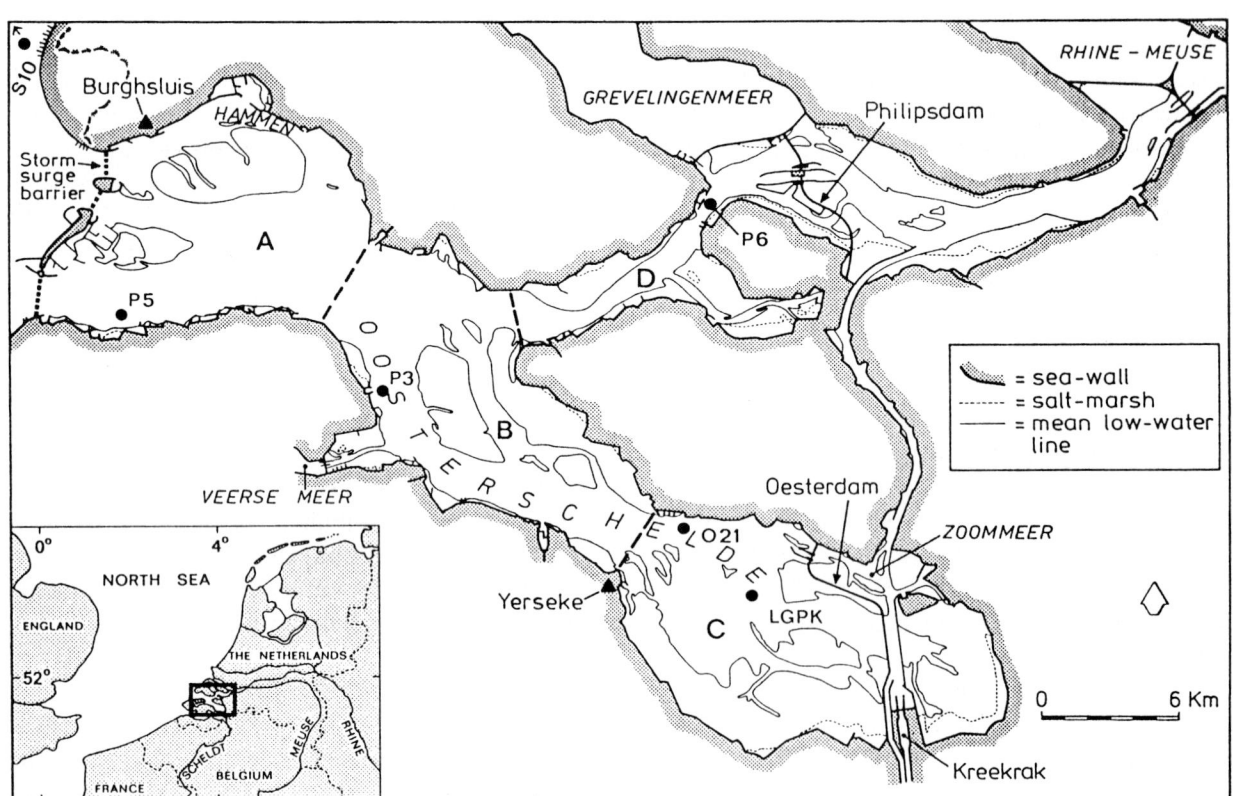

Fig. 1. Location of the Oosterschelde compartments (A = western part, B = central part, C = eastern part, D = northern part) and sampling stations.

evaluate whether or not these characteristics were influenced by the civil engineering works.

Materials and methods

Study area

The Oosterschelde is situated in the south-west of The Netherlands (Fig. 1). The area is extensively used for mussel culture (Smaal & van Stralen, 1990). As a result of the strong tidal currents the Oosterschelde was well mixed during the pre-barrier period (Dronkers & Zimmerman, 1982). From temperature profiles in the Hammen (Fig. 1; H. Haas, unpubl. data) it can be concluded that also the post-barrier Oosterschelde is well mixed (W. ten Brinke, pers. comm.); only in the northern branch some salt-stratification appeared (K. B. Robaczewska, pers. comm.). Four geographical compartments have been distinguished: a western, a central, an eastern and a northern compartment, each containing (at least) one sampling station (Fig. 1); their morphometrical and hydrographical characteristics for the pre- and post-barrier periods are given in Wetsteyn et al. (1990) and Wetsteyn & Bakker (1991).

At present most freshwater arrives from shipping locks in the Philipsdam (Fig. 1), polderwater discharges, precipitation and from Lake Veere (Veerse Meer) (Fig. 1), an eutrophic brackish lake.

Water sampling, light measurements and nutrient analysis

Water samples for nutrients, suspended matter, chlorophyll-a and primary production at stations P5, P3, LGPK and P6 (Fig. 1) were taken at a depth of ca 1 m. Samples were taken randomly in the tidal period. Although tidal variation severely can influence suspended matter concentrations (Cadée, 1982), the tidal variation in suspended matter concentrations in the Oosterschelde (see e.g. Ten Brinke, 1991) is much smaller than the annual variation. Furthermore it is shown in Bakker & Vink (1994) for samples taken during half-tide at maximum current velocities that the trends for transparency (and thus also for suspended matter concentrations) and nutrients are the same as will be described here.

At station O21 (Fig. 1) samples for chlorophyll-a and primary production were obtained by mixing water taken at depths of 0, 1, 2, 4 and 8 m. At all sampling stations the water temperature, salinity and Secchi disc visibility were measured.

The vertical attenuation coefficient at stations P5, P3, LGPK and P6 was measured with an energy-cell constructed by TFDL (Technical and Physical Service of the Agricultural University, Wageningen), mounted on a vertically moveable measuring device; at station O21 with a Licor quantum sensor (LI-192SB) connected to a Licor quantum meter (LI-185B). Daily irradiation (PAR: 400–700 nm) was measured at Burghsluis (1980–1984, Fig. 1), Kreekrak (1980–1984, Fig. 1) and Yerseke (1980–1990, Fig. 1) with TFDL energy-cells and Kipp CC2 Solar Integrators.

Filtered water samples (using Whatman GF/C filters) were analysed with Technicon Auto-Analyzers for inorganic nutrients (phosphate, nitrite + nitrate, ammonium and silicate) according to Grasshoff et al. (1983). No release of silicate from the GF/C filters could be detected (A.G.A. Merks, pers. comm.).

Suspended matter was measured gravimetrically.

Sampling frequencies are given in Table 1.

Phytoplankton

Total primary production (particulate + dissolved production) at stations P5, P3, LGPK and P6 was measured in a laboratory incubator using the ^{14}C method (see Wetsteyn et al., 1990); the incubation period was always 2 hours. Particulate primary production at station O21 was also measured in a laboratory incubator using the ^{14}C method (see Vegter & de Visscher, 1984a; 1987); in general, the incubation period lasted 5 hours. An intercomparison of the two different incuba-

Table 1. Sampling frequencies at the Oosterschelde stations during the period 1980–1990. 1: Monthly; 2: April–September: weekly; October–March: once or twice a month; 3: January–April: as 1; May–December: as 2; 4: not sampled.

Year	P5	P3	O21	LGPK	P6
1980	1	2	1	1	1
1981	3	2	2	1	1
1982	2	2	2	1	1
1983	2	2	2	1	1
1984	2	2	2	1	1
1985	2	4	2	1	1
1986	2	2	2	1	1
1987	2	2	2	2	1
1988	2	2	2	2	2
1989	2	2	2	2	2
1990	2	2	4	2	2

tors with a phytoplankton culture (*Thalassiosira rotula*) revealed no significant differences between the measured photosynthetic parameters. The photosynthesis-light (P-I) curves were fitted with the relation $P = I/(aI^2 + bI + c)$, where P denotes production rate in mg C m^{-3} h^{-1}, I = light intensity in μE m^{-2} s^{-1} and a, b, c are curve parameters (Eilers & Peeters, 1988). From the curve parameters the physiological photosynthetic parameters P_{max} (in mg C mg^{-1} Chl-*a* h^{-1}) and α (in mg C mg^{-1} Chl-*a* h^{-1} μE^{-1} m^2 s) were calculated. P_{max}, the photosynthetic capacity per unit chlorophyll-*a*, is the maximum rate of photosynthesis and α, the photosynthetic efficiency per unit chlorophyll-*a*, is the initial slope of the P-I curve. Integral primary production was calculated as daily column production, using a time-depth integrating procedure as described by Fee (1973) and as 'basin' production, taking into account the exponential surface-depth relation (cf. Klepper *et al.*, 1988, see also Wetsteyn *et al.*, 1990). In the original fitting procedure used to fit P-I curves for station O21 the coefficients a, b and c could become negative. Theoretically this is not possible (see Eilers & Peeters, 1988). The P-I curves for station O21 were therefore recalculated with a better fitting procedure from 1982 onwards. Hence, the data differ to the data for this station published by Vegter & de Visscher (1987) and in Wetsteyn & Bakker (1991). Recalculated produc-

tion values were higher for 1983 (44%), 1986 (47%) and 1987 (23%) and lower for 1988 (10%). All primary production measurements were done within 1 hour after sampling, at the temperature of the sampling site.

Chlorophyll-*a* was measured fluorometrically according to Strickland & Parsons (1972) during the period 1980–1985; from 1986 onwards by reversed-phase HPLC analyses. Both methods were compared in 1986. Although individual differences were observed between both methods, on average both methods were not statistically different (P. Herman, unpubl. data).

Phytoplankton carbon-specific growth rates (P/B ratio, d^{-1}) were calculated as daily 'basin' production (g C m^{-2} d^{-1}) divided by biomass (g C m^{-2}). A carbon/chlorophyll-*a* ratio of 30 (phytoplankton carbon data from station LGPK, sampled at mid-depth during half-tide, combined with chlorophyll-*a* data from station O21) was used to convert chlorophyll-*a* data into phytoplankton carbon, because this was about the mean ratio during the period 1982–1990 (Table 2).

Phytoplankton cells were counted with an inverted microscope. Algal biomass as carbon was calculated from cell counts and volume measurements using Eppley's equations (in Smayda, 1978).

Table 2. Cumulative (see text) annual concentrations of phytoplankton carbon (g C m^{-3}) and chlorophyll-*a* (mg m^{-3}) and C/Chl-*a* ratios in the eastern (LGPK, O21) compartment of the Oosterschelde.

Year	LGPK Phyto-C	O21 Chl-*a*	LGPK, O21 C/Chl-*a*
1982	54.7	1478	37
1983	41.4	1840	23
1984	63.4	1328	48
1985	51.5	3118	17
1986	93.1	2788	33
1987	64.1	3050	21
1988	55.9	2599	21
1989	86.6	3335	26

Results

Salinity and water temperature

The average salinity in the western compartment remained constant at *ca* 31 during the period 1980–1989. There was a slight increase in the average salinity in the central part from *ca* 29 to *ca* 30. However, due to the reduced freshwater load the minimum salinity in the eastern compartment increased from *ca* 25 before to *ca* 30 after the completion of the civil engineering works. A similar increase (from *ca* 22 to *ca* 25) was observed in the northern compartment. For more detailed information on salinity (and water temperature) see Wetsteyn & Bakker (1991: Table 2).

Seston and turbidity

Suspended matter concentrations were highest in the western compartment and lowest in the northern compartment (Fig. 2). Due to the reduced flow velocities there was a decrease in the amount of suspended matter, which was observed earlier in the eastern and central compartments of the basin than in the western compartment (Fig. 2).

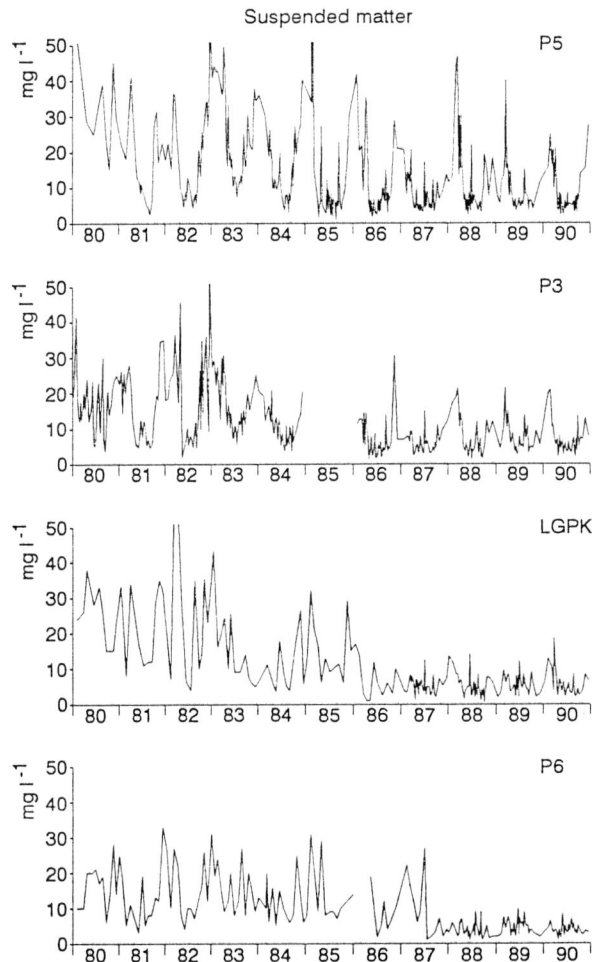

Fig. 2. Time-series of suspended matter concentrations (in mg l^{-1}) at four stations in the Oosterschelde.

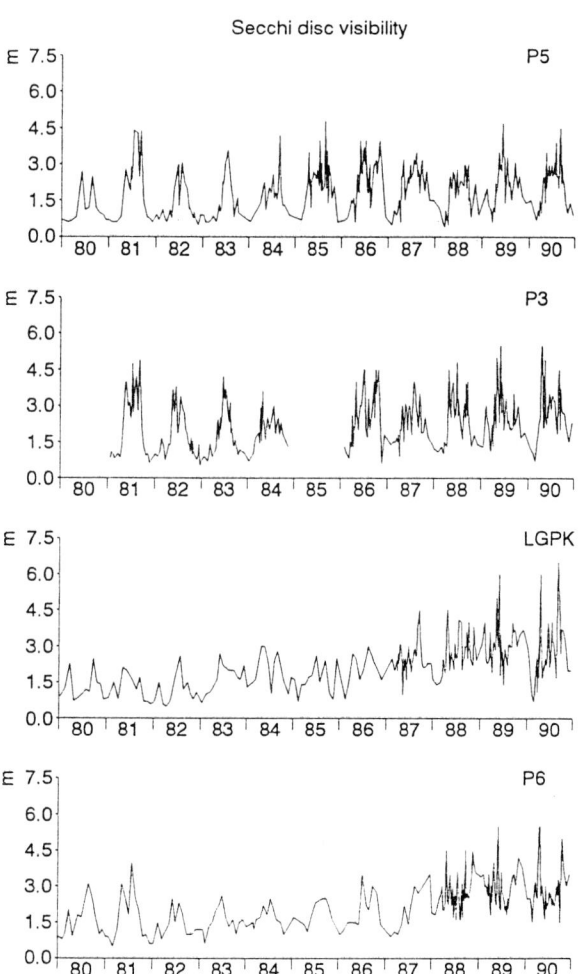

Fig. 3. Time-series of Secchi disc visibilities (in m) at four stations in the Oosterschelde.

Due to the decrease in suspended matter concentrations, the Secchi disc visibilities increased (Fig. 3). Note the disappearance of the annual pattern of suspended matter concentrations and transparencies in the eastern and northern compartments from 1987 onwards. Specific details on the behaviour of the suspended matter concentrations and Secchi disc visibilities in relation to the building of the storm-surge barrier and the compartment dams were already reported in Wetsteyn & Bakker (1991). Data on vertical attenuation coefficients are not shown because they give the same information as Fig. 3.

Nutrients

Nutrient concentrations showed characteristic seasonal cycles and a spatial gradient in autumn/winter values with highest values in the northern compartment.

Maximum phosphate concentrations in autumn-winter remained more or less constant in all compartments during the period 1980–1989 (Fig. 4). However, from 1990 onwards phosphate concentrations began to decrease. Minimum phosphate concentrations in May were less than found previously, especially in the eastern and northern compartments from 1984 onwards.

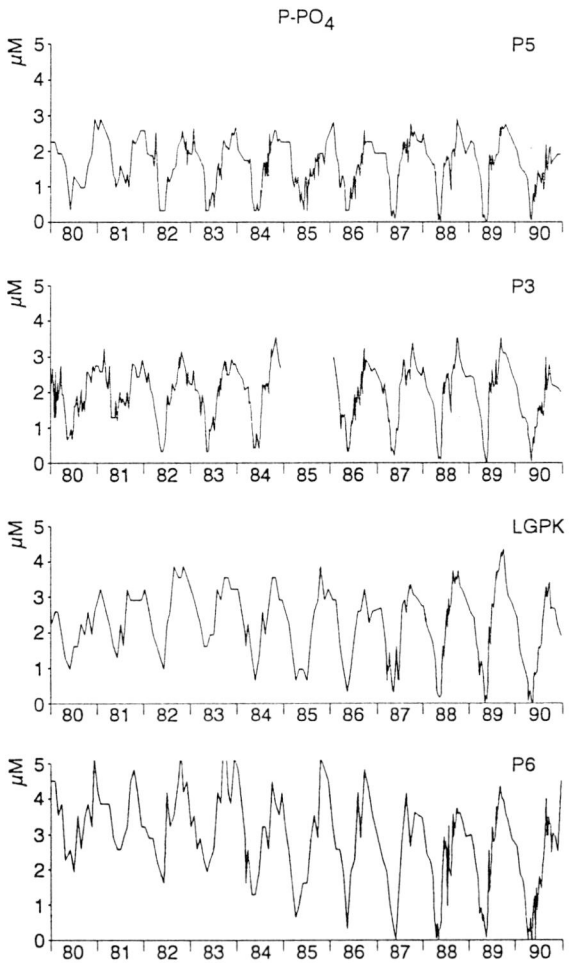

Fig. 4. Time-series of phosphate concentrations (in μM $P\text{-}PO_4$) at four stations in the Oosterschelde.

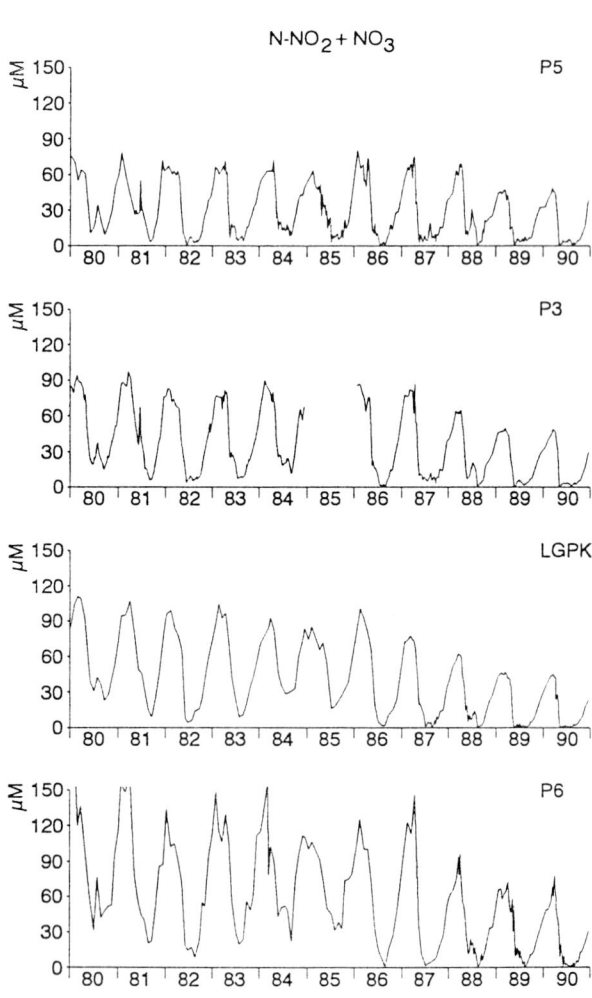

Fig. 5. Time-series of nitrite + nitrate concentrations (in μM $N\text{-}NO_2 + NO_3$) at four stations in the Oosterschelde.

Winter nitrite + nitrate concentrations decreased after the closure of the Philipsdam in 1987, especially in the central, eastern and northern compartments (Fig. 5). In the eastern compartment this happened a year earlier than in the other compartments, because the Oesterdam was closed in 1986 before the closure of the Philipsdam in 1987. Winter concentrations in the central and eastern compartments reached similar values as in the western, most seaward compartment, indicating a greater influence of the North Sea in the post-barrier situation. This is obvious in the salinity distribution too (see Wetsteyn & Bakker, 1991: Table 2). During the

period 1986–1990, minimum summer concentrations in the central, eastern and northern compartments were lower than before.

Winter ammonium concentrations remained more or less similar in the western and central compartments during the period 1980–1990, while those in the eastern and northern compartments tended to decrease after 1987 (Fig. 6).

Winter silicate concentrations decreased after the closure of the Philipsdam in 1987, especially in the eastern and northern compartments (Fig. 7). As with the nitrite + nitrate concentrations, the decrease of silicate concentrations in the eastern compartment started a year earlier.

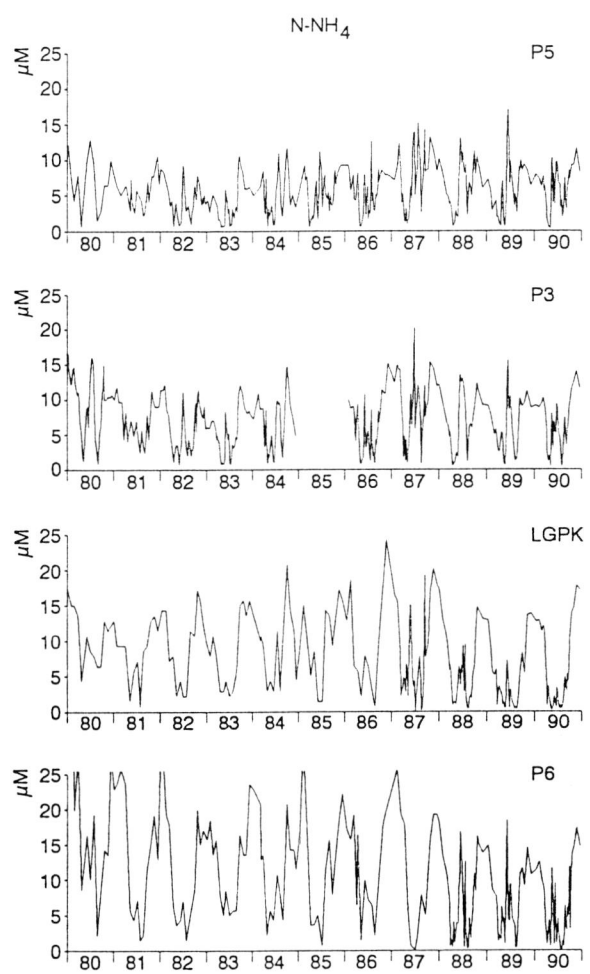

Fig. 6. Time-series of ammonium concentrations (in μM N-NH$_4$) at four stations in the Oosterschelde.

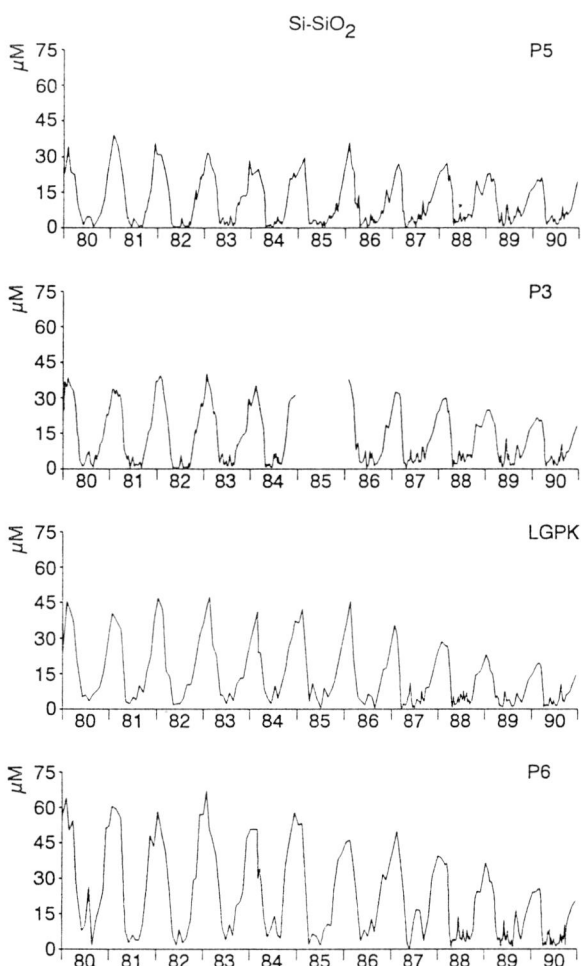

Fig. 7. Time-series of silicate concentrations (in μM Si-SiO$_2$) at four stations in the Oosterschelde.

Winter concentrations in the central and eastern compartments reached the same values as in the western compartment.

From the dissolved nutrient concentrations we calculated the molar N/P-, Si/P- and Si/N-ratios (N as nitrite + nitrate + ammonium) (Figs 8–10). During the post-barrier period higher N/P-, Si/P- and Si/N-ratios were calculated. The higher N/P- and Si/P-ratios are caused by the relatively more decreased P concentrations in comparison with the N- and Si concentrations; the higher Si/N-ratios can be explained by the relatively more decreased N concentrations when compared with the Si concentrations. During the growth season, the molar N/P- and Si/P-ratios were much smaller

than the Redfield ratio of 16, indicating a relative shortage of N and a surplus of P. The Si/N-ratio was far below the Redfield ratio of 1, suggesting a relative shortage of silicate (see Discussion).

Information on nutrients within a more detailed time-scale is presented in Bakker & Vink (1994).

Phytoplankton biomass

From the annual patterns in chlorophyll-*a* concentrations in the western and central compartments no obvious trend is visible; in the eastern and northern compartments higher chlorophyll-*a* concentrations were measured from 1985 on-

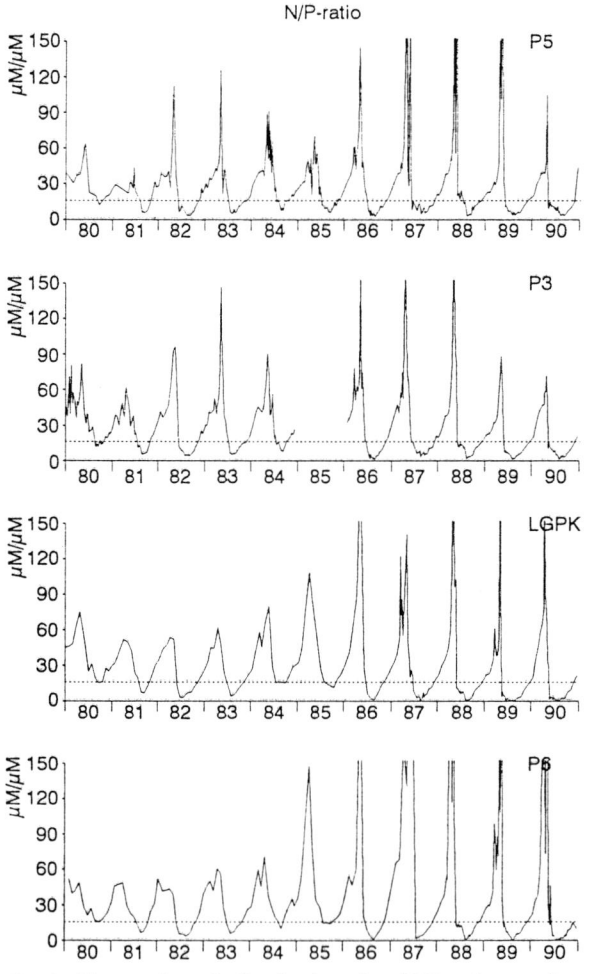

Fig. 8. Time-series of dissolved molar N/P-ratios at four stations in the Oosterschelde. Dashed lines indicate N/P = 16.

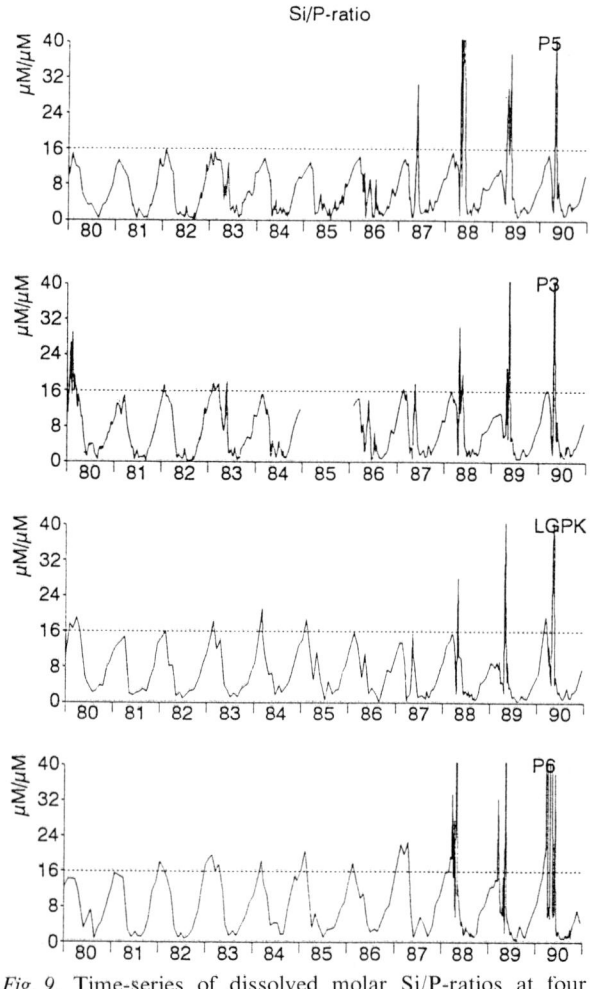

Fig. 9. Time-series of dissolved molar Si/P-ratios at four stations in the Oosterschelde. Dashed lines indicate Si/P = 16.

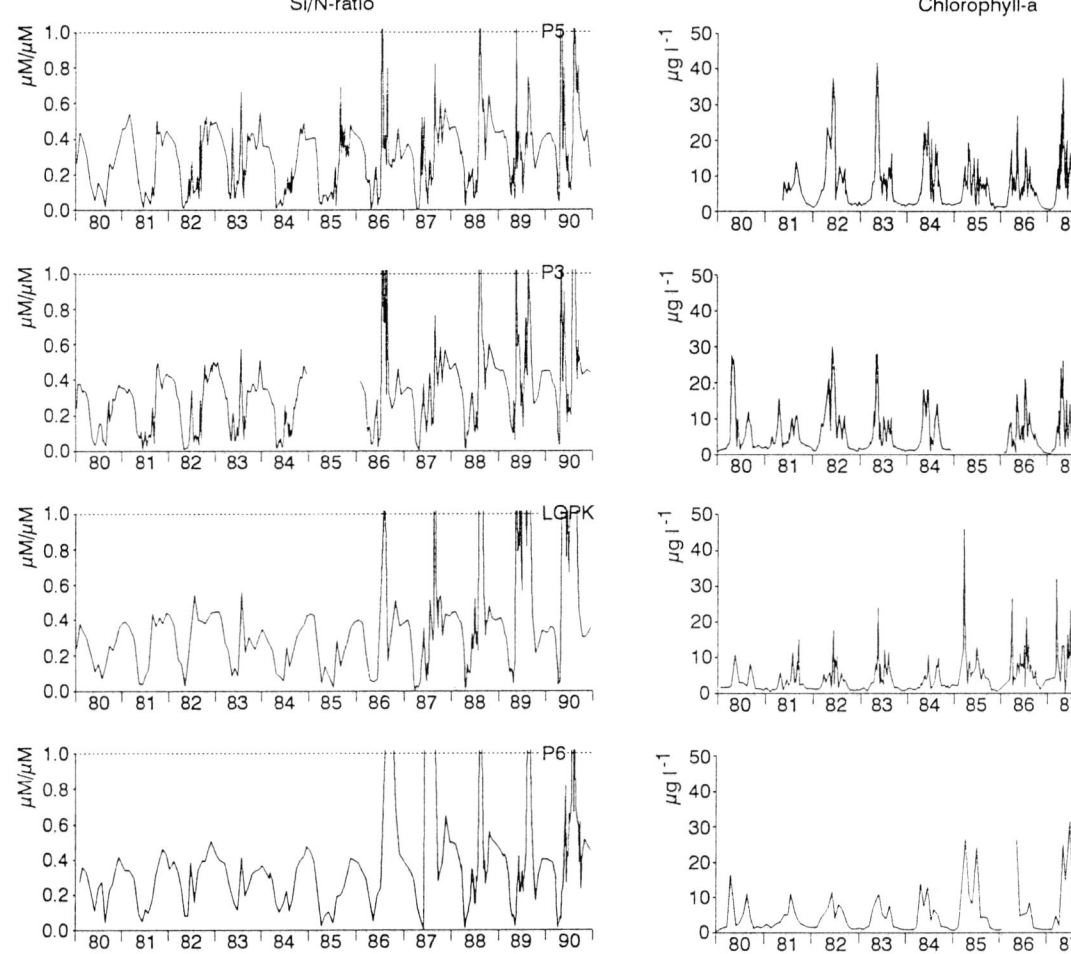

Fig. 10. Time-series of dissolved molar Si/N-ratios at four stations in the Oosterschelde. Dashed lines indicate Si/N = 1.

Fig. 11. Time-series of chlorophyll-*a* concentrations (in $\mu g\, l^{-1}$) at four stations in the Oosterschelde.

wards (Fig. 11). Looking at the cumulative annual phytoplankton carbon concentrations in the eastern compartment, the period 1986–1990 seems to have higher values (Table 2). This is even more striking for the cumulative annual chlorophyll-*a* concentrations (*i.e.* the total chlorophyll-*a* concentration (the area) under the graph of date (day) against chlorophyll-*a* concentration) in the period 1985–1990 (Table 2). Hence, from the cumulative annual yields, it is obvious that in the post-barrier period the phytoplankton biomass in the eastern compartment increased substantially.

Phytoplankton primary production

Daily column ('basin' production is not reported in this paper, but see Wetsteyn *et al.*, 1990) production is given in Fig. 12 (stations LGPK and P6 excluded). Column production maxima in the Oosterschelde compartments in general reached values between *ca* 4 g C m^{-2} d^{-1} and 6 g C m^{-2} d^{-1}. Annual column production has been summarized in Table 3 and data from completely measured years during the pre-barrier (1980–1984) and post-barrier (1988–1990) periods were used to produce Fig. 13. Figure 13 shows that during the pre-barrier period a gradient existed

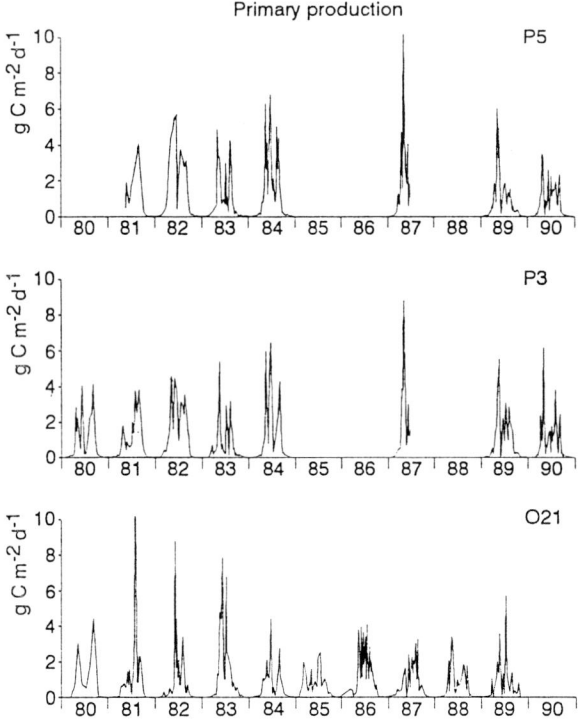

Fig. 12. Time-series of daily primary production (in g C m^{-2} d^{-1}) at three stations in the Oosterschelde.

with highest values at the western, most seaward side (P5). During the post-barrier period annual

Table 3. Annual column production (g C m^{-2}) in the Oosterschelde. Stations P5, P3, LGPK (1987–1990) and P6: particulate + dissolved production; Stations O21 and LGPK (1984): particulate production. Data for O21 in 1980 and 1981 and LGPK in 1984 from Vegter & de Visscher (1987, and unpublished). nm: not measured; *: based on monthly measurements; **: from 20 May onwards; ***: measurements until July.

Year	P5	P3	O21	LGPK	P6
1980	nm	327	345*	nm	nm
1981	280**	275	338	nm	nm
1982	540	406	264	nm	nm
1983	283	240	380	nm	nm
1984	425	373	176	176	nm
1985	nm	nm	198	nm	nm
1986	nm	nm	460	nm	nm
1987	260***	233***	250	169***	nm
1988	nm	nm	230	nm	nm
1989	277	319	229	294	550
1990	223	242	nm	237	502

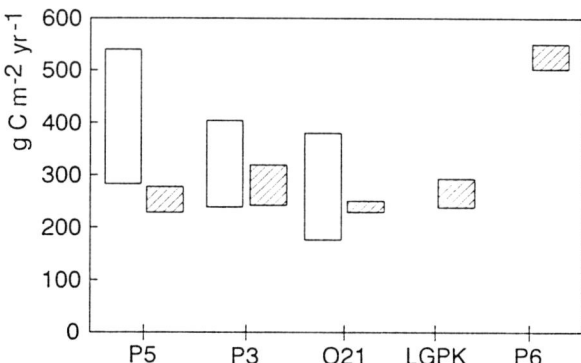

Fig. 13. Ranges of measured annual column production values at five stations in the Oosterschelde; open bars: pre-barrier period (1980–1984), hatched bars: post-barrier period (1987–1990).

production at P5 fell beneath the range measured during the pre-barrier period, while at stations P3 and O21 all annual production values fall within the range measured during the pre-barrier period. The highest annual production was measured at

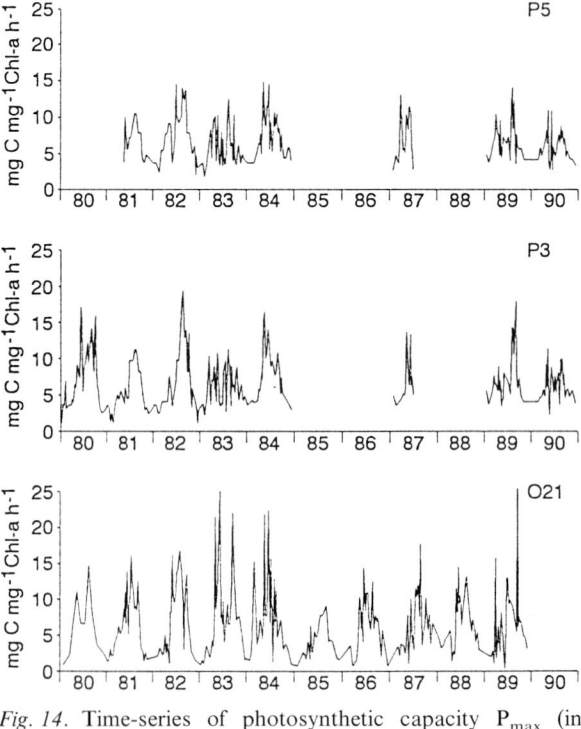

Fig. 14. Time-series of photosynthetic capacity P$_{max}$ (in mg C mg^{-1} Chl-a h^{-1}) at three stations in the Oosterschelde.

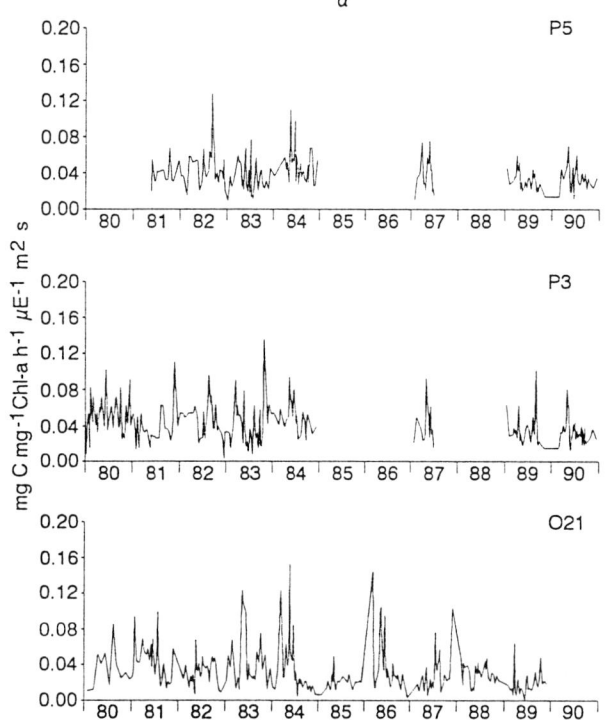

Fig. 15. Time-series of photosynthetic efficiency α (in mg C mg^{-1} Chl-*a* h^{-1} μE^{-1} m^2 s) at three stations in the Oosterschelde.

station P6; from this station no pre-barrier production data are available. During the post-barrier situation annual primary production values at all sampling stations, except station P6, fall within the same range.

Physiological photosynthetic parameters

The annual trends of photosynthetic capacity P_{max} at stations P5, P3 and O21 are shown in Fig. 14. As expected, highest values were observed in summer, with maxima greater than 10 mg C mg^{-1} Chl-*a* h^{-1} (see Discussion). During the pre-barrier period, mean values in spring and summer seem higher than during the barrier and post-barrier period (values exceeding 15 mg C mg^{-1} Chl-*a* h^{-1} are more frequent during the pre-barrier period).

The photosynthetic efficiency α at stations P5, P3 and O21, too, seem to decrease from 1985 onwards (Fig. 15). The decrease in photosynthetic capacity and efficiency becomes more obvious when the average seasonal values (*i.e.* April–September) of P_{max} and α are calculated

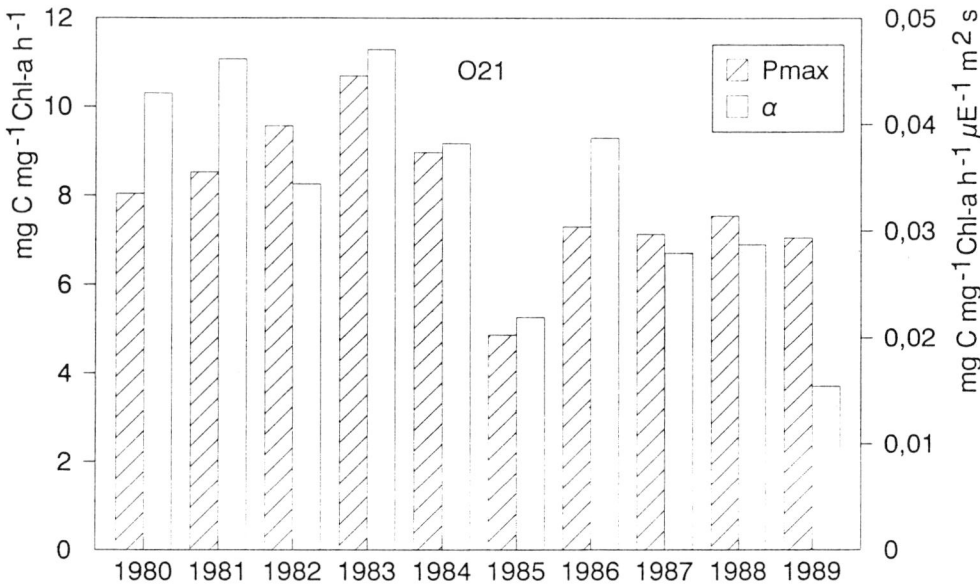

Fig. 16. Average seasonal (April–September) values of the photosynthetic capacity P_{max} (hatched bars, in mg C mg^{-1} Chl-*a* h^{-1}) and the photosynthetic efficiency α (open bars, in mg C mg^{-1} Chl-*a* h^{-1} μE^{-1} m^2 s) at station O21 in the Oosterschelde.

(Fig. 16; only data from station O21). Annually averaged P_{max} and α values at station O21 were 22%, respectively 43% lower; P_{max} and α values at station P5 decreased with 16 and 26% and at station P3 with 16 and 24%.

Phytoplankton carbon-specific growth rates

Phytoplankton carbon-specific growth rates at station O21 varied between < 0.1 d^{-1} in winter to more than 3 d^{-1} during summer; at stations P5 and P3 maximum carbon-specific growth rates were two to three times lower (Fig. 17). The few exceptionally high values in 1983 and 1984 at station O21 are likely due to measuring errors. Average seasonal values (April–September) during the pre- and post-barrier periods were all in the same range.

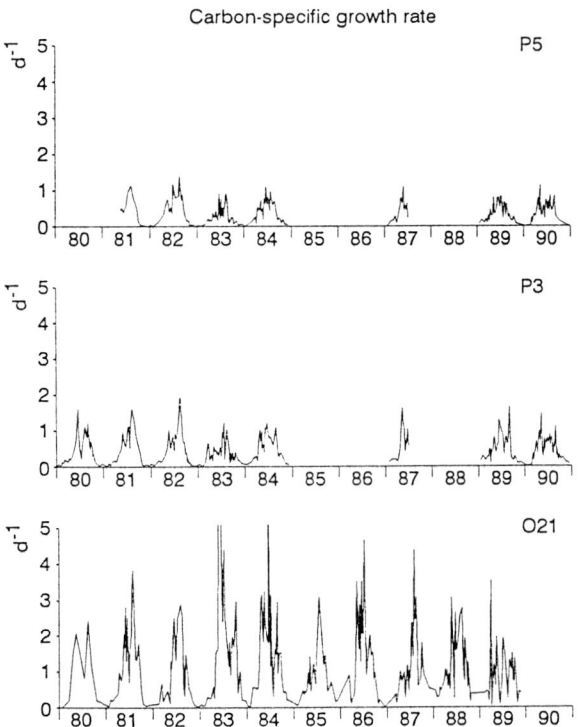

Fig. 17. Time-series of carbon-specific growth rates (in d^{-1}) at three stations in the Oosterschelde.

Discussion

Turbidity

The maximum flow velocities in the Oosterschelde have decreased by 25–86%, depending on the location in the basin (for more specific information see Wetsteyn & Bakker, 1991: Table 3 and discussion). As mentioned before, these reduced velocities resulted in increased water transparencies.

Nutrients

The total freshwater loads were reduced after April 1987 by *ca* 64% (Wetsteyn & Bakker, 1991; Smaal & Boeije, 1991: Fig. 3.8). This decrease of freshwater discharges reduced the nutrient loads, especially to the northern compartment. The nutrient loading to the northern compartment severely influenced the nutrient concentrations in this compartment and in the central and eastern compartments as well. This is because water from the northern compartment enters the central compartment during ebb-tide. During flood-tide this water flows into the eastern compartment via ebb and flood tidal gullies.

After the reduction of nutrient loads, the winter concentrations of nitrite + nitrate (Fig. 5, Table 4) and silicate (Fig. 7, Table 4) decreased, especially in the central, eastern and northern compartments.

The reduction of the nutrient loads did not produce immediately lower autumn phosphate concentrations (Fig. 4), a probable effect of phosphate release from the bottom sediments. In 1990 the autumn phosphate concentrations in the Oosterschelde started to decrease, at first in the western compartment; this decrease was continued in 1991 and 1992 (not shown). Summer phosphate concentrations at stations P5 and P3 did not increase as fast as in former years, which is an indication that phosphate release from the bottom sediments is decreasing. The importance of phosphate release from bottom sediments was also reported by De Jonge & Essink (1991) for

Table 4. Maximum nutrient concentrations at four stations in the Oosterschelde and at one station in the adjacent coastal waters (S10, Fig. 1) during the pre-barrier period 1980–1984 (given as ranges) and in the post-barrier year 1990. N-NO$_2$ + NO$_3$, P-PO$_4$ (station S10) and Si-SiO$_2$: mean January + February values; P-PO$_4$ (other stations): mean October + November values.

	N-NO$_2$ + NO$_3$ (μM)		P-PO$_4$ (μM)		Si-SiO$_2$ (μM)	
	1980–1984	1990	1980–1984	1990	1980–1984	1990
S10	30.5–45.8	22.1	1.5–1.9	1.3	13.8–22.3	9.3
P5	59.2–69.6	38.7	1.8–2.4	1.7	23.2–35.2	19.0
P3	72.0–86.9	39.4	2.5–3.2	2.2	31.0–36.3	20.5
LGPK	68.2–100.0	38.8	2.3–3.7	2.6	29.6–44.3	17.9
P6	114.9–155.6	51.1	3.6–4.8	2.8	50.9–58.4	24.5

the Dutch Wadden Sea and by Kelderman (1984) for Lake Grevelingen (Grevelingenmeer) (Fig. 1).

The observed decrease in winter nutrient concentrations cannot be explained by the civil engineering works only. As can be seen in Table 4, nutrient concentrations in the neighbouring coastal regions of the North Sea decreased in this period too. This may be of consequence for the nutrient concentrations in the Oosterschelde which are now to a large extent influenced by the nutrient situation of the North Sea coastal waters; only in the northern compartment the freshwater influence is still measurable.

The average molar nutrient content of phytoplankton is P:N:Si = 1:16:16 (Gillbricht, 1988). These ratios, together with the half-saturation constants for nutrient uptake by natural phytoplankton populations (phosphate 0.1–0.5 μM, dissolved inorganic nitrogen 1–2 μM, silicate 1–5 μM, Fisher *et al.*, 1988) can be used to estimate the growth limiting factors. As in nearly all cases phosphate concentrations (Fig. 4) were higher than 1 μM and the N/P- and Si/P-ratios (Figs 8–10) during the growth season far below 16, it seems justified to conclude that phosphate was never a growth-limiting nutrient for phytoplankton. In contrast, both dissolved inorganic nitrogen and silicate reached growth limiting concentrations. This is supported by the results of bio-assays in 1989 with the diatom *Skeletonema costatum* and the flagellate *Rhodomonas sp.* (unpublished data Wetsteyn & Rijstenbil). As the Si/N-ratio (Fig. 10) in nearly all cases was below

1 (except during the winter), it is likely that Si will finally be the limiting nutrient for diatom growth. As a consequence, the diatom population will collapse and non-silicate containing species will take over. This is consistent with the observed situation in the Oosterschelde (Bakker *et al.*, 1990) and in other Dutch coastal waters (e.g. Van Bennekom *et al.*, 1975; Gieskes & Kraay, 1975; Cadée, 1986a; Veldhuis *et al.*, 1986). For the non-Si containing algae, nitrogen is likely to be the growth-limiting nutrient during the post-barrier period. On the other hand, we have to be careful to draw these conclusions on the basis of nutrient concentrations alone. Due to the microbial loop activity, nutrient fluxes may be high, in spite of low concentrations. It might be better to compare ratios on basis of total contents in the water (*i.e.* on dissolved and particulate nutrients) as the microbial loop activity can cause a rapid transition from one phase to another (Sherr & Sherr, 1984). However, this will not seriously affect our conclusions, because silicate turns over more slowly than nitrogen (Officer & Ryther, 1980).

De Jong *et al.* (1994) found a strong increase of microphytobenthic biomass during post-barrier years and it is well documented that phytoplankton (especially diatoms) may compete for nutrients with benthic diatoms (Bakker & de Vries, 1984; De Vries & Hopstaken, 1984). This aspect of competition between phytoplankton and microphytobenthos for nutrients in the water column is discussed in Bakker & Vink (1994).

Light limitation of the phytoplankton is another

possibility which has to be considered. The photosynthetic parameters, however, suggest that light was not a limiting factor. This will be discussed in detail below.

Phytoplankton

As light and available nutrients are the most important factors governing phytoplankton growth and primary production, their changes in the Oosterschelde must have affected the phytoplankton. As shown above, the light conditions for phytoplankton have improved due to an increase in transparency. A Canonical Correspondence Analysis showed that the phytoplankton assemblage adopted (on taxonomic and population dynamical level) summer characteristics both earlier and later in the year and that light was the driving factor (Bakker *et al.*, 1990, 1994).

During the barrier period, the annual phytoplankton biomass increased in the eastern compartment (Fig. 11, Table 2). Because this increase already appeared in 1985, before the switch from fluorometry to HPLC, these higher cumulative annual chlorophyll-*a* concentrations are not likely to be caused by methodological differences, at least on an annual basis (see Materials and methods). Apparently, in the eastern compartment the reduction in water velocity, causing an increase in transparency, was of greater importance for algal growth than the reduction in nutrient input in 1987 (but see below). In the northern compartment higher chlorophyll-*a* concentrations from 1985 onwards can be explained by the improved light conditions, while nutrient concentrations were high enough to sustain phytoplankton growth.

During the post-barrier period annual primary production values at all stations, except station P6, reached comparable values. The total primary production at station O21 may have been underestimated, because here only particulate production was measured. If dissolved production (extracellular release) is taken into account (10–15% of annual total production; Iturriaga & Zsolnay, 1983; Lancelot, 1983; Vegter & de Visscher, 1984b) the similarity between the measured post-barrier values is even more striking.

During the post-barrier period annual primary production values at stations P3 and O21 fell within the ranges measured during the pre-barrier period (Fig. 13). This was not the case at station P5, where the post-barrier values were beneath the range measured during the pre-barrier period (Fig. 13). Yet it is difficult to decide whether or not this is a real decrease because of the great variability in annual production values. The lower annual production values at this station during the post-barrier situation were caused by the absence of summer peaks, which coincides with the period of low nutrient concentrations. We conclude that at least in the central and eastern compartments annual primary production has not changed and that a certain degree of functional homeostasis exists (cf. Klepper, 1989 and Scholten *et al.*, 1990). This homeostasis makes it difficult to compare different coastal ecosystems by means of primary production only and therefore this parameter alone is not suited for biomonitoring purposes. Rather large-scale changes with respect to phytoplankton structure have to be taken into account. Many tables with annual primary production values have been published for reasons of comparison (Colijn, 1983; Flint, 1984; Vegter & de Visscher, 1984a; Pennock & Sharp, 1986; Quéguiner & Tréguer, 1986), comprising a broad range (from 40 to 820 g C m^{-2} yr^{-1}) of different values for estuaries (including the riverine and marine parts) and coastal areas of North-America and Europe. Large year-to-year variations may be the result of natural variability, caused by variations in environmental conditions. Also increasing production, due to eutrophication (e.g. the Marsdiep tidal inlet of the western Wadden Sea (Cadée, 1986b)), may give rise to a large range of annual production values. Just as in the Oosterschelde (176–550 g C m^{-2} yr^{-1}, this study), large long-term ranges of annual primary production values have also been observed in Central Bay (337–782 g C m^{-2} yr^{-1}: Boynton *et al.*, 1982), in the Marsdiep tidal inlet of the western Wadden Sea (150–400 g C m^{-2} yr^{-1}: Cadée & Hegeman, 1991: Fig. 2a) and in the Delaware estuary (190–400 g C m^{-2} yr^{-1}:

Pennock & Sharp, 1986). As can be seen, despite the fact that these are rather different ecosystems, the primary production of the systems is comparable.

Our analyses also show that physiological changes in photosynthetic parameters occurred at station O21, but also at stations P5 and P3. The photosynthetic capacity P_{max} showed a decrease in the barrier and post-barrier period compared to the pre-barrier period (Fig. 14). In the pre-barrier period values between 10 and 25 mg C mg^{-1} Chl-a h^{-1} were common whereas in the barrier and post-barrier period most values were below 15 mg C mg^{-1} Chl-a h^{-1}. During and after the barrier period photosynthetic efficiency α had decreased (in many cases below 0.02 mg C mg^{-1} Chl-a h^{-1} μE^{-1} m^2 s, Fig. 15)

Light-shade adaptation theory (see e.g. Dubinsky, 1980; Falkowski, 1980; Berner et al., 1989) predicts that the photosynthetic capacity P_{max} and the photosynthetic efficiency α (both expressed per unit chlorophyll-a) will increase when algae receive more light, or when a nutrient-limitation is alleviated. The increase in α is primarily due to a decrease in the package effect. Hence, our data suggest that, as a result of the civil engineering works, especially the phytoplankton in the eastern compartment became more shade adapted. The post-barrier data are similar to P-I curve parameters observed for phytoplankton in the nearby, very turbid and eutrophic Westerschelde estuary (Van Spaendonk et al., 1993). However, an adaptation to a deteriorated light climate is very unlikely because the water transparency increased and the availability of nutrients decreased. This apparent discrepancy may be explained by the species shift which occurred as a result of the improved light climate (Bakker et al., 1990, 1993). Smaller and more fragile algal forms (especially diatoms) occurred after the construction of the storm-surge barrier and the two compartment dams. This suggests an adaptation to lower nutrient (especially silicate in the case of diatoms) conditions and to a reduced turbulence (see also Van der Tol & Scholten, 1992).

As shown earlier (Wetsteyn et al., 1990: Fig. 8) the carbon-specific growth rates were higher in the eastern than in the western compartment. This can be explained by the large differences in depth of the compartments (Wetsteyn et al., 1990). The western compartment is much deeper and hence has a lower euphotic to mixing depth ratio with concomitant higher respiratory losses. No significant changes in carbon-specific growth rates were observed (Fig. 17).

As the biomass in the eastern compartment increased (Table 2) regeneration of nutrients is likely to be of greater importance during the post-barrier period. Unfortunately, the ratio of new versus regenerated production (cf. Dugdale & Goering, 1967, 1986) was not measured in the different periods.

From model simulations it appears that the Oosterschelde is relatively insensitive to changes in nutrient loadings and that suspension-feeders substantially contribute to a functional stability (Herman & Scholten, 1990; Van der Tol & Scholten, 1992). However, benthic suspension feeders are not very important in the eastern compartment of the Oosterschelde. Copepod biomass in the eastern compartment increased during the post-barrier years and thus will exert a stronger grazing pressure on the phytoplankton (Bakker & van Rijswijk, 1993). Model simulation and experiments showed a slight increase in the rate of zooplankton grazing in the post-barrier period (Van der Tol & Scholten, 1992; C. Bakker, pers. comm.). To explain the increase in biomass in the eastern compartment, total loss processes have to be smaller in the post-barrier period than before. Sedimentation losses are likely to increase due to the reduction in flow velocities. On the other hand, the phytoplankton composition changed in such a way (more smaller forms) that losses by sedimentation might actually have decreased. Another explanation may be the reduced flushing time: phytoplankton biomass produced in the eastern compartment is washed out with a lower rate.

Conclusions

The result of the civil-engineering works was a levelling (i.e. a loss of gradients within the Oosterschelde) of salinities, nutrient concentrations, chlorophyll-*a* concentrations and annual primary production. Because of the loss in estuarine gradients, which were already small, the Oosterschelde became more and more a marine tidal basin. The annual primary production did not change significantly, despite the large changes in the chemical and physical environment; this homeostasis was due to a change in phytoplankton species composition.

Acknowledgements

We thank C. Bakker for providing phytoplankton carbon data from station LGPK. We wish to thank the crews of the R. V. Argus, Delta, Ventjager, Jan Verwey and Luctor for sampling and the personnel of the laboratories in Middelburg and Yerseke for the chemical analyses. The comments of two referees helped to improve the manuscript. Communication No. 650 of the Netherlands Institute of Ecology, Yerseke, The Netherlands.

References

Bakker, C. & P. van Rijswijk, 1994. Zooplankton biomass in the Oosterschelde (SW Netherlands) before, during and after the construction of a storm-surge barrier. Hydrobiologia 282/283: 127–143.

Bakker, C. & M. Vink, 1994. Nutrient concentrations and planktonic diatom-flagellate relations in the Oosterschelde (SW Netherlands) during and after the construction of a storm-surge barrier. Hydrobiologia 282/283: 101–116.

Bakker, C. & I. de Vries, 1984. Phytoplankton- and nutrient dynamics in saline Lake Grevelingen (SW Netherlands) under different hydrodynamical conditions in 1978–1980. Neth. J. Sea Res. 18: 191–220.

Bakker, C., P. M. J. Herman & M. Vink, 1990. Changes in seasonal succession of phytoplankton induced by the storm-surge barrier in the Oosterschelde (SW Netherlands). J. Plankton Res. 12: 947–972.

Bakker, C., P. M. J. Herman & M. Vink, 1994. A new trend in the development of the phytoplankton in the Oosterschelde (SW Netherlands) during and after the construction of a storm-surge barrier. Hydrobiologia 282/283: 79–100.

Berner, T., Z. Dubinsky, K. Wyman & P. G. Falkowski, 1989. Photoadaptation and the 'package' effect in Du-

naliella tertiolecta (Chlorophyceae). J. Phycol. 25: 70–78.

Boynton, W. R., W. M. Kemps & C. W. Keefe, 1982. A comparative analysis of nutrients and other factors influencing estuarine phytoplankton production. In V. S. Kennedy (ed.), Estuarine comparisons. Academic Press, New York: 69–90.

Cadée, G. C., 1982. Tidal and seasonal variation in particulate and dissolved organic carbon in the western Dutch Wadden Sea and Marsdiep tidal inlet. Neth. J. Sea Res. 15: 228–249.

Cadée, G. C., 1986a. Recurrent and changing seasonal patterns in phytoplankton of the westernmost inlet of the Dutch Wadden Sea from 1969 to 1985. Mar. Biol. 93: 281–289.

Cadée, G. C., 1986b. Increased phytoplankton primary production in the Marsdiep area (western Dutch Wadden Sea). Neth. J. Sea Res. 20: 285–290.

Cadée, G. C. & J. Hegeman, 1991. Phytoplankton primary production, chlorophyll and species composition, organic carbon and turbidity in the Marsdiep in 1990, compared with foregoing years. Hydrobiol. Bull. 25: 29–35.

Colijn, F., 1983. Primary production in the Ems-Dollard estuary. Thesis, University of Groningen.

De Jong, D. J., P. H. Nienhuis & B. J. Kater, 1994. Microphytobenthos in the Oosterschelde estuary (The Netherlands), 1981–1990; consequences of a changed tidal regime. Hydrobiologia, 282/283: 183–195.

De Jonge, V. N. & K. Essink, 1991. Long-term changes in nutrient loads and primary and secondary producers in the Dutch Wadden Sea. In M. Elliott & J.-P. Ducrotoy (eds), Estuaries and Coasts: Spatial and Temporal Intercomparisons. Proceedings of the ECSA 19 Symposium, Caen 1989. Olsen & Olsen, Fredensborg, Denmark: 307–316.

De Vries, I. & C. F. Hopstaken, 1984. Nutrient cycling and ecosystem behaviour in a salt-water lake. Neth. J. Sea Res. 18: 221–245.

Dronkers, J. & J. T. F. Zimmerman, 1982. Some principles of mixing in tidal lagoons with examples of tidal basins in the Oosterschelde. In: P. Lasserre & H. Postma (eds), Coastal Lagoons. Proc. Int. Symp. Coastal Lagoons, Gauthier-Villars: 460–474.

Dubinsky, Z., 1980. Light utilization efficiency in natural phytoplankton communities. In P. G. Falkowski (ed.), Primary productivity in the sea. Plenum Press, New York: 83–97.

Dugdale, R. C. & J. J. Goering, 1967. Uptake of new and regenerated forms of nitrogen in primary production. Limnol. Oceanogr. 12: 196–206.

Dugdale, R. C. & J. J. Goering, 1986. The use of ^{15}N to measure nitrogen uptake in eutrophic oceans; experimental considerations. Limnol. Oceanogr. 31: 673–680.

Eilers, P. H. C. & J. C. H. Peeters, 1988. A model for the relationship between light intensity and the rate of photosynthesis in phytoplankton. Ecol. Modell. 42: 199–215.

Falkowski, P. G., 1980. Light-shade adaptation in marine phytoplankton. In P. G. Falkowski (ed.), Primary productivity in the sea. Plenum Press, New York: 99–119.

Fee, E. J., 1973. A numerical model for determining integral primary production and its application to Lake Michigan. J. Fish. Res. Bd Can. 30: 1447–1468.

Fisher, T. R., L. W. Harding, D. W. Stanley & L. G. Ward, 1988. Phytoplankton, nutrients and turbidity in the Cheasapeake, Delaware, and Hudson estuaries. Estuar. coast. Shelf Sci. 27: 61–93.

Flint, R. W., 1984. Phytoplankton production in the Corpus Christi Bay estuary. Contributions in Marine Science 27: 65–83.

Gieskes, W. W. C. & G. W. Kraay, 1975. The phytoplankton spring bloom in Dutch coastal waters of the North Sea. Neth. J. Sea Res. 9: 166–196.

Gillbricht, M., 1988. Phytoplankton and nutrients in the Helgoland region. Helgoländer Meeresunters. 42: 435–467.

Grasshoff, K., M. Ehrhardt & K. Kremling, 1983. Methods of seawater analysis. Verlag Chemie, Weinheim.

Herman, P. J. & H. Scholten, 1990. Can suspension-feeders stabilise estuarine ecosystems? In M. Barnes & R. N. Gibson (eds), Trophic Relationships in the Marine Environment. Proc. 24th Europ. Mar. Biol. Symp. Aberdeen University Press: 104–116.

Iturriaga, R. & A. Zsolnay, 1983. Heterotrophic uptake and transformation of phytoplankton extracellular products. Bot. Mar. 26: 375–381.

Kelderman, P., 1984. Phosphate budget and sediment-water exchange in Lake Grevelingen (SW Netherlands). Neth. J. Sea Res. 14: 229–236.

Klepper, O., J. C. H. Peeters, J. P. G. van de Kamer & P. Eilers, 1988. The calculation of primary production in an estuary. A model that incorporates the dynamic response of algae, vertical mixing and basin morphology. In A. Marani (ed.), Advances in environmental modelling. Elsevier, Amsterdam: 373–394.

Klepper, O., 1989. A model of carbon flows in relation to macrobenthic food supply in the Oosterschelde estuary (S.W. Netherlands). Thesis, Landbouwuniversiteit Wageningen.

Knoester, M., J. Visser, B. A. Bannink, C. J. Colijn & W. P. A. Broeders, 1984. The Eastern Scheldt Project. Wat. Sci. Tech. 16: 51–77.

Lancelot, C., 1983. Factors affecting phytoplankton extracellular release in the Southern Bight of the North Sea. Mar. Ecol. Prog. Ser. 12: 115–121.

Nienhuis, P. H. & A. C. Smaal, 1994. The Oosterschelde estuary, a case study of a changing ecosystem: an introduction. Hydrobiologia 282/283: 1–14.

Officer, C. B. & J. H. Ryther, 1980. The possible importance of silicon in marine eutrophication. Mar. Ecol. Prog. Ser. 3: 83–91.

Pennock, J. R. & J. H. Sharp, 1986. Phytoplankton production in the Delaware estuary: temporal and spatial variability. Mar. Ecol. Prog. Ser. 34: 143–155.

Quéguiner, B. & P. Tréguer, 1986. Freshwater outflow effects in a coastal, macrotidal ecosystem as revealed by hydrological, chemical and biological variabilities (Bay of Brest, Western Europe). In S. Skreslet (ed.), The role of freshwater outflow in coastal marine ecosystems. NATO ASI Series, Springer-Verlag, Berlin: 219–230.

Scholten, H., O. Klepper, P. H. Nienhuis & M. Knoester, 1990. Oosterschelde estuary (S.W. Netherlands): a self-sustaining ecosystem? Hydrobiologia 195: 201–215.

Sherr, B. F. & E. B. Sherr, 1984. Role of heterotrophic protozoa in carbon and energy flow in aquatic ecosystems. In M. J. Klug & C. A. Reddy (eds), Current perspectives in Microbial Ecology. American Society for Microbiology, Washington D.C.: 412–423.

Smaal, A. C. & M. van Stralen, 1990. Average annual growth and condition of mussels as a function of food source. Hydrobiologia 195: 179–188.

Smaal, A. C. & R. C. Boeije, 1991. Veilig getij, de effecten van de waterbouwkundige werken op het getijdemilieu van de Oosterschelde. Nota GWWS 91.088 DGW/Directie Zeeland, Middelburg (in Dutch).

Smaal, A. C., M. Knoester, P. H. Nienhuis & P. M. Meire, 1991. Changes in the Oosterschelde ecosystem induced by the Delta works. In M. Elliott & J.-P. Ducrotoy (eds), Estuaries and Coasts: Spatial and Temporal Intercomparisons. Proceedings of the ECSA 19 Symposium, Caen 1989. Olsen & Olsen, Fredensborg, Denmark: 375–384.

Smayda, T., 1978. From phytoplankters to biomass. In A. Sournia (ed.), Phytoplankton Manual. Unesco, Paris: 273–279.

Strickland, J. D. H. & T. R. Parsons, 1972. A practical handbook of seawater analysis. Bull. Fish. Res. Bd Can. 167: 1–310.

Ten Brinke, W. B. M., 1991. Quantifying mud exchange between the Eastern Scheldt tidal basin and the North Sea. In N. C. Kraus, K. J. Gingerich & D. L. Kriebel (eds), Coastal Sediments '91. Volume 1. American Society of Civil Engineers, New York: 760–774.

Van Bennekom, A. J., W. W. C. Gieskes & S. B. Tijssen, 1975. Eutrophication of Dutch coastal waters. Proc. R. Soc. Lond. B. 189: 359–374.

Van der Tol, M. W. M. & H. Scholten, 1992. Response of the Oosterschelde ecosystem to a changing environment: adaptive or functional? Neth. J. Sea Res. 30: 175–190.

Van Spaendonk, J. C. M., P. R. M. de Visscher & J. Kromkamp, 1993. Primary production of phytoplankton in a turbid coastal plain estuary, The Westerschelde (The Netherlands). Neth. J. Sea Res. (in press).

Vegter, F. & P. R. M. de Visscher, 1984a. Phytoplankton primary production in brackish Lake Grevelingen (SW Netherlands) during 1976–1981. Neth. J. Sea Res. 18: 246–259.

Vegter, F. & P. R. M. de Visscher, 1984b. Extracellular release by phytoplankton during photosynthesis in Lake Grevelingen (SW Netherlands). Neth. J. Sea Res. 18: 260–272.

Vegter, F. & P. R. M. de Visscher, 1987. Nutrients and phytoplankton primary production in the marine tidal Ooster-

78

schelde estuary (The Netherlands). Hydrobiol. Bull. 21: 149–158.

Veldhuis, M. J. W., F. Colijn & L. A. H. Venekamp, 1986. The spring bloom of Phaeocystis pouchetii (Haptophyceae) in Dutch coastal waters. Neth. J. Sea Res. 20: 37–48.

Vroon, J., 1994. Hydrodynamic characteristics of the Oosterschelde in recent decades. Hydrobiologia 282/283: 17–27.

Wetsteyn, L. P. M. J., J. C. H. Peeters, R. N. M. Duin, F. Vegter & P. R. M. de Visscher, 1990. Phytoplankton primary production and nutrients in the Oosterschelde (The Netherlands) during the pre-barrier period 1980–1984. Hydrobiologia 195: 163–177.

Wetsteyn, L. P. M. J. & C. Bakker, 1991. Abiotic characteristics and phytoplankton primary production in relation to a large-scale coastal engineering project in the Oosterschelde (The Netherlands): a preliminary evaluation. In M. Elliott & J.-P. Ducrotoy (eds), Estuaries and Coasts: Spatial and Temporal Intercomparisons. Proceedings of the ECSA 19 Symposium, Caen 1989. Olsen & Olsen, Fredensborg, Denmark: 365–373.

Hydrobiologia **282/283**: 79–100, 1994.
P. H. Nienhuis & A. C. Smaal (eds), The Oosterschelde Estuary.
© 1994 *Kluwer Academic Publishers. Printed in Belgium.*

A new trend in the development of the phytoplankton in the Oosterschelde (SW Netherlands) during and after the construction of a storm-surge barrier

C. Bakker, P. M. J. Herman & M. Vink
Netherlands Institute of Ecology, Centre for Estuarine and Coastal Ecology, Vierstraat 28, 4401 EA Yerseke, The Netherlands

Key words: phytoplankton species, diatoms, flagellates, light, nutrients, pre- and post-barrier Oosterschelde

Abstract

During the pre-barrier period (1982–83), the Oosterschelde phytoplankton were a diatom-dominated community, comprising a species-rich assemblage throughout the year. Assemblages of spring, early summer and summer, developed in response to a gradually evolving turbidity-light gradient during the course of the year.

During the barrier-construction period (1984–87), characterized by decreasing current velocities, increasing sedimentation of suspended matter, increasing water transparencies and unchanged nutrient conditions, the growth season for the phytoplankton started earlier and lasted longer. Some flagellate species responded by much higher biomass than before. The impact of short-term climatic factors during this period, notably severe winters, could be illustrated with examples of clear responses of some species (e.g. *Biddulphia aurita*).

In the post-barrier years (1987–90) a changed light-nutrient-salinity regime (i.e. much light, limitation of nitrate, high salinity) was demonstrated and an extended summerseason developed, without the original gradual transitions. This was reflected in an a-seasonal trend of the phytoplankton assemblage, where summer species were already observed in spring and spring species decreased in abundance. In summer small flagellates increased and some weakly silicified diatom species made their appearance. In the eastern compartment no colony formation of *Phaeocystis* occurred in summer and this was thought to be due to nitrate limitation. Changes in abundance of some species (*Phaeocystis, Ditylum brightwellii, Skeletonema costatum*), occurring during the entire period of investigation (1982–90), could be explained using field observations compared with experimental evidence from the literature.

The relationship between species composition and biomass on the one hand and environmental variables on the other hand, was analysed in a Canonical Correspondence Analysis, for both compartments separately.

Introduction

The Oosterschelde is a tidal inlet (surface area ± 350 km²) of the Southern North Sea (Fig. 1).

Preliminary data on the seasonal succession of the phytoplankton in this basin were presented by Bakker (1964; 1978). Rijstenbil (1987) studied the development of dominant diatom species in relation to salinity, nutrients, light and turbulence in 1980. Peperzak (1988, 1989, 1991; unpublished reports) made a qualitative analysis of the phytoplankton ($> 30 \mu m$ fraction) in the entire estuary during 1987–90.

During the years 1984–86 a storm-surge barrier was constructed in the western inlet together with two auxiliary dams in the rear ends of the estuary (Nienhuis & Smaal, 1994) (Fig. 1). A comprehensive long-term ecological study started in 1981 to analyse the situation before, during and after this period. The construction of the barrier and the accompanying works in the Eastern (Markiezaatsdam and Oesterdam) and Northern (Philipsdam) area (Fig. 1), represented a unique natural experiment: the hydrodynamics of the estuary were seriously influenced, particularly with respect to tidal amplitude and current velocities

(Vroon, 1994). Current velocities decreased, especially in the shallow Eastern compartment, suspended matter contents declined and water transparency rose concomitantly (Bakker *et al.*, 1990; Wetsteyn & Bakker, 1991). For further data on hydrography, nutrient concentrations and primary production, especially with regard to the pre-barrier and barrier construction periods: cf. Vegter & De Visscher (1987), Klepper (1989) and Wetsteyn *et al.* (1990). During the same periods Bakker *et al.* (1990) followed the seasonal succession of the phytoplankton and analysed the relationship between species composition, biomass and environmental variables. During the barrier construction period the phytoplankton assemblage responded clearly to the improved light conditions, obtaining more and more a summer character within an extended growth season.

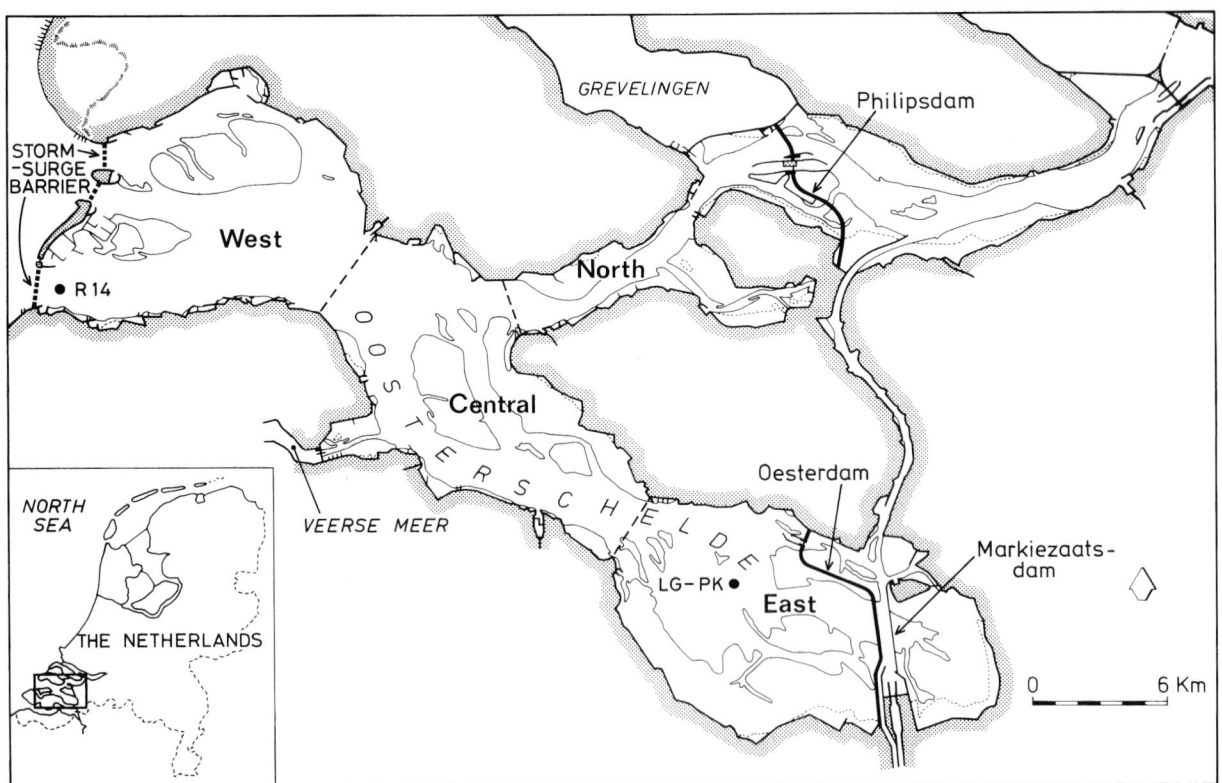

Fig. 1. Map of the Oosterschelde with indication (broken lines) of the 4 compartments and the sampling localities in West (R14) and East (LG-PK). Note the position of the storm-surge barrier (West), of the Oester- and Markiezaatsdam (East) and of the Philipsdam (North).

The seasonal succession of phytoplankton, as well as the relation of successional patterns with physico-chemical changes and biotic interferences, has been well documented, especially in temperate coastal waters (e.g. Guillard & Kilham, 1977; Margalef, 1978; Smayda, 1980; Harris, 1986). Less is known of responses of phytoplankton to major changes in the environment.

In this paper we analyse the results obtained from the phytoplankton investigations carried out during the post-barrier period (1987–90). Data from the preceding periods have been incorporated in an evaluation of the phytoplankton development during the entire period of investigation (1982–90). The ultimate aim is to reveal the response of the phytoplankton species composition and succession to the growth conditions in the present-day Oosterschelde, characterized by constantly high water transparencies and strongly decreased nutrient concentrations during the growing season.

Materials and methods

Environmental data

Only the eastern and western compartments of the Oosterschelde were sampled (Fig. 1); these will be further referred to as E and W respectively.

In W, samples were taken at buoy R14 (main channel in the vicinity of the storm-surge barrier, Fig. 1), depth ± 35 m; in E at the centrally located buoy LG-PK (Fig. 1), depth ± 20 m. All samples were taken at mid-depth during half-tide periods when maximum current velocities caused vertical distributions of the phytoplankton as homogeneous as possible (Bakker, unpubl. data). Sampling was carried out weekly from March to October (1982–90) and incidentally during the rest of the year. In 1982 W was sampled biweekly, in 1985 only E was sampled.

Water transparency was measured using a Secchi disc. Data on column light intensity, water temperature, suspended matter concentrations and macronutrients (silicate, ammonium, nitrate and phosphate) were derived from Vegter & De

Visscher (1987), Klepper (1989), Wetsteyn et al. (1990), Wetsteyn & Bakker (1991); Wetsteyn & Kromkamp (1994); Bakker & Vink (1994). Wind velocity data were obtained from the Royal Dutch Meteorological Institute (Station Vlissingen).

Phytoplankton analysis

Water for phytoplankton analysis was sampled in 11 glass bottles and preserved directly with Lugol's solution. Phytoplankton was counted in aliquots of 1–10 ml using an inverted microscope.

For the nomenclature of diatoms mainly Hustedt (1930) and Drebes (1974) were followed. Following Sournia (1988) the haptophycean flagellate *Phaeocystis* was not identified to species level, but very probably all blooms observed in Dutch coastal waters refer to *P. globosa* (Cadée, 1991).

Cell counts and volume measurements were converted to carbon using the equations of Eppley (in Smayda, 1978). For the presentation and the preprocessing of the species data, logarithmic values of the carbon content of each species population per sampling date were calculated and plotted symmetrically as kite diagrams around the horizontal (time) axis (Figs 2 and 3). Kite diagrams, first constructed by Lohmann (1908), are still considered to present adequately phytoplankton species assemblages and succession. In these diagrams time is the horizontal axis. Species are arranged vertically in a sequence that approximates their seasonal succession. For 1982–86 this sequence was derived from the divisive hierarchical clustering method TWINSPAN (Hill, 1979a), which performs a simultaneous ordering of samples and species. The 1982–86 data, although published earlier (Bakker et al., 1990), are presented once again together with those for the years 1987–1990 for an adequate evaluation of the entire data set. For practical reasons the sequence of the species in the latter years is similar to that in the former, which facilitates comparison with the previous results (Bakker et al., 1990, Figs 4 and 5). The additional species for the new period have not been placed in between but at the bottom.

A number of diatoms and flagellates of different dimensions was not identified to species level. In order to estimate the carbon contents more precisely, these species were distinguished in size classes, indicated in Table 1.

Ordination of phytoplankton species and environmental variables

Multivariate analysis techniques such as classification and ordination are used more and more to explore the combined data sets of environmental variables and of species. These methods, however, are indirect: ordination or classification is performed first; a relationship with environmental variables is established only by correlative approaches afterwards. This contrasts with direct (regression) methods, in which a statistical model is built directly with the environmental variables. Recently, regression and ordination have been integrated into techniques of multivariate direct

gradient analysis, called Canonical Correspondence Analysis (CCA) (Jongman *et al.*, 1987, Ter Braak, 1986, 1988, 1989). In CCA the axes of the ordination are directly derived under the constraint that they must be linear functions of the environmental variables. The resulting ordination diagrams represent 'biplot scores' of environmental factors, species and samples. We have represented species scores and environmental factors in the same biplot (Fig. 6), and sample scores in separate plots (Fig. 7). These, however, can also be represented in a biplot with environmental factors or with species. For the interpretation it is important to take into account the three different sets.

The length of the arrow of an environmental factor in the species-environment biplot indicates the degree of correlation between the environmental factor and the ordination plane shown. The angle between environmental arrows indicates correlation between the factors: the smaller the angle between two arrows of environmental factors is, the higher the correlation between them. Arrows at an angle of $90°$ indicate no correlation, at $180°$ strong negative correlation. The perpendicular projection of a species' position on an environmental arrow approximates the optimum of this species' response curve with respect to that environmental factor. The biplot score of a sample is a weighted average of the scores of the species occurring in that sample. The scores of the environmental variables have to be used also for the interpretation of the ordination diagram of the samples.

In our study the relation between species composition and environmental variables was analyzed via CCA using the program package CANOCO (Ter Braak, 1988, 1989). Separate analyses were performed for both compartments E and W. All species were used in the analysis after a log $(X + 1)$ transformation of carbon content to avoid overweighting of abundant species due to dominance effects.

'Downweighting of rare species' was performed according to Hill (1979b): if A_{max} is the frequency of the commonest species, then the abundances of species rarer than $A_{max}/5$ are reduced in pro-

Table 1. Size clases of some diatoms and flagellates indicated as sp. 1, sp. 2, etc. in Figs 2 and 3. Pennate diatoms: length; discoid diatoms and globular forms: diameter. Derived from Bakker *et al.* (1990).

Pleurosigma	$0 \ \mu m < \text{sp. } 1 < 100 \ \mu m$
	$100 \ \mu m < \text{sp. } 2 < 200 \ \mu m$
Discoid diatoms (small)	$0 \ \mu m < \text{sp. } 1 < \ 20 \ \mu m$
	$20 \ \mu m < \text{sp. } 2 < \ 40 \ \mu m$
	$40 \ \mu m < \text{sp. } 3 < \ 60 \ \mu m$
	$60 \ \mu m < \text{sp. } 4 < 100 \ \mu m$
Discoid diatoms (large cf. *Coscinodiscus*)	$100 \ \mu m < \text{sp. } 1 < 200 \ \mu m$
	$200 \ \mu < \text{sp.} 2$
Chaetoceros	$0 \ \mu m < \text{sp. } 1 < \ 20 \ \mu m$
	$20 \ \mu m < \text{sp. } 2 < \ 40 \ \mu m$
	$40 \ \mu m < \text{sp. } 3 < \ 60 \ \mu m$
	$60 \ \mu m < \text{sp. } 4$
C. socialis and related forms	*ca.* $6 \ \mu m : \text{sp. } 5$
Mesodinium cf. rubrum	$0 \ \mu m < \text{sp. } 1 < \ 20 \ \mu m$
	$20 \ \mu m < \text{sp. } 2 < \ 40 \ \mu m$
	$40 \ \mu m < \text{sp. } 3 < \ 60 \ \mu m$
	$60 \ \mu m < \text{sp. } 4$
Cryptomonas	$0 \ \mu m < \text{sp. } 1 < \ 10 \ \mu m$
	$10 \ \mu m < \text{sp. } 2 < \ 20 \ \mu m$
Flagellates (autotrophic species)	$0 \ \mu m < \text{sp. } 1 < \ 10 \ \mu m$
Small round (non-flagellate, pigmented) cells	*ca.* $4 \ \mu m$

portion to their frequency. This step effectively reduced the effect of species which occur rarely, but have large biomasses when they do occur.

Environmental factors were subsequently deleted from the analysis to assess their importance for the results.

CCA is essentially a correlative study in which a statistical model, incorporating explicitly environmental factors, is fitted to the data. Ter Braak (1986) gives details about the characteristics of this model. Interpretation of the results can yield qualitative insight into the relations between species and environment.

Results

Phytoplankton composition and seasonal succession (Figs 2 and 3)

In the following the term 'maximum' is used for the biomass peaks a species developed during its temporal evolution. The term 'bloom' is reserved when a species reached a carbon content $> 100 \mu g \ C \ l^{-1}$.

Pre-barrier period (1982–83)

Diatoms. Marine diatoms predominated in the Oosterschelde throughout the year and demonstrated large species diversity. The majority of species were pelagic but also several littoral-benthic forms could be found in the plankton samples.

The seasonal succession of diatoms comprised the following species:
– Species starting their development in late winter, often with extended maxima in spring, were: *Thalassiosira nordenskioeldii, Plagiogramma brockmannii, Asterionella kariana, Thalassionema nitzschioides* (notably in 1982), *Skeletonema costatum.* Some of these also occurred in summer (*Plagiogramma, Skeletonema,* and *Thalassionema*) and formed small blooms in autumn (*Plagiogramma*). *Thalassiosira nordenskioeldii* was not observed in 1984.
– Benthic-pelagic winter/spring species were

Biddulphia aurita (with a large spring bloom), *Nitzschia closterium* (with small biomass only), *Paralia sulcata* (no spring bloom).
– A number of species showed an irregular seasonality, occurring in spring and summer with blooms of varying strength and duration: *Asterionella glacialis* (more numerous in W), *Streptotheca thamensis, Ditylum brightwellii, Rhizosolenia setigera, Thalassiosira rotula. Rhizosolenia delicatula* was nearly perennial, with intermittent strong ($> 500 \mu g \ C \ l^{-1}$) blooms (oscillating species).
– Dominant species in early summer were: *Rhizosolenia stolterfothii* (especially in W), *R. shrubsolei* (W), *Cerataulina pelagica* (notably in 1982, during 1983 not developing in E), *Lauderia annulata, Eucampia zoodiacus, Guinardia flaccida* (more observations in W), *Nitzschia* cf. *seriata.*
– Full-summer species with extended maxima were: *Biddulphia sinensis* blooming strongly in both compartments, small *Chaetoceros* (sp. 5), *Coscinodiscus granii.*
– Benthic-pelagic summer species without clear maxima were: *Lithodesmium undulatum, Bellerochea malleus* (more observations in the shallow E).
– Benthic pennate species with irregular occurrence and abundance were mainly composed of: *Raphoneis amphiceros* (attached to sand grains and empty diatom frustules), several unidentified pennate diatoms (notably sp. 2 and *Pleurosigma*). The presence of such species in the plankton samples indicates that resuspension from the bottom is a commonly occurring phenomenon in a shallow coastal environment.

Non-diatom phytoplankton. Cryptophycean flagellates represented a persistent and abundant component of the microphytoplankton without a clear seasonality (from 1983 onwards *Cryptomonas* spp. were counted in two size classes). Small round non-flagellate cells could nearly always be found, in varying densities.
– The small unarmored dinoflagellate *Katodinium rotundatum* occurred in the majority of the samples (although less numerous than *Cryptomonas* spp.). The thecate dinoflagellates *Exuviella* sp., *Ceratium fusus* and *Prorocentrum micans* were

84

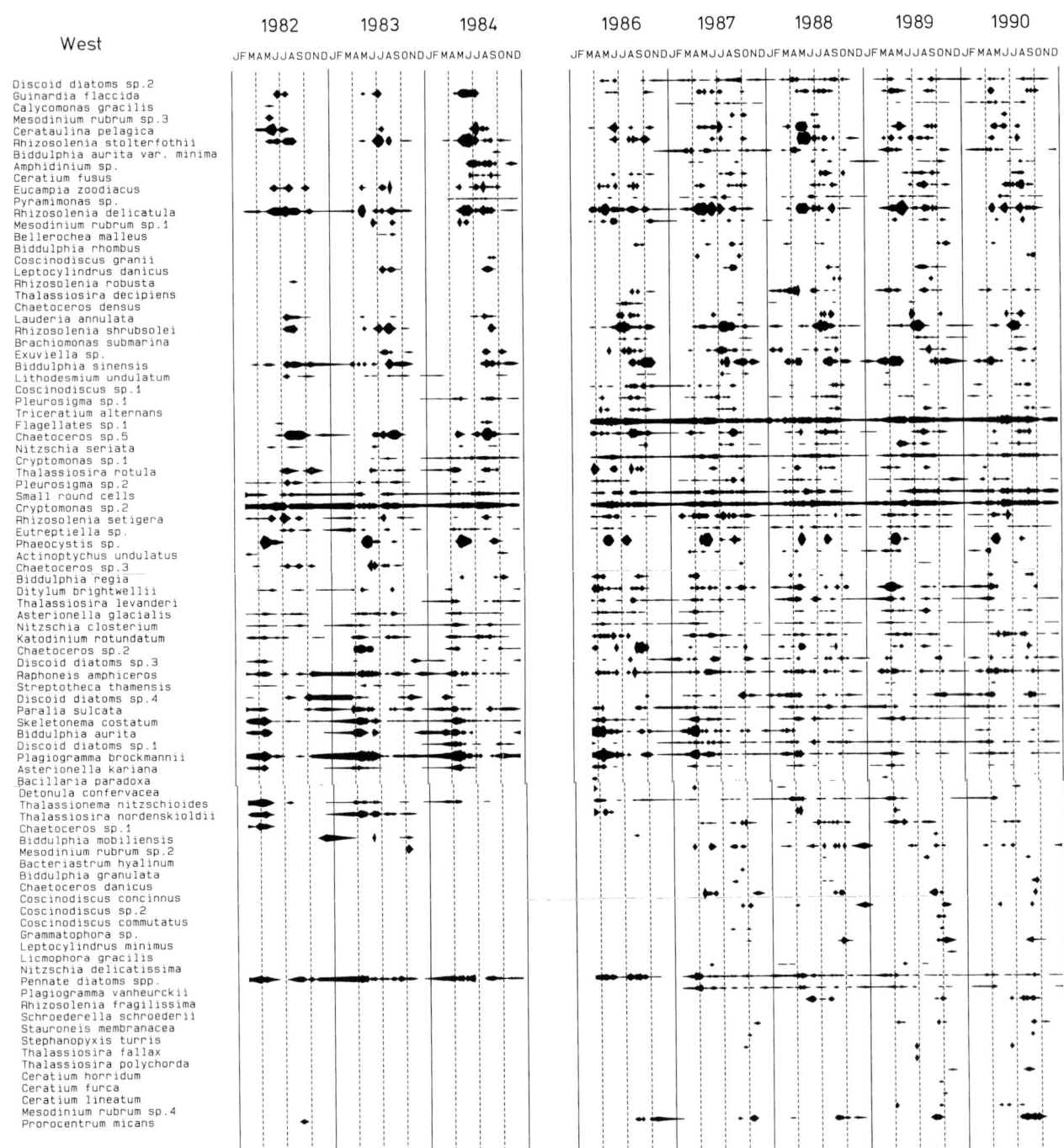

Fig. 2. Kite diagram of the phytoplankton composition in the Western compartment of the Oosterschelde during the period 1982–90. The horizontal axis represents time, with quarters separated by dashed lines. The amplitude of the black blocks for each species is proportional to the logarithm of the organic carbon concentrations of that species' population. For further explanation see material and methods and Table 1.

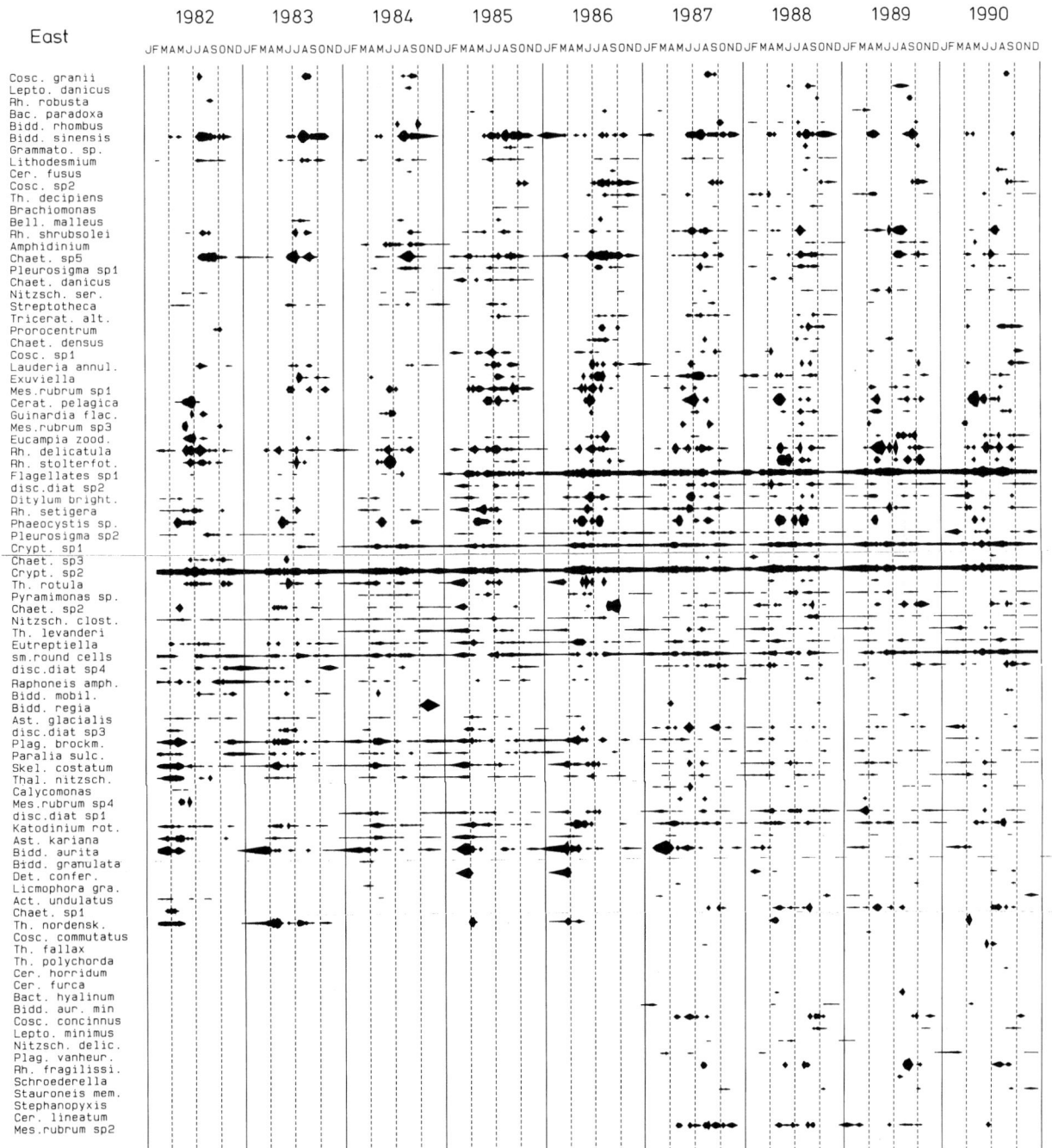

Fig. 3. Kite diagram of the phytoplankton composition in the Eastern compartment of the Oosterschelde during the period 1982–90. For further legends and the complete species names: see Fig. 2.

present in summer in varying densities, *Ceratium* mainly in W.

– *Eutreptiella* sp. (Euglenophyceae) was found regularly from spring onwards.

– *Mesodinium rubrum* (autotrophic ciliate, with cryptophycean zooxanthellae) occurred irregularly during short periods, with differing cell sizes.

– *Phaeocystis* sp. (Haptophyceae) (see also

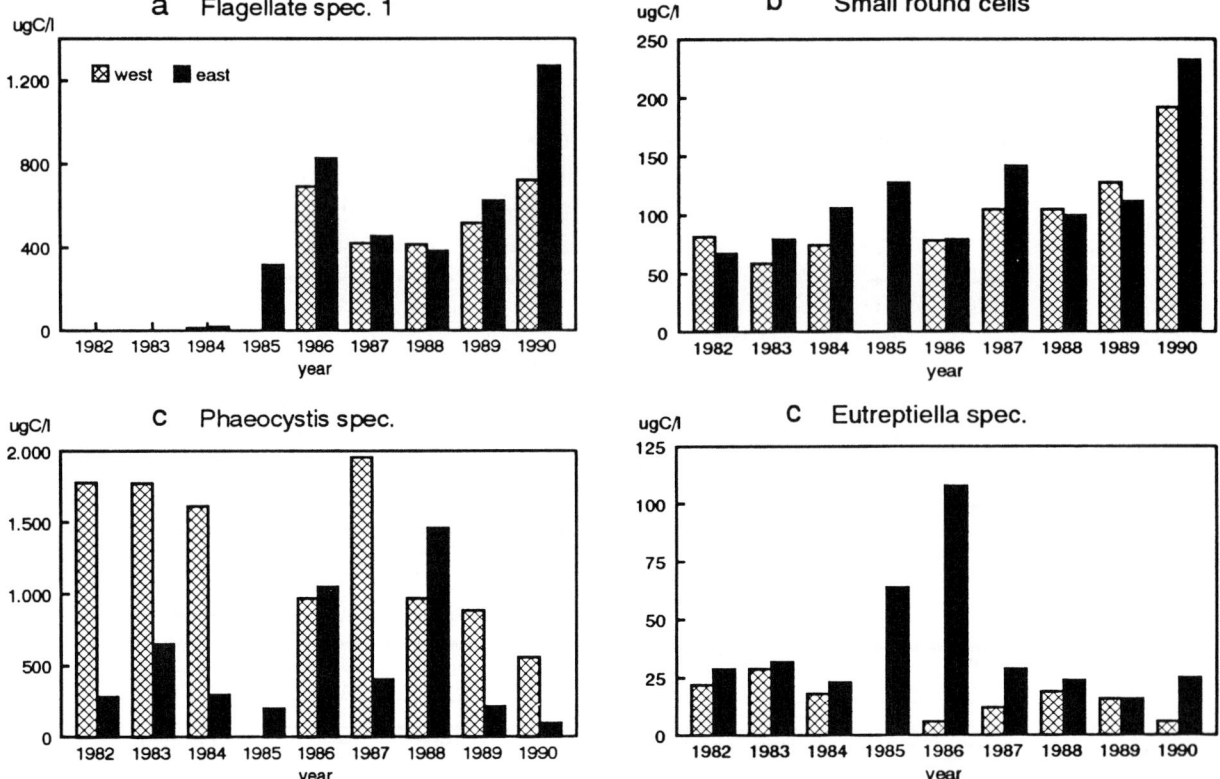

Fig. 4. Total accumulated annual carbon concentrations from 1982 to 1990 in the Western and Eastern compartments for the populations of a number of non-diatom species. Note: 1. the different vertical scales; 2. no observations in West during 1985.

Fig. 5) bloomed strongly in May, notably in W, occurring in gelatinous colonies of 1,0–1,5 cm diam. Its occurrence was highly predictable as the development started during, and the peak appeared immediately after, the diatom spring bloom.

The clear seasonality of a large number of spring- and summer species was a marked feature of the phytoplankton assemblages in the pre-barrier period.

Barrier-construction period (1984–spring 1987)
We have to note that three severe winters just occurred in this period. Consequently, it was to be expected that some species would respond to the very low temperatures prevailing in spring 1985–87. This holds especially for the diatom *Biddulphia aurita*, demonstrating earlier and heavier spring blooms than before in both com-

partments. The stenothermal species *Detonula confervacea* was observed abundantly during the same time (except 1987) in E, but hardly in W.

– Some species, found only incidentally and in very low abundance earlier, became numerically important from 1985 onwards. From these species especially *Coscinodiscus* sp. 2 (replacing *C. g-ranii* in E, 1985–86), *Triceratium alternans* (1985–86), *Thalassiosira decipiens* (1986) and *Brachiomonas submarina* (1985–86) can be mentioned.

– Especially in E some species developed higher maxima than before, notably the diatom *Ditylum brightwellii* (1986) (Fig. 8b) and the flagellates *Katodinium rotundatum* (1986), *Eutreptiella* sp. (1985–86) (Fig. 4d) and *Exuviella* sp. (1986). In addition to the spring bloom *Phaeocystis* developed summer maxima too (Fig. 5), thus reaching higher total carbon concentrations than be-

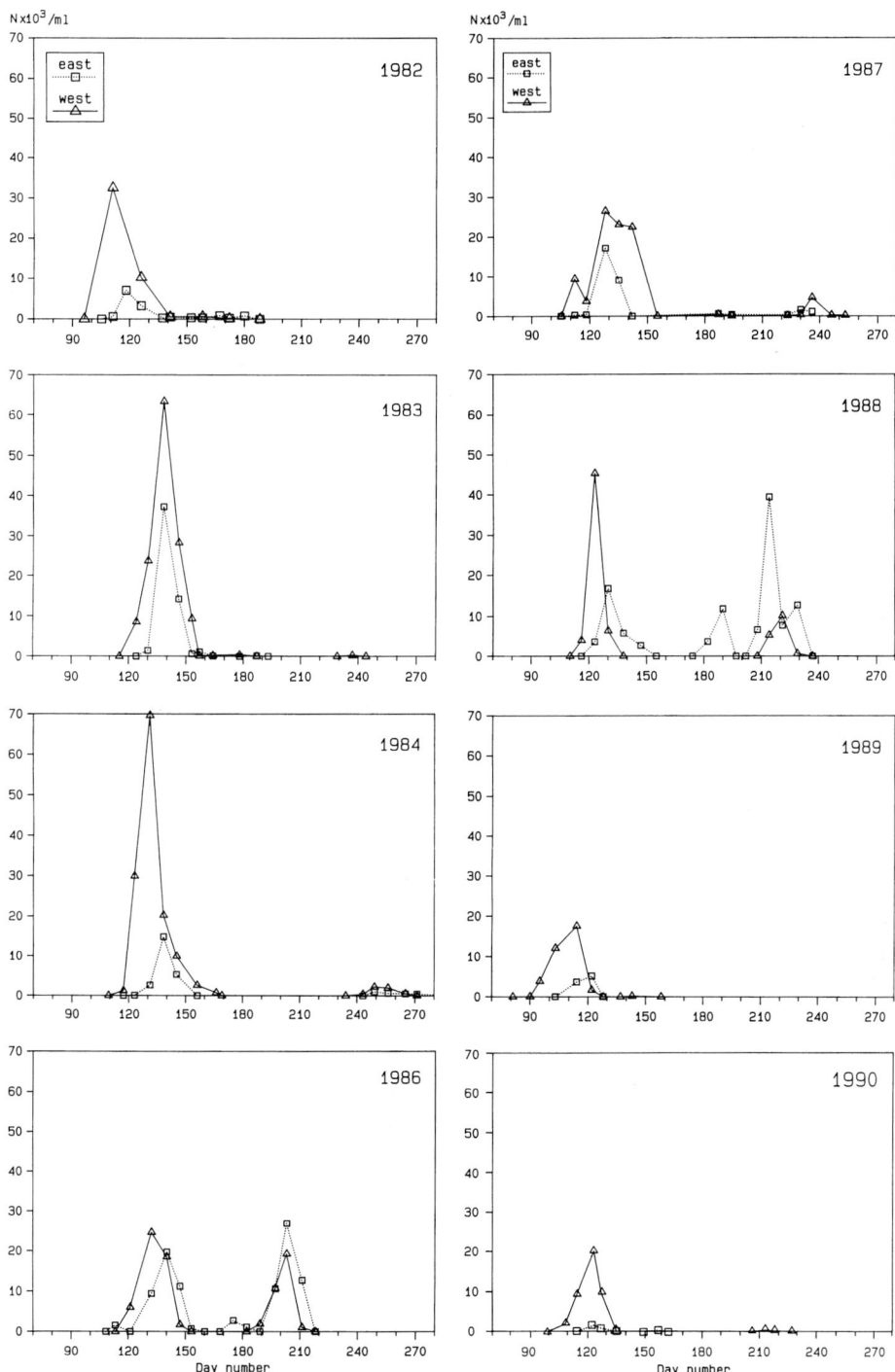

Fig. 5. Annual abundance curves of *Phaeocystis* (cells ml l^{-1}) in the Western and Eastern compartments of the Oosterschelde during the years 1982–84 and 1986–90.

fore (Fig. 4c) and equalling (for the first time) the annual value in W (1986). Unidentified small flagellates (sp. 1) increased clearly, both in W and E (Fig. 4a).

– Pennate diatoms and *Raphoneis*, on the other hand, decreased in E after 1984 (Fig. 8a).

– Many species were recorded for a longer period (from summer to late autumn) during the barrier period: this was the case for *Biddulphia sinensis*, *Chaetoceros* sp. 5, *Lauderia annulata*, *D. brightwellii*, *Rhizosolenia setigera*, *Thalassionema nitzschioides*, *M. rubrum* sp. 1. Autumnal developments were noticed also for the 'new' species mentioned above. Some species, on the other hand, were found earlier in spring during the barrier period: *Biddulphia sinensis* (1986), *Thalassiosira rotula* (1985–86). Both tendencies, prolonged presence and earlier appearance, were more conspicuous in E than in W.

Post-barrier period (summer 1987–90)

– The tendency of extending the seasonal presence still continued in this period, as can be seen for the following species: *Biddulphia sinensis*, *Rhizosolenia shrubsolei*, small *Chaetoceros* sp. 5, *Eucampia zoodiacus*, *Nitzschia seriata*, *Prorocentrum micans* and *Exuviella* sp. Another group of spring species, on the contrary, showed decreasing abundances: the diatoms *Skeletonema costatum*, *Asterionella kariana*, *Plagiogramma brockmannii*, *Thalassiosira nordenskioeldii*. A decrease of *Biddulphia aurita* in the plankton was observed, probably mainly influenced by the mild winters of this period. The occurrence of summer species in spring and the decreasing abundance of spring species represents an important change in the phytoplankton from the pre-barrier to the post-barrier years.

– Just as in the barrier construction period, some species were observed for the first time in the Oosterschelde: the diatoms *Thalassiosira fallax*, *Rhizosolenia fragilissima* (prevailing in E) and *Leptocylindrus minimus* (much more abundant in W). These species are characterized by thin, fragile, weakly silicified frustules.

The diatoms *Plagiogramma vanheurckii* and *Stauroneis membranacea*, in earlier years found in

net plankton samples only (unpubl. obs.), could be observed in the sedimented 1 litre samples during the last years and must therefore have increased in abundance. This holds too for *Coscinodiscus concinnus* in 1987 (spring) and 1988–90 (summer–autumn).

– In 1990 some small non-diatom species increased strongly in E. Small flagellates (Fig. 4a) were observed in numbers from 1985 onwards. During the preceding years these cells, if present, may have been overlooked owing to the prevailing high suspended matter concentrations. An increase was observed during the period 1988–90 in both compartments but culminating in E. A similar rise during the same period was seen for small round cells (Fig. 4b).

Figures 4a and 4b contrast with Fig. 4c, giving the data for *Phaeocystis* occurring as colonies and strongly declining from 1988–90 in E. The species did not occur in colonial form during the summer of 1990. Because *Phaeocystis* is a notorious nuisance bloom alga along the Dutch Coast (Cadée, 1990; Riegman *et al.*, 1992), this species deserves special attention with regard to its occurrence and bloom development in the Oosterschelde (see discussion).

– With the year 1990 a new period was started, due to a likely nitrogen limitation becoming manifest from May onwards, notably in E, which was unprecedented for the Oosterschelde (Bakker & Vink, 1993). Although silica was sufficiently available, much lower total diatom carbon was demonstrated in summer (Bakker & Vink, *op. cit.*). Figures 2 and 3 indicate that a number of diatom species had decreased compared to the previous years, especially *Biddulphia sinensis* in E, and *Chaetoceros* sp. 5 in W; *Triceratium alternans* was not at all observed in E, 1989–90; *Rhizosolenia delicatula* formed a number of modest blooms only, in June/July; further some small species (the already mentioned *Rh. fragilissima*, *L. minimus* and *Th. fallax*) were present. Some flagellate species, on the other hand, reached much larger densities than before: *Ceratium fusus*, *Prorocentrum micans* (in E), *Katodinium rotundatum*.

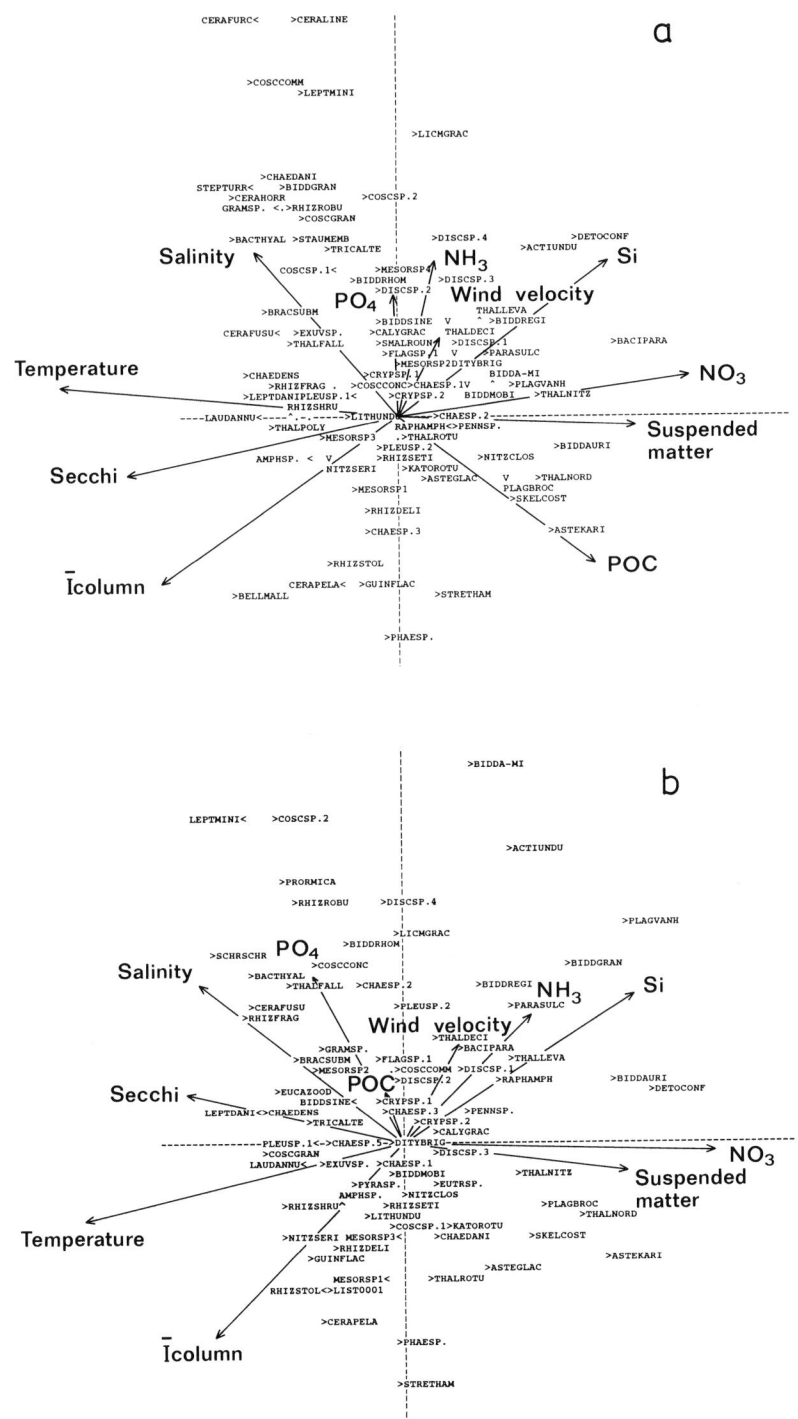

Fig. 6. Canonical Correspondence Analysis of the phytoplankton in the Oosterschelde from 1982–90: scores of environmental variables and species in the plane of the first two axes of the CCA ordination for (a) the Western and (b) the Eastern compartment. For the species codes: see the complete species names in Figs 2 and 3.

Results of the Canonical Correspondence Analysis (Figs 6 and 7)

The biplot diagram of environmental variables, comprising an analysis of all data (Fig. 6), shows the following features:

– The upper right quadrant shows high silica- and ammonium concentrations while irradiance is low; in lower left low nutrient concentrations and high irradiance are found: so a main seasonal gradient is visible, running from top right (spring) to bottom left (summer). Generally, higher wind

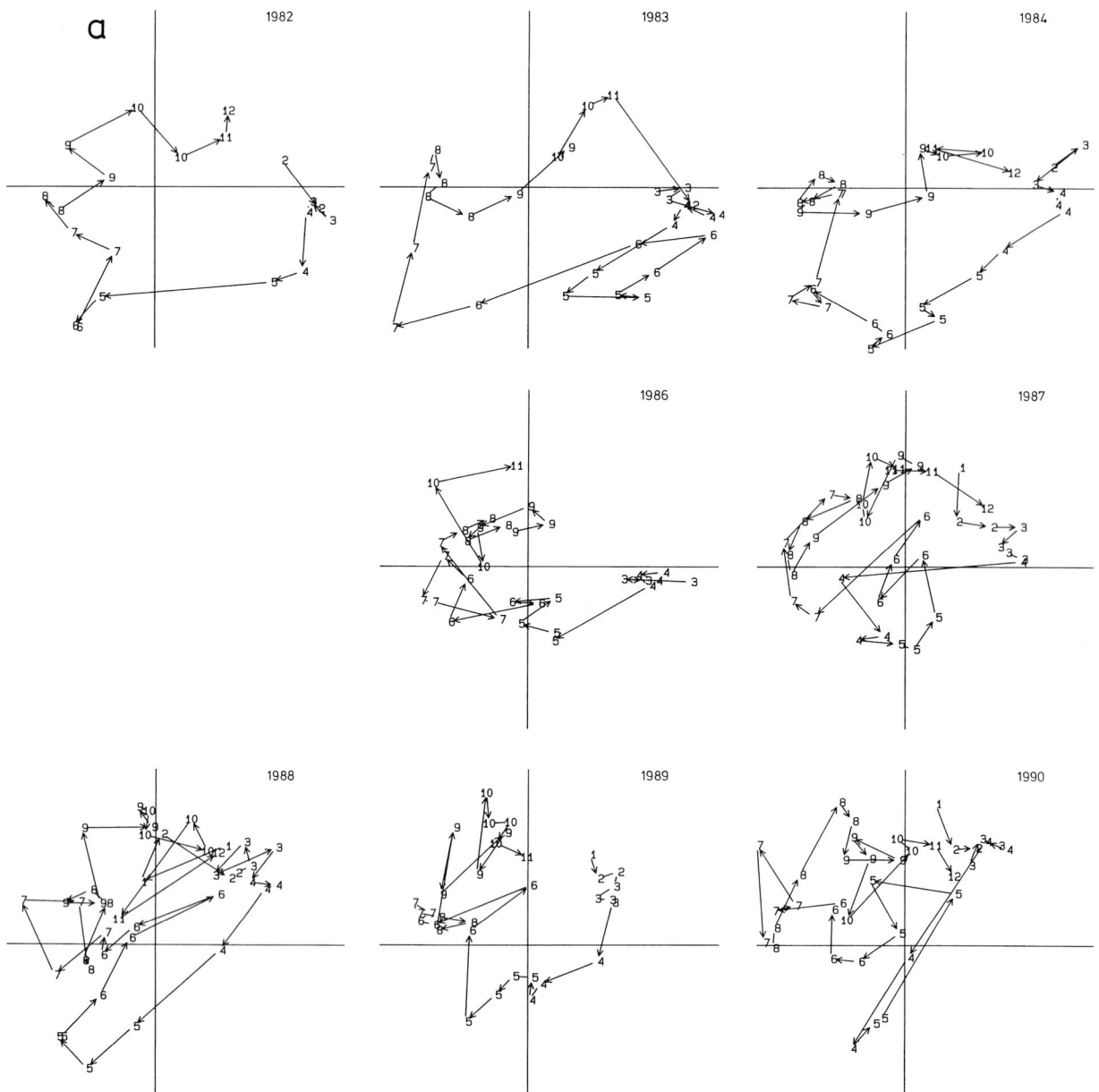

Fig. 7. Canonical Correspondence Analysis of the phytoplankton in the Oosterschelde from 1982–90: position of the samples in the plane of the first two axes of the CCA ordination for (a) the Western and (b) the Eastern compartment. The axes in the yearly subplots have the same scaling. The arrows represent the time course, the numbers are month numbers.

velocities occur during the winter–spring period (top right).

– Another axis is presented by the suspended matter (seston)- water transparency (Secchi disc visibility) parameters, in E diverging *ca* 45–60° from the above mentioned main seasonal gradient. In W this deviation is smaller than in E.

– The salinity arrow, showing an angle of *ca* 90° with the seasonal gradient, indicates that salinity is not seasonally determined. This is a consequence of the diminished influence of riverine discharge via the Philipsdam sluices (Fig. 1). In the pre-barrier period, without the Philipsdam, high river discharges caused decreasing salinities,

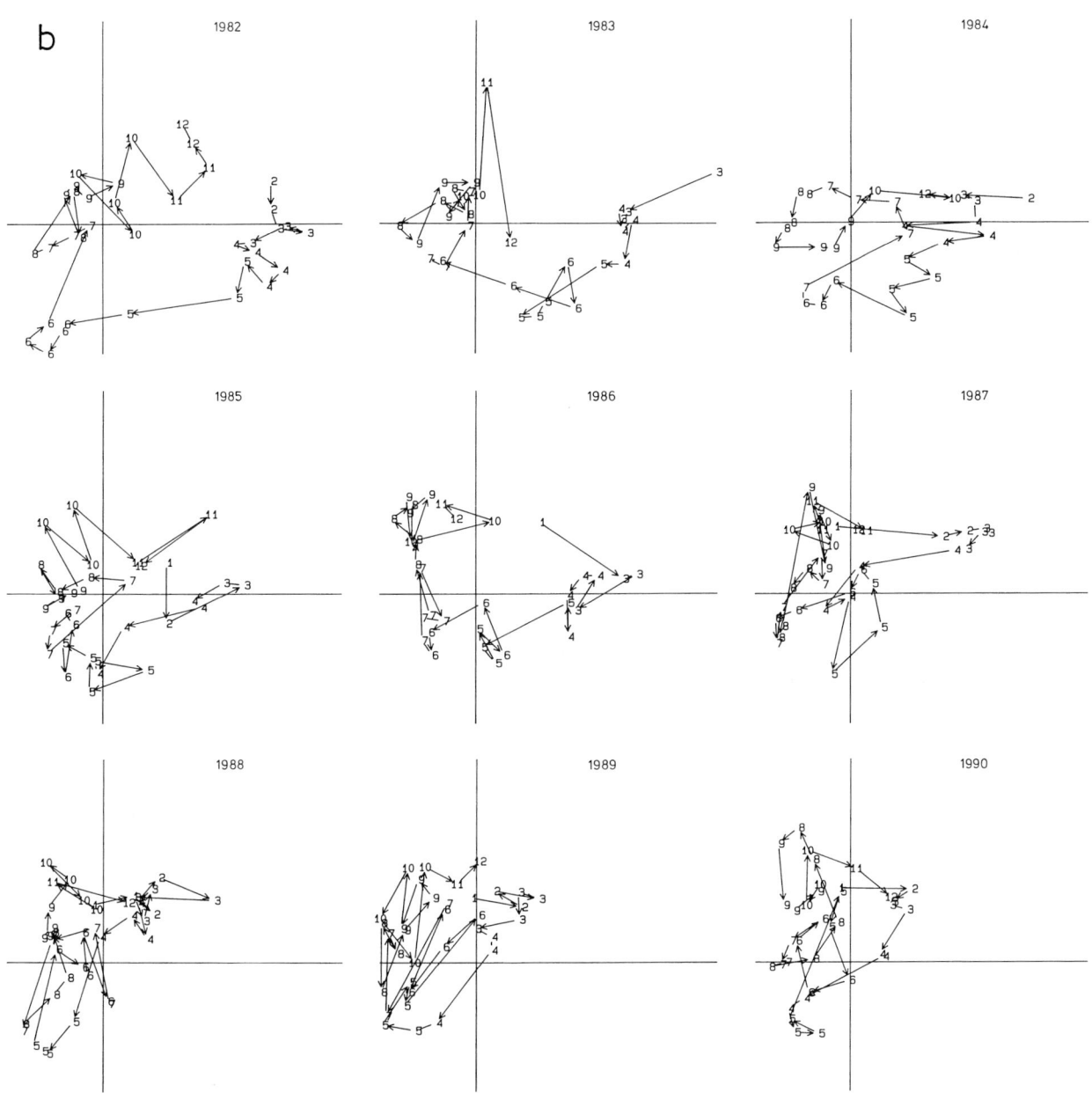

Fig. 7b.

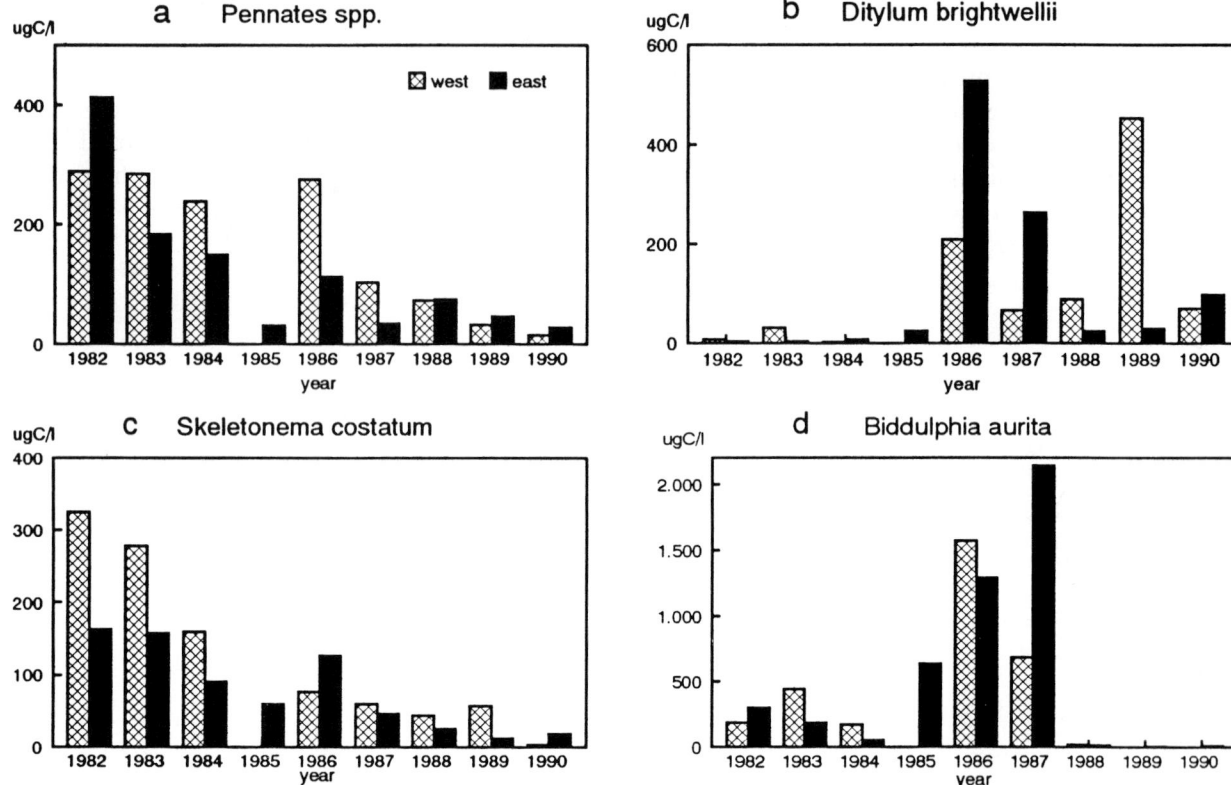

Fig. 8. Total accumulated annual carbon concentrations from 1982 to 1990 in the Western and Eastern compartments for the populations of a number of diatom species. For further legends see Fig. 4.

especially in E during winter and spring, and therefore salinity was more or less seasonally determined during that time. Also dissolved phosphate has no connection with the main seasonal gradient, especially in E. Nitrate appears to be situated near the turbidity gradient. Temperature in E shows an intermediate position between Secchi and irradiance. In W, temperature shifts to the upper left quadrant and a strong negative correlation exists with suspended matter. In W, POC (occurring in higher concentrations than in E) is strongly negatively correlated with salinity, which is not the case in E.

In Fig. 6 the species scores have also been plotted. Corresponding to the main seasonal gradient of cold, less transparent and nutrient-rich spring water, early spring phytoplankton species (like e.g. *Thalassiosira levanderi*) are found in the upper right quadrant and summer species (like e.g. *Rhizosolenia stolterfothii* and *Nitzschia seriata*)

accordingly occur in the lower left quadrant. Many other species, however, are not found along the seasonal gradient and are situated elsewhere in the plot (lower right and upper left): these species apparently do not show a clear seasonal character, although their occurrence in lower right and upper left indicates affinity to spring and summer-autumn conditions respectively. An intriguing difference between both compartments is the position of *Ditylum brightwellii*, in E located at the origin of the ordination plane indicating indifference to seasonal environmental parameters, in W situated in upper right according to its prevailing occurrence in spring in this area.

Figure 7 gives the sample scores of the entire period of investigation per compartment, separately for each year. Generally, it can be seen that the seasonal cycle is moving round from top right through bottom right and bottom left to top left. This pattern is clear during pre-barrier and barrier periods and becomes less well-marked in post-barrier years. During this period the spring samples moved in the direction of lower left, especially in E (Fig. 7b). This is the expression of the above mentioned changed temporal distributions of summer species extending now into spring, and the decreasing importance of a number of spring species (Figs 2 and 3).

Discussion

Pre-barrier period

Before the construction of the storm-surge barrier the Oosterschelde could be considered a well mixed estuary (Dronkers & Zimmermann, 1982). During the pre-barrier period a clear seasonality existed in the combination of the factors light and nutrients in the course of the growth season (Bakker & Vink, 1994). The phytoplankton responded by clearly different spring-, early summer- and full summer assemblages (Bakker et al., 1990). Tidal movements and small average depth (ca 8 m. in the entire basin) caused a strong influence of bottom and shores on the suspended matter content of the water column. Monthly averaged suspended matter concentrations varied from 10 to over 50 mg l^{-1} (Wetsteyn & Bakker, 1991), and repeated resuspension of benthic diatoms gave the phytoplankton assemblage a littoral aspect (Bakker, 1978).

Barrier-construction period

From 1984 to April 1987 inclusive the basin changed hydrodynamically as a consequence of the large infrastructural works carried out. Current velocities decreased (Vroon, 1994), sedimentation of suspended matter increased (Wetsteyn & Bakker, 1991), and abundance of benthic diatoms in the plankton decreased (Fig. 8a), as a consequence of the diminished resuspension of bottom material. A rise in water transparency was observed during a prolonged period of the year, notably in E (Bakker et al., 1990).

A number of species increasing in E, notably

the diatom *Ditylum* (Fig. 8b) and the flagellate *Eutreptiella* (Fig. 4d), are common in Grevelingen (Bakker & de Vries, 1984). Grevelingen is a tideless saline lagoon, a former tidal estuary, previously connected to the Oosterschelde (Fig. 1). After its closure from the North Sea, tidal influence disappeared completely and water transparencies rose strongly (Bakker & Vegter, 1978). As the eastern compartment of the Oosterschelde, in the barrier-construction period, revealed a comparable trend with regard to water transparency, it was not surprising that the above mentioned 'Grevelingen' species increased their densities. Moreover, during the barrier-construction period, there was a substantial discharge (15 m^3 s^{-1}) of Grevelingen water into the Oosterschelde (Nienhuis & Smaal, 1994). This water must have contained an important inoculum of the relevant species and these were subsequently distributed throughout the entire basin. Especially flagellates, being able to perform small-scale vertical migrations, could be expected to grow more rapidly in the less turbulent and more transparent water of E. Obviously, *Ditylum* also found better growing conditions in the Oosterschelde during the barrier-construction period than before. In W such developments did not take place, which can be attributed to the less favourable depth- and transparency conditions in this compartment.

Finally, using a Canonical Correspondence Analysis, it could be demonstrated that the phytoplankton assemblage during the barrier period more and more attained a summer character, extending its growth season both earlier and later in the year. The transition of spring to summer phytoplankton proceeded parallel to the changed light-turbidity gradient. The light factor not only explained the seasonal pattern, but also the long-term trend from pre-barrier to barrier construction period (Bakker et al., 1990).

Post-barrier period.

In the course of the post-barrier period the growth season became increasingly more uniform with respect to the underwater light climate as transparencies were high throughout the year with small differences only between spring and summer. Also the nutrient regime underwent a clear change: only during winter and early spring were nutrients present in excess, but the prolonged summer season was characterized by increasingly lower nutrient concentrations, especially in E and notably with regard to nitrate-nitrogen. Silica levels did, however, increase in E and were generally higher during the post-barrier years than before, without the periods of undetectable Si-concentrations observed earlier (Bakker & Vink, 1994). In E the nitrogen limitation resulted in strongly decreased diatom biomass in summer. This Oosterschelde summer situation resembles the nutrient conditions observed in Grevelingen lagoon with the important distinction however that in the latter stagnant basin stratification occurs during summer. Incidental increase of diatom biomass in summer could be explained (in Grevelingen) by sudden increases of silica and nitrogen concentrations caused by release from the bottom and turnover of the stratification (Bakker & de Vries, 1984).

In the post-barrier Oosterschelde, E, water transparency demonstrates a clear deviation from the former seasonal character. This is accompanied by high salinities throughout the year (Wetsteyn & Bakker, 1991), without the fluctuations of the past when the freshwater discharges were larger. In summer these high levels of transparency and salinity coincide with the very low concentrations of nitrate in E. The combined set of environmental parameters including also the reduced tidal movement and the larger residence time, accentuate the environment as a marine tidal bay, less estuarine and more constant than before. Consequently, a divergence could be noticed between the old (and still existing) seasonal gradient of high nutrient concentrations and low irradiance in spring *vs* low nutrient contents and high irradiance in summer during the prebarrier years, and the new suspended matter/transparency gradient originating in the barrier-construction and post-barrier periods (Fig. 6). This divergence represents a striking difference from the old situation when transparency and irradiance ran parallel (Bakker *et al.*, 1990 cf.

Fig. 7). This means that in the former periods transparency still followed the seasonal trend caused by the coupling of high suspended matter contents and low irradiance in spring, and low turbidity and high irradiance in summer. In the new (post-barrier) period high transparencies and low suspended matter concentrations can be observed nearly throughout the year. Because irradiance is a seasonal factor being primarily influenced by day length, the all year-round high transparency tends to diverge from irradiance. During the post-barrier years suspended matter contents in E decreased much stronger than in W (cf. Wetsteyn & Kromkamp, 1994). Wind influence on the declining suspended matter concentrations will have decreased too, more in E than in W. This explains the increased divergence between the arrows for both variables in E (Fig. 6b) compared to the small distance in the preceding years (Fig. 7b in Bakker *et al.*, 1990).

The observed phenomena caused the phytoplankton assemblage to evolve into a direction different from that previously governed by the strictly seasonal development of the environmental conditions.

The changed light-nutrient regime, without the gradual transitions existing in the earlier years, was reflected in an increasing a-seasonality of the phytoplankton. True spring species became restricted to the early spring period only and a variety of summer species were observed during the rest of the growing season. When nitrate-nitrogen was present in very low concentrations only, from summer 1990 onwards, notably in E, a decrease in diatom biomass was accompanied by an increase of flagellates (Bakker & Vink, 1994). This can be considered a further shift in the phytoplankton species composition, mainly guided by increasing limiting nutrient conditions. Very small cells, particularly (mobile) flagellates, seem to be privileged by these conditions rather than (immobile) diatoms, except some small and weakly structured species.

Primary production and biomass in summer have become more dependent on prevailing nutrient conditions. Wetsteyn & Kromkamp (1994) determined primary production during the post-barrier period and compared their results with those of Vegter & De Visscher (1987) and Wetsteyn *et al.* (1990) for the preceding years. They established a decreased slope of the P/I-curve for which the observed shift in the phytoplankton species assemblage may be held responsible. When the phytoplankton species composition does not undergo further essential changes, future primary production levels may reach lower maximum values than before, notably in W (Van der Tol & Scholten, 1992).

With regard to the significance of the phytoplankton as food for the suspension feeding zooplankton, it is not primarily the changed species composition that is of importance but rather the increased phytoplankton carbon/POC-ratio, resulting from the enhanced sedimentation of the suspended matter (Bakker & Van Rijswijk, 1994). For further discussions about the functional response of the (changed) phytoplankton cf. Van der Tol & Scholten (1992); Klepper *et al.* (1994); Wetsteyn & Kromkamp (1994).

Year-to-year variation

In studies of marine diatoms it is well documented that species composition, -dominance and timing of the periods of maxima vary widely from year to year (Hagmeier, 1978; Matta & Marshall, 1984; Reid *et al.*, 1990; Cadée, 1992; Kat, 1992). Sometimes exceptional blooms could be explained by the occurrence of anomalous hydrographic (Reid *et al.*, 1983) or meteorological (Boalch, 1987) conditions. Changes in seasonal patterns over larger periods in open sea systems could be related to changes in climate, cf. the results of phytoplankton investigations in the Plymouth area (Maddock *et al.*, 1989) and the results of the Continuous Plankton Recorder Surveys in the North Sea since 1931 (Robinson, 1983). Long-term changes in coastal sea water, on the other hand, were observed for non-diatoms and could be related to anthropogenically influenced eutrophication (Cadée, 1990, for *Phaeocystis*; see further Radach *et al.*, 1990, for other flagellates). Also in the Oosterschelde large interannual variations of several species of the diatom assemblage (Figs 2 and 3) could be recorded, not demonstrating a clear relationship with the changes in the environment. Considering the significance of the diatom species recently found in the Oosterschelde, we consulted the available data about the composition and distribution of the phytoplankton in the coastal sea water during recent years. *Leptocylindrus minimus* was already observed *ca* 75 years ago in the Southern Bight of the North Sea (cf. Meunier, 1915, describing the species as *L. belgicus*) and can often be found together with *L. danicus* (Drebes, 1974). The species was dominant for the first time during summer 1989 and 1990 in the Marsdiep, entrance Dutch Wadden Sea (Cadée, 1992). The species was also observed during summer 1990 by Koeman *et al.* (1991) for the Westerschelde estuary, several locations in the coastal stretch of the North Sea and the Dutch Wadden Sea. In the Oosterschelde *L. minimus* was more abundant in W than in E and, consequently, its occurrence bears no relation to the changed environmental conditions but depends only on its present-day abundance in the coastal sea water and subsequent transport into the Oosterschelde.

The newcomer *Rhizosolenia fragilissima* was not observed by Cadée (1992) in the Marsdiep, but later on still demonstrated incidentally (Cadée, unpubl.). Koeman *et al.* (1991) found the species sparsely in one North Sea locality (10 km off Terschelling) only. In Grevelingen lagoon, however, a bloom was observed in summer 1990 (in earlier years no observations were made). We assume that the occurrence in Grevelingen enabled its appearance in the Oosterschelde, although discharge of Grevelingen water was negligible during the last years. *Rh. fragilissima* was flowering in the Oosterschelde every year since 1987, more regularly and abundantly in E than in W. This illustrates the suitability for this species of a marine environment with reduced water movements.

Environmental changes

A clear response to the environmental changes in the Oosterschelde is given by the diatoms *Skeletonema costatum* (Fig. 8c) and *Ditylum brightwellii*

(Fig. 8b). *Skeletonema* undoubtedly is one of the best known and most widespread euryhaline marine diatom species since Curl & McLeod (1961) studied its physiological ecology from cultures and used these results for comparison with the development of *Skeletonema* in the shelf waters of Long Island Sound. In The Netherlands the species occurs in stagnant and tidal waters. In brackish Veersemeer (Fig. 1) the diatom reached very high maxima: $\pm 10^5$ cells ml^{-1} (Bakker & De Pauw, 1974) under mesohaline-eutrophic conditions, while in saline Grevelingen lagoon the species was rare under conditions of high salinities and limiting nutrients (N, Si) in summer (Bakker & De Vries, 1984). In the Marsdiep, receiving continuously nutrient-rich Rhine water, *Skeletonema* was dominating from 1980–89, always in spring and early summer, but often also in summer (Cadée, 1992). In the Dutch coastal North Sea, Kat (1992) noted strongly varying densities: long-lasting blooms in 1973 and near-absence in 1974. These field data fit well in the results of the culture experiments of Rijstenbil (1988), who established that under strongly fluctuating mesohaline conditions *Skeletonema* competed more succesfully with other marine diatoms than under more stable polyhaline-marine conditions. The latter situation is preferred by *Ditylum*, only tolerating slow variations in salinity (Rijstenbil, 1987; 1988; Rijstenbil *et al.*, 1989). In competition experiments in the laboratory with both species, after a shift from higher to lower salinity, *Skeletonema* had a greater affinity for ammonium and outcompeted *Ditylum* (Rijstenbil, 1988). In the Oosterschelde the increasing salinities since 1985 might have influenced the competitive abilities of both species, leading to the observed shifts in annual abundance. This may explain the decrease of *Skeletonema* during the barrier- and post-barrier periods (Fig. 8c) as well as the contrasting increase of *Ditylum* (Fig. 8b) during the same period.

Phaeocystis

Among the non-diatoms we pay special attention to the haptophycean flagellate *Phaeocystis* sp., a polymorphic species with a remarkable life cycle (Kornmann, 1955; Veldhuis, 1987; Verity *et al.*, 1988; Sournia, 1988; Cadée, 1991). Peperzak (in prep.) unravelled the life cycle. He distinguished small solitary flagellated cells (microflagellates), developing into large solitary flagellated cells (macroflagellates) when total daily irradiance exceeds 100 Wh m^{-2} d^{-1} (Peperzak, 1993). Under nutrient-sufficient conditions the macroflagellates transform into large non-motile cells to form colonies. Under conditions of nutrient depletion and reduced total daily irradiance, the colonial cells develop into colonial macroflagellates and decrease in size to form colonial microflagellates. These cells are released, closing the cycle (Peperzak, in prep.). The microflagellates are present in high percentages in the decline phase of the *Phaeocystis* spring bloom along the Dutch Coast (Veldhuis *et al.*, 1986). Generally, *Phaeocystis* colonial blooms occur after the spring bloom of diatoms. This phenomenon is well-known and in a review of historical phytoplankton data of the Western Dutch Wadden Sea, Cadée & Hegeman (1991) revealed that the species was observed blooming here at least as long ago as 1897. As a probable consequence of increasing eutrophication caused by man, it was demonstrated that the species increased in cell numbers and in length of the blooming period from 1973 to 1985, not only in spring but also in summer (Cadée & Hegeman, 1986). In the former Oosterschelde estuary *Phaeocystis* blooms were restricted to spring but from the barrier period onwards also summer peaks could be observed (Fig. 5). We assume that these summer blooms were developed when the water transparencies increased, enhancing daily irradiance that governs the growth rate and colony formation of *Phaeocystis* (Peperzak, in press). Nutrients were not limiting in summer during the barrier-construction period. During the post-barrier years, however, the nutrient loads from the northern compartment were strongly reduced (total N: 49,3%; Si–SiO$_2$: 65,2%; Wetsteyn & Bakker, 1991). This initiated a trend of oligotrophication, notably clear in 1990 with regard to nitrogen. In 1990 the spring development of *Phaeocystis* was poor in W, compared to previous years (Fig. 5) and in E no cells were counted but only some disintegrating colonies could be observed in live samples (Wetsteyn & Peperzak, pers. comm.), probably remains of the tidal transport from W. Absence of local development in E may be directly related to the low nitrate concentration during that year, from May onwards. Moreover, Riegman *et al.* (1992) indicated that colony formation of *Phaeocystis* is induced when nitrate is available as the dominant N-source and that this process does not take place when ammonia has to be consumed under N-limiting conditions. The findings of Peperzak (in press) and of Riegman *et al.* (1992) may explain the periodicity of *Phaeocystis* colonies in the Oosterschelde and the absence of the colonies in E since summer 1989 (Fig. 5). Blooming of *Phaeocystis* in colonial form along the Dutch Coast in spring or/and in summer, is only possible by the regular discharge of nitrate-rich Rhine water. A phytoplankton species can be determined in its strategy by the nature of the limiting resource (Harris, 1986) and *Phaeocystis*' (non-)ability to form colonies is a nice example of such a dependence.

Severe winters

Not only the hydrodynamic changes and the decrease of nutrient concentrations (caused by the storm-surge barrier and the accompanying works respectively) can be considered driving forces for the (changes in) phytoplankton development of the Oosterschelde. With regard to climatic effects, a severe winter may exert an immediate influence on the phytoplankton development in coastal sea water: a rapid stabilization of the water column occurs, better light conditions arise and cause the spring bloom to start earlier (Gillbricht, 1964). Coinciding with the barrier-construction period, three successive severe winters occurred (1984–85, 1985–86 and 1986–87). The benthic-pelagic diatom *Biddulphia aurita* started its growth earlier in spring and was much more abundant during that period than before and afterwards (Fig. 8d). An explanation for its abundance in such condi-

tions was given by Baars (1986). He determined generation times in culture and demonstrated that these were shortest at temperatures from -1.5 to *ca* 6 °C. Moreover, the early participation of *B. aurita* in the spring plankton was facilitated by a temperature-dependent non-adherence behaviour: at 0 °C (and below) almost no adherence of this tychopelagic species to the sediment could be observed while at higher temperatures the species' colonies increasingly stuck to the substrate, causing *B. aurita* to lead a completely benthic existence in summer. Consequently, because most other diatom species are prevented from growing vigorously under these low-temperature conditions, *B. aurita*, once it becomes detached from the bottom, builds up a dense initial population without having to compete with other species. Also during the extremely cold and long winter 1962–63 massive blooms of *B. aurita* were recorded in the Elbe mouth (Kühl, in: Gillbricht, 1964) and in the Oosterschelde (Bakker, 1963; unpubl.).

While *B. aurita* is always present in temperate coastal sea water during spring, the neritic diatom *Detonula confervacea* is found at irregular times. According to Smayda (1969) the species is Arctic, frequently being observed at temperatures below 0 °C, although growth does not depend on ice cover. In the Netherlands the species was recorded earlier during spring, in the brackish (meso-polyhaline) Veersemeer (Bakker & De Pauw, 1974), in (polyhaline) Grevelingen lagoon (Bakker & De Vries, 1984) and in the small (polyhaline) Schelphoek lagoon (Rijstenbil, 1987), all in non-tidal waters. Cadée (pers. comm.) never observed the species in the Marsdiep area. The coincidence of severe winters and decreased tidal water movements during the barrier period was probably favourable for the development of *Detonula* in the Oosterschelde, especially in E where high abundance was reached.

Acknowledgements

We thank Drs Jacco C. Kromkamp, Pieter H. Nienhuis, Jan W. Rijstenbil and Lambertus

P. M. J. Wetsteyn as well as the referees Drs Gerhard C. Cadée and Linda Maddock for critical comments on an earlier version of the ms. We are grateful to Mr. Alfons A. Bolsius who refined and completed the (computer) figures and to Mrs Elly van Hulsteijn en Miss Joke D. M. Tibbe for producing patiently several successive p.c. versions of the text. Communication No. 690 of NIOO-CEMO, Yerseke, The Netherlands.

References

Baars, J. W. M., 1986. Autecological investigations on marine diatoms. 4: *Biddulphia aurita* (Lyngb.). Brébisson et Godey – A succession of spring diatoms. Hydrobiol. Bull. 19: 109–116.

Bakker, C., 1964. Plankton untersuchungen in einem holländischen Meeresarm vor und nach der Abdeichung. Helgoländer Wiss. Meeresunters. 10: 456–472.

Bakker, C. & N. De Pauw, 1974. Comparison of brackish water plankton assemblages of identical salinity ranges in an estuarine tidal (Westerschelde) and stagnant (Lake Veere) environment (S.W.-Netherlands). I. Phytoplankton. Hydrobiol. Bull. 8: 179–189.

Bakker, C., 1978. Some reflections about the structure of the pelagic zone of the brackish Lake Grevelingen (S.W.-Netherlands). Hydrobiol. Bull. 12: 67–84.

Bakker, C. & F. Vegter, 1978. General tendencies of phyto- and zooplankton development in two closed estuaries (Lake Veere and Lake Grevelingen) in relation to an open estuary (Eastern Scheldt), S.W.-Netherlands. Hydrobiol. Bull. 12: 226–245.

Bakker, C. & I. De Vries, 1984. Phytoplankton- and nutrient dynamics in saline Lake Grevelingen (S.W.-Netherlands) under different hydrodynamical conditions in 1978–1980. Neth. J. Sea Res. 18: 191–220.

Bakker, C., P. M. J. Herman & M. Vink, 1990. Changes in seasonal succession of phytoplankton induced by the storm-surge barrier in the Oosterschelde (S.W.-Netherlands). J. Plankton Res. 12: 947–972.

Bakker, C. & M. Vink, 1994. Nutrient concentrations and planktonic diatom-flagellate relations in the Oosterschelde (SW Netherlands) during and after the construction of a storm-surge barrier. Hydrobiologia 282/283: 101–116.

Bakker, C. & P. Van Rijswijk, 1994. Zooplankton biomass in the Oosterschelde (SW Netherlands) before, during and after the construction of a storm-surge barrier. Hydrobiologia 282/283: 127–143.

Boalch, G. T., 1987. Changes in the phytoplankton of the Western English Channel in recent years. Br. phycol. J. 22: 225–235.

Cadée, G. C., 1990. Increase of *Phaeocystis* blooms in the westernmost inlet of the Wadden Sea, the Marsdiep since

1973. In: C. Lancelot, G. Billen & H. Barth (eds.). Water pollution research report 12. Commission of the European Communities, Luxembourg, p. 105–112.

Cadée, G. C., 1991. *Phaeocystis* colonies wintering in the water column? Neth. J. Sea Res. 28: 227–230.

Cadée, G. C., 1992. Phytoplankton variability in the Marsdiep, 1980–1990. ICES. Variability Symp.: in press.

Cadée, G. C. & J. Hegeman, 1986. Seasonal and annual variation of *Phaeocystis pouchetii* (Haptophyceae) in the westernmost inlet of the Wadden Sea during the 1973 to 1985 period. Neth. J. Sea Res. 20: 29–36.

Cadée, G. C. & J. Hegeman, 1991. Historical phytoplankton data for the Marsdiep. Hydrobiol. Bull. 24: 111–118.

Curl, H. & G. C. McLeod, 1961. The physiological ecology of a marine diatom *Skeletonema costatum* (Grev.) Cleve. J. Mar. Res. 19: 70–88.

Drebes, G., 1974. Marines Phytoplankton. Thieme Verlag, Stuttgart.

Dronkers, J. & J. T. F. Zimmerman, 1982. Some principles of mixing in tidal lagoons with examples of tidal basins in the Oosterschelde. In P. Lasserre & H. Postma (eds), Coastal Lagoons. Proc. Int. Symp. Coastal Lagoons, Gauthier-Villars: 460–474.

Gillbricht, M., 1964. Einwirkungen des kalten Winters 1962–63 auf die Phytoplanktonentwicklung bei Helgoland. Helgoländer wiss. Meeresunters. 10: 263–275.

Guillard, R. R. L. & P. Kilham, 1977. The ecology of marine planktonic diatoms. In D. Werner (ed.). The Biology of Diatoms. Botanical Monographs, Vol. 13, Blackwell Scientific Publications, Oxford.

Hagmeier, E., 1978. Variations in phytoplankton near Helgoland. Rapp. P.-v. Réun. Cons. int. explor. Mer 172: 361–363.

Harris, G. P., 1986. Phytoplankton Ecology. Structure, Function and Fluctuation. Chapman and Hall, London and New York.

Hill, M. O., 1979a. TWINSPAN – a FORTRAN program for arranging multivariate data in ordered two-way table by classification of individuals and attributes. Cornell University, Ithaca, NY.

Hill, M. O., 1979b. DECORANA – a FORTRAN program for detrended correspondence analysis and reciprocal averaging. Cornell University, Ithaca, NY.

Hustedt, F., 1930. Die Kieselalgen. In Dr L. Rabenhorst's Kryptogamen-Flora von Deutschland, Österreich und der Schweiz, 8(1). Akad. Verlagsges., Leipzig.

Jongman, R. H. G., C. J. F. Ter Braak & O. F. R. Van Tongeren, 1987. Data Analysis in Community and Landscape Ecology. Pudoc, Wageningen.

Kat, M., 1992. Year-to-year variation in the occurrence of some dominant phytoplankton species in Dutch coastal waters, 1973–1984. Hydrobiol. Bull. 25: 225–231.

Klepper, O., 1989. A model of carbon flows in relation to macrobenthic food supply in the Oosterschelde estuary (S.W.-Netherlands). Thesis, Wageningen.

Klepper, O., M. W. M. van der Tol, H. Scholten &

P. M. J. Herman, 1994. SMOES: a simulation model for the Oosterschelde ecosystem. Part I: Description and uncertainty analysis. Hydrobiologia 282/283: 437–451.

Koeman, R., M. Rademaker & W. Gremmen, 1991. Biologische monitoring van fytoplankton in de Nederlandse zoute en brakke wateren, 1990. TRIPOS Rijswijk-Haren.

Kornmann, P., 1955. Beobachtungen an *Phaeocystis*-Kulturen. Helgoländer Wiss. Meeresunters. 5: 218–233.

Lohmann, H., 1908. Untersuchungen zur Feststellung des vollständigen Gehalts des Meeres an Plankton. Wiss. Meeresunters. Kiel 10: 129–370.

Maddock, L., D. S. Harbour & G. T. Boalch, 1989. Seasonal and year-to-year changes in the phytoplankton from the Plymouth area, 1963–1986. J. mar. biol. Ass. U.K. 69: 229–244.

Margalef, R., 1978. Life-forms of phytoplankton as survival alternatives in an unstable environment. Oceanol. Acta 1: 493–509.

Matta, J. F. & H. G. Marshall, 1984. A multivariate analysis of phytoplankton assemblages in the Western North Atlantic. J. Plankton Res. 6: 663–675.

Meunier, A., 1915. Microplancton de la Mer Flamande. 2me partie, Les Diatomacées (suite) (Le genre *Chaetoceros* excepté). Mém. Mus. Roy. Hist. Belg. 7: 1–118.

Nienhuis, P. H. & A. C. Smaal, 1994. The Oosterschelde estuary, a case study of a changing ecosystem: an introduction. Hydrobiologia 282/283: 1–14.

Peperzak, L., F. Colijn & J. C. H. Peeters, in prep. A life cycle model of *Phaeocystis*.

Peperzak, L., 1993. Daily irradiance governs growth rate and colony formation of *Phaeocystis*. J. Plankton Res. 15: 809–821.

Radach, G., J. Berg & E. Hagmeier, 1990. Long–term changes of the annual cycles of meteorological, hydrographic, nutrient and phytoplankton time series at Helgoland and at LV ELBE I in the German Bight. Cont. Shelf. Res. 10: 305–328.

Reid, P. C., H. G. Hunt & T. D. Jonas, 1983. Exceptional blooms of diatoms associated with anomalous hydrographic conditions in the Southern Bight in early 1977. J. Plankton Res. 5: 755–765.

Reid, P. C., C. Lancelot, W. W. C. Gieskes, E. Hagmeier & G. Weichert, 1990. Phytoplankton of the North Sea and its dynamics: a review. Neth. J. Sea Res. 26: 295–331.

Riegman, R., A. A. M. Noordeloos & G. C. Cadée, 1992. *Phaeocystis* blooms and eutrophication of the continental coastal zones of the North Sea. Mar. Biology 112: 479–484.

Rijstenbil, J. W., 1987. Phytoplankton composition of stagnant and tidal ecosystems in relation to salinity, nutrients, light and turbulence. Neth. J. Sea Res. 21: 113–123.

Rijstenbil, J. W., 1988. Selection of phytoplankton species by gradual salinity changes. Neth. J. Sea Res. 22: 291–300.

Rijstenbil, J. W., J. A. Wijnholds & J. J. Sinke, 1989. Implications of salinity fluctuations for growth and nitrogen metabolism of the marine diatom *Ditylum brightwellii* in comparison with *Skeletonema costatum*. Mar. Biol 101: 131–141.

Robinson, G. A., 1983. Continuous plankton records: phytoplankton in the North Sea, 1958–1980, with special reference to 1980. Br. phycol. J. 18: 131–139.

Smayda, T. J., 1969. Experimental observations on the influence of temperature, light and salinity on cell division of the marine diatom *Detonula confervacea* (Cleve) Gran. J. Phycol. 5: 150–157.

Smayda, T. J., 1978. From phytoplankters to biomass. In A. Sournia (ed.), Phytoplankton Manual. Unesco, Paris: 273–279.

Smayda, T. J., 1980. Phytoplankton species succession. In I. Morris (ed.), The Physiological Ecology of Phytoplankton. Blackwell Scientific Publications, Oxford: 493–570.

Sournia, A., 1988. *Phaeocystis* (Prymnesiophyceae): how many species? Nova Hedwigia 47: 211–217.

Ter Braak, C. J. F., 1986. Canonical correspondence analysis: a new eigenvector method for multivariate direct gradient analysis. Ecology 67: 1167–1179.

Ter Braak, C. J. F., 1988. CANOCO – a FORTRAN program for canonical community ordination by (partial) (detrended) (canonical) correspondence analysis, principal components analysis and redundancy analysis (version 2.1). Agricultural Mat. Group, Ministry of Agriculture and Fisheries.

Ter Braak, C. J. F., 1989. CANOCO, an extension of DECORANA to analyze species-environment relationships. Hydrobiologia 184: 169–170.

Van der Tol, M. W. M. & H. Scholten, 1992. Response of the foodweb of the Oosterschelde to a changing environment: functional or adaptive? Neth. J. Sea Res. 30: 175–190.

Vegter, F. & P. R. M. de Visscher, 1987. Nutrients and phytoplankton primary production in the marine tidal Oosterschelde estuary (The Netherlands). Hydrobiol. Bull. 21: 149–158.

Veldhuis, M. J. W., F. Colijn & L. A. H. Venekamp, 1986. The spring bloom of *Phaeocystis pouchetii* (Haptophyceae) in Dutch coastal waters. Neth. J. Sea Res. 20: 37–48.

Veldhuis, M. J. W., 1987. The eco-physiology of the colonial alga *Phaeocystis pouchetii*. Ph.D. Thesis State Univ. Groningen.

Verity, P. G., T. A. Villareal & T. J. Smayda, 1988. Ecological investigations of blooms of colonial *Phaeocystis pouchetii*. II. The role of life-cycle phenomena in bloom termination. J. Plankton Res. 10: 749–766.

Vroon, J., 1994. Hydrodynamic characteristics of the Oosterschelde in recent decades. Hydrobiologia 282/283: 17–27.

Wetsteyn, L. P. M. J., J. C. H. Peeters, R. N. M. Duin, F. Vegter & P. R. M. De Visscher, 1990. Phytoplankton primary production and nutrients in the Oosterschelde (The Netherlands) during the pre-barrier period 1980–1984. Hydrobiologia 195: 163–177.

Wetsteyn, L. P. M. J. & C. Bakker, 1991. Abiotic characteristics and phytoplankton primary production in relation to a large-scale coastal engineering project in the Ooster-schelde (The Netherlands): a primary evaluation. In M. Elliott & J. P. Ducrotoy (eds.). Estuaries and Coasts: Spatial and Temporal Intercomparisons. Proceedings of the ECSA 19 Symposium, Caen 1989. Olsen & Olsen, Fredensborg, Denmark: 365–373.

Wetsteyn, L. P. M. J. & J. Kromkamp, 1994. Turbidity, nutrients and phytoplankton primary production in the Oosterschelde (The Netherlands) before, during and after a large-scale coastal engineering project (1980–1990). Hydrobiologia 282/283: 61–78.

Hydrobiologia **282/283**: 101–116, 1994.
P. H. Nienhuis & A. C. Smaal (eds), The Oosterschelde Estuary.
© 1994 *Kluwer Academic Publishers. Printed in Belgium.*

Nutrient concentrations and planktonic diatom-flagellate relations in the Oosterschelde (SW Netherlands) during and after the construction of a storm-surge barrier

C. Bakker & M. Vink
Netherlands Institute of Ecology, Centre for Estuarine and Coastal Ecology, Vierstraat 28, 4401 EA Yerseke, The Netherlands

Key words: nutrient concentrations, nutrient ratios, phytoplankton biomass, diatoms, flagellates, pre- and post-barrier Oosterschelde

Abstract

Nutrients

The inflow of Rhine water into the Oosterschelde was strongly reduced from 1987 onwards. This caused the winter concentrations of silicate and nitrate to decrease in the Eastern compartment, while those in the deeper Western compartment, more dependent on North Sea concentrations, hardly changed. The result was a levelling of the former East-West gradients for these nutrients. In East, summer concentrations of nitrate reached limiting levels in the post-barrier period and molar nitrate/ammonium ratios became < 1, indicating that any release of nitrogen must be important to stimulate phytoplankton growth in this area. Silicate summer concentrations in East, on the other hand, were higher in the new situation. In West, differences in summer nutrient concentrations between the old and new situation were smaller than in East, due to the still continuing exchange with the North Sea.

Phytoplankton diatoms and flagellates

In East during summer, N-depletion and longer residence times caused the phytoplankton to become strongly dependent on nutrient regeneration processes and increased zooplankton grazing. Average diatom biomass declined, but flagellate biomass rose during summer. Spring conditions for phytoplankton development in this area improved due to the increased water transparency, nutrients being present in excess, and this resulted in a higher 'new' production of diatoms than before.

In West, summer biomass of diatoms decreased, probably due to increased consumption by mussels under conditions of longer residence times; nutrients were not limiting, due to important benthic mineralization processes and exchange with the North Sea. The previously existing West-East biomass gradients disappeared, or sometimes reversed.

Experimental (mesocosm studies) as well as field data, reported in the literature, give evidence for the given explanations.

Introduction

The Oosterschelde is a tidal bay (*ca* 350 km²) of the Southern North Sea (Fig. 1). Strong tidal ex-

change with the North Sea via relatively deep (40 m max.) channels is causing complete mixing of the system (Dronkers & Zimmerman, 1982). Suspended matter contents, nutrient concentrations and phytoplankton biomass in the Western compartment, therefore, closely resemble those of the nearby North Sea (Bakker, unpubl. data). Before 1970 when most estuaries of the Dutch Delta Area were still not closed off from the sea, the rivers Rhine and Meuse discharged periodically high volumes of fresh water via the Volkerak into the Northern compartment of the Oosterschelde (Fig. 1). Examples of high river discharge strongly influencing the salinity of the estuary have been given by Korringa (1940), Drinkwaard (1960) and Bakker (1967). When the Volkerak was closed (1969), the irregular inflow of riverwater changed into a fixed and regular discharge, amounting to ca 70 m^3 s^{-1} during the period 1980–87 (Wetsteyn & Bakker, 1991). High peaks during wet winters (such as in 1965 and 1966), which caused strong salinity decreases in the system (Bakker,

1967), were no longer possible. After the closure of the Philipsdam in 1987, the freshwater load to the Northern compartment did not exceed 10 m^3 s^{-1}. The concomitant reduction in nutrient loads amounted to 44–82% (Wetsteyn & Bakker, 1991).

During the years 1984–1986, a storm-surge barrier was constructed in the western area (Fig. 1). A comprehensive long-term ecological study was started in 1981 to analyse the situation before, during and after this period. During the construction of the barrier and the accompanying works in the Eastern (Markiezaatsdam and Oesterdam) and Northern (Philipsdam) area (Nienhuis & Smaal, 1994), the hydrodynamics of the estuary were seriously influenced, particularly with respect to tidal amplitude and current velocities (Vroon, 1994). When current velocities

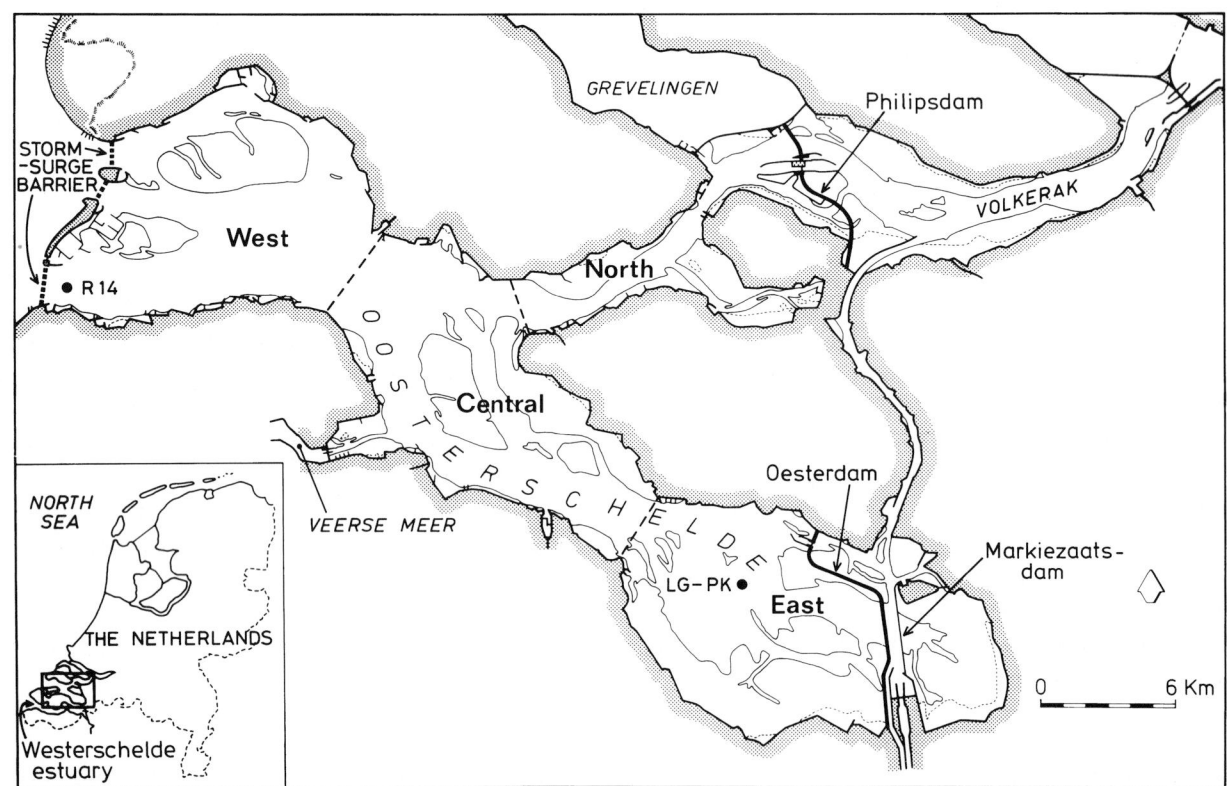

Fig. 1. Map of the Oosterschelde with indication (broken lines) of the 4 compartments and the sampling localities in West (R14) and East (LG-PK). Note the position of the storm-surge barrier (West), of the Oester- and Markiezaatsdam (East) and of the Philipsdam (North).

decreased, especially in the shallow Eastern compartment, suspended matter concentrations declined and water transparency rose concomitantly (Bakker *et al.*, 1990; Wetsteyn & Bakker, 1991; Bakker *et al.*, 1994). For additional data on hydrography, nutrient concentrations and primary production, *cf* Vegter & De Visscher (1987), Wetsteyn *et al.* (1990), Wetsteyn & Kromkamp (1994).

In this paper, we discuss the nutrient concentration and phytoplankton biomass data obtained during the entire period of investigation (1982–1990). The aim is to describe the response of the phytoplankton in terms of biomass of the main species groups, diatoms and flagellates, to the changed hydrographical conditions in the present-day Oosterschelde, caused by the barrier construction works. For the response of the phytoplankton in terms of species composition and -succession, see Bakker *et al.* (1994) and for zooplankton developments, Bakker & Van Rijswijk (1994) and Tackx *et al.* (1994).

Materials and methods

Only the Eastern and Western compartments of the Oosterschelde (Fig. 1) were sampled. These areas will be further referred to as E and W respectively.

In W, samples were taken at buoy R14 (main channel in the vicinity of the storm-surge barrier, Fig. 1), depth $\pm$ 35 m; in E at the centrally located buoy LG-PK (Fig. 1), depth $\pm$ 20 m. All samples were taken at mid-depth during half-tide periods when maximum current velocities caused vertical distributions of the phytoplankton to be as homogeneous as possible. Sampling was carried out weekly from March to October (1982–1990) and periodically during the rest of the year. In 1982, W was sampled biweekly. In 1985 only E was sampled.

Concentrations of Si-SiO_2, N-NO_3, N-NH_4 were determined as in Parsons *et al.* (1984). Water samples for phytoplankton analysis were kept in 1 l glass bottles, preserved before sedimentation with Lugol's iodide solution. Phytoplankton cells

were counted with an inverted microscope. Diatoms and non-diatoms were identified to species level. For the species list, *cf.* Bakker *et al.* (1994). Cell counts and volume measurements were converted to carbon using the equations of Eppley (in Smayda, 1978). Total biomass values were estimated for diatoms and flagellates separately, in both compartments. The curves representing nutrient concentrations (Figs 2–4) as well as phytoplankton biomass (Figs 5–6) have been smoothed by calculating moving averages of 3 successive weekly observations. To emphasize the relation between nutrient concentrations and phytoplankton biomass, the presentation of Figs 2–6 covers mainly the growth season. To avoid too much detail, we chose 4 years to represent an adequate comparison of the periods of investigation, *i.e.* 1983 (pre-barrier year, former hydrodynamic conditions, *cf.* Dronkers & Zimmerman, 1982); 1986 (year of barrier construction, transition period characterized by strongly decreased tidal movements, due to the hydraulic works, *cf.* Vroon, 1994); and the years 1988 and 1990 (post-barrier period, new situation: with reduced tidal movement compared to the pre-barrier situation but showing enhanced tidal amplitude compared to the barrier construction period, *cf.* Vroon, 1994). Intentionally we chose two years from the last period to demonstrate that some trends became increasingly clear in the course of these years. Finally, seasonal averages (spring: water temperature < 15 °C and summer: $\geq$ 15 °C), have been given for each period (Fig. 7) to summarize the observed trends over the entire period of observation.

Results

Nutrients

The following main tendencies can be distinguished:

Winter levels of Si- and NO_3-N concentrations (Figs 2–3)
In E, winter levels amounted to *ca* 1 mg l^{-1} during the pre-barrier period. After the Philipsdam

104

(Fig. 1) was finished (April 1987), the influx of
river water decreased strongly. Declining nutrient
concentrations were observed during the next
winter (1988) and became evident in 1990 (Fig. 3)
when Si- and NO₃ values did not exceed 0.6 mg
l⁻¹. During the barrier period (1986; Fig. 2) winter levels of Si and NO₃, respectively, were similar
to, respectively slightly higher than, those of the
preceding years.

In W, winter levels of Si and NO₃ of pre- and
post-barrier years hardly differed (range 0.50–
0.70 mg l⁻¹ in 1983; 0.55–0.65 mg l⁻¹ in 1990).
Winter values during the barrier period (1986)
showed higher NO₃ and similar Si maxima, as in
E. In 1988 relatively high Si- and NO₃ values
(0.7–0.9 mg l⁻¹) were measured for the last time.

Within the barrier period 3 subsequent severe
winters occurred (1984–1985; 1985–1986; 1986–

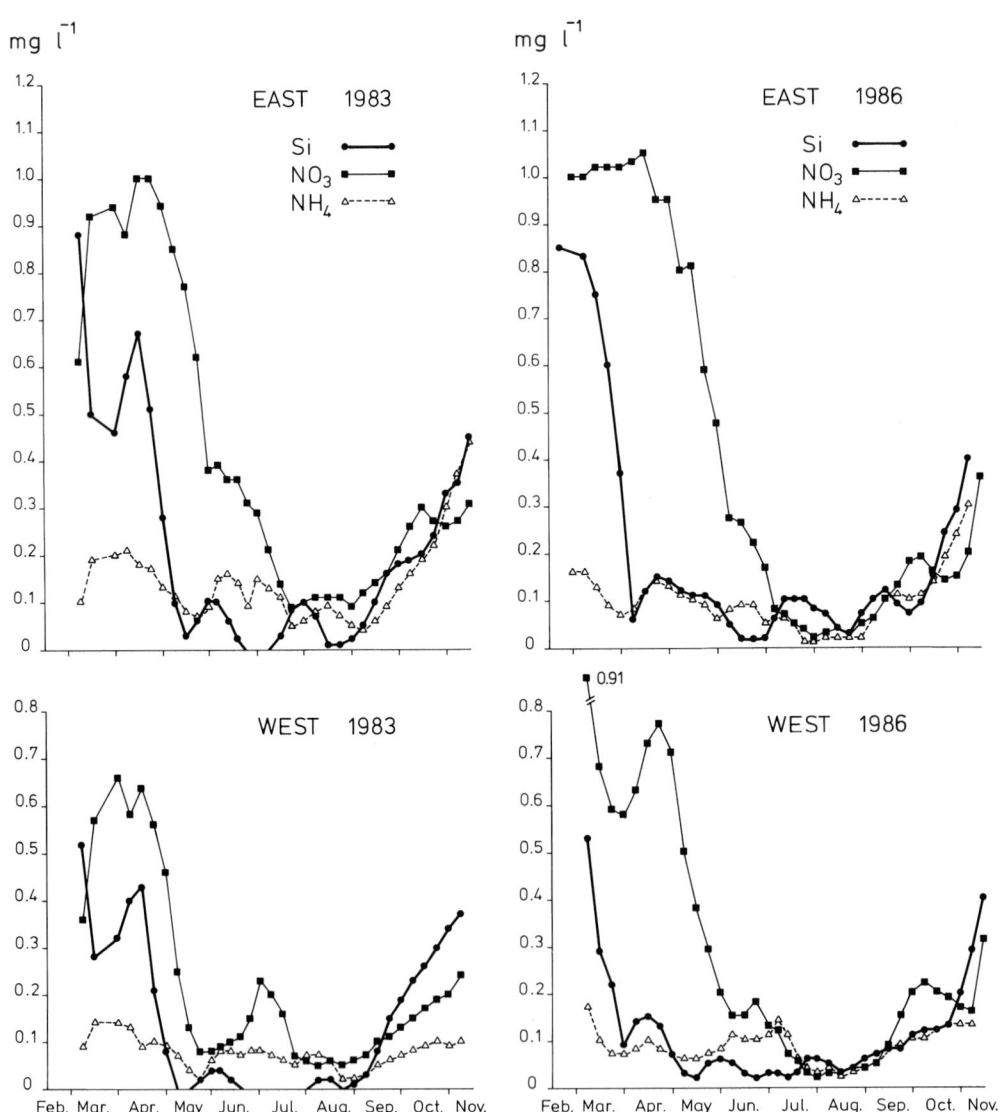

Fig. 2. Moving averages (over 3-weekly periods) of Si-SiO₂, N-NO₃ and N-NH₄ concentrations (mg l⁻¹) in the Oosterschelde,
Eastern (top) and Western (bottom) compartments; pre-barrier year 1983 (left) and barrier-construction year 1986 (right).

1987). During these winters the Si decline in spring (Fig. 2) was observed one month earlier than before and afterwards and was coupled with earlier diatom blooms (*cf.* Fig. 5).

Summer levels of Si- and NO-N₃ concentrations (Figs 2–3)
Si concentrations regularly approached zero from mid June to mid July during the pre-barrier period (1983), in both compartments; this was not found during all other years of investigation. Si levels in summer were generally higher during the post-barrier period than before, especially in E.

For NO_3 concentrations in summer, the picture was reversed compared to Si. During the pre-barrier and barrier years average concentrations in W were lower than in E; during the post-barrier years 1988–1990 concentrations in W were consistently higher than in E. In 1990, in E potentially limiting conditions (≤ 0.01 mg l^{-1}) were reached from May to the middle of August (Fig. 3).

Concentrations of NH₄-N (Figs 2–3)
NH_4 concentrations fluctuated irregularly, not showing a clear seasonality, and mostly did not exceed 0.2 mg l^{-1}. Consistent differences in winter concentrations between the different periods, as observed for NO_3, did not occur. Levels always declined in spring. Summer increases of NH_4, due to mineralization, did not reach NO_3 peak levels in the pre-barrier period, but exceeded NO_3 concentrations during 1990 (Fig. 3) which represented a new phenomenon in the post-barrier period.

N/Si-ratios (Fig. 4)
Variations in the molar ratios of the ambient concentrations of dissolved Si and inorganic N ($NO_3 + NH_4$-N; NO_2-N neglected) can lead to strong deviations from the Redfield ratio for these macro-nutrients. An intracellular ratio N/Si = 1 is indicative for nutrient-sufficient diatom populations (Brzezinsky, 1985). Consequently, potential N-limitation may occur at N/Si < 1. Figure 4 shows that in 1983 a ratio < 1 was never observed; some ratios from May to August were infinitely high due to non-detectable Si-concentrations (Fig. 2). This occurred earlier and lasted longer in W where lower winter Si-values (Fig. 2) but stronger diatom blooms (Fig. 5) were observed. In 1986 too, periods with non-detectable Si-concentrations occurred but these were shorter than in 1983. This may be influenced by the continuous supply of nutrient-rich waters from the Northern compartment, combined with a longer residence time of the water during the barrier period. In 1988 under the changed conditions of increased (compared to 1986) tidal movements, but nevertheless reduced (compared to 1983) tidal amplitude, N/Si-ratios < 1 occurred several times in E (August–September) while in W, this was a less frequent observation. In 1990 the differences with the pre-barrier period became evident. Periods of non-detectable Si-concentrations were no longer observed, but an extended period of potential N-limitation (from May to August inclusive) occurred. In W only incidental cases of a N/Si-ratio < 1 were seen.

NO₃/NH₄-ratios (Fig. 4)

In 1983 maximum NO_3/NH_4-ratios in April–May indicated a preferred NH_4-uptake by the phytoplankton. In 1986 (severe winter, barrier period) the maximum ratio was observed earlier (March), notably high in E, synchronously with an infinitely high N/Si-peak, indicating rapid uptake by diatoms (Fig. 5). In 1988 and 1990, the NO_3/NH_4-spring peaks were again recorded in April. Due to the diminished influx of nutrients from the Northern compartment after the closure of the Philipsdam, these peak ratios were highest in W, while in the previous years the values in E exceeded those in W.

Minimum NO_3/NH_4-ratios were found in summer and autumn. In 1983 when NO_3 was regularly supplied from the Northern compartment, this ratio seldom was < 1. During the barrier period, with comparable nutrient supply and a prolonged residence time, NO_3/NH_4-ratios showed a number of peaks, not seen before; in some cases the ratio declined to < 1. In 1988, post-barrier situation, the NO_3 surplus decreased

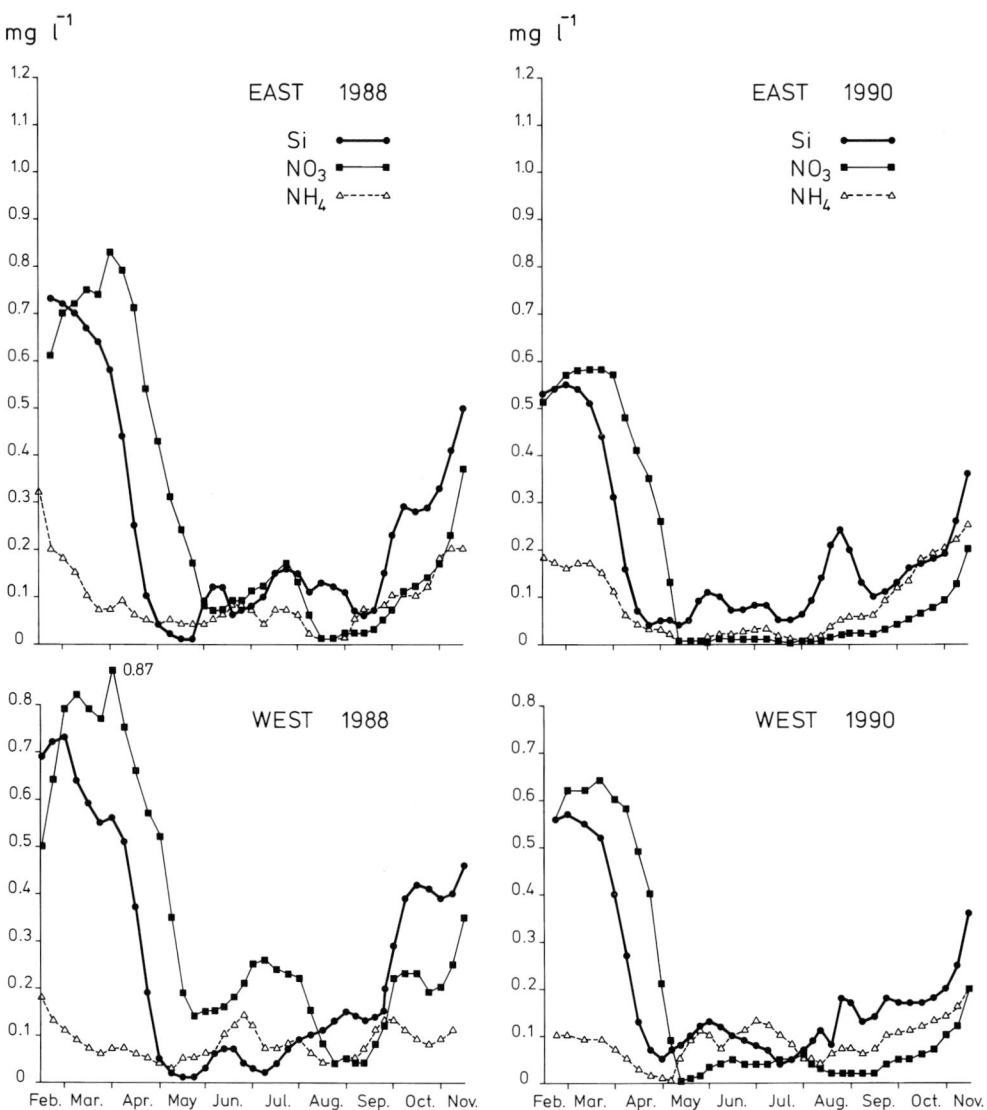

Fig. 3. Moving averages (over 3-weekly periods) of Si-SiO$_2$, N-NO$_3$ and N-NH$_4$ concentrations (mg l^{-1}) in the Oosterschelde, Eastern (top) and Western (bottom) compartments; post-barrier years 1988 (left) and 1990 (right).

but only incidentally values < 1 were reached. In 1990 NO$_3$/NH$_4$-ratios < 1 were observed from May onwards.

Phytoplankton biomass, selected years (Figs 5–6)

Diatoms

Pre-barrier period (1983). In spring, W developed a much higher carbon peak than E. In summer, however, large blooms were observed in both compartments (Fig. 5).

Barrier period (1986). After a severe winter an early (March) spring bloom developed, with highest biomass in E (Fig. 5). In April–May a large maximum was found in W, while in E diatoms did not at all develop during that period. Repeated summer blooms in E produced higher total biomass than in W.

Post-barrier period (1988, 1990). The main biomass peaks were reached later in spring (May). In 1988, this peak was largest in W (Fig. 6), as usual,

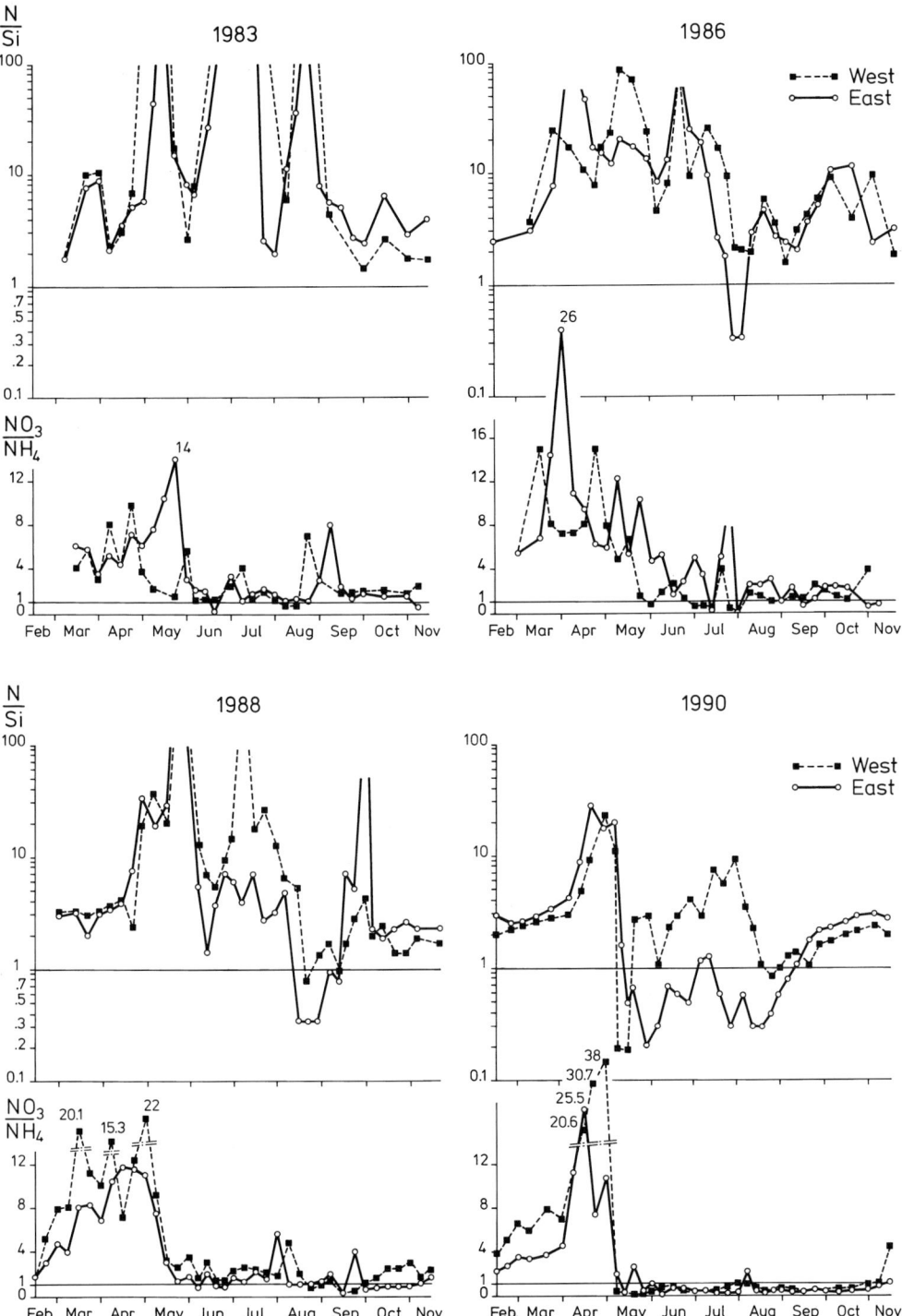

Fig. 4. Molar ratios of N to Si (top) and of NO_3- to NH_4-N (bottom) in the Oosterschelde, Eastern and Western compartments; pre-barrier year 1983, barrier-construction year 1986, post-barrier years 1988 and 1990.

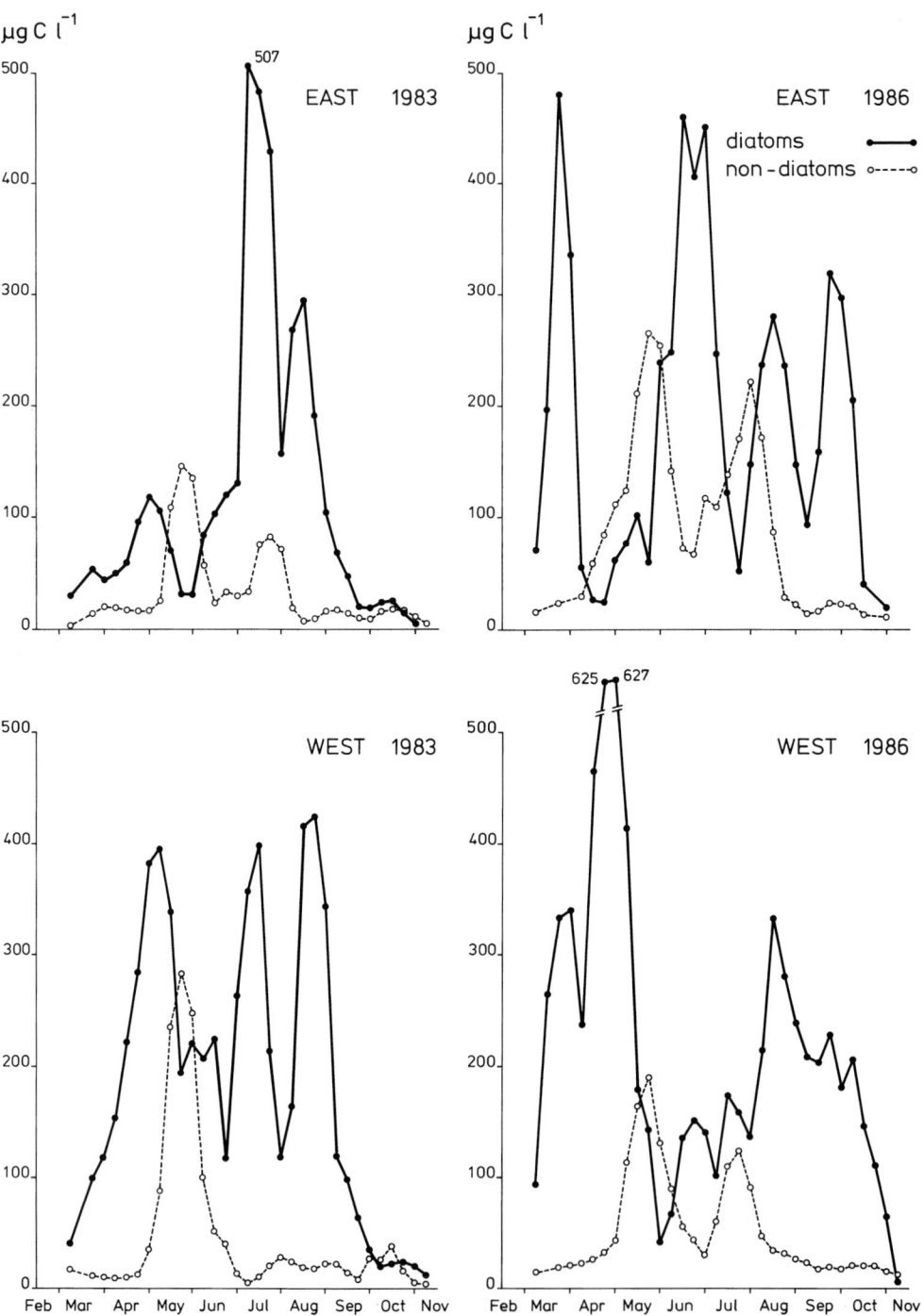

Fig. 5. Moving averages (over 3-weekly periods) of diatom- and non-diatom biomass (μg C l^{-1}) in the Oosterschelde, Eastern (top) and Western (bottom) compartments; pre-barrier year 1983 (left) and barrier-construction year 1986 (right).

but in 1990 E produced the highest spring peak recorded during the entire period of investigation, exceeding strongly the bloom in W, which was the smallest ever measured (Fig. 6). (These data correspond in detail with those for chlorophyll-a, measured by Wetsteyn & Kromkamp (1994)). The subsequent summer blooms were smaller than before and decreased in size. As a consequence, the previous spring biomass gradient from W towards E, still existing in 1988, was reversed in 1990. In summer diatom biomass gradients did not at all occur during the post-barrier period.

Non-diatoms

Pre-barrier period (1983). The spring bloom in W exceeded that in E, (Fig. 5) and mainly consisted of *Phaeocystis*, occurring as colonies and accounting for 40% of total average phytoplankton biomass. Summer development resulted in a peak (of other species) in E only (Fig. 5). Total flagellate carbon in summer amounted to 11–16% of total phytoplankton carbon only.

Barrier construction period (1986). The foregoing picture changed drastically during 1986 when a second colonial *Phaeocystis* bloom developed in summer, in both compartments and the blooms in E were higher and broader than those in W (Fig. 5).

Post-barrier period (1988, 1990). Spring blooms of *Phaeocystis* in 1988 in both compartments resembled those of the pre-barrier period, and the largest were observed in W (Fig. 6). In summer a second colonial *Phaeocystis* bloom developed, in both compartments. In 1990 *Phaeocystis* colonies were observed in W during spring, as usual, but the peak was lower than before, as for the diatoms. In E only a few disintegrating colonies could be found (Peperzak & Wetsteyn, pers. comm.) and the major bloom, occurring in May–June, consisted of cryptomonads accompanied by flagellated *Phaeocystis*-like cells only. This phenomenon was repeated in August.

Alternation of diatom and non-diatom blooms
The seasonal distributions of diatoms and non-diatoms for both compartments (Figs 5–6) show that the blooms of both speciesgroups were more or less alternating, except in spring 1988.

Phytoplankton biomass, averages for each period (Fig. 7)

Diatoms

In spring, diatom biomass in E increased from pre- to post-barrier periods. Spring diatom biomass in W reached much higher values than in E during pre-barrier and barrier periods but this difference disappeared in the post-barrier years.

In summer, diatom biomass in E did not change in pre-barrier and barrier periods but decreased strongly in the post-barrier year 1990. Summer diatom biomass in W decreased from pre- to post-barrier periods and biomass gradients W-E disappeared.

Non-diatoms
In spring large biomass differences between W and E were observed during the prebarrier period, *Phaeocystis* biomass being 2–4 times larger in W than in E. During barrier and post-barrier years biomass in W decreased, but in E increased, reducing the differences between both compartments.

In summer, flagellate biomass increased from pre- to post-barrier periods, notably in E.

Discussion

Nutrients and phytoplankton biomass

Phosphorus
Reviewing marine nutrient limitation, Howarth (1988) presented several lines of evidence that net primary production is mainly nitrogen-limited, while phosphorus may limit production in some ecosystems. In the presentation of our results we omitted data on P because arguments have been given by Vegter & De Visscher (1987), Wetsteyn

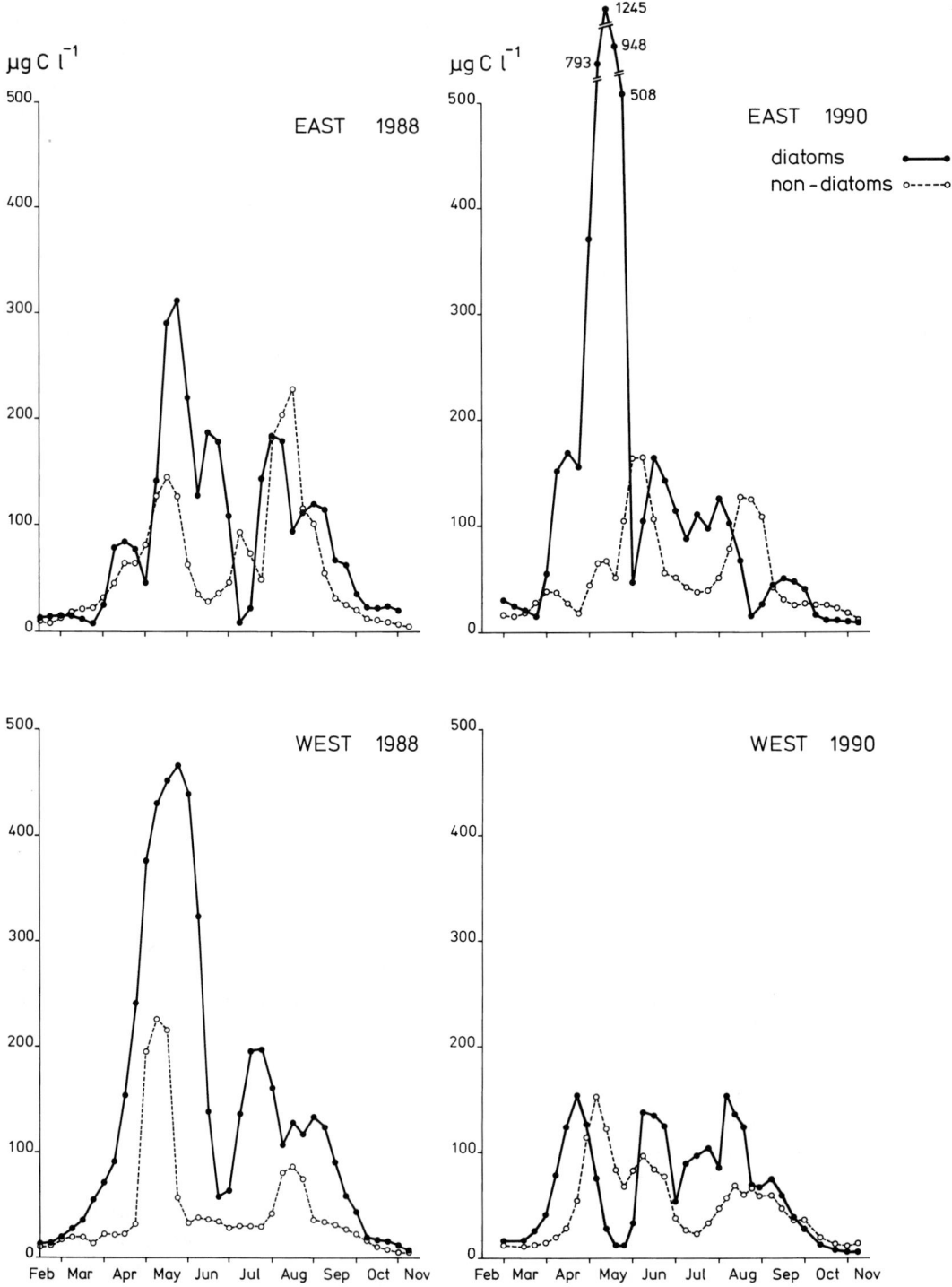

Fig. 6. Moving averages (over 3-weekly periods) of diatom- and non-diatom biomass (μg C l^{-1}) in the Oosterschelde, Eastern (top) and Western (bottom) compartments; post-barrier years 1988 (left) and 1990 (right).

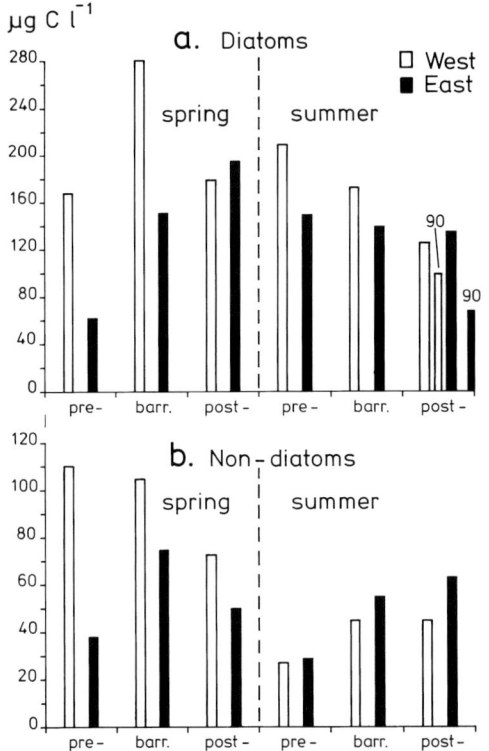

Fig. 7. Diatom (a) and non-diatom (b) biomass (μg C l^{-1}) during spring (water temperature < 15 °C) and summer (water temperature ≥ 15 °C) in the Oosterschelde, Western and Eastern compartments; pre-barrier, barrier-construction and post-barrier periods.

et al. (1990), Wetsteyn & Bakker (1991) and Bakker *et al.* (1990), that it could not be considered a limiting factor for phytoplankton growth in the Oosterschelde, neither in pre-barrier nor in following periods. However, indications can now be given (Wetsteyn & Kromkamp, 1994) that P-release from the bottom sediments is decreasing. Nevertheless, we do not discuss this nutrient and focus the discussion on Si and N. Also in Grevelingen lagoon (Fig. 1) the latter nutrients, rather than P, are considered limiting factors for phytoplankton growth (De Vries & Hopstaken, 1984; Bakker & De Vries, 1984).

Freshwater discharge
Discharges of river water via the Philipsdam (Fig. 1) and via other sluices (e.g. in the Oesterdam, Fig. 1) are responsible for the influx of nu-

trients in the Oosterschelde. Van Geldermalsen (1985) stressed the importance of this discharge for the dissolved Si concentrations in the basin during the pre-barrier period. Wetsteyn & Bakker (1991) found a strong correlation between load and winter concentrations in the Northern compartment. The strong reduction of the nutrient loads (44–82%), effectuated since April 1987, resulted in a decrease of the winter concentrations of NO$_3$ and Si in E but less so in W owing to the short distance of the sampling location R14 from the North Sea. The decreased river discharge into the Oosterschelde resulted in a levelling of the pre-barrier differences in winter levels of Si and NO$_3$ concentrations between W and E. Presently, post-barrier values in the entire sea-arm closely approach the pre-barrier conditions in W.

Daily exchange with the North Sea
During the pre-barrier period, the continuous tidal exchange with North Sea water, having lower Si- and NO$_3$ concentrations than the Oosterschelde (Stortelder *et al.*, 1984) resulted in a decrease of these nutrients in W. Wetsteyn & Kromkamp (1994) point out that the concentrations in neighbouring coastal regions of the North Sea decreased in the period of investigation. Consequently, the nutrient supply and thus -concentrations in W decreased too. Supposing direct Si-uptake by the developing diatom spring bloom and reduced uptake of NO$_3$ when NH$_4$ is present in concentrations exceeding *ca* 0.015 mg l^{-1}, then the Si-decline depends on exchange and uptake and the NO$_3$ decline at first mainly on exchange. The lag period between the Si and NO$_3$ curves can then be explained. Residence times in the basin increased with 75% (E) to 100% (W) during the barrier period (Wetsteyn *et al.*, 1990), due to a similar decrease in the rate of exchange with the North Sea. This is well illustrated by the enlarged time lag between Si- and NO$_3$ curves in spring 1986 (Fig. 2): the time shift of 1–2 months during the pre-barrier year 1983 (Fig. 2) increased to 2–3,5 months. In the post-barrier years, residence times shortened again and also the time shift between Si- and NO$_3$ curves in spring shortened, to *ca* 1 month (Fig. 3).

Uptake of nutrients by phytoplankton in the water phase

N uptake by marine phytoplankton is characterized by the preferential incorporation of NH_4 compared to NO_3 (McCarthy *et al.*, 1977). Only in case of serious depletion of NH_4, when concentrations decrease to $<2\,\mu g$ at l^{-1} ($= ca\ 0.03$ mg l^{-1}), phytoplankton shifts to the uptake of NO_3. For the majority of marine phytoplankton species growth is still unlimited at concentrations not lower than $1\,\mu g$ at N l^{-1} ($= ca\ 0,015$ mg N l^{-1}). In Figs 2, 3 and 4 it can be seen that phytoplankton growth during the years to 1988 inclusive, could hardly be limited by NH_4 and NO_3 shortages.

Si versus diatoms

Generally, the spring decline of Si was already in full progress before the start of the spring outburst of the planktonic diatoms. During the pre-barrier period (1983) this phenomenon could be observed in both compartments, the curves for W and E demonstrating a parallel course (Fig. 2). It has to be noted, however, that in E where a very modest first diatom bloom developed in 1983 (Fig. 5) starting from a high Si (winter) concentration, Si declined to values which were not as low as in W where much larger diatom blooms grew, starting from lower Si (winter) concentrations.

In the severe winters of the barrier-construction period (1986) the Si decline occurred earlier (Fig. 2) than before, and in fact, occurred at the same time as the development of an early diatom bloom (Fig. 5).

During the post-barrier year 1988, the spring decline of Si was quantitatively comparable in both compartments (Fig. 3), starting from similar winter concentrations, in spite of a much stronger diatom growth in W (Fig. 6). Consequently, other mechanisms than Si consumption by planktonic diatoms alone have to be taken into account for the explanation of the steep spring decline. (See further: microphytobenthos). Si availability in summer 1988 was greater than in 1983 (*cf.* Figs 3 and 4 with Figs 2 and 4), but diatom blooms were smaller (Fig. 6 *vs.* Fig. 5), most probably caused by increased grazing pressure

(Bakker & Van Rijswijk, 1994; see further). In 1990, E, the April bloom continued with a large peak in May (Fig. 6), but nevertheless Si did not reach zero values (Figs 3 and 4). In W the April bloom was replaced by *Phaeocystis* without a successive diatom development in May (Fig. 6). There was hardly any difference, however, with regard to the accompanying Si-curves and N/Si-ratios in both compartments (Figs 3 and 4).

Silicate and nitrogen versus total phytoplankton biomass

In early spring, the Si decline proceeded synchronously with the decrease of NH_4 and when the latter was consumed, NO_3 concentrations decreased (Figs 2 and 3). This explains the lag between the decrease of Si and NO_3 in spring. In summer, when benthic mineralization processes in shallow coastal waters prevail, the role of NO_3 for phytoplankton development can be taken over, again, by NH_4 (regenerated production).

Table 1 suggests a possible relation between abundance of diatoms *vs.* flagellates and nutrient availability in summer. For May–July 1990, the period of limiting N concentrations in E, the average biomass of diatoms and flagellates was calculated together with similar data from the same period in 1983 and 1986. Diatom contributions to total phytoplankton biomass decreased from pre-barrier-, to barrier-construction and post-barrier

Table 1. Average diatom- and flagellate biomass from May to July inclusive in pre-barrier- (1983), barrier- (1986) and post-barrier (1990) Oosterschelde, Eastern compartment.

	1983 pre-barrier	1986 barrier	1990 post-barrier
	Tidal regime undisturbed (100%)	Strongly reduced tidal movement ($\pm40\%$)	Reduced tidal movement (60%)
	N in excess	N in excess	N limiting
Diatom-C ($\mu g\ l^{-1}$)	178(72%)	226(60%)	91(48%)
Flagellate-C ($\mu g\ l^{-1}$)	69(28%)	149(40%)	97(52%)

periods and, consequently, relative flagellate biomass increased. The flagellate increase in 1986 *vs.* the high carbon biomass of both diatoms and flagellates may be related to the strongly decreased tidal amplitude, while still large nutrient concentrations were present. The further increase of the flagellates *vs.* the low absolute concentration of diatoms in 1990 is probably caused by N depletion, more severely limiting growth of diatoms than of flagellates. Also size-selective zooplankton grazing is supposed to have played a role; see further.

The foregoing is in agreement with results obtained in experimental ecosystems with mixed diatom-flagellate assemblages competing for nutrients (Harrison & Turpin, 1982; Takahashi *et al.*, 1977; 1982). In these studies it was demonstrated that centric diatoms dominated at high specific nutrient fluxes, while flagellates were more successful at a low flux. Takahashi *et al.* (1982) stated that conditions of low nutrient concentrations, generally, are not suitable for diatoms. Field evidence from estuaries where macronutrients were continuously supplied in summer, revealed that under these conditions (characteristic for the pre-barrier Oosterschelde) large phytoplankton (especially centric diatoms) may grow faster than small phytoplankton, diatoms as well as flagellates (Parsons & Takahashi, 1973).

In conclusion: experimental and field evidence from the literature confirm the explanations given for the Oosterschelde results: a large and frequent nutrient supply during the pre-barrier period resulted in diatom dominance ('new' production) also in summer; the low nutrient concentrations in the post-barrier Oosterschelde during summer since 1990, led to a decline of diatoms and to an increase of flagellates ('regenerated' production). The storm-surge barrier project represented a unique large-scale (macrocosm) field experiment leading to similar phytoplankton responses as observed in mesocosm studies.

Two other mechanisms have to be considered in relation to the observed changes in absolute and relative nutrient concentrations, *viz.* mineralization/excretion phenomena and nutrient uptake by microphytobenthos.

Mineralization and excretion of nutrients

Nutrients are recycled by the release from bacteria, zooplankton and zoobenthos. In summer, when algal uptake rates are high and nutrient concentrations in the water column are low, fluxes from the sediment are a significant source of nutrients to the water column (Boynton *et al.*, 1980). Release of nutrients from bottom organisms into the water column is characteristic for this benthic-pelagic coupling (Nixon, 1981). Also planktonic animals, notably copepods, excrete nutrients and may be of considerable importance in supplying nitrogen on a micro-spatial scale in nutrient-depleted waters (McCarthy & Goldman, 1979). Given the simultaneous decrease in river discharge and the increase in residence time (in the post-barrier-compared to the pre-barrier Oosterschelde) the relative contributions of the zoobenthos and the zooplankton to the nutrient supply of the water column, must have increased during the last years.

In W (and in the Central area) extended beds of cultured mussels occur (Van Stralen & Dijkema, 1994). It was demonstrated by Prins & Smaal (1994) that the biodeposition on the musselbed is an important source of regenerated nutrients: its contribution to total N-mineralization was in the range of 20–36% and 43–60% in spring and summer respectively. The NH_4 concentrations in W in summer exceeded those in E during the post-barrier period, especially in 1990 (Fig. 3). This may be explained by the location of the musselbeds. When mussels occur over the entire depth range of an estuary, such as in the Dutch Wadden Sea, NH_4 concentrations are higher in the shallow water (Helder, 1974.)

In E mussels are not cultured, but wild cockles are present, in numbers (Coosen *et al.*, 1994) and will mainly be responsible for the nutrient fluxes from the bottom in this compartment. These fluxes, however, are much smaller than those from the mussel beds (Prins & Smaal, 1994). Evidently, the increased zooplankton densities in E during the post-barrier period, even exceeding those of W (Bakker & Van Rijswijk, 1994), did not substantially help to raise the low NH_4 concentrations in summer in this compartment. This points again to the significance of musselbeds, lacking in E, for the release of nutrients.

NO_3 is not directly excreted by mussels but is released from the sediment and may represent 20% of the total DIN-flux (Prins & Smaal, 1994). Consequently, in summer most nitrogen is released as NH_4, leading to a decrease in the NO_3/NH_4-ratio in the water column. A decrease in the NO_3/NH_4-ratio became obvious in 1990 in both compartments (Fig. 4), but limiting concentrations were only reached in E (0–10 μg N l^{-1}, Fig. 3). This implies that any release of nitrogen must be important to stimulate phytoplankton growth in this area. In the Central area, Prins & Smaal (1993) measured a total DIN-flux of *ca* 34 mmol N m^{-2} hr^{-1}, *i.e. ca* 1.5 μg N l^{-1} d^{-1}. For the shallower E, these fluxes may be of the same order of magnitude. Such figures illustrate that the tendency towards oligotrophication in the present-day Oosterschelde (E) leads to a stronger dependency of the phytoplankton growth on nutrient regeneration processes. Very probably, the continuous low NO_3 concentrations in E during summer 1990 suppressed diatom growth and therefore the available Si could not be used. Also increased zooplankton biomass (Bakker & Van Rijswijk, 1994), leading to enhanced grazing pressure (Tackx *et al.*, 1994) will have reduced the diatom standing stock in summer. Diatom biomass in W was also low during spring and summer 1990, in spite of the higher NO_3 and NH_4 concentrations as well as the non-limiting Si concentrations. Here the relatively increased grazing pressure of the mussel population might have been responsible for the decreased phytoplankton biomass (*cf.* Herman & Scholten, 1990).

Nutrient uptake by microphytobenthos in the bottom
The strong decline of Si in spring, still before substantial consumption by planktonic diatoms could have taken place, may be partially caused (just as for NO_3) by exchange with the North Sea. Another mechanism of Si removal might be the consumption by bottom diatoms. In the saline Grevelingen lagoon a spring decline of Si occurred, without an accompanying spring in-

crease of planktonic diatoms (Bakker & De Vries, 1984). Decreasing Si concentrations were ascribed to flowerings of benthic diatoms in the shallow bottoms in early spring. During that time the flux of nutrients from the bottom is smaller than the nutritional requirements of the microphytobenthos. Nutrients in the watercolumn are available in excess then, and under favourable light conditions for the diatoms on the shallow bottoms of this stagnant basin, uptake by these benthic forms may substantially reduce nutrient concentrations (De Vries & Hopstaken, 1984). In the Oosterschelde, however, tidal hydrodynamics play an important role, restricting to a large extent the opportunities for benthic micro-algae to photosynthesize during the immersion period, especially during the pre-barrier years (De Jong *et al.*, 1990) characterized by high suspended matter concentrations. The latter workers, therefore, mainly measured photosynthesis during the emersion period and ignored under water photosynthesis. Nevertheless, rather high microphytobenthic biomass (measured as chlorophyll-a) was formed during early spring in the entire Oosterschelde and De Jong *et al.* (1990) measured high daily productions, beginning already in February and peaking in March. During the post-barrier period De Jong *et al.* (1994) demonstrated enhanced chlorophyll-a contents in the shallow bottoms of E as compared to the pre-barrier period. This can be explained by the improved light conditions in this area. The data available are inadequate to demonstrate that bottom diatoms are mainly responsible for the early decrease of Si in the Oosterschelde, but might indicate the role of this factor.

Interaction between microphytobenthos and nutrients in the water column might also occur during summer in the post-barrier years. De Jong *et al.* (1994) found a strong increase of benthic diatom biomass in the entire estuary during this season. This coincides with the continuous low phytoplankton biomass in summer. The increased mean irradiance in the water combined with the decreased water movements can be considered the main causative factors, but the question can be raised whether the microphytobenthos could again act as a successful competitor with the phytoplankton for nutrients in the water column. Again, decisive evidence cannot be given, but results of both data sets do not exclude this. In any case, we may state, in agreement with the conclusions of D. J. De Jong *et al.* (1993) that the importance of the interactions between water column and bottom increased during barrier and post-barrier periods.

Final remarks

Because residence times in the post-barrier Oosterschelde increased compared to the pre-barrier years, the combined total grazing pressure of zooplankton and zoobenthos will have increased, leading to stronger phytoplankton control especially in summer. In E, where no mussel culturing takes place, zooplankton grazing exceeded benthic consumption already in the pre-barrier period (Tackx *et al.*, 1990). As zooplankton biomass increased strongly in the post-barrier period (Bakker & Van Rijswijk, 1994), zooplankton grazing will have increased too (Tackx *et al.*, 1994). The combined effect of high zooplankton grazing pressure and potentially limiting dissolved inorganic nitrogen concentrations in the post-barrier period, will continuously keep phytoplankton summer biomass at low levels. Finally, the increased zooplankton biomass and -grazing in E may also have contributed to the observed change in the relation between diatom- and flagellate biomass in summer (Table 1): size selection by the copepods (Tackx *et al.*, 1989) favours the grazing of diatoms and this promotes a further shift from diatoms to flagellates in the phytoplankton assemblage. In W, high losses of phytoplankton biomass must be mainly caused by the grazing activity of both zoobenthos and zooplankton, as potentially limiting nutrient concentrations were found less frequently in this area.

Acknowledgements

We are much indebted to Drs Pieter H. Nienhuis, Theo C. Prins, Jan W. Rijstenbil, Michèle L. M. Tackx and Lambertus P. M. J. Wetsteyn as well as the referees Drs Joop Ringelberg and A. Davies for critical comments on earlier versions of the m.s. We thank Mr. Alfons A. Bolsius for carefully drawing the figures and Mrs. Elly van Hulsteijn for repeatedly typing the corrected texts. Communication No. 691 of NIOO-CEMO, Yerseke, The Netherlands.

References

Bakker, C., 1967. Veranderingen in milieu en plankton van het Oosterscheldegebied. Vakbl. v. Biologen 47: 181–192.

Bakker, C. & I. de Vries, 1984. Phytoplankton- and nutrient dynamics in saline Lake Grevelingen (SW Netherlands) under different hydrodynamical conditions in 1978–1980. Neth. J. Sea Res. 18: 191–220.

Bakker, C., P. M. J. Herman & M. Vink, 1990. Changes in seasonal succession of phytoplankton induced by the storm-surge barrier in the Oosterschelde (SW-Netherlands). J. Plankton Res. 12: 947–972.

Bakker, C., P. M. J. Herman & M. Vink, 1994. A new trend in the development of the phytoplankton in the Oosterschelde (SW Netherlands) during and after the construction of a storm-surge barrier. Hydrobiologia 282/283: 79–100.

Bakker, C. & P. van Rijswijk, 1994. Zooplankton biomass in the Oosterschelde (SW Netherlands) before, during and after the construction of a storm-surge barrier. Hydrobiologia 282/283: 127–143.

Boynton, W. R., W. M. Kemp & C. G. Osborne, 1980. Nutrient fluxes across the sediment-water interface in the turbid zone of a coastal plain estuary. In: V. S. Kennedy (ed): Estuarine perspectives. Academic Press, New York: 93–109.

Brzezinsky, M. A., 1985. The Si:C:N ratio of marine diatoms: interspecific variability and the effect of some environmental variables. J. Phycol. 21: 347–357.

Coosen, J., F. Twisk, M. W. M. van der Tol, R. H. D. Lambeck, M. R. van Stralen & P. M. Meire, 1994. Variability in stock assessment of cockles (*Cerastoderma edule* L.) in the Oosterschelde (in 1980–1990), in relation to environmental factors. Hydrobiologia 282/283: 381–395.

De Jong, D. J., P. H. Nienhuis & B. J. Kater, 1994. Microphytobenthos in the Oosterschelde estuary (The Netherlands), 1981–1990; consequences of a changed tidal regime. Hydrobiologia 282/283: 183–195.

De Jong, S. A., P. A. G. Hofman, A. J. J. Sandee & E. J. Wagenvoort, 1990. Primary production of benthic microalgae in the Oosterschelde estuary (SW-Netherlands). Balans nota 43, DIHO-RWS.

De Vries, I. & C. F. Hopstaken, 1984. Nutrient cycling and ecosystem behaviour in a salt-water lake. Neth. J. Sea. Res. 18: 221–245.

Drinkwaard, A. C., 1960. The quality of oysters in relation to environmental conditions in the Oosterschelde in 1958. Ann. Biol. 15: 224–233.

Dronkers, J. & J. T. F. Zimmerman, 1982. Some principles of mixing in tidal lagoons with examples of tidal basins in the Oosterschelde. In: P. Lasserre & H. Postma (eds), Coastal Lagoons. Proc. Int. Symp. Coastal Lagoons, Gauthier-Villars, pp. 460–474.

Harrison, P. J. & D. H. Turpin, 1982. The manipulation of physical, chemical and biological factors to select species from natural phytoplankton communities. In: G. D. Grice & M. R. Reeve (eds), Marine mesocosms, biological and chemical research in experimental ecosystems: Springer-Verlag, New York/Heidelberg/Berlin: 275–289.

Helder, W., 1974. The cycle of dissolved inorganic nitrogen compounds in the Dutch Wadden Sea. Neth. J. Sea Res. 8: 154–173.

Herman, P. M. J. & H. Scholten, 1990. Can suspension-feeders stabilise estuarine ecosystems? In: M. Barnes & R. N. Gibson (eds): Trophic Relationships in the Marine Environment, Proc. 24th Europ. Mar. Biol. Symp., Aberdeen Univ. Press: 104–116.

Howarth, R. W., 1988. Nutrient limitation of net primary production in marine ecosystems. Ann. Rev. Ecol. Syst. 19: 89–110.

Korringa, P., 1941. Experiments and observations on swarming, pelagic life and settling in the European flat oyster, Ostrea edulis L. Arch. Néerl. Zool. 5: 1–249.

McCarthy, J. J., W. R. Taylor and L. J. Taft, 1977. Nitrogenous nutrition of the plankton in the Chesapeake Bay. 1. Nutrient availability and phytoplankton preferences. Limnol. Oceanogr. 22: 996–1011.

McCarthy, J. J. & J. C. Goldman, 1979. Nitrogenous nutrition of marine phytoplankton in nutrient-depleted waters. Science 203: 670–672.

Nienhuis, P. H. & A. C. Smaal, 1994. The Oosterschelde estuary, a case study of a changing ecosystem: an introduction. Hydrobiologia 282/283: 1–14.

Nixon, S., 1981. Remineralization and nutrient cycling in coastal marine ecosystems. In: B. J. Neilson & L. E. Cronin (eds): Estuaries and nutrients, Humana Press, Clifton, New Jersey: 111–138.

Parsons, T. R. & M. Takahashi, 1973. Environmental control of phytoplankton cell size. Limnol. Oceanogr. 18: 511–515.

Parsons, T. R., Y. Maila & C. M. Lalli, 1984. A manual of chemical and biological methods for seawater analysis. Pergamon Press, Oxford Frankfurt.

Prins, T. C. & A. C. Smaal, 1994. The role of the blue mussel Mytilus edulis in the cycling of nutrients in the Oosterschelde estuary (The Netherlands). Hydrobiologia 282/283: 413–429.

Smayda, T. J., 1978. From phytoplankters to biomass. In: A. Sournia (ed.), Phytoplankton Manual. Unesco, Paris, pp. 273–279.

Stortelder, P. B. M., J. S. L. Vink & P. F. Havermans, 1984. Waterkwaliteitskenmerken van de Oosterschelde en Schelde-Rijn Verbinding. RWS-note 83.13.

Tackx, M. L. M., C. Bakker, J. W. Francke & M. Vink, 1989. Size and phytoplankton selection by Oosterschelde zooplankton. Neth. J. Sea Res. 23: 35–43.

Tackx, M. L. M., C. Bakker & P. van Rijswijk, 1990. Zooplankton grazing pressure in the Oosterschelde (The Netherlands). Neth. J. Sea Res. 25: 405–415.

Tackx, M. L. M., P. M. J. Herman, P. Van Rijswijk, M. Vink & C. Bakker, 1994. Plankton size distributions and trophic relations before and after the construction of the storm-surge barrier in the Oosterschelde estuary. Hydrobiologia 282/283: 145–152.

Takahashi, M., D. L. Seibert & W. H. Thomas, 1977. Occasional blooms of phytoplankton during summer in Saanich Inlet, B. C., Canada. Deep-Sea Res. 24: 775–780.

Takahashi, M., I. Koike, K. Iseki, P. K. Bienfang & A. Hattori, 1982. Phytoplankton species' responses to nutrient changes in experimental enclosures and coastal waters. In: G. D. Grice & M. R. Reeve (eds): Marine mesocosms, biological and chemical research in experimental ecosystems. Springer Verlag, New York/Heidelberg/Berlin: 333–340.

Van der Tol, M. & H. Scholten, 1992. Response of the food-web of the Oosterschelde to a changing environment: functional or adaptive? Neth. J. Sea Res. 30: 175–190.

Van Geldermalsen, L. A., 1985. Evaluation of seasonal nutrient dynamics in the Oosterschelde (SW-Netherlands). Neth. J. Sea Res. 19: 207–216.

Van Stralen, M. R. & R. D. Dijkema, 1994. Mussel culture in a changing environment: the effects of a coastal engineering project on mussel culture (Mytilus edulis L.) in the Oosterschelde (SW Netherlands). Hydrobiologia 282/283: 359–379.

Vegter, F. & P. R. M. De Visscher, 1987. Nutrients and phytoplankton primary production in the marine tidal Oosterschelde estuary (The Netherlands). Hydrobiol. Bull. 21: 149–158.

Vroon, J., 1994. Hydrodynamic characteristics of the Oosterschelde in recent decades. Hydrobiologia 282/283: 17–27.

Wetsteyn, L. P. M. J., J. C. H. Peeters, R. N. M. Duin, F. Vegter & P. R. M. de Visscher, 1990. Phytoplankton primary production and nutrients in the Oosterschelde (The Netherlands) during the pre-barrier period 1980–1984. Hydrobiologia 195: 163–177.

Wetsteyn, L. P. M. J. & C. Bakker, 1991. Abiotic characteristics and phytoplankton primary production in relation to a large-scale coastal engineering project in the Oosterschelde (The Netherlands): a preliminary evaluation. In: M. Elliott & J. P. Ducrotoy (eds). Estuaries and coasts: spatial and temporal inter-comparisons. Proc. ECSA 19 Symp., 1989. Olsen & Olsen, p. 365–373.

Wetsteyn, L. P. M. J. & J. C. Kromkamp, 1994. Turbidity, nutrients and phytoplankton primary production in the Oosterschelde (The Netherlands) before, during and after a large-scale coastal engineering project (1980–1990). Hydrobiologia 282/283: 61–78.

Hydrobiologia **282/283**: 117–126, 1994.
P. H. Nienhuis & A. C. Smaal (eds), The Oosterschelde Estuary.
© 1994 *Kluwer Academic Publishers. Printed in Belgium.*

Zooplankton species composition in the Oosterschelde (SW Netherlands) before, during and after the construction of a storm-surge barrier

C. Bakker
Netherlands Institute of Ecology, Centre for Estuarine and Coastal Ecology, Vierstraat 28, 4401 EA Yerseke, The Netherlands

Key words: zooplankton species, *Acartia tonsa, A. clausi, Synchaeta* spp. Oosterschelde, storm-surge barrier, lagoon-like character

Abstract

During the period of construction of a storm-surge barrier current velocities decreased strongly and the Eastern compartment of the basin obtained a lagoon-like character. The rotifer *Synchaeta* spp., already abundant in the neighbouring salt and brackish lakes, profited from this condition. Higher and less fluctuating salinities caused the estuarine character of this compartment to disappear in the post-barrier years. This was reflected in the obscured succession of some *Acartia* species: the estuarine *A. tonsa* dominated in pre-barrier- and barrier years during summer, while the marine *A. clausi* was abundant in the post-barrier Oosterschelde during that time. Also the changed food conditions (Bakker & Vink, 1993) may have played a role in this phenomenon.

Further changes in species composition were hardly observed. The main changes were of a quantitative nature (Bakker & Van Rijswijk, 1993; Tackx *et al.*, 1993). A list of the commonly occurring species is given.

Introduction

The Oosterschelde (Fig. 1) is a tidal inlet (*ca* 350 km^2) of the southern North Sea. The first (qualitative) plankton analysis of the basin (Eastern part) was carried out by Redeke (1902). Much later, the plankton of the central part was (quantitatively) studied by Bakker (1964). A mainly taxonomical study was devoted to the tintinnid protozoans (Bakker & Phaff, 1976). General tendencies of zooplankton development in the Oosterschelde were described and compared with those in two nearby closed estuaries (Lakes Veere and – Grevelingen: Fig. 1) (Bakker & Vegter, 1978). An inventory was made of the

aquatic plants and animals (including zooplankton) occurring in the Oosterschelde (Elgershuizen *et al.*, 1979).

A large-scale hydraulic engineering project was carried out in the Oosterschelde, involving the storm-surge barrier in the mouth (1982–86), the Oesterdam in the Eastern part (1982–86) and the Philipsdam in the Northern compartment (1985–87) (Fig. 1). For details cf. Nienhuis & Smaal (1994). During the construction of the barrier and the accompanying works, the hydrodynamics of the estuary were seriously influenced, particularly with respect to tidal amplitude and current velocities (Wetsteyn *et al.*, 1990). When current velocities decreased, especially in the shallow Eastern

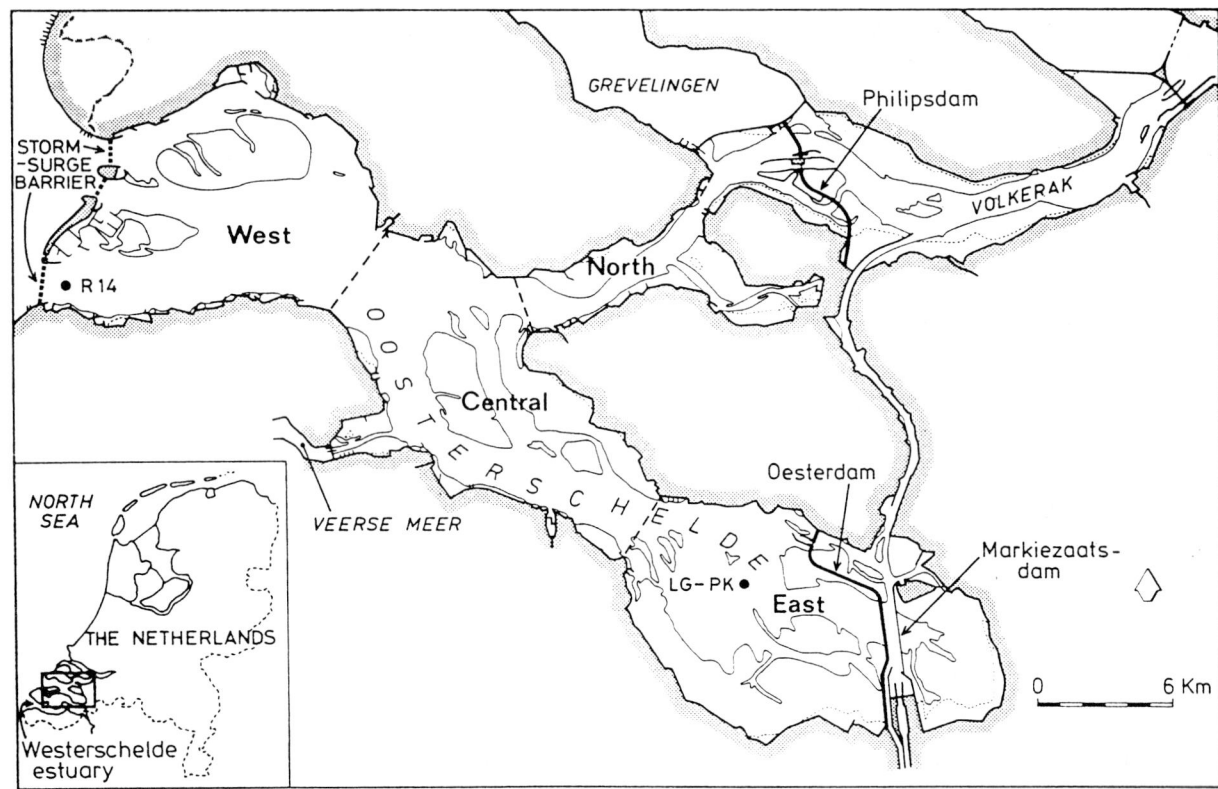

Fig. 1. Map of the Oosterschelde with indications (broken lines) of compartments and sampling localities; inset: location of the sea-arm in The Netherlands.

area, suspended matter concentrations declined (Wetsteyn & Bakker, 1991) and water transparency rose concomitantly. This influenced the species composition and seasonal succession of the phytoplankton (Bakker *et al.*, 1990, 1994) and phytoplankton biomass (Bakker & Vink, 1994).

The aim of the present paper is to discuss a possible response of the species composition of the dominant zooplankton species to the changing environmental conditions in the Oosterschelde during and after the barrier construction. A separate paper is devoted to the changes in zooplankton biomass (Bakker & Van Rijswijk, 1994) from pre- to post-barrier periods.

Attention is focussed on the Eastern part of the sea-arm as all changes were most pronounced in this area. Eastern and Western compartments are further indicated as E and W, respectively.

Methods

Samples were taken at locations LG-PK (depth 20 m, shallow E) and R14 (depth 40 m, deeper W) in the Oosterschelde (Fig. 1). During the growth season (March–October) sampling was carried out weekly, during the rest of the year less frequently. Mesozooplankton was collected by pumping with a Pleuger submersible system, capacity 200 l min^{-1}. Generally, 100 l of water were filtered through a net of 63 μm mesh gauze, retaining all copepod stages (smallest nauplii and eggs included), cladocerans, appendicularians and the majority of meroplankton (larvae of mollusks, polychaete worms, barnacles). The 63 μm mesh did not retain the smallest rotifers and protozoans: these were obtained with water bottles of 1 l, used for the sampling of the phytoplankton. Larger planktonic animals (jelly-fish, decapod

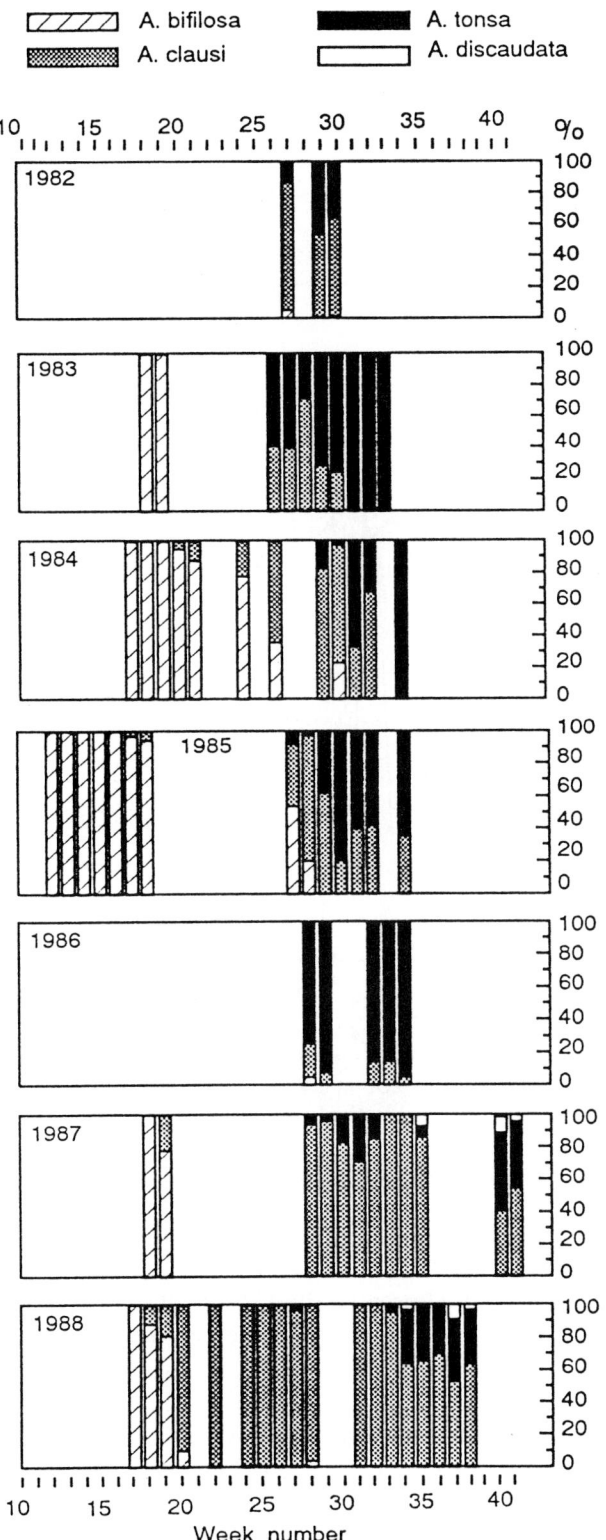

larvae, mysids) were caught with a high-speed sampler, gauze 200 μm mesh width, during tows of *ca* 5 minutes, filtering 50–100 m³ of water. Fish larvae were not identified.

Results

During the entire period of observation, the following zooplankton species were found.

Copepoda Calanoida (representing the predominating group in the Oosterschelde zooplankton): *Acartia clausi* Giesbrecht; *A. hudsonica* Pinhey; *A. discaudata* Giesbrecht; *A. tonsa* Dana; *A. bifilosa* (Giesbrecht); *Anomalocera patersoni* Templeton; *Calanus helgolandicus* (Claus); *Centropages hamatus* (Lilljeborg); *Eurytemora affinis* (Poppe); *E. americana* Williams; *Labidocera wollastoni* (Lubbock); *Paracalanus parvus* (Claus); *Pseudocalanus* sp.; *Temora longicornis* (O. F. Müller).

Highest abundances were reached by *Temora longicornis* in W and by *Acartia* spp. in E. From the *Acartia* species, in W only *A. clausi* was regularly seen (in summer). In E a seasonal succession of 3 *Acartia* species could be observed. In Fig. 2 each species is expressed as a percentage of total numbers of adult *Acartia*. In spring and early summer, an estimation of the percentual contribution often was impossible, due to a lack of adult animals in the samples. Generally, a succession was observed from *A. bifilosa* in spring to *A. clausi* and *A. tonsa* in summer. *A. bifilosa* densities were never high; some individuals were still observed in summer. The times and degree of replacement of *A. clausi* by *A. tonsa* were very different in the years of investigation. During the pre-barrier and barrier periods (1982–86) succession of *A. clausi* by *A. tonsa* was observed; in 1983 and 1986 *A. tonsa* rapidly reached dominance. During the post-barrier years 1987–88, however, summer succession of *Acartia* did hardly occur. The adult copepod population nearly exclusively

Fig. 2. Seasonal distribution of *Acartia* spp., expressed as percentages of total adult numbers, in the Eastern compartment of the Oosterschelde, 1982–88.

consisted of *A. clausi* (Fig. 2) and absolute numbers of this species were high (Bakker & Van Rijswijk, 1994).

Copepoda Harpacticoida: the pelagic species *Euterpina acutifrons* (Dana); the benthic species: *Alteutha interrupta* (Goodsir), *Harpacticus flexus* (Brady & Robertson); *Tachidius discipes* (Giesbrecht), and unidentified species of the genera *Longipedia, Microsetella* and *Tisbe*.

Copepoda Cyclopoida: only the littoral form *Cyclopina* spp. was frequently observed. The pelagic *Oithona* spp. (*O. nana*) was rarely found.

From the holoplanktonic species other than copepods the most common species are:

Cladocera: *Pleopis polyphemoides* (Leuckart) Gieskes; *Evadne nordmanni* Lovén.

Appendicularia: *Oikopleura dioica* Fol; *Fritillaria borealis acuta* Lohmann.

Scyphozoa: *Aurelia aurita* (L.); *Chrysaora hysoscella* (L.); *Cyanea capillata* (L.); *Cyanea lamarckii* (Peron & Lesueur); *Rhizostoma octopus* (L.).

Ctenophora: *Pleurobrachia pileus* (O. F. Müller); *Beröe cucumis* Fabricius.

Chaetognatha: *Sagitta setosa* O.F. Müller.

Turbellaria: *Alaurina composita* Metschnikow

Rotatoria: *Synchaeta litoralis* Rousselet; *S. vorax* Rousselet; *S. triophthalma* Lauterborn; *S. grimpei* Remane; *Trichocerca marina marina* (Daday). During the pre-barrier situation (1982–84) *Synchaeta* spp. were observed in low numbers, but during the barrier period when tidal movements slowed down, their densities increased strongly. Two high peaks developed in summer and autumn 1987, under the new (reduced) tidal regime, but numbers decreased again in 1988 (Fig. 3).

Protozoa: *Mesodinium rubrum* (Lohmann) Hamburger and Buddenbrock; *Noctiluca scintillans* (McCartney) Ehrenberg; *Favella ehrenbergii* (Claparède & Lachmann); *Helicostomella subulata* (Ehrenberg); *Stenosemella ventricosa* (Claparède and Lachmann); *Tintinnopsis beroidea* Stein; *T. campanula* (Ehrenberg) Daday; *T. cylindrica* Daday.

From the animal (sub)phyla inhabiting the sublittoral bottoms, intertidal mudflats and (artifi-

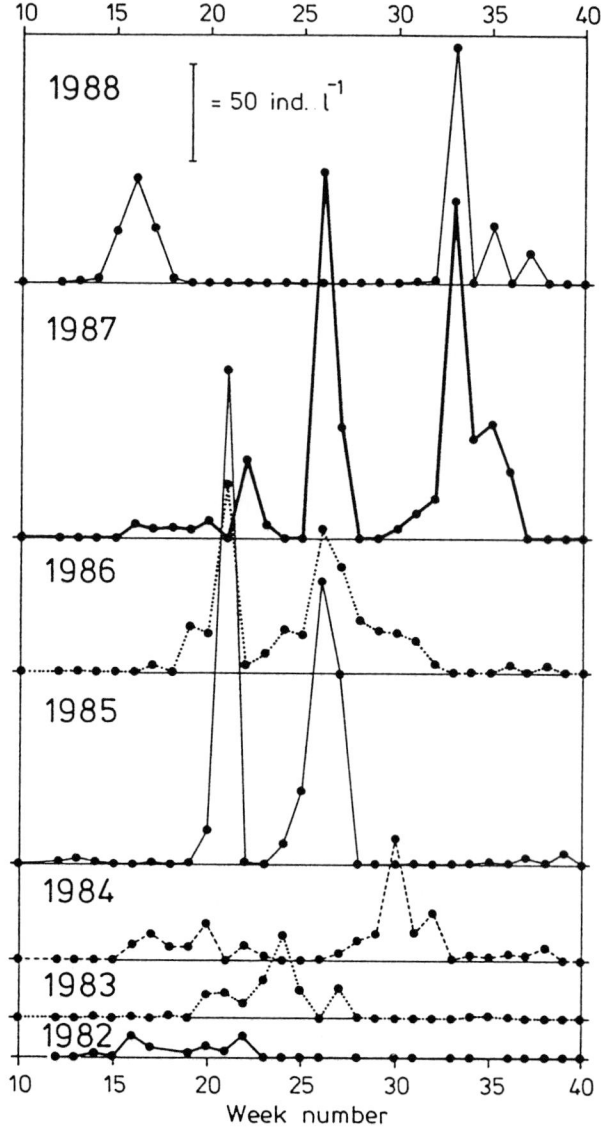

Fig. 3. Seasonal distribution of *Synchaeta* spp. (Rotatoria, N l⁻¹) in the Eastern compartment of the Oosterschelde, 1982–88.

cial) rocky shores of the Oosterschelde, the following characteristic species produce larvae spending their lives in the water column (meroplankton):

Hydromedusae: *Bougainvillea ramosa* (Whright); *Amphinema rugosum* (Mayer); *Ectopleura dumortieri* (van Beneden); *Leuckartiara octona* (Fleming); *Lovenella clausa* Hincks; *Margelopsis haeckeli* Hartlaub; *Obelia* sp.; *Phialidium* hemisphaericum (L.); *Rathkea octopunctata* (Sars); *Sarsia tubulosa* (Sars); *Steenstrupia nutans* (Sars); *Turritopsis nutricula* (McCrady); *Eucheilota maculata* Hartlaub.

Prosobranchia: *Crepidula fornicata* (L.); *Hydrobia ulvae* (Pennant); *Littorina littorea* (L.).

Nudibranchia: unidentified veliger larvae.

Lamellibranchia: *Crassostrea gigas* (Thunberg); *Cerastoderma edule* (L.); *Mytilus edulis* L.; *Ostrea edulis* L.

Polychaeta: *Anaitides mucosa* (Oersted); *Autolytus prolifer* O.F. Müller; *Eteone longa* (Fabricius); *Harmothöe impar* (Johnston); *Lanice conchilega* (Pallas); *Magelona papillicornis* O.F. Müller; *Nephtys* spp.; *Nereis* spp.; *Polydora* spp.; *Pygospio elegans* (Claparède); *Scolelepis foliosa* (Audouin & Edwards); several other spionid larvae.

Cirripedia: *Balanus balanoides* (L.); *B. crenatus* (Bruquière); *Elminius modestus* Darwin.

Decapoda: *Carcinus maenas* (L.); *Crangon crangon* L.; *Pagurus bernhardus* (L.).

Ectoprocta: *Conopeum-, Electra-* and *Membranipora* spp. Echinodermata: *Asterias rubens* (L.); *Echinocardium cordatum* (Pennant); *Ophiothrix fragilis* (Abildgaard).

Tunicata: *Styela clava* Hardmann; *Ciona intestinalis* (L.).

Mysidacea: *Gastrosaccus spinifer* (Goës), *Mesopodopsis slabberi* Van Beneden; *Praunus flexuosus* O.F. Müller; *Schistomysis kervillei* Sars; *Sch. spiritus* Norman.

Cumacea: *Diastylis rathkei* (Kröyer).

Isopoda: *Oniscus asellus* L.

Amphipoda: (*Marino*)*gammarus* spp.

Discussion

We only listed the characteristic and/or common zooplankton species of the Oosterschelde observed during the entire period of investigation. For a more complete inventory of zooplankton species occurring in the sea-arm before the construction of the storm-surge barrier we refer to Elgershuizen *et al.* (1979).

Copepods

Temora longicornis is the most abundant copepod species in the Oosterschelde, dominating in W during pre-barrier years in spring and early summer. Population dynamics resp. food selection and grazing were studied by Bakker & Van Rijswijk (1987) resp. Tackx *et al.* (1989, 1990). The copepod is still dominant in W; it increased its abundance in E during post-barrier years, due to prolonged residence times of the water and increased phytoplankton content of suspended material (Bakker & Van Rijswijk, 1994).

In a revision of the *Acartia* subgenus *Acartiura*, Bradford (1976) redescribed a number of species closely related to *A. clausi*. One of these, *A. hudsonica*, entered Lake Veere (Fig. 1) from the Oosterschelde (Revis & Bakker, 1988). The specimens of *Acartia* in the Oosterschelde samples of 1982–88, previously recorded as *A. clausi* (*sensu lato*), were not re-identified to establish the relative abundance of *A. clausi* (*sensu stricto*) and *A. hudsonica* in the Oosterschelde. A clear distinction was made, however, between *A. clausi* (*s.l.*) and *A. tonsa,* subgenus *Acanthacartia,* in E succeeding each other during summer. This succession was obvious during the pre-barrier and barrier periods (Fig. 2). For the pre-barrier period Tackx *et al.* (1989) demonstrated that *Acartia* spp., in naturally occurring seston particle distributions with distinct peaks in the $>20\ \mu$m size range, performed a clear peak tracking but could switch its feeding activity towards seston particles $<20\ \mu$m in case of flattened seston distributions extending from $4-100\ \mu$m. Although detritus was prevailing in the size class $<20\ \mu$m, during this period, it could not be shown unequivocally that *Acartia* had a detritivorous behaviour. In their experiments, Tackx *et al.* (1989) made no distinction between *A. tonsa* and *A. clausi*, but on the basis of the data presented in Fig. 2 we can state that the majority of the *Acartia* population in 1983 consisted of *A. tonsa*.

Succession of *A. clausi* to *A. tonsa* became obscure during the post-barrier years when *A. clausi* maintained dominance during the entire summer (Fig. 2). This may be explained firstly by the higher salinity of the Oosterschelde (E, Fig. 4) after the construction of the Philipsdam (Fig. 1), (see also Wetsteyn & Bakker, 1991). Water temperatures in summer were more dependent on seasonal meteorological conditions than on the hydrodynamical changes. In 1983 and 1986 most salinities were found between 15 and 16‰ Cl⁻ (Fig. 4) and temperatures rose to 22.5 °C. In 1987–88 salinities increased to 17‰ Cl⁻ (Fig. 4) with a maximum temperature of 20,1 °C. These conditions tend towards the different salinity-temperature ranges to which both species are adapted. *A. tonsa* is an eurythermal and strongly euryhaline estuarine summer species from southerly origin, *A. clausi* a less eurythermal and euryhaline early summer species from a more northern origin (Conover, 1956; Jeffries, 1967; Lee & McAlice, 1979; Turner, 1981). *A. tonsa* is also abundantly found in semi-enclosed brackish basins (Bakker *et al.*, 1977; Ambler, 1986) while *A. clausi* also occurs in marine inlets without a significant inflow of freshwater (Landry, 1978; Uye, 1982). Jeffries (1962) found a linear relationship between temperature and salinity in the field during the interaction periods of both species: the higher the salinity, the higher the temperature necessary to eliminate *A. clausi*. Temperature is the main factor responsible for replacement of *A. clausi* by *A. tonsa* (see also Sullivan & McManus, 1986). *A. tonsa* can better reproduce than *A. clausi* under conditions of lowered salinities. Consequently, interaction between

both species starts in the least saline region of an estuary permitting reproduction; in the more saline mouth the period of interaction is delayed (Jeffries, 1962). The same occurred in the Oosterschelde where succession was only observed in E under the fluctuating salinities and (incidentally) high summer temperatures of pre-barrier to barrier periods; in W *A. clausi* persisted during that time, although in small numbers (Bakker & Van Rijswijk, 1994). At the increased salinities during the post-barrier years *A. clausi* occurred in the entire basin but reached dominance in E only (Bakker & Van Rijswijk, 1994); complete replacement by *A. tonsa* did not occur (Fig. 2) under the prevailing water temperatures to *ca* 20 °C. Only during summers with higher water temperatures for long periods complete succession by *A. tonsa* can be expected.

However, beside salinity also food conditions in the Oosterschelde changed, especially in E. Suspended matter concentrations declined (Wetsteyn & Bakker, 1991), microplankton: POC ratios increased (Bakker & Van Rijswijk, 1994), diatom biomass in summer decreased while flagellates increased (Bakker & Vink, 1994) and smaller phytoplankton species replaced larger ones (Bakker *et al.*, 1994). Literature is equivocal with respect to the question whether the feeding characteristics of both species are different. As is usual for coastal-estuarine species, both are able to ingest detritus (Chervin, 1978), although we may expect that *A. tonsa* due to its occurrence in estuarine sites with high detritus concentrations will be better adapted to detritus-rich diets than *A. clausi*. *A. tonsa* increases its growth rate when detritus is added to a phytoplankton diet (Roman, 1984). But *A. tonsa* is also adapted to high phytoplankton concentrations, decreasing its clearance rate when food concentration declines below certain limits, as Paffenhöfer & Stearns (1988) demonstrated in a comparative study of the clearance rates of nearshore *A. tonsa* and offshore *Paracalanus* sp. Future studies are needed to learn if similar differences exist between *A. tonsa* and *A. clausi*. However, according to Sullivan & McManus (1986), competition for food did not appear to be an important factor controlling sea-

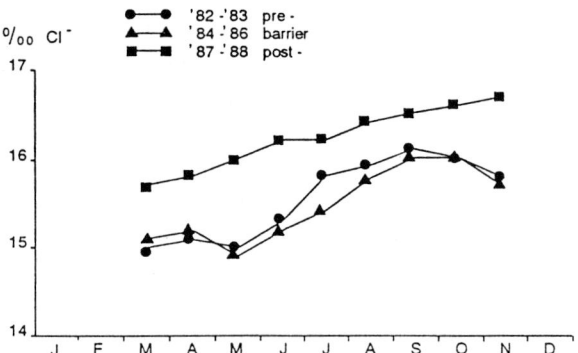

Fig. 4. Salinity (‰ Cl⁻) in the Eastern compartment of the Oosterschelde, during pre-barrier (1982–83), barrier (1984–86) and post-barrier (1987–88) periods.

sonal succession of *A. clausi* and *A. tonsa* in enclosure experiments. Nevertheless, we do not exclude that *A. clausi* may be better adapted than *A. tonsa* to the low summer concentrations of small algae (diatoms as well as flagellates) in the post-barrier Oosterschelde, food being an important factor beside temperature-salinity combination.

The largest calanoid copepod species, *Calanus helgolandicus*, occasionally enters the Oosterschelde from the North Sea; it was mainly found in the samples obtained with the high-speed sampler. *Eurytemora americana* may enter the estuary in small numbers from saline Lake Grevelingen (Fig. 1), where it can develop rather dense populations in early summer (Bakker, 1978). Some *E. affinis* are introduced from the northern branch (Fig. 1) with inflowing riverine water: during pre-barrier years regularly and abundantly, during post-barrier years incidentally and in small numbers. Introduction of *E. affinis* may also take place from the sea side (via the storm-surge barrier, Fig. 1), originating from the river Rhine plume in the coastal sea water (after continued N.W. winds) or from the Westerschelde estuary (Fig. 1), where it is the dominating estuarine copepod species (Bakker *et al.*, 1977; Soetaert & Van Rijswijk, 1993).

The adult benthic harpacticoids were mainly sampled from deeper water or were represented by their naupliar stages (notably *Longipedia*).

The rare occurrence of the cyclopoid copepod *Oithona nana* in the Oosterschelde nowadays bears no relation with the changed hydrodynamics of the sea-arm. Oldest records (Van Breemen, 1903; 1905) mention *O. nana* as a species developing in the southern North Sea, but only after being transported into this region by the Channel Current. During that time *O. nana* was found commonly indeed in the Oosterschelde (Redeke, 1902). Preliminary observations of Oosterschelde zooplankton (Bakker, unpubl. data from the sixties and seventies), revealed an incidental occurrence of the species, while Fransz (1983) noted a more regular occurrence indeed, but in small numbers only in the Dutch and German Wadden Sea. During the years of investigation of this paper

(1982–89) the species was not found at all. Strikingly, Soetaert & Van Rijswijk (1993), studying spatial and temporal patterns of the zooplankton in the Westerschelde (situated immediately south of the Oosterschelde, Fig. 1) report *O. nana* in numbers during late summer and fall 1990 in the marine reach of this estuary. In 1990 no plankton samples were taken in the Oosterschelde and therefore a possible return of the copepod in the seaarm could not be observed. As a cause of this irregular temporal pattern of occurrence, the changing intensity of the Channel Current has been suggested (Lücke, 1912). *O. nana* is an oceanic species and, thus, real autochthonic stocks are lacking in the southern North Sea. When dense summer populations have developed in the Atlantic, introduction via the English Channel can take place in case of sufficient strength of the current and then this leads to a retarded occurrence of non-reproductive populations in late-summer and fall. A comparable history can be sketched for the northern counterpart of *O. nana*: *O. similis*. This species mainly lives in the central North Sea in summer (Fransz *et al.*, 1991) and its populations are drifting into the German Wadden Sea (Martens, 1980) and the (Danish) Limfjord (Blanner, 1982) in late summer-fall.

As to the rotifers, a large number of *Synchaeta* species lives in the southern North Sea, but the majority of these can be found in coastal areas and the marine parts of estuaries (Berzins, 1960). In the Oosterschelde and the saline-brackish lakes Grevelingen and Veere 5 species occur regularly, but largest densities are always found in the lakes (Bakker & Vegter, 1978). In the Oosterschelde the abundance of *Synchaeta* increased during the barrier period (Fig. 3) when considerable quantities of Lake Grevelingen water were discharged on the Oosterschelde (Nienhuis & Smaal, 1994). The strongly reduced current velocities and increased transparencies of the water, especially in E, gave these animals better conditions of development than before. In fact, the eastern part of the sea-arm obtained a lagoon-like character, resembling Lake Grevelingen (Fig. 1), where rotifers form a major contribution to the zooplankton assemblage, preferably in spring

(Bakker, 1978; Bakker & Vegter, 1978). The transitional years of the barrier period illustrated clearly the increased possibilities of development for planktonic species such as *Synchaeta*, characteristic for brackish-marine waters without tidal movement (Veerse Meer and Grevelingen). Rotifers again decreased in abundance when current velocities and tidal volume rose after the barrier construction and the lagoon-like character of East diminished.

Systematics and ecology of the species of the neritic genus *Tintinnopsis* (pelagic Protozoa) occurring in the S.W. Netherlands, were discussed by Bakker & Phaff (1976). The 'unique photosynthetic ciliate' (Lindholm, 1985) *Mesodinium rubrum* is also found in the Oosterschelde but largest densities were again registrated in the abovementioned saline-brackish lakes (Bakker & Vegter, 1978) where the discolored water may look maroon (which never occurred in the Oosterschelde).

From the Hydromedusae a remarkable immigrant species was not yet mentioned because its occurrence in the Oosterschelde was not established with certainty: *Gonionemus vertens* A. Agassiz. This littoral species, endemic in the coastal North Pacific ocean, was discovered abundantly in the extended eelgrass beds of Lake Grevelingen (Bakker, 1981) and could be expected to occur (and spread) in the Oosterschelde too, especially during the barrier period, when prognoses were made of increasing *Zostera* in the clearer water with reduced water movement. These predictions were not fulfilled for several reasons (Smaal & Boeije, 1991) and therefore the occurrence of *Gonionemus* in the Oosterschelde is dubious.

The Polychaeta fauna of the Oosterschelde is diversified and counts considerably more taxa, also species releasing planktotrophic larvae, than mentioned in this paper. For an extended overview of the polychaets we refer to Wolff (1973). It is not known whether species shifts within this group occurred during the last decennium.

Lamellibranch larvae. There was a strong decrease of the cultured stocks of *Ostrea edulis* in the Oosterschelde, mainly due to the extremely severe winter of 1962/63. Before that time larvae of this species were abundant, especially in East (where the oyster plots were situated) during summer. In 1965 seedlings of *Crassostrea gigas*, the Pacific oyster, were introduced in the Oosterschelde. This species, replacing *O. edulis*, thrived well in the sea-arm and was common on tidal dike slopes and sandflats after about ten years (Dijkema, 1994). This means that most oyster larvae encountered nowadays in the plankton samples originate from *C. gigas*.

Mussels are introduced on a large scale from the Dutch Wadden Sea as spat and cultured in the central and western parts of the Oosterschelde (Van Stralen & Dijkema, 1994). Mussel larvae are peaking during spring and early summer. In summer, peaks of bivalve larvae mainly originate from oysters and, notably, from cockles (Bakker & Van Rijswijk, 1994).

We conclude that the zooplankton species composition of the Oosterschelde underwent some changes. Partly, these changes were primarily caused by the hydraulic works. *viz.* the abundant occurrence of the rotifer *Synchaeta* spp. in E during the period of barrier construction, as well as the retarded, obscured or even omitted seasonal succession of the copepods *Acartia clausi* and *A. tonsa* during the post-barrier period. Partly changes came from other factors *f.i.* the severe winter 1962/63 (causing strong decrease of *Ostrea edulis*) and the introduction of species (*Crassostrea gigas*), both species releasing their larvae in the water (meroplankton). Generally, the species composition was rather similar during the period of investigation. Differences between pre- and post-barrier periods were mainly of a quantitative nature (Bakker & Van Rijswijk, 1994; Tackx *et al.*, 1994).

Acknowledgements

I thank Drs Peter M. J. Herman, Pieter H. Nienhuis and Lambertus P. M. J. Wetsteyn as well as the referees Drs H. George Fransz and Michèle L. M. Tackx for valuable comments on an earlier version of the manuscript. I am much indebted to

Mr Pieter van Rijswijk who analyzed numerous samples, skillfully identified many species (particularly those of *Acartia*) and constructed the Figs 2 and 3; to Mr Alfons A. Bolsius who draw the definite figures and to Mrs Elly van Hulsteijn for typing the text. Communication No. 692 of NIOO-CEMO, Yerseke, The Netherlands.

References

Ambler, J. W., 1986. Effect of food quantity and quality on egg production of *Acartia tonsa* Dana from East Lagoon, Galveston, Texas. Estuar. coast. Shelf Sc. 23: 183–196.

Bakker, C., 1964. Planktonuntersuchungen in einem holländischen Meeresarm vor und nach der Abdeichung. Helgoländer wiss. Meeresunters. 10: 456–472.

Bakker, C. & W. J. Phaff, 1976. Tintinnida from coastal waters of the SW-Netherlands. I. The genus *Tintinnopsis* Stein. Hydrobiologia 50: 101–111.

Bakker, C., W. J. Phaff, M. v. Ewijk-Rosier & N. De Pauw, 1977. Copepod biomass in an estuarine and a stagnant brackish environment of the S.W. Netherlands. Hydrobiologia 52: 3–13.

Bakker, C., 1978. Some reflections about the structure of the pelagic zone of the brackish Lake Grevelingen (S.W. Netherlands). Hydrobiol. Bull. 12: 67–84.

Bakker, C. & F. Vegter, 1978. General tendencies of phyto- and zooplankton development in two closed estuaries (Lake Veere and Lake Grevelingen) in relation to an open estuary (Eastern Scheldt) SW-Netherlands. Hydrobiol. Bull. 12: 226–245.

Bakker, C., 1981. On the distribution of *Gonionemus vertens* A. Agassiz (Hydrozoa, Limnomedusae) a new species in the eelgrass beds of Lake Grevelingen (SW-Netherlands). Hydrobiol. Bull. 14: 186–195.

Bakker, C. & P. Van Rijswijk, 1987. Development time and growth rate of the marine calanoid copepod *Temora longicornis* as related to food conditions in the Oosterschelde estuary (southern North Sea). Neth. J. Sea Res. 21: 125–141.

Bakker, C., P. M. J. Herman & M. Vink, 1990. Changes in seasonal succession of phytoplankton induced by the storm-surge barrier in the Oosterschelde (SW-Netherlands). J. Plankton Res. 12: 947–972.

Bakker, C., P. M. J. Herman & M. Vink, 1994. A new trend in the development of the phytoplankton in the Oosterschelde (SW Netherlands) during and after the construction of a storm-surge barrier. Hydrobiologia 282/283: 79–100.

Bakker, C. & P. Van Rijswijk, 1994. Zooplankton biomass in the Oosterschelde (SW Netherlands) before, during and after the construction of a storm-surge barrier. Hydrobiologia 282/283: 127–143.

Bakker, C. & M. Vink, 1994. Nutrient concentrations and planktonic diatom-flagellate relations in the Oosterschelde (SW Netherlands) during and after the construction of a storm-surge barrier. Hydrobiologia 282/283: 101–116.

Berzins, B., 1960. Rotatoria I. *Synchaeta*. C.P.I.E.M. Zooplankton Sheet 84: 1–7.

Blanner, P., 1982. Composition and seasonal variation of the zooplankton in the Limfjord (Denmark) during 1973–1974. Ophelia 21: 1–40.

Bradford, J. M., 1976. Partial revision of the *Acartia* subgenus *Acartivra* (Copepoda: Calanoida: Acartiidae). N.Z.J. Mar. Freshw. Res. 10: 159–202.

Chervin, M., 1978. Assimilation of particulate carbon by estuarine and coastal copepods. Mar. Biol. 49: 265–275.

Conover, R. J., 1956. Oceanography of Long Island sound, 1952–1954. VI. Biology of *Acartia clausi* and *A. tonsa*. Bull. Bingham Oceanogr. Coll. 15: 156–233.

Dijkema, R., 1994. Fishery on bivalves and gastropods and culture of mussels and oysters in the Netherlands. In: McKenzie, C. *et al.*, (eds). The history, present condition and future of the molluscan fisheries of North America and Europe. Mar. Fish. Rev. (In press).

Elgershuizen, J. H. B. W., C. Bakker & P. H. Nienhuis, 1979. Inventarisatie van aquatische planten en dieren in de Oosterschelde. DIHO Rapp. en Versl. 1979–3, 105 pp.

Fransz, H. G., 1983. Zooplankton species of the Wadden Sea. In W. J. Wolff (ed.). Marine Zoology, Ecology of the Wadden Sea. 4. Invertebrata. Balkema, Rotterdam: 12–23.

Fransz, H. G., J. M. Colebrook, J. C. Gamble & M. Krause, 1991. The zooplankton of the North Sea. Neth. J. Sea Res. 28: 1–52.

Jeffries, H. P., 1962. Succession of two *Acartia* species in estuaries. Limnol. Oceanogr. 7: 354–364.

Jeffries, H. P., 1967. Saturation of estuarine zooplankton by congeneric associates. In: Lauff, G. H. (ed.). Estuaries. A.A.A.S. publ. no 83; Washington D.C., p. 500–508.

Landry, M. R., 1978. Population dynamics and production of a planktonic marine copepod, *Acartia clausi*, in a small temperate lagoon on San Juan Island, Washington. Int. Revue ges. Hydrobiol. 63: 77–119.

Lee, W. Y. & B. J. McAlice, 1979. Seasonal succession and breeding cycles of three species of *Acartia* (Copepoda: Calanoida) in a Maine estuary. Estuaries: 2: 228–235.

Lindholm, T., 1985. *Mesodinium rubrum* – a unique photosynthetic ciliate. Adv. aquat. Microbiol. 3: 1–48.

Lücke, F., 1912. Quantitative Untersuchungen an dem Plankton bei dem Feuerschiff 'Borkumriff' im Jahre 1910. Wiss. Meeresunters. (Kiel) 14: 101–128.

Martens, P., 1980. Contributions to the mesozooplankton of the Northern Wadden Sea of Sylt, German Bight, North Sea. Helgoländer Meeresunters. 34: 41–54.

Nienhuis, P. H. & A. C. Smaal, 1994. The Oosterschelde estuary, a case study of a changing ecosystem: an introduction. Hydrobiologia 282/283: 1–14.

Paffenhöfer, G.-A. & D. E. Stearns, 1988. Why is *Acartia tonsa* (Copepoda: Calanoida) restricted to near-shore environments? Mar. Ecol. Progr. Ser. 42: 33–38.

126

Redeke, H. C., 1902. Overzicht van de samenstelling van het plankton der Oosterschelde. In P. P. C. Hoek (ed.). Rapport over de oorzaken van den achteruitgang in hoedanigheid van de Zeeuwsche oester. Min. v. Waterstaat, Den Haag. Drukk. C. de Boer Jr., Helder: 115–145.

Revis, N. J. P. & C. Bakker, 1988. Zooplankton van het Veerse Meer in 1987 (with English summ.). Rep. 1988-5 DIHO, Yerseke, 1–78.

Roman, M. R., 1984. Utilization of detritus by the copepod *Acartia tonsa*. Limnol. Oceanogr. 29: 949–959.

Smaal, A. C. & R. Boeije, 1991. Veilig getij. Nota GWWS 91.088: 1–132.

Soetaert, K. & P. van Rijswijk, 1993. Spatial and temporal patterns of the zooplankton in the Westerschelde estuary. Mar. Ecol. Progr. Ser., 97: 47–59.

Sullivan, B. K. & L. T. McManus, 1986. Factors controlling seasonal succession of the copepods *Acartia hudsonica* and *A. tonsa* in Narragansett Bay, Rhode Island: temperature and resting egg production. Mar. Ecol. Progr. Ser. 28: 121–128.

Tackx, M. L. M., C. Bakker, J. W. Francke & M. Vink, 1989. Size- and phytoplankton selection by Oosterschelde zooplankton. Neth. J. Sea Res. 23: 35–43.

Tackx, M. L. M., C. Bakker & P. van Rijswijk, 1990. Zooplankton grazing pressure in the Oosterschelde (The Netherlands). Neth. J. Sea Res. 25: 405–415.

Tackx, M. L. M., J. W. Francke, P. Van Rijswijk, M. Vink & J. Rijk, 1991. Size distributions and trophic relationships of the pelagic ecosystem in the Oosterschelde (S.W. Netherlands). Hydrobiol. Bull. 25: 9–14.

Tackx, M. L. M., P. M. J. Herman, P. Van Rijswijk, M. Vink & C. Bakker, 1994. Plankton size distributions and trophic relations before and after the construction of the storm-surge barrier in the Oosterschelde estuary. Hydrobiologia 282/283: 145–152.

Turner, J. T., 1981. Latitudinal patterns of calanoid and cyclopoid copepod diversity in estuarine waters of eastern North America. J. Biogeogr. 8: 369–382.

Uye, S.-I., 1982. Population dynamics and production of *Acartia clausi* in inlet water. J. exp. mar. Biol. Ecol. 57: 58–83.

Van Breemen, P. J., 1903. Über das Vorkommen von *Oithona nana* (Giesbrecht) in der Nordsee, 1902–1908. Cons. Int. Explor. Mer, Publ. de Circonst., No. 7

Van Breemen, P. J., 1905. Plankton van Noordzee en Zuiderzee. Ph.D. thesis, Amsterdam, 180 pp.

Van Stralen, M. R. & R. Dijkema, 1994. Mussel culture in a changing environment: the effects of a coastal engineering project on mussel culture (*Mytilus edulis* L.) in the Oosterschelde (SW Netherlands). Hydrobiologia 282/283: 359–379.

Vroon, J., 1994. Hydrodynamic characteristics of the Oosterschelde in recent decades. Hydrobiologia 282/283: 17–27.

Wetsteyn, L. P. M. J., J. C. H. Peeters, R. N. M. Duin, F. Vegter & P. R. M. de Visscher, 1990. Phytoplankton primary production and nutrients in the Oosterschelde (The Netherlands) during the pre-barrier period 1980–1984. Hydrobiologia 195: 163–177.

Wetsteyn, P. M. J. & C. Bakker, 1991. Abiotic characteristics and phytoplankton primary production in relation to a large-scale coastal engineering project in the Oosterschelde (The Netherlands): a preliminary evaluation. In: M. Elliott & J. P. Ducrotoy (eds). Estuaries and coasts: spatial and temporal inter-comparisons. Proc. ECSA 19 Symp., 1989. Olsen & Olsen: 365–373.

Wolff, W. J., 1973. The estuary as a habitat. Ph.D. thesis Leiden, 242 pp.

Hydrobiologia **282/283**: 127–143, 1994.
P. H. Nienhuis & A. C. Smaal (eds), The Oosterschelde Estuary.
© *1994 Kluwer Academic Publishers. Printed in Belgium.*

Zooplankton biomass in the Oosterschelde (SW Netherlands) before, during and after the construction of a storm-surge barrier

C. Bakker & P. van Rijswijk
Netherlands Institute of Ecology, Centre for Estuarine and Coastal Ecology, Vierstraat 28, 4401 EA Yerseke, The Netherlands

Key words: zooplankton biomass, copepods, *Temora*, *Pleurobrachia*, benthic larvae, phytoplankton, seston food quality, Oosterschelde, retention time

Abstract

The hydrodynamic consequences of large coastal engineering (barrier-construction) works in the Oosterschelde were: prolonged residence times of the water, increased sinking of particulate material, and higher water transparencies. This strongly influenced the phytoplankton (Bakker *et al.*, 1990; 1994) and phytoplankton biomass increased in the shallow Eastern compartment of the Oosterschelde (Bakker & Vink, 1994) while phytoplankton concentration of the seston rose.

Zooplankton biomass, especially of copepods (*Temora*) and meroplankton (barnacle larvae) increased during the post-barrier period in the Eastern compartment. It is hypothesized that this is caused by the improved feeding conditions and the increased retention times in this area.

The barrier years 1985 and 1986 were characterized by low current velocities. In the Eastern compartment, this may have favoured the development of the rotifer *Synchaeta* (Bakker, 1994) and of the important copepod predator *Pleurobrachia* (Ctenophora).

In the Western compartment, zooplankton developments in the post-barrier years were rather similar to those in the pre-barrier period. This led to the disappearance of the previously existing biomass gradients West-East (maxima in West). At present a trend in the opposite direction (maxima in East) is observed.

Introduction

In the Oosterschelde (Fig. 1), a tidal inlet (*ca* 350 km²) of the southern North Sea, a large-scale hydraulic engineering project was carried out. This involved the construction of a storm-surge barrier in the mouth (1982–86), and dams in the Eastern part (Oesterdam, 1982–86) and in the Northern compartment (Philipsdam, 1985–87) (Fig. 1). For further details cf. Nienhuis & Smaal (1994).

The construction of the barrier and the accompanying works seriously affected tidal amplitude and current velocities (Wetsteyn *et al.*, 1990; Smaal & Nienhuis, 1992). Current velocities decreased, especially in the shallow Eastern area, suspended matter concentrations declined (Wetsteyn & Bakker, 1991) and water transparency rose concomitantly (Bakker *et al.*, 1990). This caused the phytoplankton assemblage to extend its growth season both earlier and later in the year (Bakker *et al.*, 1990).

During the post-barrier years a changed light-nutrient-salinity regime (*i.e.* high water trans-

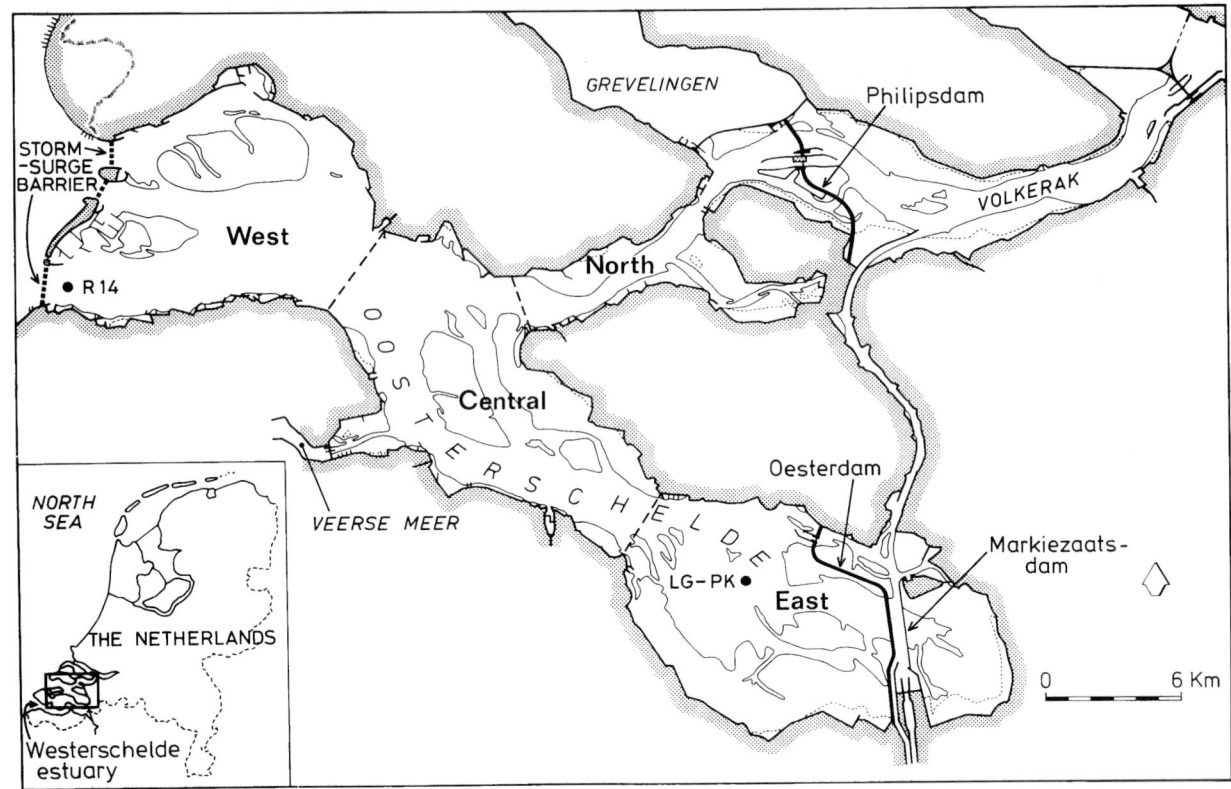

Fig. 1. Map of the Oosterschelde with indications (broken lines) of the 4 compartments and the sampling localities in West (R14) and East (LG-PK).

parency, limitation of nitrate, high salinity) was established during an extended summer season, without the gradual transitions characterising the pre-barrier period (Bakker *et al.*, 1994). In summer, biomass of diatoms decreased but biomass of flagellates increased in the Eastern compartment (Bakker & Vink, 1994). Annual primary production of the pre-barrier years, as reported by Vegter & De Visscher (1987), did hardly change in the post-barrier period (Wetsteyn & Kromkamp, 1994): slightly decreased values were observed for the Western- and similar values for the Eastern compartment.

Data on the zooplankton species composition of the Oosterschelde are given by Bakker (1994). Population dynamics and trophic relations were studied in the eighties, with special reference to the copepods. Bakker & Van Rijswijk (1987) determined the development time and growth

rate of *Temora longicornis* as related to the local food conditions. At the same time Tackx *et al.* (1989, 1990) performed grazing experiments using the copepods *Acartia* spp., *Temora longicornis* and *Centropages hamatus* as well as the nauplius larvae of *Balanus* spp., dominating in the Eastern area. They demonstrated size- and species-selective feeding in the copepods, and quantified the grazing pressure in the Western and Eastern parts of the Oosterschelde. Van Rijswijk *et al.* (1989) continued the *Temora* studies with the estimation of daily fecundity as influenced by temperature and food. Tackx *et al.* (1991, 1994) studied size distributions of phyto- and zooplankton for the analysis of the trophic relations in the pelagic subsystem in the basin. Finally, Klepper (1989) and Van der Tol & Scholten (1992) integrated all field data of the Oosterschelde in a dynamical ecosystem model in an attempt to quantify the impact of the hydraulic works on the foodweb.

The aim of the present paper is to detect a possible response of the dominant zooplankton species, notably the copepods, to the environmental conditions in the present-day Oosterschelde.

As in the phytoplankton papers (Bakker *et al.*, 1990, 1994; Bakker & Vink, 1994) the attention is focussed on the Eastern part of the sea-arm as changes were most pronounced in that area. Eastern and Western compartments are further indicated as E and W, respectively.

Methods

Samples were taken at two locations: station LG-PK in E (depth 20 m) and station R14 in W (depth 40 m, near the storm-surge barrier) (Fig. 1). During the period March to October sampling was carried out weekly, during the rest of the year less frequently.

Sampling and counting

Zooplankton was collected by pumping with a Pleuger submersible system, capacity 200 l min⁻¹. Generally, 100 l of water were filtered, through a net of 63 μm mesh gauze, retaining all copepod stages (smallest nauplii and eggs included), cladocerans, appendicularians and the majority of meroplankton (larvae of mollusks, polychaete worms, barnacles). The 63 μm mesh did not retain quantitatively rotifers and protozoans. Sampling was always done around half tide, when maximum current velocities prevailed, causing approximately homogeneous vertical distributions of zooplankton in the water column (unpublished data of a precedhng pilot study). During the years 1982–84 all samples were taken at mid depth, in the period 1985–88 samples were collected at depths of 2.5, 7.5, 12.5, 17.5 and 22 m and in W also at 27.5 and 32 m depth. At both stations, samples taken at all depths were pooled. Zooplankton was preserved with borax-buffered

formalin to a final concentration of 4% exactly (Steedman, 1976). In the laboratory a subsample was taken, representing 100 l of the original field sample. Adult copepods and copepodid stages were counted under a binocular microscope. For the copepod nauplii and the rest of the zooplankton a subsample was taken representing 10 l of water of the original field sample and the organisms were counted using an inverted microscope.

Pleurobrachia pileus (Ctenophora) was sampled at mid depth with a 200 μm mesh-sized high-speed sampler during tows of *ca* 5 minutes, filtering 50–100 m³ of water. Total numbers of comb-jellies were counted on board the ship (without determining the size), in order to obtain an impression of the abundance of this predator, potentially of importance in controlling copepod densities.

Biomass determination and calculations

Copepods were sorted and divided in the following groups: small copepodid stages (I–III, 3 × 40 individuals, large copepodids (IV–V, 3 × 30 ind.), adult females (3 × 10 ind.) and males (3 × 10 ind.). Cephalothorax lengths were measured. The animals were put in aluminum combustion vessels of 10 × 4 × 3 mm (Perkin-Elmer). Vessels were pre-dried and vessels plus animals dried in an oven at 50 °C for 24 hours. A Cahn electrobalance (model 29, accuracy 0.1 μg) was used for the determination of copepod dry weight. For corrections of dry weight values and further details: cf. Bakker & Van Rijswijk, 1987). For *Oikopleura* the length-weight relation according to King *et al.* (1980) was used. For the remaining species and groups weights were determined after calculation of volume and conversion of wet- to dry weight.

Further calculations

Weekly averages of the biomass of the two main components of the zooplankton (copepods and meroplankton) were calculated for the 3 periods of investigation: pre-barrier- (1982–83), barrier- (1984–86) and post-barrier- (1987–88) period

130

Copepods

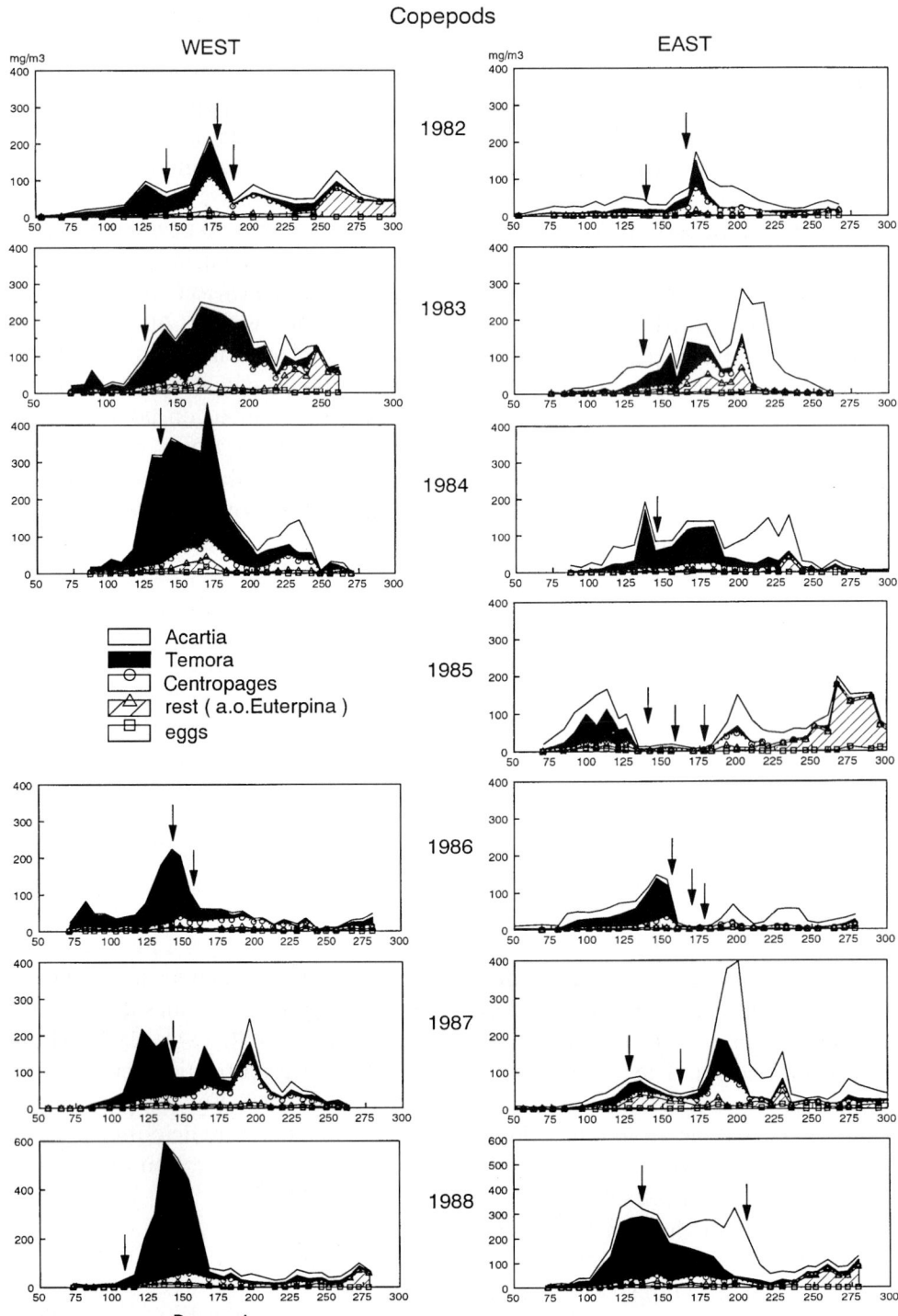

Fig. 2. Seasonal distribution of copepod biomass (mg drw m^{-3}) in the Western and Eastern compartments of the Oosterschelde, 1982–88. Arrows indicate peaks of *Pleurobrachia*. (Note the different scale in 1988).

(Fig. 8). Bakker *et al.* (1990) argued the inclusion of 1984 in the barrier construction period, on the basis of the already increasing transparencies.

Although phytoplankton is considered to be the main food for suspension-feeding (meso)-zooplankton, microzooplankton like ciliates and rotifers may also be consumed by these animals (Stoecker & Egloff, 1987). Therefore total microplankton carbon (*i.e.* phytoplankton- and microzooplankton carbon added) was calculated (based on the data of Bakker & Vink, 1994, and on unpublished data of these authors), as monthly averages for the 3 periods of investigation (Fig. 10).

In the Oosterschelde suspended matter concentrations and particulate organic carbon concentrations decreased, from the barrier period onwards (Wetsteyn & Bakker, 1991). Consequently, the ratio of microplankton carbon to POC seemed of importance as a measure of (changing) food quality. This ratio was also cal-

culated as monthly averages for the 3 periods of investigation (Fig. 11).

Results

Copepod biomass (Figs 2–3)

In spring, the most abundant copepod genera were *Temora* and *Acartia*; during summer *Centropages*, *Acartia* and *Euterpina* were found, the latter however only abounding during the barrier year 1985 in E (Fig. 2). Generally, *Pseudocalanus*, *Paracalanus* and *Euterpina* did not reach high densities and were lumped to the rest group in Figs 2–3. The seasonal distributions of copepod biomass at both stations (Fig. 2) and the average annual biomass values (Fig. 3) show *Temora longicornis* to be the dominating copepod in W throughout the entire period of investigation with *ca* 50–75% of total biomass. In E, *Acartia* spp. were more abundant, representing *ca* 40–60% of

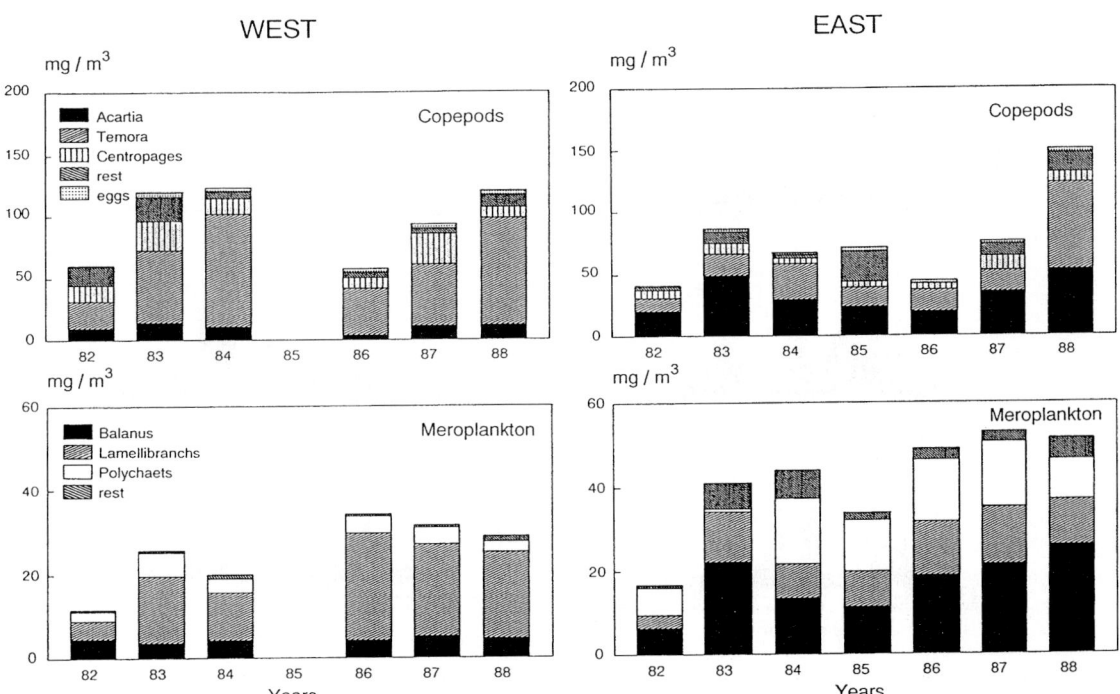

Fig. 3. Annual average copepod (top) and meroplanktonic (bottom) biomass (mg drw m⁻³) in the Western and Eastern compartments of the Oosterschelde, 1982–88.

132

Benthic larvae

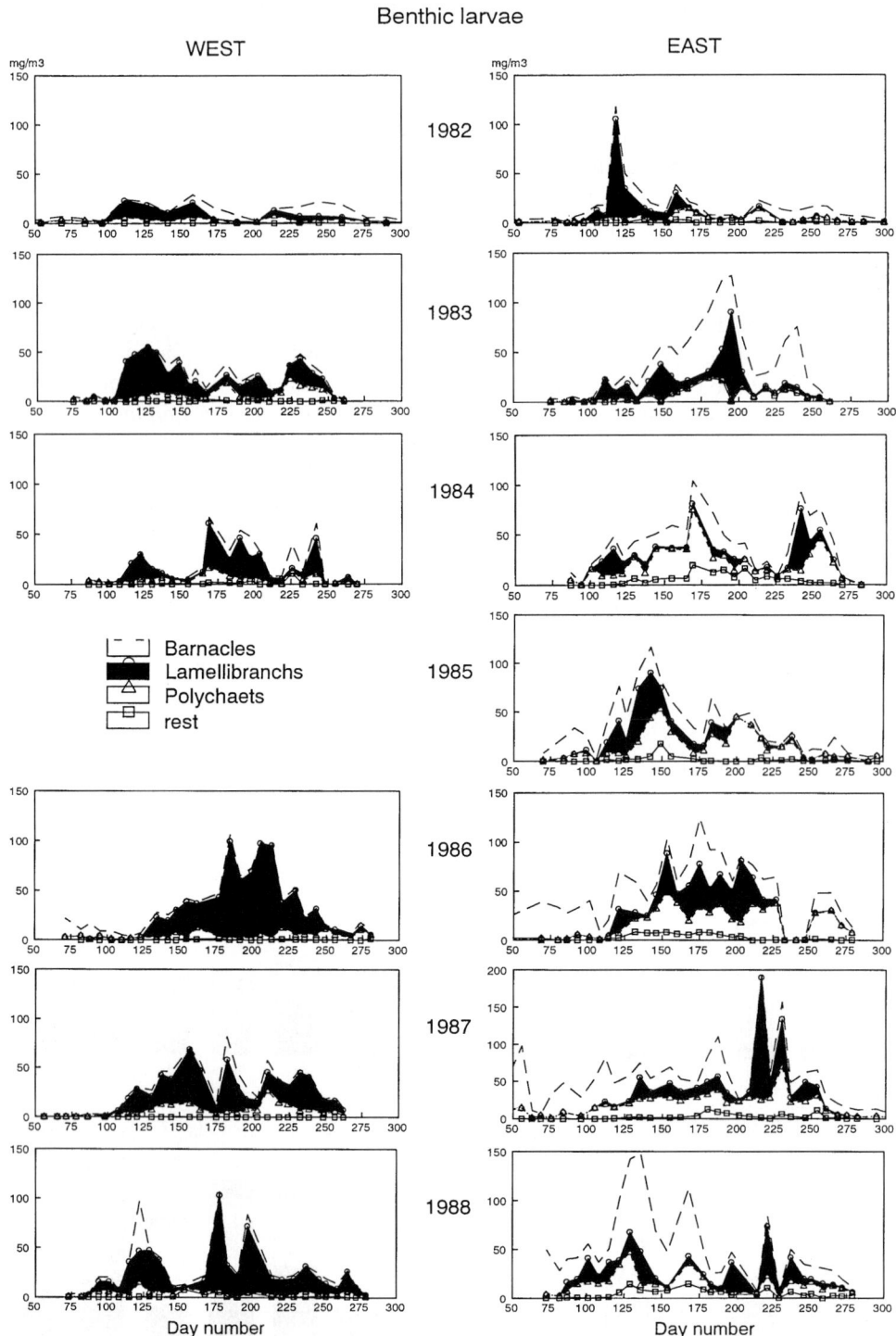

WEST · EAST

Barnacles
Lamellibranchs
Polychaets
rest

1982
1983
1984
1985
1986
1987
1988

Day number · Day number

Fig. 4. Seasonal distribution of meroplanktonic biomass (mg drw m^{-3}) in the Western and Eastern compartments of the Oosterschelde, 1982–88.

total biomass. During the postbarrier year 1988 *Temora* reached an unprecedentedly high biomass in E. Consequently, total copepod biomass rose considerably in this area in 1988 (Fig. 3). In W such an increase did not occur: average biomass of 1988 was similar to that of 1983–84 (Fig. 3). Late winter–early spring densities of *Temora* in W were larger than those in E in 1982 till 1986, but this tendency disappeared in 1987–88. Consequently, *Temora* tended to peak earlier in W than in E, but this trend was reversed in 1988.

Acartia always demonstrated much larger late-winter biomass and earlier peaks in E than near the barrier. Here, *Acartia* development was often very poor.

Peaks of *Centropages* biomass were modest, if present, and times of occurrence of peaks were similar in both areas. Late winter–early spring densities were always very small or negligible. *Centropages* generally followed *Temora*, but always had much smaller biomass.

All copepod species showed a more or less fixed pattern of seasonal occurrence, except for *Euterpina*. In W during 1982–83, this species developed well, but it could hardly be observed in 1984–87 (Fig. 2). In E a dense *Euterpina* population occurred only in 1985 (Fig. 2). In this barrier year, when current velocities began to decrease, copepod developments were strikingly different from those in all other years of investigation in this area. *Temora* peaks, for instance, were already seen around day 100, while the usual maxima around days 150–175 were completely lacking in 1985 (Fig. 2).

Eggs (of all copepod species together) always contributed very sparsely to total copepod biomass.

Meroplankton biomass (Figs 3–4)

Meroplankton (benthic larvae) in the Oosterschelde consisted mainly of barnacle larvae (*Balanus* spp.), bivalve larvae (mussels, cockles, oysters) and polychaete larvae. Other groups recorded were gastropods (periwinkles, mud snails, slipper limpets) and tunicates (tadpole larvae).

The meroplankton group composition differed considerably between the two compartments. Barnacle- and polychaete larvae were found in a 2–5 times higher biomass in E than in W. Barnacle larvae could reach high densities in E already in the pre-barrier period (1983), decreased during 1984–85 and continuously increased again from 1986 onwards (Figs 3, 4). Polychaetes, mostly comprising a substantial biomass in E, had a remarkably low density in 1983 (Fig. 3). Bivalve larvae accounted for *ca* 25% of total benthic biomass in E. In W extended mussel beds occur, causing the major benthic contribution to total larval biomass in that area, especially from 1986 onwards, when biomass values amounted to > 20 mg m^{-3}, equalling those of barnacle larvae in E (1983, 1988). Other groups of minor importance were stronger represented in E. In W the first larvae peak, in spring, was produced by mussels; the summer peaks (Fig. 4) originated from cockle larvae. In this area in 1986, exceptionally, summer biomass strongly exceeded the spring values here (Fig. 4). In E high cockle larvae peaks were observed during summer 1987 (Fig. 4).

Ctenophora (Pleurobrachia) densities (Figs 5–6)

Figure 5 gives the *Pleurobrachia* densities (N m^{-3}) from 1982 onwards at both stations. As a rule, highest densities were observed in W (1982, 1983, 1984, 1987). During 1985 and 1988 highest maxima were registrated in E, while in 1986 (barrier construction year) E started and finished higher than W, although the absolute peak was found in the latter area. In 1984, 1985 and 1988 W demonstrated relatively low maxima (≤ 10 m^{-3}) while during the other years numbers fell within the range of $> 20–> 50$ m^{-3}). *Pleurobrachia* development became visible from March–April onwards and most maxima were situated in May.

During the pre-barrier years a W–E gradient in *Pleurobrachia* density was observed which disappeared in the barrier period but returned in the post-barrier years. In 1988 (May) however, this gradient reversed due to a steady increase of

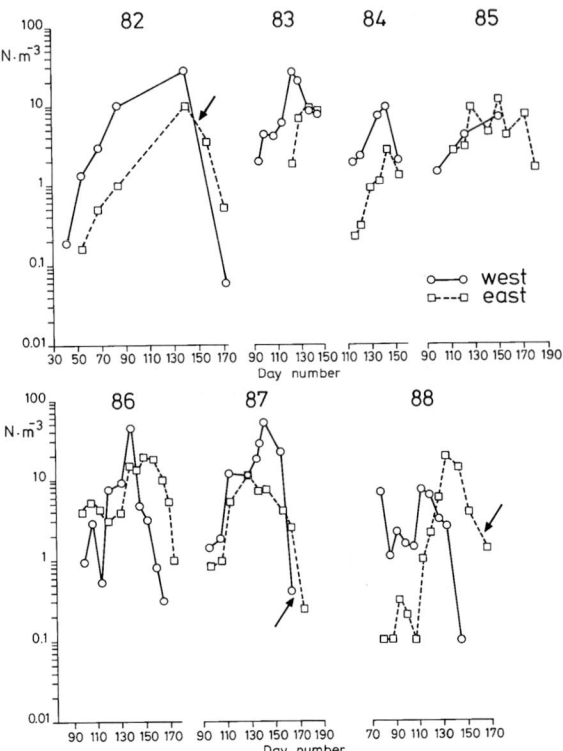

Fig. 5. Densities of *Pleurobrachia pileus* (N m⁻³) in Western and Eastern compartments of the Oosterschelde during the years 1982–88. Arrows indicate presence of *Beroe gracilis*.

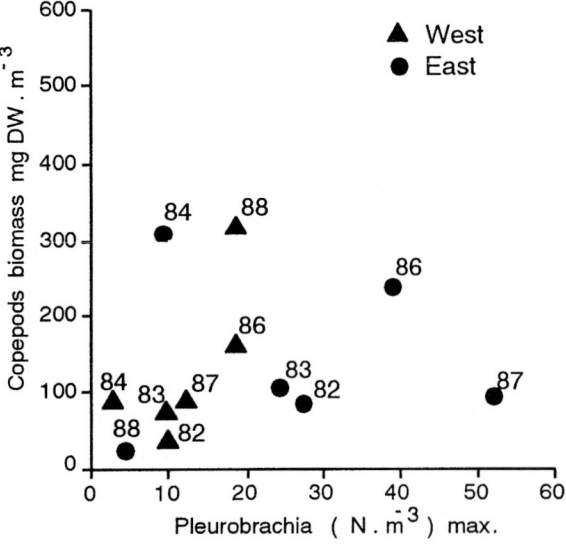

Fig. 6. Maxima (N m⁻³) of *Pleurobrachia* and co-occurring copepod biomass (mg drw m⁻³) in the Oosterschelde (1982–1988) in the Western (circles) and Eastern (crosses) compartments.

Pleurobrachia numbers in E, coinciding with a decline in West. Figure 6 shows that no correlation existed between the maximum density of *Pleurobrachia* and the co-occurring total copepod biomass. However, comparing Fig. 5 with Fig. 2, the following details can be noticed. In 1984 *Pleurobrachia* reached low peak numbers in W while at the same time (day 142) copepod (*Temora*) biomass was very large and still rose further after the definite wane of the predator. In E, 1985 was characterized by scarcity of copepods from day 130 onwards which coincided with a relatively high and long-lasting (although fluctuating) abundance of *Pleurobrachia*. From day 170 onwards when *Pleurobrachia* decreased definitely, summer copepods (mainly *Acartia*) developed. In E, 1986, the possibly predatory influence of *Pleurobrachia* shifted to June when its numbers still rose and copepod biomass strongly declined. In W, 1986, copepod biomass did not increase further when *Pleurobrachia* was at its maximum (day 140). In W, 1987, the drop of *Temora* on day 140 may be influenced by the *Pleurobrachia* maximum at the same time. In E, 1987, it was not earlier than day 170 when *Pleurobrachia* (as well as other gelatinous predators: *Aurelia*) had disappeared completely, that a strong copepod development started. In E, May 1988, copepod biomass levels decreased or flattened (*Temora*) when *Pleurobrachia* reached maximum abundance. In W, 1988, the high peak in copepod biomass developed when *Pleurobrachia* was nearly absent.

Biomass during pre-barrier-, barrier- and post-barrier periods (Figs 7–8)

Annual average total zooplankton biomass (Fig. 7). Biomass in W during the pre-barrier period (*ca* 75–*ca* 150 mg drw m⁻³) was generally higher than in E (*ca* 60–*ca* 100 mg drw m⁻³). During the years 1985–86 biomasses in W and E were similar and amounted to *ca* 100 mg drw m⁻³. During the post-barrier period E reached the largest values measured in 1988 (>200 mg drw m⁻³), while the values of W maintained the 1983–84 level (*ca* 150 mg drw m⁻³). During

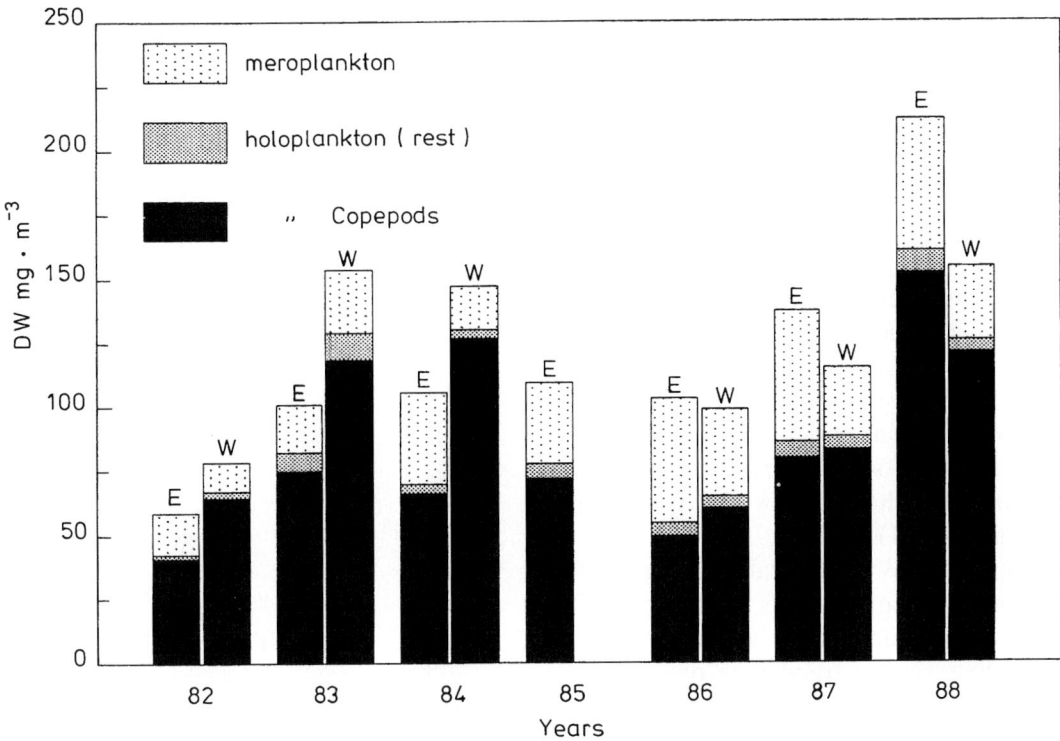

Fig. 7. Annual average values of total zooplankton biomass (mg drw m^{-3}, main contributing groups indicated) in Western (right bar) and Eastern (left bar) compartments of the Oosterschelde, 1982–88.

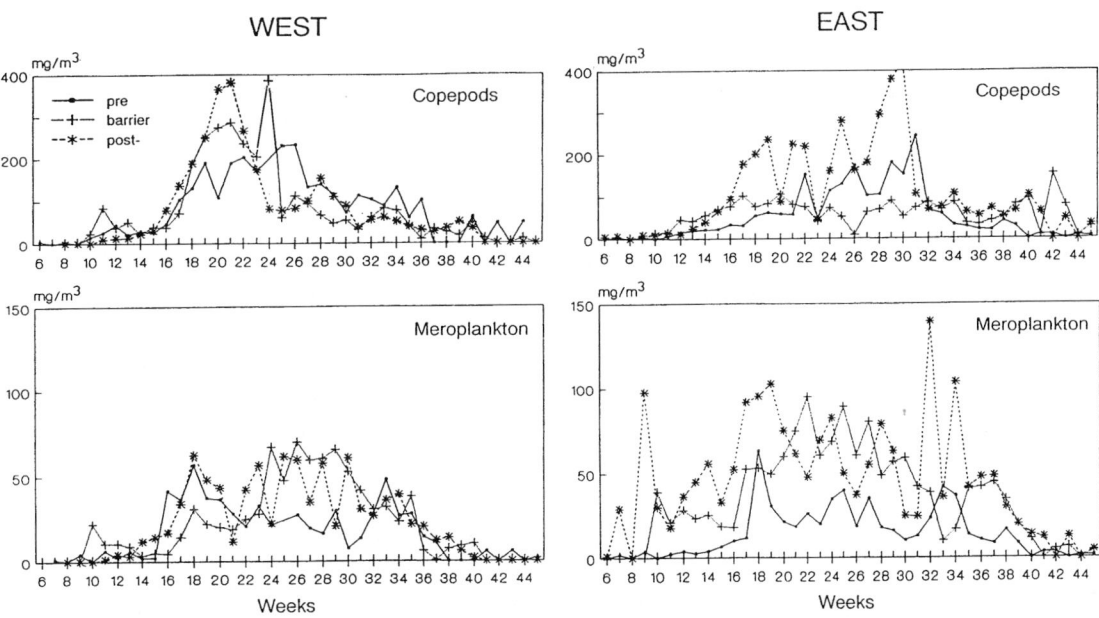

Fig. 8. Seasonal distribution of copepod (top) and meroplanktonic (bottom) biomass (mg drw m^{-3}) averaged for pre-barrier-, barrier- and post-barrier years in the Oosterschelde, Western and Eastern compartments.

1987–88 total annual biomass in W was smaller than in E, so that the previous biomass gradient W–E was reversed (on a per m³ basis).

Seasonal distribution patterns for copepods and meroplankton are shown in Fig. 8, where average weekly biomass values are given during the three periods separately, for the two compartments.

Copepods. In the post-barrier period in W, copepods demonstrated a series of higher spring but lower summer values than previously, leading to approximately similar annual averages, except 1982 (Fig. 7). In E on the other hand, copepods in the post-barrier years reached higher values than before, nearly all over the year. The years of barrier construction showed very low values in summer (weeks 24–31).

Meroplankton. In W higher values in barrier- as well as in post-barrier years were reached during summer, as compared to the pre-barrier situation. In E a similar trend was observable during nearly the whole year.

Discussion

The main change observed in this study is the important increase in zooplankton biomass during the post-barrier period in E. This led to the reversal of the previous biomass gradient W–E during the major part of the growth season in these years (1987–88). However, we face the objection that the complete zooplankton data for the last period only comprise 1988, the first complete post-barrier year, and not 1989, a year still included indeed in the phytoplankton studies. And as zooplankton biomass, especially of *Temora longicornis*, was unprecedentedly high in E during 1988, the question arises whether this increase was an incidental phenomenon only. Zooplankton samples of 1989 have been collected and analysed with regard to *Temora*. Results of the counting of the numbers of the adult animals are presented in Fig. 9, for both 1988 and 1989 (after Darboe, 1992). Maximum *Temora* densities of 1989 were as high as those of the foregoing year. The duration of the entire period of occur-

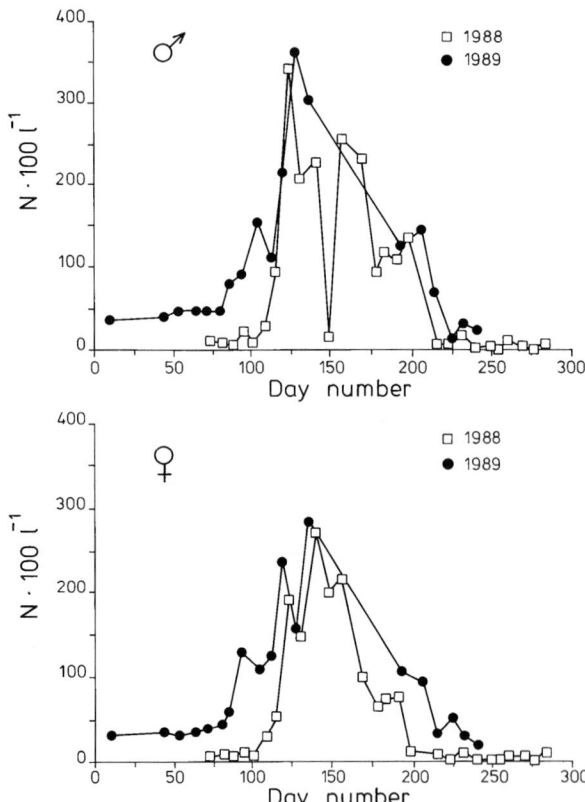

Fig. 9. Seasonal distribution of adult *Temora longicornis* densities in the Eastern compartment of the Oosterschelde in 1988 and 1989 (after Darboe, 1992).

rence was even longer in 1989 (Fig. 9). Moreover, the 1989 (and 1988) *Temora* adults reached heavier individual weights than during the pre-barrier period (Bakker & Van Rijswijk, unpubl.; Darboe, 1992). This supports the impression that the increase of copepod (notably *Temora*) biomass is a characteristic phenomenon for the post-barrier years. Also the model calculations by Van der Tol & Scholten (1992) indicate an increased zooplankton biomass and an overall increase in carbon fluxes with regard to the zooplankton during the post-barrier period.

Environmental factors in the Oosterschelde influencing the annual fluctuations in zooplankton abundance comprise both the hydrodynamical changes (specific for this basin) and the cyclic physical, chemical and biological alterations, partly of general nature, partly strengthened or weakened by the infrastructural works. Because several factors demonstrated the most conspicuous changes in E, the phytoplankton in this area were influenced most distinctly too (Bakker *et al.*, 1994; Bakker & Vink, 1994).

In relation to the zooplankton biomass increase in E, the following factors are of importance:

1. Increased residence times during barrier- and post-barrier years (cf. Wetsteyn *et al.*, 1990; Vroon, 1994). Larger residence times lead to stronger retention of zooplankton which is of importance during the growth season as the populations are allowed to stay and reproduce for a prolonged time in a smaller area than before. This may result in higher population densities during the propagative period of a species. For copepods this can also lead to higher numbers of resting eggs sedimenting on a certain bottom area and, consequently, to higher naupliar densities in the next spring period. The strongly decreased current velocities during the barrier years were most favourable for the rotifer *Synchaeta* spp., reaching

peak abundances during this period (Bakker, 1994).

2. Increased phytoplankton biomass. A shift in phytoplankton composition during the barrier period was demonstrated (Bakker *et al.*, 1990), continuing in the post-barrier period (Bakker *et al.*, 1994). It is difficult to evaluate a possible influence of a change in phytoplankton composition *per se* on the zooplankton, but when also phytoplankton biomass increases (Bakker & Vink, 1994; Wetsteyn & Kromkamp, 1994), the food conditions for the zooplankton will improve, especially when the enlarged biomass of the phytoplankton is composed of preferred species in favourable size classes (Tackx *et al.*, 1989). There are good reasons to assume that this is true indeed, as flagellate biomass increased from 1985 onwards (Bakker & Vink, 1994) and also a number of smaller diatom species appeared (Bakker *et al.*, 1994). For the copepods also microzooplankton (rotifers, ciliates) represent good food (Stoecker & Egloff, 1987). Therefore total micro-

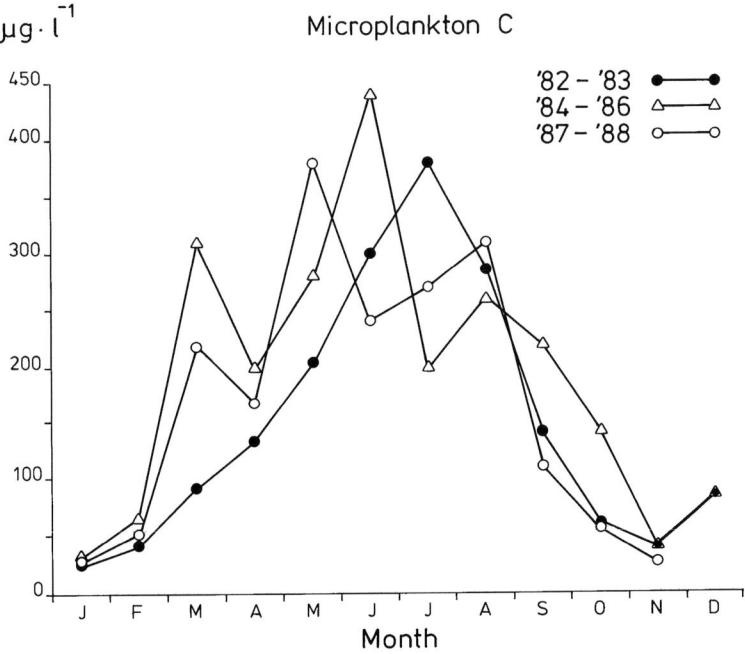

Fig. 10. Microplankton carbon (*i.e.* phytoplankton- plus microzooplankton carbon) in the Eastern compartment of the Oosterschelde, calculated as average monthly values during pre-barrier- (1982–83), barrier- (1984–86) and post-barrier (1987–88) periods.

138

plankton carbon (including microzooplankton) was calculated to illustrate the changed food quantities in the Oosterschelde during the 3 periods (Fig. 10). For the pre-barrier period a regular increase of microplankton carbon can be seen from winter to summer, with a distinct summer peak, followed by a (steeper) decrease in late summer and autumn. The barrier- and post-barrier years, characterized by higher transparencies in the water column enabling an earlier start of phytoplankton development, demonstrated much larger biomass in late winter and spring. In summer, however, carbon biomass values were smaller than previously: this may be partly due to nitrogen limitation (Bakker & Vink, 1994), partly to increased grazing pressure exerted by the larger zooplankton biomass (Fig. 8).

3. Increased food quality during barrier- and post-barrier years. Suspended matter concentrations decreased significantly from the barrier period onwards (Wetsteyn & Kromkamp, 1994; Bakker & Vink, 1994). Particulate organic carbon concentration was found to behave similarly as suspended matter concentration (Bakker & Wetsteyn: unpubl.). During the major part of the growth season in barrier- and post-barrier periods, the microplankton carbon/POC ratio changed in favour of the living component of the POC (Fig. 11). (Only during a short summer period (July) the situation was similar to the pre-barrier period). This improvement of the food value of the seston, in combination with the increased microplankton biomass *per se*, may result in a more favourable energy balance as the animals have to spend less energy and/or time to select preferred food items. This may have led to a larger productivity, of the holoplankton as well as of the benthos, and, consequently to the increased biomass and total zooplankton densities. The decrease of indigestible material in the zooplankton diet was not explicitly modelled by van der Tol & Scholten (1992).

4. Increased sinking rate during the barrier period, when current velocities and tidal amplitude continuously decreased to minimum values at the end of 1986 and the first months of 1987 (cf. Vroon, 1994). The demonstrated increased sedimentation of total suspended matter might also have comprised living organisms, zooplankton included. *Temora* abundance was rather low during 1986. Barnacle larvae, on the other hand, were more abundant than previously (Fig. 4) and notably these animals are expected to respond strongly to decreased current velocities with increased sinking, due to slower swimming movements (compared to copepods) according to their littoral character. For the copepods, an intensified vertical migration rather than increased sink-

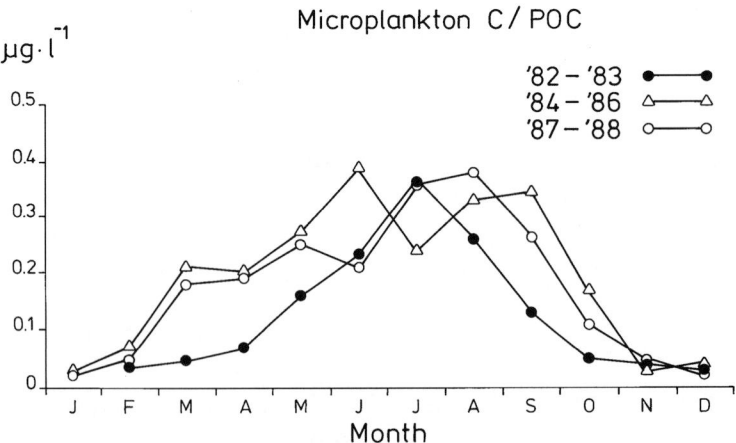

Fig. 11. The ratio microplankton carbon: particulate organic carbon in the Eastern compartment of the Oosterschelde, calculated as average monthly values during pre-barrier- (1982–83), barrier- (1984–86) and post-barrier (1987–88) periods.

ing rates might be held responsible for the smaller biomass observed, as the decreased tidal water movements were accompanied by increased transparencies (cf. Dodson, 1990).

5. Predatory control of mesozooplankton populations by invertebrate macrozooplankton. In the open North Sea, Daan (1989) demonstrated that predation, as a regulating factor, is only of importance when feeding conditions for the copepods are sub-optimal. In the coastal zone, however, other situations may prevail and predators like *Ctenophora* can reduce copepod population densities to a high extent (a.o. Greve, 1971; Deason & Smayda, 1982). (See further).

6. Water temperature, especially during winter. Influence of water temperature in zooplankton reproduction, notably of copepods, has been studied thoroughly. In our region a significant correlation was established between *Temora* fecundity and temperature in spring (Van Rijswijk *et al.*, 1989; Fransz *et al.*, 1989). Colebrook (1982, 1985) was the first to document that winter periods are often crucial for the variation between years, through differential survival of the stocks. After mild winters copepod development starts earlier and may proceed more rapidly, while after extreme winters development is strongly retarded (Fransz *et al.*, 1991). Remarkably, during the barrier period three colder winters occurred successively. Air temperature during the months January–March were 1–3 °C lower than the long-term mean (Data Dutch Meteorological Inst., unpubl.). (See further).

We assume that the effects of increased residence times, increased phytoplankton biomass and improved food quality all have influenced the zooplankton development positively during the barrier construction period. Moreover, larger volumes of Grevelingen (Fig. 1) water were discharged into the Oosterschelde in those years (Nienhuis & Smaal, 1993) and several phytoplankton- and microzooplankton organisms were introduced that found favourable conditions for development, especially in E (Bakker *et al.*,

1994; Bakker, 1994). Increased sinking rates of large phytoplankton and stronger predatory control, on the other hand, may have counteracted the favourable effects. Also the lower winter temperatures may have had negative influences. However, when we account for the strong early development of *Temora* in E in 1985 (Fig. 2), it seems that the winter effect can be neglected here and that the positive effects of food quality and -quantity prevailed.

Copepod production of resting eggs

The persistence of a neritic copepod population over successive years may depend on the ability of producing resting eggs. Overwintering eggs of *Temora longicornis* were discovered by Lindley (1986) who found that these eggs could remain viable for more than a year (Lindley, 1990). *Temora* in E probably developed autochthonous populations via increasing densities of resting eggs when exchange between the compartments decreased and retention of animals plus eggs increased. Moreover, resuspension and subsequent transport of eggs will have mainly occurred during flood, via the deeper channels in eastern direction, because the strength of the ebb current in E was reduced more strongly in barrier- and post-barrier years than the strength of the flood current (Vroon, 1994) and thus transport of eggs in seaward direction might have decreased, probably leading to accumulation of eggs in E.

Copepods (Temora) *and Ctenophora* (Pleurobrachia)

Pleurobrachia abundance in coastal embayments, at the start of its growth season, is often dependent on import (immigration) from the sea (Van der Veer & Sadée, 1984). The Oosterschelde data (Fig. 5) suggest a similar picture, cf. the W–E gradient of *Pleurobrachia* densities in the pre-barrier years.

The sudden appearance of high numbers of the comb-jellies in coastal areas is ascribed to migration of already developed populations from else-where (Reeve & Walter, 1978), or to massive re-suspension of overwintering stages on the sea floor (De Wolf, pers. comm.). *Pleurobrachia* adopts the planktonic phase when a water temperature of *ca.* 10 °C is exceeded (De Wolf, idem) and at the same time food is present in sufficient quantities. The massive recruitment from bottom stages partly explains the patchy occurrence of *Pleurobrachia*, although many other plankton organisms show patchiness phenomena (De Wolf, 1989). Once in the water phase and in favourable food conditions, propagation of the species proceeds rapidly and in the adjacent waters of the southern North Sea the peak is generally reached in May (Greve, 1971; Van der Veer & Sadée, 1984; Kuipers *et al.*, 1990; this paper).

Fraser (1970) found 80–97% of the diet to consist of crustaceans with copepods as dominating constituent. Depletion of copepod standing stocks by Ctenophora is observed regularly (Greve, 1971; Kremer, 1976; Deason & Smayda, 1982; Frank, 1986; Suthers & Frank, 1990). However, also in coastal systems the predatory impact of *Pleurobrachia* on copepod abundance is sometimes very small (Kuipers *et al.*, 1990; Dutch Wadden Sea, 1983). Moreover, negative correlations between *Pleurobrachia*- and copepod abundance, as reported in the literature, often are not significant (Miller & Daan, 1989). But when a patchy Ctenophora distribution would coincide with a uniform copepod distribution, correlations although existing, might not be demonstrated (Suthers & Frank, 1990).

We did not find a correlation between *Pleurobrachia* maxima and copepod densities at the same time (Fig. 6), but for the W compartment there were strong indications that the height of the *Pleurobrachia* maximum was influenced by either the preceding copepod maximum, or might influence (in its turn) the following copepod maximum (Fig. 12). A choice, however, between cause and consequence cannot be made. Moreover, an exact correlation can hardly be expected as the timings of the start of both prey and predator are independent of each other (see also Kuipers *et al.*, 1990). When *Pleurobrachia* comes first and food

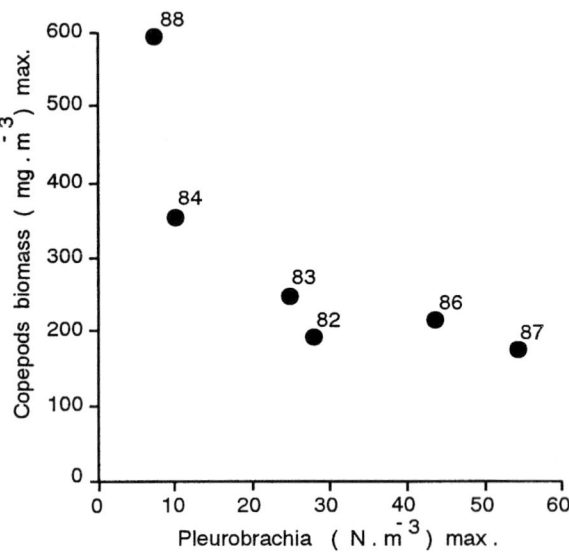

Fig. 12. Maxima (N m^{-3}) of *Pleurobrachia* and copepod biomass maxima (mg drw m^{-3}) before, during or after the *Pleurobrachia* peak in the Oosterschelde, Western compartment, 1982–1988.

is still lacking, mortality of the Ctenophora will be high and subsequent copepod development may be prosperous. A good example of this sequence may be the situation of W in 1988 (Fig. 2). In case of the reverse, *i.e.* when *Temora* starts first and develops rapidly, the later arriving *Pleurobrachia* finds a well provided table and may quickly graze down the copepod stock. This could have occurred in E, 1985, when after a cold winter, phytoplankton started its growth very early and *Temora* responded immediately, in spite of the lower water temperature. Afterwards, continued high abundance of *Pleurobrachia* may have caused the unusual disappearance of the copepod in May. In many cases development of predator and prey may run more or less parallel, e.g. in 1986, in both compartments; this was also observed by Kuipers *et al.* (1990). In W, *Temora* biomass of 1984 was the largest measured in this area during the entire period of observation and the combination of high densities of suitable phytoplankton (Bakker & Van Rijswijk, 1987) and low abundance of *Pleurobrachia* (Fig. 5) can be held responsible.

The (ctenophoran) predator on *Pleurobrachia*, often causing the final decline of the latter, is

Beroe gracilis (e.g. Greve, 1971). We paid no special attention to *Beroe*, but in case this species was observed, this always occurred during the wane of *Pleurobrachia* (cf. the arrows in Fig. 5 for 1982, 1987 and 1988).

Meroplankton

Copepods use to dominate in W (to *ca* 70% of total zooplankton biomass; Bakker & Van Rijswijk, 1987; Tackx *et al.*, 1990) and benthic larvae demonstrate higher biomass values in East, as follows from the detailed comparisons of Figs 2, 3 and 4.

Generally, tidal estuaries and sea-arms show an increase of depth of the channels in seaward direction. In the Oosterschelde a large difference between the compartments exists in depth-volume ratios: average depth in W is *ca* 12 m and in E *ca* 4 m. (For the main hydrographical features cf. Vroon, 1994). Consequently, in E a relatively larger bottom surface provides a relatively small water volume with many benthic larvae, especially those of barnacles and polychaetes, while in W comparable numbers of larvae are continuously distributed over a much larger water volume. Nevertheless, the abundance of a particular group of meroplankton, the lamellibranch larvae, is still larger in W, which is due to the localization of the majority of the mussel culture plots in this (and the central) area of the sea-arm (van Stralen & Dijkema, 1994).

In coastal marine and estuarine areas suspension-feeding zoobenthos may largely control phytoplankton biomass (Cloern, 1982; Herman & Scholten, 1990). This holds also for the Oosterschelde with its dense beds of cockles and (cultivated) mussels, implying high filtering activity (Prins & Smaal, 1994; Van Stralen & Dijkema, 1994; Herman & Scholten, 1990, Van der Tol & Scholten, 1992). The increased residence times during barrier- and post-barrier years will have been responsible for a stronger retention than before of the numerous larvae released and this explains the increased biomass of lamellibranch larvae during these periods in W (Figs 3, 4). The copepods have to compete with the mussels for phytoplankton food and did not succeed to increase their biomass during these years (Figs 2, 3).

In E, however, there is virtually no mussel culture, and it was here, that the copepods (Figs 2, 3, 7, 8) as well as the non-selective (Tackx *et al.*, 1989) barnacle larvae (Figs 3, 4, 7) could profit completely from the improved food conditions. Consequently, total zooplankton biomass here increased during nearly the entire growth season of the post-barrier period (Figs 7, 8).

During the pre-barrier period, as a rule, mussel veligers were present from the beginning of April onwards (Fig. 4). After the colder winters of the barrier years, however, development of larval populations started later. In 1988 the larvae could be observed again in April. It seems therefore that the lower water temperatures in early spring retarded mussel reproduction. For barnacle reproduction, on the other hand, a retarding effect of low temperatures during spring 1985–87 could not at all be discovered: a large larval stock was already present in March, while during the same period in 1982–84 only a few larvae were found. In 1988 at the same time many larvae occurred. For barnacles, therefore, food availability seems decisive for the early development, even in case of low winter temperatures. During the pre-barrier years, when suspended matter concentration was high due to a large contribution of non-living particulate matter, barnacle larvae, contrary to copepods not able to efficient phytoplankton selection (Tackx *et al.*, 1989), did not occur in very high abundances. It seems therefore that these organisms, even more than the copepods, profited from the new situation with increased phytoplankton concentration of the seston, making selection less necessary.

Final remarks

The increased zooplankton biomass in E (post-barrier years) will exert a stronger grazing pressure on the phytoplankton. During summer phytoplankton biomass was much smaller than previously (Bakker & Vink, 1994) indeed. Beside the role of zooplankton, the increasing nutrient limitation (especially of nitrogen) can be held responsible for the phytoplankton decline in summer (Bakker & Vink, 1994).

Tackx *et al.* (1994) calculated phyto- and zooplankton biomass distributions in E during the entire period of observation. They used these data to test the model of Sheldon *et al.* (1977; predicting biomasses at higher from those at lower trophic levels) and showed that observed standing stock ratios of zooplankton to phytoplankton agreed well with the model predictions in pre-barrier- and barrier periods. In 1988, however, the model predictions became invalid because of important changes in the zooplankton, *i.e.* the increased biomass of mainly *Temora* and *Balanus* nauplii.

We concluded that the new hydrographic regime in the Oosterschelde (via reduced current velocities, increased sinking of suspended matter and increased transparencies) influenced phytoplankton composition (Bakker *et al.*, 1990; 1994), phytoplankton biomass (Bakker & Vink, 1994) and -primary production (Wetsteyn & Kromkamp, 1994) directly. Correspondingly, we hypothesize that increased retention time, increased phytoplankton biomass and improved food quality of the seston have resulted in an increase of zooplankton biomass from 1988 onwards. These phenomena were most clearly observed in E. In this area also some changes in species composition were observed (Bakker, 1994). Future research in the Oosterschelde will reveal if this new trends of the zooplankton development will persist.

Acknowledgements

We are grateful to Drs Michèle L. M. Tackx and Lambertus P. M. J. Wetsteyn, as well as to the referees J. A. Lindley and (notably) Bouwe R. Kuipers for critical remarks greatly improving the manuscript. We thank Mr Alfons A. Bolsius who perfected the original computer figures and Mrs Elly van Hulsteijn who carefully accomplished successive P.C. versions of the text.

Communication No. 693 of NIOO-CEMO, Yerseke, The Netherlands

References

Bakker, C. & P. van Rijswijk, 1987. Development time and growth rate of the marine calanoid copepod *Temora longicornis* as related to food conditions in the Oosterschelde estuary (Southern North Sea). Neth. J. Sea Res. 21: 125–141.

Bakker, C., P. M. J. Herman & M. Vink, 1990. Changes in seasonal succession of phytoplankton induced by the storm-surge barrier in the Oosterschelde (SW-Netherlands). J. Plankton Res. 12: 947–972.

Bakker, C., P. M. J. Herman & M. Vink, 1994. A new trend in the development of the phytoplankton in the Oosterschelde (SW Netherlands) during and after the construction of a storm-surge barrier. Hydrobiologia 282/283: 79–100.

Bakker, C. & M. Vink, 1994. Nutrient concentrations and planktonic diatom-flagellate relations in the Oosterschelde (SW Netherlands) during and after the construction of a storm-surge barrier. Hydrobiologia 282/283: 101–116.

Bakker, C., 1994. Zooplankton species composition in the Oosterschelde (SW Netherlands) before, during and after the construction of a storm-surge barrier. Hydrobiologia 282/283: 117–126.

Cloern, J. E., 1982. Does the benthos control phytoplankton biomass in South San Francisco Bay? Mar. Ecol. Progr. Ser 9: 191–202.

Colebrook, J. M., 1982. Continuous plankton records: seasonal variations in the distribution and abundance of plankton in the North Atlantic Ocean and the North Sea. J. Plankton Res. 4: 435–462.

Colebrook, J. M., 1985. Continuous plankton records: overwintering and annual fluctuations in the abundance of zooplankton. Mar. Biol. 84: 261–265.

Daan, R., 1989. Predation and cannibalism as regulating processes in North Sea copepod populations. Ph. D. Thesis Univ. Amsterdam, 62 pp.

Darboe, F. S., 1992. Comparison of the *Temora longicornis* (Copepoda Calanoida) population in pre- and post-barrier Oosterschelde, the Netherlands. M. Sc. Thesis Free Univ. Brussels, 43 pp.

Deason, E. E. & T. J. Smayda, 1982. Ctenophore-zooplankton-phytoplankton interactions in Narragansett Bay, Rhode Island, USA, during 1972–1977. J. Plankton Res 4: 203–217.

De Wolf, P., 1989. The price of patchiness. Helgoländer Meeresunters. 43: 263–273.

Dodson, S., 1990. Predicting diel vertical migration of zooplankton. Limnol. Oceanogr. 35: 1195–1200.

Frank, K. T., 1986. Ecological significance of the ctenophore *Pleurobrachia pileus* off southwestern Nova Scotia. Can. J. Fish. aquat. Sci. 43: 211–222.

Fransz, H. G., S. R. Gonzalez & W. C. M. Klein Breteler, 1989. Fecundity as a factor controlling the seasonal population cycle in *Temora longicornis* (Copepoda, Calanoida). In J. S. Ryland & P. A. Tyler (eds), Reproduction, genetics and distributions of marine organisms. Olsen & Olsen: 83–90.

Fransz, H. G., J. M. Colebrook, J. C. Gamble & M. Krause, 1991. The zooplankton of the North Sea. N. J. Sea Res. 28: 1–52.

Fraser, J. H., 1970. The ecology of the ctenophore *Pleurobrachia pileus* in Scottish waters. J. Cons. Int. Explor. Mer 33: 149–168.

Greve, W., 1971. Ökologische Untersuchungen an *Pleurobrachia pileus* I. Freilanduntersuchungen. Helgoländer wiss. Meeresunters. 22: 303–325.

Herman, P. M. J. & H. Scholten, 1990. Can suspension-feeders stabilise estuarine ecosystems? In M. Barnes & R. N. Gibson (eds.). Trophic Relationships in the Marine Environment, Aberdeen University press: 104–116.

King, K. R., J. T. Hollibaugh & F. Azam, 1980. Predation-prey interactions, between the larvacean *Oikopleura dioica* and bacterioplankton in enclosed water columns. Mar. Biol. 56: 49–57.

Klepper, O., 1989. A model of carbon flows in relation to macrobenthic food supply in the Oosterschelde estuary (S.W. Netherlands). Thesis, Wageningen.

Kremer, P., 1976. Population dynamics and ecological energetics of a pulsed zooplankton predator, the ctenophore *Mnemiopsis leidyi*. In M. Wiley (ed.), Estuarine processes I. Uses, stresses and adaptation to the estuary. New York Academic Press: 197–215.

Lindley, J. A., 1986. Dormant eggs of calanoid copepods in sea-bed sediments of the English channel and southern North Sea. J. Plankton Res. 8: 399–400.

Lindley, J. A., 1990. Distribution of overwintering calanoid copepod eggs in sea-bed sediments around Southern Britain. Mar. Biol. 104: 209–217.

Miller, R. J. & R. Daan, 1989. Planktonic predators and copepod abundance near the Dutch coast. J. Plankton Res. 11: 263–282.

Nienhuis, P. H. & A. C. Smaal, 1994. The Oosterschelde estuary, a case study of a changing ecosystem: an introduction. Hydrobiologia 282/283: 1–14.

Prins, Th. & A. Smaal, 1994. The role of the blue mussel *Mytilus edulis* in the cycling of nutrients in the Oosterschelde estuary (The Netherlands). Hydrobiologia 282/283: 413–429.

Reeve, M. R. & M. A. Walter, 1978. Nutritional ecology of ctenophores – a review of recent research. Adv. Mar. Biol. 15: 249–287.

Sheldon, R. W., W. H. Sutcliffe & M. A. Paranjape, 1977. Structure of pelagic food chain and relationship between plankton and fish production. J. Fish. Res. Bd Can. 34: 2344–2353.

Smaal, A. C. & P. H. Nienhuis, 1992. The Eastern Scheldt (The Netherlands), from an estuary to a tidal bay: a review of responses at the ecosystem level. Neth. J. Sea Res. 30: 161–173.

Steedman, H. F., 1976 (ed.). Zooplankton fixation and preservation. The Unesco Press, Paris.

Stoecker, D. K. & D. A. Egloff, 1987. Predation by *Acartia tonsa* Dana on planktonic ciliates and rotifers. J. exp. mar. Biol. Ecol. 110: 53–68.

Suthers, I. M. & K. I. Frank, 1990. Zooplankton biomass gradient off south-western Nova Scotia: nearshore ctenophore predation or hydrographic separation? J. Plankton Res. 12: 831–850.

Tackx, M. L. M., C. Bakker, J. W. Francke & M. Vink, 1989. Size- and phytoplankton selection by Oosterschelde zooplankton. Neth. J. Sea Res. 23: 35–43.

Tackx, M. L. M., C. Bakker & P. van Rijswijk, 1990. Zooplankton grazing pressure in the Oosterschelde (The Netherlands). Neth. J. Sea Res. 25: 405–415.

Tackx, M. L. M., J. W. Francke, P. Van Rijswijk, M. Vink & J. Rijk, 1991. Size distributions and trophic relationships of the pelagic ecosystem in the Oosterschelde (S.W. Netherlands). Hydrobiol. Bull. 25: 9–14.

Tackx, M. L. M., P. M. J. Herman, P. Van Rijswijk, M. Vink & C. Bakker, 1994. Plankton size distributions and trophic relations before and after the construction of the storm-surge barrier in the Oosterschelde estuary. Hydrobiologia 282/283: 145–152.

Van Rijswijk, P., C. Bakker & M. Vink, 1989. Daily fecundity of *Temora longicornis* (Copepoda, Calanoida) in the Oosterschelde estuary (SW-Netherlands). Neth. J. Sea Res. 23: 293–303.

Van Stralen, M. R. & R. Dijkema, 1994. Mussel culture in a changing environment: the effects of a coastal engineering project on mussel culture (*Mytilus edulis*) in the Oosterschelde (SW Netherlands). Hydrobiologia 282/283: 359–379.

Van der Tol, M. W. M. & H. Scholten, 1992. Response of the Eastern Scheldt ecosystem to a changing environment: functional or adaptive? Neth. J. Sea Res. 30: 175–190.

Van der Veer, H. W. & C. F. M. Sadée, 1984. Seasonal occurrence of the ctenophore *Pleurobrachia pileus* in the western Dutch Wadden Sea. Mar. Biol. 79: 219–227.

Vegter, F. & P. R. M. De Visscher, 1987. Nutrients and phytoplankton primary production in the marine tidal Oosterschelde estuary (The Netherlands). Hydrobiol. Bull. 21: 149–158.

Vroon, J., 1994. Hydrodynamic characteristics of the Oosterschelde in recent decades. Hydrobiologia 282/283: 17–27.

Wetsteyn, L. P. M. J., J. C. H. Peeters, R. N. M. Duin, F. Vegter & P. R. M. de Visscher, 1990. Phytoplankton primary production and nutrients in the Oosterschelde (The Netherlands) during the pre-barrier period 1980–1984. Hydrobiologia 195: 163–177.

Wetsteyn, P. M. J. & C. Bakker, 1991. Abiotic characteristics and phytoplankton primary production in relation to a large-scale coastal engineering project in the Oosterschelde (The Netherlands): a preliminary evaluation. In: M. Elliott & J. P. Ducrotoy (eds.). Estuaries and coasts: spatial and temporal inter-comparisons. Proc. ECSA 19 Symp., 1989. Olsen & Olsen, p. 365–373.

Wetsteyn, L. P. M. J. & J. Kromkamp, 1994. Turbidity, nutrients and phytoplankton primary production in the Oosterschelde (The Netherlands) before, during and after a large-scale coastal engineering project (1980–1990). Hydrobiologia 282/283: 61–78.

Hydrobiologia **282/283**: 145–152, 1994.
P. H. Nienhuis & A. C. Smaal (eds), The Oosterschelde Estuary.
© 1994 *Kluwer Academic Publishers. Printed in Belgium.*

Plankton size distributions and trophic relations before and after the construction of the storm-surge barrier in the Oosterschelde estuary

M. L. M. Tackx[1], P. M. J. Herman[2], P. van Rijswijk[2], M. Vink[2] & C. Bakker[2]
[1] *Ecology Laboratory, Free University of Brussels, Pleinlaan 2, 1050 Brussels Belgium;*
[2] *Netherlands Institute of Ecological Research, Vierstraat 28, 4401 EA Yerseke, The Netherlands*

Key words: plankton, size distribution, Sheldon, storm-surge barrier

Abstract

Phytoplankton and zooplankton biomass distributions were calculated on a carbon basis for the inland part of the Oosterschelde, in the period before (1983), during (1984, 1986) and after (1987, 1988) the construction of the storm- surge barrier. In all years studied, both phytoplankton and zooplankton distributions are very irregular, and little consistent patterns emerge. The data were used to test the model of Sheldon *et al.* (1977). The observed standing stock ratios of zooplankton to phytoplankton agree with the model predictions in 1983, and are slightly higher during the period 1984–1987. In 1988, the model predictions are very different from the observed values, because of important changes in the zooplankton species abundance occurring in this year.

Introduction

The observation of Sheldon *et al.* (1972) that particulate material (in casu plankton) in oceanic systems is roughly equally distributed over logarithmic size intervals, has led to a number of models which relate plankton biomass in one size class to that in another (Kerr, 1974; Sheldon *et al.*, 1977; Borgmann, 1982; Platt & Denman, 1978). These models provide a tool to predict fish stocks from biomass data on lower trophic levels (Sheldon *et al.*, 1977; Moloney & Field, 1985), to compare the structure of pelagic ecosystems (Sprules & Munamar, 1986) and to study the influence of perturbations on planktonic biomass spectra (Minns *et al.*, 1987).

This study uses the size spectrum approach to compare planktonic biomass size distributions and trophic interactions between mesozooplankton and phytoplankton in the period before

(1983), during (1984, 1986) and after (1987, 1988) the construction of the storm- surge barrier in the Oosterschelde.

Materials and methods

From 1983 until 1988, phyto- and zooplankton samples were sampled at the station LGPK in the inland part of the Oosterschelde at weekly intervals during the growing season (April–September), and at biweekly intervals during winter. For zooplankton sampling, 100 l of water was pumped with a Pleuger submersible pump, having a capacity of 200 l min^{-1} and filtered through a 63 μm net. During 1983–1984, all samples were taken at mid depth, from 1984 until 1988, samples were taken at 2.5, 7.5, 12.5, 17.5 and 22 m. Samples taken at all depths were pooled and a subsample corresponding to an original volume of 100 l was taken for counting under binocular microscope. For phytoplankton sampling, a 5 l sample was taken with a Hydrobios bottle at mid depth. A 1 liter subsample was taken and preserved with lugol's solution. In the laboratory, the samples were decanted to a final volume of 100 ml and pooled. A subsample of a few ml was analyzed by inverted microscope for species composition and abundance, and size measurements were taken of all species. Zooplankton samples were analyzed for species and developmental stage composition and abundance under a binocular microscope. Individual dry weight of the various developmental stages of the dominant species were determined using an Cahn electrobalance. The reader is referred to Bakker, 1994, and Bakker *et al.*, 1994 for details on the sampling and sample analysis procedures.

Phytoplankton cell volume was calculated for each species from the microscopic size measurements using the most appropriate geometric form. Cell carbon content was calculated using the formula of Eppley (in Smayda, 1978). Carbon mass of each species and developmental stage was calculated as 45% of the individual dry mass.

Based on its carbon content, each planktonic species was placed in a given size class, in a series of size classes ranging from 0.45 to 15, 100, 000 pg C. For each size class in this series, the lower boundary is double the previous one. Total carbon concentration (in mg C m^{-3}) of each species in each sample was calculated by multiplying its individual volume by its numerical abundance. Total carbon concentration in each size class was calculated as the sum of the carbon concentrations of the species placed in this size class. Mean monthly distributions of plankton carbon concentrations in the different size classes were first calculated from the distributions obtained at each sampling date. Finally, mean annual distributions were calculated using these monthly distributions. Standing stock values of phytoplankton and zooplankton were calculated by summing the mean average biomass values (expressed in carbon) over all the size classes of the distribution in which the component occurred.

The feeding activity of the copepod species

Acartia spp., *Temora longicornis*, *Centropages hamatus* and the cirriped *Balanus* spp. on particulate matter in the 3–100 μm size range has been quantified in the pre-barrier period using the Coulter application of the counting method (Fuller & Clarcke, 1936). These organisms (called zooplankton in the following) strongly dominate the zooplankton in the Oosterschelde, and exert selective feeding on phytoplankton (Tackx *et al.*, 1989, 1990; Bakker *et al.*, 1991). The 1983 dataset was used to calculate monthly averaged daily rations (on a carbon basis) for the various zooplankton and developmental stages because this year provides the most complete information (Tackx, 1987). To calculate the zooplankton grazing activity in other years, these 1983 daily ration values were applied to the monthly averaged biomass values of the various species and developmental stages measured in each year. It was assumed that the zooplankton grazing activity during winter (October–March) is negligible.

Mean annual primary production values were obtained from experimental measurements carried out in the inland part of the Oosterschelde, using the ^{14}C and O_2 method (Wetsteyn and Kromkamp, 1994).

The obtained standing stock values for phytoplankton and zooplankton and the predation pressure values were used to test the model of Sheldon *et al.* (1977):

$$Sc/Sp = (Dc/Dp)^{0.72} \, Ge \cdot Ce , \qquad (1)$$

in which: Sc = standing stock of the predator;
Sp = standing stock of the prey;
Dc = spheric equivalent diameter of the predator;
Dp = spheric equivalent diameter of the prey;
Ge = growth efficiency (= quantity of predator tissue formed to the amount of prey ingested);
Ce = predation efficiency (= proportion of the prey production taken by the predator).

Ce values were calculated by dividing the mean annual ingestion rate of the total zooplankton population by the mean annual primary production (all in terms of mg C m^{-3} d^{-1}).

The Dc/Dp ratio was calculated as the ratio of the median size of the mean annual zooplankton and the phytoplankton standing stocks. For the zooplankton, the class corresponding to the medium carbon biomass was converted to fresh weight, by multiplying with a factor 5/0.45. The corresponding body volume was calculated assuming a density of 1 g cm^{-3} and the corresponding Spheric Equivalent Diameter (S.E.D.) was deduced from this body volume.

For phytoplankton, the biomass class, in which 50% of the cumulative biomass (expressed in carbon) was reached, was converted to volume, using the formula of Eppley (in Smayda, 1978). This results in two estimates of cell sizes, one for diatoms and one for non-diatom species. These two values were used to obtain a minimum and a maximum estimate of Dp in the right part of equation (1).

Results

Figure 1 a–e show the biomass distributions obtained. In all years, the phytoplankton, and zooplankton show a jagged distribution over the various classes in which they occur. The resulting total plankton distributions are also irregular. Except for a consistent trough in classes 18 and 19 no explicit features can be recognised. In 1984, a phytoplankton peak is observed in class 12. From 1984 till 1987, peaks in phytoplankton biomass occur irregularly spread over classes 10–17. Phytoplankton and zooplankton standing stocks and the S.E.D. values corresponding to the classes in which the median biomass is situated are given in Table 1. Phytoplankton biomass varies between 129 and 212 mg C m^{-3} without showing any specific trends with time. Zooplankton biomass varies between 16 and 35 mg C m^{-3}, the minimum value being observed in 1986, the maximum in 1988. The range of the median S.E.D. of the phytoplankton lies at 23–35 μm in 1983, 1984 and 1987. In 1986 it decreases to 18–26 μm, and in 1988 to 14–19 μm. The median

S.E.D. of the zooplankton lies at 272 μm in all years, except 1988, when it is situated at 342 μm.

Table 1 also lists the zooplankton grazing activity calculated for each year, in mg C m^{-3} d^{-1}, and expressed as percentage of the primary production (= Ce), the minimum and maximum estimates of the right part of equation (1) and the observed values of Sc/Sp. The observed Sc/Sp values fall within the range calculated by the model for 1983. From 1984 till 1987, the observed values fall slightly below (1984, 1986) or exactly on the minimum value calculated by the model (1987). In 1988, the Sc/Sp value of 0.26 is considerably lower than the minimum value of 0.58 calculated by the model.

Discussion

In a previous paper, Tackx et al. (1991) examined plankton and total particulate matter distributions in the inner part of the Oosterschelde in 1983 on a volume basis. It was shown that non of the components considered (phytoplankton, zooplankton and total particulate matter) is distributed equally over the size ranges in which they occur. The data presented here for 1983 show that, on a carbon basis, phytoplankton and zooplankton are irregularly distributed over the range in which they occur and that zooplankton standing stock is roughly 6 times lower than phytoplankton standing stock. The irregular shape of the phytoplankton and zooplankton distribution is found throughout the years of investigation. The occurrence of a phytoplankton peak in classes 10–17 in 1984–1987 corresponds to the changes in phytoplankton composition and abundance reported by Bakker et al., 1990 (see below). In 1986, the presence of this peak results in a decrease in median S.E.D. of the phytoplankton standing stock in comparison with the previous years and with 1987. The phytoplankton peak occurring in class 5 in 1988 corresponds to the increase in flagellates reported by Bakker et al., 1994 to be typical for the period after the storm surge barrier construction. In 1988, this abundance of flagellate species led to a decrease of the

148

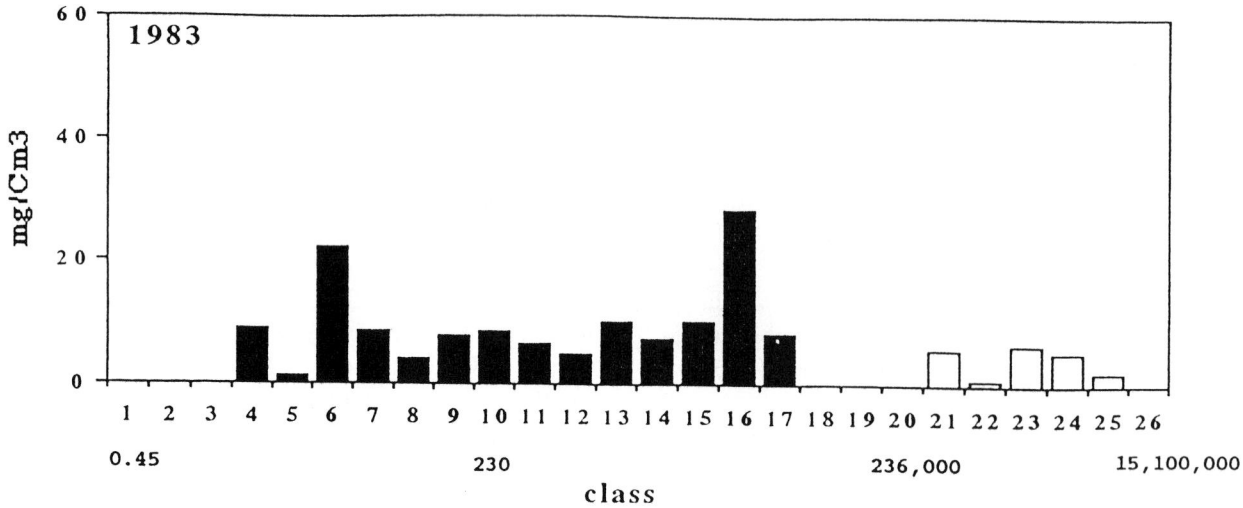

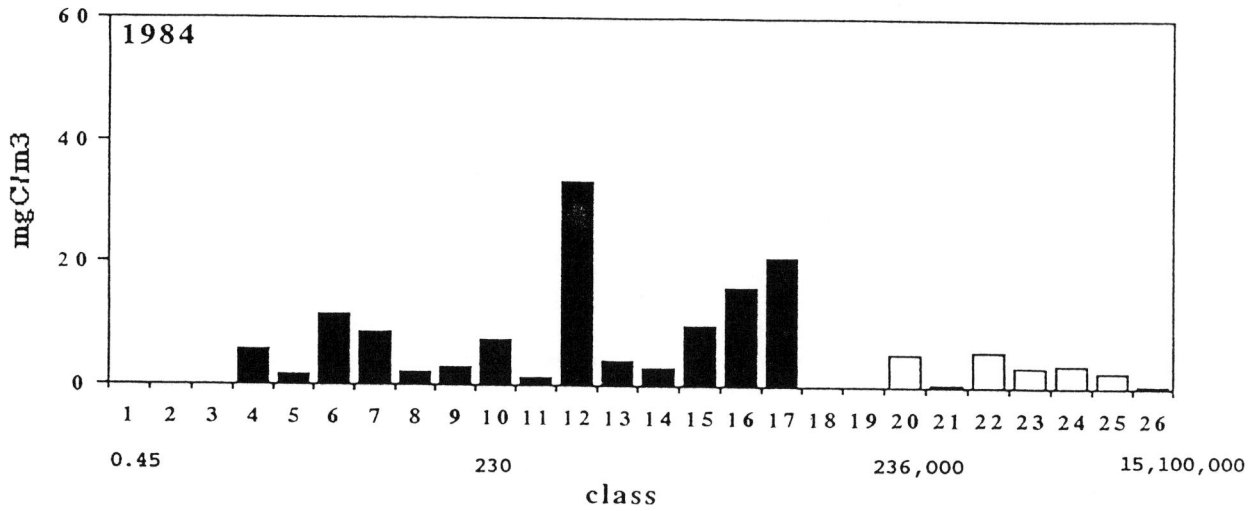

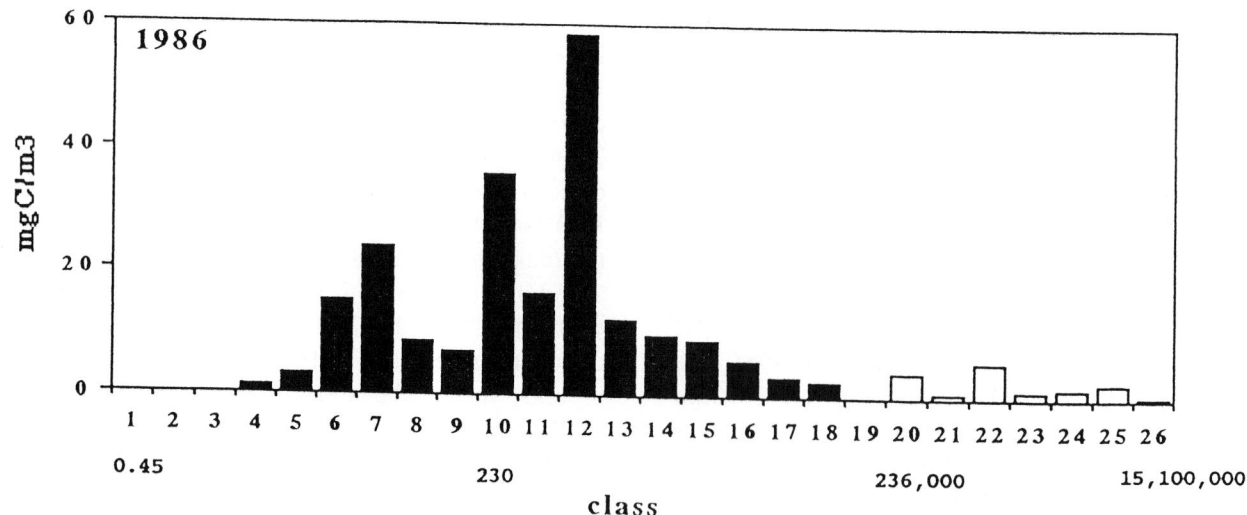

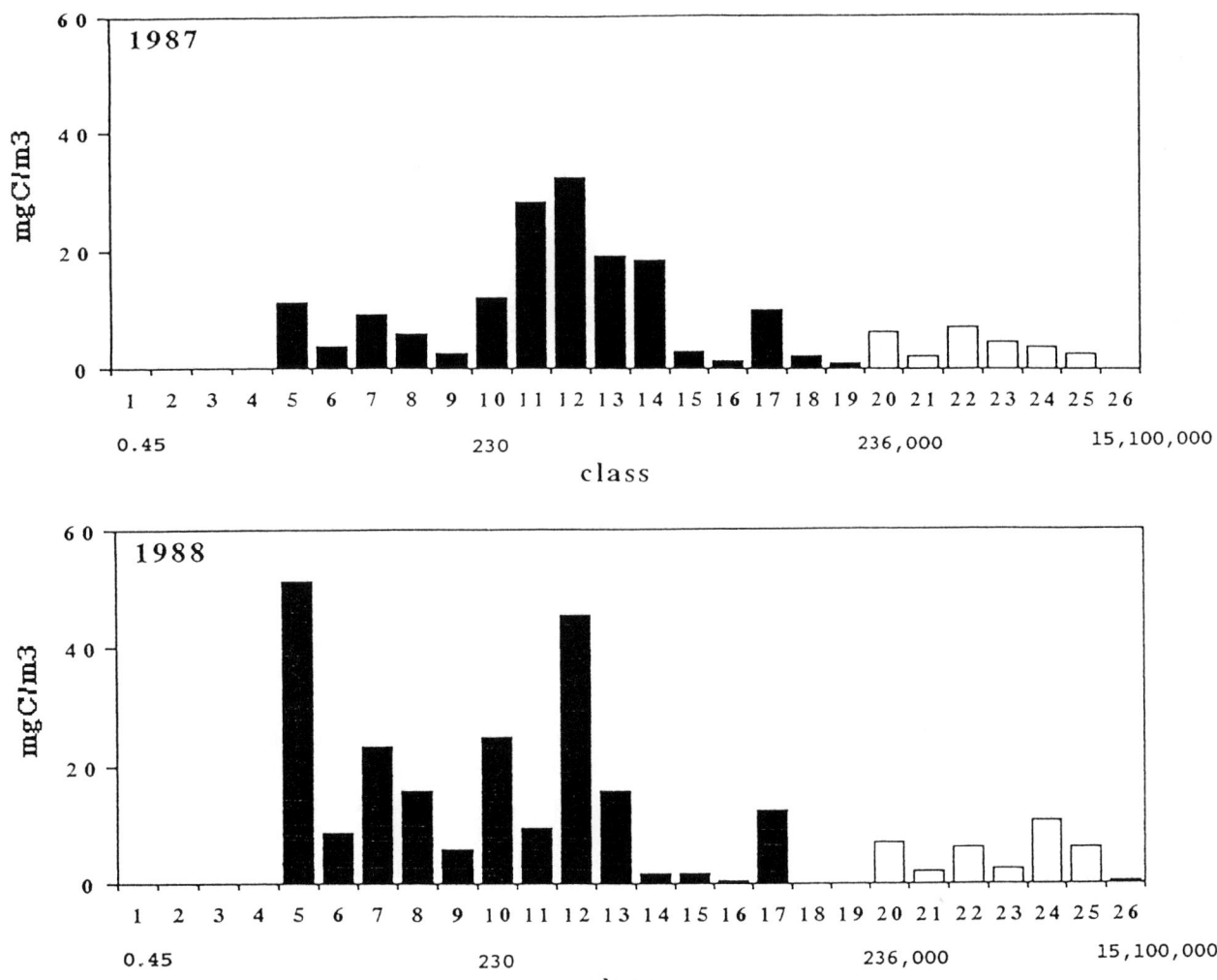

Fig. 1. Size distribution of phytoplankton (black) and zooplankton (white).

median S.E.D. of the phytoplankton population to 14–19 μm.

Considering mean annual biomass values of the various plankton components from 1983 till 1987, no differences or trends were observed between the period before, during or after the construction of the storm-surge barrier. In 1988, phytoplankton biomass stays in the same range as observed in previous years, while mesozooplankton biomass increases. The increase in zooplankton biomass is mainly caused by an increasing development of the calanoid copepod *Temora longicornis*, and also of nauplii of the cirriped *Balanus* spp. (Bakker & Van Rijswijk, 1994).

Tackx *et al.* (1991) have also tested the equation of Sheldon *et al.* (1977) for the 1984 data, on a volume basis. They calculate an *Sc/Sp* ratio of 0.08, which is just below the lower value of the 0.09–0.14 range they calculate for the right part of equation (1).

The present data show that, on a carbon basis, the observed ratio of zooplankton to phytoplankton standing stocks falls within the range of values calculated by the model of Sheldon *et al.* (1977) for 1983. From 1984 till 1987, observed zooplankton biomass to phytoplankton biomass ratios are generally somewhat lower than the model prediction. A substantial difference be-

Table 1. Standing stocks of phytoplankton (*Sp*), zooplankton (*Sc*) in the inland part of the Oosterschelde during the years studied. Maximum (*Dc* max) and minimum (*Dc* min) values of the spheric equivalent diameter of the phytoplankton and S.E.D. of the zooplankton (*Dp*). Zooplankton grazing activity, in mg C m^{-3} d^{-1} and as percentage of primary production (*Ce*). Minimum and maximum estimate of the right part of equation (1), using a *Ge* value of 0.15. Observed ratio standing stock zooplankton to phytoplankton (*Sc/Sp*). Between brackets: range of *Sc/Sp* calculated from odd and even data. All values are annual means.

Year	*Sp* mg C m^{-3}	*Sc* mg C m^{-3}	*Dc* min µm	*Dc* max µm	*Dp* µm	Zooplankton grazing mg C m^{-3} d^{-1}	*Ce*	$\left(\dfrac{Dc}{Dp}\right)^{0.72} \cdot Ge \cdot Ce$ min	max	*Sc/Sp* (range)
1983	143	22	23	35	272	20	0.18	0.12	0.16	0.15 (0.10–0.22)
1984	129	21	23	35	272	16	0.26	0.18	0.24	0.16 (0.13–0.20)
1986	212	16	18	26	272	10	0.12	0.10	0.13	0.08 (0.07–0.08)
1987	158	27	23	35	272	17	0.27	0.17	0.24	0.17 (0.16–0.17)
1988	131	35	14	19	342	31	0.49	0.58	0.74	0.26 (0.25–0.28)

tween the observed ratio and the model prediction occurs only in 1988.

As only one sampling station was used in the study, no estimate of the spatial variability of the phytoplankton and zooplankton abundances can be given. To evaluate the variability due to timing of the sampling, the annual mean values were calculated with odd and even data separately. The resulting range in *Sc/Sp* values is consistent with the trends in correspondence between model and observations described above (Table 1).

The discrepancy between the observed and the predicted zooplankton to phytoplankton biomass ratio in 1988 reflects the occurring change in phytoplankton- and zooplankton species abundance:

– *T. longicornis*, being larger than the other dominant calanoid species in the area, increased its abundance in 1988, which caused an increase of the medium size of the zooplankton standing stock as compared to the previous years. Combined with the observed decrease in the medium size of the phytoplankton standing stock caused by the increase of flagellate species, this results in an increase of the *Sc/Sp* ratio.

– Daily ration of *Balanus* spp. nauplii, as calculated from the grazing experiments performed in 1983, is very high (314% d^{-1}) in May. The high daily ration calculated in May is mainly due to the fact that in spring most phytoplanktonspe-cies occurring are non-diatoms, or diatoms with a S.E.D. between 20–30 µm. These algae have a higher carbon content per unit cell volume as the larger sized diatoms which dominate the phytoplankton later in the growing season. In 1983–1987, these high daily rations obtained by *Balanus* spp. nauplii and, to a smaller extent, by *T. longicornis*, do not represent a high total grazing activity, because the number of individuals in this month is limited. In 1988, the important increase in number of *Balanus* spp. nauplii and *T. longicornis* individuals results in a very important grazing activity. The resulting *Ce* value is 0.49, whereas this value is between 0.12 and 0.27 in the previous years. Combination with the increased *Dc/Dp* ratio, results in a high value (0.58–0.70) of the right part of equation (1). Although an increase in the zooplankton standing stock is observed in 1988, it does not match the values calculated by the model. A possible explanation for the overestimation made by the model is that *Balanus* spp. nauplii may not be completely herbivorous (Tackx *et al.*, 1989). If *Balanus* spp. feed to a considerable degree on detritus, the calculated grazing pressure is overestimated to a considerable degree in 1988, and to a smaller degree in the previous years.

The fit of the observations to the 1983 model predictions shows that the inland part of the Oosterschelde, before the construction of the storm-surge barrier, is consistent with the steady state implied by the model of Sheldon *et al.*, (1977).

Discrepancy between the model predictions and the observed zooplankton to phytoplankton biomass ratio could be a reflection of a deviation from the basic assumptions of the model taking place since the construction of the storm-surge barrier. The model of Sheldon *et al.* (1977) is based on the fact that log P/B of marine organisms decreases with body size: the smaller organisms living in an ecosystem have a higher production per unit biomass than the bigger ones. This relationship holds for organisms which are found living in a given ecosystem, with all populations adapted to the circumstances prevailing in this system. The construction of the storm-surge barrier in the Oosterschelde has led to changes in environmental conditions. Changes of importance to the phytoplankton were a more favourable light climate on the one hand and a decreased nutrient concentration on the other. As a result, flagellates and small diatom species have increased in abundance (Bakker *et al.*, 1994). In spite of the decrease in median S.E.D. of the phytoplankton population in 1988, carbon specific growth rates in the inland part of the Oosterschelde have remained in the same range as in the pre-barrier period (Wetsteyn & Kromkamp, 1994). From 1984 till 1987, changes taking place in phytoplankton species composition and abundance did not seem to result in important changes in the zooplankton species composition and abundance. Only a moderate increase in *T. longicornis* abundance is observed. *T. longicornis* is a herbivorous species, known to feed mainly on phytoplankton species $> 20 \, \mu m$ S.E.D. (Tackx *et al.*, 1989; Van Rijswijk *et al.*, 1989). Its increase could be a consequence of the increase in phytoplankton to detritus ratio (Bakker, 1994). In 1988, the shift towards small phytoplankton species became more pronounced. Simultaneously, a strong increase in abundance of *T. longicornis* and especially of *Balanus* spp. nauplii was observed. It is at present difficult to evaluate in how far the 1988 development in the zooplankton community is a direct response to the changes taking place in the phytoplankton community. The fact that *Balanus* sp. nauplii reach such high abundances in 1988 could be a result of improved feeding conditions for this organisms. If the copepod population is not well adapted to feed efficiently on the phytoplankton population in the present day Oosterschelde, more phytoplankton would be left available for *Balanus* nauplii. These organisms are less capable of phytoplankton selection (Tackx *et al.*, 1989), but could profit from the abundance of small phytoplankton cells which are not, or little eaten by copepods. Grazing pressure by benthic populations might also have decreased as a consequence of the decrease in phytoplankton size. This would add to the above described favourable conditions for *Balanus* nauplii.

The present dataset does not allow to evaluate in how far the 1988 situation is a temporary disturbance of the system, or the onset of a new equilibrium. Zooplankton samples, collected in 1989 and 1992, are presently being analyzed in order to evaluate the response of the zooplankton community on a longer term basis. In addition to these species abundance study, grazing experiments carried out with dominant zooplankton organisms on the natural particulate matter in the new occuring situations would be necessary to solidify trophic explanations of occuring trends in plankton composition.

Acknowledgements

The authors are indebted to J. Kromkamp and L. P. M. J. Wetsteyn for supplying primary production data. N. Daan and an unknown referee gave valuable criticism on this paper. C. Van Ongevalle helped with the typing.

Communication No. 694 of NIOO-CEMO, Yerseke, The Netherlands.

References

Bakker, C., P. J. M. Herman & M. Vink, 1990. Changes in seasonal succession of phytoplankton induced by the storm-surge barrier in the Oosterschelde (S.W. Netherlands). J. Plankton Res. 12: 947–972.

Bakker, C. & P. Van Rijswijk, 1994. Zooplankton biomass in the Oosterschelde (SW Netherlands) before, during and after the construction of a storm-surge barrier. Hydrobiologia 282/283: 127–143.

Bakker C., P. J. M. Herman & M. Vink, 1994. A new trend in the development of the phytoplankton in the Oosterschelde (SW Netherlands) during and after the construction of a storm-surge barrier. Hydrobiologia 282/283: 79–100.

Borgmann, U., 1982. Particle-size conversion efficiency and total animal production in pelagic ecosystems. Can. J. Fish. aquat. Sci. 40: 2010– 2018.

Fuller, J. C. & G. L. Clarke, 1936. Further experiments on the feeding of *Calanus finmarchicus*. Biol. Bull. Woodshole: 308–320.

Kerr, S. R., 1974. Theory of size distributions in ecological communities. J. Fish. Res. Bd Can. 31: 1859–1862.

Minns, E. S., E. S. Millard, J. M. Cooley, M. G. Johns, D. A. Hurley, K. H. Nicholls, G. W. Robinson, G. E. Owen & A. Crowder, 1987. Production and biomass size spectra in the Bay of Quinte, a eutrophic ecosystem. Can. J. Fish. aquat. Sci. 44: 148–155.

Platt, T. & K. Denman, 1978. The structure of pelagic marine ecosystems. Rapp. P.-V. Reun. Cons. Int. explor. Mer 173: 60–65.

Sheldon, R. W., A. Prakash & W. H. Sutcliffe, 1972. The size distribution of particles in the ocean. Limnol. Ocean. 17: 327–340.

Sheldon, R. W., W. H. Sutcliffe & M. A. Paranjape, 1977. Structure of pelagic food chain and relationship between plankton and fish production. J. Fish. Res. Bd Can., 34: 2344–2353.

Sprules, W. G., J. M. Casselman & B. J. Shuter, 1983. Size distribution of pelagic particles in lakes. Can. J. Fish. aquat. Sci. 40: 1761–1769.

Sprules, W. G. & M. Munamar, 1986. Plankton size spectra in relation to ecosystem productivity, size and perturbation. Can. J. Fish. aquat. Sci. 43: 1789–1794.

Sprules, W. G., S. B. Brandt, D. J. Stewart, M. Munamar, E. H. Jin & J. Love, 1991. Biomass size spectrum of the Lake Michigan pelagic food web. Can. J. Fish. aquat. Sci. 48: 105–115.

Tackx, M. L. M., C. Bakker, J. W. Francke & M. Vink, 1989. Size and phytoplankton selection by Oosterschelde zooplankton. Neth. J. Sea Res. 23: 35–43.

Tackx, M. L. M., C. Bakker & P. Van Rijswijk, 1990. Zooplankton grazing pressure in the Oosterschelde (The Netherlands). Neth. J. Sea Res. 25: 405–415.

Tackx, M. L. M., J. W. Francke, P. Van Rijswijk, M. Vink & J. Rijk, 1991. Size distributions and trophic relationship of the pelagic ecosystem in the Oosterschelde (S.W. Netherlands). Hydrobiol. Bull. 25: 9–14.

Van Rijswijk, P., C. Bakker & M. Vink, 1989. Daily fecundity of *Temora longicornis* (Copepoda, Calanoida) in the Oosterschelde estuary (S.W. Netherlands). Neth. J. Sea Res. 23: 293–303.

Wetsteyn, L. P. M. J. & J. C. Kromkamp, 1994. Turbidity, nutrients and phytoplankton primary production in the Oosterschelde (The Netherlands) before, during and after a large scale coastal engineering project (1980–1990). Hydrobiologa 282/283: 61–78.

Hydrobiologia **282/283**: 153–156, 1994.
P. H. Nienhuis & A. C. Smaal (eds), The Oosterschelde Estuary.
© 1994 *Kluwer Academic Publishers. Printed in Belgium.*

Theme III:
The structure of the benthic system

Patrick M. Meire
Institute of Nature Conservation, Ministry of the Flemish Community, Kiewitdreef 5, B3500 Hasselt, Belgium

The benthic compartment forms a very important link within the estuarine ecosystem. It depends to a large extent on the import of food (plankton or detritus) from the pelagic system, and it serves as food for many predators, of which fish and birds are best known. However, both the primary production within the benthic compartment by macro- and microphytobenthos as well as the consumption by infauna predators may also be important energy pathways.

This section deals mainly with some structural aspects of the benthic system and possible effects caused by the changed environmental conditions in the Oosterschelde after the construction of a storm-surge barrier and secondary dams.

Within the benthic system of the Oosterschelde a distinction must be made between two very different benthic habitats: hard and soft bottoms.

The seawalls and many other constructions provide an artificial area of hard substrate, estimated at about 1150 ha in the Oosterschelde (Smaal & Boeije, 1991). The benthic communities on these hard substrates in the Oosterschelde estuary were known to be very rich and are covered in this section by the papers of De Kluijver & Leewis (1994) and Meijer & Waardenburg (1994) for the sublittoral and littoral zone respectively. The results between both studies differ however strikingly. In the littoral zone the environmental changes did not seem to have affected the communities strongly. The reduction in tidal amplitude had a minor effect on the zonation of communities. Especially in the upper part of the

intertidal zone, a vertical shift downward by about 15–30 cm of the communities was observed, corresponding to the 22 cm decrease of the high-water line. Species composition and richness did not really change. Sublittoral, however, community structure changed rapidly according to changes in current velocities and transparency of the water. Some faunal elements disappeared, while species characteristic for the tideless Lake Grevelingen became dominant, indicating the reduced hydrodynamical impact.

The sedimentation of fine sediments, due to the lower current velocities and erosion of the tidal flats, affected the hard substrate communities in both the lower part of the littoral and sublittoral zone.

The ecology of soft sediment benthos in the Oosterschelde was studied in more detail than the hard substrate fauna. Species composition of microphytobenthos (consisting mainly of diatoms) was studied but not included in this book (see e.g. Vos, 1989) but De Jong *et al.* (1994) show clearly that the biomass of microphytobenthos increased substantially after completion of the storm-surge barrier compared to the pre-barrier period. Average annual biomass has increased by about 70%; the same increase has been calculated for primary production. One third of the primary production of the Oosterschelde comes now from microphytobenthos. The changes are ascribed to a general decrease in the dynamic forces on the intertidal flats (lower current velocities), increased water transparency and inorganic carbon flow from the open water to the shoals. This increase in biomass did not seem, until now, to be reflected in the zoobenthos.

Macrozoobenthos is dealt with in three different papers describing field data and one paper by Hummel *et al.* (1994) reviewing mainly experimental work on the effects of prolonged periods of emersion and submersion on some selected species. Where submersion does not seem to affect strongly benthic organisms, emersion can cause serious mortality in most species studied depending on temperature and other factors.

Seys *et al.* (1994) describe the temporal pattern of some overall parameters of the benthic populations (density, biomass and species composition), while Coosen *et al.* (1994) focus on the temporal pattern, distribution and habitat preference of some individual species. Meire *et al.* (1994a) describe the spatial distribution of intertidal macrobenthic communities and some temporal changes (the cockle and mussel populations of the Oosterschelde are treated separately by Coosen *et al.* (1994b) and Van Stralen & Dijkema (1994)).

The changes in the hydrodynamics and the morphology of the Oosterschelde after the completion of the storm-surge barrier do not seem to have influenced the normal patterns in macrobenthic populations, as late summer values of total biomass, total density, species richness, diversity and abundance-biomass ratio showed no overall significant trend between 1981 and 1989. The occurrence of severe versus mild winters, rather than hydrodynamic changes seemed to influence macrobenthic populations. The same conclusion was drawn by Coosen *et al.* (1994a) analysing the data for some individual species. Only at the most elevated station there is evidence that the short-term increase in exposure time, caused by the manipulation of the storm-surge barrier in 1985 influenced both macrobenthic biomass and density, especially these of *Scoloplos armiger* and juvenile *Arenicola marina*, species found to be sensitive to emersion by Hummel *et al.* (1994).

The spatial pattern of macrozoobenthos is analysed in detail by Meire *et al.* (1994a). Several clusters of sampling stations, each with a different faunal composition, were identified but did not form distinct zones on the tidal flats. The relationship between density and biomass of individual species and of different trophic groups and the mud content of the sediment and the depth was very weak and sometimes differing clearly between years due to a broad tolerance of the species to the range of the environmental variables found in our study area. A further comparison of the 1985 (pre-) and 1989 (post-barrier) data showed that the faunal changes are not necessarily linked to changes in the measured environmental parameters, suggesting that the macrobenthic populations are probably more towards the nonequilibrium end of the continuum between nonequilibrium and equilibrium communities as defined by Wiens (1984). Although the impact of the construction of the barrier on the macrobenthic community seems at present to be rather small, this does not mean that on the long-term there will be no effect. Indeed the transport of larvae can change due to the reduced current velocities and this together with e.g. increased biomasses of microphytobenthos could gradually change the benthic communities. The fact, however, that the presence of many species is to a large extent determined by factors extrinsic to the tidal flat ecotone, decrease the predictability of the ecological events and the resulting pattern of abundances and distribution (Reise, 1985).

Data on the macrozoobenthic populations in the subtidal areas of the Oosterschelde are very scarce, an important gap in our knowledge. Locally there is an important sedimentation of fine sediments, and one might expect significant changes in the macrozoobenthic populations of these sites.

Meiobenthos was sampled only in 1981, 1984 and 1985. The results, summarised by Smol *et al.* (1994) clearly point, however, at the importance of these organisms. The meiofauna was strongly dominated by nematodes, and abundance, biomass and diversity of these animals were significantly higher intertidally than subtidally and also showed a decreasing trend from east to west, gradients possibly reflecting the positive relation between numbers and silt content of the sediment and the negative relation between numbers and median grain size. Although the abundance and diversity of copepods was highest subtidally, their biomass was highest on the tidal flats. Due to a close coupling between silt content and grain size of the sediment and meiofauna Smol *et al.* (1994) predict that meiofauna will become more important in terms of abundance and biomass, mainly due to increasing numbers of nematodes, as the current velocities will alter the distribution and accumulation of the sediment particles. This could increase bioturbation, nutrient mineralisation and sustaining bacterial growth. A general decrease in meiofauna diversity is predicted. The number of copepods is expected to decrease and interstitial species will be replaced by epibenthic species, the latter being more important in terms of biomass and as food for the epibenthic macrofauna and fishes. Unfortunately no data are available to test this hypothesis but based on present data on the sediment characteristics we would expect the changes in meiofauna to occur mainly in the subtidal compartment as here the siltcontent of the sediment increases significantly compared to the tidal flats where it does not really change. Unfortunately no data are available.

From all the above mentioned studies we can conclude that the structure of the benthic system changed significantly in the subtidal hard substrate communities. It is unknown, however, what happened to the subtidal soft substrate populations of meio- and macrozoobenthos. Both on hard and soft substrates the changes of benthic populations due to the environmental changes in the intertidal area seems to be rather small. It must be stressed, however, that important changes in the intertidal hard-substrate communities occurred due to asphalting of dikes, an irreversible impact (Meijer & Waardenburg, 1994). This fact points to the important role of anthropogenic influences on the system. Shellfisheries, especially the cockle fishery which increased dramatically in recent years (see e.g. Meire *et al.*, 1994) might add to this.

Although during the entire 'Oosterschelde-project' the benthic populations were studied in detail still many questions remain unsolved. This is due partly to a lack of data but also to the fact that not all different measurements were collected in an integrated way. Further studies should aim therefore at an integrated approach sampling the different abiotic and biotic variables on some selected stations. This will greatly benefit the interpretation of the results and our insight into the structure and functioning of the benthic system. Furthermore the studies should run for a long time as the response of the different compartments to changes might take years.

References

Coosen, J., J. Seys, P. M. Meire & J. A. Craeymeersch, 1994a. Effect of sedimentological and hydrodynamical changes in the intertidal areas of the Oosterschelde estuary (SW Netherlands) on distribution, density and biomass of five common macrobenthic species: *Spio martinensis* (Mesnil), *Hydrobia ulvae* (Pennant), *Arenicola marina* (L.), *Scoloplos armiger* (Muller) and *Bathyporeia* sp. Hydrobiologia 282/283 (Dev. Hydrobiol. 97): 235–249.

Coosen, J., F. Twisk, M. W. M. van der Tol, R. H. D. Lambeck, M. R. van Stralen & P. M. Meire, 1994b. Variability in stock assessment of cockles (*Cerastoderma edule* L.) in the Oosterschelde (in 1980–1990), in relation to environmental factors. Hydrobiologia 282/283 (Dev. Hydrobiol. 97): 381–395.

De Jong, D. J., P. H. Nienhuis & B. J. Kater, 1994. Microphytobenthos in the Oosterschelde estuary (The Netherlands), 1981–1990; consequences of a changed tidal regime. Hydrobiologia 282/283 (Dev. Hydrobiol. 97): 183–195.

De Kluijver, M. J. & R. J. Leewis, 1994. Changes in the sublittoral hard substrate communities in the Oosterschelde estuary (SW Netherlands), caused by changes in the environmental parameters. Hydrobiologia 282/283 (Dev. Hydrobiol. 97): 265–280.

Hummel, H., A. W. Fortuin, R. H. Bogaards, A. Meijboom & L. De Wolf, 1994. The effects of prolonged emersion and submersion by tidal manipulation on marine macrobenthos. Hydrobiologia 282/283 (Dev. Hydrobiol. 97): 219–234.

Meire, P. M., H. Schekkerman & P. Meininger, 1994b. Consumption of benthic invertebrates by waterbirds in the Oosterschelde estuary, SW Netherlands. Hydrobiologia 282/283 (Dev. Hydrobiol. 97): 525–546.

Meire, P. M., J. Seys, J. Buijs & J. Coosen, 1994a. Spatial and temporal patterns of intertidal macrobenthic populations in the Oosterschelde: are they influenced by the construction of the storm-surge barrier? Hydrobiologia 282/283 (Dev. Hydrobiol. 97): 157–182.

156

Meijer, A. J. M. & H. W. Waardenburg, 1994. Tidal reduction and its effects on intertidal hard-substrate communities in the Oosterschelde estuary. Hydrobiologia 282/283 (Dev. Hydrobiol. 97): 281–298.

Reise, K., 1985. Tidal Flat Ecology: an experimental approach to species interactions. Springer-Verlag, Berlin.

Seys, J. J., P. M. Meire, J. Coosen & J. A. Craeymeersch, 1994. Long-term changes (1979–89) in the intertidal macrozoobenthos of the Oosterschelde estuary: are patterns in total density, biomass and diversity induced by the construction of the storm-surge barrier? Hydrobiologia 282/283 (Dev. Hydrobiol. 97): 251–264.

Smaal, A. C. & R. C. Boeije, 1991. Veilig getij, de effecten van de waterbouwkundige werken op het getijdemilieu van de Oosterschelde. Nota WS 91.088, Rijkswaterstaat, Middelburg (in Dutch).

Smol, N., K. A. Willems, J. C. R. Govaere & A. J. J. Sandee, 1994. Composition, distribution and biomass of meiobenthos, in the Oosterschelde estuary (SW Netherlands). Hydrobiologia 282/283 (Dev. Hydrobiol. 97): 197–217.

Vos, P. C., 1989. Benthische diatomeeën celtellingen in de Oosterschelde en Voordelta. RWS nota GWAO 89.001 (internal report in Dutch).

Wiens, J., 1984. On understanding a non-equilibrium world: myth and reality in community patterns and processes. In: Strong, D. R., D. Simberloff, G. Abele & A. B. Thistle (eds), Ecological Communities. Conceptual Issues and the Evidence. Princeton University Press, Princeton: 439–457.

Hydrobiologia **282/283**: 157–182, 1994.
P. H. Nienhuis & A. C. Smaal (eds), The Oosterschelde Estuary.
© 1994 *Kluwer Academic Publishers. Printed in Belgium.*

Spatial and temporal patterns of intertidal macrobenthic populations in the Oosterschelde: are they influenced by the construction of the storm-surge barrier?

Patrick M. Meire [1,2,3], Jan Seys [1], John Buijs [2] & Jon Coosen [4]

[1] *University of Ghent, Laboratory of Animal Ecology, Ledeganckstraat 35, B9000 Gent, Belgium;*
[2] *Netherlands Institute of Ecology, Centre for Estuarine and Coastal Ecology, Vierstraat 28, 4401 EA Yerseke, The Netherlands;* [3] *Institute of Nature Conservation (Ministry of the Flemish Community), Kiewitstraat 5, B3500 Hasselt, Belgium;* [4] *National Institute for Coastal and Marine Management/RIKZ, P.O. Box 8039, 4330 AE Middelburg, The Netherlands*

Key words: macrobenthos, storm-surge barrier, benthic community structure

Abstract

The construction of a storm-surge barrier in the mouth of the Oosterschelde caused important hydro-dynamical and morphological changes that could influence the macrobenthic populations. This paper is one in a series of five all dealing with the effects of the storm-surge barrier on macrozoobenthos and analyses the spatial and temporal distribution of macrozoobenthos in the Oosterschelde and its relationship with some environmental parameters, based on two large scale sampling campaigns, one before and one after the completion of the barrier.

The sediment of the sampling stations was fine, well sorted sand, with an average mud content of about 2.5%. Only in the Krabbenkreek the sediment was coarser in 1989. The tidal elevation of the sampling sites decreased significantly in 1989.

The density of macrozoobenthos was significantly lower, the biomass higher in 1989. The density was dominated by deposit feeders, the biomass by filter feeders. The difference in biomass between both years was mainly due to a substantial increase of the biomass of filter feeders in 1989. The number of species per station was significantly smaller in 1989 than in 1985. Between 1985 and 1989, frequency of occurrence decreased in 34 versus 13 which increased, density increased in 13 species and decreased in 34 species, biomass increased in 18 species and decreased in 29 species.

Based on TWINSPAN several clusters of stations, each with a different faunal composition, were identified. These clusters did not form distinct zones on the tidal flats but were dispersed widely.

The relationship between density and biomass of different trophic groups and the mud content of the sediment and the depth was analysed. This relationship sometimes differed clearly between years. The correlation coefficient of a multiple regression between density and biomass of individual species and environmental factors, although significant in most cases, was very low, indicating that only a small proportion of the species variability was explained. The relationship between benthos and environmental factors was further analysed by canonical correlation analysis and multivariate discriminant analysis that gave different results for the 1985 and 1989 data. This is probably due to the broad tolerance of the species to the range of the environmental variables found in our study area. From a TWINSPAN of the density data of 1985 and 1989 together we could conclude that, although the environmental parameters in a group of stations, showing a large faunal similarity in one year, did not change, the faunal composition did. This indicates that faunal changes are not necessarily linked to changes in the measured environmental parameters.

In the discussion the different factors affecting macrobenthic populations are situated and it is suggested that the macrobenthic populations are probably more towards the nonequilibrium end of the continuum between nonequilibrium and equilibrium communities as defined by Wiens (1984).

Although the impact of the construction of the barrier on the macrobenthic community seems at present to be rather small this does not mean that on the long-term there will be no effect.

Introduction

In the estuarine ecosystem, macrozoobenthos is an important group of animals both in terms of species richness, abundance or biomass and in the estuarine food chain. The different species can be divided into several trophic groups (e.g., Fauchald & Jumars, 1979). Filter feeders such as mussels (*Mytilus edulis*) and cockles (*Cerastoderma edule*) consume a substantial part of the primary production (Prins & Smaal, 1994). Deposit feeders, which form the majority of the species, largely rely on organic material present on or within the sediment. Some species (e.g., *Nephtys hombergii*) are predators or scavengers. The macrozoobenthic invertebrates in turn are an important food source for fish and birds (Meire *et al.*, 1994), cockles and mussels are also consumed by men (Van Stralen & Dijkema, 1994).

Several studies indicate that macrozoobenthic populations show large variations both in space and time (see e.g., papers in Elliot & Ducrotoy, 1991). Long-term studies provide a good basis for unraveling the underlying causes of this temporal variability (e.g., Dörjes *et al.*, 1986; Desprez *et al.*, 1986; Beukema, 1989), whereas large scale surveys aim to analyse factors causing spatial variation (e.g., Warwick *et al.*, 1991). Man-induced changes of the environment (pollution, coastal engineering works etc.) offer an interesting opportunity to study the factors influencing the occurrence of macrobenthic populations given a carefully designed monitoring scheme before and after the changes. The construction of the storm-surge barrier in the Oosterschelde was such an event. This paper is one in a series of five papers all dealing with the effects of the storm-surge barrier on macrozoobenthos. Seys *et al.* (1994)

describe the temporal pattern of some overall parameters of the benthic populations (density, biomass and species composition). Coosen *et al.* (1994) focus on the temporal pattern, distribution and habitat preference of some individual species. The cockle and mussel populations of the Oosterschelde are treated separately (Coosen *et al.*, 1994b; Van Stralen & Dijkema, 1994).

This paper analyses the spatial and temporal distribution of macrozoobenthos in the Oosterschelde and the relationship between macrozoobenthos and some environmental parameters, based on two large scale sampling campaigns, one before and one after the completion of the barrier. To what extent the benthic populations depend upon the measured environmental variables and whether or not this relationship was similar in both campaigns is the central question of this paper. In the discussion we try to situate the macrobenthic populations in the nonequilibrium-equilibrium continuum concept of Wiens (1984).

Material and methods

Database

For a general description of the Oosterschelde and the engineering works we refer to Nienhuis & Smaal (1994).

For the study of the macrozoobenthos we had access to three different databases, two which were focused on temporal patterns (Seys *et al.*, 1994; Coosen *et al.*, 1994a) and one on spatial patterns. The 'Interecos' campaign aimed at a description of the large scale spatial distribution before and after the construction of the barrier and formed the basis for this paper. In August

1985 and 1989, each time a total of 305 locations were sampled, on three different intertidal flats of the Oosterschelde (Fig. 1). At each location 10 samples were taken with a core (4.5 cm diameter) to a depth of 10 cm. All samples were fixed in the field with 40% neutralized formaline and sieved in the laboratory on a 1 mm mesh sieve. All organisms were picked out after staining with Rose Bengale for at least 24 h, and identified to the species level when possible. Additionally 5 samples with a core of 15 cm diameter were taken to a depth of 30 cm. These samples were sieved in the field on a 3 mm mesh sieve and the animals fixed in 7% neutral formaline. The data from the large cores were used to estimate the abundance of three species: *Arenicola marina*, *Mya arenaria* and *Mytilus edulis*. The first two are known to occur much deeper than 10 cm. Mussels were not well sampled by the small cores due to their size.

Data for all other species came from the small cores. Ash free dry weight (AFDW) is the weight after drying for 3 days at 72 °C, minus the weight after burning at 520 °C for 3.5 hours.

The 305 sampling sites were distributed over different, predefined strata according to a stratified random sampling strategy (Van Der Meer *et al.*, 1989). Averages and confidence intervals were calculated according to a simple random sampling design. In both years the same sites were sampled, however in 1989, 5 stations could not be sampled, so all comparisons are based on 300 sites.

At each site several environmental parameters were measured: median grain size, sorting coefficient (both measured in ϕ units), mud content (% of sediment smaller than 53 μ), tidal elevation (further called depth) bulk density (g/cm$_3$), chlorophyll content (μg/g), moisture content (%)

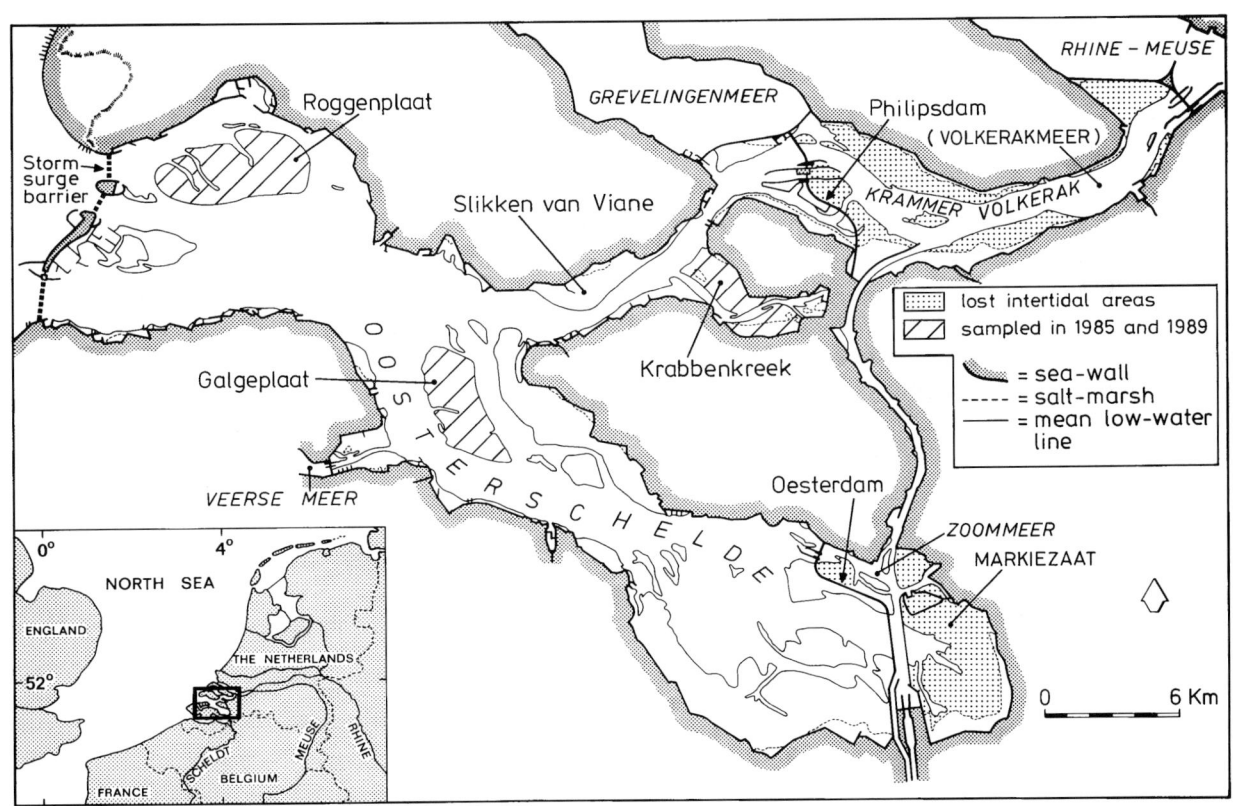

Fig. 1. Map of the Oosterschelde with the location of the storm-surge barrier, the secondary dams and the intertidal areas that were sampled.

and vegetation cover (seaweeds and eelgrass). For more details about the measurement of these parameters we refer to Ten Brinke *et al.* (1994) and De Jong *et al.*, (1994).

The identification of the species was not exactly similar in the two years. Species of genera difficult to identify, such as *Polydora*, *Nereis*, *Microphthalmus* etc. were not identified to species level in 1985. In order to compare both years the different species of these genera were lumped to the genus level in the 1989 database. In that way we retained 47 taxa that could be compared between years.

Analyses

The different species were classified into different trophic groups based on Fauchald & Jumars (1976) (for more details see Meire *et al.*, 1991a, b). The mud content of the sediment and depth were grouped in 7 and 12 classes respectively (0–1%; 1–2%; 2–3%; 3–4%; 4–5%; 5–10% and > 10% for mud and NAP −2.0–−1.75 m; −1.75–−1.5 m; ... NAP 1–1.25 m for depth) (NAP is the Dutch ordnance level and is situated at about mid tidal level).

As in both years the same stations were sampled, the results were compared with the Wilcoxon matched-pairs Signed rank test on untransformed data (Siegel, 1956).

In order to study the relation between sediment parameters and macrozoobenthos we used several types of multivariate analysis. TWINSPAN (Hill, 1979) was performed on the original data, with the following cut levels: analysis with density data: 0, 64, 512, 1028, 4112, 16440, 65792, 99999; analysis with biomass data: 0, 0.5, 1, 5, 10, 25, 100, 9999. For comparing the 1985 and 1989 density data, a TWINSPAN analysis was run after eliminating the species occurring in less than 25 stations. The same cut levels were used. To identify whether the different TWINSPAN clusters could be distinguished based on sediment parameters, a multiple discriminant analysis was used (see e.g., Weston, 1988).

Multivariate discriminant analysis (MDA) was also used to determine the linear combination of environmental parameters discriminating the 20 dominant benthic species. Discriminant analysis is viewed as a good method for defining the multidimensional habitat characteristics of the benthic fauna in order to identify species differences (Flint & Kalke, 1986; James & McCulloch, 1990). A data matrix was constructed with 20 sample groups each representing a species and 8 abiotic variables (mud content, median grain size, sorting coefficient, depth, bulk density, chlorophyll content, vegetation cover and moisture content) (only the 20 most abundant species were used in the analysis). The number of cases per sample group (or the number of observations per species) varied between 22 and 267. Only the stations where the abundance of the species was higher than the overall average were used.

The relation between sediment characteristics and benthos was further analysed by canonical correlation analysis (CCA). The analysis was similar to that described by Van Der Meer (1991), who did the analysis on the 1985 data. The same analysis was done for 1989 and for both years together. For the MDA and CCA the data were transformed according to Van Der Meer (1991).

To perform a multiple regression on both the 1985 and the 1989 data, the species data of each year were converted to Z-scores and then combined in order to eliminate between year variation. Of all environmental variables both the original value and its square was used, as a quadratic function was expected.

SYSTAT (Wilkinson, 1988) was used for all statistical analyses.

Results

Ecological parameters

The means of some important environmental parameters on the sampling sites are given in Table 1. The sediment in both campaigns can be characterized as fine, well sorted sand. Median grain size varied between 2.18 and 3.45 ϕ with a mean of 2.78 ϕ in 1985 and between 2.17 and 3.52 ϕ with a mean of 2.75 ϕ units in 1989. The mud content varied between 0% and 21.4% and

Table 1. Sediment parameters of the sampling sites. The mean and Standard Error of the parameters in both years with the Z-value and associated probability of the Wilcoxon matched pairs signed rank test (D) are given for the whole dataset and for the three flats separately (OS Oosterschelde; RP Roggenplaat; GP Galgeplaat; KB Krabbenkreek) (Med: median grain size; Sort sorting coefficient).

Parameters		Changes in sediment parameters							
		OS (*N* = 300)		RP (*N* = 120)		GP (*N* = 105)		KB (*N* = 75)	
Depth	85	− 0.14	0.03	− 0.12	0.05	− 0.32	0.04	0.07	0.07
	89	− 0.24	0.03	− 0.19	0.05	− 0.60	0.04	0.18	0.07
	D	− 7.15	<0.001	− 4.04	<0.001	− 8.87	<0.001	5.75	<0.001
Mud	85	2.63	0.21	2.25	0.36	1.77	0.19	4.47	0.52
	89	2.47	0.33	3.39	0.73	1.13	0.1	2.85	0.61
	D	− 6.32	<0.001	0.35	ns	− 5.47	<0.001	− 5.97	<0.001
Med	85	2.78	0.014	2.63	0.03	2.86	0.02	2.89	0.03
	89	2.75	0.015	2.61	0.02	2.82	0.02	2.85	0.03
	D	− 5.62	<0.001	− 3.14	<0.01	− 3.56	<0.001	− 3.23	<0.01
Sort	85	0.42	0.005	0.42	0.009	0.40	0.006	0.44	0.01
	89	0.40	0.003	0.40	0.005	0.40	0.005	0.40	0.01
	D	− 3.69	<0.001	− 3.32	<0.01	0.82	ns	− 3.75	<0.001

44.9% in 1985 and 1989 respectively, with mean values of 2.6% and 2.4%. Differences between flats existed: the Krabbenkreek was muddier than the other two. The significant decrease in mud content, median grain size and sorting coefficient between 1985 and 1989 (Table 1) was widespread over the stations but the absolute value of the decrease was very small. Only the changes in the Krabbenkreek can be considered as meaningful (Ten Brinke *et al.*, 1994). The variability of most sediment parameters, especially median grain size was very low (coefficient of variation 8.8 and 9.6% in 1985 and 1989).

The tidal elevation of the sampling sites varied between NAP −1.6 and +1.2 m in both years with means of NAP −0.14 and NAP −0.24 in respectively 1985 and 1989. The decrease in 1989 was significant (Table 1).

A description of the vegetation on the intertidal flats is given by De Jong *et al.* (1994).

Characteristics and spatial distribution of the macro-zoobenthos of the Oosterschelde in 1985 and 1989

General parameters (Density, biomass and species composition)
The density of macrozoobenthos varied between 630 and 138 150 ind. m^{-2} in 1985, between 480

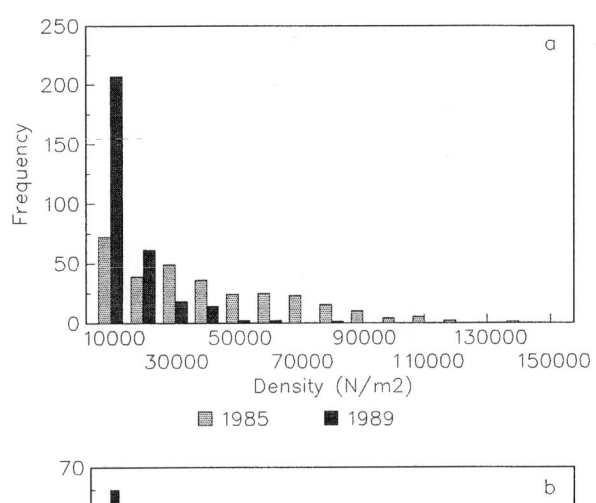

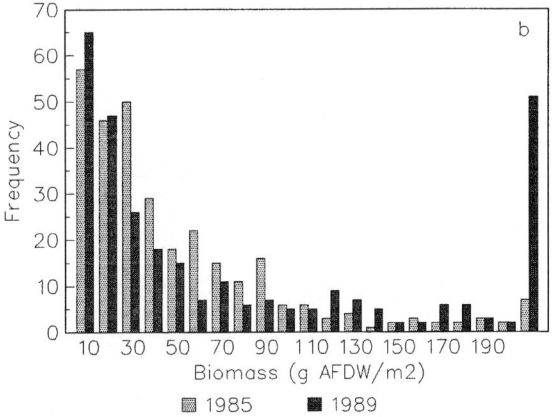

Fig. 2. Frequency distribution of density (a) and biomass of macrozoobenthos (b) in 1985 and 1989.

and 71 600 ind. m^{-2} in 1989 (Fig. 2a). It was significantly lower in 1989 (Wilcoxon matched-pairs Signed-rank test $Z = -13.6$, $N = 300$, $p < 0.001$). The majority of sites had densities lower than 30 000 ind. m^{-2}. Differences in densities between tidal flats existed, but densities were in the same order of magnitude (Fig. 3a). The biomass varied between 0.43 and 343 g AFDW m^{-2} in 1985 and 0.23 and 864.9 g AFDW m^{-2} in 1989 (Fig. 2b) and was, contrary to the density, higher in 1989 (Wilcoxon matched-pairs Signed-rank test $Z = -5.15$, $N = 300$, $p < 0.001$). In most stations biomass was lower than 100 g AFDW m^{-2}. Remarkable was the large number of stations in 1989 with biomass values of more than 200 g AFDW m^{-2}. Biomass was lowest in the Krabbenkreek in both years (Fig. 3b). The spatial variability of density and biomass was very high. The coefficient of variation for density, biomass and different trophic groups is given in Table 2.

Table 2. Coefficient of variation of density and biomass of different trophic groups in both 1985 and 1989. (DF: burrowing deposit feeders; SDF: surface deposit feeders; SF; suspension feeders; O: omnivores and predators).

Group	Spatial variability of the macrozoobenthos of the Oosterschelde			
	Density		Biomass	
	1985	1989	1985	1989
Total	79.4	109.0	110.1	132.3
DF	135.4	119.5	78.7	78.0
SDF	96.4	172.4	116.8	126.0
SF	126.6	121.8	145.3	150.9
O	106.4	386.3	214.9	184

The density was dominated by deposit feeders, mainly surface deposit feeders (Fig. 4a), the bio-

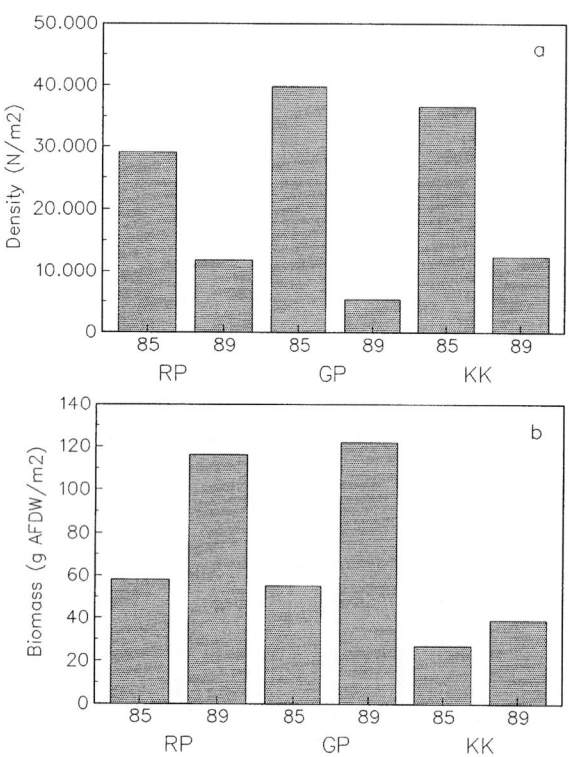

Fig. 3. Average density (a) and biomass (b) of macrozoobenthos in 1985 and 1989 at three intertidal flats. (RP Roggenplaat; GP Galgeplaat; KK Krabbenkreek).

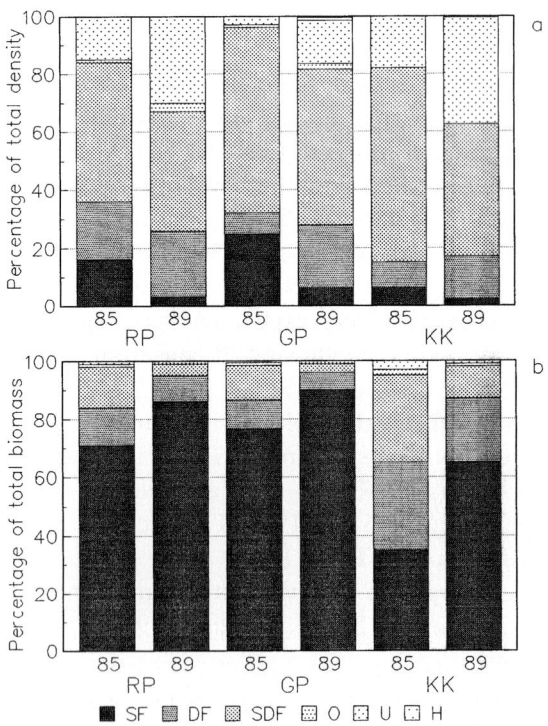

Fig. 4. The proportion of different trophic groups in the total density (a) and biomass (b) in both 1985 and 1989 on each of the three tidal flats. (RP Roggenplaat; GP Galgeplaat; KK Krabbenkreek) (SF suspension feeders; DF burrowing deposit feeders; SDF surface deposit feeders; O omnivores; U unknown; H herbivores).

mass by filter feeders, especially on the Roggen- and Galgeplaat (Fig. 4b). The difference in bio-mass between both years was mainly due to a substantial increase of the biomass of filter feed-ers in 1989 (Fig. 4b).

The total number of species found was 65, namely 32 polychaetes, 14 crustacea, 16 molluscs, 2 echinoderms and 1 nemertinea and did nearly not differ between both campaigns. Only a few species were found in only one year. Remarkable was the appearance of *Ensis directus*, an Ameri-can species that is colonizing European waters since 1979 (Essink, 1988) (see also Seys *et al.*, 1993). The number of species per station is given in Fig. 5a and was significantly smaller in 1989 than in 1985 (Wilcoxon matched-pairs Signed-rank test $Z = -5.15$, $N = 300$, $p < 0.001$). Differ-ences between flats were small (Fig. 5b).

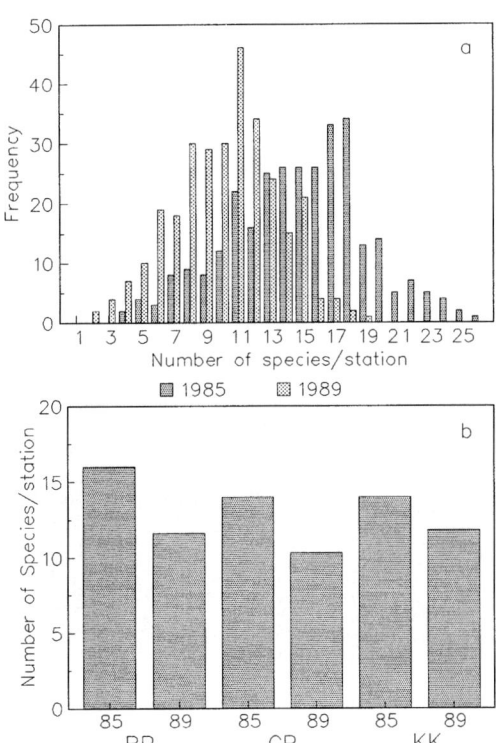

Fig. 5. Frequency distribution of the number of species per station in 1985 and 1989 (a) and the average number of spe-cies per station in 1985 and 1989 on each of the three tidal flats (b). (RP Roggenplaat; GP Galgeplaat; KK Krabben-kreek).

Individual species

In Table 3 the frequency of occurrence, mean density and mean biomass of the individual spe-cies is given for both sampling years together with the results of the Wilcoxon matched pairs Signed-rank test. Frequency of occurrence decreased in most species (34 versus 13 which increased). Density increased in 13 species and decreased in 34 species, the change being significant in 8 and 28 species respectively. Biomass increased in 18 species, decreased in 29 species, the change being significant in 11 and 22 species respectively.

Details of the distribution of each individual species are given by Meire *et al.* (1991a, b) and the relationship with environmental parameters in Seys *et al.* (1992). The distribution of some individual species is discussed in detail by Coosen *et al.* (1994a) and Seys *et al.* (1994).

Communities

In order to investigate whether different faunal groupings ('communities') could be detected, a classification method (TWINSPAN) was run on the density and biomass data for both years. This resulted in several clusters of stations, each char-acterized by some typical species. As an example, the results of the analysis of the 1985 biomass data are summarized in Fig. 6a. Seven station clusters could be delineated. The first division separated stations with high biomass, density and number of species per sample from the others. Cluster 1 consists of stations situated on mussel-beds. They are characterised by a high density, biomass and number of species. Besides mussels, some typical species are *Littorina littorea*, *Lepido-chiton cinereus* and *Lanice conchilega*. In the sta-tions of cluster 2 species like *Arenicola marina*, *Heteromastus filiformis*, *Hydrobia ulvae*, *Mya arenaria*, *Tharyx marioni*, *Nereis sp.* and Oligochae-etes have high biomass values. The species com-position of cluster 3 is more or less similar to the former cluster but biomass of many species is much lower except for cockles, whose average biomass is highest in this cluster. In cluster 4 den-sity, biomass and diversity are also lower, espe-cially for species like *Heteromastus filiformis*, *Ma-coma balthica*. In clusters 5, 6 and 7 density,

Table 3. Frequency of occurrence of different species and Wilcoxon. (F = Frequency; B = Biomass [g AFDW/m^2]; D = Density [number/m^2]; P = significance of the Wilcoxon matched pairs signed rank test).

Species	1985			1989			Difference (1985–1989)				
	F	B	D	F	B	D	F	B	P	D	P
Ampharete acutifrons	5.9	0.027	14.0	1.6	0.003	1.3	4.2	0.025	0.001	12.6	0.001
Anaitides mucosa	57	0.121	164.5	42.6	0.066	60.4	14.3	0.055	0.004	104.0	0.000
Arenicola marina	79.3	4.418	36.6	81.6	6.839	27.4	− 2.3	− 2.421	0.000	9.2	0.003
Aricidea sp.	2.2	0.001	5.7	0.6	0.000	0.6	1.6	0.001	0.046	5.0	0.074
Bathyporeia sp.	43.6	0.078	373.2	23	0.075	513.3	20.6	0.003	0.004	− 140.0	0.002
Capitella capitata	73.1	0.146	1089.9	56.6	0.033	226.2	16.4	0.113	0.000	863.6	0.000
Carcinus maenas	35.7	0.286	47.2	7.3	0.104	5.5	28.4	0.182	0.000	41.7	0.000
Cerastoderma edule	86.2	25.402	5565.2	66.6	77.117	259.5	19.5	− 51.714	0.000	5305.6	0.000
Corophium sp.	45.9	0.060	368.41	35.6	0.037	172.1	10.2	0.023	0.006	196.2	0.000
Crangon crangon	40.9	0.079	63.9	14.6	0.124	14.0	26.3	− 0.045	0.000	49.9	0.000
Crepidula sp.	0.3	0.001	0.4	1.6	0.010	1.3	− 1.3	− 0.009	0.172	− 0.9	0.279
Eteone sp.	62.6	0.056	114.0	30.6	0.020	37.0	31.9	0.037	0.000	77.0	0.000
Eumida sanguinea	0.9	0.000	0.6	3.3	0.003	6.8	− 2.3	− 0.003	0.005	− 6.2	0.014
Eurydice sp.	0.9	0.000	1.1	1.6	0.001	1.8	− 0.6	− 0.000	0.499	− 0.7	0.726
Gammarus sp.	5.2	0.006	6.4	11.3	0.047	105.0	− 6	− 0.041	0.029	− 98.6	0.001
Harmothoe sp.	0.6	0.005	0.9	2.3	0.002	1.8	− 1.6	0.003	0.441	− 0.9	0.126
Heteromastus filiformis	64.2	0.693	1622.7	43.3	0.383	132.4	20.9	0.311	0.000	1490.3	0.000
Hydrobia ulvae	84.9	3.218	13998.9	32.6	0.844	1058.3	152.2	2.374	0.000	2940.6	0.000
Lanice conchilega	12.1	0.105	34.9	25.6	0.548	71.0	− 13	− 0.443	0.000	− 36.0	0.000
Lepidochitona cinerea	5.2	0.026	8.1	1.3	0.002	1.8	3.9	0.024	0.007	6.3	0.010
Littorina littorea	9.5	0.394	67.8	0.3	0.001	0.4	9.1	0.393	0.000	67.3	0.000
Macoma balthica	86.8	2.125	727.1	65	2.224	120.4	21.8	− 0.099	0.979	606.7	0.000
Magelona papillicornis	6.2	0.007	5.2	2.6	0.010	2.0	3.5	− 0.003	0.345	3.2	0.031
Malacoceros sp.	3.2	0.002	8.4	2.6	0.002	11.2	0.6	0.000	0.683	− 2.7	0.776
Melita sp.	0.9	0.001	2.2	3	0.004	8.0	− 2	− 0.003	0.041	− 5.7	0.066
Microphthalmus sp.	2.2	0.002	1.5	10.6	0.002	17.4	− 8.3	− 0.001	0.002	− 15.8	0.000
Mya arenaria	31.1	1.131	17.3	5	0.271	2.5	26.1	0.860	0.000	14.7	0.000
Mysella bidentata	7.8	0.004	8.5	5	0.004	4.5	2.8	− 0.000	0.792	3.9	0.151
Mytilus edulis	19.3	6.485	55.7	5.6	7.258	5.5	13.6	− 0.773	0.023	50.1	0.000
Nemertinae sp.	8.1	0.014	9.1	4.3	0.008	4.5	3.8	0.006	0.442	4.5	0.098
Nephtys sp.	53.1	0.267	76.8	56.6	0.633	58.1	− 3.5	− 0.366	0.000	18.6	0.032
Nereis sp.	63.9	0.573	424.9	57.3	0.828	235.6	6.6	− 0.255	0.242	189.2	0.000
Oligochaeta	78	0.403	3771.4	69.6	0.223	2790.5	8.3	0.180	0.000	980.8	0.000
Ophelia sp.	3.6	0.003	17.1	1.3	0.001	8.0	2.2	0.002	0.084	9.1	0.094
Polydora sp.	29.1	0.055	222.9	19.3	0.014	36.0	9.8	0.041	0.000	186.8	0.000
Pygospio elegans	89.5	0.299	2955.3	64.6	0.093	724.5	24.8	0.207	0.000	2230.7	0.000
Retusa sp.	34.4	0.019	58.9	10.3	0.003	11.2	25.1	0.016	0.000	47.7	0.000
Scoloplos armiger	90.4	1.092	808.9	87.3	0.886	1357.2	3.1	0.207	0.067	− 548.3	0.004
Scolelepis sp.	8.1	0.025	30.5	2	0.012	1.8	6.1	0.012	0.001	28.7	0.000
Scrobicularia plana	11.4	0.860	13.1	10.3	0.258	12.4	1.1	0.602	0.043	0.7	0.392
Spiophanes bombyx	17.3	0.042	57.3	16.6	0.021	30.8	0.7	0.021	0.578	26.5	0.019
Spio filicornis	31.8	0.022	157.5	57.3	0.089	662.2	− 25	− 0.066	0.000	− 504.7	0.000
Spisula sp.	3.6	0.003	9.7	0.3	0.000	0.2	3.2	0.003	0.008	9.4	0.007
Streptosyllis sp.	2.9	0.005	22.6	0.6	0.000	0.4	2.2	0.005	0.018	22.2	0.018
Tellina sp.	3.9	0.061	5.2	5.6	0.024	5.0	− 1.7	0.037	0.216	0.1	0.933
Tharyx marioni	60.3	0.278	1623.3	47	0.151	818.3	13.3	0.127	0.000	804.9	0.000
Urothoe sp.	14.4	0.011	37.9	17.3	0.013	43.6	− 2.9	− 0.002	0.056	− 5.7	0.171

biomass and diversity are lower. The stations from cluster 7 consist mainly of *Bathyporeia sp.* This species is also characteristic for clusters 5 and 6. However, in the latter clusters some other species are abundant as well: *A. marina* in cluster 5 and *C. edule* in cluster 6.

The spatial distribution of these clusters is shown in Fig. 6b. Almost all clusters were found

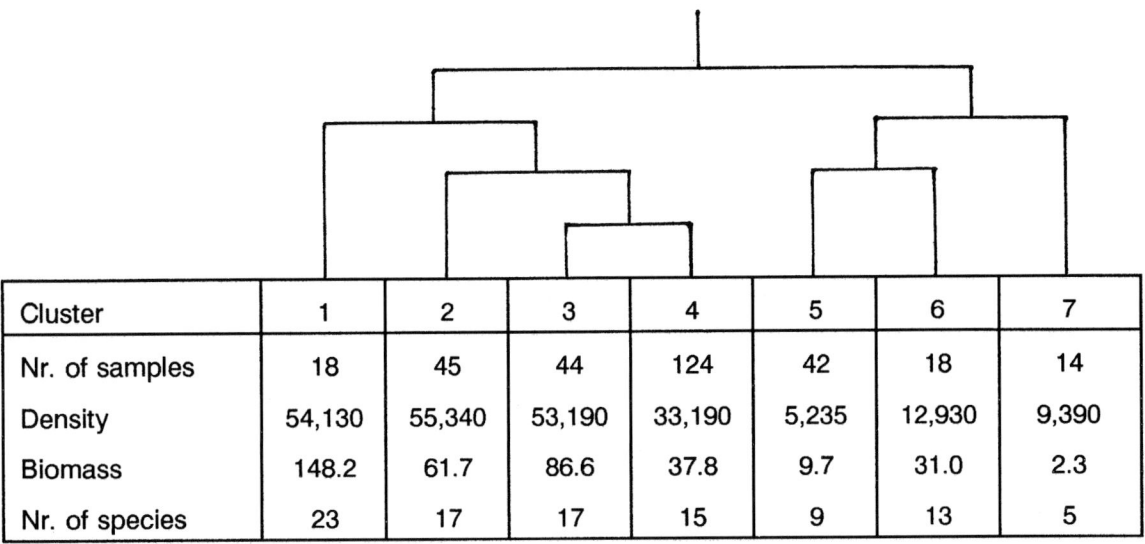

Cluster	1	2	3	4	5	6	7
Nr. of samples	18	45	44	124	42	18	14
Density	54,130	55,340	53,190	33,190	5,235	12,930	9,390
Biomass	148.2	61.7	86.6	37.8	9.7	31.0	2.3
Nr. of species	23	17	17	15	9	13	5

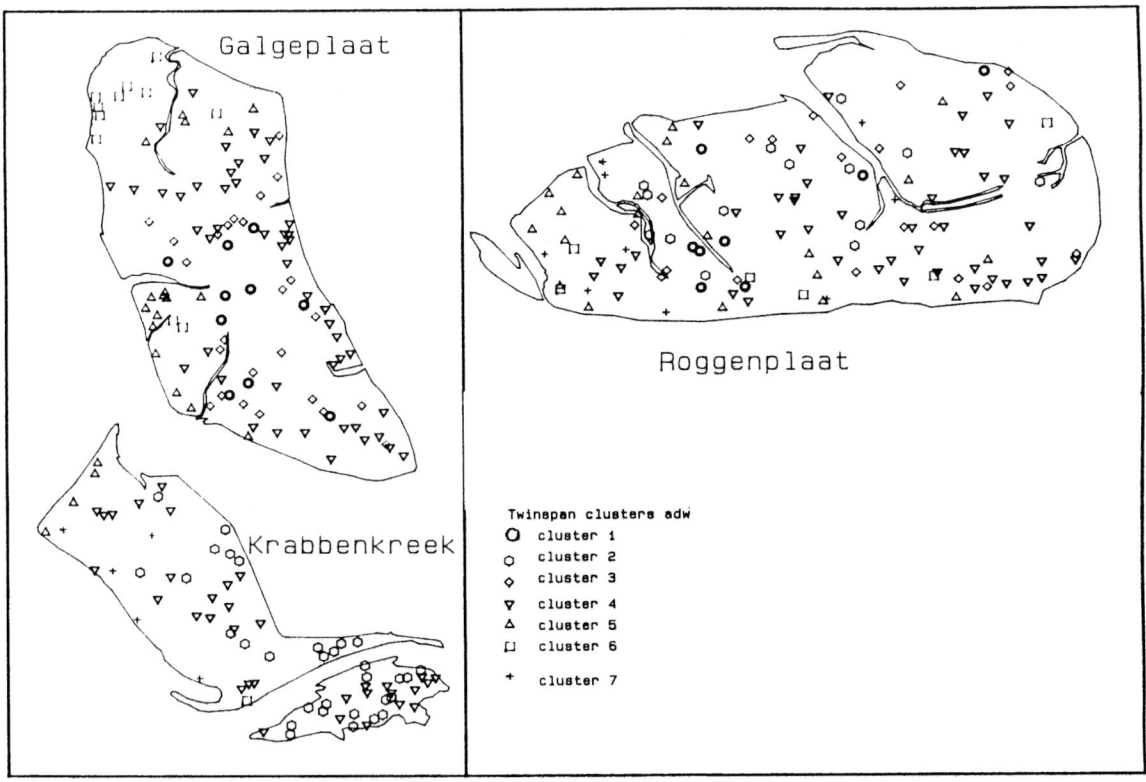

Fig. 6a. Dendrogram of the TWINSPAN divisions based on the biomass data of 1985. The number of samples, density (N/m^2), biomass (g AFDW/m^2) and the total number of species per cluster are plotted. *Fig. 6b.* Map showing the spatial pattern of the different TWINSPAN clusters based on the biomass data of 1985.

on each of the three intertidal flats. They did not form distinct zones but were dispersed over the intertidal area. The clustering of stations based on density or biomass differed within one year as

Total density and biomass

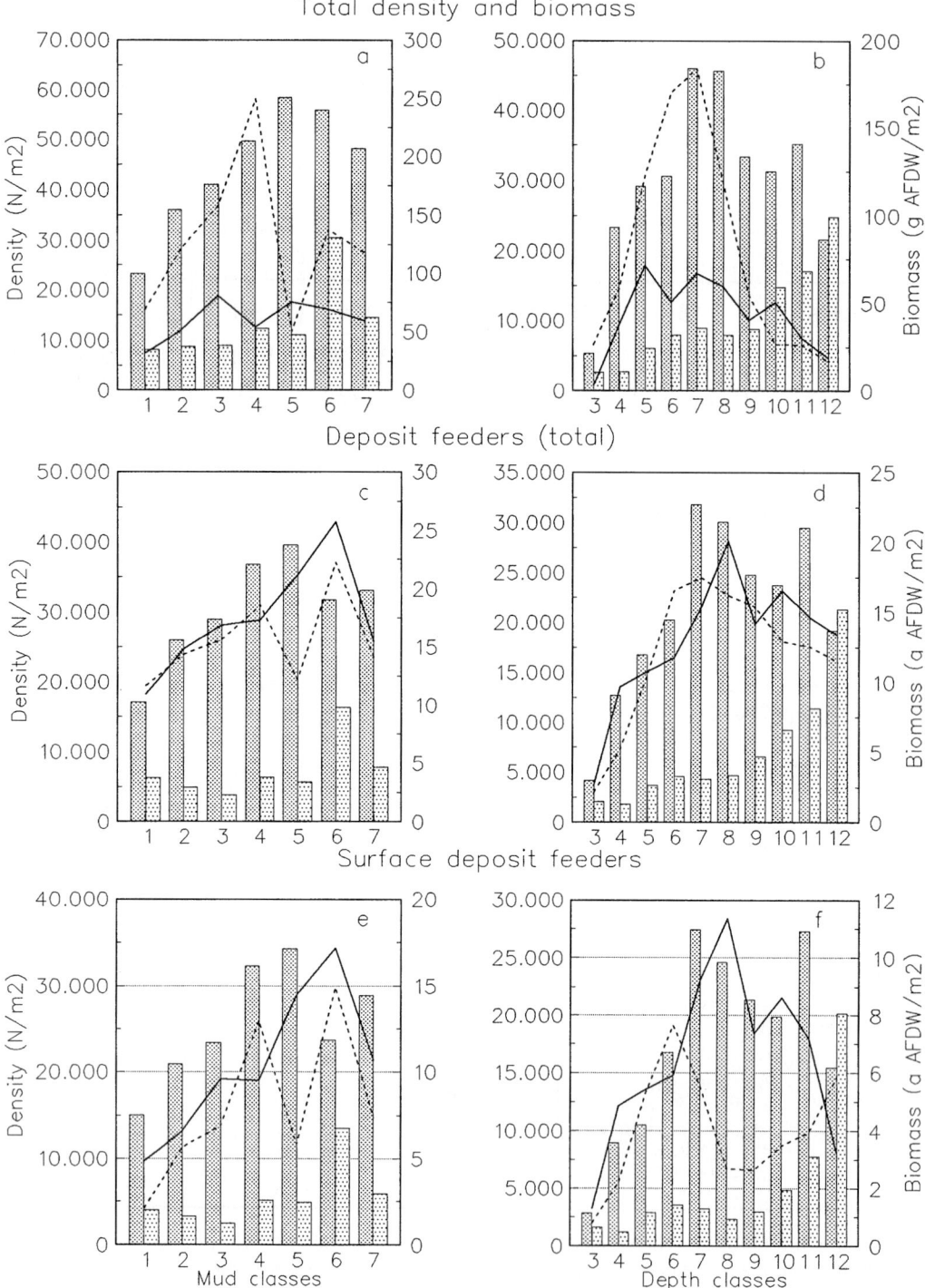

Deposit feeders (total)

Surface deposit feeders

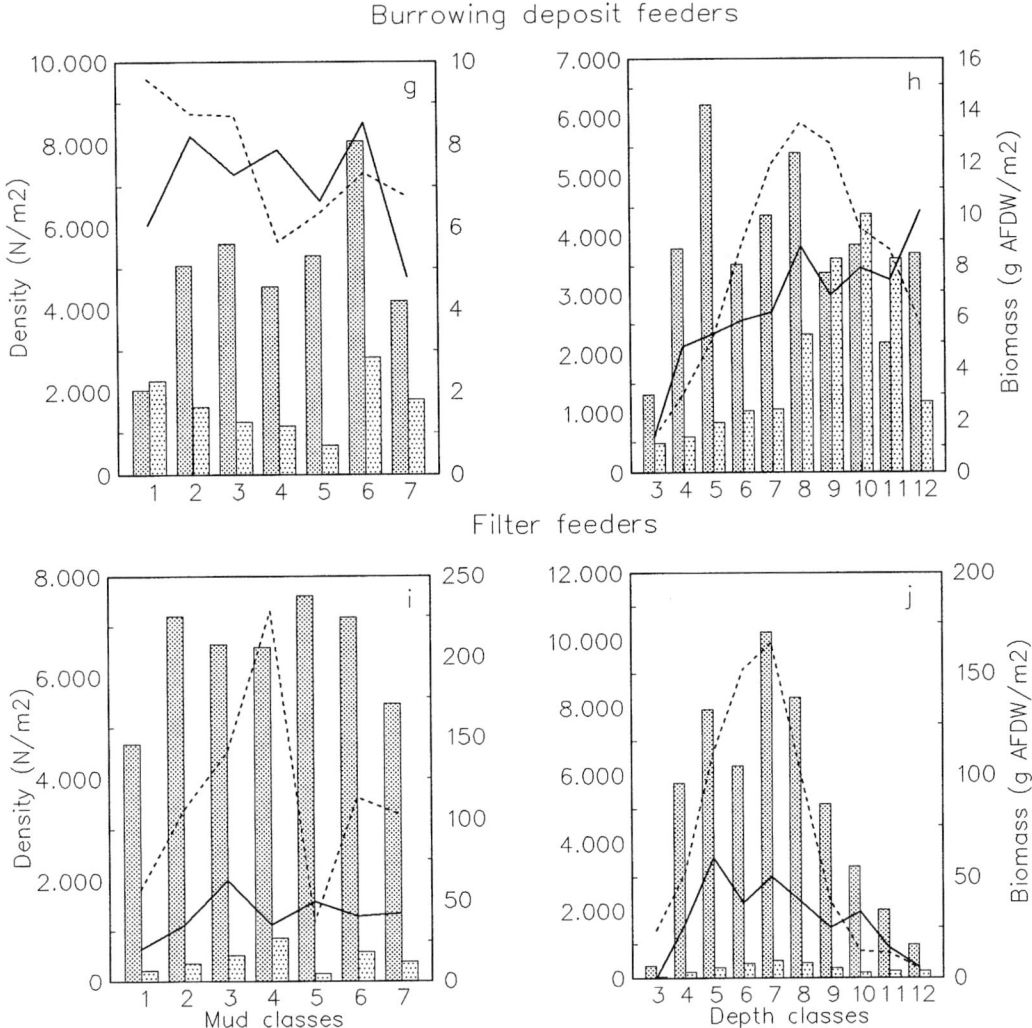

Fig. 7. Density (bar) and biomass (line) of all macrozoobenthic animals (a–b) and of the most important trophic groups (c–j) as a function of mud content of the sediment and depth (tidal elevation). (for the definition of the mud and depth classes see text) (1985 data: full line and left bar; 1989 data: broken line and right bar).

well as between years but the general pattern was similar to the one described above.

Relation between macrozoobenthos and environmental parameters

The average values of total density and biomass, and of the most important trophic groups, are plotted as a function of the different mud and depth classes in Fig. 7. In 1989, the highest biomass values occurred at or just below mid-tidal level (Fig. 7b), a pattern less pronounced but also present in 1985. In 1989 density clearly increased with tidal elevation, whereas in 1985 the highest densities occurred near mid tidallevel. Density increased, especially in 1985, with mud content of the sediment and decreased again at high mud content (Fig. 7a). Biomass, in 1989, was maximal at intermediate mud content.

In both years, the biomass of all deposit feeders clearly increased with increasing mud content of the sediment, except at the highest values (Fig. 7c). The density pattern was less clear. The

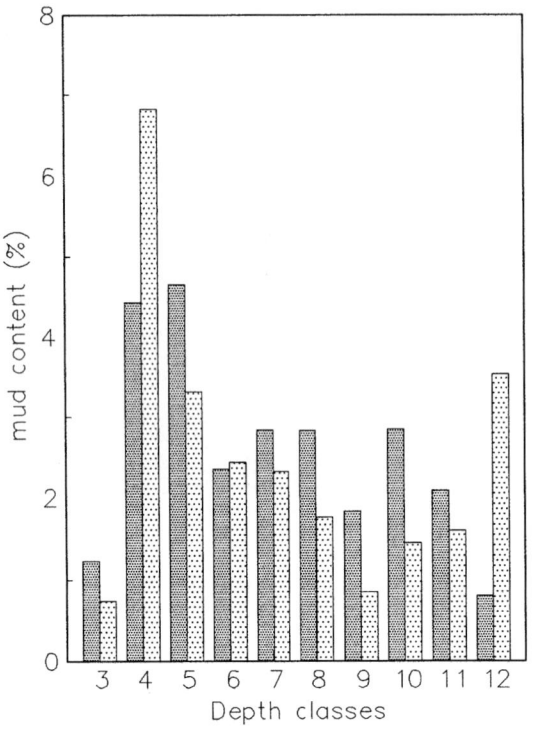

Fig. 8. Mud content of the sediment at different tidal elevations in both 1985 (left bar) and 1989 (right bar). (for the definition of the depth classes see text).

biomass of all deposit feeders also increased towards the upper part of the intertidal area (Fig. 7d). The biomass of surface deposit feeders showed a marked difference in the distribution in relation to depth between years (Fig. 7f). Compared to the mud content of the sediment in the same depth classes (Fig. 8) it is obvious that where the biomass of surface deposit feeders was lower in 1989 there was also a decrease in the mud content of the sediment. This pattern does not hold for the burrowing deposit feeders (Fig. 7h).

Suspension feeders were not clearly related to the mud content (Fig. 7i) but showed highest values of both density and biomass in the lower part (around and below NAP) of the intertidal area (Fig. 7j).

The number of species per sample increased somewhat with the mud content (Fig. 9a). There was no clear pattern with depth (Fig. 9b).

A stepwise multiple regression was applied to analyse further these relations (Table 4). The dummy variable year was significant for three species, indicating that the relation between dis-

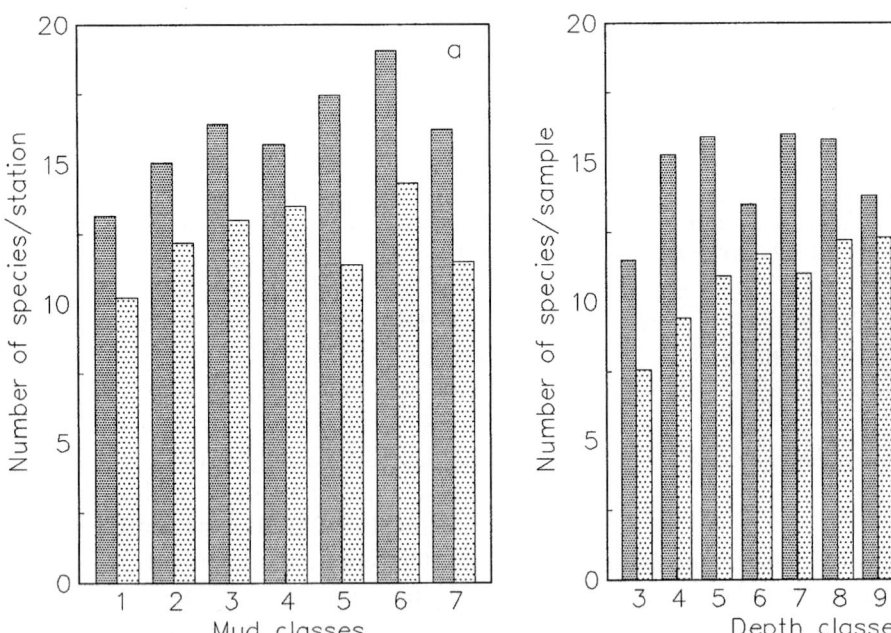

Fig. 9. Average number of species per sample in relation to tidal elevation (a) and mud content of the sediment (b). (for the definition of the mud and depth classes see text) (1985 data: left bar; 1989 data: right bar).

Table 4. Results of a stepwise multiple regression analysis based on the data of 1985 and 1989. The standardized regressions coefficients for all variables in the model together with the coefficient of determination are given. (D density; B Biomass; Dep Depth; Mud Mud content of the sediment; Med Median grain size; Sort Sorting coefficient; Oxl depth of oxygenated layer; Vege vegetation cover; *2 square of variable).

Species	D/B	Year	Dep	Dep*2	Mud	Mud*2	Med	Med*2	Sort	Sort*2	Oxl	Oxl*2	Vege	Vege*2	R2
Arenicola marina	D		0.202	-0.165		0.097					-0.08		0.069		0.109
	B			-0.283							-0.092				
Aniatides sp.	D						0.159								0.017
	B			-0.069					-0.153	0.060					0.005*
Bathyporeia sp.	D		0.241	0.122				-0.119			0.631				0.178
	B		0.272	0.130				-0.148	-0.598	0.538	0.311	-0.314			0.203
Capitella capitata	D		-0.150	-0.086											0.021
	B				0.682	-0.480			0.080					0.073	0.039
Cerastoderma edule	D		-0.230	-0.228			0.189	-0.126		-0.172					0.097
	B		-0.256	-0.259									0.064		0.071
Corophium sp.	D		0.445	0.271			0.140		-0.060						0.160
	B		0.448	0.253					-0.061						0.162
Heteromastus filiformis	D		-0.099	-0.082	0.444	-0.304		0.139	0.554						0.098
	B				0.790	-0.518		0.126		-0.571				0.122	0.272
Hydrobia ulvae	D		0.208				0.450	0.125	-0.081				0.477	-0.414	0.235
	B		0.257				0.381							0.361	0.167
Lanice conchilega	D	-0.060	-0.197			-0.084									0.148
	B	-0.071	-0.202					-0.061						0.330	0.098
Macoma Balthica	D		-0.068	-0.162	0.368	-0.373	0.124						0.086		0.081
	B		-0.122	-0.177					0.734	-0.661			-0.219	0.396	0.035
Mya arenaria	D				0.125		0.144		0.099						0.030
	B							0.468	-0.390						
Nepthys hombergii	D		-0.398		-0.331	0.142							-0.353	0.241	0.180
	B		-0.236	-0.077	-0.104				0.168						0.045
Nereis sp.	D		0.215	-0.089	0.671	-0.409		-0.168					-0.379	0.197	0.129
	B	0.086	0.082	-0.117	0.527	-0.225		-0.226					-0.148		0.087
Oligochaeta	D	0.083	0.125	-0.069	0.842	-0.641	0.123	0.091	0.690	-0.663					0.236
	B	0.073	0.086	-0.095	0.694	-0.492									0.155
Pygosipi elegans	D		0.378	0.221				-0.065							0.116
	B		0.331	0.179				-0.074							0.088
Scoloplots armiger	D		0.346	0.115				-0.095	0.070						0.133
	B		0.316								-0.069	-0.078			0.129
Scrobicularia plana	D				0.929	-0.702	-0.171		1.071	-1.187					0.044
	B							0.138	0.850	-0.752				-0.110	0.096
Spio sp.	D		-0.124				-0.253		-0.612	0.543			-0.068		0.058
	B						-0.227	-0.430	0.373						0.076
Tharyx marioni	D		0.711				0.124		0.225				-0.252	0.366	0.120
	B			-0.595											

Table 5. Results of a Canonical correlation analysis on density data of 1985 and 1989 separateley and together (Med, Median grain size).

Wilks' Lambda	Canonical correlation analysis macrozoobenthos											
	1985				1989				1985–1989			
	0.044 $F = 4.926$, $df = 207,2266$ $p < 0.001$				0.026 $F = 5.846$, $df = 207,2232$ $p < 0.001$				0.046 $f = 0.046$, $df = 207,4748$ $p < 0.001$			
	Axis	1	2	3	Axis	1	2	3	Axis	1	2	3
Canonical correlations		0.853	0.794	0.611		0.869	0.794	0.541		0.833	0.76	0.539
		1	2	3		1	2	3		1	2	3
Intra-set correlations	mud	0.783	−0.065	0.413	depth	0.979	0.026	−0.053	depth	0.695	0.639	−0.027
	depth	0.123	0.832	−0.454	mud	−0.157	0.769	0.403	mud	0.381	−0.679	−0.558
	med	0.763	−0.139	0.231	med	−0.081	0.757	−0.339	med	0.365	−0.651	−0.321

tribution and environmental variables was different between both years. Depth and/or its square was the most commonly selected environmental parameter. For most species, the results for density or biomass are often in agreement, although some clear differences exist as e.g., for *T. marioni*.

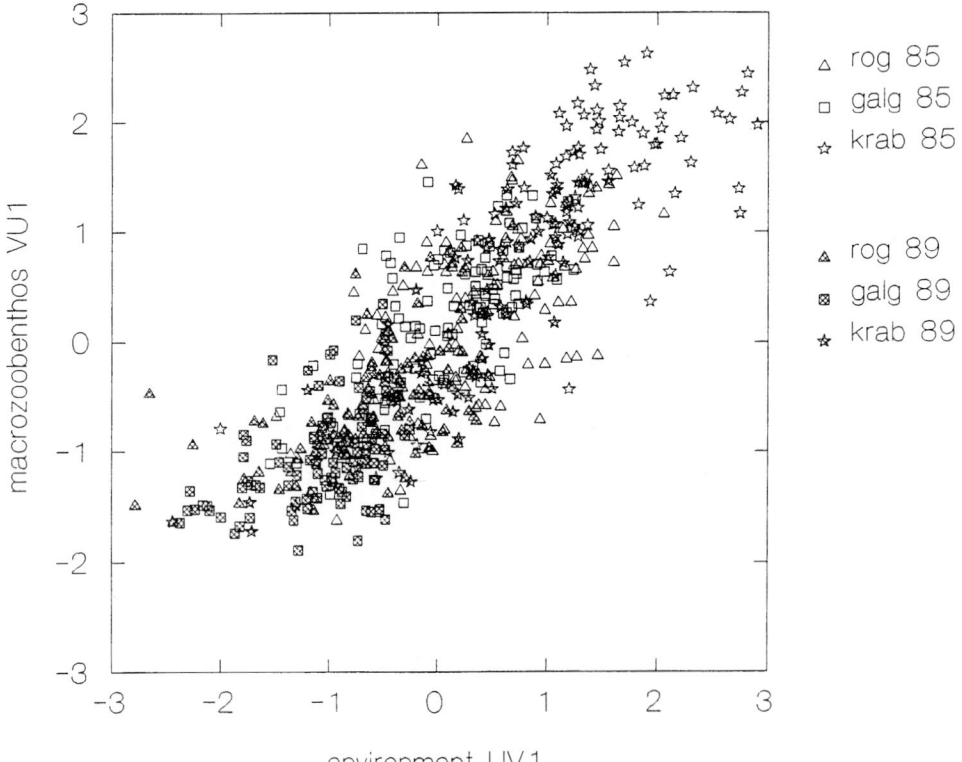

Fig. 10. Plot of the first (standardized) canonical variable environment (UV1) against macrobenthos (VU1) (rog: Roggenplaat; galg: Galgeplaat; krab: Krabbenkreek).

The multiple correlation coefficient is however in nearly all cases very low, indicating that only a small proportion of the species variability is explained.

To gain more insight into the overall relationship between the distribution of the species and the environmental variables a canonical correlation analysis was performed separately on the data of both individual years and on all data (see also Van Der Meer, 1991). The canonical correlation coefficients, the canonical coefficients and the intra-set correlations are given in Table 5. A clear linear relationship between macrobenthos and environmental variables emerged (Fig. 10). From both the canonical coefficients and the intraset correlations it is obvious that in 1985 the first canonical environmental axis was related to the sediment characteristics, mainly mud content and median grain size. The second axis was related to depth. In 1989 the first canonical environmental axis was related to depth, the second axis to mud content and median grain size. The same holds for the analysis of both years. On the second axis however all variables had high coefficients. The results of this analysis are also summarized in Fig. 11. From the confidence ellipses it is clear that the major shift between years occurred along the first axis which is related to depth. This agrees with the major change in this environmental variable.

The importance of depth and mud content of the sediment is also reflected in the TWINSPAN analysis, described before. Figure 12 shows that the mud content of the sediment was highest in the first two clusters, lower in cluster 3 and 4 and very low in the cluster 5 to 7. Median grain size also decreased from cluster 1 to 7. Depth differed between clusters but no clear pattern is obvious. Vegetation mainly occurred in clusters 2.

To test whether the different clusters, as de-

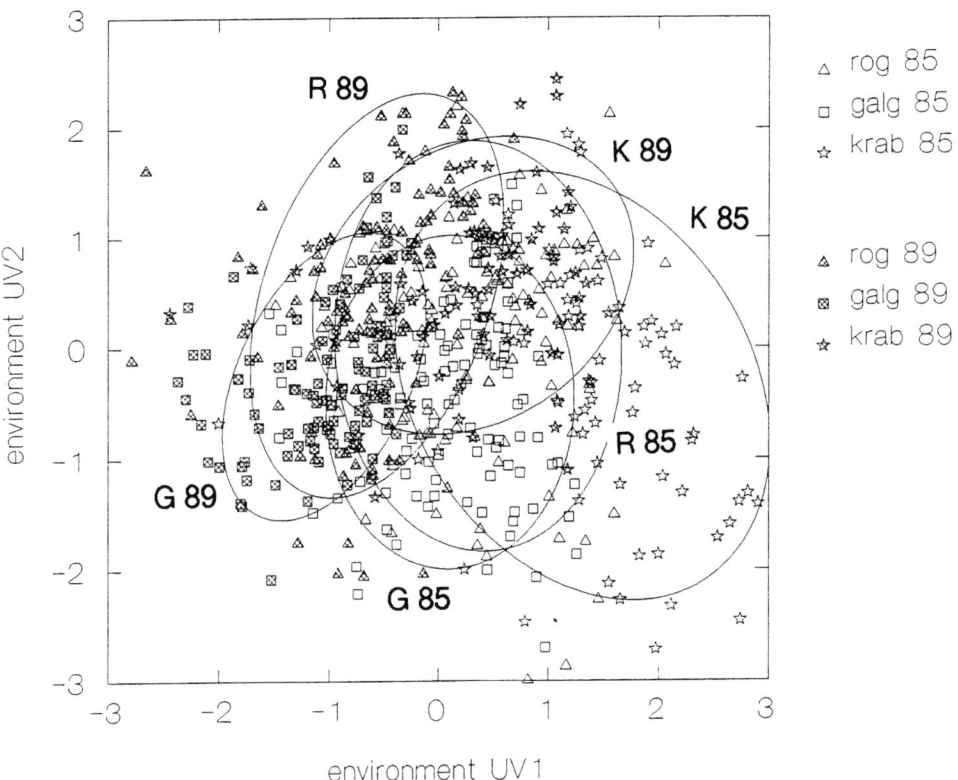

Fig. 11. Plot of the first two canonical variables (UV1 and UV2) plotted against each other. For each intertidal area the 80% confidence ellips for each year is given. (rog, R: Roggenplaat; galg, G: Galgeplaat; krab, K: Krabbenkreek).

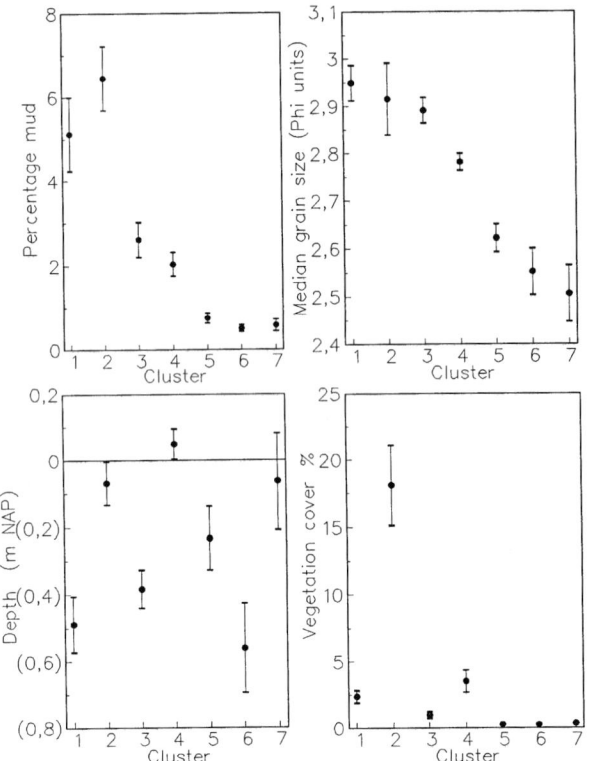

Fig. 12. Average value ($\pm$ SE) of some environmental parameters per TWINSPAN cluster based on the analysis of the biomass data of 1985.

fined by TWINSPAN, could be distinguished based on the environmental parameters measured, a Multivariate Discriminant Analysis (MDA) was performed. The different TWINSPAN clusters were the groups and the sediment parameters (mud content, median grain size sorting coefficient), depth and vegetation cover the discriminating variables. The results are given in Table 6. The discriminant functions differ quite substantially between both years. The results of the analysis on the 1989 density and biomass data are very similar, depth being the most important factor on axis one, vegetation cover on axis 2 and median grain size on axis 3. The results of the analysis of the 1985 samples are different for density and biomass. Based on the obtained discriminant functions, each station was again classified. 49, 65, 46 and 61% of the stations were classified in the correct TWINSPAN cluster for density

1985, 1989, and biomass 1985, 1989 respectively. These results indicate that the measured environmental parameters explained only partly the distribution of the species.

The previous finding could result from a broad tolerance of the species to the environmental variables. Therefore we tried to estimate 'the habitat' of each species by means of a MDA. The results for both years are summarized in Table 7. The first two discriminant functions were highly significant in differentiating between species according to the chi-square test. In 1985 the predominant environmental variable on DF1 was the mud content of the sediment on DF2 the depth. In 1989 depth was the predominant environmental variable on DF1 and median grain size and mud content on DF2. In Fig. 13 the 95% confidence ellipses for the species discriminant scores are plotted for the 1985 data showing a very large overlap between species. Although some species are clearly segregated in the discriminant space most of them are overlapping very much.

Change of the benthic community between 1985 and 1989

The previous results indicate clear differences between both sampling periods. Density and biomass did change significantly and the effect of different environmental parameters varied between years. To get an idea of the overall changes in the benthic community a TWINSPAN on the density data of all stations in both 1985 and 1989 was carried out. The results are summarized in Table 8. Seven different clusters were retained for further analysis, cluster 7 having only 11 stations. Four of the remaining six clusters consist almost exclusively of samples from one year, two clusters had samples from both years. Average values of environmental parameters per cluster are also given in Table 8. To get more insight into the changes, Table 9 shows to what TWINSPAN cluster the 1985 and 1989 data of each station belonged to. It is clear that the 1989 data of stations, whose 1985 data clustered together into 1 cluster, are found in several clusters. This means

Table 6. Results of a Multivariate Discriminant Analysis. For each TWINSPAN (on biomass or density data of 1985 or 1989) the analysis was run. Each TWINSPAN cluster was a group and the discriminating variables were the measured environmental variables.

	Biomass 1985			Biomass 1989		
	Function 1	Function 2	Function 3	Function 1	Function 2	Function 3
df	30	20	12	25	16	9
X2	212	98	50	387	134	20
p	<0.001	<0.001	<0.001	<0.001	<0.001	<0.05
Canonical cor	0.566	0.387	0.367	0.762	0.569	0.247
Mud	0.992	0.06	0.09	0.084	0.026	0.058
Med	0.181	0.363	−0.901	0.028	0.301	−0.970
Depth	0.346	−0.794	−0.505	0.926	0.383	0.131
Sort	−0.331	0.153	0.114	0.165	0.488	0.131
Vegetation	0.245	−0.263	0.701	−0.445	0.764	0.267

	Density 1985			Density 1989		
	Function 1	Function 2	Function 3	Function 1	Function 2	Function 3
df	20	12	6	20	12	6
X2	169	61	20	393	107	35
p	<0.001	<0.001	<0.001	<0.001	<0.001	<0.001
Canonical cor	0.552	0.361	0.233	0.791	0.469	0.334
Mud	0.475	−0.393	0.567	0.320	−0.123	0.083
Med	0.495	−0.699	−0.381	−0.237	0.335	0.807
Depth	0.582	0.370	−0.017	1.001	0.228	−0.095
Sort	0.306	−0.260	0.506	0.145	0.607	0.144
Vegetation	0.632	0.350	−0.074	−0.270	0.729	−0.527

Table 7. Results of a Mulivariate Discriminant Analysis to discriminate between species based on environmental parameters. For explanation see text.

	1985		1989		
	Discriminant function 1	Discriminant function 2	Discriminant function 1	Discriminant function 2	Discriminant function 3
Degrees of freedom	152	126	152	126	102
Chi-square test for significance of the discriminant function	330.8	183.9	644.4	279.4	122.6
P	<0.001	<0.001	<0.001	<0.001	<0.1
Canonical correlation	0.203	0.187	0.353	0.236	0.138
Standardized discriminant function coefficients					
mud(%)	0.738	−0.426	0.186	0.512	0.138
median grain size	0.533	−0.550	0.200	0.729	−0.500
depth	0.371	0.724	−0.937	0.297	0.051
sorting coef.	−0.455	−0.270	0.206	0.483	0.124

Table 8. Results of TWINSPAN analysis of the density of macrozoobenthos of 1985 and 1989. For each cluster the total number of stations, and the number from each sampling year is given togehter with the average value of some environmental parameters.

Cluster	Characteristics of the different TWINSPAN groups						
	1	2	3	4	5	6	7
N of stations	47	97	137	100	131	82	11
85 stations	1	92	47	99	54	9	3
89 stations	46	5	90	1	77	73	8
Depth	− 0.69	− 0.29	0.13	− 0.02	− 0.17	− 0.73	0.57
Med	2.87	2.88	2.81	2.84	2.58	2.68	2.6
Mud	5.2	5.01	2.9	2	0.69	1.2	0.42
Vegetation	32	7	4	4	0.8	9	0.3

that stations with a large faunal similarity in one year evolved in different ways, thereby reducing faunal similarity. This could be due to the fact that the environmental parameters of the stations that form one cluster in 1985 changed in different ways; in 1985 the stations could have all a similar sediment but by 1989 the sediment in these stations could have changed in different ways, e.g., some stations becoming more sandy, others

Table 9. Evolution fo some stations from 1985 to 1989 in the TWINSPAN. The first row and colum give the TWINSPAN group. Row 2 consists of all stations whose 1985 data were classified in group 2. The columns show to what TWINSPAN group these stations belonged to in 1989 (of the 92 stations of 1985 belonging to group 2, in 1989, 30 belonged to group 1, 4 to group 2 etc.). The same is given for the plots belonging in 1985 to group 3 and 4.

Cluster 1985	Cluster 1989						
	1	2	3	4	5	6	7
2	30	4	31	0	16	10	0
3	0	0	19	0	18	0	0
4	11	0	34	1	25	26	0

more muddy. To check this possibility we investigate what happened to the stations belonging in 1985 to cluster 2. Cluster 2 consists of 97 stations, 92 of them being sampled in 1985. The 1989 data of these 92 stations are found in TWINSPAN cluster 1, 3, 5 and 6 (30, 31, 16 and 10 stations respectively) (Table 9). These four subgroups of stations are called group 2a to 2d. The environmental parameters of these 4 subgroups were compared with a Kruskal Wallis analysis of Variance and the Mann Withney U test both within and between years (Fig. 14). In 1985 group 2a and 2b differed in depth. This difference is still present in 1989 and additionally vegetation cover is higher in group 2a than 2b in 1989. Between 1985 and 1989 vegetation cover increased only in group 2a, no changes were found in 2b. Whereas only 1 species (*H. filiformis*) differed in

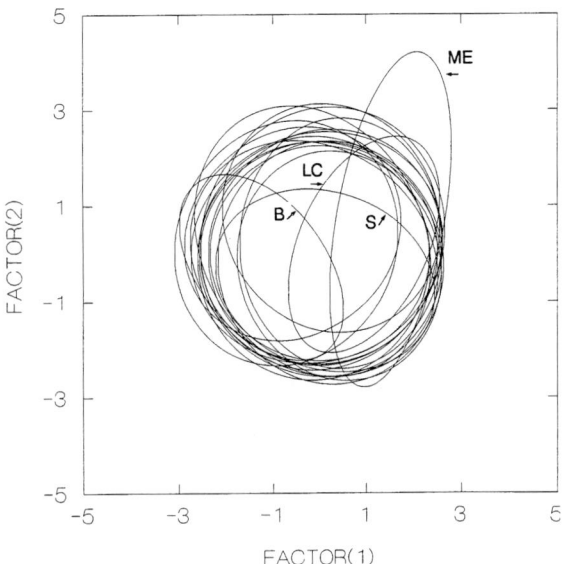

Fig. 13. Ninety-five percent confidence ellipses around the mean scores on discriminant functions (axis) 1 and 2 of the 20 most common macrobenthic species (based on a MDA on the density data of 1989). The ellipses of some species are indicated (ME, *Mytilus edulis*; LC, *Lanice conchilega*; S, *Spio sp.*; B; *Bathyporeia sp.*).-

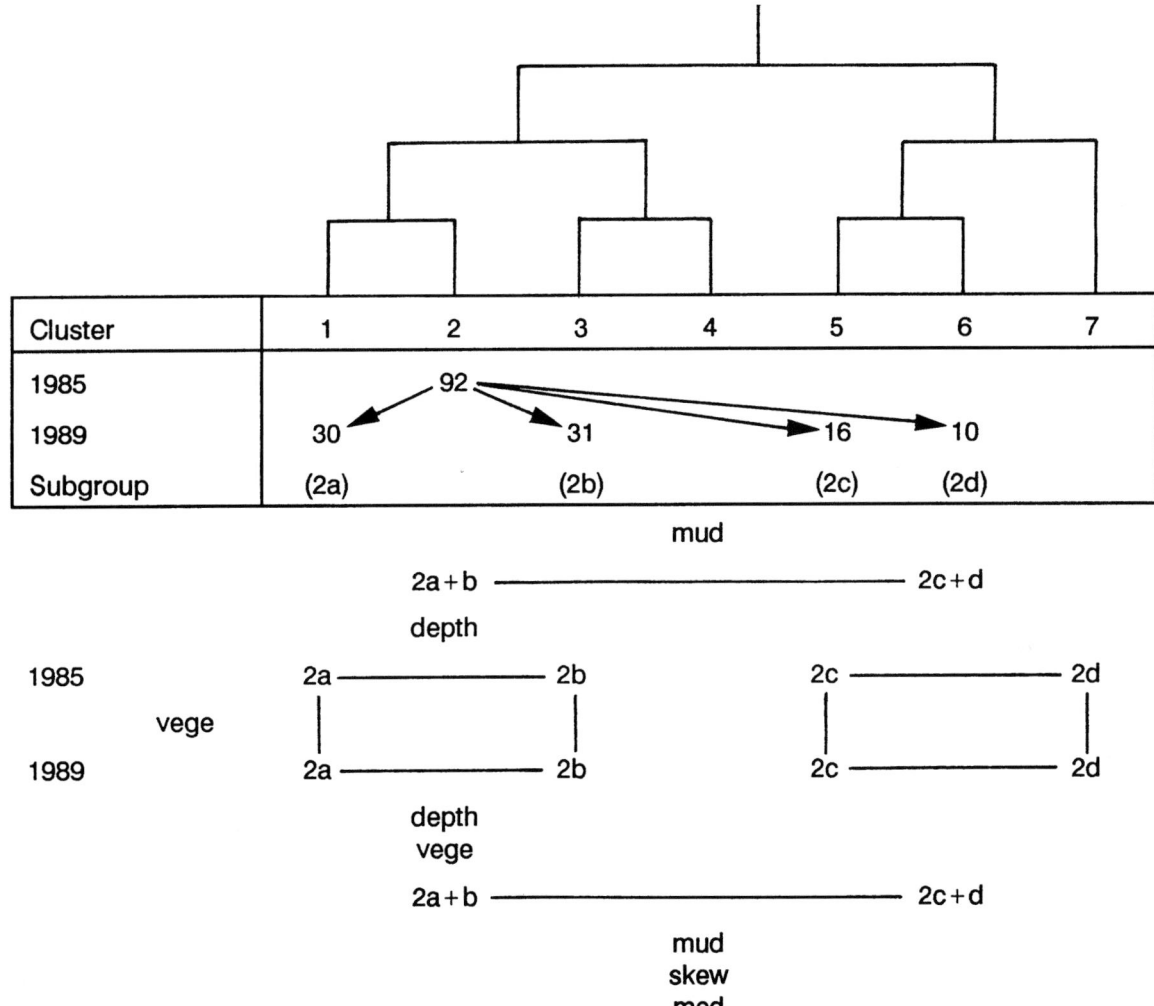

Fig. 14. Dendrogram of a TWINSPAN on density data of both 1985 and 1989 together with an analysis of what happened in 1989 to the stations whose 1985 data clustered together in cluster 2. The 1989 data of these stations were found in 4 different clusters and labelled as subgroup 2a–2d. The environmental variables that differed significantly between these subgroups both within and between years are indicated. For further explanation see text.

density between groups 2a and 2b in 1985 4 species (*A. marina, Gammarus* spec, *L. conchilega* and *S. bombyx*) did in 1989. Between years significant changes in density, number of species per sample and density of several species were found. The same holds for group 2c and d, with the exception that no differences in environmental variables were found within and between years. Within years also densities did not differ, between years it did for several species. Mud content differed between group 2(a + b) and group 2(c + d)

in 1985. In 1989 both groups differ not only in mud content but also in median grain size and sorting coefficient. Whereas no difference in density of individual species were found in 1985, in 1989 density of *H. filiformis* and Oligochaeta were significantly higher in group 2(a + b), density of *S. filicornis* significantly lower. This change in sediment characteristics is reflected in a larger faunal change since TWINSPAN cluster 5 and 6 are separated at an early stage from cluster 1 to 4. *H. filiformis* and Oligochaeta are known to prefer muddier sediments, *S. filicornis* coarser sediments. Similar results are obtained for the other groups.

These results could be an artifact if the different subgroups (e.g., 2a–d) were already separated in the TWINSPAN at a lower level of division than the one considered here. This was however not the case. The stations making up groups 2a–d were found in completely different combinations in the further subdivisions of cluster 2. The same is true for the other clusters.

From this analysis we can conclude that, although the environmental parameters in a group of stations, showing a large faunal similarity in one year, did not change, the faunal composition did. This indicates that faunal changes are not necessarily linked to changes in the measured environmental parameters.

Discussion

Observed changes in environmental parameters and benthos

Changes in environmental parameters
The coastal engineering works caused some important hydrodynamical changes in the Oosterschelde estuary (Vroon, 1994), influencing the morphology of the estuary (Mulder & Louters, 1993; Ten Brinke *et al.*, 1994), and the sediment composition of the tidal flats. On average, sediments became coarser with a lower mud content. According to Ten Brinke *et al.* (1994), the observed changes on the Roggen- and Galgeplaat can not be seen independently of the natural seasonal variation in these parameters. Only on the Krabbenkreek the changes in sediment composition could be attributed to the construction of the storm-surge barrier.

The erosion and hence the decrease in depth of the tidal flats is significant all over the Oosterschelde (Philippart, 1991; Mulder & Louters, 1994). Height of a site proper is of no importance to benthic animals but the emersion period is, which is of course function of the height (e.g., Hummel *et al.*, 1994). Due to the reduced tidal amplitude, above mid tidal level the immersion period decreased, below it increased. Thus major changes in emersion time occurred only in the upper and lower part of the intertidal zone. In between, the changes in emersion time due to the construction of the barrier are small, especially compared to the effects wind may have on the tides.

It can be concluded that on average the measured site characteristics (sediment and emersion) did not change markedly and that the range of observed values is similar before and after the construction of the barrier.

In the water column important changes occurred. The current speed was, at most sites, reduced by 30 to 40%, in the northern branch even by 70% (Vroon, 1994). Due to the manipulation of the barrier the wave attack on the tidal flats, especially during storms, increased in the zone NAP +0.5 m to NAP −1.0 m up to 30%. Primary productivity remained constant (Wetsteyn & Kromkamp, 1994).

Changes in benthic populations

Between both years significant differences in density, biomass and number of species per sample were found. Density and biomass of some species remained fairly constant while those of others changed dramatically. Comparing different years, the major benthic communities ranging from very muddy musselbeds with a high number of species to very sandy sites with a limited number of species (see Meire & Kuijken, 1984) remained present. However, within these larger habitats, we see important between year variability in the relative abundance of species etc.. The top dominant species within each community varies largely and the Canberra dissimilarity index for one study plot between two successive years amounts to about 60–70%, both for density or biomass of species (Meire unpublished data).

Based on the analyses presented above we want to investigate whether these changes are related to the environmental changes in the Oosterschelde or to other factors. We will firstly focus on the relation between the benthic invertebrates and sediment characteristics, afterwards on other environmental factors.

Environmental parameters influencing benthic populations

Relations with sediment parameters
The different methods used to describe the relationship between the distribution of benthic invertebrates and environmental parameters, all clearly indicated the importance of depth and sediment characteristics, especially mud content. This is in agreement with earlier studies showing that depth and mud content of the sediment are very important factors influencing benthic populations (e.g., Beukema, 1976; Dankers & Beukema, 1981; Gray, 1974). However, the analyses indicate that, although these parameters clearly influence the benthic invertebrates, the coupling between the species and these variables seems to be rather loose. The coefficients of determination of the multiple regression analyses are all very low (mostly lower than 0.1!) and the classification of stations by MDA showed a large degree of error. The relative importance of the different parameters also differed between years. In most analyses mud content is the dominant parameter in 1985, depth in 1989. As the range of both parameters was the same in both years the reason for this difference is not clear. The MDA results on the species classification shows a very large overlap between most species. These results contrast with those presented by Flint & Kalke (1986). In a similar analysis they found a very clear separation of most benthic species in the discriminant space. The difference probably can be explained by the fact that the range of sediment parameters in our study is very small, relative to the tolerance of the different species. This is corroborated by the TWINSPAN of both years. If the change in sediment parameters of a station is rather large, then species composition changes in a predictable way. These results indicate the importance of the sediment parameters but also indicate that they are insufficient to explain the observed patterns and that other factors are important in structuring benthos populations.

Relations with other environmental parameters
Next to sediment parameters the most important environmental variables influencing benthic populations are weather conditions. Severe winters are known to have a very important impact on benthic populations. Several species such as *Nephtys hombergii*, *Lanice conchilega* etc. are very sensitive to frost (Beukema, 1985) and their populations are likely to be decimated in severe winters. Additionally several species (*Arenicola marina*; most bivalves) are known to have very successful spatfalls after a severe winter (Beukema, 1985).

Comparison of the data presented in this paper with the long term patterns observed in the Oosterschelde indicates that the density of most species peaked in 1985 (Seys *et al.*, 1994; Coosen *et al.*, 1994a). This was probably a response to the very severe winter 1984/85 in which a high mortality of benthos occurred. In the 1985 samples, large numbers of small individuals (juveniles) were found. The cockle population collapsed resulting in low values for the total biomass, but high numbers of spat. After the severe winter of 1986/87 also much cockle spat was found. In 1989 these second year cockles made up an important part of the biomass (Coosen *et al.*, 1994a). The decrease in density and the increase in biomass do therefore not indicate long term trends but fluctuations around a long term mean, that did not seem to change until 1989, three seasons after the completion of the barrier (Seys *et al.*, 1994). The difference between years found in this study are therefore more likely to be caused by the weather conditions than by changes in sediment parameters.

Structure of the benthic communities: abiotic versus biotic control of populations

Biotic control
In general, the structure of macrobenthic communities and the concomitant processes have been studied intensively during the last decade;

some of the evidence has recently been reviewed by Wilson (1991). Although both inter- and intraspecific competition occur on tidal flats (e.g., Bonsdorff *et al.*, 1986; see Wilson, 1991 for review), predation is thought to be the most important form of biotic interaction (Reise, 1985; Wilson, 1991). Ambrose (1984) proposed a three level interactive model for soft-sediment marine systems, the first level being epibenthic predators (birds, fish, crabs and shrimps), the second being predatory infauna (mainly polychaetes like *Nereis diversicolor* and *Nephtys hombergii*) and the non-predatory infauna as third level. The importance of epibenthic predators is well known (Baird *et al.*, 1985; Meire *et al.*, 1994; Reise, 1985; Sanchez-Salazar *et al.*, 1987). The importance of infauna predators is only recognized since a decade and has recently been reviewed by Ambrose (1991). Estimates of feeding rates indicate that many predatory taxa have the potential to reduce the size of prey populations. Based on manipulative field experiments it was demonstrated that infaunal predators have indeed a significant effect on infaunal densities and affect the spatial and temporal distribution of their prey.

Abiotic control

Severe winters have an important direct effect on benthos as many individuals may die. They have also very important indirect effects (Kneib, 1991). Severe winters can be considered as a natural predator removal experiment (Reise, 1985): some predators are killed, others arrive later in spring on the flats. This can have pronounced effects. The increase of *Scoloplos armiger* after a severe winter, when *Nephtys hombergii*, its main predator, is decimated is a clear example (Coosen *et al.*, 1994a; Beukema, 1987). After a severe winter the water temperature needs more time to increase, causing juvenile crabs and shrimps, important predators of (juvenile) benthos, to move up the flats significantly later (Beukema 1985). At that time many juvenile benthos species are large enough to escape predation, and can survive, giving rise to high density populations. If large enough, they will then be prey of larger epibenthic predators as fish and birds, most of which are able to switch easily between different prey species.

Macrobenthos as a non-equilibrium community

Lying at the foundation of contemporary views of community patterns and the processes producing them is the presumption that these systems are at or close to equilibrium (by equilibrium is meant the stability or 'steadiness' of the community components) (Wiens, 1984). They are ecologically saturated, resource limited and governed by biotic interactions, especially competition. Fluctuations in natural communities are attributed to variations in resource levels which are closely tracked by members of the community. Several predictions can be made on ecomorphological patterns, diet niche relationships, patterns of habitat occupancy etc.

Estuarine benthic populations seem however not to be in any readily definable equilibrium. Species vary rather independently of one another (Beukema & Essink, 1986; Elliot & Ducrotoy, 1991; Coosen *et al.*, 1994a); the community seems not to be saturated. Several exotic species became established without any or small observable effect on native species. Examples are *Crepidula fornicata*, *Crassostrea* sp. and *Ensis directus*. The polychaete *Tharyx marioni* became established in the Wadden Sea in the early 1970's and is at present one of the most abundant species. No clear effect on other species was noticed (Reise, 1985). Also no clear effects of the introduction of *Marenzellaria viridis* on other species were found (Essink, pers. comm). Intertidal benthic populations also do not appear to be resource-limited very often. Biomass of suspension feeders varies between years with several orders of magnitude, their food supply (phytoplankton) does not (e.g., Wetsteyn & Kromkamp, 1994). A low biomass of cockles do not coincide with low values of primary production, suggesting, at that time, they will probably not be resource limited. Food supply for deposit feeders is difficult to measure but in predator exclusion experiments densities of several deposit feeders can increase substantially, which could indicate they are not resource limited (Reise, 1985). This is however in contrast to the results by Beukema & Cadee (1986) who attribute an increase in macrobenthic populations in the Wadden Sea to eutrophication. Infauna predation is also important in determining the observed patterns. Diet overlap occurs in many species (Ambrose, 1991) and the results presented above indicate there is a large overlap in habitat preference between species. Predation is probably more important than competition, which seems to occur more widely within than between species (Reise, 1985; Wilson, 1991).

Loose community structuring can be expected in highly variable or harsh environments. Under such conditions, resource levels and environmental conditions may at times be severely constraining. Such 'ecological crunches' (e.g., severe winters) may act in a major way to determine the ecological adaptations of species and the observed patterns of community composition. Much of the time the environment may be more benign and resource levels essentially non-limiting. Under these conditions populations and communities may be freed of close direct biotic or even abiotic control and vary in manners that erode the clean patterns expected from equilibrium theory (Wiens, 1984). These communities are nonequilibrium communities. They should be characterized by a general 'decoupling' of close biotic interactions, and the species should respond to environmental variations largely independent of one another. Habitats may not be fully saturated with individuals and species may be under represented. Populations and communities may be more strongly influenced by abiotic agents than by the imposition of ceilings on resource abundance, and population dynamics may be governed by effects that are largely independent of density. Collectively these factors produce communities that are only loosely structured, and clear consistent patterns may be largely lacking (Wiens, 1984). Equilibrium and nonequilibrium community structure are not discrete states of ecological communities, but are the opposite ends of a continuum (Wiens, 1984). The results presented in this paper, together with the evidence presented in the discussion, in our opinion, indicates that intertidal macrobenthic populations should be considered rather as nonequilibrium communities. Boecklen & Price (1991) came to the same conclusion for a population of sawflies on arrayo willow.

On the tidal flats different broad habitat types can be distinguished as muddy musselbeds and more exposed sites characterized by sandy sediments and large ribble marks. Most species, however, have a large tolerance for sediment parameters, especially relative to the range found in our study area. Within the range of conditions within specific habitats, e.g., musselbeds, most species can occur. Which species occur, and in which densities, is probably mainly determined by differences in settlement and subsequent predation. Differences in settlement are caused by variations in the number of offspring produced by the adults and, to a large extent, by the settlement conditions. The timing of settlement in relation to other species and especially to the invasion of the flats by shorecrabs will determine the population. This means that within each major habitat many different combinations of dominant species are possible. If the environmental parameters change drastically then a more pronounced change in benthic composition occurs. Within a habitat the benthic system can be seen in neighbourhood stability (Gray, 1977). The existence of many stable points could be seen as a consequence of the nonequilibrium community.

Conclusion

Although the impact of the construction of the barrier on the macrobenthic community seems at present to be rather small. This does not mean that on the long term there will be no effect. During 1985/1986 when the reduction in tidal amplitude was more pronounced (see Nienhuis & Smaal, 1994), clear effects on benthos were found in the upper part of the intertidal area, where the emersion time decreased drastically (Craeymeersch *et al.*, 1988; Seys *et al.*, in press). Furthermore it is possible that the important changes in the water column will have an effect on the transport or settlement of larvae, and the transport of food to the bottom. This was not studied until now. Also the availability of benthos as food for birds may decrease (Meire *et al.*, 1994). The expected long term change in the erosion-sedimentation equilibrium might over time have a pronounced effect on the whole ecosystem. Communication No. 695 of NIOO-CEMO, Yerseke, The Netherlands.

Acknowledgements

This paper could not have been written without the help of many people. The first campaign in 1985 was planned and organized by Jaap Van Der Meer who gave also many ideas for the data analysis. Dick De Jong and many other people from Rijkswaterstaat carried out the sampling. Part of the samples were analysed by Fred Twisk, Anette Van Den Dool and Ed Stickvoort. Rob Lambeck, Aad Smaal, Johan Craeymeersch, Anny Anselin, Greet Deguelder, Eckhart Kuijken, Martin Hermy, Dirk Bauwens, Carlo Heip and André Dhondt provided many valuable comments. Communication No. 695 of NIOO-CEMO, Yerseke, The Netherlands.

References

Ambrose, W. G., 1984. Role of predatory infauna in structuring soft-bottom communities. Mar. Ecol. Prog. Ser. 17: 109–115.

Ambrose, W. G. Jr., 1991. Are infaunal predators important in structuring marine soft-bottom communities?. Am. Zool. 31: 849–860.

Baird, D., P. R. Evans, H. Milne & N. W. Pienkowski, 1985. Utilization by shorebirds of benthic invertebrate production in intertidal areas. Oceanogr. Mar. Biol. annu. Rev. 23: 573–597.

Beukema, J. J., 1976. Biomass and species richness of the macro-benthic animals living on the tidal flats of the Dutch Wadden-Sea. Neth. J. Sea Res. 10: 236–261.

Beukema, J. J., 1985. Zoobenthos survival during severe winters on high and low tidal flats in the Dutch Wadden Sea. In J. S. Gray & M. E. Christiansen (eds), Marine Biology of Polar regions and effects of stress on marine organisms. Wiley & Sons, New York: 351–361.

Beukema, J. J., 1987. Influence of the predatory polychaete *Nephtys hombergii* on the abundance of other polychaetes. Mar. Ecol. Prog. Ser. 40: 95–101.

Beukema, J., 1989. Long-term changes in macrobenthic abundance on the tidal flats of the Western Wadden Sea. Helgoländer Meeresunters. 43: 405–415.

Beukema, J. J. & G. C. Cadee, 1986. Zoobenthos responses to eutrophication of the Dutch Wadden Sea. Ophelia 26: 65–76.

Beukema, J. J. & K. Essink, 1986. Common patterns in the fluctuations of macrobenthic species living at the different places on tidal flats in the Wadden Sea. Hydrobiologia 142: 199–207.

Boecklen, W. J. & P. W. Price, 1991. Nonequilibrium community structure of sawflies on arroya willow. Oecologia 85: 483–491.

Bonsdorff, E., J. Mattila, C. Rönn & C-S. Oosterman, 1986. Multidimensional interactions in shallow soft-bottom ecosystems: testing the competitive exclusion principle. Ophelia Suppl. 4: 37–44.

Coosen, J., J. Seys, P. M. Meire & J. A. M. Craeymeersch, 1994a. Effect of sedimentological and hydrodynamical changes in the intertidal areas of the Oosterschelde estuary (SW Netherlands) on distribution, density and biomass of five common macrobenthic species: *Spio martinensis* (Mesnil), *Hydrobia ulvae* (Pennant), *Arenicola marina* (L.), *Scoloplos armiger* (Muller) and *Bathyporeia* sp. Hydrobiologia 282/283: 235–249.

Coosen, J., F. Twisk, M. W. M. van der Tol, R. H. D. Lambeck, M. R. van Stralen & P. M. Meire, 1994b. Variability in stock assessment of cockles (*Cerastoderma edule* L.) in the Oosterschelde (in 1980–1990), in relation to environmental factors. Hydrobiologia 282/283: 381–395.

Craeymeersch, J. A., J. Coosen & A. Van den Dool, 1988. Trendanalyse van densiteit- en biomassawaarden van bodemdieren in het getijdengebied in de Oosterschelde (1983–1986). Rapporten en Verslagen 1988–7, DIHO, Yerseke.

Dankers, N. & J. J. Beukema, 1981. Distributional patterns of macrozoobenthic species in relation to some environmental factors. In N. Dankers, H. Kühl & W. J. Wolff (eds), Invertebrates of the Wadden Sea. Balkema, Rotterdam: 4/69–4/103.

De Jong, D. J., P. H. Nienhuis & B. J. Kater, 1994. Microphytobenthos in the Oosterschelde (The Netherlands), 1981–1990; consequences of a changed tidal regime. Hydrobiologia 282/283: 183–195.

Desprez, M., J-P. Ducrotoy & B. Sylvand, 1986. Fluctuations naturelles et évolution artificielle des biocénoces macrozoobenthiques intertidales de trois estuaires des côtes françaises de la Manche. Hydrobiologia 142: 217–232.

Dörjes, J., H. Michaelis & B. Rhode, 1986. Long-term studies of macrozoobenthos in intertidal and shallow subtidal habitats near the island of Norderney (East Frisian coast, Germany). Hydrobiologia 142: 217–232.

Elliot, M. & J. P. Ducrotoy, 1991. Estuaries and Coasts: spatial and temporal intercomparisons. Olson & Olson, International Symposium Series, Fredensborg.

Essink, K., 1988. Recent introduction of North American marine invertebrates in the North Sea and the Baltic Sea. Rijkswaterstaat, Dienst Getijdewateren, Notitie GWAO-88.2240, Haren.

Fauchald, K. & P. A. Jumars, 1979. The diet of worms. A study of polychaete feeding guilds. Oceanogr. Mar. Biol. annu. Rev. 17: 193–284.

Flint, R. W. & R. D. Kalke, 1986. Niche characterisation of dominant estuarine benthic species. Estuar. coast. shelf Sci. 22: 657–674.

Gray, J. S., 1974. Animal-sediment relationships. Oceanogr. Mar. Biol.: an annu. Rev. 12: 223–161.

Gray, J. S., 1977. The stability of benthic ecosystems. Helgoländer wiss. Meeresunters. 30: 427–444.

Hill, M. O, 1979. TWINSPAN, a fortran program for arraging multivariate data in an ordered two-way table by classification of individuals and attributes. Cornell, Ithaca; N.Y.

Hummel, H., A. W. Fortuin, R. H. Bogaards, A. Meijboom & L. De Wolf, 1994. The effects of prolonged emersion and submersin by tidal manipulation on marine macrobenthos. Hydrobiologia 282/283: 219–234.

James, F. C. & C. E. McCulloch, 1990. Multivariate analysis in ecology and systematics: panacea or pandora's box? Annu. Rev. Ecol. Syst. 21: 129–166.

Kneib, R. T., 1991. Indirect effects in experimental studies of marine soft-sediment communities. Am. Zool. 31: 874–885.

Meire, P. M. & E. Kuijken, 1984. Relations between the distribution of waders and the intertidal benthic fauna of the Oosterschelde, Netherlands. In P. R. Evans, J. D. Goss-Custard & W. G. Hale (eds), Coastal waders and wildfowl in winter. Cambridge University Press, Cambridge: 57–68.

Meire, P. M., H. Schekkerman & P. L. Meininger, 1994. Consumption of benthic invertebrates by waterbirds in the Oosterschelde estuary, SW Netherlands. Hydrobiologia 282/283: 525–546.

Meire, P. M., J. Buijs, J. Seys, F. Twisk, E. Stikvoort, D. de Jong, 1991b. Macrozoobenthos in de Oosterschelde in het najaar 1989: een situering van de gegevens. Rapport RUG-WWE nr. 30, Universiteit Gent, Gent.

Meire, P. M., J. Buijs, J. Van Der Meer, A. van den Dool, A. Engelberts, F. Twisk, E. Stikvoort, D. de Jong, R. H. D. Lambeck & J. Polderman, 1991a. Macrozoobenthos in de oosterschelde in het najaar 1985: een situering van de gegevens. Rapport RUG-WWE nr. 27, Universiteit Gent, Gent.

Mulder, J. P. M. & T. Louters, 1994. Changes in basin geomorphology after implementation of the Oosterschelde project. Hydrobiologia 282/283: 29–39.

Nienhuis, P. & A. Smaal, 1994. The Oosterschelde estuary, a case study of a changing ecosystem: an introduction. Hydrobiologia 282/283: 1–14.

Philipart, M. E., 1991. Bodemligging Oosterschelde: de invloed van de werken. Rapport 1991–12, Geografisch Instituut Rijksuniversiteit Utrecht, Utrecht.

Prins, T. & A. Smaal, 1994. The role of the blue mussel *Mytilus edulis* in the cycling of nutrients in the Oosterschelde estuary (The Netherlands). Hydrobiologia 282/283: 413–429.

Reise, K., 1985. Tidal Flat Ecology: an experimental approach to species interactions. Springer-Verlag, Berlin.

Sanchez-Salazar, M. E., C. L. Griffiths & R. Seed, 1987. The interactive roles of predation and tidal elevation in structuring populations of the edible Cockle, *Cerastoderma edule*. Estuar. coast. Shelf. Sci. 25: 245–260.

Seys, J. & P. Meire, 1992. Long-term fluctuations (1979–1989) of the intertidal macrozoobenthos in the Oosterschelde: before, during and after the construction of the storm-surge barrier. Rapport WWE nr. 33, Universiteit Gent, Gent.

Seys, J. J., P. M. Meire, J. Coosen & J. A. Craeymeersch, 1994. Long-term changes (1979–89) in the intertidal macrozoobenthos of the Oosterschelde estuary: are patterns in total density, biomass and diversity induced by the construction of the storm-surge barrier? Hydrobiologia 282/283: 251–264.

Seys, J. J., P. M. Meire, J. Coosen, J. A. Craeymeersch, R. H. D. Lambeck, A. van den Dool, in press. The intertidal macrozoobenthos in the Oosterschelde before, during and after the construction of the storm-surge barrier: some preliminary results. In N. V. Jones (ed.), The changing coastline: in press.

Siegel, S., 1956. Nonparametric Statistics for the behavioural sciences. McGraw-Hill, Tokyo.

Ten Brinke, W. B. M., J. Dronkers & J. P. M. Mulder, 1994. Fine sediments in the Oosterschelde estuary before and after partial closure. Hydrobiologia 282/283: 41–56.

Van Der Meer, J., 1991. Exploring macrobenthos-environment relationship by canonical correlation analysis. J. exp. mar. Biol. Ecol. 148: 105–120.

Van Der Meer, J., A. van den Dool, A. Engelberts, D. de Jong, R. H. D. Lambeck & J. Polderman, 1989. Het macrozoöbenthos van enkele intergetijdengebieden in de Oosterschelde. Resultaten van een inventarisatie in de nazomer van 1985. GWAO 89.307, Rijkswaterstaat, Middelburg.

Van Stralen, M. R. & R. D. Dijkema, 1994. Mussel culture in a changing environment: the effects of a coastal engineering project on mussel culture. (*Mytilus edulis* L.) in the Oosterschelde estuary (SW Netherlands). Hydrobiologia 282/283: 359–379.

Vroon, J., 1994. Hydrodynamic characteristics of the Oosterschelde in recent decades. Hydrobiologia 282/283: 17–27.

Warwick, R. M., J. D. Goss-Custard, R. Kirby, C. L. George, N. D. Pope & A. A. Rowden, 1991. Static and dynamic environmental factors determining the community structure of estuarine macrobenthos in SW Britain: why is the Severn estuary different? J. appl. Ecol. 28: 1004–1026.

Weston, D. P., 1988. Macrobenthos-sediment relationships on the continental shelf of Cape Hatteras, North Carolina. Continental Shelf Res. 8: 267–286.

Wetsteyn, L. P., M. J. & J. C. Kromkamp, 1994. Turbidity, nutrients and phytoplankton primary production in the Oosterschelde (The Netherlands) before, during and after a large-scale coastal engineering project (1980–1990). Hydrobiologia 282/283: 61–78.

Wiens, J., 1984. On understanding a non-equilibrium world: myth and reality in community patterns and processes. In D. R. Strong, D. Simberloff, L. G. Abele & A. B. Thistle (eds), Ecological communities. Conceptual issues and the evidence. Princeton University Press, Princeton: 439–457.

Wilkinson, L., 1988. SYSTAT, the system for statistics. SYSTAT, Evanstin, Illinois.

Wilson, W. H., 1991. Competition and predation in marine softsediment communities. Annu. Rev. Ecol. Syst. 21: 221–241.

Hydrobiologia **282/283**: 183–195, 1994.
P. H. Nienhuis & A. C. Smaal (eds), The Oosterschelde Estuary.
© 1994 *Kluwer Academic Publishers. Printed in Belgium.*

Microphytobenthos in the Oosterschelde estuary (The Netherlands), 1981–1990; consequences of a changed tidal regime

D. J. de Jong[1], P. H. Nienhuis[2] & B. J. Kater[1]
[1] *National Institute for Coastal and Marine Management/RIKZ, P.O. Box 8039, NL-4330 EA Middelburg, The Netherlands;* [2] *Netherlands Institute of Ecological Research, Centre for Estuarine and Marine Ecology, Vierstraat 28, NL-4401 EA Yerseke, The Netherlands*

Key words: microphytobenthos, biomass, time-series, tidal change, Oosterschelde estuary, depth distribution, hydrodynamics

Abstract

During the period 1981–1990 the functioning of microphytobenthos in the carbon cycle was studied in the Oosterschelde, a mesotidal, euhaline estuary (SW Netherlands), both before and after completion of a storm-surge barrier in the sea ward entrance of the estuary in 1986, which reduced the tidal range to 88% and current velocities to 70% of their former values on average.

The annual biomass cycle has changed from small spring and autumn peaks into a much larger summer peak. The average biomass during summer has increased from 70 to 170 mg Chlorophyll-a m^2. The average annual biomass has increased from 115 to 195 mg Chlorophyll-a m^2. As a consequence the (calculated) primary production by microphytobenthos has increased also, from 150 to 242 gC m^{-2} y^{-1} (14 045 to 22 265 tonnes C y^{-1}), and its share in the total primary production of the Oosterschelde has increased from 16 to 30%. These increases in biomass and primary production are mainly ascribed to a decrease in the dynamic forces (water current velocities) over the intertidal flats in most parts of the basin. Increased water transparency in parts of the estuary and increased import of inorganic carbon from the water column towards shoals may have contributed as well.

The rate of reworking of the top layer of the soil (0–10 cm) has not changed significantly, as the decrease in Chlorophyll-a biomass with depth has hardly changed.

Introduction

This paper looks at the microphytobenthos (unicellular algae) on intertidal sediments, mainly consisting of diatoms, (Colijn & Nienhuis, 1977; Colijn & Koeman, 1975; Colijn & Dijkema, 1981; Vos *et al.*, 1988). As they are dependent on photosynthesis for growing, the active cells are confined to the upper few millimeters of the sediment, the 'photosynthetically active' layer (Admiraal, 1980). However, they are found in substantial quantities to depths of 10 cm and more in the sediment, due to active migration (Cadée & Hegeman, 1974), hydrodynamic forces (De Jonge, 1992) and bioturbation (Cadée, 1976).

In this way, both in the western Wadden Sea (Cadée & Hegeman, 1974) and the Oosterschelde (1993), on average 25% of the biomass present in the 0–10 cm layer is present in the 0–1 cm layer versus 75% in the 1–10 cm layer. On the one hand microphytobenthos can be resuspended by hydrodynamic forces (tidal currents and waves)

(De Jonge, 1985; De Jonge & van den Bergs, 1987; Delgado et al., 1991; Baillie & Welsh, 1980; Grant, 1981; Gabrielson & Lukatelich, 1985), becoming temporarily part of the 'phytoplankton', after which they are deposited again at the sediment surface. Consequently, a considerable percentage of the microphytobenthos may be suspended permanently in the water column (De Jonge & van Beusekom, 1992). On the other hand, the cells may be buried in the sediment by hydrodynamic forces, where the diatoms remain inactive or change to heterotrophic growth (Admiraal & Pelletier, 1979; Cadée & Hegeman, 1974) until they are reworked again to the sediment surface, or die. In this way they are an important stock of primary producers, able to start to photosynthesise immediately when tidal flats are eroded during gales (Cadée & Hegeman, 1974).

Microphytobenthos plays an important role in the carbon cycle of estuarine systems (Klepper, 1989), being a significant food source for zoobenthos, both macro- and meiobenthos (e.g. Admiraal et al., 1983; H. Asmus, 1982; Bianchi & Rice, 1988). However, birds may feed on it as well: e.g. Shelduck (*Tadorna tadorna*) (Meininger & Snoek, 1992; D. J. de Jong, pers. obs.) and Brent Goose (*Branta bernicla*) (Cadée, 1972; D. J. de Jong, pers. obs.). A significant aspect of microphytobenthos occurrence is that it is present in substantial quantities the whole year round (Colijn & Dijkema, 1981), in contrast with phytoplankton which is almost absent during winter. A major physical aspect is that microphytobenthos plays an important role in the sediment-binding processes on intertidal shoals (Vos et al., 1988; Grant et al., 1986).

Much research has been carried out into spatial and seasonal distribution patterns of microphytobenthos, often related to hydrodynamic factors (e.g. R. Asmus, 1982; Baillie & Welsh, 1980; Cadée & Hegeman, 1974, 1977; Colijn & Dijkema, 1981; Colijn & Nienhuis, 1977; Colijn & Koeman, 1975; De Jonge, 1992). However, little is known about the effects of a significant change in hydrodynamics on the ecology of these microalgae.

The construction of large civil engineering works in the Oosterschelde (a storm-surge barrier in the mouth of the basin and two auxiliary dams, completed in 1987), leading to significant changes in the hydrodynamic forces (Nienhuis & Smaal, 1994; Vroon, 1994), offered ample opportunity to investigate their impact. For that purpose, research into changes in biomass of microphytobenthos was carried out both before and after the construction of these civil engineering works. Although this research was carried out in different programmes (1981–1982: Daemen et al., 1985; 1983–1984: Vos, 1986; 1985 onwards: D. J. de Jong, unpubl.), yet the field sampling methods used are comparable, which enables us to include all three programmes.

The developments observed in biomass of the microphytobenthos before and after completion of the barrier will be discussed in this paper and some remarks will be made on the consequences with respect to the primary production of these microalgae.

Materials and methods

Research area

The Oosterschelde is an euhaline mesotidal basin (Fig. 1). It has an area of about 30 000 ha, about 11 500 ha of which are sandy tidal flats, with a clay content of $< 2\%$; in sheltered places and on (cultured) mussel plots the clay content may increase to 5–10% and occasionally up to 20–30%. The water is quite clear (transparency between 0.5 m in winter and 3 m in summer). For more details see e.g. Knoester et al. (1984), Nienhuis & Smaal (1994), Mulder & Louters (1994) and Vroon (1994).

A storm-surge barrier was constructed in the seaward mouth of the estuary (completed spring 1987) reducing the tidal range by about 12% (from 3.6 to 3.2 m) and tidal currents by 30–40%, and locally by as much as 70–90%. Wave impact on the shoals changed little in general, except in the most western part, where due to the presence of the barrier it decreased to some extent. See also Fig. 1 and Nienhuis & Smaal (1994), Vroon (1994).

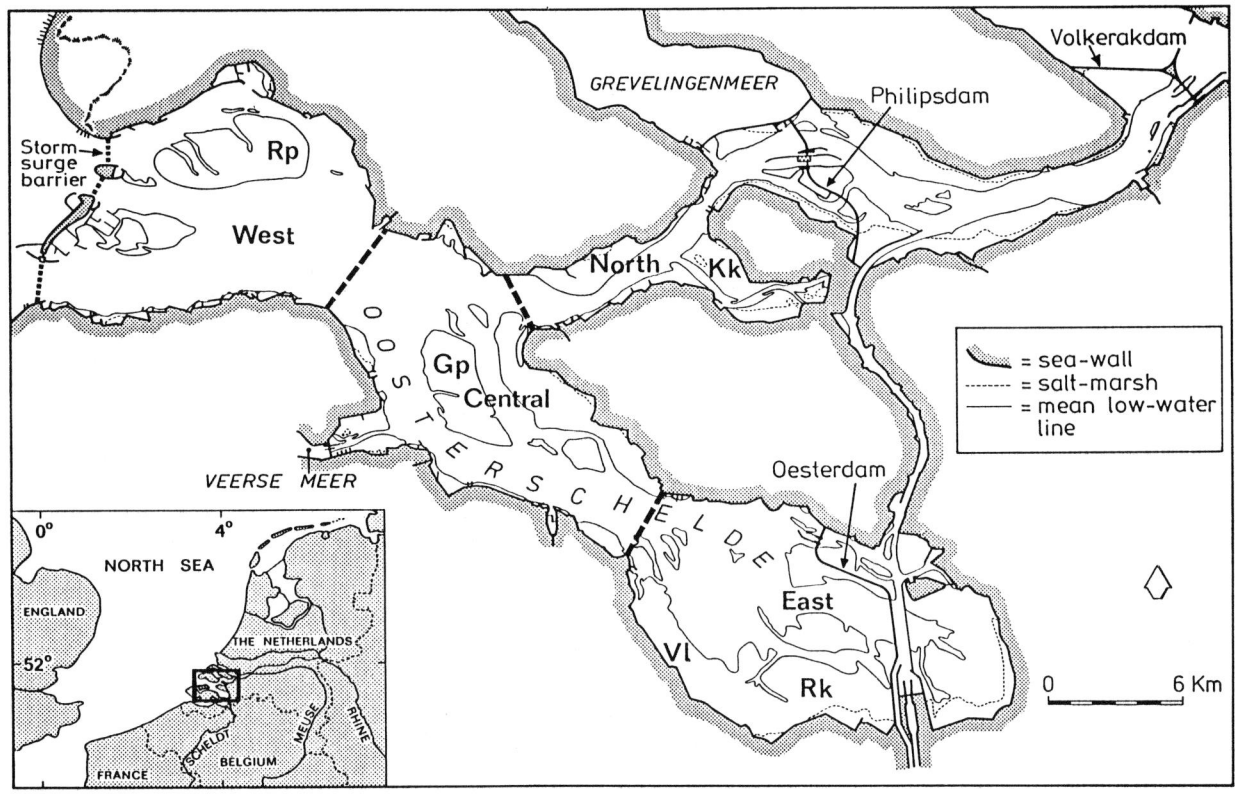

Fig. 1. Map of the Oosterschelde estuary showing the shoals and tidal flats investigated: Gp = Galgeplaat, Kk = Krabbenkreek, Rk = Rattekaai, Rp = Roggenplaat, Vl = Verdronken land; east = eastern basin, west + central + north = remaining basin.

Biomass surveys

Biomass sampling was carried out by mixing 5–10 samples from the topmost centimeter of the sediment, using a perspex corer with a diameter of 2.3 cm. The samples were either processed immediately or deep-frozen ($-20\ °C$) before further laboratory processing. The mixed samples were analyzed for chlorophyll-*a* in the laboratory. For the period 1981–82 the spectrophotometric method according to Lorenzen (Lorenzen, 1967; Daemen, 1986) was applied and from 1983 onwards the HPLC-method (Daemen, 1986). A conversion factor of 0.7 was necessary to make spectrophotometer and HPLC results comparable: spectrophotometer value * 0.7 = HPLC value (Daemen, 1986; Daemen & De Leeuw Vereecken, 1985). All results published in this paper are expressed as HPLC values (analysed by or converted to the HPLC method). The re-

sults of the chemical analyses in $\mu g\ g^{-1}$ dry sediment are converted to mg m^{-2} by multiplying by 15.5, based on an average bulk density of the Oosterschelde sediment of 1.55 g cm^{-3}.

Two major field sampling programmes have been carried out:

1. a monthly sampling programme in 1981–1984 and in 1989–1990 at 25 permanent plots on 4 intertidal areas (Roggenplaat, Galgeplaat, Rattekaai, Verdronken Land; Fig. 1), which were selected from the programmes in 1981–82 and 1983–84;
2. a summer (August) sampling programme at 300 permanent plots distributed stratified randomly over 3 intertidal areas (Roggenplaat, Galgeplaat, Krabbenkreek; Fig. 1) from 1985–1990, which were compared to 17 plots on the same intertidal areas sampled in August 1981–82.

In 1982–83 and in 1990 at a number of plots additional biomass samples were taken to a depth of 10 cm; in 1982–83 in 5 subsequent layers of 1 cm (0–1, 1–2, 2–3, 3–4, 4–5 cm) and 2 layers of 2.5 cm (5–7.5, 7.5–10 cm) and in 1990 of 10 subsequent layers of 1 cm.

As the sample plots of the monthly sampling programme have not been chosen at random, they have to be considered as individual plots which react individually to the changed situation. Therefore the biomass data of this programme has had to be analysed with a 3-way ANOVA (Sokal & Rohlf, 1981; statistical packet SPSS) to determine the significance of the observed changes. Tests were run to check that all conditions for an

Table 1. Year-averaged biomass mirophytobenthos (upper 1 cm, mg Chl-*a* m^{-2}); the plots in 1981/84 are the same as in 1989/90. (Oosterschelde: taking number of plots per area into account).

	1981/84	1989/90	N
Roggenplaat	86.9	149.8	8
Galgeplaat	112.5	134.1	7
Rattekaai	144.3	206.6	4
Verdronken Land	157.2	341.4	6
'Oosterschelde'	115	195	25

ANOVA were met; the ANOVA was carried out with year, month and plot as independent parameters and biomass as a variable.

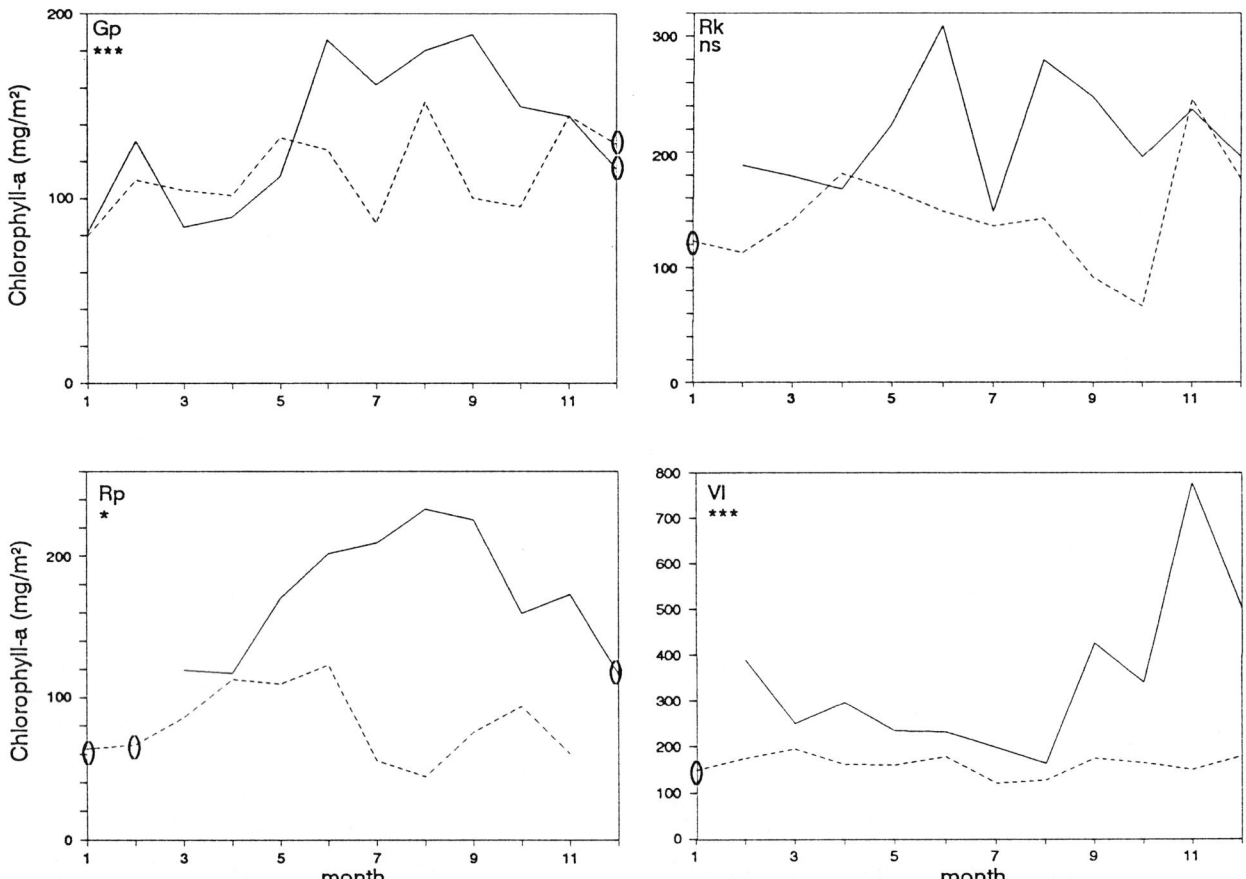

Fig. 2. Average annual variation in the biomass of microphytobenthos in 1981–84 (dashed line) and 1989–90 (solid line). Months between parentheses are not used in the ANOVA because of lack of data in the corresponding period. Gp: Galgeplaat, Rk: Rattekaai, Rp: Roggenplaat, Vl: Verdronken land. ns: not significant, *: $0.05 < p \leq 0.01$, **: $0.01 < p \leq 0.001$, ***: $p < 0.001$ (2nd order interaction year * month).

Due to the stratified random sampling strategy, the samples of the 1985–1990 August sampling programme can be considered as subsamples of the three areas investigated and thus mean and standard error may be calculated. For the sake of comparison, the same has been done with the data sampled in August 1981–82; since no data were available for some plots for August 1981 for these the mean biomass values of July and September 1981 were used.

Primary production

Gross primary production is calculated as a function of the average biomass per year according to Cadée & Hegeman (1977) and Colijn & de Jonge (1984). Based on their data of average biomass per year versus primary production two regression equations were derived:

$$\text{Cadée:} \quad P = 1.22B + 1.77 \quad (R^2 = 0.86)$$

$$\text{Colijn:} \quad P = 1.06B + 13.91 \quad (R^2 = 0.99)$$

in which:
P = gross primary production in g C m^{-2} y^{-1},
B = average biomass in mg chlorophyll-a m^{-2} y^{-1} in the 0–1 cm-layer.

As both relations are almost similar they are combined into one general equation:

$$P = 1.13B + 8.23 \quad (R^2 = 0.92)$$

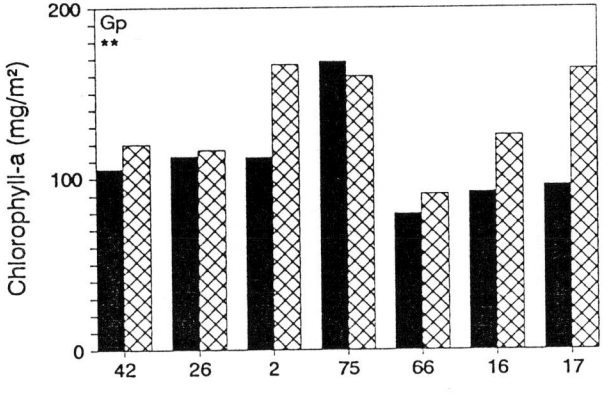
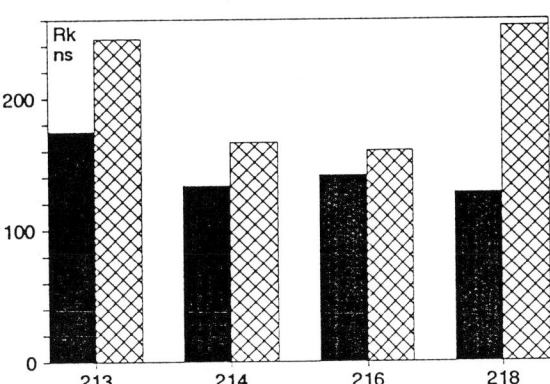
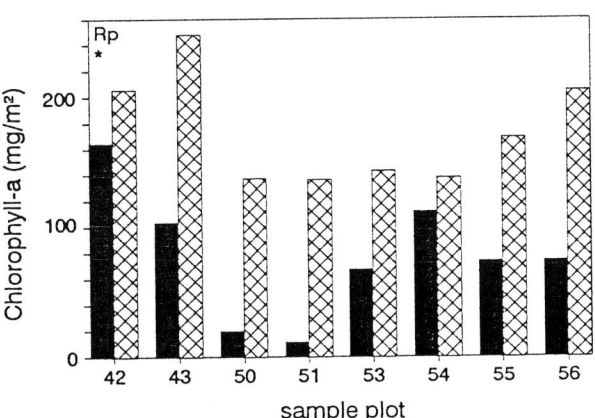
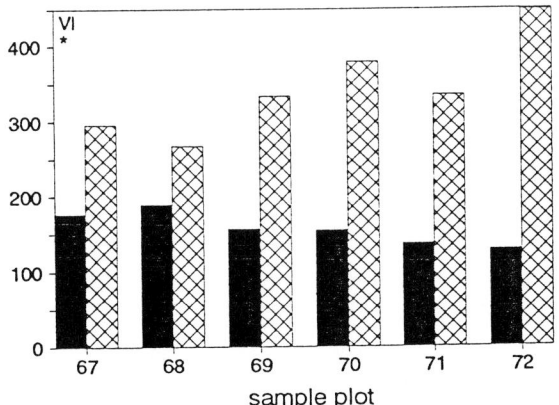

Fig. 3. Mean biomass of microphytobenthos in the pre-barrier (dark) and post-barrier (hatched) period for the individual plots; further explanation see Fig. 2. (numbers of sample plots have no specific meaning in this context) (2nd order interaction year * plot).

188

The following conversions have been made:

- biomass figures of Colijn & de Jonge (1984) are converted from a layer of 0–0.5 cm to a layer of 0–1 cm by multiplying by 1.5; as no data were available for the Ems estuary, this factor was established empirically for Oosterschelde data;
- C/Chlorophyll-*a* ratio = 40 (for HPLC-chlorophyll-*a* values); although de Jonge (1980) mentions large variations in this ratio for the Ems estuary, yet one ratio was chosen based on Oosterschelde data from Daemen & de Leeuw-Vereecken (1985).

Results

Biomass development

Table 1 indicates the year-averaged biomass in the 4 areas investigated both before and after the construction of the storm-surge barrier (= pre-barrier and post-barrier period respectively). The annual course of the biomass in each area is indicated for both periods in Fig. 2. Figure 3 shows the year-averaged biomass of each plot. Figure 4 shows the biomass in August in both periods. Table 2 gives the statistics concerning the 3-way ANOVA.

From these Tables and Figures it can be concluded that in the post-barrier period in most areas the biomass values have increased significantly, the year-averaged values by 1.2–2.2 × (average 1.7 ×) and the August values by 2.2–2.8 × (average 2.5). A striking fact is that the increase in August occurs as early as in 1985 (Fig. 4), *i.e.* before any significant change in the tidal range can be observed. In addition, the variations in biomass in the three areas are strikingly similar. It should also be noted that the samples were taken each year within a period of about 4 weeks, sampling the areas 'at random'.

Table 2 shows that 3rd order interactions never occur, but 2nd order interactions do; except for

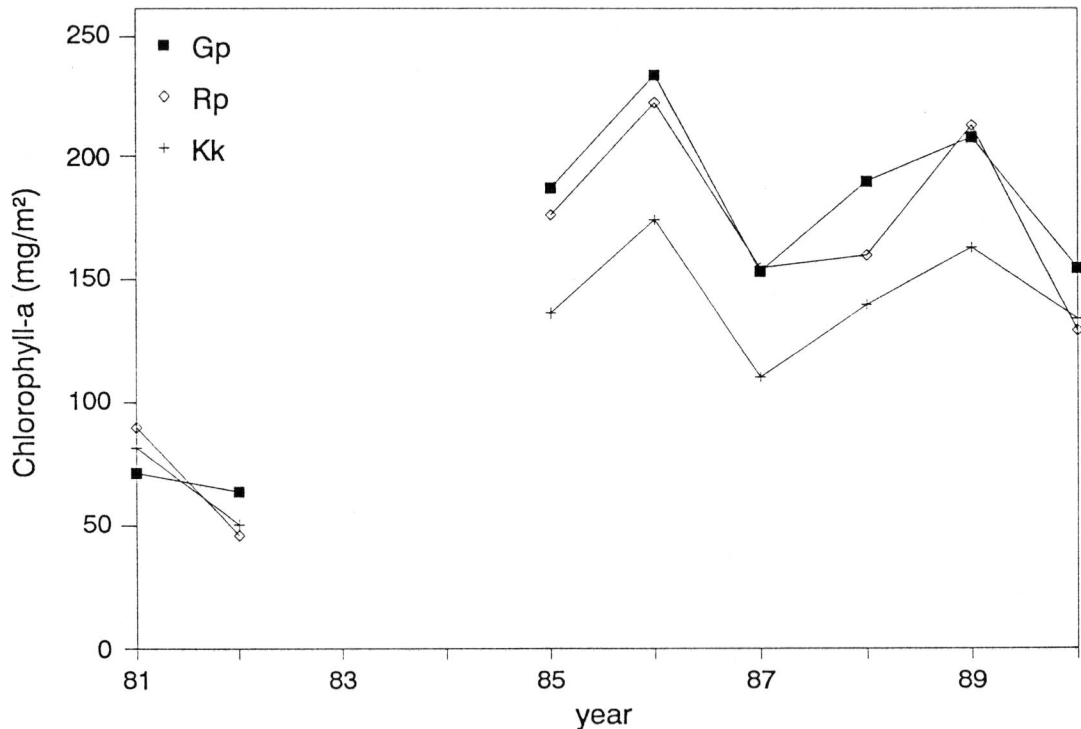

Fig. 4. August biomass of microphytobenthos at Galgeplaat (Gp), Krabbenkreek (Kk) and Roggenplaat (Rp). Standard error variations: 1981–82: 6-16; 1985–90: 4-10. Number of plots: 1981–82: Gp-5, Kk-4, Rp-8; 1985–90: Gp-110, Kk-75, Rp-120.

Table 2. Three way ANOVA (*p*-values) with plot, month and year as independent variables and biomass as the dependent variable.

	Rattekaai	Galgeplaat	Roggenplaat	Verdronken Land
1st order interaction				
Plot	0.008	0.000	0.000	0.642
Month	0.134	0.000	0.223	0.000
Year	0.000	0.000	0.000	0.000
2nd order interaction				
Plot * month	0.970	0.379	0.965	1.000
Year * month	0.236	0.000	0.012	0.000
Plot * year	0.098	0.003	0.046	0.044
3rd order interaction				
Plot * month * year	0.996	0.507	0.957	0.998

Rattekaai with only 1st order interactions. From the 2nd order interactions it can be concluded that in most areas the annual course in biomass has changed significantly (see also Fig. 2), generally from a small spring and autumn peak in the pre-barrier period into a large summer peak now. However, at the Verdronken Land, at present a large autumn/winter peak appears. At Rattekaai, the change in the annual course is not significant (Fig. 2), which might be due to the low value in July 1989/90 and the high peak in November 1984, but this has not been tested any further. It can be concluded from the 2nd order interactions that the annual increase occurs for each point individually (see also Fig. 3); again with exception of Rattekaai. However, the results of the 1st order interaction show that at Rattekaai the increase has been significant for the entire area.

Figure 5 presents the decrease in biomass in the 0–10 cm zone: although the absolute amounts have changed, the course in biomass with depth in the layer 0–10 cm has not changed.

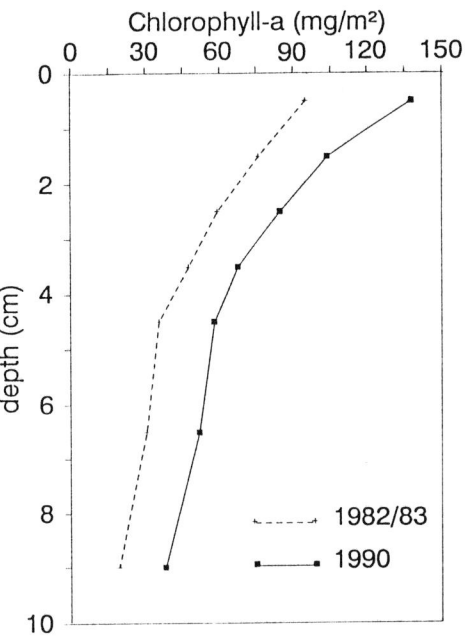

Fig. 5. Decrease in biomass with depth in the sediment in 1982/83 (dashed line) and 1990 (solid line). Data 1982/83: average of 2 plots monthly sampled + 2 plots sampled in June and November; data 1990: average of 9 plots sampled in April and June.

Primary production

In order to calculate the primary production of the microphytobenthos the Oosterschelde has been divided into two areas: the Eastern compartment (Rattekaai + Verdronken Land) and the Western + Middle compartment (Galgeplaat +

Roggenplaat). As from the Northern compartment only data from the pre-barrier period were available this compartment is not taken into account for the primary production calculations; however, this compartment is only about 14% of the total Oosterschelde, and it may be regarded

Table 3. Gross primary production of the microphytobenthos on the intertidal flats in the Oosterschelde (compartment east and west + middle resp.), calculated according to different relations between biomass and primary production. (East = average of Verdronken Land + Rattekaai; west + middle = average of Galgeplaat + Roggenplaat).

	East		West + middle		Total east + middle + west	
	< 1985	≥ 1985	< 1985	≥ 1985	< 1985	≥ 1985
Biomass						
Year-average (mg Chl-*a* m^{-2})	150	275	100	140		
Total (tonnes C)	240	440	229	320	469	760
Primary production						
g C m^{-2} y^{-1}	178	319	121	166		
Tonnes C Y^{-1}	7125	12770	6920	9495	14045	22265

as far as the benthic primary production is concerned as about similar to the Western and Middle compartment. According to Table 1 (and also to Daemen & de Leeuw-Vereecken, 1985) within these areas in both periods the yearly-averaged biomass remains in the same order of magnitude, while between these areas the biomass differs considerably. In Table 3 figures are presented for both the pre-barrier and the post-barrier period. Average gross production on the tidal flats varied roughly between 120 and 180 gC m^{-2} y^{-1} in the pre-barrier period, and increased to 165–320 gC m^{-2} y^{-1} in the post-barrier period.

Discussion

Biomass developments

The average annual amount of biomass has increased strongly when the pre-barrier and the post-barrier periods are compared. This increase is manifested mainly during the summer period, in the sediment down to a depth of 10 cm. This increase of biomass may be caused by a number of factors. A change in the method of analysis may have played a role in the first place; but also changes in one or more growth-limiting factors could be responsible, such as available quantity of light, hydraulic influences above shoals and tidal flats, grazing by zoobenthos and a changed availability of nutrients. These factors will be discussed briefly.

Change in the method of analysis

Different methods of analysis were used in 1981–1982 and from 1983 onwards: Spectro-Photo-Meter (SPM) and High Pressure Liquid Chromatography (HPLC) respectively. The total chlorophyll content is measured when SPM is used, while with HPLC the separate chlorophyll components are measured, of which chlorophyll-*a* is regarded as representative for the biomass. Consequently HPLC gives lower values than SPM and therefore both methods are only comparable when a conversion factor is used (0.7, based on Daemen, 1986). This might give a systematic difference between the analysis results. However, the fact that the increase has not been the same in all months and that the biomass has also increased significantly at Rattekaai (only HPLC) and at the Galgeplaat (except for plot 16 and 17 also only HPLC) contradicts this. Therefore, it is not thought likely that a change in the method of analysis has affected the changes in biomass.

Growth-limiting factors
= Light availability (emergence period, water transparency)

If, as De Jong *et al.* (1987, 1990) claim, production of microphytobenthos occurs mainly during emergence and stops when the sediments are flooded, the period of emergence is important. Because of the changed tidal range the emergence period has increased in the upper part of the tidal flats and decreased in the lower part. However, as these changes are small (at maximum 10% in the highest and the lowest part) and on average cancel each other out, they may be considered as insignificant.

If, as Colijn (1982), Cadée (1978) and Nienhuis *et al.* (1986) claim, production of microphytobenthos continues after flooding, water transparency is a limiting factor. In parts of the Oosterschelde water transparency has increased (Fig. 6; Wetsteyn & Bakker, 1991); in the western and central compartment only a small increase can be observed in winter and early spring, but in the northern and eastern compartment transparency has increased the whole year round. This increased light penetration may have a positive effect on biomass growth in the northern and eastern compartment. This effect may be concluded from the plots at the Verdronken Land (see Fig. 3) which represent a transect from about mean sea level (plot 67) to mean low water (plot 72). These plots show a changed trend in the course of biomass with depth: in 1981/82 decreasing with depth, in 1989/90 increasing. This

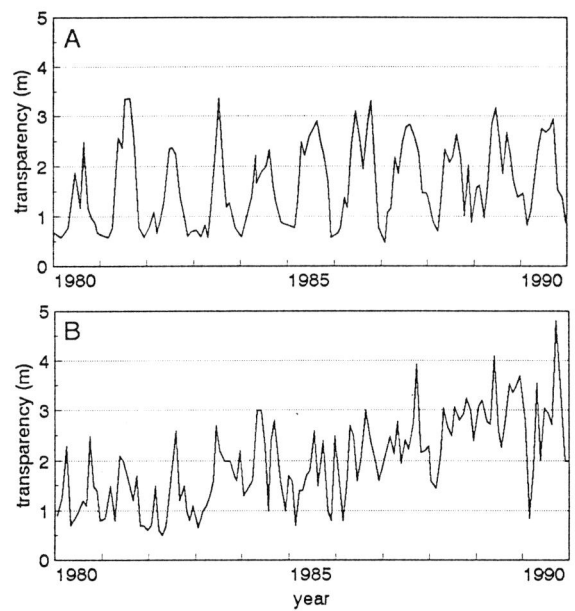

Fig. 6. Monthly averaged water transparency from 1980–1990 in the western part (A) and eastern part (B) of the Oosterschelde estuary (after Wetsteyn & Bakker, 1991).

change of trend could have been a result of the increased water transparency in this area (in summer from average 1.4 to 3 m).

= Hydrodynamics on the shoals (waves, currents)

According to de Jonge (1992) and Baillie & Welsh (1980), waves and currents strongly affect resuspension of microphytobenthos and subsequent transport to the adjacent tidal gullies. Vos *et al.* (1988) described how transplant biomass of microphytobenthos decreased considerably after a stormy period. Therefore, any permanent change in wave-impact and/or current velocities in the intertidal zone may have a considerable influence on diatom biomass.

Because of the presence of the barrier, wave impact is likely to have decreased in the western part of the Oosterschelde. In the other parts of the estuary, however, the fetch has remained unaltered and it is therefore not likely that the barrier has affected wave strength here.

Due to decreased tidal range, the intertidal zone in which wave impact occurs has narrowed. This has led to an increase in the duration of the wave impact which is estimated (from model calculations) at about 10%. From Fig. 5 it can be seen that this small increase in duration of wave impact is not reflected in a more gradual trend in biomass with depth in the sediment.

Water current velocities over the shoals have decreased. From model calculations this decrease can roughly be estimated at, on average 25%, with large variations from place to place (Vroon, 1994; Mulder & Louters, 1994). This decrease will have led to a decrease of the lateral transport of microphytobenthos, e.g. as floating mats.

The combined result could be that, although eddying up by waves has probably hardly changed, lateral transport, and thus export, of microphytobenthos (e.g. to the tidal gullies), is much less than before.

= Exposure

The more exposed shoals of Roggenplaat and Galgeplaat show a smaller biomass than the more sheltered tidal flats of Verdronken Land and Rattekaai (Figs 2, 3) of which the Verdronken Land has very high biomass values. This might be explained by the very sheltered situation of this area due to the presence of an anthropogenic reef of mussels on the 'seaward' side of the shoal (mussels growing on the brick walls of former oyster pits), which diminishes both wave impact and export of sediment. An observation (D. J. de Jong), supporting this hypothesis is that in this area a large amount of colony-forming diatoms (representative of sheltered situations) are present at the sediment surface.

The exposed Galgeplaat plots 26, 75 and 66 show hardly any increase in biomass, which is probably the result of minor changes in hydro-dynamic forces (especially waves). In contrast, the Roggenplaat plots 50 and 51 show a very large increase in biomass, which will be the conse-quence of a considerable decrease in current velocities (morphologically a shift from mega-ripples to small wave ripples).

= Changed grazing pressure

Microphytobenthos is strongly grazed by zoo-benthos like worms, amphipods and especially *Hydrobia* species. *Hydrobia* species in high den-sities are able to decrease diatom biomass con-siderably (R. Asmus, 1982; Cadée, 1980; Fenchel & Kofoed, 1980; Morrisey, 1988a, b).

As to the Oosterschelde Seys *et al.*(1994) and Coosen *et al.* (1994) mention that for worms and amphipods no significant changes in biomass have occurred during the last decade. *Hydrobia* densities were higher in 1983–85 than from 1986–87 onwards. However, in 1985, when micro-phytobenthos biomass had already increased sig-nificantly, *Hydrobia* density was still at the level of 1981–84. Therefore it is not likely that the decline of *Hydrobia* from 1986 onwards has had a sig-nificant influence on the increase of the micro-phytobenthos.

Besides, microphytobenthos is strongly grazed by meiobenthos. However, little is known about changes within this group of organisms, due to the presence of the storm-surge barrier.

= Nutrients and inorganic carbon

The availability of nutrients and inorganic carbon is assumed to be an important factor in primary production of microphytobenthos (e.g. Colijn & de Jonge, 1984). Important sources for the micro-phytobenthos are import from the water column and mineralization in the sediment.

Since 1986 the import of nutrients into the Oosterschelde has decreased significantly, lead-ing to a decrease of the nutrient concentrations in the water (Fig. 7, Wetsteyn & Bakker, 1991; Wetsteyn & Kromkamp, 1994). This is not

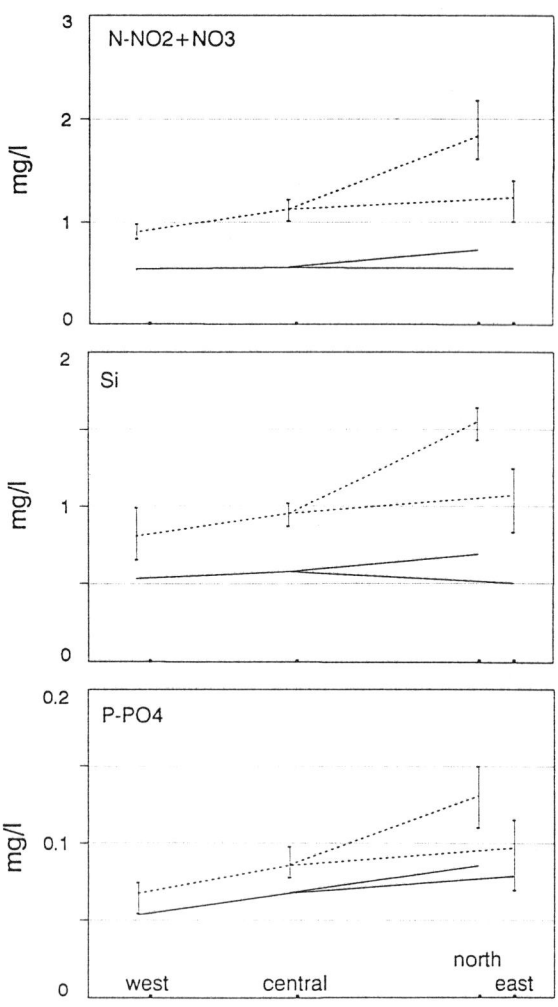

Fig. 7. Maximum nutrient concentrations in the Ooster-schelde in 1980–84 (dashed line, mean and range) and 1990 (solid line) (after Wetsteyn & Bakker, 1991).

reflected in the microphytobenthos biomass. Furthermore, it is often claimed that primary production and biomass of estuarine benthic diatoms are not limited by nutrients, but by the total amount and availability of inorganic carbon in the sediment (e.g. Admiraal, 1977; Admiraal et al., 1982; Colijn & de Jonge, 1984; Rasmussen et al., 1983; Van den Hoek et al., 1979). Assuming carbon limitation of the primary production of the microphytobenthos in the Oosterschelde in the pre-barrier period, for the present, significantly higher production and biomass of microphytobenthos a significant increase in the carbon pool in the upper sediment layer has to be presumed, at least during summer. The source of this inorganic C might be increased sedimentation of dead phytoplankton on the shoals after the spring bloom, due to decreased current velocities above the shoals.

Primary production

In relation to the increased biomass, the calculated gross primary production of the microphytobenthos has increased from 14 045 to 22 265 tonnes C y^{-1} (Table 3). For the entire Oosterschelde, with the exception of the Northern compartment, this means that the gross primary production of the microphytobenthos has increased from 46 to 73 gC m^{-2} y^{-1}. When these figures are compared with those of the phytoplankton – before 1985 ≈ 237g C m^{-2} y^{-1} and after 1986 ≈ 174 gC m^{-2} y^{-1} the share of the microphytobenthos in the primary production of the Oosterschelde has increased significantly from 16% to 30% of the total primary production. This phytoplankton primary production is based on gross column primary production figures from Wetsteyn & Kromkamp (1994) and a ratio basin production/column production from Wetsteyn (pers. comm.) to take production in shallow areas into account.

Comparison with other areas

The data for the Oosterschelde will be compared with two similar areas, the Wadden Sea and the Ems estuary. The depth distribution of the chlorophyll in all three areas is comparable: the upper 1 cm of the soil contains about 25% of the total chlorophyll-a present in the 0–10 cm layer, while the average rate of decrease over the 0–10 cm layer follows the same trend (Western Wadden Sea: Cadée & Hegeman, 1974, Ems estuary: De Jonge, 1992).

The annual average biomass in the Oosterschelde, however, is much higher than in the Western Wadden Sea and the Ems estuary, especially in the post-barrier period: in the Oosterschelde before 1985 100-150 mg chlorophyll m^{-2} y^{-1} and from 1985 onwards 150–275 mg m^{-2} y^{-1} versus 25–90 mg m^{-2} y^{-1} (Van den Hoek et al., 1979) and 35–120 mg m^{-2} y^{-1} (Cadée & Hegeman, 1977) in the western Waddensea and 23–130 mg m^{-2} y^{-1} in the Ems estuary (Colijn & de Jonge, 1984). Consequently, the same holds for gross primary production: in the Oosterschelde before 1985 120–180 g C m^{-2} y^{-1} and from 1985 onwards 165–320 g C m^{-2} y^{-1} versus in the Wadden Sea 40–180 g C m^{-2} y^{-1} (Cadée & Hegeman, 1977) and in the Ems estuary 60–100 g C m^{-2} y^{-1} with an extreme value of 250 g C m^{-2} y^{-1} in a very eutrophicated area in the innermost part of the estuary (Colijn & De Jonge, 1984, see also Van den Hoek et al., 1979).

It is clear from this that in the past in the Oosterschelde, biomass and production values of microphytobenthos were in the same order of magnitude as in the sheltered parts of the Wadden Sea/Ems estuary. However, in the present situation, the values in the Oosterschelde are much higher, especially in the most sheltered eastern part.

Acknowledgement

Communication No. 696 of NIOO-CEMO, Yerseke, The Netherlands.

References

Admiraal, W., 1977. Experiments with mixed populations of benthic estuarine diatoms in laboratory microecosystems. Bot. mar. 20: 479–485.

Admiraal, W., 1980. Experiments on the ecology of benthic diatoms in the Eems-Dollard estuary. BOEDE publications and reports nr 3-1980/Thesis State University Groningen.

Admiraal, W. & H. Pelletier, 1979. Influence of organic compounds and light limitation on the growth rate of estuarine benthic diatoms. Br. phycol. J. 14: 197–206.

Admiraal, W., H. Pelletier & H. Zomer, 1982. Observations and experiments on the population dynamics of epipelic diatoms from an estuarine mudflat. Estuar. coast. mar. Shelf Sci. 14: 471–487.

Admiraal, W., L. A. Bouwman, L. Hoekstra & K. Romeyn, 1983. Qualitative and quantitative interactions between microphytobenthos and herbivorous meiofauna on a brackish intertidal mudflat. Int. Revue ges. Hydrobiol. 68: 175–191.

Asmus, H., 1982. Field measurements on respiration and secondary production of a benthic community in the northern Wadden Sea. Neth. J. Sea Res. 16: 403–413.

Asmus, R., 1982. Field measurements on seasonal variation of the activity of primary producers on a sandy tidal flat in the northern Wadden Sea. Neth. J. Sea Res. 16: 389–402.

Baillie, P. W. & B. L. Welsh, 1980. The effect of tidal resuspension on the distribution of intertidal epipelic algae in an estuary. Estuar. coast. Mar. Sci. 10: 165–180.

Bianchi, T. S. & D. L. Rice, 1988. Feeding ecology of Leitoscoloplos fragilis; II. Effects of worm density on benthic diatom production. Mar. Biol. 99: 123–131.

Cadée, G. C., 1972. Diatomeeën als rotgans-voedsel (Diatoms as food for Brent Goose; in Dutch). Levende Natuur 75: 119–120.

Cadée, G. C., 1976. Sediment reworking by Arenicola marina on tidal flats in the Dutch Wadden Sea. Neth. J. Sea Res. 10: 440–460.

Cadée, G. C., 1978. Voedselproduktie in de Waddenzee (Food production in the Waddensea; in Dutch). Natuur & Techniek 46: 730–749.

Cadée, G. C., 1980. Reappraisal of the production and import of organic carbon in the western Wadden Sea. Neth. J. Sea Res. 14: 305–322.

Cadée, G. C. & J. Hegeman, 1974. Primary production of the benthic microflora living on tidal flats in the Dutch Wadden Sea. Neth. J. Sea Res. 8: 260–291.

Cadée, G. C. & J. Hegeman, 1977. Distribution of primary production of the benthic microflora and accumulation of organic matter on a tidal flat area, Balgzand, Dutch Wadden Sea. Neth. J. Sea Res. 11: 24–41.

Colijn, F. C., 1982. Light absorption in the waters of the Ems-Dollard estuary and its consequences for the growth of phytoplankton and microphytobenthos. Neth. J. Sea Res. 15: 196–216.

Colijn, F. C. & V. N. de Jonge, 1984. Primary production of the microphytobenthos in the Ems-Dollard estuary. Mar. ecol. Prog. Ser. 14: 185–196.

Colijn, F. C. & K. S. Dijkema, 1981. Species composition of benthic diatoms and distribution of chlorophyll-a on an intertidal flat in the Dutch Wadden Sea. Mar. ecol. Prog. Ser. 4: 9–21.

Colijn, F. C. & R. P. T. Koeman, 1975. Das Mikrophytobenthos der Watten, Strände und Riffe um den Hohen Knechtsand in der Wesermündung. Forschungsstelle Norderney, Jahresbericht 1974, 26: 53–83.

Colijn, F. C. & H. Nienhuis, 1977. The intertidal microphytobenthos of the 'Hohe Weg' shallows in the German Wadden Sea. Forschungsstelle Norderney, Jahresbericht 1977, 29: 149–174.

Coosen, J., J. Seys, P. Meire & J. A. M. Craeymeersch, 1994. Effect of sedimentological and hydrodynamical changes in the intertidal areas of the Oosterschelde estuary (SW Netherlands) on distribution, density and biomass of five common macrobenthic species: Spio martinensis (Mesnil), Hydrobia ulvae (Pennant), Arenicola marina (L.), Scoloplos armiger (Muller) and Bathyporeia sp. Hydrobiologia 282/283: 235–249.

Daemen, E. A. M. J., 1986. Comparison of methods for the determination of chlorophyll in estuarine sediments. Neth. J. Sea Res. 20: 21–28.

Daemen, E. A. M. J. & M. T. T. de Leeuw Vereecken, 1985. Kwalificering en kwantificering van het microfytobenthos in de Oosterschelde (Qualification and quantification of the microphytobenthos in the Oosterschelde. Delta Institute for Hydrobiological Research Yerseke/Rijkswaterstaat, Tidal Waters Division Middelburg, Final report Balans-project 1985–3 (internal report in Dutch).

De Jong, S. A., P. A. G. Hofman, A. J. J. Sandee & H. A. P. M. Jansen, 1987. Oxygenic photosynthesis coupled to the microdistribution of diatoms and bacteria on intertidal sediments. In: Progr. Rep. DIHO 1986: 21–24.

De Jong, S. A., P. A. G. Hofman, A. J. J. Sandee & E. J. Wagenvoort, 1990. Primary production of benthic microalgae in the Oosterschelde estuary (SW Netherlands); Delta Institute for Hydrobiological Research Yerseke/Rijkswaterstaat, Tidal Waters Division Middelburg, Final report Balans-project 1990–43 (internal report in Dutch).

De Jonge, V. N., 1980. Fluctuations in the organic carbon to chlorophyll-a ratios for estuarine benthic diatom populations. Mar. ecol. Prog. Ser. 2. 345–353.

De Jonge, V. N., 1985. The occurrence of 'epipsammic' diatom populations: a result of interaction between physical sorting of sediment and certain properties of diatom species. Estuar. coast. Shelf Sci., 21: 607–622.

De Jonge, V. N., 1992. Physical processes and dynamics of microphytobenthos in the Ems estuary (the Netherlands). Thesis State University Groningen. 176 pp.

De Jonge, V. N. & J. van den Bergs, 1987. Experiments on the resuspension of estuarine sediments containing benthic diatoms. Est. coast. Shelf Sci. 24: 725–740.

De Jonge, V. N. & J. E. E. van Beusekom, 1992. Wind and tide induced resuspension of sediment and microphytobenthos in the Ems estuary. In: de Jonge, V. N., 1992. Physical processes and dynamics of microphytobenthos in the Ems estuary (the Netherlands). Thesis State University Groningen: 139–155.

Delgado, M., V. N. de Jonge & H. Peletier, 1991. Effect of sand movement on the growth of bentic diatoms. J. exp. mar. Biol. Ecol. 145: 221–231.

Fenchel, T. & L. H. Koefoed, 1976. Evidence for exploitative interspecific competition in mud snails (Hydrobiidae). Oikos 27: 367–376.

Gabrielson, J. O. & R. J. Lukatelich, 1985. Wind-related resuspension of sediments in the Peel-Harvey estuarine system. Estuar. coast. Shelf Sci. 20: 135–145.

Grant, J., 1981. Sediment transport and disturbance on an intertidal sandflat: infaunal distribution and recolonization. Mar. ecol. Prog. Series 6: 249–255.

Grant, J., U. V. Batthmann & E. L. Mills, 1986. The interaction between benthic diatom films and sediment transport. Estuar. coast. Shelf Sci. 23: 225–238.

Klepper, O., 1989. A model of carbon flows in relation to macrobentic food supply in the Oosterschelde estuary (SW Netherlands). Thesis Agricultural University Wageningen (ISBN 90-9002944-3).

Knoester, M., J. Visser, B. A. Bannink, C. J. Colijn & W. P. A. Broeders, 1984. The Eastern Scheldt Project. Wat. Sci. Tech. 16: 51–78.

Lorenzen, C. J., 1967. Determination of chlorophyll and phaeopigments: spectrophotometric equations. Limnol. Oceanogr. 12: 343–346.

Meininger, P. L. & H. Snoek, 1992. Non-breeding shelduck (*Tadorna tadorna*) in the SW Netherlands: effects of habitat changes on distribution, numbers, moulting sites and food. Wildfowl 43: 139–151.

Morrisey, D. J., 1988a. Differences in effects of grazing by deposit-feeders Hydrobia ulvae and Corophium arenarium on sediment microalgal populations. I Qualitative differences. J. exp. mar. Biol. Ecol. 118: 33–42.

Morrisey, D. J., 1988b. Differences in effects of grazing by deposit-feeders Hydrobia ulvae and Corophium arenarium on sediment microalgal populations. II Quantitative effects. J. exp. mar. Biol. Ecol. 118: 43–53.

Mulder, J. P. M. & T. Louters, 1994. Changes in basin geomorphology after implementation of the Oosterschelde project. Hydrobiologia 282/283: 29–39.

Nienhuis, P. H., E. A. M. J. Daemen, S. A. de Jong & P. A. G. Hofman, 1986. Biomass and production of microphytobenthos. In: Progr. Rep. DIHO 1985: 11–14.

Nienhuis, P. H. & A. C. Smaal, 1994. The Oosterschelde estuary, a case study of a changing ecosystem: an introduction. Hydrobiologia 282/283: 1–14.

Rasmussen, M. B., K. Henriksen & A. Jensen, 1983. Possible causes of temporal fluctuations in primary production of the microphytobenthos in the Danish Wadden Sea. Mar. Biol. 73: 109–114.

Seys, J. J., P. M. Meire, J. Coosen & J. A. M. Craeymeersch, 1994. Long-term changes (1979–89) in the intertidal macrozoobenthos of the Oosterschelde estuary: are patterns in total density, biomass and diversity induced by the construction of the storm-surge barrier? Hydrobiologia 282/283: 251–264.

Sokal, R. R. & F. J. Rohlf, 1981. Biometry. 2nd edn. W. A. Freeman & Co, New York.

Van den Hoek, C., W. Admiraal, F. C. Colijn & V. N. de Jonge, 1979. The role of algae and saegrasses in the ecosystem of the Wadden Sea: a review. Chapter 3 in: W. J. Wolff (ed.) Flora and vegetation of the Wadden Sea; vol I in: W. J. Wolff (ed.), Ecology of the Wadden Sea. A. A. Balkema, Rotterdam 1983.

Vos, P. C., 1986. De sediment stabiliserende werking van benthische diatomeeen in het intergetijde-gebied van de Oosterschelde (The sediment stabilizing effect of benthic diatoms on the shoals of the Oosterschelde); Rijkswaterstaat Tidal Division Middelburg/State University Utrecht, RWS-86-03 (internal report in Dutch).

Vos P. C., P. L. de Boer & R. Misdorp, 1988. Sediment stabilization by benthic diatoms in intertidal sandy shoals; qualitative and quantitative observations. In: P. L. de Boer *et al.* (eds), Tide-influenced sedimentary environments and facies. D. Reidel Publ. Comp.: 511–526.

Vroon, J., 1994. Hydrodynamic characteristics of the Oosterschelde in recent decades. Hydrobiologia 282/283: 17–27.

Wetsteyn, L. P. M. J. & C. Bakker, 1991. Abiotic characteristics and phytoplankton primary production in relation to a large-scale coastal engineering project in the Oosterschelde (the Netherlands). A preliminary observation. In: M. Elliot, & J-P. Ducrotoy (eds), Estuaries and coasts: Spatial and temporal intercomparisons. Proc. ECSA19, Symposium Caen 1989. Olsen & Olsen, Fredensborg, Denmark.

Wetsteyn, L. P. M. J. & J. C. Kromkamp, 1994. Turbidity, nutrients and phytoplankton primary production in the Oosterschelde (The Netherlands) before, during and after a large scale coastal engineering project (1980–1990). Hydrobiologia 282/283: 61–78.

Hydrobiologia **282/283**: 197–217, 1994.
P. H. Nienhuis & A. C. Smaal (eds), The Oosterschelde Estuary.
© 1994 *Kluwer Academic Publishers. Printed in Belgium.*

Composition, distribution and biomass of meiobenthos in the Oosterschelde estuary (SW Netherlands)

N. Smol[1], K. A. Willems[1,3], J. C. R. Govaere[2] & A. J. J. Sandee[3]
[1]*Marine Biology Section, State University of Ghent, Ledeganckstraat 35, 9000 Gent, Belgium;*
[2]*Koninklijk Belgisch Instituut voor Natuurwetenschappen, Vautierstraat 29, 1040 Brussel, Belgium;*
[3]*Netherlands Institute of Ecology, Vierstraat 28, 4401 EA Yerseke, The Netherlands*

Key words: meiofauna, distribution, biomass, seasonal variation, Oosterschelde estuary

Abstract

Meiofauna composition, abundance, biomass, distribution and diversity were investigated for 31 stations in summer. The sampling covered the whole Oosterschelde and comparisons between the subtidal – intertidal and between the western-central – eastern compartment were made.

Meiofauna had a community density ranging between 200 and 17 500 ind 10 cm^{-2}, corresponding to a dry weight of 0.2 and 8.4 gm^{-2}. Abundance ranged between 130 and 17 200 ind 10 cm^{-2} for nematodes and between 10 and 1600 ind 10 cm^{-2} for copepods. Dry weight biomass of these taxa was between 0.5–7.0 gm^{-2} and 0.008–0.3 gm^{-2} for nematodes and copepods respectively.

The meiofauna was strongly dominated by the nematodes (36–99%), who's abundance, biomass and diversity were significantly higher intertidally than subtidally and significantly higher in the eastern part than in the western part. High numbers were positively correlated with the percentage silt and negatively with the median grain size of the sand fraction. The abundance and diversity of the copepods were highest in the subtidal, but their biomass showed an inverse trend being highest on the tidal flats.

The taxa diversity of the meiofauna community and species diversity of both the nematodes and the copepods were higher in subtidal stations than on tidal flats. In the subtidal, the meiofauna and copepod diversity decreased from west to east, whereas nematode diversity increased.

The vertical profile clearly reflected the sediment characteristics and could be explained by local hydrodynamic conditions.

Seasonal variation was pronounced for the different taxa with peak abundance in spring, summer or autumn and minimun abundance in winter.

Changes in tidal amplitude and current velocity enhanced by the storm-surge barrier will alter the meiofauna community structure. As a result meiofauna will become more important in terms of density and biomass, mainly due to increasing numbers of nematodes, increasing bioturbation, nutrient mineralisation and sustaining bacterial growth. A general decrease in meiofauna diversity is predicted. The number of copepods is expected to decrease and interstitial species will be replaced by epibenthic species, the latter being more important in terms of biomass and as food for the epibenthic macrofauna and fishes.

Introduction

Meiofauna is defined here as the benthic Metazoa that can pass through a sieve with a mesh size of 1 mm and comprises the benthic animals intermediate between the microfaunal organisms (bacteria, ciliates, foraminiferans etc.) and the macrofaunal organisms (polychaetes, bivalves, crustaceans etc.).

Meiobenthos occur in all types of sediment and occupy a wide variety of habitats. In general, grain size of the sediment is a primary factor affecting the abundance and species composition of meiobenthic organisms.

Meiofauna has been investigated in different estuaries: e.g. Blyth, England (Capstick, 1959), Elbe, Germany (Riemann, 1966), New England, USA. (Tietjen, 1969), Exe, England (Warwick, 1971), Weser, Germany (Skoolmun & Gerlach, 1971), Grevelingen, the Netherlands (Heip et al., 1977; Willems & Sandee, 1978, 1979; Willems et al., 1984), Tigris & Euphrate, Iraq (Saad & Arlt, 1977; Arlt & Saad, 1977), Swartskop, S. Africa (Dye & Furstenberg, 1978), Lynher, England (Warwick & Price, 1979), Westerschelde, the Netherlands, Belgium (Heip et al., 1979; Van Damme et al., 1980; Van Damme et al., 1984), Eems-Dollard, the Netherlands (Heip et al., 1979; Van Es et al., 1980; Bouwman, 1983; Van Damme et al., 1984), the Wadden Sea, the Netherlands (Witte & Zijlstra, 1984), Wellington, New Zealand (Coull & Wells, 1981), Ythan, Scotland (Baird & Milne,1981), Tamar, England (Warwick & Gee, 1984; Austen & Warwick, 1989), Hunter, Australia (Hodda & Nicholas, 1985, 1986).

Meiobenthos of the Oosterschelde was extensively sampled during the period 1976–1985, the period before and during the construction of the storm-surge barrier. Due to the construction of different barriers the estuarine character of the Oosterschelde is mainly lost and the salinity is fairly constant: 28–30‰. (Wolff, 1973; Duursma et al., 1982). The Oosterschelde is nowadays characterized as a polyhaline or mixo-euhaline sea arm or tidal bay (Nienhuis & Smaal, 1994).

Preliminary results on the Oosterschelde meiofauna are given by Heip et al., 1979 and Smol,

1986. This paper deals with extensive information on the spatial and seasonal distribution, density, biomass and diversity of the whole meiofauna with special focus on the two dominant taxa: nematodes and copepods.

Material and methods

Sampling

The meiofauna composition of 31 stations was investigated, covering both the subtidal and the intertidal habitats of the whole Oosterschelde area and covering different substrate types (Fig. 1). Most stations, 01 to 024, were sampled in August–September 1981. Station 017 was sampled in August 1981; station 035 in June 1984 and the stations 036a, 036b, 036c, 037a, 037b, 037c, 038a, 038b and 038c were sampled in May, August, November 1984 and February 1985 to investigate temporal changes in meiofauna community. The coordinates and depth at the moment of sampling of the stations are represented in Table 1. The first three samples were located west of the storm-surge barrier.

Sublittoral samples were collected with a modified 'Reineck box-corer' (Farris & Crezee, 1976) of which 5 replicate subsamples were taken with a 10 cm^2 perspex corer of 40 cm length. Stations 36b, 37b and 38b at −1 m depth below low water level were sampled with a hand held 10 cm^2 perspex corer using SCUBA-diving; the same corer was also used in collecting intertidal samples at low tide. At each station 4 replicate samples were analyzed for meiofauna and 1 for sediment. On some tidal flats (e.g. 017, 024, 028) samples were taken on 4 different places situated along a transect perpendicular to the water line; those data were pooled together and a mean value of the 4 samples is taken as representative for that intertidal station.

Sediment granulometry was determined using a graded series of standard sieves suited to the intervals of the Wenthworth scale (Buchanan & Kain, 1971). The degree of sorting was classified according to Wolff (1973). The silt fraction was determined as the amount of sediment passing

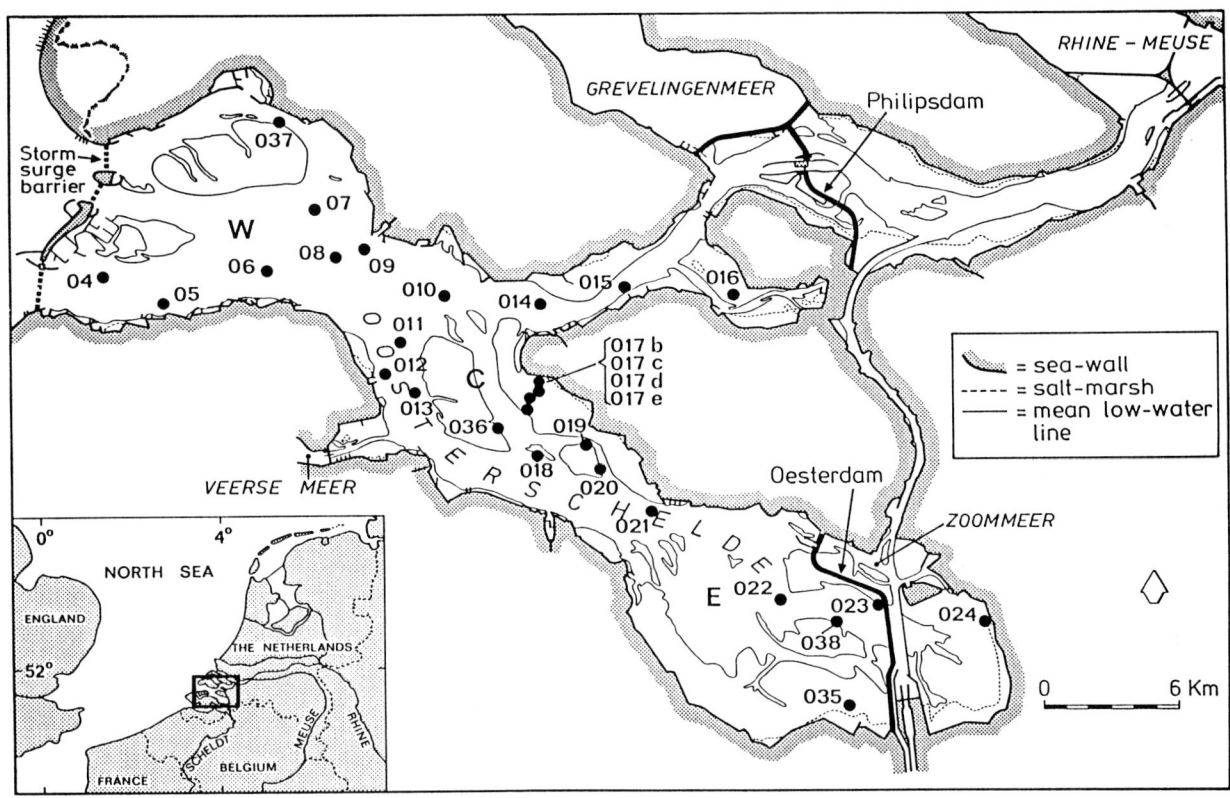

Fig. 1. Location of the stations in Oosterschelde estuary.

through the 62 μm sieves and the gravel fraction as the amount of material retained on the 1 mm sieve.

Copepods and nematodes were elutriated from the sand on a 38 μm sieve using a combination of the trough-method (Barnett, 1968) and a density gradient centrifugation technique with ludox HS 40% (Bowen *et al.*, 1972). All copepods were identified. Because of the high numbers of nematodes, only 100 chosen at random were identified to species level.

For the stations 036a, 036b, 036c, 037a, 037b, 037c, 038a, 038b, 038c mean individual dry weight of the nematodes and copepods was determined by means of a Mettler microbalance (accuracy 0.1 μg). Two hundred nematodes or twenty copepods, randomly picked out and rinsed with distilled water, were transferred to an aluminium vial and weighed after two hours of drying and 30 minutes of cooling.

For the other stations a value of 0.38 μg (mean individual dry weight in summer) was used for the nematodes and individual dry weight of the copepods was determined by the method of Willems (1989). The individual dry weight of the other meiofauna taxa was based on values presented by Faubel (1982) and Van Damme *et al.* (1980); a mean value of the different size classes was used.

The following individual dry weight values were used to calculate the total biomass of the different groups: Turbellaria: 3.7 μg, Ostracoda: 7.8 μg, Gastrotricha: 0.43 μg, Archiannelida: 4.6 μg, Hydrozoa: 3.0 μg, Halacarida: 1.8 μg, Tardigrada: 0.7 μg, Kinorhyncha: 2.1 μg, Polychaeta: 4.6 μg, Oligochaeta: 3.6 μg, Nemertini: 14.1 μg, Cumacea: 10 μg, Tanaidacea: 10 μg and Bivalvia: 5.4 μg.

Statistics

For statistical analysis only the values of the stations sampled in August and early September

Table 1. Geographic coordinates, depth and sediment characteristics of the stations.

Station	Latitude North	Longitude East	Depth m	Sediment characteristics			
				Grain size (μm)	Sorting	% silt	% gravel
O1	51°37′42″	3°19′02″	− 10	308	0.350	0.15	0.00
O2	51°40′13″	3°31′14″	− 5	275	0.340	0.20	0.00
O2	51°37′58″	3°26′33″	− 11	394	0.410	0.28	0.00
O4	51°38′28″	3°43′16″	− 23	202	0.360	0.24	0.00
O5	51°36′25″	3°45′45″	− 24	257	0.310	0.00	0.00
O6	51°37′02″	3°49′48″	− 10	180	0.320	0.96	0.00
O7	51°38′47″	3°51′37″	− 16	260	0.350	0.15	0.00
O8	51°37′29″	3°52′33″	− 6	210	0.360	0.39	0.00
O9	51°37′51″	3°53′32″	− 55	240	0.360	0.33	0.00
O10	51°36′62″	3°56′32″	− 21	246	0.340	0.12	0.00
O11	51°35′29″	3°54′59″	− 11	251	0.520	1.43	0.00
O12	51°34′17″	3°54′05″	− 36	333	0.390	0.48	4.51
O13	51°34′09″	3°55′04″	− 11	219	0.330	0.57	0.00
O14	51°36′13″	4°00′01″	− 16	242	0.340	0.00	2.93
O15	51°36′56″	4°03′09″	− 20	278	0.320	0.19	0.56
O16	51°36′28″	4°07′26″	− 6	161	0.450	8.65	1.91
O17	51°34′49″	4°00′25″	0	159	0.355	1.80	0.00
O18	51°33′15″	4°00′00″	− 3	157	0.240	0.28	0.00
O19	51°33′25″	4°02′02″	− 10	127	0.300	16.94	0.00
O20	51°32′22″	4°02′43″	− 8	116	0.360	16.38	2.24
O21	51°31′20″	4°04′36″	− 41	278	0.400	0.21	0.00
O22	51°29′12″	4°09′02″	− 18	151	0.400	2.05	0.00
O23	51°29′01″	4°13′27″	− 6	186	0.760	20.52	0.00
O24	51°27′29″	4°16′07″	0	111	0.550	29.25	0.00
O35	51°26′22″	4°10′06″	0	125	0.375	5.40	0.00
O36a	51°33′24″	3°58′28″	0	137	0.270	3.20	
O36b	51°33′24″	3°58′28″	− 1	131	0.300	2.90	
O37a	51°40′24″	3°50′14″	0	147	0.330	6.00	
O37b	51°40′24″	3°50′14″	− 1	129	0.460	16.10	
O38a	51°28′52″	4°11′15″	0	139	0.390	24.20	
O38b	51°28′52″	4°11′15″	− 1	131	0.270	7.40	

were used. Diversity of the taxonomical units was measured using the Hill's diversity numbers (Hill, 1973), the evenness indices of Heip (1974) and Alatalo (1981) and the formerly used H, H′, J and SI to allow comparison with literature. The whole set of biological data, sediment characteristics, geographic position and depth were submitted to correlation analysis using the non parametric Spearman rank correlation coefficient. Abundance, biomass and diversity of the total meiofauna, the nematodes and the copepods were analyzed either by the non parametric Kruskal-Wallis or the parametric 1-way ANOVA (for homogenous data). A detailed list of the data are given by Smol (1986).

Results

Sediment characteristics

Sediment characteristics of the investigated stations are represented in Table 1 and the mean value of the main characteristics is plotted for the subtidal and intertidal zone and the 3 parts of the Oosterschelde in Fig. 2.

Table 2. Mean density and biomass of the total meiofauna, the nematodes and the copepods for the subtidal and intertidal (per 10 cm²)

	Meiofauna		Nematoda		Copepoda	
	Density	Biomass	Density	Biomass	Density	Biomass
Subtidal	2000 ind.	0.9 g	1600 ind.	0.6 g	303 ind.	0.09 g
Intertidal	5400 ind.	2.7 g	5000 ind.	2.0 g	126 ind.	0.2 g

Median grain size. The sand fraction of the sediment consisted of very fine to medium sand, the median grain size ranging between 111 μm–394 μm. These minimum and maximum scores were found in station 024 and 03 respectively, the latter being located outside the Oosterschelde.

The subtidal stations had a mean median grain size of 220 μm. Most of them were characterized by fine and medium sands. Very fine sand was restricted to station 020 only. The coarsest sediments were found in the western region with a mean value of 246 μm. Towards the inner basin the mean value decreased, being 214 μm and 117 μm in the central and eastern region respectively.

In the intertidal most stations were characterized by fine sands. No medium sands were found. The overall mean is 116 μm, with mean values of 147 μm in the western area, 148 μm in the middle area and 83 μm in the inner basin.

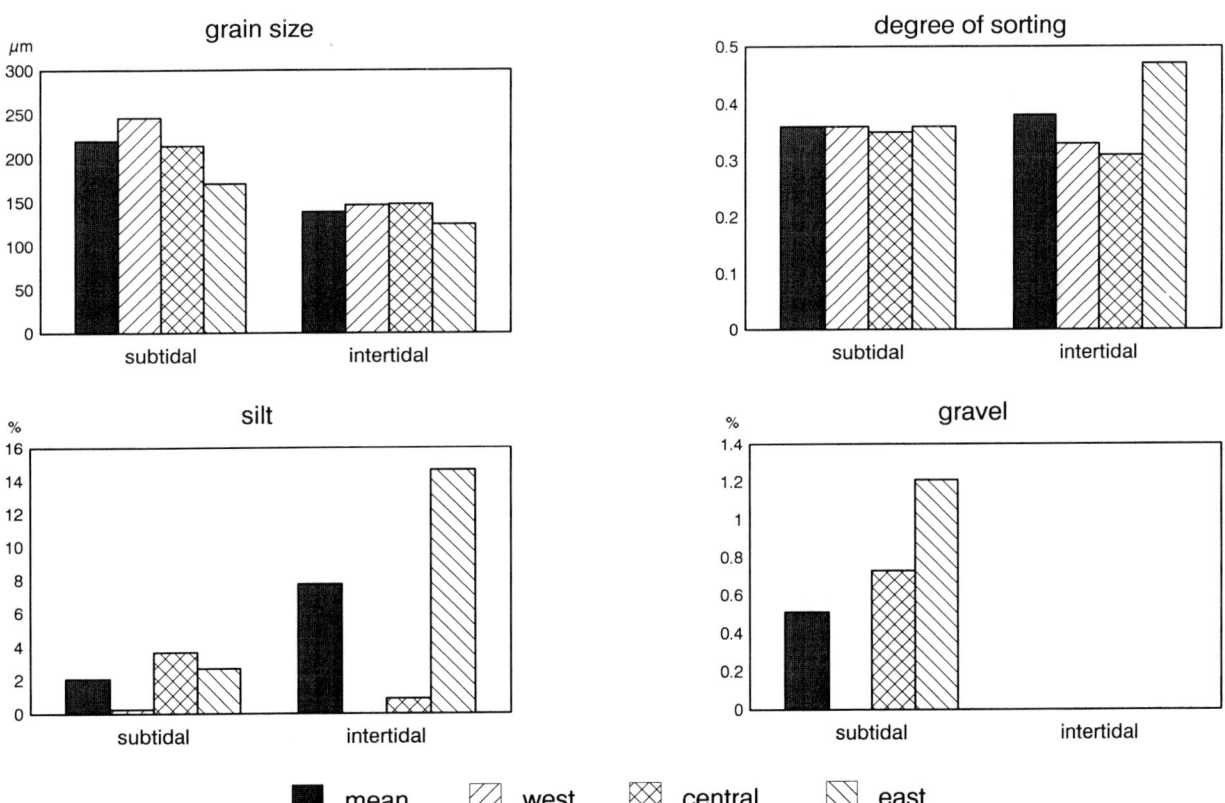

Fig. 2. Mean value of the main sediment characteristics for the subtidal and the intertidal habitat and for the 3 regions of the Oosterschelde.

The mean median grain size of all the subtidal stations was 220 µm. The coarsest sediments were found in the western region: a mean value of 246 µm, and the finest in the eastern region: 117 µm; the middle region had an intermediate position: 214 µm.

Degree of sorting. The sediments were mostly very well ($n = 15$) or well ($n = 13$) sorted. The sorting coefficients were fairly constant throughout the Oosterschelde, except for several intertidal stations, located in the inner basin.

Silt-clay. In the subtidal most stations consisted of clean sand whereas most intertidal stations were characterized by muddy sands. Increased amounts of silt-clay were found in sheltered zones of the northern area (016), the middle region (019, 020) and the inner basin.

Gravel. Occasionally small amounts of gravel were found, all of it being of biogenic origin (= shell debris).

Finally the Oosterschelde was dominated by clean, fine to medium sands which are well to very well sorted. In the shallow sheltered zones such as the inner basin the median grain size decreased paralleled by an increase of the silt-clay content.

Statistically significant differences were found between the subtidal and intertidal for both median grain size ($p < 0.011$) and silt-clay content ($p < 0.002$). However the west-east gradient could not be confirmed by statistical analysis.

The stations could be classified into following sediment types: clean fine sand, very well sorted; clean fine sand, well sorted; slightly mixed fine sand, well sorted; clean medium sand, well sorted; slightly mixed fine sand, very well sorted. All other sediment types were less important and scored less than 5% of all samples.

Meiofaunal composition and abundance

The meiobenthos community of the Oosterschelde was very diverse: 12 higher taxa were identified: Nematoda (176 species), Copepoda Harpacticoida (97 species), Turbellaria (95 species), Ostracoda (17 species), Gastrotricha (10 species), Archiannelida (8 species), Oligochaeta, Hydrozoa (4 species), Halacarida, Tardigrada, Kinorhyncha (1 nov. species), and Rotifera (9 species). A species list for each taxon is provided by Smol (1986).

The meiofauna community was dominated by nematodes throughout the Oosterschelde, followed by copepods, gastrotrichs and turbellarians. Nematodes and copepods together comprised about 90% of the meiofauna (Fig. 3). The average relative abundance of the nematodes was 88% for the intertidal and 64% for the subtidal. Although copepods were ranked second, they were particularly important in the subtidal where they represented almost one fourth (24%) of the total meiofauna density. The other taxa made up only 3% and 1% in the sub- and intertidal habitats respectively.

The density of the total meiofauna community ranged between 200 and 17500 ind 10 cm^{-2}. On tidal flats the mean total density was much higher (5400 ind 10 cm^{-2}) than in the subtidal (2000 ind 10 cm^{-2}) (Fig. 4). This figure shows the overwhelming abundance of nematodes in both habitats. The copepods and gastrotrichs ranked second depending on the community. Differences between the intertidal and subtidal were significant ($p < 0.002$).

Nematode density fluctuated between 100 ind 10 cm^{-2} (012) and 7100 ind 10 cm^{-2} (037a). A mean density of 1500 ind 10 cm^{-2} and 5000 ind 10 cm^{-2} was representative for the sub- and intertidal respectively, these scores were statistically different ($p < 0.002$). In 10 out of 24 subtidal stations the importance of the nematode group was < 50% of the meiofauna, but on tidal flats they represented mostly > 90% of the meiobenthos, occasionally decreasing to 70%.

An inverse trend was observed for the copepods, their relative abundance became important in subtidal stations, usually exceeding 50% of the total meiofauna, and with minimal scores (< 1%) in the intertidal.

The mean copepod density was 300 ind 10 cm^{-2} and 120 ind 10 cm^{-2} for the sub- and

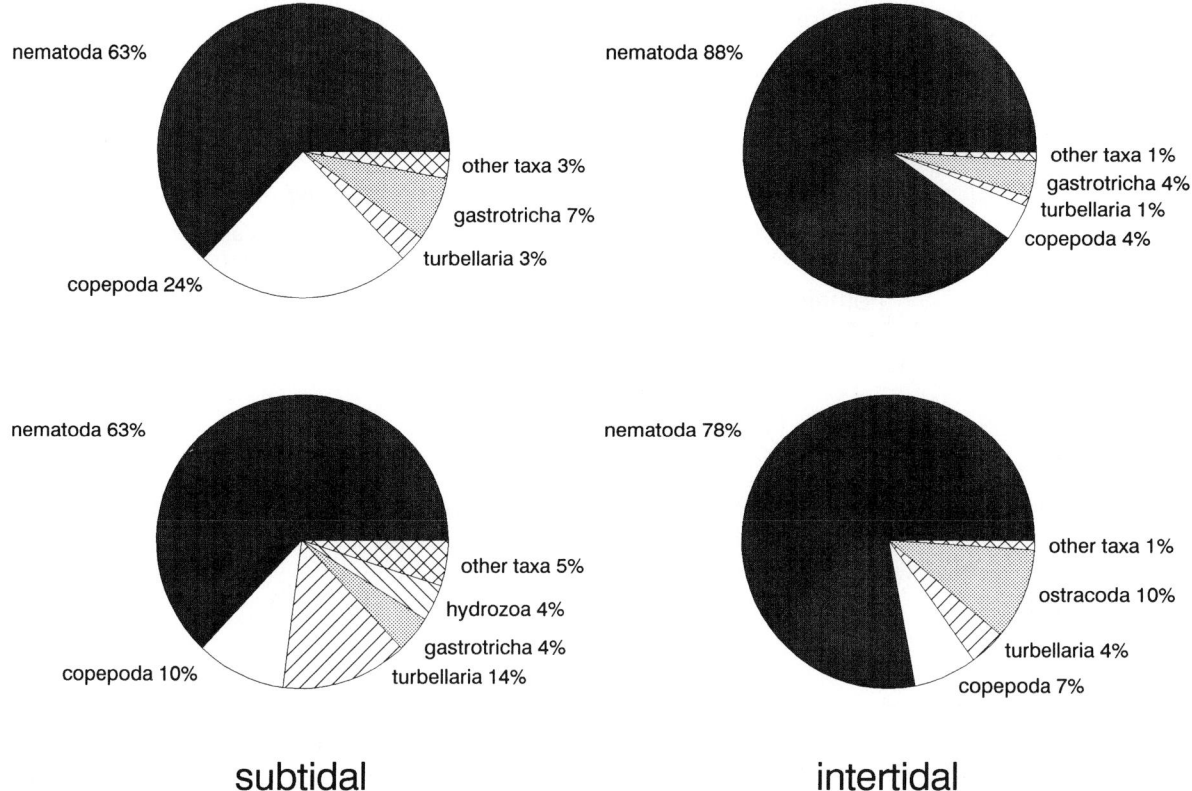

Fig. 3. Relative abundance (upper circles) and biomass (lower circles) partitioning of the meiofauna: comparison between subtidal and intertidal.

intertidal area respectively. No significant differences were observed. Station densities ranged between 10 ind 10 cm^{-2} (035b) and 1500 ind 10 cm^{-2} (015) with a mean density of 214 ind 10 cm^{-2}. Only clean medium sands showed densities > 500 ind 10 cm^{-2} due to the presence of interstitial fauna.

Turbellarians did occur in all but one station (absent in the intertidal 017). The population density reached a maximum of 114 ind 10 cm^{-2} (05, situated near the storm-surge barrier). Mean densities in the sub- and the intertidal were 34 and 25 ind 10 cm^{-2} respectively, although this difference seemed less obvious it was significant ($p < 0.000$). The relative abundance of the turbellarians was at most 12% (04).

Gastrotrichs only occured in 20 out of 31 stations and showed great variability: 1–1100 ind 10 cm^{-2}. Their mean density was significantly

($p < 0.000$) higher on tidal flats (288 ind 10 cm^{-2}) than in the sublittoral zone (109 ind 10 cm^{-2}). Their excessive numbers at particular stations resulted in a relative important ranking within the meiofauna community.

The other taxa were less important or even rare in terms of abundance although significant differences in density were observed for hydroids ($p < 0.000$) and tardigrades ($p < 0.012$), being more abundant in the subtidal and for the ostracods ($p < 0.000$) and oligochaetes ($p < 0.000$) being more abundant on tidal flats.

In Fig. 4 mean density of the most important meiofauna groups are given. From west to east mean subtidal densities of the 3 main regions showed an increasing trend for nematodes and a decreasing trend for copepods and gastrotrichs. In the intertidal an increasing trend was only found for copepods. In general the total meio-

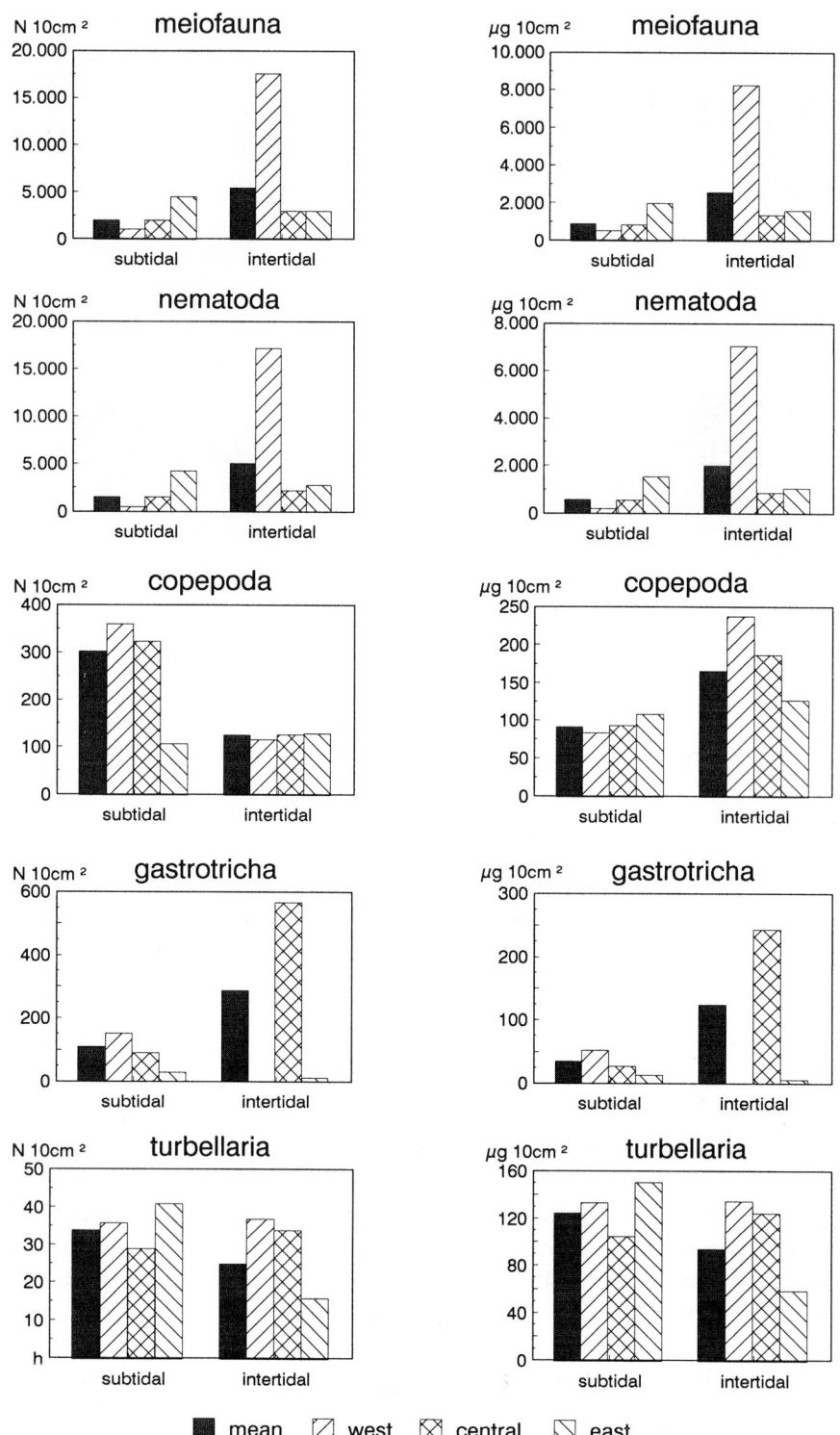

Fig. 4. Mean density (left graph) and biomass (right graph) of the meiofauna taxa: comparison between subtidal – intertidal and between the 3 main regions.

fauna and nematode density were significantly higher in the eastern part than in the western part ($p < 0.000$).

The Nematode/Copepod-ratio

The mean nematode/copepod-ratio (N/C-ratio) per station varied between 0.3 (015) and 184 (016); both these extreme values occurred in the sublittoral region. The mean N/C-ratio for this zone is 21.4, whereas the mean N/C-ratio for the intertidal is 56.1, this difference was significant ($p < 0.000$).

In the subtidal the N/C-ratio showed an increasing trend from west to east (Fig. 5), which was not found in the intertidal due to an extreme value (147) in station 037a on the Roggeplaat, located in the western region. This station was characterized by a median grain size of 147 μm and 6% silt and was located at the sheltered side of the sand flat. In summer dense algal mats, diverse macrofauna species and many egg cocoons of *Scoloplos armiger* occurred, which made this site quite unique in offering heterogeneous micro habitats. However, the algal mats reduced the oxygen in the underlying sediment, as is noted for station 037b, although not affecting the N/C-ratio.

Biomass

The standing stock dry weight biomass of the meiofauna community ranged between 200–8300 μg dwt 10 cm^{-2}. Within the subtidal the biomass was increasing from west to east, but on the intertidal flats this trend was reversed (Fig. 4). A significant difference in the mean biomass between the sub- and intertidal was found, being 900 and 2600 μg dwt 10 cm^2 respectively.

Mean individual dry weight of the nematodes in the Oosterschelde as determined by direct measurements was 0.38 μg in summer (August). Variation in time and depth occurred (Table 3), reflecting differences in species composition and population structure. The summer values were slightly lower than those in autumn and winter,

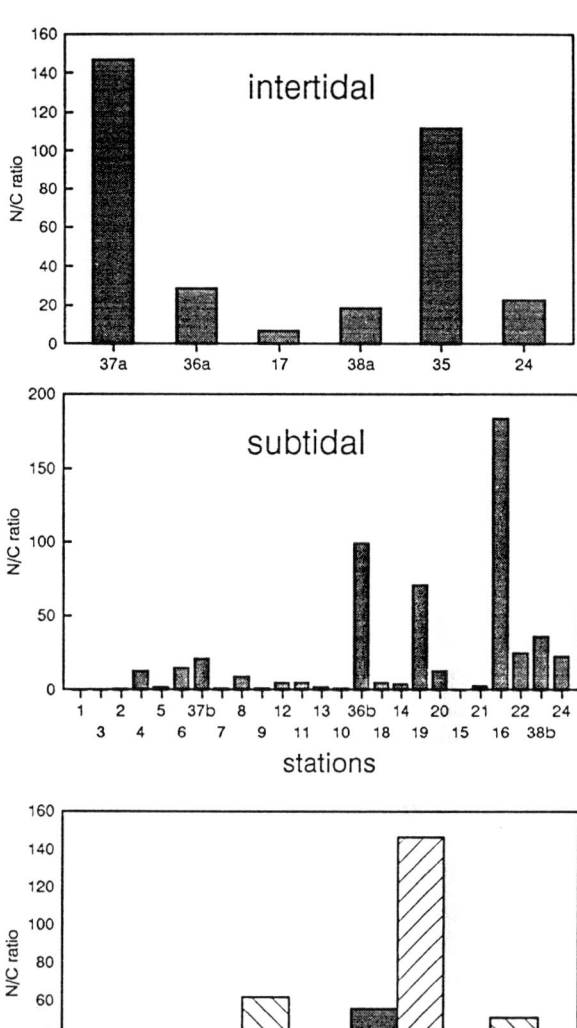

Fig. 5. The N/C-ratio per station for the subtidal and intertidal and a comparison between the mean values per habitat and region (stations are ranked from west (left) to east (right)).

probably due to an important number of juveniles. The individual dry weight of nematodes in the top 5 cm was about double as high as that of the nematodes in the deeper layers. Most species inhabiting the lower parts of the sediment were indeed characterized as long and thin species.

Although the overall mean individual dry weight of the nematodes (0.38 μg) was about 3

Table 3. Seasonal and depth variation of the mean individual dry weight (μg DW ind^{-1}) of nematodes and copepods (value between brackets = standard error, n = number of weighings).

Time

	May	Aug	Nov	Feb
Nematoda	0.45 (0.03) $n = 45$	0.38 (0.02) $n = 14$	0.46 (0.02) $n = 11$	0.41 (0.03) $n = 11$
Copepoda	1.59 (0.16) $n = 17$	1.36 (0.24) $n = 9$	1.58 (0.17) $n = 7$	1.65 (0.12) $n = 3$

Depth

	0–5 cm	5–10 cm	10–15 cm	15–20 cm
Nematoda	0.54 (0.04) $n = 30$	0.26 (0.04) $n = 11$	0.24 (0.04) $n = 3$	0.28 $n = 1$

times lower than that of the copepods (1.36 μg), this taxon again was the dominant component of the total meiofauna biomass (Fig. 3). In the subtidal station they reached 60%, followed by the turbellarians (14%) and the copepods (10%). Hydroids as well as gastrotrichs made up 4% each of the total biomass. On tidal flats the dominance of the nematodes was even more pronounced (78%). Subdominant were the ostracods (10%), followed by the copepods (7%) and the turbellarians (4%).

Nematode biomass per station ranged in summer between 49 μg dwt 10 cm^{-2} (012) and 7044 μg dwt 10 cm^{-2} (037a). The subtidal environment was characterized by a mean biomass of 575 μg dwt 10 cm^{-2} differing significantly ($p < 0.000$) from the intertidal by 1999 μg dwt 10 cm^{-2} (Fig. 4).

Copepod biomass ranged between 8.4 μg dwt 10 cm^{-2} (012) and 282.9 μg dwt 10 cm^{-2} (038b) per station.

The overall mean was 114.8 μg dwt 10 cm^{-2} with a mean value of 92 μg dwt 10 cm^{-2} and 166 μg dwt 10 cm^{-2} for the sub- and intertidal respectively, this difference was not significant. The biomass followed to a certain extend the pattern of abundance but this pattern was distorted by extreme differences in body size and thus individual dry weights of the species.

Both Paramesochridae and Cylindropsyllidae were the most abundant but as typical interstitial types their importance was strongly reduced as far as biomass goes. The same held for the mesopsammic component of the fauna. This was explained by the fact that these families only consisted of mesopsammic species *i.e.* small, vermiform copepods with an extremely low individual dry weight (range: 0.1–0.5 μg ind^{-1}).

Except for the Ectinosomatidae, the other families showed a reverse trend, since many members had a relatively high individual dry weight (range: 2–5 μg ind^{-1}). Thus, many more families shared dominance. Among the ecological groups only the psammophilous and euryoecious faunas were of any importance. Dominance in biomass was evenly shared by the mesopsammic, epi-endopsammic and euryoecious species.

Turbellarian total biomass ranged between 0.9 μg 10 cm^{-2} (018) and 421.8 μg 10 cm^{-2}. The mean value of 125 μg 10 cm^{-2} for the subtidal stations was somewhat higher than that of the intertidal: 94 μg 10 cm^{-2}, which made them subtidally more important in terms of energy flux than the copepods (Fig. 3). However it must be reminded that turbellarians are predators and one cannot compare their importance in the system on the basis of weight alone.

The total biomass of the gastrotrichs varied between 0.4 μg 10 cm^{-2} (016) and 488.4 μg 10 cm^{-2} (036a). High biomass values for the other taxa were noted in station 014: 303 μg 10 cm^{-2} (hydroids), 06: 135 μg 10 cm^{-2} (hydroids), 038b: 196 μg 10 cm^{-2} (polychaetes). On the tidal flats ostracods became an important component of the biomass: 360 μg 10 cm^{-2} (017), 140 μg 10 cm^{-2} (035), 636 μg 10 cm^{-2} (037a) and 400 μg 10 cm^{-2} (038a).

Variation in time and space

Vertical distribution
The depth distribution of the meiofauna according to the geographic location is given in Fig. 6, both for the intertidal and subtidal environment (–1 mm below low water level and > 10 m deep).

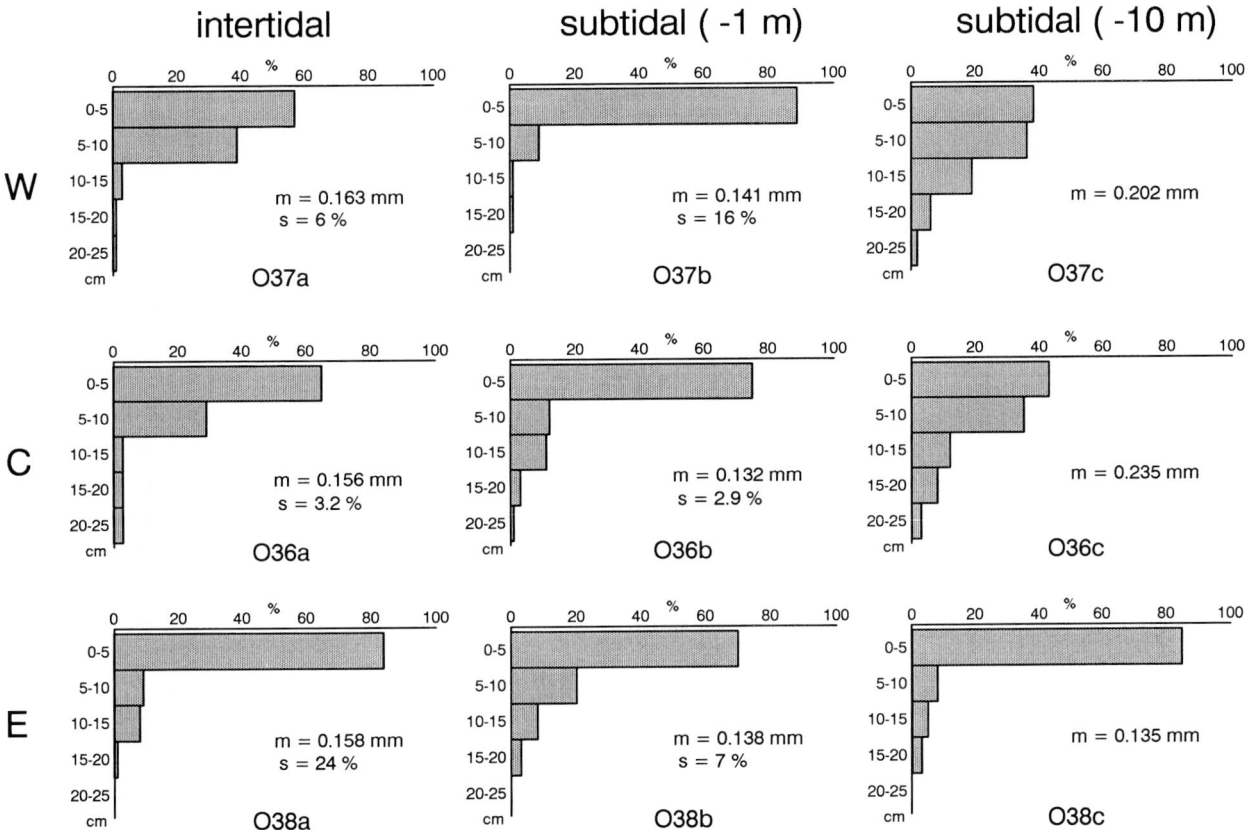

Fig. 6. Vertical distribution of the meiofauna according to the geographic location: 9 stations distributed over the subtidal and intertidal habitat and the 3 regions (W = west; C = central; E = east) of the Oosterschelde (m = median grain size, s = silt-clay fraction).

Meiofauna occurred to a depth of 25 cm (maximum sampling depth) into the sediment. In all stations the bulk of the meiofauna ($> 70\%$) inhabited the upper 10 cm, on tidal flats more than 90% were restricted to that layer. Towards the east the decreasing median grain size superimposed by an increasing silt-clay content resulted in a concentration of more than 80% of the meiofauna in the top 5 cm of the sediment. This clearly reflected a response of the meiofauna to two main sediment characteristics *i.e.* the amount of silt-clay and the median grain size.

Seasonal changes: Fig. 7
A pronounced seasonality was found to exist for most meiofauna groups, generally with a maximum abundance in the warm seasons and a minimum in winter. Highest total meiofauna densities were observed in summer: 18 000 ind 10 cm^{-2} (037a) and lowest in winter: 744 ind 10 cm^{-2} (037b).

Nematodes, being the dominant taxon, fluctuated according to the same pattern. In the intertidal the density sometimes increased up to 20 times the winter values (037a).

In the western part the density of the nematodes at the subtidal station was lower in summer than in spring. The presence of dark granules in the intestine of more than 50% of the nematodes indicated a detoxification system for sulphide ions (and oxygen depletion, cfr. Nuss, 1984). This was confirmed by a strong H_2S-smell. Most probably this H_2S-stress had its repercussion on the density of the nematodes, the copepods and other groups.

The copepod population reached its highest

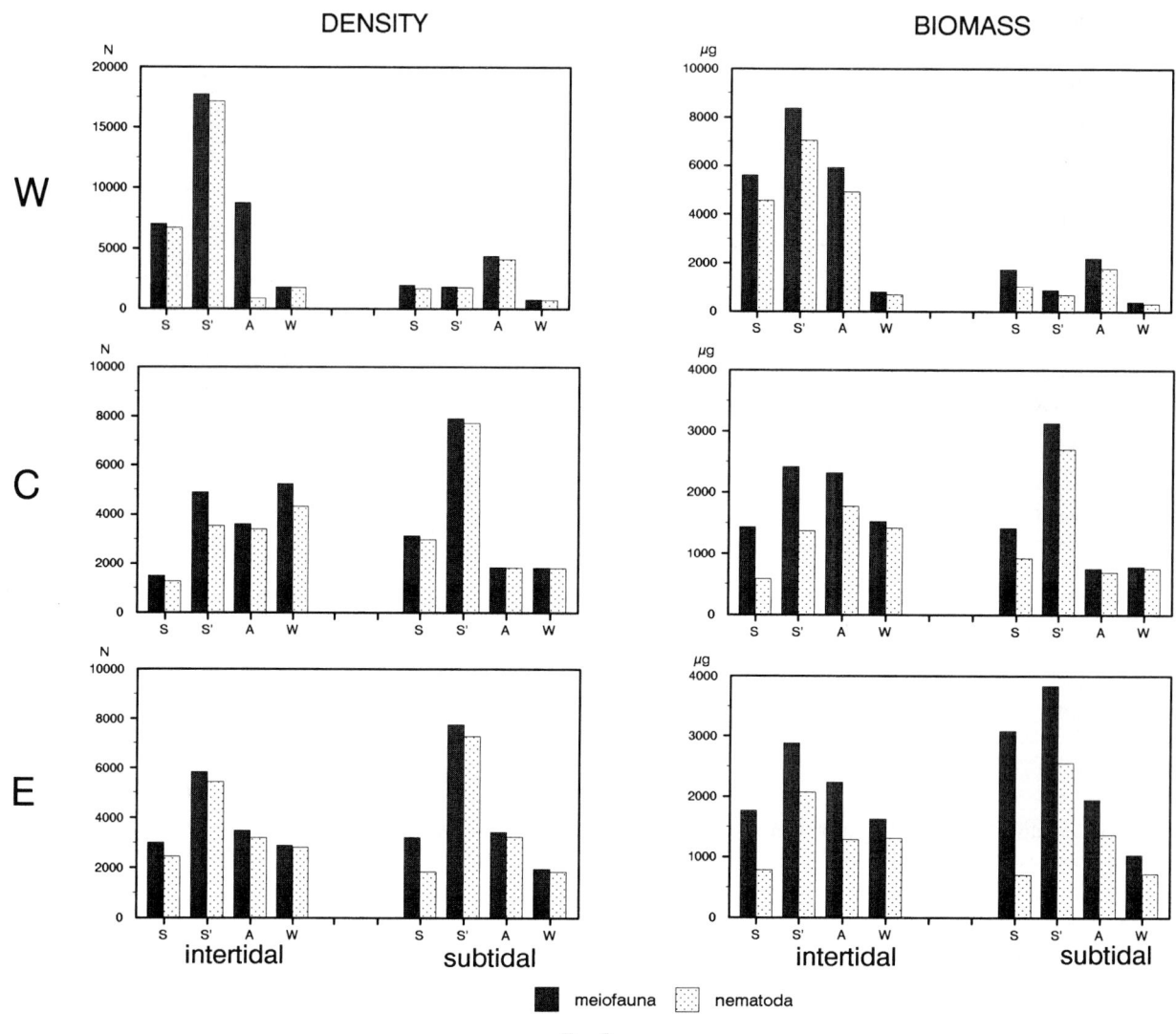

DENSITY

BIOMASS

Fig. 7a.

density in spring at the subtidal station 038b. The stations situated in the western and central part had a peak in autumn and in summer respectively. A peak abundance for the turbellarians, the ostracods and the tardigrades was noted either in spring, summer or autumn; their minimum was always observed in winter.

The total biomass of the meiofauna was highest in summer with values of 8.4 mg 10 cm^{-2} (037a), 2.4 mg 10 cm^{-2} (036a), 2.9 mg 10 cm^{-2} (038a) for the intertidal and 2.21 mg 10 cm^{-2} (037b), 3.1 mg 10 cm^{-2} (036b), 3.8 mg 10 cm^{-2} (038b) for the subtidal. The seasonal pattern was most pronounced in station 037a, the summer value being 10 times higher than in winter, due to the dramatic increase in numbers ($\times 20$) and in biomass ($\times 10$) of the nematodes. The variation of the biomass of the nematode population was in accordance with the fluctuations in density. Minimum biomass values for the copepods as well as for the other groups were observed in winter and maximum peaks occurred in the other seasons, often reaching 15 times the winter values.

The abundance and biomass of the ostracods sometimes exceeded that of the turbellarians and

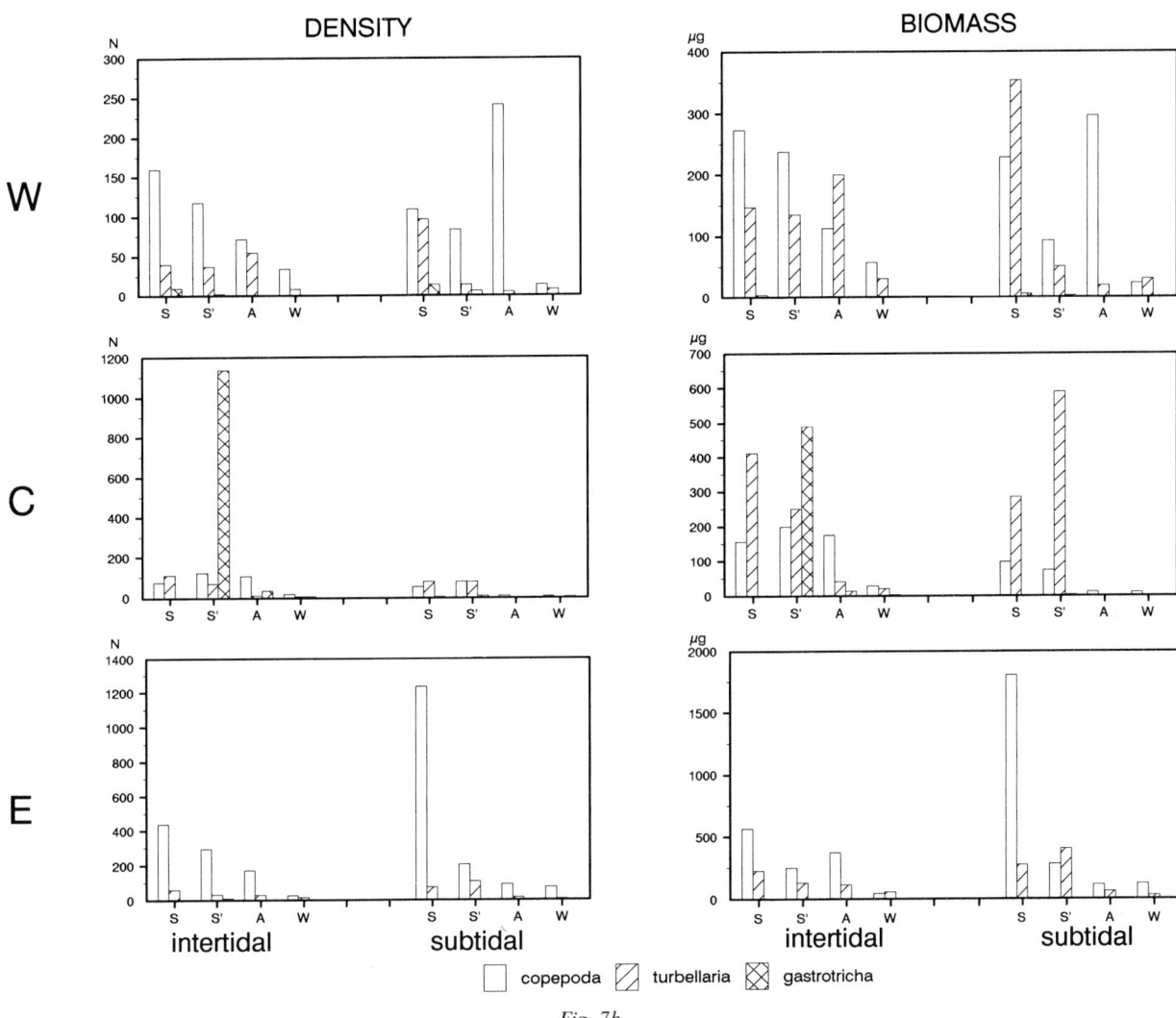

Fig. 7b.

in the central region the biomass of the latter exceeded that of the copepods in spring and summer.

Diversity

The number of meiofauna taxa was lowest in station 017: 4 and highest in 037a and 038b: 11 taxa (Fig. 8). One could expect a mean of 7 taxa in the subtidal stations and a mean of 8 on the tidal flats, although the mean diversity measured by the set of indices as indicated in Table 4 was higher in the subtidal stations than in the intertidal stations.

The number of nematode species varied between 20 and 33 per station. In the intertidal there was a clear increase in number of species from west to east and the diversity and evenness were higher intertidally than subtidally.

The species richness of the copepod community varied between 6–21 species per station. The highest scores were found in subtidal clean medium sands and subtidal muddy fine sands of the outer area and the inner basin respectively. The lowest number of species was observed in clean fine sands in the intertidal zone of the middle area.

A mean of 9 and 7 species was calculated for

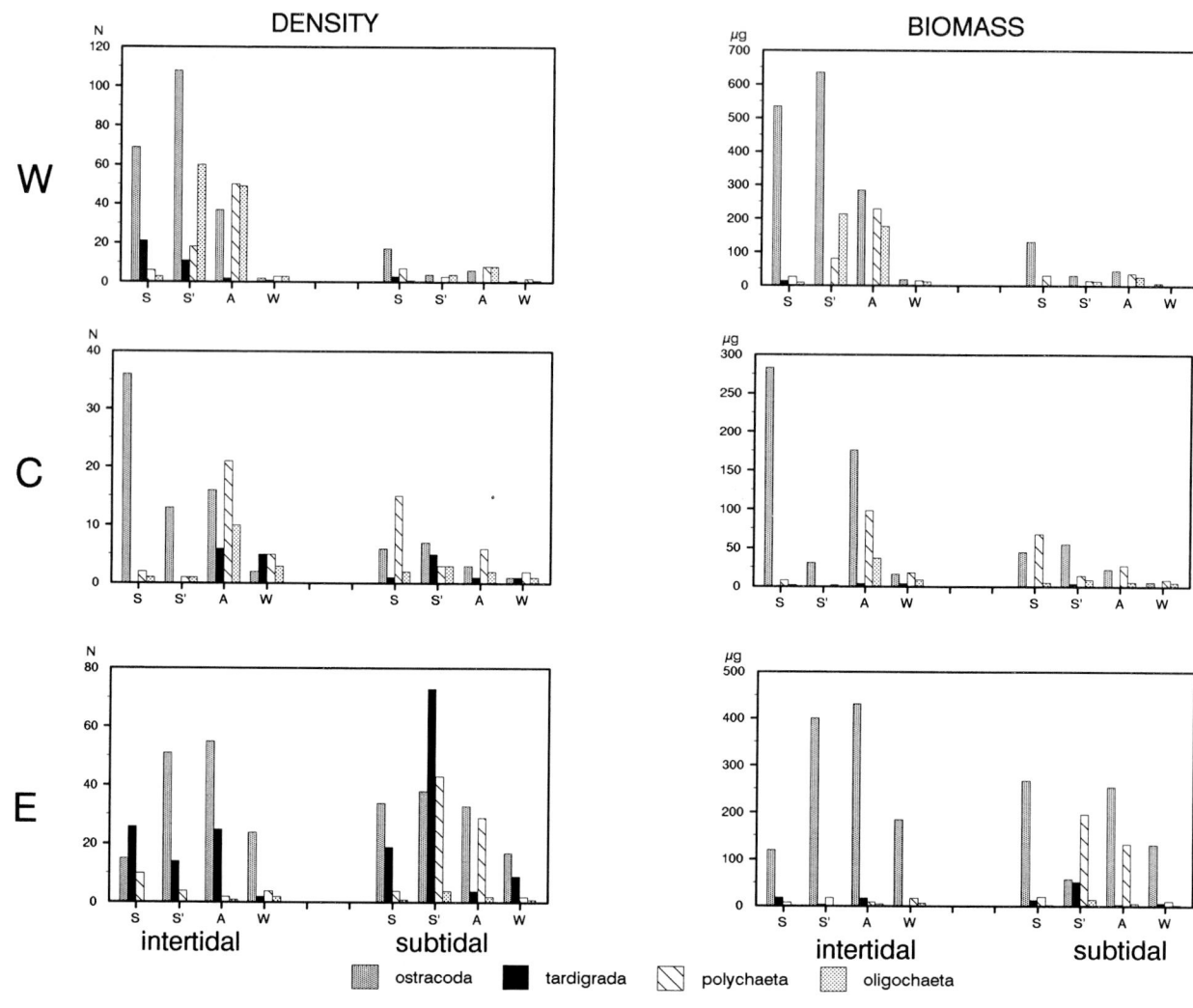

Fig. 7c.

Fig. 7. Seasonal changes in density (ind 10 cm^{-2}) and biomass (μg DW 10 cm^{-2}) of the meiofauna taxa (S = spring; S′ = summer; A = autumn; W = winter; W = west; C = central; E = east).

7a: Total meiofauna and Nematoda

7b: Copepoda, Turbellaria, Gastrotricha

7c: Ostracoda, Tardigrada, Polychaeta, Oligochaeta

subtidal and intertidal respectively. For each of the diversity indices the copepod community was more diverse in the subtidal than in the intertidal.

Comparing the diversity of the copepods between the subtidal-intertidal stations on the one hand and the 3 main regions of the Oosterschelde on the other hand, significant differences for different indices were observed as shown in Table 4.

In both cases the number of species was significantly different. The discrepancy between the sub- and intertidal was mainly caused by differences in evenness (or abundance) of the species and the difference between the western and eastern region of the Oosterschelde was mainly due to the diversity (or number of species). A classical 1-way ANOVA had the same results.

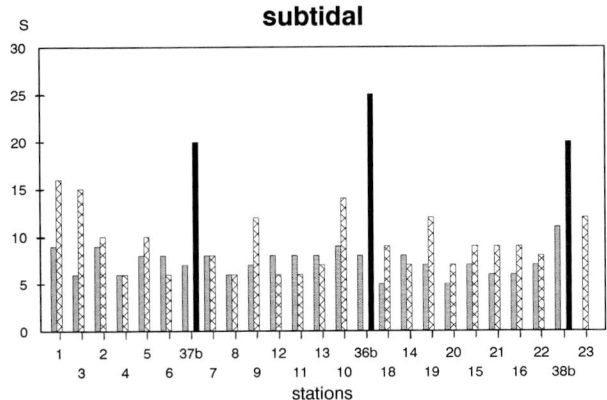

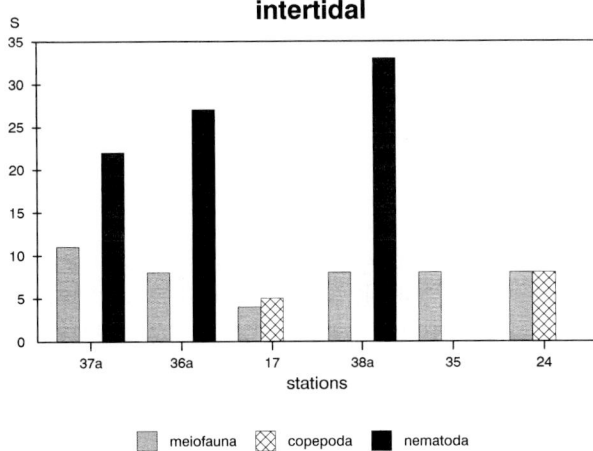

Fig. 8. Number of meiofauna taxa, and number of nematode and copepod species per station (stations ranked from west (left) to east (right).

Table 4. Diversity of the meiofauna, the nematodes and the copepods: comparison of the diversity of the copepods between subtidal–intertidal (= habitat) and between the 3 regions of the oosterschelde (= area); probabilities of the Kruskall-Wallis analysis are given (H′ = Shannon-Wiener-index, H = Brillouin-index, SI = Simpson-index, J = Pielou's index, N1, N2, E1,0, E2,1 = Hill's numbers, E′1,0 = Heip-index, E′2,1 = Alatalo-index, N∞ = dominance index).

	Habitat	Area
H′	0.650	0.006 **
H	0.750	0.008 **
SI	0.391	0.03 *
J	0.001 ***	0.169
N1	0.545	0.007 **
N2	0.356	0.032 *
E1,0	0.000 ***	0.334
E′1,0	0.001 **	0.249
E2,1	0.099	0.359
E′2,1	0.039 *	0.609
S	0.021 *.	0.048 *
N∞	0.237	0.362

Nematode community

A total of 176 species belonging to 97 genera and 20 families were identified. The number of species per station ranged between 15 to 48. Dominant families (≥ 10%) were in descending order Xyalidae, Comesomatidae, Oncholaimidae, Desmodoridae, Axonolaimidae, Enoplidae and Chromadoridae.

The subtidal stations were dominated by *Daptonema riemanni*, *Sabatieria vulgaris*, *S. breviseta*, *Viscosia viscosa*, *Enoplolaimus propinquus*, *Ascolaimus elongatus*, *Daptonema fallax*, and *Metadesmolaimus varians*.

The tidal flats were mainly occupied by *Metachromadora vivipara*, *Spirinia parasitifera*, *Oncholaimellus sp.*, *Sabatieria vulgaris*, *Ascolaimus elongatus*, *Daptonema riemanni*.

Particularly the genera *Sabatieria* and *Daptonema* were represented by several species.

Among the ecological feeding groups the nematode community was dominated in terms of abundance by non-selective deposit feeders. The predators/omnivores ranked second, followed by the epistratum feeders and the selective deposit feeders.

Copepod community

A total of 1 cyclopoid and 70 harpacticoid copepod species were found, the latter belonging to 14 families and 47 genera.

Seven families represented 56 species (78.8%); these were in order of their importance: Ectinosomatidae, Diosaccidae, Laophontidae, Cylindropsyllidae, Cletodidae, Tachidiidae and Paramesochridae. Only 2 species could be characterized as very common (> 50%) based on their occurrence in the total data-set: *Asellopsis intermedia* and *Harpacticus flexus*. Twenty species were common (25–49%); these were in order of their frequency: *Pseudobradya beduina, Tachidius discipes, Euterpina acutifrons, Arenosetella tenuissima, Canuella perplexa, Paraleptastacus espinulatus, Enhydrosoma propinquum, Pseudobradya minor, Evansula pygmaea, Kliopsillus constrictus s.str., Halectinosoma herdmani, Leptastacus laticaudatus, Kliopsyllus paraholsaticus, Canuella furcigera, Arenosetella germanica, Arenocaris bifida, Paronychocamptus nanus, P. curticaudatus, Stenhelia palustris, Halectinosoma gothiceps.* Rare (5–24%) and very rare (<5%) species constituted together 49 species (23 and 26 respectively).

In terms of their relative abundance the family Paramesochridae strongly dominated. Sub-dominance was shared between the Cylindropsillidae and Ectinosomatidae. The dominance of the Paramesochridae and Cylindropsillidae was reflected in the dominance of the psammophilous, particularly the mesopsammic species. Only the epi-endopsammic and euryoecious groups were subdominant and of equal importance.

Among the ecological groups only psammophilous and euryoecious groups were of any importance followed by the limicolous and phytophilous groups.

The groups of psammophilous and euryoecious species together comprised 51 species, 71.9% of the total number. Twenty of these species accounted for 91.0% of the very common and common species groups. Within the psammophilous fauna 20 species (62.5%) were mesopsammic, 8 of which made up 36.4% of the very common and common species. Epi-endopsammon (6 species) and euryoecious (6 species) each accounted for 27.3%.

Turbellarian community

The density of the turbellarians and their species diversity was much higher in the Oosterschelde than in lake Grevelingen or in the Westerschelde (Martens & Schockaert, 1981). The same authors found that in the Delta as a whole, Proseriata and Acoela were about equally well represented and approximately one fifth of the turbellarian fauna consisted of Neorhabdocoela and 90% of these were Kalyptorhynchia.

A list of species determined at the stations 013 and 018 in 1979 is represented in Martens and Schockaert (1981).

Abundance and environmental parameters

In the Oosterschelde the variation in space of meiofaunal abundance was significantly correlated with environmental parameters such as: median grain size of the sand fraction, amount of silt-clay, amount of gravel, sorting, depth, and longitude. Significant correlation coefficients are presented in Table 5 ($p < 0.01$ for $n = 29$, $r_s = 463$).

The density and relative abundance of the nematodes and to a lesser extent the density of the ostracods and oligochaetes increased significantly with decreasing median grain size and increasing silt content.

Turbellarian density increased with a higher degree of sorting and the same holds for the tardigrades.

The number of gastrotrichs was negatively correlated with the silt content and their importance in the meiofauna composition responded to a coarser sand and lesser mud.

Highest hydroid abundance was found in the clean coarser sands, their numbers decreasing to the east.

For the copepods only the relative importance was highly significant being positively correlated with the median grain size and negatively with the silt-clay fraction.

The N/C-ratio exhibited a similar correlation with grain size and silt-clay content as the nematodes and increased with increasing sorting and decreased with depth.

The relative abundance of nematodes was significantly correlated with longitude, becoming more important to the east. Both the percentage of nematodes and oligochaetes increased with increasing silt-clay content and decreasing median grain size and depth.

Copepods, turbellarians, gastrotrichs and hydroids became more important with increasing median grain size and decreasing silt-clay content.

Among the environmental factors the median

Table 5. Significant correlations ($p < 0.001$) between different biotic and abiotic parameters: Spearman rank correlation coefficients.

	Grain size	Sorting	Silt-clay	Gravel	Depth	Longitude
Density meiofauna	− 0.609		0.534		− 0.486	
Density nematoda	− 0.824		0.704		− 0.629	
Density turbellaria		− 0.554				
Density gastrotricha			− 0.496			
Density tardigrada		− 0.492				
Density ostracoda	− 0.574		0.527	− 0.538	− 0.691	
Density hydrozoa	0.678		− 0.721		0.647	− 0.472
Density oligochaeta	− 0.617		0.592			
N/C-ratio	− 0.824	− 0.049	0.823		− 0.575	
% nematoda	− 0.837		0.845		− 0.589	0.476
% copepoda	0.809		− 0.792	0.548		
% turbellaria	0.587		− 0.644		0.661	
% gastrotricha	0.544		− 0.566		0.491	
% ostracoda				− 0.528	− 0.628	
% hydrozoa	0.759		− 0.733		0.718	− 0.475
% polychaeta					0.540	
% oligochaeta	− 0.533		0.527			
Biomass meiofauna	− 0.657		0.505		− 0.522	
Biomass nematoda	− 0.817		0.700		− 0.635	
Biomass copepoda					− 0.509	
Biomass ostracoda	− 0.576		0.540	− 0.529	− 0.655	
Biomass turbellaria		− 0.549				
Diversity taxa	0.700		− 0.700		0.600	
Depth	0.651					
Longitude	− 0.495		0.483			
Silt-clay	− 0.844					

grain size showed a positive relation with depth and a negative relation with longitude and silt-clay content, the latter increasing from west to east and decreasing with depth.

The diversity of the meiofauna community increased in coarser sand and lower amount of silt-clay and is positively correlated with depth (below approximately known sea level). No relation was detected between the diversity of the copepods and the sediment characteristics.

Discussion

The Oosterschelde was strongly influenced by tidal currents causing a heterogeneous and unstable biotope characterized by environmental gradients. Especially the tidal flats were daily subjected to greater fluctuations of temperature and salinity, and different faunal assemblages in the subtidal and intertidal habitat are therefore expected.

The sediment composition, was primarily determined by the hydrodynamical forces and offered a variety of different biotopes within the Oosterschelde.

The meiofauna was very diverse with 12 permanent meiofauna taxa. In the nearby Westerschelde Van Damme *et al.* (1980) recorded 10 taxa in subtidal transects. In the Dutch Wadden Sea only 4 and 5 taxa were noted by Witte & Zijlstra (1984) and Bouwman (1981) respectively, but not all meiofauna groups were taken into account.

The abundance of the meiofauna was very high in both sub- and intertidal habitats, which was partly due to the sampling season (summer). Similar high densities in subtidal areas were observed by Faubel *et al.* (1983) in the Fladen

Ground, Bodin (1984) in the bay of Douarnenez, Herman *et al.* (1984) in the Belgian coastal zone, Ansari *et al.* (1980) in Goa, India; and in tidal areas by Ellison (1984) in Plymouth and Dye (1983) in South Africa.

Meiofauna abundance on the tidal flats of the Wadden Sea and the Westerschelde was generally lower.

The subtidal abundance and biomass were considerably lower and were consistent with the density of the subtidal populations in the North Sea (Heip *et al.*, 1979), of the fine sand populations in the Bay of Morlaix, France (Boucher, 1980) and of Firemore Bay, Scotland (McIntyre & Murison, 1973).

In the Oosterschelde nematode abundance and biomass differed significantly between the subtidal and intertidal. No such difference was noted in the literature, since all studies are restricted to one of these habitats only.

On the tidal flats nematodes predominated as in most estuaries. The observed abundance and standing stock were high and similar to those of the salt marsh Saaftinghe in the Westerschelde estuary (Van Damme *et al.*, 1980), of the Lynher estuary (Warwick & Price, 1979), estuaries of Georgia, USA (Teal & Wieser, 1966) and the Eems estuary, although nematode density was slightly lower there (40–10 000 ind 10 cm^{-2}, Bouwman, 1983). These high scores are typical for intertidal muddy sands.

Contradictory to our data very high numbers of copepods were found in other intertidal environments e.g. at the Galapagos a peak value of 6000 ind 10 cm^{-2} (Schmidt, 1978) and of 3400 ind 10 cm^{-2} at South Africa (McLachlan, 1977).

The number of turbellarians were in good agreement with those observed by McIntyre (1968) for an Indian estuary, but were lower than the values presented by Reise (1983) for a sand-flat in Sylt and by McIntyre & Murison (1973) for a sublittoral habitat in Firemore Bay, Scotland.

From these data it is obvious that the N/C-ratio as a monitor of pollution has to be used with precaution. Within the unpolluted Oosterschelde the N/C-ratio exhibited an important variation and reached peak values (184), which according

to Raffaelli & Mason (1981) are characteristic for polluted situations. In the Oosterschelde the N/C-ratio rather reflected more the sediment characteristics as was demonstrated by its correlation with median grain size and silt-clay fraction. This implies that the use of the N/C-ratio in monitoring pollution is only valid when comparing similar sediment types, if having any value at all. This was well illustrated by station 037a, where extremely high values of nematodes distorted the N/C-ratio without any obvious relation to perturbation. Furthermore as a ratio it is very sensitive to the aggregated pattern of both its constituent elements *i.e.* copepods and nematodes. One could expect large variation in the N/C-ratio as both the nematode and copepod populations displayed an aggregated distribution pattern.

In temperate regions intertidal and shallow subtidal meiobenthos are known to vary seasonally. In the Oosterschelde all taxa reached maximum abundance and biomass in the warmer months of the year. Whereas peak values occurred in spring, summer or autumn, according to the taxa, minimum scores were consistently found in winter. A similar seasonal pattern was observed by Little (1986), who noted maxima for the tardigrades, ostracods and some oligochaetes in spring.

In the intertidal of the Oosterschelde seasonality was more pronounced, which could be explained by the response of the meiofauna community to increased temperature (Heip & Smol, 1976).

The nematode community closely resembled that of other north-west European estuaries. At the genus level resemblance was very close with the Lynher estuary (Warwick & Price, 1979 and the Eems-Dollard (Bouwman, 1983). Although to species level the similarity was only half. Differences in microhabitat, biological interactions, food supply, structural heterogeneity of the sediment caused by macrofauna, predation etc. determined the species composition and the peculiarity of each estuary.

Although sorting has been suggested by Jansson (1967) as a relevant factor to the distribution of the meiofauna, in the Oosterschelde grain size was the primary factor controlling the meiofauna abundance pattern (next to temperature). Only the abundance of turbellarians and tardigrades responded positively to a higher degree of sorting.

Among the meiofauna taxa, the distribution of nematodes, gastrotrichs and oligochaetes demonstrated a similar affinity to high silt content and small particle size, nematodes scoring the best.

Interstitial groups e.g. gastrotrichs, tardigrades and hydroids prefered clean sand with coarser grain size.

Conclusions

It is clear that in the Oosterschelde both the grain size and the silt content were co-controlling the distribution and diversity of the meiofauna. Consequently, changes in tidal amplitude and current velocity enhanced by the storm-surge barrier, will alter the distribution and accumulation of the sediment particles. Clean sands will be most affected. As a result meiofauna will become more important in terms of abundance and biomass, mainly due to increasing numbers of nematodes. One can predict a general decrease in meiofauna diversity.

The number of copepods is expected to decrease and interstitial species will be replaced by burrowing and epibenthic species. The latter are more important in terms of biomass and thus as food for the epibenthos. Among the meiofauna taxa, differences in activity and availability result in harpacticoids as the preferred food for the epibenthos, with ostracods making up the rest (Pihl, 1985; Gee, 1987). This is more evident in muddy intertidal sediments, where almost the total population of the prey species is concentrated in the top 3 mm (Gee, 1987).

Recently it has been shown that epibenthic crustacea, particular shrimps (*Crangon crangon*), are the most important epibenthic consumers of meiofauna from intertidal and shallow subtidal habitats (review in Hicks & Coull, 1983; Pihl, 1985; Gee, 1987). According to Hostens & Hamerlynck (1993) in the Oosterschelde *Crangon crangon* belongs to the top 5 epibenthos species in terms of density as well as biomass. The crab *Carcinus maenas*, also a dominant species in the Oosterschelde is known to feed on meiofauna (Sherer & Reise, 1981). Gee (1989) reviewed the importance of fish as predators on meiofauna taxa, and notes that flatfish and gobies appear to be the main groups of fish feeding on meiofauna; the former dominating in sandy habitats and the latter being more prevalent in muddy habitats. In the Oosterschelde flatfish such as *Pleuronectes platessa*, *Limanda limanda* and the goby *Pomatoschistus minutus* score within the top 5 dominant epibenthic species. Both flatfishes also dominate the biomass of the epibenthos (Hostens & Hamerlynck, 1993). An increase in the numbers of nematodes will increase bioturbation (Cullen, 1973), nutrient remineralization (Warwick & Price, 1979, Platt & Warwick, 1980) and sustain bacterial growth (Gerlach, 1978). As the abundance of the meiofauna might increase, its role in the 'small foodweb' will become more important.

Acknowledgements

The authors acknowledge the crews of the research vessels Welsinghe and Wijtvliet of the Dutch Rijkswaterstaat and Marris Stella of the Netherlands Institute of Ecology at Yerseke for their assistance in the field. Part of the research was supported by the Balans-project of Rijkswaterstaat. The second author acknowledges a grant from the Beyerinck-Popping fund. Contribution No. 697 of NIOO-CEMO, Yerseke, The Netherlands.

References

Alatalo, R. V., 1981. Problems in the measurement of evenness in Ecology. Oikos 37: 199–204.

Ansari, A. A., A. H. Parulekar & T. G. Jagtap, 1980. Distribution of sublittoral meiobenthos off Goa Coast, India. Hydrobiologia 74: 209–214.

Arlt, G. & M. A. H. Saad, 1977. Investigations on the meiofauna and sediment of the Shatt Al-Arab near Basrah (Iraq). Freshwat. Biol. 7: 487–494.

Austen, M. C. & R. M. Warwick; 1989. Comparison of univariate and multivariate Aspects of estuarine meiobenthic community Structure. Estuar. coast. Shelf Sci. 29: 23–42.

Baird, D. & H. Milne, 1981. Energy flow in the Ythan estuary, Aberdeenshire, Scotland. Estuar. coast. Shelf Sci. 13: 455–472.

Barnett, P. R. O., 1968. Distribution and ecology of harpacticoid copepods of an intertidal mudflat. Int. Revue ges. Hydrobiol. 53: 177–209.

Bodin, Ph., 1984. Densité de la méiofaune et peuplements de copépodes harpacticoïdes en baie de Douarnenez (Finistère). Ann. Inst. océanogr. 60: 5–17.

Boucher, G., 1980. Facteurs d'équilibre d'un peuplement de nématodes des sables fins sublittoraux. Mém. Mus. natn. Hist. nat. 114: 1–81.

Bouwman, L. A., 1981. A survey of nematodes from the Ems estuary. Part I. Systematics. Zool. Jb. Syst. 108: 335–385.

Bouwman, L. A., 1983. A survey of nematodes from the Ems estuary. Part II. Species assemblages and associations. Zool. Jb. Syst. 110: 345–376.

Bowen, R. A., J. M. Onge, J. B. Colton & C. A. Price, 1972. Density-gradient centrifugation as an aid to sorting planktonic organisms I. gradient materials. Mar. Biol. 14: 242–247.

Buchanan, J. B. & J. M. Kain, 1971. Measurements of the physical and chemical environment. In N. A. Holme & A. D. McIntyre (eds), Methods for the study of marine Benthos, IBP-handbook nr. 16; Blackwell Oxford: 30–58.

Capstick, C. K., 1959. The distribution of free-living nematodes in relation to salinity in the middle and upper reaches of the River Blyth estuary. J. anim. Ecol. 28: 189–210.

Coull, B. C. & J. B. J. Wells, 1981. Density of mud-dwelling meiobenthos from the sites in the Wellington region. N.Z. J. Mar. Freshw. Res. 15: 411–415.

Cullen, D. J. 1973. Bioturbation of superficial marine sediments by interstitial meiobenthos. Nature 242: 323–324.

Duursma E. K., H. Engel & TH. J. H. Martens, 1982. De Nederlandse Delta. Natuur & Techniek: 1–511.

Dye, A. H., 1983. Vertikal and horizontal distribution of meiofauna in mangrove sediments in Transkei, Southern Africa. Estuar. coast. Shelf Sci. 16: 591–598.

Dye, A. H. & J. P. Furstenberg, 1978. An ecophysiological study of the meiofauna of the Swartkops estuary. 2. The meiofauna: composition, distribution, seasonal fluctuation and biomass. Zool. Afr. 13: 19–32.

Ellison, R. L., 1984. Foraminifera and meiofauna in an intertidal mudflat, Cornwall, England: populations, respiration and secundary production and energy budget. Hydrobiologia 109: 131–148.

Farris, R. A. & H. Crezee, 1976. An improved Reineck box for sampling coarse sand. Int. Revue ges. Hydrobiol. 66: 703–705.

Faubel, A., 1982. Determination of individual meiofauna dry weight values in relation to definite size classes. Cah. Biol. Mar. 23: 339–345.

Faubel, A., E. Hartwig & H. Thiel, 1983. On the ecology of the benthos of sublittoral sediments, Fladen Ground, North Sea. I. Meiofauna standing stock and estimation of production. 'Meteor' Forsch.-Ergebnisse 36: 35–48.

Gee, J. M., 1987. Impact of epibenthic predation on estuarine intertidal harpacticoid copepod populations. Mar. Biol. 96: 497–510.

Gee, J. M., 1989. An ecological and economic review of meiofauna as food for fish. Zool. J. Linn. Soc. 96: 243–261.

Gerlach, S. A., 1953. Die biozönotische Gliederung der Nematodenfauna am der Deutschen Kusten. Z. morph. Ökol. Tiere 41: 411–512.

Gerlach, S. A., 1978. Food chain relationships in subtidal silty sand, marine sediments and the role of meiofauna in stimulating bacterial productivity. Oecologia 33: 55–69.

Heip, C., 1974. A new index measuring evenness. J. mar. biol. Ass. U.K. 54: 555–557.

Heip, C., R. Herman & M. Vincx, 1984. Variability and productivity of meiobenthos in the Southern Bight of the North Sea. Rapp. P.-v. Réun. Cons. int. Explor. Mer 1983: 51–56.

Heip, C., R. Herman, G. Bisschop, J. C. R. Govaere, M. Holvoet, D. Van Damme, C. Vanosmael, K. A. Willems & L. De Coninck, 1979. Benthic studies of the Southern Bight of the North Sea and its adjacent continental estuaries. Progress Report I. International Report. International Council of the exploration of the Sea. C.M. 1979/L:9 Biological and Oceanography Committtee: 1–30.

Heip, C. & N. Smol, 1976. The influence of temperature on the Reproductive Potential of two brackish-water Harpacticoids (Crustacea: Copepoda). Mar. Biol. 35: 327–334.

Heip, C., K. A. Willems & A. Goossens, 1977. Vertical distribution of meiofauna and the efficiency of the Van Veen grab on sandy bottoms in Lake Grevelingen (the Netherlands). Hydrobiol. Bull. 11: 35–45.

Herman, R., L. K. H. Thielemans & C. Heip, 1984. Benthic studies of the Southern Bight of the North Sea. VIII. Evolution of the meiofauna in the belgian coastal waters from 1977 till 1983. Geconcerteerde Onderzoeksacties Oceanografie: progress report 1983: 1–19.

Hicks, G. R. F. & B. C. Coull, 1983. The ecology of marine meiobenthic harpacticoid copepods. Oceanogr. Mar. Biol. annu. Rev. 21: 67–175.

Hill, M. O., 1973. Diversity and evenness: a unifying notation and its consequences. Ecology 54: 427–432.

Hodda, M. & W. L. Nicholas, 1985. Meiofauna associated with Mangroves in the Hunter River Estuary and Fullerton Cove Southeastern Australia. Aust. J. mar. Freshwat. Res. 36: 41–50.

Hodda, M. & W. L. Nicholas, 1986a. Temporal changes in littoral Meiofauna from the Hunter River Estuary. Austr. J. mar. Freshwat. Res. 37: 729–741.

Hodda, M. & W. L. Nicholas, 1986b. Nematode diversity and industrial pollution in the Hunter River Estuary, NSW, Australia. Mar. Pollut. Bull. 17: 251–255.

Hostens, K. & O. Hamerlynck, 1994. The mobile epifauna of the soft bottoms in the subtidal Oosterschelde estuary:

structure, function and impact of the storm-surge barrier; Hydrobiologia 282/283: 479–496.

Jansson, B. O., 1966. Microdistribution of factors and fauna in marine sandy beaches. Veröff. Inst. Meeresforsch. Bremerh. 2: 77–56.

Little, C., 1986. Fluctuations in the meiofauna of the Aufwuchs community in a brackish-water lagoon. Estuar. coast. Shelf Sci. 23: 263–276.

Martens, P. M. & E. R. Schockaert, 1981. Sand-dwelling Turbellaria from the Netherlands Delta Area. Hydrobiologia 84: 113–127.

McIntyre, A. D., 1969. Ecology of marine meiobenthos. Biol. Rev. 44: 245–290.

McIntyre, A. D. & D. J. Murison, 1973. The meiofauna of a flatfish nursery ground. J. mar. biol. Ass. U.K. 53: 93–118.

McLachlan, A., 1977. Studies on the psammolittoral meiofauna of Algao Bay, South Africa. II. The distribution, composition and biomass of the meiofauna and macrofauna. Zool. Afr. 12: 33–60.

Nienhuis, P. H. & A. C. Smaal, 1994. The Oosterschelde estuary, a case study of a changing ecosystem: an introduction. Hydrobiologia 282/283: 1–14.

Nuss, B. 1984. Ultrastrukturelle und ökophysiologische untersuchungen an kristalloiden einschlüssen der muskeln eines sulfidtoleranten limnischen Nematoden (*Tobrilus gracilis*). Veröff. Inst. Meeresforsch. Bremerh. 20: 3–15.

Pihl, L., 1985. Food selection and consumption of mobile epibenthic fauna in shallow marine areas. Mar. Ecol. Progr. Ser. 22: 169–179.

Platt, H. M. & R. M. Warwick, 1980. The significance of free-living nematodes to the littoral ecosystem. In: J. H. Price, D. E. G. Irvine & W. F. Farnham, (eds) The Shore Environment 2. Ecosystems. New York, Academic Press: 729–759.

Raffaelli, D. & F. Mason, 1981. Pollution monitoring with meiofauna, using the ratio of nematodes to copepods. Mar. Pollut. Bull. 12: 158–163.

Reise, K., 1983. Experimental removal of lugworms from marine sand affects small zoobenthos. Mar. Biol. 74: 327–332.

Riemann, F., 1966. Die interstitielle Fauna im Elbe-aestuar. Verbreitung und Systematik. Arch. Hydrobiol., Supp. 31: 1–279.

Saad, M. A. H. & G. Arlt, 1977. Studies on the bottom deposits and the meiofauna of Shatt-al Arab and the arabian Gulf. Cah. Biol. mar. 18: 71–84.

Sherer, B. & K. Reise, 1981. Significant predation on micro- and meiofauna by the crab *Carcinus maenas* L. in the Wadden Sea. Kieler Meeresforsch. 5: 490–500.

Skoolmun, P. & S. A. Gerlach, 1971. Jahreszeitliche Fluktuationen der Nematodenfauna im Gezeitenbereich des Weser-aestuar. Veröff. Inst. Meeresforsch. Bremerh. 13: 119–138.

Schmidt, P., 1978. Die quantitative Verteilung und Populationsdynamik des Mesopsammons am Gezeiten-Sandstrand der Nordseeinsel Sylt. I. Faktorengefüge und biologisch Gliederung des Lebensraumes. Int. Revue ges. Hydrobiol. 53: 723–779.

Smol N., 1986. Rol van het meiobenthos in de Oosterschelde. Balans report, Delta Institute for hydrobiological Research: 1–151 (in Dutch).

Teal, J. M. & W. Wieser, 1966. The distribution and ecology of nematodes in a Georgia salt marsh. Limnol. Oceanogr. 11: 217–222.

Tietjen, J. H., 1969. The ecology of shallow water meiofauna in two New England estuaries. Oecologia 2: 251–291.

Van Damme, D., C. Heip & K. A. Willems. Influence of pollution on the harpacticoid copepods of two North Sea estuaries. Hydrobiologia 112: 143–160.

Van Damme, D., R. Herman, Y. Sharma, M. Holvoet & P. Martens, 1980. Fluctuations of the meiobenthos communities in the Westerschelde estuary. Ices-report CM/L 23: 131–170.

Van Es F. B., M. A. van Arkel, L. A. Bouwman, H. G. J. Schröder, 1980. Influence of organic pollution on bacterial, macrobenthic and meiobenthic populations in intertidal flats of the Dollard. Neth. J. Sea Res. 14: 288–304.

Warwick, R. M., 1971. Nematode associations in the Exe estuary. J. mar. biol. ass. U.K. 51: 439–454.

Warwick, R. M. & J. M. Gee, 1984. Community structure of estuarine meiobenthos. Mar. Ecol. Progr. Ser. 43: 213–219.

Warwick, R. M. & R. Price, 1979. Ecological and metabolic studies on free-living nematodes from an estuarine mudflat. Estuar. coast. mar. Sci. 9: 259–271.

Willems, K. A., 1989. Verspreiding, ecologie en gemeenschapsstructuur van benthische copepoden in het Delta gebied en de Eems-Dollard (Nederland). Ph.D.-thesis, State University Gent: 1–440.

Willems, K. A. & A. J. J. Sandee, 1978. Working group carbon cycle in the Grevelingen. The role of meiozoobenthos in the carbon cycle. In: E. K. Duursma (ed.), Progress Report 1977. Delta Insitute for Hydrobiological Research, Yerseke. Verh. kon. Ned. Akad. Wetensch. Natuurkunde 2: 28–30.

Willems, K. A. & A. J. J. Sandee, 1979. Working group carbon cycle in the Grevelingen. Zoobenthos investigations. Meiozoobenthos: density and biomass. In: E. K. Duursma (ed.), Progress Report 1978. Delta Institute for Hydrobiological Research, Yerseke. Verh. Kon. Ned. Wetensch. Natuurkunde 2: 168–170.

Willems, K. A., Y. Sharma, C. Heip & A. J. J. Sandee, 1984. Long-term evolution of the meiofauna at a sandy station in lake Grevelingen, the Netherlands. Neth. J. Sea Res. 18: 418–433.

Witte, J. IJ. & J. J. Zijlstra, 1984. The meiofauna of a tidal flat in the western part of the Wadden Sea and its role in the benthic ecosystem. Mar. Ecol. Prog. Ser. 14: 129–138.

Wolff, W. J., 1973. The estuary as a habitat. An analysis of data on the soft bottom macrofauna of the estuarine area of the rivers Rhine, Meuse and Scheldt. Zool. Verh. 126: 1–242.

Hydrobiologia **282/283**: 219–234, 1994.
P. H. Nienhuis & A. C. Smaal (eds), The Oosterschelde Estuary.
© *1994 Kluwer Academic Publishers. Printed in Belgium.*

The effects of prolonged emersion and submersion by tidal manipulation on marine macrobenthos

Herman Hummel, Anne W. Fortuin, Roelof H. Bogaards, Andre Meijboom* & Lein de Wolf
*Netherlands Institute of Ecology, Vierstraat 28, 4401 EA Yerseke, The Netherlands; *Research Institute for Nature Management Texel, The Netherlands*

Key words: tidal manipulation, effects of emersion, submersion, marine macrozoobenthos, Oosterschelde estuary

Abstract

Effects of tidal manipulation, resulting in prolonged periods of emersion and submersion or in protracted tidal cycles, on estuarine benthic animals are reviewed.

Prolonged submersion periods did not show effects on mortality of most benthic animals tested, with the exception of the crumb-of-bread sponge *Halichondrea panicea,* which, at low water-flow rates, was covered with a layer of bacteria and subsequently died.

Protracted low-water periods of 18 hours during several weeks hardly caused any mortality. However, protracted low-water periods of 30 hours during some weeks or emersion during several days caused a strong increase in mortality, depending on: the duration of emersion, temperature, condition of the animals, species and age. At temperatures below $-1\,^\circ$C and above 24 $^\circ$C mortality was generally high. Animals with a low glycogen content were more sensitive to emersion than those with a high content. Species with a shell and those that are relatively big were less sensitive than those without a shell or of small size.

The reproductive cycle of benthic animals could be delayed or accelerated by both emersion and submersion.

Introduction

Benthic organisms in coastal and estuarine areas are adapted to tidal fluctuations, *i.e.* they are adjusted to changing water currents, and, in the eulittoral zone, to alternating periods of submersion and exposure to air (Yonge, 1976; Newell, 1979). The major demand all these marine animals make to their environment is the presence of streaming salt water during at least part of the day. Tidal manipulation by means of dams and storm-surge barriers may influence this vital condition drastically. In this way, the impact of man on coastal and estuarine ecosystems has been considerable during the last decades.

Closure of coastal areas by means of barriers, such as the Thames barrier in England or the Oosterschelde barrier in the Netherlands results in temporary changes of the tidal cycle. Tidal currents and the tidal amplitude will be reduced (Elgershuizen, 1981; Smies & Huiskes, 1981), resulting in a lengthening of the periods of submersion or emersion, depending on the level of closure, and consequent hampering of the food and oxygen supply. A storm-surge barrier will be closed as long as a storm is threatening the protected areas; this will not be longer than 1 to 3 days for the barriers mentioned above. However, in cases, e.g. a threatening oil-spill, the barrier might be used too and longer periods of closure may occur. Closure may result in minor to irreversible and lethal effects on animals and ecosystems, depending on the time scale of the changes and the sensitivity of the animals.

In this treatise, the effects of tidal manipulation by means of barriers on marine benthic organisms are reviewed. Besides the studies carried out in relation to the Dutch barrier, only a few references are directly related to this topic. In relation to the Dutch Oosterschelde barrier the species with the highest biomass or numbers were studied in both field and laboratory experiments. These species are the gastropods *Hydrobia ulvae* (Pennant) and *Littorina littorea* (L.), the bivalves *Cerastoderma edule* (L.), *Macoma balthica* (L.), *Mytilus edulis* L. and *Ostrea edulis* L., the polychaetes *Arenicola marina* (L.), *Heteromastus filiformis* (Claparède), *Malacoceros fuliginosus* (Claparède), *Nephtys hombergii* Savigny, *Nereis diversicolor* O.F. Müller, *Pygospio elegans* Claparède, *Scoloplos armiger* (O.F. Müller), and *Tharyx marioni* (de St. Joseph), the anemones *Diadumene cincta* Stephenson and *Sagartia troglodytes* (Price), and the sponge *Halichondria panicea* (Pallas).

The characteristics determined in these species are mortality, condition (*i.e.* weight per volume or glycogen content), reproduction and biochemical constitution.

Because of its parallels with effects of starvation and anaerobiosis relevant references on these topics have been incorporated.

Water content, salinity and temperature of the sediment

Animals living in the sediment experience the buffering capacity of the sediment. Fluctuations of temperature, salinity and water content decrease with depth (Wilde & Berghuis, 1979; Hummel *et al.*, 1986a). Nevertheless, temperatures in emerged sediments were shown to follow the mean air temperature. Then, although fluctuations will be reduced. the mean temperature in the sediment can reach lethal limits during emersion (above 24 °C).

Changes in water content and salinity were only found in the upper 10 cm of emerged sediment, both during short-term experiments (Hummel *et al.*, 1986a; 14 days) as during long-term emersion in the field (Hummel *et al.*, 1986b; 3 months). No changes occurred in the deeper layers of emerged sediments or in the submerged sediments.

In summer, in outdoor laboratory experiments with rained (daily addition of tap water on top of the sediment) and non-rained emerged sediments, the salinity in the upper layer of both sediment-types increased, whereas the water content decreased. This may be due to evaporation (Capstick, 1957; Oglesby, 1969; Anderson & Howell, 1984; Hummel *et al.*, 1986a, b). In contrast to the summer, probably due to a lower evaporation rate, in the other seasons a decrease of the salinity was found in the rained sediment and a decrease of the water content only in the non-rained sediment.

The increased salinity in the top layer of the emerged sediment did, even after 14 days, not reach the upper limits for salinity tolerance, which lies above 35‰ for most marine animals (Kinne, 1971; Wolff, 1973).

The mortality rate of benthic animals in rained and non-rained emerged sediment showed no differences (Hummel *et al.*, 1986b). Obviously, the fluctuations in salinity and water content of the emerged sediment are too small, and not of importance to the survival of benthic animals.

It is concluded that the factors absence of overlying water and ambient temperature are more important than changes in salinity and water content of the emerged sediment.

Mortality

General effects of submersion and emersion

In general, the impact of prolonged submersion and prolonged exposure to air is quite distinct.

This was best shown by the ultimate effect: the death of the animals subjected to it.

During prolonged (up to 2 weeks) submersion in stagnating (but not anoxic) water, most of the benthic animal species tested showed no mortality (Hummel *et al.*, 1986b). One exception was found: the sponge *Halichondria panicea* was very susceptible to a reduction of the water flow rate (Hummel *et al.*, 1988b). Benthic animals subjected in the field to submersion in completely stagnant water for periods longer than one month showed a high mortality (Bogaards *et al.*, 1980; Hummel *et al.*, 1986a).

Animals exposed to air showed a considerable mortality, depending on the species, age, temperature and duration of the emersion (Hummel *et al.*, 1986a, b, 1988a; Fortuin *et al.*, 1989a).

Submersion

Lack of food and oxygen
Submersion for two weeks in oxic stagnant water did not influence the species studied (Hummel *et al.*, 1986b), with the only exception of the crumb-of-bread sponge *Halichondria panicea*. However, under anoxic conditions during submersion benthic animals show a fast mortality (Dries & Theede, 1974, 1976). Yet, in the areas protected by means of storm-surge barriers periods of anoxia by (temporary) closure are not expected to occur on the short term and on a wide scale; oxygen depletion is only expected to occur locally in summer near the sediment as a result of high mineralization activity (Smies & Huiskes, 1981), and near the surface of dense mussel beds, as a consequence of mussel respiration (Dankers *et al.*, 1986). In an experimental set-up the oxygen levels above a mussel bed in stagnant water decreased almost to zero within 2 days. The mussels survived this 4-days experiment, but all crabs and starfishes died. This corresponds with the results of LD_{50} experiments in anoxic running sea water; echinoderms and crustaceans reach 50% mortality in 2 hours to $3^1/_2$ days, bivalves in 8 to 80 days (Dries & Theede, 1974).

During long periods of closure (months) anoxia

may occur in the whole system, as is known from permanently closed areas, such as the Grevelingen estuary and the Markiezaat area in the Netherlands. The Grevelingen estuary was closed in May 1971 at air-temperatures up to 28 °C. Within a month the first benthic animals died and initiated a chain reaction in the submerged areas (Bogaards *et al.*, 1980). The bacterial mineralization of the dead animals resulted in oxygen depletion. The oxygen depletion, in its turn, increased the mortality of the benthic animals during the following months. In submerged areas of the Markiezaat, closed in 1983 at temperatures around 5 to 20 °C, an intensive growth and decay of macroalgae intensified the lack of food and oxygen for benthic animals and initiated the death of animals between the third and sixth month after closure (Hummel *et al.*, 1986a). The difference in temperature during closure may have caused the difference in time after which the mortality of benthic animals increased drastically. In the Markiezaat area, *Nereis diversicolor* appeared to be the most resistant to anoxic conditions, and still lived after 6 months.

It is concluded that, stagnation of the water column, and the consequent lack of oxygen, may severely disturb tidal benthic animals, at least on the long term, but most probably also on the short term (weeks).

Reduced water flow rates and sponges
The only species found to die due to reduced water flow rates in oxic water was the common crumb-of-bread sponge *Halichondria panicea*. Healthy specimens are yellow or olive-yellow. Submersion in water with reduced water-flow rates resulted in a darkening of the sponge as a result of bacterial growth. This was often succeeded by the appearance of white or grey bacterial layers on the surface of the sponge (Hummel *et al.*, 1988b). The bacterium *Alcaligenes faecalis* was isolated from the primary infection, and the bacterium *Acinetobacter calcoaceticus* subsp. *anitratis* from the white or grey layers. Higher temperatures and lower water flow rates promoted the growth of the bacteria and the bacterial coverage of the sponge (Figs. 1, 2). Sponges that were

222

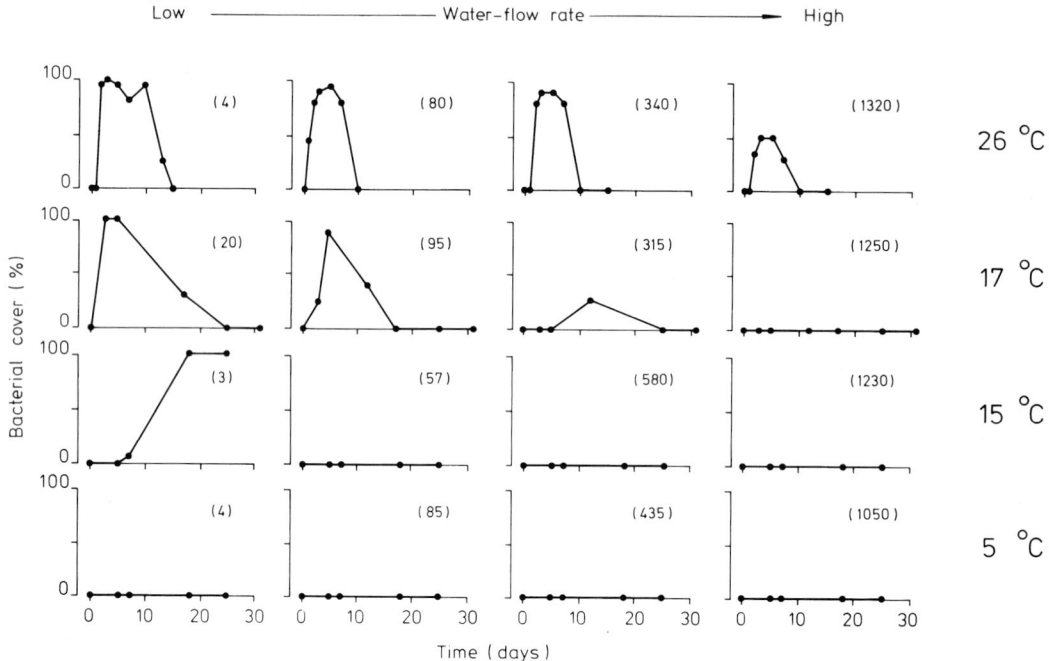

Fig. 1. Changes in bacterial cover of the crumb-if-bread sponge *Halichondria panicea* during experiments with lowered water-flow rates (water-flow rate in ml min⁻¹ are indicated in brackets) (after Hummel *et al.*, 1988b). The disappearance of the bacterial cover, after it reached its maximum, is due to a complete decay of the sponge.

infected with bacteria for more than 70% of their surface could not recover; after the bacterial cover reached its maximum a patch of rotten sponge tissues remained.

Consequently, tidal manipulation might lead to a mass mortality of sponges in the enclosed areas. Mass mortalities of sponges, and their associated fauna, do occur regularly in the open sea, and are

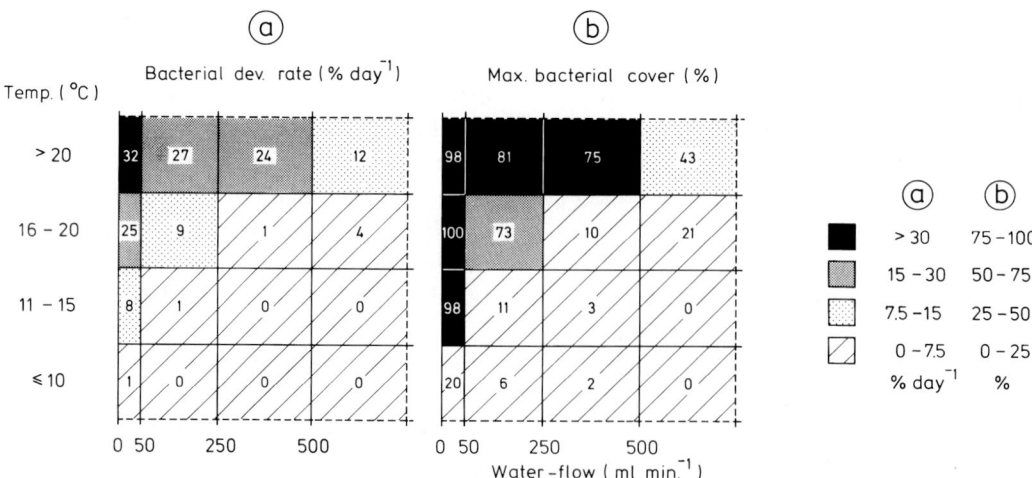

Fig. 2. Development rate (a) and maximal cover (b) of bacteria on the crumb-of-bread sponge *Halichondria panicea* in relation to temperature and water-flow rate (after Hummel *et al.*, 1988b).

mostly ascribed to bacterial and fungal infections (Lauckner, 1980). The etiology of these sponge diseases is hardly investigated and far from being understood. Moreover, a relation with water-flow rates has never before been noticed. No other references on the impact of tidal manipulation or reduction of the water-flow rates on sponges are available. Yet, it is known that the internal ventilation of sponges is partly effected by the beating of the flagella of the choanocyte cells, and partly by passive ventilation from the incurrent pores to the higher, towering, excurrent osculum, caused by velocity gradients in the water-flow near the sponge surface (Vogel & Bretz, 1972). The passive ventilation will decrease to the same degree the water-flow rate is reduced. The more the water-flow rate is reduced, the more bacterial substrate will accumulate inside as well as on the top of the sponge. Thus, the increased bacterial growth at reduced water-flow rates probably was a consequence of a hampered removal of sponge exudates and excretion products from the surface of the sponge. It has to be kept in mind that the crumb-of-bread sponge has, in relation to other species, low concentrations of antimicrobial substances (Thompson et al., 1985). Therefore, it can not be ascertained that after reduction of the water-flow rates other species of sponges will be covered by bacteria in the same manner as the crumb-of-bread sponge.

It can be concluded that sensitive species, such as the crumb-of-bread sponge, may be infected, die and deteriorate soon after reduction of water-flow rates below a certain threshold level. In that case a chain of cumulating adverse reactions (through anaerobiosis) can start.

Emersion

Prolonged emersion and protracted tidal cycles

The mortality of benthic animals during prolonged emersion is strongly dependent on the species, the duration of emersion and temperature (Fig. 3) Hummel et al., 1988a). The longer an emersion period lasted the higher the mortality (Hummel et al., 1986a, b, 1988a: Boyden 1972). The mortality includes the, often considerable, mortality during a subsequent period of 2 to 4 weeks in which the animals had been returned to a normal tidal cycle (Hummel et al., 1986a).

In experiments with protracted tidal cycles the animals were subjected to an extended low-water period (= exposure to air) of 18 or 30 hours, instead of normal 6 hours, followed by 6 hours submersion. If the protracted tidal cycles, with a low-water period of 18 hours, lasted at least 2 weeks, the species known to be sensitive to prolonged emersion showed a considerable mortality (Fortuin et al., 1989b). At high (above 20 °C) or low temperatures (below 0 °C) the less sensitive species showed also a higher mortality. Protracted tidal cycles with low-water periods of 30 hours resulted within 3 weeks in high mortality of sensitive as well as less sensitive animals.

Interspecific differences, a relation with morphology

The animal species tested showed different sensitivity to emersion (Fig. 3) (Hummel et al., 1988a; Fortuin et al., 1989a). The gastropods showed the strongest resistance to emersion. No significant mortality occurred for at least 14 days of emersion at the temperatures tested. The polychaetes Nereis diversicolor and Pygospio elegans also belonged to that emersion-resistant group. Dries & Theede (1974) showed that during anaerobiosis, among several bivalves and polychaetes, N. diversicolor also belonged to the longest living species. The group of emersion resistant species was followed by 2 groups, including all bivalves tested and the polychaete Arenicola marina, which showed mortality rates around 5 and 15% per day respectively, especially at the higher and lower temperatures and at periods of emersion longer than 5 days. The next group, formed by the polychaetes Nephtys hombergii, Heteromastus filiformis and Malacoceros fuligonosus, and the anemones Diadumene cincta and Sagartia troglodytes, showed already a clear mortality (around 5% per day) at temperatures around 5 °C and during short periods of emersion. Maximal mortality rates were similar to those of the previous group (up to 15% per day). The polychaetes Scoloplos armiger and Tharyx marioni were the most sensitive to emersion; already after 2 days of emersion,

224

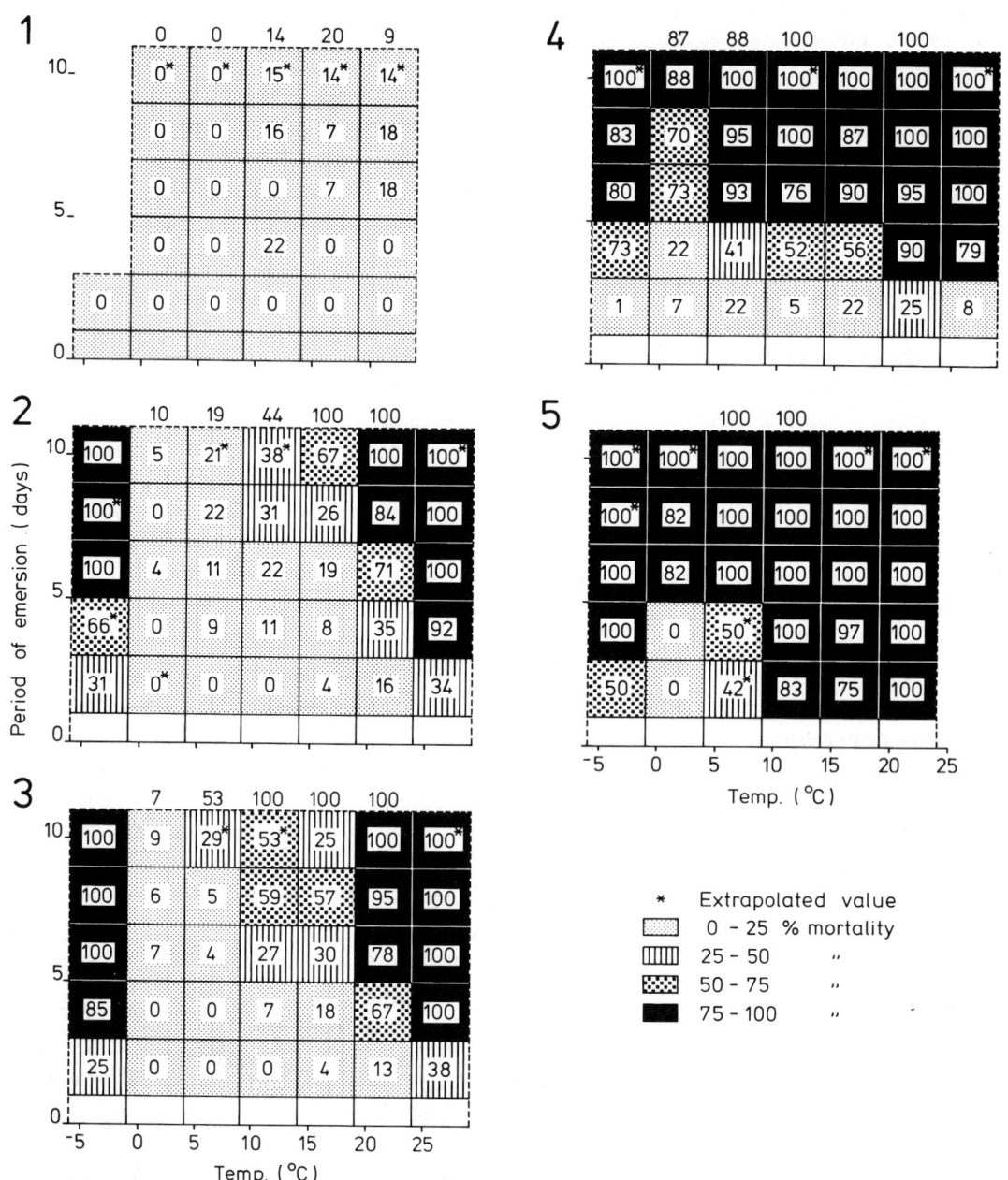

Fig. 3. Percent mortality of five groups of macrobenthic species in relation to temperature and period of emersion. In the upper row the percent mortality of animals emerged for 14 days as found by Hummel *et al.,* (1986a, b) is shown (after Hummel *et al.,* 1988a) Group 1: *Hydrobia ulvae* (Pennant) and *Littorina littorea* (L.), *Nereis diversicolor* O. F. Müller, *Pygospio elegans* Claparède; Group 2: *Ostrea edulis* L., *Arenicola marina* (L.); Group 3: *Cerastoderma edule* (L.), *Macoma balthica* (L.), *Mytilus edulis* L.; Group 4: *Diadumene cincta* Stephenson, *Heteromastus filiformis* (Claparède), *Malacoceros fuliginosus* (Claparède), *Nephtys hombergii* Savigny, *Sagartia troglodytes* (Price); Group 5: *Scoloplos armiger* (O. F. Müller), *Tharyx marioni* (de St. Joseph).

and at almost all temperatures investigated, a high mortality rate was found of up to 50% per day.

The interspecific differences coincided largely with taxonomic and morphological differences. Gastropods with a shell were the least sensitive, bivalves and the bigger polychaetes were intermediate, small polychaetes and anemones were the most sensitive. Similarly, intraspecific differences were observed, smaller (or juvenile vs adult) individuals of *Macoma balthica* and *Nereis diversicolor* were more sensitive to prolonged emersion than the larger individuals (Fortuin *et al.*, 1989a). Only *Pygospio elegans* (small but very resistant) did not conform to this generalization. The lower loss of water in the bigger animals and the animals with a shell might be responsible for the observed differences in the resistance to emersion.

From the results mentioned above it can be concluded that morphological differences (volume, shell) are a prime factor in determining a species resistance to emersion.

Effects of temperature
The lowest mortalities were found at temperatures around 2 °C (Fig. 3) (Hummel *et al.*, 1988a). With increasing temperature the mortality during emersion increased progressively. The mortality of emerged animals kept under constant temperature did not differ from that of animals kept under fluctuating (daytime 5 °C above, night-time 5 °C below average) temperatures.

At temperatures above 24 °C, most species reached a mortality of approximately 100% within 3 to 4 days. At temperatures below −1 °C a sudden strong increase of mortality occurred in almost all species. Thus, starting from 2 °C, a temperature decrease of 5 °C gave the same effect as an increase of 20 °C.

A high mortality at high temperatures was also found in field studies (Hummel *et al.*, 1986a). During the closure of the Markiezaat area an increase in temperature by 10 °C, up to 28 °C, in 4 days was observed, which was followed by a high mortality of all macrobenthic species found in that area.

A greater mortality at temperatures below −1 °C and above 24 °C is in agreement with previous studies on the thermal tolerance of macrobenthic species (Kristensen, 1958; Williams, 1970; Kennedy & Mihursky, 1972; Bayne, 1976; Beukema, 1979, 1985; Murphy, 1979, 1983; Murphy & Johnson, 1980; Ansell *et al.*, 1981; Aarset

& Zachariassen, 1982). The lower limits for thermal tolerance, at exposition periods longer than 1 day, are in between −2 and −10 °C, the upper limits between 24 and 33 °C (if exposition periods at extreme temperatures were shorter than 1 day, more extreme temperatures were reported). The limits found by us correspond with the least extreme values mentioned, thus the animals can only survive prolonged emersion within narrow temperature-limits. This indicates that emerged animals are very vulnerable to extra stress (from temperature) in addition to the stress from emersion.

The upper and lower temperature, −1 and 24 °C resp., correspond well with the minimal and maximal mean monthly air temperatures in the region of research (Fortuin *et al.*, 1989a). This suggests that beyond these natural limits deviations from normal living conditions by emersion are hardly survived by most species. From this, one can predict to a certain extent the lethal temperature limits for benthic species living in other areas subjected to tidal manipulation.

Intraspecific differences
Intraspecific differences in mortality were only shown in species that are sensitive to emersion, at least do show a mortality of more than 50% within 2 to 4 days (Fortuin *et al.*, 1989a). In a comparison of animals from two intertidal stations, one with an average emersion period of 1 1/4 hours and another of 4 3/4 hours per tidal cycle, intraspecific differences were found for the sensitive species *H. filiformis* and *S. armiger*. Animals from the lower (shorter emerged) station died earlier during prolonged emersion than those from the higher station.

Moreover, as stated before, an intraspecific difference was found for juvenile (small) vs adult (large) specimens of *M. balthica* and *N. diversicolor* (Fortuin *et al.*, 1989a). The smaller specimens were more sensitive to emersion than the larger ones.

Relations with tidal zonation
Inter- and intraspecific differences in the sensitivity to emersion have often been related to the

zonation of a species or an organism in the intertidal zone (gastropods: Broekhuysen, 1940; Brown, 1960; Kensler, 1967; Davies, 1969; bivalves: Boyden, 1972; Kennedy, 1976; crustaceans: Barnes et al., 1963; Foster, 1971; several groups: Hummel et al., 1988a). Animals living higher in the intertidal zone are more resistant to exposure to air. Possible causes are differences in the ability to prevent evaporation or loss of water, differences in the tolerance to dehydration and extreme temperatures, and differences in the ability to reduce the metabolic energy expenditure during emersion (Beukema, 1985; Boyden, 1972; Newell, 1979; Hummel et al., 1988a).

The last statement is hard to verify, because for most species studied in our exposure experiments no information is available on the metabolic expenditure during normal and stress situations. Moreover, the information available on *Mytilus edulis* and *Cerastoderma edule* indicates that they have strongly diverging adaptations to reduce their energy expenditure (Boyden, 1972; Widdows et al., 1979; Bayne & Newell, 1983; Akberali & Trueman, 1985; Widdows & Shick, 1985).

Both the ability to prevent the evaporation of water and the ability to reduce the energy expenditure are obviously related to the morphological characteristics that were of importance in determining the interspecific differences in mortality during the emersion experiments. Species with a shell will be more protected against evaporation than un-shelled, as will be burrowing species versus epibenthic species. Moreover, bigger animals have a lower surface to volume ratio and by that a relatively lower evaporation. Indeed, the species of which the tidal zonation in the Dutch Delta area was known (Coosen, pers. comm.) followed this pattern: animals with a shell (*Cerastoderma edule, Hydrobia ulvae, Macoma balthica, Mytilus edulis*) and the bigger animals without shell (*Arenicola marina, Nereis diversicolor*) are found higher in the intertidal zone than the small animals without shell (*Diadumene cincta, Heteromastus filiformis, Sagartia troglodytes, Scoloplos armiger*). The animals of the last group all belong to the species sensitive to experimental prolonged emersion.

Condition and glycogen content

The relation between glycogen and mortality

When benthic invertebrate animals are unable to find enough food or oxygen, they are able to use biochemical reserves, such as glycogen and lipids or protein, and as a consequence, show a reduction in their condition (weight per unit of volume; glycogen content) (Dales, 1958; Pora et al., 1969; Zwaan & Zandee, 1972a, b; Gabbott & Bayne, 1973; Dries & Theede, 1976; Riley, 1976; Nagabushanam & Talikhedar, 1977; Zwaan, 1977; Bayne et al., 1982; Gabbott, 1983; Gäde, 1983; Carr & Neff, 1984; Rodhouse & Gaffney, 1984; Akberali & Trueman, 1985; Gaffney & Diehl, 1986; Lane, 1986). This situation may occur during the winter season, or during sudden artificial environmental changes (see reviews of Gabbott, 1983; Gäde, 1983; Bayne et al., 1985).

In marine bivalves a low condition index and a low glycogen content coincide (Hummel et al., 1989b; Mann, 1978). Consequently, bivalves with a low glycogen content have an overall low (physiological) condition. During stress their reserves are finished within a shorter time than animals with a high glycogen content and as a result are more vulnerable.

Indeed, a relation was found between the glycogen content and the mortality of emerged animals (Hummel et al., 1986b). In autumn as well as in spring, moderate temperatures occurred during the experiments, which should result in similar, rather low, mortalities. However, the average mortality and mortality rate differed significantly (Table 1). The much higher mortality during spring was explained by the much lower glycogen content during this season.

In addition to this, a much higher mortality was found in mussels *Mytilus edulis* with an initial glycogen content below 4% (dry weight), than in mussels with a higher glycogen content (Fig. 4) (during the first 8 to 10 days of emersion at temperatures between 0 and 17 °C; Hummel et al., 1989a).

Thus a clear distinction has to be made be-

Table 1. Seasonal averages of all data on the mortality and mortality rate of benthic animals after 14 days of emersion (see list of species in Fig. 3), of all data on the glycogen content at the beginning of the experiments (per species indexed on the basis of the annual highest glycogen content (g g^{-1}) in the animals), and of all data on the average temperature during the experiments (Hummel *et al.*, 1986b).

	Autumn	Winter	Spring	Summer
Mortality (%)	48.0	48.8	77.7	86.4
Mortality rate (% day^{-1})	2.9	2.3	6.5	16.3
Glycogen (index)	0.79	0.50	0.47	0.76
Temperature (°C)	7	4	10	21

tween animals with a high and those with a low glycogen content or condition.

A latent period and a sudden decrease in glycogen

In the reviewed studies changes in the condition and glycogen content during stress are mostly gradual in time. The disturbances, in the field but also in experiments, mostly were caused by a lack of food or oxygen.

Emersion would bring the animals in a similar condition as they experience during starvation.

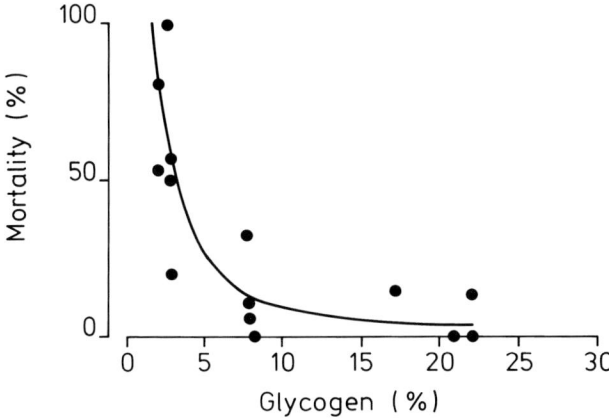

Fig. 4. Percentage mortality of mussels emerged for 8 to 10 days in relation to the glycogen content (data of experiments carried out at temperatures between −1 and 18 °C) (after Hummel *et al.*, 1989a).

Glycogen is known to be the main energy substrate during exposure to air in marine bivalves (Zwaan, 1983). Thus, at least as a result of emersion the glycogen content, and with it the condition, is expected to decrease. However, no decrease of the condition or glycogen content was found during experiments with emerged animals when measured at intervals of 7 to 14 days (Hummel *et al.*, 1986a, b). A significant decrease in glycogen content was sometimes only observed at the very last days before death. On the other hand, an increase in the condition or glycogen content of emerged and submerged animals was also absent, whereas, in spring and summer, it did occur in simultaneously sampled animals from the field (Hummel *et al.*, 1986a, b, 1988c). This indicated that the animals certainly were stressed.

A significant decrease in condition and glycogen content indeed was found in more detailed experiments with *Mytilus edulis* and *Cerastoderma edule*, during which the glycogen content and condition was measured at intervals smaller than 7 days (Hummel *et al.*, 1989b). Changes in the condition were very small, but statistically significant. Changes in the glycogen content depended on the level of glycogen at the start of the experiment. In animals with a low glycogen content at the start of emersion, the glycogen content decreased from the beginning. In specimens with a high glycogen content the glycogen content did not decrease until after 3 to 7 days, at high (20 °C) and low (5 °C) temperatures, respectively. At that moment a high mortality occurred. A latent period of some days before glycogen is used, was not notified before. An acclimatization period, as used in other studies, might obscure such a latent period. However, the absence of a decrease during the first 3 to 7 days may also be obscured by a high individual variation by which it is difficult to obtain in such a short period significant changes.

As stated before, a low condition index mostly coincides with a low glycogen content (Hummel *et al.*, 1989b; Mann, 1978). From this it is not amazing that a possible latent period was found primarily in animals with a high glycogen content, and not always in those with a lower glycogen content. The animals with a low glycogen content have an overall low (physiological) condition, and as a result cannot cope with much stress. They have to use their reserves from the beginning.

Disregarding the first days, the results with emerged animals resemble those of other studies; the glycogen content decreases during stress. The decrease of glycogen in emerged animals, however, may occur suddenly, mostly coinciding with a high mortality. This might explain that a decrease in glycogen content was not found when glycogen was measured at intervals of 7 to 14 days (Hummel et al., 1986a, b). This sudden decrease in glycogen content may be a (post-) mortem phenomenon; during necrosis of tissues, hydrolysis of polymers, such as glycogen, may appear.

The formation of acetic acid

The strong mortality and decrease in glycogen content in both mussels and cockles coincided with a sudden increase in the concentration of acetic acid. It was calculated that at low temperature 37% and at high temperature more than 100% of the glycogen finally was transformed into acetic acid. When glycogen is used during stress several organic acids are formed (Zwaan & Zandee, 1972a; Gäde, 1975; Zwaan, 1977; Zandee et al. 1980; Gäde & Meinardus, 1981; Gäde, 1983). At the moment glycogen is used at high rates a major part is transformed into acetic acid, especially at the higher temperatures. A slow accumulation of acetic acid and the excretion of acetic acid at the moment animals died has been reported before for *M. edulis* kept under anoxic conditions (Kluytmans et al., 1977, 1978; Thillart & Vries, 1985). This phenomenon was also observed in *in vitro* experiments with isolated excised tissues of the mussel, which produced acetic acid at high rates caused by tissue damage (Zurburg, pers. comm.). The high (calculated) amounts (up to more than 100%) of glycogen transformed, do indicate other additional sources for acetic acid. Most probably higher microbial activities in dying animals are the cause.

Thus, a lack of energy reserves does not seem to be the cause of death, as the condition (weight per volume) hardly decreased. The accumulation of organic acids might cause a fatal acidification, while the onset of a high acetic acid production (by microbes) is accompanied with death.

In the above experiments no difference between mussels and cockles was found. The absence of a clear difference was unexpected. Mussels are able to decrease their metabolic expenditure during stress periods to 4% of their normal level, whereas, cockles can only reduce it to 28% (Boyden, 1972; Widdows et al., 1979; Meinardus & Gäde, 1981; Bayne & Newell, 1983; Akberali & Trueman, 1985; Widdows & Shick, 1985). The mussel is able to keep its valves closed during aerial exposure and respire anaerobically; the cockle uses aerial oxygen and respires anaerobically to a much lesser extent or not. In spite of these differences both species showed the same response towards stress by emersion. Similarly, no differences in the mortality of the two species were found after prolonged (2 to 14 days) emersion (Hummel et al., 1988a). Probably, anaerobic respiration and the reduction of metabolism in the mussel have not evolved as adaptation to prolonged emersion but to anaerobic periods which might occur during weeks or months in stratified (subtidal) water types. Then, cockles would be evolutionary more adapted to intertidal areas and mussels to subtidal areas.

Conclusions from the use of reserves during emersion

The general picture which arises from the above is that bivalves are affected differently by stress from emersion than by stress from anaerobiosis or starvation. First, the condition (measured as weight per volume) of the emerged animals hardly decreases. Second, a distinction between animals with a higher and lower glycogen content has to be made. Third, at least in animals with a high glycogen content, a latent period of 3 to 7 days occurs, during which the use of glycogen generally can not be measured. Fourth, in the animals with a high glycogen content the decrease in glycogen was not gradual, but sudden and strong after the first 3 to 7 days (probably due to hydrolysis during necrosis). Fifth, a sudden and high mortality during emersion is not due to a lack of energy reserves, but, most probably, due to acidification by accumulation of organic acids (probably caused by microbial activity).

Effects on reproduction

During gametogenesis glycogen is reallocated from the soma to the gonads (see reviews of Riley, 1976; Sastry, 1979; Bayne *et al.*, 1985). As was shown before, tidal manipulation, at least emersion, may alter normal metabolism and stimulate the use of glycogen in benthic animals. Therefore, an impact of tidal manipulation on reproduction is to be expected. Indeed, the reproductive cycle of some benthic animals is disturbed after prolonged emersion, but, surprisingly, also after a period of submersion in stagnant water (Table 2) (Hummel & Bogaards, 1989; Hummel *et al.*, 1989a). The effects of a period of emersion or submersion on reproduction could arise several months after the period of stress. The disturbance was most distinct at the end of gametogenesis and during the period of spawning (spring). In the

mussel gametogenesis was delayed with $1/2$ to 6 months, whereas, in the cockle it was accelerated with 1 to 2 months. The cause of these opposite effects in the 2 species studied is unknown.

Similar effects in benthic animals were found in studies of effects of starvation (Coe & Turner, 1938; Bayne, 1975; Bayne *et al.*, 1978, 1982; Pipe, 1985), extreme temperatures (Bayne *et al.*, 1978; Maung-Myint & Tyler, 1982), low salinities (Butler, 1949) and contamination by metals (Maung-Myint & Tyler, 1982; Akberali & Trueman, 1985). Minimal effects in those studies were found at the end of summer or in autumn and were attributed to a good condition, to high concentration of biochemical reserves and to gametogenesis being in the resting stage. Maximal effects were found halfway gametogenesis, *i.e.* in early winter because of a reduced condition and low glycogen content. At the end of gametogenesis no change in the period of spawning would occur anymore; at that time, stress would result, as a consequence of reallocation of energy reserves, in fewer and smaller oocytes produced and even some oocytes resorbed. The results with emerged and submerged cockles and mussels fit this pattern only partly, *i.e.* maximal effects if the animals were stressed halfway or at the end of gametogenesis. However, a disturbance at the end of gametogenesis also caused a change in the period of spawning, and no resorption of oocytes was found (Hummel & Bogaards, 1989; Hummel *et al.*, 1989a).

Surprisingly, no differences in the effects of emersion or submersion were found. This means that submersion in stagnant water, although not being lethal, must have evoked a comparable condition in the animals as emersion. Starvation might be this common ground; in both treatments the animals were unfed.

Because glycogen is reallocated from the soma to the gonads, and subsequently lost during spawning, a lowered glycogen content can be found at the end of the spawning season. In most benthic animals, mentioned in this review, the spawning season coincides with the end of the winter season or start of the growing (spring) season (Hummel *et al.*, 1988c). In such species a

Table 2. The number of months for which the spawning period in spring or sumer was delayed ($+ n$) or accelerated ($- n$) after a period of emersion or submersion as measured by means of changes in gonadal stages, oocyte-volume, or oocyte-diameter in *Cerastoderma edule* (C) and *Mytilus edulis* (M). The reproduction of the disturbed animals was followed during 1 year (abstracted from Hummel & Bogaards, 1989; Hummel *et al.*, 1989a).

Start of experiment (stress-period)	Emersion		Submersion			
	7 days		7 days		14 days	
	C	M	C	M	C	M
Winter (January)	0	$+6$	-1	0	-1	$+1/2$
Spring (April)	0	$+2$	-3	$+1$		$+3$
Summer (July)	$+$[a]	$+$[a]	-0[b]	0	-0[b]	0
Autumn (October)	-1	$+1$	-1	0	-1	0

[a] All animals died as a consequence of the stress.
[b] The second spawning period in summer disappeared as a consequence of the stress.

decrease in glycogen can be attributed to nutritional stress (during winter) as well as to gametogenesis and spawning. The ultimate result of both phenomena is a lowered glycogen content and consequently higher sensitivity to stress during spring (table 1).

Seasonal differences

Strong seasonal differences were found in the sensitivity of the animals to prolonged emersion or submersion (Hummel & Bogaards, 1989; Hummel *et al.*, 1986a, b, 1989a; Fortuin *et al.*, 1989a) (Table 1, 2). The seasonal differences, however, can be explained from differences in temperature or physiological condition (glycogen content and reproductive stage; Table 1). During summer high temperatures during prolonged emersion result in a high mortality, in spite of a high glycogen content. In autumn moderate temperatures and a high glycogen content keep mortality at a low level during disturbances. In winter the mortality is higher again because of a decrease in the glycogen content and occasional temperatures below −1°C. In spring mortality can be very high, irrespective of the moderate temperatures, because the glycogen content is often very low as a consequence of nutritive stress during winter and reallocation of glycogen from the soma into gametes.

Effects on the benthic community and ecosystem

If the species studied are assumed to be representative for the benthic community as a whole, a 50% community mortality (average of the mortality of the 16 most studied species) is not reached within 11 days of emersion between temperatures of −1 and 4 °C, whereas, it is reached after only 3 days at temperatures below −1 °C and above 19 °C (Fig. 5) (Hummel *et al.*, 1988a). Assuming that the species studied are representative for the benthic community may have major drawbacks (Cairns & Niederlehner, 1987). The response of the species tested might not correspond with the response of a wide array of organisms in the natural system. Vulnerable species might have escaped the attention of researchers, because they could have died as a consequence of sampling or during acclimatization in the experimental systems. Moreover, the responses chosen (mortality, condition, reproduction) may not be the only one changing due to prolonged emersion. Competition and other interactions might change too. Therefore, a graph with average mortalities for a benthic community can only be applied with much care, taking into account that some species might have died, whereas a community-LD_{50} is not reached. Because of this differential, species-specific, mortality, inferences about effects on the other links of the food chain, *i.e.* their food spe-

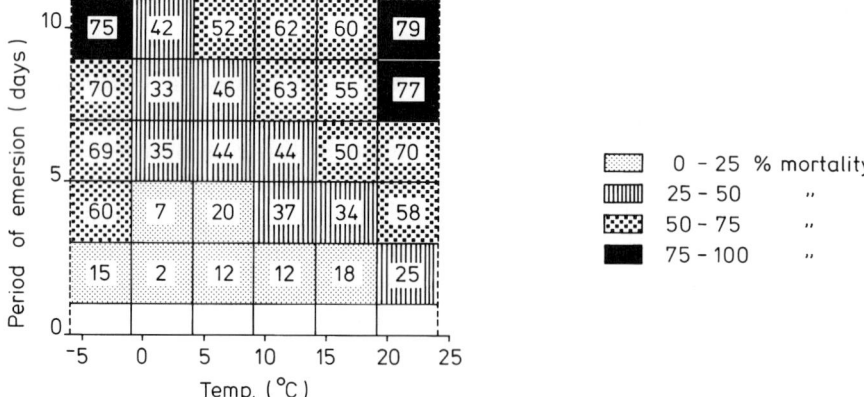

Fig. 5. Mortality of the macrobenthos community in the Oosterschelde sea-arm (SW Netherlands) in relation to temperature and period of emersion (calculated from the weighted averages of the five groups in Fig. 3) (after Hummel *et al.*, 1988a).

cies and their predators, are difficult, not to say impossible.

Post-experiment mortality and recovery

During the experiments on the effects of emersion a number of animals, emerged for some days and then replaced in sea-water with a normal tidal cycle, showed a delayed mortality (Hummel *et al.*, 1986a, b). The strongest post-experiment mortality occurred in the first 2 weeks following emersion; but occasionally even in the third and fourth week animals were found dying. This means that carrying out experiments without a recovery period, as is common practice in laboratory and response-dose (LD_{50}) experiments, will result in an underestimate of the effects of a treatment. For a proper evaluation of the effects of emersion or submersion, data on the post-experiment mortality are indispensable.

During the experiments on the effects of emersion or submersion a recovery of the animals, *i.e.* an increase of the condition, was never noticed; the experimentation period might have been to short (mostly only 4 to 6 weeks).

Data on the recovery of defaunated areas after prolonged emersion are not available. Investigations dealing with comparable situations, *i.e.* a locally defaunated area close to an undisturbed area (e.g. due to dredging or anoxic conditions) show recolonization rates of 2 to 8 weeks for small and mobile species, such as polychaetes and amphipods (Dauer & Simon, 1976; Dauer, 1984; Jones, 1986). Recolonization by less mobile or sessile species depends on the reproduction period and may take more than half a year. Total recovery of species numbers and biomass may take several years (Kaplan *et al.*, 1975; Bonsdorff, 1980; Hily, 1983).

Acknowledgement

This publication is Communication No. 698 of NIOO-CEMO, Yerseke, The Netherlands.

References

Aarset, A. V. & K. E. Zachariassen, 1982. Effects of oil pollution on the freezing tolerance and solute concentration of the blue mussel *Mytilus edulis*. Mar. Biol. 72: 45–51.

Akberali, H. B. & E. R. Trueman, 1985. Effects of environmental stress on marine bivalve molluscs. Adv. Mar. Biol. 22: 101–198.

Anderson, F. E. & B. A. Howell, 1984. Dewatering of an unvegetated muddy tidal flat during exposure-desiccation or drainage. Estuaries 7: 225–232.

Ansell, A. C., P. R. O. Barnett, A. Bodoy & H. Masse, 1981. Upper temperature tolerance of some European molluscs. III. *Cardium glaucum* and *C. edule*. Mar. Biol. 65: 177–183.

Barnes, H., D. M. Finlayson & J. Piatigorsky, 1963. The effect of desiccation and anaerobic conditions on the behaviour, survival and general metabolism of three common cirripedes. J. anim. Ecol. 32: 233–252.

Bayne, B. L., 1975. Reproduction in bivalve molluscs under environmental stress. In: J. Vernberg (ed.), Physiological ecology of estuarine organisms. University of south Carolina Press, Columbia, South Carolina, pp. 259–277.

Bayne, B. L., 1976. Marine mussels: their ecology and physiology. Cambridge University Press, Cambridge, 506 pp.

Bayne, B. L., D. A. Brown, K. Burns, D. R. Dixon, A. Ivanovici, D. R. Livingstone, D. M. Lowe, M. N. Moore, A. R. D. Stebbing & J. Widdows, 1985. The effects of stress and pollution on marine animals. Praeger, New York, 384 pp.

Bayne, B. L., A. Bubel, P. A. Gabbott, D. R. Livingstone, D. M. Lowe & M. N. Moore, 1982. Glycogen utilisation and gametogenesis in *Mytilus edulis* L.. Mar. Biol. Lett. 3: 89–105.

Bayne, B. L., D. L. Holland, M. N. Moore, D. M. Lowe & J. Widdows, 1978. Further studies on the effects of stress in the adult on the eggs of *Mytilus edulis*. J. mar. biol. Ass. U.K. 58: 825–841. Bayne, B. L. & R. C. Newell, 1983. Physiological energetics of marine molluscs. In: K. M. Wilbur (ed.), The Mollusca. Vol. 4. Physiology. Part 1. Academic Press, New York, 407–515 pp.

Beukema, J. J., 1979. Biomass and species richness of the macrobenthic animals living on a tidal flat area in the Dutch Wadden Sea: effects of a severe winter. Neth. J. Sea Res. 13: 203–223.

Beukema J. J., 1985. Zoobenthos survival during severe winters on high and low tidal flats in the Dutch Wadden Sea. In J. S. Gray & M. E. Christiansen (eds), Marine biology of polar regions and effects of stress on marine organisms. Wiley, Chicester: 351–361.

Bogaards, R. H., J. W. Francke, R. H. D. Lambeck & C. H. Borghouts–Biersteker, 1980. De afsluiting van de Grevelingen en de gevolgen voor de aan het substraat gebonden macrofauna. De Levende Natuur 3: 109–118.

Bonsdorff, E., 1980. Macrozoobenthic recolonization of a

232

dredged brackish water bay in SW Finland. Ophelia, suppl. 1: 145–155.

Boyden, C. R., 1972. The behaviour, survival and respiration of the cockles *Cerastoderma edule* and *C. glaucum* in air. J. mar. biol. Ass. U. K. 52: 661–680.

Broekhuysen, C. J., 1940. A preliminary investigation of the importance of desiccation, temperature and salinity as factors controlling the vertical distribution of certain marine gastropods in False bay, South Africa. Trans. R. Soc. S. Afr. 28: 255–292.

Brown, A. C., 1960. Desiccation as a factor influencing the vertical distribution of some South African Gastropoda from intertidal rock shores. Port. Acta biol. 7: 11–23.

Butler, P. A., 1949. Gametogenesis in the oyster under conditions of depressed salinity. Biol. Bull. 96: 263–269.

Cairns, J. & B. R. Niederlehner, 1987. Problems associated with selecting the most sensitive species for toxicity testing. Hydrobiologia. 153: 87–94.

Capstick, C. K., 1957. The salinity characteristics of the middle and upper reaches of the river Blyth estuary. J. anim. Ecol. 26: 295–315.

Carr, R. S. & J. M. Neff, 1984. Field assessment of biochemical stress indices for the sandworm *Neanthes virens* (sars). Mar. Envir. Res. 14: 267–280.

Coe, W. R. & H. J. Turner, 1938. Development of the gonads and gametes in the soft-shell clam (*Mya arenaria*). J. Morphol. 62: 91–111.

Dales, R. P., 1958. Survival of anaerobic periods by two intertidal polychaetes, *Arenicola marina* (L.) and *Owenia fusiformis* Delle Chiaje. J. mar. biol. Ass. U.K. 37: 521–529.

Dankers, N., K. Kersting, M. Binsbergen & K. Zegers, 1986. De effecten van het stoppen van de stroming op een mosselbank. Research Institute for Nature Management, Texel, RIN-report, 86/2, 24 pp.

Dauer, D. M., 1984. High resilience to disturbance of an estuarine polychaete community. Bull. Mar. Sc: 34: 170–174.

Dauer, D. M. & J. L. Simon, 1976. Habitat expansion among polychaetous annelids repopulating a defaunated marine habitat. Mar. Biol. 37: 169–177.

Davies, P. S., 1969. Physiological ecology of *Patella*. III. Desiccation effects. J. mar. biol. Ass. U.K. 49: 291–304.

De Zwaan, A., 1977. Anaerobic energy metabolism in bivalve molluscs. Oceanogr. Mar. Biol. Annu Rev. 15: 103–187.

De Zwaan, A., 1983. Carbohydrate catabolism in bivalves In P. W. Hochachka (ed.), The Mollusca. Vol. 1.. Metabolic biochemistry and molecular biomechanics. Academic Press, New York: 137–175.

De Zwaan, A. & D. I. Zandee, 1972a. The utilization of glycogen and accumulation of some intermediates during anaerobiosis in *Mytilus edulis* L.. Comp. Biochem. Physiol. 43B: 47–54.

De Zwaan, A. & D. I. Zandee, 1972b. Body distribution and seasonal changes in the glycogen content of the common mussel *Mytilus edulis*. Comp. Biochem. Physiol. 43A: 53–58.

Dries, R. R. & H. Theede, 1974. Sauerstoffmangelresistenz mariner Bodenevertebraten aus der Westlichen Ostsee. Mar. Biol. 25: 327–333.

Dries, R. R. & H. Theede, 1976. Stoffwechselintensität und Reservestoffabbau einiger mariner Muscheln bei herabgesetzter Sauerstoffsättigung des Mediums. Kieler Meeresforsch., Sonderh. 3: 37–48.

Elgershuizen, J. H. B. W., 1981. Some environmental impact of a storm surge barrier. Mar. Pollut. Bull. 12: 265–271.

Fortuin, A. W., H. Hummel, A. Meijboom & L. de Wolf, 1989a. Expected effects of the use of the Oosterschelde storm surge barrier on the survival of the intertidal fauna. Part 1 – The effects of prolonged emersion. Mar. Envir. Res. 27: 215–227.

Fortuin, A. W., H. Hummel, A. Meijboom & L. de Wolf, 1989b. Expected effects of the use of the Oosterschelde storm surge barrier on the survival of the intertidal fauna. Part 2 – The effects of protracted tidal cycles. Mar. Envir. Res. 27: 229–239.

Foster, B. A., 1971. Desiccation as a factor in the intertidal zonation of barnacles. Mar. Biol. 8: 12–29.

Gabbott, P. A., 1983. Developmental and seasonal metabolic activities in marine molluscs. In P. W. Hochachka (ed.), The mollusca, Vol. 2, Environmental biochemistry and physiology. Academic Press, New York: 165–217.

Gabbott, P. A. & B. L. Bayne, 1973. Biochemical effects of temperature and nutritive stress on *Mytilus edulis* L.. J. mar. biol. Ass. U.K. 53: 269–286.

Gäde, G., 1975. Anaerobic metabolism of the common cockle, *Cardium edule*. I. The utilization of glycogen and accumulation of multiple end products. Arch. Int. Phys. Biochem. 83: 879–886.

Gäde, G., 1983. Energy metabolism of arthropods and mollusks during environmental and functional anaerobiosis. J. exp. Zool. 228: 415–429.

Gäde, G. & G. Meinardus, 1981. Anaerobic metabolism of the common cockle *Cardium edule*. V. Changes in the level of metabolites in the foot during aerobic recovery after anoxia. Mar. Biol. 65: 113–116.

Gaffney, P. M. & W. J. Diehl, 1986. Growth, condition and specific dynamic action in the mussel *Mytilus edulis* recovering from starvation. Mar. Biol. 93: 401–410.

Hily, C., 1983. Macrozoobenthic recolonization after dredging in a sandy mud area of the Bay of Brest enriched by organic matter. Oceanol Acta 4: 113–120.

Hummel, H. & R. H. Bogaards, 1989. Changes in the reproductive cycle of the cockle *Cerastoderma edule* after disturbance by means of tidal manipulation. In: J. S. Ryland & P. A. Tyler (eds), Reproduction, genetics and distributions of marine organisms. Olsen & Olsen, Fredensborg, pp. 133–136.

Hummel, H., E. B. M. Brummelhuis & L. de Wolf, 1986a. Effects on the benthic fauna of embanking an intertidal flat area (the Markiezaat, Eastern Scheldt estuary, the Netherlands). Neth. J. Sea Res. 20: 397–406.

Hummel, H., A. W. Fortuin, R. H. Bogaards, L. de Wolf &

A. Meijboom, 1989b. Changes in *Mytilus edulis* in relation to short-term disturbances of the tide. In R. Z. Klekowski, E. Styczynska–Jurewicz & L. Falkowski (eds), Proceedings 21st EMBS. Ossolineum, Gdansk: 77–89.

Hummel, H., A. W. Fortuin, L. de Wolf & A. Meijboom, 1988a. Mortality of intertidal benthic animals after a period of prolonged exposure. J. exp. mar. Biol. Ecol. 121: 247–254.

Hummel, H., A. Meijboom & L. de Wolf, 1986b. The effects of extended periods of drainage and submersion on condition and mortality of benthic animals. J. exp. mar. Biol. Ecol. 103: 251–266.

Hummel, H., A. B. J. Sepers, L. de Wolf & F. W. Melissen, 1988b. Bacterial growth on the marine sponge *Halichondria panicea* induced by reduced waterflow rate. Mar. Ecol. Prog. Ser. 42: 195–198.

Hummel, H., L. de Wolf & A. W. Fortuin, 1988c. The annual cycle of glycogen in estuarine benthic animals. Hydrobiol. Bull. 22: 199–202.

Hummel, H., L. de Wolf, W. Zurburg, L. Apon, R. H. Bogaards & M. van Ruitenburg, 1989b. The glycogen content in stressed marine bivalves: the initial absence of a decrease. Comp. Biochem. Physiol. 94B: 729–733.

Jones, A. R., 1986. The effects of dredging and spoil disposal on macrobenthos, Hawkesbury estuary, NSW. Mar. Pollut. Bull. 17: 17–20.

Kaplan, E. H., J. R. Welker, M. G. Kraus & S. McCourt, 1975. Some factors affecting the colonization of a dredged channel. Mar. Biol. 32: 193–204.

Kennedy, V. S. 1976. Desiccation, higher temperatures and upper intertidal limits of three species of sea mussels (Mollusca: Bivalvia) in New Zealand. Mar. Biol. 35: 127–137.

Kennedy, V. S. & J. A. Mihursky, 1972. Upper temperature tolerances of some estuarine bivalves. Chesapeake Sci. 12: 193–204.

Kensler, C. B., 1967. Desiccation resistance of intertidal crevice species as a factor in their zonation. J. anim. Ecol. 36: 391–406.

Kinne, O., 1971. Salinity, Animals, Invertebrates. In O. Kinne (ed.), Marine Ecology. Vol. 1.2.. Wiley Interscience, London: 821–995.

Kluytmans, J. H., A. M. T. de Bont, J. Janus & T. C. M. Wijsman, 1977. Time dependent changes and tissue specificities in the accumulation of anaerobic fermentation products in the sea mussel *Mytilus edulis* L.. Comp. Biochem. Physiol. 58B: 81–87.

Kluytmans, J. H., M. van Graft, J. Janus & H. Pieters, 1978. Production and excretion of volatile fatty acids in the sea mussel *Mytilus edulis* L.. J. Comp. Physiol. 123: 163–167.

Kristensen, I., 1958. Differences in density and growth in a cockle population in the Dutch Wadden Sea. Arch. Neerl. Zool. 12: 351–453.

Lane, J. M., 1986. Allometric and biochemical studies on starved and unstarved clams, *Rangia cuneata* (Sowerby, 1831). J. exp. mar. Biol. Ecol. 95: 131–143.

Lauckner, G., 1980. Diseases of Porifera. In O. Kinne (ed.), Diseases of marine animals. Vol. I. General aspects. Protozoa to Gastropoda. Wiley, Chicester: 139–165.

Mann, R., 1978. A comparison of morphometric, biochemical, and physiological indexes of condition in marine bivalve molluscs. In: J. E. Thorpe & J. W. Gibbons (eds), Energy and environmental stress in aquatic systems. DOE Symposium Series 48: 484–497.

Maung–Myint, U. & P. Tyler, 1982. Effects of temperature, nutritive and metal stressors on the reproductive biology of *Mytilus edulis*. Mar. Biol. 67: 209–223.

Meinardus, G. & G. Gäde, 1981. Anaerobic metabolism of the cockle, *Cardium edule*. IV. Time dependent changes of metabolites in the foot and gill tissue induced by anoxia and electrical stimulation. Comp. Biochem. Physiol. 70B: 271–277.

Murphy, D. J., 1979. A comparative study of the freezing tolerances of the marine snails *Littorina littorea* and *Nassarius obsoletus*. Physiol. Zool. 52: 219–230.

Murphy, D. J., 1983. Freezing resistance in intertidal invertebrates. Ann. Rev. Physiol. 45: 289–299.

Murphy, D. J. & L. C. Johnson, 1980. Physical and temporal factors influencing the freezing tolerance of the marine snail *Littorina littorea*. Biol. Bull. 158: 220–232.

Nagabushanam, T. & P. M. Talikhedar, 1977. Seasonal variation in proteins, fat and glycogen of the wedge clam *Conax cuneatus*. Indian J. Mar. Sci. 6: 85–87.

Newell, R. C., 1979. Biology of intertidal animals. Marine Ecological Surveys Ltd, Faversham, 781 pp.

Oglesby, L. C., 1969. Salinity–stress and desiccation in intertidal worms. Am. Zool. 9: 319–331.

Pipe, R. K., 1985. Seasonal cyles in and effects of starvation on egg development in *Mytilus edulis*. Mar. Ecol. Prog. Ser. 24: 121–128.

Pora, E. A., C. Wittenberger, G. Suarez & N. Portilla, 1969. The resistance of *Crassostrea rhizoporea* to starvation and asphyxia. Mar. Biol. 3: 18–23.

Riley, R. T., 1976. Changes in the total protein, lipid, carbohydrate, and extracellular body fluid free amino acids of the Pacific oyster, *Crassostrea gigas*, during starvation. Proc. natl. Shellfish Assoc. 65: 84–90.

Rodhouse, P. G. & P. M. Gaffney, 1984. Effect of heterozygosity on metabolism during starvation in the American oyster *Crassostrea virginica*. Mar. Biol. 80: 179–187.

Sastry, A. N., 1979. Pelecypoda (excluding Ostreidae). In A. C. Giese & J. S. Pearse (eds), Reproduction of marine invertebrates. Vol. 5. Academic Press, New York: 113–292.

Smies, M. & A. H. L. Huiskes, 1981. Holland's Eastern Scheldt estuary barrier scheme: some ecological considerations. Ambio 10: 158–165.

Thillart, G. van den & I. de Vries, 1985. Excretion of volatile fatty acids by anoxic *Mytilus edulis* and *Anodonta cygnea*. Comp. Biochem. Physiol. 80B: 299–301.

Thompson, J. E., R. P. Walkner & D. J. Faulkner, 1985. Screening and bioassays for biologically-active substances

234

from forty marine sponge species from San Diego, California, USA. Mar. Biol. 88: 11–21.

Vogel, S. & W. Bretz, 1972. Interfacial organisms: passive ventilation in the velocity gradients near surface. Science 157: 210–211.

Widdows, J., B. L. Bayne, D. R. Livingstone, R. C. Newell & P. Donkin, 1979. Physiological and biochemical responses of bivalve molluscs to exposure to air. Comp. Biochem. Physiol. 62A: 301–308.

Widdows. J. & J. M. Shick, 1985. Physiological responses of *Mytilus edulis* and *Cardium edule* to aerial exposure. Mar. Biol. 85: 217–232.

Wilde, P. A. W. J. de & E. M. Berghuis, 1979. Cyclic temperature fluctuations in tidal mud-flat. In E. Naylor & R. G. Hartnol (eds), Cyclic phenomena in marine plants and animals. Pergamon Press, Oxford: 435–441.

Williams, R. J., 1970. Freezing tolerance in *Mytilus edulis*. Comp. Biochem. Physiol. 35: 145–161.

Wolff, W. J., 1973. The estuary as a habitat. An analysis of data on the soft-bottom macrofauna of the estuarine area of the rivers Rhine, Meuse and Scheldt. Zool. Verh., Leiden, 126, 242 pp.

Yonge, C. M., 1976. Origin and nature of bivalves. In C. M. Yonge & T. E. Thompson (eds), Living marine molluscs. Williams Collins Sons and Co Ltd Glasgow: 142–153.

Zandee, D. I., J. H. Kluytmans, W. Zurburg & H. Pieters, 1980. Seasonal variations in biochemical composition of *Mytilus edulis* with reference to energy metabolism and gametogenesis. Neth. J. Sea Res. 14: 1–29.

Hydrobiologia **282/283**: 235–249, 1994.
P. H. Nienhuis & A. C. Smaal (eds), The Oosterschelde Estuary.
© 1994 *Kluwer Academic Publishers. Printed in Belgium.*

Effect of sedimentological and hydrodynamical changes in the intertidal areas of the Oosterschelde estuary (SW Netherlands) on distribution, density and biomass of five common macrobenthic species: *Spio martinensis* (Mesnil), *Hydrobia ulvae* (Pennant), *Arenicola marina* (L.), *Scoloplos armiger* (Muller) and *Bathyporeia* sp.

J. Coosen[1], J. Seys[2,3], P. M. Meire[2,3] & J. A. M. Craeymeersch[4]
[1]*National Institute for Coastal and Marine Management/RIKZ, P.O. Box 8039, 4330 EA, Middelburg, The Netherlands;* [2]*Laboratory of Animal Ecology, Zoogeography and Nature Conservation, University of Ghent, K.L. Ledeganckstraat 35, 9000 Gent, Belgium; Present address:* [3]*Ministry of the Flemish Community, Institute of Nature Conservation, Kiewitdreef 5, 3500 Hasselt, Belgium;* [4]*Netherlands Institute of Ecology – C.E.M.O., Vierstraat 28, 4401 EA Yerseke, The Netherlands*

Key words: Oosterschelde, tidal flats, macrozoobenthos, population dynamics

Abstract

In order to evaluate the impact of the construction of the storm-surge barrier and secondary dams on macrobenthos of the tidal flats in the Oosterschelde (SW Netherlands), changes in distribution, density and biomass of five common species (*Spio martinensis, Hydrobia ulvae, Arenicola marina, Scoloplos armiger* and *Bathyporeia sp*) were analysed.

Data on macrobenthos were collected from 1979 to 1989 on five different tidal flats. Changes in sediment texture and hydrodynamic factors during the construction and after the completion of the coastal engineering project were taken into account.

Three severe winters in a row caused more disturbance in the population of the main predator of *S. armiger* than did the hydrodynamical changes. A temporary prolongation of the emersion time (in 1986 and 1987) caused a temporary decrease in juvenile *A. marina*. But afterwards they still occupy the same 'nursery grounds'. Increased wave action on the edges of the flats probably created new niches for *Bathyporeia sp.* and *Spio martinensis*, replacing other benthic species.

It is not yet clear what has caused the decline of *H. ulvae* in many places in the Oosterschelde estuary. Parasitic infestation is one of the possibilities.

Introduction

The distribution, abundance and biomass of intertidal macrobenthos species are related to various environmental factors such as height in the intertidal zone (immersion time) and sediment characteristics (Wolff, 1973; Gray, 1974; Beukema, 1982; Dankers & Beukema, 1983; Forbes & Lopez, 1990; Fortuin *et al.*, 1989; Hummel *et al.*, 1994).

Due to the construction of the storm-surge barrier and additional dams, some environmental conditions in the Oosterschelde have changed (Nienhuis & Smaal, 1994). For some tidal flats it has been shown that the amount of silt in the upper layer has decreased (Ten Brinke *et al.*,

1994) and the bottom height has changed, because of changes in erosion/sedimentation processes (Mulder & Louters, 1994).

During construction, increased wave dynamics were observed on the edges of the tidal flats and a 'washing-out' of fine sediments during storms occurred, without replacement by sedimentation, because of the low percentage of silt in the water under the new conditions (Ten Brinke *et al.*, 1994). A slow erosion of the tidal flats, resulting in a net loss of 15% intertidal area within 30 years, is the dominant process now (Smaal & Nienhuis, 1992).

Apart from the impact of the coastal engineering project, the effect of severe winters or high summer temperatures on benthic populations can be substantial (Beukema, 1979; Beukema, 1985; Dörjes, 1980; Ziegelmeier, 1964). Seys *et al.* (1994) found the observed patterns in total biomass, total density and diversity to be largely determined by the alternation of severe and mild winters.

In order to evaluate the impact of the hydrodynamic and sedimentological changes on macrobenthos of the tidal flats in the Oosterschelde a monitoring programme was executed. Three approaches have been used to treat the data. This paper deals with analysis of the population dynamics at the species level. Meire *et al.* (1994) studied the same dataset from a community viewpont and Seys *et al.* (1994) assessed trends and patterns in total density, total biomass and diversity of the benthic community.

Material and methods

Sampling stations and methods

The analysis is based on three datasets: INTER-ECOS, COST and VIANE.

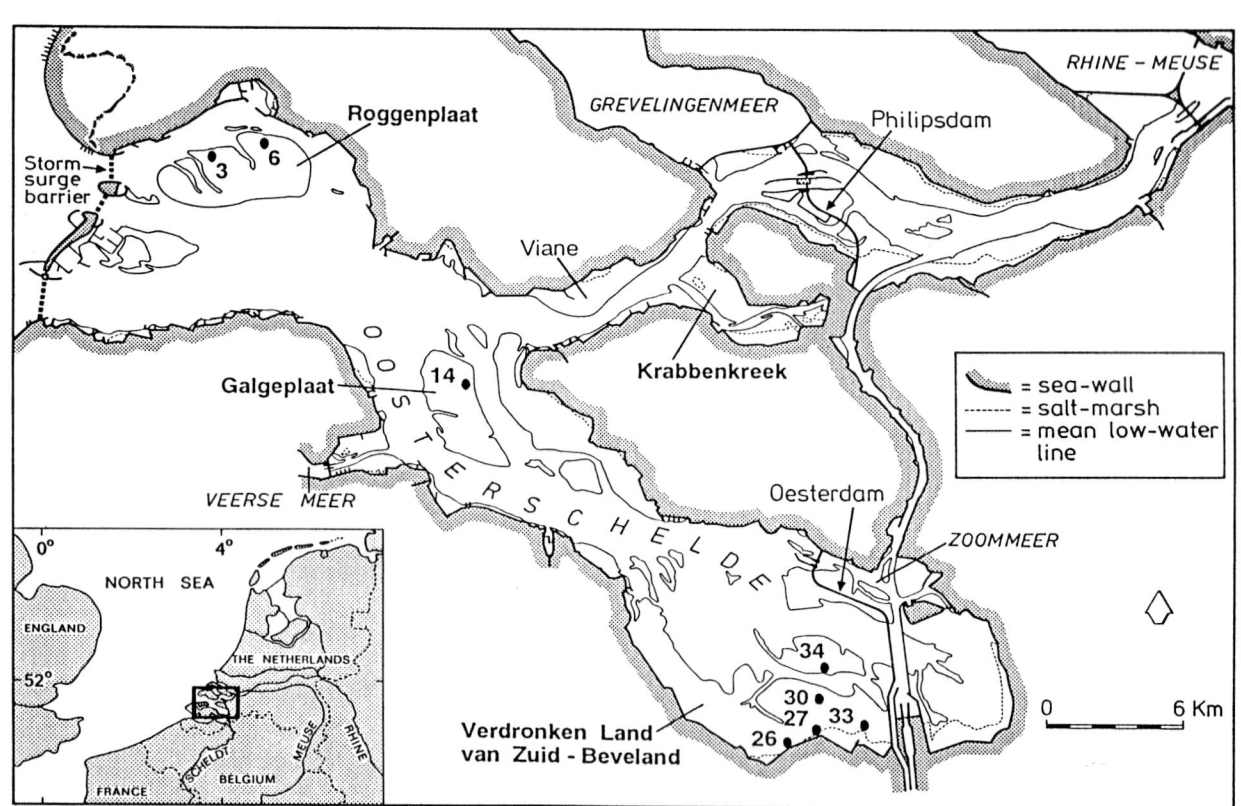

Fig. 1. Map of the Oosterschelde indicating sampling stations mentioned in the text.

The INTERECOS dataset consists of a macro-zoobenthic survey in August/September 1985 (pre-barrier situation) and a similar one in 1989 (post-barrier situation) with 305 sampling stations distributed over three tidal flats, *i.e.* in the western (Roggenplaat), central (Galgeplaat) and northern (Krabbenkreek) part of the Oosterschelde (Fig. 1). To compare the results of both years the Wilcoxon matched-pairs signed rank test (Siegel, 1956) was used. The sediment texture of the 3 large tidal flats does show some variation in average values (Table 1). Sediments were very sandy with the fines content generally less than 5%. Roggenplaat and Galgeplaat showed little difference in percentages of silt and median grain size from 1985 to 1989. Only the Krabbenkreek shows a significant decrease in both (Table 1). The tidal elevation of the sampling sites varied between NAP (Dutch Ordnance Level) −1.6 m and NAP +1.2 m in both years with means of NAP −0.14 m and NAP −0.24 m in 1985 and 1989 respectively. The decrease in 1989 was significant (Meire *et al.*, 1994: Table 1). The higher zones of the tidal flats (up to NAP +0.75 m) were less silty in 1989 than in 1985 (Fig. 2).

The COST dataset comes from a monitoring programme in the period 1983–1989 based on two stations from the Roggenplaat in the west of the Oosterschelde and five at the eastern tidal flats (Verdronken land van Zuid Beveland [Fig. 1]). Samples were initially (1983 and 1984) taken four times a year, and twice since then (1985–1989). The sediment in most of the permanent stations was silty sand with a median grain-size of 2.5–3.2 $\emptyset$ (Seys *et al.*, 1993). Two stations higher in the intertidal zone in the east-

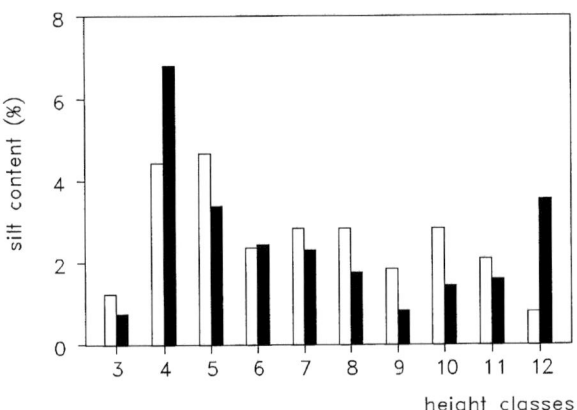

Fig. 2. Silt content in the sediment in relation to height in the intertidal zone of the tidal flats of the Oosterschelde in 1989. White bars = 1985; black bars = 1989. Height classes, legend: 3 = NAP-150 cm/NAP-125 cm; 4 = −125/−100; 5 = −100/−75; 6 = −75/−50; 7 = −50/−25; 8 = −25/NAP; 9 = NAP/+25; 10 = +25/+50; 11 = +50/+75; 12 = +75/= 100.

ern part were more silty before the construction of the dams. The tidal elevation of the stations varied between NAP −0.70 and NAP +1.1 m before the construction of the dams. In 1989 one station showed sedimentation (30 cm), three showed erosion (10–15 cm) (Seys *et al.*, 1994).

The VIANE dataset was derived from a monitoring programme (1979–1989) on 6 stations on a tidal flat (The Slikken van Viane), situated in the northern part of the Oosterschelde (Fig. 1). Samples were taken (almost) yearly in late summer. Until 1986 the Viane stations had a higher percentage of silt (fraction < 53 μ), in the order of 5 to 10% (Seys *et al.*, 1994). In 1987 the percentage dropped to an average of 2.5 (except for station V-39). The tidal elevation of the stations

Table 1. Mean values of percentage silt and median grain size of 3 intertidal areas in the Oosterschelde in 1985 and 1989 (excluding mussel plots and clay banks) {after Ten Brinke *et al.*, 1994}. * Difference 1989–1985 statistically significant according to Wilcoxon signed ranks test (Siegel, 1956).

Area	% silt fraction < 53 μm				Median grain size			
	1985	1989	Diff.	Test	1985	1989	Diff.	Test
Roggenplaat	0.86	1.14	0.28		2.56	2.54	−0.02	*
Galgeplaat	1.01	0.76	−0.25	*	2.80	2.78	−0.02	
Krabbenkreek	4.47	2.86	−1.61	*	2.89	2.86	−0.03	*

varied between NAP −0.80 and NAP + 0.60 before the construction of the dams. In 1988 most stations showed erosion (10–15 cm) (Seys *et al.*, 1994).

In all the datasets, core-samples were taken and sieved on a 1 mm-mesh sieve. Numbers of animals are counted and identified and ash free dry weight (AFDW) is determined by drying, weighing, ashing and reweighing. For more details on sampling and laboratory methods see Meire *et al.* (1994), Craeymeersch *et al.* (1988) & Seys *et al.* (1993) and Meire & Dereu (1990) for the INTERECOS, COST and VIANE datasets respectively.

Results

In the Oosterschelde tidal flats, biomass and density are dominated by 10 to 15 species. Of these species very few reach mean densities above 1000 m^{-2} or mean biomass above 10 g AFDW m^{-2} (Meire *et al.*, 1994). The species in Figs 3 and 4 are the most abundant and dominant in biomass. For the purpose of this paper five species have been chosen from that list. *Cerastoderma edule* and *Mytilus edulis* are discussed in separate papers in this volume (Coosen *et al.*, 1994; Van

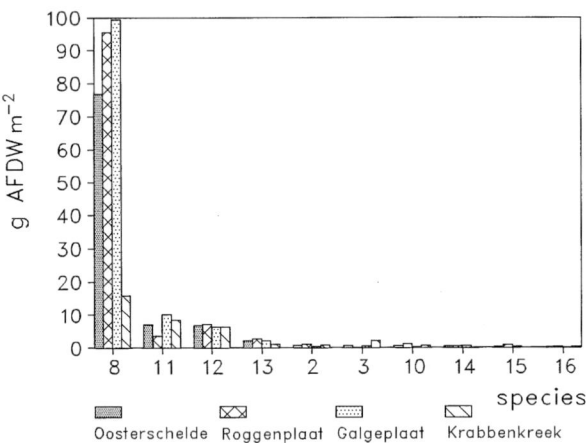

Fig. 4. Mean biomass of 10 most dominant species in the Oosterschelde, per sub-area; (2, 3, 8 & 10 = see figure 3; 11 = *Mytilus edulis*; 12 = *Arenicola marina*; 13 = *Macoma balthica*; 14 = *Nephthys hombergii*; 15 = *Lanice conchilega*; 16 = *Heteromastus filiformis*.

Stralen & Dijkema, 1994). From the group of small opportunistic polychaetes like *Pygospio, Capitella, Tharyx* and *Spio*, the last has been selected because of its remarkable density peaks in 1986 and 1989. The deposit-feeding lugworm *A. marina* is the polychaete species contributing most to the biomass. It is a very common inhabitant of the tidal flats, important as prey for wading birds and flatfish and known to have a life-cycle on different levels of the tidal flats (Reise, 1985). The herbivorous gastropod *H. ulvae* is very common on the tidal flats and serves as food for Shelduck (*Tadorna tadorna*) and Redshank (*Tringa totanus*) and is a consumer of organic debris or the micro-organisms attached to it (Newell, 1965). Together with *S. armiger* (one of the most common species in the Oosterschelde) its distribution might be influenced by a change in sediments. *Bathyporeia sp.*, an inhabitant of more exposed sites, can give an indication of the extent to which the new hydrodynamic properties of the Oosterschelde have influenced that habitat for benthic organisms.

Spio martinensis

The two INTERECOS surveys showed a remarkable increase in frequency of occurrence, mean

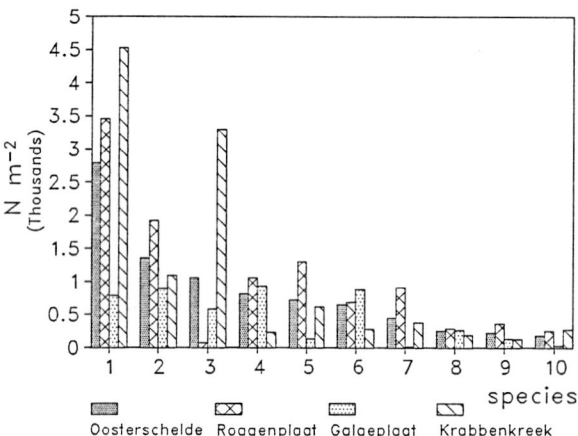

Fig. 3. Mean density of 10 most abundant species in the Oosterschelde, per area (1 = *Oligochaeta*; 2 = *Scoloplos armiger*; 3 = *Hydrobia ulvae*; 4 = *Tharyx marioni*; 5 = *Pygospio elegans*; 6 = *Spio martinensis*; 7 = *Bathyporeia pilosa*; 8 = *Cerastoderma edule*; 9 = *Capitella capitata*; 10 = *Nereis diversicolor*).

Table 2. INTERECOS data (number of samples, frequency of occurrence, mean density and mean biomass) of 5 species. ↑* = significant increase; ↓* significant decrease. Number of samples (1985 & 1989) Roggenplaat: 120; Galgeplaat: 110; Krabben-kreek: 75.

	Roggenplaat			Galgeplaat			Krabbenkreek		
	1985	1989	Sign	1985	1989	Sign	1985	1989	Sign
(a) *Spio martinensis*									
Freq. of occurr. (%)	42	61		36	77		9	25	
Mean density $N \cdot m^{-2}$	150	691	↑*	200	870	↑*	270	430	↑*
Mean biomass g $AFDW \cdot m^{-2}$	0.02	0.114	↑*	0.027	0.113	↑*	0.027	0.039	
(b) *Hydrobia ulvae*									
Freq. of occurr. (%)	72	18		95	27		92	65	
Mean density $N \cdot m^{-2}$	3550	75	↓*	23000	580	↓*	17000	3000	↓*
Mean biomass g $AFDW \cdot m^{-2}$	1.4	0.9	↓*	4.5	4	↓*	4	2	↓*
(c) *Arenicola marina*									
Freq. of occurr. (%)	79	89		75	68		87	89	
Mean density $N \cdot m^{-2}$	31	30		22	15		68	41	↓*
Mean biomass g $AFDW \cdot m^{-2}$	4.9	7.4	↑*	3.8	6.5	↑*	4.6	6.5	
(d) *Scoloplos armiger*									
Freq. of occurr. (%)	97	89		87	85		85	88	
Mean density $N \cdot m^{-2}$	770	1920	↑*	600	900		1190	1090	
Mean biomass g $AFDW \cdot m^{-2}$	1.07	1.24		0.88	0.46	↓*	1.13	0.92	
(e) *Bathyporeia sp.*									
Freq. of occurr. (%)	52	18		44	4		31	29	
Mean density $N \cdot m^{-2}$	420	1020		140	25	↓*	640	390	
Mean biomass g $AFDW \cdot m^{-2}$	0.08	0.15		0.05	0.013	↓*	0.117	0.05	

biomass and mean density on all three tidal flats between 1985 and 1989 (Table 2a). Results from the permanent stations gave no evidence for a gradual pattern of increase: numbers at most permanent stations were only high in 1986 and 1989, shown for stations 3, 30 & 34 in Fig. 5.

A preference of *Spio* for sandy sediments (<1% silt) is obvious from the INTERECOS data (Fig. 6a). The vertical distribution of *Spio*, however, has drastically changed from 1985 to 1989 (Fig. 6b). In 1985 it was most abundant in the zones below NAP, but in 1989 the higher zones were occupied too by this small polychaete.

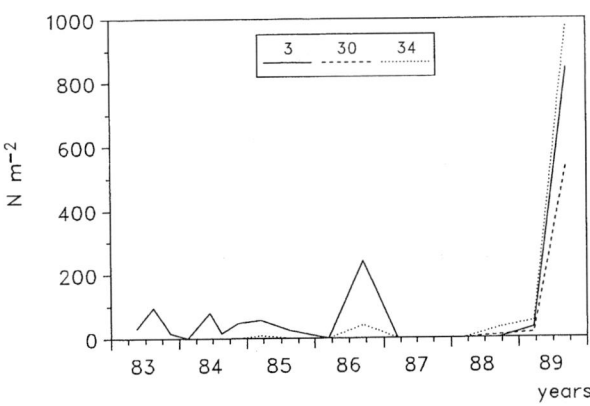

Fig. 5. Density (in $N\ m^{-2}$) of *Spio martinensis* at COST stations 3, 30 and 34.

Hydrobia ulvae

Comparing 1985 and 1989 (INTERECOS), frequency of occurrence, mean biomass and mean density decreased dramatically on the Roggen-plaat and Galgeplaat (Table 2b), a pattern also found at most of the permanent stations there (Fig. 7a). The decrease in density and biomass occurs in all types of sediment (Fig. 6c) and in

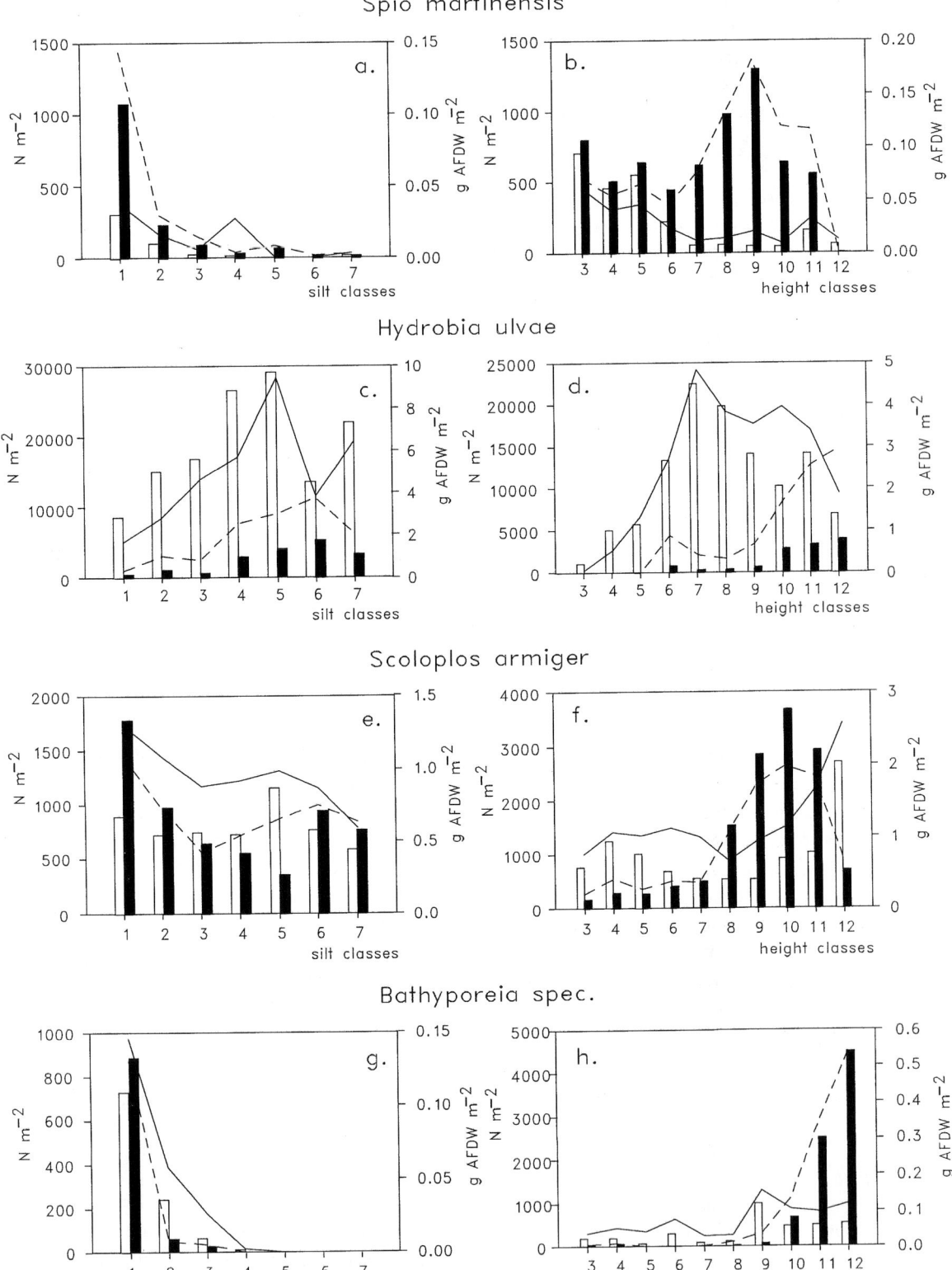

Spio martinensis

Hydrobia ulvae

Scoloplos armiger

Bathyporeia spec.

most tidal areas (up to NAP + 0.5 m: Fig. 6d). At the Krabbenkreek the change in frequency of occurrence, mean biomass and mean density was less pronounced (Table 2b), but still significant. At the Slikken van Viane the mudsnail is dominant in the higher zones, with a distinct downward trend in biomass in station V32 (Fig. 7b). In the eastern part, the three permanent stations where *H. ulvae* is common show a mixed development in biomass variation from year to year. At two stations a seasonal pattern with low spring biomass values and higher autumn values is found during the whole period of investigation, except in autumn 1985 when low values were recorded (Fig. 7c). At the third station (34) *H. ulvae* almost disappeared from autumn 1987 onwards.

Arenicola marina

The surveys of 1985 and 1989 do not show a significant change in the distribution pattern of this species. It was present in 70 to 90% of the samples in both years (Table 2c). On the Roggenplaat mean biomass increased between 1985 and 1989, mean densities remained stable. The results from the permanent stations on the Roggenplaat (3 & 6) showed a normal seasonal pattern, with lower biomass values in spring and higher values in autumn, except in 1985 and 1988 at station 6 (Fig. 8a). Densities were fairly constant: (25–30 m^{-2}) in 1983 to 1984 and again in 1988 to 1989, but were three times as high in autumn 1985 and 1987. In the central part overall mean biomass values increased from 1985 to 1989 (Table 2c); densities decreased. In the eastern part of the Oosterschelde densities were higher at the higher, more silty stations (27 and 33) than in

the lower stations (26, 30 and 34) (Figs 8b and 8c). The former represent typical 'nursery grounds', since autumn values were mostly high, together with low individual weights in stations 27 and 33. In contrast, individual weights in stations 30 and 34 showed higher values in autumn and numbers are fairly constant compared to stations 27 and 33.

Scoloplos armiger

The surveys of 1985 and 1989 show no significant change in the distribution pattern of *S. armiger*. The species is present in almost 90% of the samples taken at the three tidal flats (Table 2d). Mean biomass was usually low and stayed the same, except on the Galgeplaat, where it decreased in 1989. Mean density only increased at the Roggenplaat (Table 2d).

Time series at the permanent stations show in general low numbers from 1980 to 1984, an increase in 1985 in the eastern part, followed in the western and central part with a peak in 1986. At most stations 1988 and 1989 showed high numbers too (Fig. 9). Biomass values usually remain around 1 g AFDW m^{-2}, but occasionally reach 2 g AFDW m^{-2}. On the higher zone in the eastern part of the Oosterschelde (stations 27, 33 and 34) the seasonal pattern was interrupted in autumn 1986: density and biomass decreased and the species was absent until spring 1988 (Fig. 9b). *S. armiger* does not seem to have a real preference for one type of sediment (Fig. 6e), although it is slightly more abundant in less silty sediments in 1989. Data from the same survey showed that densities on the higher parts of the intertidal zones increased (Fig. 6f).

Fig. 6. Biomass and density of four species in relation to different silt classes of the sampling plots in the INTERECOS surveys of 1985 and 1989. Silt = fraction < 53 μ. Class 1 = ≤ 1%; 2 = 1–2%; 3 = 2–3%; 4 = 3–4%; 5 = 4–5%; 6 = 5–10%; 7 = > 10%. Left vertical axis: White bars = density 1985; black bars = density 1989. Right vertical axis: solid line = biomass 1985; broken line = biomass 1989. a = *Spio martinensis*; c = *Hydrobia ulvae*; e = *Scoloplos armiger*; g = *Bathyporeia sp.* Biomass and density of four species in relation to different height classes of the sampling plots in the INTERECOS surveys of 1985 and 1989. Height classes, legend: 3 = NAP-150 cm/NAP-125 cm; 4 = − 125/ − 100; 5 = − 100/ − 75; 6 = − 75/ − 50; 7 = − 50/ − 25; 8 = − 25/NAP; 9 = NAP/ + 25; 10 = + 25/ + 50; 11 = + 50/ + 75; 12 = + 75/ + 100. Left vertical axis: White bars = density 1985; black bars = density 1989. Right vertical axis: solid line = biomass 1985; broken line = biomass 1989. b = *Spio martinensis*; d = *Hydrobia ulvae*; f = *Scoloplos armiger*; h = *Bathyporeia sp.*

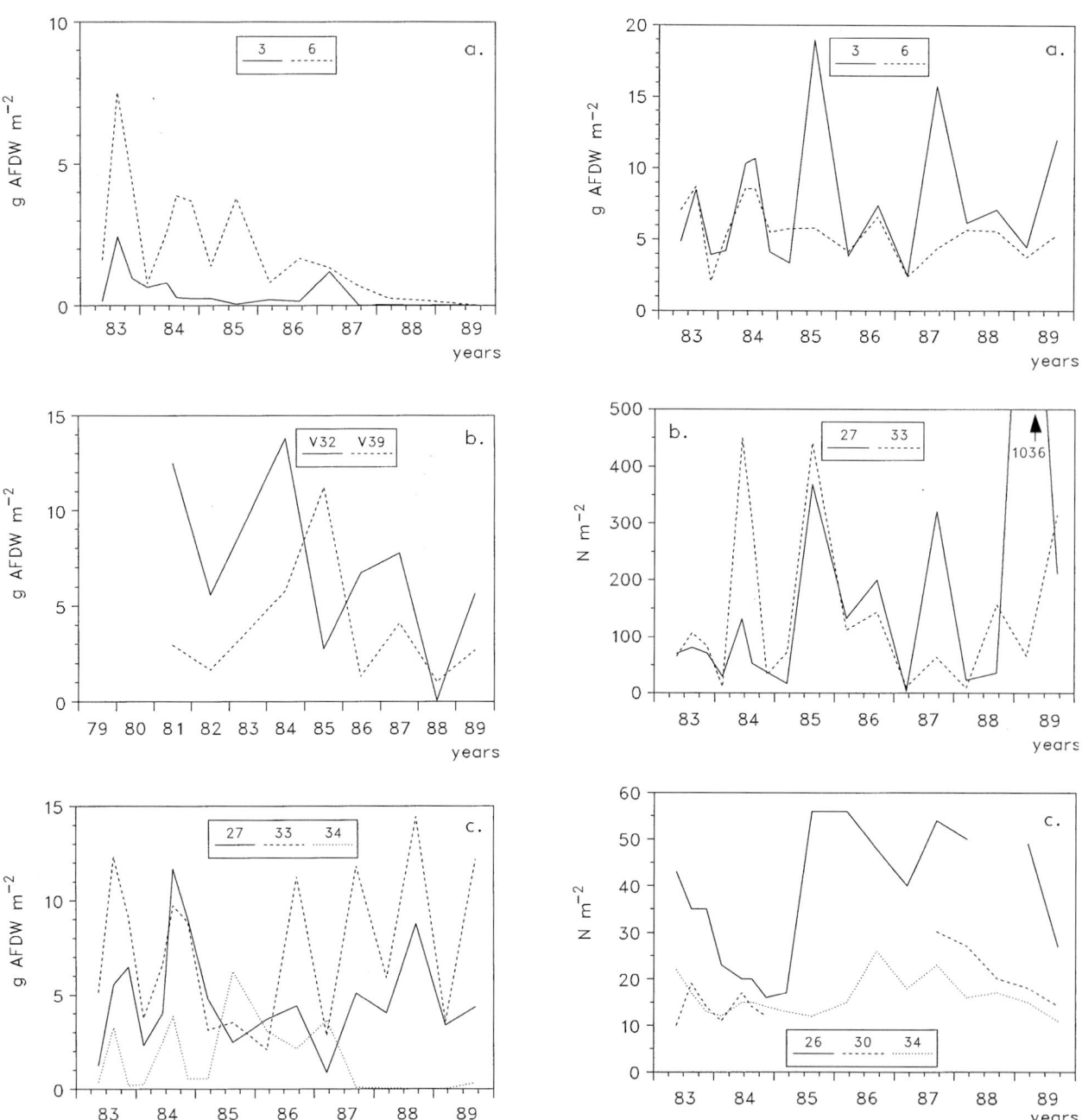

Fig. 7. Biomass (in g AFDW m^{-2}) of *Hydrobia ulvae* at COST stations 3 and 6 (a); at Viane stations V32 and V39 (b); at COST stations 27, 33 and 34 (c).

Fig. 8. Biomass (in g AFDW m^{-2}) of *Arenicola marina* at COST stations 3 and 6 (a); Density (in N m^{-2}) of *Arenicola marina* at COST stations 27 and 33 (b); at COST stations 26, 30 and 34 (c).

Bathyporeia sp.

The Amphipods *Bathyporeia pilosa* (Lindström) and *B. sarsi* (Watkin) are both more or less abun-

dant on the tidal flats of the Oosterschelde. Since identification problems can arise, all animals were grouped as *Bathyporeia sp.* The survey of 1989

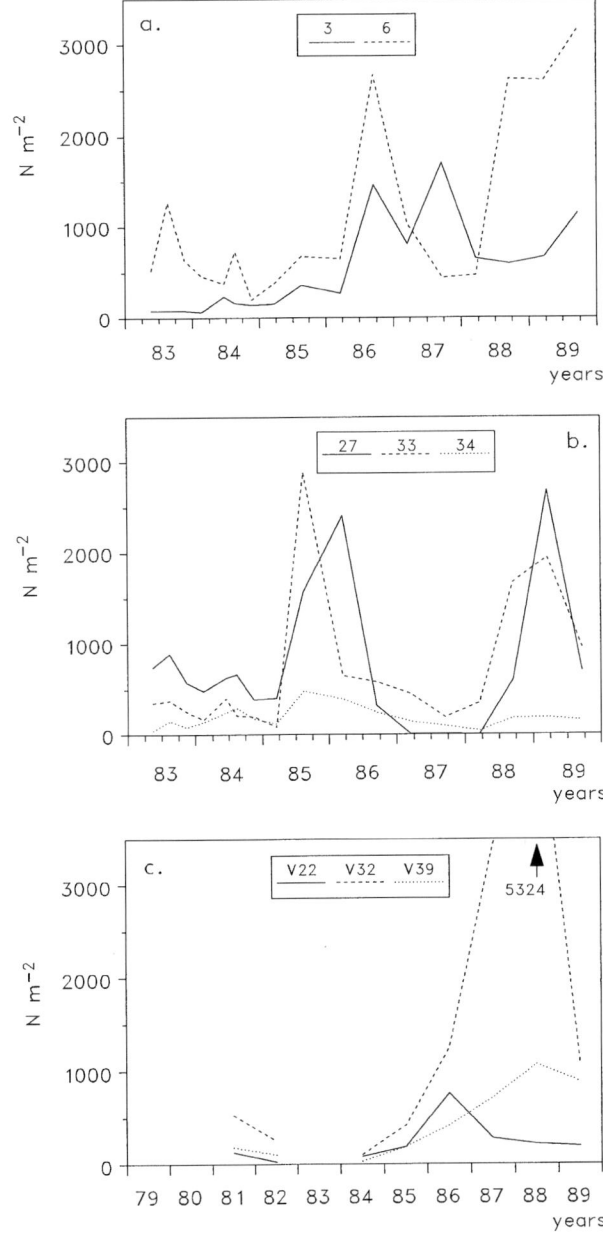

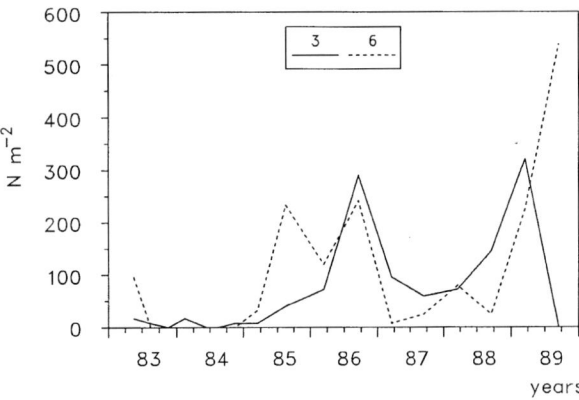

Fig. 10. Density (in N m^{-2}) of *Bathyporeia sp.* in COST stations 3 and 6.

(Table 2e). In 1989 *Bathyporeia sp.* was found in only 18% (Roggenplaat) and 4% (Galgeplaat) of the samples. In the Krabbenkreek it was still present in 30% of the samples. Mean biomass was usually very low on all three locations, not more than 0.15 g AFDW m^{-2} and mean density varies between 25 and 1000 m^{-2}. They only decreased on the Galgeplaat. The preference for sandy sediments is obvious in both years (Fig. 6g). Overall, the change in distribution between 1985 and 1989 indicates a change of biomass and density towards the higher intertidal zones (Fig. 6h). In the permanent stations where this amphipod was found, rather low numbers were recorded in 1980–1984. Density increased temporarily in 1985–1986 and again in 1988–1989 (Fig. 10).

Discussion

Sediments and height

Ten Brinke *et al.* (1994) assume that the decrease in percentage silt and median grain size found in the Krabbenkreek is representative of all landward tidal flats (also for the Verdronken Land van Zuid Beveland). This assumption corresponds with the decrease at the higher stations 27 and 33, but not with the observations at the low lying station 30 (Seys *et al.*, 1994). So, a general pattern of change in sediment texture is hard to

Fig. 9. Density (in N m^{-2}) of *Scoloplos armiger* at COST stations 3 and 6 (a); at COST stations 27, 33 and 34 (b); at Viane stations V22, V32 and V39 (c);

showed a negative change in the distribution pattern of *Bathyporeia sp.* on the Roggenplaat and the Galgeplaat compared with the survey of 1985. The species was present in almost 50% of the samples taken at these two tidal flats in 1985 give on the basis of the available data. The decrease in height between 1985 and 1989 is significant for all tidal flats examined (Mulder & Louters, 1994). Consequently, the emersion periods for most sites have changed. Due to the reduced tidal amplitude emersion period also changed, resulting in a decrease above mid tidal level and an increase below. But overall, the range of observed emersion periods on the tidal flats of the Oosterschelde did not change, so benthic organisms still have to deal with the same variation in exposure and submersion.

The ongoing shift of sediment from the higher to the lower intertidal parts (Mulder & Louterse, 1994) will lead to a substantial decrease in exposed total tidal flat area during low tide.

Macrobenthos

The density-peaks of *Spio martinensis* in 1986 and 1989, especially on the higher dynamic zones of the tidal flats, cannot both be explained as a response to the hydrodynamic changes that took place in 1986. The opportunistic life style with two to four reproductive periods in a year, together with a short life span of one year (Gudmundsson, 1985) makes an immediate reaction possible, but the factor that caused the 3-fold increase in 1989 was not of hydrodynamic nature.

Although the higher zones of the tidal flats became less silty after the completion of the engineering works, *S. martinensis* did not increase immediately. The settlement in 1989 in the higher zones of *Spio* may be explained by lack of competition or predators. At the Krabbenkreek and the Galgeplaat, where the decrease in silt fraction was significant, the frequency of occurrence increased from 9 to 25% and from 36 to 77% respectively, but also on the Roggenplaat this species became more common.

The high spatial variability in density and biomass is quite common in *Hydrobia* populations. Beukema & Essink (1986) and Dörjes *et al.* (1986) found almost exclusively non-coinciding density fluctuations in the Wadden Sea; this was explained by the active migration of the species.

Four possible explanations for the decline of the mudsnail in the Oosterschelde will be discussed:

(a) hydrodynamic changes due to the engineering works;
(b) influence of cold winters on recruitment;
(c) parasitic infestation.

(a) Since *H. ulvae* is known to favour silty sediments (Newell, 1965; Fenchel, 1975, Barnes & Greenwood, 1978), a lowering of the percentage of silt in the upper layers could be one of the reasons for the observed decline in 1989. During the higher wave impact of 1986 (Mulder & Louters, 1994) the sediment of the lower intertidal zone, including the mudsnails, could have been 'washed out'. That habitat would than be less suited to larvae of the mudsnail.

On the tidal flats examined, the silt fraction was lower in 1989 than in 1985, but only on the Krabbenkreek were the differences in percentage silt and median grain size significant (Table 1; Ten Brinke *et al.*, 1994), compared with the temporal variability. On that intertidal flat, the decrease in silt in the sampling stations was not correlated with the decrease in biomass of the mudsnail in the same stations (Fig. 11). Other taxa that are favoured by silty sediments, like *Heteromastus filiformis* and *Oligochaeta*, did not show any decrease from 1985 to 1989. The de-

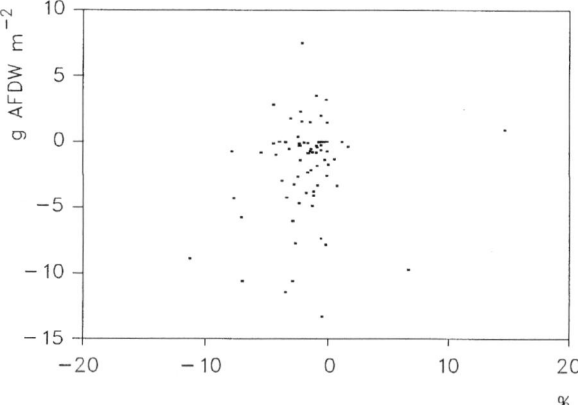

Fig. 11. Decrease/increase in biomass of *Hydrobia ulvae* in relation to decrease/increase of silt content in sediments from the Krabbenkreek (1989 *vs* 1985).

cline of *H. ulvae* is just as strong in the parts of the tidal flats that still have a high percentage of silt (Fig. 6c).

There are more facts that refute the 'washing out'-hypothesis. If the animals are 'washed away' by hydrodynamic forces, the chance of this happening must have been greatest in the period October 1986, since at that time the storm surge barrier was closed during a severe storm. No collapse of the population (in numbers and/or biomass) was found in spring 1987, however (Fig. 7). At three of the stations (3, 6 and 34) the decrease starts in autumn 1987 and at the Slikken van Viane autumn 1988 gave the lowest values of the study period. The observed low value in spring 1987 at station 27, high in the intertidal of the eastern part of the Oosterschelde, is probably related to the preceeding period of reduced immersion times (Hummel *et al.*, 1994), because in autumn 1987 numbers were high again and biomass was back to previous levels. The higher zones of the INTERECOS surveys also showed a minimum change in biomass (Fig. 6d). The ability of *Hydrobia* to disperse easily points towards a non-hydrodynamical factor to explain the decrease in numbers and biomass on many locations, but it is possible that the permanent increased wave impact has an effect on the settlement of larvae on a small zone of the tidal flats.

It seems that the disappearance of *H. ulvae* did not happen in the entire Oosterschelde and where it happened it was not simultaneous.

(b) *H. ulvae* is not very sensitive to cold winters (Beukema, 1979), but it could well be that the cold winter of 1984/1985 caused a decrease in numbers of its predators (*i.e.* shore crabs), so allowing the new settlement of 1985 of *H. ulvae* to spread all over the estuary. On the other hand, warm summers can cause low egg production and hence a small recruitment (Barnes & Greenwood, 1978). The summer temperature of 1989 was indeed above normal (KNMI – weather reports). Recruitment success is one of the determining factors that govern density fluctuations. It could well be that the INTERECOS survey of 1985 coincided with a remarkable strong settlement.

The mudsnail is known for its variable pattern of strong and weak year classes (Fish & Fish, 1974, 1977). Data from the Slikken van Viane and one of the eastern stations (34) support this view, but other stations show highest numbers of juveniles in other years. Again, the dispersion of *Hydrobia* is an important phenomena explaining density differences. A comparison of the mean individual weights of the mudsnails in 1985 and 1989 (Table 3) indicates smaller specimens in 1985. The observed biomasses (including shell) in 1985 (0.20 to 0.45 mg AFDW) suggest a strong dominance of 0 -year individuals, if compared with the weights found by Wolff & de Wolf (1977) in the former Grevelingen estuary. Dekker (1979) also found comparable weights (0.20–0.60 mg AFDW) in the Western Dutch Wadden Sea.

(c) Parasitic infestation.

H. ulvae is an important intermediate host for several trematodes (Kinne, 1980; Rothschild, 1936; Ankel, 1962; Morrisey, 1990; Jensen & Mouritsen, 1992). Infestation rates can vary from place to place, from only a few individuals to over 90% of the population (Kinne, 1980). There are no data on parasitic infestation of *H. ulvae* in the Oosterschelde. At several stations, however, high individual weights were recorded in 1987 to 1989. This may point to infestation. The destruction of the gonads results in changes in physiological and hormonal mechanisms, causing the specimens to grow bigger than normal (gigantism) and make them sterile. A mortality of 40% in the Danish Wadden Sea in 1990 was attributed to infestation with trematodes. High temperatures were suggested as a triggering factor for the mass development of the trematodes in *Hydrobia* (Jensen & Mouritsen, 1992).

Table 3. Mean individual biomass (mg AFDW, including shell) of *H. ulvae* on three tidal flats in the Oosterschelde in 1985 and 1989.

Tidal flats	1985	1989
Roggenplaat	0.45	1.14
Galgeplaat	0.20	1.06
Krabbenkreek	0.23	0.73

The conclusion must be, with Barnes (1988), that population limiting factors and their differential effects on juveniles and adults are still contentious or unknown. The occurrence of a strong year class in 1985, coinciding with different factors that caused a decline in the population at many places, does not allow a proper assessment of the impact of the engineering works.

Living rather deep in the sediment, the adult *Arenicola marina* is not very sensitive to hydrodynamic disturbances. Its ecological amplitude is broad, but very soft sediments are not favoured, certainly not by adults (De Wilde & Farke, 1981). Severe winters have no influence on the size of the population (Dörjes *et al.*, 1986; Beukema, 1982). Large density fluctuations from year to year do not occur (Reise, 1985; Beukema, 1982), presumably because the number of juveniles that settles between the adults is limited by the number of adults (Farke & Berghuis, 1979). Only on typical juvenile settlement places ('nursery flats') on the higher shore do large numbers occur. The developments of *A. marina* in the Oosterschelde over the period 1983–1989 do not deviate from studies elsewhere. Only on the higher zones, juvenile settlement was interrupted in the period of reduced tidal amplitude.

Longbottom (1970) has shown that the abundance of *A. marina* can be correlated with the abundance of organic matter in the sediment. The changes in fine sediment content that occurred on some places did not influence its distribution on the Oosterschelde, however.

Scoloplos armiger is widespread in subtidal and intertidal areas in all sorts of sediments (Wolff, 1973), but particulary common in fine to muddy sands (Gibbs, 1968). The increase in numbers and biomass from 1985 onwards occurred in other coastal areas too, such as the Wadden Sea (Beukema & Essink, 1986). This indicates that the development in the Oosterschelde is probably not caused by local environmental changes, but by larger scale events such as weather or climate. *S. armiger* is a northern species, not very sensitive to severe winter temperatures (Beukema, 1989), whereas one of its major predators *Nephtys hombergii* is then subject to a high mortality

(Beukema, 1987). In the permanent plots *N. hombergii* declined sharply between 1984–1987 (severe winters 1984–1985, 1985–1986 and 1986–1987). Recolonisation did not start until 1989 (Seys *et al.*, 1994). Moreover, although *S. armiger* is known to be sensitive to longer exposure times (Fortuin *et al.*, 1989) and it is therefore not likely to occur at the highest intertidal level (Schöttler & Grieshaben, 1988), it was most abundant in 1989 in the higher zones (Fig. 6f). At the same time *N. hombergii* became re-established higher on the tidal flats (Fig. 12). This is an indication that the pattern of distribution of *S. armiger* in the Oosterschelde is temporally and spatially (under the normal immersion regime) influenced by the occurrence of the predator *N. hombergii*.

B. pilosa and *B. sarsi* live in the upper layers of unstable, sandy tidal flats (<2% silt content, median grain size 150–220 μ; Kharyallah & Jones, 1980). There they find enough water movement (oxygen supply) and the sediment structure that enables them to dig and feed in their typical way (Nicolaisen & Kanneworff, 1969).

The temporal density pattern of *Bathyporeia sp.* in the Oosterschelde is neither related to severe winter temperatures (1984–1985 to '86–'87), nor to manipulations of the storm surge barrier

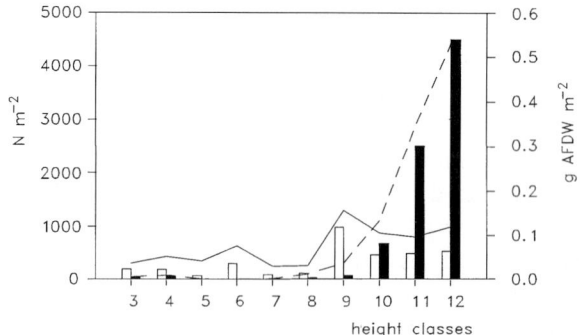

Fig. 12. Biomass and density of *Nephtys spec.* in relation to different height classes of the sampling plots in the INTER-ECOS surveys of 1985 and 1989. legenda: 3 = NAP-150 cm/NAP-125 cm; 4 = −125/−100; 5 = −100/−75; 6 = −75/−50; 7 = −50/−25; 8 = −25/NAP; 9 = NAP/+25; 10 = +25/+50; 11 = +50/+75; 12 = +75/+100. Left vertical axis: White bars = density 1985; black bars = density 1989. Right vertical axis: solid line = biomass 1985; broken line = biomass 1989.

(1986–1987). However, the shift towards higher tidal levels on a spatial scale and the increase in density and biomass at the Roggenplaat indicate that local changes in hydrodynamics may play a role. Nicolaisen & Kanneworff (1983) found rather large seasonal and year to year variations in absolute and relative numbers in a Danish coastal area, and they mention water movement as one of the most important environmental factors for the distribution of the species. In the Westerschelde *Bathyporeia sp.* is one of the most common species of the high dynamic sand flats (Ysebaert & Meire, 1991). In the Oosterschelde current velocities decreased on average by 33% (Ten Brinke *et al.*, 1994) while in a limited range (NAP + 50 cm to NAP + 100 cm) of the tidal area the impact of wave-stress increased (Mulder & Louters, 1994). This range is just the range where *Bathyporeia sp.* is still present in high or even higher densities.

Conclusions

Various fluctuations in the distribution, density and biomass of some common species of the tidal flats have been observed before, during and after the construction and presence of the storm-surge barrier and secondary dams, but only little of that variation could be explained by the changes in hydrodynamic factors or sediment characteristics. It seems to be restricted to a temporal decline in numbers on the higher zones of the flats during the period of reduced immersion (1986–1987). Tidal conditions and sediment variation are still of the same order as before, but the current velocities are lower in many places. The effect of the new hydrodynamical conditions on transport and settlement of pelagic larvae was not studied. It is however unlikely that the mechanical properties of the sediment have altered in such a way that the growth of juveniles is restricted. Transport of food to the bottom may be affected on the long run. The erosion process of the tidal flats may eventually cause a shift in benthic communities: further observations on a large scale of all benthic organisms can give answers to the many questions that still remain.

Acknowledgements

We wish to thank A. van den Dool, J. Dereu, E. C. Stikvoort, E. Brummelhuis and E. Wessel who assisted in the sampling, the determination and the recording of the data. J. Buijs and E. C. Stikvoort carefully handled the data and prepared the figures. We would also like to thank R. H. D. Lambeck for his helpful comments on the draft. Communication No. 699 of NIOO-CEMO, Yerseke, The Netherlands.

References

Ankel, F., 1962. *Hydrobia ulvae* Pennant und *Hydrobia ventrosa* Montagu als Wirte larvaler Trematoden. Vidensk. Meddr Dansk Naturh. Foren. 124: 1–100.

Barnes, R. S. K., 1988. On reproductive strategies in adjacent lagoonal and intertidal-marine populations of the Gastropod *Hydrobia ulvae*. J. mar. biol. Ass. U.K. 68: 365–375.

Barnes, R. S. K. & J. G. Greenwood, 1978. The response of the intertidal gastropod *Hydrobia ulvae* Pennant to sediments of differing particle size. J. exp. mar. Biol. Ecol. 31: 43–54.

Beukema, J. J., 1979. Biomass and species richness of the macrobenthic animals living on a tidal flat area in the Dutch Wadden Sea: effects of a severe winter. Neth. J. Sea Res. 13: 203–223.

Beukema, J. J., 1982. Annual variation in reproductive success and biomass of the major macrozoobenthic species living in a tidal flat area of the Wadden Sea. Neth. J. Sea Res. 16: 37–45.

Beukema, J. J., 1985. Zoobenthos survival during severe winters on high and low tidal flats in the Dutch Wadden Sea. In J. S. Gray & M. E. Cristiansen (eds), Marine Biology of Polar Regions and Effect of Stress on Marine Organisms. John Wiley & Sons Ltd: 351–361.

Beukema, J. J., 1987. Predatory effects of *Nephtys hombergii* on other polychaetes in tidal flat sediments. Mar. Ecol. Prog. Ser. 40: 95–101.

Beukema, J. J., 1989. Long-term changes in macrozoobenthic abundance on the tidal flats of the western part of the Dutch Wadden Sea. Helgoländer wiss. Meeresunters. 43: 405–415.

Beukema, J. J. & K. Essink, 1986. Common patterns in the fluctuations of macrozoobenthic species living at different places on tidal flats in the Wadden Sea. Hydrobiologia 142: 199–207.

Coosen, J., F. Twisk, M. W. M. van der Tol, R. H. D. Lambeck, M. R. van Stralen & P. M. Meire, 1994. Variability in stock assessment of cockles (*Cerastoderma edule* L.) in the Oosterschelde (in 1980–1990), in relation to environmental factors. Hydrobiologia 282/283: 381–395.

Craeymeersch, J., J. Coosen, A. v. d. Dool, 1988. Trendanalyse van densiteit- en biomassawaarden van bodemdieren in het getijdengebied van de Oosterschelde (1983–1986). DIHO Rapporten en verslagen nr. 1988–7 (in Dutch).

Dankers, N. & J. J. Beukema, 1983. Distributional patterns of macrozoobenthic species in relation to some environmental factors. In N. Dankers, H. Kühl & W. J. Wolff (eds), Invertebrates of the Wadden Sea. Balkema, Rotterdam: 69–103.

Dekker, R., 1979. Numbers, growth, biomass and production of organic and calcareous matter of Hydrobia ulvae (Gastropoda: Prosobranchia) in the western Dutch Wadden Sea. Interne verslagen NIOZ Texel 1979–15, 27 pp.

De Wilde, P. A. W. J. & H. Farke, 1981. The Lugworm Arenicola marina. In N. Dankers, H. Kühl & W. J. Wolff (eds), Invertebrates of the Wadden Sea. Balkema, Rotterdam: 111–113.

Dörjes, J., 1980. Auswirkungen des kalten Winters 1978/79 auf das marine Macrobenthos. Natur und Museum 110: 109–115.

Dörjes, J., H. Michaelis & B. Rhode, 1986. Long-term studies of macrozoobenthos in intertidal and shallow subtidal habitats near the island of Norderney (East Frisian coast, Germany). Hydrobiologia 142: 217–232.

Farke, H. & E. M. Berghuis, 1979. Spawning, larval development and migration of Arenicola marina under field conditions in the western Wadden Sea. Neth. J. Sea Res. 13: 529–535.

Fenchel, T., 1975. Factors determining the Distribution Patterns of Mud Snails (Hydrobiidae). Oecologia 20: 1–17.

Fish, J. D. & S. Fish, 1974. The breeding cycle and growth of Hydrobia ulvae in the Dovey estuary. J. mar. biol. Ass. U.K. 54: 685–697.

Fish, J. D. & S. Fish, 1977. The effects of temperature and salinity on the embryonic development of Hydrobia ulvae Pennant. J. mar. biol. Ass. U.K. 57: 213–218.

Forbes, V. E. & G. R. Lopez, 1990. The role of sediment type in growth and fecundity of mud snails (Hydrobiidae). Oecologia 83: 53–61.

Fortuin, A. W., H. Hummel, A. Meijboom & L. de Wolf, 1989. Expected Effects of the Use of the Oosterschelde Storm Surge Barrier on the Survival of the Intertidal Fauna: part 1 – The Effects of Prolonged Emersion. Mar. Envir. Res. 27: 215–227.

Hummel, H., A. W. Fortuin, R. H. Bogaards, A. Meijboom & L. de Wolf, 1994. The effects of prolonged emersion and submersion by tidal manipulation on marine macrobenthos. Hydrobiologia 282/283: 219–234.

Gibbs, P. E., 1968. Observations on the populations of Scoloplos armiger at Whitstable. J. mar. biol. Ass. U.K. 48: 225–254.

Gray, J. S., 1974. Animal sediment relationships. Oceanogr. Mar. Biol. annu. Rev. 12: 223–261.

Gudmundsson, H., 1985. Life history patterns of Polychaete species of the family Spionidae. J. mar. biol. Ass. U.K. 65: 93–111.

Jensen, K. T. & K. N. Mouritsen, 1992. Mass mortality in two common soft-bottom invertebrates, Hydrobia ulvae and Corophium volutator – the possible role of trematodes. Helgoländer Meeresunters 46: 329–339.

Kharyallah, N. H. & A. M. Jones, 1980. The ecology of Bathyporeia pilosa (Amphipoda: Haustoriidae) on the Tay Estuary. II. Factors affecting the micro-distribution. Proc. r. Soc. Edinb. 78: 121–130.

Kinne, O., 1980. Diseases of marine animals. Volume 1: General aspects, Protozoa to Gastropoda. J. Wiley & Sons, New York, 466 pp.

Longbottom, M. R., 1970. The distribution of Arenicola marina (L.) with particular reference to the effects of particle size and organic matter of the sediments. J. exp. mar. Biol. Ecol. 5: 138–157.

Meire, P. M., J. Seys, J. Buijs & J. Coosen, 1994. Spatial and temporal patterns of intertidal macrobenthic populations in the Oosterschelde: are they influenced by the construction of the storm-surge barrier? Hydrobiologia 282/283: 157–182.

Meire, P. M. & J. Dereu, 1990. Use of the abundance/biomass comparison method for detecting environmental stress: some considerations based on intertidal macrozoobenthos and bird communities. J. appl. Ecol. 27: 210–223.

Mulder J. P. M. & T. Louters, 1994. Changes in basin geomorphology after implementation of the Oosterschelde estuary project. Hydrobiologia 282/283: 29–39.

Morrisey, D. J., 1990, Factors affecting individual body weight in field populations of the mudsnail Hydrobia ulvae. J. mar. biol. Ass. U.K. 70: 99–106.

Newell, R., 1965. The role of detritus in the nutrition of two marine deposit feeders, the prosobranch Hydrobia ulvae and the bivalve Macoma balthica. Proc. zool. Soc. Lond. 144: 25–45.

Nicolaisen, W. & E. Kanneworff, 1969. On the burrowing and feeding habits of the Amphipods Bathyporeia pilosa Lindström and Bathyporeia sarsi Watkin. Ophelia 6: 231–250.

Nicolaisen, W. & E. Kanneworff, 1983. Annual variations in vertical distribution and density of Bathyporeia pilosa Lindström and Bathyporeia sarsi Watkin at Julebaek (North Sealand, Denmark). Ophelia 22: 237–251.

Nienhuis, P. H. & A. C. Smaal, 1994. The Oosterschelde estuary, a case study of a changing ecosystem: an introduction. Hydrobiologia 282/283: 1–14.

Reise, K., 1985. Tidal flat ecology. An experimental approach to species interactions. Ecological studies, vol. 54. Springer-Verlag, Berlin, Heidelberg, New York, Tokyo, 191 pp.

Rothchild, M., 1936. Gigantism and variation in Peringia ulvae (Pennant, 1977), caused by infection with larval trematodes. J. mar. biol. Ass. U.K. 20: 537–546.

Schöttler, U. & M. Grieshaber, 1988. Adaptation of the polychaete worm Scoloplos armiger to hypoxic conditions. Mar. Biol. 99: 215–222.

Seys, J., P. M. Meire, J. Coosen, J. A. Craeymeersch, R. H. D. Lambeck & A. Wielemaker-van den Dool, 1993.

Long term changes in the intertidal macrobenthic fauna on eight permanent stations in the Oosterschelde – effects of the construction of the Storm Surge Barrier: first results. Proceedings ECSA 20.

Seys, J., P. M. Meire, J. Coosen & J. A. Craeymeersch, 1994. Long-term changes (1979–1989) in the intertidal macrozoobenthos of the Oosterschelde estuary: are patterns in total density, biomass and diversity induced by the construction of the storm-surge barrier? Hydrobiologia 282/283: 251–264.

Siegel, S., 1956. Nonparametric statistics for the behavioural sciences. MacGraw-Hill, Tokyo.

Smaal, A. C. & P. H. Nienhuis, 1992. The Eastern Scheldt (The Netherlands), from an estuary to a tidal bay: a review of responses at the ecosystem level. Neth. J. Sea Res. 30: 161–173.

Ten Brinke, B. M., J. Dronkers & J. P. M. Mulder, 1994. Fine sediments in the Oosterschelde tidal basin before and after partial closure. Hydrobiologia 282/283: 41–56.

Van Stralen, M. R. & R. D. Dijkema, 1994. Mussel culture in a changing environment: the effects of a coastal engineering project on mussel culture (*Mytilus edulis* L.) in the Oosterschelde estuary (SW Netherlands). Hydrobiologia 282/283: 359–379.

Wolff, W. J., 1973. The estuary as a habitat. An analysis of data on the soft bottom macrofauna of the estuarine area of the rivers Rhine, Meuse and Scheldt. Zoologische verhandelingen, 126: 1–242.

Wolff, W. J. & L. de Wolf, 1977. Biomass and production of the zoobenthos in the Grevelingen estuary, The Netherlands. Estuar. coast. mar. Sci. 5: 1–24.

Ysebaert, T. & P. M. Meire, 1991. Het macrozoobenthos van de Westerschelde en de Beneden Zeeschelde. Rapport WWE 12 (in Dutch).

Ziegelmeier, E., 1964. Einwirkungen des kalten Winters 1962/63 auf das Macrobenthos im Ostteil der Deutschen Bucht. Helgoländer wiss. Meeresunters. 10: 276–282.

Hydrobiologia **282/283**: 251–264, 1994.
P. H. Nienhuis & A. C. Smaal (eds), The Oosterschelde Estuary.
© 1994 *Kluwer Academic Publishers. Printed in Belgium.*

Long-term changes (1979–89) in the intertidal macrozoobenthos of the Oosterschelde estuary: are patterns in total density, biomass and diversity induced by the construction of the storm-surge barrier?

Jan J. Seys [1,2], P. M. Meire [1,2], J. Coosen [3,4] & J. A. Craeymeersch [4]
[1] *Laboratory of Animal Ecology, Zoogeography and Nature Conservation, University of Ghent, K. L. Ledeganckstraat 35, 9000 Gent, Belgium;* [2] *Ministry of the Flemish Community, Institute of Nature Conservation, Kiewitdreef 5, 3500 Hasselt, Belgium;* [3] *National Institute for Coastal and Marine Management/RIKZ, P.O. Box 8039, NL-4330 AE Middelburg, The Netherlands;* [4] *Netherlands Institute of Ecology, Centre for Estuarine and Marine Ecology, Viersraat 28, 4401 EA Yerseke, The Netherlands*

Key words: Oosterschelde, macrozoobenthos, storm-surge barrier, long-term changes, monitoring

Abstract

To evaluate the effects of the construction of a storm surge barrier in the Oosterschelde, long-term patterns (1979–89) in abundance and biomass of the intertidal macrozoobenthos were studied at 14 permanent stations. Additionally, data of a large-scale survey in late summer 1985 and 1989 were analysed. In this paper, patterns in general parameters are discussed.

Late summer values of total biomass, total density, species richness, diversity and abundance- and biomass ratio show no overall significant trend during the study period. The changes in the hydro-dynamics and the morphology of the Oosterschelde after the completion of the storm surge barrier do not seem to have influenced the normal patterns in benthic populations. The observed patterns are determined by the occurrence of severe versus mild winters, rather than by hydrodynamic changes caused by the construction of the barrier. Low biomasses, high densities (particularly of opportunistic species) and higher 'stress-values" (abundance- and biomass ratio) in 1985(−87) indicate a temporal disturbance by severe winter weather. At the elevated COST-station 27, total biomass decreased sharply in 1985, due to a short-term increase in exposure time, caused by the manipulation of the storm surge barrier.

Introduction

When the decision was made to build a storm surge barrier in the mouth of the Oosterschelde estuary, an ecological monitoring program was developed to assess the impact of these engineering works. The interest in biomonitoring programs has been widely recognised. This seems to be part of a maturation process in ecology, resulting from the need for integration on a larger temporal and spatial scale (Reise, 1989). The recent success of

ecological monitoring has different reasons. First of all, it improves our understanding of natural patterns and processes; by statistical modelling, long-term data sets can be used to assess the appropriateness of sampling schedules (spatial and temporal) and as a predictor of future trends (Coull, 1986; Gray & Christie, 1983). Secondly, departures from a common pattern may indicate local effects of pollutants or other factors of disturbance (Gray & Christie, 1983; Beukema & Essink, 1986). It is a necessary precondition for translating the principle of anticipatory action into practical policy (Reise, 1989). With an increasing anthropogenic impact on the natural environment, the development of biomonitoring studies is necessary.

Because of the crucial importance of the intertidal macrozoobenthos in the food-web of the intertidal system (as consumers of plankton, detritus; as prey for higher trophic levels: birds, fish), intensive studies on this group were started as part of the monitoring program in the Oosterschelde (Nienhuis & Smaal, 1994).

In this paper, trends and patterns of some general characteristics (total biomass, density, species richness, diversity, abundance- and biomass ratio) of the macrozoobenthos at 14 permanent stations during the period 1979–89 and at 300 stations in late summer 1985 and 1989 are presented and the effects of the changed environmental conditions in the Oosterschelde (cf. Nienhuis & Smaal, 1994) on the macrozoobenthic system is discussed. Structural aspects and changes in the distribution and biomass of some important macrobenthic species are treated respectively in Meire et al. (1994) and Coosen et al. (1994).

Material and methods

Sampling methods and -stations

Three data-sets were available for the analysis presented in this paper. For all of them, core-samples were taken and sieved on a 1 mm-mesh size. All animals were identified, counted and the ash-free dry weight (AFDW) was determined by drying, weighing and ashing. For more details on sampling and laboratory methods, we refer to [1]Meire et al. (1991a & b; 1994), [2]Seys et al. (1993b) and [2]Craeymeersch et al. (1988) and [3]Meire et Dereu (1989) for the INTERECOS-[1], COST-[2] and VIANEN-[3] dataset respectively.

The INTERECOS data-set consists of a survey in late summer 1985 (pre-barrier situation) and 1989 (post-barrier situation) at 305 (resp. 300) sampling stations distributed over three tidal flats, i.e. in the western, central and northern part of the estuary (Fig. 1).

The COST data-set comes from a monitoring program in the period 1983–1989 based on two stations from the Roggenplaat in the west of the Oosterschelde, one from the Galgeplaat (central part) and five at the eastern tidal flats (Verdronken Land van Zuid-Beveland; Fig. 1). Samples were initially (1983 and 1984) taken 4 times a year, and twice since then (1985–1989). No data were available from station 14 in 1983 and 1986, from station 26 in 1988 and from station 30 in 1985–1986. For two of the stations (3 and 27), all species were picked out and identified; for the other six stations, only data for the 11 biomass-dominant species (*Arenicola marina* L., *Cerastoderma edule* L., *Heteromastus filiformis* Claparède, *Hydrobia ulvae* Pennant, *Lanice conchilega* Pallas, *Macoma balthica* L., *Mya arenaria* L., *Nephtys hombergi* Savigny, *Nereis diversicolor* O. F. Müller, *Scoloplos armiger* O. F. Müller and *Scrobicularia plana* Da Costa) were available. For these stations, 'total biomass' means the sum of biomass of these 11 species. Total density, species richness, diversity, and abundance- and biomass ratio were not calculated for these stations.

The VIANEN-dataset comes from a monitoring program at 6 stations on the 'Slikken van Vianen' (Meire & Dereu, 1989), a tidal flat in the northern branch of the Oosterschelde (Fig. 1). Samples were taken annually in late summer 1981, 1982 and 1984–1989. Additional samples of late summer 1979 were taken at station 60, 10 and 13.

This paper is primarily based on late summer data from the COST- and VIANEN-set, supplemented with INTERECOS-data where necessary. For data on seasonal variability, we refer to Coosen et al. (1994) and Seys et al. (1993a, b, c). For some stations, data from one or more years are missing. Therefore not all stations could be used in all statistical analyses. Three different combinations of permanent stations were used for analysis (Table 1). The feuer stations are included in the combination the more year data are available. The CVO-combination includes data from all 14 COST/VIANEN-stations. The

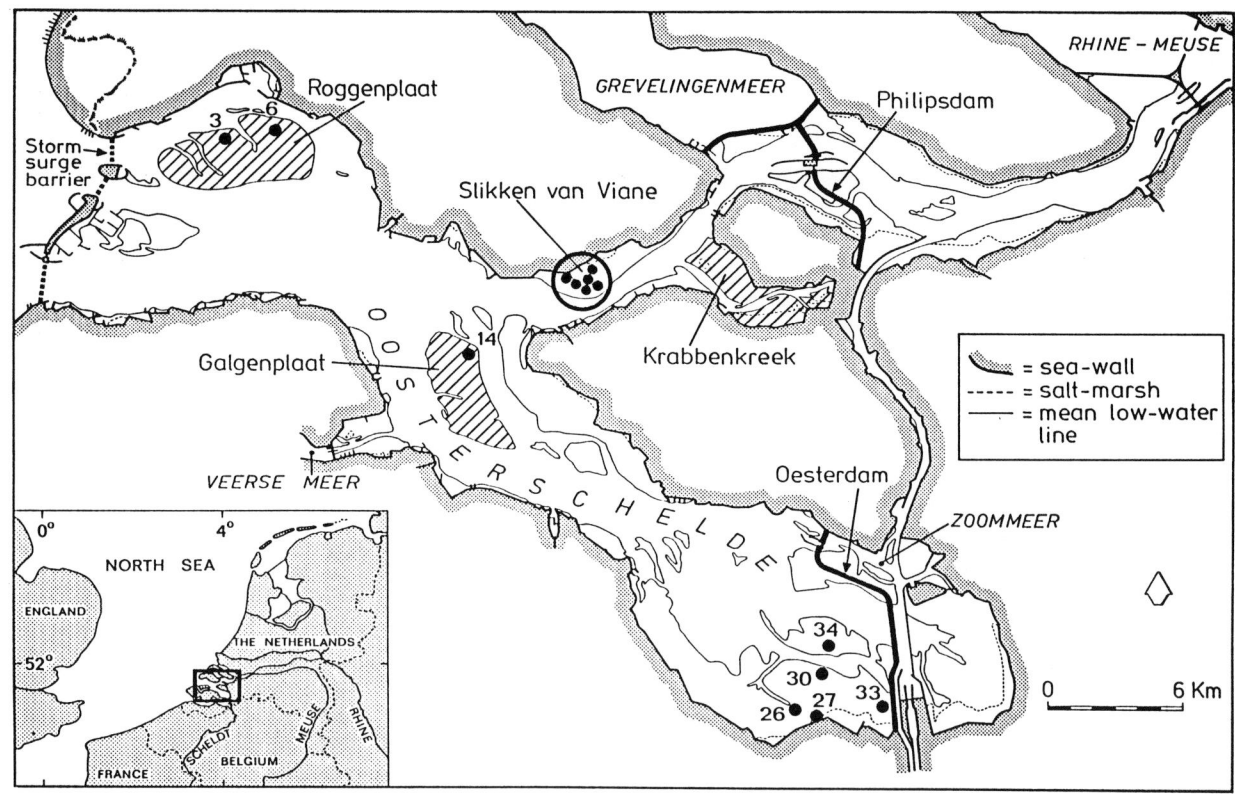

Fig. 1. Study area and sampling stations of macrozoobenthos. The 8 COST-stations (3, 6, 14, 26, 27, 30, 33 and 34) are spread over the Oosterschelde intertidal flats, the 6 VIANEN-stations are enclosed by a circle. The three tidal flats sampled during the INTERECOS-surveys 1985 and 1989 (resp. 305 and 300 sampling points) are shaded.

COST-stations 14, 26 and 30, of which less than 7 late summer data are available are omitted in the CV1-combination. The COST-stations with incomplete data on smaller species (all COST-stations except 3 and 27) were excluded from combination CV2. Calculations of total density, species richness, diversity and abundance – and biomass ratio's were done only with the CV2 combination.

Data-analysis

Biomass was expressed as g ash free dry weight (AFDW) m^{-2} and density as number of individuals m^{-2}. The abundance ratio, Abundance/number of species (A/S), and biomass ratio, Biomass/abundance (B/A) were used as stress-indicators (Gray *et al.*, 1988). For the division into feeding guilds, we refer to Seys *et al.* (1993c). Species richness and diversity was expressed as Hill-indices N_0–N_2 (Hill, 1973).

Since the time series available is short and contains gaps for most stations no statistical time series analysis could be performed on the data. To detect if any long term change was present in the data a simple non-parametrical test, the Spearman's rank correlation, was calculated. This test enables us to detect gradual increases or decreases in the selected variables during the study-period, but can not reveal any other pattern.

A Detrended Correspondence Analysis (Hill, 1979) on density and biomass data of all (314) INTERECOS-, COST- and VIANEN sampling stations in 1989, was performed with all species present in more than 5 samples.

Table 1. Available data on abiotic characteristics of the CV0-stations in the period 1979–1990 (TL = tidal level; MGS = median grain size (in φ units); SC = silt ($<53\,\mu$) content). The classification of stations in the CV0, CV1 and CV2 groups is also indicated. (V = Vianen, C = Cost data set).

CV2						
CV1						
CV0						
Location	V	V	V	V	V	V
Variable/station	Year	10	22	60	13	32	39

Variable/station	Year	10	22	60	13	32	39
TL (MTL + cm)	79	−81		−70	−41		
	84	−78	−78	−82	−51	+63	−10
	85	−68	−69	−78	−57	+59	−13
	86	−92	−94	−98	−80	+43	−18
	87	−91	−94	−97	−79	+42	−18
	88	−93	−88	−85	−72	+47	−15
	89	−97	−92	−88	−83	+42	−19
	90	−97	−94	−90	−96	+42	−20
MGS (φ units)	79	3.17		3.17	3.07		
	81	2.91	2.99	2.89	2.66	3.02	2.83
	84	2.85	2.76	2.95	2.88	2.94	2.96
	85	2.92	2.87	2.84	2.78	2.95	2.89
	86	2.73	2.69	2.69	2.84	2.97	2.93
	87	2.75	2.82	2.74	2.78	2.89	2.95
	88	2.50	2.53	2.55	2.80	2.63	2.61
SC (%)	79	8.5		9.2	3.9		
	81	7.6	16.6	6.0	7.0	5.0	6.0
	84	4.5	4.0	10.0	11.0	4.0	7.0
	85	5.7	5.6	5.2	4.5	3.2	6.2
	86	5.6	7.0	5.1	13.6	5.3	5.2
	87	2.0	2.8	2.5	2.8	0.8	5.6
	88	3.4	1.54	3.2	4.7	2.1	2.3

CV2								
CV1								
CV0								
Location	C	C	C	C	C	C	C	C

Variable/station	Year	3	27	6	33	34	14	26	30
TL (MTL + cm)	83	+36	+110	+50	+60	−35	−50	−50	−70
	85					−50			−25
	89			+40		−50	+0	−25	−40
	90	+25		+40				−50	
MGS (φ units)	83	2.65	3.39	2.74	3.47	3.04	3.00	3.09	3.12
	87	2.50	3.03						
	89	2.31	3.22	2.48	3.26	2.73	2.31	2.85	2.82
SC (%)	83	4.8	3.9	0.5	12.0	0.7	1.2	2.2	0.5
	87	2.8	7.6						
	89	0.9	2.7	0.1	7.8	0.4	0.3	2.1	1.1

Environmental changes in the Oosterschelde

For a detailed description of the Oosterschelde and the engineering works we refer to Nienhuis & Smaal (1994). The major changes are shortly summarized. Hydrodynamic changes started medio 1985 and resulted, after the completion of the storm surge barrier (October 1986) and the compartmentalization dams (Oesterdam, October 1986; Philipsdam, April 1987) in a reduction of the tidal range (-13%); current velocities were reduced by 30% in the western sector and 70% in the northern sector; tidal volume decreased by 28%, mean fresh water input was 64% less and nitrogen input decreased by 58%. Water residence time increased by 100% and chlorinity by 14%. Primary production has slightly increased ($+5\%$) whereas zooplankton is now much more abundant ($+60\%$). In the period end 1986–April 1987, the tidal reduction was more pronounced, caused by the manipulation of the storm surge barrier for the completion of the compartmentalization dams. This resulted in significantly longer exposure periods in the upper part and significantly longer immersion periods on the lower part of the intertidal areas (see Nienhuis & Smaal, 1994; Seys *et al.*, 1993c).

Results

Representativity of the monitoring stations

The representativity of the COST/VIANEN monitoring stations for the Oosterschelde can be tested by comparing the 1985 and 1989 COST/VIANEN data with the results of the two large scale surveys (INTERECOS) made in both years. The results are summarized in Table 2. The biomass estimates are in close agreement in both years. The biomass of the CV2 combination is higher as the proportion of sampling stations on musselbeds is rather high in this group. The densities are also very comparable, although in 1989 they are higher in the monitoring stations than in the INTERECOS set. In the INTERECOS-survey 1989, 65 species were found. It is not possible to compare this value with the results of

1985 since not all organisms were identified to species level in the INTERECOS survey 1985. In all CV2 stations 57 species were found.

To see whether the benthic communities found at the monitoring stations are representative for the Oosterschelde, a Detrended Correspondence Analysis (Hill, 1979) both on density and biomass data of the COST-, VIANEN-, and INTERECOS stations of late summer 1989 was done. In both analyses the first two DCA-axes were responsible for most of the variation and all COST/VIANEN monitoring stations are very well spread within the cloud of INTERECOS points as shown in Fig. 2 for the density data. Nearly all main benthic community types found in the INTERECOS-set (Meire *et al.*, 1994) were represented by one or more monitoring stations. The results of the COST/VIANEN stations can therefore be used for further analysis of the temporal patterns of the Oosterschelde macrozoobenthos.

Changes in environmental parameters

Available data on some environmental parameters of the stations are summarized in Table 1. In the period 1979–83, all stations were situated between Mean Tidal Level (MTL) -80 cm and MTL $+110$ cm. In 1988–89, nine stations had eroded with 5–30 cm, while at three stations sedimentation (25–50 cm) occurred. The sediments at all COST/VIANEN-stations, except one, became coarser between 1979–83 and 1987–89: median grain size changed from 2.65–3.47 φ to 2.31–3.26 φ, while the range in silt-content ($<53\ \mu$) shifted from 0.5–16.6% to 0.1–7.8%.

Temporal variations in some general parameters

The temporal patterns of total biomass, biomass of different feeding guilds, density, species richness, diversity and abundance- and biomass ratio's are illustrated in Fig. 3. As most of the patterns observed in the different VIANEN or COST-stations were found to be significantly concordant (based on the Kendall Coefficient

Table 2. Mean biomass, density and number of species from the different data sets in 1985 and 1989 (I INTERECOS, CV0, CV1 and CV2 see Table 1).

Parameter	Data set	1985 Mean	1989 Mean
Biomass	I	49.3	99.3
(g AFDW m^{-2})	CV0	48.3	88.2
	CV1	52.5	105.8
	CV2	64.1	134.2
Density	I	34715	9673
(numbers m^{-2})	CV2	31838	15254
Number of species	I		65
	CV2		57

of Concordance), average values for the six VIANEN-stations (1979–89) and the eight COST-stations were used to produce Fig. 3. To

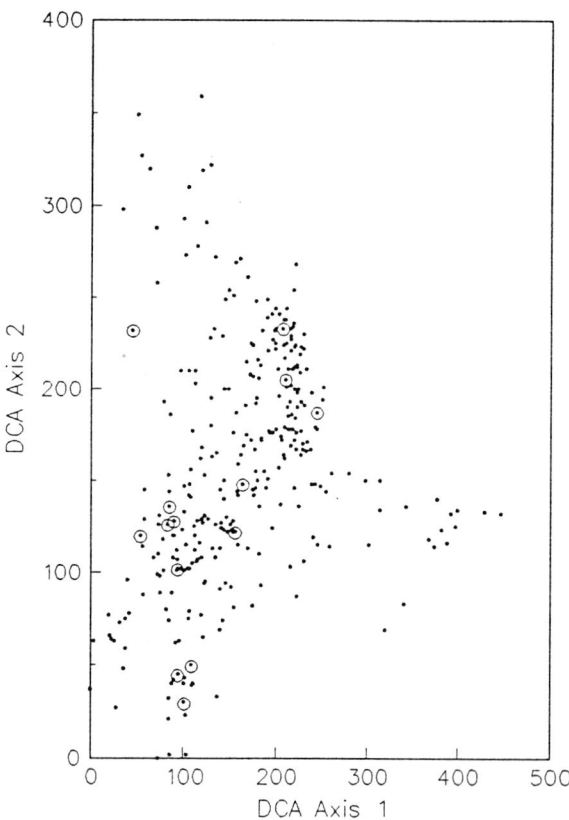

Fig. 2. Detrended Correspondence Analysis on the density data of the INTERECOS- and COST/VIANEN-stations in late Summer 1989 (314 stations). Stations are presented according to their position on the first and second DCA-axis. The CV-stations are encircled.

exclude overruling effects of some individual stations in the figures absolute values per station were transformed to percentages of the long-term average at this station, before different stations were combined. The results of the Spearman Rank correlation coefficient are summarized in Table 3 and 4.

Biomass

The average tota l biomass in the INTERECOS-stations changed from 49.3 in 1985 to 99.3 g AFDW m^{-2} and 1989 respectively. For the CV0-stations, total biomass varied between 5.5 and 531 g AFDW m^{-2} with an overall mean value of 76.2 g (Table 5). The highest values occur at the VIANEN-musselbed stations 10, 22 & 60 (Table 5).

Total biomass is dominated by the taxonomically diverse group of deposit feeders and the filter feeders, essentially comprised of *C. edule* and *M. edulis*. Particularly this last group fluctuated rather strongly, with low values in 1985–86 (Fig. 3) and high values in 1987–89 especially at the VIANEN-stations (Fig. 3). The low values are due to mass mortality after the severe winter 1984/85–1985/86, the high values to the abundant spatfall, especially of *C. edule*, after the severe winters and their subsequent survival. Without *C. edule* and *M. edulis*, an overall mean biomass of only 19.3 g AFDW m^{-2} was found with a much smaller range (8.54–28.93 g AFDW

Table 3a. Trends in total biomass (TB), biomass excluding *C. edule*/*M. edulis* (TE) and biomass of feeding guilds (FF = filter feeders; DF = deposit feeders; OP = omnivores/predators; GR = grazers) at the CV1-stations. The Spearman Rank correlation coefficients are based on late summer values 1979–1989 (station 60, 10, 13), 1981–1989 (station 22, 32, 39) or 1983–1989 (station 3, 6) (N = number of cases; * $p < 0.05$; ** $p < 0.01$; ns = not significant).

Station	N	TB	TE	FF	DF	OP	GR
C3	7	− 0.679 ns	− 0.750 ns	− 0.679 ns	− 0.357 ns	− 0.750 ns	+ 0.099 ns
C6	7	− 0.821 *	− 0.786 *	− 0.714 ns	− 0.786 *	+ 0.179 ns	+ 0.408 ns
C27	7	− 0.571 ns	− 0.571 ns	+ 0.107 ns	− 0.500 ns	+ 0.321 ns	+ 0.089 ns
C33	7	+ 0.143 ns	+ 0.071 ns	− 0.000 ns	+ 0.214 ns	− 0.143 ns	+ 0.408 ns
C34	7	+ 0.107 ns	+ 0.214 ns	+ 0.179 ns	+ 0.143 ns	+ 0.321 ns	+ 0.450 ns
V60	9	− 0.133 ns	− 0.333 ns	− 0.133 ns	− 0.400 ns	− 0.400 ns	+ 0.367 ns
V10	9	+ 0.150 ns	+ 0.033 ns	+ 0.150 ns	+ 0.183 ns	+ 0.233 ns	+ 0.500 ns
V13	9	+ 0.950 **	+ 0.583 ns	+ 0.950 **	+ 0.433 ns	+ 0.283 ns	+ 0.800 *
V22	8	− 0.381 ns	− 0.690 ns	− 0.333 ns	− 0.952 **	− 0.119 ns	+ 0.071 ns
V32	8	+ 0.667 ns	− 0.000 ns	− 0.714 ns	− 0.167 ns	+ 0.571 ns	+ 0.910 **
V39	8	+ 0.667 ns	− 0.190 ns	− 0.667 ns	− 0.286 ns	− 0.310 ns	+ 0.333 ns

m^{-2}). In the INTERECOS-surveys 1985 and 1989 biomass without *C. edule* and *M. edulis* was 17.0 and 14.9 g AFDW m^{-2} respectively. The temporal pattern of biomass excluding *C. edule* and *M. edulis*, is rather stable, with slightly lower values in 1985, mainly due to a decrease of deposit feeders and omnivores/predators.

No significant trend was found at 9 of the 11 CV1 stations (Table 3a). The positive trend at station 13 (due to a re-establishment of filter feeders – in casu *C. edule* – in the second half of the study period) disappeared if *C. edule* and *M. edulis* were omitted. The decrease in biomass at station 6 resulted from lower biomasses of deposit feeders. Also station 22 had lower bio-

masses of deposit feeders at the end of the study period (Table 3b). Although no overall trend in total biomass was found at station 13 and 39, the grazers, which were of minor importance for total biomass, showed a negative trend.

Density

The average density at the INTERECOS stations in 1985 was 34 715 m^{-2} (29 094 m^{-2} excluding *C. edule*/*M. edulis*), while in 1989 much lower values were found (resp. 9673 and 9409 m^{-2}). The average value for the CV2-stations in the period 1984–89 was 21 426 m^{-2} (without

Table 3b. Trends in total density (DT), density excluding *C. edule*/*M. edulis* (DE) and diversity (Hill numbers N_0, N_1, N_2) at the CV2-stations. The Spearman Rank correlation coefficients calculated for these variables are based on available late summer values of the period 1979–1989 (stations 60, 10, 13, 22, 32, 39) or 1983–1989 (stations 3, 27) (N = number of cases; * $p < 0.05$; ** $p < 0.01$; ns = not significant).

Station	N	DT	DE	N_0	N_1	N_2
C3	7	− 0.786 *	− 0.857 *	− 0.673 ns	+ 0.321 ns	+ 0.321 ns
C27	7	− 0.857 *	− 0.857 *	− 0.611 ns	+ 0.536 ns	+ 0.429 ns
V60	9	+ 0.017 ns	− 0.033 ns	− 0.600 ns	− 0.533 ns	− 0.517 ns
V10	9	+ 0.267 ns	+ 0.333 ns	− 0.641 ns	− 0.883 **	− 0.817 *
V13	9	+ 0.650 ns	+ 0.600 ns	− 0.025 ns	− 0.283 ns	− 0.250 ns
V22	8	− 0.524 ns	− 0.524 ns	+ 0.036 ns	+ 0.714 ns	+ 0.286 ns
V32	8	− 0.571 ns	− 0.690 ns	− 0.626 ns	+ 0.881 **	+ 0.786 *
V39	8	− 0.429 ns	− 0.476 ns	− 0.182 ns	− 0.190 ns	− 0.595 ns

258

COST

VIANEN

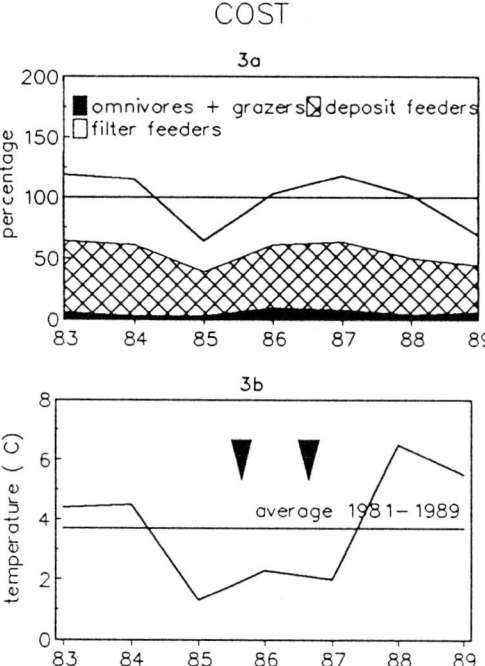

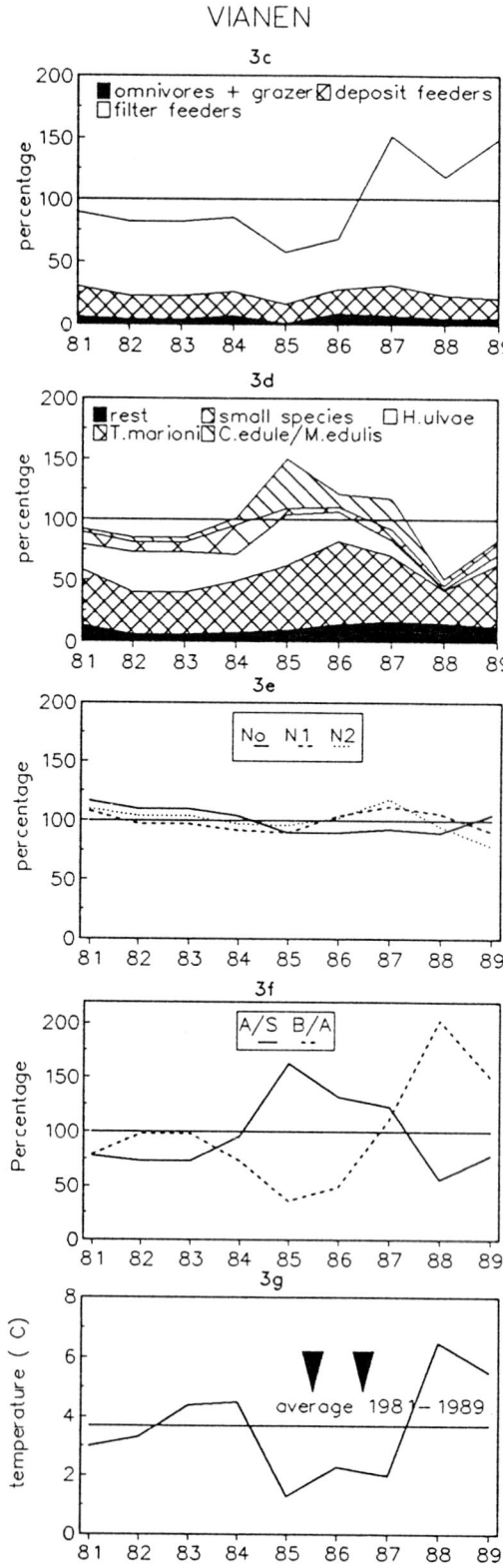

Fig. 3. Change in total biomass (3a, c), total density (3d), diversity ($N_0 N_1$ and N_2)(3e) and abundance- (A/S) and biomass ratio (B/A) (3f) at 8 COST-, resp. 6 VIANEN-stations in the period 1983–1989, resp. 1981–1989. Values are converted to percentages of the long-term average, *i.e.* the average of the whole study period. For the COST-stations, total biomass is the biomass of the 11 biomass-dominant species and density, diversity and A/S, B/A are not calculated (not all species sorted at 6 of the 8 stations). The 'small species' in 3d are: *Anaitides* spec., *Bathyporeia* spec., *Boccardia redeki, Capitella capitata, Corophium* spec., *Eteone* spec., *Heteromastus filiformis,* Oligochaeta, *Polydora* spec. and *Pygospio elegans.* In 3b, g, the average winter air temperature (average of mean monthly temperature December–February) and the period of maximal manipulation of the storm surge barrier (arrows) are indicated.

Table 4. Trends in abundance ratio (A/S) and biomass ratio (B/A) at the CV2-stations. The Spearman Rank correlation coefficients calculated for these variables are based on available late summer values of the period 1979–1989 (station 60, 10, 13, 22, 32, 39) or 1983–1989 (station 3, 27) (N = number of cases; * $p < 0.05$; ** $p < 0.01$; ns = not significant).

Station	N	A/S		B/A	
3	7	− 0.857	*	+ 0.357	ns
27	7	− 0.893	**	+ 0.714	ns
60	9	+ 0.533	ns	− 0.283	ns
10	9	+ 0.333	ns	− 0.233	ns
13	9	+ 0.683	*	+ 0.750	*
22	8	− 0.571	ns	+ 0.286	ns
32	8	− 0.357	ns	+ 0.643	ns
39	8	− 0.500	ns	+ 0.690	ns

C. edule and *M. edulis*: 19 317 m^{-2}). The absolute values of total density were quite different between these stations: they ranged from 6058 m^{-2} at station 13 to 38 256 m^{-2} at station 39 (Table 5).

No significant trend was found in 6 of the 8 CV2 stations, a decrease in 2 stations (Table 3b). Peak values (= more than average) were found in 1985–87 (Fig. 3b). These values are caused by an increase in the abundance of (small) *C. edule*/ *M. edulis* and of a number of other small species (*Anaitides mucosa, Bathyporeia pilosa/sarsi, Boccardia redeki, Capitella capitata, Corophium volutator/arenarium, Eteone longa, Heteromastus filiformis, Oligochaeta, Polydora ligni, Pygospio elegans*). For all these invertebrates, 1988 showed low abundances. Two species with high densities in the first half of the study period (*Tharyx marioni* and *Hydrobia ulvae*) became less common after 1985–86. The overall density of the other species was more or less stable.

Diversity

In the INTERECOS-survey 1989, 65 species were found. Although it was not possible to compare this value with the results of 1985 (not all organisms were identified to species level in the INTERECOS survey 1985), no species seemed to have (dis)appeared. For all CV2 stations an average value of 22 species per station was found, with a range of 18–26 species (Table 5).

The number of species (N_0) did not change significantly in the study period (Table 3b). The species richness N_0 and the Hill-indices of diversity N_1 and N_2 (Hill, 1973) were rather constant over time, with a small increase in 1987 (Fig. 3e). No significant trend was found for N_1 and N_2 at six CV2-stations, a negative significant trend at station 10 and a positive, significant trend at station 32. The negative sign of R_s for N_0, in contrast with positive R_s-values for N_1 & N_2 at station 32 illustrates the different rationale behind these Hill numbers. While there was apparently no significant change in the total number of species (N_0), a significant decrease in numbers of *H. ulvae* (*cf.* Coosen *et al.*, 1994) had increased the (N_2)-diversity (expressed as the reciprocal of the probability to take two individuals of the same species, if one samples at random and without replacement).

Abundance/number of species − Biomass/ abundance ratio's

Except for station 13, none of the CV2-stations had increasing A/S values over the study period 1979–89 (Table 4). At stations 3 & 27, A/S decreased significantly, because of lower numbers of *H. ulvae* in the second half of the study period. The biomass ratio B/A showed an increasing trend at station 13, due to significant higher biomasses in the second half of the study period. Lowest values were observed in 1985–86 (less than 50% of the long-term average), and high values (above average) in the period 1987–89. The abundance ratio showed an opposite pattern (Fig. 3f).

Discussion

The monitoring plots on the Slikken van Vianen were selected to represent the major macrobenthic habitats. The selection of the six stations was based on a larger number of plots that were

Table 5. Average late summer values (1984–1989) of total biomass (TB), total density (TD) and diversity (DI) indices N_0, N_1 and N_2 at the CV0-stations (including overall average), compared to total biomass and -density values of all INTERECOS-samples 1985 and 1989 (IE). For total biomass and -density, 'all' means all species included, 'excl' means all species excluding *C. edule* and *M. edulis*.

	Station	V10	V22	V60	V13	C30	C26	C14	C34	V39
TB	All	184.0	134.7	218.5	88.9	16.4	48.4	43.9	48.7	79.9
	Excl	27.6	28.9	27.7	8.5	12.2	14.7	17.9	13.6	21.5
TD	All	19741	25643	32143	6058					38256
	Excl	15300	25187	28109	3979					34504
DI	N_0	25	26	26	18					22
	N_1	6.78	4.64	5.57	7.40					4.82
	N_2	4.28	2.47	3.34	4.95					3.10

	Station	C3	C6	V32	C33	C27	Mean CV0	IE 1985	IE 1989
TB	All	102.3	27.2	27.7	30.4	15.4	76.2	49.3	99.3
	Excl	24.8	16.6	16.8	23.5	15.2	19.3	17.0	14.9
TD	All	8033		16884		24651	21426	34715	9673
	Excl	6415		16497		24547	19317	29094	9409
DI	N_0	24		18		18	22		65
	N_1	8.69		3.14		3.25			
	N_2	6.14		1.93		2.18			

sampled in previous years (Meire & Dereu, 1989). The selection of the COST stations is based on a large scale mapping of the major macrobenthic habitats of the Oosterschelde (Coosen *et al.*, 1988). The good resemblance between the results of the monitoring stations in late summer 1985 and 1989 and the INTERECOS large scale surveys indicate that the results from the monitoring stations can be seen as representative for the Oosterschelde. Also the changes in the environmental parameters of the different monitoring stations are similar to the observed changes in the rest of the Oosterschelde (Mulder & Louters, 1994; Ten Brinke *et al.*, 1994).

The biomass of the Oosterschelde intertidal macrozoobenthos is high, compared to other Dutch estuaries and brackish lakes: in the Dutch Wadden Sea, a very productive area, a year-average biomass of 38.5 g AFDW m^{-2} (28.1 g AFDW m^{-2}, excluding *C. edule/M. edulis*) was found in 1987 (Beukema, 1989). Corrected for the differing sampling period (biomass in late summer is about 20% higher than year-average bio-

mass; Beukema, 1974) and the incalculation of organic matter in bivalve shells (measured by not removing the flesh from the shell in our study compared to Beukema, 1989) (biomass 17% higher if organic matter in shells is included: Beukema, 1974), a late summer biomass of 54 g AFDW m^{-2} for the Dutch Wadden Sea can be calculated. This value is intermediate between the biomass of the INTERECOS-surveys 1985 (49.3 g AFDW m^{-2}) and 1989 (99.3 g AFDW m^{-2}) and lower than the overall CV0 biomass for the period 1984–89 (76.2 g AFDW m^{-2}). In the marine part of the estuaries Westerschelde and Ems, lower values have been found: in 1987, a late summer biomass of 33 g AFDW m^{-2} was observed in the Westerschelde and 22 g AFDW m^{-2} in the Ems (Meire *et al.*, 1991c). Wolff & De Wolf (1977) found an average of 37.5 g AFDW m^{-2} for the Grevelingen estuary in September, with minima of 12.2 g AFDW m^{-2} in December. Lower biomasses were also found in the saline lake Grevelingen (24.8–38.6 g AFDW m^{-2} in spring 1985–1989: Fortuin & Altena, 1990) and in the brackish Lake Veere (22.2 g AFDW m^{-2} in late summer 1987: Seys & Meire, 1988).

The general characteristics of the macrobenthic system in the Oosterschelde did not yet change markedly after the construction of the storm-surge barrier. Notwithstanding some variation, especially in the period 1985–1987, no overall increasing or decreasing trends in biomass, density, species richness, diversity, abundance- or biomass ratio were observed. These results suggest that hydrodynamical (Vroon, 1994) and sedimentological (Ten Brinke et al., 1994) changes in the Oosterschelde after the completion of the storm surge barrier (Nienhuis & Smaal, 1994) had, by now, only a small effect on the general characteristics of the benthic fauna, compared to the impact of natural phenomena, like the occurrence of severe winters.

In most long-term data-sets (*Wadden Sea*: Beukema, 1989; Essink & Beukema, 1986; Dörjes et al., 1986; Reise et al., 1989; *Northumberland coast*: Buchanan & Moore, 1986; *Scottish lochs*: Pearson et al., 1986; *Baie de Morlaix*: Dauvin & Ibanez, 1986; Ibanez & Dauvin, 1988; *Baltic Sea*: Rosenberg & Loo, 1988; Cederwall & Elmgren, 1980; Persson, 1987; Joseffson & Widbom, 1988) there is quite a large natural seasonal and year to year variation. This makes it often difficult to separate these fluctuations from effects of gradual man-induced changes in the system, particularly because species react in different ways. Only when the system is affected abruptly and severely (heavy pollution: Dauvin & Ibanez, 1986; Pearson et al., 1986; complete closure of an estuary: Lambeck, 1981; dredging activities: Lopez-Jamar et al., 1986), general parameters of the macrozoobenthic community react clearly upon these changes. In two Scottish lochs, Pearson et al. (1986) found the organic input from a paper mill to determine the major changes in total density and biomass. However, this input was on average 4–14 times larger than the natural carbon input from planktonic sources. Neither the changes in the sediments of the Oosterschelde nor the change in primary production were so drastic (Wetsteyn & Kromkamp, 1994) although the plankton communities did change (Bakker et al., 1994).

It was shown that the most pronounced changes in some general parameters of the benthic system occurred in 1985(–87). This can result from effects of the first (1984–1985) of three successive severe winters (Fig. 3g), but also from the major hydrodynamic changes, started medio 1985. If hydrodynamic changes would have an effect on the general characteristics of the Oosterschelde macrozoobenthos, we would expect a gradual change after 1985 when the hydrodynamic changes occurred. Although the time series is rather short no other long-term trend was observed before or after the changes. Secondly, at the COST-stations low biomass- and density values were found already in spring 1985 (Seys et al., 1993b; Craeymeersch et al., 1988), i.e. before the major hydrodynamic changes. Moreover, different species known as winter sensitive collapsed immediately after the winter 1984–1985 and/or 1985–1986, 1986–1987 (Coosen et al., 1993; Seys et al., 1993b). Indeed a closer examination of the long-term pattern of 25 important species in the data-set (Seys et al., 1993c) revealed that 17 species were directly or indirectly affected by the severe winter periods: the populations of winter-sensitive species, such as *C. edule*, *L. conchilega*, *S. plana* and *N. hombergii*, were decimated during one or more of the severe winters 1984–85, 1985–86 and 1986–87. *C. edule* and some other large species (*A. marina*, *M. balthica*) may have a good recruitment after severe winters, while others (*S. armiger*, *H. filiformis*) can benefit from the high mortality in the populations of one of their predators: *N. hombergii* (Beukema, 1989). In the Oosterschelde, there was a good spatfall of *C. edule* after the winter 1984–1985 and *S. armiger* did indeed increase significantly in numbers where *N. hombergii* was decimated (Smaal et al., 1991; Coosen et al., 1993; Seys et al., 1993c). A group of small species with a large ecological spectrum and a short generation time, was particularly abundant immediately after the severe winters 1984–85 and 1985–86. Since these opportunists can recolonise empty niches in any 'stress' situation (after pollution, dredging, severe winters, etc.) they are often used as 'stress' indicators. The rationale behind the use of the biomass ratio B/A and the abundance ratio A/S (Gray *et al.*, 1988) is based on this mass occurrence of opportunists following any kind of disturbance. In the Oosterschelde, maximal stress values (high A/S, low B/A) were concentrated in 1985, *i.e.* after the first severe winter (1984/85).

Although there were no clear effects of hydrodynamic changes in the Oosterschelde on the general characteristics of the benthic system, some changes could be observed if we looked in more detail. On a small spatial scale, some stations were clearly influenced by the hydrodynamic changes. At one elevated COST-station (station 27: NAP + 110 cm), total biomass, total density and the abundance of individual species declined sharply between late summer 1986 and spring 1987, due to a considerable temporal decrease in inundation time (Coosen *et al.*, 1994; Seys *et al.*, 1993c). At this tidal level the mudflat was exposed for several successive days, due to the manipulation of the barrier. Experimentally it was demonstrated that most estuarine organisms show a considerable mortality after four days of exposure (Hummel *et al.*, 1994).

Although by now there is not much evidence of major changes in the macrozoobenthic populations of the Oosterschelde this does not mean that on the long run there will be no effects. Indeed changed hydrodynamic conditions may have an impact on the distribution of larvae, on the sedimentation of food particles to the bottom etc. Furthermore the clear erosion of the intertidal flats (Mulder & Louters, 1994) will in the end certainly have an impact on the total benthic populations. Therefore it is of crucial importance to continue the monitoring of the benthic populations in order to detect the long term effect of the construction of the storm surge barrier on the benthos.

Acknowledgements

This study was financially supported by the Dutch Ministry of Transport and Public Works, Tidal Waters Division. We are grateful to Anette van den Dool for sorting many samples. We received many constructive comments on earlier versions of this manuscript from Eckhart Kuijken, André Dhondt, Anny Anselin, Dirk Bauwens, Rob Lambeck, Mike Elliot and Piet Nienhuis. Communication No. 700 of NIOO-CEMO, Yerseke, The Netherlands.

References

Bakker, C., P. M. J. Herman & M. Vink, 1994. A new trend in the development of the phytoplankton in the Oosterschelde (SW Netherlands) during and after the construction of a storm-surge barrier. Hydrobiologia 282/283: 79–100.

Beukema, J. J., 1974. Seasonal changes in the biomass of the macrobenthos of a tidal flat area in the Dutch Wadden Sea. Neth. J. Sea Res. 8: 94–107.

Beukema, J. J., 1989. Long-term changes in macrozoobenthic abundance on the tidal flats of the western part of the Dutch Wadden Sea. Helgoländer Meeresunters. 43: 405–415.

Beukema, J. J. & K. Essink, 1986. Common patterns in the fluctuations of macrozoobenthic species living at different places on tidal flats in the Wadden Sea. Hydrobiologia 142: 199–207.

Buchanan, J. B. & J. J. Moore, 1986. Long-term studies at a benthic station off the coast of Northumberland. Hydrobiologia 142: 121–127.

Cederwall, H. & R. Elmgren, 1980. Biomass increase of benthic macrofauna demonstrates eutrophication of the Baltic Sea. Ophelia suppl. 1: 287–304.

Coosen, J., P. Meire & A. Van Den Dool, 1988. Kartering bodemdieren intergetijdengebieden Oosterschelde. Rapport Balans 1988–15. RWS, Middelburg, KNAW-DIHO, Yerseke.

Coosen, J., J. Seys, P. Meire & J. A. M. Craeymeersch, 1994. Effect of sedimentological and hydrodynamical changes in the intertidal areas of the Oosterschelde estuary (SW Netherlands) on distribution, density and biomass of five common macrobenthic species: *Spio martinensis* (Mesnil), *Hydrobia ulvae* (Pennant), *Arenicola marina* (L.), *Scoloplos armiger* (Muller) and *Bathyporeia* sp. Hydrobiologia 282/283: 235–249.

Coull, B. C., 1986. Long–term variability of meiobenthos: value, synopsis, hypothesis generation and predictive modelling. Hydrobiologia 142: 271–279.

Craeymeersch, J. A., J. Coosen & A. van den Dool, 1988. Trendanalyse van densiteit- en biomassawaarden van bodemdieren in het getijdengebied van de Oosterschelde (1983–1986). Rapporten en verslagen DIHO 1988–7.

Dauvin, J-C. & F. Ibanez, 1986. Variations à long-terme (1977–1985) du peuplement des sables fins de la Pierre Noire (baie de Morlaix, Manche occidentale): analyse statistique de l'évolution structurale. Hydrobiologia 142: 171–186.

Dörjes, J., H. Michaelis & B. Rhode, 1986. Long-term studies of macrozoobenthos in intertidal and shallow subtidal habitats near the island of Norderney (East Frisian coast, Germany). Hydrobiologia 142: 217–232.

Essink, K. & J. J. Beukema, 1986. Long-term changes in intertidal flat macrozoobenthos as an indicator of stress by organic pollution. Hydrobiologia 142: 209–215.

Fortuin, A. W. & H. C. Altena, 1990. Macrozoöbenthos in het Grevelingenmeer: bestandsopname in voorjaar 1989. Rapporten en verslagen DIHO, Yerseke, 1990–15: 41 p.

Gray, J. S. & H. Christie, 1983. Predicting long-term changes in marine benthic communities. Mar. Ecol. Progr. Ser. 13: 87–94.

Gray, J. S., M. Aschan, M. R. Carr, K. R. Clarke, R. H. Green, T. H. Pearson, R. Rosenberg & R. M. Warwick, 1988. Analysis of community attributes of the benthic macrofauna of Frierfjord/Langesundfjord and in a mesocosm experiment. Mar. Ecol. Prog. Ser. 46: 151–165.

Hill, M. O., 1973. Diversity and eveness: a unifying notation and its consequences. Ecology 54: 427–432.

Hill, M. O., 1979. DECORANA – A Fortran program for detrended correspondence analysis and reciprocal averaging. Ecology and Systematics, Cornell University Ithaca, New York.

Hummel, H., A. W. Fortuin, R. H. Bogaards, A. Meijboom & L. de Wolf, 1994. The effects of prolonged emersion and submersion by tidal manipulation on marine macrobenthos. Hydrobiologia 282/283: 219–234.

Ibanez, F. & J-C. Dauvin, 1988. Long-term changes (1977 to 1987) in a muddy fine sand Abra alba – Melinna palmata community from the Western English Channel: multivariate time-series analysis. Mar. Ecol. Progr. Ser. 49: 65–81.

Joseffson, A. B. & B. Widbom, 1988. Differential response of benthic macrofauna and meiofauna to hypoxia in the Gullmar Fjord basin. Mar. Biol. 100: 31–40.

Lambeck, R. H. D., 1981. Effects of closure of the Grevelingen estuary on survival and development of macrozoobenthos. In N. V. Jones & W. J. Wolff (eds), Feeding and survival strategies of estuarine organisms. Plenum Publishing Corporation: 153–158.

Lopez-Jamar, E., G. Gonzalez & J. Mejuto, 1986. Temporal changes of community structure and biomass in two subtidal macroinfaunal assemblages in La Coruna bay, NW Spain. Hydrobiologia 142: 137–150.

Meire, P. M. & J. Dereu, 1989. Use of the Abundance/ Biomass Comparison method for detecting environmental stress: some considerations based on intertidal macrobenthos and bird communities. J. appl. Ecol. 27: 210–223.

Meire, P. M., J. Buys, J. Van Der Meer, A. Van Den Dool, A. Engelberts, F. Twisk, E. Stikvoort, D. De Jong, R. H. D. Lambeck & J. Polderman, 1991a. Macrozoobenthos in de Oosterschelde in het najaar 1985. een situering van de gegevens. Rapport RUG-WWE nr. 27, Gent; RWS, Middelburg, DIHO, Yerseke.

Meire, P. M., J. Buys, J. Seys, F. Twisk, E. Stikvoort & D. De Jong, 1991b. Macrozoöbenthos in de Oosterschelde in het najaar van 1989: een situering van de gegevens. Rapport RUG-WWE nr. 30, Gent; RWS, Middelburg; DIHO, Yerseke.

Meire, P. M., Seys, J., T. J. Ysebaert & J. Coosen, 1991c. A comparison of the macrobenthic distribution and community structure between two estuaries in SW Netherlands. In M. Elliott & J-P. Ducrotoy (eds), Estuaries and Coasts: Spatial and Temporal Intercomparisons, Proc. ECSA 19 Symposium. Olsen & Olsen, Fredensborg: 221–230.

Meire, P. M., J. Seys, J. Buijs & J. Coosen, 1994. Spatial and temporal patterns of intertidal macrobenthic populations in the Oosterschelde: are they influenced by the construction of the storm-surge barrier? Hydrobiologia 282/283: 157–182.

Mulder, J. P. M. & T. Louters, 1994. Changes in basin geomorphology after the implementation of the Oosterschelde estuary project. Hydrobiologia 282/283: 29–39.

Nienhuis, P. H. & A. C. Smaal, 1994. The Oosterschelde estuary, a case study of a changing ecosystem: an introduction. Hydrobiologia 282/283: 1–14.

Pearson, T. H., G. Duncan & J. Nuttall, 1986. Long term changes in the benthic communities of Loch Linnhe and Loch Eil (Scotland). Hydrobiologia 142: 113–119.

Persson, L-E., 1987. Baltic eutrophication: a contribution to the discussion. Ophelia 27: 31–42.

Reise, K., 1989. Monitoring the Wadden Sea – an introduction. Helgoländer Meeresunters. 43: 259–262.

Reise, K., E. Herre & M. Sturm, 1989. Historical changes in the benthos of the Wadden Sea around the island of Sylt in the North Sea. Helgoländer Meeresunters. 43: 417–433.

Rosenberg, R. & L-O. Loo, 1988. Marine eutrophication induced oxygen deficiency: effects on soft bottom fauna, Western Sweden. Ophelia 29: 213–225.

Seys, J. & P. M. Meire, 1988. Macrozoobenthos van het Veerse Meer. Rapport RUG-WWE nr. 4, Gent.

Seys, J., P. M. Meire, J. Coosen & J. Craeymeersch, 1993a. Macrozoöbenthos op acht stations in de Oosterschelde 1983–1989: presentatie data. Rapport RUG-WWE (in press).

Seys, J., P. Meire & J. Coosen, 1993b. Macrobenthic populations in the Oosterschelde before, during and after major coastal engineering works. Rapport RUG-WWE, Gent (in press).

Seys, J., P. M. Meire, J. Coosen & J. Craeymeersch, 1993c. Longterm changes in the macrobenthic fauna at eight intertidal permanent stations in the Oosterschelde estuary – effects of the construction of the storm surge barrier: some preliminary results. Proceedings ECSA-Hull sept 1990 (in press).

Smaal, A. C., M. Knoester, P. Nienhuis & P. M. Meire, 1991. Changes in the Oosterschelde ecosystem induced by the Delta works. In M. Elliott & J-P. Ducrotoy (eds), Estuaries and Coasts: Spatial and Temporal Intercom-

264

parisons, Proc. ECSA 19 Symposium. Olsen & Olsen, Fredensborg: 375–384.

Ten Brinke, W. B. M., J. Dronkers & J. P. M. Mulder, 1994. Fine sediments in the Oosterschelde tidal basin before and after partial closure. Hydrobiologia 282/283: 41–56.

Vroon, J., 1994. Hydrodynamica characteristics of the Oosterschelde in recent decades. Hydrobiologia 282/283: 17–27.

Wetsteyn, L. P. J. M. & J. C. Kromkamp, 1994. Turbidity, nutrients and phytoplankton primary production in the Oosterschelde (The Netherlands) before, during and after a large-scale coastal engineering project (1980–1990). Hydrobiologia 282/283: 61–78.

Wolff, W. J. & L. De Wolf, 1977. Biomass and production of zoobenthos in the Grevelingen estuary, The Netherlands. Estuar. coast. mar. Sci. 5: 1–24.

Hydrobiologia **282/283**: 265–280, 1994.
P. H. Nienhuis & A. C. Smaal (eds), The Oosterschelde Estuary.
© 1994 *Kluwer Academic Publishers. Printed in Belgium.*

Changes in the sublittoral hard substrate communities in the Oosterschelde estuary (SW Netherlands), caused by changes in the environmental parameters

M. J. de Kluijver [1] & R. J. Leewis [2,*]
[1] *Institute of Taxonomic Zoology, University of Amsterdam, P.O. Box 4766, 1009 AT Amsterdam, The Netherlands;* [2] *National Institute for Coastal and Marine Management/RIKZ, P.O. Box 20907, 2500 EX The Hague, The Netherlands;* *Present address: RIVM (LWD), P.O. Box 1, 3720 BA Bilthoven, The Netherlands*

Key words: sublittoral hard substrate communities, storm-surge barrier, environmental parameters, Oosterschelde estuary

Abstract

In order to assess the effects of the execution of the Delta Project, the sessile sublittoral communities on hard substrates in the Oosterschelde estuary and the environmental parameters were quantitatively investigated from 1985 till 1990. During the construction period of the barrier, three communities were sampled in the photic zone and four in the aphotic zone. The distribution of the communities in the photic zone seemed to be determined by the exposition to water movement and depth, while the communities in the aphotic zone were restricted to geographic areas, with differences in tidal current velocities: the mouth of the estuary, the Hammen, the central part and the Zijpe. Two years after the completion of the enclosure works, the community structure changed rapidly, caused by decreases of tidal current velocities, increases of the amounts of sedimentation, especially of fine sediments, and an increase of the transparency of the water. Changes within the associated vagile animals showed the same tendency as the sessile communities: under less exposed conditions the number of organisms remained the same or increased, while at some locations this increase was nullified by increasing amounts of sedimentation.

Introduction

Recently a growing interest is shown in the use of the structure of sublittoral benthic communities on hard substrate for ecological monitoring. Partly this interest was stimulated by the development of SCUBA techniques, which enables marine biologists to survey the communities in their natural habitat and partly because of the increasing acceptance of the importance of the benthic hard substrate communities for the entire ecosystem. Even on the artificial rocky shores in the Oosterschelde estuary, benthic hard substrate organisms stand for 32.7% of the total benthic biomass (Leewis & Waardenburg, 1989). To use these hard substrate communities for ecological monitoring, a reproducible description of both communities and environmental parameters should be available. Because of this reproducibility, the communities should be studied as a whole, including both floral and faunal components, and, if possible, the associated vagile fauna as well. Secondly, the data must be quantified as much as possible. A regulation model for sublittoral communities by the environmental parameters is proposed by De Kluijver (1989). However, testing these models brings forth a problem, because most studies contributing to the development of these models were carried out in stable marine environments (e.g., De Kluijver, 1991), wherein it is difficult to assess the impact of the individual environmental parameters. The large scale engineering works in the southern Delta area (SW Netherlands) provide a unique situation in which it is possible to follow the effects of changing parameters.

The engineering works in the Oosterschelde estuary were completed in 1987. The effects of the construction of the storm-surge barrier in the mouth of the estuary and of two compartmentalisation dams in the eastern part on the marine environment are revealed by Wetsteyn *et al.* (1990) and Nienhuis & Smaal (1994). Before the completion of the Delta works, the sublittoral communities were studied by Leewis & Waardenburg (1989; 1990) and De Kluijver (1989). To evaluate the effects of the engineering works,

the quantitative data of De Kluijver (1989) were used.

Material and methods

During the years of 1985 and 1986 (construction period) 79 stations were investigated along the dikes of the southern Delta area (De Kluijver, 1989). During the post-barrier period (1987–1990) another 175 stations were investigated in the same area. The locations of the stations are shown in Fig. 1.

The communities were sampled using quadrats, in which the percentages of coverage in vertical projection of all sessile macro-organisms were estimated. At each station a minimal sampling area (Weinberg, 1978) of 31 dm², as determined in the construction period, was sampled. Vagile organisms, which are able to move around,

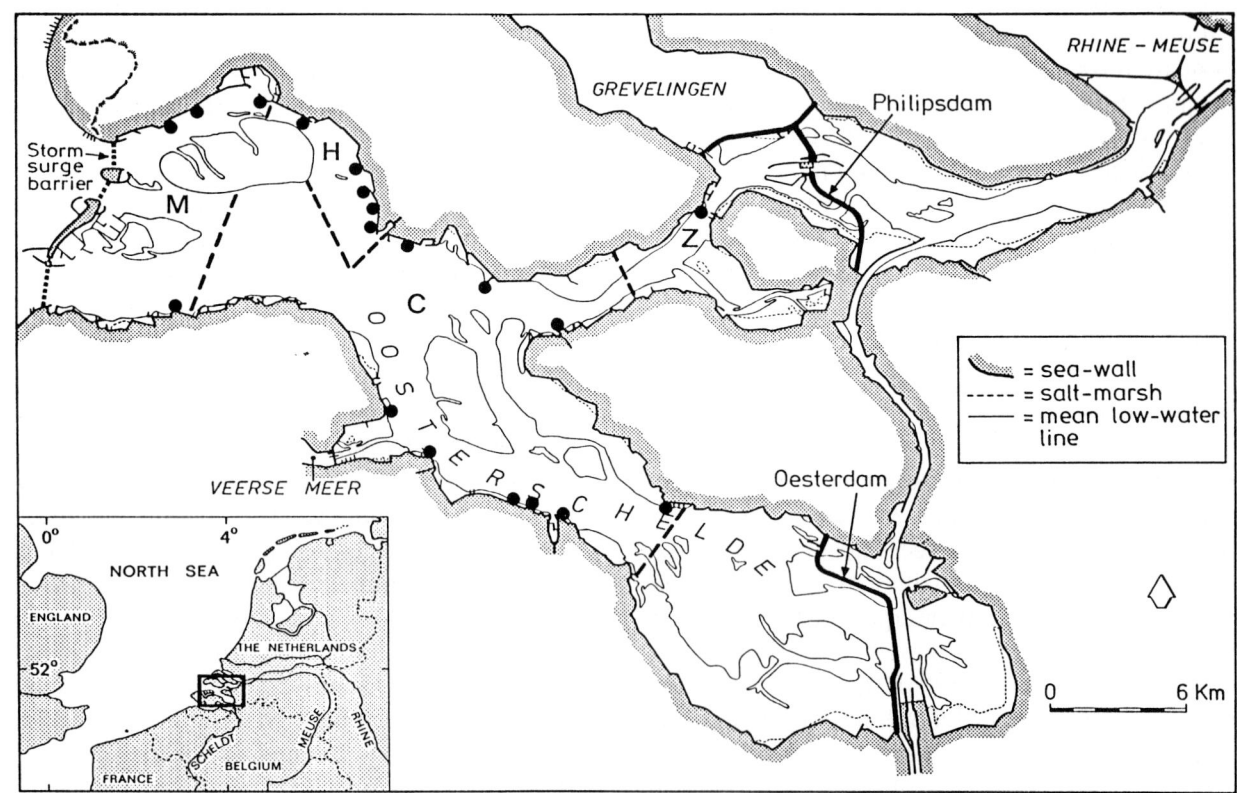

Fig. 1. Map of the study area showing the sampled locations (dots) and the four different regions found in the construction period: M (mouth), H (Hammen), C (central part) and Z (Zijpe).

were scored qualitatively within the stations, but were not used for cluster analysis. Cluster analysis was carried out using the computerprogram CLUSTAN1 C2 (Wishart, 1978), with logarithmically transformed data. The average-linkage method (Sokal & Michener, 1958) was used in combination with the Bray-Curtis coefficient. With the results of the normal analysis an inverse analysis was performed, which procures information concerning the species composition of the clusters. With the program SRTORD (Kaandorp, 1986) the distribution of the quantities of the species over the clusters was calculated. Characteristic species were distinguished at a concentration level of 90% within a community. To be distinguished as dominant or characteristic species, the species had to be present in at least 67% of the stations in the community concerned. Characteristic species were restricted to just one community, while dominant species were abundant (5% percentage cover or more) and occurred in more communities. The communities, here defined as assemblages of species which inhabit a specific set of environmental parameters and remain stable in time and space, consist of different structural layers. In the studied area a distinguishable top layer (TL) of thalli of large brown algae of the genera *Laminaria* and *Sargassum* may be developed. A middle layer (ML) is formed by organisms growing erect from the substrate, but which do not reach the canopy of the top layer. Also epiphytic and epizoic organisms belong to this layer. The encrusting layer (EL) consists of organisms totally adhering to the substrate, e.g., algae and bryozoans. As a result of the changing environmental parameters, the assemblages of species changed in such a way that either new communities could be distinguished or the assemblages could be regarded as a variant of an existing community. The difference between a community and its modified variant is frequently related to the abundance of species.

The most important environmental parameters were measured.

– Depth was referred to NAP, the Dutch Ordnance Level, which approximately corresponds to mean sea level. The tidal difference increases towards the east. In the central part, during the construction period, it reached 3.5 m at springtide.
– The submarine daylight was measured using a relative 'Underwater Hemispherical Irradiance Meter' (UHIM). The spectral sensitivity of this meter (peak value 480 nm; band-width 60 nm) roughly corresponds to the transmission characteristics of the water. By these measurements, the vertical extinction coefficient (k in m^{-1}) was calculated.
– The exposure of the communities to water movement was determined using erosion of gypsum blocks and the composition of the bottom sediments. The erosion of the gypsum blocks is expressed as the weight loss ($g\,h^{-1}$) during one lunar day (24.45 hours). The erosion values obtained in the construction period were statistically tested on the current velocities measured by the Tidal Waters Division (current velocities (in $cm\,s^{-1}$) = 806.1, erosion values ($g\,h^{-1}$) – 41.5; correlation coefficient = 0.99). Sediment characteristics were determined using 7 graded sieves (2.8–0.053 mm). The characteristics were expressed as the proportional contribution of the dry weights of the different sieved fractions.
– The potential sedimentation in the communities was measured using sediment traps. Simple cylinders were used with an inner diameter of 11.7 cm and a length to diameter ratio of 5:1. Sedimentation was expressed in $g\,m^{-2}\,d^{-1}$ dry weight, measured over one month.
– Temperature was measured using a modified mercury thermometer.

During 1988 and 1989 monthly colonization and succession experiments were carried out on artificial concrete substrates at two locations in the mouth and in the eastern part of the estuary.

Results

Based on the distribution of the sublittoral communities during the construction period (De Kluijver, 1989; Leewis & Waardenburg, 1989;

268

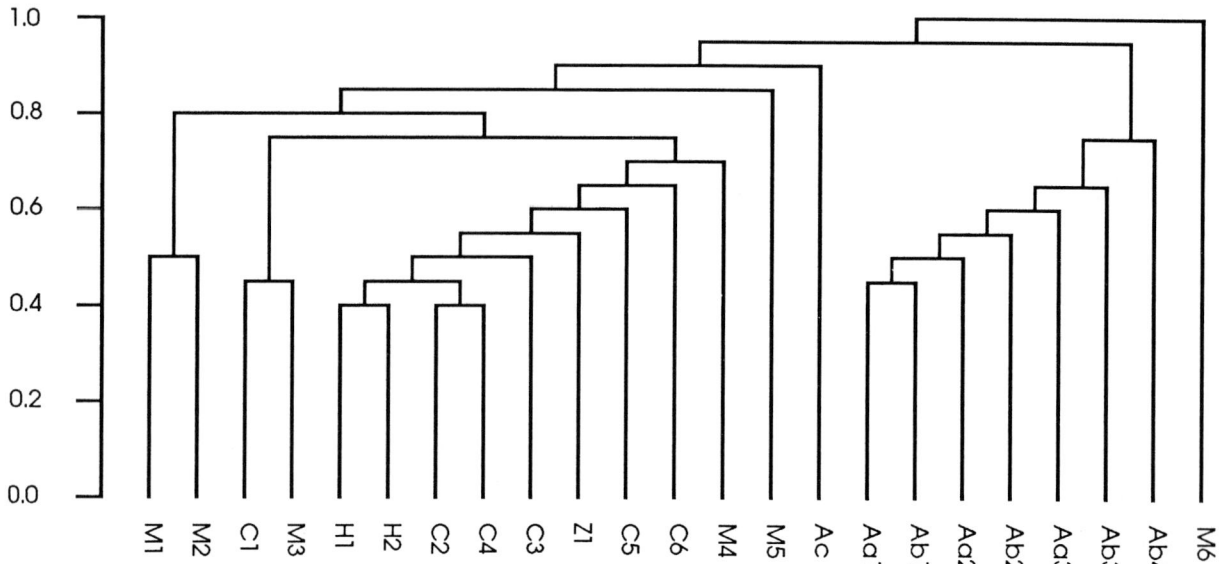

Fig. 2. Simplified dendrogram resulting from the cluster analysis of the data collected during 1985–1990. The communities and variants distinguished are marked on the horizontal axis (explanation see text).

1990), the Oosterschelde estuary was divided into four different regions (Fig. 1). A simplified dendrogram of the 254 sampled stations during 1985–1990 is shown in Fig. 2. The communities restricted to the photic zone are marked Aa1 through Ac. These communities are dominated by algae in the middle structural layer. The communities restricted to the aphotic zone are marked M, H, C or Z, corresponding with the main distribution in the regions of Fig. 1. Communities distributed within more regions of the eastern part of the estuary are marked C as well.

Communities restricted to the photic zone

The photic zone is the depth zone in which the communities are dominated by foliose algae in the middle structural layer. In addition to this, a top layer of thalli of large brown algae of the genera *Laminaria* and *Sargassum* and an encrusting layer of red algae may be developed. This zone extends from the tidemark during the lowest water during spring tide to depths where 10% of the surface light intensity is available (measured by 480 nm). In the construction period (1985–1986) three dif-

ferent communities were found in the Oosterschelde estuary (Figs 3, 4, 5 & 6). Dominant species, which were present in at least 67% of the

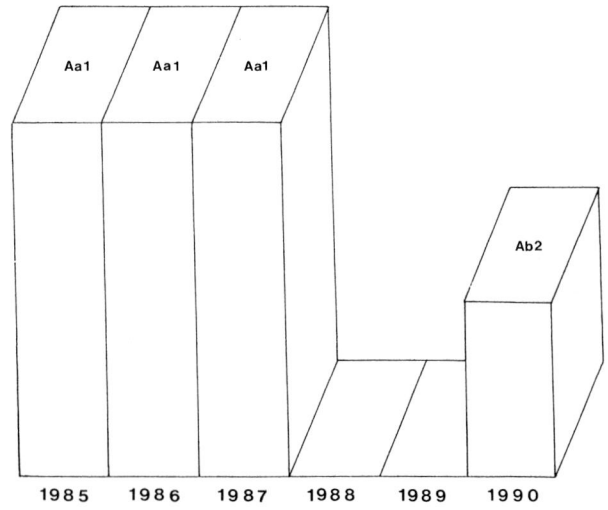

Fig. 3. Distribution of the communities in the photic zone of the mouth of the Oosterschelde estuary. On the x-axis the different years are shown (1985–1990), on the y-axis the relative richness of all sessile species present in at least 67% of the stations and on the z-axis the proportional contribution of the communities per year (length of axis is 100%).

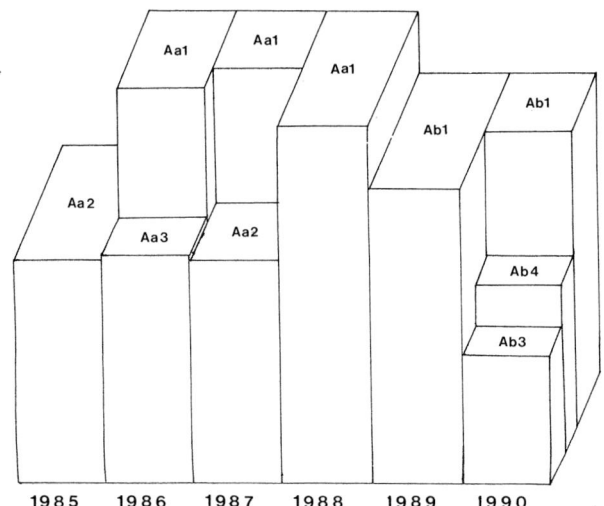

Fig. 4. Distribution of the communities in the photic zone of the Hammen. For an example see Fig. 3.

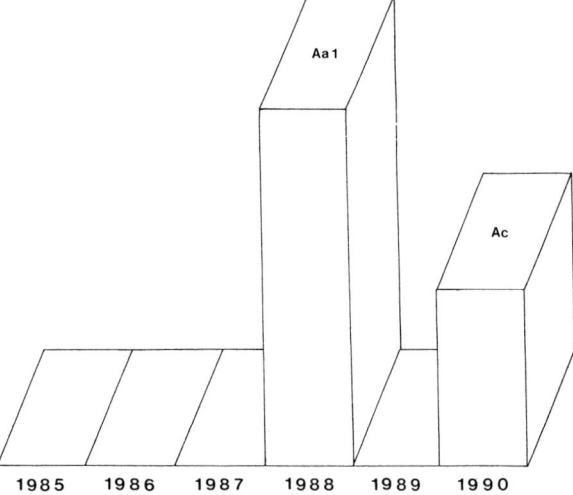

Fig. 6. Distribution of the communities in the photic zone of the Zijpe. For an explanation see Fig. 3.

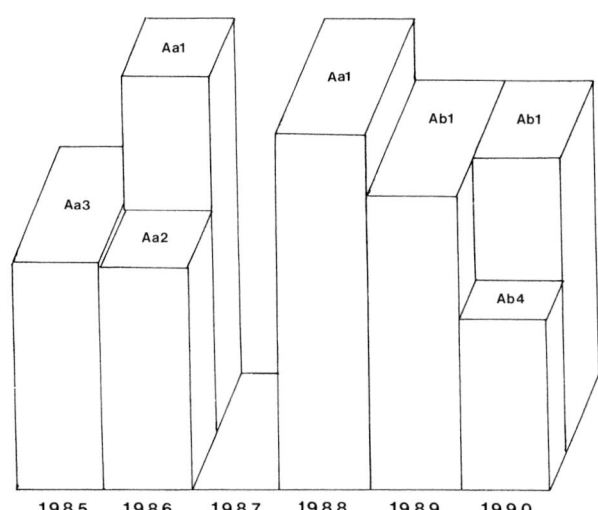

Fig. 5. Distribution of the communities in the photic zone of the central part. For an explanation see Fig. 3.

stations of one of the communities, are listed in Table 1.

Community Aa1 was found in all four regions, between 2 and 3.5 m minus NAP. The middle structural layer of this community was dominated by the red algae *Ceramium rubrum*, *Phyllophora pseudoceranoides* and *Polysiphonia nigrescens* and the brown alga *Dictyota dichotoma* and the tenuous form of the green alga *Codium fragile*. The

bryozoan *Electra pilosa* was found epiphytic within this structural layer. The top layer was mainly formed by *Sargassum muticum* and sporadically by *Laminaria saccharina*. An encrusting layer, formed by *Phymatolithon lenormandi*, was hardly developed.

Community Aa2 was found at exposed locations in the Hammen and the central part, between 2.7 and 3.5 m minus NAP. The community possessed a well developed canopy of *Laminaria saccharina*. The middle structural layer was dominated by the red algae *Ceramium rubrum*, *Phyllophora pseudoceranoides*, *Polysiphonia nigrescens* and *Gracilaria verrucosa* and the tunicate *Styela clava*. Like in community Aa1, *Electra pilosa* was found epiphytic again on these algae.

Community Aa3 was also sampled in the Hammen and in the central part, between 2.7 and 4.0 m minus NAP. A top layer was not developed within this community. The middle structural layer was, besides the red alga *Hypoglossum woodwardii*, dominated by heterotrophic elements from the aphotic zone (e.g., the tunicate *Styela clava*, the anthozoan *Diadumene cincta*, the sponge *Halichondria panicea* and barnacles). The bryozoan *Electra pilosa* is found epiphytic and epizoic on the other species.

During the first two years of the post-barrier

Table 1. Percentage coverage of the dominant species in the communities in the photic zone. Species which are found in at least 67% of the stations of one community are marked with an asterisk.

Community	Aa1	Aa2	Aa3	Ab1	Ab2	Ab3	Ab4	Ac
Antithamnion plumula (Ellis) Thur. in Le Jolis	1.7*	–	0.1	1.1*	0.3*	13.3*	–	3.3*
Ceramium rubrum (Huds.) C. Ag.	15.2*	35.0*	2.4*	15.6*	14.1*	–	2.4*	–
Dumontia contorta (Gmel.) Rupr.	0.1	–	–	–	0.1	–	8.6*	–
Gracilaria verrucosa (Huds.) Papenf.	0.1	5.8*	0.1	0.3	–	–	–	–
Hypoglossum woodwardii Kützing	4.4*	4.4*	6.7*	2.0*	0.7*	0.7*	–	0.6
Phyllophora pseudoceranoides (Gmel.) N. et T.	4.6*	28.7*	0.3*	16.5*	0.7*	0.7*	17.5*	–
Polysiphonia nigrescens (Huds.) Grev.	13.0*	4.6*	2.5*	7.1*	17.5*	1.3*	4.2*	1.0
Dictyota dichotoma (Huds.) Lamour.	7.4*	–	–	6.8*	0.2	1.3*	0.1	–
TL-*Laminaria saccharina* (L.) Lamour.	0.1	14.7*	–	–	–	–	–	–
Bryopsis hypnoides Lamour.	0.6	0.4	0.7	3.1*	–	22.7*	0.3*	1.4
Bryopsis plumosa (Huds.) C. Ag.	2.3*	0.5	0.1	1.1	–	7.3*	0.7*	0.2*
Codium fragile (tenuous) Hariot	5.1*	1.9*	0.6*	0.4	–	–	–	–
Ulva lactuca L.	1.2*	1.0*	0.4	5.6*	11.6*	–	11.5*	0.6
Diadumene cincta Stephenson	0.9*	1.6*	8.8	0.7*	0.2	5.0*	0.2*	0.1
Styela clava Herdman	3.2	4.8*	5.1*	0.3	–	–	–	3.7*
Halichondria panicea (Pallas)	0.7	2.4*	7.7	0.1	0.1	–	–	–
Electra pilosa (L.)	7.9*	16.3*	11.4*	7.2*	1.0*	1.0*	0.6*	0.3
Crassostrea gigas (Thunberg)	2.9	–	0.1	4.4	0.2	28.7*	7.7*	32.5*
Cirripedia	2.8*	2.5*	8.3*	0.3	2.7	–	5.4	1.6*
Bare substrate	28.2	31.7	29.1	40.0	53.3	28.3	30.9	29.2
Number of stations	12	3	4	11	2	1	2	2
Number of sessile species	69	43	44	57	34	25	33	35
Number of vagile species	13	7	5	23	4	2	5	5

period (1987–1988) no changes in the community structure were noticeable. Communities Aa1 and Aa2 showed the same distribution pattern.

After 1988 the species composition of the communities changed quickly, the Aa communities were no longer sampled, instead a new community (Ac) and a number of variants (Ab) of community Aa1 were found. Many species strongly decreased in percentage cover (e.g., the sponge *Halichondria panicea*, the bryozoan *Electra pilosa* and the hydrozoans *Sertularia cupressina* and *Calicella syringa*), while other species, characteristic for the tideless Lake Grevelingen communities (see de Kluijver, 1989), became more abundant (e.g., the sponge *Halichondria bowerbanki*, the tunicates *Botryllus schlosseri* and *Ascidiella aspersa* and the algae *Bryopsis hypnoides*, *Dumontia contorta* and *Callithamnion byssoides*).

Variant Ab1 was found in the Hammen and the central part, between 1.7 and 3.1 m minus NAP. This variant is dominated by the same algae as in

community Aa1 and the green alga *Ulva lactuca*. The main difference between Aa1 and Ab1 is a reduction in species richness of 17% and an increase of 42% silt covered substrate.

Variant Ab2 was found in the mouth of the estuary, between 1.7 and 1.9 m minus NAP. This variant is dominated by the red algae *Polysiphonia nigrescens* and *Ceramium rubrum* and the green alga *Ulva lactuca*. The variant must be regarded as a further decline of variant Ab1. Species richness, compared to variant Ab1, is reduced 40% and the silt covered substrate is increased with 33%, up to 53% of the total available hard substrate surface.

Variant Ab3 was found in the Hammen at 3.7 m minus NAP. The middle structural layer was dominated by the algae *Bryopsis hypnoides*, *Bryopsis plumosa*, *Antithamnion plumula*, the anthozoan *Diadumene cincta* and the bivalve *Crassostrea gigas*.

Variant Ab4 was found in the Hammen and the

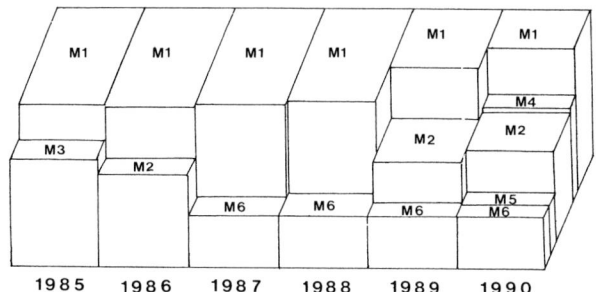

Fig. 7. Distribution of the communities in the aphotic zone of the mouth of the Oosterschelde estuary. For an explanation see Fig. 3.

central part at 1.8 m minus NAP. The middle structural layer is dominated by the algae *Ulva lactuca*, *Phyllophora pseudoceranoides*, *Dumontia contorta*, the bivalve *Crassostrea gigas* and by barnacles. Species composition indicates that this variant must be considered as an impoverished derivative of community Aa1.

Community Ac was found in the Zijpe at 3.9 m minus NAP. Algae were not dominant in the middle structural layer, but this layer was totally dominated by the bivalve *Crassostrea gigas*. Besides many algae with low percentage covers, this layer consists of many animals which are characteristic for the tideless Lake Grevelingen (e.g., the tunicates *Ascidiella aspersa*, *Botryllus schlosseri* and *Ciona intestinalis*, the sponge *Haliclona xena* and scyphistomae of the jelly-fish *Aurelia aurita*).

Communities restricted to the aphotic zone

The aphotic zone is the depth zone which is limited by a 10% illumination boundary and extends to depths where no hard substrate is available. A top layer is absent in these communities and an encrusting layer may consist of bryozoans. The communities consist of heterotrophic animals, independent of the available light. In the construction period four different communities were found.

During the construction period of the barrier, in the mouth of the estuary especially community M1 was abundant (Fig. 7). The middle layer of this community was dominated by the anthozoan *Metridium senile* and the sponge *Halichondria panicea* (Table 2). During this period two variants of community M1 were found. In addition to the dominant species of community M1, variant M3 was also dominated by the polychaete *Lanice conchilega*. Most of the substrate was covered with a thick layer of sediment. Variant M2 was totally dominated by the anthozoan *Metridium senile* and compared to community M1 the species richness was decreased by 35%.

In the Hammen, during the construction period, community H1 was abundant (Fig. 8). This community was dominated by the sponge *Halichondria panicea*, the anthozoan *Diadumene cincta* and the hydrozoan *Halecium halecinum* (Table 3). Near its eastern boundary, community C1 was

Table 2. Percentage coverage of the dominant species in the communities in the mouth of the estuary in the aphotic zone. For explanation see Table 1.

Community	M1	M2	M3	M4	M5	M6
Diadumene cincta Stephenson	3.5*	0.5*	–	6.2*	9.7*	0.2
Metridium senile (L.)	43.2*	45.0*	21.0*	0.3*	51.7*	60.5*
Cliona celata Grant	1.3	–	–	–	5.3*	–
Halichondria panicea (Pallas)	12.0*	0.8*	16.8*	–	–	1.8
Lanice conchilega (Pallas)	0.1	–	13.5*	–	–	–
Bare substrate	33.6	49.4	41.3	85.0	40.0	36.5
Number of stations	36	7	1	1	1	5
Number of sessile species	28	18	21	19	10	10
Number of vagile species	27	13	7	8	3	9

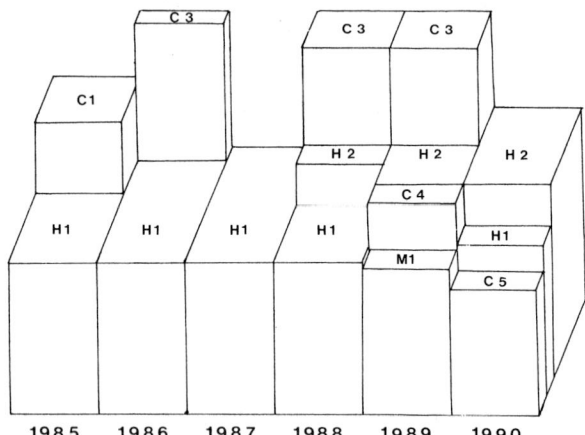

1985 1986 1987 1988 1989 1990

Fig. 8. Distribution of the communities in the aphotic zone of the Hammen. For an explanation see Fig. 3.

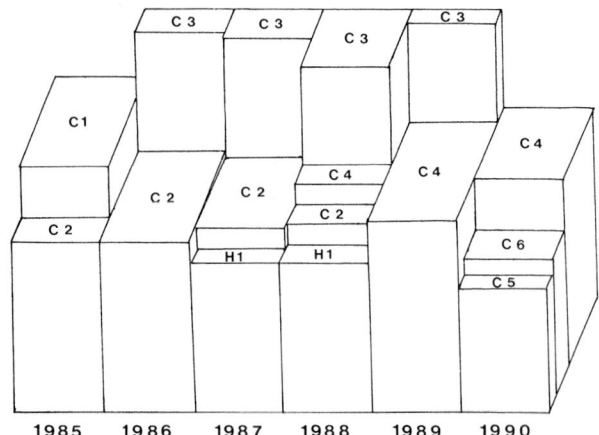

1985 1986 1987 1988 1989 1990

Fig. 9. Distribution of the communities in the aphotic zone of the central part. For an explanation see Fig. 3.

sampled in 1985 and in 1986 community C3 was found, which became common in the post-barrier period.

In the central part community C1 was common (Fig. 9). The middle structural layer was dominated by the anthozoan *Diadumene cincta*, the tunicate *Styela clava* and the sponges *Halichondria panicea* and *Haliclona oculata* (Table 4). *Electra pilosa* was found epizoic. Other variants of community C1, sampled in this period were C2 and C3, both becoming more important during the post-barrier period.

In the Zijpe only community Z1 was sampled during the construction period (Fig. 10). The middle structural layer was dominated by the bivalve *Crassostrea gigas*, the tunicate *Styela clava*, the anthozoan *Diadumene cincta* and the sponges *Halichondria panicea* and *Haliclona oculata* (Table 5).

During the first two years of the post-barrier period (1987–1988), changes in community structure were of minor importance, comparable with the photic zone.

In the mouth of the estuary 80% of the sampled stations still belonged to community M1, while

Table 3. Percentage coverage of the dominant species in the communities in the Hammen in the aphotic zone. For explantion see Table 1.

Community	H1	C1	C3	H2	C4	C5
Haliclona oculata (Pallas)	2.7*	5.1*	2.9*	0.9*	1.1*	0.6
Diadumene cincta Stephenson	14.2*	15.1*	12.2*	29.6*	11.0*	14.4*
Ciona intestinalis (L.)	2.7	0.1	1.3*	0.4	5.6*	1.1*
Styela clava Herdman	2.7*	13.0*	7.3*	1.9*	3.3*	2.2
Halichondria panicea (Pallas)	33.3*	7.2*	5.9*	7.2*	3.4*	2.5*
Electra pilosa (L.)	1.0*	14.8*	0.7*	0.6*	0.5*	0.6*
Eudendrium spec. Ehrenberg	0.3	0.1	0.5	5.4*	0.2	–
Halecium halecinum (L.)	4.9*	1.6	4.7*	6.9*	1.1	0.1
Crassostrea gigas (Thunberg)	–	–	0.1	0.7	9.9*	4.8*
Bare substrate	32.6	30.2	34.1	34.8	41.8	59.2
Number of stations	21	8	12	7	18	2
Number of sessile species	29	43	56	37	37	24
Number of vagile species	23	7	23	20	26	7

Table 4. Percentage coverage of the dominant species in the communities in the central part in the aphotic zone. For explantion see Table 1.

Community	C1	C2	C3	C4	C5	C6
Haliclona oculata (Pallas)	5.1*	5.2*	2.9*	1.1*	0.6	0.1
Diadumene cincta Stephenson	15.1*	26.6*	12.2*	11.0*	14.4*	2.5*
Stylea clava Herdman	13.0*	16.2*	7.3*	3.3*	2.2	1.6*
Ostrea edulis L.	0.1	2.4*	2.6*	1.8	–	7.7
Ciona intestinalis (L.)	0.1	0.2	1.3*	5.6*	1.1*	6.8*
Cirripedia	2.2*	6.6*	2.2*	0.5*	–	–
Halichondria panicea (Pallas)	7.2*	14.2*	5.9*	3.4*	2.5*	0.2
Electra pilosa (L.)	14.8*	2.6*	0.7*	0.5*	0.6*	0.6*
Crassostrea gigas (Thunberg)	–	2.5*	0.1	9.9*	4.8*	0.1
Bare substrate	30.2	23.5	34.1	41.8	59.2	70.0
Number of stations	8	12	12	18	2	2
Number of sessile species	43	33	56	37	24	27
Number of vagile species	7	18	23	26	7	6

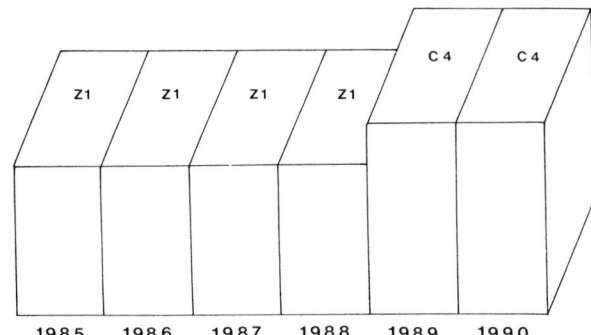

Fig. 10. Distribution of the communities in the aphotic zone of the Zijpe. For an explanation see Fig. 3.

Table 5. Percentage coverage of the dominant species in the communities in the Zijpe in the aphotic zone. For an explanation see Table 1.

Community	Z1	C4
Haliclona oculata (Pallas)	5.6*	1.1*
Diadumene cincta Stephenson	10.9*	11.0*
Styela clava Herdman	15.3*	3.3*
Ciona intestinalis (L.)	0.1	5.6*
Halichondria panicea (Pallas)	15.7*	3.4*
Crassostrea gigas (Thunberg)	28.2*	9.9*
Bare substrate	21.4	41.8
Number of stations	7	18
Number of sessile species	29	37
Number of vagile species	10	26

the other stations belonged to variant M6. This variant displayed an even higher dominance of the anthozoan *Metridium senile* than variant M2 and species richness had decreased another 44%.

In the Hammen community H1 remained dominant, but in 1988 variants C3 and H2 were sampled, especially H2 became important later in the post-barrier period.

In the central part variants C2 and C3 became more important. Variant C2 was dominated by the anthozoan *Diadumene cincta*, the tunicate *Styela clava*, barnacles and the sponges *Halichondria panicea* and *Haliclona oculata*. Compared to community C1 the species richness decreased

with 23%. Variant C3 was dominated by the anthozoan *Diadumene cinta*, the tunicate *Styela clava* and the sponge *Halichondria panicea*. Besides these species, the middle structural layer consisted of many floral elements general in the photic zone (e.g., the algae *Griffithsia devoniensis*, *Polysiphonia nigra*, *Ceramium deslongchampsii*, *Dictyota dichotoma* and *Callithamnion byssoides*). In the western part community H1 was sampled and in 1988 variant C4 was sampled.

In the Zijpe, only community Z1 was sampled during the years of 1987 and 1988.

After 1988 the species composition of the communities changed quickly. In the mouth of the estuary community M1 was mainly found at greater depths, while at shallow places the variants M2, M4, M5 and M6 were found.

Variant M5 was dominated by the anthozoans *Metridium senile* and *Diadumene cincta* and the sponge *Cliona celata*. Variant M4 was dominated by *Diadumene cincta* while *Metridium senile* became sparse. In both variants the sponge *Halichondria panicea* was totally absent.

In the Hammen, community H1 was found only once, mostly this community was replaced by variant H2. Compared to community H1, the percentage cover of *Diadumene cincta* had doubled, while the cover of *Halichondria panicea* had strongly decreased. The hydrozoans *Halecium halecinum* and *Eudendrium* spec. were dominant. Other variants sampled in this period were C3, C4 and C5. Variant C5 was dominated by *Diadumene cincta* and the bivalve *Crassostrea gigas*. This variant had the lowest species richness of all variants in the eastern part. In the western part of the Hammen, at the bottom of the tideway, community M1 was sampled.

In the central part variant C3 was still found in 1989. In 1990 only the variants C4, C5 and C6 were sampled. In addition to *Diadumene cincta*, variant C4 was dominated by the tunicate *Ciona intestinalis* and the bivalve *Crassostrea gigas*. Variant C6 was dominated by *Ciona intestinalis* and the bivalve *Ostrea edulis*.

In the Zijpe community Z1 was totally replaced by the variant C4, caused by strong decreases in the percentage covers of the tunicate *Styela clava*, the sponges *Halichondria panicea* and *Haliclona oculata* and the bivalve *Crassostrea gigas*.

The numbers of vagile species found in the different communities are listed in Tables 1, 2, 3, 4 and 5.

In the photic zone the shore crab *Carcinus maenas* was common during both construction and post-barrier period. The starfish *Asterias rubens*, the crustacean *Caprella linearis* and the seaspider *Pycnogonum littorale* were especially common during the construction period, while the brittle-star *Ophiotrix fragilis*, the velvet swimming crab *Necora puber*, the mollusc *Crepidula fornicata* and

the black goby *Gobius niger* were more common during the post-barrier period. With the exception of variant Ab1, all variants in the post-barrier period accommodated less vagile animals than the variants in the construction period did.

The same holds for the variants in the apothic zone in the mouth of the estuary. *Ophiotrix fragilis* and *Necora puber* became more common during the post-barrier period, while nudibranchs (e.g., *Janolus cristatus*, *Dendronotus frondosus* and *Aeolidia papillosa*) became rare in this period. In the Hammen the number of vagile species did not change during both periods. In most variants *Caprella linearis*, *Asterias rubens* and the crustacean *Pagurus bernhardus* were common, but *Ophiotrix fragilis*, *Necora puber*, *Gobius niger*, and the crustacean *Macropodia* spec. became more abundant in the post-barrier period. In the central part the number of vagile species slightly increased. *Asterias rubens*, *Ophiotrix fragilis*, *Pagurus bernhardus*, *Macropodia* spec. and *Necora puber* became more common in the post-barrier period, while *Caprella linearis*, *Carcinus maenas* and nudibranchs became less abundant. In the Zijpe the number of vagile species strongly increased during the post-barrier period. *Asterias rubens* and *Carcinus maenas* were common in both periods, while *Ophiotrix fragilis*, *Pagurus bernhardus*, *Macropodia* spec., *Crepidula fornicata* and *Necora puber* became more abundant in the post-barrier period and the abundance of *Caprella linearis* and *Homarus gammarus* declined.

Environmental parameters

Changes in the sublittoral environment are connected to human activities in the southern Delta area. Since the start of the execution of the Delta Project, the Oosterschelde estuary became isolated from river influences and from adjoining estuaries (Nienhuis & Smaal, 1994). The successive operations gradually changed the water exchange with the North Sea, the mean velocity of the tidal currents and the fresh-water load of the estuary, which led to changes in the other environmental parameters (e.g., submarine daylight,

Table 6. Comparison of environmental parameters in the construction- and post-barrier period. Erosion values marked with asterisk are obtained through extrapolation of data of Tidal Waters Division. Sediment characteristics are given of bottom sediments, sampled at 10 m minus NAP, particle size given in mm.

	mouth	Hammen	central part	Zijpe	Grevelingen
Construction period:					
Erosion value (g h^{-1})*	0.17	0.17	0.14	0.20	–
k (in m^{-1})	0.78	0.57	0.57	–	–
Sediment characteristics	0.30–0.15	0.15–0.09	0.15–0.09	0.30–0.15	–
Post-barrier period:					
Erosion value (g h^{-1})	0.13	0.12	0.11	0.08	0.07
k (in m^{-1})	0.52	0.45	0.40	0.39	0.32
Sediment load (g m^{-2} day^{-1})	430–3130	270–1450	230–760	30–1150	10–1160
Sediment characteristics	0.15–0.09	< 0.09	< 0.09	0.15–0.09	0.30–0.09

sediment load and sediment characteristics). In Table 6 the changes of the most important environmental parameters are summarized.

Although the current velocities had already changed during the construction period, the data obtained using gypsum blocks corresponded with the flow velocities given by Wetsteyn *et al.* (1990): maximum current velocities were found in the mouth of the estuary, in the Hammen and in the Zijpe and less severe conditions were found in the central part. In the post-barrier period a decrease of 31% was found in the mouth of the estuary and a diminution of 25% in the central part (Wetsteyn *et al.*, 1990). Due to the construction of the Philipsdam, the current velocity in the Zijpe decreased even more, comparable to values measured in tideless water basins. Within the estuary, a periodicity in current velocities existed, with maximum current velocities in autumn and minimal values during winter. This periodicity is also displayed by the sedimentation and composition of the bottom sediments (Fig. 11).

All through 1980–84, turbidity was measured by Wetsteyn *et al.* (1990) using a Secchi disc and different energy cells. The highest vertical extinction coefficients (k in $^{-1}$) were measured during autumn and winter, the lowest during spring and summer. During 1986–90, the UHIM was used only during spring and summer. During the construction period the highest values were measured in the western part of the estuary and lower values in the central and eastern parts. In the post-barrier period the vertical extinction coefficient decreased in the whole estuary, but shortterm fluctuations became larger, especially in the western part.

At shallow locations, the particle size of the deposited material in the construction period was characterized by a dominance of silt and clay fractions (<0.05 mm). At most deeper locations in the tidal channels in the mouth of the estuary and in the Zijpe, the sediments were dominated by the sand fraction between 0.3–0.15 mm, and in the Hammen and the central part the sediments were dominated by the fine-grained sand fraction (0.15–0.05 mm). Since 1988 an increase of the smaller fractions took place. In the mouth of the estuary, the original sand fraction was found below 15 m minus NAP, while at most shallow places fine-grained sand, silt and clay were dominant. The Hammen was totally dominated by the fine-grained sand, silt and clay fractions. In the central part and in the Zijpe the sediments were dominated by the fine-grained sand fraction, but silt and clay fractions started to become more dominant. The already noted periodicity in current velocities was also found in the particle size (Fig. 11).

The disturbance of the estuary is most clearly shown by the sediment load. The gradient in sediment load ranges from west to east from 430–3130 till 30–1150 g m^{-2} d^{-1} dry weight, while the gradient in erosion values ranges from 0.07–0.20 until 0.06–0.14 g h^{-1}. The variation in sedi-

276

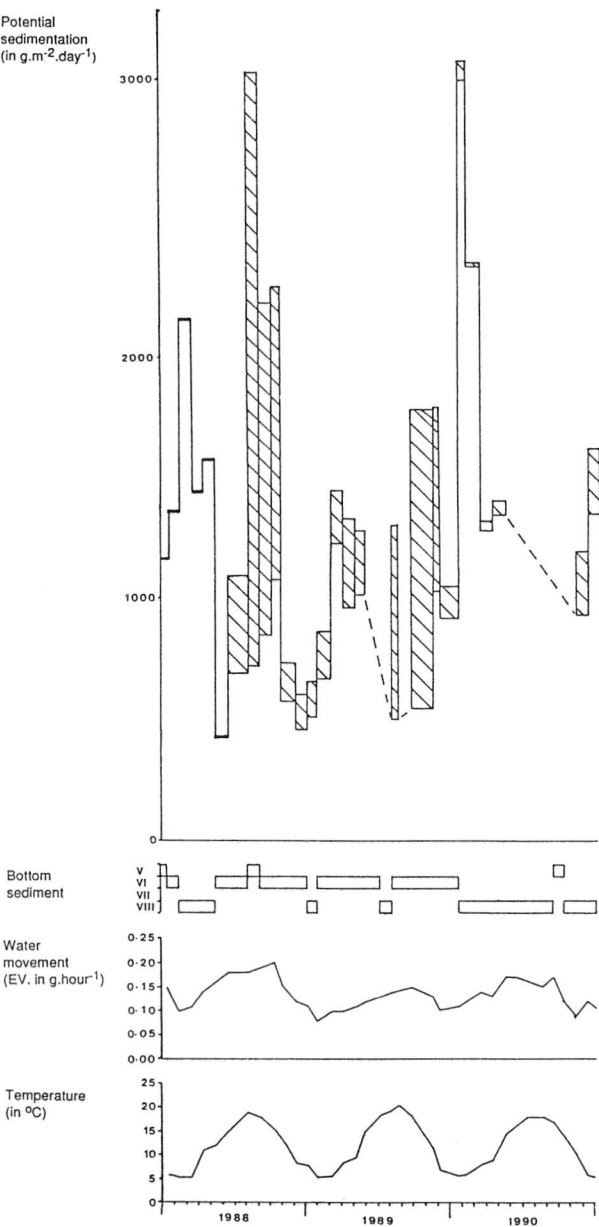

ment load was mainly caused by the periodicity of the current velocities (Fig. 11). Periods with strong sedimentation, coincided with periods with strong and weak water movement, with maximal sedimentation during weak tidal currents. Compared to periods with weak tidal currents, coarse-grained sediments were deposited during strong tidal currents. The contradiction between current velocities and sedimentation was partly caused by the reduction of the tidal volume in relation to the cross section of the different regions and partly by the supply of sediments (see Mulder & Louters, 1994). Sand import from the North Sea has been measured to be $1.10^6 \, \mathrm{m}^3 \, \mathrm{y}^{-1}$, while the total sediment deficit is calculated to the amount of $400–600 \; 10^6 \, \mathrm{m}^3$ (Kohsiek et al., 1987). A deficit of 40% of this amount exists in the mouth of the estuary, leading to bottom sediments characterized by sand fractions. The fine-grained sand, clay and silt fractions, deposited in areas with weak current velocities, probably originated from the erosion of tidal flats.

Temperature displayed a gradient in the Oosterschelde estuary. In the period 1988–1990 the temperature in the mouth of the estuary (measured at 4.5 m minus NAP) ranged between 5.5 °C and 20.5 °C. In the same period in the Zijpe (measured at 6.0 m minus NAP) temperature reached higher values during summer (21.0 °C) and lower values during winter (4.5 °C).

Discussion

The three sublittoral communities in the photic zone, found during the construction period (1985–1986), did not change during the first two years of the post-barrier period (1987–1988). Community Aa1 was common in all regions, while community Aa2 was found under exposed conditions. Measurements, using gypsum blocks during south-west winds, showed erosion values between 0.10 and 0.12 g h^{-1} for community Aa1 and 0.25 g h^{-1} for community Aa2. At deeper places community Aa3 occurred. When, after 1988, the species composition of these commu-

Fig. 11. Environmental parameters in the mouth of the estuary at the location Plompetoren, 4.5 m minus NAP, during the post-barrier period. The upper graph shows the potential sedimentation (g m^{-2} d^{-1}) and the input of Ophiotrix fragilis (shaded). Underneath the composition of the bottom sediments is given. Type V is dominated by the 0.3–0.15 mm fraction, type VI by 0.15–0.09 mm fractions, VII by 0.09–0.05 mm fractions and type VIII is dominated by fractions ≤0.05 mm. The next graph shows the total amount of water movement (EV. in g h^{-1}). The bottom graph shows the water temperature (T in °C).

nities changed, the algal composition was unaffected in variants Ab1, Ab2 and Ab4. On the other hand, some of the faunal components, especially sponges, hydrozoans and bryozoans, suddenly disappeared. Only the bivalve *Crassostrea gigas* became more abundant in the variants Ab3 and Ab4 and community Ac. In variant Ab3 algae and the in community Ac faunal components characteristic for the tideless Lake Grevelingen, became more abundant.

It is not likely that changes in current velocities are directly responsible for the changes in community Aa1, because this community was found in all regions on both north and south shores of the estuary. The strength of the water movement in the photic zone is strongly dependent on the geographical position relative to prevailing winds. It is unlikely as well, that these changes are caused by predation or grazing activities of the vagile associated fauna. Except in variant Ab1, the number of vagile organisms found in the Ab-variants was less then found within the Aa-communities.

The decrease of sessile species richness was probably caused by the increased amounts of fine-grained sediments on the substrate. Especially sponges (e.g., *Prosuberites epiphytum* and *Halichondria panicea*), hydrozoans and bryozoans seem to be sensitive to strong sedimentation. The increase of elements characteristic for the tideless Lake Grevelingen, in variant Ab3 and community Ac, is probably a direct result of decreasing current velocities. The tunicates (*Ciona intestinalis*, *Botryllus schlosseri* and *Ascidiella aspersa*) are characteristic for other sheltered environments in the North Sea (De Kluijver, 1991).

Although the most important changes in the communities in the aphotic zone took place after 1988, since 1985 (construction period) minor changes in species composition were already detectable. Initially, community M1 dominated the mouth of the estuary, H1 the Hammen, C1 and partly C2 the central part and Z1 the Zijpe. In all these communities the sponge *Halichondria panicea* was dominant. In the mouth of the estuary *Metridium senile* was a dominant species, while in

all communities eastwards *Diadumene cincta* was dominant. A dominant species in the Hammen community was the hydrozoan *Halecium halecinum*. The tunicate *Styela clava* was dominant in both central part and the Zijpe. The difference between C1 and C2 was a domination of *Electra pilosa* and *Lanice conchilega* in community C1, and barnacles besides *Haliclona oculata* in variant C2. The community Z1 in the Zijpe was further dominated by *Crassostrea gigas* and *Haliclona oculata*. The species richness (only those species were considered which were present in at least 67% of the stations of a community) in this initial stage was 28 species in the mouth of the estuary, 29 in the Hammen, 43 in the central part and 29 in the Zijpe (Tables 2, 3, 4 and 5). In 1986 variant M2 was sampled in the mouth of the estuary. This variant, only dominated by *Metridium senile*, showed a low species diversity (species richness = 18). In the Hammen and the central part, community C3 was found (species richness = 56), which consisted, besides the dominant species of H1, of many algae general in the photic zone. The penetration of algae into the aphotic zone is only possible in case of a lower vertical extinction coefficient in the Hammen and central part. In preceding studies it was found that algal dominance stopped at a depth of a 10% illumination level (De Kluijver, 1989, 1991).

In 1987, shortly after completion of the compartmentalization dams, variant M6 was found in the mouth of the estuary, mainly dominated by *Metridium senile* (species richness = 10). In the other regions the same communities were found as during the construction period. In 1988 a new community in the Hammen (H2) was sampled. Compared to H1 *Eudendrium* spec. also became dominant and the species richness increased to 37. In the central part, variant C4 was found dominated by *Diadumene cincta*, *Ciona intestinalis* and *Crassostrea gigas* (species richness = 37).

In 1989 in the mouth of the estuary 50% of the sampled stations belonged to the derivative variants of community M1, variant C4 was found eastwards, in the Hammen, the central part and in the Zijpe. In the Zijpe, it totally replaced the existing community Z1. In 1990 in the mouth of the estuary mainly variants were found, dominated by *Metridium senile* or *Diadumene cincta* or both. In the Hammen, mainly H2 was found, dominated by *Diadumene cincta*, *Halichondria panicea*, *Halecium halecinum* and *Eudendrium* spec.. In the central part and the Zijpe, mainly variant C4 was found, dominated by *Diadumene cincta*, *Ciona intestinalis* and *Crassostrea gigas*.

In the construction period the mouth of the estuary was an exposed environment with a residence time of 10 tides. In the post-barrier period, tidal currents decreased with 30% and the tidal residence increased 100% (Wetsteyn *et al.*, 1990). At shallow places (<15 m) the increased sedimentation fragmented the original M1 community into derivative variants. The bottom sediments in the new variants were dominated by fractions <0.15 mm (broken line in Fig. 12). At greater depths (>15 m), community M1 was still found, but variant M2 was abundant as well. The bottom sediments were dominated by the sand fraction (0.3–0.15 mm). Colonization experiments during 1988 and 1989 showed that settlement mainly occurred in the period from April till September. Species colonizing artificial substrates in the shallow aphotic zone were barnacles, the anthozoan *Sagartia troglodytes* and the bivalve *Mytilus edulis*. The dominant *Metridium senile*, only colonized new substrates by asexual reproduction and migration. Under these stress

conditions (high sedimentation and strong tidal currents) it is not likely that other species are able to establish themselves at shallow places in this region. Furthermore most of the hard substrates are covered with thick layers of sediment. Whether community M1 is able to maintain itself at greater depths, will depend on the particle size of deposited sediments and the availability of hard substrates at greater depths.

In the Hammen, the central part and the Zijpe, a heterogeneous community will remain for a longer period, dominated by *Diadumene cincta* and *Crassostrea gigas*, with local differences: hydrozoans and bryozoans especially in the western part and tunicates in the central and eastern part. Colonization experiments in the Zijpe already showed the settlement of the tunicates *Botryllus schlosseri*, *Ascidiella aspersa* and *Ciona intestinalis*. Compared to the initial communities in the Hammen and the Zijpe, this community has a higher species richness but remains the same in the central part. In the post-barrier period tidal currents are less severe and thus potentially more species are able to settle, but increased sedimentation, of especially the fine-grained silt and clay fractions, prevents many species from doing so. Species, like *Halichondria bowerbanki*, which can survive sedimentation (Vethaak *et al.*, 1982) will become more abundant in this community.

It is remarkable that changes in most commu-

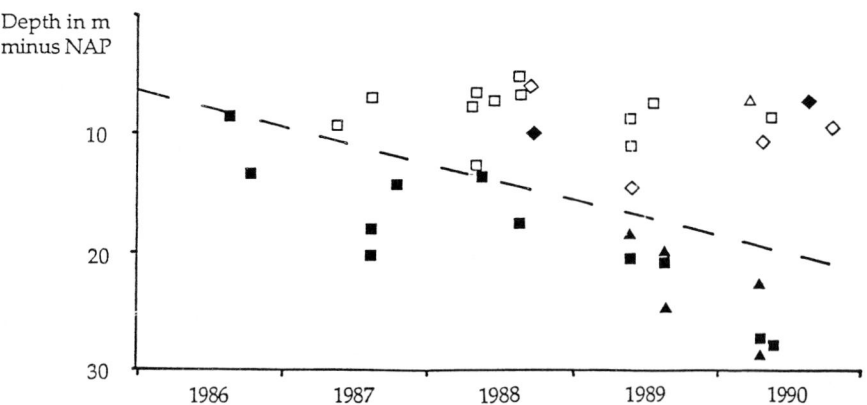

Fig. 12. Depth distribution of community M1 (squares), variant M2 (triangles) and the other M variants (diamonds) in the mouth of the estuary during the consecutive years. Black symbols indicate a dominance of the sand fractions (0.15–0.3 mm) in the bottom sediment. Open symbols indicate a dominance of fractions ≤0.15 mm.

nities after completion of the Delta Project occur with a considerable time-lag. This is also found in the total biomass (Leewis *et al.*, 1994). As already stated, three executive parameters were responsible for these changes: 1) the decrease of the vertical extinction coefficient, which enabled some algae to penetrate into the aphotic zone; 2) the decrease of tidal current velocities, which enabled other species, not adapted to extreme conditions, to settle in the mouth of the estuary, the Hammen and the Zijpe; 3) the increase of sedimentation, which prevented settlement and survival of species in certain regions.

Due to the execution of the Delta Project, parameter 3 was already effective in 1985, covering hard substrates in the mouth of the estuary, including the benthic communities, with a thick layer of sediment (see variant M3). Parameter 1 expressed itself in 1986, showing some algae to penetrate into the aphotic zone (see variant C3). In 1989 parameters 2 and 3 cooperated, causing a deterioration of sessile animals in both photic and aphotic zone and the introduction of new species into the eastern part of the Oosterschelde estuary. It took the communities two years to react on the most important changes (2 and 3). Partly this might be explained by the slow response of a biological system. Colonization is only possible during a short period of the year (April-September) and the existing community must deliver possibilities for the settlement of new species. On the other hand it is possible that it took some time for the redistribution of the fine-grained sediments to reach a lethal effect on the species which have disappeared.

The distribution of vagile animals does not indicate any regulation by these organisms. Together with the richness of sessile organisms in the photic zone and aphotic zone in the mouth of the estuary, the richness in vagile organisms decreased. In the central part species richness remained the same and in the Hammen and Zijpe richness increased. The increase in abundance of *Ophiotrix fragilis*, *Necora puber* and *Macropodia* spec. is caused by mild winters since 1987. In Fig. 12 the increase of *Ophiotrix fragilis* during summertimes is clearly visible. The increases in abundance of *Crepidula fornicata* and *Gobius niger* are caused by the decrease of tidal current velocities. These species were characteristic for the tideless Lake Grevelingen. The increase of the *Ophiotrix fragilis* population is not responsible for the changes in the benthic communities, because the new variants were already established before the brittlestar reached its maximum density. Furthermore, this kind of population explosions occurred frequently in the past, without lasting changes in benthic community structure. The role of vagile organisms in the regulation of the community structure is probably negligible in the southern Delta area, and it is likely that their distribution is also determined by the environmental parameters (L.A.M. Aerts, pers. comm.).

Conclusions

The sublittoral hard substrate communities of the southern Delta area are mainly governed by the environmental parameters. After the completion of the Delta Project in 1987, these environmental parameters changed and subsequently the communities changed dramatically. A number of executive parameters caused changes in the communities in the photic and aphotic zone:

- A decrease of tidal volume of 30%, leading to decreases of tidal current velocities, enabled species, not adapted to extreme tidal currents, to colonize other regions. Examples are some algae and tunicates which were characteristic for the tideless Lake Grevelingen, but became dominant in the post-barrier period in some communities in the Oosterschelde estuary.
- The decrease of tidal volume also disturbed the relation between the cross section of the tideways and tidal currents in the different regions, causing a sediment deficit within the Oosterschelde estuary. Increased sedimentation and a redistribution of the sediments caused a deterioration of some animals, while other organisms came in a more favourable position.
- A decrease of the vertical extinction coefficient

enabled some algae to penetrate into the aphotic zone.

After a time-lag of two years the seven initial communities changed rapidly. In the photic zone, some faunal elements disappeared, while elements characteristic for the tideless Lake Grevelingen, became dominant. The net result of the changes was a decrease in species richness. Correlated with increased sedimentation, the original community in the aphotic zone in the mouth of the estuary, fragmented into a number of derivated variants, all with a low species richness. In the three regions eastwards a new variant is in development. In the Hammen and in the Zijpe species richness increases, correlated with a severe decrease of tidal current velocities, while in the central part the species richness remains equal or decreases.

Acknowledgements

Dr R. W. M. van Soest and Dr J. A. Kaandorp of the Institute of Taxonomic Zoology (Amsterdam) and Miss N. V. P. Groot are thanked for reviewing the first drafts of the manuscript and Dr J. A. Kaandorp for his advice during cluster analysis; Drs L. A. M. Aerts and Drs M. A. E. Leloup for the co-operation during various parts of the field work; Mr J. M. Villerius and Miss M. Eland for the accommodation and support during the field work.

This study was supported by grants from the 'Stichting Bouwstenen voor de Dierenbescherming' (Amsterdam), the 'Beijerinck-Popping Fonds' (Amsterdam), the 'Van Tienhoven Studiefonds (Vereniging tot Behoud van Natuurmonumenten)' ('s-Graveland) and the 'Ministry of Transport and Public Works, Tidal Waters Division' (Den Haag). 'Aqua Diving' (Amsterdam), 'De Grevelingen' and 'De Kabbelaar' (Scharendijke) are thanked for their material support.

References

Aerts, L. A. M., in prep. Seasonal distribution and ecology of the Dutch nudibranchs (Gastropoda; Opistobranchia) in the southern Delta area, SW Netherlands.

De Kluijver, M. J., 1989. Sublittoral hard substrate communities of the southern Delta area, SW Netherlands. Bijdr. Dierk. 59: 141–158.

De Kluijver, M. J., 1991. Sublittoral hard substrate communities off Helgoland. Helgoländer Meeresunters. 45: 317–344.

Kaandorp, J. A., 1986. Rocky substrate communities of the infralittoral fringe of the Boulonnais coast, NW France: a quantitative survey. Mar. Biol. 92: 255–265.

Kohsiek, L. H. M., J. P. M. Mulder, T. Louters & F. Berben, 1987. De Oosterschelde naar een nieuw onderwaterlandschap. Nota DGW. AO 87.029, Rijkswaterstaat, 48 p.

Leewis, R. J. & H. W. Waardenburg, 1989. The flora and fauna of the sublittoral part of the artificial rocky shores in the south-west Netherlands. Progress in Underwater Science 14: 109–122.

Leewis, R. J. & H. W. Waardenburg, 1990. Flora and fauna of the sublittoral hard substrate in the Oosterschelde (The Netherlands) – interactions with the North Sea and the influence of a storm surge barrier. Hydrobiologia 195: 189–200.

Leewis, R. J., H. W. Waardenburg & M. W. M. van der Tol, 1994. Biomass and standing stock on sublittoral hard substrates in the Oosterschelde estuary (SW Netherlands). Hydrobiologia 282/283: 397–412.

Mulder, J. P. M. & T. Louters, 1994. Changes in basin geomorphology after implementation of the Oosterschelde estuary project. Hydrobiologia 282/283: 29–39.

Nienhuis, P. H. & A. C. Smaal, 1994. The Oosterschelde estuary, a case study of a changing ecosystem: an introduction. Hydrobiologia 282/283: 1–14.

Sokal, R. R. & C. D. Michener, 1958. A statistical method for evaluating systematic relationships. Univ. Kansas Sci. Bull. 38: 1409–1438.

Vethaak, A. D., R. J. A. Cronie & R. W. M. van Soest, 1982. Ecology and distribution of two sympatic, closely related sponge species, *Halichondria panicea* (Pallas, 1766) and *H. bowerbanki* Burton, 1930 (Porifera, Demospongiae), with remarks on their speciation. Bijdr. Dierk. 52: 82–102.

Weinberg, S., 1978. The minimal area problem in invertebrate communities of Mediterranean rocky substrata. Mar. Biol. 49: 33–40.

Wetsteyn, L. P. M. J., J. C. H. Peeters, R. N. M. Duin, F. Vegter & P. R. M. de Visscher, 1990. Phytoplankton primary production and nutrients in the Oosterschelde (The Netherlands) during the pre-barrier period 1980–1984. Hydrobiologia 195: 163–177.

Wishart, D., 1978. CLUSTAN user manual. Program Library Unit Edingburgh University: 1–175. (Edingburgh).

Hydrobiologia **282/283**: 281–298, 1994.
P. H. Nienhuis & A. C. Smaal (eds), The Oosterschelde Estuary.
© 1994 *Kluwer Academic Publishers. Printed in Belgium.*

Tidal reduction and its effects on intertidal hard-substrate communities in the Oosterschelde estuary

A. J. M. Meijer & H. W. Waardenburg
Bureau Waardenburg BV Consultants for Environment and Ecology, P.O. Box 365, 4100 AJ Culemborg, The Netherlands

Key words: hard substrates, hard-substrate communities, benthic algal communities, species-distribution, Oosterschelde estuary, storm-surge barrier, intertidal zone

Abstract

Developments of intertidal hard-substrate communities in the Oosterschelde estuary were examined in perpendicular transects between high-water line and low-water line in the period 1982–1992. Prior to the beginning of the Oosterschelde estuary works a typology of communities was established and an overall survey of the estuary was carried out. The communities contain flora (algae) as well as fauna. Due to asphalting of dikes in 1986, much of the surface of several communities has been destroyed. The originally well developed communities with large species-richness have not returned. The small reduction in tidal amplitude due to the construction of the storm-surge barrier had a minor effect on the zonation of communities. In the upper part of the intertidal zone the boundaries of the communities moved 0.5–1.0 m downward in the transects along the dike-slopes. At an average inclination of 18° this means a vertical shift of about 15–30 cm. This reflects the reduction of the tidal amplitude: the high-water line shifted *ca* 22 cm downward. In a number of places sedimentation has caused a reduction in the number of smaller seaweed species in the lower eulittoral zone. At monitoring locations presence of the original communities is rather unchanged. Rare species like *Pelvetia canaliculata*, *Actinia equina* and *Gelidium pusillum* have been able to maintain quite successfully.

Introduction

Hard substrates on sea walls, civil-engineering constructions, dikes and stone deposits bordering the Oosterschelde estuary can be regarded as artificial rocky shores. Due to gradients in environmental circumstances in the intertidal area, a zonation of communities has developed. These communities are composed of different types of plants and animals belonging to many taxonomic groups. The position of the zones (belts) is especially determined by the tides and the type of substratum. The degree of sedimentation is one of the factors determining the species composition (Den Hartog, 1959; Nienhuis, 1980).

Our research started in 1982. The main objective was to establish the original situation before completion of the storm-surge barrier and auxiliary dams. Possible changes in the distribution and structure of the communities could then be determined afterwards. After making a classification of the intertidal communities of hard substrates an overall survey was made in the period 1983–1985, covering most of the total area. In the period 1986–1987 attention was given to developments during completion of the storm-surge barrier. At eight locations permanent transects were surveyed during a monitoring research until 1991, which followed the developments after completion of the storm-surge barrier. Recently, also attention has been paid to the composition of biomass inside the transects. Due to many changes to revetments of dikes in 1986, repeated research was carried out in the period 1988–1992 dealing with the recolonisation of a number of locations. This paper mainly deals with a description of the intertidal communities in the pre-barrier and post-barrier period. Changes in the subtidal communities have been described by De Kluijver & Leewis (1994).

Study area

Hard substratum in the intertidal zone of the Oosterschelde estuary

Along the Oosterschelde estuary, sea walls and dikes are enforced with stony materials, originally mainly limestone and basalt columns. These materials protect the dike-body against waves and the erosive effects of the water. During this century, several modern materials as concrete blocks, asphalt, stone-asphalt, rubble-stone, mine-stone, phosphorus-slag copper-slag and furnace-slag were used. The presence of hard materials offers opportunities for settlement of communities usually found on rocky coasts. Many organisms can only occur because the use of hard materials in the past by man. The Oosterschelde estuary provides the larger part of hard substratum in the intertidal zone along the Dutch coast. This emphasises the value of this area. Figure 1 shows the area of hard substratum on dikes of the Oosterschelde estuary. Apart from many seaweeds (Chlorophyta, Phaeophyta as well as Rhodophyta) various animal taxa occur: Porifera, Anthozoa, Hydrozoa, Crustacea, Gastropoda, Bivalvia, Bryozoa, Echinodermata and Tunicata. In the intertidal zone communities have been distinguished. A community is regarded as an organized complex with a typical floristic or faunistic

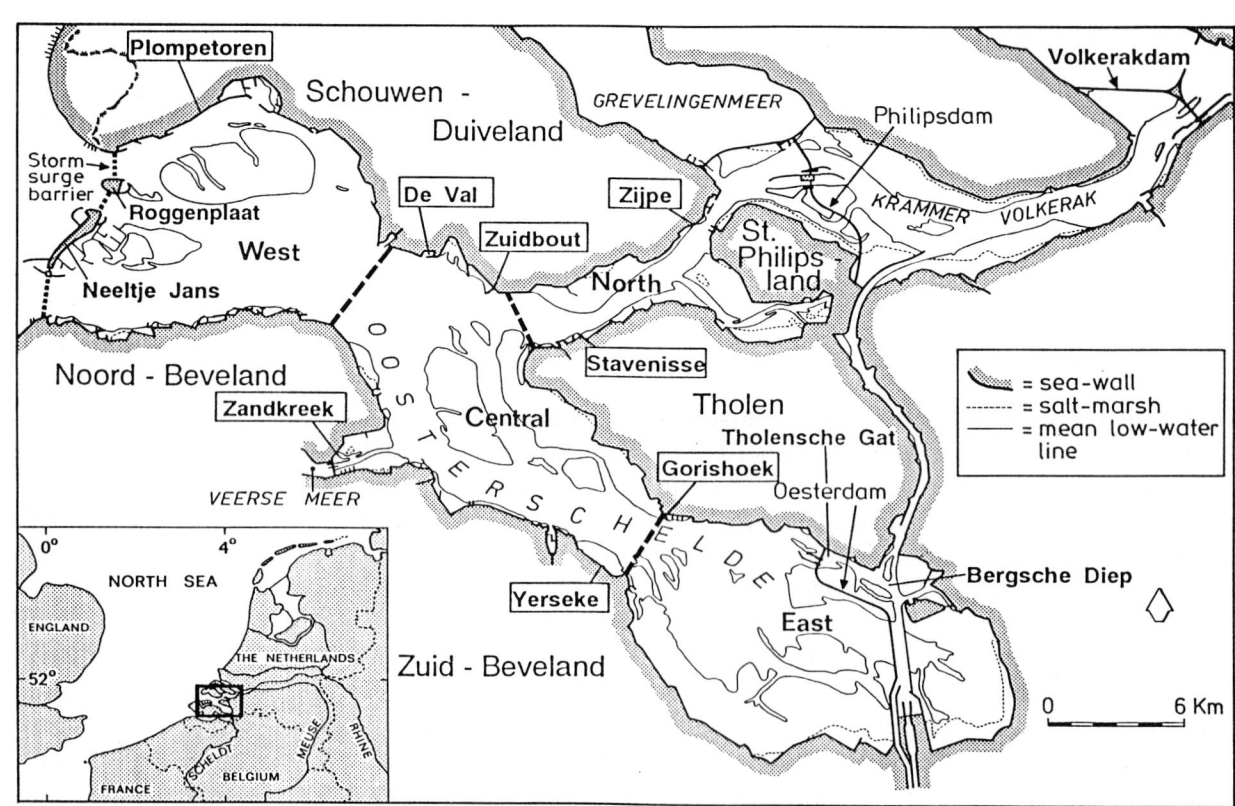

Fig. 1. Distribution of hard substratum in the intertidal zone of the Oosterschelde estuary and transect-location (in boxes).

Table 1. Length and surface data of hard substrates and hard substrate communities in the intertidal zone of the Oosterschelde estuary (after Meijer & Van Beek, 1988; with addition of new data).

Total length coastline around Oosterschelde estuary	230 km
Total length with intertidal hard substrate on dike-slopes	200 km
Total length with hard-substrate in lower eulittoral zone	150 km
Total surface hard substrates on dike-slopes	250 hectares
Total surface hard substrates in lower eulittoral zone	90 hectares
Total surface communities on dike-slopes	200 hectares
Total surface communities in lower eulittoral zone	90 hectares

composition and morphological structure (Shimwell, 1971). The communities often form belts.

Table 1 presents information on hard substratum surface in the intertidal zone of the Oosterschelde estuary.

Profile of a sea-wall along the Oosterschelde estuary

The body of an Oosterschelde-estuary dike is made of sand and clay (Fig. 2). The top (1) is often covered with grass, on which sheep graze. The zone below the top is tightly set with stones of a more or less regular form (2), such as basalt columns or concrete blocks, sometimes covered with asphalt. Around the low-water line there is a slightly sloping zone, usually irregularly covered with rubble-stone and stone debris (3). Below the low-water line, the dike-slope is covered with fascines, kept in place by blocks of natural stone and sometimes furnace-slag of different sizes usually mixed with silt (4). Zone 2 and zone 3 are dealt with in this paper. These zones correspond with the supra littoral zone and the eulittoral zone according to Den Hartog (1959).

Zonation of communities influenced by environmental factors

In a perpendicular transect along a sea-wall the submergence time increases downward, implying

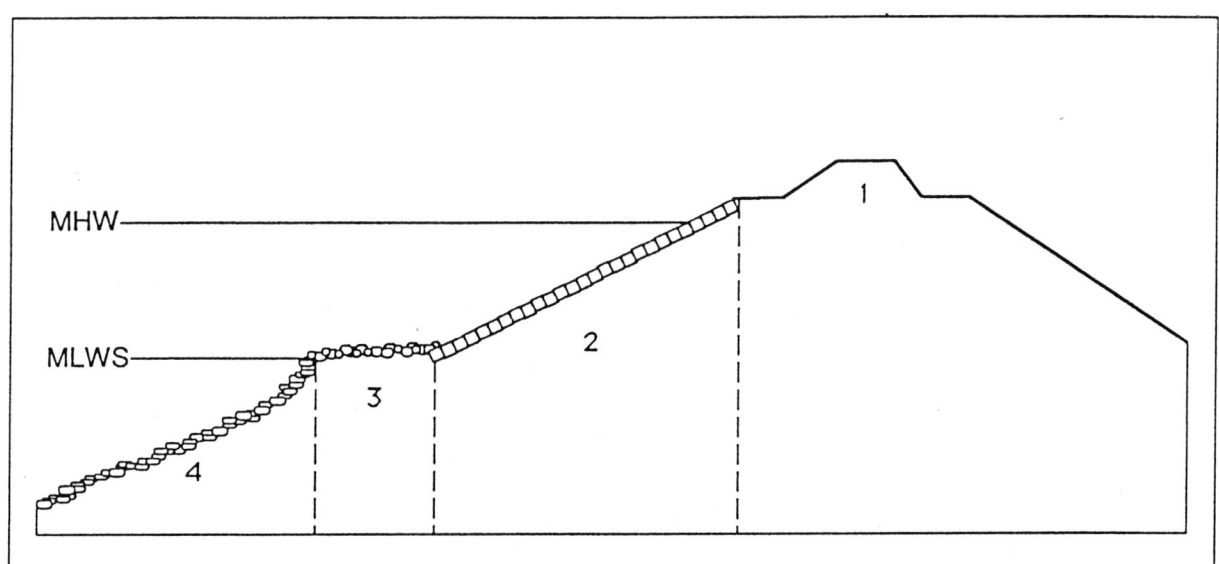

Fig. 2. Transect of an Oosterschelde estuary dike. MHW = mean high water. MLWS = mean low water at springtide. 1 = top of dike overgrown with grass. 2 = lower supralittoral and upper and middle eulittoral zone. 3 = lower eulittoral zone. 4 = subtidal zone.

vertical gradients in abiotic circumstances such as the availability of water, influence of light, desiccation, temperature, wave action and salinity. Due to this phenomenon a zonation pattern of clearly defined horizontal belts (Fig. 3) is formed, containing littoral organisms, which occurs almost universally on all rocky shores around the world (Mathieson & Nienhuis, 1991).

The development of the zonation in the Oos-

terschelde estuary depends on inclination of the sea-wall, exposure to wind, waves, current speed, sediment contents of the water, degree of sedimentation and several properties of the hard substratum (see e.g. Den Hartog, 1959; Nienhuis, 1969: Meijer & Van Beek, 1988; Leewis et al., 1989). The type of substratum is usually the only factor that can be changed by man.

Table 2. Synoptic table (after Meijer & Van Beek, 1988).

Community number (c.n.)		1	2	3	4	5	6	7	8	9	10	11	12	13	Total number of samples: 4,811	Total number as percentage of 4,811 samples
Number of quadrat samples		1,034	354	39	160	731	245	151	481	177	492	657	80	210		
c.n.	Differential species															
1	Lichens	100	19	3	7	1	0.4								1,120	23
2	Entophysalis deusta	1	100	18	21	1	3	1	0.2						424	9
3	Pelvetia canaliculata		1	100	1			4			0.4				52	1
3	Catenella caespitosa			3		0.3					1				6	0.1
4	Blidingia spec.	0.4	36	18	100	2	7	10			0.2				343	7
6	Enteromorpha spp.	0.1	2	10	8	5	100	60	45	69	42	9	14	11	1,040	22
6	Porphyra spec.		0.3		6	2	32	13	23	43	12	9	8	0.5	434	9
6	Porphyra umbilicalis		0.3		7	1	20	11	2						96	2
6	Ulothrix/Urospora				1		5	3		2					23	0.5
7	Fucus spiralis		2	15	13	7	24	100	5			3	4		1	360	7
7	Actinia equina		0.3					3							6	0.1
8	Fucus vesiculosus				1	11	18	17	100	67	54	39	48	37	1,391	29
9	Fucus serratus					1	11		6	100	25	3	34		406	8
9	Chondrus crispus					0.4			9	32	12	12	33	5	278	6
9	Ectocarpaceae					0.1	8		9	21	12	1	3		175	4
9	Cladophora rupestris						6	2	6	21	13	4			173	4
9	Actinaria/Anthozoa						1	3	4	26	8	2	11		135	3
9	Cladophora spp.						2	1	1	11	5	1			67	1
9	Chaetomorpha aerea									4					7	0.1
9	Bryopsis hypnoides									2		0.3			5	0.1
10	Ascophyllum nodosum		0.3	8		7		6	24	37	100	15	9	2	843	18
10	Hildenbrandia rubra					1		11	13	10	29	7			293	6
10	Ralfsia spec.					1	0.4	5	8	11	24	8		3	256	5
10	Rhodochorton purpureum							1		2	11	2			72	1
10	Gelidium pusillum					0.4			0.2		2				14	0.3
10	Diadumene cincta										2				11	0.2
10	Codium spec.										2				10	0.2
5/11	Elminius modestus	0.1	8	69	12	83	64	57	85	75	84	84	90	90	2,695	56
5/11	Littorina litorea	0.4	5	15	11	72	16	54	86	79	83	100	99	100	2,607	54
5/11	Balanus balanoides		0.3	3		35	19	8	57	59	49	75	65	82	1,650	34
5/11	Littorina saxatilis	1	8	82	19	70	20	52	26	10	29	36	13	20	1,319	27
5/11	Cirripedia		1		4	3		25	11	11	10	13	10	10	302	6
12	Crassostrea gigas			3		5	1		52	63	29	69	100	83	1,255	26
12	Gigartina stellata					1		1	16	16	17	23	38	14	406	8
12	Polysiphonia nigrescens								3	1		2	9		38	1
12	Crepidula fornicata										3		16		28	1
12	Callithamnion roseum									1	3		10		24	0.5
12	Styela clava					0.4							10		9	0.2
12	Haliclona panicea					0.4							9		8	0.2
12	Nereis spec.											0.5	3		5	0.1
13	Mytilus edulis					20	12	25	60	82	50	93	100	100	1,799	37
	Non differential species															
8	Nucella (Thais) lapillus								1			0.5			7	0.1
9	Placophora									3	5			3	24	0.5
9	Asterias rubens									3		1	1		12	0.2
9	Callithamnion spp.						1			2	0	0.2	1		10	0.2
8/9	Electra spp.								3	3	0.2	1			26	1
9/10	Littorina litoralis					0.3		15	29	58	47	4	20		540	11
9/10	Phymatolithon lenomandii							1	6	37	30	4	23	3	297	6
9/10	Dynamena pumila							1	3	26	23	1	1		186	4

Table 2. (Continued)

Community number (c.n.)	1	2	3	4	5	6	7	8	9	10	11	12	13	Total number of samples: 4,811	Total number as percentage of 4,811 samples
Number of quadrat samples	1,034	354	39	160	731	245	151	481	177	492	657	80	210		
9/10 *Polysiphonia* spp.									12	12	0.2	9		86	2
9/10 Polychaeta										3				16	0.3
9/10 *Acrochaetium/ Goniotrichum*								0.2	1	2	0.2	1		15	0.3
9/10 *Chaetomorpha melagonium*						1			1	2				11	0.2
9/11 *Petalonia fascia*									1		1			7	0.1
9/12 *Ulva* spec.					1	11	9	9	25	11	4	34	2	244	5
9/12 *Ceramium deslongchampsii*					0.1	6	1	8	24	11	5	33	13	239	5
9/12 *Ceramium rubrum*								1	12	3	0.5	19		62	1
n.d. *Pseudendoclonium marinum*				2	15	2	23	26	3	40	26	26	8	690	14
n.d. *Lipura maritima*	0.3	7	31	13	12	6	33	16	5	29	5	16	7	505	10
n.d. Amphipoda		1	3	2	0.3	4	35	31	25	43	4	1		504	10
n.d. *Carcinus maenas*				1	0.1	13	12	25	37	35	5	41		476	10
n.d. *Hydrobia ulvae*		2		2	2	1	5	3		3	2			77	2
n.d. Isopoda	0.1	0.3		1				3	2		0.5			24	0.5
n.d. *Sagartia troglodytes*								2	1		0.4	1	1	17	0.4
n.d. *Scytosiphon lomentaria*						2		1						11	0.2
Total number of species found	10	18	15	20	30	32	34	44	50	48	45	38	20	74	

Species occurring in < 5 samples: *Prasiola stipitata, Caprella linearis, Bryopsis plumosa, Scypha* spec., *Cystoclonium purpureum, Pycnogonum littorale, Tubularia* spec., *Diatomeae, Dumontia contorta, Hypoglossum woodwardii.*

The numbers in the centre of this table indicate the frequency, expressed as percentages, in the total number of samples per community (on top of table).
For example: Lichens were found in 7% of the samples which have been characterized as community nr. 4 (Blidingia community).
The differentiating species per community are printed bold and underlined. Explanation of community numbers:

1 Lichens community
2 *Entophysalis* community
3 *Pelvetia* community
4 *Blidingia* community
5 Cirripedia/Littorinidae community
6 *Enteromorpha* community
7 *Fucus spiralis* community

8 *Fucus vesiculosus* community
9 *Fucus serratus* community
10 *Ascophyllum* community
11 Cirripedia/Littorinidae/*Crassostrea/Mytilus* community
12 *Crassostrea* community
13 *Mytilus* community

Material and methods

During low tide transect surveys were made by using quadrats of 0.5 m × 0.5 m laid down contiguously along the slope of the sea-wall between the upper side of the stony slopes and the low-water line. The abundance and cover of species was determined in each quadrat. In this way an accurate view of the zonation of communities in the transects is obtained, but also of the vertical distribution of the separate types of organisms. Sessile and semi-mobile organisms have been surveyed according to the Braun-Blanquet (1964) cover-abundance scale, adapted by Barkman *et al.* (1964) and ourselves (Meijer & Van Beek, 1988). Mobile species (such as *Carcinus maenas*) have been recorded as present or absent.

Results survey 1983-1985 (pre-barrier period)

Typology of the communities

In the period 1983–1985 155 transects have been investigated on Schouwen-Duiveland, St. Philipsland, Tholen, Zuid-Beveland, Noord-Beveland and Neeltje Jans. In total 5468 quadrat samples were made. Marine organisms were found in 4811 samples. The typology of the communities has been based on these 4811 samples. They have been ordered according to the Braun-Blanquet method (as described by Den Held & Den Held, 1973). Thirteen groups of quadrat samples were distinguished, which could be considered representative of separate communities with specific dominant species ($> 50\%$ cover). Table 2 gives a summary: the synoptic table. A number of sub-types has been distinguished within the communities (e.g. poor and rich in species respectively). These sub-types will not be dealt with in this paper. With regard to seaweeds, a large number of the communities distinguished corresponds to those identified by Den Hartog (1959) and Nienhuis (1980) respectively, but the communities in this study are determined by seaweeds as well as fauna, however.

The communities are described below. The numbers refer to those in the top of Table 2.

Lichens community
The Lichens community is especially found in the supra littoral zone above the high-water line. The lower limit is determined by the 'splash zone'. Lichens were normally found on all hard substratum above the high water line. Hardly any other marine organisms are found in this Lichens community. The most abundant Lichens are *Caloplaca* spp. and *Xanthoria* spp. These yellow Lichens give the belt its characteristic colour. Other species are *Verrucaria* spp. and *Lecanora* spp., which are black and grey respectively. The number of species per quadrat sample shows almost no variation (Table 4).

Entophysalis *community*
This community is found in the transition of the supra littoral zone and eulittoral zone and lies during high tide in the 'splash zone'. This community, poor in species, consists mainly of *Entophysalis deusta*, a Cyanobacteria community distinguished by Nienhuis (1980). The community can easily be identified as a black belt between the yellow Lichens belt and the green seaweed belt of *Blidingia* spp. and/or *Enteromorpha* spp. at the top of the eulittoral zone. The number of species per quadrat sample shows almost no variation (Table 4).

Pelvetia *community*
The *Pelvetia* community was only found on a small number of dike-slopes, almost exclusively in the eastern half of the Oosterschelde estuary area, in Zuid-Beveland particularly. The community is characterised by the rare brown seaweed *Pelvetia canaliculata* and it is found in a narrow belt just above the high-water line. Only in a few places *Pelvetia canaliculata* forms a closed vegetation. Usually, there are scattered groups of individuals. The number of species per quadrat sample shows slight variation (Table 4).

Blidingia *community*
On many places near the high water line, a narrow belt dominated by the green seaweed *Blidingia* spp. is found. This seaweed belongs together with *Enteromorpha* spp. to the pioneering species colonising a bare substratum. Monitoring research indicated that this belt is continually found in many places. The circumstances in this belt exclude settlement of seaweeds such as Fucoids apparently. The *Blidingia* community forms a narrow green belt between the *Entophysalis* community and the *Fucus spiralis* community. When Fucoids occur only in low density, the bottom part of the belt merges into the Cirripedia/Littorinidae community. The number of species per quadrat sample shows some variation (Table 4).

Cirripedia/Littorinidae community
This species-poor community is mainly found in the upper part of the eulittoral zone. The Cirripedia are represented by *Elminius modestus* and *Balanus balanoides*. The Littorinidae are *Littorina littorea* and *L. saxatilis*. The community is found on a large number of different types of substratum. The numbers of species per quadrat sample show variation (Table 4), depending on the presence of algae. Although the total number of species found can amount to 30, an average of only 4 species is found per quadrat sample (Table 4).

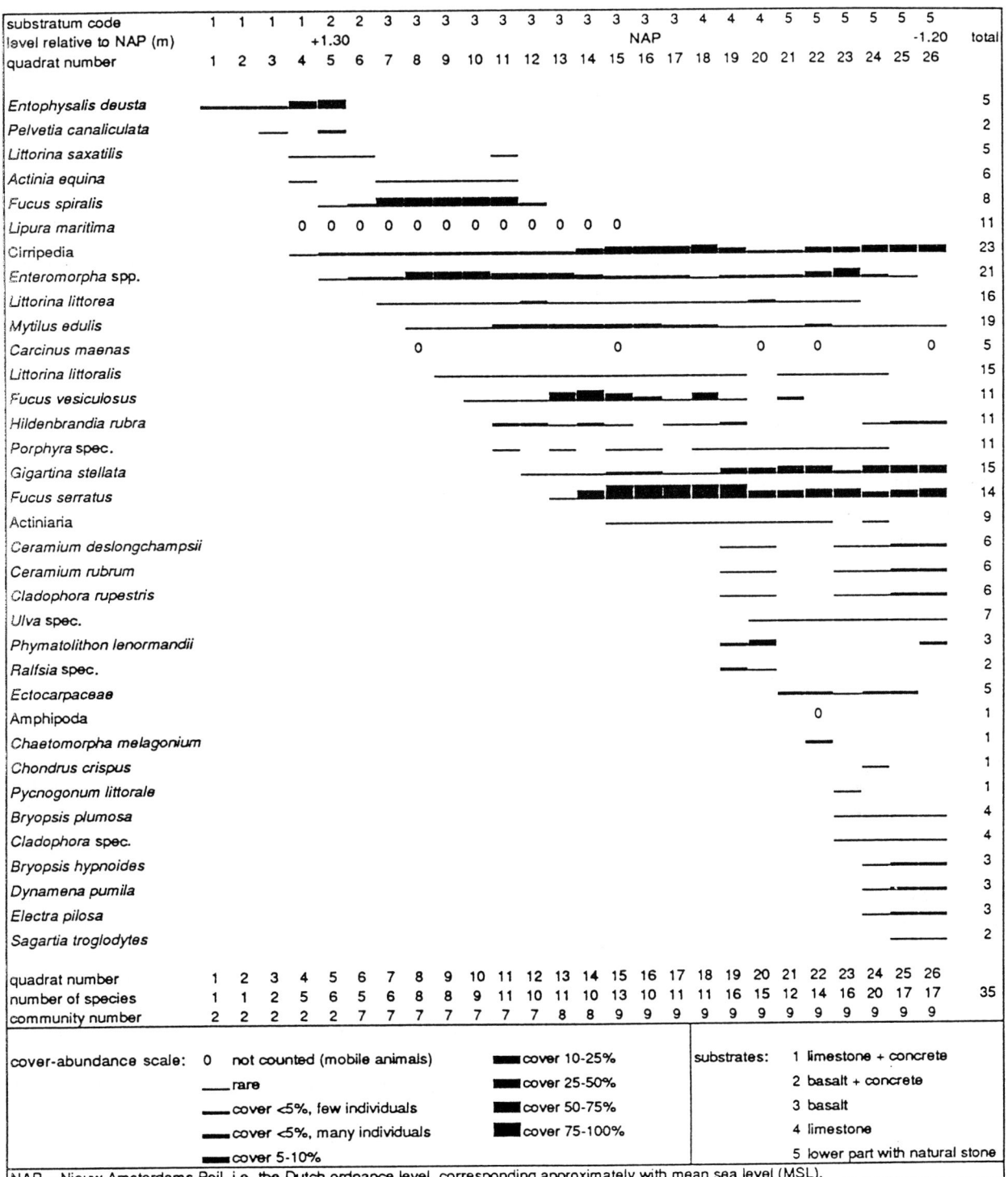

Fig. 3. The vertical distribution of a number of dominant and accompanying species in the intertidal zone at Plompetoren, summer 1982. After Meijer & Van Beek (1988).

Enteromorpha *community*

The green seaweed *Enteromorpha* spp. may be considered as pioneering species. When bare substratum is available, often a vegetation develops of almost exclusively *Enteromorpha*. In many places this is only temporarily, as a result of damage by storm and frost (drift ice) destroying the existing brown seaweed vegetations. Later on, *Enteromorpha* spp. is usually superseded by the originally present brown seaweeds. Under circumstances such as strong wave action and asphalt-substrate *Enteromorpha* spp. seem to maintain remarkably longer. Within the community *Enteromorpha compressa* and *Enteromorpha intestinalis* are abundant. The number of species per quadrat shows variation. Although the total number of species found amounts to 32, an average of only 4 species is found per quadrat sample (Table 4).

Fucus spiralis *community*

In many places, just below the high-water line, a rather narrow belt is found, dominated by *Fucus spiralis*. This community lies between the *Blidingia* and the *Fucus vesiculosus* or the *Ascophyllum* community. The cover of *Fucus spiralis* varies strongly from place to place. The belt seems very sensitive to strong wave action (storm) and frost (drift ice). In general *Fucus spiralis* forms a narrow belt but in places where the slope is less steep, the belt can be remarkably wider. The species richness in the community is related to the cover by *Fucus spiralis*. When density increases, the vegetation offers a suitable habitat for more organisms. Although the total number of species found amounts to 34, an average of only 6 species is found per quadrat sample (Table 4).

Fucus vesiculosus *community*

In a large part of the vertical transect the *Fucus vesiculosus* community can be found, usually between the *Fucus spiralis* community and the low-water line. The number of species increases towards the low-water line but can vary strongly (Table 4). At the upper limit species of the *Fucus spiralis* community such as *Littorina saxatilis* are found. Further down the transect *Ceramium deslongchampsii, C. rubrum, Cladophora rupestris, Chondrus crispus, Gigartina stellata, Phymatolithon lenormandii* and *Polysiphonia nigrescens* are found. Within a quadrat sample several layers can be distinguished. A crusty layer is formed by *Phymatolithon lenormandii, Hildenbrandia rubra* and/or *Ralfsia* spec. Cirripedia, *Mytilus edulis* and *Crassostrea gigas* can also be found in this layer. A bushy layer is formed mainly by *Gigartina stellata* and *Chondrus crispus*. The uppermost layer is formed by *Fucus vesiculosus* and possibly other Fucoids (with a low cover). The total number of species found amounts to 44, on average only 8 species are found per quadrat sample (Table 4).

Fucus serratus *community*

This community, characterised by the dominance of *Fucus serratus*, is found in the lower part of the eulittoral zone. In general, the community can be found between the *Fucus vesiculosus* community or *Ascophyllum* community and the low-water line at the bottom side. As is in the *Fucus vesiculosus* community, layers can be distinguished. A crusty layer is formed by *Phymatolithon lenormandii, Hildenbrandia rubra* and/or *Ralfsia spec.* Cirripedia. *Mytilus edulis* and *Crassostrea gigas* can also be included. A bushy layer is especially formed by *Gigartina stellata* and/or *Chondrus crispus*. The top layer is formed by *Fucus serratus* sometimes in combination with *Fucus vesiculosus*. The highest number of species was found in this community. However a large number of species has a low frequency (Table 2). These are species which are specifically found along the low-water line and which belong in fact to the upper part of the sublittoral zone. The number of species per quadrat sample increases towards the low-water line. The total number of species found is 50, on average 13 species per quadrat sample (Table 4). It can be concluded that the *Fucus serratus* community is one of the most species-rich communities.

Ascophyllum *community*

A wide belt of *Ascophyllum nodosum* is found in moderately exposed areas. In the widest belts this community is found between *Fucus spiralis* community near the high-water line and the lower eulittoral zone. In a number of places the community is found in a narrower belt: for instance between the *Fucus vesiculosus* community in the upper part of the eulittoral zone and the *Fucus serratus* community on the lower eulittoral zone. Most authors consider *Ascophyllum nodosum* and *Fucus vesiculosus* as replacing one another: on more quiet places the former can be found. When more waves and/or current occur the latter can be found (Den Hartog, 1959; Nienhuis, 1980; Meijer & Van Beek, 1988). In this community too, layers can be distinguished. The crusty layer is formed by *Phymatolithon lenormandii*, *Hildenbrandia rubra*, *Pseudendoclonium marinum*, *Mytilus edulis* and/or *Ralfsia* spec., together with Cirripedia, *Mytilus edulis* and/or *Crassostrea gigas*. The bushy layer is formed by *Gigartina stellata* and/or *Chondrus crispus*. The upper layer is formed by *Ascophyllum nodosum* and possibly other Fucoids. The total number of species found amounts to 48, an average of 13 species is found per quadrat sample (Table 4). A large number of species has a very low frequency. It concerns particularly the species found along the low-water line. The numbers of species per quadrat sample vary strongly (Table 4). In general, the number increases approaching the low-water line. It can be concluded that this community and the Fucus serratus community are the most species-rich communities. Both communities can be considered the final stage of succession.

Cirripedia/Littorinidae/Crassostrea/Mytilus *community*

Some of the lower parts of the dike-slopes and some of the lower eulittoral zones have no seaweed vegetations. Here the Cirripedia/Littorinidae community seems present, but *Mytilus edulis* and often *Crassostrea gigas* determine the aspect. A new community can be distinguished: the Cirripedia/Littorinidae/Crassostrea/Mytilus community. The four species mentioned above are found in large numbers, but neither *Mytilus edulis* nor *Crassostrea gigas* reaches a cover higher than 25%. A large number of other species can be found in this community (Tables 2 and 4). None of them reaches a high frequency (Table 2). The number of species per quadrat sample increases approaching the low-water line. The number of species per quadrat sample varies, with an average of 7 species (Table 4).

Crassostrea *community*

In this community *Crassostrea gigas* dominates (>50% cover). The community is often found on the lower eulittoral zone, near the low-water line. The distribution is probably restricted to the less exposed places (ports, unexposed dike-slopes). *Crassostrea gigas* is accompanied by *Mytilus edulis*, *Littorina littorea*, *Elminius modestus* and *Balanus balanoides*. *Fucus vesiculosus* is also frequently found, with a sparse cover though. Other species are remarkably less frequently found, with exception of *Ceramium deslongchampsii*, *C. rubrum* and *Ulva* spec. (Table 2). The numbers of species per quadrat sample vary strongly (Table 4).

Mytilus *community*

In several places, at the foot of the dike-slope, or in the lower eulittoral zone, a high cover of *Mytilus edulis* is found. The aspect deviates in such a way that a separate community can be distinguished. *Littorina littorea*, *Elminius modestus*, *Balanus balanoides* and *Crassostrea gigas* are accompanying species. The total number of found species is 20 (Table 4). This indicates a relatively species-poor community compared to the previously described communities. The average number of species per quadrat sample is 7 (Table 4).

Survey of the communities in the Oosterschelde estuary

Applying our typology, communities have been surveyed on most sea-walls in the area. Using the results, detailed surface cover calculations have been made, as shown in Table 3.

Table 3. Surface data of hard substrate communities in the intertidal zone of the Oosterschelde estuary (after Meijer & Van Beek, 1988).

The investigated area consists of Schouwen-Duiveland, Sint-Philipsland, Tholen, Zuid-Beveland and Noord-Beveland.

a. Dike-slopes

Total investigated coast-length	169 km
Total length with intertidal hard substrate on dike-slopes	137 km
Total surface of hard substrates on dike-slopes	166 hectares
Total surface of communities on dike-slopes (survey 1983–1985)	139 hectares
Total surface of species-rich communities (survey 1983–1985)	30 hectares
Total surface of dike-slopes reinforced in 1986	10 hectares

Community	Total surface per community 1983/1985 (hectares)	Percentage of total surface 1983/1985 (%)	Surface affected in 1986 (hectares)	Percentage of community surface (%)
1. Lichens	39.0	28	0.7	2
2. *Entophysalis*	16.4	12	0.5	3
3. *Pelvetia*	1.4	1	0.3	21
4. *Blidingia*	4.1	3	0.4	10
5. Cirripedia/Littorinidae	26.6	19	1.6	6
6. *Enteromorpha*	4.7	3	0.1	2
7. *Fucus spiralis*	5.1	4	0.8	16
8. *Fucus vesiculosus*	8.9	6	0.5	6
9. *Fucus serratus*	3.0	2	0.9	30
10. *Ascophyllum*	15.8	11	3.4	22
11. *Cirr/Litt/Crass/Myt*	12.5	9	1.1	9
12. *Crassostrea*	0.0	0	0.0	0
13. *Mytilus*	1.2	1	0.01	1
Total surface communities	138.7 hectares	100%	10.3 hectares	7%

b. Lower part of eulittoral zone

Total length lower part of eulittoral zone	104 km
Total surface lower part of eulittoral zone	48 hectares
Total surface communities on lower part of eulittoral zone	48 hectares

Community	Total surface per community 1983/1985 (hectares)	Percentage of total surface 1983/1985 (%)
5. Cirripedia/Littorinidae	4.5	10
6. *Enteromorpha*	1.3	3
8. *Fucus vesiculosus*	3.9	8
9. *Fucus serratus*	3.1	6
10. *Ascophyllum*	3.2	7
11. Cirr/Litt/Crass/Myt	21.5	45
12. *Crassostrea*	2.6	5
13. *Mytilus*	7.6	16
Total surface communities	47.7 hectares	100%

Results surveys during building period and post-barrier period

Consequences of extra reduction in tidal amplitude during 1986 and 1987

The building of the Oesterdam and Philipsdam caused additional reductions in tidal amplitude in the period 1986–1987 (Nienhuis & Smaal, 1994). An increase in mortality of algae (a.o. *Pelvetia canaliculata* and *Fucus spiralis*), Cirripedia and Littorinidae was expected in the upper intertidal zone, which would stay dry during high tide (Waardenburg & Meijer, 1985). To assess this mortality certain locations were surveyed several times during 1986–1987. The expected mortality was indeed observed (Waardenburg & Meijer, 1986, 1987). In summer and autumn of 1987 natural recovery took place.

During the closure of the Tholensche Gat in October 1986, the storm-surge barrier was closed temporarily. During this period of maximum tide reduction a storm caused strong wave attack, which destroyed the larger algae (Fucoids) at many places. This damage was more severe than effects of the temporary closure itself.

During the closure of the Krammer in April 1987 only small changes in species composition or cover were found. In the upper part of the eulittoral zone tidal reduction caused dehydration of algae. In certain areas Cirripedia, *Mytilus edulis, Fucus spiralis, Blidingia* spp. and *Actinia equina* died. In the days around the closure at some places a considerable deposition of silt was observed at the lower part of the dike-slope and in the lower eulittoral zone (Zijpe, Zuidbout, Plompetoren). This deposition varied from several millimetres or centimetres on the dike-slope to 20–50 centimetres between stones at the lower eulittoral zone. Particularly the cover of smaller algae in the lower part of the intertidal zone was affected by this silt deposition. One week after re-opening of the storm-surge barrier this material had disappeared again in many areas.

Changes as a result of the gradual tidal reduction caused by the storm-surge barrier

The completion of the storm-surge barrier and the auxiliary dams caused changes in the abiotic environment, which directly and indirectly affected the communities on hard substratum. The following changes were expected (Waardenburg & Meijer, 1985). (1) The reduction in current-velocity in certain areas might result in additional deposits of silt gradually covering the hard-substrate communities. (2) Changes in current-velocity, visibility, and sediment characteristics will, in general, be beneficial for certain types of organisms, but harmful to others. This might change the biomass distribution. (3) Tidal reduction might affect the zonation of communities between the levels of high and low-water line. (4) The distribution of communities along the Oosterschelde estuary might also change. (5) Management of the storm-surge barrier possibly influences the adaptation of the ecosystem to the changes in environment.

To establish developments surveys of transects were made several times a year in the period 1985–1988 in seven locations, in some already since 1982. As several locations were covered with asphalt due to the dike reinforcements in 1986 some new locations had to be chosen. For an undisturbed sequence only the results of the locations Plompetoren, Zuidbout, Zijpe, Stavenisse en Gorishoek (Fig. 1) could be used. Since 1989 some new locations were studied too: De Val, Yerseke and Zandkreek.

The scope of this paper does not allow us to mention all the detailed developments at all locations. Only the most remarkable observations will be discussed. General conclusions drawn from this monitoring research programme are found in the last paragraph.

The gradual decrease in tidal amplitude might result in gradual shifts of upper and lower borders of communities. To distinguish these shifts the quadrat samples of zones where steady species as Lichens and Fucoids were dominant, were presented in time series (as shown in Table 4). Pioneer species such as *Blidingia* spp., *Enteromorpha*

Table 4. Numbers of species found per quadrat sample (0.5 × 0.5 m) (after Meijer & Van Beek, 1988).

Investigated area: dike-slopes and lower parts of the eulittoral zone on Schouwen-Duiveland, Sint-Philipsland, Tholen, Zuid-Beveland and Noord-Beveland.

Community	Minimum number of species found per sample	Maximum number of species found per sample	Average number of species found in all samples within community	Total number of species found in all samples within community
1. Lichens	1	3	1	10
2. *Entophysalis*	1	6	2	18
3. *Pelvetia*	2	6	3	15
4. *Blidingia*	1	10	2	20
5. Cirripedia/Littorinidae	1	11	4	30
6. *Enteromorpha*	1	9	4	32
7. *Fucus spiralis*	4	12	6	34
8. *Fucus vesiculosus*	5	18	8	44
9. *Fucus serratus*	4	21	13	50
10. *Ascophyllum*	2	20	13	48
11. Cirr/Litt/Crass/Myt	3	13	7	45
12. *Crassostrea*	6	19	10	38
13. *Mytilus*	3	9	7	21

Total number of species found in 4.811 samples: 74 (Table 2). Species within the groups of Lichens, Amphipoda, Isopoda, Polychaeta and within the genera of *Enteromorpha*, *Cladophora* and *Polysiphonia* have not been distinguished.
In the surveys at the monitoring locations other species have been noted.
The total number examined in this research-programme approximates 100 species.

spp. and Cirripedia, as well as mobile species were considered less representative, their cover shows stronger variation to season and storms. Nevertheless their borders were studied too. Only some locations could be studied for Fucoids (Plompetoren, De Val, Stavenisse, Gorishoek, Yerseke and Zandkreek). At the other locations no Fucoids were dominant or present at all.

Plompetoren (monitored during 1982–1991)
During 1986/1987 several species showed a temporary decrease in presence in the transect: *Actinia equina*, Actinaria, *Blidingia* spp., *Carcinus maenas*, Cirripedia, *Enteromorpha* spp., *Entophysalis deusta*, *Fucus spiralis*, *Fucus vesiculosus*, *Porphyra umbilicalis*, *Prasiola stipitata*. In contrast, *Chondrus crispus*, *Crassostrea gigas* and *Ulva spec.* were temporarily more abundant in the middle part of the eulittoral zone. *Actinia equina* can still be found, but in a narrow zone compared to the original situation (Table 6).

The inclination at this location is 22°, a shift of one quadrat in the transect means a vertical shift of *ca* 19 cm. Compared to 1982 the borders of the Lichens-, *Entophysalis*- and *Blidingia*-communities show a vertical shift of *ca* 19–38 cm. The *Fucus spiralis* community shifted at the upper side with *ca* 19 cm downward, the lower border remained unchanged (Table 6). In the middle and lower part of the eulittoral zone no shifts occurred.

The shifting of the different belts did not co-occur. The *Entophysalis*, *Blidingia* en *Fucus spiralis* belts moved between April and August 1986. The new lower border of the Lichens was first found in November 1987 (Table 5).

Summarizing, compared to the original situation, no major changes have taken place. The rarer species can still be found, whereas the zonation pattern is rather unchanged. In the upper part a vertical shift of *ca* 19 cm downward occurred.

Table 5. Time series of dominant species per quadrat, location Plompetoren.

Date/Quadrat nr.	1	2	3	4	5	6	7	8	9	10	11	12	13	14	15	16	17	18	19	20	21	22	23	24	25	26	27	28	29	30	31	32	33	34	35
1982-09-01	1	1	2	2	2	2	2	7	7	7	7	7	8	8	8	8	9	9	9	9	9	9	9	9	9	9	9	9	n.i.	n.i.	n.i.	n.i.	n.i.	n.i.	n.i.
1984-04-18	1	1	2	2	2	2	4	7	7	7	7	7	8	8	8	9	9	9	9	9	9	9	9	9	9	9	9	9	9	9	9	9	9	9	9
1984-10-16	1	1	2	2	2	2	6	7	7	7	7	7	8	8	8	9	9	9	9	9	9	9	9	9	9	9	9	9	9	9	9	9	9	n.i.	n.i.
1985-04-23	1	1	2	2	2	2	6	7	7	7	7	7	8	9	9	9	9	9	9	9	9	9	9	9	9	9	9	9	9	9	9	9	n.i.	n.i.	n.i.
1985-08-09	1	1	2	2	2	2	6	7	7	7	7	7	8	8	9	9	9	9	9	9	9	9	9	9	9	9	9	9	9	9	9	9	n.i.	n.i.	n.i.
1985-11-04	1	1	–	2	2	2	4	7	7	7	7	8	8	8	9	9	9	9	9	9	9	9	9	9	9	9	9	9	9	9	9	9	n.i.	n.i.	n.i.
1986-04-30	1	1	1	2	2	2	4	4	7	7	7	7	9	9	9	9	9	9	9	9	9	9	9	9	9	9	9	9	9	9	9	9	n.i.	n.i.	n.i.
1986-05-26	1	1	1	2	2	2	2	4	7	7	7	7	9	9	9	9	9	9	9	9	9	9	9	9	9	9	9	9	9	9	9	n.i.	n.i.	n.i.	n.i.
1986-08-08	1	1	–	2	2	2	4	4	7	7	7	7	9	9	9	9	9	9	9	9	9	9	9	9	n.i.	n.i.	n.i.	n.i.	n.i.	9	n.i.	n.i.	n.i.	n.i.	n.i.
1986-10-06	1	1	–	–	–	–	–	–	5	7	7	7	9	9	9	9	9	9	9	9	9	9	n.i.	n.i.	n.i.	9	9	9	n.i.	9	n.i.	n.i.	n.i.	n.i.	n.i.
1986-11-03	1	1	–	–	–	–	–	–	5	7	7	7	8	9	9	9	9	9	9	9	9	9	n.i.	n.i.	n.i.	9	9	9	9	9	n.i.	n.i.	n.i.	n.i.	n.i.
1986-11-21	1	1	–	–	–	–	–	–	5	7	7	7	9	9	9	9	9	9	9	9	9	9	n.i.	9	9	9	9	9	9	9	n.i.	n.i.	n.i.	n.i.	n.i.
1987-03-20	1	1	–	–	–	4	4	4	5	5	7	7	9	9	9	9	9	9	9	9	9	9	9	9	9	9	9	9	9	9	9	9	n.i.	n.i.	n.i.
1987-04-13	1	1	–	–	–	4	4	4	5	7	7	7	9	9	9	9	9	9	9	9	9	9	9	9	9	9	9	9	9	9	9	9	n.i.	n.i.	n.i.
1987-04-18	1	1	–	–	–	4	4	4	5	7	7	7	9	9	9	9	9	9	9	9	9	9	9	9	9	9	9	9	9	9	9	9	n.i.	n.i.	n.i.
1987-04-27	1	1	–	–	–	4	4	4	5	7	7	7	9	9	9	9	9	9	9	9	9	9	9	9	9	9	9	9	9	9	9	9	n.i.	n.i.	n.i.
1987-05-25	1	1	–	–	4	4	4	4	5	7	7	7	9	9	9	9	9	9	9	9	9	9	9	9	9	9	9	9	9	9	9	9	n.i.	n.i.	n.i.
1987-08-12	1	1	–	–	–	2	2	2	2	5	7	7	7	9	9	9	9	9	9	9	9	9	9	9	9	9	n.i.	n.i.	9	9	9	9	n.i.	n.i.	n.i.
1987-09-28	1	1	1	1	–	2	2	2	6	4	7	7	7	7	9	9	9	9	9	9	9	9	9	9	9	9	n.i.	n.i.	9	9	9	9	n.i.	n.i.	n.i.
1987-11-10	1	1	1	2	2	2	2	4	6	7	7	8	8	8	9	9	9	9	9	9	9	9	9	9	9	9	9	9	n.i.	9	n.i.	n.i.	n.i.	n.i.	n.i.
1988-04-05	1	1	1	2	2	4	6	6	7	7	7	8	8	8	8	9	9	9	9	9	9	9	9	9	9	9	9	9	n.i.	9	9	9	9	9	n.i.
1988-08-02	1	1	1	2	1	2	6	6	7	7	7	8	8	8	8	8	9	9	9	9	9	9	9	9	9	9	9	9	9	9	9	9	n.i.	n.i.	n.i.
1988-11-14	1	1	1	2	1	2	4	4	7	7	7	7	8	8	9	9	9	9	9	9	9	9	9	9	9	9	9	9	9	9	9	n.i.	n.i.	n.i.	n.i.
1989-03-31	1	1	1	2	2	2	4	4	7	7	7	7	8	9	9	9	9	9	9	9	9	9	9	9	9	9	9	9	9	n.i.	n.i.	n.i.	n.i.	n.i.	n.i.
1989-08-01	1	1	1	2	2	2	4	2	7	7	7	7	8	9	9	9	9	9	9	9	9	9	9	9	9	9	9	9	9	9	9	9	n.i.	n.i.	n.i.
1989-10-19	1	1	1	2	–	2	2	2	7	7	7	7	9	9	9	9	9	9	9	9	9	9	9	9	9	9	9	9	9	9	9	9	n.i.	n.i.	n.i.
1990-11-23	1	1	1	2	2	2	2	4	7	7	7	7	9	9	9	9	9	9	9	9	9	9	9	9	9	9	9	9	9	9	9	9	9	n.i.	n.i.
1991-12-12	1	1	1	2	2	2	2	4	7	7	7	7	9	9	9	9	9	9	9	9	9	9	9	9	9	9	9	9	9	9	9	9	n.i.	n.i.	n.i.

The dominant species per quadrat sample is presented, using the following numbers:

1 = Lichens, 2 = *Entophysalis deusta* and/or *Prasiola stipitata*, 4 = *Blidingia* spec., 5 = Cirripedia and/or Littorinidae, 6 = *Enteromorpha* spp., 7 = *Fucus spiralis*, 8 = *Fucus vesiculosus*, 9 = *Fucus serratus* (not found are communities nr. 3, 10, 11, 12 and 13).

n.i. = not investigated due to low-water level, – = no living organisms, originally found organisms have died.

Quadrat nr. 1 is located at the upper limit of the revetment, quadrat nr. 35 is located at the lowest part of the eulittoral zone.

Table 6. Time series of *Actinia equina*, location Plompetoren.

Date/Quadrat nr.	1	2	3	4	5	6	7	8	9	10	11	12	13	14	15	16	17	18	19	20	21	22	23	24	25	26	27	28	29	30	31	32	33	34	35
1982-09-01	–	–	–	–	–	r	r	r	r	r	r	r	r	r	r	–	–	–	–	–	–	–	–	–	–	–	–	–	n.i.	n.i.	n.i.	n.i.	n.i.	n.i.	n.i.
1984-04-18	–	–	–	–	–	–	–	–	r	r	r	1	1	1	r	r	r	–	–	–	–	–	–	–	–	–	–	–	–	–	–	–	–	–	–
1984-10-16	–	–	–	–	–	–	r	r	r	r	1	2m	1	r	r	x	x	–	–	–	–	–	–	–	–	–	–	–	–	–	–	–	–	n.i.	n.i.
1985-04-23	–	–	–	–	–	–	–	–	–	–	–	r	r	r	r	x	–	–	–	–	–	–	–	–	–	–	–	–	–	–	–	–	n.i.	n.i.	n.i.
1985-08-09	–	–	–	–	–	–	–	–	r	r	r	r	r	1	x	–	–	–	–	–	–	–	–	–	–	–	–	–	–	–	–	–	n.i.	n.i.	n.i.
1985-11-04	–	–	–	–	–	–	–	–	–	–	r	r	r	r	r	–	–	–	–	–	–	–	–	–	–	–	–	–	–	–	–	–	n.i.	n.i.	n.i.
1986-04-30	–	–	–	–	–	–	–	–	r	r	1	1	r	r	r	–	–	–	–	–	–	–	–	–	–	–	–	–	–	–	–	–	n.i.	n.i.	n.i.
1986-06-25	–	–	–	–	–	–	–	–	r	r	1	2m	2m	1	r	–	–	–	–	–	–	–	–	–	–	–	–	–	–	–	–	n.i.	n.i.	n.i.	n.i.
1986-08-08	–	–	–	–	–	–	–	–	r	r	1	1	1	1	1	–	–	–	–	–	–	–	–	–	n.i.	n.i.	n.i.	n.i.	n.i.	n.i.	n.i.	n.i.	n.i.	n.i.	n.i.
1986-10-06	–	–	–	–	–	–	–	–	r	r	–	r	r	r	r	–	–	–	–	–	–	–	–	–	n.i.	n.i.	n.i.	n.i.	n.i.	n.i.	n.i.	n.i.	n.i.	n.i.	n.i.
1986-11-03	–	–	–	–	–	–	–	–	–	–	–	r	r	r	r	–	–	–	–	–	–	–	–	–	–	–	–	n.i.	–	–	n.i.	n.i.	n.i.	n.i.	n.i.
1986-11-21	–	–	–	–	–	–	–	–	–	–	r	r	r	r	–	–	–	–	–	–	–	–	n.i.	n.i.	–	–	–	–	–	n.i.	n.i.	n.i.	n.i.	n.i.	n.i.
1987-03-20	–	–	–	–	–	–	–	–	–	–	–	r	x	–	–	–	–	–	–	–	–	–	n.i.	n.i.	n.i.	n.i.	n.i.	n.i.	n.i.	n.i.	n.i.	n.i.	n.i.	n.i.	n.i.
1987-04-13	–	–	–	–	–	–	–	–	–	–	–	x	x	r	–	–	–	–	–	–	–	–	–	–	n.i.	n.i.	n.i.	n.i.	n.i.	n.i.	n.i.	n.i.	n.i.	n.i.	n.i.
1987-04-18	–	–	–	–	–	–	–	–	–	–	–	x	x	–	–	–	–	–	–	–	–	–	–	–	–	–	–	–	–	–	n.i.	–	n.i.	n.i.	n.i.
1987-04-27	–	–	–	–	–	–	–	–	–	–	–	–	x	–	–	–	–	–	–	–	–	–	–	–	–	–	–	–	–	–	n.i.	–	n.i.	n.i.	n.i.
1987-05-25	–	–	–	–	–	–	–	–	–	–	–	–	x	–	–	–	–	–	–	–	–	–	–	–	–	–	–	–	–	–	–	n.i.	n.i.	n.i.	n.i.
1987-08-12	–	–	–	–	–	–	–	–	–	–	–	x	x	–	–	–	–	–	–	–	–	–	–	–	–	–	–	–	–	–	–	n.i.	n.i.	n.i.	n.i.
1987-09-28	–	–	–	–	–	–	–	–	–	–	–	–	x	–	–	–	–	–	–	–	–	–	–	–	–	–	–	–	–	n.i.	n.i.	n.i.	n.i.	n.i.	n.i.
1987-11-10	–	–	–	–	–	–	–	–	–	–	–	–	x	–	–	–	–	–	–	–	–	–	–	–	–	–	–	–	–	–	n.i.	n.i.	n.i.	n.i.	n.i.
1988-04-05	–	–	–	–	–	–	–	–	–	–	–	–	x	x	x	–	–	–	–	–	–	–	–	–	–	–	–	–	–	–	n.i.	n.i.	n.i.	n.i.	n.i.
1988-08-02	–	–	–	–	–	–	–	–	–	–	–	r	1	r	–	–	–	–	–	–	–	–	–	–	–	–	–	–	–	–	–	–	n.i.	n.i.	n.i.
1988-11-14	–	–	–	–	–	–	–	–	–	–	x	x	x	x	–	–	–	–	–	–	–	–	–	–	–	–	–	–	n.i.	n.i.	n.i.	n.i.	n.i.	n.i.	n.i.
1989-03-31	–	–	–	–	–	–	–	–	–	x	–	x	x	x	–	–	–	–	–	–	–	–	–	–	–	–	–	–	n.i.	n.i.	n.i.	n.i.	n.i.	n.i.	n.i.
1989-08-01	–	–	–	–	–	–	–	–	–	–	–	x	x	x	–	–	–	–	–	–	–	–	–	–	–	–	–	–	–	–	n.i.	n.i.	n.i.	n.i.	n.i.
1989-10-19	–	–	–	–	–	–	–	–	–	–	–	r	r	r	r	–	–	–	–	–	–	–	–	–	–	–	–	–	–	–	n.i.	n.i.	n.i.	n.i.	n.i.
1990-11-23	–	–	–	–	–	–	–	–	–	r	1	1	1	r	–	–	–	–	–	–	–	–	–	–	–	–	–	–	–	–	–	–	n.i.	n.i.	n.i.
1991-12-12	–	–	–	–	–	–	–	–	–	1	1	1	x	x	–	–	–	–	–	–	–	–	–	–	–	–	–	–	–	–	–	–	n.i.	n.i.	n.i.

Presence of *Actinia equina* per quadrat sample is presented, using the following codes:

x = absent in transect, found sparsely on same position nearby; r = rare (1–5 ind./quadrat); 1 = 5–10 ind./quadrat, total cover <5%;

2m = >10 ind./quadrat, total cover <5%.

– = no *Actinia equina* found.

n.i. = not investigated due to low-water level.

Quadrat nr. 1 is located at the upper limit of the revetment, quadrat nr. 35 is located at the lowest part of the eulittoral zone.

De Val (monitored in 1983 and 1989–1991)
The inclination at this location is 20°, a shift of one quadrat in the transect means a vertical shift of *ca* 17 cm. The original zonation pattern has somewhat changed. The lower border of the lichens moved *ca* 34 cm downward in the transect. The upper border of the Fucoid-zone moved *ca* 17 cm downward. The zone of *Entophysalis deusta* was reduced with one quadrat. The presence of *Catenella caespitosa* and *Gelidium pusillum* in the latest years is remarkable high. *Ascophyllum nodosum* has reduced during the last years, *Fucus spiralis* and *Fucus vesiculosus* became more important. The total number of species per quadrat sample in the middle eulittoral zone is reduced compared to 1983.

Zuidbout (monitored during 1985–1988)
The inclination at this location is 18°, a shift of one quadrat in the transect means a vertical shift of *ca* 16 cm downward. In the upper part of the eulittoral zone *Blidingia* spp. shifted *ca* 30 cm downward. During August 1987–August 1988 the upper borders of Cirripedia, *Littorina littorea* and *Mytilus edulis* show a vertical shift of 40–50 cm downward. Since April 1988 *Chondrus crispus* and *Gigartina stellata* increased upward. Conclusions about permanent changes in species composition and total numbers of species per quadrat sample cannot be made for the lower eulittoral zone.

Zijpe (monitored during 1985–1988)
The inclination at this location is 18°, a shift of one quadrat in the transect means a vertical shift of *ca* 16 cm. The lower border of the Lichens zone has shifted *ca* 16 cm downward since April 1988. Cirripedia were temporarily less abundant during 1986–1987, now the upper border is found *ca* 16 cm down the original border. *Fucus spiralis* was found temporarily in 1986 and 1987, but disappeared again. *Mytilus edulis* shows migration downward the transect. Total numbers of species per quadrat sample show no great changes in the lower eulittoral zone.

Stavenisse (monitored during 1985–1988)
The inclination at this location is 16°, a shift of one quadrat in the transect means a vertical shift of *ca* 13 cm. The upper borders of *Chondrus crispus,* Cirripedia, *Fucus spiralis, Fucus vesiculosus,* and Lichens are found *ca* 13–39 cm downward compared to the former situation. Total numbers of species per quadrat sample show no great changes in the lower eulittoral zone.

Gorishoek (monitored during 1984–1991)
At this location vertical shifts could not be found for sure. This location showed the largest number of changes at species level. *Fucus vesiculosus* has disappeared almost since November 1986. *Fucus serratus* became less abundant and has disappeared completely in the middle eulittoral zone. Accompanying species such as *Littorina littoralis,* Actinaria and hydroids disappeared in this zone too. The belt of *Fucus serratus* at the lower part of the eulittoral zone is declining. *Bryopsis hypnoides* and *Gelidium pusillum* show a strong increase in the latest years. *Dumontia contorta* is found since November 1987. Cover of *Gigartina stellata* increased, this species shifted its upper border upwards the transect. The year 1991 showed a large species diversity. *Gelidium pusillum* reached a cover up to 50% in the middle eulittoral zone, higher as seen before.

Yerseke (monitored in 1985 and during 1989–1991)
The inclination at this location is 15°, a shift of one quadrat in the transect means a vertical shift of *ca* 13 cm. On this location the lower border of the Lichens zone moved *ca* 65 cm downward. The upper border of the Fucoids moved *ca* 26 cm downward. This means that the intermediate zone of *Entophysalis deusta* and *Blidingia* spp. became less wide. The number of species is reduced compared to 1985. In 1985 in the lower eulittoral zone 15 species could be found per quadrat sample, in 1991 only 8 species were found: the small green and red seaweeds disappeared (*Cladophora* spp., *Ceramium deslongchampii, Polysiphonia* spp.), possibly due to sedimentation.

Zandkreek (monitored in 1985 and during 1989–1991)

The inclination at this location is 19°, a shift of one quadrat in the transect means a vertical shift of *ca* 16 cm. The original zonation pattern has slightly changed. The lower border of the Lichens zone shifted *ca* 16 cm vertical downward. The upper border of the Fucoid zone moved *ca* 16 cm downward, a new zone is formed by *Fucus spiralis*. The rare alga *Pelvetia canaliculata* is still present, the cover varies. In 1991 cover was relatively high (up to 25%) compared to former years (<10%). Total numbers of species per quadrat sample have increased since 1985.

Other locations

As mentioned below many dike-slopes were reinforced in 1986. In most cases the upper parts were not affected. The lower borders of the Lichens zone in these parts could be studied for indications of downward shifts in the perpendicular transects. At 29 dike-slopes the lower border shifted downward, variering from 1 to 6 quadrats, with an average of 1.9 quadrats. These dike-slopes have an average inclination of about 15°. This means an average vertical shift of *ca* 25 cm downward.

Disturbance by dike reinforcements in 1986

In 1986 the Dutch Water Authorities reinforced the dike-slopes along the Oosterschelde estuary in order to prevent damage of the dike-slopes caused by continually strong wave attack in periods with a closed storm-surge barrier. These reinforcements consisted mainly of asphalting the dike-slopes. Some slopes were covered completely with asphalt. On a number of places rubble-stone and granite (10–15 cm diametre) was applied followed by asphalt injection. Also, on a number of dike-slopes, parts of the basalt or limestone slopes were replaced by concrete blocks.

As a result of the reinforcements a number of communities were affected. Using the survey of the period 1983–1985 the extent of the affected communities could be calculated (Table 3). On

Schouwen-Duiveland 45% of the *Fucus serratus* community disappeared. On Zuid-Beveland the *Pelvetia*-, *Blidingia*-, *Fucus spiralis*-, *Fucus serratus*-, *Ascophyllum*- and the Cirripedia/Littorinidae/*Crassostrea*/*Mytilus* community were severely affected (46%, 22%, 27%, 34%, 42% and 21% disappeared respectively). On Zuid-Beveland, a large part of the *Pelvetia* community was destroyed (46%). The disappearance of a large surface of the *Pelvetia* community, due to dike reinforcements, threathened *Pelvetia canaliculata* in its existence in the Netherlands.

In order to gain information of the recovery, the reinforced dike-slopes were surveyed every year in the period 1988–1992 (Meijer, 1993). The most important results will be mentioned here briefly.

With the adaptations of the dike-slopes, large areas of richly developed communities were lost. After 6 years, one can hardly speak of any development into a climax stage. In many places pioneering species still dominate. In sheltered areas, more or less closed *Fucus* communities occur. However, they only consist of a minimal number of species, while undergrowth is lacking. It can be concluded that either the largest part of the adapted dike-slopes has become unsuitable for the development of species-rich communities, or that recolonization has slowed down severely. In places where the original substrate appears again (due to the breaking up of the asphalt) or on the applied rubble-stone (providing it emerges above the asphalt sufficiently), further colonisation occurs. On a number of dike-slopes, more or less closed vegetations of *Fucus vesiculosus* appear, where seaweed vegetations were not found before 1986. At these sites too, the total number of species per quadrat sample is low.

Conclusions

– The temporary extra tidal-reduction in 1986 and 1987 caused some small effects, the communities recovered in a relatively short period. Most species occur again in their original belts from 1988 onwards, although the upper or lower border of the belts have shifted slightly downward for some species.

– The construction of the storm-surge barrier and auxiliary dams results in a gradual reduction in tidal movements, to which the hard-substrate communities have adapted. An exact comparison in numbers of the original situation with the present-day situation cannot be made since only few locations have been surveyed and some locations have been disturbed by dike reinforcements. Without a repeated sampling of the same transects as made in 1983–1985 it is not possible to calculate precisely the ultimate effects of the tidal reduction, because developments differ per location (dike-slope).

– At the investigated locations the communities of the upper eulittoral zone have only slightly shifted downward in the transects. The lower border of the Lichens belt has shifted downward, on most locations 1–2 quadrats, which means an increase of the total surface area for this community. The zone consisting of *Entophysalis deusta* and *Blidingia* spp. communities is reduced in wide at several locations. The upper borders of the Fucoid communities shifted at most locations 1–2 quadrats. Permanent changes in the middle and lower eulittoral zone cannot be found, due to variation in species diversity.

– In general it is concluded that the gradual reduction of the tidal amplitude has resulted in a shift of 1–2 quadrats downward in most transects. This means a vertical shift of *ca* 15–30 cm, depending on inclination (in most locations between 15° and 20°). This reflects the reduction of the tidal amplitude: the high-water line shifted *ca* 22 cm downward (Nienhuis & Smaal, 1993).

– The changes depend on location, exposure and inclination of the dike-slope. For places where the lower eulittoral zone is situated alongside a gully it was expected that the upward shift of the low-water line would benefit species diversity. The occurrence of sedimentation, however, prevented improvement of circumstances locally. A temporary explosive growth of *Ophiotrix fragilis* in the winters 1988/89 and 1989/90 had a negative effect on the presence of small seaweeds. The results do not permit the conclusion that the upward shift of the low-water line benefits species diversity in the lower eulittoral zone in general.

– The adaptations of the *Entophysalis*, *Blidingia* en *Fucus spiralis* communities seemed to occur during 1986–1987. The Lichens community has been changing its lower border since spring 1988.

– In general, it is concluded that the building of the storm-surge barrier did not cause dramatic changes in species composition and zonation of communities. Locally, sedimentation causes a reduction of species-diversity. On the other hand, some rare species have become more common, like *Gelidium pusillum* and *Catenella caespitosa*.

– Rare species, such as *Actinia equina* and *Pelvetia canaliculata* are still found. The number of locations were the latter can be found is strongly reduced compared to the pre-barrier period. Mortality caused by the temporary tidal reductions in 1986/1987 and dike reinforcements in 1986 caused an overall decrease in occurrence of the *Pelvetia* community. At present this community must be considered very rare. Only at some undisturbed locations *Pelvetia canaliculata* has maintained its position.

– The reconstruction of dike revetments in 1986 had considerable negative effects on some communities. Although only 7.4% of the total area of dike-slopes has been re-surfaced this had a considerable effect for certain rare communities (*Pelvetia* community), or those rich in species (*Fucus vesiculosus*-, *Fucus serratus*- and *Ascophyllum* community).

– The local dike reinforcements in 1986 have had much greater effects on distribution and surface area of several communities than the building of the storm-surge barrier.

298

Acknowledgements

Research mentioned in this article was carried out by Bureau Waardenburg in commission of the Delta Department and the Tidal Waters Division, both of the Ministry of Public Works and Transport. We thank R. J. Leewis, J. Leentvaar and J. Coosen being project managers of the Ministry. Mrs R. Guicherit and P. de Joode did translating work. We also thank P. H. Nienhuis for his useful comments.

References

Barkman, J. J, H. Doing & S. Segal, 1964. Kritische Bemerkungen und Vorschläge zur quantitativen Vegetationsanalyse. Acta. bot. neerl. 13: 394–419.

Braun-Blanquet, J., 1964. Pflanzensociologie, Grundzüge der Vegetationskunde, 3rd edn. Springer, Vienna/New York.

De Kluijver, M. J. & R. J. Leewis, 1994. Changes in the sublittoral hard substrate communities in the Oosterschelde estuary (SW Netherlands), caused by changes in the environmental parameters. Hydrobiologia 282/283: 265–280.

Den Hartog, C., 1959. The epilithic algal communities along the coast of The Netherlands. Wentia 1: 1–241.

Den Held, J. J. & A. R. den Held, 1973. Beknopte Handleiding voor Vegetatiekundig Onderzoek. Wetenschappelijke Mededelingen K.N.N.V. nr. 97.

Leewis, R. J., H. W. Waardenburg & A. J. M. Meijer, 1989. Active management of an artificial rocky coast. Hydrobiol. Bull. 23: 91–99.

Lewis, J. R., 1964. The Ecology of Rocky Shores. English Univ. Press, London.

Mathieson, A. C. & P. H. Nienhuis (eds), 1991. Intertidal and littoral ecosystems. Ecosystems of the World 24: 1–564, Elsevier, Amsterdam.

Meijer, A. J. M. & A. C. van Beek, 1988. De levensgemeenschappen op harde substraten in de getijdezone van de Oosterschelde. Report Bureau Waardenburg, Culemborg: 141 pp + appendices (Unpubl. report).

Meijer, A. J. M., 1993. Aangroei en ontwikkeling van levensgemeenschappen op aangepaste en recent aangelegde dijkglooiingen in de getijdezone van de Oosterschelde. Resultaten inventarisatie 1988 t/m 1992; Report Bureau Waardenburg, Culemborg (Unpubl. report).

Nienhuis. P. H., 1969. The significance of the substratum for intertidal algal growth on the artificial rocky shore of The Netherlands. Int. Revue ges. Hydrobiol. 54: 207–215.

Nienhuis, P. H., 1980. The epilithic algal vegetation of the S.W. Netherlands. Nova Hedwigia, 33: 1–94.

Nienhuis, P. H. & A. C. Smaal, 1994. The Oosterschelde estuary, a case study of a changing ecosystem: an introduction. Hydrobiologia 282/283: 1–14.

Shimwell, D. W., 1971. Description and classification of vegetation. Sidgwick & Jackson, London. 1–322.

Waardenburg, H. W. & A. J. M. Meijer, 1985. Hardsubproject; Notities betreffende mogelijke effecten van tijdelijke extra reductie van het getijdeverschil op de levensgemeenschappen van het harde substraat. Report Bureau Waardenburg, Culemborg (Unpubl. report).

Waardenburg, H. W. & A. J. M. Meijer, 1986. Resultaten onderzoek naar effecten sluiting Tholense Gat op hardsubstraatlevensgemeenschappen in litoraal en sublitoraal van de Oosterschelde. Report Bureau Waardenburg, Culemborg (Unpubl. report).

Waardenburg, H. W. & A. J. M. Meijer, 1987. Beknopte bespreking resultaten onderzoek effecten sluiting Krammer op hardsubstraatlevensgemeenschappen in de Oosterschelde. Report Bureau Waardenburg, Culemborg (Unpubl. report).

Hydrobiologia **282/283**: 299–301, 1994.
P. H. Nienhuis & A. C. Smaal (eds), The Oosterschelde Estuary.
© 1994 *Kluwer Academic Publishers. Printed in Belgium.*

Theme IV:
Ecology of the salt marshes

Jan De Leeuw
International Institute for Aerospace Surveys and Earth Sciences (ITC), P.O. Box 6, 7500 AA Enschede, The Netherlands

It is generally accepted that biotic and abiotic aspects of salt-marsh ecosystems are strongly determined by the frequency and duration of flooding by the tidal waters. This would imply that a change of the tidal regime would affect both the abiotic and biotic conditions in salt-marsh ecosystems. The decision to construct a semi-permeable storm-surge barrier in the mouth of the Oosterschelde estuary provided a unique opportunity to study the impact of tidal change on salt-marsh ecosystems. Long time before completion of the barrier it was predicted that the reduction in cross section of the mouth of the estuary would lead to a reduction of the tidal range (Nienhuis & Smaal, 1994) and hence affect salt marsh ecosystems (Smies & Huiskes, 1980; Saeijs & Al, 1982; De Jong & De Kogel, 1983). With this in mind several research groups decided in the early 1980's to initiate research programmes or intensify existing ones to predict and monitor the response of the Oosterschelde salt-marsh ecosystems to the anticipated change of the tidal regime.

The chapters written by De Jong *et al.* (1994), De Jong & Van der Pluijm (1994) and De Leeuw *et al.* (1994) describe the observed response of sediments and vegetation to these tidal changes. The following changes have been reported with respect to the abiotic environment. Inundation frequencies decreased from September 1985 onwards. The higher and middle salt-marsh environment was not flooded by the tides in most of 1986 and the first months of 1987. From April 1987 onwards the marshes became inundated again, but the frequency of flooding was lower than before the construction of the barrier. The marsh sediments, which used to be waterlogged, were drained from 1985 onwards. Oenema (1990) reported in 1985 and 1986 aeration of the soil, oxidation of pyrites to sulphates and a subsequent decrease of the pH in response to the lack of flooding. Low pH values were locally observed even after April 1987 (De Jong *et al.*, 1994). Besides, De Jong *et al.* (1994) observed ripening and settling of the soil, leading to a steepening of the height gradients between levees and back-marshes. In many places erosion rates at the salt-marsh edge increased since 1986 (De Jong *et al.*, 1994) leading to increased loss of the area of salt marsh in the Oosterschelde estuary.

Before construction of the barrier most researchers expected that the distribution of salt-marsh species would move to lower elevations (Smies & Huiskes, 1980; Saeijs & Al, 1982; De Jong & De Kogel, 1983). This expectation was confirmed by the observation that by 1990–1991 most species had moved to lower elevations in the marsh which corresponded to their pre-barrier inundation frequencies (De Jong & Van der Pluijm, 1994; De Leeuw *et al.*, 1994). De Leeuw *et al.* (1994) investigating the relationship between species composition and inundation frequency could not detect differences in this relationship before (1984) and after (1990) the reduction of the tides. They concluded that species composition had re-equilibrated with the newly established tidal conditions in 1990.

However, other vegetation responses were observed as well. The cover of some species (e.g.

300

Spartina anglica, Halimione portulacoides) was reduced in 1985, prior to the reduction of the tides. This was most likely a result of frost damage in the winter of 1984–1985. This initial reduction in cover was followed by a further reduction and locally by large scale die back of the original perennial vegetation in 1986. The die back was accompanied by the appearance of glycophytes and annual species. The cover of the annual species was greatly reduced in 1987, but returned to higher cover values in 1988. Various hypotheses were put forward to explain these ephemeral changes (De Leeuw *et al.*, 1994; De Jong & Van der Pluijm, 1994). The perennial vegetation recovered after the new tidal regime had been established.

Another unexpected development since the reduction of the tides was the spread of *Halimione portulacoidces* and *Elymus pycnanthus*, species normally observed on well drained levees, into waterlogged back marshes. De Jong & Van der Pluijm (1994) attribute the appearance and persistence of these species in the back marshes to irreversible and thus persisting changes in soil conditions. De Jong *et al.* (1994) describe these altered soil conditions as follows: since the reduction of the tides the soils in the basins in the middle high salt marsh have the dry nature of the levees. De Leeuw *et al.* (1994) in contrast, hypothesized that these levee species established under more favourable conditions in the interim period, and have persisted since. This hypothesis is based on previous observations that *Halimione portulacoides* persisted for more than ten years in waterlogged back marsh, after it had been established after artificial disturbance of the original vegetation and ephemeral aeration of the soil (De Leeuw *et al.*, 1992).

While the distribution of most species shifted down the marsh, several other species did not. The abundance of these species declined (De Leeuw *et al.*, 1994; De Jong & Van der Pluijm, 1994). This caused a considerable reduction of the diversity in the marshes. The species involved are generally of small stature such as *Limonium vulgare, Triglochin maritima, Juncus gerardi* and *Plantago maritima*. Both De Jong & Van der Pluijm (1994) and De Leeuw *et al.* (1994) raise the question whether the disappearing species will return at lower levels in the marsh. This depends to a large extent on the factors which are responsible for the reduction of these species. One possible explanation could be that these species have slower dispersal mechanisms. An alternative hypothesis could be that the species of smaller stature tend to be outcompeted by more competitive species like *Elymus pycnanthus* and *Halimione portulacoides*. These two latter species do frequently form species-poor stands, particularly in non-grazed salt marshes. This is attributed to their superior ability to compete for light (De Leeuw *et al.*, 1992; Bakker, 1990). When grazed however, these species-poor stands may be converted in more species-rich plant communities (Bakker, 1990). Grazing might therefore be a suitable management option to counterbalance the loss of species diversity. However, although at short-term it may be the most important one, it is just one of the changes in the Oosterschelde salt marshes caused by the storm-surge barrier. Erosion may, at longer time scales, form an even more important threat for the salt marshes. In the light of the foregoing it becomes clear that continued monitoring of both erosion, particularly the retreat of the marsh edge, and species composition remains necessary for sound management of these marshes.

References

Bakker, J. P., 1990. Nature management by grazing and cutting. On the ecological significance of grazing and cutting regimes applied to restore former species-rich grassland communities in the Netherlands. Kluwer Academic Publishers, Dordrecht, 400 pp.

De Jong, D. J. & T. J. de Kogel, 1985. Salt-marsh research with respect to the civil-engineering project in the Oosterschelde estuary, the Netherlands. Vegetatio 62: 425–432.

De Jong, D. J. & A. M. van der Pluijm, 1994. Consequences of a tidal reduction for the salt-marsh vegetation in the Oosterschelde (The Netherlands). Hydrobiologia 282/283 (Dev. Hydrobiol. 97): 317–333.

De Jong, D. J., Z. de Jong & J. P. M. Mulder, 1994. Changes in area, geomorphology and sediment nature of salt marshes in the Oosterschelde (SW Netherlands) due to tidal changes. Hydrobiologia 282/283 (Dev. Hydrobiol. 97): 303–316.

De Leeuw, J., L. P. Apon, P. M. J. Herman, W. de Munck & W. G. Beeftink, 1992. Vegetation response to experimental and natural disturbance in two salt marsh plant communities in the southwest Netherlands. Neth. J. Sea Res. 30: 279–288.

De Leeuw, J., L. P. Apon, P. M. J. Herman, W. de Munck & W. G. Beeftink, 1994. The response of salt marsh vegetation to tidal reduction caused by the Oosterschelde storm-surge barrier. Hydrobiologia 282/283 (Dev. Hydrobiol. 97): 335–353.

Nienhuis, P. H. & A. C. Smaal, 1994. The Oosterschelde estuary, a case study of a changing ecosystem: an introduction. Hydrobiologia 282/283 (Dev. Hydrobiol. 97): 1–14.

Oenema, O., 1990. Pyrite accumulation in salt marshes in the Eastern Scheldt, southwest Netherlands. Biogeochemistry 9: 75–98.

Saeijs, H. L. F. & J. P. Al, 1982. Environmental impact assessment for design and management of the Oosterschelde barrier. In: H. L. F. Saeijs, Changing estuaries, PhD thesis, University of Leiden: 279–320.

Smies, M. & A. H. L. Huiskes, 1981. Holland's Eastern Scheldt Estuary barrier scheme: some ecological considerations. Ambio 10: 158–165.

Hydrobiologia **282/283**: 303–316, 1994.
P. H. Nienhuis & A. C. Smaal (eds), The Oosterschelde Estuary.
© 1994 *Kluwer Academic Publishers. Printed in Belgium.*

Changes in area, geomorphology and sediment nature of salt marshes in the Oosterschelde estuary (SW Netherlands) due to tidal changes

D. J. de Jong, Z. de Jong* & J. P. M. Mulder
*National Institute for Coastal and Marine Management/RIKZ, P.O. Box 8039, NL-4330 EA Middelburg, The Netherlands; * P.O. Box 8358, NL-3503 RJ Utrecht, The Netherlands*

Key words: tidal reduction, soil ripening, soil acidification, areal changes, cliff erosion, accretion

Abstract

As a result of the construction of a storm-surge barrier across the mouth of the Oosterschelde (SW Netherlands) in 1987, the tidal range and mean high water level in the estuary have been reduced permanently to about 88% of their original values. During the final stage of construction (1985–1987) the tidal range and mean high water level were reduced even further for more than 18 months, by up to about 65% of their original values. This paper describes the consequences of these reductions for some abiotic aspects of the salt marshes.

Strong ripening of the soil, especially in basins of the middle high salt marshes, resulted in the soils in these basins having more or less the dry nature of levees. This may cause moisture deficits for the vegetation during dry periods locally, and may lead locally to acidification of the soil as a result of oxidation of pyrite.

Erosion of the edges of the salt marshes has increased in many places since 1986, both due to lowering of the surface level of the foreland, causing wave action to affect the marsh cliff more strongly than before, and weakening of cliff strength as a consequence of desiccation of the salt marsh soil and subsequent withering of plants and plant roots. In addition, the gradual salt marsh gradients have decreased on a large scale, as a consequence of increased wave attack and frost damage to *Spartina*. Finally, also due to desiccation and plant withering, levees have degraded and eroded, forming shoulders in the creeks.

Settling, especially in the basins, has steepened and narrowed the height gradients between basin and levee.

Introduction

In order to protect the SW Netherlands against extreme storm surges, a storm-surge barrier has been constructed across the mouth of the Oosterschelde, in combination with two dams in the eastern and northern part of the basin (Fig. 1). These structures mean that tidal range as well as Mean High Water (MHW) in the Oosterschelde have been reduced to about 88% of their original values (in the *pre-barrier period* = period until

autumn 1985) from April 1987 onwards (= the *post-barrier period*). During the final construction stage (here called the *interim period* = the period autumn 1985 - April 1987) the tidal range was reduced even further, up to a maximum of about 65% of the original situation in the period of summer 1986 to April 1987 (see Fig. 2a). More detailed information on the Oosterschelde project is given in Nienhuis & Smaal (1994) and Vroon (1994).

Important parts of tidal systems are salt

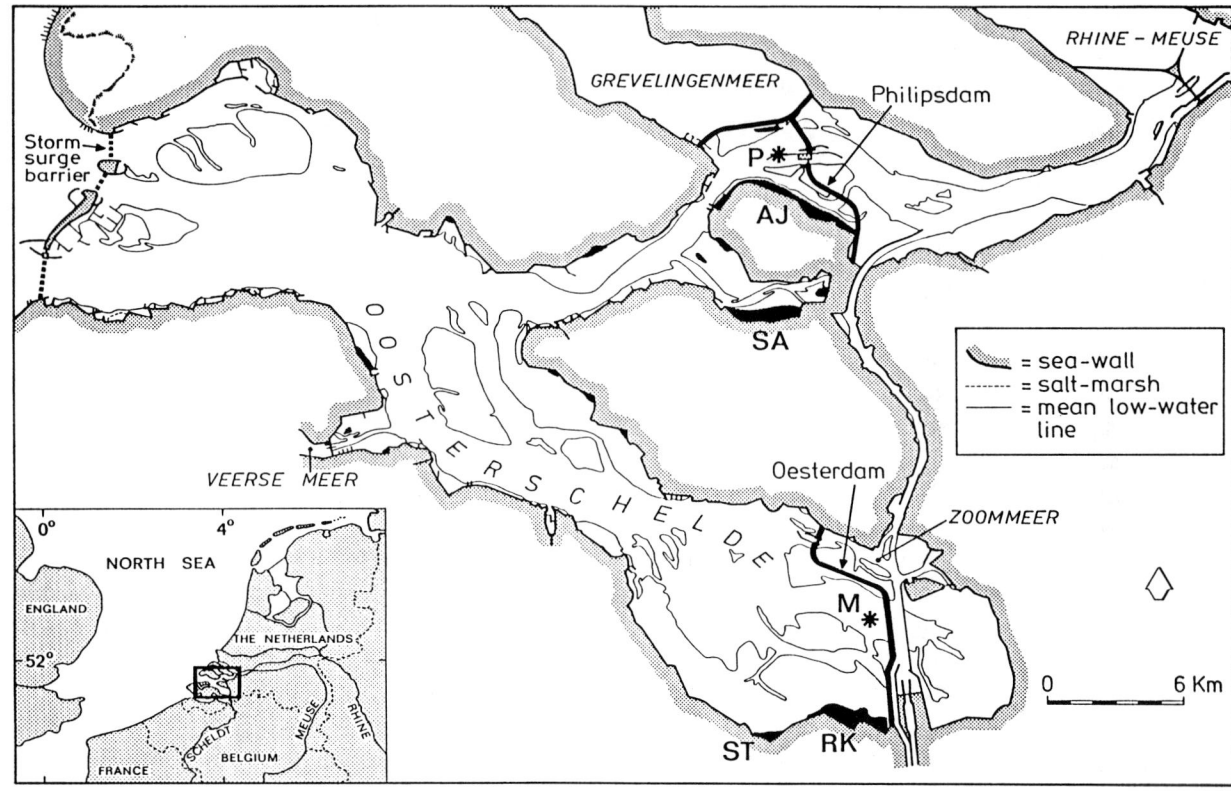

Fig. 1. The Oosterschelde and its salt marshes; salt marshes researched: AJ = Anna Jacobapolder, RK = Rattekaai, SA = St Annaland, ST = Stroodorpepolder; self-registering gauges: M = Marollegat, P = Philipsdam-west.

marshes, areas overgrown with vascular plants and situated in the higher parts of the tidal zone. They are dissected by a network of creeks which are bordered by relatively high levees (or creek banks), enclosing relatively low basins (or back marshes). Dominant factors in the development and erosion of salt marshes are the action of tide and waves, determining the balance between net supply or removal of sediment. Net sedimentation may occur in the less dynamic parts of a tidal basin. As soon as these parts have been built up sufficiently high, pioneer salt marsh plants will be able to colonise the areas. The interaction between tide and plants leads to further accretion on the salt marsh. Both the tidal amplitude and the plants strongly determine the zoning of sedimentation on the salt marsh, and by this geomorphological and pedological structure as well. The result of these interactions is a complex system of geomorphological, hydraulical, hydrological and pedological gradients, which is reflected in the distinct zoning and patterns of vegetation that are generally to be found on salt marshes. See also Adam (1990), Beeftink (1965), Long & Mason (1983), Van Diggelen (1988).

In view of the dominant effects of the tidal amplitude and flooding frequency on the salt marsh system, the significant changes in tidal range in the Oosterschelde may have substantial consequences for the salt marshes in this estuary. Several predictions have been made about possible consequences for the abiotic aspects: cliff erosion might decrease as a result of decreased wave attack on the cliffs (Schoot & Van Eerdt, 1985), significant expansion of salt marshes might be possible (Saeijs & Al, 1982), accretion rate at the top of the salt marshes might decrease as a result of the decreased flooding frequency (Vranken *et al.*, 1990; Oenema & DeLaune, 1988); and oxidation of pyrite (FeS_2) might occur

Table 1. General data concerning the Oosterschelde (A) and the salt marshes researched (B). (Mean elevation and elevation of cliff base are averaged figures; NAP = Dutch Ordnance Level = appr. Mean Sea Level).

A		≤ 1985		≥ 1987	
Surface area total (ha)		45,200		35,000	
Intertidal area (ha)		17,000		11,300	
Salt marsh (ha)		1,725		645	
Average tidal range Yerseke (m)		3.6		3.2	

B	Surface (ha)	MHW (m + NAP)		Mean elevation (m + NAP)	Cliff base (m + NAP)
		≤ 1985	≥ 1987		
Rattekaai	100	2.05	1.85	1.9	1.5–2.2
Stroodorpepolder	25	2.05	1.85	2.1	1.5–2.1
St Annaland	195	1.80	1.55	1.5	0.6–1.0
Anna Jacobapolder	160	1.80	1.55	1.6	0.4–1.0

locally on the middle salt marshes in dry summers and lead to acidification of the soil (Vranken *et al.*, 1990).

In order to monitor the consequences of the tidal reduction on the salt marshes and to test these predictions, research has been carried out into ripening of the soil (including settling, decalcification and acidification), accretion on the salt marsh and changes in surface area. Besides, a sudden reduction in tidal range and flooding frequency is a very rare event, and experience with the Oosterschelde salt marshes might benefit researchers and policy-makers concerned with similar civil engineering projects elsewhere. However, this unusual situation makes comparison with analogous situations difficult.

Materials and methods

Description of the area

The Oosterschelde is a euhaline (average chlorinity 16 g Cl$^-$ l^{-1}), mesotidal estuary with a total surface area of about 45000 ha before 1985 and about 35000 ha since 1987. The salt marshes are situated mainly in the eastern and northern parts of the estuary. The measurements for this research were concentrated on four salt marshes: Ratte-kaai, St Annaland, Anna Jacobapolder and Stroodorpepolder (Fig. 1, Table 1). The elevation of these salt marshes until 1985 was roughly between 0.5 m below and 0.5 m above MHW with flooding frequencies ranging from about 600 to 20 times a year. The soil consists of silt and silty clay, varying in thickness from 0.2 m to 1.5 m, deposited on the sandy subsoil of the tidal flat. On one side the salt marshes are confined by a dike and on the other side by intertidal flats. In 1980, 80% of the salt marsh edges consisted of eroding cliffs. Horizontal expansion to the tidal flat occurred only sporadically. The vegetation in the basins varied from monostands of *Spartina anglica* [1] on the primary salt marsh and in the lower basins to a mix of *Spartina*, *Puccinellia maritima*, *Triglochin maritima* and *Limonium vulgare* in the higher basins. On the lower levees *Halimione portulacoides* dominates as a rule and on higher levees *Festuca rubra* or *Elymus pycnantus*. More extensive descriptions of the Oosterschelde in general and the salt marshes are given by Beeftink (1965), Oenema & DeLaune (1988), Oenema (1988a), Schoot & Van Eerdt (1985) and De Kogel & De Jong (1980).

---

[1] Nomenclature of vascular plants follows Tutin *et al.* (1964–1980).

Measurements

Tidal data from the Rijkswaterstaat's self-registering gauges at Marollegat and Philipsdam-west (Fig. 1) have been used to calculate flooding frequencies for the salt marshes in the south-eastern and north-eastern parts respectively.

The *soil* has been investigated in a total of 27 transects with a length of about 10 to 20 m on the Rattekaai and St Annaland salt marsh, each transect extending from the basin over the levee to the opposite side of the creek. In field observations of these transects, the following features of the soil were described at intervals of 1 m:

- stratification of the soil down to the underlying tidal sands (by augering);
- presence of calcium carbonate at various depths (qualitative estimation of liberated CO_2 after adding HCl to the soil sample in 3 classes: 1 = no, 2 = little and 3 = much CO_2 liberation);
- oxydation-reduction boundary zone (visually determined);
- degree of ripening in 5 classes: 1 = unripened, 2 = almost unripened, 3 = half ripened, 4 = almost ripened and 5 = fully ripened (after Pons & Zonneveld, 1965);
- settling (by levelling the elevation of the surface twice a year).

Research into *acidification* of the soil was carried out in the post-barrier situation on the four salt marshes. To that end, 61 borings (6 cm diameter auger) were made in basins, down to the underlying tidal sands or into the reduced zone. Besides field measurements on texture, calcium carbonate content, soil ripening and aeration depth (for methods see above), laboratory analyses were carried out on calcium carbonate content (Scheibler method), actual pH (pH-H_2O) and potential pH (oxidation with H_2O_2, not cooled) (Houtekamer, 1991).

Sedimentation was measured on the Rattekaai and St Annaland salt marshes. For this 37 'kaoline fields' (0.3 × 1.0 m) were laid out in 7 transects perpendicular to a creek in which a tracer layer of fine white kaoline clay was applied to the surface. By means of a small auger (diameter 12 mm) the thickness of the sediment on top of the tracer layer was determined; see also Oenema & DeLaune (1988).

Lateral changes in the salt marsh cliffs were determined in 14 transects (20 m length, perpendicular to the cliffs) on 3 salt marshes: Rattekaai, Anna Jacobapolder and St Annaland. The elevation of the marsh cliffs and the foreland was levelled 2 to 4 times a year.

An overall view of the surface area of the salt marshes before and after the finishing of the barrier has been obtained from aerial photographs. Using these, the position of the edges of all salt marshes has been mapped precisely. Salt marsh was defined as the area with a vegetation cover $\geq 50\%$. Although the 'photographic method' only provided information for a measurement interval of about 7 years altogether ($\pm$ 1983– $\pm$ 1990), it has been possible to assess the areal changes for seperate periods (1983–86, 1986–87, 1987–88, 1988–90) by combining the 'photographic method' with the more frequent measurements of the cliffs.

Results

Vertical tide

Figure 2A presents the tidal range in the central part of the Oosterschelde for the period 1984 to 1989. This has been transposed into flooding frequencies for the three important elevation zones in the salt marshes in Fig. 2B. From this it is clear that on the middle and high salt marsh zones flooding was (almost) absent during the interim period. In the post-barrier period all parts of the salt marshes have been reflooded, albeit with a largely decreased frequency. The salt marsh cliffs were reached by the water at high water considerably less frequently during the interim period than they were during the pre-barrier period. In the post-barrier period high water reaches the cliffs more frequently again, but less frequently than before.

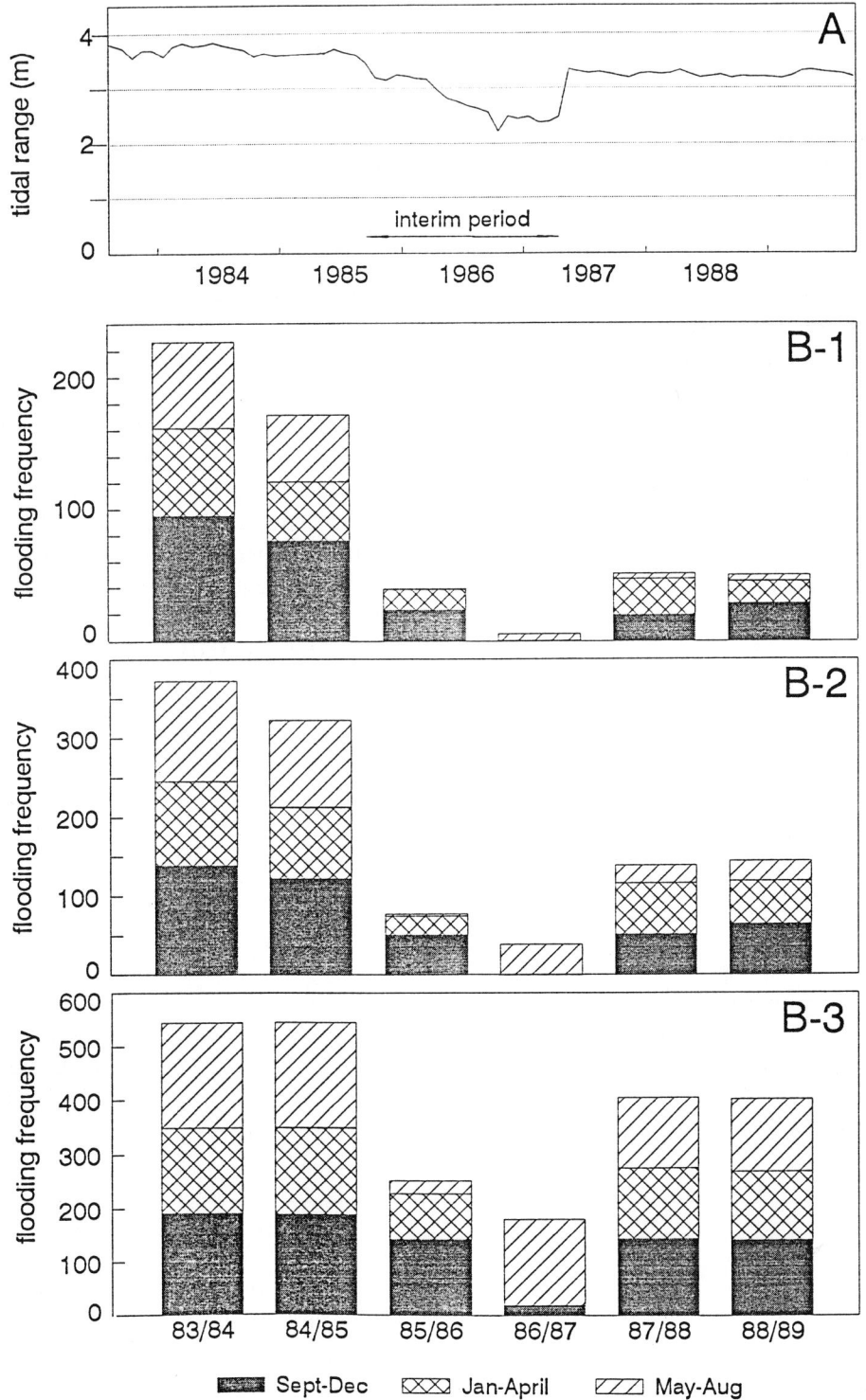

Fig. 2. Tidal range and flooding frequency in the Oosterschelde from 1984 to 1989. (A) Tidal range in the central part of the Oosterschelde (Yerseke). (B) Flooding frequency per 'vegetation year' for three salt marsh elevations. I: high salt marsh (elevation about MHW + 0.2 m during the pre-barrier period); II: middle salt marsh (elevation about MHW during the pre-barrier period); III: low salt marsh (elevation about MHW – 0.3 m during the pre-barrier period). ('vegetation year': period of importance for the growing season; e.g. 'vegetation year' '84/'85: September 1984–August 1985).

308

Soil characteristics

In 1986 and 1987 the soils of the salt marshes were desiccated and in many areas shrinkage cracks with depths up to over 0.3–0.4 m developed, causing strong aeration of the soil. Aeration was increased in many places by the strong burrowing activities of mice and rabbits, which invaded large parts of the salt marshes during 1986.

During the interim period, in the autumn of 1986, the lower boundary of the aerated zone at the Rattekaai salt marsh dropped by an average of 0.2 m relative to NAP (Normaal Amsterdam Peil = about Mean Sea Level) (Oenema, 1988a; Vranken *et al.*, 1990). Our measurements in the transects in 1990 showed that this boundary was about 0.05–0.1 m lower relative to NAP as compared to the pre-barrier period. However, due to settling of the soil (about 0.05–0.1 m in the basins) the depth of the aerated zone below surface level hardly changed in the post-barrier period (Fig. 3).

Oenema (1988a) mentions increased physical ripening of the soil in the basins of the Rattekaai salt marsh due to a downward movement of the water table in 1986. Our measurements in the transects in both salt marshes also showed that physical ripening of the soil had taken place, predominantly in the clayey basins of the middle salt marsh and less on the sandier levees and on the low (primary) salt marsh. The mean ripening stage in the transects had increased from almost unripened (class 2) in the pre-barrier situation to half ripened (class 3) in the post-barrier situation.

Elevation measurements in the transects at both salt marshes showed that settling had occurred, varying from about 0.01 to 0.1 m, in the basins especially and, moreover, mostly during the interim period (Fig. 3, 4). These observations are consistent with the findings of Oenema (1988a) and Vranken *et al.* (1990) who measured a settling of 0.01–0.05 and 0.01–0.08 m at Rattekaai as early as autumn 1986.

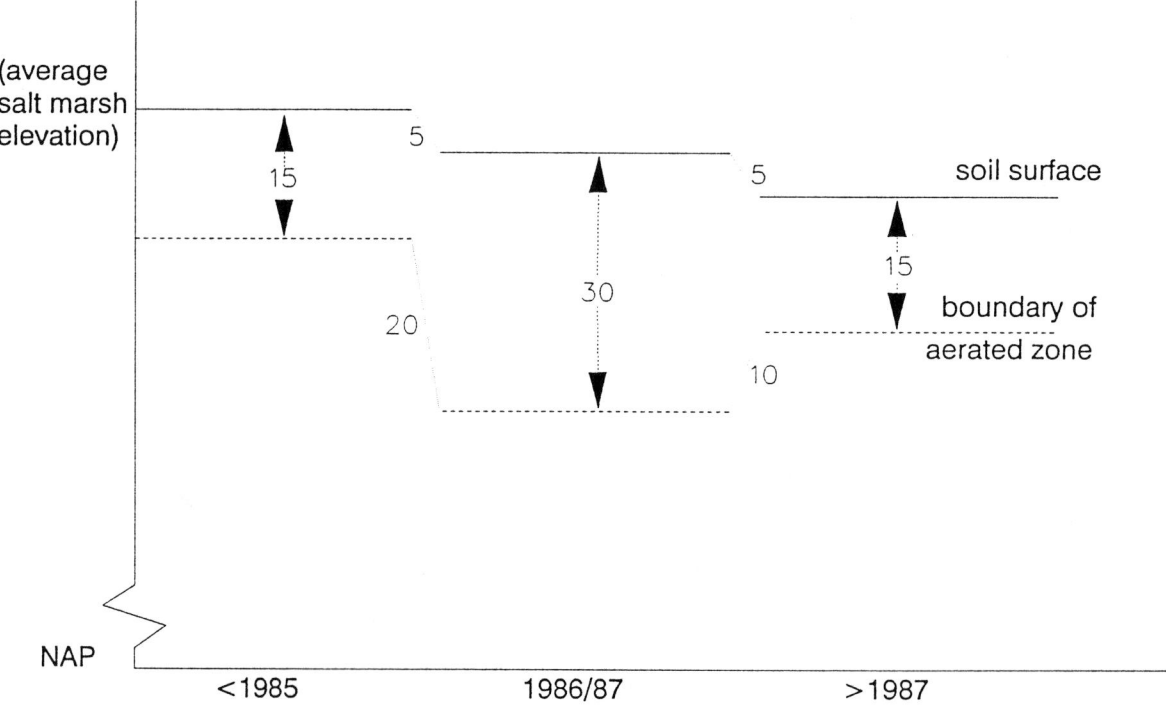

Fig. 3. Average changes in the elevation of the salt marsh surface and of the depth of the aerated zone during the interim period and the post-barrier period (1986/87 after Vranken *et al.*, 1990) (as it is a general indication no actual values are indicated on the Y-axis; however, average salt marsh elevation concerned may be stated at MHW and above).

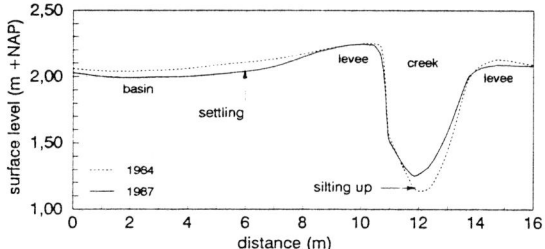

Fig. 4. Changes in surface level (m + NAP) in a transect at Rattekaai from the pre-barrier situation, 1984, (dashed line) to the post-barrier situation, 1987 (solid line). The significant settling in the basin and also the partly silted-up creek can be distinguished. In the post-barrier period the silt in the creek is largely eroded (not shown).

A strong red colouring of the water and an 'oil-like film' on the water have been observed in many smaller ramifications of the creeks in the basins since the start of the interim period, espe-

cially in Rattekaai; these are indications respectively of the oozing out of iron and elementary sulphur. Although less than during the interim period, this oozing out is still being observed to a stronger degree than in the pre-barrier period.

Low pH values, with 3 as an extreme, were measured in three basins during the interim period in 1986 (Vranken *et al.*, 1990). In the post-barrier period, low pH values were measured in a larger number of basins (Houtekamer (1991): in 1990 the minimum was pH 5.9 and in 1991 pH 5.5, compared to an average pH of 7–8 in the pre-barrier period. In many basins the calcium carbonate content appeared to be very low, with values down to about 0.1% $CaCO_3$, compared with average values of 3–6% in basins in Rattekaai and 0.5–2% in St Annaland in the pre-barrier period. Laboratory experiments, using

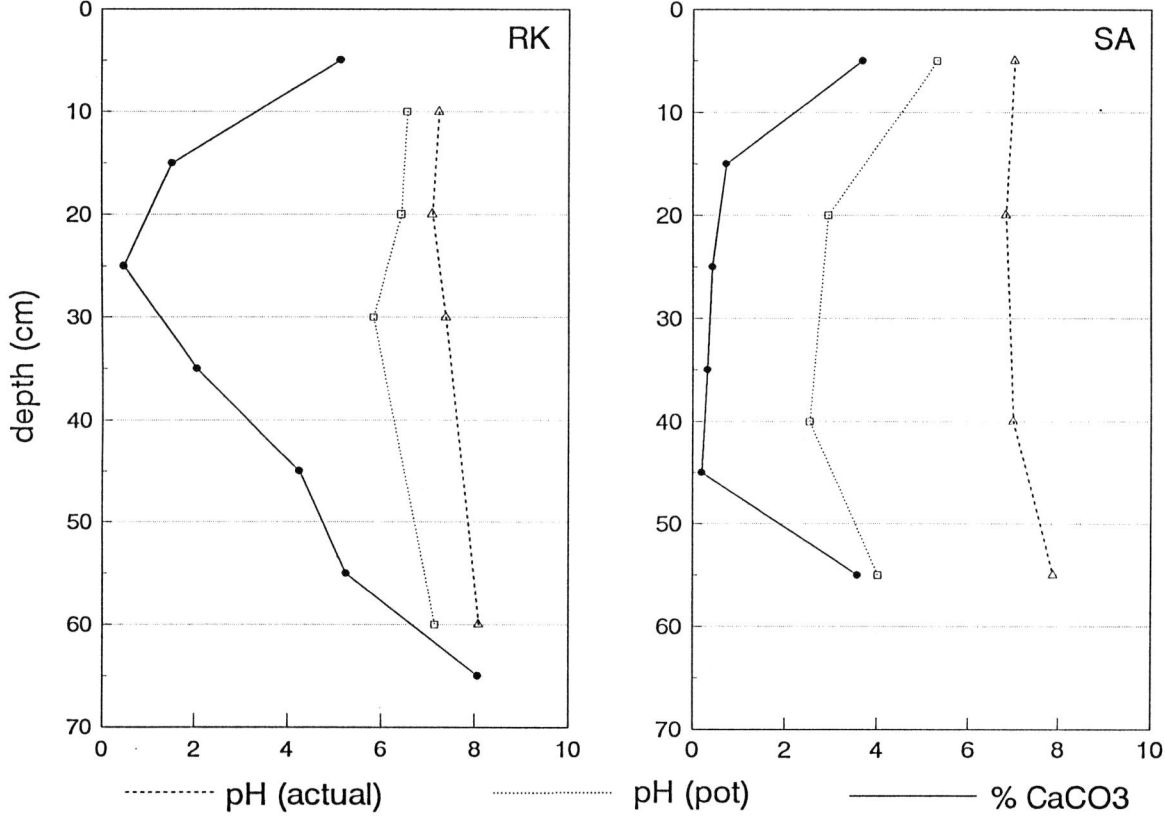

Fig. 5. Depth-profile (in cm) of calcium carbonate (solid line), pH (H_2O) (dashed line) and the potential minimum pH (after oxidation with H_2O_2) (dotted line) in three salt marsh basins. SA: St Annaland, RK: Rattekaai. X-axis: pH and % calcium-carbonate. (after Houtekamer, 1991).

310

H$_2$O$_2$ as an oxidant, indicated that the pH of these soils could potentially decrease to very low values of pH 2.5 to 3. Two examples of the profile of calcium content and pH with depth are shown in Fig. 5.

Accretion

At Rattekaai, accretion on the salt marsh surface remained the same during the first two years of the post-barrier period compared with the long term average in the pre-barrier period; after that it decreased to about half the amount and finally it turned into net erosion in 1991/92. At St Annaland accretion first increased but it has decreased sharply since 1990; see Fig. 6.

Besides accretion at the top of the salt marsh, there was also a marked silting up of the creeks in the interim period, especially on the Rattekaai salt marsh. This varied from an average of 0.1–0.5 m with peaks up to 1 m locally (Fig. 4). During the first two years of the post-barrier period (1987/1988) this sediment was partly eroded. In

the north-eastern salt marshes hardly any silting up of creeks was observed.

The salt marsh edge

During the interim period little erosion of the cliffs occurred. Field observations showed that the soil of the cliffs was strongly desiccated, while the vegetation on top had declined sharply even disappearing locally due to withering.

In the post-barrier period cliff erosion increased sharply in most places compared with the pre-barrier period. Generally, this increased erosion decreased after some time on cliffs with a more elevated cliff base (e.g. Rattekaai), while increased erosion continued on cliffs with a less elevated cliff base (e.g. St Annaland) (Fig. 7).

The effects on the total area of the salt marshes, derived from the aerial photograph analysis, are shown in Table 2. This analysis showed that many salt marsh edges characterised by a gradual transition had retreated as well. Unfortunately, no data from transects are available to specify the

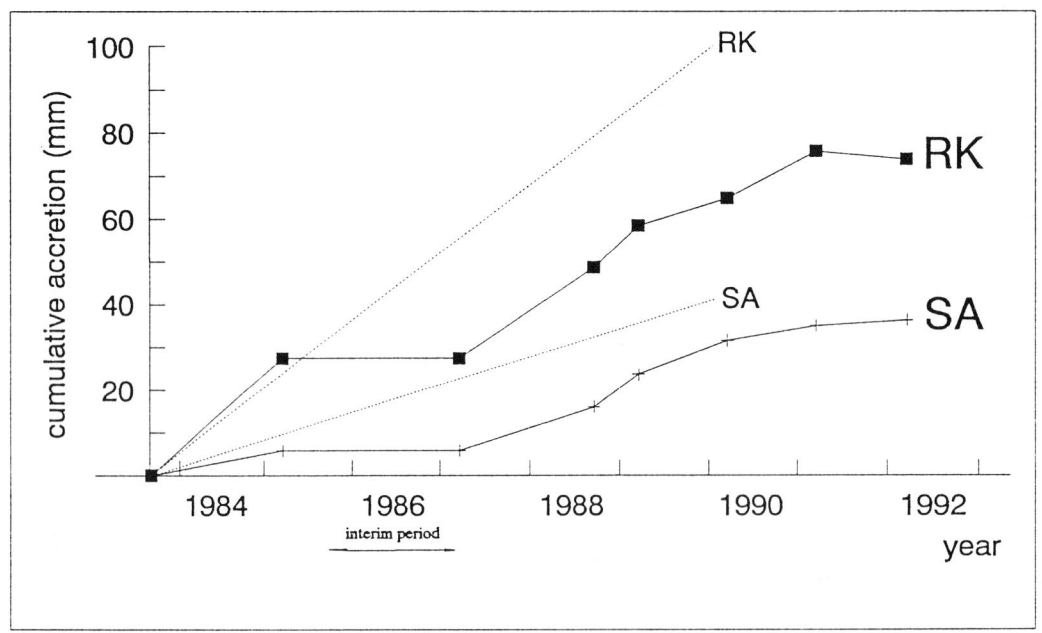

Fig. 6. Cumulative net sedimentation at Rattekaai (RK) and St Annaland (SA) in the period 1984–1992. In addition, mean sedimentation as measured for the period 1954–1985 has been indicated (dashed lines) (after Oenema & DeLaune, 1988).

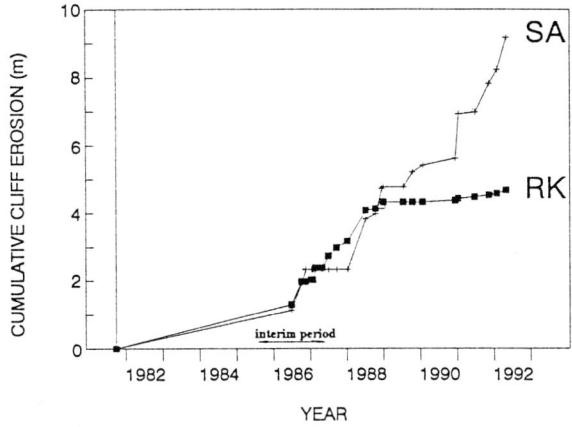

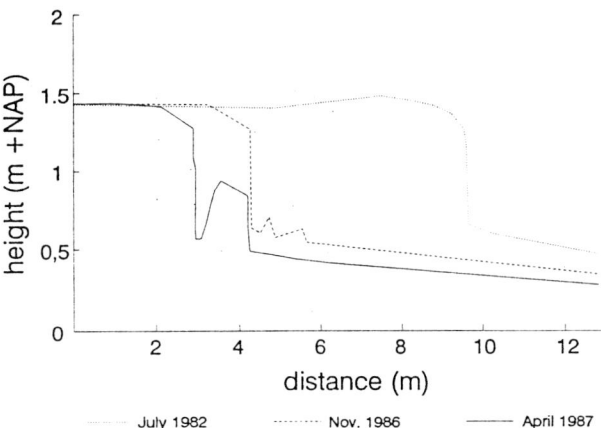

Fig. 7. Cumulative cliff erosion (recession of cliff base) of two cliffs for 1982–1992. Cliff RK (= Rattekaai), with a high elevation cliff-base, shows first a significant increase in erosion after the interim period followed by a decrease in erosion; cliff SA (= Sint Annaland), with a low elevation cliff-base, shows a continuing increased erosion.

Fig. 8. Cross section of a salt marsh cliff (Anna Jacobapolder), showing a lowering of the surface level of the foreland. The 'peaks' in front of the cliff are lumps of clay dislodged from the cliff.

period(s) over which this retreat has taken place. In the period analysed no expansion of salt marshes has been observed. Taken together, the continuing erosion of the salt marsh cliffs and transitional salt marsh edges, and the lack of newly generated salt marshes mean that salt marshes in the Oosterschelde are declining at present at a net rate of 4–6 ha y^{-1}, compared with about 1–2 ha y^{-1} (Schoot & Van Eerdt, 1985) in the last years of the pre-barrier period.

Levelling surveys of the salt marsh cliffs (Fig. 8) and field observations show that since the start of the interim period the elevation of the foreland has been lowered in many places, causing the boundary of the clayish salt marsh soil and the underlying tidal sands to become an outcrop. Consequently hollows have occurred at this boundary ('wave-cut notches'), changing the cliff profile from a straight cliff into an undermined one (Fig. 9).

Observations confirm that many levees show a similar development as cliffs. During the interim

Table 2. Changes in salt marsh surface area (in h y^{-1}). Subdivision of the changes into periods is estimated using data from cliff transect measurements. # assuming erosion of transitional salt marsh edges to have taken place mainly since 1987; # # assuming erosion of transitional salt marsh edges to have occurred uniformly over the entire period of measurement. * mean value for the entire period.

| | 1983–87 | | 1987–88 | 1988–90 |
	1983–86	1986–87		
Cliff with high elevation	− 0.3	0	− 0.4	− 0.4
Cliff with low elevation	− 0.9	0	− 4.1	− 2.1
Transitional salt marsh edge (#)	0	0	− 3.8	− 3.8
or (# #)	< -------------------------- − 1.6 -------------------------->			
Horizontal accretion *	< -------------------------- + 0.5 -------------------------->			
Net change of surface area (#)	− 0.7		− 7.8	− 5.8
or (# #)	− 2.3		− 5.6	− 3.6

312

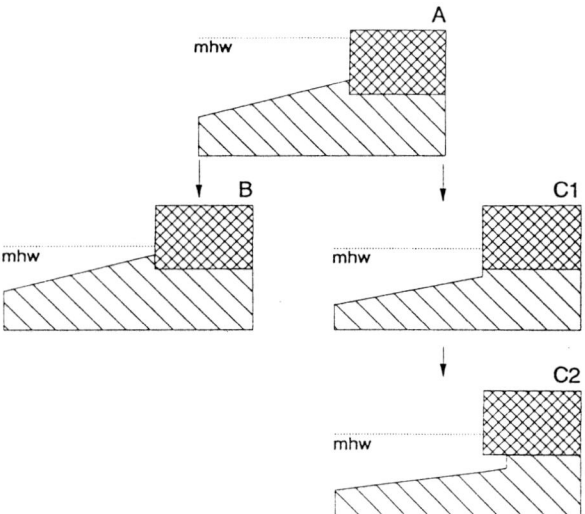

Fig. 9. Schematic representation of the effect on wave height of lowering of the tidal flat lying in front of the salt marsh cliff. A: base situation; B: high elevation cliff base: when MHW is lowered, the mean water depth in front of the cliff decreases proportionally; C: low elevation cliff base: if MHW and the elevation of the foreland are lowered simultaneously the water depth does not decrease or at least decreases less significantly. C2 illustrates the genesis of an undermined cliff ('wave-cut notch').
(single hatched: tidal flat, double hatched: salt marsh)

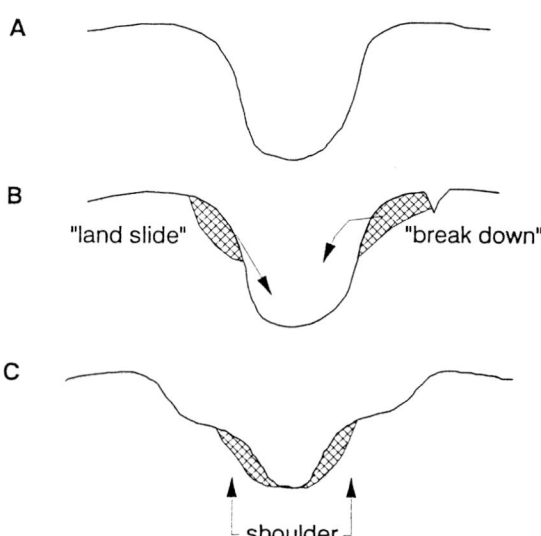

Fig. 10. Schematic representation of levee erosion leading to 'shoulders' in the creek A: original creek profile, B: 'solifluction' or break down of lumps, C: new creek profile with 'shoulders'.

period desiccation of the soil occurred as well as die-back of the withered vegetation on the top. Serious erosion occurred locally in the period immediately afterwards, either by breaking off, or slumping down of large lumps. As a result, at many places there are lowered strips, so-called 'shoulders' or 'creek benches', in the creeks (Fig. 10).

Discussion

Soil

Soil features observed in 1986 and 1987, -shrinkage cracks, settling, increased aeration and ripening- all indicate a strong drying out of the soil (see also Pons & Zonneveld, 1965), the most marked effects occurring during the 65% tidal reduction during the interim period. We attribute this to the (near) absence of flooding for about 1.5 years, as also confirmed by the observations of Vranken et al. (1990) and Oenema (1988a). The vegetation also showed strong indications of soil ripening during 1986 and 1987 (De Jong & Van der Pluijm, 1994): glycophytes indicated desalination of the soil and a sharp increase in the number of nitrophytes indicated release of nutrients.

The salt marsh soils have been flooded more frequently again since 1987. However, the ripening of the soil which has occurred is irreversible (Pons & Zonneveld, 1965), causing the soils of the salt marshes to remain riper than may be expected from their flooding frequency in the post-barrier period. This may have permanent consequences for some abiotic aspects of the salt marshes.

Desiccation

Both the ripening of the soil and the shrinkage cracks may have disturbed the original layered structure of the soil, resulting in an increased vertical transport of water in the soil, especially in basins. Because of this the soil may dry out faster in the post-barrier period after flooding and during dry spells. As a consequence the basins have a drier character than would be expected from their elevation and geomorphology; in soil structure and geohydrology they resemble a levee more than a basin. This is confirmed by De Jong & Van der Pluijm (1994), who observed a continuing invasion of 'levee species' into many basins since 1987.

Settling

As a consequence of the variations in settling between basins and levees height differences between these two morphological units have increased; this manifests itself as a steeper and narrower height gradient between basins and levees (Fig. 4).

Acidification

The measured decrease of the pH in 1986 (Vranken *et al.*, 1990) is ascribed to oxidation of pyrite (FeS_2). The observed heavy oozing out of iron and elementary sulphar is also an indication of this process. When pyrite is oxidized the calcium carbonate present in the soil is used to neutralize the sulphuric acid freed (after Van Breemen, 1973):

$$4FeS_2 + 15O_2 + 10H_2O \rightarrow 4FeOOH + 8H_2SO_4 ,$$

$$H_2SO_4 + CaCO_3 \rightarrow CaSO_4 + H_2O + CO_2 .$$

For basin soils rich in calcium carbonate this is not a problem. However, basin soils with a low calcium carbonate content may become decalcified by pyrite oxidation, eventually causing low pH values. Tests with H_2O_2 have shown that in some of the basins researched the pH may reach very low values (2–3). These low pH values may occur in dry basins, whether only temporarily during dry spells in summer when the water table drops and the soil temporarily oxidizes deeper (as indicated by Vranken *et al.*, 1990), or permanently. This means that in dry basins with a calcium carbonate deficit there is a serious risk of low pH values ('cat clay') at present. De Jong & Van der Pluijm (1994) deal with possible consequences for the vegetation.

Vertical accretion

It was predicted that the accretion rate on the salt marshes would decrease markedly after completion of the Oosterschelde barrier, because of the large decrease in flooding frequency and the reduction of suspended matter content in the water flooding the salt marsh (Oenema & DeLaune, 1988). However, the accretion rate actually measured in the post-barrier period initially remained at the pre-barrier level, and only declined after some years.

These successive developments are probably connected with the general process of vertical accretion on the salt marshes of the Oosterschelde (Oenema, 1988b; Oenema & DeLaune, 1988): the major source of the accumulated sediment most likely is the tidal flat in front of the salt marsh. Sediment accumulates on the tidal flats during less dynamic periods, from where it is resuspended and transported towards the adjacent salt marsh by wave action during storms. In that way the dimension of the foreland plays an important part in the amount of sediment available to a salt marsh. This is confirmed by Schoot & Van Eerdt (1985) (see also Oenema & DeLaune, 1988), who observed that in the pre-barrier situation the sediment concentration in the water over the tidal flat at Rattekaai, with a relatively large foreland, was 3–4 × higher than at St Annaland, with a relatively small foreland. Besides, seasonal dynamics of silt contents on shoals and tidal flats have been observed by Oenema (1988b) and Ten Brinke *et al.* (1994).

Taking this into account we suggest that the initial continuation of the former accretion rate may be due to the fact that accumulated silt from the foreland does not disappear instantly. Besides, at St Annaland a marked lowering of the foreland occurred during the interim period and the first few years afterwards, which may even have led to an increased supply of silt to this salt marsh. The subsequent drop in accretion rate will be caused by the large decrease in sedimentation on the foreland due to the much clearer water of the Oosterschelde under present conditions (Wetsteyn & Kromkamp, 1994), and erosion of the foreland of St Annaland has almost stopped.

Owing to the decrease in sediment supply it is probable that in future the accretion rate on the salt marsh will remain permanently reduced, as predicted. This may have consequences in the long run in connection with sea level rise as accretion on the salt marsh may not be able to keep up with this.

Changes of surface area

It was predicted that the rate of cliff erosion would diminish in the post-barrier period, as a result of a decline in eroding wave action against the cliffs (Schoot & Van Eerdt, 1985). However, after a period of hardly any cliff erosion during the interim period, from 1987 onwards cliff erosion initially increased sharply compared with the pre-barrier period, after which, depending on the elevation of the cliff base, it remained at this high rate or resumed the old rate of erosion.

Due to lack of flooding during the interim period, the cliff soil dried out and vegetation on top withered. As plant roots play a significant part in the strength of cliffs (Van Eerdt, 1985), this combination of withered plants and dried out soil seriously weakened the cliffs. In this way they became susceptible to increased wave attack from April 1987 onwards. The hypothesis is proposed therefore that the initial increase in cliff erosion was caused by weakening of the cliffs during the interim period.

The subsequent continuation of accelerated erosion of cliffs having a less elevated cliff base (St Annaland and Anna Jacobapolder) may be caused by a lowering of the foreland by 0.1–0.2 m locally (Figs. 8, 9). Erosion up to 0.1 m on occasion could be observed after closures of the barrier during heavy storms, both in the interim and the post-barrier periods. Erosion of the foreland during heavy storms may also have occurred in the pre-barrier period, but in the post-barrier period there is no longer any compensation as, due to the reduction of current velocities in the channels, sediment supply from the channels has de-

creased drastically (Mulder & Louters, 1994). The lowered foreland undermined the cliffs ('wave-cut notches', Fig. 9C2), with the overhang able to break off quite easily. Moreover, the lowering of the tidal flats means that water depth in front of the salt marsh has not diminished (Fig. 9-C1), and so the impact of wave action has not decreased. This lowering of the foreland was not taken into account by Schoot & Van Eerdt (1985) in their predictions.

At cliffs with a relatively high elevation cliff base (Rattekaai) (Fig. 9-B) erosion of the foreland was much less marked and had a relatively small effect on wave impact on the cliffs. Consequently, cliff erosion has now roughly resumed the low rates of the pre-barrier situation.

The strong recession of transitional salt-marsh edges was not expected, as these zones were considered to be potential areas of expansion of the salt marsh. The recession which occurred may be ascribed to two factors: increased wave attack and severe frost. In the pre-barrier period the zone with transitional gradients was at low elevation relative to MHW. However, along with the lowering of the MHW, the zone of maximum wave attack has now also been lowered; wave attack in the transitional salt marsh zone has therefore increased and with it erosion of bed material, including the uprooting of plants.

In addition, severe frosts during three successive winters (1985, 1986, 1987) is presumed to be the cause of a strong die-back of *Spartina anglica* in this zone. *Spartina* is a species susceptible to frost (Beeftink, 1979; De Leeuw *et al.*, 1994). With the frequent absence of flooding during these winters, the frost might have had a greater impact on the plants than would usually be the case.

In view of the foregoing we expect that the recession rate of the transitional edges will diminish in the future until no such edges will remain in the zone where wave action is concentrated most.

Very little generation of new salt marshes or expansion of existing salt marshes has been observed since 1986. Obviously the balance between erosive and depositional forces prevents net sedimentation and the establishment of pioneer vegetation. In addition, the very low sediment content in the flooding water will prevent further accretion on places where salt marsh plants might become established. This means that any new salt marshes will remain stuck at an elementary stage.

Consequences of the interim period

The 65% reduction in tidal range during the interim period raises the question whether this phenomenon has had specific and lasting consequences for abiotic aspects of the salt marshes (see also De Jong & De Jong, 1993).

One important feature, significantly influenced by the large reduction in tidal range, is the strong ripening of the soil, resulting in temporary or permanent acidification of the soil locally, 'levee-like' basins and steeper height gradients between levee and basin. It seems plausible that if there had been a direct shift in tidal range to 88%, the soil would have ripened less, more gradually, and without the severe consequences mentioned earlier.

Another feature, that can be ascribed largely to the interim period, is the (temporary) increased erosion of the salt marsh cliffs in 1987/88 as a result of the weakening of cliff strength. The continuing increased erosion rate of the cliffs with a less elevated cliff base may be caused by the lowering of the foreland. However, this process is not due to the interim period alone, as it would probably have occurred sooner or later as a result of the changed erosion and sedimentation processes on the foreland in the post-barrier period (see also Mulder & Louters, 1994). The increased wave impact on the transitional salt marsh edges would also have occurred without the interim period.

In general, it can be concluded that the temporary 65% reduction in the tidal range over 18 months has had lasting consequences for some abiotic aspects of the salt marshes. A direct shift to an 88% reduction of the tidal range would probably have had far less irreversible consequences.

Acknowledgements

The research for this paper has only been possible thanks to the efforts of many people. Of these we would mention particularly Mrs Houtekamer and Messrs Joosse, Boer, Siereveld, Vonk, Bleijenberg, Jonkers and Apon.

References

Adam, P., 1990. Saltmarsh Ecology. Cambridge Studies in Ecology, Cambridge University Press; ISBN 0-521-24508-7 (461 pp).

Beeftink, W. G., 1965. De zoutvegetatie van zuidwest Nederland beschouwd in europees verband (Salt marsh communities of the SW-Netherlands in relation to the European halophytic vegetation, in Dutch), thesis. Mededelingen Landbouwhogeschool 65-1, Wageningen (167 pp).

Beeftink, W. G., 1979. The structure of salt marsh communities in relation to environmental disturbances. In: Jefferies, R. L. & A. J. Davy (eds), Ecological processes in coastal environments. Blackwell, Oxford: 77–93.

De Jong, D. J. & Z. de Jong, 1993. The consequences of a one-year tidal reduction of 35% for salt marshes in the Oosterschelde (SW Netherlands). Proceedings ECSA.

De Jong, D. J. & A. M. van der Pluijm, 1994. Consequences of a tidal reduction for the salt-marsh vegetation in the Oosterschelde estuary (The Netherlands). Hydrobiologia 282/283: 317–333.

De Kogel, T. J. & D. J. de Jong, 1980. Vegetatiekartering van de schorren in de Oosterschelde en het Krammer-Volkerak, 1978 (Vegetation mapping of the salt-marshes in the Oosterschelde and the Krammer-Volkerak, 1978; in Dutch). Rijkswaterstaat, Deltadienst. Nota DDMI-80.20.

De Leeuw, J., L. Apon, P. Herman, W. de Munck & W. G. Beeftink, 1994. The response of salt-marsh vegetation to tidal reduction caused by the Oosterschelde storm-surge barrier. Hydrobiologia 282/283: 335–353.

Houtekamer, N. L., 1991. Schorren in de Oosterschelde: onderzoek naar het verzuringsrisico van schorbodems door de oxidatie van pyriet (salt marshes in the Oostershelde: research into acidification risks of the salt marsh soil by oxidation of pyrite, in Dutch). State University Utrecht, Institute of Geographical Research, Report GEOPRO 1991.05 (IRO).

Long, S. P. & C. F. Mason, 1983. Saltmarsh ecology. Blackie, Glasgow; ISBN 0-216-91438-8, 160 pp.

Mulder, J. P. M. & T. Louters, 1994. Changes in basin geomorphology after implementation of the Oosterschelde estuary project. Hydrobiologia 282/283: 29–39.

Nienhuis, P. H. & A. C. Smaal, 1994. The Oosterschelde estuary, a case study of a changing ecosystem: an introduction. Hydrobiologia 282/283: 1–14.

Oenema, O., 1988a. Pyrite accumulation in salt marshes in the Eastern Scheldt. In: Oenema, O. 1988: Early diagenesis in recent fine-grained sediments in the Eastern Scheldt. State University Utrecht, Institute of Earth Sciences, Thesis.

Oenema, O., 1988b. Distribution and cycling of fine-grained sediment in the Eastern Scheldt. In: Oenema, O., 1988: Early diagenesis in recent fine-grained sediments in the Eastern Scheldt. State University Utrecht, Institute of Earth Sciences, Thesis.

Oenema, O. & R. D. DeLaune, 1988. Accretion rates in salt marshes in the Eastern Scheldt, SW Netherlands. Estuar coast. Shelf Sci. 26: 379–395.

Pons, L. J. & I. S. Zonneveld, 1965. Soil ripening and soil classification. Int. Inst. for Land Reclamation and Improvement, publ. 13, Veenman, Wageningen, 128 pp.

Saeijs, H. L. F. & J. P. Al, 1982. Environmental impact assessment for design and management of the Oosterschelde barrier. In: Saeijs, H. L. F., Changing estuaries; State University Leiden, thesis/Rijkswaterstaat communications nr 32, Government Publishing Office, The Hague 1982. (ISBN 90-12-03921-5), 414 pp.

Schoot, P. M. & M. M. van Eerdt, 1985. Toekomstige ontwikkelingen van de schor gebieden in de Oosterschelde, procesonderzoek schorsystemen (Future developments of salt marshes in the Oosterschelde, process research on salt marsh systems; in Dutch). Rijkswaterstaat Tidal Waters Division, Report DDMI-85.23, Middelburg

Ten Brinke, W. B. M., J. Dronkers & J. P. M. Mulder, 1994. Fine sediments in the Oosterschelde tidal basin before and after partial closure. Hydrobiologia 282/283: 41–56.

Tutin, T. G., V. H. Heywood, N. A. Burges, D. M. Moore, D. H. Valentine, S. M. Walters & D. A. Webb, 1964–1980. Flora Europaea. Cambridge University Press, Cambridge. 5 Volumes.

Van Breemen, N., 1973. Soil forming processes in acid sulphate soils. Proc. Int. Symp. on Acid Sulphate Soils. Volume I. Wageningen.

Van Diggelen, J. 1988. A Comparative study on the ecophysiology of salt marsh halophytes. Thesis, Free University Press, Amsterdam. ISBN 90-6256-622-7 CIP, 208 pp.

Van Eerdt, M. J., 1985. The influence of vegetation on the processes of erosion and sedimentation in the salt marshes of the Oosterschelde. Vegetatio 62: 367–373.

Vranken, M., O. Oenema & J. P. M. Mulder, 1990. Effects of tide range alterations on salt marsh sediments in the Eastern Scheldt, SW Netherlands. Hydrobiologia 195: 13–20.

Vroon, J., 1994. Hydrodynamic characteristics of the Oosterschelde in recent decades. Hydrobiologia 282/283: 17–27.

Wetsteyn, L. P. M. J. & J. C. Kromkamp, 1994. Turbidity, nutrients and phytoplankton primary production in the Oosterschelde (The Netherlands) before, during and after a large-scale coastal engineering project (1980–1990). Hydrobiologia 282/283: 61–78.

Hydrobiologia **282/283**: 317–333, 1994.
P. H. Nienhuis & A. C. Smaal (eds), The Oosterschelde Estuary.
© 1994 *Kluwer Academic Publishers. Printed in Belgium.*

Consequences of a tidal reduction for the salt-marsh vegetation in the Oosterschelde estuary (The Netherlands)

D. J. de Jong & A. M. van der Pluijm
National Institute for Coastal and Marine Management/RIKZ, P.O. Box 8039, NL-4330 EA Middelburg, The Netherlands

Key words: salt marsh, vegetation, tidal reduction, Oosterschelde (NL), desiccation

Abstract

A storm-surge barrier was constructed in the mouth of the Oosterschelde, a euhaline mesotidal estuary in the SW Netherlands (mean tidal range 3.6 m). As a consequence, the tidal range and the Mean High Water in the estuary have been reduced to about 88% of their original values.

During the final construction stage of this barrier (1986–87) both were reduced to a maximum of 65% for more than 18 months. During this period, large-scale die-back of the vegetation occurred in vast areas on the salt marshes; locally, a complete die-back of the vegetation took place. Glycophytes and disturbance indicating species appeared on a large scale and grew abundantly. After the new tidal regime had been established, the vegetation recovered. The species characteristic of disturbance, are gradually being replaced by perennial salt marsh species. In addition, most species are shifting into zones of lower elevation, which correspond (in 1990/1991) more or less with the original flooding frequencies. Moreover, in many basins the 'levee'-species *Halimione portulacoides* and *Elymus pycnanthus* are far more prominent than before, probably as a result of the strong ripening of the soil that has occurred in these basins during the extra tidal reduction. In 1991, four years after the establishment of the new tidal regime, the salt marsh vegetation had still not been stabilized.

Introduction

Generally, the vegetation of salt marshes shows a zonation, which primarily seems to be related to elevation with respect to MHW (or to flooding frequency). However, a basin and a levee with the same flooding frequency have different vegetation, and in addition, many plant species may be regarded as more or less typical of levees or of basins. This implies that, as well as flooding frequency, other abiotic aspects also play an important role, e.g. morphology, soil and drainage. However, these other abiotic aspects may also be affected by flooding, its frequency as well as its duration. (For further details, see Chapman,

1977; Adam, 1990; Long & Mason, 1983; Beeftink, 1965, 1977a/b; van Diggelen, 1988).

The major impact of flooding on the salt marsh system means that changes in flooding will be reflected in changes in the soil and the vegetation. Recently, such an alteration in the flooding frequency occurred in the salt marshes of the Oosterschelde, an estuary in the SW Netherlands (Fig. 1). This alteration, a lowering of the MHW, was the result of the construction of the Oosterschelde works, consisting of a storm-surge barrier (SSB) in the mouth and two partitioning dams in the eastern part of the area (Fig. 1).

Owing to the presence of this SSB, since April 1987, (the present situation, hereafter referred to

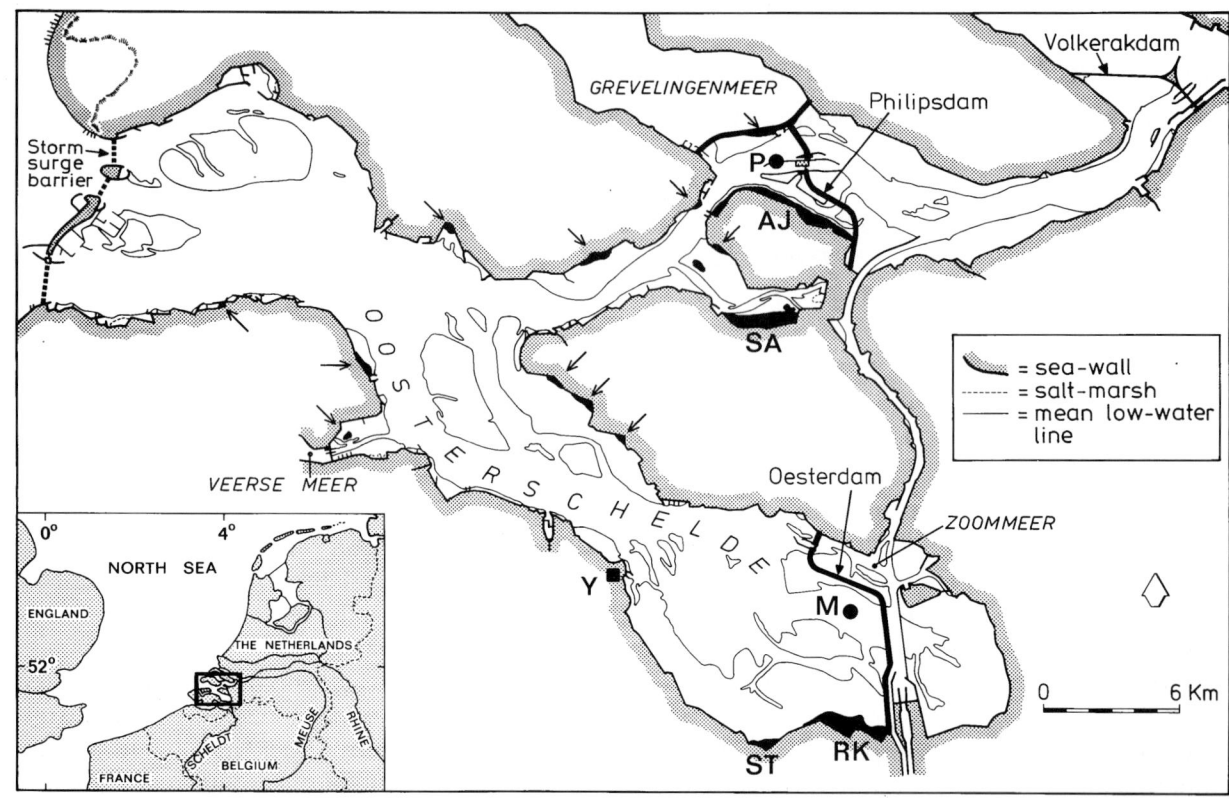

Fig. 1. Situation of the Oosterschelde and the salt marshes; salt marshes researched: RK = Rattekaai, SA = St Annaland, ST = Stroodorpepolder, AJ = Anna Jacobapolder (arrows: other salt marshes). Y = Yerseke; self-registering gauges: M = Marollegat, P = Philipsdam-west.

as the post-barrier period), both the tidal range and the MHW in the Oosterschelde have been reduced to 88% of the baseline situation (hereafter referred to as the pre-barrier period, that is the period up to the end of 1985).

During the final construction stage of the SSB (the interim period) the tidal range and MHW were temporarily additionally reduced to a maximum of 65% of their level during the pre-barrier period, from summer 1986 to April 1987 (Fig. 2). It is evident from a more detailed analysis of the annual number of floodings in three main elevation zones (Fig. 3) that, at the middle and high salt marsh elevations, flooding was almost completely absent during the interim period, while in the low salt marshes, flooding was quite drastically reduced during this period. Figure 3 also shows that flooding in the high salt marsh was

almost absent and in the middle salt marsh remained very low up to the summer of 1987. Therefore in this paper the interim period is defined as the period autumn 1985 to summer 1987.

At present, in the post-barrier period, all parts of the salt marsh are flooded again, but the frequency has dropped. See Nienhuis & Smaal (1994) and Vroon (1994) for more detailed information on the Oosterschelde works and alterations to the tidal regime.

The near or complete absence of flooding in the interim period, as well as the decreased flooding in the post-barrier period, have caused a strong ripening of the soil, especially in the middle and high salt marshes. This has had a number of important consequences (De Jong *et al.*, 1994; De Jong & De Jong, 1994; Vranken *et al.*, 1990):

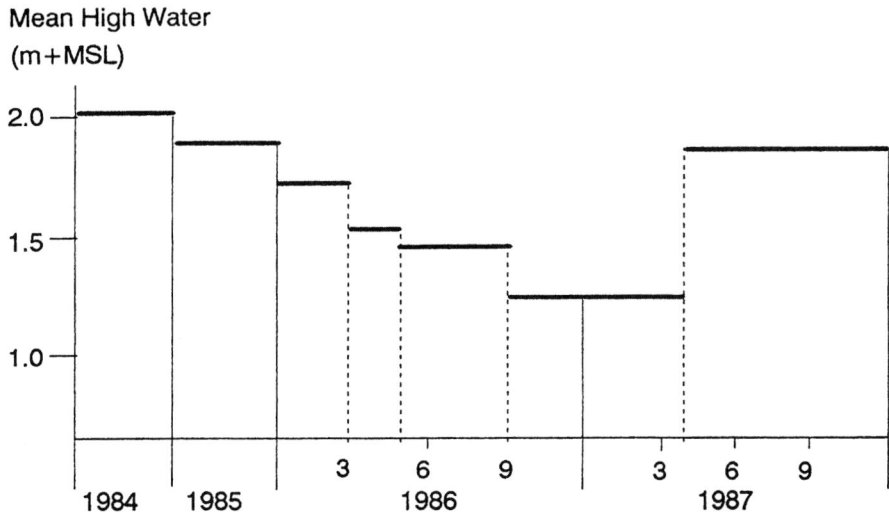

Fig. 2. Course of Mean High Water (MHW) (in m + Mean Sea Level, near Yerseke) in the period 1984–1987. Since 1988, MHW has been comparable to that of 1987.

= the horizontal stratification of the soil has been broken, as a consequence of which the water in the basins is able to drain faster. This has resulted in the basins being drier than they used to be. As a consequence, a considerable number of the basins, especially in the middle-high salt marsh at Rattekaai, have acquired the character of a levee with regard to soil structure and geohydrology.

= owing to the oxidation of pyrite in dry basins with low calcium carbonate levels, the pH decreases in a number of basins to values of about 5–6 in summer, in contrast with previous levels of 7–8. Moreover, laboratory analyses have shown that, in a number of basins, the pH may decrease much more, to values of 3–4.

= the surface level of the soil has subsided up to 10 cm owing to settling, a result of which is that flooding frequency has been reduced somewhat less than it would have been without settling. In addition, the transitional height gradient between basin and levee has become steeper and narrower (Fig. 5).

In the past, several predictions were made concerning the possible reactions of the vegetation to the decrease in flooding frequency (De Jong & De Kogel, 1985; Groenendijk, 1987; Saeijs & Al, 1982; Smies & Huiskes, 1981). Apart from differences in the details, the general expectation has been that the vegetation will shift to lower positions, in accordance with changes in flooding frequency.

The subject of this paper is to describe the actual changes in the vegetation caused by the tidal reduction. Attention is paid to both developments resulting from the temporarily sharp reduction during the interim period and the recovery and development of a new equilibrium in the present situation.

Materials and methods

Description of the area

The Oosterschelde is a euhaline (average chlorinity 16 g $Cl^- \cdot l^{-1}$), mesotidal basin with various salt marshes, mainly in the eastern part (Fig. 1, Table 1). Until 1985, the elevation of these salt marshes was between about 0.5 m below and 0.5 m above MHW (flooding frequency 600–20 times a year). The salt marshes are bordered on the seaward side by a tidal flat, generally via an eroding salt marsh cliff, and on the landward side

320

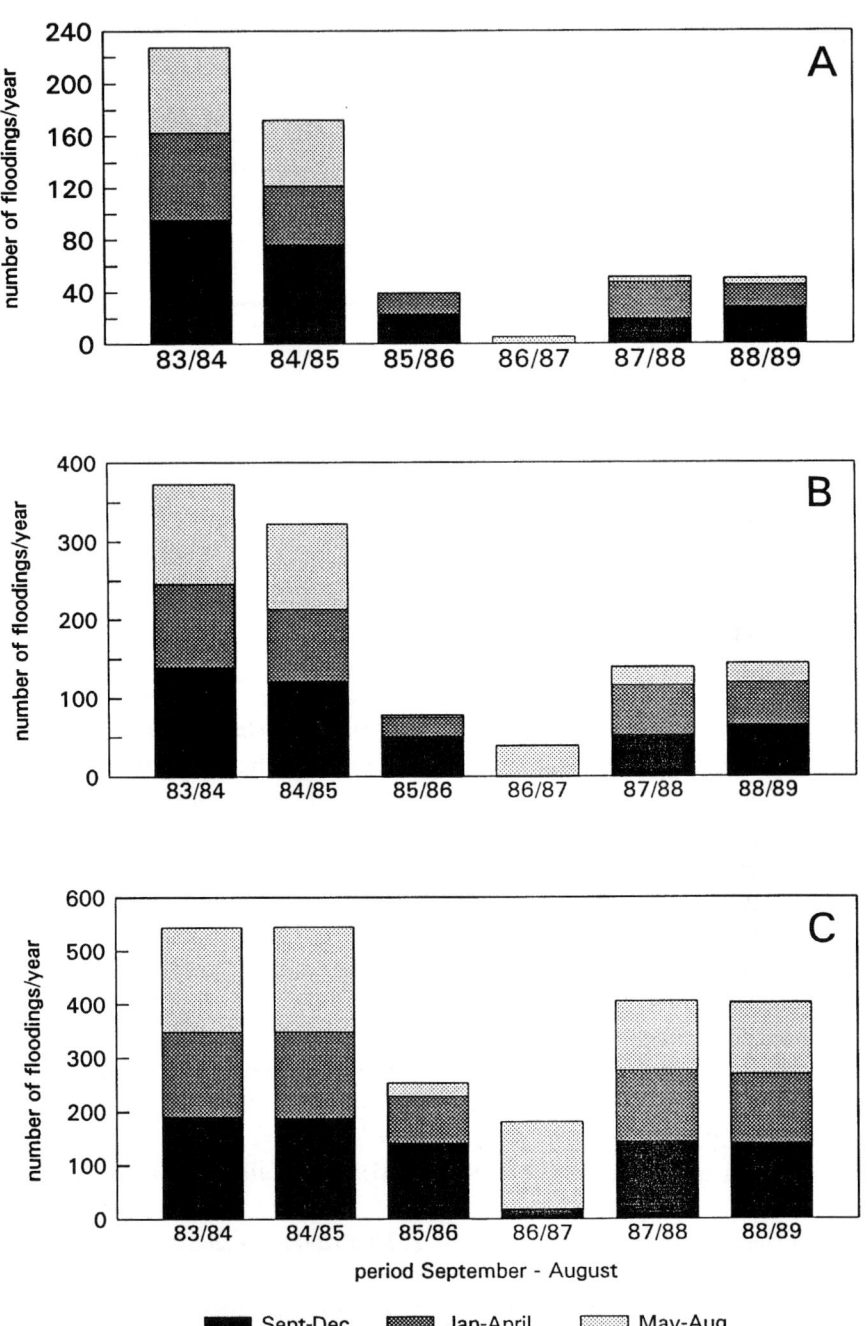

Fig. 3. Number of floodings in six subsequent periods of September to August *) for three elevation-zones of the salt marsh; A: high marsh (= aprox. 0.2 m above MHW in the pre-barrier period), B: middle marsh (= approx. around MHW), C: low marsh (= approx. 0.3 m below MHW).

* the reactions of the vegetation during the growing season are determined by the number of floodings since the previous September.

Table 1. Some general characteristics of the Oosterschelde (A) and of the salt marshes investigated (B).

A	≤ 1985	≥ 1987
Surface area total (ha)	45,200	35,000
intertidal area (ha)	17,000	11,300
salt marsh (ha)	1,725	645
Average tidal range Yeseke (m)	3.6	3.2
Mean High Water Yerseke (m + NAP)	2.1	1.8

B	Surface (ha)	Mean elevation (m + NAP)	Av. high water (m + NAP)	
			≤ 1985	≥ 1987
Rattekaai	100	1.9	2.05	1.85
St Annaland	195	1.5	1.80	1.55
Stroodorpepolder	25	2.1	2.05	1.85
Anna Jacobapolder	160	1.6	1.80	1.55

by a dike. Their soil consists of silt to silty clay, varying in thickness from 0.3 to 1.5 m, deposited at the sandy soil of the tidal flat.

The surveys were carried out on two salt marshes, called Rattekaai and St Annaland. Additional observations were made in most of the other salt marshes, Stroodorpepolder and Anna Jacobapolder being the most important (Fig. 1, Table 1). The four salt marshes mentioned together comprise about 75% of the total salt marsh area still present.

The salt marshes are divided into three 'elevation' types (based on the pre-barrier tidal range):

= high salt marsh: the zone with basins flooded less than 300 times a year;
= middle salt marsh: the zone with basins flooded 300–500 times a year;
= low salt marsh: the zone with basins flooded more than 500 times a year.

Rattekaai (and Stroodorpepolder) comprised all three elevation zones and St Annaland (and Anna Jacobapolder) mainly middle and low salt marshes.

The vegetation in the salt marshes studied consisted mostly of plant communities characteristic of low and middle salt marshes. The vegetation in the basins varied from monostands of *Spartina anglica** (often accompanied by *Aster tripolium*) on the initial salt marsh and in the lower basins, to mixed vegetations of *Spartina, Puccinellia maritima, Triglochin maritima, Plantago maritima* and *Limonium vulgare* in the middle basins. On the middle levees *Halimione portulacoides* dominated, and on the higher ones, *Festuca rubra* (sometimes accompanied by *Artemisia maritima* and/or *Elymus pycnanthus*). *Elymus* occurred in some high basins as well. More extensive descriptions of the Oosterschelde salt marshes are given by Beeftink (1965), Oenema (1988a), De Kogel & De Jong (1980) and De Jong & De Kogel (1988).

Measurements of the vegetation

28 transects (with a length of 10 to 20 m) were marked out with poles in the two salt marshes, 13 at Rattekaai and 15 at St Annaland. Each transect starts at a point in the centre of the basin and crosses the levee to the other side of the adjoining creek. The transects are distributed from the seaward edge to the dike, over one creek-system. Together, they comprise the most important vegetation zones. The vegetation in these transects has been described in adjacent plots of 0.5 × 0.5 m annually since 1984, always in the same month. Besides general data, such as total cover and height of the plants, this description accounts for the individual species with their percentage cover, subdivided into the following classes: 0.1, 0.2, ..., 0.5, 1, 2, ., 5, 10, 15, .., 95, 100%.

The results of these measurements are shown by means of histograms (Fig. 4) in which the horizontal axis depicts the length of the transect and the height of the bars represents the extent (in % cover) to which the various species occur. In these diagrams the original 'dominant' perennial species are black, and the species that appeared afterwards are grey. Height profiles of the transects are shown in Fig. 5.

---

* Nomenclature of vascular plants follows Tutin *et al.* (1964–1980).

Tidal data to calculate flooding frequencies are taken from selfregistering gauges in the Oosterschelde (Marollegat, Philipsdam-west), as provided by Rijkswaterstaat. Weather data are used from the Royal Dutch Meteorological Institute in De Bilt, as measured in their weather-station in Vlissingen.

Results

The history of the vegetation will be described chronologically, and illustrated by means of five transects (Fig. 4) reflecting the main vegetation changes. As both salt marshes studied show differences in developmental trends they are described here separately.

Rattekaai

Pre-barrier period
In 1985, *Halimione* and *Spartina* declined in cover by 10 to 30 percent in most of the salt marsh except for the initial salt marsh (Fig. 4: A, B, C).

Interim period
In 1986 most perennial species declined in cover to a greater or lesser extent, and annual disturbance indicating species invaded to a large extent: *Atriplex prostrata* in the higher parts and *Suaeda maritima* and *Aster* in the lower parts; locally, *Atriplex* and *Suaeda* almost formed 'monostands' (Fig. 4: A). In 1987 not only did the original species die back in large parts of this salt marsh, but the disturbance indicating species also showed an

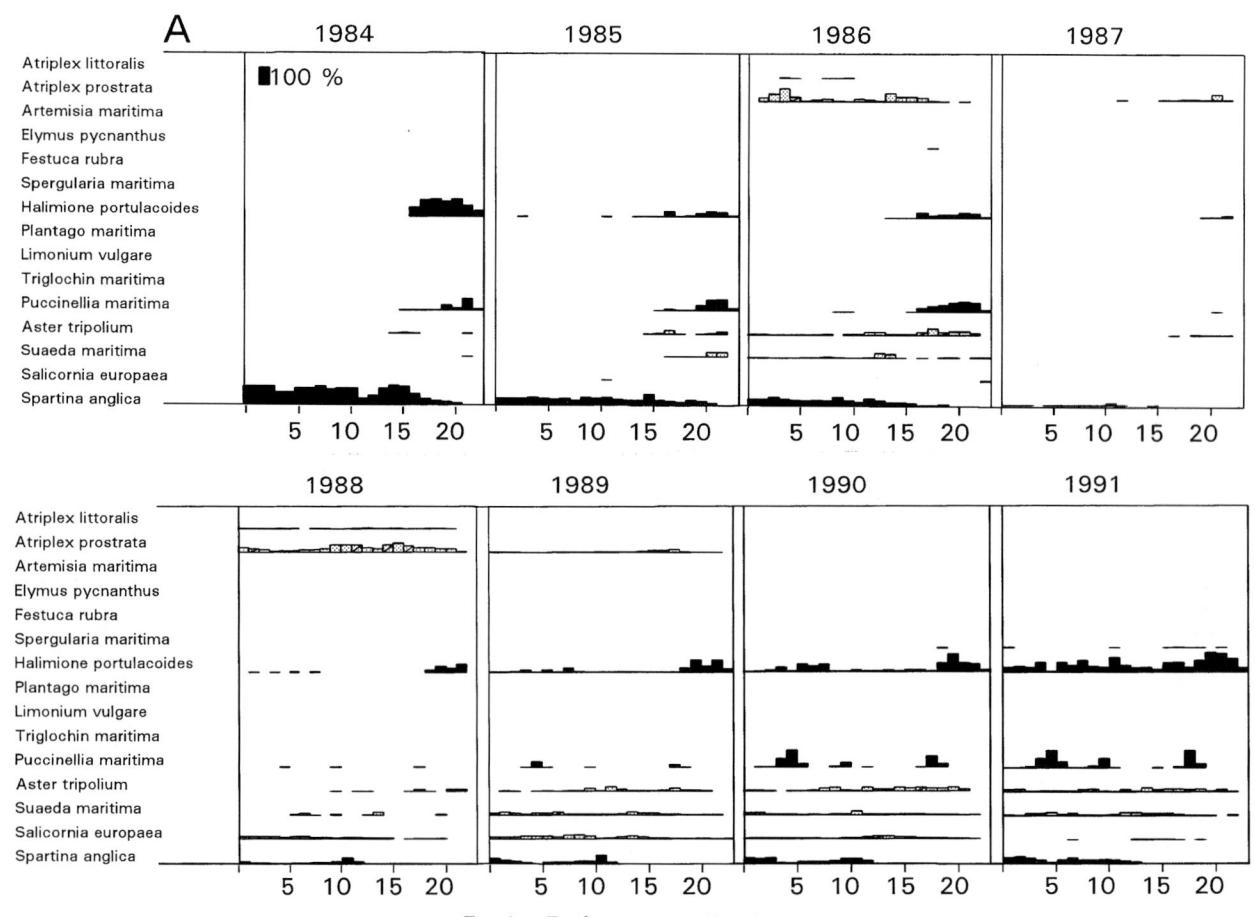

Fig. 4a. Explanation, see Fig. 4e.

almost complete die-back; this resulted in extensive bare areas (about 40% of the marsh) in the middle salt marsh zone (Fig. 4: A, B, C). At other places, the original species and/or the disturbance indicating species held their own reasonably well (Fig. 4: C).

Nitrophylous species such as *Aster*, *Suaeda* and *Atriplex* grew 2–3 × higher than normal in 1986; however, in 1987, the latter two remained very small, only a few centimetres, while *Aster* reached normal size. Furthermore many glycophytes were found on the middle and high levees in 1986 and 1987; these included *Matricaria maritima ssp. inodora*, *Solanan nigrum*, *Sonchus arvensis*, *Plantago major*, *Cirsium arvense* and even *Sambucus nigra*. The density of this group varied between 0.1 and 1 specimen per m[1] levee.

Post-barrier period
In 1988, the glycophytes had disappeared completely. The disturbance indicating species held their own strongly or re-appeared and often became dominant, determining the overall appearance (Fig. 4: A, B, C-partly). Since 1989 onwards, the disturbance indicating species have gradually disappeared.

The perennial salt marsh species have re-appeared since 1988 as well. On the one hand, this has been a return of species to their original zone, and on the other hand, a gradual shift or invasion into the lower basins by 'levee' species which originally occurred in higher zones, especially *Halimione* (Fig. 4: A, C) and *Elymus* (Fig. 4: B). At Rattekaai there is a development towards 'monostands' of these 'levee' species. The original

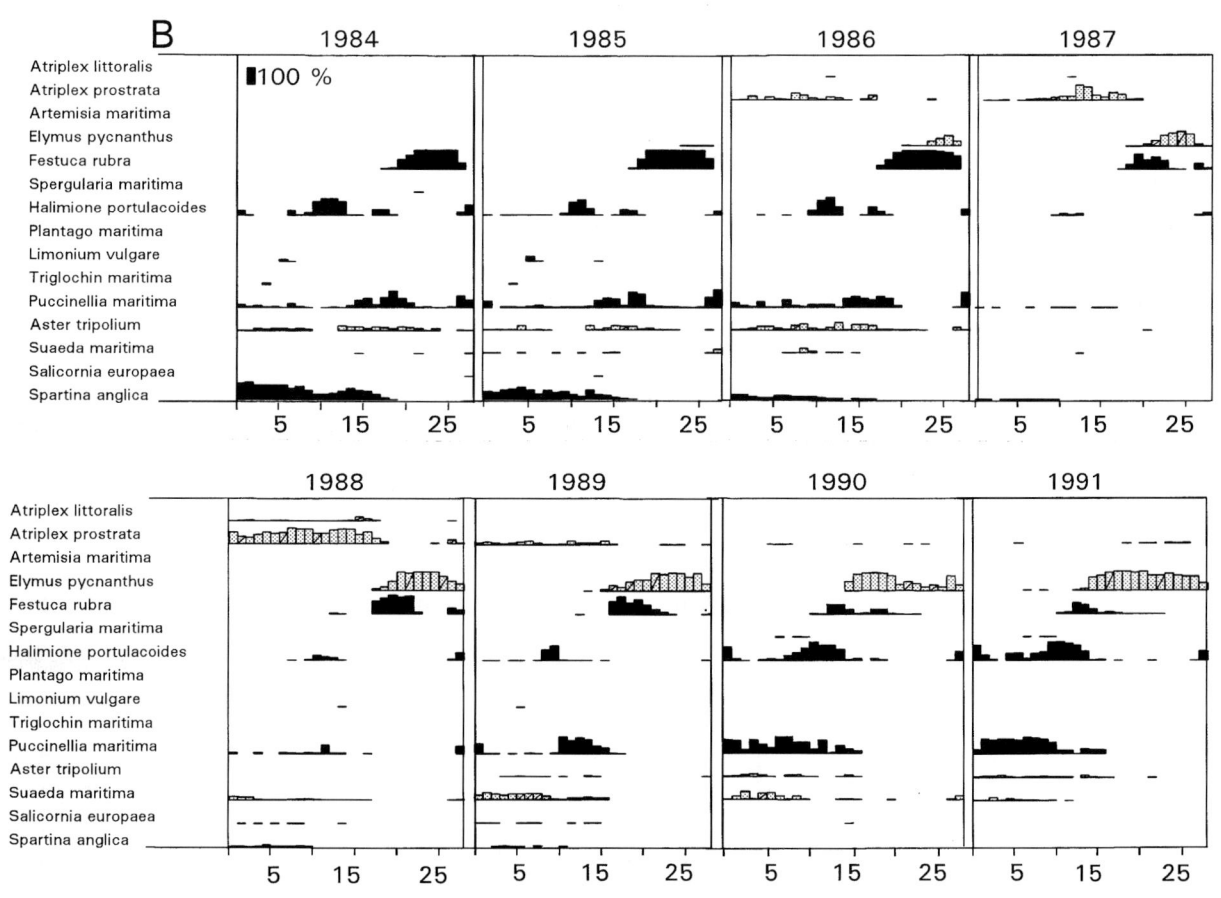

Fig. 4b. Explanation, see Fig. 4e.

324

'basin' species, *Limonium* and *Puccinellia*, are thriving better than before, in contrast with *Triglochin* and *Plantago maritima* which are often doing less well. In 1991, these developments were still continuing.

St Annaland

Pre-barrier period
In 1985, *Halimione* and *Spartina* showed a decline over the larger part of the salt marsh, varying from 10 to 30 percent (Fig. 4: D, E).

Interim period
In 1986 and 1987, disturbance indicating species increased, but the original vegetation was more or less able to hold its own and it continued to dominate the general picture (Fig. 4: D, E). In 1987, no bare spots occurred, neither 'giant' growth nor 'dwarf' growth of disturbance indicating species in 1986 and 1987 resp.. Glycophytes were hardly observed.

Post-barrier period
In 1988, the sporadically present glycophytes vanished completely. The disturbance indicating species, however, held their own or re-appeared, only to disappear gradually in the years after. However, they were not dominantly present in this salt marsh in 1988. The perennial species have re-appeared since 1988 as well. Here too, 'new' species are appearing besides the ones originally present, but to a far lesser extent than at

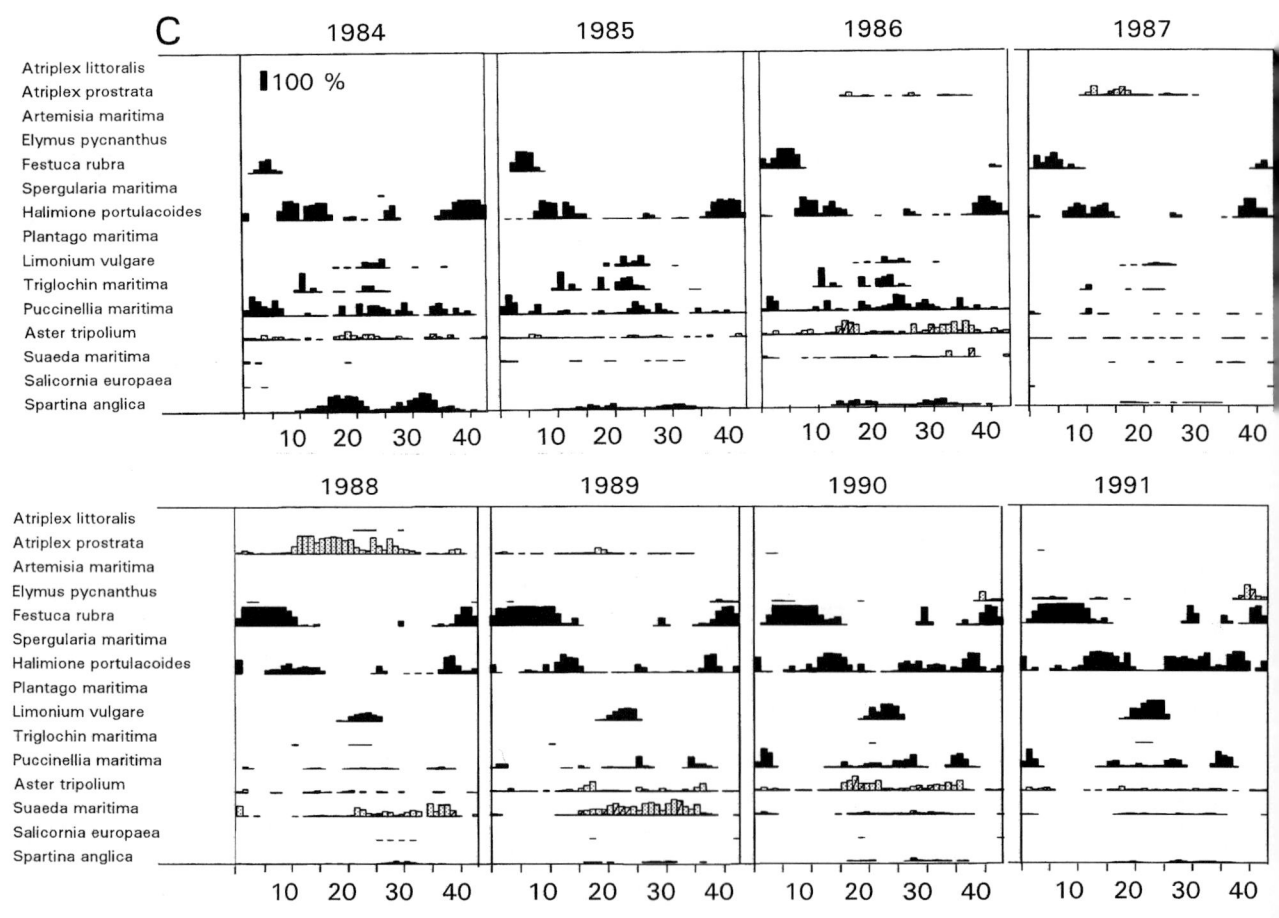

Fig. 4c. Explanation, see Fig. 4e.

Rattekaai. Finally, there is also often a gradual shift or invasion into the lower parts of the salt marsh by species that occurred originally at higher elevations. However, here not only 'levee' species are concerned, but also 'basin' species of the middle basins, such as *Puccinellia*, *Plantago maritima*, *Triglochin* and *Limonium* (Fig. 4: D, E) (all four species are doing better than before on the northeastern salt marshes). In many originally low *Spartina* basins this development has led to the growth of the varied vegetation characteristic of middle basins (Fig. 4: D). Low *Spartina/Aster* levees are developing a middle 'levee' vegetation of *Halimione* (Fig. 4: D). Developments towards vegetations dominated by one species have hardly occurred up to 1991.

Figure 6 indicates the overall trend for the dominant species/vegetation types from the pre-barrier to the post-barrier period (1984 to 1991).

Basins with *Spartina* as the dominant species have generally undergone a sharp change in character, with the exception of the lowest salt marsh zone (the initial salt marsh zone) in which the *Spartina* vegetation held its own fairly well during the whole period. In the low and middle-high basins, *Spartina* has generally been replaced by *Puccinellia* and/or *Halimione*, especially at Rattekaai. At St Annaland the low *Spartina* basins have changed in many cases into a mixed 'basin' vegetation of *Spartina* with species such as *Limonium*, *Puccinellia*, *Triglochin* and *Plantago*.

Basins with a mixed vegetation of *Spartina*, *Puccinellia*, *Triglochin*, *Limonium* and *Plantago* partly stayed mixed, but with less or no *Spartina*,

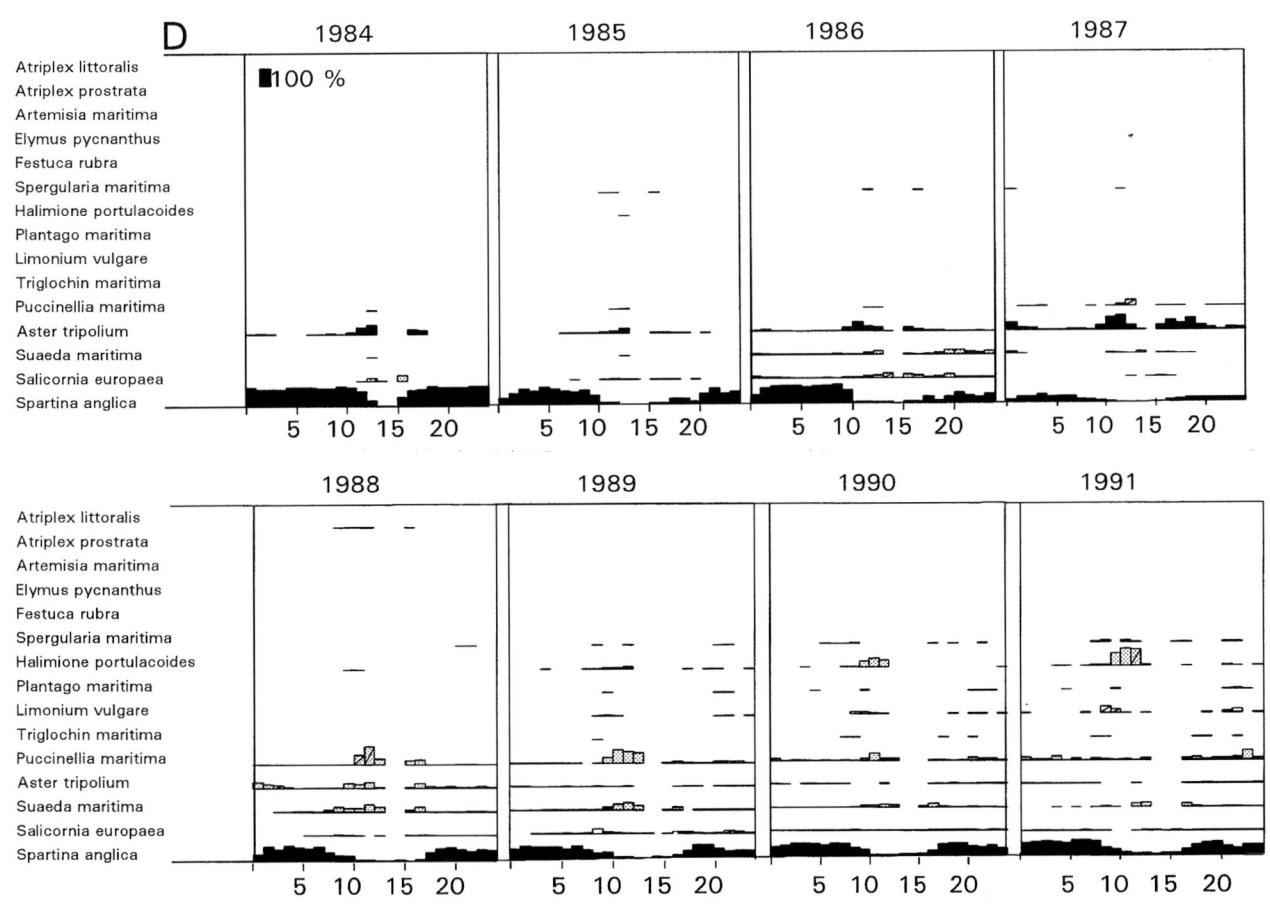

Fig. 4d. Explanation, see Fig. 4e.

326

while *Limonium* and *Puccinellia* thrived better than before; at St Annaland, *Triglochin* and *Plantago* also often increased. In other mixed basins, the vegetation has shifted to a strong dominance by *Puccinellia* or *Halimione* to which *Elymus* has been added recently in some places.

The levees with *Halimione* have held their own in many cases in the post-barrier period, independently of a decline in 1985 and/or during the interim period. At present, this species is extending quite vigorously towards the adjoining basins. On some of these levees, however, *Halimione* is being replaced or succeeded at a high rate by *Festuca* or *Elymus*. At St Annaland, new *Halimi-*

one levees are developing alongside small creeks in the low *Spartina* basins.

On levees with a dominance of *Festuca rubra* the development of this species was two-fold. Either it has been replaced by *Elymus* (especially when *Elymus* was already present in the vicinity in the pre-barrier period), or it has been able to hold its own and even to extend towards the adjoining basins.

Levees with a dominance of *Elymus pycnanthus* have held their own well and are often extending in the direction of the adjoining basins in the post-barrier period.

Figure 7 shows the shift in elevation for three

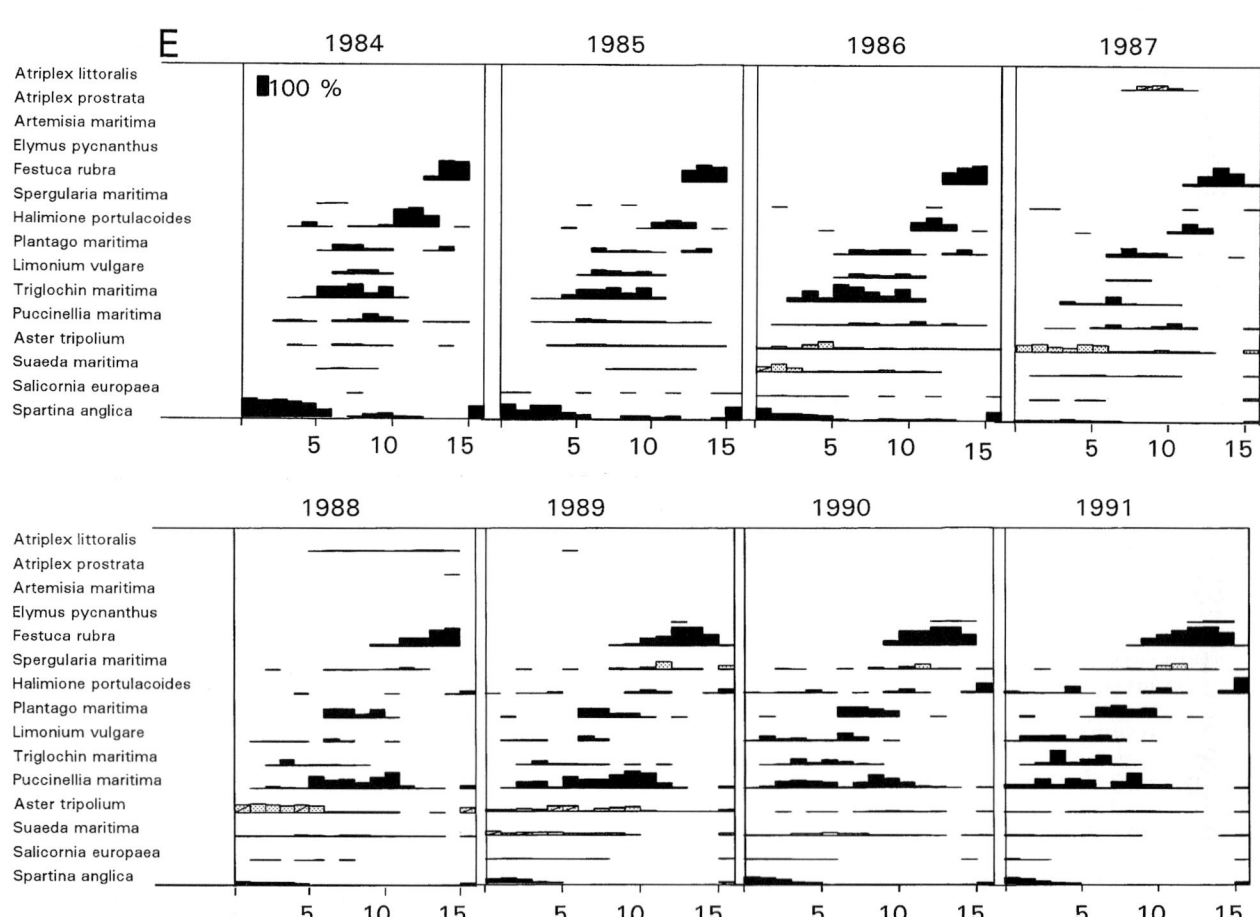

Fig. 4. Vegetation development in five representative transects. A: Rattekaai low marsh, B: Rattekaai middle marsh, C: Rattekaai high marsh, D: St Annaland low marsh, E: St Annaland middle marsh; 1984 and 1985: pre-barrier period, 1986 and 1987: interim period, 1988–1991: post-barrier period.

Horizontal axis is distance in m. Each bar represents a distance in the transect of 0.5 m.; height of the bars represents plant cover (in %); original dominant species are black, species appearing afterwards are grey.

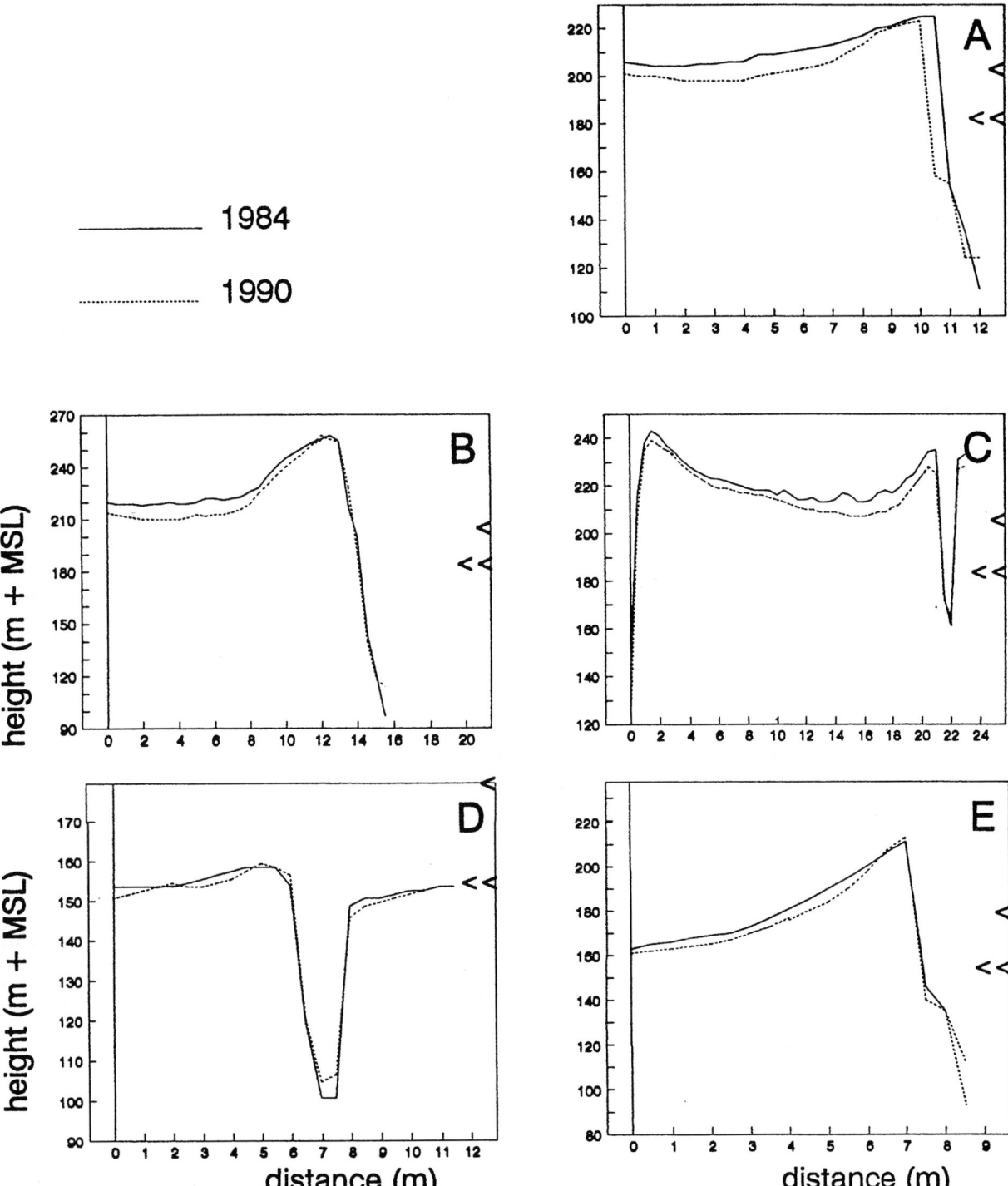

Fig. 5. Elevation profile of the vegetation transects in Fig. 4 in both 1984 (pre-barrier period) and 1990 (post-barrier period). The figure also shows the subsidence due to shrinkage of the soil in the interim period. A, B, etc. see Fig. 4. Arrows indicate MHW in pre-barrier (<) and post-barrier period (< <).

328

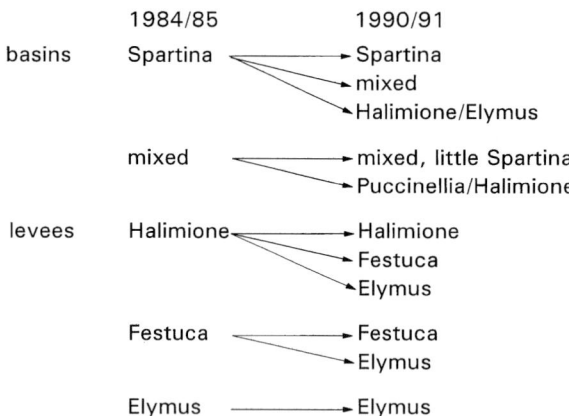

Fig. 6. Scheme with the overall changes in dominance of the species from 1984/85 (pre-barrier period) to 1990/91 (post-barrier period).

Table 2. Monthly mean temperature (°C) and number of ice days (= days with max. temp. < 0 °C) in January and February in the period 1985-1991, as well as the averages for the period 1950-1990 for station Vlissingen. (data after Royal Dutch Meteorological Institute (KNMI), De Bilt)

	Average temperature (°C)		Number of icedays	
	January	February	January	February
1985	− 1.5	− 0.3	13	10
1986	3.4	− 3.6	1	12
1987	− 1.3	2.5	13	0
1988	6.6	5.3	0	0
1989	5.3	5.6	0	0
1990	5.9	7.9	0	0
1991	4.1	0.5	0	6
Long-term average	3.1	3.1	3	2

dominant species. This figure makes clear that all three species have shifted downwards with some 20 cm, approximately to the same extent as the downward shift in MHW. It is also clear that *Halimione*, especially, is still extending downwards, even further than the 20 cm lowering of MHW. At present *Halimione* can be found at a relatively lower level (= higher flooding frequency) than in the pre-barrier period, a process that still seems to be continuing.

Discussion

Die-back of Spartina and Halimione in 1985

As early as 1985, even before flooding frequencies in the middle and low salt marsh decreased significantly, *Halimione* and *Spartina* declined in many of the transects. Given the severe winter of 1985 (the winters of 1985, 1986 and 1987 rank among the severest winters in recent decades, see Table 2) and the susceptibility of both species to sharp frost (Beeftink, 1979, 1987; Beeftink *et al.*, 1978; De Leeuw *et al.*, 1992), it may be assumed that the decline of these species in 1985 was due to the severe winter of that year. This assumption is confirmed by the analyses of De Leeuw *et al.* (1994).

Tidal reduction during the interim period

The reactions of the vegetation observed in 1986 and 1987 (die-back of original species, intense invasion of disturbance indicators and glycophytes) are quite similar to those that occur in salt marshes in the first two years after being completely closed off from the tide; e.g. such as was observed in Lake Veere and Lake Grevelingen, two estuaries in SW Netherlands that were closed off from the tide in 1961 and 1970 respectively (Beeftink, 1975, 1979, 1987; De Jong & De Kogel, 1979). The marsh developments in these lakes depend on the ripening of the soil and desalinisation, which both start as soon as flooding stops. Vranken *et al.* (1990), De Jong & De Jong (1994) and De Jong *et al.* (1994) observed a fast and strong ripening of the salt marsh soil in 1986 and]987. Moreover, the process of ripening caused by the significant decline in the tide was probably intensified by the sharp frost in 1986 and 1987, since frost affects the ripening of soil positively. The changes in the vegetation of the salt marshes in the Oosterschelde in 1986 and 1987 also indicate a strong ripening of the soil as well as a rapid desalinisation in large parts of the salt marshes in this period. Given the speed and extent by which the vegetation reacted, as early as 1986 (many glycophytes and an intensified growth of nitro-

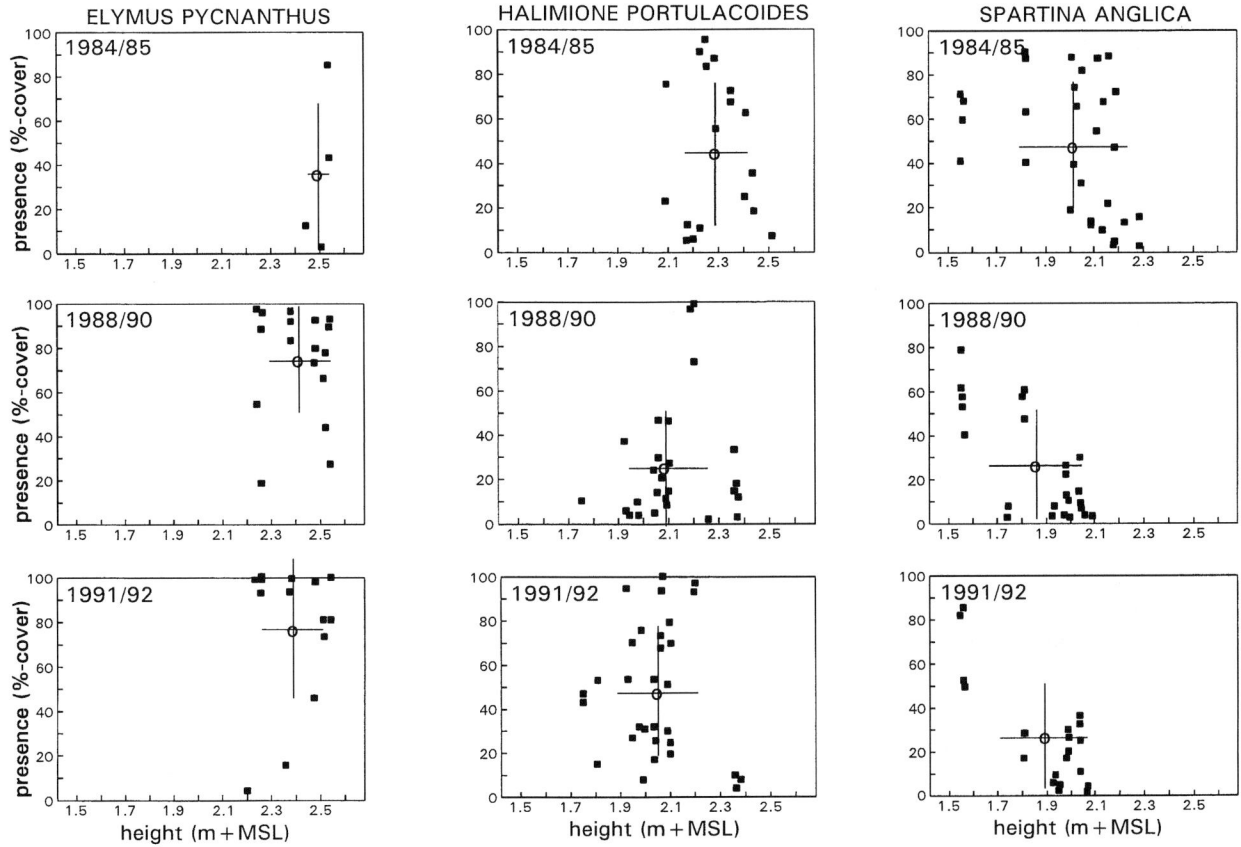

Fig. 7. Relation to height with respect to Mean Sea Level of the three most important species in the Oosterschelde salt marshes in the pre- and post-barrier period; Mean and standard deviations are indicated.

phyllous species), after only a few months with hardly any flooding, these processes in the soil are clearly present. Moreover, plants may have been weakened by the frost in 1986 and 1987, and as a consequence, they may have reacted faster and more strongly to the occurring ripening of the soil than without the frost.

Many glycophytes occurred on the higher levees of the middle and high salt marsh, especially at Rattekaai in 1986, owing to the desalinisation of the soil as a consequence of the almost complete absence of flooding. It is striking that in 1987, when the tidal range increased from 65% to 88%, the glycophytes occurred on a similar scale to that in 1986 (this observation is in contrast with the claim of De Leeuw *et al.* (1994)). This continuing presence can be explained by the low number of floodings in the middle and high salt

marshes before September 1987 (Fig. 3), as a result of which resalinisation of the soil was delayed.

A noticeable fact is that there was an extreme growth in height of some disturbance indicating species (especially *Suaeda*, *Aster* and *Atriplex*) (2–3 × higher than normal) in 1986, but that in the years thereafter, the nitrophyllous disturbance indicating species (*Suaeda* and *Atriplex*) remained extremely small (5–10 cm). This might indicate an almost complete consumption by these species of the nutrients released during the initial ripening of the soil in 1986, so that in the years afterwards there was a shortage of them. No data are available on this subject, however.

A second striking phenomenon is the large-scale die-back of all vegetation at Rattekaai in 1987, both of the original species and the disturbance indicating species. Both *Spartina* in the basins and *Halimione* on the levees in the middle salt marsh died in about 40% of this zone. There are two possible explanations for this. In the first place, a massive layer of dead plant debris was found in this area in 1987, since material that had died off earlier had not been washed away by lack of flooding in the preceding winter. This massive layer probably prevented the germination of seeds completely and/or suffocated possible seedlings; a process that is similar to the local washing up of massive rafts of tidal litter, owing to which the vegetation dies off temporarily (e.g. Adam, 1990; Long & Mason, 1983). A second explanation could be that germination did occur in the early spring, but that the recurrence of flooding since mid-April was a too great an osmotic shock (re-salinisation of the soil) to the seedlings which caused their die-off (Adam, 1990) (see also De Leeuw *et al.*, 1994).

Tidal reduction in the post barrier period

As soil ripening is essentially an irreversible process (De Jong *et al.*, 1994), the strong soil ripening during the interim period will have effects in the post-barrier period. For the vegetation this means that the soil of the basins in large areas of the salt marshes have acquired the character of that of a levee, with a better aerated and faster drained soil, probably leading to moisture deficiency temporarily. In addition to that acidification phenomena might occur in basins (De Jong *et al.*, 1994). This is at present expressed in the development of the vegetation by way of a sharp increase in more or less typical 'levee' species (*Halimione*, *Festuca* and *Elymus*) in the middle and low basins. They occur in basins with a relative elevation that used to be the domain of 'basin'-species such as *Spartina*, *Puccinellia* and *Limonium*.

Consequences of possible severe moisture deficiency in the basins have been observed on a small scale up to 1991. In the vegetation these patches (of several m^2) are characterized by a low vegetation density of mainly *Suaeda*, *Salicornia*

and *Aster*, the plants remaining low, deformed and being very susceptible to fungi and insects.

It appears from the data from the transects that the shift of 'levee' species towards the basins is a gradual process that has manifested itself since 1988. On the one hand, the species concerned establish themselves more or less randomly in the basin after which they spread out gradually, until eventually a dominance of one species occurs. On the other hand, the species concerned gradually invade from the levee. Both processes indicates that this is a development towards a new equilibrium, in which the 'levee' species colonize the basins on a broad scale. This contradicts De Leeuw *et al.* (1993) who claim that *Halimione* should have colonized the basins only temporarily after a large invasion in 1986. From our measurements (see also Fig. 4) it appears that this species was hardly present if at all, in the basins in 1986 and 1987, but that it has invaded them on a broad scale since 1988.

Regional differences between Rattekaai and St Annaland

The two salt marshes studied, Rattekaai and St Annaland, have reacted quite differently to the decrease in flooding frequency: Rattekaai has shown much more intense reactions than St Annaland. Moreover, the general observations showed that Stroodorpepolder showed tendencies similar to those in Rattekaai, and Anna Jacobapolder similar ones to St Annaland. Consequently, a distinction can be made between the reactions of the salt marshes in the SE part of the Oosterschelde and those in the NE part: in the first area the reactions are more intense.

This contrast is probably connected to differences in history between the two salt marshes with respect to tidal range. As a result of the construction of the Volkerakdam in 1969/70, the St Annaland salt marsh was subject to a larger tidal range and a higher MHW (about 0.2 m) in the period 1970–1985 than until 1970; since 1987, the tidal range and MHW have returned to about the level of before 1970 (Vroon, 1994; Beeftink,

1979); at Rattekaai, no significant tidal changes took place at that time. The consequence of the increase in flooding frequency was that the salt marsh at St Annaland became a low-to-middle salt marsh during the period 1970–1985 instead of the middle-to-high salt marsh that it used to be. This expressed itself in such manifestations as a shift in the higher salt marsh vegetation types to middle salt marsh vegetation types. The vegetation in the middle and low salt marsh changed little (Beeftink, 1979).

The soil in the St Annaland salt marsh had to adjust to the higher flooding frequency in the period 1970–1985. However, as ripening of the soil is basically an irreversible process (De Jong et al., 1994), the degree of ripening was hardly reversed; instead, the soil in the salt marsh became more or less inundated. The temporary absence of flooding in 1986 and the return to the pre-1970 MHW level from 1987 onwards meant that the new ripening led to a return to the original pre-1970 ripening state. The soil in the St Annaland salt marsh, therefore, had much less of a shock than that in Rattekaai, and this far smaller shock has resulted in a far less intense reaction by the vegetation.

Conclusion

The character of the vegetation in the salt marshes of the Oosterschelde has been changed by the tidal reductions. The greatest changes have occurred in the salt marshes where the soil has changed (ripened) most. The extra reduction in tidal range during the interim period, and the severe frost, have had an intensifying effect on the ripening of the soil and on the reaction of the vegetation to it.

This leads to the conclusion that species like *Spartina* and *Halimione* have declined more in vitality owing to the combination of an extra tidal reduction (of 65%) and sharp frost immediately before and during the interim period than would have occurred with a 88% tidal reduction only. Owing to this, the frost may have played an important part in the creation of bare patches in the

basins dominated by *Spartina*, and in the invasion of salt marsh species from the middle salt marsh, such as *Puccinellia*, *Limonium* and *Halimione*, in this normally persistent vegetation.

Future developments in the salt marsh vegetation

Since between 1990 and 1991 still significant shifts in the vegetation took place, it may be concluded that in 1991 there was no evidence of a new equilibrium in the vegetation (in contrast with the view of De Leeuw et al., 1994). Although most species seem to have returned to their former relative elevation (flooding frequency) (Fig. 7; see also De Leeuw et al., 1994), important shifts are still occurring, for some species even to lower relative elevations (see *Halimione* in Fig. 7). How the vegetation will develop in the near future may be deduced from the recent developments in the vegetation in the transects, taking into account the developments in the soil. Given the differences between Rattekaai and St Annaland, a separate prognosis will be made for each salt marsh.

The basins at *Rattekaai* are fairly strongly ripened and have become better aerated and drained so that, with regard to soil structure and geohydrology, they resemble a well-aerated and drained levee more than a reduced, wet basin. Besides, during dry periods, they have become susceptible to strong dehydration and probably locally also to acidification. This means that the vegetation in the basins will obtain the character of a levee, with *Halimione* in the lower and middle basins and *Elymus* in the higher ones as the main species. Only in the initial salt marsh *Spartina* will probably dominate the vegetation. Moreover, the lower high water levels might lead to a decreased export of dead plant material and to more frequent deposition of plant debris on the higher levees and in the higher basins. This might enhance the settlement of *Elymus*, as well as of *Atriplex prostrata* and *A litoralis* in these areas. Eventually, the vegetation at Rattekaai will mainly be characterized by three types poor in species: *Elymus* (and *Atriplex*) on the high and middle levees and in the high basins, *Halimione* on the low levees and in the middle and low basins, and *Spartina* in the initial salt marsh. Whether species like *Limonium* and *Puccinellia* are able to withstand the invasion of *Halimione* in the middle basins is still unclear. If there are basins that seriously dehydrate periodically, possibly combined with acidification, the vegetation in the basins concerned will die off partly or completely during such a period and will be replaced temporarily by annuals like *Suaeda* and *Salicornia* and a species like *Aster*. If dehydration occurs frequently, or if the acidification becomes more permanent, these basins will be permanently covered with these species or even remain permanently bare.

The vegetation on *St Annaland* has generally held its own up to now. In the lower parts, there is even an increasing diversity. Given the history of the salt marsh, it is not likely that the soil in the basins has acquired a typical 'levee' structure, although they might be somewhat drier than before. A massive invasion of *Halimione* in the basins and of *Elymus* on the levees has not occurred here (yet). The problems connected with a decreased export of dead plant material and frequent deposition of wracked material will be less as this salt marsh is situated lower according to MHW. Eventually, at St Annaland the middle basins will probably keep their 'basin' character, with a vegetation rich in species in which *Puccinellia* will play an important part. The low *Spartina*-basins will acquire a vegetation rich in species characteristic of the middle basins. On the high levees, a *Festuca* or *Elymus* vegetation will develop, on the middle levees a *Festuca* vegetation, and on the low levees a *Halimione* vegetation. In general, this salt marsh will probably acquire a character that is richer in species and more varied than at Rattekaai.

The period necessary to reach a new equilibrium in the vegetation may be derived from Beeftink (1979). He mentions a period of six years after sharp frost and four to seven years after a sudden increase in tidal range; for the latter, he found that when the increase in the tidal range (disturbance) is smaller, the period of recovery becomes longer (three to four years at an increase of 30 cm and seven years at an increase of 10 cm). As a result,

the equilibrium in Rattekaai (where the shock was bigger than in St Annaland) would be restored sooner than in St Annaland. At the moment, the vegetation has had four years to recover from the shock, and an equilibrium has not yet been reached at all. A few years would have to be added for Rattekaai and a few years more for St Annaland. This suggests that in Rattekaai the equilibrium of the vegetation may be reached after some eight to ten years later than 1987 (in 1995–1997) and in St Annaland after ten to twelve years (in 1997–1999).

References

Adam, P., 1990. Saltmarsh ecology. Cambridge University Press, Cambridge.

Beeftink, W. G., 1965. De zoutvegetatie van ZW Nederland beschouwd in europees verband (Salt marsh communities of the SW Netherlands in relation to the European halophytic vegetation; in Dutch). Ph. D. Thesis. Mededelingen Landbouwhogeschool 65–1, Wageningen, 167 pp.

Beeftink, W. G., 1975. The ecological significance of embankment and drainage with respect to the vegetation of the South-West Netherlands. J. Ecol., 63: 423–458.

Beeftink, W. G., 1977a. Salt-marshes. In: R. S. K. Barnes (ed.), The Coastline, Chapter 6, John Wiley & Sons, London: 93–121.

Beeftink, W. G., 1977b. The coastal salt marshes of Western and Northern Europe: An ecological and phytosociological approach. In V. J. Chapman (ed.), Wet coastal ecosystems, Elsevier, Amsterdam: 109–155.

Beeftink, W. G., 1979. The structure of salt marsh communities in relation to environmental disturbances. In: R. L. Jefferies & A. J. Davy (eds), Ecological processes in coastal environments. Blackwell, Oxford: 77–93.

Beeftink, W. G., 1987. Vegetation responses to changes in tidal inundation of salt marhes. In: J. van Andel, J. P. Bakker & R. W. Snaydon (eds), Disturbance in grasslands. Junk publishers, Dordrecht: 77–117.

Beeftink, W. G., W. de Munck & J. Nieuwenhuize, 1978. Aspects of population dynamics in Halimione portulacoides communities. Vegetatio 36: 31–43.

Chapman, V. J., 1977. Introduction. In: V. J. Chapman (ed.), Wet Coastal Ecosystems. Elsevier, Amsterdam: 1–29.

De Jong, D. J. & T. J. de Kogel, 1979. Vegetatieontwikkeling op de Slikken van Flakkee van 1972 tot en met 1978 (Vegetation development at the Slikken van Flakkee from 1972–1978; in Dutch). Rijkswaterstaat, Middelburg (NL) report 79–11.

De Jong, D. J. & T. J. de Kogel, 1985. Salt-marsh research with respect to the civil-engineering project in the Oosterschelde, The Netherlands. Vegetatio 62: 425–432.

De Jong, D. J. & Z. de Jong, 1994. The consequences of a one-year tidal reduction of 35% for salt marshes in the Oosterschelde (SW Netherlands). In Jones, N. V. & J. S. Pethick (eds), The Changing Coastline. Coastal Zone Topics: Process, Ecology and Management, Vol. I. Joint Nature Conservation Council, in press.

De Jong, D. J., Z. de Jong & J. P. M. Mulder, 1994. Changes in area, geomorphology and sediment nature of salt marshes in the Oosterschelde estuary (SW Netherlands) due to tidal changes. Hydrobiologia 282/283: 303–316.

De Kogel, T. J. & D. J. de Jong, 1980. Vegetatiekartering van de schorren in de Oosterschelde en het Krammer-Volkerak, 1978 (Vegetation mapping of the salt marshes in the Oosterschelde and Krammer-Volkerak, 1978; in Dutch). Rijkswaterstaat, Middelburg (NL). Nota DDMI 80.20.

De Leeuw, J., L. Apon, P., M. J. Herman, W. de Munck & W. G. Beeftink, 1992. Vegetation response to experimental and natural disturbance in two saltmarsh plant communities in the SW Netherlands. Neth. J. of Sea Res. 30: 279–288.

De Leeuw, J., L. Apon, P. Herman, W. de Munck & W. G. Beeftink, 1994. The response of salt-marsh vegetation to tidal reduction caused by the Oosterschelde storm-surge barrier. Hydrobiologia 282/283: 335–353.

Groenendijk, A. M., 1987. Ecological consequences of a storm-surge barrier in the Oosterschelde: the salt marshes. Ph. D. Thesis, State University Utrecht/Delta Inst. f. Hydrob. Res. Comm. Nr. 359, 177 pp.

Long, S. P. & C. F. Mason, 1983. Saltmarsh ecology. Blackie, Glasgow, 160 pp.

Nienhuis, P. H. & A. C. Smaal, 1994. The Oosterschelde estuary, a case study of a changing ecosystem: an introduction. Hydrobiologia 282/283: 1–14.

Oenema, O., 1988. Pyrite accumulation in salt marshes in the Eastern Scheldt. In: Oenema, O., 1988: Early diagenesis in recent fine-grained sediments in the Eastern Scheldt. State University Utrecht, Institute of Earth Sciences, Ph. D. Thesis, 222 pp.

Saeijs, H. L. F. & J. P. Al, 1982. Environmental impact assessment for design and management of the Oosterschelde Barrier. In: Saeijs, H. L. F., 1982: Changing estuaries. Rijkswaterstaat Communications nr 32; Min of Transport and Public Works, The Hague, The Netherlands: 279–320.

Smies, M. & A. H. L. Huiskes, 1981. Holland's Eastern Scheldt Estuary Barrier Scheme: Some ecological considerations. Ambio 10: 158–165.

Tutin, T. G., V. H. Heywood, N. A. Burgess, D. M. Moore, D. H Valentine, S. M. Walters & D. A. Webb, 1964–1980. Flora Europaea. Cambridge University Press, Cambridge, 5 Volumes.

Van Diggelen, J., 1988. A comparative study on the ecophysiology of salt marsh halophytes. Ph. D. Thesis, Free University Press, Amsterdam, 208 pp.

Vranken, M., O. Oenema & J. P. M. Mulder, 1990. Effects of tide range alterations on salt marsh sediments in the Eastern Scheldt, SW Netherlands. Hydrobiologia 195: 13–20.

Vroon, J., 1994. Hydrodynamic characteristics of the Oosterschelde in recent decades. Hydrobiologia 282/283: 17–27.

Hydrobiologia **282/283**: 335–353, 1994.
P. H. Nienhuis & A. C. Smaal (eds), The Oosterschelde Estuary.
© 1994 *Kluwer Academic Publishers. Printed in Belgium.*

The response of salt marsh vegetation to tidal reduction caused by the Oosterschelde storm-surge barrier

Jan de Leeuw[1], Leo P. Apon[2], Peter M. J. Herman, Wim de Munck & Wim G. Beeftink
Netherlands Institute of Ecology, Centre for Estuarine and Coastal Ecology, Vierstraat 28, 4401 EA Yerseke, The Netherlands; [1]*Present address: Division of Vegetation and Agricultural Sciences, I.T.C., P.O. Box 6, 7500 AA, Enschede, The Netherlands;* [2]*Present address: Zuiveringsschap Hollandse Eilanden en Waarden, P.O. Box 469, 3300 AL Dordrecht, The Netherlands*

Key words: salt marsh vegetation, species composition, response to tidal reduction, ecological modelling, multivariate statistics, prediction of environmental impact

Abstract

In 1986 a sluice gate barrier was completed in the mouth of the Oosterschelde estuary. The barrier has been partially or completely closed during 1986 and the first months of 1987. Consequently the high tides were reduced to such a level that the salt marshes were scarcely flooded. Since April 1987 the barrier has been closed on average during two out of 706 high tides a year. Although the barrier allows tidal exchange the tidal flow has been restricted as a result of the reduced width of the mouth of the estuary. This restriction of the tidal flow caused a 26 cm decrease of mean high water in the eastern part of the estuary. As a result the inundation frequencies of the salt marshes decreased.

The response of salt-marsh vegetation to this tidal reduction was analyzed using annual records (1982 till 1990) of species composition in 57 permanent plots in two marshes at the southern shore of the estuary. Analysis of the response of individual species to marsh elevation in the pre and the post-barrier situation revealed that most species moved down the marsh elevation gradient. The first axis of an ordination (DCA-1) was significantly negatively related to inundation frequency. Between 1984 and 1990 all plots were displaced towards a significantly higher ordination score, indicating a trend towards a species composition from higher up the marsh. The position of most plots along DCA-1 remained stable until 1985 and started to increase in 1986 or 1987. The vegetation in plots dominated either by *Halimione portulacoides** or by *Spartina anglica*, started to change in 1985. This premature change was attributed to frost damage in January 1985. The initially high rate of change along DCA-1 decreased in 1989 and 1990. This would suggest that the vegetation re-equilibrated with the newly established tidal conditions. Further analysis revealed no significant difference in the relation between inundation frequency and sample score along DCA-1 between 1984 and 1990. This corroborates the view that species composition had re-equilibrated with the tidal conditions. Along DCA-2 the samples were displaced towards a significantly higher score as well. The change was attributed to an increase of the perennial *Halimione* and the annual *Suaeda maritima*. The annual *Atriplex hastata* displayed an increase from 1986 till 1988, but strongly declined in 1990. The transient response of this annual was described by DCA-3.

During the pre-barrier phase several attempts had been made to predict the vegetation response to tidal reduction. We developed a multivariate model (CCA) with inundation frequency as a constraining factor to describe the relation between inundation frequency and species composition in the pre-barrier phase in 1984. Next we used this model to predict the score of the samples along CCA-1 in the post-barrier phase from the inundation frequencies in 1990. The actual ordination score in 1990 was calculated from the observed species composition. The predicted ordination scores did not differ significantly from the observed ones. In retrospect it is concluded that the response of the vegetation to the reduction of the tides, as far as the fraction of species composition which is related to inundation frequency is concerned, could have been predicted by our model.

Introduction

Salt-marsh vegetation generally displays a zonation along the marsh elevation gradient. This zonation has attracted the attention of ecologists for nearly a century and resulted in a great number of descriptions of the distribution pattern of salt-marsh species along the elevation gradient (Adam, 1990; Long & Mason, 1983; Chapman, 1974). The vertical distribution of salt marsh species has been attributed to the differential response of species to abiotic parameters such as soil salinity, sulphide and reduced iron (Rozema et al., 1985, Van Diggelen, 1988). These stress parameters are strongly determined by the degree of tidal flooding (Armstrong et al., 1985; De Leeuw et al., 1991) which in turn depends on the elevation of the marsh with respect to the tidal waters.

When the zonation is determined by the tides it follows that salt marsh vegetation would be sensitive to changes in tidal inundation. The response of salt marsh vegetation to changing tidal conditions has been documented in several studies. Beeftink (1987) and Olff et al. (1988) related fluctuations in species composition to short term variation of sea level. Beeftink (1979, 1987) described the response of salt marsh vegetation to a man-made increase of estuarine water levels. Ericson & Wallentinus (1979) and Cramer & Hytteborn (1987) related long-term changes of Baltic shore vegetation to a gradual decrease of flooding frequency caused by geotectonic uplift of the land.

The construction of a semi-permeable barrier in the mouth of the Oosterschelde estuary provided a unique opportunity to monitor the re-

sponse of salt marsh vegetation to tidal reduction. This storm surge barrier remains open during normal circumstances, but it is closed when high water levels (> 300 cm NAP) are predicted. Such water levels are exceeded on average once or twice a year, generally during north-westerly storms. Before the completion of the barrier it was predicted that the reduction in cross-section of the mouth of the estuary would lead to a reduction of the tidal range (Nienhuis & Smaal, 1994). This implied that the salt marshes would be flooded less frequently. With this in mind Smies & Huiskes (1981), Saeijs & Al (1982) and De Jong & De Kogel (1985) predicted strong changes of the salt-marsh vegetation, while Groenendijk et al. (1987) did not expect such a response.

In this article we present the results of an investigation of the response of salt marsh vegetation to the tidal reduction caused by the construction of the Oosterschelde storm surge barrier. First we analyze the intensity and the causes of the tidal changes. Secondly we compare the vertical distribution of species in the pre- and the post-barrier situation in order to evaluate whether individual species moved down the marsh elevation gradient. Next we investigate the temporal aspects of the changes in the vegetation. Then we address the question whether species composition has re-equilibrated with the newly established tidal conditions. Finally we present and evaluate a model which could have been used to predict the response of the vegetation to the reduction of the tides.

Material and methods

Study area

The present study was executed in two non-grazed salt marshes along the Oosterschelde es-

---

* Plant nomenclature according to Tutin et al. (1964–1980).

tuary (Fig. 1). The Oosterschelde forms a tidal bay with nearly constant water salinity since 1969 when the Volkerakdam closed it off from fresh water input from the rivers Rhine and Meuse (De Leeuw *et al.*, 1991). Until 1986 tides were semi-diurnal with a range of 395 cm at Razernijpolder (average 1971–1980; De Leeuw *et al.*, 1992). Both marshes are situated along the southern shore of the estuary and bordered by dikes at their land-ward side. The Stroodorpepolder marsh (26 ha) is a relatively high marsh with a cliff at its seaward side. In 1978 the vegetation consisted mainly of middle and high marsh plant communities (De Kogel & De Jong, 1980). The central part of the Rattekaai marsh (145 ha) is situated at relatively lower elevations and has a more gradual transition from low marsh to bare mudflats. Prior to 1986 most of this central part of the marsh was dominated by middle and low marsh plant communities. The soil in the Oosterschelde marshes consists of up to 2 m thick clay horizons which

overlay a more sandy substrate (Kooistra, 1978). The vegetation of the salt marshes in the SW Netherlands has been described by Beeftink (1965) and De Kogel & De Jong (1980).

Chronology of the construction of the storm surge barrier

In 1976 it was decided to construct a storm surge barrier in the mouth of the Oosterschelde estuary. Two additional structures, the Philipsdam and the Oesterdam, were projected in the rear of the estuary (Fig. 1). One of the conditions for the design of the barrier was a mean tidal range near Yerseke of at least 2.70 m. Therefore the barrier was designed with a 99% probability of exceeding a mean tidal range of 2.70 m. The barrier consists of 65 prefabricated piers which support sliding steel gates that can be closed when necessary. In 1984 the piers were placed, while the

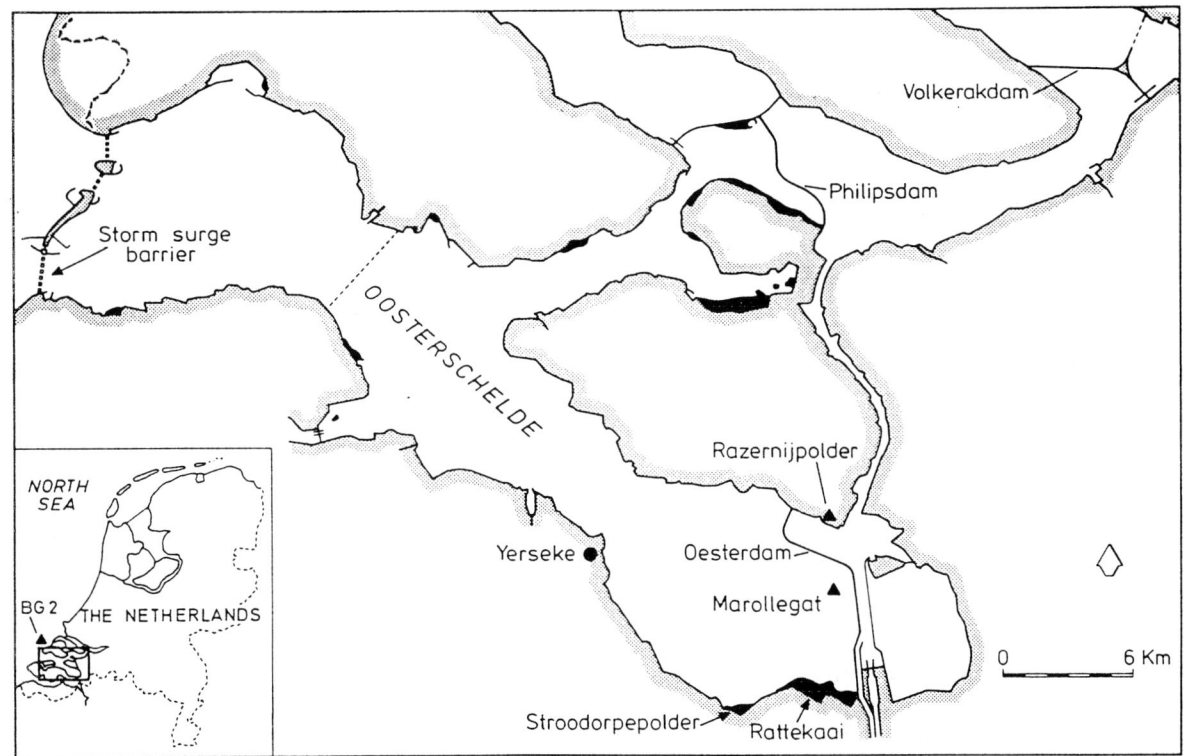

Fig. 1. Map of the Oosterschelde indicating all salt marshes (black), the Stroodorpepolder and the Rattekaai marshes, the tide gauge stations referred to in this study (triangles), the storm-surge barrier and various dams.

sill beams and upper beams were installed between 1984 and 1986. The continuing placement of these structures in the mouth of the estuary gradually reduced its cross section from 80 000 m^2 in the pre-barrier situation to 17 550 m^2 in April 1986. The tidal exchange started to be influenced by mid 1985 when the cross section was reduced below 35 000 m^2. A variable number of gates of the storm surge barrier have been closed during most of 1986 and the first months of 1987 for various reasons: to facilitate the construction of the barrier and the closure of the dams in the rear of the estuary and to reduce tidal currents which hindered shipping. This partial closure of the barrier lasted until the closure of the Philipsdam in April 1987. From then on the barrier has been operated as intended. For further information on the construction of the storm surge barrier we refer to Watson & Finkl (1990) and Smaal & Boeije (1991).

Vegetation data

In the present study we used annual records of species composition in permanent quadrates over the period 1982–1990. Species cover in 34 permanent quadrates (12 to 32 m^2) in the Stroodorpepolder marsh, established by Beeftink and co-workers in the 1960's and 1970's, has been recorded using the combined scale of Doing-Kraft (1954). The estimates were transformed to numerical values (0–100%) according to Van Der Maarel (1979). Species composition in 23 permanent quadrates (1 m^2) at the Rattekaai marsh, established in 1984 by D. De Jong and co-workers, was estimated to the nearest percent.

Estuarine water level, marsh level and inundation frequencies

Daily high water records for the Marollegat tide gauge station, situated north of the marshes studied (Fig. 1), and for BG 2, a station offshore in the North Sea, were obtained from the tidal water division of Rijkswaterstaat. The Marollegat high water data were analyzed for long-term trends in

MHW using a running average of two lunar cycles (113 high waters) to filter out the short term variation. We calculated a running average (113 high waters) of the difference of high water levels between the Oosterschelde and the North Sea high water levels to analyze the influence of the storm surge barrier on high water levels in the Oosterschelde. The level of the permanent quadrates with respect to Dutch Ordnance Level (NAP) was measured in 1984 and 1990 using a theodolite. Inundation frequencies were calculated from elevation of the plots and recorded water levels, assuming that water levels in the marsh attained the same height as the water levels recorded at the tide gauges.

Vegetation analysis

Pre- and post-barrier distribution of species

First we investigated whether individual species had moved down the marsh as a result of tidal reduction. The distribution of species along the marsh elevation gradient before (1984) and after (1990) the completion of the barrier was analyzed using the CURVE module of VEGROW (Fresco, 1990; Huisman et al., 1993). A hierarchical set of five regression models was fitted according to

$$ y = M * \frac{1}{1 + e^{a + bx}} * \frac{1}{1 + e^{c + dx}} $$

where y is the predicted cover of the species, M the maximal cover of the species in the dataset used to establish the regression, x the elevation and a, b, c and d the regression parameters to be estimated. The equation can describe, depending on the number of parameters involved, constant abundance, a one sided sigmoid curve increasing to a maximum equal to M, a one sided curve increasing to a maximum lower than M, a symmetric distribution and a skewed distribution. The best fitting of the five models was selected using an F test.

Year-to-year change in species composition

To analyze the year-to-year changes of the vegetation we ordinated the complete dataset. We used detrended correspondence analysis (Hill & Gauch, 1980) since preliminary analysis had revealed a strong arch effect without detrending. The axis produced by the ordination were interpreted using the linear correlation between inundation frequency and the ordination score of the 1984 and the 1990 samples.

When the tidal conditions which are thought to determine the species composition are modified, one would expect to observe a general change of the vegetation in one direction. Such general change may either be permanent (trend) or transient (fluctuation). In order to detect such responses we first analyzed the displacement of plots along the first four DCA axis by eye. Next we used the following statistical procedures to analyze the change of the vegetation along each of these ordination axis. We investigated whether the position along an axis of the 1990 samples differed significantly from that of the 1984 samples using a geometric mean regression between the position of the samples in 1984 and 1990. We tested (t-test) the significance of the deviation of the slope and the intercept of these regressions from one respectively zero. When the regression model was not statistically significant we tested whether the mean difference between the paired observations differed from zero (paired t-test).

Since we assumed that species composition of salt-marsh vegetation is determined by inundation frequency we hypothesized that an abrupt change of the mean high water level would disequilibrate the relation between species composition and inundation frequency. Furthermore we suggested that species composition would eventually re-equilibrate with the newly established tidal conditions. This raised the question whether the same relationship between species composition and inundation frequency existed in the post-barrier situation of 1990 as in the pre-barrier situation. We used analysis of covariance (Sokal & Rohlf, 1981) to test whether the relation between ordination score and inundation frequency (calculated over the 12 months preceding the sampling of the vegetation) in 1990 differed from the 1984 relation.

Prediction of the response of the vegetation

Prior to the construction of the storm surge barrier several attempts have been made to predict the response of the vegetation to tidal reduction. A downward migration of the vegetation zones was expected by Smies & Huiskes (1981), Saeijs & Al (1982) and De Jong & De Kogel (1985). These predictions were based on the assumption that the zonation of the vegetation was related to the tidal factor. However, inundation frequency was, in none of these models, explicitly related to species composition.

We are able to evaluate whether the actual change of the vegetation could have been predicted with a model describing the relation between inundation frequency and species composition in the pre-barrier situation. Only a fraction of the variation in species composition will statistically be related to inundation frequency. We hypothesize that the fraction of species composition which is related to inundation frequency would re-equilibrate towards a relation between inundation frequency and species composition similar to that observed in the pre-barrier situation. With respect to the fraction of the variation in species composition not related to inundation frequency we have no *a priori* expectations whether the vegetation will change, nor in what direction. The best we can do is to assume that there will be no systematic change between the pre- and the post-barrier situation.

We used canonical correspondence analysis (CCA, Ter Braak, 1986) of the 1984 samples with inundation frequency as the sole environmental parameter to describe the vegetation in 1984. Since the 1984 dataset did not contain high marsh samples above 250 cm NAP we added four high marsh samples from elevations above 250 cm NAP published by Van Diggelen (1988). The first axis of this ordination which is related to inundation frequency, catches all variation in species composition related to inundation frequency. Higher order axes describe the variation not related to inundation frequency. We used this model to predict the vegetation after reduction of the tides (1990). These predictions were evaluated through comparison with actual observations.

For the first axis we established the linear regression between inundation frequency and sample score of the 1984 vegetation records. This regression served as a model to predict the ordination score of the samples in 1990 from the inundation frequencies over the preceding 12 months. The actual ordination score of the samples in 1990 was calculated by weighted averaging of the species composition in 1990 and the species scores derived from the 1984 ordination. We used a geometric mean regression (Model II regression) to evaluate the relation between predicted and observed CCA scores, since both the predicted and observed CCA scores are random variables.

For the higher order ordination axes we calculated the sample scores in 1990 by weighted averaging of 1984 species scores by the 1990 species composition. We used a geometric mean regression to test whether these 1990 sample scores deviated significantly from the 1984 sample scores. When the regression model was not statistically significant we tested whether the mean difference between the paired observations differed from zero (paired t-test).

Results

Changes in tide level and inundation

Mean high water (running average over two lunar cycles) recorded at Marollegat station in the eastern part of the estuary fluctuated from 1982 till the first half year of 1985 between 1.95 to 2.25 m NAP (Fig. 2a). Before the second half of 1985 the difference between MHW at the North Sea and MHW in the Oosterschelde estuary remained fairly stable (Fig. 2b). This indicates that high water in the estuary followed the fluctuations of high water on the North Sea. MHW in the estuary started to decrease in the second half year of 1985 (Fig. 2a). The difference between MHW at the North Sea and in the estuary started to decline in the same period (Fig. 2b). This shows that the 1985 reduction of MHW in the estuary was

not related to a decrease of water levels on the North Sea. We therefore conclude that the 1985 reduction of MHW at Marollegat Station has to be attributed to the impact of the storm surge barrier on high water levels in the estuary.

Water levels in the estuary continued to fall over 1986 and early 1987. At the Marollegat station a minimum mean MHW of 1.08 m + NAP was recorded in early 1987. The decrease of the high water level over 1986 and 1987 was caused by the partial closure of the storm surge barrier to allow construction works in the eastern part of the Oosterschelde basin. From April 1987 the storm surge barrier has been operated as intended and since then gates have only been closed when predicted high water levels exceeded 300 cm NAP. As a result the running average of high water rapidly rose to higher values. From April 1987 on MHW fluctuated between 1.70 and 2.00 m above NAP. MHW in the post-barrier situation (average April 1987–April 1989) has been reduced by 26 cm (Table 1) when compared to the pre-barrier situation (average 1983 and 1984).

Figure 3 demonstrates the influence of the above mentioned variations of the tides on the inundation frequency at elevations of 1.85, 2.25 and 2.50 m NAP, which represents the average pre-barrier elevation for low, middle and high marsh respectively. On average 12.5% of the tides exceeded 2.50 m NAP in 1982, 1983 and 1984. None of the tides reached this level in 1986 and the first two quarters of 1987, while later on 2.50 m was reached by only 3.0% of the high waters. High waters did not pass this level of 2.50 m NAP in the second quarter of 1988, 1989 and 1990. On average 33.8% of tides exceeded 2.25 m NAP before 1985, while this level has been exceeded by 10.7% of the tides since the second quarter of 1987. While 72.9% of the tides exceeded a level of 1.85 m NAP before 1985 only a small percentage of the tides did so during 1986 and the first quarter of 1987. Since the second quarter of 1987 50.0% of the tides exceeded a level of 1.85 m NAP. The inundation frequency of the marsh is determined both by the level of high water and the elevation of the marsh.

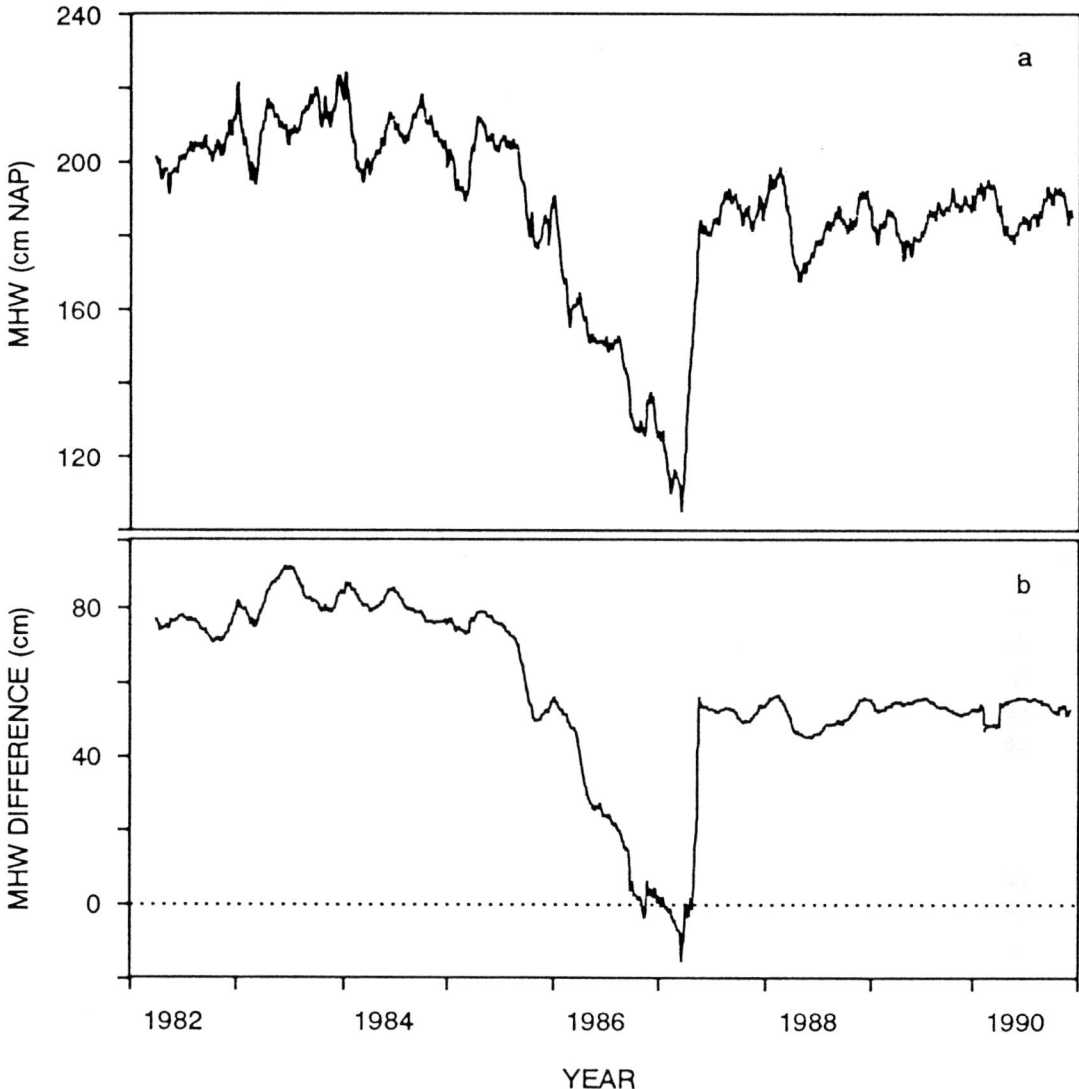

Fig. 2. A: Running average (two lunar cycles) of high water (cm NAP) as recorded at Marollegat tide gauge station, the Oosterschelde, The Netherlands, between 1982 and 1990 and B: running average (two lunar cycles) of the difference in high water recorded at the North Sea (BG 2) and in the estuary (Marollegat).

The 57 permanent plots used in the analysis of the vegetation were situated in 1990 on average 2.25 cm lower than in 1984 (Fig. 4). There were however, pronounced differences between the plots. Several plots were lowered by 10 cm while the elevation of other plots was not changed. This lowering of the marsh has been attributed to drainage caused by the decreased inundation frequencies (De Jong *et al.*, 1994). The differences between plots has to be attributed to local geomorphological differences within the marsh. Plots situated in marsh depressions, where the soil consisted of thicker layers of unsettled clay, were lowered by 5 to 10 cm. Plots situated either in the low marsh, where the clay layer was very thin, or on high and middle marsh creek banks, where the soil consisted of a coarser and more settled substrate, were not or only slightly lowered.

Table 1. Expected reduction of tidal range used for predictions with respect to the impact of tidal reduction on salt-marsh vegetation and actual reduction of tidal range and mean high water as recorded at Yerseke and Marollegat. Sources: 1: Rijkswaterstaat 1978, 2: Unpublished data Rijkswaterstaat, 3: Unknown, 4: Rijkswaterstaat 1989, 5: this study.

	Tidal range		Mean high water	
	Yerseke	Marollegat	Yerseke	Marollegat
Predicted				
Smies and Huiskes, 1981	− 80[1]			
Saeijs and Al, 1982	− 70[2]			
De Jong and De Kogel, 1985	− 70[2]			
Groenendijk *et al.*, 1987	− 40[3]			
Observed (April 1987–April 1989)				
Without barrier versus observed	− 48[4]	− 58[4]	− 25[4]	− 28[4]
MHW 1983 & 1984 versus observed				− 26[5]

Response of the vegetation

Pre- and post-barrier distribution of species

The distribution of the major species along the marsh elevation gradient in 1984 is shown in Fig. 5. *Spartina anglica* dominated the low marsh, *Halimione portulacoides*, *Puccinellia maritima*, *Triglochin maritima* and *Limonium vulgare* were the most abundant species in the middle marsh, *Festuca rubra* displayed a peak somewhat higher, while *Elymus pycnanthus* dominated the high marsh. Most species have moved down the marsh elevation gradient between 1984 and 1990 (Fig. 6). *Elymus* has become more abundant in plots which used to be vegetated by middle marsh species. Similarly *Festuca* has moved down the marsh gradient. Some middle marsh species (e.g. *Halimione*, *Puccinellia*, *Limonium*) have shifted down the gradient as well. Other middle marsh species have declined strongly (*Triglochin*) or have disappeared (*Juncus gerardi*, *Plantago maritima*; not shown). The low marsh species *Spartina*, which dominated in 1984 up to above 200 cm NAP, has also moved down the gradient. However, its cover in 1990 was inferior to that in 1984. The annual species *Atriplex hastata* and *Suaeda maritima* have moved down the gradient as well. Besides, their maximum cover in 1990 was higher than in 1984. Overall, the figure shows that between 1984 and 1990 individual species have moved down the marsh elevation gradient. The figure does however not show when these changes have occurred.

Year-to-year changes in species composition

The temporal aspects of change were analyzed in more detail by ordination (DCA). On the first ordination axis (Fig. 7) low marsh species like *Spartina* and *Salicornia europaea* have been placed at the lower end while high marsh species such as *Elymus* and *Festuca* occurred at the high end of the ordination axis. Middle marsh species have been situated in between. The position of the samples along the first ordination axis was highly significantly related to inundation frequency in both 1984 and 1990 (Table 2).

Over time there were pronounced changes in the position of the plots along the first axis (Fig. 8). We divided all plots into three groups, according to their ordination score on the first axis in 1984. The position of the plots with a 1984 score > 2.8 started to change in 1985, prior to a reduction of MHW. This was related to a strong decline from 1984 to 1985 of *Halimione* (Fig. 9), most likely as a result of frost damage in January 1985. However, the trend along the first ordination axis which started in 1985 (Fig. 8) continued over the following year, and all plots moved to a score of 4 or higher, indicating that *Elymus* came to dominance in these plots.

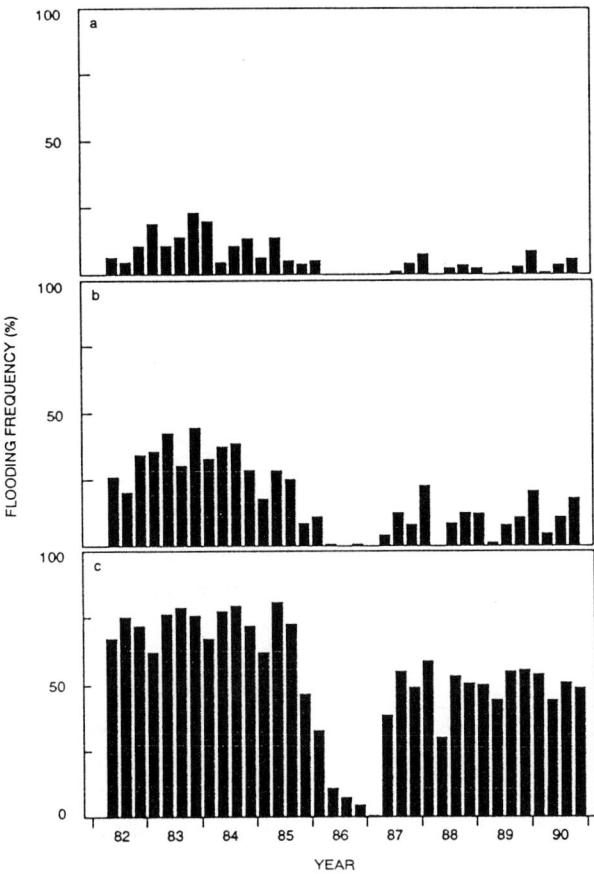

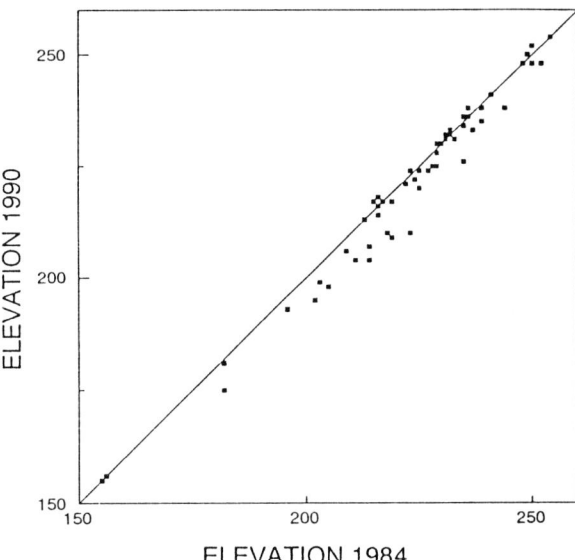

Fig. 4. Elevation (cm above NAP) of the permanent plots in the Stroodorpepolder and the Rattekaai marsh in 1984 and 1990.

Fig. 3. Change in inundation frequency (per three month period) from the second quarter of 1982 till the last quarter of 1990. Inundation frequencies were calculated from recorded high water data at Marollegat station for elevations of a. 250 cm, b. 225 cm and c. 185 cm above NAP (Dutch Ordnance Level). Before 1986 these elevations corresponded to the average level of occurrence of high, middle and low salt-marsh plant communities respectively.

The position of middle marsh plots, with a score in 1984 between 0.9 and 2.8, remained remarkably stable before 1986. The ordination score started to increase in 1986 in some plots, while the remaining plots started to change in 1987 (Fig. 8). The increase was most pronounced between 1986 and 1988 and levelled off in 1989 and 1990. The change showed a synchronous pattern, which indicates that some external factor affecting all plots should be responsible for the change.

The low marsh plots with a 1984 score <0.9 showed a more complicated pattern of change

(Fig. 8). The score of most plots increased in 1985. Compared to 1984 the cover of *Spartina* had decreased in 1985 in many plots (Fig. 9). We suggest that this reduction was the result of frost damage to *Spartina* in January 1985. The score of the plots increased from 1985 till 1986, mainly due to increased abundance of the annuals *Atriplex*, *Suaeda* and *Salicornia* (Fig. 10). In 1987 many plots displayed a sharp drop of the ordination score, followed by a quick recovery in 1988 (Fig. 8). This drop and the subsequent increase of the ordination score was caused by a strong decrease followed by an increase of the above-mentioned annual species (Fig. 10).

These results indicate that the three groups of plots behaved different along the first ordination axis. However, the following two generalizations apply to the development of all plots. First of all they had moved to a higher ordination score in 1990 compared to 1984: the displacement of the plots along the first ordination axis was statistically significant (Table 3). Secondly, the rate of change along the first ordination axis seems to have slowed down by 1990. This suggests that the vegetation has re-equilibrated with the newly established environmental conditions. As the first

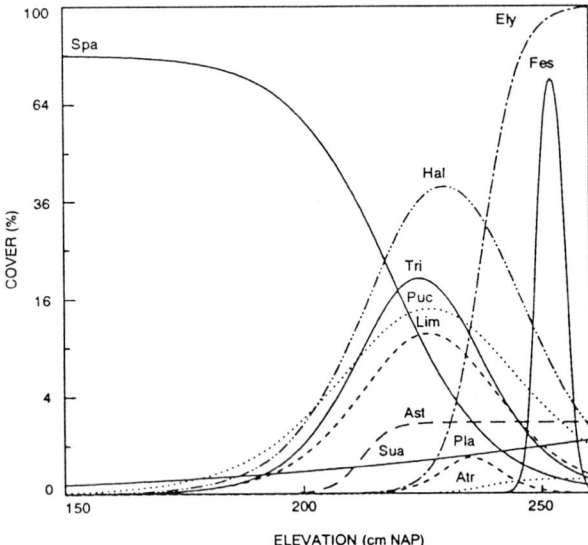

Fig. 5. Cover percentage along the marsh elevation gradient of the eleven most common species in 1984 (cm above Dutch Ordnance Level, NAP). Based on 57 permanent quadrates. Curves were fitted by the CURVE module of VEGROW (Fresco 1990). Ast = *Aster tripolium*, Atr = *Atriplex hastata*, Ely = *Elymus pycnanthus*, Fes = *Festuca rubra*, Hal = *Halimione portulacoides*, Lim = *Limonium vulgare*, Pla = *Plantago maritima*, Puc = *Puccinellia maritima*, Spa = *Spartina anglica*, Sua = *Suaeda maritima*, Tri = *Triglochin maritima*.

axis was related to inundation frequency, it seems relevant to question whether the vegetation has re-equilibrated with inundation frequency along this first ordination axis. Analysis of covariance revealed that the relation between inundation frequency and score along the first DCA axis did not differ significantly between 1984 and 1990. Apparently, in 1984 and 1990 the same relationship held between inundation frequency and sample score along the first ordination axis.

The second ordination axis was weakly correlated with inundation frequency in 1984 but not in 1990 (Table 2). The extreme ends of the axis were determined by the shrub *Artemisia maritima* and the grass *Festuca* respectively (Fig. 7). A limited number of plots moved towards or from these extremes (Fig. 11). These fluctuations were determined by changes in abundance of *Artemisia* or *Festuca*. The majority of the plots were situated within a more limited range (Fig. 7). The dominant marsh species (Fig. 6) *Elymus*, *Spartina* and

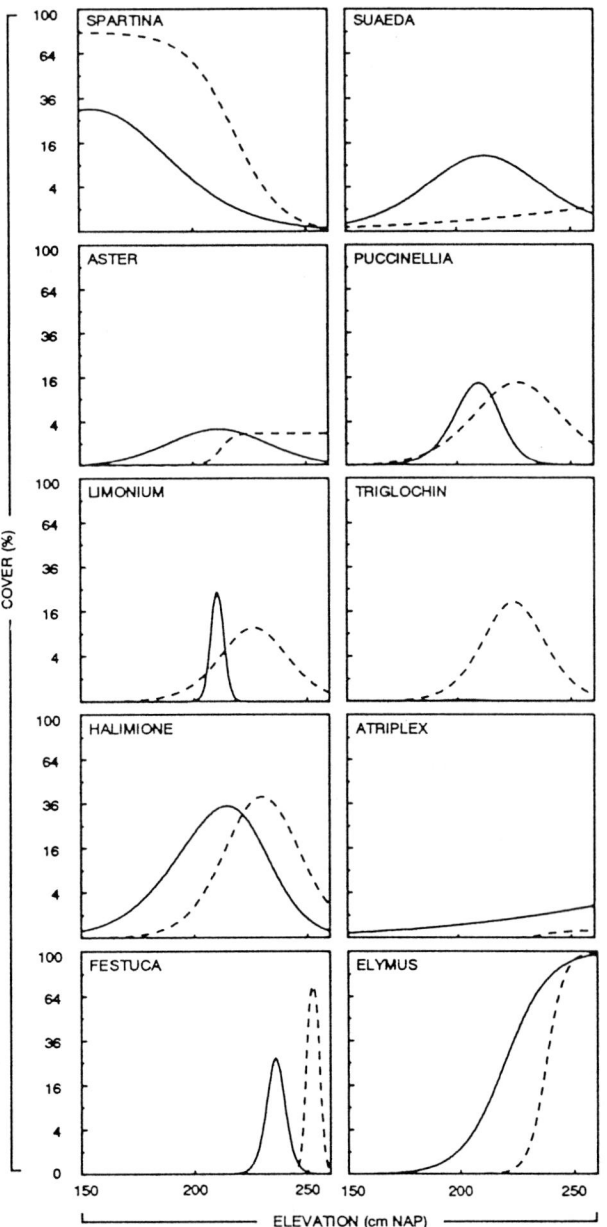

Fig. 6. Change in distribution of ten salt-marsh species along the marsh elevation gradient between 1984 (dashed line) and 1990 (solid line). Most species moved downward, most likely as a result of tidal reduction. Abbreviations of species names as in Fig. 5.

Halimione, with a score of 2–2.5, were situated near the centre of this range (Fig. 7). Somewhat lower, around a score of 1, occurs a cluster of species from waterlogged marsh depressions, e.g.

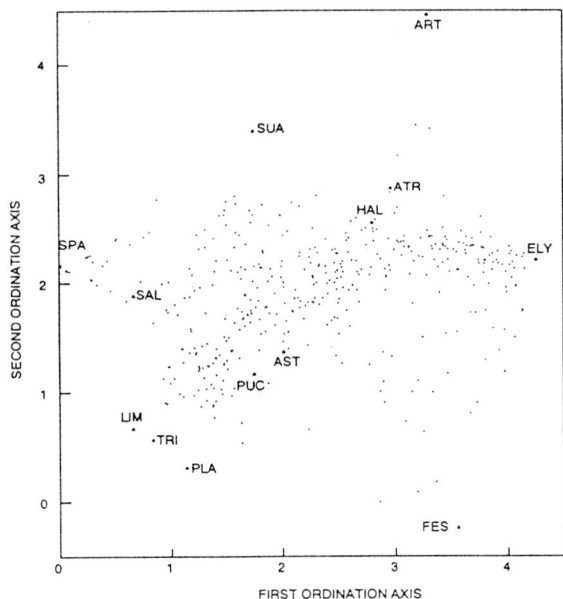

Fig. 7. Ordination biplot for the complete dataset (1982–1990); first and second axis. Abbrevation of species names as in Fig. 5.

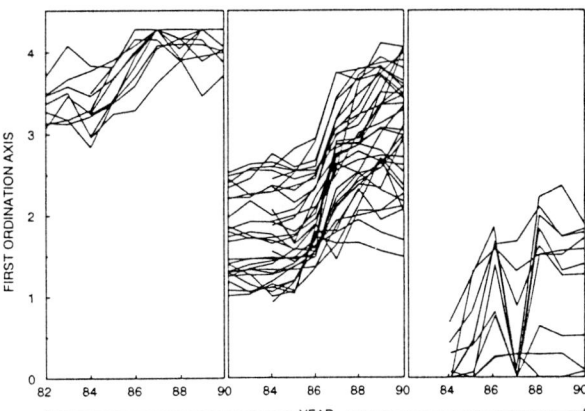

Fig. 8. Displacement of permanent plots along the first ordination axis. Permanent plots were divided into three groups, according to the sample score on the first axis in 1984. The first group (a) consisted in 1984 of high marsh and well drained middle marsh vegetation dominated by *Elymus*, *Festuca* or *Halimione*, the second group (b) was vegetated in 1984 by waterlogged middle marsh communities, while the third group (c) was dominated in 1984 by the low marsh species *Spartina anglica*.

Plantago, *Triglochin*, *Puccinellia* and *Limonium*. Annual species *Suaeda* and *Atriplex* were situated at a higher score of approximately 3. Within this more restricted range the following displacements were observed. High marsh samples were not displaced (Fig. 11a). Middle marsh plots increased from 1987 onward, and seem to have settled down at a higher position in 1990. Low marsh plots increased to 1986, then decreased in 1987 and increased once more in 1988. This pattern resembles the pattern described on the first axis. The figure suggests that the plots, particularly the middle marsh plots, reached a higher score in 1990 than in 1984. The paired t-test revealed that the samples displayed a significant positive dis-

placement along the second ordination axis between 1984 and 1990 (Table 3). The middle marsh plots responsible for this significant displacement were situated at elevations between 200 and 235 cm NAP. Most of the original species had in 1990 been replaced by *Halimione* and *Suaeda* (Fig. 6). These have a much higher species score along the second axis than the original species.

Table 2. Linear correlation coefficient (r) between inundation frequency and sample score on the first three ordination (DCA) axis in 1984 and 1990 ($n = 57$).

DCA axis	r_{1984}	r_{1990}
I	− 0.82***	− 0.84***
II	+ 0.31*	+ 0.12[n.s.]
III	+ 0.45***	+ 0.20[n.s.]

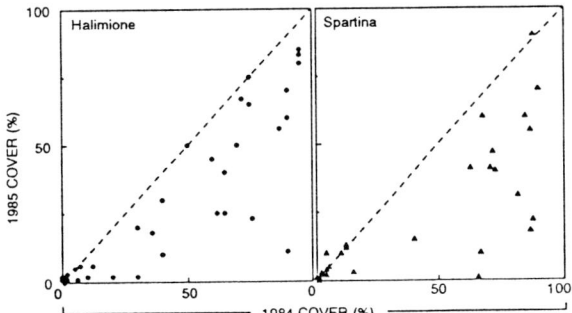

Fig. 9. Relationship between 1984 and 1985 cover of A. *Halimione portulacoides* and B. *Spartina anglica* as recorded in 55 permanent quadrats in the Stroodorpepolder and the Rattekaai marshes. Cover of both species declined most likely as a result of frost damage in the winter of 1984/1985.

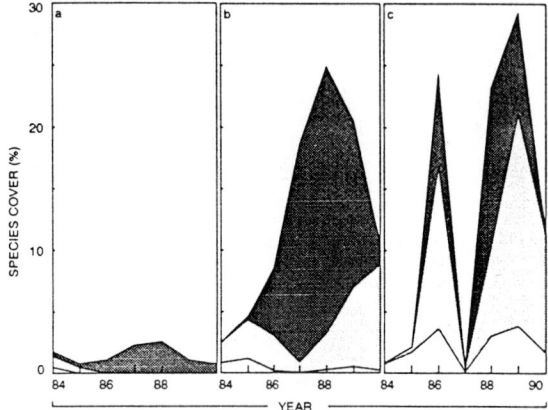

Fig. 10. Change in average cover (%) of the annual species *Salicornia europaea* (white), *Suaeda maritima* (light shading) and *Atriplex hastata* (dark shading) in the three groups of permanent quadrates (See Fig. 8).

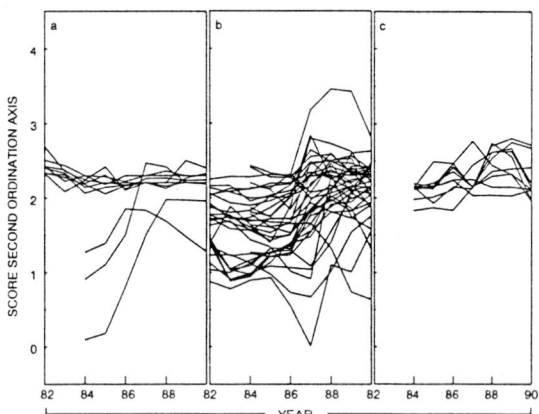

Fig. 11. Displacement of permanent plots along the second ordination axis. Permanent plots were divided in three groups (See Fig. 8).

The third ordination axis contrasted the annual *Atriplex* on the higher side and the shrub *Halimione* at the lower end (Fig. 12). There was a significant correlation between inundation frequency and the score of the samples on this third axis in 1984 but not in 1990 (Table 2). This may be related to the fact that *Atriplex* was restricted to the highest positions in the marsh in 1984 while it occurred all over the marsh in 1990 (Fig. 6). Many plots displayed an increase followed by a sharp decrease along this third ordination axis (Fig. 13). We suggest that this has to be attributed to the transient dominance of *Atriplex* in many plots (Fig. 10). The majority of the samples had in 1990 returned to their pre-barrier ordination score. We were unable to detect a statistically significant difference in the position of the plots between 1984 and 1990 (Table 3).

Prediction of the response of the vegetation

The regression between inundation frequency and the sample score along the first axis of the canonical ordination explained 76% of the variation of the samples along this axis (Fig. 14). The figure shows that the samples were displaced between 1984 and 1990 more or less parallel to this regression line. The geometric mean regression revealed that the predicted sample score did not differ significantly from the observed scores (Table 4). This indicates that the response of the fraction of species composition related to inundation frequency could have been predicted accurately before the reduction of the tides. The 1984 sample scores differed significantly from the observed 1990 score along the second and the fourth axis but not along the third axis (Table 4). This indicates that the response of the fraction of

Table 3. Parameters describing the relation between the ordination score of the samples ($n = 57$) in 1984 and 1990 for the first four DCA axis: correlation coefficient (r), slope (B_{GM}) of the geometric mean regression and intercept (A_{GM}) of this regression, and mean difference (D) between paired observations. The standard errors of the estimated parameters are given by figures between brackets.

	Axis 1	Axis 2	Axis 3	Axis 4
r	0.84***	0.21[n.s.]	0.47***	0.43***
B_{GM}	1.04 (0.07)[n.s.]	–	0.93 (0.12)[n.s.]	0.96 (0.12)[n.s.]
A_{GM}	1.11 (0.18)***	–	0.07 (0.17)[n.s.]	0.19 (0.18)[n.s.]
D	–	0.41 (0.07)***	–	–

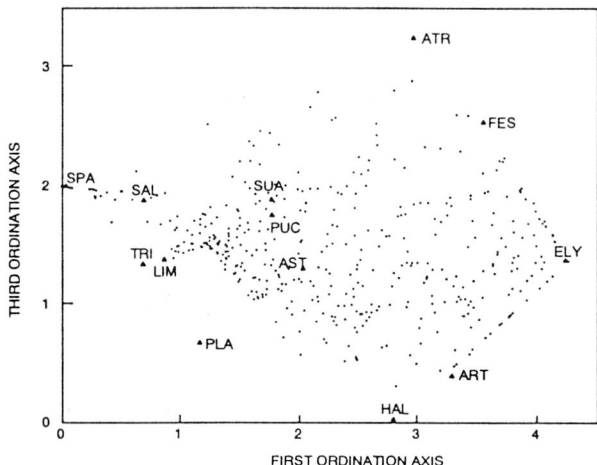

Fig. 12. Ordination biplot of species and samples for the complete dataset (1982–1990); first and third axis. Abbrevations of species names as in Fig. 5.

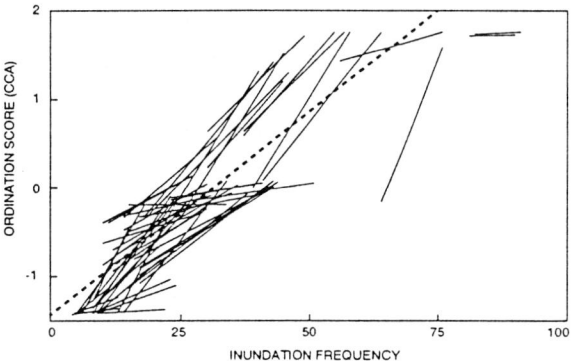

Fig. 14. Linear regression ($R^2 = 0.76$, $n = 57$) between inundation frequency and ordination score in 1984 (interrupted lines) and displacement between 1984 and 1990 of the permanent quadrates (solid lines) in a biplot of inundation frequency and the ordination score (first CCA axis).

species composition not related to inundation frequency could not have been predicted prior to the reduction of the tides.

Discussion

In the present study we describe the changes of the vegetation in two salt marshes along the Oosterschelde estuary during the 1980's. By the end of the 1980's MHW had decreased 26 cm compared with the beginning of the 1980's and inundation frequencies, particularly in the middle and high marsh environment had strongly decreased. Apart from this permanent tidal reduction the marsh had scarcely or not been flooded in 1986 and 1987. Besides, we suggested that several species had suffered from frost damage in 1985. This implies that the observed changes in the vegetation may have been determined by more than one disturbance. Below we will discuss the various observed responses of the vegetation and try to identify the factors responsible.

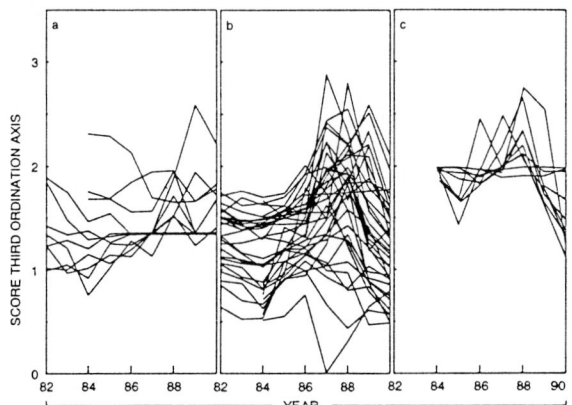

Fig. 13. Displacement of permanent plots along the third ordination axis. Permanent plots were divided in three groups (See Fig. 8).

The response of annual species

First of all we note a transient increase in abundance of several annual species (*Suaeda*, *Atriplex* and to a lesser extent *Salicornia*) in 1986 and subsequent years. It has frequently been observed that natural and man-made disturbances increase the cover of annual species to values greatly exceeding their abundance in undisturbed salt-marsh vegetation (Beeftink *et al.*, 1978; Beeftink, 1979, 1987; Bertness & Ellison 1987; De Leeuw *et al.*, 1992). This response has been related to the capacity of annuals to rapidly colonize the gaps created. Over time, however, these early coloniz-

Table 4. Evaluation of a multivariate model to predict the response of the vegetation to tidal reduction. Parameters describe the relation between the predicted and observed ordiantion score of the samples in 1990 for the first four CCA axes: correlation coefficient (r), slope (B_{GM}) of the geometric mean regression and intercept (A_{GM}) of this regression, and mean difference (D) between paired observations. The standard error of the estimated parameters are given by figures between brackets. The prediction for the first axis was derived from inundation frequency. The prediction for the higher order axis was equal to the score observed in 1984.

	Axis 1	Axis 2	Axis 3	Axis 4
r	0.86***	0.09[n.s.]	0.07[n.s.]	0.52***
B_{GM}	0.93 (0.07)[n.s.]	–	–	0.74 (0.12)*
A_{GM}	0.14 (0.07)[n.s.]	–	–	– 0.04 (0.07)[n.s.]
D	–	0.64 (0.15)***	– 0.19 (0.13)[n.s.]	–

ers are overgrown and replaced by perennial species (Bertness & Ellison, 1987; De Leeuw *et al.*, 1992).

In the present study we recorded a similar pattern in the middle marsh plots. Annual species, particularly *Atriplex*, first increased and later decreased again. The gaps created by frost damage to *Halimione* did apparently not contribute to the success of the annuals, since the real increase started in 1986 only. We suggest the following explanation for this increase: the extreme reduction of the tides in 1986 and 1987 reduced the salinity of the soils, so that the middle marsh became a suitable habitat for *Atriplex*. The open structure of the vegetation in depressions in the middle marsh, did probably further contribute to the success of the annuals.

Suaeda was by far the most important annual in the low marsh plots. The abundance of annuals in the low marsh plots displayed a more complicated pattern. A first an increase in 1986 was followed by a strong decline in 1987 and a return in 1988, followed by a gradual decline thereafter. We suggest the following explanation for the 1987 decline. The low marsh had not been flooded for most of 1986 and the first months of 1987. Infrequently flooded marsh sediments display a low salinity in spring, mainly because rainfall exceeds evapotranspiration in autumn, winter and spring (De Leeuw *et al.*, 1991). Because the low marsh had not been flooded for such a long period we assume that seeds of annual species germinated in spring 1987 (March, April) under low salinity

conditions. Then, all of a sudden in April, the low marsh was flooded by sea water again. Consequently the young seedlings were exposed to high salinity. We suggest that this caused an osmotic shock which decimated the populations of these annuals. Higher up the marsh the annuals did not show such a dramatic decrease of their populations in 1987. The observation that the middle and high marsh were rarely flooded in the second quarter of 1987 might explain why annual populations higher up the marsh were not so strongly affected.

Frost damage interfering with downward migration of the vegetation

The ordination of the entire dataset showed that the vegetation was relatively stable before 1985. We would not have expected a detectable response of the vegetation before 1986 because the water levels started to decline at the end of the growing season of 1985 only. The initiation in 1986 of a trend along the first ordination axis of all the middle marsh plots agreed with this expectation. The synchrony of the changes suggest that the response of the vegetation has to be attributed to a common external factor (such as tidal reduction), which affected all plots. However, the vegetation in the *Halimione* and *Spartina* dominated quadrates started, contrary to the expectation expressed above, to change in 1985. We suggest that this premature change was not related to changes in hydrological conditions caused by the storm surge barrier but instead has to be attributed to the extreme cold in January 1985.

Frost damage in *Halimione* in the SW Netherlands has been described by Beeftink *et al.* (1978) and by De Leeuw *et al.* (1992) while Kamps (1962) reported frost damage in *Spartina* in the northern Netherlands. The interference of frost damage with the reduction of the tides complicated the analysis of the changes of the vegetation in these plots. Beeftink *et al.* (1978) and De Leeuw *et al.* (1992) reported that *Halimione* vegetation after being damaged by frost reverted back to the species composition of undisturbed plots within a couple of years. However, in the present study *Halimione* plots affected by frost damage did not return to their original species composition or ordination score. Instead they moved to a higher score due to increasing dominance of *Elymus*. We argue that this trend towards a higher ordination score has to be attributed to the modified hydrology. First of all because *Elymus* has never been observed after disturbance in *Halimione* vegetation and secondly because *Elymus* does normally occur higher up the marsh. Frost damage only advanced the change of species composition in the *Halimione* vegetation. The same seems to hold for the low marsh formerly dominated by *Spartina*. Frost damage advanced the change of species composition in this vegetation; later on, however, the vegetation stabilized at a higher DCA score than before.

Re-equilibration of the vegetation with the new tidal conditions

Together our results demonstrate that frost damage interfered with the influence of the permanent reduction of the tides on species composition at a short time scale. However, the first ordination (DCA) axis revealed that at the longer time scale the samples were significantly displaced. Following the reduction of the tides the vegetation in all plots moved to a species composition which used to be characteristic of a position higher up the marsh. The decreasing rate of change along the first ordination axis and the fact that the first ordination axis was related to inundation frequency led to the suggestion that the vegetation was approaching equilibrium with the newly established tidal regime. We used analysis of covariance to test for differences in the relation between inundation frequency and sample score along the first DCA axis between 1984 and 1990. The analysis revealed no significant differences. Hence, the same relationship seems to hold in 1984 and 1990. This corroborates the view that the vegetation reached equilibrium with the newly established tidal conditions.

The position of the samples along the second ordination axis changed significantly between 1984 and 1990. The change observed along this axis was particularly caused by samples from depressions. The original vegetation in these depressions, which contained species indicating waterlogged conditions, was replaced by the annual *Suaeda* and the perennial *Halimione*. The increased abundance of the annual may be interpreted as a consequence of the disturbances in the second half of the 1980's. We expect that *Suaeda* will decline over time. *Halimione* normally prefers the better drained conditions than the original vegetation in the depressions. It is tempting to attribute the increase of *Halimione* to improved drainage of the soil caused by the reduced inundation frequencies. However, the weak correlation of the second axis with inundation frequency in 1984 and the absence of such relation in 1990 suggests that this is not a very likely explanation.

An alternative hypothesis would be that *Halimione* was able to colonize the depressions during the excessive drainage of the marsh in 1986 and 1987. The improved drainage conditions, however, persisted for one and a half year only. When the inundation frequencies increased again in 1987 the soil most likely became more waterlogged. Hence one would expect a return of species indicating more waterlogged conditions. Up till 1990 this has not been observed. De Leeuw *et al.* (1992) reported that *Halimione*, once established, persisted for many years in a similar waterlogged marsh depression. This suggests that the abundance of *Halimione* will not easily decrease. This raises the question whether the change along the second axis will persist, or alternatively, that the vegetation returns to pre-barrier conditions. Continued monitoring of the vegetation will have to reveal whether the observed change along this axis was transient or permanent.

The drainage of the marsh in 1986

So far we described several consequences of the one year of tidal restriction in 1986 and 1987. It may be questioned whether the scarce flooding in 1986–1987 exerted additional influence on the vegetation. Glycophytic species were observed in 1986 but their presence was transient because they have not been observed in 1987 and subsequent years. These species most likely invaded the marsh because the soil was temporarily desalinated.

De Jong *et al.* (1994) describe the changes of several soil parameters caused by the drainage of the marshes in 1986 and 1987. They question whether these changed physical and chemical soil parameters would exert a lasting influence on the vegetation. As a result of the tidal reductions the elevation with respect to NAP of the permanent plots was reduced in 1990 between 0 and 10 cm. These values agree with Oenema (1990) who recorded 1 to 5 cm subsidence of the Rattekaai back marshes in 1986. We suggest that this soil shrinkage exerted and still exerts some influence on the vegetation because it changed the inundation frequencies of the plots. As such the shrinkage of the soil partly counterbalanced the influence of the tidal reductions on the decrease of the inundation frequencies. This counterbalancing influence will have been larger in the backmarsh depressions than either the low marsh or the creek banks, since the backmarsh sites were lowered more strongly.

Prediction of the response of the vegetation to tidal reduction

In this study we present a prediction of the response of the vegetation to tidal reduction. Our model distinguishes variation in species composition related to inundation frequency and unrelated variation. Evaluation of the predictions with the observations reveals that our model accurately predicted the response of the fraction of the species composition related to inundation frequency. However, our model failed to predict the response of the variation in species composition not related to inundation frequency. Probably this unpredictable response is related to the changes in species composition described by the second DCA axis.

The possible influence of tidal reduction on salt marsh vegetation has been investigated before the construction of the storm surge barrier (Smies & Huiskes, 1981; Saeijs & Al, 1982; De Jong & De Kogel, 1985; Groenendijk *et al.*, 1987). The predictions with regard to the response of the vegetation were all related in some way to expectations with respect to tidal reduction (Table 1). All authors referred to the predicted reduction of tidal range instead of MHW. The latter parameter was considered more relevant for salt marsh vegetation (D. de Jong, pers. comm.) but predictions were not available. In practice it was assumed that MHW would be reduced by approximately half the predicted reduction of the tidal range (D. de Jong, pers. comm.). The table shows that over the years the predicted tidal reduction decreased. It should therefore be borne in mind that part of the differences between the studies with respect to the predicted response of the vegetation, which will be discussed in more detail below, may be attributed to the expected tidal reduction (Table 1).

The predictions of the response of the vegetation to tidal reduction can be classified in three groups: (i) a possible invasion and subsequent dominance by non-halophytic species as a result of desalinization in the upper marsh, (ii) a downward shift of the existing zonation of halophytic vegetation and (iii) a seaward extension of low marsh plant species into the mudflats leading to an increase of the total marsh area.

Saeijs & Al (1982) predicted for 0 and 2 closures of the barrier per year 3.2 and 11.3 ha of desalinated marsh respectively. Actually the barrier has been closed on average 2 times per year between April 1987 and the end of 1991. So far invasion of the marshes by non-halophytes has not been observed (De Jong, pers. comm.). However, the predictions by Saeijs & Al (1982) were based on a 70 cm reduction of the tidal range with the barrier opened, whereas the actual reduction turned out to be 48 cm only (Table 1). Because this implies that the highest zones of the marshes are more frequently flooded than expected by Saijs and Al (1982) it may well explain why halophytes continue to dominate.

Completely divergent views were expressed with regard to the likelihood of a down-shore movement of the vegetation zones in response to the expected tidal reduction. Saeijs & Al (1982) referring to Beeftink (1979) expected a downward movement of the major vegetation zones. This view was supported by De Jong & De Kogel (1985) who predicted the post-barrier distribution of plant communities along the marsh elevation gradient with a model describing the pre-barrier relationship between inundation frequency and plant communities. In contrast Smies & Huiskes (1981) expected that the pre-barrier vegetation zones would not simply move down the marsh elevation gradient but instead be replaced by an impoverished post-barrier zonation. They argued that the species-rich high marsh communities (referred to as middle marsh communities in this article) would most likely disappear, because of the supposed inability to outcompete the existing *Spartina* vegetation. The observation that the cover of middle marsh species such as *Triglochin* and *Limonium* declined while *Plantago* disappeared altogether agrees with this prediction. However, the disappearance of these species can not be attributed to the inability to compete with *Spartina*, because the cover of this species had been reduced by frost damage in 1985. We suggest that either competition with annual species (e.g. *Suaeda*) which filled the gaps in the low marsh vegetation or the poor reproductive capacity of the mentioned middle marsh species are more likely explanations for the lack of recovery of these species at lower elevations.

Groenendijk *et al.* (1987) considered it unlikely that the zonation of the vegetation would show a large scale change. This prediction was based on two experiments with respect to the influence of tidal reduction on the competitive interactions between three middle marsh species, and another two investigations of the influence of tidal manipulation on the competitive interactions between three low marsh species. However, the influence of tidal reduction on the competitive interactions between low, middle and high marsh species was not investigated in these experiments. Two of the species from higher up the marsh, which as we know by now displayed a very strong response to tidal reduction, *Elymus* and *Halimione*, were not included. The method used in these experiments was suitable for the purpose of predicting the influence of tidal reduction on zonation. However, the experimental set up did not allow to predict the influence of tidal reduction on the zonation of salt marsh vegetation because the species interactions investigated formed an incomplete and biased sample of all possible (relevant) species interactions along the marsh elevation gradient.

With regard to a possible seaward extension of the marsh Saeijs & Al (1982) expected formation of 200 to 500 ha of new salt marshes, mainly because tidal reduction would move intertidal flats to within the range of inundation frequencies where colonization by salt marsh species, particularly *Spartina anglica*, becomes possible. Groenendijk (1986) in contrast expected no large scale invasion of the tidal mudflats by *Spartina*, because no change in wave action, which was considered the main factor controlling the natural spread of *Spartina*, was expected. The developments since 1986 follow the predictions put forward by Groenendijk (1986); so far no large scale spread of *Spartina* has been observed.

Acknowledgements

Thanks are due to D. J. de Jong (Rijkswaterstaat, Tidal Waters Division) for the use of vegetation data of the Rattekaai saltmarsh. Communication No. 641 of NIOO-CEMO, Yerseke, The Netherlands.

352

References

Adam, P., 1990. Saltmarsh ecology. Cambridge University Press, Cambridge, 461 pp.

Armstrong, W., E. J. Wright, S. Lythe & J. T. Gaynard, 1985. Plant zonation and the effects of the spring neap tidal cycle on soil aeration in a Humber salt marsh. J. Ecol. 73: 323–339.

Beeftink, W. G., 1965. Salt marsh communities of the SW-Netherlands in relation to the European halophytic vegetation (In Dutch). PhD Thesis, Wageningen University, 167 pp.

Beeftink, W. G., M. C. Daane, W. de Munck & J. Nieuwenhuize, 1978. Aspects of population dynamics in *Halimione portulacoides* communities. Vegetatio 36: 31–43.

Beeftink, W. G., 1979. The structure of salt marsh communities in relation to environmental disturbances. In R. L. Jefferies & A. J. Davy (eds), Ecological processes in coastal environments. Blackwell, Oxford: 77–93.

Beeftink, W. G., 1987. Vegetation responses to changes in tidal inundation of salt marshes. In: J. van Andel, J. P. Bakker & R. W. Snaydon (eds), Disturbance in grasslands. Dr W. Junk Publishers, Dordrecht: 97–117.

Bertness, M. D. & A. M. Ellison, 1987. Determinants of pattern in a New England salt marsh plant community. Ecol. Monogr. 57: 129–147.

Chapman, V. J., 1974. Salt marshes and salt-deserts of the world. Cramer, Lehre, 2nd edn. 392 pp.

Cramer, W. & H. Hytteborn, 1987. The separation of fluctuation and long-term change in vegetation dynamics of a rising seashore. Vegetatio 69: 157–167.

De Jong, D. J., Z. de Jong & J. P. M. Mulder, 1993. Changes in area, geomorphology and sediment nature of salt marshes in the Oosterschelde (SW Netherlands) due to tidal changes. Hydrobiologia, in press.

De Jong, D. J. & T. J. de Kogel, 1985. Salt-marsh research with respect to the civil-engineering project in The Oosterschelde, The Netherlands. Vegetatio 62: 425–432.

De Kogel, T. J. & D. J. de Jong, 1980. Vegetatiekartering van de schorren in de Oosterschelde en het Krammer-Volkerak, 1978. Rijkswaterstaat, Dienst Getijdewateren. Nota DDMI-80.20.

De Leeuw, J., A. van den Dool, W. de Munck, J. Nieuwenhuize & W. G. Beeftink, 1991. Factors influencing the soil salinity regime along an intertidal gradient. Estuar. coast. Shelf Sci. 32: 87–97.

De Leeuw, J., L. P. Apon, P. M. J. Herman, W. de Munck & W. G. Beeftink, 1992. Vegetation response to experimental and natural disturbance in two salt-marsh plant communities in the south-west of the Netherlands. Neth. J. Sea Res. 30: 279–288.

Doing-Kraft, H. 1954. L'analyse des carrés permanents. Acta Bot. Neerl. 3: 421–425.

Ericson, L. & H. G. Wallentinus, 1979. Sea-shore vegetation around the Gulf of Bothnia. Wahlenbergia 5: 1–142.

Fresco, L. F. M., 1990. VEGROW: Processing of vegetation data. Shorthand manual V.4.0. S-BEES, Haren.

Groenendijk, A. M., 1986. Establishment of a *Spartina anglica* population on a tidal mudflat: a field experiment. J. Envir. Mg 22: 1–12.

Groenendijk, A. M., J. G. J. Spieksma & M. A. Vink-Lievaart, 1987. Growth and interactions of salt marsh species under different flooding regimes. In A. H. L. Huiskes, C. W. P. M. Blom & J. Rozema (eds), Vegetation between land and sea. Dr W. Junk Publishers, Dordrecht: 236–258.

Hill, M. O. & H. G. Gaugh, 1980. Detrended correspondence analysis, an improved ordination technique. Vegetatio 42: 47–58.

Huisman, J., H. Olff & L. F. M. Fresco, 1993. A hierarchical set of models for species response analysis. J. Veg. Sci. 4: 37–46.

Kamps, L. F., 1962. Mud distribution and land reclamation in the eastern Wadden shallows. Rijkswaterstaat Communication 4: 1–73.

Kooistra, M. J., 1978. Soil development in recent marine sediments of the intertidal zone in the Oosterschelde the Netherlands. Netherlands Soil Survey Institute, Wageningen. Soil Survey Papers 14: 1–183.

Long, S. P. & C. F. Mason, 1983. Saltmarsh ecology. Blackie, Glasgow, 160 pp.

Nienhuis, P. H. & A. C. Smaal, 1994. The Oosterschelde estuary, a case-study of a changing ecosystem: an introduction. Hydrobiologia 282/283: 1–14.

Oenema, O., 1990. Pyrite accumulation in salt marshes in the Eastern Scheldt, southwest Netherlands. Biogeochemistry 9: 75–98.

Olff, H., J. P. Bakker & L. F. M. Fresco, 1988. The effects of fluctuations in tidal inundation frequency on a salt marsh vegetation. Vegetatio 78: 13–19.

Rijkswaterstaat, 1978. Aanstaande veranderingen in de getijdebeweging op het Oosterschelde-bekken. Driemaandelijks Bericht Deltawerken 86: 284–290.

Rijkswaterstaat, 1989. Nieuwe randvoorwaarden voor het Oosterscheldesysteem. EOS-RVW rapportage, Middelburg.

Rozema, J., P. Bijwaard, G. Prast & R. Broekman, 1985. Ecophysiological adaptations of coastal halophytes from foredunes and salt marshes. Vegetatio 62: 499–521.

Saeijs, H. L. F. & J. P. Al, 1982. Environmental impact assessment for design and management of the Oosterschelde barrier. In: H. L. F. Saeijs, Changing estuaries, PhD Thesis, University of Leiden: 279–320.

Smaal, A. C. & R. C. Boeije, 1991. Veilig getij, de effecten van de waterbouwkundige werken op het getijmilieu van de Oosterschelde. Nota GWWS 91.088 DGW/Directie Zeeland, Middelburg, 132 pp. (in Dutch).

Smies, M. & A. H. L. Huiskes, 1981. Holland's Eastern Scheldt Estuary barrier scheme: some ecological considerations. Ambio 10: 158–165.

Sokal, R. R. & F. J. Rohlf, 1981. Biometry. Freeman, New York, 859 pp.

Ter Braak, C. J. F., 1986. Canonical correspondence analysis: a new eigenvector technique for multivariate direct gradient analysis. Ecology 67: 1167–1179.

Tutin, T. G., V. H. Heywood, N. A. Burges, D. M. Moore, D. H. Valentine, S. M. Walters & D. A. Webb, 1964–1980. Flora Europaea. Cambridge University Press, Cambridge. 5 Volumes.

Van Diggelen, J., 1988. A comparative study on the ecophysiology of salt marsh halophytes. PhD Thesis, Free University, Amsterdam, 208 pp.

Van der Maarel, E., 1979. Transformation of cover-abundance values in phytosociology and its effects on community similarity. Vegetatio 39: 97–114.

Watson, I. & C. W. Finkl Jr., 1990. State of the art in storm-surge protection: the Netherlands Delta Project. J. coast. Res. 6: 739–764.

Hydrobiologia **282/283**: 355–357, 1994.
P. H. Nienhuis & A. C. Smaal (eds), The Oosterschelde Estuary.
© 1994 *Kluwer Academic Publishers. Printed in Belgium.*

Theme V:

The response of benthic suspension feeders to environmental changes

Aad C. Smaal
National Institute for Coastal and Marine Management/RIKZ, P.O. Box 8039, 4330 EA Middelburg, The Netherlands

Introduction

In many estuaries, benthic suspension feeders generally occur in high densities and biomass, often concentrated in beds and reefs. In the Oosterschelde estuary the main species are cockles (*Cerastoderma edule*), mussels (*Mytilus edulis*) and the (subtidal) rocky shore populations of e.g. oysters (*Crassostrea gigas*), Ascidians, sponges and brittle stars (*Ophiothrix fragilis*). Biomass, averaged over the whole area in the post-barrier period, was estimated for cockles, mussels and rocky shore populations as 13.1, 10.7 and 2.5 g ash-free dry weight (AFDW) m^{-2}, respectively (Smaal & Nienhuis, 1992). In comparison with the mean biomass over the whole area of the other benthic populations (mainly deposit feeders) of 5 g AFDW m^{-2}, the dominance of the suspension feeders is obvious. Biomass data can also be presented per unit of habitat surface: in that case mean biomass of cockles and benthic deposit feeders on the tidal flats is 40 and 15 g AFDW m^{-2} respectively, mean mussel biomass on mussel cultivation plots is 170 g AFDW m^{-2} and mean biomass of the rocky shore community is 88 g AFDW per m^{-2} hard substrate (Seys *et al.*, 1994; Leewis *et al.*, 1994).

Biomass and densities of benthic suspension feeders in temperate estuaries such as the Oosterschelde, are extremely variable. As a result of storm events intertidal populations may be decimated, severe winters cause mass mortality of sensitive species such as e.g. *Ophiothrix fragilis*, and mild winters seem to induce recruitment failure for the bivalves and hence large year-to-year variation in year-class strength. This phenomenon is extensively discussed for cockle biomass estimations by Coosen *et al.* (1994).

The high densities of benthic suspension feeders and their feeding activity result in a dominant and potentially controlling role in the main nutrient fluxes in estuarine ecosystems (Dame, 1993). Feeding is performed by pumping and filtering large volumes of water through the gills, thereby creating a large flux of particulate material towards the bottom. In various ecosystems depletion of particulate material in the overlying water owing to biofiltration has been observed. Only a small part of the filtered material is ingested and absorbed by the animals, and the majority is biodeposited on the benthic suspension feeder beds or reefs. Depending on local hydrography, accumulation or resuspension of biodeposition will occur. The release of large amounts of dissolved inorganic nutrients has been observed on mussel beds and oyster reefs and is ascribed to remineralisation of biodeposits. The benthic suspension feeders have a large impact on the benthic-pelagic coupling, and enhance the cycling of nutrients (Smaal & Prins, 1993).

Another important feature of estuariene systems is the ability to cultivate bivalve suspension feeders. In the Oosterschelde estuary there is an extensived cultivation of mussels on bottom plots, which started more than 150 years ago. The actual mussel population is for more than 95% controlled by the mussel farmers, and one of the main objectives to protect the Oosterschelde area from flooding by a storm-surge barrier instead of a closed dam, was the maintenance of the mussel culture.

Impact of the civil-engineering project

The civil-engineering project has resulted in reduction of current velocities, freshwater and nutrient loadings, increase of the water residence time, and geomorphological changes. These changes interfered with other factors such as severe and mild winters. The main conclusions from the Oosterschelde studies with respect to the dynamics of benthic suspension feeders, presented in this section, can be summarized as follows:

There is no relation between the dynamics of the cockle population and the civil-engineering project; despite methodological difficulties in quantitative sampling of variable stocks, biomass fluctuations between years could be ascribed to climate effects on recruitment success and consequent year-class strength (Coosen et al., 1994).

The rocky shore suspension feeder communities show a considerable variation in biomass per m^{-2} values over the period 1979–1990, with a minimum in 1988. No direct relation could be established with the civil engineering project, and the variation could partly be explained by fluctuations in brittle star densities; in years after mild winters this species occurred in extremely high densities that may have had negative effects upon other rocky shore populations. On a local scale deposition of silt has reduced the availability of hard substrate habitat in the post-barrier period, but on other sites new substrate has been created owing to the engineering works. A shift in sponges and ascidian populations has been observed owing to reduced current velocities: passive filter feeding sponges standing stock has decreased and actively filter feeding ascidians have increased (Leewis et al., 1994).

The mussel farmers have anticipated the new hydrodynamic conditions by shifting culture plots to new, previously unsuitable areas. As a consequence the yields of the mussel culture could be maintained per unit standing stock (Smaal & Nienhuis, 1992). The average growth of mussels has not changed in the post-barrier situation, although the original West-East gradient in growth has disappeared (Van Stralen & Dijkema, 1994). In general, the growth of mussels correlates with phytoplankton primary production. This correlation is also true in the post-barrier period where the primary production pattern has maintained on average; primary production decreased in the western part and increased in the northern part of Oosterschelde estuary (Wetsteyn & Kromkamp, 1994). As a consequence of increased residence time and decreased nutrient loadings, internal nutrient cycling has become more important in the post-barrier period. The regeneration of nitrogen by mussel beds now contributes significantly to the nutrient availability in some periods of the year. It is postulated that the filtration by benthic suspension feeders stimulates the turnover of phytoplankton (Prins & Smaal, 1994).

Consequences for the Oosterschelde's carrying capacity

In general, benthic suspension feeder populations were not negatively affected by the engineering project. Adaptations have occurred, partly man-induced such as extension of hard substrate surface and exploitation of mussel culture plots on new locations, partly as a natural response such as the disappearance of the West-East gradient in mussel growth and the increased benthic-pelagic coupling. The consequences for the ecosystem have been evaluated with respect to the carrying capacity of the ecosystem for exploitation of benthic suspension feeders by man and by birds. Van Stralen & Dijkema (1994) have shown a clear relationship between growth of mussels, standing stock of mussels and cockles, and primary production: in years with a high primary production, there is a relatively high growth rate of mussels, unless there is a high standing stock of suspension feeders. There is competition for food between the suspension feeders, and extension of suspension feeder stocks will reduce the growth rate, unless the primary production increases. This is, however, very unlikely to occur because nutrient concentrations have reduced and are extremely low in summer. A model simulation has shown that extension of mussel culture by a factor 2 results in a reduction of mussel growth to negative values. The carrying capacity of the Oosterschelde ecosystem for mussel culture is limited by the actual level of primary production and does not allow extension of the cultures.

The exploitation of cockles has changed in the course of the engineering project. Fisheries of cockles increased from 8% to 53% of the available standing stock from the pre- to the post barrier period, respectively. The consumption of cockles by oystercatchers increased in the same period from 25% to 46% of the available stock (Smaal & Nienhuis, 1992; Schekkeman et al., 1994; Meire et al., 1994): the competition between man and waders is obvious. The carrying capacity of the Oosterschelde for waders has decreased in the post-barrier period by a combination of factors influencing food availability: low standing stocks of cockles due to recruitment failure and fisheries and decreased exposure time of foraging areas, being a consequence of the execution of the engineering project. In conclusion, the basis of the food chain, including the benthic suspension feeders, has shown a resilient response to the engineering project; yet, the top of the food chain is affected by the geomorphological changes.

References

Coosen, J., F. Twisk, M. W. M. van der Tol, R. H. D. Lambeck, M. R. van Stralen & P. M. Meire, 1994. Variability in stock assessment of cockles (*Cerastoderma edule L.*) in the Oosterschelde (in 1980–1990), in relation to environmental factors. Hydrobiologia 282/282 (Dev. Hydrobiol. 97): 381–395.

Dame, R. F., 1993. The role of bivalve filter feeder material fluxes in estuarine ecosystems. In R. F. Dame (ed.), Bivalve Filter Feeders in Estuarine and Coastal Ecosystem Processes. NATO ASI Series G, Vol. 33, Springer Verlag, Berlin: 245–270.

Leewis R. J., H. W. Waardenburg & M. W. M. van der Tol, 1994. Biomass and standing stock on sublittoral hard substrates in the Oosterschelde estuary (SW Netherlands). Hydrobiologia 282/283 (Dev. Hydrobiol. 97): 397–412.

Meire P. M., H. Schekkerman & P. L. Meininger, 1994. Consumption of benthic invertebrates by waterbirds in the Oosterschelde estuary, SW Netherlands. Hydrobiologia 282/283 (Dev. Hydrobiol. 97): 525–546.

Prins T. C. & A. C. Smaal, 1994. The role of the blue mussel (*Mytilus edulis*) in the cycling of nutrients in the Oosterschelde estuary (The Netherlands). Hydrobiologia 282/283 (Dev. Hydrobiol. 97): 413–429.

Schekkerman, H., P. L. Meininger & P. M. Meire, 1994. Changes in the waterbird populations of the Oosterschelde (SW Netherlands) as a result of large-scale coastal engineering works. Hydrobiologia 282/283 (Dev. Hydrobiol. 97): 509–524.

Seys, J. J., P. M. Meire, J. Coosen & J. A. Craeymeersch, 1994. Long-term changes (1979–89) in the intertidal macrozoobenthos of the Oosterschelde estuary: are patterns in total density, biomass and diversity induced by the construction of the storm-surge barrier? Hydrobiologia 282/283 (Dev. Hydrobiol. 97): 251–264.

Smaal, A. C. & P. H. Nienhuis, 1992. The Eastern Scheldt (The Netherlands), from an estuary to a tidal bay: a review of responses at the ecosystem level. Neth. J. Sea Res. 30: 161–173.

Smaal A. C. & T. C. Prins, 1993. The uptake of organic matter and the release of inorganic nutrients by bivalve suspension feeder beds. In R. F. Dame (ed.), Bivalve Filter Feeders in Estuarine and Coastal Ecosystem Processes. NATO ASI Series G, Vol. 33. Springer Verlag, Berlin: 271–299.

Van Stralen M. R. & R. D. Dijkema, 1994. Mussel culture in a changing environment: the effects of a coastal engineering project on mussel culture (*Mytilus edulis* L.) in the Oosterschelde estuary (SW Netherlands). Hydrobiologia 282/283 (Dev. Hydrobiol. 97): 359–379.

Wetsteyn L. P. M. J. & J. Kromkamp, 1994. Turbidity, nutrients and phytoplankton primary production in the Oosterschelde (The Netherlands) before, during and after a large-scale coastal engineering project (1980–1990). Hydrobiologia 282/283 (Dev. Hydrobiol. 97): 61–78.

Hydrobiologia **282/283**: 359–379, 1994.
P. H. Nienhuis & A. C. Smaal (eds), The Oosterschelde Estuary.
© 1994 *Kluwer Academic Publishers. Printed in Belgium.*

Mussel culture in a changing environment: the effects of a coastal engineering project on mussel culture (*Mytilus edulis* L.) in the Oosterschelde estuary (SW Netherlands)

M. R. van Stralen & R. D. Dijkema
Netherlands Institute for Fisheries Research (RIVO-DLO), P.O. Box 77, 4400 AB, Yerseke, The Netherlands

Key words: mussel, mussel culture, growth, standing stock, food, carrying capacity, hydrodynamics

Abstract

To evaluate the effects of a large scale coastal engineering project on the mussel (*Mytilus edulis*) bottom culture in the Oosterschelde estuary (S.W. Netherlands), mussel growth and production in the period 1980–1990 are studied in relation to food supply and the hydrodynamic conditions. Due to the construction of a storm-surge barrier and two additional dams, the risk that mussels are swept away by high current velocities decreased, resulting in an increase of the area in the Oosterschelde potentially suitable for mussel culture and in food availability now being more important as a limiting factor. For the Oosterschelde, a clear relation between mussel growth, stock sizes, and phytoplankton dynamics has been demonstrated. The meat yield of mussels landed in autumn – which is an index for growth rate – seems to be determined by the phytoplankton production in the preceding summer. In years with dense bivalve stocks, phytoplankton production and meat yields are relatively low. It is concluded that an increase of the mussel biomass cultured can result in a reduction of the primary production and, consequently, in a deterioration of the growing conditions for suspension-feeders in the estuary. This conclusion is supported by model calculations. An expansion of mussel culture in the new Oosterschelde is therefore dissuaded. Apart from primary production and stock sizes, food supply for mussels on culture lots appeared to be controlled by the horizontal advection of phytoplankton between and within the tidal channels. An observed decline in mussel landings from certain areas is attributed to the reduced mixing energy of the estuary in relation to the present distribution of the lots over the estuary. Production figures from the experimental lots, established in 1988 in the newly available areas, demonstrate that the yield of mussels can be enhanced by relaying culture lots towards the areas where the phytoplankton is produced. It is expected that by redistributing the culture lots, without expanding the biomass cultured, the carrying capacity of the Oosterschelde for mussel culture can be maintained.

Introduction

In the Netherlands, mussels are cultured on bottom in the Oosterschelde estuary (S.W. Netherlands) and in the Wadden Sea (Fig. 1). The average annual yield is 100 000 metric tons fresh weight, one third of which is produced in the Oosterschelde. To protect the Delta region in the southwestern part of the Netherlands from flooding, a large scale coastal engineering project has been carried out (Huis in 't Veld, 1984; Knoester, 1984). This resulted in the closure of various

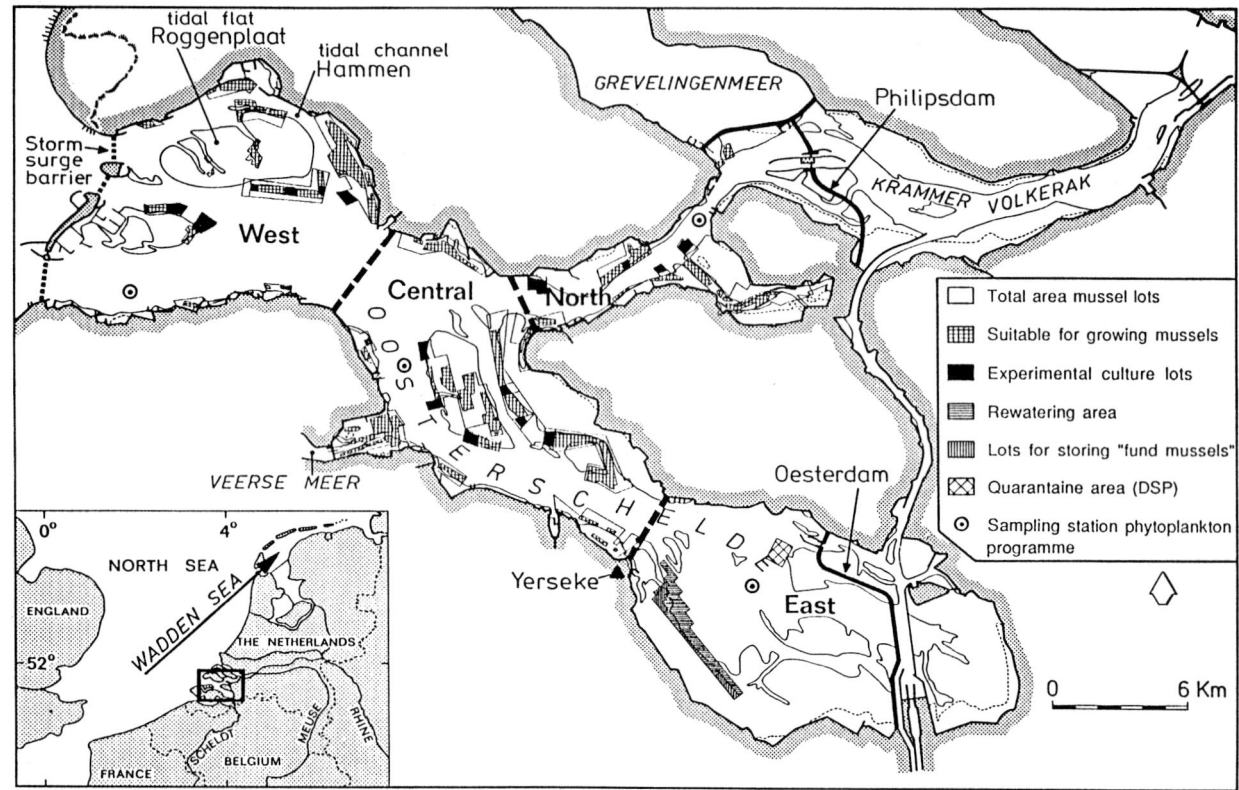

Fig. 1. The Oosterschelde after the completion of the barrier in 1987. The estuary is divided in four sections: the Western, Central, Eastern and Northern area. The locations of the Oosterschelde in the Delta area and of the Wadden Sea is depicted in the inset, showing a general map of the Netherlands. For the bottom culture lots, the area let to the farmers (including the experimental lots) and the area actually utilized for growing mussels (post-barrier situation) are indicated. Lots in the Eastern area of the estuary are used for storing landed mussels. Indicated are the sampling stations of the phytoplankton programme.

estuaries from the North Sea by dams or sluices, followed by the disappearance of shellfish fisheries and culture where the newly created lakes became brackish or fresh (Dijkema, 1988). Originally, a similar closure of the Oosterschelde estuary was planned. Due to pressure from the oyster and mussel growers and from nature conservation groups, followed by an extensive political debate, the plans were changed into the construction of a storm surge barrier, which is closed only under severe storm conditions (Knoester *et al.*, 1984). The design of the barrier aimed at maintaining the natural values of the area and preserving the shellfish culture and fisheries (Saeijs, 1982). The barrier in the Oosterschelde was completed in 1986. Two additional dams were completed in 1987, cutting off the newly

created lake 'Zoommeer' from the Oosterschelde. The impacts of these works on the ecosystem have been extensively studied from 1981 to 1990 (reviews by Smaal & Nienhuis, 1992; Smaal *et al.*, 1991). A specific study, dealing with the impact of the project on mussel cultivation in the Oosterschelde, is presented here.

The building of the barrier and adjacent dams resulted in a reduction of the current velocities in the estuary (Vroon, 1994). Before the barrier was completed, it became obvious that the area potentially suitable for mussel culture increased considerably, since the risk that mussels are swept away by high currents is reduced as well (Dijkema, 1988). It was expected that the capacity of the Oosterschelde to feed suspension-feeders is not significantly changed by the engineering works (Scholten *et al.*, 1990). The carrying capacity of the new Oosterschelde for the mussel culture is therefore studied with emphasis on mussel growth in relation to stock sizes and food supply.

Growth of mussels often depends on the availability of food (Boje, 1965; Pieters *et al.*, 1980; Wallace, 1980; Page & Hubbard, 1987; Mallet *et al.*, 1987), which is extracted from the water column by filtration. From the filtered seston, edible particles are selected from the gills and ingested (Kiørboe & Møhlenberg, 1981; Shumway *et al.*, 1985). The remaining material, mainly inorganic silt, is deposited. Phytoplankton is often an important food source (Rodhouse *et al.*, 1984). For a number of areas it has been shown that the feeding capacity for filter-feeders is related to plankton dynamics (Tenore & Gonzalez, 1976; Incze & Lutz, 1980; Rosenberg & Loo, 1983) and hence to factors influencing the production and biomass of phytoplankton. Also for the Oosterschelde it is shown that phytoplankton is the most important food source for mussels, and the growth rate of mussels increases when the plankton production is high (Smaal & Van Stralen, 1990). In this paper the growth of mussels in the Oosterschelde in the period 1980–1990 is related to phytoplankton production, chlorophyll concentrations and the biomass of mussels and cockles (*Cerastoderma edule*). The results of a statistical analysis of time-series are compared with results of model calculations, obtained with the ecosystem model 'Simulation Model Oosterschelde EcoSystem' (abbr. SMOES). In this model, the major carbon fluxes in the ecosystem have been integrated for the first and second trophic level (Klepper, 1989; Scholten & Van der Tol, 1994).

Study area

Based on hydrographic conditions, the Oosterschelde is divided into four sections: the Western, Central, Eastern and Northern area (Fig. 1). Most lots for growing mussels are located in the Western and Central area. For the West, special

attention is given to the tidal channel 'Hammen', in which two-third of the landed market-sized mussels are produced. After the completion of the barrier, however, conditions for growing mussels in the Hammen deteriorated and the landings from this area dropped dramatically. Due to the construction of the barrier, the hydrodynamic conditions have changed since 1985. The engineering works were completed in April 1987, resulting in a reduction of current velocities of 30% near the barrier up to 70% in the Northern section and a reduction of the tidal range by 12% to 3.25 m at station Yerseke. Nienhuis & Smaal (1994) defined the construction period from 1985 until April 1987. In this paper a time lag is incorporated, because the landing data of mussels (meat yields, landed amounts in autumn) reflect the production conditions in the preceding growing season(s). Mussel landed in 1985 are considered to be produced in the pre-barrier period. Production figures obtained in 1986 and 1987 are considered representative for the construction period, and, consequently, the post-barrier period starts in 1988.

A detailed overview of the characteristics of the Oosterschelde and changes due to the completion of the barrier is given by Nienhuis & Smaal (1994). The average biomass of mussels and cockles in the pre-barrier period (1982–1983) is 7.2 and 5.6 g C m^{-2} respectively. For the post-barrier period (1988–1989), these figures are 5.6 and 5.0 g C m^{-2}. The biomass of other macrobenthic species (tunicates, sponges, hydroids, oysters) decreased from 3.2 to 2.7 g C m^{-2}. The average annual biomass of zooplankton increased from 0.3 g C m^{-2} in the pre-barrier situation to 0.5 g C m^{-2} after the completion of the barrier. Primary production of phytoplankton ranges from 200–400 g C m^{-2} yr^{-1}, which is 85% of the total primary production in the estuary; the average annual chlorophyll-a concentration is about 5 μg l^{-1} (Wetsteijn & Bakker, 1991). These food parameters are considered representative for large areas (compartments as distinguished in Fig. 1), because the water in the estuary is well-mixed (Dronkers & Zimmerman, 1982; Bakker *et al.*, 1990; Smaal & Van Stralen, 1990).

The practice of mussel culture

The cultivation of mussels in the Netherlands dates from the end of the last century, when wild stocks of market-sized mussels became overfished (Korringa, 1976). Mussel culture is based on catching one-year old mussels of 20 mm shell length (called 'mussel seed') from natural beds and redistributing these on culture lots in the Oosterschelde and the Wadden Sea. Generally, the seeded mussels grow up to a marketable size of 50–60 mm within two years. Most culture lots are situated on the banks of tidal channels. Culture lots in the Central area of the Oosterschelde are traditionally used for the production of half-grown mussels (shell length 35–45 mm). During winter, these half-grown mussels were originally relaid mainly to culture lots in the Western area, where food conditions are better. With the completion of the engineering-works, however, the difference in growth conditions have disappeared. Culture lots in the Central area are now used more frequently for the production of market-sized mussels. The mussel farmers rent their lots from the state. Of the total surface of culture lots in the Oosterschelde (4000 ha), 56% is actually in use for growing mussels, which is 7% of the total surface area of the estuary. The remaining area is unsuitable for mussel culture because current velocities are too high (pre-barrier situation) or – for intertidal lots – the area is submerged too short. In the Wadden Sea the total surface of the culture lots is 6000 ha, of which 62% is in use. Due to the building of the 'Philipsdam', 662 ha of lots in the Krammer and Volkerak (North) were lost. Apart from the experimental lots laid out in 1988, this reduction is partly compensated by an increased suitability of the remaining lots in the Oosterschelde. In Table 1, the surface area actually in use for mussel culture before and after the completion of the barrier in 1987 is represented.

Since 1985, the mussels seeded on lots in the Oosterschelde originate almost completely from the Wadden Sea due to the failure of spatfall in the Oosterschelde for unknown reason. Wild Oosterschelde-mussels contribute only 5% of the total stock and are mainly found intertidally on rocky shores. Since 1985, the transfer of seed and half-grown mussels from the Wadden Sea to the Oosterschelde is not longer restricted to spring, when wild mussel beds are opened for the fishery. Especially in years with dense stocks of mussels in the Wadden Sea, in autumn and winter considerable quantities of half-grown mussels are dredged on lots in the Wadden Sea and brought to the Oosterschelde. The reason for re-seeding these mussels, despite the generally better growing conditions in the Wadden Sea, is the relatively high risk in the Wadden Sea to loose mussels during storms in winter. After the very successful settlement of mussels in the Wadden Sea in 1987,

Table 1. Total surface area actually utilized for growing and storing mussels in the Oosterschelde before and after 1987.

Surface area (ha)	West	Central	East	North	Total 1991	Total before 1987
Utilised for:						
producing half-grown	45	405	–	90	540	1081
producing market-size	510	610	–	190	1310	1010
experimental lots	162	193	–	48	403	–
storing and rewatering	–	10	330	–	340	250
storing fund mussels	–	–	110	–	110	175
quarantine	–	–	80	–	80	–
Total culture lots	717	1208	–	328	2253	2091
Total lots	717	1218	520	328	2783	2516

in 1988 and 1989 the biomass of mussels seeded in the Oosterschelde (respectively 63000 and 39000 tonnes fresh weight) even exceeded the annual production in this area.

All harvested mussels are sold at the auction in Yerseke. This auction is managed by the Commodity Board for Fish and Fish Products and is financed by the growers and traders who pay a levy per tonne of landed mussels. To inform the traders, every load of mussels is sampled by the auctioneers to determine size, meat yield and the total net weight of the landed mussels. The price paid for the mussels landed depends strongly on their meat yield (Fig. 3). The landing season starts in July, when the mussels have recovered from spawning, and lasts until the beginning of April next year. In Fig. 2 the annual landings from the Oosterschelde and Wadden Sea in the period 1950–1990 are presented. After being sold, the mussels are stored several weeks on special lots near Yerseke. When the supply of mussels is too high or their quality is judged too poor (meat yield must be above 16%, and at least 40% of the landed mussels must have a shell length of 50 mm or more), mussels are sold to an intervention fund which is financed through a levy on landed mussels as well. These mussels are stored on special lots and after the season are sold back to the mussel farmers, to be relaid on culture lots. The

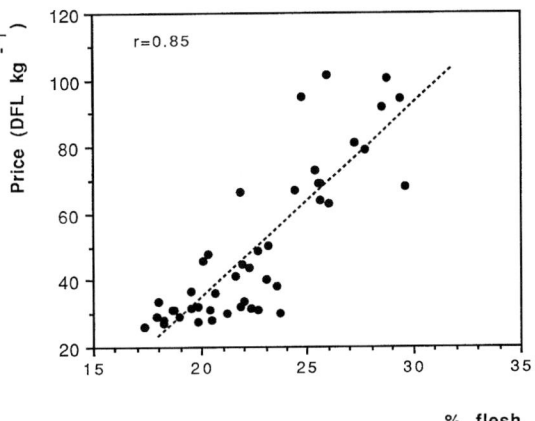

Fig. 3. Price per kg in Dutch guilders in January 1990 in relation to the meat yield of the landed mussels (1DFL = 0.55 US$).

location of the different types of lots in the Oosterschelde is given in Fig. 1. About 80 firms are culturing mussels and provide labour for 240 people (Mes, 1991). Respectively 110 and 50 people are employed in cockle and oyster fisheries. In the whole shellfish industry, including additional activities (packing for the fresh market, cookeries and canning, etc.), 1500 people are employed. Figure 4 shows the economic value of the landed mussels since 1982.

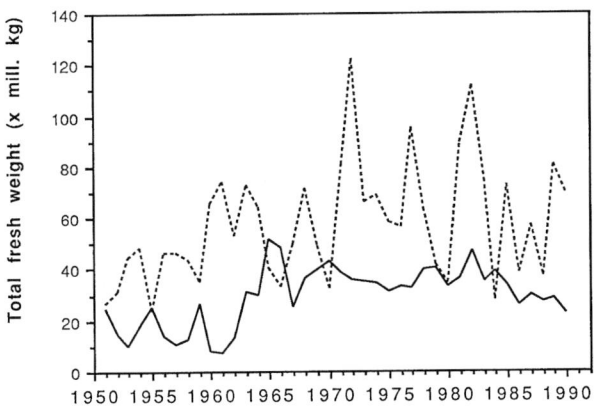

Fig. 2. The landings of mussels in the period 1950–1990 from the Delta region (solid line) and in the Wadden Sea (dotted line). All mussels landed from the Delta after 1972 originated from the Oosterschelde estuary.

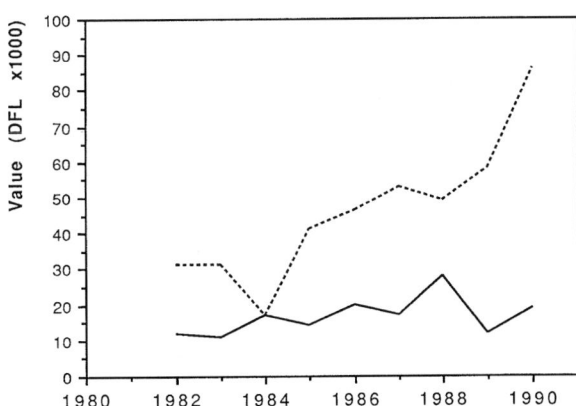

Fig. 4. Total value of landed mussels from the Oosterschelde (solid line) and the Wadden Sea (dotted line) in Dutch guilders (1DFL = 0.55 US$).

Material and methods

Meat yields of landed mussels as index for mussel growth

A preliminary study made clear that the meat yield of mussels (= amount of steamed flesh as percentage of the fresh weight) landed in autumn is strongly related to the growth of mussels in mg ash-free dry weight (ADW) per day in the preceding growing season (June–November) (Smaal & van Stralen, 1990). The growth rate of mussels is calculated from monthly samples during 1985–1987 on 20 to 40 culture lots, using regression techniques, and normalized for mussels of a length of 45 mm in May by analysis of covariance (Sokal & Rohlf, 1981). Averages of these normalized growth rates correlate very well with the average meat yield of mussels landed from the same area between August and November in the same year (Fig. 4; $r = 0.84$, $n = 13$).

Food supply

The meat yields of landed mussels are compared to the average chlorophyll concentrations and primary production in: (a) April–May, when plankton concentrations are generally high due to blooms, and (b) June–November, the mussel growing season. The averages are based on data from Wetsteyn et al. (1990), Wetsteyn & Bakker (1991).

Stock size of mussels and cockles and filtration activity

The estimation of the standing stock of mussels between 1980–1990 is based on (a) records from the Commodity Board of Fish and Fish Products of half-grown mussels re-seeded from lots in the Wadden Sea to the Oosterschelde, (b) on auction records of landed market-sized mussels and (c) on information from fishermen about the amounts of mussel seed fished on wild beds. The calculations were carried out as a part of the development of the model SMOES (Scholten &

Van der Tol, 1994). Figures on the biomass of cockles in the period 1980–1989 are calculated from data collected during a variety of field surveys (Coosen et al., 1994). For 1990 cockle biomass was estimated from landing figures and from data on spatfall (Van Stralen, unpublished data). For 1980 cockle biomass is probably seriously overestimated, and is therefore excluded from the statistical analysis presented in this study. Cockles biomass in 1980 was calculated from a field study on cockle dynamics in areas which are important for cockle fishery, but where the very abundant year-class 1979 had not yet been fished (Coosen et al., 1994).

The filtration activity of mussels and cockles is defined as the − area average − amount of water pumped daily per square meter by these species, and is calculated from (a) the biomass curves, (b) the age distribution and (c) the average annual growth of the different age groups (year classes) of the mussel and cockle stocks. The filtration (F in m³ d⁻¹) of individual mussels and cockles of a certain body weight (W in g ADW) is described by the relation $F = aW^b$. The values used for the parameters a and b (for mussels 0.062 and 0.55 and for cockles 0.058 and 0.65) originate from SMOES (Scholten & Van der Tol, 1994). Further, for cockles it is assumed they are submerged for 75% of the time.

Production of mussels on the experimental culture lots

To test wether the newly available areas are more suitable for growing mussels than the existing lots, a study with 40 experimental lots, spread over the Oosterschelde (Fig. 1), was started in 1988. The surface area of the experimental lots is about 400 ha, which represents a 22% increase of the area in the Oosterschelde used for mussel culture. The lots were let to the mussel growers for a period of three years. The growers were free how to use the lots, but were obliged to keep detailed records on seeded and fished amounts of mussels, their size and meat yield and on other activities like fishing starfish and impressions

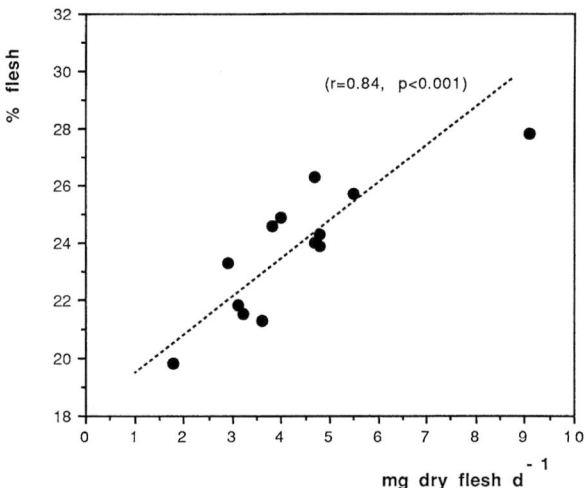

Fig. 5. Relation between growth rate of cultured mussels during summer (in mg dry tissue weight per day) and the meat yield of mussels (= steamed flesh as percentage of the fresh weight) landed in autumn from these production areas.

from sampling. From these data, production and growth statistics were derived and compared to figures from the existing ('old') lots.

The amount of mussels produced on a certain lot strongly depends on its surface area and the availability of mussels to seed on it. The productivity of lots is therefore expressed as the production efficiency, defined as the biomass of mussels present after one year as percentage of the amount of mussels seeded. For the experimental lots, production efficiencies are calculated from the biomass of mussels fished and seeded by fitting an exponential production curve. For the existing culture areas, production efficiencies are based on time-series of mussel biomass, collected by 'Van Veen' bottom grab sampling (Van Stralen, in prep.). The production efficiency is the final result of growth and mortality of mussels and therefore strongly depends on the size of the mussels seeded. Young mussels are generally more productive since they grow faster. To avoid bias from the size of the seeded mussels, only the production of market-sized mussels from half-grown mussels (40 mm) is considered here. The production of half-grown mussels from seed is not investigated, because, due to the scarcity of mussel

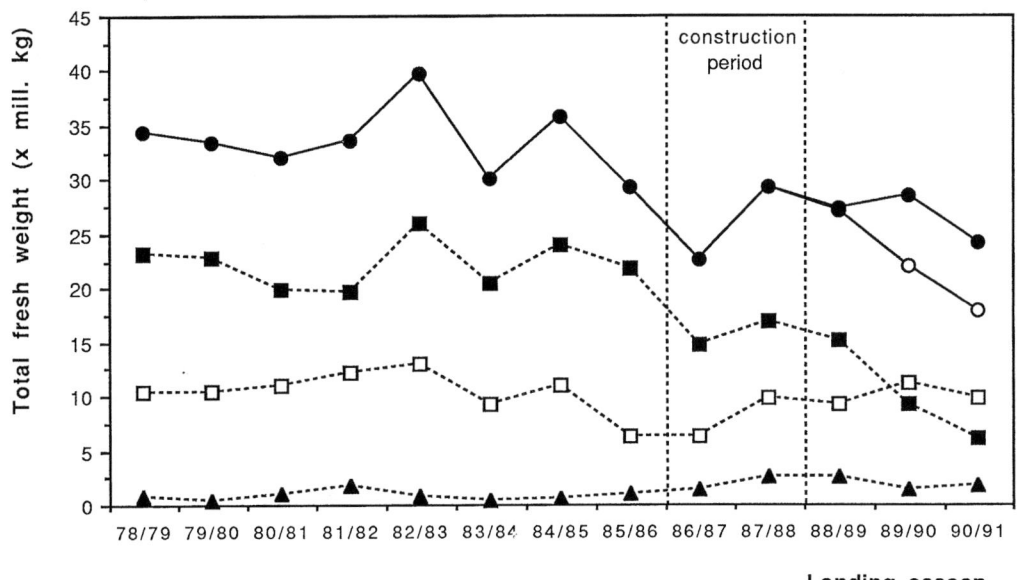

Fig. 6. Landings of market-sized mussels from the Oosterschelde (× 1000 tonnes total fresh weight) between 1978 and 1990. The landing season starts in summer and ends on the first of April next year. Mussels landed in the seasons '86/'87 and '87/'88 are considered to be cultured during the construction period. For the whole estuary (solid lines) for 1989 and 1990 total landings are given including (solid dots) and excluding (open dots) the experimental lots. The dotted lines refer to landings from culture lots in the Western (closed squares), Central (open squares) and Northern area (closed triangles) of the Oosterschelde.

seed on wild beds in 1989 and 1990, most lots were seeded with half-grown mussels.

Results

Production of market-sized mussels

The annual landings of market-sized mussels from 1978 until 1991 are presented in Fig. 6. The total landings from the Oosterschelde declined from 33 500 metric tons fresh weight in the pre-barrier period (1978–1986) to 26 400 tons (= 78%) after 1986 (*t*-test, *p* = 0.0021). In 1989 and 1990 25% of these landings originates from the experimental culture lots. Without these lots, the average landing of mussels after the completion of the barrier is 22 300 tons, which is 67% of the pre-barrier production. The decline in mussel landings carries back to the reduced production of market-sized mussels in the Western area, while the landings from the Central and Northern area remain about the same. In the Western area, the mussel production decreased from 22 200 tons

fresh weight in the pre-barrier years to 15 800 ton (= 71%) in 1986 and 1987 and further to 11 800 ton (= 53%) in the post-barrier situation. In the Western area the landings from the eastern section of the tidal channel 'Hammen' dropped from 21 300 tons before 1987, which was almost two-hird of the total production, to 1700 tons (= 8%) in 1989 and 1990.

Meat yield of landed mussels

In Fig. 7, the average meat yield of mussels landed between August and November is presented as a function of the distance between the production areas and the North Sea. For the pre-barrier period, relatively high meat yields (max. up to 30%) are found closer to the North Sea. This trend has disappeared during the building of the barrier due to a decrease in meat yields of mussels from the Western area and an increase in the Central area. After 1987, meat yields in the Western area, and especially in the eastern section of the Hammen, decreased further (down to 20%), while in the

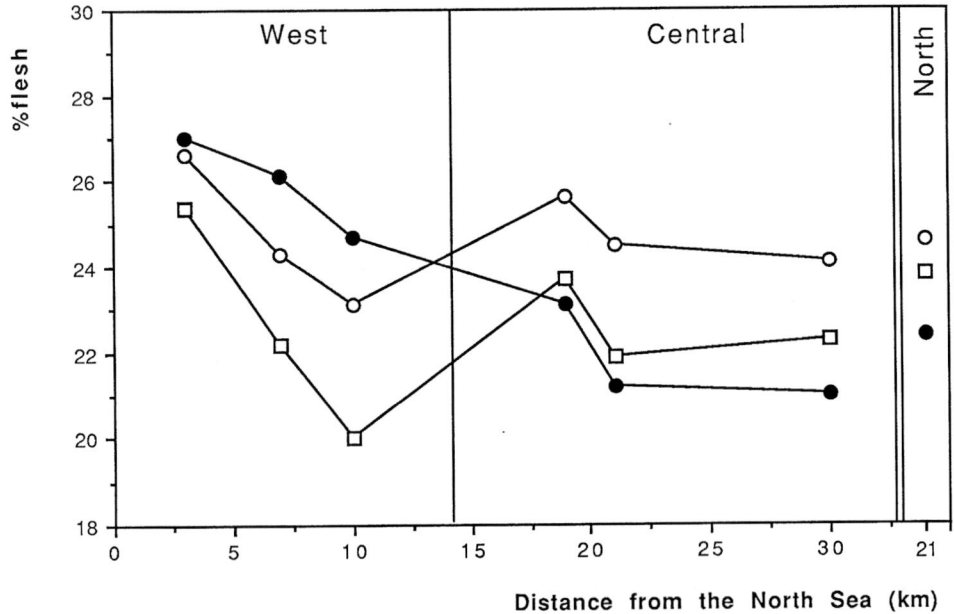

Fig. 7. Average meat yield of landed mussels in relation to the distance of the culture areas from the North Sea in the pre-barrier situation (1978–1985, closed dots), during the construction period (1986 and 1987, open dots) and in the postbarrier situation (1988–1990, open squares).

367

Central area the average meat yield decreased to pre-barrier values (23%). The low meat yields in the Hammen resulted in a change in the use of the lots in this area. Instead of growing market-sized mussels, the lots are now used more frequently for the production of half-grown mussels, which are later relaid onto lots with better growing conditions elsewhere in the Oosterschelde. The reduction in mussel production in the Hammen is therefore lower then the landing figures might suggest. Exact figures on the local production of half-grown mussels are, however, not available.

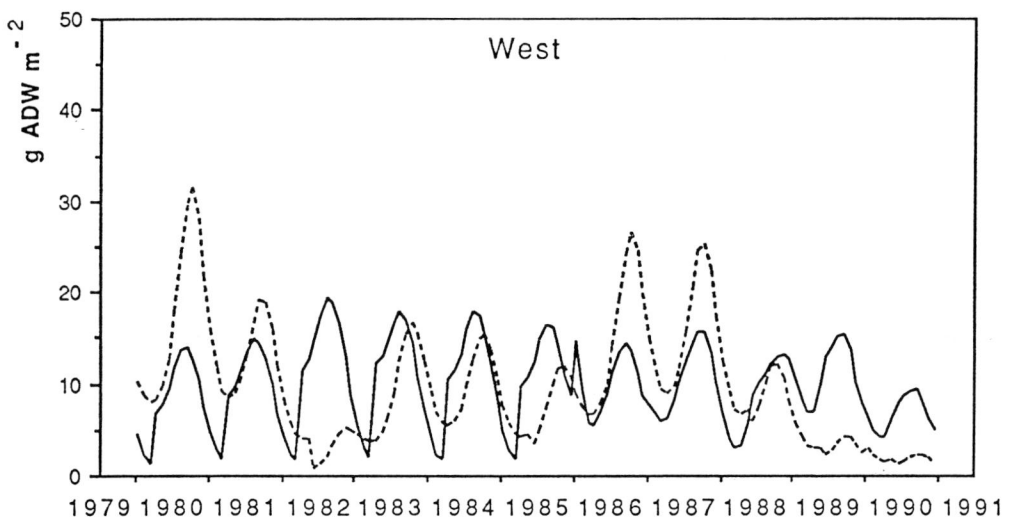

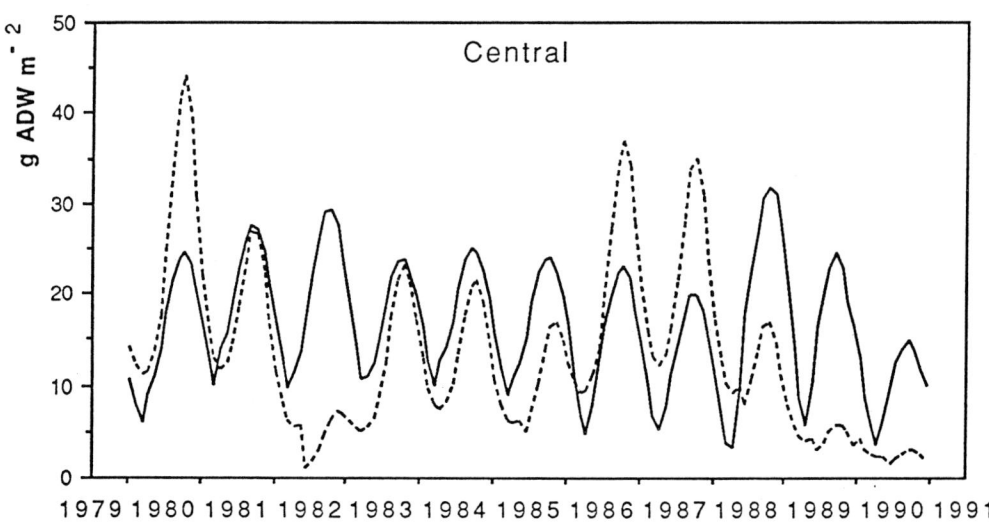

Fig. 8. Biomass of mussels (solid line) and cockles (dotted line) in the Western and Central area of the Oosterschelde (in g ADW m^{-2}) between 1980 and 1990.

Biomass and filtration activity of mussels and cockles

In Fig. 8 the biomass of mussels and cockles in the Western and Central area in the period 1980–1990 is presented. The filtration activity of the mussel and cockle stocks is shown in Fig. 9. Averaged over the whole Western area, mussels and cockles potentially filter 1 to 2 m³ water per square meter per day, with maxima up to 3 m³ m⁻² d⁻¹ during summer, when stock sizes are high. Taking into account the mean water depth of this area (12.2 m), this means that half of the water volume in the entire area is cleansed from particles every 3 to 9 days. The time in which the volume of the Central area is filtered is even shorter, since the average biomass of mussels and cockles per unit surface area is higher (Fig. 8) and water depth is less (10.4 m). During summer, half of the water volume in this area can be filtered within about 2 days. Variation in filtration activity from year to year is mainly determined by differences in the stock size of cockles, as is illustrated in Fig. 8.

Standing stock, food availability and mussel growth

For the *Western area* in Fig. 10.1 the meat yield of mussels landed from the existing culture lots is presented in relation to the average daily primary production of phytoplankton from June until November. The correlation between both variables is statistically significant, when calculated over the whole study period (1981–1990, $r = 0.904$, $n = 6$, $p < 0.05$). Striking are the relatively low meat yields and primary production observed for the post-barrier years. The time-series is too short for separate calculations for the pre- ($r = 0.837$, $n = 4$, not significant) and post- ($n = 2$) barrier period. It seems, however, that the response of the mussels to primary production has not been influenced by the engineering works. In Fig. 10.2, meat yields in the Western area are plotted as a function of the average chlorophyll-a concentration from June till November. The correlation between these variables (whole period: $r = 0.827$, $n = 10$, $p < 0.01$) confirms the relation found for meat yield and food availability expressed as primary production. In years with low chlorophyll concentrations, meat yields are low too, and again the relation between mussel condition and food availability appeared not to be influenced by the construction of the barrier. Because phytoplankton production and concentration are related as well ($r = 0.938$, $p < 0.01$, $n = 6$) it is not yet clear which of both factors is primarily determining mussel growth. Based on partial correlation coefficients (Sokal & Rohlf,

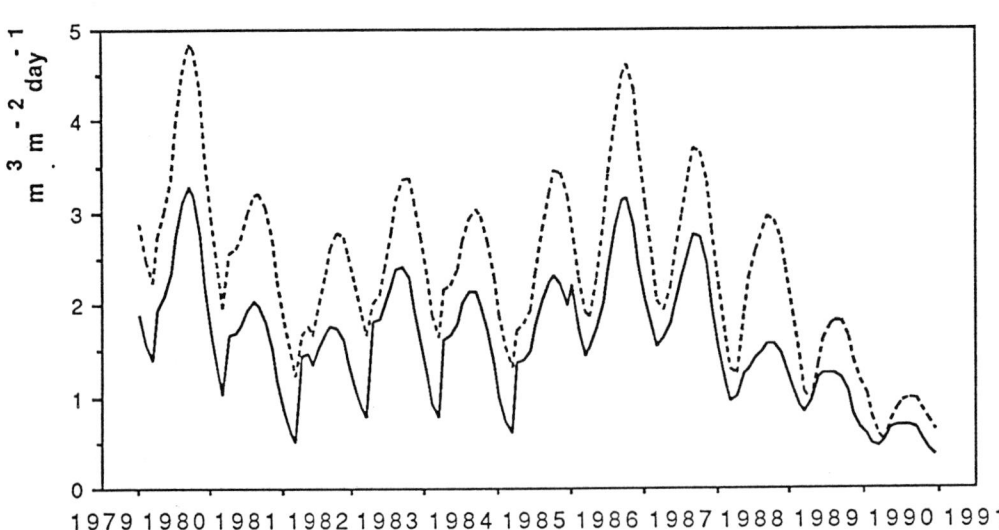

Fig. 9. The filtration activity of mussels and cockles in m³ m⁻² d⁻¹ in the Western (solid line) and Central area (dotted line).

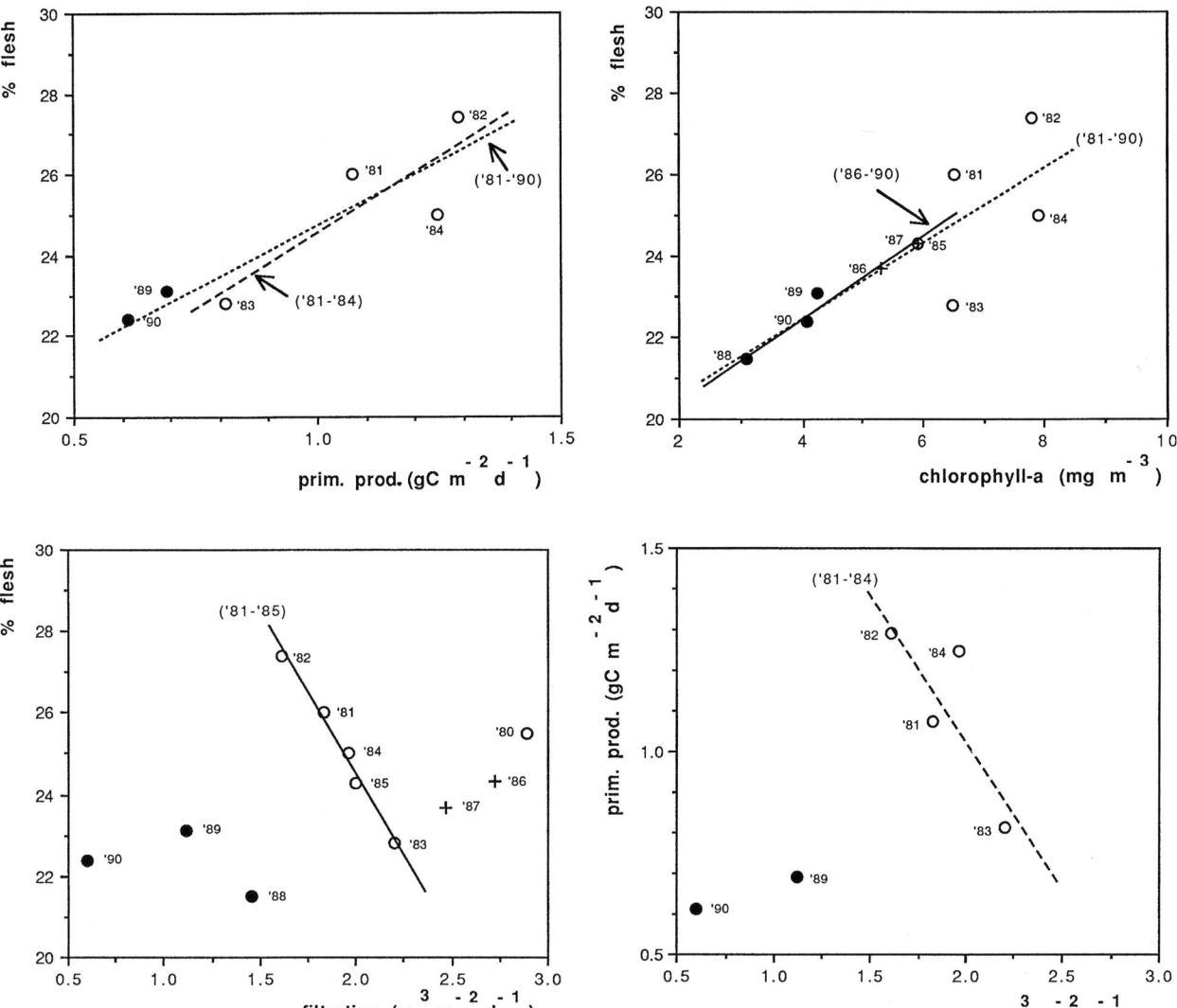

Fig. 10. Correlations between meat yields of mussels in autumn, food availability and filtration activity of mussels and cockles in the Western area of the Oosterschelde. Solid lines refer to statistically significant correlations for the pre- or for the post-barrier situation. Dashed lines refer to relations which appeared to be present in these periods but which are not statistically significant. Dotted lines refer to significant correlations over the whole period 1980–1990. The used symbols refer to the pre-barrier period (open dots), the construction period (+) and to the post-barrier situation (closed dots). (* = $p < 0.05$, ** = $p < 0.01$).

(1) Primary production [g C m^{-2} d^{-1}] *vs* meat yield of mussels
(2) Chlorophyll-*a* concentration in [mg m^{-3}] *vs* meat yield of mussels
(3) Filtration by mussels and cockles [m^3 m^{-2} d^{-1}] *vs* meat yield of mussels
(4) Filtration [m^3 m^{-2} d^{-1}] *vs* primary production [g C m^{-2} d^{-1}]

1981), it seems that the primary production is steering mussel growth ($r = 0.916$, $n = 6$, $p < 0.05$, Table 2). The correlation between chlorophyll concentrations and the meat yield of mussels in autumn vanishes when the calculations are based on year-average chlorophyll concentrations ($r = 0.568$, $n = 6$). This seems to be related to the high chlorophyll concentrations during phytoplankton blooms in spring. The low correlation between chlorophyll concentrations in April and

Table 2. Correlations and partial correlations between meat yields, primary production and chlorophyll concentrations in the Western and Central area. For the Western area correlation coefficients are calculated for the whole study-period (1981–1990) as well as for the pre-barrier situation (1981–1984). For the Central area only correlation coefficients for the pre-barrier situation are calculated, since the relation between food availability and mussel growth appeared to be influenced by the completion of the engineering works (ANCOVA, $P<0.01$). For the post-barrier situation for both areas time series is to short ($n=2$) to calculate correlations.

		Correlations			Partial correlations (first order)		
		chl-*a*	PP	Con	chl-*a*	PP	Con
WEST	chl-*a*	1		(df = 2)	1		(df = 1)
pre-barrier	PP	0.866	1		0.932	1	
(1981–1984)	Con	0.511	0.837	1	− 0.712	0.918	1
WEST	chl-*a*	1		(df = 4)	1		(df = 3)
whole period	PP	0.938**	1		0.916*	1	
(1981–1990)	Con	0.752'	0.904*	1	− 0.647	0.869'	1
CENTRAL	chl-*a*	1		(df = 2)	1		(df = 1)
pre-barrier	PP	0.989**	1		0.964*	1	
(1981–1984)	Con	0.833'	0.833'	1	0.112	0.112	1

chl-*a* = Chlorophyll-*a* concentration between June and November.
PP　 = Primary production between June and November.
Con 　= Average meat yield of mussels between August and November.
df　 = degrees of freedom.
'　 = $0.05<p<0.10$ (not significant).
*　 = $0.01<p<0.05$.
**　 = $p<0.01$.

May and the meat yields of mussels in autumn ($r=0.031$, $n=6$) indicates that mussels do not utilize these blooms for tissue growth.

In Fig. 10.3 the meat yield of mussels in autumn is presented as a function of the standing stock of mussels and cockles. In the pre-barrier period, a high filtration due to dense stocks coincides with relatively low meat yields ($r=-0.995$, $n=5$, $p<0.01$). Primary production and chlorophyll concentrations are negatively related to the filtration activity of mussels and cockles as well (Fig. 10.4, $r=-0.813$ resp. $r=-0.507$, $n=4$). Apart from the increased competition for food when stocks are high, mussel growth is apparently reduced due to a decrease of the production of phytoplankton. The low primary production observed in the post-barrier period, however, cannot be attributed to dense stocks of filter-feeders. In contrary, compared to the pre-barrier situation, mussel and cockle stocks were relatively low too in these years.

As observed for the Western area, in the *Central area* a high primary production of phytoplankton coincides with a high meat yield of mussels in autumn (Fig. 11.1). In contrast to the Western area, the relation between phytoplankton production and meat yields (pre-barrier period: $r=0.833$, $n=5$, $0.05<p<0.10$; post: $n=2$) seems to have changed since the completion of the engineering works. Despite the relatively low primary production, in 1989 and 1990, meat yields of mussels reached the same meat yields as observed in the pre-barrier period. The relation between meat yield and food availability, expressed as chlorophyll concentration (Fig. 11.2; pre-barrier: $r=0.833$, $0.05<p<0.10$, $n=6$, post: $r=0.936$, $p<0.05$, $n=3$) has changed statistically significantly (ANCOVA, $p<0.01$). Apparently, in the new situation, mussels on culture lots in the Central area are utilising the available food more efficiently.

Primary production and chlorophyll concentrations are highly correlated too ($r=0.984$,

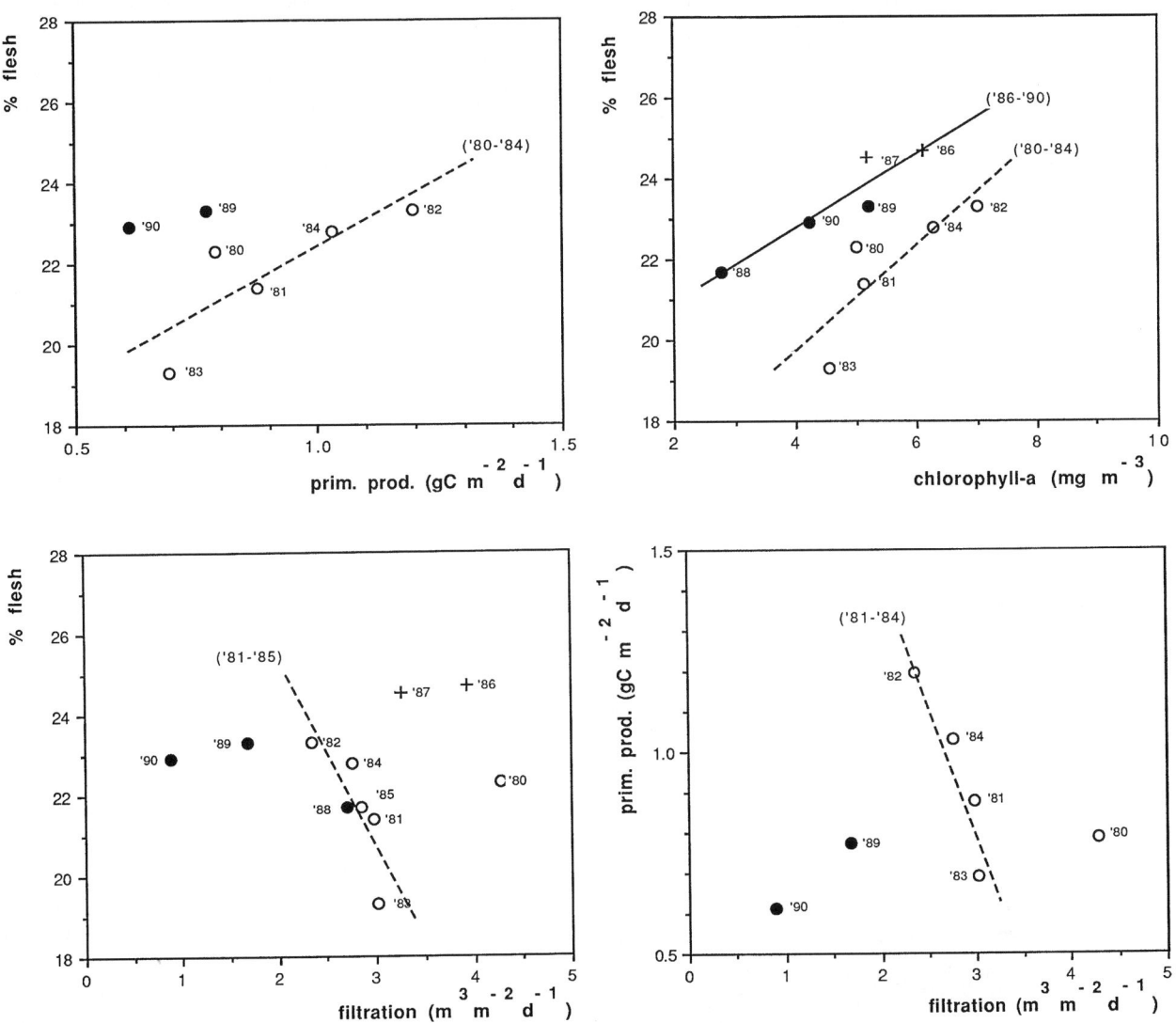

Fig. 11. Central area of the Oosterschelde: see legend of Fig. 10.

$p < 0.001$, $n = 6$). For the Central area, however, it is not yet clear which of both factors is primarily determining mussel growth. Partial correlations between meat yield and primary production respectively chlorophyll concentration are both low (Table 2). It has to be noticed, however, that in calculating partial correlations coefficients, hardly any degrees of freedom are left, and therefore the sensitivity to detect relations is low. In Figs 11.3 and 11.4, the meat yield of mussels and the primary production are plotted as a function of the filtration activity of mussels and cockles. The re-

lations found (pre-barrier period, excluding 1980: $r = -0.820$, $p < 0.1$, $n = 5$ resp. $r = -0.929$, $p < 0.05$, $n = 4$) confirm those found for the Western area.

In the *Northern area*, chlorophyll concentrations almost doubled: from 5.3 mg m^{-3} in the period 1980–1986 to 11.2 mg m^{-3} in 1987 until 1990 (mean of year averages). Primary production figures are not available, except for 1989 and 1990, but it is assumed that primary production has increased as well (Smaal & Nienhuis, 1992). Mussel growth in the Northern area, however,

372

cannot be explained by these figures. A relation between the average chlorophyll concentration (June–November) and meat yield of mussels landed appeared to be absent (pre-barrier period: $r = -0.21$, $n = 6$; post-barrier: $r = -0.36$, $n = 3$; whole period 1981–1990: $r = 0.09$, $n = 9$).

The risk that mussels are lost by high current velocities

With the validation of the predicted changes on the hydrodynamic circumstances in the Ooster-schelde (Vroon, 1994) a definitive map of areas suitable for growing mussels could be compiled (Fig. 12). As predicted, only in the deeper parts of the larger tidal channels current velocities are still too high for culturing mussels (> 60 cm s^{-1}).

Suitability of the new areas: the experimental lots

In Figs 13 and 14 average meat yields of landed mussels and production efficiencies (annual yield per unit of seeded biomass) are compared for the experimental and existing lots. Special attention is given to the eastern section of the Hammen-channel because of its high production in the pre-barrier period and the dramatic drop of the landings from this area in the new situation. The meat yield of mussels from the experimental lots is generally higher than from the existing lots. The largest differences are found for the Western area. Meat yields of mussels produced on the experimental lots were on average 25%, on the existing lots 21%. This difference reflects the difference in growing conditions in general between the Hammen-channel, where most of the existing lots

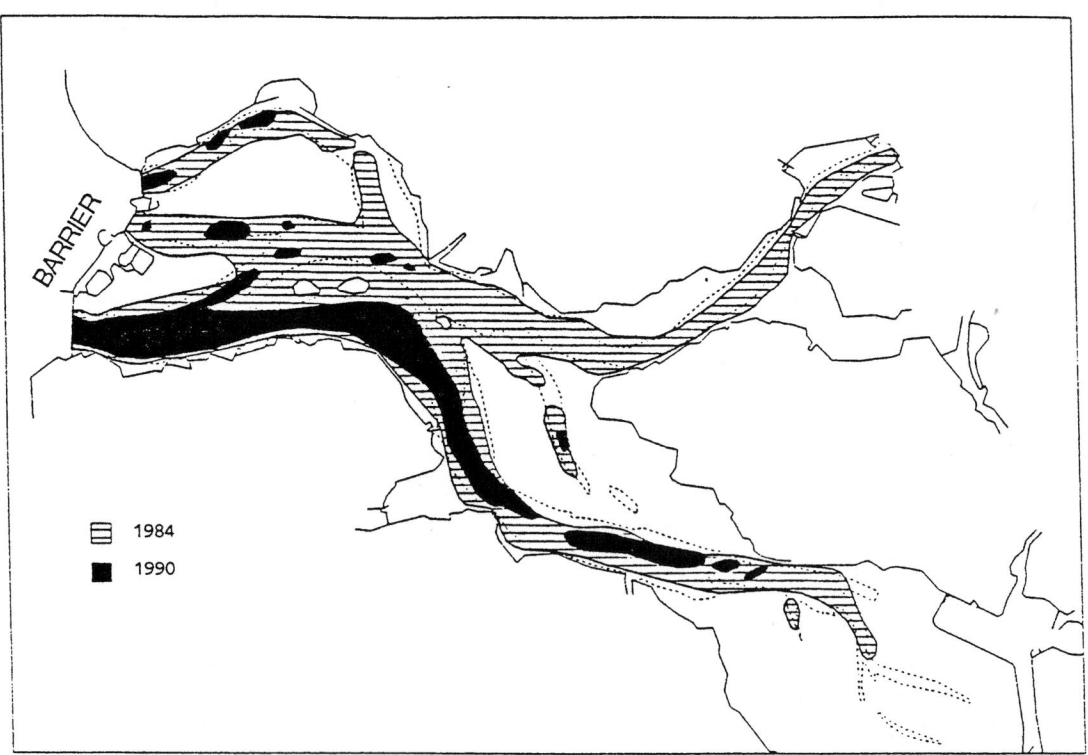

Fig. 12. Risk of mussels being swept away due to high current velocities. Before 1987 in the hatched and black areas current velocities were over 60 cm s^{-1} and therefore unsuitable for musselculture. After the completion of the barrier in the black areas current are still over 60 cm s^{-1}.

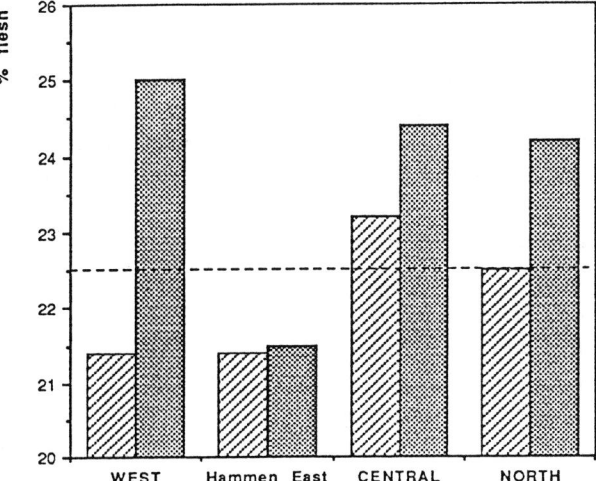

Fig. 13. Average meat yield of mussels landed from existing lots (hatched bars) and experimental lots (shaded bars) in the Western, Central and Northern area of the Oosterschelde in 1989 and 1990. Separate calculations are presented for the eastern section of the Hammen-channel, because of its high production in the pre-barrier situation and the dramatic drop of the landings from this area since the completion of the barrier. The horizontal line represents the average meat yield of all mussels landed from the Oosterschelde.

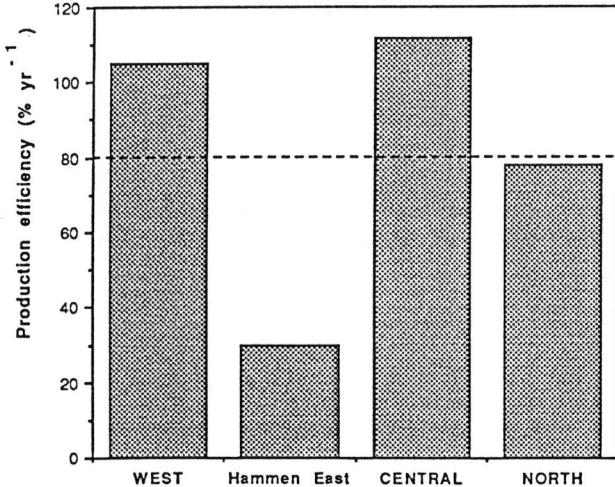

Fig. 14. Production efficiencies (annual yield per unit of seeded biomass in %) of the experimental culture lots in the West, Central and North of the Oosterschelde between 1988 and 1990. The horizontal line represents the average production efficiency for all existing lots.

are situated, and the two tidal channels to the south, where most of the experimental lots were established. The differences between the existing and experimental lots in the Central and Northern area are less, since the meat yields of mussels produced on the existing lots in these areas is maintained at pre-barrier values (on average 23%). With respect to the amount of mussels produced (Fig. 14), over the whole estuary in 1989 and 1990 the average production efficiency on the existing lots was 0.80. For the experimental lots in the Western area (excluding lots in the Hammen) and Central area production efficiencies were generally higher (on average 1.09). The productivity of the experimental lots in the Northern area (0.78) is about the same as on the existing culture lots. The production efficiency on the experimental lots in the Hammen was: 0.30.

Discussion

Mussel growth in relation to phytoplankton dynamics

In the Oosterschelde, a clear relation between plankton production and concentration during summer and the meat yield of mussels in autumn has been demonstrated. In the Western area of the estuary, primary production seems to determine mussel growth. Since primary production correlates with chlorophyll concentrations, differences in meat yield can be predicted by chlorophyll concentrations as well. Phytoplankton blooms in spring seem to be of no importance for the meat yield of mussels in autumn. The filtered phytoplankton is utilized mainly for gonad development and spawning. Moreover, the feeding activity of the mussels might be reduced during spawning, as observed by Newell & Thompson (1984). It is concluded that the growth of mussels and the meat yield of mussels in autumn are determined by food availability during summer. In this study the growing season for mussels is therefore defined as the period between the first of June and November.

For the Western area, the relation between food

availability during summer and mussel condition in autumn appears not to have been influenced by the engineering works. In the Central area, however, after the completion of the barrier, mussels utilize the available phytoplankton more efficiently, resulting in higher meat yields. With respect to the quality of the seston as food for mussels, inorganic silt concentrations in the Oosterschelde have decreased considerably (Wetsteyn & Bakker, 1990). A more efficient utilisation of the available phytoplankton by filter-feeders, when silt concentrations are low, is suggested by Winter (1973) and Theisen (1977). For the Oosterschelde this is supported by model calculations (Van der Tol & Scholten, 1992; Scholten & Van der Tol, 1994). Also zooplankton in the Oosterschelde seems to profit from the decreased silt concentrations (Tackx et al., 1990). A higher quality of the seston as food for mussels, however, does not explain why only mussels cultured in Central area utilize the produced food more efficiently, considering that silt concentrations in the Western area are reduced as well (Wetsteyn & Bakker, 1990). As explained in next section, the growing conditions for mussels cultured in the Oosterschelde appeared to be determined mainly by the distribution of the culture lots over the estuary in relation to the (changed) hydrodynamic conditions.

Site selection

The dramatic drop of mussel growth in the Hammen after the completion of the barrier indicates that the carrying capacity of this channel for mussel culture has been reduced. In the Hammen, the rate in which food is depleted by filtration is relatively high. Most of the lots for mussel culture in the Western area are situated in this channel, producing two third of the annual production of market-sized mussels from the Oosterschelde (pre-barrier situation). Phytoplankton is also depleted by cockles on the tidal flat 'Roggenplaat', of which gullies and creeks drain mainly onto the Hammen. The high filtration activity, in combination with the fact that the

Hammen transports only 20% of the total tidal volume of the estuary, stresses the importance of horizontal transport of phytoplankton from the surrounding areas towards this channel. During the pre-barrier period, the Oosterschelde estuary was well-mixed. Smaal & Van Stralen (1990) demonstrated that mussels on lots utilize the phytoplankton produced over large areas. For mussels in the Hammen, it was concluded that also phytoplankton produced in the North Sea is utilized as a food source. Under present conditions (reduced current velocities by 30%; in the eastern part of the Hammen up to 40%), water exchange with the surrounding areas is apparently too low to supply the dense stocks of filter-feeders in the area with sufficient food. Given the fact that within the Western area mussel growth in the less densely populated channels situated more southerly is still good, it is expected that mussel growth in the Hammen will recover when the mussel biomass cultured in this area is reduced. This conclusion is supported by the observed recovery of meat yields of mussels in the Hammen in 1991, when the mussel and cockle stocks were extremely low. Lots for mussel culture were seeded scarcely due to a lack of seed mussels in the Wadden Sea, and the cockle biomass on the Roggenplaat was low due to a lack of spatfall since 1988.

Due to the reduction of (high) current velocities, the area in the Oosterschelde potentially suitable for mussel culture increased considerably. In the newly available areas the experimental lots were established. Also within the boundaries of the existing lots, especially in the Central and the Northern area the surface area actually utilized for mussel culture increased considerably. Compared to the existing lots, the meat yield of mussels from the new production sites is generally high, which seems to be related to their location on the banks of the larger tidal channels. Mainly because of their depth, these channels produce most of the phytoplankton. Apart from the short distance between the food source and the cultured mussels, food supply towards the new sites is stimulated by the relatively high current velocities compared to the existing production sites. It is concluded that in the Oosterschelde not only between tidal channels – as discussed for the Hammen case – but also within the tidal channels food supply for aggregations of mussels is controlled by the advection of phytoplankton. Apart from horizontal transport, local food availability for benthic filter-feeders is controlled by vertical mixing of the water column, which is also a function of current velocities. For the pre-barrier situation, reduced food concentrations near the mussel bed could not be demonstrated and it was concluded that mussels cultured in the Oosterschelde utilize the whole water column (Smaal & Van Stralen, 1990). In the new situation, current velocities locally dropped dramatically, especially in the Northern area. Due to the construction of the 'Philipsdam', in this area current velocities are reduced by 70% to maximum values less than 20–30 cm s^{-1}. Food depletion near the mussel bed might determine now mussel growth, and explain why meat yields has not increased in accordance to the observed doubling of the phytoplankton concentrations in the centre of the main channel of this area.

For areas with similar current conditions reduced mussel growth due to food depletion near the mussel bed has been demonstrated by various authors. From a study on lots for mussel bottom culture on the coast of Maine (USA), it is shown that mussel growth on the edges of aggregations of mussels is generally higher than in the centre (Newell, 1990). For locations with relatively high current speeds (mean over 10 cm s^{-1}) this phenomenon was observed for the larger patches (patch diameter over 10 m). Food depletion through filtration by mussels resulting in reduced plankton concentrations near the mussel bed was demonstrated (Newell *et al.*, 1989). Also the growth of mussels suspended one meter above the mussel bed can be reduced by food depletion (Fréchette & Bourget, 1985). These authors found reduced food concentrations near mussel beds at current speeds varying between 10 and 25 cm s^{-1} (Fréchette *et al.*, 1989). For the Northern area it may be worthwhile to reduce mussel densities on culture lots, for instance by increasing their surface without increasing the seeded biomass, and

to improve the spreading of mussels during seeding as suggested by Newell (1990). It is not clear yet wether it is also worthwhile to produce mussels in off bottom cultures instead of on bottom lots. Some experimental long-line cultures have been started, producing annually 1–2 hundred tonnes fresh weight. Compared to bottom lots, mussels growth is relatively high. The surface area suitable for off bottom cultures, however, is restricted and labour costs per unit mussels produced are high.

The carrying capacity of the Oosterschelde for mussel culture

For the Oosterschelde, the total primary production of phytoplankton is maintained at the level observed in the pre-barrier period (Wetsteyn & Bakker, 1990; Smaal & Nienhuis, 1994). The impact on phytoplankton production of the lower nutrient concentrations, caused by a reduced fresh-water load, seems to be compensated by the reduced turbidity and therefore lower light attenuation. Changes occurred in the seasonal pattern of the primary production – it starts earlier and lasts longer – and in the succession of phytoplankton species. Within the Oosterschelde, primary production and chlorophyll concentrations increased in the Northern and Eastern area, and decreased in the Western and Central area. As shown by a number of authors, phytoplankton dynamics can be influenced locally by the consumption of phytoplankton by filter-feeders (Fréchette & Bourget, 1985; Fréchette *et al.*, 1989; Newell, 1990), as well as over large areas (Carlson *et al.*, 1984; Cloern, 1982; Nichols, 1985). Also for the Oosterschelde estuary it has been shown that a high filtration activity by mussels and cockles is accompanied by low chlorophyll concentrations and primary production. In this context, the reduction of the primary production in 1989 and 1990 in the Western and Central area is striking, considering the low stocks of mussels and cockles in these years. The low biomass of filter-feeders cannot be attributed to the completion of the engineering works. Cockle biomass is

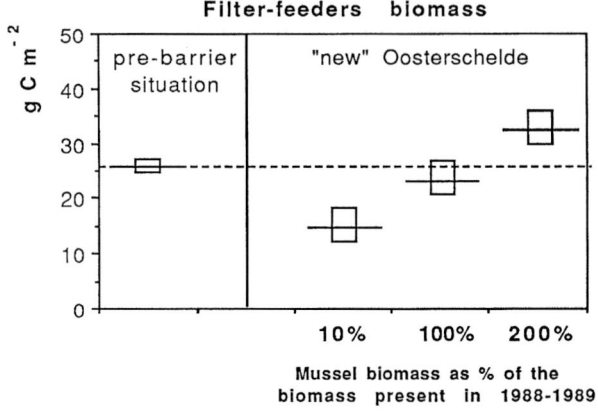

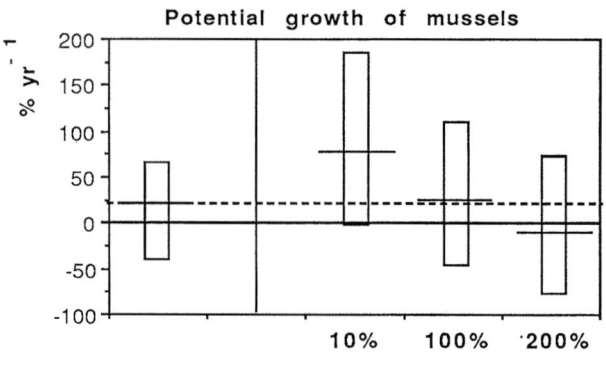

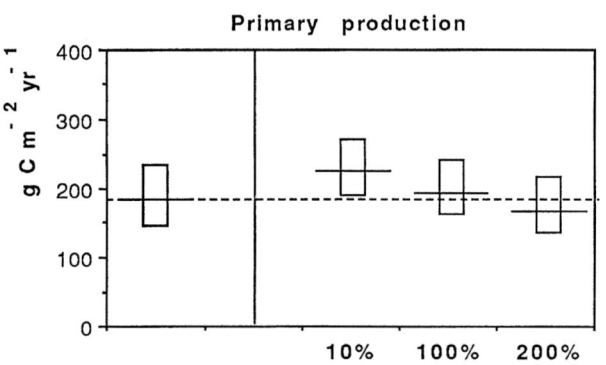

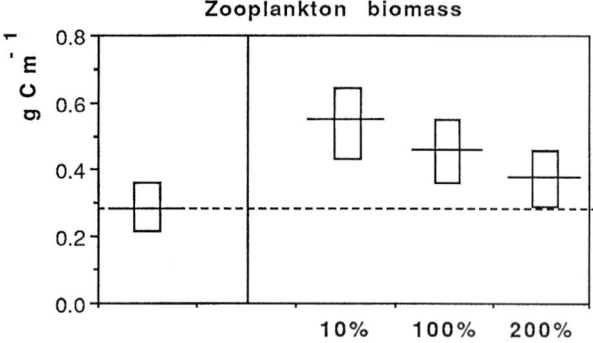

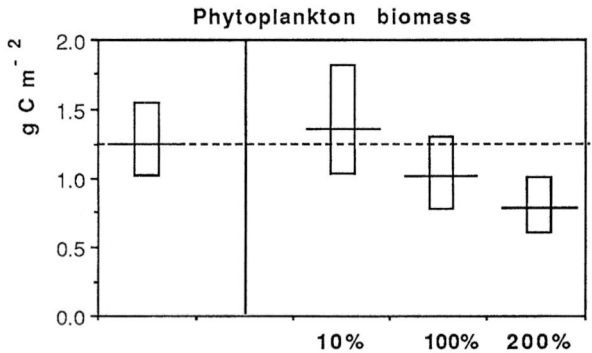

Fig. 15. Results of calculations with the mathematical ecosystem model SMOES, simulating a doubling of the biomass of mussels (200% scenario) and a reduction to 10% of the biomass level present in 1989 and 1990. Shown is also the situation for the pre-barrier period (left and dotted lines), obtained by calibrating the model for the years 1982 and 1983. Given intervals reflect model outcomes between 5 and 95% of all outcome values.

low due to the failure of spatfall, which Coosen *et al.* (1994) attribute to the very mild winters since 1988. Mussel biomass on culture lots decreased steadily since 1989 due to a lack of mussel seed on wild beds the Wadden Sea. To obtain insight in the carrying capacity of the 'new' Oosterschelde for mussel culture, calculations with the model SMOES have been carried out, simulating

different levels of mussel biomass. The simulations are made with the model version calibrated for the post-barrier situation (1988 and 1989) (Scholten & v.d. Tol, 1994). Figure 15 shows the impact of a doubling of the mussel biomass and of a reduction to 10% of the mussel biomass level in 1988 and 1989 on: (1) the total filter-feeder biomass, (2) the primary production, (3) plankton biomass, (4) mussel growth and (5) zooplankton biomass. Values for these variables for the pre-barrier situation (1982–1983) are given as well.

Doubling of the mussel biomass results in an increase of the total biomass of suspension-feeders by 41%. Compared to the pre-barrier situation (1982–1983), doubling of the mussel biomass as present in 1988 and 1989, corresponds with an increase by 26% of the filter-feeder biomass in 1982–1983. Such an increase could also occur naturally in case of strong cockle year-classes, such as in 1979, 1986 and 1987. The reduction of mussel biomass to 10% can be considered as a scenario for an Oosterschelde without mussel culture. Doubling of the mussel biomass results in a decrease in plankton biomass and primary production by 20% and 14% respectively. Primary production decreases less than plankton biomass because the increased depletion of phytoplankton by (more) mussels is accompanied by an increase of the release of nutrients from the mussel beds, stimulating the growth of the algae. The importance of mussel beds in the Oosterschelde in the regeneration of inorganic nutrients through mineralisation of deposited faeces and pseudofaeces is supported by field studies (Prins & Smaal, 1990; Prins & Smaal, 1994; Dame *et al.*, 1991). The reduced food availability when mussel biomass is doubled, and the fact that the available food is shared by more consumers, results in a reduction of the potential growth of mussels and a reduction of the biomass of zooplankton (−19%). When mussel biomass is reduced (the 10% scenario), the opposite effects have been calculated: a higher plankton production (+16%) and concentration (+35%) and consequently a better growth of mussels and zooplankton (+17%).

The model-predictions support the conclusions drawn from the analysis of time-series presented before. It is expected that in the new Oosterschelde the plankton production and consequently the productivity of the whole ecosystem, decreases when mussel culture is expanded. Strong year classes of cockles might have the same effect. In addition, also the colonisation of

subtidal areas by wild suspension-feeders is now possible, since the risk of being swept away as a result of high current velocities is reduced. Already in 1987, settlement of the bivalve *Spisula subtruncata* was observed in new areas. This stock, however, disappeared within one year due to predation by starfish.

Conclusions and consequences for mussel culture

The impact of the engineering works in the Oosterschelde has resulted in a biological instead of a physical limitation for mussel culture. The risk that mussels are swept away by high current speeds has been reduced, and in principle large areas are now suitable for growing mussels. It is expected that an increase of the mussel biomass will lead to a reduction of the primary production and to lower meat yields and consequently lower prices obtained by the grower for the mussels landed. From an economic point of view, however, a higher landing accompanied by lower meat yields, might still be more profitable. A reduction of the production of phytoplankton and the consequences for other suspension-feeders, however, conflicts with the main objective in the governmental policy for the Oosterschelde: the preservation of the natural values of the area (see Nienhuis & Smaal, 1994). An expansion of mussel culture is therefore dissuaded. It is concluded that food availability for mussels on culture lots is controlled by the horizontal transport of phytoplankton between and within the tidal channels. The decline in mussel landings from the existing culture lots seems at least partly a result of the reduced mixing energy of the estuary and the present distribution of the culture lots over the estuary. The results of mussel culture can be enhanced by redistributing the mussel lots towards the areas where the phytoplankton is produced or where water exchange with the surrounding area is relatively high. A redistribution of culture lots over the Oosterschelde is recommended. With respect to local food depletion for the Northern area it may be worthwhile to spread the mussels over a larger surface without increasing the cultured biomass. For the Hammen a reduction of the biomass cultured should be achieved. Further, in view of the high risk of loosing mussels due to storms in winter in the Wadden Sea, it is suggested to utilize the carrying capacity of the Oosterschelde for the production of market-sized mussels by on-growing half-grown mussels from the Wadden sea. Given the present primary production and the yields from the experimental culture lots, it is expected that production and meat yield of cultured mussels can be similar to those before 1987.

Acknowledgements

We are grateful to P. Dare, N. Dankers and A. Smaal for valuable comments on the manuscript; to the members of the EOS-project J. van Buuren, W. ten Brinke, H. Haas, A. Smaal, M. van der Tol and J. Vroon for their stimulating discussions and cooperation, to J. Bol, C. Brand, A. Doornekamp and the crew of the research vessel 'Schollevaar' for their help with sampling and sorting and to H. Scholten and M. van der Tol providing data on model simulations.

References

Bakker, C., P. M. J. Herman & M. Vink, 1990, Changes in seasonal succession of phytoplankton, J. exp. mar. Biol. Ecol. 148: 215–232.

Boje, R., 1965, Die Bedeutung von Nahrungsfaktoren für den Wachtstum von *Mytilus edulis* L. in de Kieler Förde und im Nord-Ostsee-Kanal. Kieler Meeresf. 21: 81–100.

Carlson, D. J., D. W. Townsend, A. L. Hyliard & J. F. Eaton, 1984. Effect of an intertidal mudflat on plankton of the overlaying water column. Can. J. aquat. Sci. 41: 1523–1528.

Cloern, J. E., 1982. Does benthos control phytoplankton biomass in South San Fransisco Bay? Mar. Ecol. Prog. Ser. 9: 191–202.

Coosen, J., F. Twisk, M. W. M. van de Tol, R. H. D. Lambeck, M. R. van Stralen and P. M. Meire, 1994. Variability in stock assessment of cockles (*Cerastoderma edule*) in the Oosterschelde (1980–1990), in relation to environmental factors. Hydrobiologia 282/283: 381–395.

Dame, R., N. Dankers, T. Prins, H. Jongsma & A. Smaal, 1991. The influence of mussel beds on nutrients in the Western Wadden Sea and Eastern Scheldt estuaries. Estuaries 14: 130–138.

Dijkema, R., 1988. Shellfish cultivation and fisheries before and after a major flood barrier construction project in the Southwestern Netherlands. J. Shellfish Res. 7: 241–252.

Dronkers, J. & J. T. F. Zimmerman, 1982. Some principles of mixing in tidal lagoons with examples of the tidal basins in the Oosterschelde. In: P. Lasserre & H. Postma (eds), Coastal Lagoons, Proc. Int. Symp. Coastal Lagoons, Gauthier-villars: 460–474.

Fréchette, M. & E. Bourget, 1985. Food limited growth of Mytilus edulis L. in relation to the benthic boundary layer. Can. J. Fish. aquat. Sci. 42: 1166–1170.

Fréchette, M., C. A. Butman & W. Rockwell Geyer, 1989. The importance of the benthic boundary layer flows in supplying phytoplankton to the benthic suspension feeder, Mytilus edulis L. Limnol. Oceanogr. 34: 19–36.

Huis in 't Veld, J. C., J. Stuip, A. W. Walter & J. M. van Westen (eds), 1984. The closure of tidal basins. Delft University Press, 450 pp.

Incze, L. S. & R. A. Lutz, 1980. Mussel culture: an east coast perspective, In: R. A. Lutz (ed.), Mussel culture and harvest; a North American perspective. Elsevier, Amsterdam: 99–140.

Kiørboe, T. & F. Møhlenberg, 1981. Particle selection in suspension feeding bivalves. Mar. Ecol. Prog. Ser. 5: 291–296.

Klepper, O., 1989. A model of carbon flows in relation to macrobenthic food supply in the Oosterschelde estuary (S.W. Netherlands). Thesis, Wageningen, 270 pp.

Knoester, M., 1984. Introduction to the Delta case studies. Wat. Sci. Tech. 16: 1–9.

Knoester, M., J. Visser, B. A. Bannink, C. J. Colijn & W. P. A. Broeders, 1984. The eastern Scheldt project. Wat. Sci. Tech. 16: 51–77.

Korringa, P., 1976. Farming marine organisms low in the food chain. Elsevier, Amsterdam.

Mallet, A. L., C. E. A. Carver, S. S. Coffen & K. R. Freeman, 1987. Winter growth of the blue mussel Mytilus edulis L.: importance of stock and site. J. exp. mar. Biol. Ecol. 108: 217–228.

Mes, C. L., 1991. De economische betekenis van de vissserij op de Oosterschelde. Prov. Griffie, Middelburg. In Dutch.

Newell, C. R., 1990. The effects of mussels (Mytilus edulis, L) position in seeded bottom patches on growth at subtidal lease sites in Maine. J. Shellf. Res. Vol. 9: 113–118.

Newell, C. R., S. E. Shumway, T. L. Cucci & R. Selvin, 1989. The effects on natural seston particle size and type on feeding rates, feeding selectivity and food resource availability for the mussel (Mytilus edulis) at bottom culture sites in Maine. J. Shellf. Res. 8: 187–196.

Newell, R. I. E. & R. J. Thompson, 1984. Reduced clearance rates associated with spawning in the mussel Mytilus edulis L. (bivalvia, Mytilidae). Mar. Biol. Lett. 5: 21–33.

Nichols, F. H., 1985. Increased benthic grazing, an alternative explanation for low phytoplankton biomass in

Northern San Fransisco Bay during the 1976–1977 drought. Estuar. coast. Shelf Sci. 21: 379–388.

Nienhuis, P. H. & A. C. Smaal, 1994. The Oosterschelde estuary, a case study of a changing ecosystem: an introduction. Hydrobiologia 282/283: 1–14.

Page, H. M. & D. M. Hubbard, 1987. Temporal and spatial patterns of growing mussels Mytilus edulis on an offshore platform: relationship to water temperature and food availability. J. exp. mar. Biol. Ecol. 111: 159–179.

Pieters, H., J. H. Kluytmans, D. I. Zandee & G. C. Cadee, 1980. Tissue composition and reproduction of Mytilus edulis in relation to food availability. Neth. J. Sea Res. 14: 349–361.

Prins, T. & A. C. Smaal, 1990. Benthic pelagic coupling: the release of inorganic nutrients by an intertidal bed of Mytilus edulis. In: M. Barnes & R. N. Gibson (eds); Trophic Relationships in the Marine Environment. Proc. 24th Eur. Mar. Biol. Symp., Aberdeen Univ. Press, Aberdeen: 89–103.

Prins, T. & A. C. Smaal, 1994. The role of the blue mussel Mytilus edulis in the cycling of nutrients in the Oosterschelde estuary (The Netherlands). Hydrobiologia 282/283: 413–429.

Rodhouse, P. G., C. M. Roden, M. P. Hensey & T. H. Ryan, 1984. Resource allocation in Mytilus edulis on the shore and in suspended culture. Mar. Biol. 84: 27–34.

Rosenberg, R. & L. O. Loo, 1983. Energy flow in a Mytilus edulis culture in Western Sweden. Aquaculture 35: 151–161.

Saeijs, H. L. F., 1982. Changing estuaries, a review and new strategy for management and design in coastal engineering. Rijkswaterstaat Communications 32, The Hague.

Scholten, H., O. Klepper, P. H. Nienhuis & M. Knoester, 1990. Oosterschelde estuary (S.W. Netherlands: a self sustaining ecosystem? Hydrobiologia 195: 201–215.

Scholten, H. & M. van der Tol, 1994. SMOES: a simulation model for the Oosterschelde ecosystem. Part II: Calibration and validation. Hydrobiologia 282/283: 453–474.

Shumway, S. E., T. L. Cucci, R. C. Newell & C. M. Yentsch, 1985. Particle selection, ingestion and absorbtion in filter-feeding bivalves. J. exp. mar. Biol. Ecol. 91: 77–92.

Smaal, A. C. & M. R. van Stralen, 1990. Average annual growth and condition of mussels as a function of food supply. Hydrobiologia 195: 179–188.

Smaal, A. C., M. Knoester, P. H. Nienhuis & P. M. Meire, 1991. Changes in the Oosterschelde ecosystem induced by the Delta Works. In: M. Elliot & J. P. Ducrotoy (eds), Estuaries and coasts: Spatial and temporal intercomparisons: 365–373. Olsen and Olsen, Fredensborg, Denmark.

Smaal, A. C. & P. H. Nienhuis, 1992. The Oosterschelde (The Netherlands), from an estuary to a tidal bay: response at the ecosystem level. Neth. J. Sea Res. 30: 161–173.

Sokal, R. R. & F. J. Rolhf, 1981. Biometry. 2nd edn. W. H. Freeman and Company, San Fransisco, 860 pp.

Tackx, M. L. M., C. Bakker & P. van Rijswijk, 1990. Zooplankton grazing pressure in the Oosterschelde. Neth. J. Sea Res. 25: 405–415.

Tenore, K. R. & N. Gonzalez, 1976. Food chain patterns in the Ria de Arosa, Spain: an area of intense mussel culture. Proc. 10th EMBS Ostend Belgium. Vol 2: 601–619.

Tol, M. W. M. van der & H. Scholten, 1992. Response of the foodweb of the Oosterschelde to a changing environment: adaptive or functional? Neth. J. Sea Res. (in press).

Theisen, B., 1977. Feeding rate of Mytilus edulis from different parts of Danish waters in water of different turbidity. Ophelia, 16: 221–232.

Vroon, J., 1994. Hydrodynamic characteristics of the Oosterschelde in recent decades. Hydrobiologia 282/283: 17–27.

Wallace, J. C., 1980. Growth rates of different populations of the edible mussel, Mytilus edulis, in North Norway. Aquaculture 19: 303–311.

Wetsteyn, L. P. M. J. & C. Bakker, 1991. Changes of abiotic characteristics and consequences for the fytoplankton development during and after a large scale coastal engineering project in the Oosterschelde (The Netherlands): a preliminary evaluation. In: M. Elliot & J. P. Ducrotoy (eds), Estuaries and coasts: Spatial and temporal intercomparisons, Olsen and Olsen, Fredensborg, Denmark.

Wetsteyn, L. P. M. J., J. C. H. Peeters, R. N. M. Duin, F. Vegter & P. R. M. de Vischer, 1990. Phytoplankton primary production and nutrients in the Oosterschelde (The Netherlands) during the pre-barrier period 1980–1984. Hydrobiologia 195: 163–177.

Winter, J. E., 1973. Growth in Mytilus edulis using different types of food. Berichte der Deutschen Wissenschaftlichen Kommission für Meeresforschung 23: 360–375.

Hydrobiologia **282/283**: 381–395, 1994.
P. H. Nienhuis & A. C. Smaal (eds), The Oosterschelde Estuary.
© *1994 Kluwer Academic Publishers. Printed in Belgium.*

Variability in stock assessment of cockles (*Cerastoderma edule* L.) in the Oosterschelde (in 1980–1990), in relation to environmental factors

J. Coosen[1,2], F. Twisk[1], M. W. M. van der Tol[1], R. H. D. Lambeck[2], M. R. van Stralen[3] &
P. M. Meire[4]

[1] *National Institute for Coastal and Marine Management/RIKZ, Middelburg, The Netherlands;*
[2] *Netherlands Institute of Ecology – Centre for Estuarine and Coastal Ecology, Yerseke, The Netherlands;*
[3] *Netherlands Institute for Fisheries Research, Yerseke, The Netherlands;* [4] *University of Ghent, Belgium;*
present adress: Institute of Nature Conservation, Hasselt, Belgium

Key words: cockle, tidal flats, Oosterschelde, macrozoobenthos, spatfall, stock assessment, mortality, growth curves

Abstract

The edible cockle (*Cerastoderma edule L.*) is a dominant suspension feeder in the Oosterschelde, a 351 km^2 tidal bay in the SW Netherlands. To establish its role in the benthic foodweb, and to assess the impact of human activities, data on density, age composition, biomass and growth were collected from several tidal flats in the Oosterschelde between 1980 and 1990.

To estimate the overall biomass development of the cockle, a simple model was used, in which three growing seasons are defined for the cockle population. A standard individual growth curve was constructed. A negative exponential mortality function was assumed to estimate the number of recruits. By combining the estimated number of recruits, the estimated specific mortality rate and the standard individual growth curve, numbers and biomass of each age group in the Oosterschelde population were estimated. Average biomass (including shell organics) per m^2 of tidal flat in August varied from 140 g AFDW in 1980 to 21 g AFDW in 1989, implying a total cockle stock on all tidal flats of 19 170 to 2350 tonnes AFDW (72×10^3 to 9×10^3 tonnes flesh), respectively.

A comparison of results from field surveys and the reconstructed stock estimations showed large deviations. However, an uncertainty analysis performed on the model showed that most field data fitted within the minimum and maximum biomass calculated.

Total biomass is largely dependent on the strength of certain year classes. In this respect, the year classes 1979, 1982, and 1985 were good. Effects of the construction of the storm-surge barrier and the compartmentalisation dams could not be demonstrated.

The year-to-year variation in cockle stocks, assessed in the way described in this paper should be regarded as relative, because a systematic survey of the intertidal flats was not performed every year, but population dynamics from selected stations were used instead.

Introduction

Next to the blue mussel (*Mytilus edulis L.*), the edible cockle (*Cerastoderma edule L.*) is the domi- nant suspension feeder in the Oosterschelde, a 351 km^2 tidal bay in the SW Netherlands. It is very abundant on most intertidal flats and is also found on subtidal slopes.

Cockles are mostly aggregated in banks, that appear at different places from time to time, depending on local conditions like the presence of low current velocity and silty sediment, which favour the settlement of spat (Wolff, 1973).

Density varies from year to year, depending on success of spatfall, mortality due to low winter temperature, fishing activity and predation by benthic invertebrates, fish and birds. Biomass depends on density and growth, which in turn is related to food availability. A Simulation Model Oosterschelde EcoSystem (SMOES) has been designed for the Oosterschelde (Klepper, 1989), describing transport processes, nutrient levels and energy fluxes between primary production, filter-feeders, deposit feeders and a number of other variables (Scholten & v.d. Tol, 1994). Since cockles may contribute up to 50% of the biomass of filter feeders and the latter is one of the forcing functions in SMOES, it is important to assess the biomass of different age groups per year.

The aim of this paper is to reconstruct the biomass of intertidal cockles from data collected by different workers on spatfall, growth and biomass of individual cockles, at several locations in the Oosterschelde during the years 1980 to 1990. A comparison will be made with extrapolations of field data to establish the sensitivity of the estimate.

In the study period (1980 to 1990) the hydrodynamics and morphology of the Oosterschelde changed due to the construction of the storm surge barrier and two additional dams. Factors such as severe winters and fishery activities could also have influenced the size of the cockle stock in the Oosterschelde. A second aim of this paper is therefore to investigate if variations in the cockle stock can be attributed to specific environmental factors.

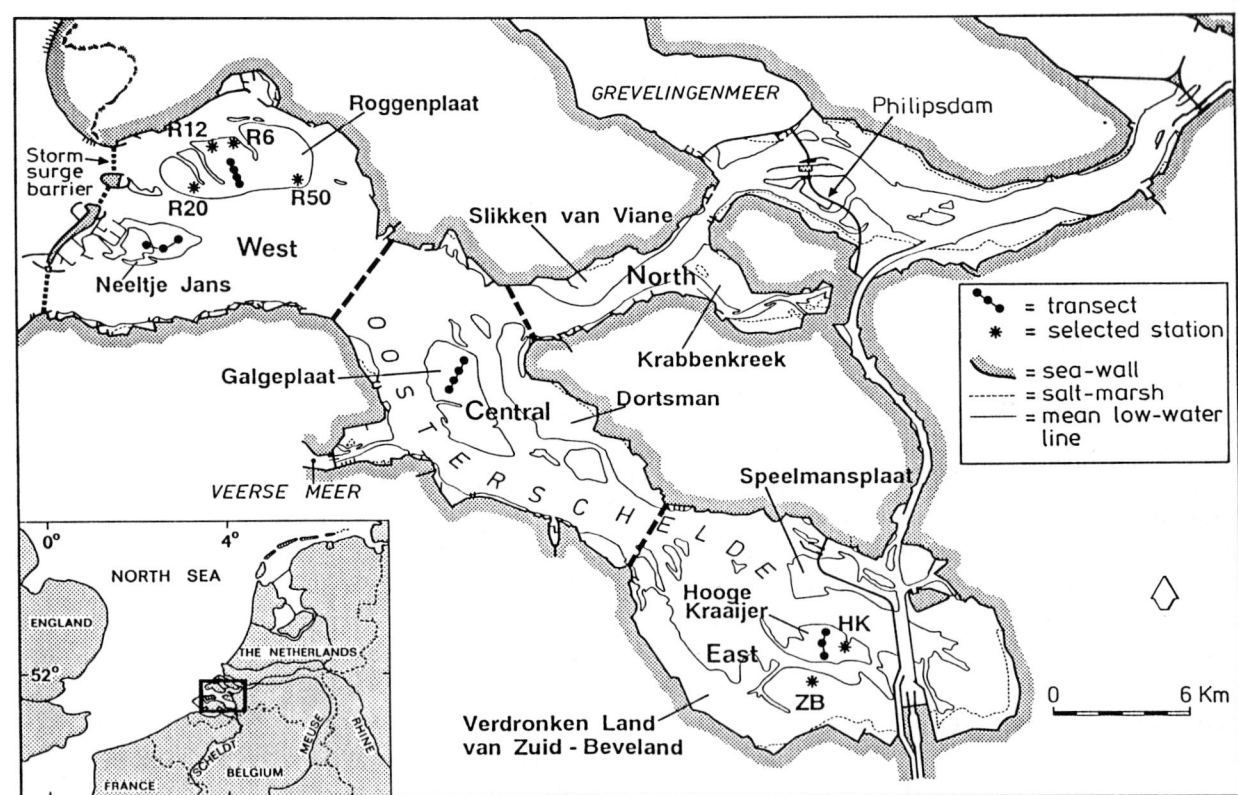

Fig. 1. Map of the Oosterschelde showing sections and sampling locations.

Materials and methods

Study area

A general description of the area and the engineering works is given by Nienhuis & Smaal (1994). The total surface area of the 'new' Oosterschelde is 351 km². The area of tidal flats, south of the Philipsdam, decreased from 137 km² in 1980–1982 to 120 km² in 1983–1986 (after closure of the Markiezaatskade) and decreased further to 114 km² from 1987 onwards (Mulder et al., 1994). The area north of the Philipsdam has not been considered.

There are four sections in the Oosterschelde, based on hydrographic characteristics: west, central, east and north (Fig. 1).

The hydrodynamics started to change in autumn 1985, two years before the completion of the storm surge barrier. From 1987 on current velocities were reduced on average by 30% in the western section and 70% in the northern section. The tidal range was reduced by about 12%. As a result, wave energy dissipation is concentrated on a smaller part of the intertidal flats and salt marshes (Vroon, 1994). Salinity is fairly constant and high in the Oosterschelde (> 30‰). Fine suspended sediment concentration and thus turbidity have reduced sharply, coinciding with the reduction of current velocities (Ten Brinke et al., 1994).

Sediments of the intertidal flats studied were very sandy with fine sediments content generally less than 5% by weight. Most flats showed little difference in sediment texture before and after the works, except the tidal flat in the northern section (the Krabbenkreek, Fig. 1) which showed a remarkable decrease in fine sediment content of 1.6% (Ten Brinke et al., 1994). Environmental conditions in the Oosterschelde are largely influenced by climatic factors like water temperature and air temperature. Severe and very mild winters did occur during the evaluation period. According to the 'IJnsen value' (IJnsen, 1981; Meininger et al., 1991) the winters of 1978/1979 and 1984/1985 were severe, 1981/1982, 1985/1986 and 1986/1987 all being cold (Fig. 2).

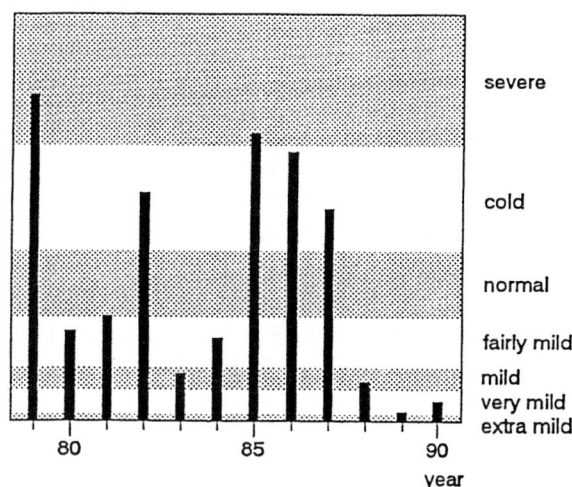

Fig. 2. The severity of the winters 1979 to 1990, characterised by the 'IJnsen value' (after IJnsen, 1981; Meininger et al., 1991).

The other winters in the study period were very mild.

Sampling of cockles

Cockles born in a certain year (called a 'year class') were divided into three age-groups (0-group, 1-group and 2 + -group). Parameters for stock assessments are the number of cockles per age group per m² and their individual weight during the year. Numbers of 0-group (spatfall) and other age groups were obtained mostly from spatial surveys, five transects and six permanent stations (Fig. 1). Data from six other selected stations were used to measure growth of individual cockles and also for determining the size of the 0-group (spatfall).

All data are from intertidal areas; so calculations only apply to that habitat. The total biomass was calculated using the mean biomass found at a limited number of stations on the main tidal flats and the total surface area of all the flats. The assumption was made that cockles occur in the entire intertidal area, irrespective of tidal elevation.

Density data

Figure 1 gives the position of the sampling locations on the tidal flats. Three datasets were available for the estimation of cockle densities:

A. Spatial surveys

Surveys were carried out on the Verdronken Land van Zuid Beveland (1984), Neeltje Jans (1985), Galgeplaat & Krabbenkreek (1985 and 1989) and Hooge Kraaijer & Roggenplaat (1984 through 1989). Details of sampling methods are given by Lambeck *et al.* (1988) and Meire *et al.* (1994).

In autumn 1989 all the intertidal flats (except the Roggenplaat, Galgeplaat and Krabbenkreek already surveyed) were surveyed by sampling over 300 transects using an adapted cockle sucker (Van Stralen *et al.*, 1991).

B. At the Slikken van Viane cockles were sampled in autumn 1979, 1981, 1982 and 1984 to the end of 1989 on six intertidal stations (Seys *et al.*, 1994).

C. Four selected transects were sampled in spring and autumn on Roggenplaat, Neeltje Jans, Galgeplaat and Speelmansplaat from 1980 to 1985.

Monitoring spatfall and growth

In addition to the density data, four stations on Roggenplaat (R6, R12, R20 and R50) and one on each of the Verdronken Land van Zuid Beveland (ZB) and Hooge Kraaijer (HK) [Fig. 1] were sampled every two to four weeks from 1984 to 1988 during the growth period (to get more detailed information on spatfall and growth).

Each sample consisted of 1 to 16 subsamples of 0.0625 m², dug out to a depth of 10 cm. The actual number of subsamples depended on the cockle density at the site. In order to obtain a reliable average individual weight, at least 100 cockles were collected if available. Samples were washed in the field on a 3 mm sieve, except in August and September when a 1 mm sieve was used. As an estimate of the number of new spat the number of 0-group cockles in August (of any given year) is taken. At that time the percentage of animals < 1 mm in the population is usually very low. In the laboratory the number of cockles per year class and per length class (class width 1 mm) was determined. The age of individuals was determined by growth ring analysis.

Biomass

Ash-free dry weight (AFDW) was obtained by weighing after drying for 12 hours (110 °C), weighing again after incinerating for 2 hours (550 °C) and substracting the two values. All AFDW data include the organics of the shell.

Construction of a standard growth curve and total biomass calculation

To obtain an overall picture of the biomass development of the cockle in the Oosterschelde the following assumptions were made:

– Each year class was presumed to have settled on June 1st, year X. The 0-group 'ended' on May 31, year $X + 1$, the 1-group on May 31, year $X + 2$, and all older animals were put in the 2-group (ending May 31, year $X + 3$).

– The growth in each group was estimated using a standard individual growth curve in which all growth after the second growing season was assumed to take place in the third growing season. To obtain this curve, results from all the stations were analysed and only the average weights based on more than 10 individuals were used. Not all stations or year classes could contribute to the data series for each year, due to lack of animals.

– The number of 0-group cockles per m² from all spatial surveys on the tidal flats in the Oosterschelde in the period from August to the end of May was used to calculate back to the number of recruits per m² at June 1st (day one) by assuming a negative exponential mortality for this group. The specific mortality rate was estimated by combining the results of a model study (Klepper, 1989) and data from selected stations, arriving at a value of $0.0036\,\mathrm{d}^{-1}$ (range 0.0029–0.0056).

By combining the estimated number of recruits, the estimated specific mortality rate and the standard individual growth curve, the number and biomass of each cohort in the Oosterschelde population was calculated, according to the following formulae:

$$N(t) = N(0) \cdot e^{-m \cdot t}$$

$$W(t) = W_{\mathrm{ind}}(t) \cdot N(t)$$

where $N(t)$ = number of individuals per m² tidal flat of age t (days); $N(0)$ = number of individuals per m² tidal flats of age zero (recruits); $W(t)$ = biomass in g AFDW per m² tidal flat with age t; Wind(t) = individual weight in g AFDW of a cockle of age t; t = time in days; m = specific mortality rate (per day).

Results

0-group cockles (spatfall)

Figure 3 shows the density of the 0-group cockles of each year class per section in the summer of settlement and in the subsequent spring. During the study period several abundant spatfalls were observed. Data from the transects at

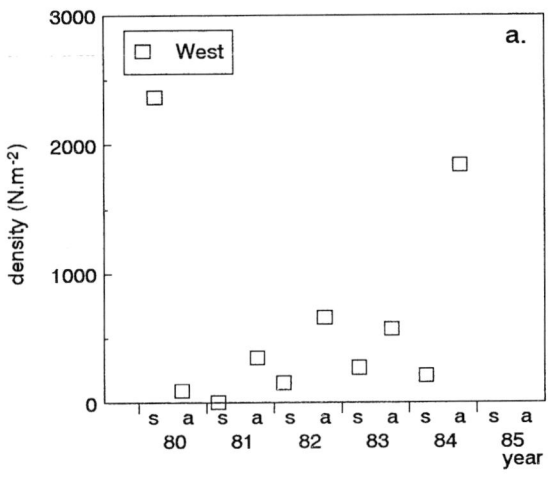

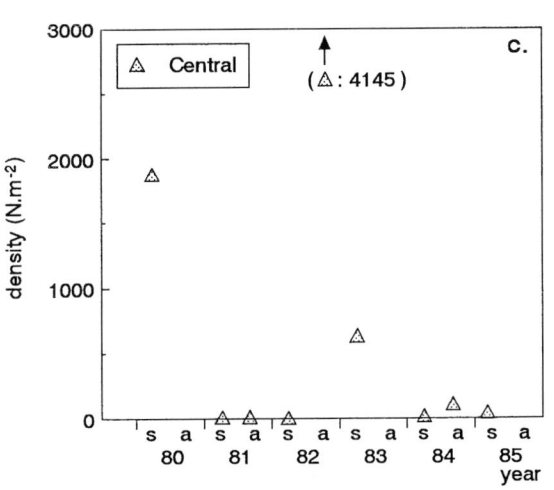

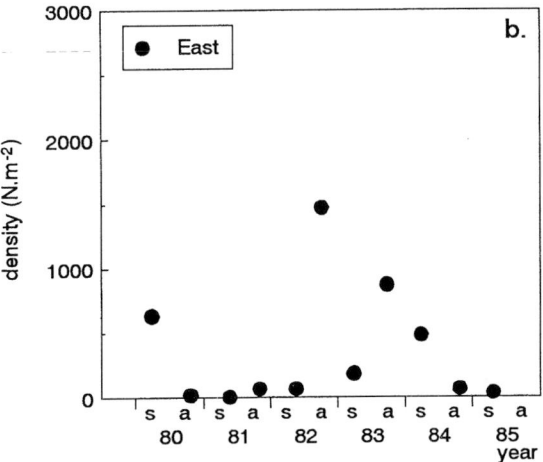

Fig. 3. Density of 0-group cockles from 3 intertidal transects in the western (a), central (b) and eastern (c) part of the Oosterschelde (1980–1985). From each year class the densities in autumn (a) of the year of recruitment and in spring (s) of the subsequent year are shown.

386

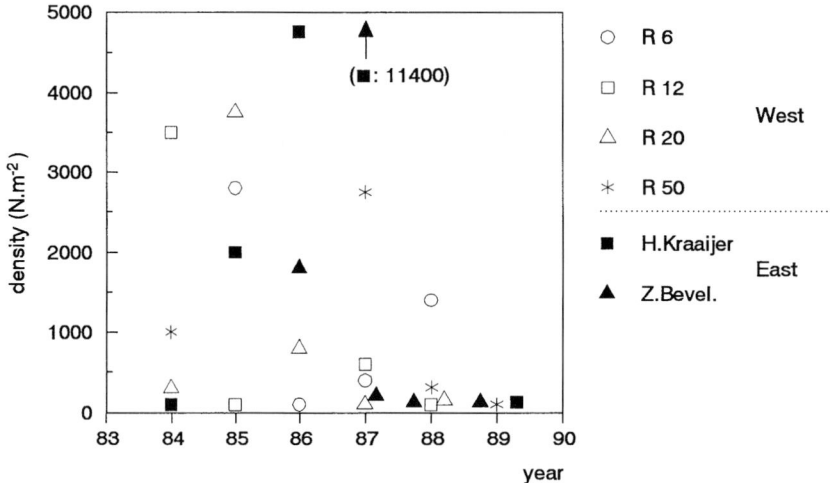

Fig. 4. Density of 0-group cockles in August in 6 intertidal stations (1984–1988).

Roggenplaat and Neeltje Jans (west); Galgeplaat (central); and Hooge Kraaijer and Speelmans-plaat (east) revealed two strong year classes: 1979 and 1982. The large spatfall of 1979 was still reflected in very high densities of 0-group cockles in the western and central part of the Oosterschelde in spring 1980. The high spatfall of year class 1982 was largely reduced in the winter '82–'83, especially in the central section. Patterns in the numbers of 0-group cockles at the 6 selected stations, sampled frequently in the period 1984–1989, are shown in Fig. 4. Densities above 1000 per m^2 in August did not usually occur at more than one of the stations in the same year, except in 1985 when high densities occurred at 4 stations. More than 5000 0-group cockles per m^2 were only found twice, in 1985 in the central part (Galgeplaat) and in 1987 in the eastern part (Hooge Kraaijer). In recent years (1988, 1989) spatfall has been very poor.

Data on 0-group cockles at the Slikken van Viane can only be derived indirectly, since the population sampled was not divided into age groups. A high density and a low mean individual biomass indicate peak numbers of juveniles in 1985. In 1988 and 1989 only older cockles were present in small numbers (Fig. 5). The August 1985 surveys of Roggenplaat, Galgeplaat and Krabbenkreek confirmed the presence of the

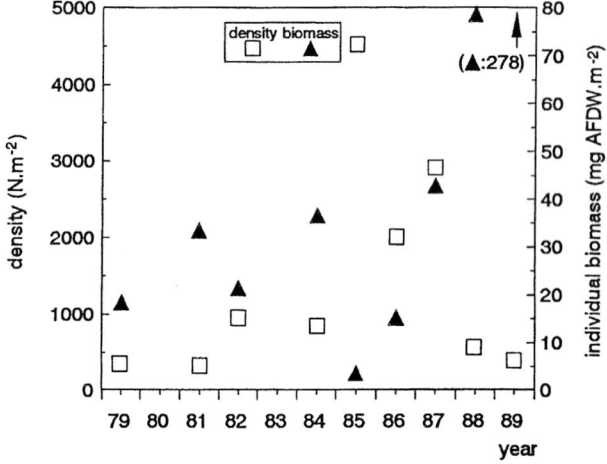

Fig. 5. Mean density per m^2 and mean individual weight of cockles in autumn from 6 intertidal stations at the Slikken van Viane (1980–1989).

strong 1985 year class (Meire *et al.*, 1994). Mean densities (with 95% confidence interval) in that month were 3600 (1020), 8600 (1510) and 1940 (436) individuals per m^2 for these three intertidal flats respectively.

Growth

Growth is discussed on the basis of typical growth curves for two selected stations in the western

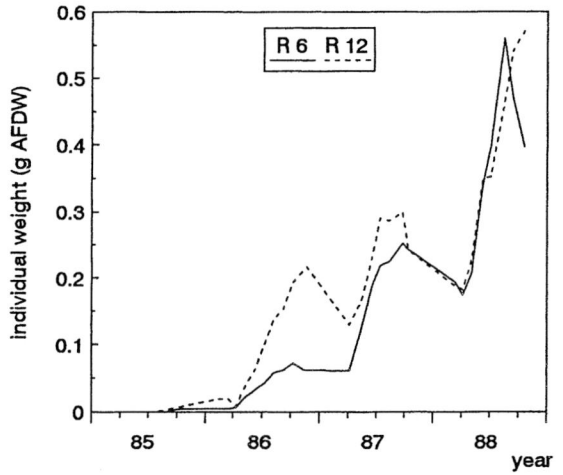

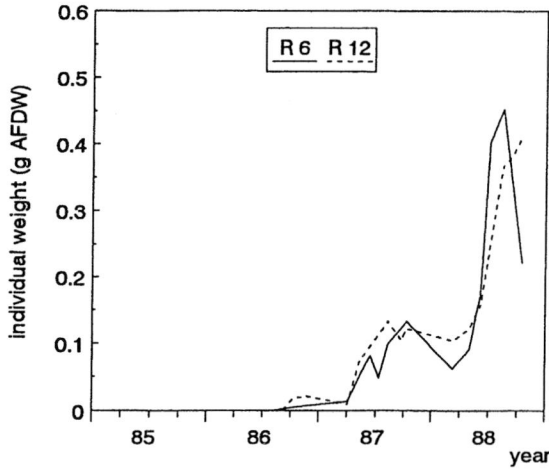

Fig. 6. Growth of cockles of year class 1985 at two Roggen-plaat stations.

Fig. 7. Growth of cockles of year class 1986 at two Roggen-plaat stations.

Table 1. Average weight of individual cockles in May and September from 6 intertidal stations in mg. AFDW (for locations: see Fig. 1). – = No data available.

	Station (high←position in tidal zone→low)						Mean of stations
	R20	R6	R50	R12	HK	ZB	
Yearclass 1985							
2nd growing season							
Weight in May	50	24	–	43	32	34	37
Weight in Sept.	140	62	–	153	249	277	177
Increase	90	38	–	110	217	243	140
3rd growing season							
Weight in May	–	124	171	165	–	–	153
Weight in Sept.	–	252	333	299	–	–	295
Increase	–	128	162	134	–	–	142
4th growing season							
Weight in May	–	207	328	228	–	–	254
Weight in Sept.	–	475	499	543	–	–	506
Increase	–	268	171	315	–	–	252
Yearclass 1986							
3rd growing season							
Weight in May	–	–	–	117	105	147	123
Weight in Sept.	–	–	–	382	–	306	344
Increase	–	–	–	265	–	159	221
Yearclass 1987							
2nd growing season							
Weight in May	42	19	66	57	–	–	46
Weight in Sept.	226	263	263	412	–	–	291
Increase	184	244	197	355	–	–	245

part of the Oosterschelde (Figs. 6 & 7) and the average individual biomass values of the cockles in May and September for all selected stations (Table 1, Fig. 8).

In the first growing season, the weight increase is usually small (10–20 mg AFDW). Most stations show a steep increase in individual weight from May to September in the following growing seasons. However, in some years growth commenced in April and growth can sometimes continue in October, and even in November. A decrease in body weight takes place during the winter months.

Most field data obtained in the present study fitted well in the standard growth curve (Fig. 8) used by Klepper (1989). However, the September values of the second growing season showed a wide variation and the curve overestimated growth in the third growing season. Despite these imperfections, the standard growth curve was used in the stock assessment.

Biomass calculations

For each age-group (cockles from the first, the second and the third + subsequent growing seasons) the seasonal patterns of the biomass per m² intertidal flat were reconstructed using the standard individual growth curves. Together with the field data (average per section [west, central, east and north]) these patterns are given in Figs. 9a to 9c.

Combining biomass calculations for the three age groups to produce temporal trends in biomass figures per m² for the cockles on the Oosterschelde intertidal flats resulted in the 'average biomass curve' given in Fig. 9d.

Highest field data were found at the Neeltje Jans Plaat (west) in the early 1980s. Sometimes the maximum biomass was reached later in the season than shown by the average biomass curve (e.g. 1981), or biomass was much higher in May (e.g. 1984).

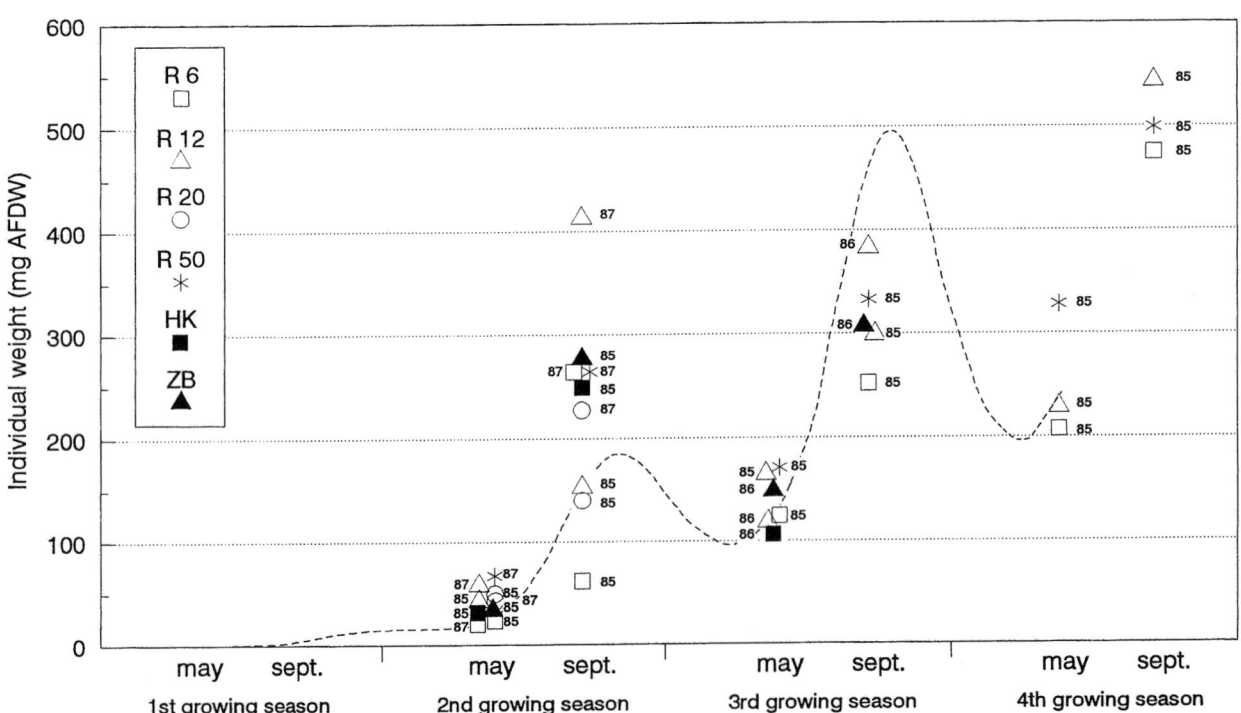

Fig. 8. Standard growth curve for individual cockles on the intertidal flats of the Oosterschelde together with empirical data from 6 intertidal stations.

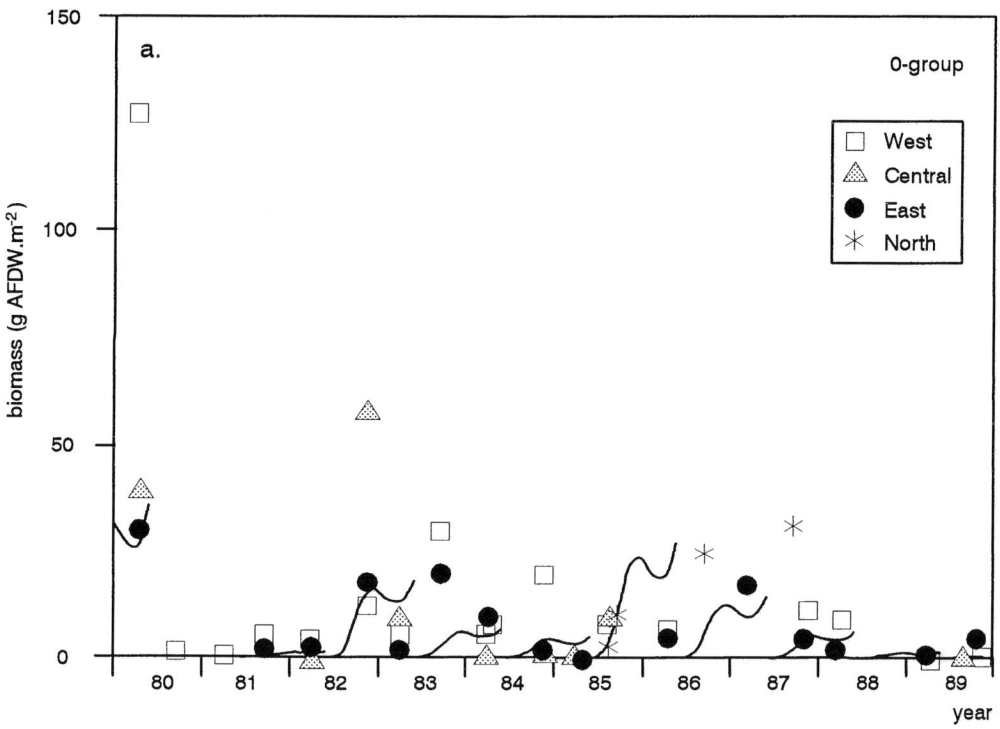

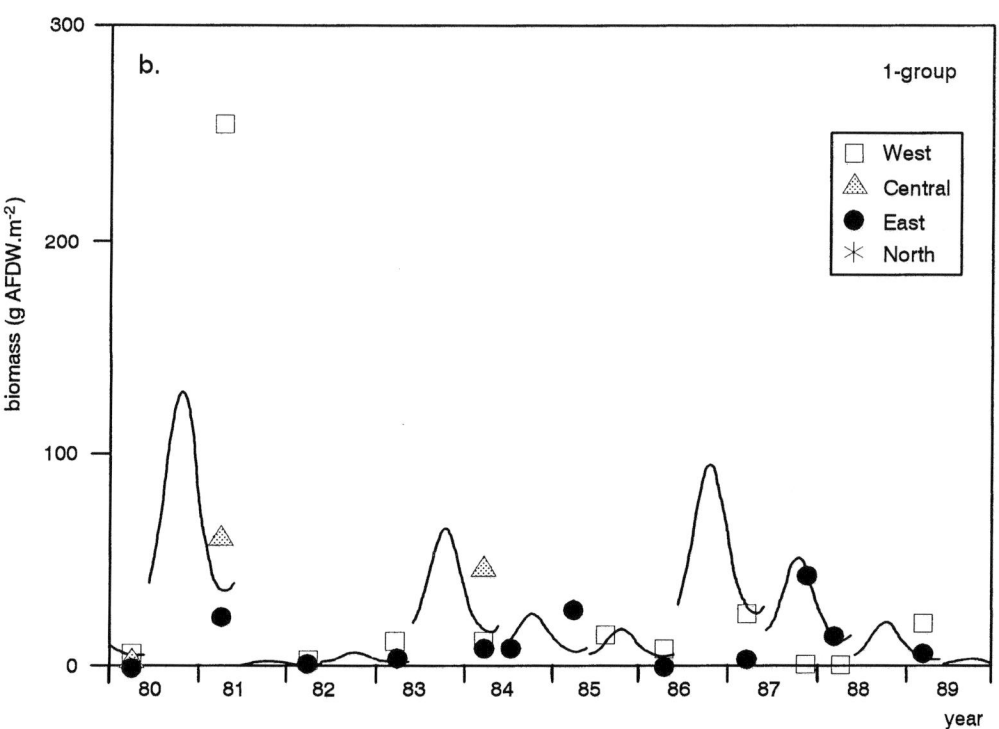

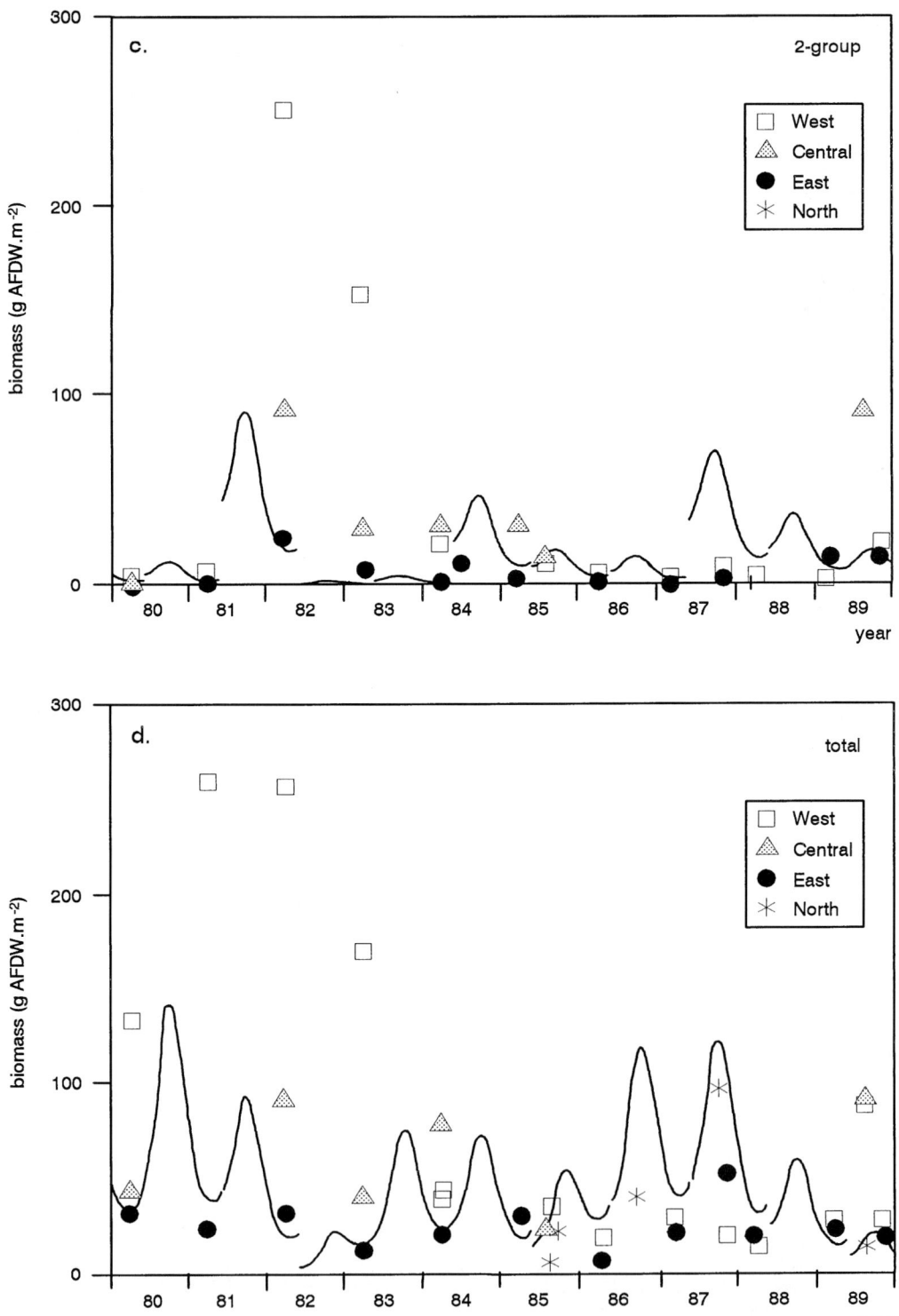

Fig. 9. Calculated biomass (g AFDW/m²) of cockles on intertidal flats from 1980 to 1989, together with empirical data from 4 sections west, central, east and north). (a) first year (0-group); (b) second year (1-group); (c) third year and older (2 + -group); (d) total cockle population.

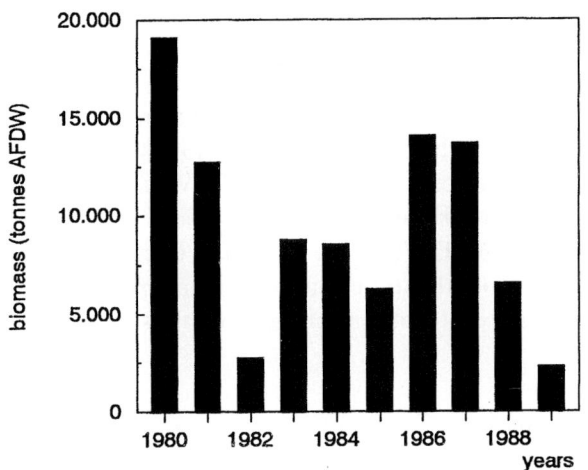

Fig. 10. Estimated total biomass of cockles (stock; in tonnes AFDW) at intertidal flats in the Oosterschelde in August/September.

The total biomass of cockles present on the Oosterschelde intertidal flats in August/September is calculated as the product of August values of the average biomass curve and the surface area (Fig. 10). The results show a decrease in the cockle stock in 1988 and 1989 to a very low value.

Two analyses were performed to test whether the stocks estimated by calculating backwards using standard growth and mortality rates, are in agreement with extrapolated biomass field data. These analyses were (1) an uncertainty analysis; (2) calculation from the large surveys (for 1985 & 1989).

Firstly, an uncertainty analysis, with the simple submodel outlined in the Methods section, was performed to assess the reliability of the estimate made in the standard way.

The estimated number of recruits not only depends on variations in mortality rate, but also on the variance in the results of the surveys. This variance is included in the uncertainty analysis by assuming a 25–175% range around the estimated mean for each cohort (age-group of

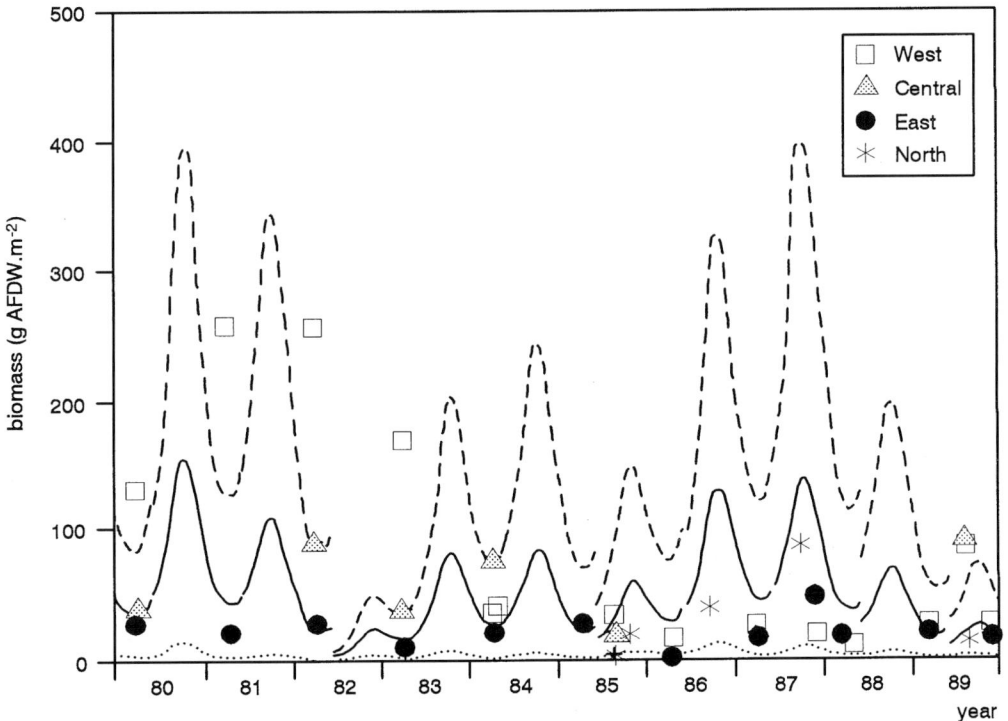

Fig. 11. Calculated biomass (g AFDW/m²) of total cockle population on intertidal flats from 1980 to 1989 with uncertainty range of 25–175%, together with field data from different sections. Legend model calculations: – – – = 175%; —— = 100%; ······ = 25%.

Table 2. Stock-assessment of cockles at the Oosterschelde intertidal flats (August, 1985). * After Nijkamp (1988).

Area	Surface-area (ha)	Biomass (g AFDW/m^2)	Stock (ton AFDW)
*West**			
Neeltje Jans	499	21.2	105.8
Roggenplaat	1625	37.5	609.4
Central	3483	23.0	801.1
North			
Viane	453	16.5	74.8
Rest	1347	5.2	70.0
East	4266	10.0	426.6
Total			2087.7

cockles). Figure 11 gives the minimum and maximum biomass of cockles per m^2 tidal flat calculated using these assumptions.

In the second test, a calculation of the standing stock in 1985 and 1989 for each tidal flat was made, using direct measured biomass data from the August surveys of those years (the results are summarised in Tables 2 and 3).

The stock in August 1985 (6300 tonnes AFDW), calculated from the average biomass curve (Fig. 9d) is much higher than the one based

Table 3. Stock-assessment of cockles at the Oosterschelde intertidal flats (August, 1989). * = After Van Stralen (1991).

Area	Surface-area (ha)	Biomass (g AFDW/m^2)	Stock (ton AFDW)
West			
Neeltje Jans	421		209.6*
Roggenplaat	1600	87.6	1401.6
Central			
Galgeplaat	1000	90.7	907.0
Dortsman	2334		340.8*
North			
Krabbenkreek	600	14.7	88.2
Viane	400	109.5	438.0
East			
Hooge Kraayer	455	34.8	158.3
Verdronken Land	3548		160.0*
Total			3703.5

on the biomass field data (2090 tonnes AFDW; Table 2). Only when biomass values from the lower part of the uncertainty range (Fig. 11) are used, estimates approach the field results.

In 1989 the opposite is true, however. The calculated stock (2500 tonnes AFDW) is lower than the figure based on biomass field data (3700 tonnes AFDW; Table 3) and a cockle biomass value from the upper part of the uncertainty range is required to obtain an acceptable similarity.

Discussion

The calculation of the stock using only field data showed that the method of stock assessment proposed by us calculating backwards using the standard growth curve and standard mortality rate, should be interpreted with care, bearing in mind the large uncertainty range. Nevertheless, even with standardised growth and mortality rates it is evident that large fluctuations in cockle stock exist from year to year.

The discussion will focus on two aspects:

– what is the effect of the use of a standard growth curve and standard mortality rate in the stock assessment?;
– Is there a master factor that explains the variation in stocks from year to year?

The SMOES submodel works with three age groups, each with a standardised growth. Combination of the third and subsequent growing seasons, leading to a high biomass for the 2 + -group, gives a higher stock in the third growing season for any given year class, and a sudden disappearance of the older age groups in the following years. Only with exceptionally strong year classes will this lead to a substantial underestimation of the stock in the fourth season after the settlement of such a strong year class. This effect may have contributed to the very low biomass in 1982 (strong year class 1979). If the density data from the third year are used to reconstruct spatfall, this may lead to an overestimation of the stock in the second year. However, the share of older age groups in the cockle population is mostly small.

The mean maximum growth in each growing season is well within the ranges observed in other estuaries (Jones, 1979; West *et al.*, 1979; Ducrotoy *et al.*, 1991). Sager (1986), who worked on data from Beukema (1975–1987), employed a growth curve based on a doubly seasonal-modified Richards function that fitted the data quite well. In that growth curve, cockles reached a weight of 0.45 g AFDW in 3 growing seasons.

By using back-calculated numbers to start each year class, and applying an uncertainty range to these figures, the result of the variation in growth and mortality is taken into account. By doing so, mortality caused by severe winters or by fishing are included. So, even if direct growth measurements for each location for each year had been available and had been applied, the net result for the Oosterschelde intertidal flats as a whole would have been quite similar.

We cannot deny the fact that a large temporal and spatial variation exists within the cockle population of the Oosterschelde, but it is also evident that a stock assessment can only be made if a number of assumptions are made. Only when large spatial surveys are carried out every year will stock assessment with less uncertainty be possible.

Some of the factors known to influence the growth of cockles have changed during the construction of the storm surge barrier and the other dams. The period of prolonged immersion in 1986 (Vroon, 1994) may have influenced the survival of the cockles lower in the intertidal zones: they were less vulnerable to cold spells (freezing) and predation by birds, and had better foods conditions, due to longer periods of filtration.

Current velocity has decreased, the amount of silt has decreased and sediment texture at some tidal flats has changed (Ten Brinke *et al.*, 1994) with different settling places as a consequence. Plankton production has changed little (Wetsteyn & Bakker, 1990), so food conditions have probably not altered very much. The differences in growth shown in Fig. 8 (or Table 1) may be related to changes in some of these factors; however, this is beyond the scope of this paper.

With the assumption of a standard growth and a standard mortality rate the variation in stock can largely be attributed to the number of recruits, calculated backwards from density values each year. So, analysis of the number of 0-group cockles gives us the best clue to a master factor regulating the size of the cockle stock. Although cockle spatfall shows a high degree of spatial variability, some years were characterised by an extraordinarily good spatfall for the entire area. Various authors (Beukema, 1979, 1982, 1985; Dörjes, 1980; Desprez *et al.*, 1990) found an enhanced recruitment of different bivalve species after severe winters, when few surviving adults are left. Physiological studies suggest that a sudden rise in temperature in spring after a cold or severe winter is a more important stimulus for spawning than absolute temperature (Boyden, 1971). But severe winters are not always followed by a large spatfall. Ducrotoy *et al.* (1991), who compared population patterns in several sites along the European coast did not find a clear relationship between cold winters and the subsequent extraordinarily high spatfall of cockles. After mild winters, high or low spatfall may occur. High numbers in the German Wadden sea were recorded for instance in 1984 after a mild winter, although in that case local storm surges had probably severely reduced the adult population in the previous months.

The winters preceding our good spatfall years (1979, 1982 and 1985) were indeed severe or cold (Fig. 2). However, the cold winter of 1985/1986 was not followed by a large spatfall in the whole area, only the eastern locations and the Slikken van Viane showed a fairly large number of juvenile cockles. It is difficult to determine whether the tidal manipulation of 1986 affected the spatfall in the other sections of the Oosterschelde or not. The absence of good recruitment as found in the Oosterschelde in 1988, 1989 and 1990 was also recorded for the Dutch Wadden sea (De Vlas, pers. comm.; Van Stralen, 1990). The exceptionally mild winters of 1987/1988, 1988/1989 and 1989/1990 may have favoured the predators of cockle spat. Such a factor seems to be more important in determining survival of spat than the number of larvae produced (Dijkema, pers. comm.).

There is probably no clear relationship between the density of adults and the spatfall success. Only if the number of adults is very high, is spatfall likely to be adversely affected (Groenewold, 1986; Brock, 1980; André & Rosenberg, 1991). The latter certainly does not apply to the conditions on the Oosterschelde tidal flats (adults densities < 100 per m^2) in 1980 to 1990.

Cockle fishery in the Oosterschelde has became increasingly important during the study period. Since only older cockles are fished, and this mortality is included in the model, no effect of fishery on cockle stocks can be detected by this method. In years with a low standing stock fishery can reduce the biomass of cockles with 20% (Smaal & Nienhuis, 1992).

So the major factor that determines the cockle stock in the Oosterschelde turns out to be the climate, a natural phenomenon. Spatial and temporal variations in growth and mortality do exist. It is very difficult to say if the engineering works during the study period had more than a local effect on the population dynamics of the cockle. Given the enormous natural variation in cockle stocks, it is almost impossible to demonstrate any effects using data collected in the two years after the completion of the barrier.

Any attempt to use these stock assessment for impact studies of fishery activities or to calculate allowable catches should be accompanied by an evaluation of the uncertainty ranges. The estimated total biomass of cockles on intertidal flats (Fig. 10) should be interpreted as relative values over the study period.

Conclusions

Total biomass on the Oosterschelde tidal flats is largely dependent on the strength of specific year classes, such as 1979, 1982 and 1985 in our study. Patterns of growth and mortality were assumed to be roughly equal all over the Oosterschelde and an average biomass per m^2 was applied on all intertidal flats. Effects of the construction of the

storm surge barrier and the compartmentation dams could not be demonstrated in this way. Cockle stocks, assessed in the way described in this paper should be regarded as relative, if the large uncertainty range is taken into account.

Acknowledgements

We would like to thank J. Bol, A. van den Dool, A. Engelberts, J. Dereu, H. Nijkamp, R. Pouwer, R. van der Sande, A. Schoenmaker, E. C. Stikvoort & C. Vroonland for their help with sampling and sorting.

E.C. Stikvoort carefully prepared the figures. Communication No. 702 of NIOO-CEMO, Yerseke, The Netherlands.

References

André, C. & R. Rosenberg, 1991. Adult-larval interactions in the suspension feeding bivalves *Cerastoderma edule* and *Mya arenaria*. Mar. Ecol. Prog. Ser. 71: 227–234.

Beukema, J. J., 1979. Biomass and species richness of the macrobenthic animals living on a tidal flat area in the Dutch Wadden Sea: effects of a severe winter. Neth. J. Sea Res. 13: 203–223.

Beukema, J. J., 1982. Annual variation in reproductive success and biomass of the major macrozoobenthic species living in a tidal flat area of the Wadden Sea. Neth. J. Sea Res. 16: 37–45.

Beukema, J. J., 1985. Zoobenthos survival during severe winters on high and low tidal flats in the Dutch Wadden Sea. In J. S. Gray & M. E. Christiansen (eds), Marine Biology of Polar Regions and Effects on Marine Organisms. Wiley & Sons, Chichester: 351–361.

Brock, V., 1980. Notes on relations between density, settling and growth of two sympatrick cockles, *Cardium edule* (L.) and *C. glaucum* (Bruguiére). Ophelia, Suppl. 1: 241–248.

Boyden, C. P., 1971. A comparative study of the reproductive cycles of the cockles *Cerastoderma edule* and *C. glaucum*. J. mar. biol. Ass. U.K. 51:605–622.

Desprez, M., J.-P. Ducrotoy & H. Rybarczyk, 1990. Fonctionnement biologique des gisements de coques (*Cerastoderma edule*) en baie de Somme (France) à la suite de recrutement massif de 1987. Rapport scientifique du GEMEL 3: 52 pp.

Dörjes, J., 1980. Auswirkungen des kalten Winters 1978/79 auf das marine Macrobenthos. Natur und Museum: 109–115.

Ducrotoy, J.-P., H. Rybarczyk, J. Souprayen, G. Bachelet, J. J. Beukema, M. Desprez, J. Dörjes, K. Essink, J. Guillou,

H. Michaelis, B. Sylvand, J. G. Wilson, B. Elkaïm & F. Ibanez, 1991. A comparison of the population dynamics of the cockle (*Cerastoderma edule L.*) in North Western Europe. In M. Elliot & J.-P. Ducrotoy (eds), Estuaries and Coasts: Spatial and Temporal Intercomparisons. Olson & Olson, Fredensborg.: 173–184.

Groenewold, A., 1986. De invloed van de plaatselijke dichtheid van oude kokkels (*Cerastoderma edule*) op vestiging en groei van kokkelbroed. NIOZ Internal Reports, Texel. 1986-4 (in Dutch).

IJnsen, F., 1981. Onderzoek naar het optreden van winterweer in Nederland. K.N.M.I., Scientific Report WR 74-2, 2nd edn. De Bilt (in Dutch).

Jones, A. M., 1979. Structure and growth of a high level population of *Cerastoderma edule* (Lamellibranchiata). J. mar. biol. Ass. U.K. 59: 277–287.

Klepper, O., 1989. A model of carbon flows in relation to macrobenthic food supply in the Oosterschelde estuary (S.W. Netherlands). Dissertation, Agricultural University, Wageningen.

Lambeck, R. H. D., C. M. Berrevoets, E. B. M. Brummelhuis, A. van den Dool & A. Hannewijk, 1988. The cockle population of an intertidal flat in the Oosterschelde: the impact of severe winters, fishery and hydraulic engineering projects. In C. H. R. Heip & E. S. Nieuwenhuize (eds), Progress Report 1987, Delta Institute for Hydrobiological Research, Yerseke: 36–38.

Meininger, P. L., A. M. Blomert & E. C. L. Marteijn, 1991. Watervogelsterfte in het Deltagebied, ZW Nederland, gedurende de drie koude winters van 1985, 1986 en 1987. Limosa 64: 89–102 (in Dutch).

Meire, P. M., J. Seys, J. Buijs & J. Coosen, 1994. Spatial and temporal patterns of intertidal macrobenthic populations in the Oosterschelde: are they influenced by the construction of the storm-surge barrier? Hydrobiologia 282/283: 157–182.

Mulder J. P. M. & T. Louters, 1994. Changes in basin geomorphology after implementation of the Oosterschelde estuary project. Hydrobiologia 282/283: 29–39.

Nienhuis, P. H. & A. C. Smaal, 1994. The Oosterschelde estuary, a case study of a changing ecosystem: an introduction. Hydrobiologia 282/283: 1–14.

Nijkamp, H., 1988. Verspreiding, populatiesamenstelling, broedval en groei van de kokkel (*Cerastoderma edule L.*) in de Oosterschelde. Student Report D 4 -1988, Delta Institute for Hydrobiological Research, Yerseke: 69 pp. (in Dutch).

Sager, G., 1986. Wachstumsspezifische Approximationen für die Herzmuschel *Cerastoderma edule L.* in der Waddenzee nach Daten von BEUKEMA (1975–78). Beitrage zur Meereskunde, Heft 55, S.: 55–66.

Scholten, H. & M. W. M. van der Tol, 1994. SMOES: a simulation model for the Oosterschelde ecosystem. Part II: Calibration and validation. Hydrobiologia 282/283: 453–474.

Seys, J., P. M. Meire, J. Coosen & J. A. Craeymeersch, 1994. Long-term changes (1979–1989) in the intertidal macrozoobenthos of the Oosterschelde estuary: are patterns in total density, biomass and diversity induced by the construction of the storm-surge barrier? Hydrobiologia 282/283: 251–264.

Smaal, A. C. & P. H. Nienhuis, 1992. The Eastern Scheldt (The Netherlands), from an estuary to a tidal bay: a review of responses at the ecosystem level. Neth. J. Sea Res. 30: 161–173.

Smaal, A. C., M. Knoester, P. H. Nienhuis & P. M. Meire, 1991. Changes in the Oosterschelde ecosystem induced by the Delta works. In M. Elliot & J.-P. Ducrotoy (eds), Estuaries and Coasts: Spatial and Temporal Intercomparisons. Olson & Olson, Fredensborg.: 375–384.

Ten Brinke, W. B. M., J. Dronkers & J. P. M. Mulder, 1994. Fine sediments in the Oosterschelde tidal basin before and after partial closure. Hydrobiologia 282/283: 41–56.

Van Stralen, M. R., 1990. Het kokkelbestand in de Oosterschelde en de Waddenzee in 1990. RIVO Report AQ 90-03 (in Dutch).

Van Stralen, M. R., J. J. Kesteloo-Hendrikse & C. M. Brand, 1991. Bestands grootte en visserijmortaliteit van kokkels in 1989. RIVO Report AQ 91-02 (in Dutch).

Vroon, J., 1994. Hydrodynamic characteristics of the Oosterschelde in recent decades. Hydrobiologia 282/283: 17–27.

West, A. B., J. K. Partridge & A. Lovitt, 1979. The cockle *Cerastoderma edule* (L.) on the South Bull, Dublin Bay: population parameters and fishery potential. Irish Fish. Invest., ser. B, 20: 3–18.

Wetsteyn, L. P. M. J. & C. Bakker, 1991. Abiotic characteristics and phytoplankton primary production in relation to a large-scale coastal engineering project in the Oosterschelde (The Netherlands): a preliminary evaluation. In M. Elliot & J.-P. Ducrotoy (eds), Estuaries and Coasts: Spatial and Temporal Intercomparisons. Olson & Olson, Fredensborg: 365–373.

Wolff, W. J., 1973. The estuary as a habitat. An analysis of data on the soft bottom macrofauna of the estuarine area of the rivers Rhine, Meuse and Scheldt. Zoologische Verhandelingen, Leiden, 126: 1–242.

Hydrobiologia **282/283**: 397–412, 1994.
P. H. Nienhuis & A. C. Smaal (eds), The Oosterschelde Estuary.
© 1994 *Kluwer Academic Publishers. Printed in Belgium.*

Biomass and standing stock on sublittoral hard substrates in the Oosterschelde estuary (SW Netherlands)

R. J. Leewis [1]*, H. W. Waardenburg [2] & M. W. M. van der Tol [1]
[1] *National Institute for Coastal and Marine Management/RIKZ, P.O. Box 20907, 2500 EX The Hague, The Netherlands;* [2] *Bureau Waardenburg, P.O. Box 365, 4100 AJ Culemborg, The Netherlands;*
*Present address: National Institute for Public Health and Environmental Protection (RIVM), P.O. Box 1, 3720 BA Bilthoven, The Netherlands

Key words: hard substrates, species, distribution, biomass, standing stock, Oosterschelde, storm-surge barrier

Abstract

From 1979 to 1991 the species composition of communities living on hard substrata (hardsub) in the Oosterschelde has been studied – in both the littoral and sublittoral zones. From 1984 onwards, biomass was also measured. This paper deals mainly with the distribution and the development of biomass on sublittoral hardsub in the Oosterschelde. Analysis has shown that the most important abiotic factors regulating the flora and fauna are: quantity and nature of the substrate; sedimentation; exposure to water movement (mainly currents); and light. The construction of the storm-surge barrier has influenced those factors. The main consequences for the flora and fauna on sublittoral hard substrata have been through the increased amount of available hard substratum by about 10% until 1984 and a further 20% from 1984 to 1987, the main barrier construction period). Within the same period (until 1987) the biomass per square metre also increased. This caused a net increase of hardsub biomass – in the sublittoral – of about 35%.

After the barrier was completed sedimentation increased; in some parts of the basin hardsub organisms were covered by sediment and have not recovered; the total quantity of available hard substratum decreased by an amount yet to be established. For the purpose of this paper it is tentatively estimated at 20%, but the process is still going on.

Tidal current velocities are smaller in the post-barrier situation, which caused a shift from more passive suspension feeders to more actively filtering species. The relative importance of suspension feeders on hard substrata has decreased by about 20% after the building of the storm-surge barrier. In 1990 and 1991 it increased again.

Overall water transparency increased, but the lower limit of macroalgal growth has not gone deeper, as nearshore turbulence and turbidity did not change significantly.

Effects on hardsub were small in the beginning. During the construction period (1985–1987) no clear effects were registered. After the completion of the barrier total species diversity increased at first, followed by a decrease from the second half of 1988 onwards. Biomass increased rather sharply, at first, but decreased very sharply in 1989. In 1990 a recovery in biomass became apparent. Developments in biomass and species composition differed per sampling location. An attempt is made to explain some of those developments, in relation to the abiotic changes brought about by the storm-surge barrier. This appeared difficult, because climatic influences obscured the effects of the barrier. The most explicit of those masking effects was brought about by a temporary, huge increase of the brittlestar (*Ophiothrix fragilis*). This animal covered the substratum in relatively thick layers (up to 5 cm) and more or less suffocated the other fauna. It was therefore difficult to quantify the effect of increased sedimentation on the fauna. The increase of *Ophiothrix* is probably not caused by the storm-surge barrier, but by a succession of several mild winters.

It is clear that a new equilibrium in the basin is still to be reached. Total effects in terms of species richness and of biomass will continue to be monitored, and the results used to advise the water authorities as to management and nature friendly dike building methods.

Introduction

The building of a storm-surge barrier and accessory dams in the Oosterschelde estuary has modified the values of environmental factors such as current velocities, sediment load and submarine light level within the basin (Wetsteyn *et al.*, 1990). These factors, together with exposure in general and the nature of the substrate, are the most important determinants for the structure of the communities on hard substrates (Warner, 1984). De Kluijver (1989) and Leewis & Waardenburg (1989; 1990) showed this to hold also for the Oosterschelde. Of course, water temperature, salinity and other water quality parameters, and topographical factors can also play a role. Hiscock (1976) and Hiscock & Mitchell (1980) described all these factors in an attempt to classify the different hard substrate communities found around the British Isles.

The expected changes in the hard substrate communities of the Oosterschelde were studied by starting a monitoring program 7 years before the completion of the barrier, and trying to relate the spatial and temporal distribution of species and biomass to environmental parameters. After completion of the barrier monitoring continued in order to compare real developments to expected changes.

Methods and materials

A general description of the Oosterschelde is given by Nienhuis & Smaal (1994). There are about 160 km of dikes and sea-walls around the basin, part of which lies at the edge of tidal mud flats, and part along deep water (up to 55 m depth).

The underwater foreshore of the latter is often protected with mattresses, kept in place by means of blocks of natural stone, and sometimes by irregular concrete blocks or furnace (e.g. phosphorus) slag. This paper deals with this sublittoral type of shore protection.

Sampling was done in early spring, summer, and late autumn by SCUBA divers at 56 stations distributed over 8 transects from Mean Low Water level (MLW) to 15 m waterdepth (see Fig. 1). Transect locations were marked exactly by means of recognition points on the dike; station locations were determined by means of readings on a depth gauge. Sampling started in the summer of 1979 and continued through 1991, although since 1988, for financial reasons, with a reduced program of one sampling series (the autumn series) per year. Therefore, in this paper only the biomass data from the autumn samplings were used. Samples were taken to the laboratory for identification, and, since 1984, also for biomass determination. Originally there were 9 transects, but one transect (Fig. 1, no. 3) was soon abandoned, because it was destroyed by dike reinforcement works, and the data is not used here. Station depths were: 1, 2, 3, 5, 7, 9 and 12 or 15 m below MLW (all depths mentioned in this paper refer to MLW). At each station, the percentage cover of each species was estimated using 50 by 50 cm quadrats, and recorded using a slightly simplified Braun-Blanquet scale (Braun-Blanquet, 1964). The minimal area to be sampled was determined for all locations, by determining the similarity of animal species (sessile as well as vagile) composition between each two squares of 625 cm^2 and multiples of that area, using the qualitative Sørensen similarity index, $Is = 2Cpq/(Cp + Cq)$, in which Cpq = the number of species

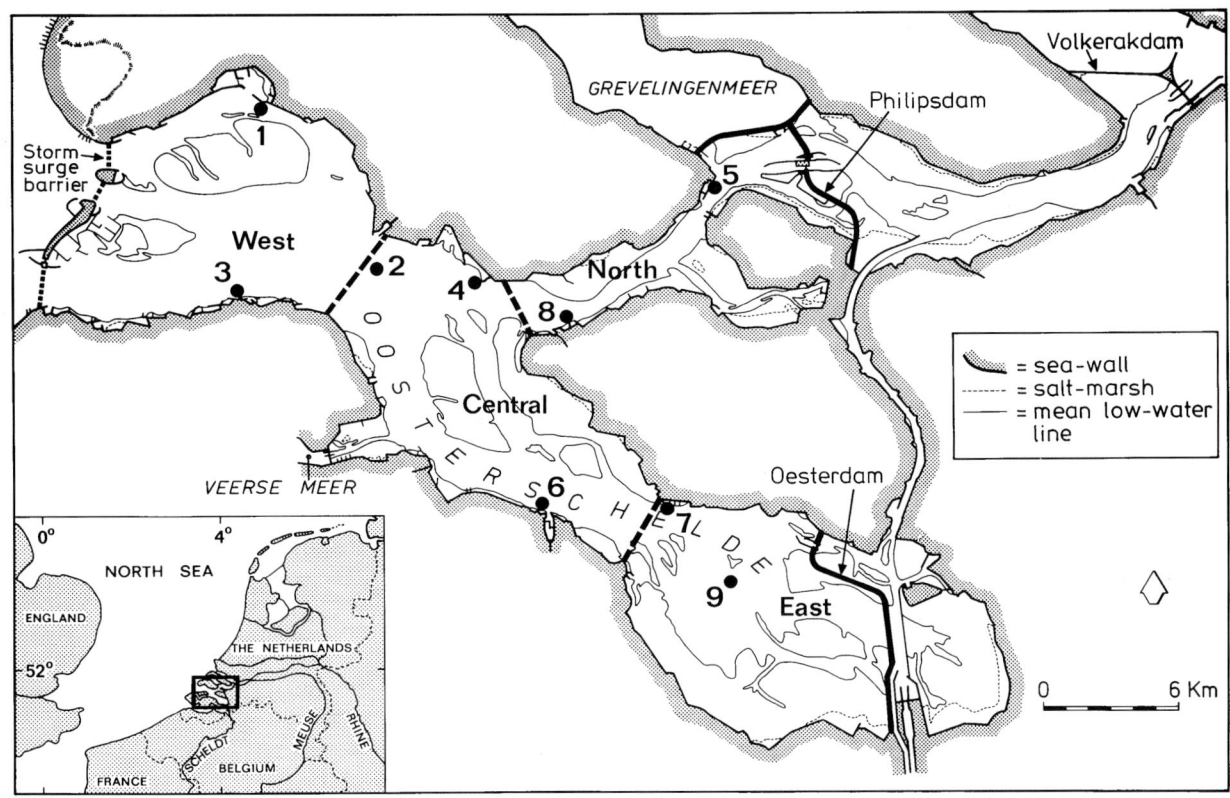

Fig. 1. The study area, with the four sub-areas, and the sampling stations. 1 = Schelphoek; 2 = Pijler Zeelandbrug; 3 = Wissenkerke; 4 = Zuidbout; 5 = Zijpe; 6 = Wemeldinge; 7 = Gorishoek; 8 = Stavenisse; 9 = Boomkil.

that occur in both squares p and q, Cp = the number of species occurring in p, and Cq = the number of species occurring in q (see Weinberg, 1978; Van Soest & Weinberg, 1981). The minimal area is assumed to be reached when $Is = 0.85$. The chosen standard sample area of 2500 cm^2 was, in most cases, sufficient to meet this criterion. In 1991 minimal area was determined once more, but now using only the sessile organisms, and using the quantitative Bray-Curtis index for similarity. Results agreed very well with those with the Sørensen index (De Meij & Van der Sloot, pers. comm.).

Biomass was determined by taking organisms from the immediate vicinity of the stations with their substratum (stones or concrete blocks) to the laboratory, where the organisms were carefully scraped off the substratum. They were then separated into 6 species groups and a 'rest' group,

dried for 4 days (for some large samples it took up to 7 days to dry completely) at 80 °C and ignited for 2 hours at 560 °C. The surface they covered on the stones was also measured. As it was not possible to collect and process biomass samples from all different depths where cover was estimated, biomass material was collected between 5 and 7 m depth only. In 1988 and 1989 only one sample per location was taken, except for Gorishoek in 1989, where two samples were taken. In 1990 and 1991 samples from the 15 m level were also taken, to add to the accuracy of the biomass estimate.

From 1979 through 1983 only percentage cover was estimated, and no biomass measurements took place. To obtain an impression of the biomass during that period, biomass was measured on a number of stations where first percentage cover was estimated. Thus, correlations and regressions could be calculated for a number of dominant species (see Table 1 and 2; Fig. 2). After testing this method on the period from which there are cover estimations as well as actual biomass measurements, the cover estimations from 1979 through 1983 were then used to calculate biomass values for that period.

Total available surface of hard substrates in the Oosterschelde was determined in three steps: on specially prepared maps (so-called criteria-kaarten) the horizontal projection of the bottom area covered with hard substrates was measured. Then the slope at each transect was measured, to determine trigonometrically the actual bottom area covered with hard substrates.

Then a relief factor was determined at each transect. This was done in a 10 by 10 m square. At every 1 m, both in horizontal and in vertical direction, a chain was draped over the substrate, and the total length of that chain measured. Sediment area between the stones was deducted from the total. The latter accounts for the fact that the relief factor sometimes is smaller than 1: in those cases there was more sediment area between the stones than surface increase by the relief of the stones (see Fig. 3). The sediment areas always appeared to be without any appreciable relief. To

Table 2. Linear regression coefficients a and b for the regression ($y = ax + b$) of biomass (y) on percentage cover (x), for selected species.

Species	a	b
Aplidium glabrum	26.51	− 0.05
Ascidiella aspersa	2.44	1.33
Ciona intestinalis	1.75	1.33
Halichondria bowerbanki	5.31	0.19
Haliclona oculata	1.56	1.62
Diadumene cincta	5.91	0.72
Metridium senile	1.41	− 0.40
Sagartia troglodytes	1.74	0.58
Crassostrea gigas	0.20	14.91
Mytilus edulis	0.99	0.38
Tubularia larynx	1.39	0.35
Bryozoa non det.	12.88	0.46

determine the relief factor a theoretical 10 by 10 m square was constructed, containing several geometric bodies (pyramids, cubes) simulating the stones. Total surface was calculated by adding

Table 1. Correlation coefficients between percentage cover and biomass for selected species. Between brackets the level of significance ($> 0.05 =$ not significant: n.s.).

Species	
Aplidium glabrum	0.845 (< 0.034)
Ascidiella aspersa	0.985 (< 0.001)
Ciona intestinalis	0.872 (< 0.001)
Styela clava	0.435 (< 0.138) n.s.
Halichondria bowerbanki	0.778 (< 0.008)
Haliclona oculata	0.878 (< 0.001)
Diadumene cincta	0.886 (< 0.001)
Metridium senile	0.958 (< 0.001)
Sagartia troglodytes	0.842 (< 0.017)
Hydrozoa non det.	0.630 (< 0.129) n.s.
Tubularia larynx	0.971 (< 0.001)
Crassostrea gigas	0.876 (< 0.001)
Mytilus edulis	0.932 (< 0.001)
Bryozoa non det.	0.926 (< 0.001)
Balanus balanoides	0.654 (< 0.029)

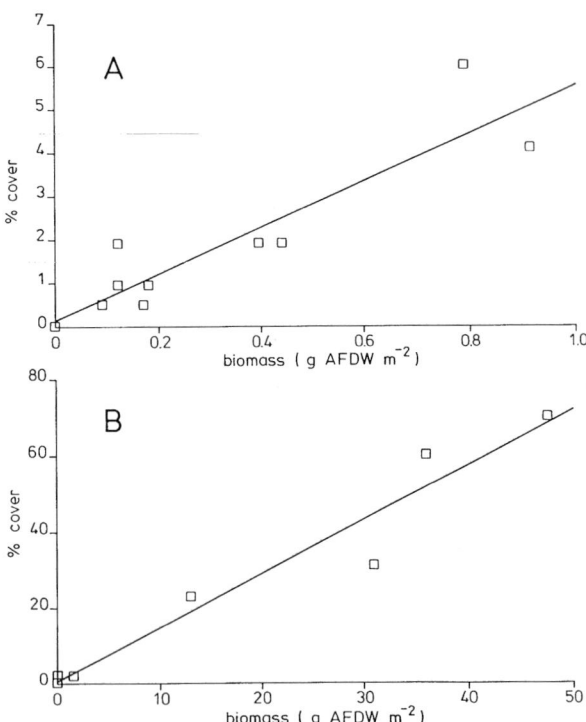

Fig. 2. Two examples of regression plots: a. *Halichondria bowerbanki*; b: *Metridium senile*.

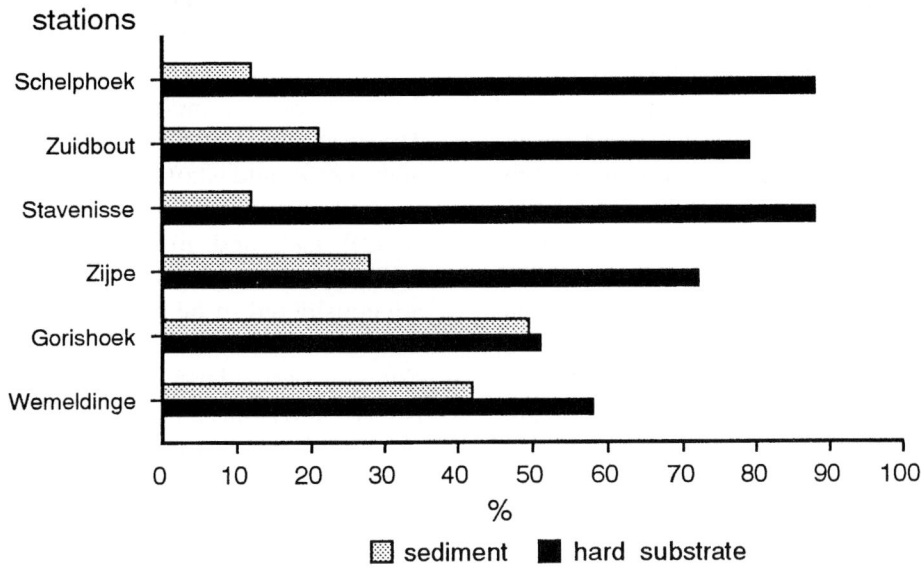

Fig. 3. Percentage of hard and soft bottom on each of the sampled stations. Unpublished data from De Mey & Van der Sloot.

the surfaces of all those bodies and flat surfaces. Result of this calculation was the figure of 144.5 m^2; this means that the available hardsub surface in an average 10 by 10 m square can be found by multiplying that surface by 1.45, after deduction of the percentage sediment area. This is the same figure as Dahl (1973) found for a partly sedimented field of coral boulders. An average 10 by 10 square has 75% of hardsub surface. Figure 3 shows the percentages of hardsub measured at the various locations. The result of the multiplication was called the relief factor, which appeared to lie between 0.74 and 1. 28 (see Table 3).

Water quality parameters, current velocities and irradiance were measured routinely by the

Table 3. Relief factor for 6 locations, in 1991 (explanation in text).

Schelphoek	1.28
Zuidbout	1.15
Gorishoek	0.74
Zijpe	1.04
Wemeldinge	0.84
Stavenisse	1.28

Tidal Waters Division at a number of locations, although not exactly at the locations of this study. As the Oosterschelde is a well-mixed basin, these measurements were assumed to be sufficiently representative to be compared with the distribution of species and biomass on the hard substrates. This assumption was shown to be correct for at least the current velocities: measurements of erosion values of gypsum blocks (clod-cards) on the bottom stations showed a correlation coefficient higher than 99% with the routine measurements by the Tidal Waters Division (De Kluijver & Leewis, 1994).

Results

Biomass

Hard substrata occur mainly along the edges of the Oosterschelde basin. Their total surface is only a fraction of the bottom area (*ca* 2% in 1979, *ca* 4% in 1990). Their faunal biomass, however, was high: Leewis & Waardenburg (1990) estimated the mean value in 1984–1986, averaged over the whole basin at 327 g ash free dry weight

(AFDW) m^{-2} ($\pm$ 141). In 1986, a total area of 13.17 km^2 of available substratum was estimated (shore protection: 8.08; pillars Zeeland bridge: 0.05; harbour piers etc.: 0.01; oyster-ponds and mussel-beds: 0.02; peat banks: 0.01; and storm-surge barrier: 5 km^2). Extrapolation of the biomass data gave 4306 tons AFDW in the whole basin, which was, in 1986, about 30% of the total benthic invertebrate biomass of the Oosterschelde (see also Coosen *et al.* 1994; Smaal & Van Straalen, 1990). Sponges accounted for 24% of total biomass, ascidians for 14%, hydroids for 10%, oysters for 10%, mussels and slipper limpets for 10%, and sea-anemones + 'rest' for 30%. Sea-anemones can locally be dominant.

For the analysis of the horizontal distribution of biomass, the Oosterschelde basin was divided into 4 areas (Fig. 1). Mean biomass m^{-2} was different in the four areas, as was the quantity of available substratum (Table 4). Therefore, the distribution of standing stock is different from the distribution of biomass m^{-2}. The development of biomass and standing stock in the 4 sub-areas is given in Fig. 4a and 4b, respectively. The large decrease in 1989 coincided with a decrease of the number of species. Only at Schelphoek there was some biomass left; at this location, the number of species did not decrease in 1989. In 1990 a re-

covery of biomass started, the strongest at location Zijpe.

The contribution to total biomass is different for each species, and can change over the years. This is shown in Tables 5 and 6: the development of biomass and standing stock of the 7 distinguished species groups for the 4 sub-areas, in g AFDW m^{-2} and in tons/sub-area, respectively.

The peat banks in the east of the Oosterschelde are considered to be part of the hard substrates. They are covered with epifauna; in the pre-barrier situation particularly with sponges (*Halichondria panicea* and *Haliclona oculata*), the ascidian *Styela clava* and the sea-anemone *Diadumene cincta*. After completion of the barrier relatively few brittlestars were found here, while in 1988 the ascidian *Styela clava* had become dominant. The peat banks have not been included in the biomass calculations, because it appeared impossible to remove the peat – representing a considerable amount of biomass – from the samples. In 1988, however, this was successfully accomplished. The biomass measured then was very high: 3.4 kg m^{-2}; more than 90% of this amount represented *Styela clava*.

Biomass in the 4 sub-areas, calculated from cover estimations for the period 1979–1983 is given in Fig. 5. The values do not deviate markedly from the values for the first years of actual biomass measurements (Fig. 4A). Figure 6 shows the development of hardsub biomass in the whole Oosterschelde from 1979 through 1991. Although it was also possible to calculate the values for the 7 distinguished species groups, the values are probably an underestimation. At least one dominant species (*Ciona intestinalis*) could not be included in the calculations, as the correlation between biomass and cover was not significant. The upward trend in the development of the mean hardsub biomass in the whole Oosterschelde between 1979 and 1984, as given in Fig. 6, will therefore in reality be a little less steep.

Table 4. Available hard substratum (km^2) in the four areas (see Fig. 1) of the Oosterschelde, from 1979 through 1990 (measured in 1986; total surface of the storm-surge barrier was estimated at 5 km^2. These figures have to be corrected for the relief factor (see Table 3).

Year	West	Central	East	Ne branch
1979	4.18	1.79	0.45	1.75
1980	4.18	1.79	0.45	1.75
1981	5.01	1.79	0.45	1.75
1982	5.85	1.79	0.45	1.75
1983	6.68	1.79	0.45	1.75
1984	7.51	1.79	0.45	1.75
1985	8.35	1.79	0.45	1.75
1986	9.18	1.79	0.45	1.75
1987	8.72	1.70	0.43	1.66
1988	8.26	1.61	0.41	1.58
1989	7.80	1.52	0.38	1.49
1990	7.34	1.43	0.36	1.40

Influence of abiotic factors

The quantity of available substrate has changed as a consequence of the building of the barrier.

Firstly, substrate was added: a lot of concrete was placed on the seabed. Earlier studies have shown that concrete is one of the best substrates for hardsub flora and fauna, although differences in growth between stone types in the sublittoral are smaller than in the littoral (Leewis *et al.*, 1989).

Secondly, substrate disappeared under sedimenting silt, although sedimentation was not evenly distributed over the Oosterschelde bottom. Distribution of sediment types changed (see De Kluijver & Leewis, 1994). It is still not possible to provide a good estimate of the quantity of hard substrate covered by sediment since the comple-

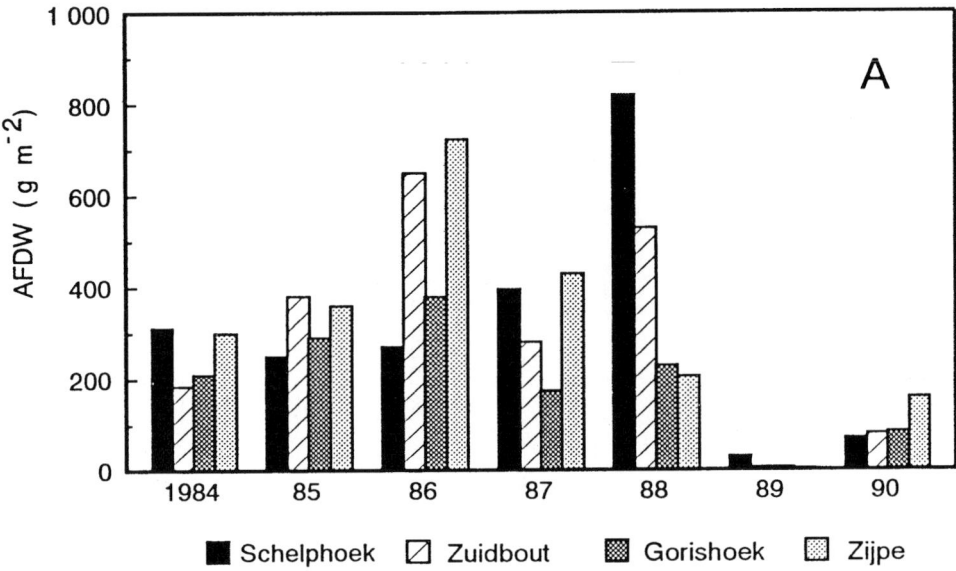

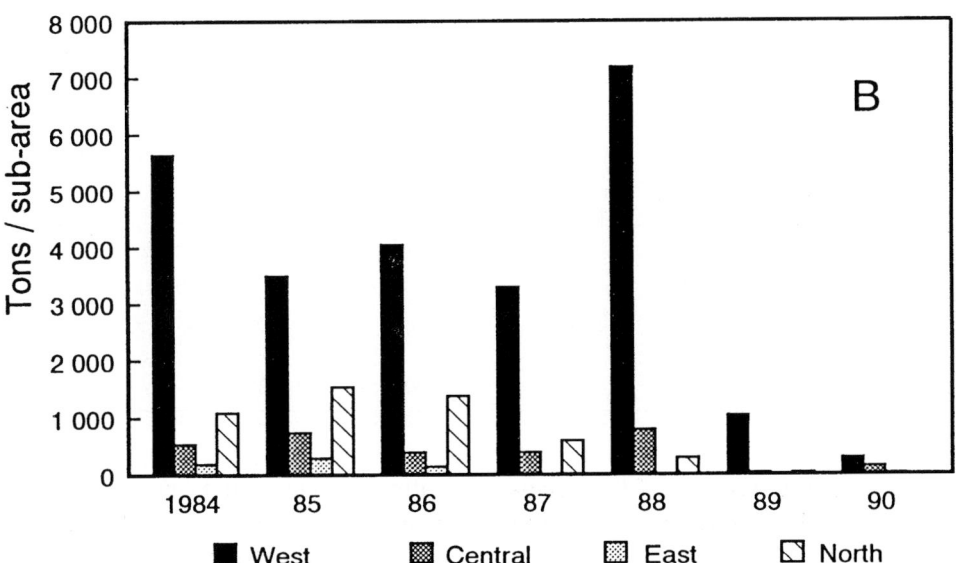

Fig. 4. a. Development of biomass in the four sub-areas – each represented by one sampling station, in g AFDW m^{-2}; b. standing stock in tons AFDW/sub-area.

Table 5. Development of biomass for selected species groups, in g AFDW m^{-2}, for the four distinguished sub-areas. Autumn values. n.s. = not sampled. With standard error of menas (s.e.m.). Where no s.e.m. is given, only one sample was taken.

	1984	1985	1986	1987	1988	1989	1990	1991
WEST (Schelphoek)								
Sponges	31 ± 31	55 ± 50	95.5 ± 172.5	42 ± 32	15	0	1 ± 2	3 ± 2.5
Ascidians	0 ± 0	0 ± 0	0 ± 0	0 ± 0	5	–	–	–
Brittlestars	1 ± 2	–	–	92 ± 94	529	27	10 ± 4	71 ± 2
Muss./Slipp.	0 ± 0	0 ± 0	–	7 ± 15	–	–	0 ± 0	1 ± 1
Oysters	21 ± 44	–	–	–	–	–	1 ± 1	8 ± 4
Hydroids	2 ± 3	40 ± 38	38 ± 40	3 ± 2.5	–	–	2 ± 2	–
Rest	323 ± 110	139 ± 106	132 ± 124	242 ± 140	261	91	48 ± 15	114 ± 19
Total	378	234	265.5	386	810	118	62	197
CENTRAL (Zuidbout)								
Sponges	14 ± 12	40 ± 57	43 ± 59	39 ± 36	4	–	–	0.5 ± 1
Ascidians	23 ± 24	36 ± 28	32 ± 15	29 ± 45	30	2	1 ± 1	109 ± 87
Brittlestars	24 ± 15	–	–	32 ± 21	190	0	64 ± 28	–
Muss./Slipp.	6 ± 7	5 ± 6	2 ± 2	6 ± 8	2	–	0.5 ± 1	–
Oysters	21 ± 40	64 ± 76	79 ± 107	40 ± 81	65	11	0 ± 0	19 ± 27
Hydroids	1 ± 7	3 ± 3	11 ± 6	0 ± 0	–	–	–	–
Rest	68 ± 18	85 ± 24	64 ± 33	102 ± 29	172	8	12 ± 7	22 ± 30
Total	157	223	231	248	463	21	77	150.5
EAST (Gorishoek)								
Sponges	30 ± 6	33 ± 25	42 ± 43	28 ± 27	n.s.	0 ± 0.5	22 ± 31	0.5 ± 0.5
Ascidians	53 ± 22	65 ± 50	27 ± 4	25 ± 29		0 ± 0	4.5 ± 2	53 ± 8
Brittlestars	18 ± 15	0 ± 0	–	4 ± 1		0 ± 0	88 ± 30.5	–
Muss./Slipp.	3 ± 7	6 ± 4	1 ± 1	6 ± 10		2 ± 3	0.5 ± 1	0.5 ± 1
Oysters	34 ± 105	146 ± 120	19 ± 26	22 ± 37		–	–	33 ± 47
Hydroids	15 ± 36	16 ± 5	18 ± 9	–		0 ± 0	1 ± 1	–
Rest	56 ± 26	77 ± 43	45 ± 18	51 ± 31		22 ± 26	12 ± 4	33 ± 2
Total	209	343	152	136		24	128	120
NORTH (Zijpe)								
Sponges	32 ± 33	285 ± 195	136 ± 101	31 ± 29	21	0	–	0.5 ± 0
Ascidians	12 ± 9	12 ± 12	25 ± 34	66 ± 72	7	3	22.5	34 ± 18
Brittlestars	0 ± 0	–	–	3 ± 1	–	6	63	–
Muss./Slipp.	125 ± 43	70 ± 52	79 ± 114	33 ± 45	42	2	4	2 ± 1
Oysters	117 ± 94	37 ± 49	90 ± 99	246 ± 269	95	8	60	52 ± 47
Hydroids	4 ± 3	1 ± 1	7 ± 7	0 ± 0	–	–	–	–
Rest	36 ± 19	41 ± 27	62 ± 54	50 ± 7	38	6	3	36 ± 16
Total	326	446	399	429	203	25	152.5	124.5
Mean whole basin	267.5	314	262	300	(492)	47	105	148

tion of the barrier. For the purpose of the calculations with SMOES, the carbon model of the Oosterschelde, an estimation of the quantity of hard substrate available (*i.e.*: not covered with sediment) was made on the basis of the sedimentation measurements in the tidal gullies. The results per year are given in Table 4. Measurements of sediment load (potential sedimentation) on a local level are reported by De Kluijver & Leewis (1994). This parameter showed a large temporal

Table 6. Development of standing stock for selected species groups, in tons AFDW m-2sub-area, for the four distinguished sub-areas. Autumn values. n.s. = not sampled.

	1984	1985	1986	1987	1988	1989	1990
WEST (Schelphoek)							
Sponges	466	825	1497	370	133	4	27
Ascidians	0	0	0	0	44	–	–
Brittlestars	15	–	–	807	4687	239	71
Muss./Slipp.	0	0	–	62	–	–	0
Oysters	315	–	–	–	–	–	9
Hydroids	30	600	570	26	–	–	–
Rest	4837	2085	1977	2130	2312	806	159
Total	5663	3510	4044	3395	7176	1049	266
CENTRAL (Zuidbout)							
Sponges	50	140	126	68	7	–	–
Ascidians	80	126	130	51	52.5	3.5	2
Brittlestars	84	–	–	56	332.5	0	112
Muss./Slipp.	21	17.5	4	10.5	3.5	–	2
Oysters	74	224	14	70	114	19	0
Hydroids	4	10.5	49	0	–	–	–
Rest	238	297.5	98	178.5	301	14	7
Total	551	766.5	421	434	810.5	36.5	123
EAST (Gorishoek)							
Sponges	27	31	40	4	n.s.	0	6
Ascidians	47	62	27	3		0	1
Brittlestars	20	0	–	0.5		0	9
Muss./Slipp.	0	6	0	1		–	0
Oysters	33	139	20	3		–	0
Hydroids	13	15	23	–		0	–
Rest	53	73	40	7		1	2
Total	193	325	150	18.5		1	18
NORTH (Zijpe)							
Sponges	112.5	997.5	475	43	29	0	1
Ascidians	42.5	42	87.5	92	10	4	61
Brittlestars	0	–	–	4	–	8	–
Muss./Slipp.	437.5	245.5	277.5	46	59	3	4
Oysters	410	129.5	315	344	133	11	93
Hydroids	15	3.5	25	0	–	–	–
Rest	125	143.5	217.5	70	53	8	64.5
Total	1142.5	1561	1397.5	599	284	34	223.5
Total basin:	7549.5	6162.5	6012.5	4446.5	8270	1120.5	630.5

variation, and a clear gradient from west to east (from mouth of the Oosterschelde: 430–3130 g $m^{-2} d^{-1}$ to location Zijpe: 30–1150 g m^{-2} d^{-1}). Some areas in the west are now covered so heavily with sediment, that only large *Metridium senile* can maintain themselves. All the other species have disappeared under the silt layer. It was shown by Leewis & Waardenburg (1989) that water quality parameters like salinity, nutrients and oxygen content are not important in deter-

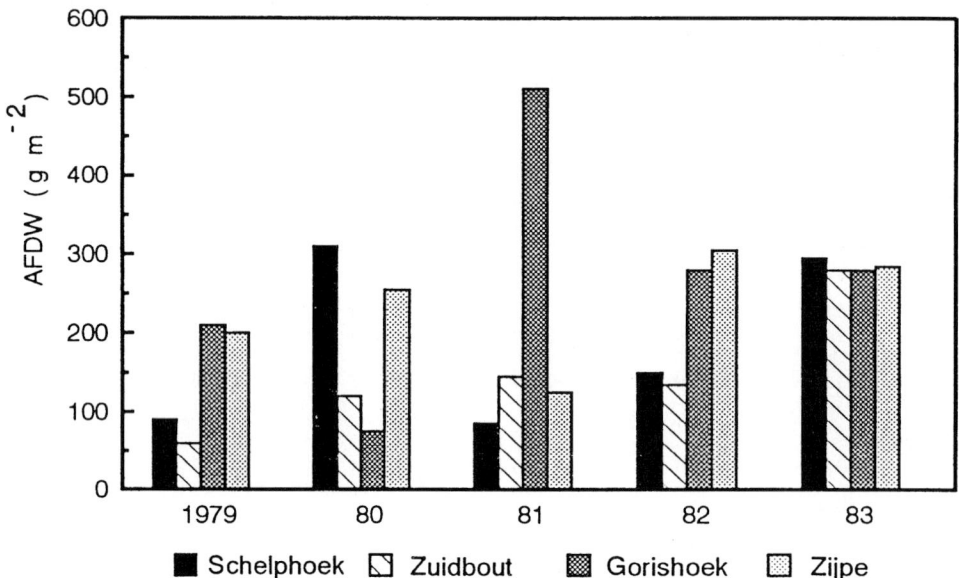

Fig. 5. Development of biomass in the four sub-areas from 1979 through 1983, based on calculation of biomass from cover estimations, in g AFDW m^{-2}.

mining the distribution of species and biomass within the Oosterschelde. The influence of other abiotic factors on the Oosterschelde system often greatly affects the biology. This can be illustrated by considering single species developments. Figure 7 shows the development of the abundance of the brittlestar *Ophiothrix fragilis* Abildgaard at Zuidbout (see Fig. 1: no. 4). This animal does not survive in water temperatures below 1 °C (Wolff,

1968). After the cold winters of 1978/'79, 1984/'85 and 1985/'86 the population of brittlestars was strongly reduced. During the following summers young brittlestars were recorded everywhere in the basin, and following mild winters (1979/'80 and 1987/'88) enormous increases in the population were found. The winter of 1988/'89 was mild too, resulting in masses of brittlestars covering large areas of hardsub in the summer of 1989 – probably smothering many other organisms by

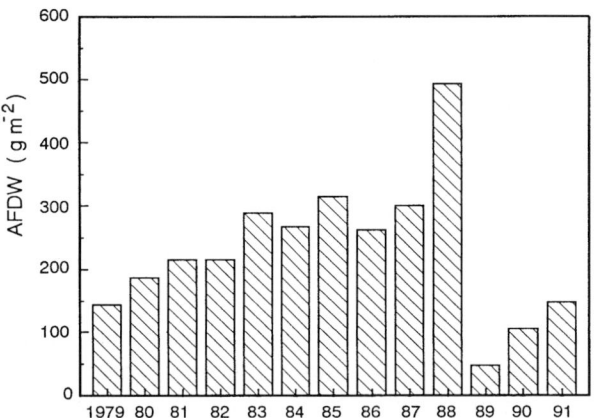

Fig. 6. Development of mean hardsub biomass in the whole Oosterschelde, from 1979 through 1991, in g AFDW m^{-2}.

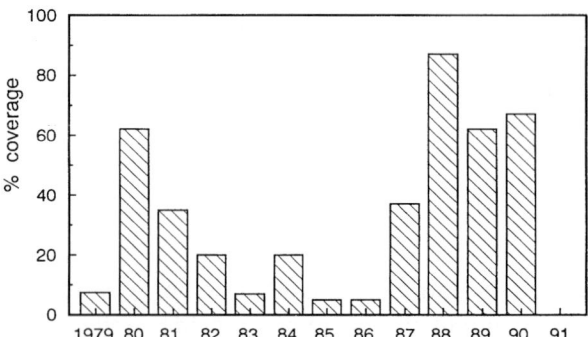

Fig. 7. Development of the percentage of coverage by the Brittlestar *Ophiothrix fragilis* at 5 to 7 m depth at Zuidbout (no. 4 in Fig. 1).

their sheer mass. The next winter showed a very cold spell during about a fortnight, and in the spring of 1990 dead brittlestars were found everywhere in the Oosterschelde. Along with them were other organisms with similar temperature preferences, e.g. the velvet swimming crab (*Lio-carcinus puber*).

Biomass of brittlestars was low before 1987 (*ca* 4% of the total in 1984; in 1985 and 1986 there were virtually no brittlestars; in 1987 the percentage was *ca* 11%; and in 1988 this percentage increased to 49%; in 1989 it went back to 17%, and in 1990 it became 54%, only to go down again to 12% in 1991. The extremely high biomass in 1988 in the western and central parts of the Oosterschelde was nearly exclusively due to brittlestars. In the preceding years sponges, oysters and anthozoa (as part of the category 'rest') made up the bulk of the biomass.

Oysters (*Crassostrea gigas*) became more important in the post-barrier period. Particularly in the eastern part of the Oosterschelde they formed a considerable part of the total biomass. In the pre-barrier period they have been present in the intertidal area for a number of years (Meijer, pers. comm.), and increasing in numbers. This increase is not to be attributed to expansion as a consequence of recent introduction. The species was introduced into the Oosterschelde (deliberately) in 1965, and expansion levelled off in the end of the seventies (Dijkema, 1993, and pers. comm.). Probably the higher low waters and the decreased current velocities in the post-barrier situation have enabled the species to expand its influence into the subtidal area. Sponges decreased markedly in biomass and coverage in the years after completion of the barrier. Ascidians, on the other hand, appear to be able to maintain themselves in about

the same numbers and biomass as in the pre-barrier period. As a result, the ratio of ascidians:sponges was increased (Table 7). After 1988 the situation for the sponges became even worse, as a consequence of ongoing sedimentation, although there is a recovery in 1990 and 1991; this appears to be mainly due to the expansion of *Halichondria bowerbanki* and of the branching sponge *Haliclona oculata*.

The majority of the hardsub organisms are filter feeders, in particular ascidians, sponges, mussels and oysters, and partly brittlestars and anthozoans. Together they constituted, in the pre-barrier period, 79% of the biomass of sessile hard substrate organisms. In 1988 this was 67% and in 1989 64%. This trend is apparent at all studied locations, the strongest at Schelphoek (Fig. 1, no. 1), while Schelphoek showed the lowest values (from 53% in 1984 to 43% in 1990) and Zijpe (Fig. 1, no. 5) the highest (from 93% in 1984 to 74% in 1990). This indicates that less filter feeders were able to maintain themselves in the post-barrier situation, which is in agreement with the expectation that the increased sedimentation (of food particles and of silt) should cause a shift in favour of deposit feeders.

Discussion

A period with cold winters preceded the completion of the barrier, while the winters just after completion of the barrier were very mild. Since several species are sensitive to low winter temperatures (e.g. *Ophiothrix fragilis*), the winter temperatures may partly mask effects of reduced current velocities and increased sedimentation following the completion of the barrier. Further-

Table 7. Biomass of sponges and ascidians, before and after completion of the barrier. Autumn, mean of all locations, in g AFDW m^{-2}. N.B.: mean max current velocity in 1988–1991 was 52% of the velocity in the period before the barrier was finished.

Year	'84	'85	'86	'87	'88	'89	'90	'91
Ascidians	22	28	21	30	14	1	7	49
Sponges	27	103	79	35	13	0	6	1
Ratio	0.82	0.27	0.26	0.85	1.08	∞	1.17	49

more, analysis of species composition on North Sea shipwrecks appears to indicate a change around the year 1987, which cannot be explained yet (Waardenburg, unpubl. data). Biomass on those shipwrecks also decreased since 1987. This suggests a more widespread influence that cannot be attributed to the Oosterschelde barrier.

A further complicating factor is the fact that during construction of the Philipsdam and the Oesterdam, the storm-surge barrier has been closed several times to reduce current velocities in the basin to facilitate construction of those dams; this happened mainly in 1986 and 1987 (Nienhuis & Smaal, 1994). The effect on hardsub and other organisms was monitored, and no severe damage was observed in the sublittoral. But it may well be that some after-effect has occurred, resulting in the two year time lag observed before the biomass (and the community composition as well) of hardsub changed considerably. Probably the decreased current velocities, in combination with the increased sedimentation, have caused the decrease of the sponges. Sponges are passive filter feeders, needing a certain external current velocity to be able to maintain a certain minimum internal velocity for the transport of food particles and waste products (Vogel, 1974). Ascidians, on the other hand, are active filter feeders, which may explain their relative increase in importance after the completion of the storm-surge barrier. Figure 8a shows the relationship between sponge biomass and the current velocities in the pre- and post-barrier situation, Fig. 8b the same for ascidian biomass. But the total biomass of ascidians did not increase greatly. The difference was nearly entirely caused by a decrease of sponge biomass. These animals most probably suffered under the increased sediment load. High growing, branching sponges like *Haliclona oculata* suffer less from sedimentation than encrusting ones. Also Haliclona has a very simple, shallow aquiferous system, which needs very little outside water pressure (Kaandorp, 1992). That may explain that in the last few years the relative importance of *Haliclona* has been increasing. The mass development of *Ophiothrix fragilis* as a consequence of several consecutive mild winters has no doubt had its

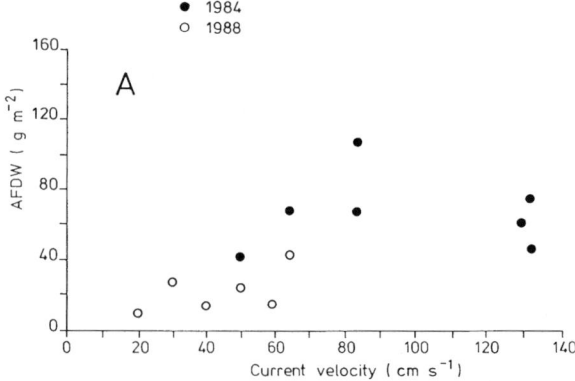

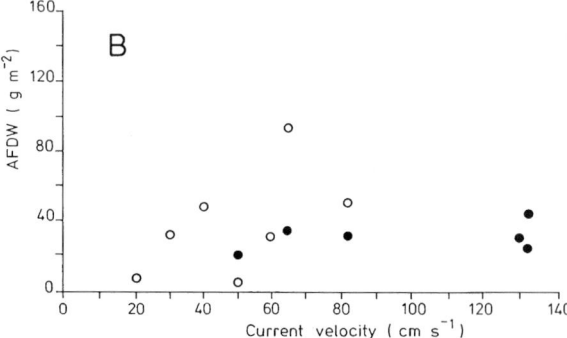

Fig. 8. Biomass of sponges (a) and ascidians (b) in Oosterschelde estuary as a function of current velocity.

influence on the sharp decrease of total biomass on hard substrata in 1989. Apart from eating away a lot of the available food, the brittlestars probably have suffocated or crushed underlying fauna by their sheer mass. The recovery in the following two years, when *Ophiothrix* started to decline, is an indication of the adverse influence of *Ophiothrix* on other species.

The biomass found in 1988 on the peat banks in the east of the Oosterschelde was extremely high (3.4 kg AFDW m^{-2}, consisting mainly of *Styela clava*). As this is only a single observation, no conclusions can be drawn from it.

In evaluating the meaning of the above biomass figures it is essential to look at the impact of the hardsub organisms on the ecosystem. The filtration capacity of hardsub organisms must be compared to that of the other important filter feeder groups (mussels, cockles and zooplankton). Based on literature (particularly Bayne & Newell, 1983; Fiala-Médoni, 1978a, b; Holmes,

1973; La Barbera, 1987; Randløv & Riisgard, 1979; Reiswig, 1971, 1974; Robbins, 1983; Shumway, 1978; Warner & Woodley, 1975; Frost, 1976; Gerodette & Flechsig, 1979) and our own research of the Tidal Waters Division a rather clear picture of this aspect has been formed. This will be reported elsewhere in detail by Leewis & Smaal. Some of the results, however, are relevant for this discussion.

Figure 9a shows the development of the total filtration of hard substrate organisms (hardsub) in the four sub-areas. Figure 9b shows the same for hardsub compared with the other important filter feeder groups (calculations with the carbon flow model of the Oosterschelde, called SMOES (Klepper, 1989; Klepper et al., 1994; Scholten & Van der Tol, 1994).

As stated above, measurements of biomass in 1990 and 1991 show a recovery of the figures for hardsub, and filtration capacity follows the same pattern. Comparison of the filtration of the hardsub organisms with that of the other groups of filter feeders (Fig. 9b) shows that the relative importance of filtration by hardsub organisms changed through the years. In 1984 it amounted to 39% of the total filtration by the benthic filter feeders. In the following years it was 34%, 26%, 17%, 33%, and 8%, respectively. In 1990 and 1991 it increased again. Other groups did not suffer such a marked decrease of biomass in that year. This appeared to be a temporary effect, probably to be attributed to the mass development of brittlestars, which occurred mainly on the hard substrates.

Total filtration of benthic filter feeders shows a maximum in 1986. Next it decreased steadily until a minimum in 1989. Gross primary production followed about the same pattern, although the range and yearly variation are rather large (Van der Tol & Scholten, 1992; Wetsteyn & Kromkamp, 1994).

The estimation of available surface for hardsub organisms is a very complicated problem. This kind of problem has only rarely been studied in mathematics (Lingeman, pers. comm.). Possibly the solution may be found in fractal geometry (Gleick, 1987).

Conclusions

Although at first an increase in faunal biomass was found after completion of the storm-surge barrier, after two years a collapse of biomass occurred. In 1990 a recovery started, which continued in the next year and of which the end result can't be established yet. Different species became dominant in the post-barrier period; e.g. *Halichondria panicea* disappeared largely, while *H. bowerbanki*, *Haliclona oculata* and some other sponge species became more important. *Crassostrea gigas* increased in numbers, particularly in the eastern part of the basin. The increased sedimentation caused the composition of communities to change. Biomass decreased locally because of the disappearance of nearly the whole community under a layer of silt.

The ratio ascidians: sponges changed in favour of the ascidians; this can be attributed to the changed current regime (smaller current velocities). But the sedimentation may also be partly responsable for the decrease of the sponges, while the mass development of brittlestars (*Ophiothrix fragilis*) as a consequence of several consecutive mild winters in a row undoubtedly plays a role. So, a climatic factor masks effects of the storm-surge barrier, and sedimentation masks the effects proper of decreased current velocities. In accordance with the expectation, the percentage of filter feeders among the hardsub organisms decreased in the post-barrier situation.

The hardsub organisms constitute an important factor in evaluating and modelling the ecosystem as a whole. Their relative impact, expressed as filtration capacity, compared with the other filter feeder groups in the Oosterschelde decreased steadily from 1984 until 1989 (with the exception of 1988). It may be expected that the hardsub filter feeders will regain pre-barrier importance when the system reaches a new equilibrium, as the years 1990 and 1991 already show a recovery, and particularly the ascidians – with a relatively high filtration capacity – have not suffered from the effects of the storm-surge barrier.

The impact of hardsub filter feeders per unit biomass was high in the northern part, and rela-

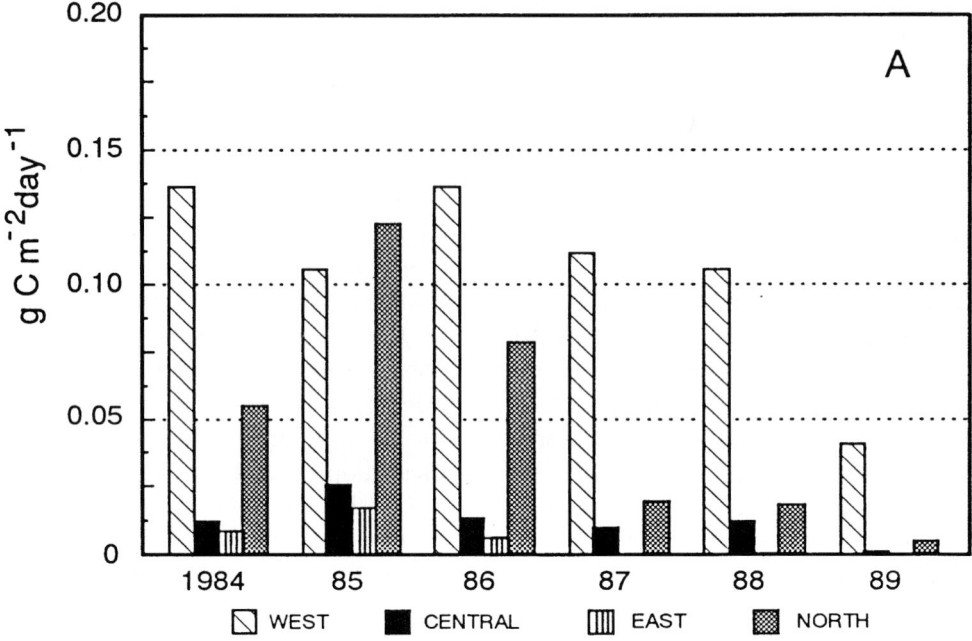

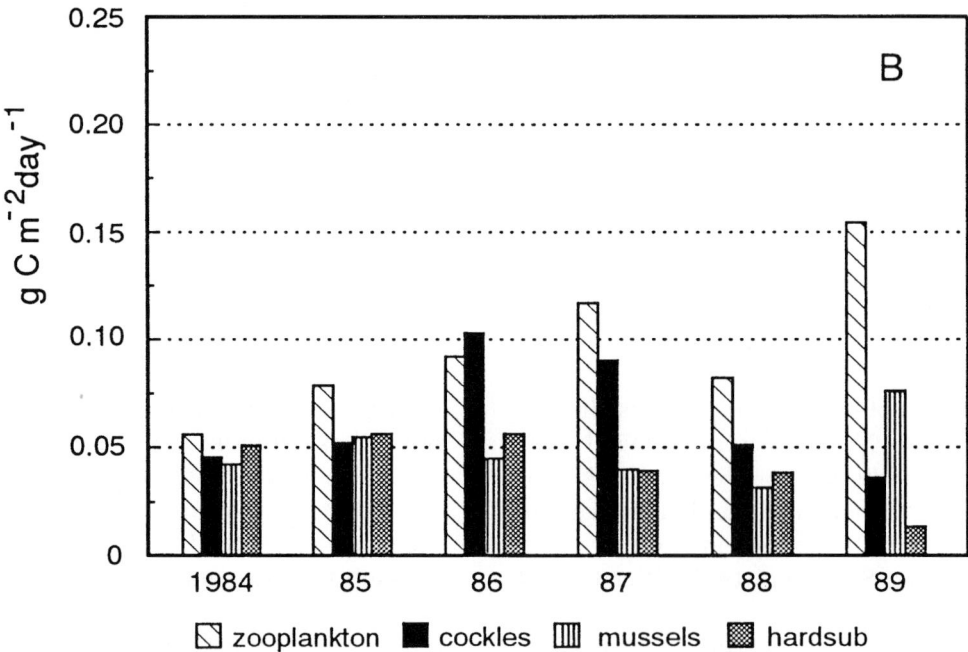

Fig. 9. a. Development of the filtration of phytoplankton by organisms on hard substrates in the four sub-areas (g C m^{-2} d^{-1}); b. development of the filtration of phytoplankton in the Oosterschelde by the organisms on hard substrates, compared to the other important groups of filter feeders (g C m^{-2} d^{-1}).

tively low in the western part of the Oosterschelde. The method of estimating biomass on the basis of estimations of percentage cover proved to be workable. It has to be tested further in order to reduce uncertainty.

Acknowledgements

The Directie Zeeland and the Meetdienst Vlissingen, both of Rijkswaterstaat are thanked for making the specially prepared maps ('criteriakaarten') available. J. v. d. Horst, R. Duyts, B. Haenen, M. Weststrate and W. Storm have prepared some of the tables, calculations and figures. H. Verkooyen, A. Engelberts and F. Stikvoort helped with the assessment of biomass. Without W. Wolters sampling would have been a lot more difficult. P. v. d. Hurk, F. Molenaar, J. Koop, D. Liebregts, J. Reijnders, B. de Mey, J.-H. v. d. Sloot and C. van Mechelen have conducted student research on several aspects of this project.

References

Bayne, B. L. & R. C. Newell, 1983. Physiological energetics of marine molluscs. In: Wilbur & Younge: The Mollusca, Vol. 4, Physiology, part 1. Academic Press.

Braun-Blanquet, J., 1964. Pflanzensociologie, Grundzuege der Vegetationskunde. 3rd edn. Springer-Verlag, Wien-New York.

Coosen, J., F. Twisk, M. W. M. van der Tol, R. H. D. Lambeck, M. R. van Stralen & P. M. Meire, 1994. Variability in stock assessment of cockles (Cerastoderma edule L.) in the Oosterschelde (in 1980–1990), in relation to environmental factors. Hydrobiologia 282/283: 381–395.

Dahl, A. L., 1973. Surface area in ecological analysis: Quantification of benthic coral-reef algae. Mar. Biol. 23: 239–249.

De Kluijver, M. J., 1989. Sublittoral hard substrate communities of the southern Delta area, SW Netherlands. Bijdragen tot de Dierkunde 59: 141–158.

De Kluijver, M. J. & R. J. Leewis, 1994. Changes in the sublittoral hard substrate communities in the Oosterschelde estuary (SW Netherlands), caused by changes in the environmental parameters. Hydrobiologia 282/283: 265–280.

Dijkema, R., 1993. Shellfish culture in the Netherlands (in press).

Fiala-Médioni, A., 1978a. Filter feeding ethology of benthic invertebrates (Ascidians). III. Recording of water current in situ. Rate and rhythm of pumping. Mar. Biol. 45: 185–190.

Fiala-Médioni, A., 1978b. Filter feeding ethology of benthic invertebrates (Ascidians). IV. Pumping rate, filtration efficiency. Mar. Biol. 48: 243–249.

Frost, T. M., 1976. Sponge feeding: a review with a discussion of some continuing research. In: Harrison F. W. & R. R. Cowden (eds), Aspects of spongebiology. Acad. Press: 283–298.

Gerodette, T. & A. O. Flechsig, 1979. Sediment-induced reduction in the pumping rate of the tropical sponge Verongia lacunosa. Mar. Biol. 55: 103–110.

Gleick, J., 1987. Chaos: making a new science. Viking, New York: 1–314.

Hill, M. O., 1979. TWINSPAN – A FORTRAN program for arranging multivariate data in an ordered two-way table by classification of the individuals and attributes. Ecology and Systematics, Cornell University, Ithaca, New York.

Hiscock, K., 1976. The influence of water movement on the ecology of sublittoral rocky areas. Ph.D.Thesis, Univ. of Wales.

Hiscock, K. & R. Mitchell, 1980. The description and classification of sublittoral epibenthic ecosystems. In: J. H. Price, D. E. G. Irvine and W. F. Farnham (eds): The Shore Environment, Vol. 2: Ecosystems. Acad. Press, London and New York: 323–370.

Holmes, N., 1973. Water transport in the ascidians Styela clava and Ascidiella aspersa. J. exp. mar. Biol. Ecol. 11: 1–13.

Kaandorp, J. A., 1992. Modelling growth forms of biological objects using fractals. Thesis, Univ. of Amsterdam: I–X + 1–158.

Klepper, O., 1989. A model of carbon flows in relation to macrobenthic food supply in the Oosterschelde estuary (S.W.Netherlands). Thesis, Univ. of Wageningen: 1–270.

Klepper, O., M. W. M. van der Tol, H. Scholten & P. M. J. Herman, 1994. SMOES: a simulation model for the Oosterschelde ecosystem. Part I. Description and uncertainty analysis. Hydrobiologia 282/283: 437–451.

LaBarbera, M., 1987. Particle capture by a pacific brittle star. Science 201: 1147–1149.

Leewis, R. J. & A. C. Smaal (in prep.). Filtration activity of several filter feeder groups in the Oosterschelde.

Leewis, R. J. & H. W. Waardenburg, 1989. The flora and fauna of the sublittoral part of the artificial rocky shores in the southwest Netherlands. Progress in Underwater Science 14: 109–122.

Leewis, R. J., H. W. Waardenburg & A. J. M. Meijer, 1989. Active management of an artificial rocky coast. Hydrobiol. Bull. 23: 91–99.

Leewis, R. J. & H. W. Waardenburg, 1990. Flora and fauna of the sublittoral hard substrata in the Oosterschelde (The Netherlands) – interactions with the North Sea and the influence of a storm surge barrier. Hydrobiologia 195: 189–200.

412

Nienhuis, P. H. & A. C. Smaal, 1994. The Oosterschelde estuary, a case study of a changing ecosystem: an introduction. Hydrobiologia 282/283: 1–14.

Reiswig, H. M., 1971. Particle feeding in natural populations of three marine sponges. Biol. Bull. 141: 568–591.

Reiswig, H. M., 1974. Water transport, respiration and energetica of three tropical marine sponges. J. exp. mar. Biol. Ecol. 14: 231–249.

Randløv, A. & H. U. Riisgard, 1979. Efficiency of particle retention and filtration rate in four species of ascidians. Mar. Ecol. Progr. series 1: 55–59.

Robbins, I. J., 1983. The effects of body size, temperature and suspension density on the filtration and ingestion of inorganic particulate suspensions by ascidians. J. exp. mar. Biol. Ecol. 70: 65–78.

Scholten, H. & M. W. van der Tol, 1994. SMOES: a simulation model for the Oosterschelde ecosystem. Part II: Calibration and validation. Hydrobiologia 282/283: 453–474.

Shumway, S. E., 1978. Respiration, pumping activity and heart rate in *Ciona intestinalis* exposed to fluctuating salinities. Mar. Biol. 48: 235–242.

Smaal, A. C. & M. van Stralen, 1990. Average annual growth and condition of mussels as function of food sources. Hydrobiologia 195: 179–188.

Van der Tol, M. W. M. & H. Scholten, 1992. Response of the Eastern Scheldt stem to a changing environment: functional or adaptive Neth. J. Sea Res. 30: 175–190.

Van Soest, R. W. M. & S. Weinberg, 1981. Preliminary quantitative assessment of the marine hard substrate communities of Roaringwater Bay. J. of Sherkin Island I: 10–26.

Vogel, S., 1974. Current-induced flow through the sponge *Halichondria bowerbanki*. Biol. Bull. 147: 443–456.

Warner, G. F., 1984. Diving and Marine Biology. The Ecology of the Sublittoral. Cambridge Univ. Press, Cambridge-New York-Melbourne: 1–210.

Warner, G. F. & J. D. Woodley, 1975. Suspension feeding in the brittle star *Ohiothrix fragilis*. J. mar. biol. Ass. U.K. 55: 199–210.

Weinberg, S., 1978. The minimal area problem in invertebrate communities of Mediterranean rocky substrata. Mar. Biol. 49: 33–40.

Wetsteyn, L. P. M. J., J. C. H. Peeters, R. N. M. Duin, F. Vegter & P. R. M. de Visscher, 1990. Phytoplankton primary production and nutrients in the Oosterschelde (The Netherlands) during the pre-barrier period 1980–1984. Hydrobiologia 195: 163–177.

Wetsteyn, L. P. M. J. & J. C. Kromkamp, 1994. Turbidity, nutrients and phytoplankton primary production in the Oosterschelde (The Netherlands) before, during and after a large-scale coastal engineering project (1980–1990). Hydrobiologia 282/283: 61–78.

Wolff, W. J., 1968. The Echinodermata of the estuarine region of the rivers Rhine, Meuse and Scheldt, with a list of species occurring in the coastal waters of The Netherlands. Neth. J. Sea Res. 4: 59–85.

Hydrobiologia **282/283**: 413–429, 1994.
P. H. Nienhuis & A. C. Smaal (eds), The Oosterschelde Estuary.
© 1994 *Kluwer Academic Publishers. Printed in Belgium.*

The role of the blue mussel *Mytilus edulis* in the cycling of nutrients in the Oosterschelde estuary (The Netherlands)

T. C. Prins [1] & A. C. Smaal [2]

[1] *Netherlands Institute of Ecology, Centre for Estuarine and Coastal Ecology, Vierstraat 28, 4401 EA Yerseke, The Netherlands;* [2] *National Institute for Coastal and Marine Management/RIKZ, P.O. Box 8039, 4330 EA Middelburg, The Netherlands*

Key words: Oosterschelde estuary, *Mytilus edulis*, nutrient cycling, phytoplankton, biodeposition, N-cycle

Abstract

The fluxes of particulate and dissolved material between bivalve beds and the water column in the Oosterschelde estuary have been measured *in situ* with a Benthic Ecosystem Tunnel. On mussel beds uptake of POC, PON and POP was observed. POC and PON fluxes showed a significant positive correlation, and the average C:N ratio of the fluxes was 9.4. There was a high release of phosphate, nitrate, ammonium and silicate from the mussel bed into the water column. The effluxes of dissolved inorganic nitrogen and phosphate showed a significant correlation, with an average N:P ratio of 16.5. A comparison of the *in situ* measurements with individual nutrient excretion rates showed that excretion by the mussels contributed 31–85% to the total phosphate flux from the mussel bed. Ammonium excretion by the mussels accounted for 17–94% of the ammonium flux from the mussel bed. The mussels did not excrete silicate or nitrate. Mineralization of biodeposition on the mussel bed was probably the main source of the regenerated nutrients.

From the *in situ* observations net budgets of N, P and Si for the mussel bed were calculated. A comparison between the uptake of particulate organic N and the release of dissolved inorganic N (ammonium + nitrate) showed that little N is retained by the mussel bed, and suggested that denitrification is a minor process in the mussel bed sediment. On average, only 2/3 of the particulate organic P, taken up by the mussel bed, was recycled as phosphate. A net Si uptake was observed during phytoplankton blooms, and a net release dominated during autumn. It is concluded that mussel beds increase the mineralization rate of phytoplankton and affect nutrient ratios in the water column. A comparison of N regeneration by mussels in the central part of the Oosterschelde estuary with model estimates of total N remineralization showed that mussels play a major role in the recycling of nitrogen.

Introduction

A dominant feature of estuaries and shallow coastal areas is the interaction between the water column and the benthic system (Zeitzschel, 1980).

This benthic-pelagic coupling affects many of the physical, chemical and biological processes in shallow marine systems. Populations of suspension feeding bivalves can attain high densities (Bahr, 1976; Wolff, 1983; Asmus, 1987; Jør-

414

gensen, 1990) and often dominate the estuarine fauna (Wolff, 1983). Dame *et al.* (1980) suggested that bivalve populations play a significant role in the coupling between the water column and the benthic system in an estuary. The bivalves increase the sedimentation of particulate material in estuaries (Verwey, 1952; Haven & Morales-Alamo, 1972; Dame *et al.*, 1984; Smaal *et al.*, 1986; Sornin *et al.*, 1983, 1986; Kautsky & Evans, 1987) and may even cause a depletion of plankton biomass (Wright *et al.*, 1982; Carlson *et al.*, 1984; Wildish & Kristmanson, 1984; Fréchette & Bourget, 1985; Nichols, 1985). Indeed, *in situ* observations on reefs of the American oyster *Crassostrea virginica* (Dame *et al.*, 1985, 1989) and on beds of the blue mussel *Mytilus edulis* (Dame & Dankers, 1988; Asmus & Asmus, 1991; Dame *et al.*, 1991) have shown that a high uptake of particulate matter occurs.

The presence of large numbers of bivalves does not merely lead to a removal of material from the water column. *In situ* measurements have shown that high amounts of inorganic nutrients (N, P, Si) are released from oyster reefs (Dame *et al.*, 1984, 1985, 1989) and mussel beds (Dame & Dankers, 1988; Asmus *et al.*, 1990; Prins & Smaal, 1990; Dame *et al.*, 1991; Asmus & Asmus, 1991). The large fluxes of inorganic nutrients from these bivalve communities into the water column are due to direct excretion by the bivalves and to mineralization processes in the adjacent sediment. The production of biodeposits by the bivalves enriches the sediment with organic material and stimulates mineralization processes (Dahlbäck & Gunnarsson, 1981; Kautsky & Evans, 1987; Dame *et al.*, 1991).

The availability of nutrients in coastal marine systems is largely dependent on benthic nutrient regeneration processes (Zeitzschel, 1980; Nixon, 1981). The regeneration of nutrients by bivalves possibly plays an important role in the cycling of nutrients in coastal systems and may influence pelagic production (Dame *et al.*, 1985; Doering *et al.*, 1986; Kautsky & Evans, 1987; Doering, 1989; Dame *et al.*, 1991). In addition to an overall increase in nutrient regeneration rates, the benthos may affect nutrient ratios as a conse-

quence of differences in the regeneration rates of N, P and Si (Doering *et al.*, 1987; Dame *et al.*, 1989).

Bivalve suspension feeders are very abundant in the Oosterschelde estuary. The biomass of the blue mussel *Mytilus edulis* (5.3 g C m^{-2}, Smaal *et al.*, 1986) is controlled by mussel culture, whereas wild stocks of the cockle *Cerastoderma edule* (3.1 g C m^{-2}, Smaal *et al.*, 1986) are present. As a consequence of the construction of a storm surge barrier in the mouth of the estuary and the building of dams on the eastern and northern boundaries of the estuary, a number of abiotic changes have taken place. Among others, the water exchange with the North Sea and the freshwater discharge have decreased (Smaal *et al.*, 1991). As a consequence of the longer residence time of the water in the estuary benthic-pelagic exchange processes will get more dominant in the ecosystem. The decreased freshwater discharge led to a significant reduction in nutrient load (Wetsteyn & Bakker, 1991). It is to be expected that the tendency to 'oligotrophication' will extend the period of nutrient limitation of primary production, and will lead to an increased dependency of phytoplankton primary production on nutrient regeneration processes.

From laboratory estimates of filtration rates it was calculated that the bivalve population in the Oosterschelde may filter the entire volume of the basin in 4–5 days (Smaal *et al.*, 1986). This is further supported by the results of *in situ* observations of chlorophyll uptake by mussel beds, from which a turnover time of phytoplankton of 5 days was calculated (Dame *et al.*, 1991). High rates of inorganic nutrient release from mussel beds in the Oosterschelde have been reported earlier. A comparison of nutrient turnover time as a consequence of nutrient release from mussel beds, and the residence time of the water in the Oosterschelde estuary suggests that the mussel beds may be a major source of regenerated nutrients, particularly nitrogen (Prins & Smaal, 1990; Dame *et al.*, 1991).

We hypothesize that nutrient regeneration by mussel beds plays a significant role in the cycling of nutrients in the Oosterschelde estuary. In the present study we present results of *in situ* observations on the uptake and release of particulate organic nutrients and dissolved inorganic nutrients by intertidal beds of *Mytilus edulis* in the Oosterschelde estuary. We will compare the size and composition of the particulate and the dissolved material fluxes between the water column and the mussel bed, and will discuss the contribution of the bivalves to the regeneration of nutrients in the estuary.

Material and methods

In situ *measurements*

The exchange of dissolved and particulate matter between beds of the blue mussel *Mytilus edulis* L. and the water column were measured with a Benthic Ecosystem Tunnel. During low water, a plexiglass tunnel (12 m long, 0.80 cm width, covering 7.8 m^2 of sediment) was placed on an intertidal bivalve bed, without disturbing the animals. During the time of submersion of the tunnel samples of inflowing and outflowing water were collected at regular intervals. Current speeds in the tunnel were recorded continuously. The maximum current speed observed was 31 cm s^{-1}, the average current speed was 9.1 cm s^{-1}. No samples were collected when current speeds were lower than 1 cm s^{-1}. The average residence time of the water between sampling points was *ca* 2 minutes.

Measurements on mussel beds were carried out in 1987 (June, 1 series; September, 1 series), 1988 (June, 3 series; September, 1 series) and 1989 (April, 2 series). All measurements were carried out on the same site, near the low-tide level at an intertidal mussel bed in the Zandkreek, in the central part of the Oosterschelde estuary. In June 1988 a control experiment was carried out on a bed made of empty mussel shells.

Sampling procedure

Each observation period lasted two tidal cycles, and water samples were taken every 30 minutes during the period of submersion (*ca* 9 h.) of the tunnel. During each tidal cycle 15–20 samples were collected. After collection the samples were transported to the laboratory in Middelburg and processed. For the determination of particulate organic carbon (POC) and particulate organic nitrogen (PON) 1 litre samples were filtered on a Whatman GF/C filter and analyzed in a Carlo-Erba CHN analyzer. Particulate organic phosphorus (POP) was determined by filtration of 0.5 litre on a Whatman GF/C filter and analysis on a Technikon auto-analyzer after destruction of the POP to orthophosphate. Chlorophyll *a* was analyzed by HPLC. Dissolved inorganic nitrogen (DIN: N_4^+, NO_2^- and NO_3^-), dissolved inorganic phosphorus (PO_4^{3-}), and silicate (H_4SiO_4) were analyzed on a Technikon auto-analyzer. Details of the sampling procedure are given in Prins & Smaal (1990) and Dame *et al.*, (1991).

After each series of measurements the macrobenthos in the mussel bed was sampled by taking six 0.0177 m^2 core samples. The samples were sieved through a 1 mm sieve. Mussels were measured to the nearest millimetre of shell-length, dried at 60 °C for 48 hours and burned at 520 °C in a muffle furnace to establish the ash-free dry weight. All other macro-organisms were enumerated, and the ash-free dry weight of each species was established.

Calculation of fluxes

Water flow through the tunnel was estimated from current velocity data that were continuously recorded with a NSW Meerestechnik magnetic-induction current speed meter. The flux of material between the water column and the bivalve bed was calculated from water flux times the difference between the inflow and outflow concentrations of a nutrient.

Particulate silicon was not measured in our experiments. The amount of biogenic silicon taken up from phytoplankton by the bivalves was calculated by combining carbon:chlorophyll ratios measured by Bakker (pers. comm.) with our chlorophyll *a* fluxes. Thus, the phytoplankton uptake was recalculated to carbon units. From the observed diatom fraction in the phytoplankton biomass (Bakker, pers. comm.) and a C:Si ratio of 106:16 (Day *et al.*, 1989) the flux of diatom-bound silicon was calculated.

Laboratory experiments

After each field measurement 12 individuals from the mussel population were transported to the Tidal Waters Division field station at Jacobahaven, where they were kept for 24 hours in flowing seawater, pumped directly from the Oosterschelde. After this acclimation period the animals were placed in small chambers with a volume of 0.3 l, and incubated for 1–2 hours. The water was sampled before and after incubation and the inorganic nutrient concentrations (phosphate, silicate, ammonium, nitrate, nitrite) were determined. The individual rates of nutrient excretion (U) were calculated from the following formula:

$$U = \frac{C_t - C_0}{t} V \text{ in mg h}^{-1}, \qquad (1)$$

where C_o is the nutrient concentration at $t = 0$, C_t is the concentration at the end of the experiment, t is the incubation time and V is the volume of the chamber. After the experiment the ash-free dry weight of the animals was determined.

Comparison of in situ *fluxes with individual excretion rates*

The amount of nutrients excreted by the bivalves depends on the biomass and the size of the animals. The individual excretion rates are an allometric function of body weight (Bayne *et al.*, 1976). In order to enable a comparison between the *in situ* rates of nutrient release from the bivalve community and the laboratory observations of individual nutrient excretion rates, the *in situ* fluxes were recalculated to rates per unit body weight:

$$a = \frac{F}{\sum n_i W_i^{0.68}} \text{ in } \mu\text{mol g AFWD}^{-1}\text{h}^{-1}, \qquad (2)$$

where F is the observed *in situ* flux, n_i is the number of mussels from sizeclass i, with an individual ash-free dry weight W_i, and 0.68 is the weight-exponent for nutrient excretion (Bayne & Widdows, 1978).

The nutrient excretion rates of individual animals, measured in the field station, were also recalculated to rates per unit body weight by dividing the excretion rate by the metabolic weight ($W^{0.68}$) of the animal.

Pooling of results

Experiments in June 1987 were carried out during a phytoplankton bloom (13.1 ± 1.2 μg chlorophyll a l^{-1}, $n = 19$), whereas all observations in June 1988 fell within a long period of very low phytoplankton concentrations (0.59 ± 0.03 μg chlorophyll a l^{-1}, $n = 92$). The inorganic nutrient fluxes observed in these two periods differed significantly (Prins & Smaal, 1990). In the subsequent processing of data the results of these two periods have been treated separately. The results of the two series of observations in April have been pooled, and the same was done with the results of both September experiments.

Table 1. Numbers and biomass (ash-free dry weight, AFDW) of mussels and total macrozoobenthos biomass (ash-free dry weight) in the tunnel experiments.

Date	Number m^{-1}	AFDW mussels g m^{-2}	Total AFDW macrozoobenthos g m^{-2}
11-6-1987	1458	891	1009
16-9-1987	1779	1273	1382
8-6-1988	3349	2187	2315
29-6-1988	2852	1448	1534
29-6-1988	3396	2147	2270
28-9-1988	2599	1721	1815
12-4-1989	6497	1206	1217
26-4-1989	4680	1569	1659

Results

Biomass of macrobenthos

The number of mussels on the mussel bed ranged between 1460 and 6500 animals m^{-2}, and the biomass between 890 and 2190 g AFDW m^{-2}. The mussels accounted for *ca* 90% of the total macrozoobenthic biomass on the mussel bed (Table 1).

Fluxes of particulate material

The average flux of particulate material between the water column and the mussel bed was calculated for each tidal cycle. The fluxes showed a large variation between measurements: POC fluxes had a range between a release from the

mussel bed of 40 mmol m^{-2} h^{-1} and an uptake by the mussel bed of 227 mmol m^{-2} h^{-1}, PON fluxes varied between 1.27 (release) and 23.14 (uptake) mmol m^{-2} h^{-1}, and POP fluxes between 1.17 (release) and 1.23 (uptake) mmol m^{-2} h^{-1}. In the control experiment with empty mussel shells a minor, non-significant, uptake of POC and PON was observed (*t*-test, $p > 0.05$). The correlation between POC and PON fluxes on the mussel bed and in the control experiment is shown in Fig. 1 ($r^2 = 0.75$, $p < 0.001$). The dashed line represents the C:N ratio (by atoms) of living phytoplankton (6.6; Redfield *et al.*, 1963). From the geometric mean regression (Ricker, 1973) a molar C:N ratio of the fluxes to the mussel bed (excluding the control measurement) of 9.4 ± 1.4 (mean $\pm$ standard error, $n = 14$) was estimated. In general, the fluxes were composed of material with a low C:N ratio compared to the seston at the inflow of the tunnel; the C:N ratios in the seston varied between experiments from 7.7 (June 1987) to 21.3 (September 1988). The C:N ratio of the suspended particulate material was higher at the outflow of the tunnel than at the inflow (Wilcoxon matched pairs-test, $p < 0.010$, $n = 164$).

Fluxes of dissolved material

There was a predominant release of dissolved inorganic nutrients from the mussel bed. The fluxes of DIN (ammonium + nitrite + nitrate) varied between an uptake by the mussel bed of 3.8 mmol m^{-2} h^{-1} and a release of 33.7 mmol m^{-2} h^{-1} (average per tidal cycle). Orthophosphate fluxes ranged from 0.08 mmol m^{-2} h^{-1} (uptake) to 0.86 mmol m^{-2} h^{-1} (release). The fluxes of DIN and phosphate are shown in Fig. 2. No significant fluxes were observed in the control experiment with empty mussel shells. One observation (night tidal cycle in June 1987) showed a very high release of DIN, and a much higher (86:1) N:P ratio than the other observations. With exclusion of this observation and the control, the fluxes showed a significant positive correlation ($r^2 = 0.69$, $p < 0.001$) and the geometric mean regression estimate of the N:P ratio fo the dissolved

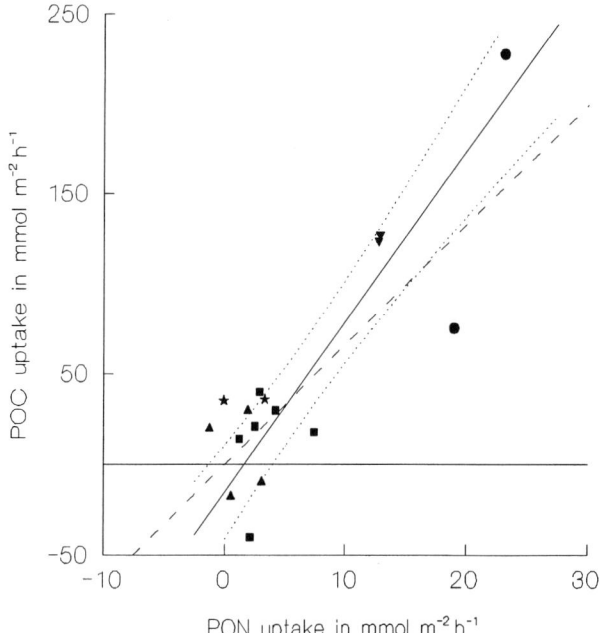

Fig. 1. The uptake of particulate organic carbon (POC) and particulate organic nitrogen (PON) by mussel beds (average per tidal cycle). Negative values indicate release of material from the mussel bed into the water column. The solid line is the geometric mean regression line between POC and PON fluxes with 95% confidence interval. The dashed line represents the Redfield ration C:N = 6.6:1.
● = *June 1987;* ▲ = *September 1987, 1988;* ■ = *June 1988;* ▼ = *April 1989;* ★ = control mussel bed.

418

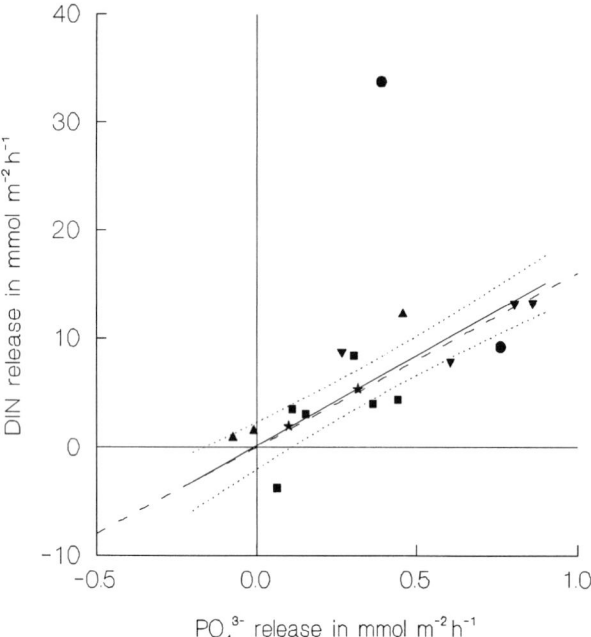

Fig. 2. The release of dissolved inorganic nitrogen (DIN, $NO_2^- + NO_3^- + NH_4^+$) and PO_4^{3-} by mussel beds (average per tidal cycle). Negative values indicate uptake of nutrients by the mussel bed. The solid line is the geometric mean regression line between DIN and phosphate fluxes with 95% confidence interval. The dashed line represents the Redfield ratio N:P = 16:1. ● = June 1987; ▲ = September 1987, 1988; ■ = June 1988; ▼ = April 1989; ★ = control mussel bed.

inorganic fluxes was 16.5 ± 2.7 ($n = 15$), which was not significantly different from the Redfield ratio.

Silicate fluxes ranged from 2.8 mmol m^{-2} h^{-1} (uptake) to 6.1 mmol m^{-2} h^{-1} (release) (Fig. 3). Silicate fluxes in the control experiment were not significant. Due to the high DIN release in the night tidal cycle of June 1987, this observation showed a very high N:Si ratio. When this observation was excluded, DIN and silicate fluxes were not significantly correlated. In general, the N:Si ratio of the fluxes was higher than the Redfield ratio (broken line).

Contribution of excretion by the bivalves to nutrient release

In the laboratory experiments excretion of ammonium and phosphate by the mussels was ob-

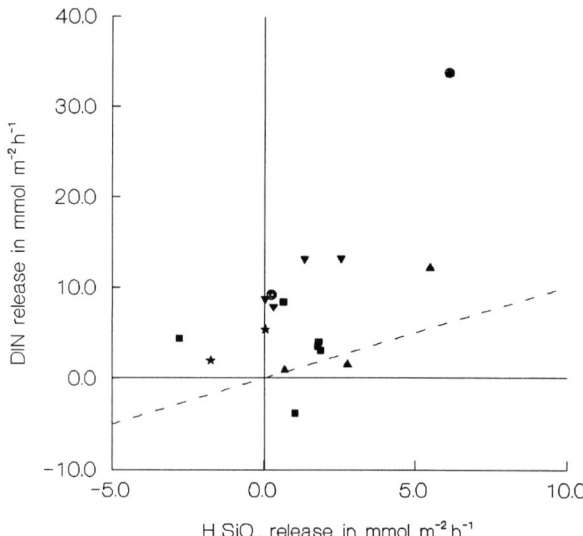

Fig. 3. The release of dissolved inorganic nitrogen (DIN, $NO_2^- + NO_3^- + NH_4^+$) and silicate by mussel beds (average per tidal cycle). The dashed line represents the ratio N:Si = 1:1. ● = June 1987; ▲ = September 1987, 1988; ■ = June 1988; ▼ = April 1989; ★ = control mussel bed.

served. No significant excretion of silicate or nitrate occurred. The excretion rates of ammonium and phosphate were correlated ($r^2 = 0.61$, $p < 0.001$), and the N:P ratio of the excretion was 7.7. The N:P ratio of the excretion was not significantly different between experiments (ANOVA, $p > 0.05$). The observed *in situ* release rates by the mussel bed of ammonium and nitrate, and of phosphate were recalculated to rates per unit body weight, and were compared to the individual excretion rates per unit body weight (Figs. 4 and 5). The estimated ammonium excretion by the mussels was equal to 41% of the *in situ* ammonium release by the mussel bed in April, 17% in June 1987, 94% in June 1988, and 28% in September, respectively. The phosphate excretion by the mussels corresponded for a larger part with the *in situ* release: 67% in April, 85% in June 1988, and 31% in September. In June 1987 phosphate excretion rates were not determined.

N, P, and Si budgets of the mussel bed

From the observed uptake of PON and POP and the release of DIN and phosphate by the mussel

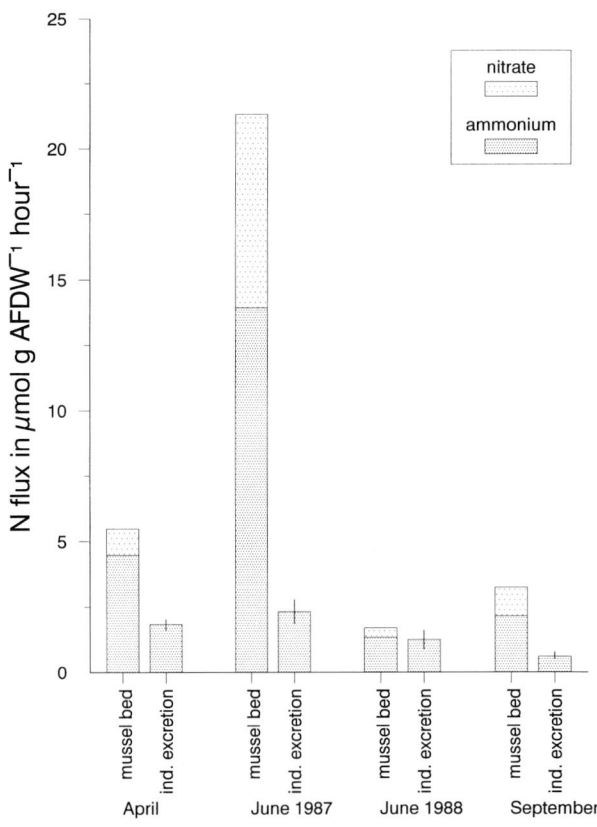

Fig. 4. The *in situ* flux of dissolved inorganic nitrogen from the mussel bed, recalculated to a rate per unit body weight, and the individual NH_4^+ excretion rate per unit body weight. The error bars indicate the 95% confidence limits.

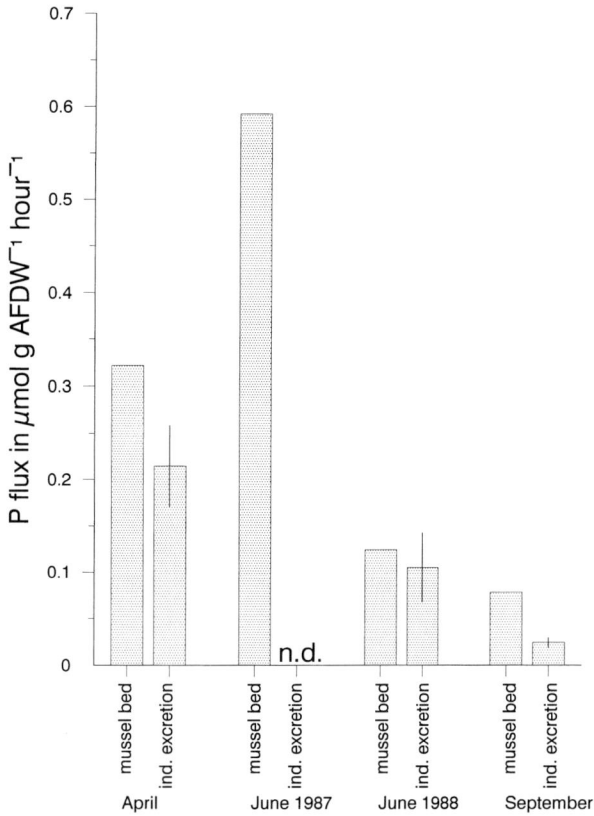

Fig. 5. The *in situ* flux of PO_4^{3-} from the mussel bed, recalculated to a rate per unit body weight, and the individual PO_4^{3-} excretion rate per unit body weight. The error bars indicate the 95% confidence limits. n.d. = not determined.

bed an estimate was made of the net budgets of nitrogen and phosphorus. Results were averaged for April, June 1987, June 1988 and September. In Table 2 the resulting nitrogen budget of the mussel bed is shown. The highest uptake of PON and the highest release of DIN was observed in April and in June 1987. PON uptake and DIN release in September were not significantly different from zero (*t*-test, $p > 0.05$). The budgets

Table 2. Uptake of particulate organic nitrogen (PON) and release of NH_4^+ and dissolved inorganic nitrogen (DIN; N $O_2^- + NO_3^- + NH_4^+$) (mean ± s.e.) by an intertidal mussel bed. A *t*-test was used to test whether fluxes were significantly different from zero: * $p < 0.050$; ** $p < 0.010$; *** $p < 0.001$; ns not significant.

Month	PON uptake		NH_4^+ release		DIN release	
	mmol m^{-2} h^{-1}	*n*	mmol m^{-2} h^{-1}	*n*	mmol m^{-2} h^{-1}	*n*
April	12.78 ± 1.68***	33	8.96 ± 0.55***	62	10.92 ± 0.72***	62
June 1987	21.68 ± 6.35**	17	13.54 ± 3.55***	17	20.72 ± 4.21***	17
June 1988	3.27 ± 0.89***	75	2.68 ± 0.92**	70	3.19 ± 1.60*	70
September	0.45 ± 1.00 ns	47	3.63 ± 1.68*	39	5.72 ± 3.52 ns	39

showed no significant differences between the up-take of PON and the release of DIN.

Table 3 shows the mean uptake of particulate organic phosphorus (POP) and the mean release of phosphate. The results showed that in most observations uptake of POP was higher than the release of phosphate, although the differences were not significant. The release of phosphate was higher than the POP uptake in June 1987. when POP uptake rates were low and non-significant (*t*-test, $p > 0.05$). Phosphate release in September was not significant (*t*-test, $p > 0.05$). On average, the phosphate release from the mus-sel bed was equal to 67% of the POP uptake by the mussel bed.

The uptake of biogenic silicon was calculated from the amount of diatoms taken up by the mus-sel bed. The estimated silicon budget (Table 4) showed that uptake of silicon was dominant in spring. Silicate release in June 1988 was low and not significant ((*t*-test, $p > 0.05$). Regeneration of silicate was higher than the estimated uptake of silicon by the mussel bed in autumn.

Discussion

Uptake of particulate matter

The *in situ* observations on exchange of material between mussel beds and water column in the Oosterschelde estuary showed that the mussel beds process high amounts of particulate mate-rial. Earlier published estimates of the amount of

Table 3. Uptake of particulate organic phosphorus (POP) and release of PO_4^{3-} (mean ± s.e.) by an intertidal mussel bed. A *t*-test was used to test whether fluxes were significantly dif-ferent from zero: * $p < 0.050$; ** $p < 0.010$; *** $p0.001$; [ns] not significant.

Month	POP uptake		PO_4^{3-} release	
	$mmol\ m^{-2}\ h^{-1}$	n	$mmol\ m^{-2}\ h^{-1}$	n
April	0.89 ± 0.25 ***	33	0.64 ± 0.09 ***	65
June 1987	0.21 ± 0.66 [ns]	17	0.58 ± 0.17 **	20
June 1988	0.56 ± 0.33 [ns]	27	0.24 ± 0.13 *	71
September	0.73 ± 0.34 *	20	0.16 ± 0.33 [ns]	42

Table 4. Uptake of biogenic silicon (POSi) and release of H_4SiO_4 (mean ± s.e.) by an intertidal mussel bed. A *t*-test was used to test whether fluxes were significantly different from zero: * $p < 0.050$; ** $p < 0.010$; *** $p0.001$; [ns] not significant.

Month	POSi uptake		H_4SiO_4 release	
	$mmol\ m^{-2}\ h^{-1}$	n	$mmol\ m^{-2}\ h^{-1}$	n
April	2.32 ± 0.23 ***	63	1.11 ± 0.27 ***	65
June 1987	3.63 ± 0.68 ***	17	3.20 ± 1.21 **	20
June 1988	0.56 ± 0.12 ***	88	0.80 ± 0.55 [ns]	74
September	0.15 ± 0.02 ***	47	3.27 ± 1.45 *	38

particulate C, N and P filtered and biodeposited by bivalves in estuaries (e.g Kuenzler, 1961; Sor-nin *et al.*, 1986; Kautsky & Evans, 1987; Dame & Dankers, 1988; Dame *et al.*, 1989; Asmus *et al.*, 1990; Dame *et al.*, 1991) show a large range, which is a consequence of differences in filter feeder biomass, concentrations of particulate ma-terial in the water column, resuspension due to turbulence etc. Our observations fall within the range of published fluxes. The net flux of particu-late matter to the mussel bed consisted of mate-rial with a low C:N ratio (9.4) compared to the composition of the seston. This low C:N ratio can be explained by resuspension and export of a carbon-rich component of the filtered material. This is supported by the fact that the C:N ratio of the seston was higher at the outflow than at the inflow of the tunnel. As a result of this process the net flux of particulate matter from the water col-umn to the mussel bed consists of material with a relatively high food quality.

Contribution of excretion to nutrient release

As the density of the mussels was high, it is ob-vious that excretion by the bivalves might con-tribute significantly to the nutrient release into the water column. In order to asses the contribution of excretion by mussels to the observed nutrient release we compared the *in situ* fluxes from the mussel bed to direct excretion rates by the ani-mals.

Estimates of the contribution of ammonium ex-cretion by bivalves to the sediment-water fluxes vary between 5–90% (Murhpy & Kremer, 1985; Kaspar *et al.*, 1985; Doering *et al.*, 1986; Dame *et al.*, 1989; Asmus *et al.*, 1990). In our experiments the excretion by the mussels accounted for 17–94% of the total ammonium flux from the mussel bed. The ammonium excretion rates in our experiments corresponded with published values (Bayne & Scullard, 1977; Bayne & Widdows, 1978; Hawkins *et al.*, 1985). We did not measure any nitrate excretion by the mussels although Boucher & Boucher-Rodoni (1988) have observed a small rate of nitrate excretion by the oyster *Crassostrea gigas*. Our results showed that in most observations direct excretion by the mussels could only partly explain the amount of inorganic nitrogen released by the mussel bed. More than 50% of the ammonium and all of the nitrate released by the mussel bed came from another source.

The excretion rates of phosphate by the mussels in our experiments were comparable to rates observed by Kautsky & Wallentinus (1980) (0.01–0.61 μmol g ADW^{-1} h^{-1}) and Asmus *et al.* (1990) (0–0.73 μmol g ADW^{-1} h^{-1}). The phosphate excretion accounted for a major part of the observed phosphate flux, and was equal to 31–85% of the total phosphate release by the mussel bed.

Mussels can excrete silicate (Asmus *et al.*, 1990). Silicate excretion is related to the feeding activity and depends on the amount of diatoms in the food ingested by the mussels (Asmus *et al.*, 1990). In our experiments silicate excretion rates were below the detection limit, and we conclude that the contribution of direct excretion by the mussels to the silicate flux from the mussel bed in our measurements was insignificant.

Our results show that the direct excretion of inorganic nutrients by the mussels contributed only partly to the release of DIN and phosphate by the mussel beds, while silicate excretion was unimportant. Other benthic fauna were a small (< 10%) fraction of the total biomass in the mussel bed, and we assume that the contribution of excretion by other organisms in the mussel bed was insignificant. Thus, we conclude that mineralization of organic matter deposited on the mus-

sel bed (biodeposition) must be an important source of inorganic nutrients in the sediment of a mussel bed.

Size and composition of dissolved inorganic nutrient flux

Our observations of nutrient release rates show agreement with published data on nutrient regeneration by bivalve communities. The fluxes of dissolved inorganic nutrients from beds of bivalve suspension feeders into the water column are significantly higher than sediment-water fluxes from other benthic habitats (Table 5). Total fluxes of ammonium, nitrate and phosphate, as well as fluxes per unit biomass, are highest from bivalve communities that attain high densities, and accumulate high amounts of biodeposition, like beds of *Mytilus edulis*, *Crassostrea virginica* and *Crassostrea gigas*. This reflects the dependence of the mineralization process in the sediment on the supply of organic matter by biodeposition. The amount of biodeposits produced is determined by the concentration of particulate matter in the water column, the clearance rate of the bivalves, and the density of the bivalves. The accumulation of biodeposition on the bivalve bed is dependent on local physical conditions (current speed, turbulence etc.). Infaunal species like the cockle *Cerastoderma edule* or the carpet clam *Ruditapes decussatus* do not accumulate large amounts of biodeposition on the sediment. Consequently the nutrient efflux from these communities is much smaller.

When compared to the biomass of the macrobenthos, relatively high silicate fluxes have been observed in experiments with *Cerastoderma edule*, *Mercenaria* and *Arenicola marina*. This may have been caused by higher rates of bioturbation (Asmus, 1986; Helder & Andersen, 1987) or higher temperatures (Helder *et al.*, 1983; Doering *et al.*, 1987) in these experiments.

Nitrogen regeneration

Ammonium was the main component of the released inorganic nitrogen, while nitrate comprised

Table 5. Minimum and maximum nutrient fluxes between bivalve beds and the water column. The period of observations and the biomass of the dominating bivalve species are included. For comparative purposes results obtained on *Arenicola marina* flats and on sediments without dense bivalve communities are included. Positive fluxes indicate a release into the water column.

NH_4^+ mmol m^{-2} h^{-1}	NO_3^- mmol m^{-2} h^{-1}	PO_4^{3-} mmol m^{-2} h^{-1}	H_4SiO_4 mmol m^{-2} h^{-1}	Period	Biomass g ADW m^{-2}	Species
−0.020/29.52	−5.46/6.37	−0.43/0.85	−1.66/8.16	Apr–Sep	890–2190	*Mytilus edulis*, Oosterschelde [1]
0.71/15.71	−5.71/−1.43*	−1.94/2.90	−0.36/2.14	Jun–Aug	820–940	*Mytilus edulis*, Wadden Sea (west)[2]
0.60/5.52	−0.06/1.07	−0.21/0.85	−0.49/2.66	Jul–Sep	1560	*Mytilus edulis*, Wadden Sea (east)[3]
4/5	−	−	−			*Mytilus edulis*, laboratory[4]
0.00/0.37	−0.02/0.01	−0.00/0.12	0.02/1.21	May–Nov	−**	*Mytilus edulis*, Mediterranean Sea[5]
−0.16/0.81	−0.20/0.34	0.06/0.22	0.04/0.98	June	20–100	*Cerastoderma edule*, Oosterschelde[6]
0.12/16.9	−1.2/0.10*	−	−	Jun–Aug	200	*Crassostrea virginica*, South Carolina[7]
1.00	0.00*	0.03	−	Year	200	*Crassostrea virginica*, South Carolina[8]
0.05/0.38	−0.11/0.05*	−	−	Year	100–200	*Crassostrea gigas*, France[9]
0.02/0.40	−0.15/0.03	−0.00/0.05	−	Year	20–140	*Mercenaria mercenaria*, California[10]
0.23	0.02*	−	0.10/0.70	Apr–Aug	16 ind m^{-2}	*Mercenaria mercenaria*, mesocosm[11]
0.02/0.04	−0.01/−0.00	0.00/0.01	0.00/0.03	Nov–Mar	90 ind m^{-2}	*Ruditapes decussatus*, Portugal[12]
−0.30/0.40	−0.25/0.00	−	−2.50/1.00	Feb–Nov	−	*Arenicola marina*, Wadden Sea (east)[13]
−0.13/1.02	−1.15/1.10	−0.02/0.05	−0.46/0.50	−	−	Other estuarine/marine sediments[14]

[1] = this study; [2] = Dame & Dankers, 1988, Dame *et al.*, 1991; [3] = Asmus *et al.*, 1990, Asmus & Asmus, 1991; [4] = Nixon *et al.*, 1976; [5] = Baudinet *et al.*, 1990; [6] = Prins & Smaal, in prep.; [7] = Dame *et al.*, 1984, 1985; [8] = Dame *et al.*, 1989; [9] = Boucher & Boucher-Rodoni, 1988; [10] = Murphy & Kremer, 1985; [11] = Doering *et al.*, 1987; [12] = Falcao & Vale, 1990; [13] = Asmus, 1986; [14] = reviews in Nixon (1981), Lerat *et al.* (1990). * NO_2^- + NO_3^-. ** Sediment under rope culture.

ca 20% of the total DIN flux. In general, nitrite fluxes were negligible. Only in one series of measurements a significant release of nitrite was observed, which accounted for 8% of the DIN flux (Prins & Smaal, 1990). Our results are in accordance with most observations on benthic nitrogen regeneration that show ammonium to be the main constituent of the DIN flux (Nixon, 1981; Nixon & Pilson, 1983).

Nitrate was predominantly released from the mussel bed in our experiments, which is unlike most observations of sediment-water fluxes, where nitrate fluxes between the water column and bivalve communities are usually small and erratic, and are often directed into the sediment (e.g. Murphy & Kremer, 1985; Doering *et al.*, 1987; Boucher & Boucher-Rodoni, 1988; Dame & Dankers, 1988; Dame *et al.*, 1989; Asmus *et al.*, 1990). A large uptake of nitrate from the water was observed on mussel beds by Dame & Dankers (1988), which was ascribed to uptake by microphytobenthos. Our results agree with the significant release of nitrate from the sediment observed in spring and summer in oyster cultivation areas (Feuillet-Girard *et al.*, 1988; Lerat *et al.*, 1990), and the increased nitrification observed in the presence of oysters (Boucher & Boucher-Rodoni, 1988; Boucher-Rodoni & Boucher, 1990). The observed nitrate release (Boucher & Boucher-Rodoni, 1988; Feuillet-Girard *et al.*, 1988; Lerat *et al.*, 1990; Boucher-Rodoni & Boucher, 1990;) seems to indicate that, in addition to ammonification, nitrification is an important process at the sediment-water interface of bivalve beds. This may be related to the high nitrification potential of bivalve faecal pellets (Henriksen *et al.*, 1984), and an increased nitrification due to resuspension of biodeposition (Owens, 1986).

Many publications on benthic nutrient regeneration mention sediment-water fluxes with anomalously low N:P ratios (< 10). The main cause for this phenomenon is assumed to be a loss of nitrogen from the sediment due to denitrification (Boynton *et al.*, 1980; Nixon, 1981; Seitzinger, 1988). Denitrification has been observed in areas with rope cultures of mussels (Kaspar *et al.*, 1985; Baudinet *et al.*, 1990). Denitrification can also occur in anaerobic microniches within faecal pellets (Jørgensen, 1977; Sayama & Kurihara, 1983), but a rapid breakdown of the faecal pellets, as was observed by Oenema (1988), would prevent the development of anoxic microsites and inhibit denitrification within the pellets. In many sediments nitrification in the aerobic layer and denitrification in the anaerobic sediment layer are tightly coupled processes (Jenkins & Kemp, 1984; Kemp *et al.*, 1990). As was hypothesized by De Vries & Hopstaken (1984), the presence of benthic epifauna may cause a spatial separation of nitrification and denitrification. A mussel bed is characterized by an impoverished infauna, due to oxygen deficiency and H_2S production (Asmus, 1987). This may result in a reduced bioturbation and hence a decreased transport of nitrate to the deeper sediment layers. Experimental results with oysters showing high rates of nitrification without immediate denitrification (Boucher & Boucher-Rodoni, 1988) support this hypothesis. Our observations show fluxes with N:P ratios close to the Redfield

ratio, and our estimates of the nitrogen budget of the mussel bed (Table 2) indicate that little nitrogen was retained by the mussel bed. From this we conclude that denitrification in the sediment of the mussel bed must have been a minor process compared to the total fluxes of nitrogen between the water column and the mussel bed.

The highest release of DIN was observed in the night tidal cycle in June 1987. As was shown earlier nutrient fluxes during the day period were significantly lower, probably as a consequence of uptake by micro-algae (Prins & Smaal, 1990). The high nitrogen release coincided with high phytoplankton concentrations, and hence a high supply of organic matter to the mussel bed. The initial degradation of mussel biodeposits concerns the mineralization of the labile fraction, and this seems to occur on a very short time scale (3–10 days) (Stuart *et al.*, 1984; Grenz *et al.*, 1990). Grenz *et al.* (1990) make the suggestion that intestinal bacteria, deposited with the faeces, are responsible for the rapid initial phase of mineralization. The fast initial degradation of biodeposits may explain the high release of DIN in a period (June 1987) with a high supply of organic matter.

Phosphorus release

The highest phosphate release rates have been observed on mussel beds in the western Dutch Wadden Sea (Dame & Dankers, 1988; Dame *et al.*, 1991). Phosphate fluxes from mussel beds in the German Wadden Sea (Asmus *et al.*, 1990) and the Oosterschelde (this study) were lower. The difference is probably caused by the relatively high discharge of phosphorus into the western Dutch Wadden Sea (Van der Veer *et al.*, 1989; Van Raaphorst & Van der Veer, 1990), compared to the German Wadden Sea (Hickel, 1989) and the Oosterschelde estuary (Wetsteyn & Bakker, 1991).

Whereas the mussel bed seemed very efficient in recycling N, the estimated P budget suggests that P may partly be retained (Table 3). Dame *et al.* (1989) also observed a retention of P by an oyster reef, and estimated that on an annual basis only 11% of the particulate P flux was recycled as phosphate. Our results showed a higher recycling of P (66%), but relative to N a larger amount of P is stored in the mussel bed. Our results from June 1987 deviate from this pattern, with a low uptake of POP and hence a higher phosphate release than POP uptake. The observations in June 1987 were done during a phytoplankton bloom (13.1 μg chlorophyll a l^{-1}) and a high uptake of PON and chlorophyll a (Prins & Smaal, 1990) was observed. The low uptake of POP does not fit with this, and may have been due to analytical or sampling errors. Often, phosphate fluxes from the sediment seem unaffected by the presence of macrofauna or are even lower than macrofaunal phosphate excretion (Nixon et al., 1980; Murphy & Kremer, 1985; Doering et al., 1987), which is probably due to adsorption of phosphate on sediment particles under aerobic conditions (Balzer et al., 1983). Thus, the retention of phosphorus on the mussel bed in our experiments may have been caused by adsorption of phosphate on sediment particles in the aerobic sediment layer (Balzer et al., 1983) and on resuspended biodeposition (Oenema, 1988).

Silicate release

Few data are available on the release of silicate from sediments. The highest release of silicate has been observed on mussel beds (Asmus et al., 1990; Dame et al., 1991; this study). Compared to other estuarine sediments *Arenicola marina* flats and sediments under mussel rope cultures show relatively high release rates of silicate, too (Asmus, 1986; Baudinet et al., 1990), probably as a consequence of bioturbation and ventilation of burrows (Aller, 1979; Asmus, 1986; Helder & Andersen, 1987).

The uptake of silicon by the mussel bed was estimated from the observed chlorophyll fluxes and from observations on the phytoplankton composition in the Oosterschelde estuary. It is clear that this conversion adds considerably to the uncertainty in the estimated particulate silicon flux. In addition, our estimates of silicon uptake by the mussel bed were probably an underestimate as we have only estimated the particulate silicon supply from diatoms, and did not take into account possible other sources, e.g. clay particles.

The silicate release rates were lower than the estimated uptake of biogenic silicon in April, and in June 1987 when a high phytoplankton concentration caused a high uptake of biogenic silicon (Table 4). Our results showed a silicon accumulation in the mussel bed during spring, probably connected to the spring phytoplankton bloom (Wetsteyn & Bakker, 1991). High silicate release was observed in the night tidal cycle of June 1987. The lower silicate fluxes during the day observations were probably due to uptake of silicate by diatoms during the day (Prins & Smaal, 1990). A rapid remineralization of silicate has been suggested by several authors (Callender & Hammond, 1982; Baudinet et al., 1990) and may explain the high silicate effluxes in the June 1987 measurement. The high silicate release in fall probably reflected the increased rate of dissolution of silicate at higher water temperatures (Helder et al., 1983).

Nutrient budget of the mussel bed

An estimate was made of the amount of nitrogen and phosphorus that the mussels need to maintain a positive growth rate. In the calculations we used the biomasses of the mussel beds in our experiments, and we assumed a constant C:N:P ratio for mussel flesh (204:5.5:1; Vonck, unpublished report). Growth rates were used that are typical for mussels of this size in the central part of the Oosterschelde estuary (0.3–1.0% day^{-1}; Smaal & Van Stralen, 1990; Van Stralen, unpublished report). Although this calculation can only give an approximation of the assimilation of N and P by the mussels, the estimates (Table 6) indicate that the amount of N and P that should have been assimilated by the mussels was relatively small compared to the observed PON and POP fluxes. This estimate makes it probable that a substantial fraction of the particulate N and P flux to the mussel bed was not assimilated by the

Table 6. Estimated assimilation of PON and POP by the mussels, necessary to enable observed growth rates.

Month	N uptake by mussels mmol m^{-2}h^{-1}	P uptake by mussels mmol m^{-2}h^{-1}
April	3.43	0.093
June 1987	1.32	0.036
June 1988	2.14	0.058
September	1.53	0.042

mussels and was deposited on the sediment as faeces or pseudofaeces. This corresponded with earlier estimates that show that 85–93% of the filtered nutrients are biodeposited by bivalves (Kuenzler, 1961; Sornin *et al.*, 1986; Feuillet-Girard *et al.*, 1988). Silicon is not assimilated by the animals, although a small fraction may be excreted by the mussels after digestion of diatoms (Asmus *et al.*, 1990). Consequently, practically all of the biogenic silicon filtered will be biodeposited. Due to mineralization of the biodeposition most of the N and Si is recycled, whereas P is partly retained by the mussel bed, probably through sediment adsorption. Hence, it can be concluded that the mussels are an important agent in the nutrient cycling in an estuary, as they transform particulate organic nutrients into inorganic nutrients which are recycled to the pelagic system, while only a small fraction of the nutrients are stored in biomass or biodeposition.

Significance on system level

The density of mussels in the Oosterschelde estuary is high, and the grazing pressure probably keeps phytoplankton biomass low (Smaal *et al.*, 1986; Herman & Scholten, 1990). Nevertheless, the mussel population not only acts as a sink for nutrients. Storage of nutrients in biomass or biodeposition seems relatively small. The main effect of the mussel population is to increase the mineralization rate of nutrients that were stored in phytoplankton biomass. Nutrient turnover rates as a consequence of regeneration on mussel beds are much higher than the rates of water re-

newal in the central part of the Oosterschelde, and it was proposed earlier that the regeneration of DIN and Si by mussel beds may be an important source of nutrients for the phytoplankton in summer (Prins & Smaal, 1990; Dame *et al.*, 1991), when primary production is nutrient-limited (Wetsteyn & Bakker, 1991).

In order to establish the role of the mussels in the cycling of nutrients in the Oosterschelde estuary, the exchange of material between the water column and the mussels should be compared with the fluxes between the other components in the ecosystem. The nitrogen regeneration on the mussel beds was estimated from the observed *in situ* DIN release (this study) and total mussel biomass in the area (Van Stralen & Dijkema, 1994). These rates were compared to estimates of the total nitrogen mineralization (pelagic + benthic) calculated with the model SMOES (Simulation Model Oosterschelde EcoSystem). This model describes the carbon and nutrient fluxes between the main functional groups in four spatial compartments of the Oosterschelde estuary and was calibrated with an extensive data set of field observations (Klepper, 1989; Scholten & Van der Tol, 1994). An uncertainty analysis with the model eventually gives a range of values for the model variables. A comparison was made for the years 1988–1989, for the central part of the Oosterschelde area. The results (Fig. 6) show that the estimated nitrogen regeneration by the mussels is equal to about 50% of the median value of the model result for total nitrogen mineralization. Of course, the error in these estimates is quite large as a consequence of the variability in the *in situ* observations and the uncertainty in the estimates of mussel biomass. Moreover, the range of values for the nitrogen mineralization calculated by the ecosystem model is large. Nevertheless, this comparison supports the hypothesis that the mussel population plays a major role in the recycling of nitrogen in the central part of the Oosterschelde ecosystem.

The suggestion has been made that, due to the high regeneration of nutrients by mussel populations, mussels may even increase eutrophication (Baudinet *et al.*, 1990). As was shown by Herman

426

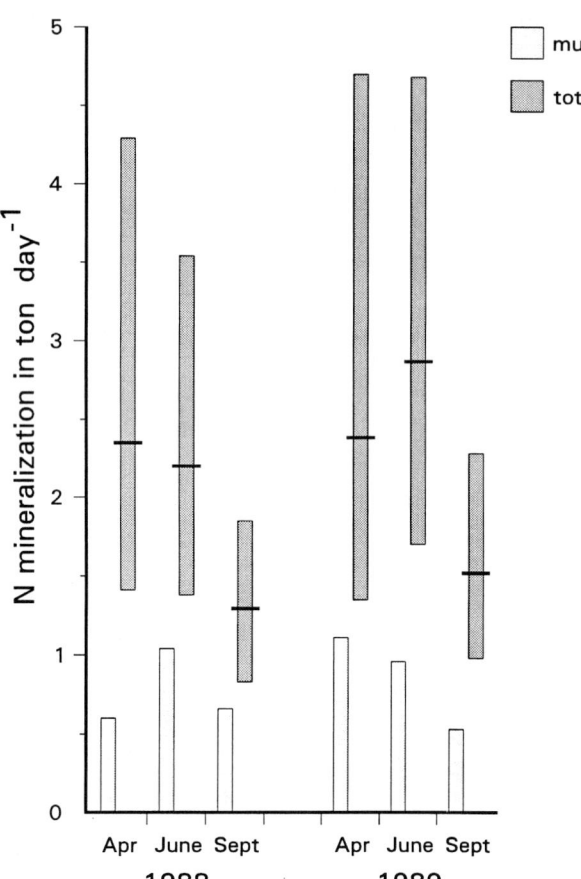

□ mussels
▨ total

Fig. 6. The amount of nitrogen mineralized by mussel beds in the central part of the Oosterschelde estuary, compared to estimates of total nitrogen mineralization (benthic + pelagic) from the Oosterschelde ecosystem model SMOES. The bar represents the 10–90% range of model estimates, the line shows the median.

& Scholten (1990) however, an effective 'top-down' control by grazing can keep phytoplankton biomass low, even when nutrient loadings to an ecosystem are high. The presence of high amounts of filter feeders merely leads to a higher turnover of phytoplankton (Sterner, 1986; Doering *et al.*, 1986; Doering, 1989; Asmus & Asmus, 1990). The control on phytoplankton biomass stabilizes the ecosystem (Herman & Scholten, 1990), as long as the primary producers do not escape filter feeder control by shifts to species that are not or less efficiently filtered, like macrophytes, small cells (Riemann *et al.*, 1988) or *Phaeocystis pouchetii* (Wolters, personal communication).

Acknowledgements

We would like to thank A. Pouwer for technical assistance, B. Boer, K. de Dreu, H. Hiemstra, C. Joosse, W. Nijsse, J. Siereveld, R. Vonck and the crews of the M. S. Agger and the M. S. Hontsloo for assistance during the field work. The chemical analyses were carried out by the Tidal Waters Division, Middelburg; We are much obliged to N. Dankers, K. Koelemaij, A. Meijboom and J. Zegers of the Institute of Forestry and Nature Management for cooperation and the supply of materials. The comments of R. F. Dame, P. H. Nienhuis and W. J. Wolff and two anonymous referees greatly improved this paper. Communication No. 703 of NIOO-CEMO, Yerseke, The Netherlands.

References

Aller, R. C., 1979. Relationships of tube-dwelling benthos with sediment and overlying water chemistry. In: K. R. Tenore and B. C. Coull (eds), Marine benthic dynamics. Univ. South Carolina Press, Columbia: 285–308.

Asmus, H., 1987. Secondary production of an intertidal mussel bed community related to its storage and turnover compartments. Mar. Ecol. Prog. Ser. 39: 251–266.

Asmus, H., R. Asmus & K. Reise, 1990. Exchange processes in an intertidal mussel bed: a Sylt-flume study in the Wadden Sea. Ber. Biol. Anst. Helgoland 6: 1–79.

Asmus, R., 1986. Nutrient flux in short-term enclosures of intertidal sand communities. Ophelia 26: 1–18.

Asmus, R. M. & H. Asmus, 1991. Mussel beds: limiting or promoting phytoplankton? J. exp. mar. Biol. Ecol. 148: 215–232.

Bahr, L. M., 1976. Energetic aspects of the intertidal oyster reef community at Sapelo Island, Georgia (USA). Ecology 57: 121–131.

Balzer, W., K. Grasshoff, P. Dieckmann, H. Haardt & U. Petersohn, 1983. Redox-turnover at the sediment/water interface studied in a large bell jar system. Oceanol. Acta 6: 337–344.

Baudinet, D., E. Alliot, B. Berland, C. Grenz, M.-R. Plante-Cuny, R. Plante & C. Salen-Picard, 1990. Incidence of mussel culture on biogeochemical fluxes at the sediment-water interface. Hydrobiologia 207: 187–196.

Bayne, B. L. & C. Scullard, 1977. Rates of nitrogen excretion by species of Mytilus (Bivalvia:Mollusca). J. mar. biol. Ass. U.K. 57: 355–369.

Bayne, B. L. & J. Widdows, 1978. The physiological ecology of two populations of *Mytilus edulis* L., Oecologia 37: 137–162.

Bayne, B. L., J. Widdows & R. J. Thompson, 1976. Physiology II. In: B. L. Bayne (ed.), Marine Mussels, their ecology and physiology. Cambridge University Press, Cambridge: 207–260.

Boucher, G. & R. Boucher-Rodoni, 1988. In situ measurement of respiratory metabolism and nitrogen fluxes at the interface of oyster beds. Mar. Ecol. Prog. Ser. 44: 229–238.

Boucher-Rodoni, R. & G. Boucher, 1990. In situ study of the effect of oyster biomass on benthic metabolic exchange rates. Hydrobiologia 206: 115–123.

Boynton, W. R., W. M. Kemp & C. G. Osborne, 1980. Nutrient fluxes across the sediment-water interface in the turbid zone of a coastal plain estuary. In: V. S. Kennedy (ed.), Estuarine perspectives. Academic Press, New York: 93–109.

Callender, E. & D. E. Hammond, 1982. Nutrient exchange across the sediment-water interface in the Potomac river estuary. Estuar. Coast. Shelf Sci. 15: 395–413.

Carlson, D. J., D. W. Townsend, A. L. Hilyard & J. F. Eaton, 1984. Effect of an intertidal mudflat on plankton of the overlying water column. Can. J. Fish. aquat. Sci. 41: 1523–1528.

Dahlbäck, B. & L. H. Gunnarson, 1981. Sedimentation and sulfate reduction under a mussel culture. Mar. Biol. 63: 269–275.

Dame, R. F. & N. Dankers, 1988. Uptake and release of materials by a Wadden Sea mussel bed. J. exp. mar Biol. Ecol. 118: 207–216.

Dame, R. F., J. D. Spurrier & T. G. Wolaver, 1989. Carbon, nitrogen and phosphorus processing by an oyster reef. Mar. Ecol. Prog. Ser. 54: 249–256.

Dame, R. F., T. G. Wolaver & S. M. Libes, 1985. The summer uptake and release of nitrogen by an intertidal oyster reef. Neth. J. Sea Res. 19: 265–268.

Dame, R. F., R. G. Zingmark & E. Haskin, 1984. Oyster reefs as processors of estuarine materials. J. exp. mar. Biol. Ecol. 83: 239–247.

Dame, R. F., R. Zingmark, H. Stevenson & D. Nelson, 1980. Filter feeding coupling between the estuarine water column and benthic subsystems. In: V. S. Kennedy. (ed.), Estuarine perspectives. Academic Press, New York: 521–526.

Dame, R. F., N. Dankers, T. Prins, H. Jongsma & A. Smaal, 1991. The influence of mussel beds on nutrients in the Western Wadden Sea and Eastern Scheldt estuaries. Estuaries 14: 130–138.

Day, J. W., C. A. S. Hall, W. M. Kemp & A. Yáñez-Arancibia, 1989. Estuarine ecology. Wiley-Interscience, New York, 558 pp.

De Vries, I. & C. F. Hopstaken, 1984. Nutrient cycling and ecosystem behaviour in a salt-water lake. Neth. J. Sea Res. 18: 221–245.

Doering, P. H., 1989. On the contribution of the benthos to pelagic production. J. mar. Res. 47: 371–383.

Doering, P. H., C. A. Oviatt & J. R. Kelly, 1986. The effects of the filter-feeding clam Mercenaria mercenaria on carbon cycling in experimental marine mesocosms. J. Mar. Res. 44: 839–861.

Doering, P. H., J. R. Kelly, C. A. Oviatt & T. Sowers, 1987. Effect of the hard clam Mercenaria mercenaria on benthic fluxes of inorganic nutrients and gases. Mar. Biol. 94: 377–383.

Falcao, M. & C. Vale, 1990. Study of the Ria Formosa ecosystem: benthic nutrient remineralization and tidal variability of nutrients in the water. Hydrobiologia 207: 137–146.

Feuillet-Girard, M., M. Héral, J. M. Sornin, J. M. Deslous-Paoli, J. M. Robert, F. Mornet & D. Razet, 1988. Eléments azotés de la colonne d'eau et de l'interface eau-sédiment du bassin de Marennes-Oléron: influence des cultures d'huîtres. Aquat. Living Resourc. 1: 251–265.

Fréchette, M. & E. Bourget, 1985. Food-limited growth of Mytilus edulis L. in relation to the benthic boundary layer. Can. J. Fish aquat. Sci. 42: 1166–1170.

Grenz, C., M.-N. Hermin, D. Baudinet & Daumas, 1990. In situ biochemical and bacterial variation of sediments enriched with mussel biodeposits. Hydrobiologia 207: 153–160.

Haven, D. S. & R. Morales-Alamo, 1972. Biodeposition as a factor in sedimentation of fine suspended solids in estuaries. Geol. Soc. Am. Mem. 133: 121–130.

Hawkins, A. J. S., P. N. Salkeld, B. L. Bayne, E. Gnaiger & D. M. Lowe, 1985. Feeding and resource allocation in the mussel Mytilus edulis: evidence for time-averaged optimization. Mar. Ecol. Prog. Ser. 20: 273–287.

Helder, W. & F. Ø. Andersen, 1987. An experimental approach to quantify biologically mediated dissolved silica transport at the sediment-water interface. Mar. Ecol. Prog. Ser. 39: 305–311.

Helder, W., R. T. P. De Vries & M. M. Rutgers van der Loeff, 1983. Behavior of nitrogen nutrients and dissolved silica in the Ems-Dollard estuary. Can. J. Fish aquat. Sci. 40 (Suppl. 1): 188–200.

Henriksen, K., A. Jensen & M. B. Rasmussen, 1984. Aspects of nitrogen and phosphorus mineralization and recycling in the northern part of the Danish Wadden Sea. Neth. Inst. Sea Res. Publ. Ser. 10: 51–69.

Herman, P. M. J. & H. Scholten, 1990. Can suspension-feeders stabilise estuarine ecosystems? In M. Barnes & R. N. Gibson (eds.), Trophic relationships in the marine environment. Aberdeen University Press, Aberdeen: 104–116.

Hickel, W., 1989. Inorganic micronutrients and the eutrophication in the Wadden Sea of Sylt (German Bight, North Sea). Proc. 21st EMBS, Polish Acad. Sciences, Inst. Limnology, Gdansk: 309–318.

Jenkins, M. C. & W. M. Kemp, 1984. The coupling of nitrification and denitrification in two estuarine sediments. Limnol. Oceanogr. 29: 609–619.

Jørgensen, B. B., 1977. Bacterial sulfate reduction within reduced microniches of oxidized marine sediments. Mar. Biol. 41: 7–17.

Jørgensen, C. B. 1990. Bivalve filter feeding: hydrodynamics,

428

bioenergetics, physiology and ecology. Olsen & Olsen, Fredensborg, Denmark. 140 pp.

Kaspar, H. F., P. A. Gillespie, I. C. Boyer & A. L. McKenzie, 1985. Effects of mussel aquaculture on the nitrogen cycle and benthic communities in Kenepuru Sound, Marlborough Sounds, New Zealand. Mar. Biol. 85: 127–136.

Kautscky, N. & S. Evans,; 1987. Role of biodeposition by *Mytilus edulis* in the circulation of matter and nutrients in a Baltic coastal ecosystem. Mar. Ecol. Prog. Ser. 38: 201–212.

Kautsky, N. & I. Wallentinus, 1980. Nutrient release from a Baltic *Mytilus*-red algal community and its role in benthic and pelagic production. Ophelia, suppl. 1: 17–30.

Kemp, W. M., P. Sampou, J. Caffrey, M. Mayer, K. Henriksen & W. R. Boynton, 1990. Ammonium recycling versus denitrification in Chesapeake Bay sediments. Limnol. Oceanogr. 35: 1545–1563.

Kuenzler, E. J., 1961. Phosphorus budget of a mussel population. Limnol. Oceanogr. 6: 400–415.

Lerat, Y., P. Laserre & P. Le Corre, 1990. Seasonal change in pore water concentrations of nutrients and their diffusive fluxes at the sediment-water interface. J. exp. mar. Biol. Ecol. 135: 135–160.

Murphy, R. C. & J. N. Kremer, 1985. Bivalve contribution to benthic metabolism in a California lagoon. Estuaries 8: 330–341.

Nichols, F. H., 1985. Increased benthic grazing: an alternative explanation for low phytoplankton biomass in northern San Francisco Bay during the 1976–1977 drought. Estuar. coast Shelf Sci. 21: 379–388.

Nixon, S. W., 1981. Remineralization and nutrient cycling in coastal marine ecosystems. In: B. J. Neilson & L. E. Cronin (eds), Estuaries and nutrients. Humana Press, Clifton, New Jersey: 111–138.

Nixon, S. W., J. R. Kelly, B. N. Furnas, C. A. Oviatt & S. S. Hale, 1980. Phosporus regeneration and the metabolism of coastal marine bottom communities. In: K. R. Tenore and B. C. Coull (eds), Marine benthic dynamics. Univ. South Carolina Press, Columbia: 219–242.

Nixon, S. W., C. A. Oviatt, J. Garber & V. Lee, 1976. Diel metabolism and nutrient dynamics in a salt marsh embayment. Ecology 57: 740–750.

Nixon, S. W. & M. E. Q. Pilson, 1983. Nitrogen in estuarine and coastal marine ecosystems. In: E. J. Carpenter & D. G. Capone (eds), Nitrogen in the marine environment. Academic Press, New York: 565–648.

Oenema, O., 1988. Early diagenesis in recent fine-grained sediments in the Eastern Scheldt. Ph. D. thesis, University of Utrecht, 222 pp.

Owens, N. J. P., 1986. Estuarine nitrification: a naturally occurring fluidized bed reaction? Estuar. coast. Shelf Sci. 22: 31–44.

Prins, T. C. & A. C. Smaal, 1990. Benthic-pelagic coupling: the release of inorganic nutrients by an intertidal bed of *Mytilus edulis*. In M. Barnes & R. N. Gibson (eds), Trophic

relationships in the marine environment. Aberdeen University Press, Aberdeen: 89–103.

Ricker, W. E., 1973. Linear regressions in fishery research. J. Fish. Res. Bd Can. 30: 409–434.

Redfield, A. C., B. H. Ketchum & F. A. Richards, 1963. The influence of organisms on the composition of sea-water. In N. M. Hill (ed.), The Sea, Vol. 2. Wiley-Interscience, New York: 26–77.

Riemann, B., T. G. Nielsen, S. J. Horsted, P. K. Bjørnsen & J. Pock-Steen, 1988. Regulation of phytoplankton biomass in estuarine enclosures. Mar. Ecol. Prog. Ser. 48: 205–215.

Sayama, M. & Y. J. Kurihari, 1983. Relationships between burrowing activity of the polychaetous annelid, *Neanthes japonica* (Izuka) and nitrification-denitrification processes in the sediments. J. exp. mar. Biol. Ecol 72: 233–241.

Scholten, H. & M. W. van der Tol, 1994. SMOES: a simulation model for the Oosterschelde ecosystem. Part II: Calibration and validation. Hydrobiologia 282/283: 453–474.

Seitzinger, S. P., 1988. Denitrification in freshwater and coastal marine ecosystems: Ecological and geochemical significance. Limnol. Oceanogr. 33: 702–724.

Smaal, A. C., M. Knoester, P. H. Nienhuis & P. M. Meire, 1991. Changes in the Oosterschelde ecosystem induced by the Delta works. In M. Elliott and J-P. Ducrotoy (eds), Estuaries and coasts: Spatial and temporal intercomparisons. Olsen & Olsen, Fredensborg, Denmark: 375–384.

Smaal, A. C. & M. R. Van Stralen, 1990. Average annual growth and condition of mussels as a function of food source. In D. S. McClusky, V. N. de Jonge and J. Pomfret (eds), North Sea-Estuaries Interactions. Hydrobiologia 195: 179–188.

Smaal, A. C., J. H. G. Verhagen, J. Coosen & H. A. Haas, 1986. Interactions between seston quantity and quality and benthic suspension feeders in the Oosterschelde, The Netherlands. Ophelia 26: 385–399.

Sornin, J. M., M. Feuillet, M. Héral & J. M. Deslous-Paoli, 1983. Effet des biodépots de l'huître *Crassostrea gigas* (Thunberg) sur l'accumulation de matières organiques dans les parcs du bassin de Marennes-Oléron, J. moll. Stud., suppl. 12A:: 185–197.

Sorning, J. M., M. Feuillet, M. Héral & J. C. Fardeau, 1986. Influence des cultures d'huîtres *Crassostrea gigas* sur le cycle du phosphore en zone intertidale: rôle de la biodéposition. Oceanol. Acta 9: 313–322.

Sterner, R. W., 1986. Herbivores' direct and indirect effects on algal populations. Science 231: 605–607.

Stuart, V., R. C. Newell & M. I. Lucas, 1984. Conversion of kelp debris and gaecal material from the mussel *Aulacomya ater* by marine microorganisms. Mar. Ecol. Prog. Ser. 7: 47–57.

Van der Veer, H. W., W. van Raaphorst & M. J. N. Bergman, 1989. Eutrophication of the Dutch Wadden Sea: external nutrient loadings of the Marsdiep and Vliestroom basin. Helgoländer Meeresunters. 43: 501–515.

Van Raaphorst, W. & H. W. van der Veer, 1990. The phosphorus budget of the Marsdiep tidal basin (Dutch Wadden

Sea) in the period 1950–1985: importance of the exchange with the North Sea. Hydrobiologia 195: 21–38.

Van Stralen, M. R. & R. D. Dijkema, 1994. Mussel culture in a changing environment: the effects of a coastal engineering project on mussel culture (*Mytilus edulis* L.) in the Oosterschelde estuary (SW Netherlands). Hydrobiologia 282/283: 359–379.

Verwey, J., 1952. On the ecology of distribution of cockle and mussel in the Dutch Waddensea, their role in sedimentation and the source of their food supply. Arch. Néerl. Zool. 10: 171–239.

Wetsteyn, L. P. M. J. & C. Bakker, 1991. Abiotic characteristics and phytoplankton primary production in relation to a large-scale coastal engineering project in the Oosterschelde (The Netherlands): a preliminary evaluation. In M. Elliott & J-P. Ducrotoy (eds), Estuaries and coasts: spatial and temporal intercomparisons. Olsen & Olsen, Fredensborg, Denmark: 365–373.

Wildish, D. J. & D. D. Kristmanson, 1984. Importance to mussels of the benthic boundary layer. Can. J. Fish. aquat. Sci. 41: 1618–1625.

Wolff, W. J., 1983. Estuarine benthos In: B. H. Ketchum (ed.), Ecosystems of the world, 26. Estuaries and enclosed seas. Elsevier, Amsterdam: 151–182.

Wright, R. T., R. B. Coffin, C. P. Ersing & D. Pearson, 1982. Field and laboratory measurements of bivalve filtration of natural marine bacterioplankton. Limnol. Oceanogr. 27: 91–98.

Zeitzschel, B., 1980. Sediment-water interactions in nutrient dynamics. In: K. R. Tenore and B. C. Coull (eds.), Marine benthic dynamics. Univ. South Carolina Press, Columbia: 195–218.

Hydrobiologia **282/283**: 431–436, 1994.
P. H. Nienhuis & A. C. Smaal (eds), The Oosterschelde Estuary.
© 1994 *Kluwer Academic Publishers. Printed in Belgium.*

Theme VI:

The analysis of a dynamic simulation model for the Oosterschelde ecosystem – a summary

Peter M. J. Herman
Netherlands Institute of Ecology, CEMO, Vierstraat 28, 4401 EA Yerseke, The Netherlands

Introduction

Modelling was an essential part of the scientific programme developed for monitoring the changes in the Oosterschelde ecosystem caused by the large engineering works. In view of the major changes expected, of the available management options and of the anticipated socio-economic problems, this modelling was focused on the major energy and matter flows in the system. The choice for the construction of a storm-surge barrier instead of a closed dam was motivated by the will to preserve the diversity of habitats and therefore of species assemblages in the system, and to preserve the economically valuable mussel cultivation and cockle fisheries.

Prediction of change in habitats and therefore in dominating species assemblages was mainly addressed in statistical models. For many intertidal habitats (where most changes were estimated to occur) there is a clear dependence between geomorphological change and occurrence of the habitat. Examples are the vegetation of salt marshes (see de Leeuw *et al.*, 1994), and the overwintering populations of wader birds dependent on intertidal mud flats for food. Where the dependence on geomorphology is the dominant force, 'black box' approaches may constitute valuable tools for prediction, even if the mechanisms for this dependence are unknown. For salt marsh vegetation, e.g., the relationship between inundation frequency and vegetation structure permits a fairly accurate prediction of the future

vegetation from predictions of the relative sea level change (de Leeuw *et al.*, 1994).

Several of the management options taken, however, influence the complex interactions by which the components of the aquatic ecosystem depend on each other. Even before the construction of the dams and barrier, internal recycling of nitrogen constituted the major source of this nutrient in the course of the season. 'New production', dependent on winter nutrient levels, was by far insufficient to explain the level of primary production in the system (Wetsteyn & Kromkamp, 1994). Recycling by heterotrophic activity and fast remineralization of detritus was an indispensable factor in this explanation. In turn, the short turnover times of the phytoplankton and the importance of the phytoplankton as food for the next trophic level made this level, at least in principle, very vulnerable to any changes in primary production and thus in nutrient fluxes. Fast turnover of the phytoplankton could e.g. be inferred from the high pressure exerted by the suspension feeders on the system. The system could therefore be characterized by high production rates, tight benthic-pelagic coupling and a high degree of biological control between components.

The management options would clearly interfere with these mechanisms of internal cycling and mutual (biological) control: external (freshwater) sources of nutrients would largely be cut off from the system; increased residence times would further increase the importance of internal cycling; reduced turbidity and sediment load in the water would favor primary production. All these changes were expected to have profound influences on the mussel and cockle stocks, and thus on the economic values in the Oosterschelde system.

Modelling is an essential tool for predicting the consequences of these management options on the structure and functioning of the ecosystem. The general objectives of carbon and nutrient modelling in ecosystems can be defined as the consistent use of available knowledge for prediction purposes. These general objectives are restricted, however, by the scale and the scope of the modelling exercise. As argued in detail by Klepper et al. (1994), a single model cannot cope with all the different scales on which processes in ecosystems take place. Spatially, these scales range from the μm scale (interactions between cells of individual microbial organisms) to the 10 km scale. Temporally, scales range from milliseconds to decades. The incorporation within a single model of all these different scales would yield insurmountable numerical problems, due to the strong non-linearity in the ecological processes. This problem is a good reason in itself for restricting the scale and scope of a model. The concept of the hierarchical structure of ecosystems also provides a theoretical basis for this approach: emergent properties of ecosystems on the system level are a consequence of the organization of lower level complexes (individuals, populations) into a system, and are not reducible to the properties of the lower level complexes themselves. It should suffice to represent essential characteristics of the lower level complexes (relevant on the scale of the larger system), in order to deduce the essential dynamics of the larger scale system.

Model formulation

In practice, models should be restricted to scales that are relevant to the problem posed: model scales are imposed by the model purpose. In the case of the Oosterschelde model SMOES, temporal scales are weeks to years; spatial scales are in the order of 10 km (Klepper et al., 1994). The scope of the model, in terms of the processes and variables incorporated into the model, is closely linked to the scale: processes which have a characteristic scale very different from the model scale will not be explicitly incorporated. The parametrization of primary production by phytoplankton in SMOES (see Klepper et al., 1994) is a good example of this choice. A submodel is constructed, based on known physiological responses of phytoplankton primary production. From this submodel, a parametrized equation describing phytoplankton primary production is derived: it yields column production from knowledge of variables known at the scale of the model: average water depth, light attenuation, daily averaged irradiation. In SMOES itself, physiological responses are not resolved. In general, the model formulations described by Klepper et al. (1994) have been carefully chosen to preserve as much as possible this homogeneity of scales.

Apart from the model scale, also the model purpose determines the scope of the model. For the Oosterschelde, the interference of the engineering works with the nutrient cycles and the shellfish fisheries were defined as the most crucial issues to be predicted by SMOES. The variables and processes were consequently chosen according to their quantitative importance for these aspects. The quantitative importance of fluxes and standing stocks to the overall nutrient and carbon flux was estimated from a budget modelling approach to the Oosterschelde. It was published elsewhere (Klepper & van de Kamer, 1987, 1988) but is relevant to the dynamic modelling in SMOES. In this study, available data on biomass of relevant groups of organisms and on fluxes between them were collected and checked for consistency. This 'inverse modelling' approach defines an assumed structure for the system, and checks whether the available data can consistently be brought into this structure. The procedure can be explained with a simple hypothetical example. If phytoplankton primary production were estimated as between 100 and 200 gC m^{-2} y^{-1}, and suspension feeder secondary production as between 350 and 500 gC m^{-2} y^{-1}, then any model assuming that the suspension feeder secondary production is entirely dependent on phytoplankton primary production is clearly inconsistent with the available data. If, on the other hand, secondary production would be measured as between 150 and 250 gC m^{-2} y^{-1}, one could conclude that the assumed structure is consistent with the measurements. In other words, the joint study of both fluxes imposes constraints on what are acceptable hypotheses on the ecosystem structure. At the same time, if the assumed structure is consistent, it imposes constraints on the measured fluxes which are tighter than the uncertainty intervals from field and lab studies. Actually, for the Oosterschelde food web structure, no inconsistencies were found. It was shown that the measured fluxes and biomasses were consistent with the hypothesized food web. The results of the budget studies were used as a basis for decisions on whether or not to include particular components or processes in the dynamical model. Moreover, application of this approach showed that the Oosterschelde is largely a self-sustained ecosystem. Total production and consumption of organic material in the system are balanced, implying that the state of the system is largely dependent on its internal biological interactions (Scholten et al., 1990).

Uncertainty analysis

In yearly averaged budget studies, mutual constraints can reduce some of the uncertainty in the description of the system. Adding time dynamics to the model structure can enhance the strength of the constraints imposed. In the hypothetical example given above, if all the primary production would occur in spring, and all suspension feeder growth in autumn, this growth could not be dependent on primary production, no matter what the yearly averages are. Therefore a model translating hypotheses concerning the time dynamics of the system produces a tighter set of mathematical constraints which should be consistent with the field data. The analysis of uncertainty in Klepper et al. (1994) and the calibration of the model in Scholten & van der Tol (1994) perform this comparison between model and data.

In general, inconsistency between a model run output and measured data in the field can result from a number of causes:

* Model formulations can be bad representations of reality. If massive mortality of mussels by extreme oxygen stress would be the dominant process in the dynamics of these animals, then a description where growth and mortality are only dependent on temperature and food would clearly not be able to describe the dynamics appropriately. Modelling, in other words, should be based on sufficient knowledge. Although this knowledge basis was enforced by embedding the present modelling exercise in a broad research project, basic process knowledge is still under development. Therefore the adequacy of the knowledge basis in ecological modelling is, and will remain for some time, a matter of debate. As argued by Klepper et al. (1994), it is difficult to devise a general formal procedure to ensure the adequacy of the model structure.
* Model descriptions use a number of parameters, which are assumed to be constant and of known value. For a number of reasons, these parameters are only approximately known. Measurements and experiments do not result in unique values for rate parameters, but in estimates with a statistical confidence interval. Measurements of the same parameter, performed in slightly different circumstances or with different methodology, can yield widely different results. Moreover, model rate parameters are often defined for a group of organisms (e.g. diatom phytoplankton), whereas measurements can only be performed on single species or on the specific assemblage present at the moment of the experiments. At best, therefore, a range of possible values can be defined for the parameters; this range may sometimes encompass more than one order of magnitude.
* Model inputs (e.g. freshwater discharge, light climate, effects of future management) are also

only approximately known, again resulting in model uncertainty.

The analysis of uncertainty performed in Klepper *at al.* (1994) mainly addresses the uncertainty resulting from parameter uncertainty. It is based on the following ideas. Not all parameters are equally important for the model outcome. Critical model outcome (as defined by the model objectives) can be mostly dependent on a limited subset of all parameters used; most emphasis should therefore be placed on a correct estimation of these most critical parameters. Further, not all parameters affect the model results in an independent way. In some cases, approximately the same model output will be generated by a high value of a first parameter and a low value of a second, or a low value of the first one and a high value of the second one. Parameters therefore form a limited number of sets of correlated groups, as far as their effect on model outcome is concerned. Identification of these groups permits the description of the model behavior in terms of a small number of parameter combinations. Identification of the most crucial parameter (where reduction of parameter uncertainty would reduce model uncertainty most) in each group, leads to an identification of the 'weakest points' in the model: the issues that would require most attention in field studies to improve the model reliability. In SMOES these weakest points turned out to be processes that were indeed not very well studied in the whole research project, demanding specific attention in further studies of the system.

Model calibration

The calibrations described in Scholten & van der Tol (1994) and Van der Tol & Scholten (1992) bring in field data in order to reduce the uncertainty in model outcome. The initial uncertainty in the model resulting from uncertainty in parameters, input and formulations can be substantially reduced by requiring the model to describe the set of field data. Certain model runs within the un-

certainty range are clearly much more inconsistent with the available data than others, and the aim is to restrict the model uncertainty range to those runs which maximize the consistency with the data. The calibration method used results in narrower parameter ranges than initially used: only that part of the initial range yielding 'acceptable' model behavior (in terms of consistency with the data) is retained.

Model validation

Scholten & van der Tol (1994) used the calibration results for the difficult problem of model validation. Broadly speaking, a model is valid if it accurately predicts system behavior under the scenarios for which it was constructed. The Oosterschelde studies offer a unique opportunity to develop concepts of model validation: the important impacts on the system were accompanied by intensive field and modelling studies both before any works were carried out, and for a considerable period after the completion of the works. It is therefore possible to actually compare model performance with a substantial data set. In fact, the Oosterschelde evolution in the eighties has been a systemscale experiment. Scholten & van der Tol (1994) have performed replicate calibrations (there is a degree of arbitrariness in the calibration results, since the methodology is based on Monte Carlo procedures) on the data set obtained before the completion of the works, and on the data set obtained after the completion. They have also made 'predictions' (in fact, these are rather hindcasts, since they were produced after the 'facts') of the situation after completion of the works, based on the knowledge available before this completion. For all sets of model runs, they have evaluated the adequacy of the model runs (are all field data falling within the model uncertainty) and the reliability of the model runs (how good is the average fit between model and data). An ideal model can maximize both adequacy and reliability: after calibration goodness of fit is maximal, while all field data will still fall within the uncertainty range. SMOES did not behave like this ideal model, and a certain trade-off between reliability and adequacy was observed: a well-calibrated model maximized reliability, at the cost of narrowing down the uncertainty range, with some observations now falling outside the range. It was important to observe, however, that this was true both for actual descriptions and for predictions. The conclusion therefore is that SMOES is less than ideal for the description of the dynamics of the system (although practically speaking its performance is quite acceptable), but it is as good (or bad) for prediction as for description. There was no dramatic loss of neither reliability nor adequacy when the model was used for prediction, compared to the level of these measures when used for description.

In addition to these overall measures describing the validation of the model, Van der Tol & Scholten (1992) give a more detailed analysis of the model performance in the pre- and post-barrier conditions. In this study they point out that for several important variables (e.g. phytoplankton biomass, grazing rate of suspension feeders on phytoplankton), a great deal of overlap could be observed between the model descriptions for the pre- and post-barrier periods. For others (e.g. zooplankton biomass), a shift in the state of the system occurred. In general, the state of the system did not change very much between the two periods. However, for phytoplankton biomass and grazing rate of suspension feeders on phytoplankton, considerable differences were found between the post-barrier calibration and the predictions based on pre-barrier parameters. Essential in this difference was a substantial shift in parameter values describing the affinity of phytoplankton for light and nutrients between the pre- and post-barrier calibrations. The authors advocate that this reflects a true, adaptive (as contrasted to functional) response in the system. The results of the model calibrations are consistent with field observations on the structure of the phytoplankton (Bakker *et al.*, 1990; 1994); a shift occurred from relatively large species with a high affinity for light and a relatively low affinity for nutrients, to smaller species with lower light

affinity and higher nutrient affinity. Most probably, this shift occurred as a response to the combined reduction in nutrient availability and increase of light availability caused by the engineering works. Other responses in the system, e.g. the relationship between suspension feeders and phytoplankton biomass are rather of the functional type, i.e. without changing the essential parameters of the system.

These observations shed a particular light on the predictions made (again using SMOES) on the relationship between suspension feeders and phytoplankton by Herman & Scholten (1990). Using the pre-barrier version of SMOES (post-barrier calibration was not yet available), these authors varied the biomass of the suspension feeders and the input of nutrients in the system, and investigated the influence on the phytoplankton biomass. They concluded that suspension feeders exert a strong stabilizing influence on the phytoplankton under changing nutrient inputs. Grazing control took over the role of nutrient limitation at high nutrient levels, while at low nutrient levels the rapid turnover of nutrients resulting from grazing maintained the primary production. These qualitative conclusions are consistent with the field results and the post-barrier calibration, but cannot be reproduced when the pre-barrier version of the model is used for post-barrier prediction. This, it turns out, may be the result of compensating errors in the approach of Herman & Scholten (1990). By taking pre-barrier rate parameters for the phytoplankton, and using pre-barrier light conditions, Herman & Scholten (1990) implicitly assumed a functional constancy for the phytoplankton dynamics. The analysis of Van der Tol & Scholten (1992) showed that indeed the production of the phytoplankton remained at a comparable level in the two conditions, but that this is also the result of adaptive changes in the phytoplankton structure, combined with an increased turnover of the phytoplankton under the post-barrier conditions. It is as yet uncertain, therefore, what the precise role of the suspension feeders in the relative functional stability of the system is. The prominent role attached to them by Herman & Scholten (1990) is probably an overemphasis. Adaptations in the phytoplankton have played, at least, an additional role.

Conclusions

With regard to the adequacy and usefulness of SMOES as a description and prediction tool for the Oosterschelde system, what can be concluded from these analyses? From the uncertainty analysis, performed for the pre-barrier conditions, Klepper *et al.* (1994) identify as major gaps in knowledge: microbiological processes, algal respiration, zooplankton food limitation and loss processes, carbon to chlorophyll ratio, transport of algae and detrital silicon, assimilation efficiency of cockles and gas exchange with the atmosphere. From the analysis of Scholten & van der Tol (1994) and Van der Tol & Scholten (1992) this list is overshadowed by the gaps in predictive power for adaptive responses in the ecosystem. These gaps are prominent weaknesses in the otherwise qualitatively good predictive capabilities of the model. The difference between the two priority lists is not surprising. The first one is concerned with what is needed to reduce the uncertainty in the model outcome, the second one with what is needed to extrapolate the model outcomes beyond the state of the system for which the model has been calibrated. If a model, similar to SMOES, were constructed for an in-depth analysis of the flows and the states in an essentially unchanging system, the first list would constitute the major priority list for research. In order to increase the predictive power of the model under changed forcing, however, incorporation of adaptive change in the system structure would constitute the utmost priority. Formulations for this adaptive change should rely on comparative studies between a number of similar habitats. It should be possible to identify the essential characters of the biota and their concomitant rate parameters, and relate them to the forcing functions to which they are submitted. Methodologically, the analysis of SMOES has shown that the explicit handling of uncertainty model and measurements, can greatly improve the insight in the model behavior. Moreover, it facilitates the identification of essential gaps in knowledge and of the possibilities for model improvement.

References

Bakker, C., P. M. J. Herman & M. Vink, 1990. Changes in seasonal succession of phytoplankton induced by the storm-surge barrier in the Oosterschelde (S.W. Netherlands). J. Plankton Res. 12: 947–972.

Bakker, C., P. M. J. Herman & M. Vink, 1994. A new trend in the development of the phytoplankton in the Oosterschelde (SW Netherlands) during and after the construction of a storm-surge barrier. Hydrobiologia 282/283 (Dev. Hydrobiol. 97): 79–100.

De Leeuw, J., L. P. Apon, P. M. J. Herman, W. de Munck & W. G. Beeftink, 1994. The response of salt-marsh vegetation to tidal reduction caused by the Oosterschelde storm-surge barrier. Hydrobiologia 282/283 (Dev. Hydrobiol. 97): 335–353.

Herman, P. M. J. & H. Scholten, 1990. Can suspension-feeders stabilise estuarine ecosystems? In: M. Barnes & R. N. Gibson (eds), Trophic Relationships in the Marine Environment, Proc. 24th Europ. Mar. Biol. Symp., Aberdeen Univ. Press: 104–116.

Klepper, O. & J. P. G. Van de Kamer, 1987. The use of mass balances to test and improve the estimates of carbon fluxes in an ecosystem. Math. Biosc. 85: 37–49.

Klepper, O. & J. P. G. Van de Kamer, 1988. A definition of the consistency of the carbon budget of an ecosystem and its application to the Oosterschelde estuary, S.W. Netherlands. Ecol. Modelling 42: 217–232.

Klepper, O., M. W. M. van der Tol, H. Scholten & P. M. J. Herman, 1994. SMOES: a simulation model for the Oosterschelde ecosystem. Part I: Description and uncertainty analysis. Hydrobiologia 282/283 (Dev. Hydrobiol. 97): 437–451.

Scholten, H., O. Klepper, P. H. Nienhuis & M. Knoester, 1990. Oosterschelde estuary (S.W. Netherlands): a self-sustaining ecosystem? Hydrobiologia 195 (Dev. Hydrobiol. 55): 201–215.

Scholten, H. & M. W. M. van der Tol, 1994. SMOES: a simulation model for the Oosterschelde ecosystem. Part II: Calibration and validation. Hydrobiologia 282/283 (Dev. Hydrobiol. 97): 453–474.

van der Tol, M. W. M. & H. Scholten, 1992. Response of the Eastern Scheldt ecosystem to a changing environment: functional or adaptive? Neth. J. Sea Res., 30: 175–190.

Wetsteyn, L. P. J. M. & J. C. Kromkamp, 1994. Turbidity, nutrients and phytoplankton primary production in the Oosterschelde. (The Netherlands) before, during and after a large-scale coastal engineering project (1980–1990). Hydrobiologia 282/283 (Dev. Hydrobiol. 97): 61–78.

Hydrobiologia **282/283**: 437–451, 1994.
P. H. Nienhuis & A. C. Smaal (eds), The Oosterschelde Estuary.
© *1994 Kluwer Academic Publishers. Printed in Belgium.*

SMOES: a simulation model for the Oosterschelde ecosystem

Part I: Description and uncertainty analysis

Olivier Klepper[1], Marcel W. M. van der Tol[2], Huub Scholten[3] & Peter M. J. Herman[4]
[1]*National Institute for Public Health and Environmental Protection, Centre for Mathematical Methods, P.O. Box 1 3720 BA Bilthoven, The Netherlands;* [2]*National Institute for Coastal and Marine Management/RIKZ, P.O. Box 20907, 2500 EX The Hague, The Netherlands;* [3]*Department of Information Science, Wageningen Agricultural University, Dreyenplein 2, 6703 HB Wageningen, The Netherlands;* [4]*Netherlands Institute of Ecology, Centre for Estuarine and Coastal Ecology, Vierstraat 28, 4401 EA Yerseke, The Netherlands*

Key words: Oosterschelde estuary, ecosystem, simulation model

Abstract

The model SMOES integrates the results of the ecological research program conducted in the Oosterschelde estuary before and during the construction of a storm surge barrier. Its aim is to provide a quantitative summary of the research findings and to provide a tool for analysis and prediction of the ecosystem in response to human manipulations. This chapter describes model background and formulations. An uncertainty analysis is used to analyze the effect of uncertainties in model parameters on model outcome. The results of the sensitivity analysis are classified by distinguishing groups of model parameters with a qualitatively different effect on model results. Within these groups, a quantitative ranking of the parameters is possible. It appears that the most sensitive parameters represent processes that are relatively little studied in the Oosterschelde, which may provide guidelines for further research.

Introduction

As part of the ecosystem study described in this volume, a dynamic mathematical simulation model was developed in order to integrate research results and to provide a management tool. SMOES (= Simulation Model Oosterschelde EcoSystem) describes the main carbon and nutrient flows with a spatial scale of 10–20 km and a temporal scale of approximately one day (Klepper, 1989; Scholten *et al.*, 1990). In this paper the background and the most important model formulations are given. For a full documentation of the model the reader is referred to Klepper (1989).

Model structure (both in terms of spatial and

trophic) resolution was designed according to the following two criteria: (1) structure should be in accordance with model objectives, and (2) quantitatively important processes should be emphasized.

As an example of the first criterium: a model that aims to predict the fate of an oil spill should have a quite different time scale from a model that aims at predicting long-term heavy metal accumulation in the sediment. For the Oosterschelde model, the human influence on which the model focuses are the presence of the storm-surge barrier (reducing exchange with the North Sea), manipulations with the number of mussels (possibly reducing their own food levels) and the freshwater input (containing nutrients for phytoplankton growth). The time scale on which these manipulations may show effects (on phytoplankton, mussel growth) is weeks to years. The spatial scale that is relevant is 10 to 20 km. This criterium also implies that the model must be able to answer the questions that it will be asked. A model that is to be used as a management tool to predict macrobenthic food supply in the Oosterschelde should include the various management options available, but management options that are *a priori* excluded, such as long-term closure of the barrier need not be included in the model.

The second criterium is sometimes difficult to meet: often the quantitative importance of a process is not apparent before it is actually included in the model. In these cases it is obviously the safest choice to include a process when in doubt, although this will lead to (large) parts of the model that influence the results only to a minor extent. In a previous study (Klepper & Van de Kamer, 1987, 1988) result of the Oosterschelde research project were also combined in a modelling study, but only in accounting terms (*i.e.* no process information) and on a yearly-averaged basis. The questions underlying this steady-state model were whether a closed carbon budget for the Oosterschelde can be found on the basis of the experimental results, and what the most important carbon-flows in this budget are. The conclusion was that the data are not inconsistent with a closed carbon budget. However, they show a considerable uncertainty. From the yearly averaged budget it can be concluded that the major carbon flows are all on the first and second trophic levels: the higher food chain (fish, birds) is quantitatively negligible, and can therefore be safely omitted in more detailed carbon flow models.

For most of the parameters in the formulations, generally only a range can be given. These ranges are the basis of the uncertainty analysis and calibration of the model. In this paper, model formulations and some results of an uncertainty analysis are presented. Model calibration is discussed by Klepper *et al.* (1991) and Scholten & van der Tol (1994).

Model description

The model simulates 11 state variables in 4 compartments (boxes). In every compartment, model structure is identical, but there are morphological parameters such as depth, surface area, etc., which are of course compartment-dependent. The actual model consists of formulating the rate of change in every state variable as a result of the state of the system itself and of (time-dependent) external conditions. External conditions include forcing functions (for example light intensity, temperature), inputs (for example nutrient discharges) and boundary conditions (concentrations of each state variable in the adjacent North Sea to the west and Volkerak to the North, see Fig. 2). After calculating all rates of change, the new state of the system can be calculated, and the simulation may move one time-step ahead.

Figure 1 gives a schematic representation of SMOES, showing the most important state variables. Included in this graph are suspension feeders (mainly cockles and mussels) which are not treated as state-variables but as forcing functions. Treatment as a state variable is only possible in a trivial fashion for this group because their number in the Oosterschelde is mainly determined by human activity (mussel culture and cockle fishery). The other bottom fauna (depositfeeders and meiobenthos) have been imposed as a forcing function too, because of their relatively minor role in the food web (Klepper & Van de Kamer, 1987).

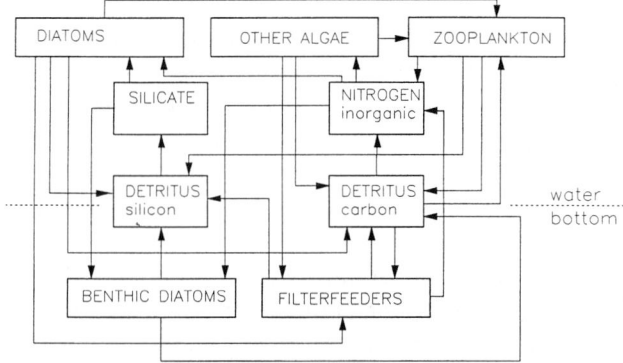

Fig. 1. Schematic representation of the simulationmodel SMOES, showing the most important state variables. Filterfeeders are treated as forcing function (Klepper, 1989).

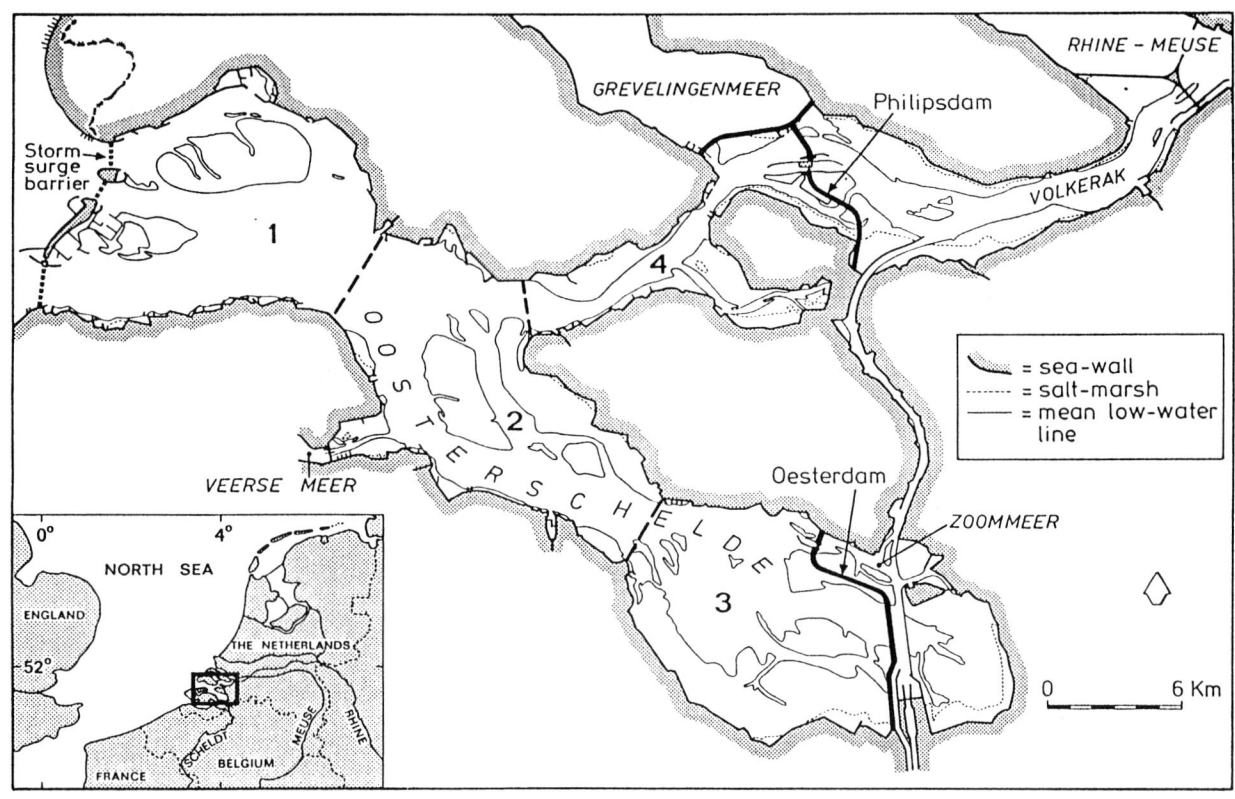

Fig. 2. Oosterschelde estuary showing the four compartments (1–4) used in the model SMOES.

The state variables salinity, refractory detritus and oxygen are not shown in fig. 1 as these do not play a role in the biological food web. Freshwater inputs are limited, so that salinity is always > 24‰ and its fluctuations have been assumed not to influence the biota. It is included as a state variable however as concentrations give valuable information for the calibration of the transport model. Refractory detritus is considered inedible for the consumers; in the model its concentration is determined by transport processes only. It was included as a state variable to enable a comparison with observed particulate organic carbon (POC) concentrations. POC measurements include all particulate organic matter (phytoplankton, labile and refractory detritus and zooplankton), but the refractory detritus is generally dominant. Oxygen concentrations are generally high in the Oosterschelde (> 80% saturation), and have been assumed not to limit aerobic pro-

cesses. The concentration of oxygen may serve as an indicator of whole-system metabolism, and calculated values (determined by exchange with the atmosphere, primary production and respiration) are useful for a judgment of model performance.

Transport model
The temporal scale of the model does not allow for an explicit description of the diurnal tide in the estuary. The tidal excursion in the Oosterschelde is approximately 10–15 km (Dronkers, 1980), but the tidally averaged displacement of water is only small. It is possible to separate the periodic tidal movement from the net movement (O'Kane, 1980). By taking into account the moment in the tidal cycle that a water sample is taken, it is possible to calculate the position that a particular water mass would have at mid-tide. This makes it possible to correct the positions of the sampling

stations for comparison with the calculated tidally averaged changes in water quality.

Although this transformation gives an excellent approximation to the actual oscillating solution (O'Kane, 1980), some problems may arise in case of the interaction between the fixed bottom (filterfeeders) and oscillating water. The present model makes two assumptions: (1) the filterfeeders have a fixed position in the moving water mass and (2) they feed only from the compartment in which they reside at mid-tide. The first assumption has been investigated by Klepper (1989) with a biologically simplified but spatially much more detailed model. In this model the actual situation could be simulated in which the water moves over a mussel bed and the mussels feed only for a short period from a certain water mass. It turns out that there is only a small difference between this dynamical calculation and a quasi steady-state approximation in which the mussel bed is spread out over the tidal path and part of the mussels feed continuously from their assigned water mass. The second assumption is potentially more critical, but still acceptable in view of the spatial resolution of the model (Klepper, 1989).

The transport of dissolved substances (salinity, nutrients) has been modelled according to a 'forward time centered space' approximation to the advection-dispersion model (O'Kane, 1980; Ruardij & Baretta, 1989). This approach is usually chosen when dispersive transport dominates advective transport as in the Oosterschelde (Dronkers, 1980). Advective transport gives the concentration changes as a result of the net flow Q:

$$T_{Ai} = 1/2 Q_{i-1} (C_{i-1} + C_i)$$
$$- 1/2 Q_i (C_i + C_{i+1}), \qquad (1)$$

where

T_A = advective transport (g s^{-1});
i = compartment index;
Q = net flow (m^3 s^{-1});
C = concentration (g m^{-3});

Dispersive transport describes the effect of mixing in the estuary, which is caused mainly by irregularities in tidal flows and topography. For the formulation of dispersive transport we first define an exchange coefficient:

$$E = \frac{DA}{L}, \qquad (2a)$$

where

E = exchange coefficient (m^3 s^{-1});
D = dispersion coefficient (m^2 s^{-1});
A = average cross-sectional area between two adjacent compartments (m^2);
L = average distance between midpoints of two adjacent compartments (m);

(see Table 1) dispersive transport can now be formulated as:

$$T_{Di} = E_{i-1} (C_{i-1} - C_i) - E_i (C_i - C_{i+1}). \qquad (2b)$$

In the two transport equations, the net flows (from freshwater discharges, extractions, precipitation and evaporation) are known, which makes it possible to determine the dispersion coefficients by comparison with measured salinity data (Table 1).

Table 1. Average (pre-barrier) flow rates, dispersion coefficients and exchange volumes in transport model. Positive sign: landward, negative sign: seaward. Compartment numbers: see Fig. 2. See text for definition of apparent flow rate.

From–to	Net water flow Q (m^3 s^{-1})	Dispersion (m^2 s^{-1})	Exchange volume (m^3 s^{-1})	Apparent flow Q' (m^3 s^{-1})
Sea-1	− 65	373 ± 11	1748 ± 52	− 530 ± 980
1–2	− 64	226 ± 6	994 ± 26	− 380 ± 750
2–3	− 8	144 ± 23	313 ± 50	− 350 ± 460
2–4	− 55	360 ± 5	379 ± 5	− 680 ± 800
4-Volkerak	− 54	216 ± 2	224 ± 2	− 120 ± 140

Part of the transport of particulate matter is similar to that of dissolved substances. For the part of the particulate matter which settles to the bottom during a tidal cycle and is later resuspended, substantial differences may occur. Because these particles do not take part in the tidal movement for part of the cycle, the path during flood may be longer (or shorter) than the path during ebb, and a net landward (seaward) displacement may occur (Postma, 1967; Dronkers, 1986) Only a small tidal asymmetry can have a large effect because the tidal excursion is much larger (20–30 km/day) than the residual displacement of water (50–100 m/day).

In the Postma-Dronkers model, the path of a single particle is described and the implicit assumption is made that the processes remain the same independent of the actual concentration of particles. In other words, total sediment flux across a compartment boundary (g s^{-1}) can be described as the product of the particle concentration (g m^{-3}) and an apparent flow rate Q' (m^3 s^{-1}). This allows us to use the same transport model for dissolved and particulate matter, with the important distinction that they may differ widely in flowrates (Q and Q', respectively). As the tidal and morphological characteristics underlying the Postma-Dronkers transport model are not time-dependent, it has been assumed that the apparent flow rate Q' is a constant.

The same dispersion coefficient for the dissolved and particulate substances have been used. Although the net displacements for the two groups may be quite different, both are negligible compared to gross (tidal) transport, which is responsible for the mixing.

There are two important remaining issues in the particulate transport model. The first is the estimation of the flow rate Q', and the second is the 'mixed' behavior of algae and POC in comparison with anorganic suspended matter.

The apparent flow rates Q' for suspended sediment have been estimated from fixed-point measurements (summing the product of flow and concentration continuously during a tidal cycle), from moving-frame measurements (recording sedimentation and resuspension in a certain water mass

to estimate the parameters in the Postma-Dronkers model) and from long term erosion data. All methods are rather unreliable as discussed by Klepper (1989); although the average results point to export of particulate matter from the Oosterschelde, the large uncertainty does not rule out the possibility of import (Table 1).

The transport of POC and chlorophyll has been estimated by using the vertical distributions of the different particulate fractions. The suspended sediment is not distributed homogeneously over the water column, but concentration increases approximately twofold towards the bottom. For dissolved substances, there is no such gradient, and POC and chlorophyll show intermediate behavior. The POC and chlorophyll gradients can be described as if these substances behaved partly as dissolved matter (no gradient) and partly as suspended sediment (twofold increase in vertical direction). It has been assumed that the horizontal transport behavior can be described by the same fractions: partly as dissolved matter (flow rate Q), partly as suspended sediment (apparent flow rate Q'). For chlorophyll the 'dissolved' fraction could be estimated as 0.79 ± 0.09, for POC as 0.48 ± 0.15.

The continuous sedimentation and resuspension of particulate matter means that there is a strong coupling between detritus in water column and bottom. In fact, it is not useful to consider the two as separate pools on the time scale of the model, and they have been lumped into a single state variable in the model. Part of the total labile detritus pool is in suspension (and thus available as food for suspension feeders and copepods), part is buried in the upper sediment layer and may be eaten by depositfeeders, meiobenthos, etc.

The fraction in suspension depends on water turbulence, caused by wind and North Sea waves. This fraction has been modelled empirically by using observed suspended sediment concentrations. We may assume that the total amount of sediment available for resuspension is constant: during a prolonged storm, the amount in suspension initially increases but levels off after some time (Ruardij & Baretta, 1989). This total amount has been estimated from peak sediment concentrations in each compartment. The observed concentrations directly allow an estimate of the suspended fraction by subtracting the amount in the water column from this total. For the pre-barrier situation, historical records of suspended sediment concentrations could be used, for predictions an estimate of future suspended sediment concentrations is required.

In the pre-barrier situation (most of) the measurements indicated an export of sediment from the Oosterschelde. In this situation, the bottom is slowly eroding. As the sediment contains approximately 5% POC (almost exclusively detritus, model parameter QPOCSED), this buried detritus is continuously introduced into the system. Whether this detritus is degradable or refractory is not known, but it is most probably refractory. To indicate the possibly degradable fraction of eroded detritus the parameter SEDLABQ (range 0–0.05) is introduced.

In the post-barrier situation, sediment concentrations are considerably lower (Ten Brinke *et al.*, 1993), although peak concentrations are comparable. There are some measurements of sediment transport, which give similar values for the apparent flow rate as those presented in the Table 1, *i.e.* not significantly different from zero. The lower concentrations cause an increased dispersive transport of sediment from the North Sea. As a result, the Oosterschelde is now a sediment-importing system.

Phytoplankton model

The phytoplankton model is mainly based on the work of Eilers & Peeters (1988 and in press) who describe algal photosynthesis in response to light intensity as a dynamic process. During illumination, photosynthetic units may be inactivated by excess light, which makes the response to light time-dependent. The vertical mixing in the water column causes light-to-dark transitions that stimulate photosynthesis by continuously supplying 'fresh' (not yet inactivated) cells into the bright surface layer. This stimulation could already be considerable (5–40% increase) in the pre-barrier

Oosterschelde (Klepper *et al.*, 1988), and increased in the post barrier period due to higher light intensities. Calculating column-averaged production therefore requires modelling of this mixing effect (Klepper *et al.*, 1988). However, the time scale for these calculations is not suitable for the present model. Therefore, column-averaged daily production has been calculated for a wide range of environmental conditions and algal properties with a detailed mixing model, and a response surface fitted to the results (Klepper *et al.*, 1988). This 'meta-model' can now be easily incorporated into the present model (Klepper, 1989).

The Eilers-Peeters model requires three properties of the algae as input: maximum production rate (P_{max}), the initial slope of the production *vs* light curve (s) and optimal light intensity (I_{opt}).

Under nutrient-saturated conditions, P_{max} is a function of temperature (Eppley, 1972), which can be modelled by a standard Q10 formulation (*i.e.* an approximate doubling of P_{max} for 10 °C temperature increase). On theoretical grounds it can be reasoned that the optimal light intensity of the algae is determined by temperature only (Eilers & Peeters, 1988); this prediction is supported by Oosterschelde measurements (Wetsteyn, unpublished results). In the model a Q10 formulation has been used.

Only inorganic nitrogen (the sum of NO_3 and NH_4) and silicate are included in the model. Phosphorus was not considered as a potentially limiting nutrient in the pre-barrier Oosterschelde (Wetsteyn *et al.*, 1990). If the nitrogen to phosphorus ratio of the inputs remains the same in future or nitrogen inputs become relatively lower, there is no need to change this model assumption, especially in view of the possible increase in denitrification (see below). If phosphorus inputs would show a relative decrease compared to nitrogen or silicon, it could be necessary to incorporate this nutrient into the model.

Nitrogen, although not limiting in the pre-barrier situation either (Wetsteyn *et al.*, 1990) was included in the model to be able to incorporate the effect of increased denitrification after the barrier (see below).

The effect of nutrients on P_{max} was modelled using a hyperbolic saturation function (Dugdale, 1967), taking into account the effect of the most limiting nutrient only (De Groot, 1981). Half-saturation concentrations between 0.005–0.2 g m^{-3} (nitrogen) and 0.002–0.1 g m^{-3} (silicate) were used.

The initial slope of the production curve is proportional to the chlorophyll content of the cells, which determines their capacity to capture light (Bannister & Laws, 1980). The chlorophyll content (chlorophyll: carbon ratio) of algae is very variable, and is related to growing conditions. In the present model a slightly simplified version of the Bannister-Laws model of chlorophyll content in relation to nutrient- and light-limitation has been incorporated (Bannister & Laws, 1980; Chalup & Laws, 1990):

$$Chlf:C = Chlf_{max} F_n (1 - F_1)^{0.6} \qquad (3)$$

where

$Chlf:C$ = chlorophyll to carbon ratio (W/W);

$Chlf_{max}$ = maximal $Chlf:C$ (between 1:8 and 1:20);

F_n = nutrient limitation function, giving actual production as a fraction of P_{max};

F_1 = light limitation function (idem).

Net population increase of algae equals gross production minus respiration and excretion. These processes are very important for the model results, as they are of the same order of magnitude as gross production. Unfortunately, they are much less studied and very imperfectly known.

Respiration can be divided into maintenance and growth respiration. The first is determined by protein turnover and the costs of osmoregulation and ion exchange with the environment. These loss processes are influenced by temperature, with a Q10 of approximately 2. The actual level of maintenance respiration is very variable (0–10% day^{-1}), depending on species and physiological state (cellular composition). Better known is the respiration associated with growth, which varies with the relative amounts of protein, fat and car-

bohydrates in the cell (Penning de Vries, 1973). Depending on this composition, growth respiration fraction varies between 0.30 and 0.55 of gross production.

Phytoplankton may excrete a considerable fraction of production as dissolved organic carbon. In the model this is added to the labile detritus pool. Excretion rates are negligible in exponentially growing populations and under light-limiting conditions. Under conditions of nutrient limitation however, an excretion of 10–40% of the carbon fixation is reported (Sharp, 1977; Fogg. 1983). This has been incorporated into the model by making excretion a function of nutrient limitation:

$$E = E_{max} (1 - F_n)P_g, \qquad (4)$$

where

E = excretion rate (day^{-1});

E_{max} = maximal excretion rate as a fraction of gross production (–);

F_n = nutrient limitation function (see above);

P_g = gross production rate (day^{-1}).

The algae in the Oosterschelde can be divided into two broad groups: diatoms and non-diatoms, of which the first is dominant (Wetsteyn et al., 1990; Bakker et al., 1990). Only the diatoms require silicon for growth, and this nutrient generally reaches growth-limiting concentrations in summer (Wetsteyn et al., 1990). Apparently this handicap is offset by some other physiological advantage. Although this advantage is not known (literature ranges in all physiological parameters show a wide overlap between the groups), the model assumes that it is a combination of a flatter temperature response (making diatoms less sensitive to low temperatures) and higher efficiency (lower respiration and excretion rates).

Microphytobenthos

The production of microphytobenthos on the tidal flats of the Oosterschelde takes place in an extreme environment, in which a biomass per square meter of the same order of magnitude as the phytoplankton biomass is concentrated in a few millimeters of sediment. As a result, the pools of nutrients are almost negligible and -on the relevant temporal scale of a day- the question of nutrient-limitation is not determined by nutrient concentrations (as in the water column) but rather directly by nutrient supply: if demand exceeds production, the concentration will drop rapidly (within hours) and phytobenthos production will also drop until an equilibrium is reached between supply and demand.

The microphytobenthos model is described by Scholten (pers. comm.). Briefly, potential (light-limited) production is calculated by the Eilers-Peeters model (also used for the phytoplankton, see above), taking into account the (biomass-dependent) vertical distribution of the algae. This potential production is then compared with total mineralization plus CO_2 diffusion (for potential CO_2 limitation), net N-mineralization (N-mineralization minus denitrification) and Si-dissolution (the phytobenthos consists of diatoms). The realized production equals the most limiting of these factors. This situation implies that microphytobenthos production consists only for a minor part of 'new' (i.e. not regenerated) production.

The respiration of phytobenthos is modelled in the same way as phytoplankton (see above). Observations have shown that the phytobenthos normally does not go into suspension on flooding in the Oosterschelde (Scholten et al., 1990), but during storms, a considerable fraction may be suspended into the water column. In the model this process is described using the empirical suspended sediment time series (see above). It is assumed that this phytobenthos does not return to a suitable environment, and its biomass is added to the labile detritus pool. Losses of phytobenthos by grazing are described below.

Zooplankton

The zooplankton in the Oosterschelde has been studied by Tackx, Bakker and coworkers (Tackx, 1987; Tackx et al., 1989, 1990; Bakker & Van

Rijswijk, 1987). The present model is an adaptation of the model by DiToro et al. (1971) based on the results by Tackx and coworkers.

In the Oosterschelde two groups of zooplankton are about equally importance: the true zooplankton (i.e. having their whole life cycle in the water, mainly the copepods *Acartia spp.*, *Centropagus hamatus* and *Temora longicornis*), and the larvae of benthic animals (mainly from Barnacles *Balanus spp.*). The first group is a state variable in the model, the biomass of the second group is given as a time-series, with a summer (April-September) value increasing from 0.02 gC m^{-3} in the west to 0.05 gC m^{-3} in the eastern part of the Oosterschelde. The grazing activity of both groups is modelled identically.

The food source for the zooplankton is primarily phytoplankton which the animals are able to select from the suspended particles (dominated by detritus). If phytoplankton biomass is low, the zooplankton is able to supplement its diet with detritus, although part of the detritus (28% -Tackx, 1987) is too small (< 3 μm) to be eaten. There is no constant size distribution for the phytoplankton, and the fraction that is too small to be captured is variable. Generally the phytoplankton population tends to be smaller sized under nutrient limitation (Parsons & Takahashi, 1973; Laws, 1975). In the model this has been expressed by using the nutrient-limitation function of the phytoplankton for its edible fraction, i.e. 100% under nutrient saturation tending to O under nutrient limitation.

Below a certain threshold, the food intake of zooplankton is proportional to food concentration. Above the threshold, a maximum intake is attained. This is expressed by the following function (Kremer and Nixon, 1978):

$$R_t = R_{\max_t} \min \left\{ 1, \frac{F}{F_{\lim}} \right\}, \qquad (5)$$

where

R_t = (temperature dependent) ingestion per unit of biomass ('daily ration') in gC gC^{-1} day^{-1};

R_{max_t} = (temperature dependent maximum daily ration gC gC^{-1} day^{-1};
F = food concentration (gC m^{-3});
F_{lim} = threshold food concentration (gC m^{-3}).

The value of F_{lim} has been estimated on the basis of a literature review (Klepper, 1989) to be in the range 0.2–0.3 gC m^{-3}. The maximum daily ration of zooplankton at 15 °C was estimated on the basis of the work by Tackx (1987) between 0.5–2.0 day^{-1}. The effect of temperature was modelled by a Q10 between 2 and 3.

Respiration is also temperature dependent, with a Q10 between 1.5 ad 2.5. The respiration at 15 °C varies between 0.05 and 0.25 day^{-1} (review by Klepper, 1989). Assimilation efficiency is assumed constant, independent of temperature or food-intake (Klepper, 1989). In the model a range of 0.4–0.9 is used.

Mortality of zooplankton is caused mainly by grazing by fish, jellyfish and other macrozooplankton and by zooplankton itself (juveniles eaten by adults). The grazing activity of these higher trophic levels has not been studied in the Oosterschelde and could not be explicitly incluced in the model. DiToro *et al.* (1971) used a constant mortality rate. This is not followed in the present model, as in spring and early summer, the biomass of predators will be highest, more or less coinciding with the zooplankton biomass peak. This coincidence has been exploited in the present model, by making zooplankton mortality rate proportional to zooplankton biomass:

$$Z_{\mathrm{mort}} = mqq\,B \qquad (6)$$

where

Z_{mort} = mortality rate (day^{-1});
mqq = coefficient (day^{-1} (gC m^{-3})$^{-1}$);
B = zooplankton biomass (gC m^{-3});

the proportionality constant as obtained by calibration is approximately 5; with a summer biomass between 0.025 and 0.15 gC m^{-3} giving a mortality rate of approximately 0.12–0.75 day^{-1}.

Macrobenthos

Macrobenthos may be divided into two groups: the depositfeeders (grazing on bottom algae and labile detritus) and the suspension feeders (who feed on pelagic labile detritus and phytoplankton). Both groups have been described as a time-series, *i.e.* are not modelled as a state variable. For the first group this choice has been made because they play a relatively minor part in the Oosterschelde food chain (Klepper & van de Kamer, 1987, 1988): they graze on phytobenthos and benthic labile detritus, but for the detritus the microbial degradation is dominant. The suspension feeders do play an important role in the food chain, both as grazers and as a food source, which would justify their inclusion as a state variable. However, their biomass is mainly determined by human activities (cockle fisheries and mussel culture for the shellfish, and the introduction of stones and concrete for the rocky substrate organisms). Biomass data of cockles (Coosen *et al.*, 1994) and rocky substrate organisms (Leewis *et al.*, 1994) were estimated from surveys. Biomass data of mussels were estimated from market statistics (Van Stralen & Dijkema, 1994).

There are three groups of depositfeeders in the model: meiobenthos (a diverse group of nematodes and harpacticoid copepods), surface deposit feeders (mainly the snail *Hydrobia* spec.) and other deposit feeders (mainly the worm *Arenicola* spec.). The activity of the three groups is described similarly, as a temperature dependent (Q10 = 2) constant rate:

$$G = c_T B_g (B_p + B_d), \qquad (7)$$

where

G = ingestion rate (gC m^{-2} day^{-1});
c_T = temperature-dependent ingestion coefficient (day^{-1} (gC m^{-2})$^{-1}$);
B_g = biomass of grazer (gC m^{-2});
B_p = biomass of phytobenthos (gC m^{-2});
B_d = amount of benthic labile detritus (gC m^{-2}).

The ingestion is assimilated with a constant efficiency, the rest is respired. In addition to direct grazing, the worms cause phytobenthos mortality by bioturbation, burying the algae in deeper sediment layers. In the model, the biomass of these buried algae is added to the labile detritus pool.

A comparable formulation is used for grazing by both mussels and cockles, but with different parameter values. The model is a considerably adapted version of the one developed by Bayne (1976). Grazing is expressed in a clearance rate, *i.e.* the volume swept clear of particles per unit time. This depends on body weight, temperature and suspended sediment concentration, and is modelled as:

$$CR = f_T g_s a W^b , \qquad (8)$$

where

CR = clearance rate (m^3 day^{-1});
f_T = exponential temperature function ($-$);
g_s = negative-exponential function of suspended sediment ($-$);
a = coefficient for clearance at 10 °C and 0 g m^{-3} suspended sediment (m^3 day^{-1} W^{-b});
W = body weight (gram dry flesh weight);
B = coefficient in allometric relation with weight.

For all parameters in the model, the range reported in the literature is wide (reviewed by Klepper, 1989). Clearance rates for an averaged-sized mussel (0.8 g) range from 1.2 to 4.5 l h^{-1}; for an averaged sized cockle (0.4 g) the range is 0.7 to 2.5 l h^{-1}. It is not clear whether this range is due to variations in temperature or suspended sediment concentrations, as the effects of both factors are uncertain. For temperature there are some papers reporting no effect (Q10 = 1), others report a fairly strong effect for both cockles and mussels (Q10 = 3). For suspended sediment concentration there is a similar uncertainty: some authors find a depression in clearance rate only at high concentrations (> 50 g m^{-3}), others find a steady decline in clearance rate with suspended matter concentration, starting already at 3 g m^{-3}.

Only part of the filtered material is actually ingested by the shellfish: with increasing concentration an increasing fraction is rejected as pseudofaeces. This is modelled following Bayne (1976):

$$PSF = C \max \{0, (S - S_{psf})\} , \qquad (9)$$

where

PSF = pseudofaeces production (g day^{-1});
C = clearance rate (m^3 day^{-1});
S = suspended matter concentration (g m^{-3});
S_{psf} = threshold concentration for pseudofaeces production (g m^{-3}).

The role of pseudofaeces is again controversial, as discussed by Klepper (1989). According to one view the pseudofaeces is rejected in the same composition as the suspended matter. The second view is that the filterfeeders are able to select all or most of the organic material before rejecting the rest. In the model it is assumed that selection of organic matter is possible up to a certain concentration, but that selection efficiency decreases above a certain threshold. This can be expressed by putting the model pseudofaeces threshold higher than the actual field threshold. In this way, ingestion would still increase linearly with filtered amount in the range between field and model threshold (*i.e.* 100% efficiency). With further increasing concentration, ingestion remains constant, which means that the relative amount that is selected from the pseudofaeces decreases. By choosing the model-PSF equal, somewhat higher or very high compared to field-PSF, we can simulate no, limited (decreasing) and full selection efficiency, respectively.

The assimilation efficiency of cockles and mussels is constant in the model. Several relations with physiological and environmental variables that have been proposed in the literature are discussed by Klepper (1989). It appears that these relations are either not consistently found or may be attributed to experimental artifacts.

The model does not distinguish phytoplankton and labile detritus as having a different food quality for macrobenthos. Although this assumption may seem unrealistic in view of the known lower affinity for detritus as compared to phytoplankton, it should be realized that a considerable part of the Oosterschelde detritus is refractory. Expressing detritus assimilation (of which only the labile part is assimilated in the model) per unit of total detritus, a much lower assimilation efficiency for detritus than for phytoplankton would result. For the actual model calculations, only a consideration of the labile detritus is required: ingestion of refractory detritus, which is not assimilated and for 100% egested as faeces can be simply left out of the mussel budget altogether.

Respiration is calculated in two ways in the model. Because biomass is described as a forcing function, there is no check whether the filterfeeder mass balance is closed. For example respiration may be always much lower than assimilation. In the case of a state-variable, this would result in a steady increase in biomass up to a new food-limited biomass level at which assimilation and respiration are again in equilibrium. In case of a forcing function for the filterfeeders however, such discrepancy would result in a permanent 'carbon sink'. To obtain a closed budget the filterfeeder respiration is therefore assumed to be equal to assimilation. For the comparison between filterfeeder food-intake and their actual energy requirements the actual respiration is calculated from:

$$R = f_T p W^q, \tag{10}$$

where

R = respiration rate (ml O_2 h^{-1});
f_T = exponential temperature function;
W = dry body weight in gram;
p, q = coefficients.

The difference between assimilation and actual respiration expressed as a fraction of body weight is termed the scope for growth, and is a useful measure of e.g. food-limitation of the cockles and mussels.

For the rocky-substrate organisms a slightly different version of the activity model is used. There is no relation between suspended matter concentration and filtration and no pseudofaeces production. Assimilation efficiency is not assumed constant for these organisms but formu-

lated as a decreasing function of the fraction of organic matter in the suspended matter.

Mineralization

The decomposition of dead organic material in the water phase is modelled as a first-order decay process. Although microorganisms are responsible for the decomposition, their potential growth rate is so high that in practice substrate availability is always the limiting factor. A decay rate at 20 °C in the range of 0.01–0.20 is used, with a Q10 between 1.5 and 2.5.

It has been assumed that all non-refractory organic matter in the model (including detritus) has a fixed N:C ratio, and that inorganic nitrogen is released in this ratio during decomposition. The N:C ratio of organic matter in the model is based on a literature review (Klepper, 1989), which showed N:C ratios in the range 0.06–0.20 (by weight).

For the release of silicate, a similar assumption can not be made and a separate state variable particulate biogenic silicon has been introduced. Dissolution rate is again modelled as a first order process, with a rate at 10 °C between 0.01–0.03 day^{-1}, and a Q10 between 2 and 4.

Anaerobic decomposition in the bottom is responsible for part of the mineralization. Some of these processes may be considered to be included in the estimate of aerobic mineralization: it may be assumed that accumulation of the products of fermentation (acetate, lactate, H_2), methanogenesis (CH_4) or sulphate reduction (H_2S, FeS) is negligible or at most a small fraction of total production, so that these products are generally oxidized as soon as they diffuse into the aerobic sediment or the water column. For denitrification this is not true and N_2 will generally escape to the atmosphere.

In contrast to the nearby saline lakes (Grevelingen, Lake Veere), the Oosterschelde N-budget is dominated by transport processes and denitrification is only of secondary importance. Nevertheless, denitrification is thought to have been a relatively important factor in the eastern part of the Oosterschelde already before the construction of the barrier (Vegter & de Visscher, 1987), and its importance has increased since then because suspended sediment concentration and aerobic decomposition have decreased in favor of a larger fraction of decomposition in the sediment.

Denitrification is a function of substrate and nitrate availability. The first is generally the dominant factor, with muddy sediments (high organic matter content) having a much higher denitrification than sandy bottoms. Therefore, denitrification is expressed as a fraction of bottom detritus, modified by the availability of nitrate. Denitrification consumes approximately 1 g of nitrate-N per gram of carbon oxidized. In the model, nitrate is not modelled separately, and it is assumed that total inorganic nitrogen may be used as a measure for the amount of available nitrate.

$$Dc = NC \, Cr_{10} \, f(T) \, \frac{N}{K_m + N} \quad (11)$$

with

NC = N consumption per unit of C consumption ($gN \, gC^{-1}$);

Dc = denitrification rate in carbon units ($gC \, m^{-2} \, day^{-1}$);

C = bottom labile detritus concentration ($gC \, m^{-2}$);

r_{10} = relative rate at 10 °C and saturating DIN (day^{-1});

$f(T)$ = exponential (Q10) temperature function (–);

N = inorganic nitrogen concentration ($g \, m^{-3}$);

K_m = half-saturation coefficient ($g \, m^{-3}$).

Summarizing the Nitrogen cycle of the model: the main factor determining inorganic nitrogen concentration is transport: input and exchange with the North Sea. In addition, there is the following biological cycle: uptake by primary production (phytoplankton and phytobenthos), release as inorganic nitrogen (respiration) and labile detritus (faeces) by grazers (zooplankton, benthos), return as inorganic nitrogen from mineralization by bacteria or loss as denitrification.

Uncertainty analysis

At several points in the foregoing discussion some uncertainty concerning the model was expressed, for example, rate constants were given as a range rather than as a single value. A listing of all parameter ranges used in the model is given in part II of this paper. Although some researcher may feel that a certain formulation and parameter value is the best, he or she will generally admit that other possibilities exist as well, and if a model (as was presently the case) is developed by a group of workers, there are frequent discussions about possible formulations and parameter choices. Clearly, it is misleading to present a single model with unique formulations and parameter values as the outcome of model development: such a model is generally a subjective choice and therefore rather arbitrary, and moreover, the user will get the impression of very exact predictions where in reality only a range of values or a general trend can be predicted.

For these reasons, the quantification of uncertainty is an important issue in model development. As model results depend on model structure, model inputs and model parameters, we have to consider uncertainty in each factor. The first factor is the hardest to quantify, as the formulation and numerical solution of only two alternative models is already quite time consuming. During the modelling process, many choices are made (number of state variables, spatial resolution, which processes to include etc.) that would lead to an unpractical number of alternative models. Although the only general solution to this problem seems to behave cautiously in model formulation (include a process if it might be important), in some cases the inclusion of alternative models is possible by means of a parameter. For example, the effect of suspended matter concentration on mussel filtration can be modelled as an exponential decrease with concentration. By putting the exponential coefficient at zero, we can easily exclude such an effect.

The uncertainty in past model inputs (boundary conditions, wheather conditions, etc.) is usually small compared to other uncertain factors, but future inputs are highly uncertain and may dominate model results. In the present analysis this uncertainty has not been included, as the aim was not to predict the actual situation in the future Oosterschelde, but rather to compare certain management strategies given a certain input.

Assuming pre-determined inputs, and having tried to put as much of the uncertainty in model formulations into parameter values as possible, we are left to quantify the effect of parameter uncertainty on model results. This was done by means of a Monte Carlo analysis, in which the parameters were varied using independent uniform distributions over their entire ranges (Klepper, 1989).

The output of the model consists of a number ($j = 1..m$) of variables at time-steps t: M_{jt}, of which the value depends on parameters p_i ($i = 1..n$), each with an uncertainty range r_i. The Monte Carlo runs allow us to calculate a linear regression between parameter values and model results:

$$M_{jt} = a + b_{ijt} \cdot p_i \qquad (12a)$$

with a and b as linear regression coefficients. An uncertainty coefficient can now be defined as:

$$S_{ijt} = \frac{b_{ijt} \cdot r_i}{M_{jt}} \qquad (12b)$$

with M_{jt} the average value of the model results.

The coefficients S_{ijt} give the average relative change in M_{jt} als a result of varying p_i over its uncertainty range. For practical purposes the number of output variables considered in the uncertainty analysis was limited to 12, and the number of time steps to 3 periods: early spring, summer and autumn, giving a total of 36 M_{jt} values.

The analysis of uncertainty coefficients is a multivariate problem: the ranking of the parameters depends not only on the magnitude of the uncertainty coefficients, but also on different possible types of behavior. For example the parameter DAYRZOO (zooplankton grazing) has a large influence on summer phytoplankton biomass, the parameter CFLUX (rate of CO_2 diffusion from atmosphere) on phytobenthos biomass and silicate concentration in autumn. As DAYRZOO does not influence phytobenthos, nor CFLUX

phytoplankton, it makes little sense to say that model sensitivity to DAYRZOO is greater or smaller than that to CFLUX.

This issue can be illustrated on a fictitious example, where the effect of 5 parameters on 6 model outputs (2 state variables at 3 time values) is examined. Consider the following sensitivity matrix S:

	M_{11}	M_{12}	M_{13}	M_{21}	M_{22}	M_{23}
p_1	0.1	0.3	0.1	0.0	0.0	0.0
p_2	0.0	0.0	0.0	-0.2	-0.4	-0.3
p_3	0.3	0.9	0.3	0.0	0.0	0.0
p_4	-0.2	-0.6	-0.2	0.0	0.0	0.0
p_5	0.1	0.3	0.1	0.2	0.4	0.3

It is clear that the parameters p_1 and p_3 have the same effect on model results; p_4 appears to have the opposite effect, but this is merely a matter of sign: a decrease in p_4 has the same effect as an increase in $p1$. Further, p_2 has an effect on the model that is quite different from the $\{p_1, p_3, p_4\}$ group. Finally, the effect of p_5 appears to be intermediate between those of the other two groups: the same effect on the M's can be obtained by a combined increase in $p1$ and a decrease in p_2. In fact, the matrix S has a rank of only two: the model behavior in this example can be described by two independent groups of parameters. By means of this classification we can now order the parameters, and for example rank p_3 as the most important uncertain parameter within its group.

In practice the rank of S cannot be determined exactly. Because of the stochastic nature of the Monte Carlo analysis, the S_{ijt} coefficients are only approximated, which would make, for example, the M_{11} and M_{13} columns only nearly equal. The approximate rank of the matrix of uncertainty coefficients of SMOES was determined by principal component analysis (Pielou, 1984) to be between 15 (95% of the variance) and 22 (99% of variance explained). To arrive at a grouping of the parameters a clustering algorithm was used on the S-matrix, leading to 21 groups (Klepper, 1989).

450

By taking the most uncertain parameter (in terms of its effect) within each group we can inventory our main gaps of knowledge. The parameters were related to: microbiological processes (carbon mineralization, dissolution of silicate, denitrification), algal respiration (both phytoplankton and phytobenthos), zooplankton food limitation and loss processes, carbon to chlorophyll ratio, transport of algae and detrital silicon, assimilation efficiency of cockles and gas exchange with the atmosphere (both CO_2 and O_2).

It appears that the most uncertain parameters generally represented processes that were little studied in the research program. This does not mean that the program has studied the wrong processes: the value of S is the product of the uncertainty in the parameter and its effect on model outcome. A well-known parameter (low r in equation 11) that represents an important process (high b) may still have a low S-value.

In view of (a particular) model application the columns in the sensitivity matrix need not be of equal importance. A particular output variable may be quite uncertain, but may be only indirectly related to the question which is asked of the model. As one of the main points of interest is the growth potential for mussels in the Oosterschelde, the various model other outputs were related to this particular one. By comparing the results of different rus of a sensitivity analysis, it turned out that some variables were strongly correlated (for example: if for a particular run chlorophyll concentration was high, mussel scope for growth was high as well and vice versa). Other variables seemed almost independent of mussel growth, in particular copepod biomass and benthic chlorophyll concentration. Therefore, if the model was meant to be applied for mussel cultivation management only, it could be considerably simplified in these areas.

The fact that only in a few cases (e.g. transport) the processes that were studied are still relatively uncertain can actually be interpreted as a generally successful research program. The present model results could be useful as a guide for possible further research.

Acknowledgement

This paper is Communication No. 704 of NIOO-CEMO, Yerseke, The Netherlands.

References

Bakker, C. & P. Van Rijswijk, 1987. Development time and growth of the marine calanoid copepod *Temora longicornis* as related to food conditions in the Oosterschelde estuary (southern North Sea). Neth. J. Sea Res. 21: 125–141.

Bakker C., P. M. J. Herman & M. Vink, 1990. Effect of infrastructural works on seasonal succession of phytoplankton in the Oosterschelde (S. W. Netherlands). J. Plankton Res.

Bannister, T. T. & E. A. Laws, 1980. Modelling phytoplankton carbon metabolism. In: P. G. Falkowski (ed.) Primary production in the sea. Plenum Press, New York: 243–258.

Bayne, B. L., 1976. Marine mussels – their ecology and physiology. International Biological Program vol. 10.

Chalup, M. S. & E. A. Laws, 1990. A test of the assumptions and predictions of recent microalgal growth models with the marine phytoplankter *Pavlova lutheri*. Limnol. Oceanogr. 35: 583–596.

Coosen, J., F. Twisk, M. W. M. van der Tol, R. H. D. Lambeck, M. R. van Stralen & P. M. Meire, 1994. Variability in stock assessment of cockles (*Cerastoderma edule* L.) in the Oosterschelde (in 1980–1990), in relation to environmental factors. Hydrobiologia 282/283: 381–395.

De Groot, W. T., 1983. Modelling the multiple nutrient limitation of algal growth. Ecol. Modelling 18: 99–119.

DiToro, D. M., D. J. O'Conner & R. V. Thomann, A dynamic model of the phytoplankton in the Sacramento-San Joaquin delta. Adv. in Chem. Series 106: 131–180.

Dronkers, J., 1980. Kwalitatieve interpretatie van zoutmetingen bij constante zoetwateraanvoer op het Volkerak. Concept nota DDWT-80. Rijkswaterstaat, Den Haag (in Dutch).

Dronkers, J., 1986. Tide induced residual transport of fine marine sediment. In: J. van de Kreeke (ed.) Physics of shallow estuaries and bays. Springer Lecture notes on coastal and estuarine studies 16. Berlin etc.

Dugdale, R. G., 1967. Nutrient limitation in the sea: dynamics, identification and significance. Limnol. Oceanogr. 12: 685–695.

Eilers, P. H. C. & J. C. H. Peeters, 1988. A model for the relationship between light intensity and the rate of photosynthesis in phytoplankton. Ecol. Modelling 42: 185–198.

Eilers, P. H. C. & J. C. H. Peeters, (in press). Dynamic behaviour of a model for photosynthesis and photoinhibition. Ecological modelling.

Eppley, R. W., 1972. Temperature and phytoplankton growth in the sea. Fish. Bull. 70: 1063–1085.

Fogg, G. E., 1983. The ecological significance of extracellular products of phytoplankton photosynthesis. Botanica Marina XXVI:3–14.

Klepper, O., 1989. A model of carbon flows in relation to macrobenthic food supply in the Oosterschelde estuary (S. W. Netherlands). Ph. D. thesis, Wageningen Agricultural University.

Klepper, O., J. C. H. Peeters, J. P. G. van de Kamer & P. H. C. Eilers, 1988. The calculation of primary production in an estuary. A model that incorporates the dynamic response of algae, vertical mixing and basin morphology. In: A. Marani (ed.) Advances in environmental modelling, Elsevier, Amsterdam, 373–394.

Klepper, O. & J. P. G. van de Kamer, 1987. The use of mass balances to test and improve the estimates of carbon fluxes in an ecosystem. Math. Biosci. 85: 37–49.

Klepper, O. & J. P. G. van de Kamer, 1988. A definition of the consistency of the carbon budget of an ecosystem and its application to the Oosterschelde estuary, S. W. Netherlands. Ecol. Modelling 42: 217–232.

Klepper, O. H. Scholten & J. P. G. van de Kamer, 1991. Prediction uncertainty in an ecological model of the Oosterschelde estuary. J. Forecasting (in press).

Kremer, J. N. & S. W. Nixon, 1978. A coastal marine ecosystem – simulation and analysis. Springer Verlag, Berlin.

Laws, E. A., 1975. The importance of respiration losses in controlling the size distribution of marine phytoplankton. Ecology 56: 419–426.

Leewis, R. J., H. W. Waardenburg & M. W. M. van der Tol, 1994. Biomass and standing stock on sublittoral hard substrates in the Oosterschelde estuary (SW Netherlands). Hydrobiologia 282/283: 397–412.

O'Kane, J. P., 1980. Estuarine water quality management. Pitman, Boston.

Parsons, T. R. & M. Takahashi, 1973. Environmental control of phytoplankton cell size. Limnol. Oceanogr. 19: 367–368.

Penning de Vries, F. W. T., 1973. Substrate utilization and respiration in relation to growth and maintenance processes in higher plants. Ph. D. Thesis, Agricultural University, Wageningen.

Pielou, E. C., 1984. The interpretation of ecological data, a primer on classification and ordination. John Wiley, New York etc.

Postma, H., 1967. Sediment transport and sedimentation in the estuarine environment. In: G. H. Lauf (ed.), Estuaries. Am. Ass. Adv. Sci. 158–179.

Ruardij, P. & J. W. Baretta, 1989. The construction of the transport sumodel. In: J. Baretta & P. Ruardij (eds), Tidal flat estuaries – Ecological studies 71. Springer Verlag, Berlin etc. p 65–76.

Scholten, H., O. Klepper, P. H. Nienhuis & M. Knoester, 1990. Oosterschelde estuary (S. W. Netherlands): a self-sustaining ecosystem? Hydrobiologia 195: 201–195.

Scholten, H. & M. W. van der Tol, 1994. SMOES: a simulation model for the Oosterschelde ecosystem. Part II: Calibration and validation. Hydrobiologia 282/283: 453–474.

Sharp, J. H., 1977. Excretion of organic matter by marine phytoplankton: do healthy cells do it? Limnol. Oceanogr. 22: 331–398.

Smaal, A. C. & Scholten, H., in press. The dominant role of shellfish in an estuarine ecosystem: a model approach. Aquaculture.

Tackx, M. L. M., 1987. Grazing door zooplankton in de Oosterschelde. Ph. D. thesis, Lab. voor ecologie en systematiek, Vrije Universiteit Brussel. (in Dutch).

Tackx, M. L. M., C Bakker, J. W. Francke & M. Vink, 1989. Size and phytoplankton selection by Oosterschelde zooplankton. Neth. J. Sea Res. 23: 35–43.

Tackx, M. L. M., C. Bakker & P. Van Rijswijk, 1990. Zooplankton grazing pressure in the Oosterschelde (The Netherlands). Neth. J. Sea Res 25: 405–415.

Ten Brinke, W. B. M., J. Dronkers & J. P. M. Mulder, 1994. Fine sediments in the Oosterschelde tidal basin before and after partial closure. Hydrobiologia 282/283: 41–56.

Van Stralen, M. R. & R. D. Dijkema, 1994. Mussel culture in a changing environment: the effects of a coastal engineering project on mussel culture (*Mytilus edulis* L.) in the Oosterschelde estuary (SW Netherlands). Hydrobiologia 282/283: 359–379.

Vegter, F. & P. R. M. de Visscher, 1987. Nutrients and phytoplankton primary production in the marine tidal Oosterschelde estuary (the Netherlands). Hydrobiological Bulletin 21: 149–158.

Wetsteyn, L. P. M. J., J. C. H. Peeters, R. N. M. Duin, F. Vegter & P. R. M. de Visscher, 1990. Phytoplankton primary production and nutrients in the Oosterschelde (the Netherlands) during the pre-barrier period 1980–1984. Hydrobiologia 195: 163–177.

Hydrobiologia **282/283**: 453–474, 1994.
P. H. Nienhuis & A. C. Smaal (eds), The Oosterschelde Estuary.
© 1994 *Kluwer Academic Publishers. Printed in Belgium.*

SMOES: a simulation model for the Oosterschelde ecosystem
Part II: Calibration and validation

Huub Scholten [1] & Marcel W. M. van der Tol [2]
[1] *Netherlands Institute of Ecology, Centre for Estuarine and Coastal Ecology, Vierstraat 28, 4401 EA Yerseke, The Netherlands (present address: Wageningen Agricultural University, Dept. of Computer Science Dreijenplein 2, 6703 HB Wageningen, The Netherlands;* [2] *National Institute for Coastal and Marine Management/RIKZ, P.O. Box 20907, 2500 EX The Hague, The Netherlands*

Key words: ecosystem model, calibration, uncertainty analysis, validation, Oosterschelde

Abstract

SMOES is an invalid model in the Popperian sense of the word. Yet it might be a useful model, which accurately simulates some of the ecosystem behavior. Our knowledge of the Oosterschelde ecosystem is empirical and consists of a comprehensive set of observed data (experiments and in situ measurements). Model usefulness will be investigated in terms of model adequacy (can a model simulate all system behavior) and reliability (does it simulate observed system behavior only).

The large impact of the building of a storm-surge barrier in the Oosterschelde Mouth (compartment West) allows us to distinguish two different systems. The quality of the simulation model, built and calibrated for the pre-barrier system, can so be tested by using it for the (new) post-barrier system. The question if the model does predict the post-barrier future will be answered.

We apply a calibration method which generates not merely a single best value for each of the many badly known parameters, but which also enables us to estimate the uncertainty in the model outcome. If we use this method to calibrate SMOES for the pre- and for the post-barrier situation, and subsequently estimate the uncertainty in the model results, we find no large (functional) shifts in the behavior of the main ecosystem properties.

Introduction

SMOES (= Simulation Model Oosterschelde EcoSystem) is one of the major products of the ecosystem study, discussed in this volume. It integrates research results and is used as management and hypotheses generating tool. The model was developed to simulate the main carbon and nutrient flows of the Oosterschelde ecosystem with a temporal resolution of approximately one day and a spatial resolution of 10–20 km (Fig. 1).

An ecological characterization of the Oosterschelde estuary is given by Nienhuis & Smaal (1994). A full description of SMOES is given by Klepper (1989). For a general introduction and a review of the processes included in SMOES the reader is also referred to Klepper *et al.* (1993). SMOES is used to evaluate the relative importance of autochthonous and imported organic carbon (Scholten *et al.*, 1990b), to determine the key-role of suspension feeders in ecosystem stability (Herman & Scholten, 1990) and to investi-

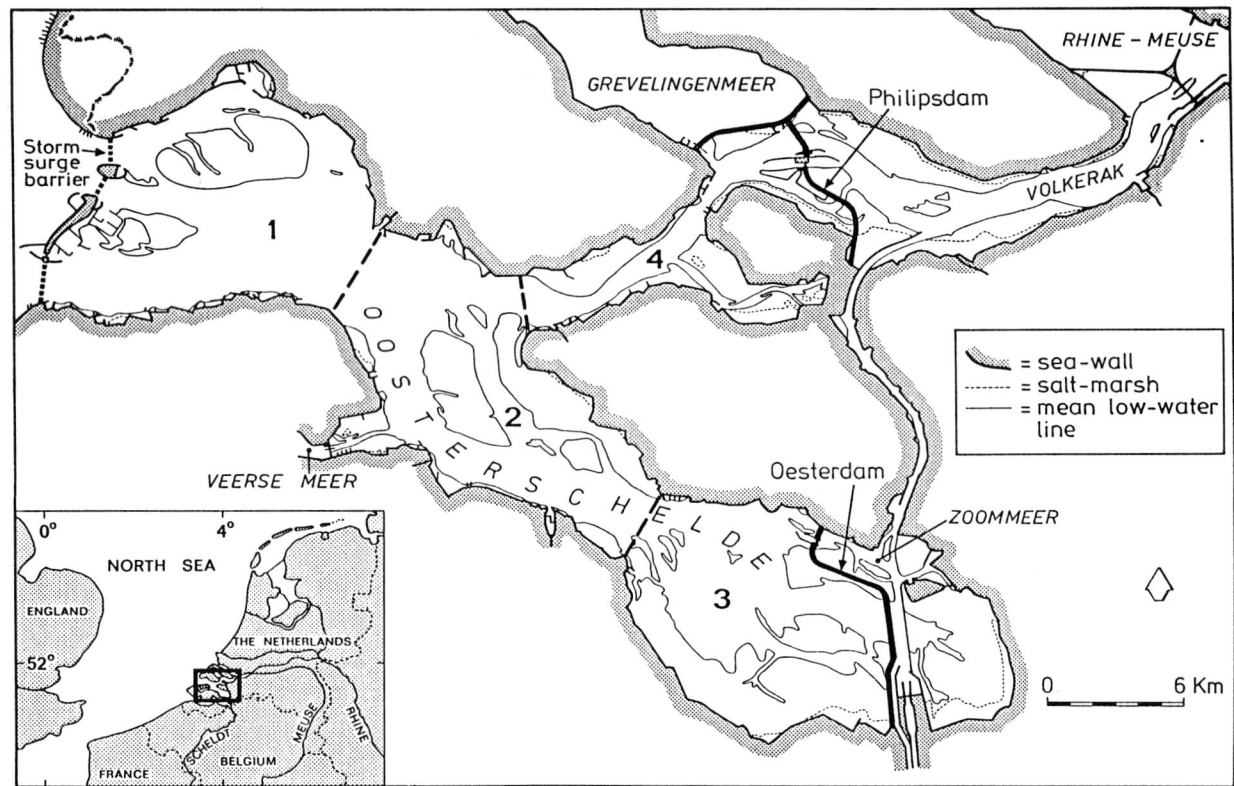

Fig. 1. The Oosterschelde estuary after the completion of the storm-surge barrier, the Philipsdam and the Oesterdam showing the four spatial compartments used in SMOES. Model compartment 1 = West, 2 = Central, 3 = East, and 4 = North.

gate adaptive changes within the ecosystem due to changing environmental conditions (Van der Tol & Scholten, 1992).

The building of a storm-surge barrier (1984–1987) caused a significant decrease in the amount of water entering and leaving the estuary during each tidal cycle with as major effects lower water flow velocities (25–80%) and lower suspended matter concentrations. The latter resulted in an increase in water transparency, which stimulates photosynthesis and thus algal growth. The coinciding lower nutrient concentrations (as a result of reduced loads of nutrient rich freshwater) counteracted the latter phenomenon and prevented an *a priori* prediction of the impact of the storm-surge barrier on the ecosystem with a qualitative approach (qualitative reasoning, qualitative models, professional judgement).

SMOES was designed as simple as possible to allow a thorough model analysis, but sufficiently detailed for an appropriate description of relevant ecological, chemical and physical processes within the constraints of the large number of field-observations collected and the very detailed approach many researchers insisted on. The ambiguity in these modelling design prerequisites produced a model which is too detailed in some ecological aspects (food consumption of mussels and cockles) while other processes are modelled in a rather coarse way (transport of particulate and dissolved substances). The resulting model includes descriptions of many processes each with several parameters of which many are not well known. The latter element requires a calibration method capable to deal with very complex optimization problems. Furthermore, any model prediction should be accompanied by an estimate of its uncertainty, especially if two model results are being compared. This requires tools to investigate these uncertainties, given the complexity and the non-linear interactions of the model at hand.

Most ecosystem models like those of Naranganset Bay (Kremer & Nixon, 1978), the Bristol Channel and Severn Estuary (Radford, 1978), the Ems-Dollard estuary (Colijn et al., 1987; Baretta & Ruardij, 1987; Baretta & Ruardij, 1988; De Jonge & De Groot; 1989), the Cumberland Basin (Keizer et al., 1987), the Western Wadden Sea (Lindeboom et al., 1989), the stagnant salt water Lake Grevelingen (De Vries et al., 1988), the stagnant brackish Lake Veere (De Vries et al., 1990), the heavily polluted Scheldt estuary (Ouboter, 1988; Van Eck & De Rooij, 1990), and a large number of fresh water systems (Los, 1980; Los et al., 1984; Bak et al., 1990) simulate ecosystems, which are not exposed to major, actually realized changes in the physical environment. SMOES has to deal with three significantly distinct states of the ecosystem due to the storm-surge barrier construction. Here we will discuss the pre- and post-barrier period only, as the transitional period (1984–1987) is not relevant for the questions we attempt to answer. The building of the storm-surge barrier induced a drastic and relatively rapid change in some Oosterschelde ecosystem properties. The many observations of these properties created a unique opportunity to compare the qualities of SMOES for both periods. In the previous mentioned ecosystems such discontinuities did not occur, which prevents a thorough validation of the other models.

A standard definition of validation is given by Schlesinger et al. (1979): 'Substantiation that a computerized model within its domain of applicability possesses a satisfactory range of accuracy consistent with the intended application of the model'. Many validation techniques have been suggested (Hermann, 1967; Wigan, 1972; Sargant, 1982, 1984, 1989a, 1989b; Lewandowski, 1982; Young, 1984; Reckhow, 1989), but most authors offer merely a terminology instead of a methodology.

SMOES could easily be invalidated or falsified, in the terminology of Popper (1959), but this is the case with most ecosystem simulation models. This does not mean that such an invalidated model would not be useful (Mankin et al., 1975). Its usefulness depends on the kind of answers we expect of the model and thus of its objective. Under certain circumstances and for a selected set of variables the model behavior can be judged as good or, preferably, useful, whereas the model can be considered unuseful under different circumstances or for other variables. As SMOES is not only intended to summarize research results, but also to predict future behavior (management tool), the model has to be validated: under which circumstances does the model predict correctly (Mankin et al., 1975)? Two aspects of validation, proposed by Mankin et al. (1975) are emphasized in this paper. First, we have to determine which part of the system behavior can be adequately simulated by the model (model adequacy). Secondly, which part of the model outcome matches system behavior (model reliability).

To validate SMOES we try to calibrate the same model for both the pre- and post-barrier period, assuming that the environmental changes in the ecosystem allow us to call this procedure a model validation (Wigan, 1972; Mankin et al., 1975; Beck, 1987). Subsequently we will try to answer several questions. Is the set of mathematical equations with its parameters, environmental forcing functions and other input, i.e. the model, an adequate and reliable description of both the pre- and post-barrier ecosystem? If we would have predicted the post-barrier ecosystem at the time of the building of the storm-surge barrier, how accurate would our forecast have been? To this type of prediction we will refer as 'historical forecasting'. What has been changed in the ecosystem and does the model outcome agree with ecological, chemical and other research results? If significant differences are found between the results of the pre- and post-barrier calibrations, what is the reason and how do we have to interpret this result? In Van der Tol & Scholten (1992) an attempt has been made to answer the latter question in more detail. Finally, the robustness of our calibration method will be discussed.

456

Methods

Concepts

Let R be the set of observable properties of the real system we are interested in. O is the set of observed data by lab experiment, direct measurement or even determined with some model, as is commonly done, this study not excepted, to 'measure' primary production of algae (Wetsteijn *et al.*, 1994). M is the set of model results, *i.e.* model outcomes for a series of output variables. Model results M are rather uncertain, because M depends on model parameters (P) which are not well known, the choice of model structure and uncertain model inputs like forcing functions (O'Neill & Gardner, 1979; Walters, 1986). The incomplete knowledge we have on the badly known parameters is collected by measurements, experiments and from literature. If a set of values for each of these badly known parameters is called a parameter vector, then there are many parameter vectors in agree with our (limited) *a priori* knowledge. Each of these parameter vectors will give different model results. In order to make statements on reality (R) we have to rely on observations of the reality (O) or on the *a priori* knowledge and general theories built in the model (M). We may expect that at least a part (but not all) of the observations match reality ($R \cap O \neq \varnothing$). In Fig. 2a model results (M) match partly with reality and with observations. Note that some model results coincide with the unobserved reality.

R, O, M and P are hyperspaces with time, space, and a series of (state) variables as dimensions. Model calibration aims to find those parameters vectors or the part of the parameter hyperspace which will produce the maximum of $M \cap O \cap R$. But, because R is unknown, calibration will be restricted to maximizing $M \cap O$ (the shaded area in Fig. 2b). Not knowing errors in O, will hamper calibration even further.

With this concept, model adequacy (which part of the ecosystem behavior can be simulated with the model?) is defined as $(M \cap O)/O$ and model reliability (fraction of model simulations matching observations) as $(M \cap O)/M$. Mankin *et al.* (1975) proposed these definitions, but did not distinguish R and O as two different sets. The ultimate goal of calibration, related to the inverse problem theory (Tarantola, 1987; Klepper & Rouse, 1991) is to find the parameter vectors (a subset of P), which realize perfectly adequate and perfectly reliable model results. Following Mankin (1975) we will call SMOES valid if $M - (O \cap M) = \varnothing$ (the model corresponds to system behavior under all conditions), but the model will be useful if $O \cap M \neq \varnothing$ (it represents some of the ecosystem behavior). Probably all ecosystem models fall into the latter, useful, but not valid class of models.

Because it was not possible to calculate the subspaces O and M in a more appropriate way, we use the number of observed data of a certain

a

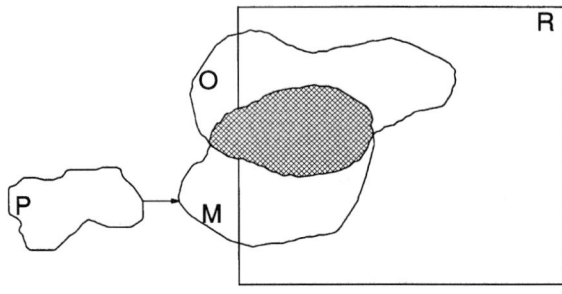

b

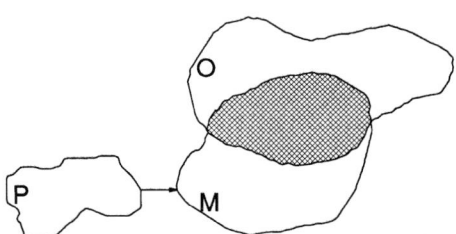

Fig. 2. Twodimensional projected hyperspaces of reality (R), observed data (O), model results (M), and, *a priori* known, parameter uncertainty (P), with known R (a) and unknown R (b). The shaded area is the overlap between O and M, which has to be maximized.

variable, within the model generated uncertainty bands, divided by the total number of observations in the simulation period as operational definition of the adequacy of the model to simulate the behavior of that variable. Instead of the reliability we use the median value of the Goodness-of-Fit of a variable (GoF V^{mean}, see (3)) as a measure for the reliability of the model with respect to that variable.

Goodness of Fit

Our automatic calibration method requires, like many optimization algorithms, an object or penalty function which has to be minimized. In general this object function (here called Goodness-of-Fit) is a quantitative measure for the discrepancy between model result and observed data. Obviously, we have to deal with a multivariate problem, but we will start with the univariate case and subsequently discuss the multivariate case (Scholten *et al.*, 1990a).

For each (output) variable, for which at least observed data are available in six output intervals within the simulation period, an average observation and normalized residuals are calculated:

$$\bar{d}_i = \frac{1}{n_i} \sum_{j=1}^{n_i} d_{ij}, \tag{1}$$

with:
$\bar{d}_i$ – weighted average of observed data of variable i within time interval j;
d_{ij} – observed data of variable i averaged over time interval j;
n_i – number of output intervals with observed data for variable i.

$$\check{e}_{ij} = \left| \left(\frac{y_{ij} - \bar{d}_i}{s_i} \right) - \left(\frac{d_{ij} - \bar{d}_i}{s_i} \right) \right| = \left| \frac{y_{ij} - d_{ij}}{s_i} \right|, \tag{2}$$

with:
$\check{e}_{ij}$ – normalized residuals of y_{ij} and d_{ij}; some variables are [10]log-transformed (Table 1);
y_{ij} – average model result of variable i for time interval j;

s_i – standard deviation of observed data of variable i.

For the calculation of the goodness of fit for variable i, the average of the normalized residuals of model result and observed data is calculated. It is often referred to as the L_1-norm (Tarantola, 1987; Klepper, 1989):

$$\mathrm{GoF}\,V_i^{\mathrm{mean}} = \frac{1}{n_i} \sum_{j=1}^{n_i} \check{e}_{ij}, \tag{3}$$

with:
GoF V_i^{mean} – time averaged Goodness-of-Fit of variable i;
n_i – number of output intervals with observed data for variable i.

In the multivariate case the goodness of fit of single variables have to be combined to one model goodness of fit which will be used in calibration:

$$\mathrm{GoF}M^{\mathrm{mean}} = \frac{1}{v} \sum_{i=1}^{v} W_i * \mathrm{GoF}\,V_{\mathrm{mean}}^i, \tag{4}$$

with:
GoFM^{mean} – time averaged Goodness-of-Fit of model;
W_i – weight of variable $i (0 \le W_i \le 1)$;
v – number of variables of which output is used for calibration.

Note that the relative importance of each variable in the calculation of the model GoF can be manipulated through the weights W_i. Low values for GoF V_i^{mean} (= good fit) mean that the behavior of variable i is simulated with high reliability.

Calibration algorithm

In the case of a linear model and a quadratic distance between model (M) and observed data (O) the calibrated parameter range can be calculated analytically. For nonlinear models the next procedure is therefore often followed (Beck & Young, 1976; Carver, 1980; Draper & Smith, 1981; Birta, 1984; Tarantola, 1987): run the model with an initial parameter vector and calculate the

Table 1. Number of observed data and attributes of output variables used for calibration.

Variable	Comp.[1]	Pre-barrier		Post-barrier		Weight[2]	Log[3]	Interpretation	Unit
		1982	1983	1988	1989				
bchlf	west	11	0	0	7[4]	1	+	benthic	$g\,m^{-2}$
	central	12	2	0	7[4]	1	+	chlorophyl	
	east	12	1	0	10[4]	1	+		
	north	11	0	0	0[4]	1	+		
chlf	west	47	54	36	34	1	+	phytoplankton	$g\,m^{-3}$
	central	68	80	43	39	1	+	chlorophyl	
	east	13	14	42	39	1	+		
	north	13	12	43	38	1	+		
cop	west	15	26	27	0	1	+	copepods	$(g\,C)\,m^{-3}$
	central	0	0	0	0	1	+		
	east	28	26	27	0	1	+		
	north	0	0	0	0	1	+		
din	west	41	44	12	12	1	−	dissolved	$g\,m^{-3}$
	central	33	45	43	39	1	−	inorganic	
	east	13	14	42	39	1	−	nitrogen	
	north	13	13	43	38	1	−		
extinc	west	28	35	0	27	0.1	+	extinction	m^{-1}
	central	31	39	0	32	0.1	+		
	east	17	19	0	33	0.1	+		
	north	0	0	0	32	0.1	+		
iopt	west	28	35	0	27	0.1	+	optimal light	$W\,m^{-2}$
	central	31	39	0	32	0.1	+	intensity for	
	east	17	19	0	33	0.1	+	phytoplankton	
	north	0	0	0	32	0.1	+	production	
ox	west	45	52	11	12	1	−	dissolved	$g\,m^{-3}$
	central	41	52	11	12	1	−	oxygen	
	east	13	14	11	12	1	−		
	north	13	13	10	12	1	−		
pmax	west	28	35	0	27	0.1	+	maximal	$(g\,C)\,(g\,Chlorophyl)^{-1}h^{-1}$
	central	31	39	0	32	0.1	+	growth	
	east	17	19	0	33	0.1	+	rate	
	north	0	0	0	32	0.1	+		
poc	west	47	54	36	28	1	+	particulate	$(g\,C)\,m^{-3}$
	central	42	53	41	32	1	+	organic	
	east	12	14	42	32	1	+	carbon	
	north	13	13	43	31	1	+		
proda	west	28	35	0	27	0.25	+	gross	$(g\,C)\,m^{-2}\,d^{-1}$
	central	31	39	0	32	0.25	+	primary	
	east	17	19	0	33	0.25	+	production	
	north	0	0	0	32	0.25	+		
silic	west	47	54	35	34	1	−	dissolved	$g\,m^{-3}$
	central	77	74	42	39	1	−	inorganic	
	east	13	14	41	39	1	−	silicon	
	north	13	13	42	38	1	−		

[1] Spatial compartment (Fig. 1).
[2] Weight used in calculating Goodness-of-Fit (Formula 4).
[3] + = ^{10}log-transformed in calculating Goodness-of-Fit and − = not transformed.
[4] Benthic chlorophyll concentrations of 1989 could not be used for calibration.

distance between M and O. Follow a path through the parameter space (P) leading to a better Goodness-of-Fit and find a best parameter vector by steepest descent or quasi-newton methods, and estimate a range for each parameter from the local behavior of the model around this best parameter vector. This method has several limitations, as the final results depend strongly on the initial guess, local minima in the distance between model results and data and, finally, local linearization may often lead to unrealistic results (Klepper *et al.*, 1991).

The calibration method we used here overcomes the disadvantages of the above mentioned, more generally applied method quite satisfactory (Klepper, 1989; Scholten *et al.* 1990a, Klepper *et al.*, 1991). The method we used, called 'Controlled Random Search', was originally proposed by Price (1979) as a constrained polynomial function-minimization and it solves complex optimization problems with many skewed or disjoint distributed parameters and with multiextremal object functions. Further it is related to optimization algorithms as 'Simulated Annealing' (Metropolis, 1953; Kirkpatrick *et al.*, 1983; Press *et al.*, 1986; Tarantola, 1987; Davis, 1987) and 'Genetic Algorithms' (Holland, 1975; Davis, 1987; Davis & Coombs, 1989; Goldberg, 1989).

To estimate n parameters the procedure starts with sampling N parameter vectors, with $N \gg n$. Here a Latin Hypercube Sampling method was used (McKay, Conover & Beckman, 1979; McKay, 1988; Iman & Helton, 1988) from the *a priori* parameter ranges and distributions. Subsequently, the model is run with each of the sampled parameter vectors, and the Goodness-of-Fit (4) is calculated and stored with the corresponding parameter vector. From a subset of k parameter vectors a centroid is calculated (5) and from the remainder of the $N - k$ stored vectors a new vector is drawn randomly and reflected in the centroid (6). If each parameter value of the newly generated parameter vector is within its initial range, the model will be run with this new parameter vector. If the Goodness-of-Fit with this new vector is better than the poorest fit of the stored vectors, the latter will be replaced by the new vector.

$$C_j = \frac{1}{k} \sum_{i=1}^{k} P_{ij}, \qquad (5)$$

with:

C_j – j^{th} element of the centroid;

P_{ij} – j^{th} element of the ith parameter vector of the N stored;

k – number of vectors used to calculate the centroid.

The jth element of the reflected parameter vector is given by:

$$P'_j = 2 * C_j - P_{(k+1)j}, \qquad (6)$$

with:

P'_j – j^{th} element of reflected parameter vector;

$P_{(k+1)j}$ – j^{th} element of a randomly drawn parameter vector from the stored parameter vectors (not used for calculating the centroid).

If some stopcriterion is met the procedure stops, but in all calibrations presented here we used a constant number of 5000 calibration runs. Here we calibrate SMOES for 88 badly known parameters, of which 500 vectors are stored, so the condition, $N \gg n$, was satisfied.

Uncertainty analysis

If a model has to be used for quantitative predictions, an estimate of its uncertainty should be carried out (Klepper, 1989; Klepper *et al.*, 1991). To demonstrate the uncalibrated model a simple Monte-Carlo-approach (Fedra, 1983) was used with a Latin Hypercube Sampling (McKay, Conover & Beckman, 1979; McKay 1988; Iman & Helton, 1988) of the parameters independently from their initial ranges (see Table 2a-e). To present the results of the calibration, SMOES was run with each of the stored parameter vectors, of which many are significantly correlated. The huge resulting files are shown at each output interval as maximum, minimum, and median variable values.

Table 2a. Ranges of the parameters of the pelagic submodel used for calibration.

Name	Initial range		Pre-barrier cal.		Post-barrier cal.		Unit	Meaning
	min.	max.	min.	max.	min.	max.		
airfac	0.5	3	1.1886	2.4760	1.0901	2.5325	m d^{-1}	reaeration coefficient
arat	1	4	1.2265	3.9206	1.1116	3.0916	–	favor coeff. other algae vs. diatoms
cchlmin	0.008	0.02	0.0091	0.0169	0.0099	0.0167	(g C) (mg CHLF)$^{-1}$	min. carbon/chlorophyll ratio
dayrzoo	0.5	2	0.8050	1.3864	0.8806	1.7618	d^{-1}	max. daily ration at 15 °C copepods
dislitf	0.4	1	0.5362	0.9005	0.518	0.9153	–	exponent in chlorofyl/carbon litfun
effzoo	0.35	0.9	0.4258	0.7140	0.4865	0.8124	–	assimilation efficiency zooplankton
exmaxq	0.01	0.7	0.1192	0.4075	0.1930	0.5805	–	max. excretion fr. of nutrient lim. prod.
foodlim	0.2	0.3	0.2333	0.2805	0.2224	0.2805	(g C) m^{-3}	treshold foodconc. copepods
iopt10	100	140	110	124	107	124	W m^{-2}	optimal light intensity algae
mrtqq	2	5	2.3717	4.8547	2.5863	4.8284	(g C)$^{-1}$ m^{-3} d^{-1}	mortality rate zooplankton
pmax10	0.2	0.6	0.2925	0.4710	0.3215	0.5195	(mg C) (mg CHLF)$^{-1}$ h^{-1}	pmax at 10 °C, no nutrient limitation
q10alg	1.5	2.5	1.7762	2.3612	1.7757	2.4184	–	q10 other algae
q10dia	1.5	2.5	1.8959	2.4620	1.6027	2.3097	–	q10 diatoms
q10zin	1.5	4	1.7373	3.2017	1.5012	2.3368	–	q10 zooplankton ingestion
q10zrs	1.5	3	1.7279	2.4846	1.6457	2.8054	–	q10 zooplankton respiration
resfrac	0.09	0.43	0.1208	0.3612	0.1208	0.3016	–	resp. fr. of algal production
resqmin	0	0.09	0.0103	0.0670	0.0264	0.0760	d^{-1}	minimal diatom resp. at 10 °C
resqz	0.05	0.25	0.0662	0.1807	0.0506	0.1817	d^{-1}	respiration rate zooplankton
slopechl	0.21	0.28	0.2210	0.2516	0.2119	0.2544	(g C) (g CHLF)$^{-1}$ h^{-1} (W m^{-2})$^{-1}$	slope phytoplankton production curve

Table 2b. Ranges of the parameters of the zoobenthos submodel used for calibration.

Name	Initial range		Pre-barrier cal.		Post-barrier cal.		Unit	Meaning
	min.	max.	min.	max.	min.	max.		
accoc	0.024	0.09	0.0344	0.0796	0.0368	0.0687	m^3 d^{-1}	a-value in a*w^b cockle clearance
acmus	0.03	0.09	0.0455	0.0823	0.0375	0.0668	m^3 d^{-1}	a-value in a*w^b mussel clearance
arcoc	0.002	0.01	0.0020	0.0055	0.0020	0.0047	(g C) d^{-1}	a-value in a*w^b cockle respiration
armus	0.0029	0.0096	0.0031	0.0055	0.0029	0.0055	(g C) d^{-1}	a value in a*w^b respiration mussels
bccoc	0.5	0.7	0.5305	0.6558	0.5537	0.6745	–	b-value in a*w^b cockle clearance
bcmus	0.3	0.7	0.3887	0.6874	0.3152	0.5903	–	b-value in a*w^b clearance mussel
brcoc	0.23	1	0.6453	0.9098	0.6170	0.9207	–	b-value in a*w^b cockle respiration
brmus	0.2	1.1	0.6338	1.0831	0.4327	0.9709	–	b value in a*w^b respiration mussels
cocfac(1)	0.25	1.75	0.4144	1.5594	0.5646	1.4267	–	factor for cockle biomass manipulation
cocfac(2)	0.25	1.75	0.4766	1.3224	0.5492	1.4628	–	
cocfac(3)	0.25	1.75	0.4583	1.6746	0.6874	1.5096	–	
cocfac(4)	0.25	1.75	0.5513	1.2061	0.6787	1.4333	–	
cockmort	0.0029	0.0055	0.0035	0.0051	0.0032	0.0048	d^{-1}	relative mortality rate cockles
crtrate	0.024	0.05	0.0249	0.0458	0.0310	0.0448	m^3 d^{-1} (g adw)$^{-1}$	hardsub filterfeeders clearance at 10 °C
effcoc	0.12	0.22	0.1452	0.2043	0.1309	0.2130	–	assimilation efficiency of cockles
effmus	0.09	0.43	0.1699	0.3734	0.1665	0.3925	–	assimilation efficiency of mussels
eftmax	0.85	0.95	0.8592	0.9024	0.8694	0.9389	–	factor for max. assimilation hardsub ff.
psfcoc	2	75	17	43	17	49	g m^{-3}	pseudofaeces treshold conc. cockles
psfmus	2	74	11	59	8	53	g m^{-3}	pseudofaeces treshold conc. mussels
q10cocc	0.9	1.5	0.9782	1.4404	1.0120	1.3374	–	q10 value cockle clearance
q10cocr	0.9	1.5	1.0374	1.3911	0.9671	1.2219	–	q10 value cockle respiration
q10musc	0.9	2.5	1.4616	2.4861	1.2242	2.1633	–	q10 value mussels clearance
q10musr	0.9	1.5	0.9206	1.4151	0.9011	1.2553	–	q10 value mussel respiration
sesqc	0	0.004	0.0014	0.0039	0.0016	0.0037	m^3 g^{-1}	reduction factor cockle clearance
sesqm	0	0.005	0.0008	0.0041	0.0012	0.0037	m^3 g^{-1}	reduction factor mussel clearance

Table 2c. Ranges of the parameters of the bottom micro-processes submodel used for calibration.

Name	Initial range min.	Initial range max.	Pre-barrier cal. min.	Pre-barrier cal. max.	Post-barrier cal. min.	Post-barrier cal. max.	Unit	Meaning
bch10coef	0.001	0.1	0.0208	0.0637	0.0262	0.0848	d^{-1}	mineralisation on channel bottoms
bioturq	0.01	0.03	0.0109	0.0235	0.0109	0.0232	$d^{-1}(g\,C\,p\,m^{-2})^{-1}$	bioturbation coefficient
bn10coef	0.001	0.05	0.0115	0.0321	0.0147	0.0399	d^{-1}	coef. for denitrification on tidal flats
bresfrac	0.01	0.2	0.0628	0.1829	0.0269	0.1672	–	respiration fr. of microfytobenthos prod.
bresmin	0.001	0.1	0.0025	0.0440	0.0350	0.0815	d^{-1}	minimal respiration microfytobenthos
cflux	0.05	3	1.1931	2.2059	0.3164	2.5332	$(g\,C)\,m^{-2}\,d^{-1}$	CO_2-flux at water/air interface
dcoef	0.0001	0.02	0.0005	0.0128	0.0004	0.0089	$d^{-1}(g\,C\,m^{-2})^{-1}$	prey eaten per g depositfeeder per day
dshapeb	0.5	5	1.9913	4.4614	1.4112	3.8212	$g\,C\,m^{-2}$	shape par. for biomass distr. in bottom
faecdfrac	0.3	0.9	0.4097	0.7731	0.5048	0.8656	–	faecal faction for deposit feeders
faechfrac	0.3	0.9	0.4905	0.8692	0.4796	0.7995	–	faecal fraction for surf.dep.feeders
faecmfrac	0.3	0.9	0.3159	0.7862	0.3222	0.8395	–	faecal fraction for meiobenthos
hcoef	0.001	0.03	0.0069	0.0203	0.0048	0.0248	$d^{-1}(g\,C\,m^{-2})^{-1}$	prey eaten per gr. surf.dep.feed. per d.
ioptb10	50	120	66	109	55	115	$W\,m^{-2}$	iopt phytb at 10 °C
kzpal	0.01	0.03	0.0138	0.0230	0.0134	0.0259	$m^{-2}\,W^{-1}$	light attenuation in upper bottom layer
maxconsb	100	400	225	326	157	328	$g\,CHLF\,m^{-2}$	max. chlf. conc. mfb. in pal
mcoef	0.001	0.03	0.0080	0.0215	0.0012	0.0206	$d^{-1}(g\,C\,m^{-2})^{-1}$	prey eaten per gram meiob. per day
pmaxb10	0.2	0.8	0.3440	0.5983	0.3802	0.6840	$(g\,C\,m^{-2}\,h^{-1})(g\,C\,m^{-2})^{-1}$	max; productivity
q10bp	0.9	3	1.5616	2.3728	1.5721	2.9670	–	q10 phytobenthos production
q10graz	1	4	1.6295	3.5184	2.1733	3.4774	–	q10 for hydrobia's, depf. and meiob.
q10pbres	0.9	2.5	1.1420	2.0995	1.4262	2.2660	–	q10 of phytobenthos respiration
slopeb	0.002	0.01	0.0042	0.0081	0.0045	0.0091	$d^{-1}(W\,m^{-2})^{-1}$	slope of production curve
zpalmax	1	3	1.0655	2.6003	1.1053	2.5408	mm	maximal photosynthetic ative layer

Pre- and post-barrier model and observed data

The parameters of which the uncertainty was used to calibrate the model, are given in Tables 2a–2e, including some of their attributes.

The output variables used for calibration, including the number of samplings for the pre- and post-barrier period, are presented in Table 1. The choice of a fixed volume transport modelling approach (Klepper, 1989; Klepper *et al.*, 1994)

Table 2d. Ranges of the parameters of the forcing function submodel used for calibration.

Name	Initial range min.	Initial range max.	Pre-barrier cal. min.	Pre-barrier cal. max.	Post-barrier cal. min.	Post-barrier cal. max.	Unit	Meaning
botrat	0.5	1	0.6027	0.9575	0.5752	0.8616	–	ratio bottom min. rate/water min. rate
bsilt	1.5	2.5	1.7388	2.3340	1.7073	2.3851	–	ratio near bottom conc./surface conc. silt
cchlb	0.02	0.06	0.0245	0.0600	0.0251	0.0535	$(g\,C)(mg\,CHLF)^{-1}$	carbon chlorophyl ratio for microphytobenthos
cwmin10q	0.004	0.14	0.0472	0.1374	0.0375	0.1174	d^{-1}	mineralisation rate carbon (10 oc)
dqalg	0.6	0.9	0.6752	0.8599	0.6416	0.8889	–	fraction algae dissolved like behaviour
dqdet	0.4	0.6	0.4127	0.5550	0.4230	0.5199	–	fraction detritus dissolved like behaviour
dqpsil	0	1	0.1310	0.8617	0.0729	0.8236	–	fr. particulate silicon dissolved like behaviour
kmdin	0.005	0.2	0.0274	0.1149	0.0179	0.1173	$(g\,N)\,m^{-3}$	saturation constant for din
kmsil	0.002	0.2	0.0345	0.1255	0.0461	0.1864	$(g\,Si)\,m^{-3}$	saturation constant for dissolved silicon
mufrac	0.1	0.4	0.1814	0.3807	0.2046	0.3540	–	fr. of poc below 3 μm: not edible for zoopl.
ncrat	0.06	0.2	0.0683	0.1981	0.1069	0.1968	$(g\,N)(g\,C)^{-1}$	N/C ration in all ecogroups
psil10	0	0.1	0.0121	0.0997	0.0319	0.0962	d^{-1}	mineralisation rate silicon at 10 °C
q10cmin	1	3	1.1563	2.9614	1.6608	2.9965	–	q10 C and N miner. in pgic and bmic
q10smin	1	3	1.1324	2.7028	1.1557	2.3926	–	q10 Si regeneration in pgic and bmic
sicrat	0.15	0.6	0.1731	0.4580	0.2488	0.4702	$(g\,Si)(g\,C)^{-1}$	Si/C ratio in all ecogroups

Table 2e. ranges of the parameters of the transport submodel used for calibration.

Name	Initial range		Pre-barrier cal.		Post-barrier cal.		Unit	Meaning
	min.	max.	min.	max.	min.	max.		
qpocsed	0	0.1	0.0145	0.0428	0.0134	0.0748	–	ratio rdet increase at surface to silt export
qpsilsed	0	0.01	0.0011	0.0093	0.0034	0.0096	–	ratio psil increase at surface to silt export
sedlabq	0	0.05	0.0086	0.0383	0.0136	0.0385	–	labile fraction of detritus in sediment
tsiltfac(1)	0.25	1.75	0.6376	1.4875	0.3637	1.6135	–	factor for manipulation of fixed amount of total silt
tsiltfac(2)	0.25	1.75	0.4829	1.3024	0.3075	1.1966	–	(bottom + suspended)
tsiltfac(3)	0.25	1.75	0.6006	1.4846	0.7192	1.6904	–	
tsiltfac(4)	0.25	1.75	0.2998	1.6840	0.8254	1.5554	–	

required a correction of the raw observed data. Using the sampling time and date, the sampling position at mid tide was calculated. Linear interpolation produced average values for the center of each compartment.

For the pre- and post-barrier model runs SMOES had not been modified, except for the following details. Different initial values of most state variables were used for both periods. Compared to the pre-barrier conditions many of the physical attributes were changed (volumes, mean water level area, tidal flat area, average emersed periods of tidal flats, depths, see Nienhuis & Smaal (1994)). Some transport aspects are different (transport of particulate matter over the compartment boundaries was set to zero for the post-barrier period). For the pre-barrier period no particulate silicon erosion was assumed, opposite to the post-barrier period. Finally we used different dispersion coefficients for both periods estimated with the calibration method, using observed salinity data.

Model experiments

SMOES is calibrated for the pre- and post-barrier period. Here we show results (including uncertainty) of the uncalibrated model, the pre-barrier calibration, the post-barrier calibration, and a test on the consistency of the calibration method. The effect of uncertain initial values was not shown, because the effect was marginal and noticeable for a short period only.

Finally we investigated what SMOES would have predicted for the post-barrier period at the time of the building of the storm-surge barrier (historical forecasting).

Software

The model was developed and investigated (calibration and uncertainty analysis) with SENECA version 1.2 (Simulation ENvironment for ECological Application; Scholten *et al.*, 1990a). Figures and tables were also produced with SENECA. Correlation matrices were calculated and Principal Component Analyses were carried out using specially developed software.

Results

Results without calibration

Without calibration the results showed wide uncertainty bands for most variables. As examples phytoplankton biomass (chlorophyll), zooplankton biomass, dissolved inorganic nitrogen, dissolved silicon, phytoplankton primary production, and particulate organic carbon in the Oosterschelde compartment West are shown (Fig. 3a–f).

The adequacy is high (Fig. 5a) and the model shows system behavior quite well, but reliability is low (Fig. 5b). Making predictions with the uncalibrated model will give unacceptable uncertain results.

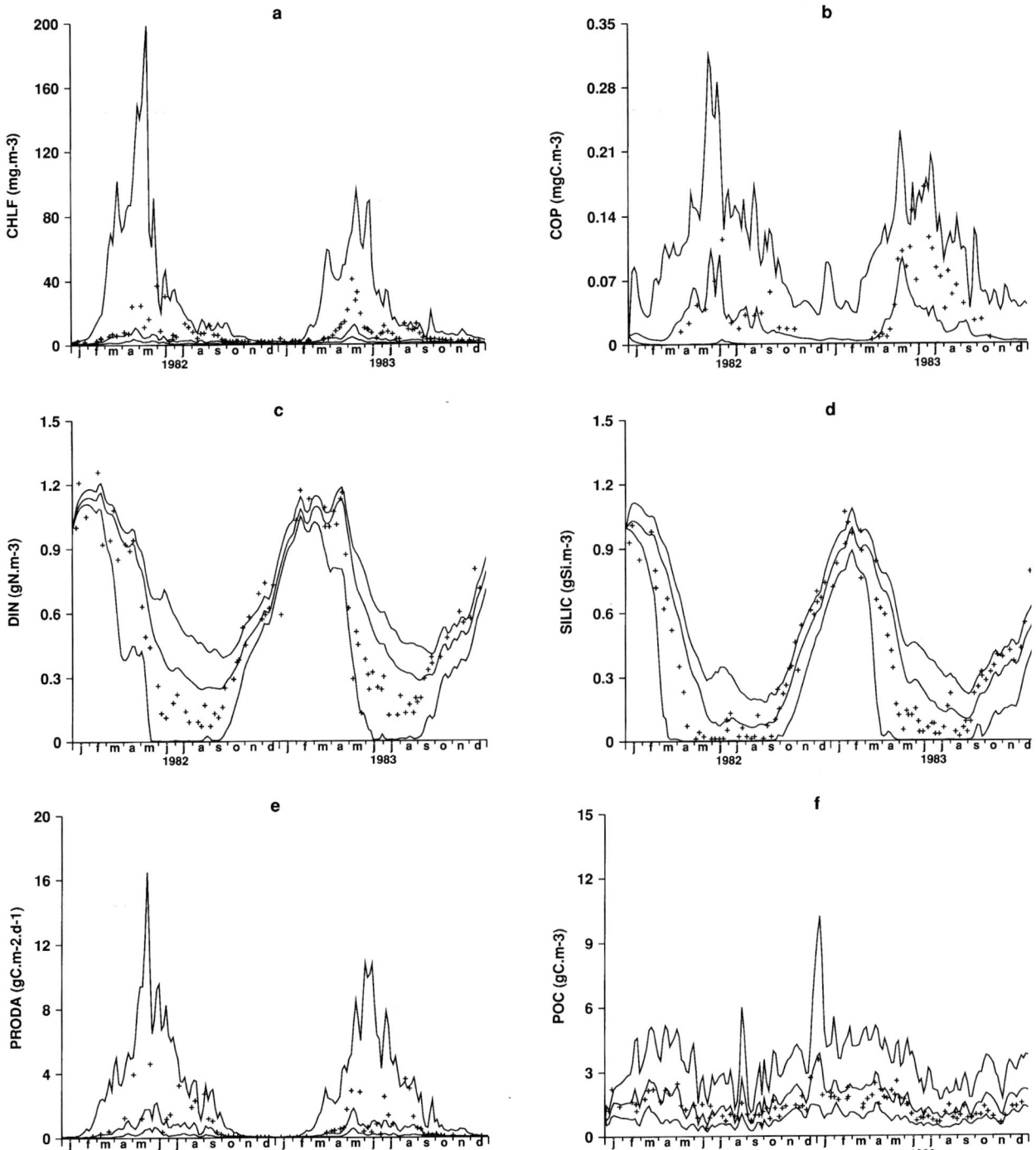

Fig. 3. Uncertainty in the model results of the uncalibrated model using the *a priori* uncertainty in 88 model parameters for the pre-barrier period in compartment West, a - Chlorophyll-*a* concentration as measure of phytoplankton biomass (CHLF), b - zooplankton biomass (COP), c - dissolved inorganic nitrogen concentration (DIN), d - dissolved silicon (SILIC), e - primary production of phytoplankton (PRODA), f - particulate organic carbon concentration (POC). Minimum, median and maximum values of the model output space at each output interval are given as lines. Observed data = + .

464

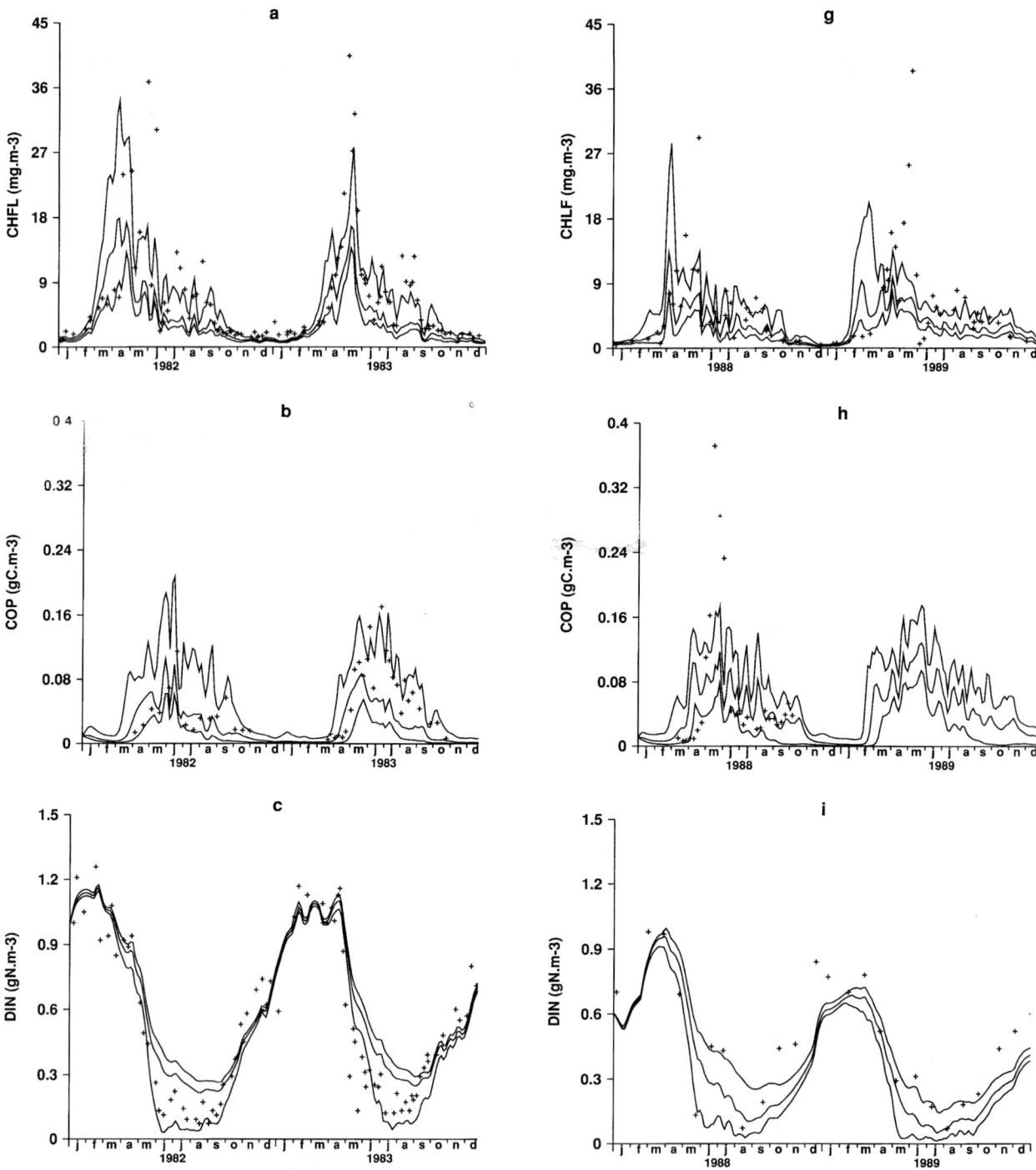

Pre-barrier and post-barrier calibration

Both in the pre- and post-barrier calibrations the uncertainty in all variables is decreased compared to the uncalibrated model (Fig. 4a–l). Goodness-

of-Fit values are decreased (which means a better fit between model and observations and thus a higher reliability) at the cost of a lower model adequacy (Fig. 5).

In 1982 and 1983 the spring bloom peaks in

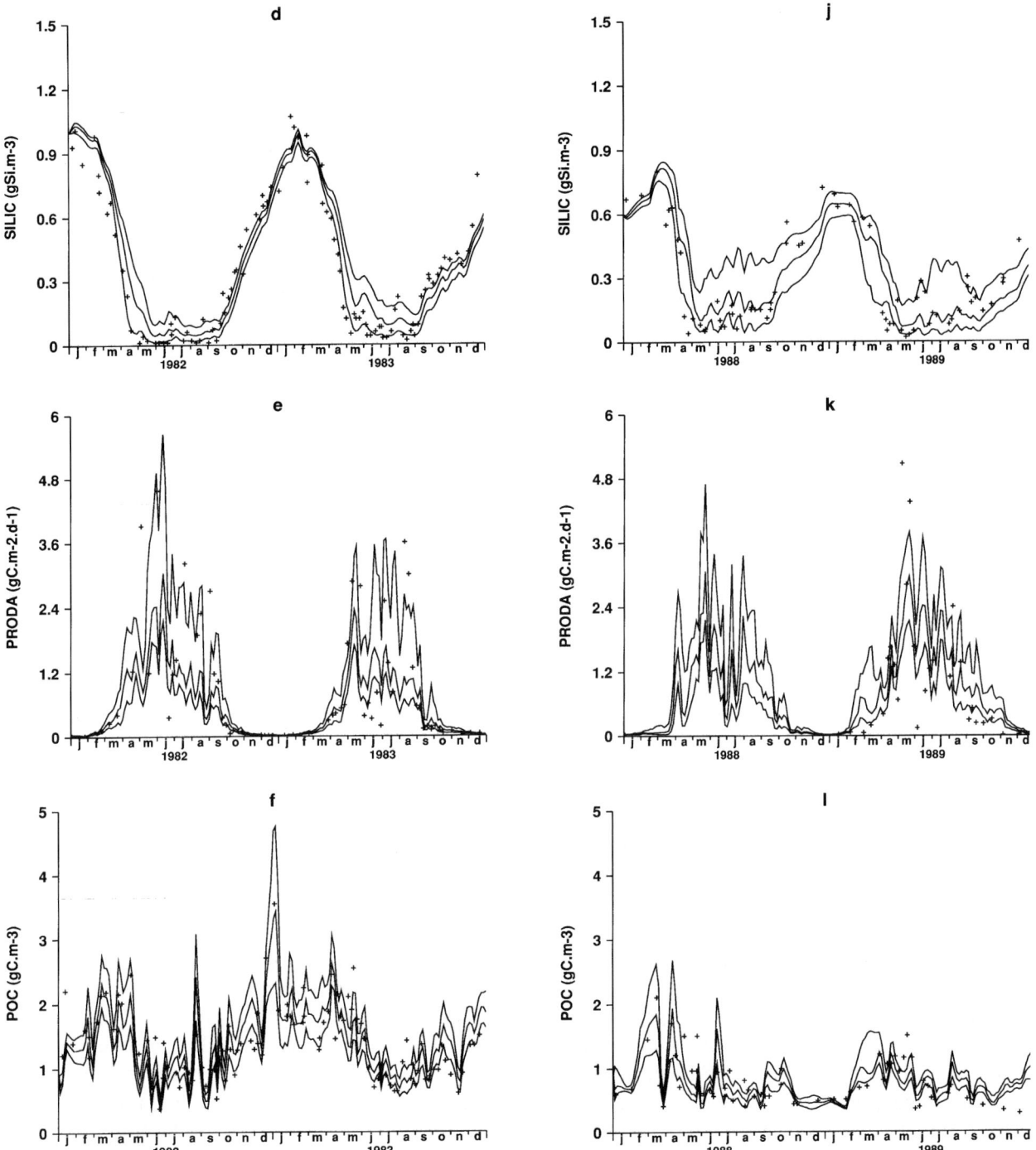

Fig. 4. Uncertainty in the model results using the 500 parameter vectors from a 5000 run calibration for the pre-barrier period (a to f) and the post-barrier period (g to l). a and g - Chlorophyll-*a* concentration as measure of phytoplankton biomass (CHLF), b and h - zooplankton biomass (COP), c and i - dissolved inorganic concentration (DIN), d and j - dissolved silicon (SILIC), e and k - primary production of phytoplankton (PRODA), f and l - particulate organic carbon concentration (POC). See also Fig. 3.

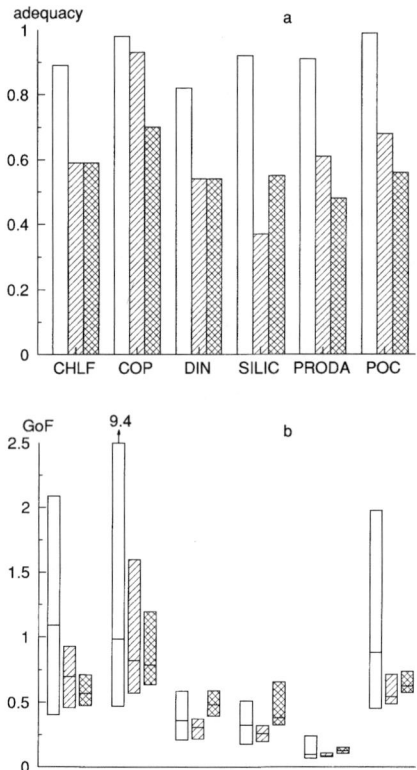

Fig. 5. Adequacies (a) and Reliabilities (b) of the uncalibrated pre-barrier results (empty), the calibrated pre-barrier results (hatched), and the calibrated post-barrier results (crosshatched) in compartment West of the Oosterschelde. Adequacies are represented by a single value and reliabilities are given as maximum, median and minimum. High Goodness of Fit values of a variable correspond with a bad fit and a low reliability. See text for definitions of Adequacy, Reliability, and Goodness-of-Fit and Fig. 3 for the variable names.

phytoplankton chlorophyll are not fully simulated by the model (Fig. 4a). The post-barrier years, 1988 and 1989 show this effect even more pronounced (Fig. 4g). Summer values are slightly too low. Zooplankton results (Fig. 4b) are correct for 1983. SMOES overestimates 1982 values and the May 1988 peak is not simulated at all (Fig. 4h). The observed zooplankton data of 1989 were not available at the time of the calibration exercise. Nutrients are simulated acceptably good in summer, but winter concentrations are less satisfactory (Fig. 4c, 4d, 4i and 4j). Primary production of phytoplankton (diatoms and other algae) was accurately simulated, both average and variance

(Fig. 4e and 4k). Particulate organic carbon (including phytoplankton and detritus) shows a good average level and an appropriate variance compared to the observed concentrations (Fig. 4f and 4l). Other output variables (not shown) produce similar results as those presented here.

Testing the calibration method

To test the consistency of the calibration method, we applied a Principal Component Analyses on the correlation matrix of the resulting parameter vectors after calibration. The giant correlation matrix showed that more than half of the parameters are significantly correlated (at a 5% level, two sided). At least 95% of the model variance was represented by the first 15 to 17 Principal Components. This means that SMOES is strongly overdimensioned in parameter numbers: 15 to 17 new parameters could generate almost all model outcome variance.

A second result of the PCA was the inconsistency of the resulting Principal Components, when carried out for several replicate calibrations (each with a different seed for the LHS random number generator). Each Principal Component is a linear combination of the (original) model parameters and we expected to find some resemblance between the Principal Components of different calibrations. Such similarities were not discovered however. This shows that the applied calibration method does not generate consistent parameter vectors.

Comparison of model results produced by several calibrations is a second way to evaluate the calibration method. We carried out four different post-barrier calibrations with four different seeds for the LHS random number generator. The results for phytoplankton biomass (Fig. 6a–d) and phytoplankton primary production (Fig. 6e–h) show quantitative differences, but are qualitatively similar.

We also may expect high resemblance in the adequacy and reliability of a variable between the four calibrations with different seeds. Adequacies of both phytoplankton biomass (chlorophyll) and phytoplankton primary production are rather consistent for each of the four seeds (Fig. 7a). The Goodness-of-Fit values for phytoplankton biomass (chlorophyll) of the four post-barrier calibrations are in similar ranges (Fig. 7b). Phytoplankton primary production shows even more consistent results (Fig. 7b). The calibration method seems to produce (qualitatively) consistent results in respect to the model output. The results of four other (historical forecasting) calibrations (Fig. 7) support our conclusion.

Historical forecasting

Besides its descriptive (research summarizing) purpose, SMOES was also intended to be used as predictive (management) tool. To test its predictive potentials we used the model as it was at the time of the completion of the storm-surge barrier, but with the real system inputs (freshwater loadings, temperature, light, etc.) to 'forecast' the post-barrier period. We used four pre-barrier calibrations with post-barrier system inputs and compared the results of these historical forecasts with four post-barrier calibrations.

The results (Fig. 6a–d) show that the historical forecasts predict consistently higher phytoplankton biomasses (chlorophyll-*a*) than the post-barrier calibrations. Reliabilities of the historical forecasts and post-barrier (normal) calibrations were more or less comparable, but slightly better (lower Goodness-of-Fit) for the normal post-barrier calibration (Fig. 7b).

Goodness-of-Fit values of phytoplankton primary production were in all cases better (with lower values) than those of phytoplankton biomasses (chlorophyll). The historical forecasts overestimated the primary production consistently (Fig. 6e–h). The reliability of the post-barrier calibration was somewhat better (lower values of the Goodness-of-Fit) than the reliability of the historical forecast primary production (Fig. 7b).

No significant differences in adequacies between the set of replicates of the historical forecasts and the set of normal calibrations for both phytoplankton biomass and primary production were found. This proves that SMOES could have been used to predict these two major variables at the time of the building of the storm-surge barrier with equally results (both for adequacy and reliability) as it can be used now, when the behavior of the Oosterschelde ecosystem is known.

Discussion

A model is valid if its behavior corresponds to system behavior under all conditions (Mankin *et al.*, 1975). SMOES is thus an invalid model. Yet, SMOES is a useful model, because for almost all variables, of which observed data are available (Table 1), there is some overlap between model output (M) and observed data (O): SMOES accurately simulates some of the system behavior. All ecosystem models are invalid, as there always exists an experiment or measurement, which can not be represented by the model. Model validity, in this sense, is an irrelevant item. We try to find out to what extent SMOES is a useful model. All simulation models aim to represent system behavior with a limited set of knowledge. SMOES was developed for two purposes: as a summary of ecosystem research and to predict future behavior. The first aim does not require any validation at all, unless an inspection of face validity (Hermann, 1967; Sargent, 1984), which has been carried out in several workshops with the researchers, whose ideas fuelled the model. How we deal with the second objective (SMOES as a prediction tool) will be discussed below.

A simulation model consists of a model structure (which state variables, process formulations, etc.), initial values of the state variables, well known parameters, badly known parameters, and system inputs. Klepper *et al.* (1994) give a description of the model structure of SMOES. We just tested a single (the present) model structure, using the initial uncertainty in the parameters (*a priori* knowledge) or the highly correlated parameter vectors, generated by calibration, and, obviously, the numerous, yet insufficient, observed data from two ecosystem studies (Nienhuis & Smaal, 1994).

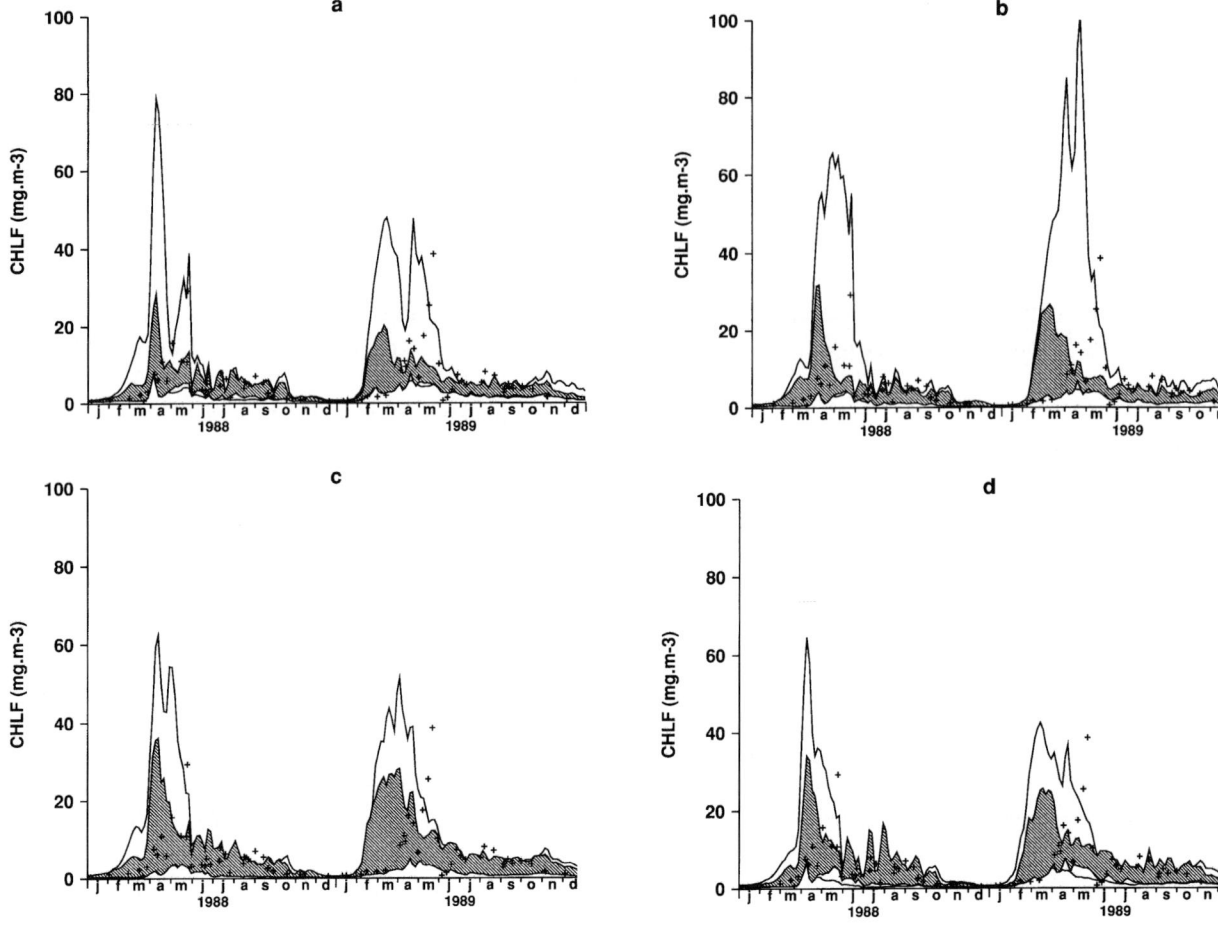

Observed data were averaged for a spatial compartment after correction for tide. This approach will introduce errors, as will sampling frequency (Jørgensen, 1986), sampling and analysis methods (especially for patchy distributed particulate matter like algae and detritus). We only have a vague idea of these errors in the data used for calibration, and therefore we assume to know observed data very precisely without errors.

Uncertainty in the initial values of state variables is a minor source of uncertainty in the Oosterschelde ecosystem. Effects are limited to the first two or three months only (Klepper, 1989). The results are not shown here.

We suppose that most system inputs are fully known, which obviously is not true. The only uncertain system inputs we included here were the biomasses of cockles and the total silt con-

centrations, because of the enormous yearly uncomprehended variances in these variables. Standing stocks of mussels, which have an equally important impact on the Oosterschelde ecosystem as cockles, are also difficult to measure. Mussels are patchy distributed, but, opposite to cockles, mussel biomass is mainly man manipulated (mussel culture), without the enormous natural variability commonly found in Oosterschelde cockle reproduction. The mismatch of winter nutrient concentrations (Fig. 4c, 4d, 4i and 4j) motivated us to investigate the effects of uncertain nutrient and particulate organic carbon loadings (from marine origin and other sources). Assuming that the associated uncertainty is time invariant, we defined two additional parameters to control these inputs. A Monte Carlo uncertainty analysis, while keeping all other parameters at

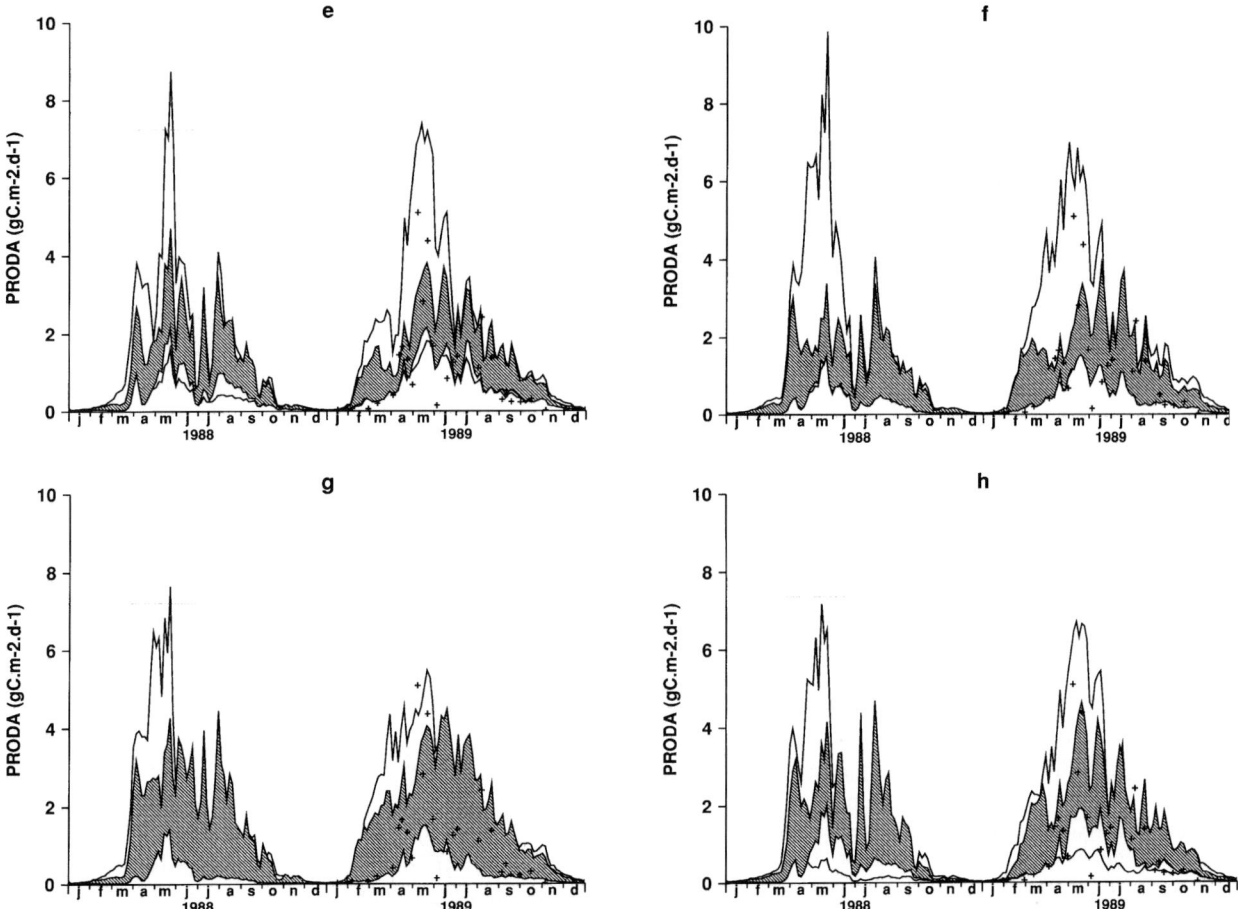

Fig. 6. Historical forecast results, generated with four different seeds of the LHS random number generator compared to four post-barrier calibration results. a–d - Chlorophyll-*a* concentration as measure of phytoplankton biomass (CHLF), e–h - primary production of phytoplankton (PRODA). The hatched area covers the minimum and maximum results at each output interval of the post-barrier calibration and the lines represent the minimum and maximum results of the historical forecast. See also Fig. 3 and text.

their best (calibrated) values, showed that winter concentrations of both dissolved nitrogen and dissolved silicon could match the observed data perfectly (not shown). Effects on other variables were directly related to nutrients and light extinction (as an effect of varying silt concentrations). Although we certainly would have attained a better match between model results (M) after calibration and observed data (O), we did not include these 'loading' parameters in our other analyses, as we intend to investigate the system itself and not to achieve just an optimal fit.

Without an appropriate stopcriterion Controlled Random Search (Price, 1979), the method

we used for model calibration, will reduce uncertainty in the model outcome infinitely, without guarantee that the observed data fall within the resulting output ranges. We can not use any information on the errors in the observed data, which could enable us to halt the calibration just in time. Therefore we use a constant number of 5000 optimization runs, which contributes an arbitrarily aspect to the procedure. The weights in the calculation of the Goodness-of-Fit (Table 1) are also subjective aspects. These weights have been chosen accordingly to the relative importance we attached to an accurate simulation of the variable under consideration. Finally, the

470

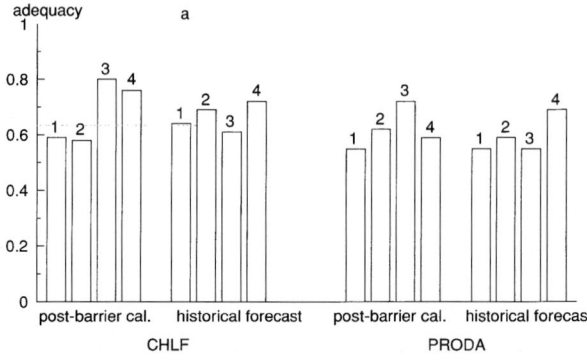

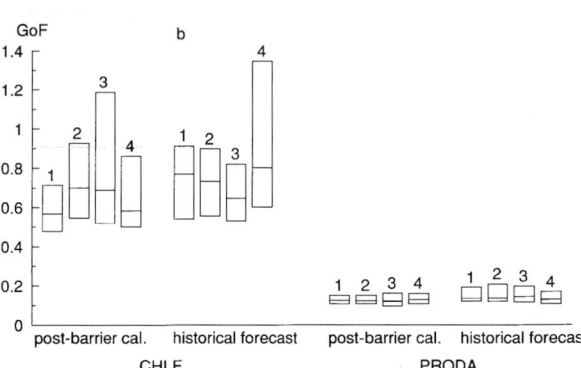

Fig. 7. Adequacy (a) and Goodness-of-Fit (b) of phytoplankton chlorophyll-*a* and phytoplankton primary production for
8 different calibrations, of which 4 historical forecasts. See
text for definitions of Adequacy, Reliability, and Goodness-
of-Fit and also Fig. 3. Numbers above bars indicate replicate.

number of stored parameter vectors is chosen
arbitrarily. Too less parameter vectors will easily
reduce the dimension of the parameter (hyper)-
space with the risk of elimination the most interesting subspace. Storing too many parameter
vectors will slow down convergence, but the parameter space will be searched more thoroughly.

The choice of the years 1982–1983 as representative for the pre-barrier period and 1988–
1989 for the post-barrier period was enforced by
the availability of observed data and, obviously,
by the building of the storm-surge barrier between
1984 and 1987. Variance in climate (water temperature and irradiance) between years is not very
large, but 1988 had a low irradiance (> 10%
below a 10 year average), which might have influenced post-barrier calibration results.

The calibration procedure yields two different
results: a set of parameter vectors and model out-

comes using this set of parameters. The absence
of consistent results in the resulting set of parameter vectors produced with different random
number seeds is due to high correlation between
many parameters, and effects of one parameter
are easily compensated by one or more other
(correlated) parameters: both parameter value
combinations have similar effects on model outcome. Comparison of results, simulated using the
parameter vector sets from different calibrations,
shows qualitative similarities. The inconsistency
of resulting parameter vectors and the lack of
quantitative similar model outcomes from different calibrations is primarily due to unidentifiabil-
ity of these type of simulation models. There is no
unique 'best' combination of parameter values
which fit the data: many combinations are equally
good (Beck, 1987). As we stated above, the set of
parameter vectors produced by calibration shows
high correlation between parameters, associated
with the lack of identifiability. The use of a (best)
single parameter vector for predictions with the
model will produce misleading results (Klepper &
Rouse, 1991). Therefore we employ the complete
set of parameter vectors, with more or less equally
good fits, to investigate model outcome uncertainty. Beck (1987) suggested a similar approach
to investigate prediction errors. The use of global
optimization methods, which aim to find the best
parameter vector (Pintér, 1988) is not appropriate in the case of an unidentifiable model as
SMOES. Additionally, these methods can handle
differential equation models with only a few (5 to
6) uncertain parameters. The Horberger-Spear-
Young-algorithm (Young, 1978, 1983; Hornberger & Spear, 1983; Beck, 1987) is appropriate
to be used in the process of model structure identification. Instead of trajectories of time-series observations it uses a definition of (past) behavior
in terms of constraints derived from a limited
number of observations. Subsequently, Monte-
Carlo analysis results in a set of acceptable and
a set of unacceptable model runs. If more observed data are available as in this Oosterschelde
study the HSY-algorithm does not use this system behavior describing information fully.

The subjective aspects mentioned above prevent that the model analysis procedure we apply (calibration, analysis of model usefulness, and uncertainty analysis for prediction) is completely objective, but our approach makes it at least explicit. We also quantify usefulness of the model in terms of adequacy and reliability. SMOES appears to be a useful model in most of its variables. Only some highly relevant variables of one spatial compartment (the Oosterschelde compartment West) are discussed here.

The uncalibrated model (Fig. 3) shows very high uncertainties (associated with *a priori* parameter uncertainty, in which state of the art knowledge of Oosterschelde ecosystem processes and characteristics are included). The uncalibrated model simulates the observed system adequately (Fig. 3 and 5a): most observed data fall within the minimum and maximum model outcome range. The reliability of the 'biological' variables (zooplankton and phytoplankton biomass, and, less pronounced, phytoplankton primary production) is rather low (= high values for GoF, Fig. 5b). Nutrients have relatively high reliabilities and their uncertainty bands hardly require any reduction by calibration. POC, which consists of living and dead algae, show a similar pattern as chlorophyll. Our calibration method uses a single GoF value for the whole model and compromises between good fits of all variables of which observed data are used for calibration. A better GoF for CHLF inevitably coincides with a worse fit of other variables. The search for a better fit with less uncertainty in chlorophyll concentrations and copepods biomasses, will also lead to a decrease in the nutrient ranges. If we compare the pre-barrier calibration results (Fig. 4a–f) with the uncalibrated results (Fig. 3), the uncertainty ranges are smaller for all variables, the nutrients included. The aim of the calibration is to reduce uncertainty in the model output in order to increase predictive capabilities (higher reliability). Calibration increases the reliabilities of the biological important variables (CHLF and COP), but the effect on nutrient reliability (Fig. 5b) is less pronounced. The increase in reliability occurred at the cost of adequacy (Fig. 5a). The set of observed data represents our knowledge of the Oosterschelde eco-

system. The calibrated model cannot simulate all observed states of the ecosystem, but we may have more confidence in predictions of the calibrated model.

The building of the storm-surge barrier has not changed the ecosystem drastically, at least for the ecological relevant variables discussed here. Differences, both in model results and data, fall within annual variances. If we compare the results of the pre-barrier and post-barrier calibration the uncertainty ranges of the output variables are approximately equally wide (Fig. 4). Some of the variables presented here show lower adequacies in the pre-barrier and others in the post-barrier calibration results (Fig. 5a). Reliabilities of CHLF and COP are better (with a lower value of the GoF) in the post-barrier calibration results, but differences are relatively small compared to differences between the uncertainties in the results of the uncalibrated model and the pre-barrier calibration results (Fig. 5b).

In order to evaluate the predictive power of the model, we carried out several historical forecasts: we predict the future with the model calibrated for the pre-barrier period, using post-barrier system inputs (as we do not want to forecast future weather conditions). We go back to the past to predict a future which we already know and use observed data to evaluate our 'predictions'. To test the calibration method we use four replicate sets of pre-barrier parameter vectors (produced by four replicate calibrations using different 'seeds' for the random number generator) to (historically) forecast the post-barrier ecosystem and compare these historical forecast results with four replicates of post-barrier calibration results and, obviously, a single set of post-barrier observed data. CHLF and PRODA, important predictors of mussel secondary production, show quantitative differences between all replicates of the post-barrier calibration, and also between the historical forecast replicates (Fig. 6). Random aspects in the calibration method must be held responsible for the variance between the replicates. But, qualitatively, differences between historical forecasts and post-barrier calibration results are equivalent. The historical forecasts overestimate phytoplankton biomass and primary production in all replicates. This indicates that the calibration method may be unable to produce quantitatively consistent results, but qualitatively we may have a reasonable confidence in the method. We expect and found (Fig. 7a) approximately equal adequacies in all cases (just as for the pre- and post-barrier calibration results of Fig. 5a). On the contrary, the reliabilities of the historical forecasts should be worse than those of the post-barrier calibration (if prediction of the future is more difficult than description of the past), but the Goodness of Fit of CHLF is only slightly lower for the post-barrier calibrations compared to those of the historical forecasts, which indicates that SMOES predicts the future approximately equally good (or bad) than it describes the past.

We try to define model usefulness in terms of adequacy and reliability. Mankin *et al.* (1975) defined model adequacy as $(M \cap O)/O$ and model reliability as $(M \cap O)/M$. Without knowing the uncertainties in the observed data we calculate model adequacy as the number of observed data within the uncertainty bands divided by the total number of observations during the simulated period and we use the GoF of a variable (Equation 3) as an estimate of its reliability. If any variable has an adequacy of 0, the model is completely inadequate and unuseful for this variable. An adequacy of 1 means that the model simulates all observed system behavior. Our measure of reliability (GoF) is less straightforward. A value of 0 means no distance between model and observations, but there is no upper limit. High reliability (low value of GoF) does not guarantee that any observation falls within the uncertainty bands. When making predictions with the model for management purposes, variables with high reliabilities can predict trends in future system behavior, but uncertainties in the predictions, necessary for risk-analysis, are not estimated. Predictions require high reliability with simultaneously optimal adequacy, which ensures an accurate account for uncertainties in the model predictions and a realistic simulation of system behavior.

Future complex model analysis studies would benefit from improvements of the method presented here. Estimated error in the observations or desired accuracy with which the model should describe the behavior of a variable can be used to determine when the calibration has to be stopped. This would make the calibration procedure more objective. Several ideas for the calculation of model adequacy and model reliability have to be evaluated. Adequacy and reliability will follow directly from error distributions of the observations and the model generated uncertainty bands. These two model quality parameters are ideal candidates to be incorporated in an object (penalty) function for calibration instead of the now used GoF.

The substantial changes in the Oosterschelde ecosystem caused by the building of the storm-surge barrier between 1984 and 1987 created a unique chance for a proper validation procedure of SMOES. Although SMOES is as invalid as most ecosystem models, it meets the two main objectives it was designed for. It summarizes a huge set of research results and it would have predict some major system properties at the time of the building of the storm-surge barrier as good or as bad as it does today, now we know its effects on the Oosterschelde ecosystem.

Acknowledgements

The authors wish to thank all members of the research teams of both Oosterschelde ecosystem studies for their stimulating discussions and the use of their data. This paper is Communication No. 705 of NIOO-CEMO, Yerseke, The Netherlands.

References

Bak, C., D. Ten Hulscher & J. Pintér, 1990. Calibration of the model system DELWAQ-IMPACT for the lake Ketelmeer, RIZA, Lelystad, Research report, RIZA, 82 pp.
Baretta, J. W. & P. Ruardij, 1987. Evaluation of the Ems Estuary ecosystem model. Continental Shelf Research 7: 1471–1476.

Baretta, J. W. & P. Ruardij, 1988. Tidal Flat Estuaries: Simulation and Analysis of the Ems Estuary. Springer-Verlag, Berlin, Heidelberg, New York, London, Paris, Tokyo, 353 pp.

Beck, M. B., 1987. Water quality modeling: a review of the analysis of uncertainty. Wat. Resour. Res. 23: 1393–1442.

Beck, M. B. & P. C. Young, 1976. Systematic identification of DO-BOD model structure. J. Env. Eng. Div., Proc. Am. Ass. Civil Eng., 102, no. EE5: 909–927.

Birta, L. G., 1984. Optimization in simulation studies. In: T. I. Oren, B. P. Zeigler & M. S. Elzas (eds), Simulation and Model-based Methodologies: an Integrative View, 10 in the series: NATO ASI Series F: Computer and System Science, Springer-Verlag, Berlin, Heidelberg: 451–473.

Carver, M. B., 1980. Parameter optimization in the continuous simulation packages FORSIM and MACKSIM. Mathematics in Computers and Simulation 22: 298–318.

Colijn, F., W. Admiraal, J. W. Baretta & P. Ruardij, 1987. Primary production in a turbid estuary, the Ems-Dollard: field and model studies. Continental Shelf Research 7: 1405–1409.

Davis, L. (ed), 1987. Genetic Algorithms and Simulated Annealing. Pitman, London, 216 pp.

Davis, L. & S. Coombs, 1989. Optimizing network link sizes with genetic algorithms. In: M. S. Elzas, T. I. Oren & B. P. Zeigler (eds), Modelling and Simulation Methodology, 4 in the series: Modelling and Simulation North-Holland, Amsterdam: 317–331.

De Jonge V. N. & E. G. de Groot, 1989. The mutual relationship between the monitoring and modelling of estuarine ecosystems. Helgolander Meeresuntersuchungen 43: 537–548.

De Vries I., F. Hopstaken, H. Goossens, M. de Vries, H. de Vries & J. Heringa, 1988. GREWAQ: an ecological model for Lake Grevelingen: documentation report and addendum: tables and figures, Delft Hydraulics/DGW, Delft, 159 (+83) pp.

De Vries, I., M. de Vries & F. Hopstaken, 1990. Development and application of VEERWAQ for management analysis of Lake Veere, Delft Hydraulics/DGW/Dir. Zeeland, Delft, T430, 87 pp (in Dutch).

Draper, N. R. & H. Smith, 1981. Applied Regression Analysis, 2nd edn. In the series: Wiley Series in Probability and Mathematical Statistics. Wiley, New York, 709 pp.

Fedra, K., 1983. A Monte Carlo approach to estimation and prediction. In: M. B. Beck & G. Van Straten (eds), Uncertainty and Forecasting of Water Quality. Springer Verlag, New York, 259–291.

Goldberg, D. E., 1989. Genetic Algorithms in Search, Optimization, and Machine Learning. Addison-Wesley Publishing Company Inc., Reading, Massachusetts etc., 412 pp.

Herman, P. M. J. & H. Scholten, 1990. Can suspension-feeders stabilise estuarine ecosystems? In: M. Barnes & R. N. Gibson (eds), Trophic Relations in the Marine Environment, in the series: Proc. 24th Europ. Mar. Biol. Symp. Aberdeen University Press, Aberdeen: 104–116.

Hermann, C. F., 1967. Validation problems in games and simulation with special reference to models of international politics. Behavior Science 12: 216–231.

Holland, J. H., 1975. Adaptation in Natural and Artificial Systems. The University of Michigan Press, Ann Arbor.

Hornberger, G. M. & R. C. Spear, 1983. An approach to the analysis of behavior and sensitivity in environmental systems. In: M. B. Beck & G. Van Straten (eds), Uncertainty and Forecasting of Water Quality Models, Springer-Verlag, Berlin, Heidelberg, New York: 101–116.

Iman, R. L. & J. C. Helton, 1988. An investigation of uncertainty and sensitivity analysis techniques for computer models. Risk Analysis 8: 71–90.

Jørgensen, S. E. (ed.), 1986. Fundamentals in Ecological Modelling, 9 in the series: Developments in Environmental modelling. Elsevier, Amsterdam, 391 pp.

Keizer P. D., D. C. Gordon Jr., P. Schwinghammer, G. R. Daborn & W. Ebenhoeh, 1987. Cumberland Basin Ecosystem Model: Structure Performance and Evaluation, 1547 in the series: Canadian Technical Report of Fisheries and Aquatic Sciences, Scotia-Fundy Region, Department of Fisheries and Oceans, Dartmouth, Nova Scotia, report, 202 pp.

Kirkpatrick, S., C. D. Gelatt & M. P. Vecchi, 1983. Optimization by simulated annealing. Science 220: 671–680.

Klepper, O., 1989. A model of carbon flows in relation to macrobenthic food supply in the Oosterschelde estuary, DGW-LUW, Wageningen, thesis, 270 pp.

Klepper, O. & D. I. Rouse, 1991. A procedure to reduce parameter uncertainty for complex models by comparison with real system output illustrated on a potato growth model. Agricultural Systems 36: 375–395.

Klepper, O., H. Scholten & J. P. G. van de Kamer, 1991. Prediction uncertainty in an ecological model of the Oosterschelde estuary, S. W. Netherlands. J. Forecast. 10: 191–209.

Klepper, O., M. W. M. van der Tol, H. Scholten & P. M. J. Herman, 1994. SMOES: a simulation model for the Oosterschelde ecosystem. Part I: Description and uncertainty analysis. Hydrobiologia 282/283: 437–451.

Kremer, J. N. & S. W. Nixon, 1978. A coastal marine ecosystem: simulation and analysis, 24 in the series: Ecological studies analysis and synthesis. Springer Verlag, Berlin, 217 pp.

Lewandowski, A., 1982. Issues in model validation. Angewandte Systemanalyse 3: 2–11.

Lindeboom, H. J., W. van Raaphorst, H. Ridderinkhof & H. W. van der Veer, 1989. Ecosystem model of the western Waddensea: a bridge between science and management. Helgolander Meeresunters. 43: 549–564.

Los, F. J., 1980. Application of an algal bloom model (BLOOM II) to combat eutrophication. Hydrobiol. Bull. 14: 116–124.

Los, F. J., N. M. de Rooij & J. G. C. Smits, 1984. Modelling eutrophication in shallow Dutch lakes. Verh. Int. Ver. Limnol. 22: 917–923.

Mankin, J. B., R. V. O'Neill, H. H. Shugart & B. W. Rust, 1975. The importance of validation in ecosystem analysis. In: G.S. Innis (ed.), New Directions in the Analysis of Ecological Systems, Part 1, vol. 5, No. 1 in the series: Simulation Councils Proc. Ser, Simulation Councils Inc., Lajolle, California, USA: 63–71.

McKay, M. D., 1988. Sensitivity and uncertainty analysis using a statistical sample of input values. In: Y. Ronen (ed.), Uncertainty analysis. CRC Press, Inc., Boca Raton, Florida: 145–186.

McKay, M. D., W. J. Conover & R. J. Beckman, 1979. A comparison of three methods for selecting values of input variables in the analysis of output from a computer code. Technometrics 21: 239–245.

Metropolis, N., A. W. Rosenbluth, M. N. Rosenbluth, A. H. Teller & E. Teller, 1953. Equation of state calculations by fast computing machines. J. Chem. Phys. 21: 1087–1092.

Nienhuis, P. H. & A. C. Smaal, 1994. The Oosterschelde estuary, a case study of a changing ecosystem: an introduction. Hydrobiologia 282/283: 1–14.

O'Neill, R. V. & R. H. Gardner, 1979. Sources of uncertainty in ecological models. In: B. P. Zeigler, M. S. Elzas, G. J. Klir & T. I. Oren (eds), Methodology in Systems Modelling and Simulation, North-Holland Publ. Co., Amsterdam: 447–463.

Pintér, J., 1988. Branch-and-bound algorithms for solving global optimisation problems with Lipschitzian structure. Optimization 19: 101–110.

Popper, K. R., 1959. The Logic of Scientific Discovery, 2nd edn. Unwin Hyman Ltd, London, 480 pp.

Press, W. H., B. P. Flannery, S. A. Teukolsky & W. T. Vetterling, 1986. Numerical Recipes: the Art of Scientific Computing. Cambridge University Press, Cambridge, 818 pp.

Price, W. L., 1977. A controlled random search procedure for global optimization. Comp. J. 20: 367–370.

Radford, P. J., 1978. Some aspects of an estuarine ecosystem model – GEMBASE. In: S. E. Jorgensen (ed.), State of the Art in Ecological Modelling, Vol. 7. Pergamon Press, Oxford: 301–322.

Reckhow, K. H., 1989. Validation of simulation models: philosophy and statistical methods of confirmation. In: M. G. Singh (ed.), Systems and Control Encyclopedia: Theory, Technology, Applications, Vol. 6. Pergamon Press, Oxford: 5011–5015.

Sargent, R. G., 1982. Verification and validation of simulation models. In: F. E. Cellier (ed.), Progress in Modelling and Simulation. Academic Press, London: 159–169.

Sargent, R. G., 1984a. Simulation model validation. In: T. I. Oren et al. (eds), Simulation and Model-based Methodologies: an Integrative View, Springer Verlag, Berlin Heidelberg: 537–555 (NATO ASI Series F10).

Sargent, R. G., 1984b. A tutorial on verification and validation of simulation models. In: S. Sheppard, U. Pooch & D. Pegden (eds), Proceedings of the 1984 Winter Simulation Conference: 115–121.

Sargent, R. G., 1989a. Validation of simulation models: general approach. In: M. G. Singh (ed.), Systems and Control Encyclopedia: Theory, Technology, Applications, Vol. 6. Pergamon Press, Oxford: 5008–5011.

Sargent, R. G., 1989b. Validation of simulation models: statistical approach. In: M. G. Singh (ed.), Systems and Control Encyclopedia: Theory, Technology, Applications, vol. 6. Pergamon Press, Oxford: 5015–5019.

Schlesinger, S., R. E. Crosbie, R. E. Gagne, G. S. Innis, C. S. Lalwani, J. Lich, R. J. Sylvester, R. D. Wright, N. Kheir & D. Bartos, 1979. Terminology for model credibility. Simulation 32: 103–104.

Scholten, H., B. J. De Hoop & P. M. J. Herman, 1990a. SENECA 1.2: a Simulation Environment for ECological Application (Manual), DIHO, Yerseke, Ecolmod report EM-4, ISBN 90-9003978-3, 150 pp.

Scholten, H., O. Klepper, P. H. Nienhuis & M. Knoester, 1990b. Oosterschelde estuary (S.W. Netherlands): a self-sustaining ecosystem? In: McLusky D. S., V. N. de Jonge & J. Pomfret (eds), North Sea–Estuaries Interactions. Developments in Hydrobiology 55. Kluwer Academic Publishers. Dordrecht: 201–215. Reprinted from Hydrobiologia 195.

Tarantola, A., 1987. Inverse problem theory. Methods for data fitting and model parameter estimation. Elsevier, Amsterdam, 613 pp.

Van der Tol, M. W. M. & H. Scholten, 1992. Response of the Eastern Scheldt ecosystem to a changing environment: functional or adaptive? Neth. J. Sea Res. 30: 175–190.

Van Eck, G. ThM. & N. M. de Rooij, 1990. Development of a water quality and bio-accumulation model for the Scheldt estuary. In: W. Michaelis (ed.), Coastal and Estuarine Studies. Springer-Verlag, Berlin, Heidelberg: 95–104.

Walters, C., 1986. Adaptive Management of Renewable Resources, in the series: Biological Resource Management. MacMillan Publ. Co., New York, 374 pp.

Wetsteyn, L. P. M. J. & J. C. Kromkamp, 1994. Turbidity, nutrients and phytoplankton primary production in the Oosterschelde (The Netherlands) before, during and after a large-scale coastal engineering project (1980–1990). Hydrobiologia 282/283: 61–78.

Wigan, M. R., 1972. The fitting, calibration and validation of simulation models. Simulation 18: 188–192.

Young, P. C., 1978. General theory for badly defined systems. In: G. C. Vansteenkiste (ed.), Modeling, Identification and Control in Environmental Systems. North-Holland Publishing Company, Amsterdam: 103–135.

Young P., 1983. The validity and credibility of models for badly defined systems. In: M. B. Beck & G. van Straten (eds.), Uncertainty and Forecasting of Water Quality Models. Springer-Verlag, Berlin, Heidelberg, New York: 69–98.

Hydrobiologia **282/283**: 475–478, 1994.
P. H. Nienhuis & A. C. Smaal (eds), The Oosterschelde Estuary.
© 1994 *Kluwer Academic Publishers. Printed in Belgium.*

Theme VII:
The higher trophic levels

Jan Mees
Marine Biology Section, Zoology Institute, University of Ghent, Ledegangckstraat 35, B-9000 Gent, Belgium

Higher trophic levels are considered to be negligible in the overall carbon balance of the Oosterschelde and were therefore not included in the SMOES model. Nevertheless fish, birds and seals are quite conspicuous members of the ecosystem, both before and after completion of the Delta plan. Especially the large numbers of waterbirds and the occasional seal attract considerable attention from the public. For bird populations, the Oosterschelde is of particular significance: it is one of the major wintering haunts for estuarine birds (especially waders, geese and ducks) in Western Europe and it is an important staging and moulting site in spring and autumn. Furthermore, the Oosterschelde is utilized by a rich fish and epibenthic fauna, which is clearly more diverse than that of neighbouring areas. Historically, the Delta area is also one of the southernmost habitats for the harbour seal.

These higher trophic levels may be important locally or episodically as consumers and they can significantly affect their environment. Their activities may influence other populations, communities, sediments, nutrient and energy flows, and they can contribute to the linking of different ecosystems (Day *et al.*, 1989). Next to this ecological role, estuarine wildlife is culturally important as objects of study and enjoyment. Because they appeal to the people, these animals can be used to focus attention on environmental problems and the management of estuarine ecosystems (Day *et al.*, 1989). The larger carnivores are also vulnerable to toxic pollution. In this context fishing, birdwatching and seal counting may be cheap and effective techniques for monitoring environmental quality.

As a consequence of the construction of the storm-surge barrier and the secondary compartmentalization dams, important changes have occurred in the hydraulics, geomorphology and ecology of the Oosterschelde system. The loss of intertidal areas – an overall reduction of 33% and, more specifically, a 63% loss of the salt marsh area – is an ongoing process: a further loss of 15% is expected in the next few decades. Current velocity and the influx of freshwater have decreased significantly, while transparency of the water has increased. Whether these changes have resulted in changes at the higher trophic levels (or if effects are to be expected in the future), is discussed in this part of the volume. The six papers under Theme VII deal with fish and mobile epifauna, birds and the harbour seal, the only marine mammal present in the area.

Hamerlynck & Hostens (1994) and Hostens & Hamerlynck (1994) attempt to evaluate the impact of the engineering works on the fish and epibenthic fauna of the Oosterschelde. The authors analyze two time series. In a first study fortnightly fyke catches taken in several locations in the Oosterschelde from 1979 through 1988 are analyzed. In a second paper, quarterly beam trawl surveys from the pre-barrier period (1984–1985) are compared with quarterly surveys from the post-barrier period (1988–1989). Furthermore, the community structure of the epibenthic fauna of the Oosterschelde is described in detail and it is compared to that of the neighbouring Voordelta and Westerschelde areas. The authors emphasize that the available data are of limited value to assess the effects of the construction works. Although a cluster grouping the samples from 1979–1984 could be separated from a cluster comprising the samples from 1985–1988 (roughly corresponding to the pre- and post-barrier periods), the effects of the storm-surge barrier on the fish community seem to have been marginal. Only the decrease in anadromous fish species (e.g. allis shad, sea trout, sea lamprey and lampern) can be attributed to the engineering works. The building of compartmentalization dams in the landward part of the bay significantly reduced communication with freshwater. For the structure and function of the fish community in the Oosterschelde the observed changes in anadromous species are probably unimportant. However, for the species themselves the marginal habitat offered by the Oosterschelde may have been one of their last refuges from which recolonization of true estuaries like the Westerschelde could have originated after restoration of the water quality. It should be noted here that a clean Westerschelde is also a prerequisite for the successful restoration of a seal population and for the health of bird populations in the Oosterschelde.

Other observed changes in individual fish species rather reflect fluctuations in yearclass strength or changes occurring on a wider geographical scale. Longer-term community changes and trends in population densities seem to predominate. The increase in gadoids (bib and whiting), for example, was probably due to stronger post-barrier year classes. Still, decreased current velocities in certain parts of the Oosterschelde may account for the local increase in certain flatfish species (mainly plaice and dab). An increased sedimentation of silt may have created the right conditions for the settlement of rich macrobenthic populations and thus an enhanced nursery function of the area for flatfish. Similarly, the decrease of sandeel (a species typical for dynamic sandy bottoms) and its predators (e.g. brill) may be linked to the reduction of current velocities in the Oosterschelde. The lower nutrient inputs and the associated changes in the phytoplankton may, in combination with an increase of predators, have contributed to the (marginal) decline of the brown shrimp in the Oosterschelde through a higher predation pressure. Still, for the populations of most species, possible effects of the construction works are difficult to separate from natural events. For example, the increase in visual predators like cod may also be partly linked to increased water transparency.

No information seems to be available on the possible effects of the loss of intertidal areas in the Oosterschelde on the distribution of toppredators. This is a pity since these habitats, especially salt marshes, are known to be important nursery areas and feeding grounds for fish and crustacean species, many of which are commercially exploited (e.g. Boesch & Turner, 1984; Cattrijsse et al., in press).

Schekkerman et al. (1994) and Meire et al. (1994) analyze an 8 year time series of bird counts and report on the consumption of, and predation pressure on populations of benthic invertebrates. The interpretation of the effects of the engineering works on bird populations is complicated by the relatively short period and by the fact that the pre-barrier study period (1978–1982) was characterized by two cold winters (78/79 and 81/82), while the post-barrier period (1987–1990) was characterized by three mild winters. For some species, mortality during severe winters may have been an important mechanism in the decrease. The total number of waterbirds (a mean maximum of some 300,000 present in the Oosterschelde and the Krammer-Volkerak area) has hardly changed, though the density peak has shifted from January to October. Still, significant shifts in the composition of the bird communities were observed. Species dependent on the intertidal areas for foraging generally decreased, while species feeding on open water remained stable or increased.

In the group of the intertidal foragers, a significant winter decrease of numbers and bird-days was observed for shelduck, pintail, teal, shoveler, oystercatcher, avocet, kentish plover, grey plover, dunlin and redshank. These decreases seem to have been mainly caused by the loss in feeding habitat. The loss of feeding area was not compensated by higher bird densities in the remaining part of the Oosterschelde, suggesting that the numbers of intertidal foragers were close to carrying capacity in the period before the coastal engineering works. Meire *et al.* (1993) present circumstantial evidence that, at least for a number of species, the decline in numbers in the Oosterschelde is due to food shortage (food is limiting at some times of the year). The authors show that the role of birds in the carbon balance of the Oosterschelde is indeed rather small, but that their impact on benthic populations may be important. During cold winters, food shortage in combination with cold stress can cause the death of many birds. Still, it can be expected that a further loss of intertidal area will result in a further reduction of bird numbers. The Oosterschelde also lost part of its significance as a moulting area for waders, due to the increased recreation and disturbance (mainly boating and surfing) and the increased accessibility of formerly isolated mudflats.

The increase of the pelagic foragers is obviously linked with increased water transparency. The authors calculate a 30-fold increase in water volume in which prey can be detected by pursuit-diving piscivores. This has resulted in a significant increase of visual predators like great crested grebe, cormorant and goldeneye. Though the total West European populations of these species are increasing, the authors could demonstrate that he rate of increase is locally higher in the Oosterschelde. This means that, for most species, changes in the Oosterschelde did not merely reflect large-scale population trends. Of the total amount of food taken by birds, less than 0.5% is taken subtidally. Species feeding there could potentially increase substantially in numbers.

Reijnders (1994) and Mees & Reijnders (1994) report on the history and the status of the harbour seal population in the Delta area. At present, less than 20 seals inhabit the Dutch Delta. Since, both before and after the construction of the storm-surge barrier, virtually no seals were present, effects of the works cannot be detected. Though, a short-term natural development of a viable population is not to be expected, the authors stress that

the Oosterschelde still is a potentially suitable habitat for seals. Through retrospective population analysis, Reijnders (1994) gives an important reference value for management: he estimates that about 4000 seals could still inhabit the Delta area. Chances for recovery to this population size depends on a wide range of environmental conditions, some of which were already mentioned above: improvement of water quality and safeguarding of resting places. Another problem is increased recreation.

In summary, it is difficult to assess the impact of the construction works on the higher trophic levels from the available data. The Oosterschelde is still a highly productive bay with an excellent water quality. The composition of the fish and epibenthic communities did not change substantially. Some fish and bird species – mainly visual predators – have increased. Marine fish which spawn in freshwater and birds dependent on intertidal areas for food have decreased, possibly as a consequence of the engineering works. Observed changes in commercially important fish species were rather related to yearclass strength of the North Sea population. Shrimp densities did not change substantially. The settling possibilities for seals were not affected by the works.

The changes induced by the construction works that may have effects for the populations of species belonging to the higher trophic levels include: (1) loss of intertidal areas due to decreased tidal volume: a reduction of the feeding area for birds and the possible loss of nursery area for fish and epibenthic crustaceans; (2) increased accessibility of the Oosterschelde mudflats and disturbance (mainly due to recreation): reduction of the potential feeding time for birds and of the potential resting time for seals; (3) reduced communication with freshwater: the decrease of anadromous fish species; (4) reduced current velocities: the alteration of the sedimentation characteristics (more silt in certain areas) may cause (local) changes in the epibenthic community structure; and (5) increased transparency of the watercolumn: a general increase of visual predators (pelagic foragers, both fish and piscivorous birds) and a possible decrease of their prey populations.

478

References

Boesch, D. F. & R. E. Turner, 1984. Dependence of fishery species on salt marshes: the role of food and refuge. Estuaries 7: 460–468.

Cattrijsse, A., E. S. Makwaia, H. R. Dankwa, O. Hamerlynck & M. A. Hemminga, in press. Nekton communities of an intertidal creek of a European estuarine brackish marsh. Mar. Ecol. Prog. Ser.

Day, J. W., C. A. S. Hall, W. M. Kemp & A. Yáñez-Arancibia (eds), 1989. Estuarine ecology. John Wiley & Sons, New York, 558 pp.

Hamerlynck, O. & K. Hostens, 1994. Changes in the fish fauna of the Oosterschelde estuary – a ten-year time series of fyke catches. Hydrobiologia 282/283 (Dev. Hydrobiol. 97): 497–507.

Hostens, K. & O. Hamerlynck, 1994. The mobile epifauna of the soft bottoms in the subtidal Oosterschelde estuary: structure, function and impact of the storm-surge barrier. Hydrobiologia 282/283 (Dev. Hydrobiol. 97): 479–496.

Mees, J. & P. J. H. Reijnders, 1994. The harbour seal, *Phoca vitulina*, in the Oosterschelde: decline and possibilities for recovery. Hydrobiologia 282/283 (Dev. Hydrobiol. 97): 547–555.

Meire, P. M., H. Schekkerman & P. L. Meininger, 1994. Consumption of benthic invertebrates by waterbirds in the Oosterschelde estuary, SW Netherlands. Hydrobiologia 282/283 (Dev. Hydrobiol. 97): 525–546.

Reijnders, P. J. H., 1994. Historical population size of the harbour seal, *Phoca vitulina*, in the Delta area, SW Netherlands. Hydrobiologia 282/283 (Dev. Hydrobiol. 97): 557–560.

Schekkerman, H., P. L. Meininger & P. M. Meire, 1994. Changes in the waterbird populations of the Oosterschelde (SW Netherlands) as a result of large-scale coastal engineering works. Hydrobiologia 282/283 (Dev. Hydrobiol. 97): 509–524.

Hydrobiologia **282/283**: 479–496, 1994.
P. H. Nienhuis & A. C. Smaal (eds), The Oosterschelde Estuary.
© 1994 *Kluwer Academic Publishers. Printed in Belgium.*

The mobile epifauna of the soft bottoms in the subtidal Oosterschelde estuary: structure, function and impact of the storm-surge barrier

K. Hostens & O. Hamerlynck
Marine Biology Section, University of Ghent, Ledeganckstraat 35, B-9000 Gent, Belgium and Netherlands Institute of Ecology – Centre for Estuarine and Coastal Ecology, Vierstraat 28, NL-4401 EA Yerseke, The Netherlands

Key words: spatial community structure, production, soft sediment, mobile epifauna, beam trawl, Oosterschelde estuary

Abstract

Data on the mobile epifauna of the Oosterschelde estuary, collected by beam trawl, were compiled from several studies. Multivariate statistical techniques brought out the fact that the Oosterschelde, when compared with neighbouring areas, has a characteristic epibenthic fauna. Diversity as measured by Hill's diversity numbers N through $N_{+\infty}$, is higher for the Oosterschelde ($N_1 = 4.5$) than for the Voordelta ($N_1 = 3.5$) and the Westerschelde ($N_1 = 2.2$).

Four epifaunal communities can be distinguished within the Oosterschelde, the two most seaward communities being the richest. Annual production is estimated at about 6 gADW m^{-2} yr^{-1}, annual consumption is estimated at over 25 gADW m^{-2} yr^{-1}. These results are highly dependent on the assumptions. Over 85% of the epibenthic production and consumption in the Oosterschelde is accounted for by only six species: starfish *Asterias rubens*, plaice *Pleuronectes platessa*, bib *Trisopterus luscus*, brown shrimp *Crangon crangon*, shore crab *Carcinus maenas* and dab *Limanda limanda*. In spite of its abundance, the sand goby *Pomatoschistus minutus* contributes little to the production.

From the available data it is difficult to assess the impact of the construction of the storm-surge barrier and the compartmentalization dams on the epibenthic fauna. The increase in flatfish in the Hammen area is probably linked to the decrease in current velocities in that area. On the other hand the increase in the gadoids bib and whiting *Merlangius merlangus* is predominantly due to the stronger year classes in the post-barrier time period. Lower nutrient inputs through the Northern branch, in combination with the increase of the gadoids, may have caused the decline of the brown shrimp in the Oosterschelde. A decrease has also been observed in the sandeel *Ammodytes tobianus* and the hooknose *Agonus cataphractus*.

Introduction

Shallow coastal areas and estuaries are dynamic, productive and economically important aquatic ecosystems (McLusky *et al.*, 1978). The high primary productivity and the relatively low grazing activity of the zooplankton are the basis for the importance of benthic heterotrophs in shallow coastal areas (Hannon & Joiris, 1989). Several studies have demonstrated the significant role of the mobile epifauna (predominantly consisting of demersal fishes, shrimps, crabs and starfish) in the matter and energy fluxes in shallow marine areas, both in the intertidal (Kuipers *et al.*, 1981) and in the shallow subtidal (Evans, 1984; Möller *et al.*, 1985). Many epibenthic predators have been shown to affect the structure of infaunal communities (review in Wilson, 1991).

Though extensive data collection has been done on the epibenthos of the soft sediments (both subtidal and intertidal) in the Oosterschelde over the last ten years, only a limited amount of data are presently available. The data were collected by several research groups for different purposes using different methodologies. Data on the subtidal macrofauna and on the bottom characteristics at most of the sampling sites are conspicuously absent. This paper can therefore only present a preliminary analysis of the structure and function of the mobile subtidal epibenthos in the Oosterschelde.

To characterise the Oosterschelde, the epibenthic fauna of three localities in the estuary in 1988 is compared to the fauna in the Westerschelde estuary and the neighbouring coastal area (Voordelta). Then the Oosterschelde is analyzed more detailed using the quarterly surveys of 1988 and 1989. The 0osterschelde is divided into four communities on the basis of multivariate statistical methods. Biomass composition in these communities and in the 'entire' Oosterschelde are analysed. Subsequently production and consumption are estimated for the main epibenthic species.

As a consequence of the construction of the storm-surge barrier and the secondary compart mentalization dams, important changes have occurred in the hydraulics (Vroon, 1994), geomorphology (Mulder & Louters, 1994) and ecology (Nienhuis & Smaal, 1994) of the Oosterschelde. An attempt is made to evaluate the impact of the construction works on the mobile epibenthos of the Oosterschelde estuary.

Material and methods

Sampling

All samples were taken from the R.V. Luctor (34 m, 500 Hp) using a 3 metre beam trawl, equipped with a 6 metre long net, a tickler chain and a chain in the groundrope. Trawling was always done with the tide parallel to the depth contours at 5 to 20 m below NAP (approximately Mean Tidal Level). Trawls were approximately 1000 m in length: trawling started from a buoy or other fixed marker and the distance covered was read from the radar screen.

Standard length (SL) was recorded for all epibenthic fish specimens. Invertebrates were counted and (wet)weighed. Net efficiency was assumed to be 20% for all size classes of fish and epibenthic invertebrates, regardless of mesh size in the cod end. The lower size limit for fishes is approximately 30 mm SL. the Echinodermata Crinoidea were considered to be sessile animals. Therefore starfish *Asterias rubens* were the only echinoderms quantified. Amongst the crustaceans caught, only postlarval shrimp and crabs were quantified. A number of crabs typical for hard substrates such as *Macropodia rostrata*, *Liocarcinus arcuatus* and *L. puber*, and rare shrimps such as *Crangon allmani* and *Pandalus montagui* were occasionally noticed but not quantified. The species *Pagurus bernhardus* is common but was not quantified.

Biomass was calculated from the size-frequency distributions using length – Ash-free Dry Weight (ADW) regressions for fish and wet weight – ADW conversions for the invertebrates (Hostens, pers. comm.).

The Oosterschelde in comparison with neighbouring areas

From November 1987 to December 1988 fortnightly beam trawl samples, using a 5×5 mm mesh in the cod end, were taken at three localities (Fig. 1) in the central part of the Oosterschelde (OEV = 20 in Fig. 1, OKT = 15 and

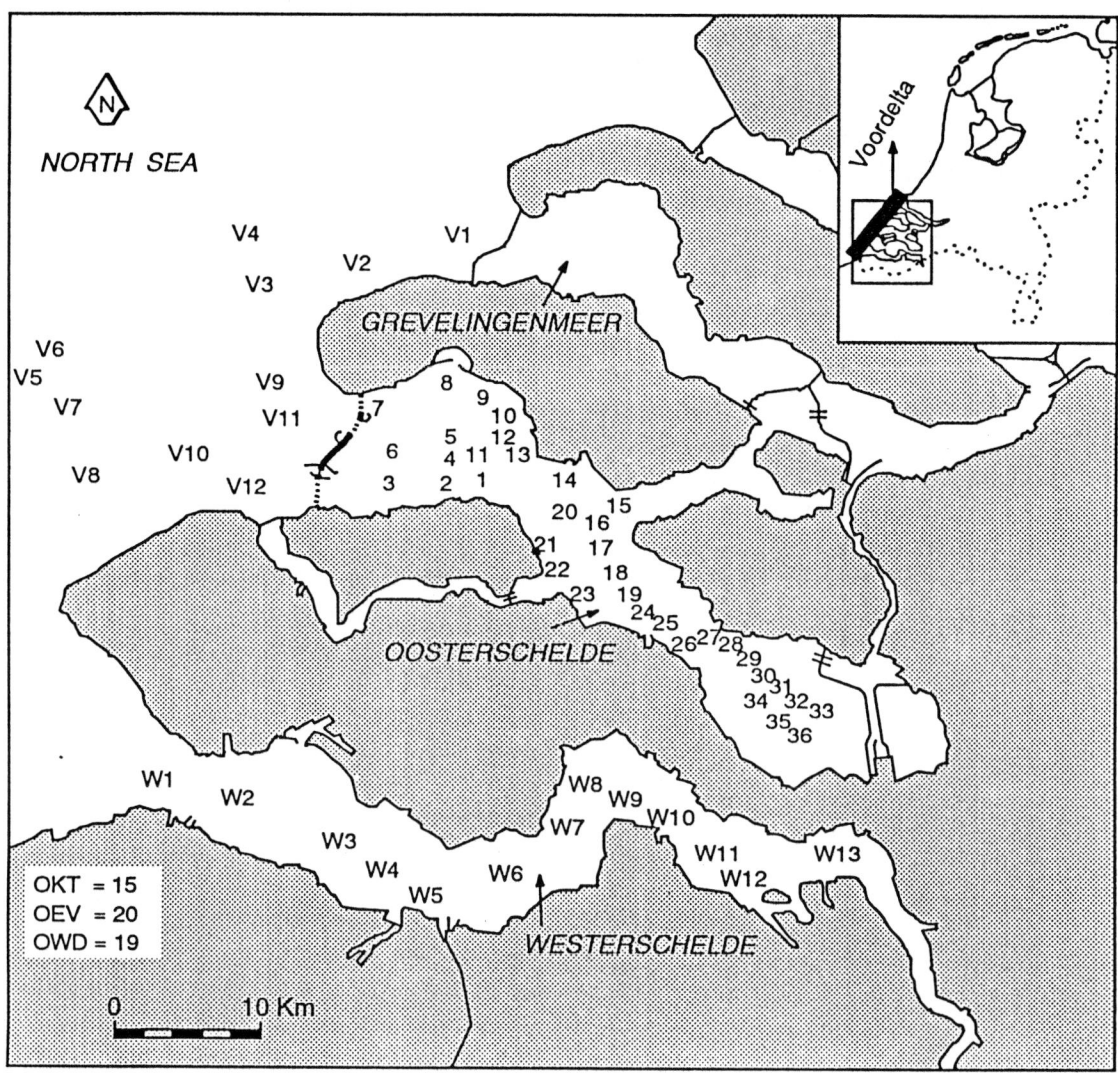

Fig. 1. Study area with the sampling localities for the different datasets. The quarterly surveys were taken in the Oosterschelde at numbers 1 through 36. The fortnightly localities in the Oosterschelde OKT, OEV and OWD correspond with the numbers 15, 20 and 19 respectively.

OWD = 19), with 3 trawls per locality (e.g. OEV1, OEV2 and OEV3). From January to December 1989 the sampling frequency was halved to a monthly basis, but one locality was added.

Only the fortnightly samples from May to December 1988 could be used for comparison with neighbouring areas. In this time period monthly samples were taken at 14 localities in the Westerschelde (W1 to W14) and 12 localities in the Voordelta (Fig. 1). In the Voordelta each locality represents two stations, one in the gully at NAP – 10 m (e.g. V1G), the other at NAP – 5 m on the sandbank slope (e.g. V1B).

In order to characterise the fish fauna of the Oosterschelde in relation to the neighbouring areas a Correspondence Analysis (CA) (Jongman *et al.*, 1987) and a Two Way Indicator Species Analysis (TWINSPAN) (Hill, 1979) were performed on the average densities (N per 1000 m^2) of fish and epibenthic invertebrates per station, after a variance-stabilising 4th root transformation (Field *et al.*, 1982). Cut-levels for the TWIN-SPAN were 0, 0.75, 0.98, 1.25, 1.74, 2.65 and 7. These were chosen in order to distribute the number of density values equally among the cut-levels. More details on the methodology of the multivariate statistical techniques are given in Hamerlynck *et al.* (in press).

Diversity within the communities, as defined by the CA, was calculated using Hill's diversity numbers of the order 0, 1, 2 and $+\infty$ (Hill, 1973). N_0 is defined as the average number of species per station, N_1 is the exponent of the Shannon-Wiener diversity index, N_2 is the reciprocal of Simpson's dominance index and $N_{+\infty}$ is equal to the reciprocal of the proportional abundance of the commonest species (Heip *et al.*, 1988).

Spatial community structure within the Oosterschelde

Between June 1983 and November 1989 22 surveys of the epibenthos of the subtidal Oosterschelde were undertaken. Only the 8 surveys conducted in 1988 and 1989 were available for the analysis of spatial structure. These surveys covered 36 stations spread out in a more or less regular way (Fig. 1). Samples were taken with a 10×10 mm mesh in the cod end. Per station average density and biomass were calculated combining the eight surveys. The 4th root transformed biomass data were used as input for a TWIN-SPAN. Seven cut-levels were used: 0, 0.21, 0.37, 0.71, 1.31, 2.35 and 5.

Annual production and consumption

Annual production for all species was calculated on the basis of a P/B ratio = 2.5. Because of the important differences between the subareas defined in the community analysis, annual mean biomass was calculated per subarea from the quarterly surveys (July 1988 through May 1989). Annual production for the 'entire' Oosterschelde was then calculated using the proportional extent of these subareas.

A P/C ratio of 0.3 for fish and 0.2 for crustaceans (Pihl, 1985) was used to estimate annual consumption of the epibenthos from the production estimates. For starfish the P/C ratio for crustaceans was used.

Changes in the epibenthos of the Oosterschelde: a comparison between 1984–1985 and 1988–1989

To evaluate the impact of the construction of the storm-surge barrier and the compartmentalization dams on the epibenthos a pairwise comparison using Wilcoxon's signed rank test for two groups was performed (Sokal & Rohlf, 1981). The two groups consist of five quarterly surveys from the period 1984–1985 and five surveys from the period 1988–1989. These were selected from all available surveys on the basis of the fact that they were conducted at similar temperatures. Thus July 1984 (16.4 °C) was paired to July 1988 (17 °C), September 1984 (14.7 °C) to September 1988 (15.4 °C), etc. This pairing was deemed necessary because, in a study of the temporal and spatial structure of the mobile epifauna in the neighbouring Voordelta, temperature was shown to be a dominant factor that strongly affected community structure (Arellano, 1991). Input was the average biomass per species or taxonomic group from the ten quarterly surveys selected. The analysis was done both for the 'entire' Oosterschelde and for the separate subareas.

Results

As it is the case for other estuarine and shallow coastal areas, the mobile epifauna makes use of the Oosterschelde in a varied way. Most species can be assigned to one of six categories (Costa & Elliot, 1991): estuarine residents *i.e.* species completing their life cycle entirely within estuaries, marine juvenile migrants (MJ) that use the area as a nursery, marine seasonal migrants (MS) whose adults make seasonal incursions into estuaries, marine occasional species (MO), fresh water occasional species (FW) and catadromous and anadromous species (CA). The species quantified in this study have been assigned to these categories in Table 2. For the purpose of this paper the conventional category 'estuarine resident' was converted into Oosterschelde resident (OR), because the Oosterschelde is not a true estuary any more and some of the OR species are not typical estuarine species. There is some degree of arbitrariness in the assignment of species to some of the categories. Most of the epibenthic MJ species enter the Oosterschelde as larval planctivorous animals in spring, become benthivorous after metamorphosis and leave the Oosterschelde as adults or subadults in winter. Some individual subadults stay in the Oosterschelde for one or more years, others that have left can have a regular cycle of entering and leaving the Oosterschelde until they reach maturity. The adults can afterwards be either seasonal or occasional visitors. The ecological types followed by a question-mark refer to species whose exact category is still in doubt.

The Oosterschelde in comparison with neighbouring areas

The Oosterschelde has a characteristic epibenthic community in comparison to neighbouring areas. In the plot of the sample scores of the Correspondence Analysis (Fig. 2a) for the first two axes (eigenvalues 0.11 and 0.11) the six Oosterschelde stations from the OWD and OKT localities form a separate subgroup towards the lower left. The three stations from the locality closest to the mouth (OEV) take up an intermediate position between this subgroup and the origin. The two other areas also form distinct clusters, the Westerschelde cluster along the first (horizontal) axis and the Voordelta cluster along the second. From the species plot (Fig. 2b) it is clear that this structure is linked to the presence of a few characteristic species in the Oosterschelde, *i.e.* the sea scorpion *Taurulus bubalis* Euphrasen 1786, the lemon sole *Microstomus kitt* Walbaum 1792, the sand smelt *Atherina presbyter* Cuvier 1829, the sea bass *Dicentrarchus labrax* Linnaeus 1758, the butterfish *Pholis gunnellus* Linnaeus 1758, the greater

pipefish *Syngnathus acus* Linnaeus 1758, the eelpout *Zoarces viviparus* Linnaeus 1758, the stickleback *Gasterosteus aculeatus* Linnaeus 1758, the bull-rout *Myoxocephalus scorpius* Linnaeus 1758 and the brill *Scophthalmus rhombus* Linnaeus 1758.

The TWINSPAN (Fig. 3) classifies the nine Oosterschelde stations (community OS) with eight stations from the Voordelta (community VD A). This cluster, together with the cluster containing the Westerschelde stations (community WS) are separated from the cluster containing all the other Voordelta stations (community VD B). As in this paper we are mainly interested in the Oosterschelde we have lumped the two Westerschelde communities (within WS) and the two Voordelta communities (within VD B) that were distinguished at the third level of division. Indicator species for the Voordelta A subcluster, dividing it from the Oosterschelde is Lozano's goby *Pomatoschistus lozanoi* de Buen 1923. Four stations of the VD A community are located in the inner part of the lagoon that has come into being in what was formerly the ebb-tidal delta of the Grevelingen estuary (Louters *et al.*, 1991). The other four stations are located in front of the storm-surge barrier. Details on the structure of the epibenthic communities in the other areas will be published elsewhere (Hamerlynck *et al.*, in press).

Table 1 shows the average densities for the four communities defined by the TWINSPAN. A number of species have similar densities in the VD A and the OS communities and are less abundant in both other communities, *i.e.* the dragonet *Callionymus lyra* Linnaeus 1758 (indicator species), Nilsson's pipefish *Syngnathus rostellatus* Nilsson 1855 and *P. gunnellus*. Species that are more abundant in the OS community than elsewhere are the plaice *Pleuronectes platessa* Linnaeus 1758, the shore crab *Carcinus maenas* Linnaeus 1758, the tub gurnard *Trigla lucerna* Linnaeus 1758, the snake pipefish *Entelurus aequoreus* Linnaeus 1758 and the 'characteristic' species as defined in the CA. These species appear as preferentials for the OS cluster in the TWINSPAN. The Oosterschelde stations are

484

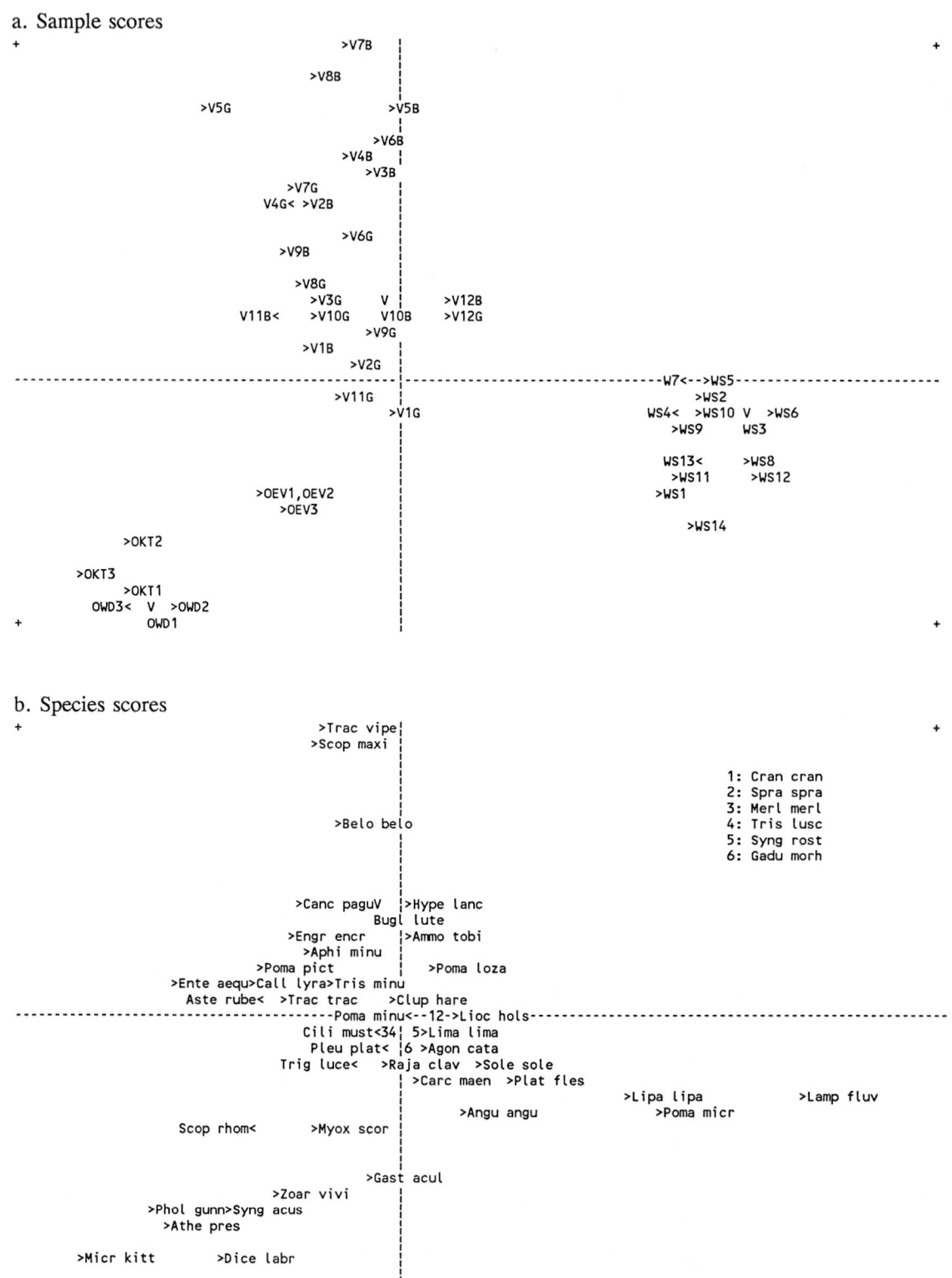

Fig. 2. Result of the Correspondence Analysis: sample (a) and species (b) scores in the plane of the first two axes. Species names are abbreviated to four letters of the genus name followed by four letters of the species name.

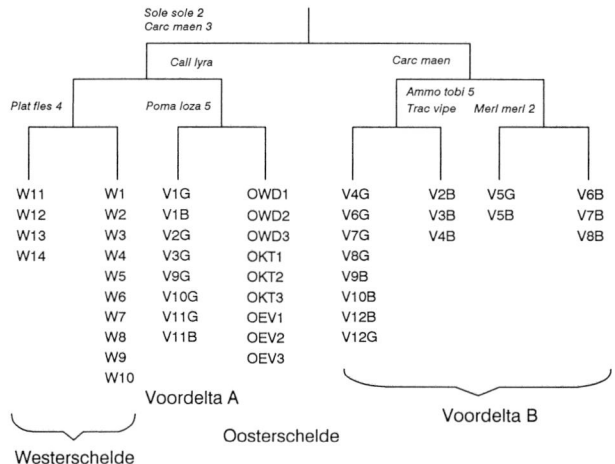

Fig. 3. Result of the TWINSPAN on the 4th root transformed density data of the monthly (fortnightly) samples. The indicator species (with cut-levels 1 or given) for each division are abbreviated.

poorer than the surrounding areas for the brown shrimp *Crangon crangon* Linnaeus 1758, the swimming crab *Liocarcinus holsatus* Fabricius 1798 and, of course, the 'negative' indicator species *P. lozanoi* (Table 1).

Diversity

In the Oosterschelde 40 species were recorded, in the Voordelta 42 species and only 32 species in the Westerschelde. However the average number of species recorded per station for each community (N_0) is significantly higher for the Oosterschelde (29.3 ± 0.7 S.E.) than for the Voordelta (22.5 ± 0.6 S.E.) or for the Westerschelde (21.5 ± 0.4 S.E.). The diversity numbers of higher order show the same pattern as N_0, but as the influence of species richness decreases with increasing order, the values for the Voordelta come closer to these of the Oosterschelde (Fig. 4). Hill's N_1 f.i. is 4.5 (± 0.5 S.E.) for the Oosterschelde, 3.5 (± 0.2 S.E.) for the Voordelta and 2.2 (± 0.1 S.E.) for the Westerschelde.

Spatial structure in the subtidal of the Oosterschelde

The TWINSPAN classifies the epibenthic fauna of the Oosterschelde into 4 communities (Figs 5

& 6). Two communities (Vuilbaard & Hammen) in the western part, a community (Central) corresponding to the central part and a community (Eastern) corresponding to the eastern part. Four stations (23, 24, 26 and 28) do not conform to the general geographical pattern and are not used for further analyses. Relative surface areas for these communities are: Hammen 22%, Vuilbaard 28%, Central 40% and Eastern 10%.

Indicator species for the two western communities are the bib *Trisopterus luscus* Linnaeus 1758, the hooknose *Agonus cataphractus* Linnaeus 1758 and *P. lozanoi*. Other indicator species are the sandeel *Ammodytes tobianus* Linnaeus 1758 for the Vuilbaard, the flounder *Platichthys flesus* Linnaeus 1758 for the Hammen and cod *Gadus morhua* Linnaeus 1758, *D. labrax* and *C. maenas* for the Eastern community (Fig. 7).

Table 2 gives the biomass composition for the different communities and for the 'entire' Oosterschelde. More than 80% of the average biomass of 2.5 gADW m^{-2} in the subtidal Oosterschelde is formed by only five species *i.e.* starfish *Asterias rubens* Linnaeus 1758 (1 gADW m^{-2}), *P. platessa* (0.6 gADW m^{-2}), *T. luscus* (0.2 gADW m^{-2}), dab *Limanda limanda* Linnaeus 1758 (0.15 gADW m^{-2}) and *C. crangon* (0.1 gADW m^{-2}). Adding the next seven species, *i.e.* *C. maenas*, *M. scorpius*, eel *Anguilla anguilla* Linnaeus 1758, *Z. viviparus*, sole *Solea solea* Linnaeus 1758, whiting *Merlangius merlangus* Linnaeus 1758 and *P. flesus*, more than 96% of the biomass is accounted for.

The faunal composition of the four communities shows that the Hammen community is much richer than the other three communities (Fig. 6). This is reflected in the highest biomass for the top 14 species in the ranking for the 'entire' Oosterschelde with the exception of *C. maenas* of *M. scorpius*, two species that have their highest biomass in the Eastern community (Table 2).

The Vuilbaard community is the second richest one but supports only about 40% of the biomass of the Hammen community. A number of species have their highest biomass in the two westernmost communities, *i.e.* *A. rubens*, *C. crangon*, *M. merlangus*, *L. holsatus*, poor cod *Trisopterus*

Table 1. Average abundance (numbers per 1000 m^2) for the four communities as defined by Twinspan, ranked according to abundance for the Oosterschelde community. OS = Oosterschelde; VD A = Voordelta A; VD B = Voordelta B; WS = Westerschelde.

Species	Community			
	OS	VD A	VD B	WS
Asterias rubens Linnaeus 1758	1400	890	360	–
Crangon crangon L. 1758	380	6200	1100	2300
Pomatoschistus minutus Pallas 1769	140	710	65	130
Pleuronectes platessa L. 1758	80	100	10	10
Limanda limanda L. 1758	60	540	30	110
Carcinus maenas L. 1758	45	30	2	15
Clupea harengus L. 1758	20	25	15	4
Trisopterus luscus L. 1758	20	60	6	4
Liocarcinus holsatus Fabricius 1798	10	320	40	160
Merlangius merlangus L. 1758	7	35	3	1
Callionymus lyra L. 1758	6	6	4	–
Syngnathus rostellatus Nilsson 1855	6	11	2	5
Zoarces viviparus L. 1758	4	0.4	–	0.1
Pomatoschistus lozanoi de Buen 1923	4	560	100	100
Myoxocephalus scorpius L. 1758	4	0.6	0.1	0.1
Sprattus sprattus L. 1758	4	4	3	9
Ammodytes tobianus L. 1758	3	1	21	2
Solea solea L. 1758	1	35	0.8	9
Gadus morhua L. 1758	0.9	4	0.4	0.6
Syngnathus acus L. 1758	0.8	–	<0.1	0.1
Agonus cataphractus L. 1758	0.5	35	1	1
Pholis gunnellus L. 1758	0.5	0.6	–	–
Trigla lucerna L. 1758	0.5	0.1	0.1	0.1
Microstomus kitt Walbaum 1792	0.4	–	–	–
Gasterosteus aculeatus L. 1758	0.3	–	–	0.1
Scophthalmus rhombus L. 1758	0.3	–	<0.1	–
Pomatoschistus pictus Malm 1865	0.2	1.0	0.4	–
Trachurus trachurus L. 1758	0.2	0.4	0.3	<0.1
Anguilla anguilla L. 1758	0.2	0.5	<0.1	0.4
Hyperoplus lanceolatus le Sauvage 1824	0.2	0.3	0.5	0.1
Atherina presbyter Cuvier 1829	0.2	–	–	<0.1
Aphia minuta Risso 1810	0.1	4	0.5	<0.1
Platichthys flesus L. 1758	0.1	10	0.1	3
Pomatoschistus microps Krøyer 1838	0.1	–	<0.1	3
Entelurus aequoreus L. 1758	0.1	–	<0.1	–
Dicentrarchus labrax L. 1758	0.1	–	–	<0.1
Trisopterus minutus L. 1758	0.1	0.2	0.1	–
Taurulus bubalis Euphrasen 1786	<0.1	–	–	–
Ciliata mustela L. 1758	<0.1	0.1	<0.1	<0.1
Liparis liparis L. 1758	<0.1	1.1	<0.1	1.2
Buglossidium luteum Risso 1810	–	–	<0.1	<0.1
Cancer pagurus L. 1758	–	0.1	0.1	–
Engraulis encrasicolus L. 1758	–	–	<0.1	–
Scophthalmus maximus L. 1758	–	–	0.1	–
Belone belone L. 1758	–	–	<0.1	–
Trachinus vipera Cuvier 1829	–	–	0.5	–
Raja clavata L. 1758	–	0.1	–	–
Lampetra fluviatilis L. 1758	–	–	–	<0.1

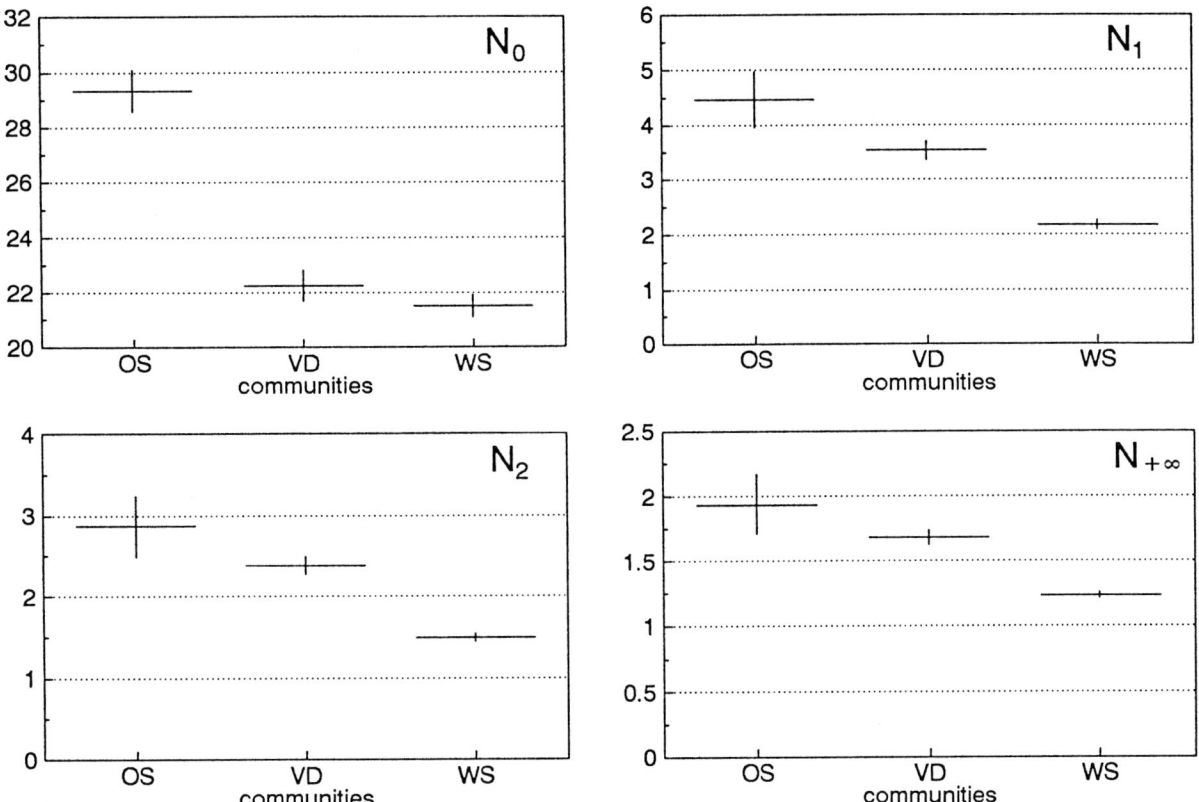

Fig. 4. Hill's diversity numbers N_0, N_1, N_2 and $N_{+\infty}$ for the different communities as defined by the CA, mean with Standard Error (S.E.). OS = Oosterschelde; VD = Voordelta; WS = Westerschelde

minutus Linnaeus 1758 and, of course, the indicator species for these two communities, *i.e. T. luscus, A. cataphractus* and *P. lozanoi*.

The biomass composition of the Central (and poorest) community is rather similar to the ranking for the total Oosterschelde. Very few species reach their highest biomass in this community, none of which is in the top 15 for the Oosterschelde. The species are *A. tobianus*, the greater sandeel *Hyperoplus lanceolatus* le Sauvage 1824, *S. rhombus, M. kitt* and scad *Trachurus trachurus* Linnaeus 1758.

The Eastern community supports the highest biomass of *C. maenas, M. scorpius*, herring *Clupea harengus* Linnaeus 1758, sprat *Sprattus sprattus* Linnaeus 1758, sand goby *Pomatoschistus minutus* Pallas 1769, black goby *Gobius niger* Linnaeus 1758 and sea-snail *Liparis liparis* Linnaeus 1758.

Annual production and consumption

The annual production of the epibenthic fauna of the Oosterschelde, using a P/B ratio of 2.5, amounts to about 6 gADW m^{-2} yr^{-1} in the production year July 1988–May 1989 (Table 3). This would represent a consumption of more than 25 gADW m^{-2} yr^{-1}.

More than half of the total production is formed by the invertebrates of which the largest share (3 gADW m^{-2} yr^{-1}) is accounted for by the starfish. Flatfish produce the bulk (70%) of the total fish production, which amounts to 3 gADW m^{-2} yr^{-1}. The fish of the Oosterschelde would consume about 10 gADW m^{-2} yr^{-1}.

The annual production for dab *L. limanda* is estimated to be 0.15 gADW m^{-2} yr^{-1} and consumption about 0.5 gADW m^{-2} yr^{-1}. For plaice *P. platessa* an annual production of 1.9 gADW

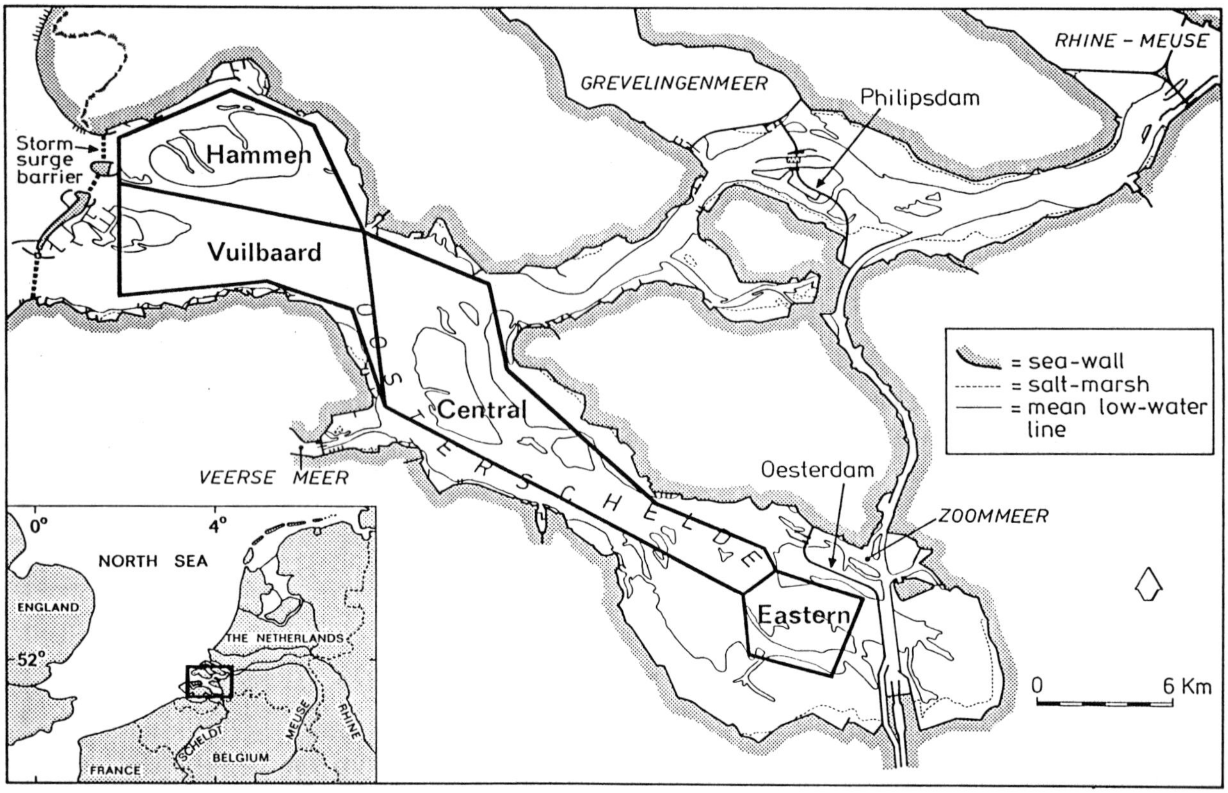

Fig. 5. The four mobile epifauna communities in the Oosterschelde estuary as defined by TWINSPAN.

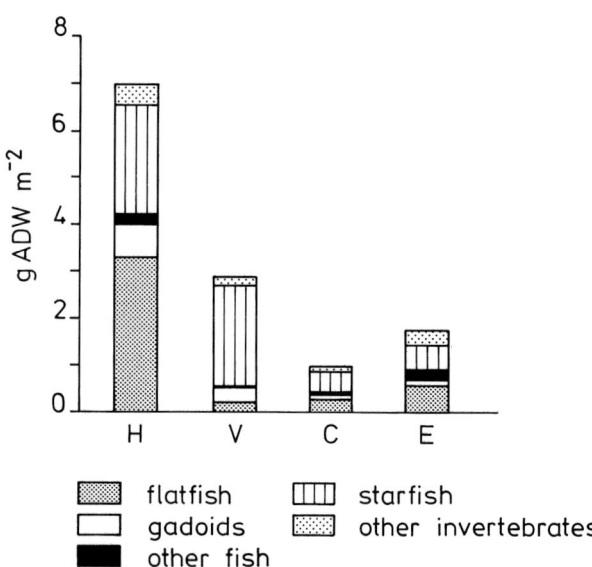

Fig. 6. The four communities as defined by the TWINSPAN (H = Hammen; V = Vuilbaard; C = Central; E = Eastern) with the annual mean biomass composition per community.

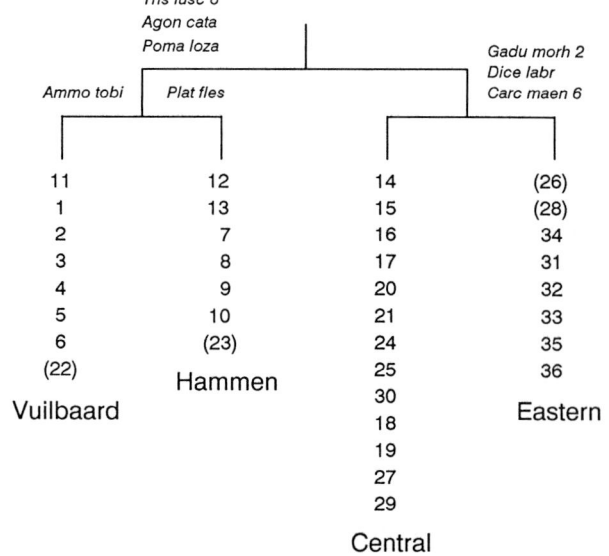

Fig. 7. Result of the TWINSPAN on the 4th root transformed biomass data of the quarterly surveys. The indicator species (with cut-levels 1 or given) for each division are abbreviated, as in Fig. 2. Numbers represent sampling localities in Fig. 1.

Table 2. Average biomass (gADW per 1000 m^2) for the four communities as defined by Twinspan, ranked according to biomass for the 'entire' Oosterschelde. For the explanation of the ecological types, see text.

Species	Community					
	Hammen	Vuilbaard	Central	Eastern	Entire	Type
Asterias rubens	2300	2100	410	470	1050	OR
Pleuronectes platessa	2500	150	210	360	610	MJ
Trisopterus luscus	330	200	60	95	160	MJ
Limanda limanda	510	40	70	95	150	MJ
Crangon crangon	285	140	55	60	110	OR
Carcinus maenas	100	25	50	260	90	OR
Myoxocephalus scorpius	95	15	35	110	65	OR
Anguilla anguilla	115	30	7	40	40	CA
Zoarces viviparus	170	5	9	6	35	OR
Solea solea	80	35	10	40	30	MJ
Merlangius merlangus	100	40	7	7	30	MJ
Platichthys flesus	90	4	5	55	30	CA
Gadus morhua	60	35	2	6	20	MJ/MO?
Liocarcinus holsatus	50	25	10	3	20	MS/OR?
Clupea harengus	8	0.7	7	35	15	MJ
Pomatoschistus minutus	8	7	5	30	10	OR
Sprattus sprattus	1	3	5	40	9	MJ
Callionymus lyra	7	0.8	2	3	3	MS
Ammodytes tobianus	–	3	4	0.3	2	OR
Scophthalmus rhombus	–	–	3	3	2	MO/OR?
Trigla lucerna	5	0.9	1	–	1	MJ
Ciliata mustela	3	–	0.8	2.0	1	MS
Trisopterus minutus	3	1	0.1	1	1	MO
Microstomus kitt	–	–	1	0.8	0.9	MO/OR?
Syngnathus acus	0.9	0.1	1	1	0.8	OR
Agonus cataphractus	2	0.5	0.3	0.0	0.6	OR
Dicentrarchus labrax	–	–	0.4	3	0.6	MJ
Pholis gunnellus	2	–	0.1	1	0.5	OR
Hyperoplus lanceolatus	–	–	0.8	–	0.3	MO/OR?
Entelurus aequoreus	–	0.4	–	–	0.1	MO/OR?
Trachurus trachurus	–	<0.1	0.3	–	0.1	MJ/MS
Taurulus bubalis	–	0.2	0.1	0.1	0.1	OR
Gobius niger L. 1758	–	–	–	0.4	0.1	OR
Syngnathus rostellatus	0.1	<0.1	0.1	<0.1	0.1	OR
Gasterosteus aculeatus	<0.1	<0.1	0.1	<0.1	<0.1	FW/OR
Liparis liparis	–	–	<0.1	0.1	<0.1	OR
Pomatoschistus lozanoi	0.1	0.1	<0.1	–	<0.1	MJ
Pomatoschistus microps	<0.1	<0.1	<0.1	<0.1	<0.1	OR
Atherina presbyter	<0.1	–	–	–	<0.1	OR
Aphia minuta	–	>0.1	–	–	<0.1	MO
Sum	6800	2800	970	1700	2500	

m^{-2} yr^{-1} leads to a consumption of 6 gADW m^{-2} yr^{-1}. Bib *T. luscus* produces approximately 0.4 gADW m^{-2} yr^{-1} and consumes 1.4 gADW m^{-2} yr^{-1}. For whiting *M. merlangus* annual production and consumption are estimated to be 0.07 gADW m^{-2} yr^{-1} and 0.25 gADW m^{-2} yr^{-1} respectively. Annual production for sand goby *P. minutus* is 0.02 gADW m^{-2} yr^{-1} and consumption is estimated to be about 0.05 gADW m^{-2} yr^{-1}.

Table 3. Annual production and consumption (gADW m^{-2} yr^{-1}) between July 1988 and May 1989 per subarea (H: Hammen, V: Vuilbaard, C: Central and E: Eastern) with P/B = 2.5 and P/C = 0.3 (fish) or 0.2 (invertebrates) ranged according to production in the 'entire' Oosterschelde (OS).

Species	Production					Consumption
	H	V	C	E	OS	OS
Asterias rubens	5.8	4.0	1.2	0.7	2.9	14.5
Pleuronectes platessa	6.7	0.4	0.6	0.8	1.9	6.3
Trisopterus luscus	0.6	0.6	0.2	0.4	0.4	1.4
Crangon crangon	0.7	0.3	0.09	0.04	0.3	1.4
Carcinus maenas	0.3	0.07	0.12	0.7	0.2	0.9
Limanda limanda	0.4	0.06	0.05	0.2	0.15	0.5
Myoxocephalus scorpius	0.2	0.04	0.06	0.09	0.09	0.3
Anguilla anguilla	0.3	0.04	0.01	0.08	0.08	0.3
Merlangius merlangus	0.2	0.09	0.02	0.02	0.07	0.2
Zoarces viviparus	0.2	0.02	0.02	0.01	0.06	0.2
Solea solea	0.15	0.04	0.02	0.05	0.06	0.2
Gadus morhua	0.05	0.15	<0.01	0.02	0.05	0.16
Liocarcinus holsatus	0.14	0.02	0.02	0.01	0.05	0.2
Platichthys flesus	0.13	<0.01	0.01	0.08	0.04	0.14
Sprattus sprattus	<0.01	<0.01	<0.01	0.16	0.02	0.06
Pomatoschistus minutus	0.02	0.02	0.01	0.05	0.02	0.05
Clupea harengus	0.02	<0.01	0.01	0.06	0.01	0.04
Callionymus lyra	0.02	<0.01	0.01	<0.01	0.01	0.03
Ammodytes tobianus	–	0.01	0.01	<0.01	0.01	0.02
Trigla lucerna	0.02	<0.01	<0.01	–	0.01	0.02
Scophthalmus rhombus	–	–	0.01	<0.01	<0.01	0.02
Other fish	0.02	0.01	0.01	0.02	0.02	0.05
Total fish	8.9	1.4	1.1	2.0	3.0	10.0
Total invertebrates	6.9	4.3	1.4	1.4	3.4	17.1
Total epifauna	16.0	5.7	2.5	3.3	6.4	27.1

Besides the four invertebrate species, the four most important flatfish species and four gadoids, only *M. scorpius* and *A. anguilla* have important individual productions. Only the top 20 epibenthic species have a production higher than 0.01 gADW m^{-2} yr^{-1}. The sum of the productions of the other 23 species is only 0.02 gADW m^{-2} yr^{-1}.

For almost all species the Hammen subarea has a production per unit area that is much higher than the other subareas, except for a few species like *S. sprattus* and *P. minutus* which have their highest production in the Eastern part or like *G. morhua*, *A. tobianus* and *S. rhombus* which have their highest production in the Central part. The Central part has the lowest productivity. Consumption in the different subareas will natu-rally reflect the different productivity levels (Table 3).

Changes in the epibenthos of the Oosterschelde: a comparison between 1984–1985 and 1988–1989

The pairwise comparison using Wilcoxon's signed rank test for two groups showed a significant increase (p < 0.05) in the flatfish and gadoid biomass for the Hammen subarea (Fig. 8). The bulk of the flat-fish increase is accounted for by plaice *P. platessa* and dab *L. limanda*. Average biomass for plaice increased more than tenfold from 0.2 to 3 gADW m^{-2}. For dab a sevenfold increase from 0.03 to 0.22 gADW m^{-2} was recorded and the biomass for flounder *P. flesus* doubled. The bulk

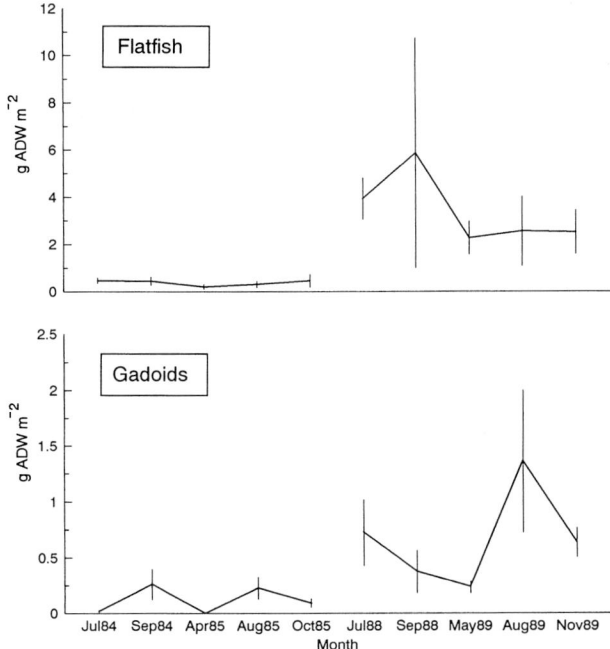

Fig. 8. Average biomass per month (± Standard Error) in gram Ash-free Dry Weight per m² for flat-fish (top) and gadoids (bottom): a comparison between 1984–'85 and 1988–'89 for the Hammen subarea.

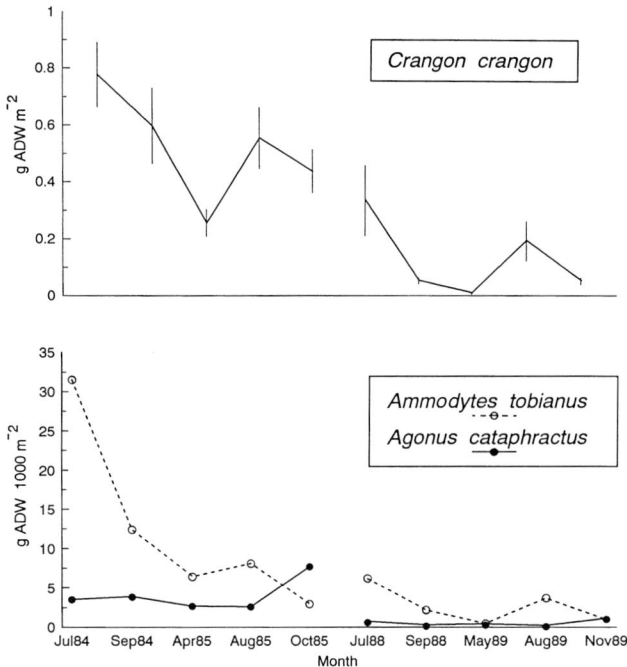

Fig. 9. Average biomass per month (± Standard Error) in gram Ash-free Dry Weight per m² for *Crangon crangon* (top) and average biomass per month per 1000 m² for sandeel and hooknose (bottom): a comparison between 1984–'85 and 1988–'89 for the 'entire' Oosterschelde.

of the gadoid increase is accounted for by bib *T. luscus* and whiting *M. merlangius*. Average biomass for bib increased more than sevenfold from 0.06 to 0.43 gADW m⁻², for whiting the increase was about fourfold from 0.03 to 0.13 gADW m⁻².

Significant decreases for the total Oosterschelde were found for shrimp *C. crangon*, sandeel *A. tobianus* and hooknose *A. cataphractus* (Fig. 9). Average biomass for shrimp decreased fourfold from 0.5 to 0.13 gADW m⁻², sandeel decreased fourfold and hooknose eightfold. Many other species showed decreases but these were not significant.

Discussion

The sampling scheme of the quarterly surveys was less than ideal for a quantification of the epibenthic fauna. In principle stratified random sampling should be the best strategy (Sissenwine et al., 1983). Sampling of the subtidal was more

or less systematic but limited both in extent (e.g. the Northern branch was not sampled) and coverage (e.g. some areas are inaccessible because of construction works, ammunition dumps, mussel culture, etc.). Data from the intertidal stratum have been collected but were not available for this study. Thus estimates for the 'entire' Oosterschelde apply only to the sampled area.

Only few authors did detailed research on net efficiency for beamtrawls (e.g. Kuipers, 1975; Doornbos & Twisk, 1984; 1987; Rogers & Lockwood, 1989). Though there is still high uncertainty about net efficiency for different bottoms (Kuipers et al., in press) there seems to be a strong consensus that the efficiency for demersal fish on 'normal' soft bottoms will be between 20 and 50% (J.J. Zijlstra, pers. com.). Elliot & Taylor (1989) use a 33% efficiency for their study of the fish community of the Forth estuary. For this study an efficiency of 20% for all length classes of all species was used. This corresponds to the efficiency for flatfish as found by Kuipers (1975). For the immobile starfish the efficiency is probably underestimated and is more likely to be around 50% (M. Fonds, pers. com.). Efficiency of a scallop-dredge for starfish was estimated at 44% (Guillou, 1990). Thus the estimates for density, biomass, production and consumption for starfish should be (at least) halved.

The use of diversity indices has been rightfully criticised (review in Washington, 1984). Hill's diversity numbers N_0 through $N_{+\infty}$ are useful (univariate) indices of community structure if, when comparing different areas, care is taken that sampling method, sample size, depth of sampling, duration of sampling, time of year and taxonomic level are similar in all areas (Hughes, 1978). Considering the number of trawls performed in the three areas (OS 142, VD 190 and WS 126) and the fact that in the Oosterschelde only three localities, all belonging to the central community, were sampled, it is clear that the Oosterschelde has a diverse epibenthic community. Besides the 'characteristic' species as defined by the CA a number of other typical species were caught when beamtrawling in the Oosterschelde in the following years, i.e. Dasyatis pastinaca Linnaeus 1758, Cyclopterus lumpus Linnaeus 1758, Mugilidae species, Raniceps raninus Linnaeus 1758, Homarus gammarus Linnaeus 1758 and Liocarcinus arcuatus Leach 1814 (Hostens, unpubl. data). These species, except for L. arcuatus which is occasionally observed in the Voordelta, were not recorded from the neighbouring areas. As very little is known on habitat preferences of most of the typical Oosterschelde species it is difficult to explain this exceptional epibenthic fauna. Some of the typical species, such as T. bubalis and P. gunnellus are dependent on the hard substrates of the Oosterschelde. The rarity of P. lozanoi in the Oosterschelde, in comparison with neighbouring areas, is probably linked to the near absence of mysids in the subtidal Oosterschelde (Mees & Hamerlynck, 1992). Mysids are the most important prey for this species (Hamerlynck et al., 1990).

The importance of pelagic fish is surely underestimated in beamtrawl catches. Still, in shallow coastal areas grazing activity of the zooplankton is relatively unimportant compared to the open sea (Hannon & Joiris, 1989). Therefore production of pelagic fish and their predators is unimportant in these areas compared to the production by benthic heterotrophs. Unfortunately there are no data on the macrobenthic fauna of the subtidal Oosterschelde, but the strong dominance of the macrobenthic component, i.e. the molluscs, in the energy fluxes in the Oosterschelde (Herman & Scholten, 1990) is probably the basis for the dominance of the demersal macrobenthos feeders in the epibenthos: i.e. plaice P. platessa, dab L. limanda, starfish A. rubens, shrimp C. crangon and shore crab C. maenas (Pihl, 1985; Degel & Gislason, 1988).

Despite the fact that dab L. limanda is the most important demersal fish species in biomass terms in the Southern Bight of the North Sea (Daan et al., 1990) only few studies on its biology have been carried out. Dab is restricted to the subtidal for most of its demersal life (Poxton et al., 1982). Density and biomass as estimated in this study are therefore more reliable than for species exhibiting strong tidal migrations or for species spending an important part of their life in the intertidal.

The three other abundant flatfish species were surely underestimated. Juvenile plaice P. platessa concentrate in the intertidal (van der Veer, 1986). Sole S. solea prefers muddy substrates and is buried rather deeply during most of the day (Boerema, 1964). Flounder P. flesus was underestimated because it spends most of its life in the shallow areas.

The Oosterschelde seems attractive to young bib T. luscus and whiting M. merlangus. These are known to feed on shrimp (Henderson & Holmes, 1989; Redant, 1982). The biomass is probably underestimated, because length frequency distributions in catches by line fishermen and in fykes (Hamerlynck & Hostens, 1994) indicate that older gadoids are certainly more abundant in the Oosterschelde than is apparent from the beamtrawl catches.

The use of a general P/B ratio of 2.5 can only yield an indication of possible production levels. There are wide discrepancies in the values found for different animal groups in the literature (see review by Redant, 1989), e.g. for *C. crangon* published P/B ratios vary from 2 to 9.3. In their study of the Forth estuary Elliott & Taylor (1989) used a P/B ratio of 2.75. Similarly there is a great need for more accurate estimates of consumption rates in the different groups. Another problem is the fact that the 'annual' production is calculated for only one year and is therefore strongly influenced by year class strength for a number of species.

Annual production (6 gADW m^{-2} yr^{-1}) and consumption (27 gADW m^{-2} yr^{-1}) for the total epibenthos of the Oosterschelde are similar to production (5 gADW m^{-2} yr^{-1}) and consumption (26 gADW m^{-2} yr^{-1}) estimates for the mobile epifauna on the Swedish coast (Möller *et al.*, 1985; Pihl, 1985). Yet, these results are highly dependent on the assumptions.

Changes in species abundance due to the construction of the storm-surge barrier and the compartmentalization dams have to be evaluated in relation to fluctuations in year class strength (see also Hamerlynck & Hostens, 1994). Year class strengths for plaice have been relatively constant since 1982, with the exception of the strong 1985 yearclass (Rijndorp *et al.*, 1991). As the bulk of plaice biomass occurring in the beam trawl catches consists of 1 + and 2 + group fish, it is unlikely that the observed increase in the Hammen is due to differences in year class strength for the two time periods. For dab no year class strength data are available, but bycatch data from the Wadden Sea do not show differences in abundance between 1984–85 and 1988 (Tiews, 1990). The increase of flatfish (mainly plaice and dab) in the Hammen is probably linked to the decrease in current velocities in this area (Mulder & Louters, 1994) with the subsequent sedimentation of silt. As the major changes in current dynamics occurred in the summer of 1985, we must hypothesize the existence of a time lag between these changes and the reaction of the epibenthic community. It also seems logical that the last two surveys of 1985 (August and October) showed no spectacular changes, because most settlement processes for macrobenthic and epibenthic animals are over by the summer. It could be argued

that these last two surveys are not independent of one another because a rich settlement is bound to propagate through all data collected for the rest of that year. The same objection could be brought against the use of both the August and November 1989 surveys. Still, the differences between the two time periods (1984–85 and 1988–89) are so obvious that counting the summer and autumn surveys as only one observation would not have altered the conclusions. In support of the hypothesis, that the increase in the flatfish in the Hammen area is linked to the changed sedimentation pattern, there is the observation from the former ebb-tidal delta of the Grevelingen, where the silting up to the tidal gullies was accompanied by an increased entrapment of passively transported elements like fish eggs and larvae, macrobenthic larvae, detritus, etc. (Hamerlynck & Mees, 1991). This created the right conditions for the settlement of rich macrobenthic populations and enhanced the nursery function of the area for flatfish (Hamerlynck *et al.*, 1992).

The gadoids, mainly bib and whiting also increased significantly, but in contrast to the situation for flatfish, this increase is probably not due to the construction works, because gadoid year classes have clearly been stronger in the postbarrier time period (see Hamerlynck & Hostens, 1994).

Sandeel *A. tobianus* and greater sandeel *H. lanceolatus* on the other hand are species typical for dynamic sandy bottoms (Hamerlynck *et al.*, in press). The decrease of sandeel with the reduction of current velocities in the Oosterschelde was therefore to be expected. As sandeel are the main prey for brill *S. rhombus* (Braber & De Groot, 1973) the decrease in brill (Hamerlynck & Hostens, 1994) may be linked to the decrease in sandeel. There is at present no explanation for the decrease in hooknose *A. cataphractus*. In the bycatch data from the Wadden Sea the hooknose was more abundant in 1985 then in 1984, but no difference was found between 1985 and 1988 (Tiews, 1990).

Judging from the Catch per Unit Effort data of the Belgian shrimp fishery (F. Redant, pers. com.) there have been no major changes in shrimp

494

abundance in the Belgian coastal area over the time period studied. The decrease in shrimp *C. crangon* in the Oosterschelde therefore seems to be a local phenomenon and is possibly due to the decrease of nutrient inputs in the Oosterschelde from the river Rhine (Nienhuis & Smaal, 1994) and the associated changes in the phytoplankton (Bakker *et al.*, 1990). It was indeed shown by Boddeke *et al.* (1986) that productivity of the brown shrimp in the Dutch coastal area was strongly influenced by nutrient inputs from the Rhine. A second factor in the decrease could be the increased predation pressure from the strong gadoid year classes.

It is a pity that no data are available on the sediment characteristics, current velocities, and macrobenthic biomass at the 36 stations sampled in the quarterly surveys. This would have allowed an analysis of the relationship between the presence of a certain type of fish assemblage and the environment. The changes in these variables as a result of the construction works could have given some indications regarding the possible causes of the observed changes in the mobile epifauna.

Conclusions

The Oosterschelde has a characteristic and diverse fish community. Four communities can be distinguished within the Oosterschelde, the two most seaward communities being the richest. More than 80% of the epibenthic biomass in the subtidal is formed by only five species: starfish, plaice, bib, dab and brown shrimp. Annual production is estimated at about 6 gADW m^{-2} yr^{-1}, annual consumption at over 25 gADW m^{-2} yr^{-1}. However these results are highly dependent on the assumptions regarding net efficiency, P/B and P/C ratios. Changes in the current velocities in the Hammen area, which occurred as a consequence of the construction of the storm-surge barrier, are probably the cause for the increase in flatfish in that area. The decreased input of nutrient rich water from the river Rhine through the northern branch of the Oosterschelde is probably the cause for the observed decrease in the brown

shrimp in the Oosterschelde. The decrease in sandeel may also be linked to the reduced current velocities in the Oosterschelde.

In future studies much could be gained from the concomitant monitoring of macrobenthic populations and bottom characteristics when monitoring mobile epibenthic animals.

Acknowledgements

Part of this study was sponsored by Rijkswaterstaat, Dienst Getijdewateren and by FKFO grant 2.0086.88. Part of the data were collected under the direction of R. H. D. Lambeck and G. Doornbos. Fieldwork was greatly facilitated by the enthousiastic help of W. Röber, P. de Koeyer, C. van Sprundel and J. W. Francke. The first author acknowledges a grant from the Instituut tot aanmoediging van het Wetenschappelijk Onderzoek in Nijverheid en Landbouw (IWONL). Contribution No. 706 of NIOO-CEMO, Yerseke, The Netherlands.

References

Arellano, R. V., 1991. Community ecology and productivity of the epibenthos of a shallow coastal area. Unpubl. M. Sc. Thesis. University Gent, 93 pp.

Bakker, C., P. M. J. Herman & M. Vink, 1990. Changes in seasonal succession of phytoplankton induced by the storm-surge barrier in the Oosterschelde (S.W. Netherlands). J. Plankton Res. 12: 947–972.

Boddeke, R., G. Driessen, W. Doesburg & G. Ramaekers, 1986. Food availability and predator presence in a coastal nursery area of the brown shrimp (*Crangon crangon*). Ophelia 26: 77–90.

Boerema, L. K., 1964. Some effects of diurnal variation in the catches upon estimates of abundance of plaice and sole. Rapp. P-v. Réun. cons. perm. int. Explor. Mer 155: 52–57.

Braber, L. & S. J. de Groot, 1973. The food of five flatfish species (Pleuronectiformes) in the southern North Sea. Neth. J. Sea Res. 6: 163–172.

Costa, M. J. & M. Elliott, 1991. Fish usage and feeding in two industrialised estuaries the Tagus, Portugal, and the Forth, Scotland. In: M. Elliott & J. P. Ducrotoy (eds), Estuaries and coasts: spatial and temporal intercomparisons: 289–297.

Daan, N., P. J. Bromley, J. R. G. Hislop & N. A. Nielsen, 1990. Ecology of North Sea fish. Neth. J. Sea Res. 26: 343–386.

Degel, H. & H. Gislason, 1988. Some observations on the food selection of plaice and dab in Øresund, Denmark. ICES CM 1988/G:50, 16 pp. (mimeo).

Doornbos, G. & F. Twisk, 1984. Density, growth and annual food consumption of plaice (*Pleuronectes platessa* L.) and flounder (*Platichthys flesus* L.) in Lake Grevelingen, The Netherlands. Neth. J. Sea Res. 18: 434–456.

Doornbos, G. & F. Twisk, 1987. Density, growth and annual food consumption of gobiid fish in the saline Lake Grevelingen. Neth. J. Sea Res. 21: 45–74.

Elliot, M. & C. J. L. Taylor, 1989. The structure and functioning of an estuarine/marine fish community in the Forth estuary, Scotland. Proceedings of the 21st European Marine Biology Symposium, Gdansk: 227–240.

Evans, S., 1984. Energy budgets and predation impact of dominant epibenthic carnivores on a shallow soft bottom community at the Swedish west coast. Estuar. coast. Shelf Sci. 18: 651–672.

Field, J. G., K. R. Clarke & R. M. Warwick, 1982. A practical strategy for analysing multispecies distribution patterns. Mar. Ecol. Progr. Ser. 8: 37–52.

Guillou, M., 1990. Biotic interactions between predators and super-predators in the Bay of Douarnenez, Brittany. In M. Barnes & R. N. Gibson (eds), Trophic relationships in the marine environment. Aberdeen University Press, Aberdeen: 141–156.

Hamerlynck, O. & K. Hostens, 1994. Changes in the fish fauna of the Oosterschelde – a ten year time series of fyke catches. Hydrobiologia 282/283: 497–507.

Hamerlynck, O & J. Mees, 1991. Temporal and spatial structure in the hyperbenthic community of a shallow coastal area and its relation to environmental variables. Oceanol. Acta. Vol. sp. 11: 205–212.

Hamerlynck, O., K. Hostens, R. V. Arellano, J.Mees & P. van Damme, in press. The mobile epibenthic fauna of soft bottoms in the Dutch Delta (south-west Netherlands): spatial structure. Neth. J. Aquat. Ecol.

Hamerlynck, O., P. Van de Vyver & C. R. Janssen, 1990. The trophic position of *Pomatoschistus lozanoi* (Pisces: Gobiidae) in the southern Bight. In M. Barnes & R. N. Gibson (eds.), Trophic relationships in the marine environment. Aberdeen University Press, Aberdeen: 183–190.

Hamerlynck, O., K. Hostens, J. Mees, R. V. Arellano, A. Cattrijsse, P. Van de Vyver & J. A. Craeymeerch, 1992. The ebb-tidal delta of the Grevelingen: a man-made nursery for flatfish? Neth. J. Sea Res. 30: 1–8.

Hannon, B. & C. Joiris, 1989. A seasonal analysis of the southern North Sea ecosystem. Ecology 70: 1916–1934.

Heip, C., P. Herman & K. Soetaert, 1988. Data processing, evaluation and analysis. In: R. P. Higgins & H. Thiel (eds), Introduction to the study of meiofauna. The Smithsonian Institution Press, Washington: 197–231.

Henderson, P. A. & R. H. A. Holmes, 1989. Whiting migration in the Bristol Channel: a predator-prey relationship. J. Fish Biol. 34: 409–416.

Herman, P. M. J. & H. Scholten, 1990. Can suspension-feeders stabilise estuarine ecosystems? In M. Barnes & R. N. Gibson (eds.), Trophic relationships in the marine environment. Aberdeen University Press, Aberdeen: 104–116.

Hill, M. O., 1973. Diversity and evenness: a unifying notation and its consequences. Ecology 54: 427–432.

Hill, M. O., 1979. TWINSPAN a FORTRAN program for arranging multivariate data in an ordered two-way table by classification of individuals and attributes. Cornell University, Ithaca, 90 pp.

Hughes, B. D., 1978. The influence of factors other than pollution on the value of Shannon's diversity index for benthic macroinvertebrates in streams. Wat. Res. 12: 359–364.

Jongman, R. H. G., C. J. F. ter Braak & O. F. R. van Tongeren, 1987. Data analysis in community and landscape ecology. Pudoc, Wageningen, 299 pp.

Kuipers, B. R., 1975. On the efficiency of a two-metre beam trawl for juvenile plaice (*Pleuronectes platessa*). Neth. J. Sea Res. 9: 69–85.

Kuipers, B. R., P. A. W. J. de Wilde & F. Creutzberg, 1981. Energy flow in a tidal flat ecosystem. Mar. Ecol. Progr. Ser. 5: 215–221.

Kuipers, B. R., H. W. van der Veer, B. MacCurrin, J. M. Miller & J. IJ. Witte, in press. On the efficiency of demersal gears in sampling juvenile flatfishes. Neth. J. Sea Res.

Louters, T., J. P. M. Mulder, R. Postma & F. P. Hallie, 1991. Changes in coastal morphological processes due to the closure of tidal inlets in the SW Netherlands. J. Coast. Res. 7: 635–652.

McLusky, D. S., M. Elliot & J. Warnes 1978. The impact of pollution on the intertidal fauna of the estuarine Firth of Forth. In: D. S. McLusky & A. J. Berry (eds), Physiology and Behaviour of Marine Organisms. Proceedings of the 12th European Marine Biology Symposium. Pergamon Press, Oxford: 203–210.

Mees, J. & O. Hamerlynck (1992). Spatial community structure of the permanent hyperbenthos of the Schelde-estuary and the adjacent coastal waters. Neth. J. Sea Res. 29(4): 357–370.

Möller, P., L. Phil & R. Rosenberg, 1985. Benthic faunal energy flow and biological interaction in some shallow marine soft bottom habitats. Mar. Ecol. Progr. Ser. 27: 109–121.

Mulder, J. P. M. & T. Louters, 1994. Changes in basin geomorphology after implementation of the Oosterschelde estuary project. Hydrobiologia 282/283: 29–39.

Nienhuis, P. H. & A. C. Smaal, 1994. The Oosterschelde estuary, a case study of a changing ecosystem: an introduction. Hydrobiologia 282/283: 1–14.

Pihl, L., 1985. Food selection and consumption of mobile epibenthic fauna in shallow marine areas. Mar. Ecol. Progr. Ser. 22: 169–179.

Poxton, M. G., A. Eleftheriou & A. D. McIntyre, 1982. The population dynamics of O-group flatfish on nursery grounds in the Clyde Sea area. Estuar. coast. Shelf Sci. 14: 265–282.

Redant, F., 1982. Caridean shrimps in the food of demersal fish of the Belgian coast. 1. Gadiformes. ICES CM 1982/K:25. 12 pp. (mimeo).

Redant, F., 1989. Productivity of epibenthic species: a review. ICES CM 1989/L:2. 33 pp. (mimeo).

Rijnsdorp, A. D., N. Daan, F. A. van Beek & H. J. L. Heessen, 1991. Reproductive variability in North Sea plaice, sole, and cod. J. Cons. int. Explor. Mer 47: 352–375.

Rogers, S. I. & S. J. Lockwood, 1989. Observations on the capture efficiency of a two-metre beam trawl for juvenile flatfish. Neth. J. Sea Res. 23: 347–352.

Sissenwine, M. P., T. R. Azarovitz & J. B. Suomola, 1983. Determining the abundance of fish. In A. G. McDonald & I. G. Priede (eds), Experimental biology at sea. Academic Press, London: 51–101.

Sokal R. R. & F. J. Rohlf, 1981. Biometry (second edition). W. H. Freeman, San Fransicco: 959 pp.

Tiews, K., 1990. 35-Jahres-Trend (1954–1988) der Häufigkeit von 25 Fisch- und Krebstierbeständen an der deutschen Nordseeküste. Arch. FishWiss. 40: 39–48.

Van der Veer, H. W., 1986. Immigration, settlement, and density-dependent mortality of a larval and early postlarval O-group plaice (*Pleuronectes platessa*) population in the western wadden Sea. Mar. Ecol. Progr. Ser. 29: 223–236.

Vroon, J., 1994. Hydrodynamic characteristics of the Oosterschelde in recent decades. Hydrobiologia 282/283: 17–27.

Washington, H. G., 1984. Diversity, biotic and similarity indices. Wat. Res. 18: 653–694.

Wilson, W. H., 1991. Competition and predation in marine soft-sediment communities. Annu. Rev. Ecol. Syst. 21: 221–241.

Hydrobiologia **282/283**: 497–507, 1994.
P. H. Nienhuis & A. C. Smaal (eds), The Oosterschelde Estuary.
© 1994 Kluwer Academic Publishers. Printed in Belgium.

Changes in the fish fauna of the Oosterschelde estuary – a ten-year time series of fyke catches

O. Hamerlynck & K. Hostens
*Marine Biology Section, State University of Ghent, Ledeganckstraat 35, B-9000 Gent, Belgium and
Netherlands Institute of Ecology – Centre for Estuarine and Coastal Ecology, Vierstraat 28, NL-4401 EA
Yerseke, The Netherlands*

Key words: fish community; fykes, storm surge barrier, Oosterschelde estuary

Abstract

Frequency of occurrence of fish species was monitored on a fortnightly basis in four fykes and a weir in the Oosterschelde estuary from 1979 through 1988. This was done in order to record changes in the fish fauna that may have occurred as a response to the construction of a storm-surge barrier in the mouth of the Oosterschelde (1984–1986) and the concomitant building of compartmentalization dams in the landward part. These compartmentalization dams reduced the freshwater inflow into the system. Principal component analysis using the annual averages in frequency of occurrence suggests a slight shift occurred in the fish community separating a cluster of years 1979–1984 from the cluster 1985–1988. Many of the changes in individual species could be attributed to fluctuations in yearclass strength or were part of changes occurring on a wider geographical scale. The only impact of the construction works seems to be the decrease in a number of anadromous fish. Fish traps seem to be useful as a monitoring tool for a number of species. The value of the data collected could be improved if catch size and length-frequency data are recorded.

Introduction

Fish traps have until recently rarely been used in fisheries research (Hinz, 1989). Their suitability for the monitoring of biological effects on the fish fauna has been emphasised by Ruth & Berghahn (1989). If fishing is strictly standardised fykes may also be used for the study of long-term trends in flatfish populations (Van der Veer *et al.*, 1992).

Fyke nets have the advantage of being relatively unselective (Ruth & Berghahn, 1989). They sample demersal fish of stony ground, *i.e.* species that are difficult to catch using towed gears. Fykes also catch pelagic species that are only caught incidentally in a beam trawl (Hinz, 1989).

In this study the frequency of occurrence of fish species has been monitored on a fortnightly basis in four fykes and a weir in the Oosterschelde from 1979 through 1988. This was done in order to record changes in the fish fauna that may have occurred as a response to the engineering works in the area. A storm-surge barrier was constructed in the mouth of the Oosterschelde and the building of compartmentalization dams in the landward part reduced the freshwater inflow into the system. In 1984 the impact of the construction works on the hydrodynamics of the system was still very limited. The biggest changes occurred in 1985 and by mid 1987 the new situation was implemented (Nienhuis & Smaal, 1994).

498

Materials and methods

Commercial fishermen were asked to monitor the catches of four fyke nets and a weir on a fortnightly basis from 1979 through 1988 at different localities in the Oosterschelde (Fig. 1). Both types of gear are described in Nédélec (1982).

The four fykes are located close to the dykes and are deployed for catching eel *Anguilla anguilla* Linnaeus 1758. Fykes are emptied on average every three days. Mesh size is 21 mm stretched.

The weir is a traditional fishery directed at the anchovy *Engraulis encrasicolus* Linnaeus 1758. There are two leaders (often several hundred metres long) of stakes set out on an intertidal flat and converging towards the gully. Fish swimming over the shallow areas at high tide are driven towards the chamber in the V-shaped point of the gear

when leaving the tidal flat. When the gear is operated actively large schools of anchovy are directed from the chamber into the net attached to the fyke opening of the chamber. Outside the anchovy season and between active catches there is a fyke net attached to this opening and the device functions more or less like a normal fyke net.

A total of 860 samples were used in the analysis. In the summer of 1986 fishing was hampered because of the presence of large quantities of coelenterates in the Oosterschelde. The time series is incomplete for the easterly stations. The locality Zandkreek was only sampled in spring and autumn. There are no data for this station for 1987. Ice floes destroyed the weir during the winters of 1984–1985, 1985–1986 and 1986–1987 and fishing could only be resumed in April.

With the closure of the Oesterdam in the au-

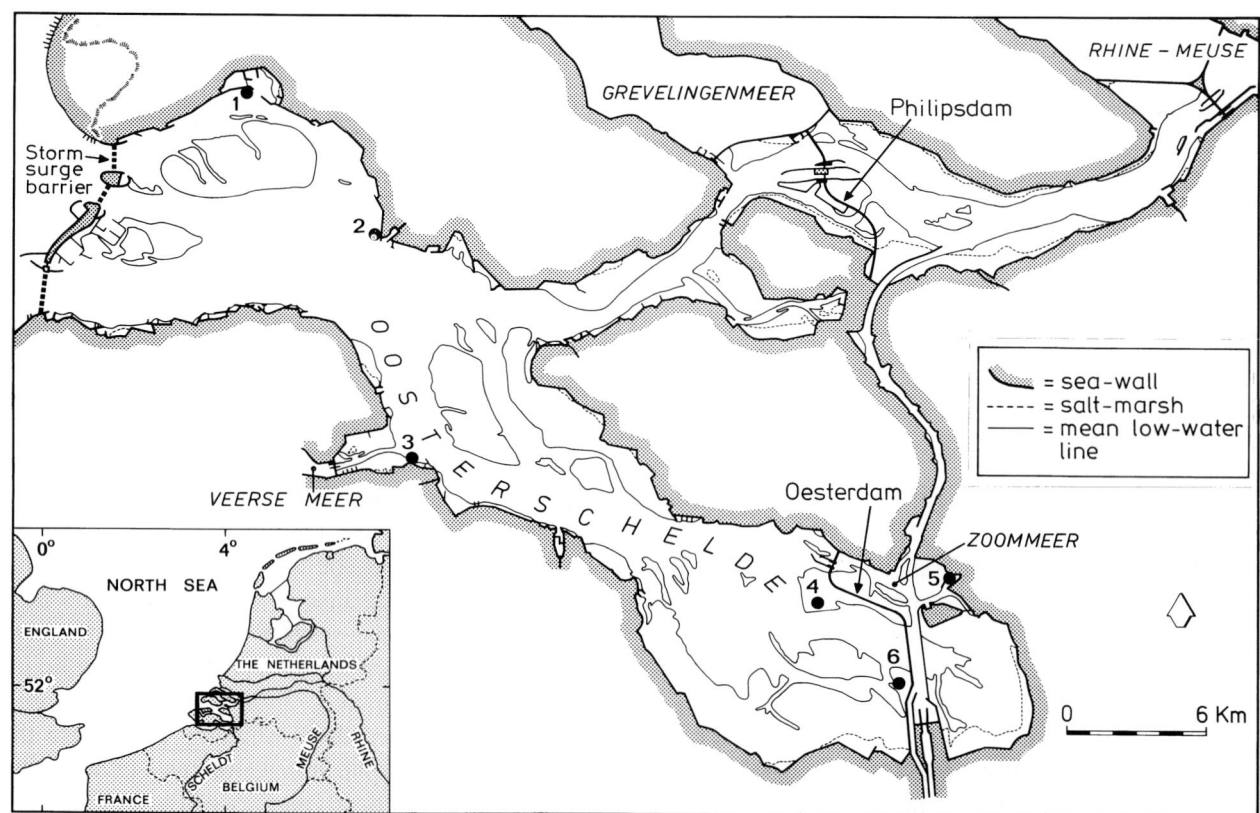

Fig. 1. Oosterschelde estuary with sampling localities; 1 = Schelphoek; 2 = Zierikzee; 3 = Zandkreek; 4 = Speelmansplaat (weir)); 5 = Bergen op Zoom (until 1986); 6 = Oesterdam (after autumn 1987); Inset: Oosterschelde estuary in The Netherlands on the North Sea.

tumn of 1986 the station Bergen op Zoom became part of the Zoommeer. The Zoommeer remained connected rather indirectly to the Oosterschelde through the Philipsdam until April 1987. In the summer of 1987 the Zoommeer became a freshwater lake. Starting in the autumn of 1987 a new locality called 'Oesterdam' was introduced into the monitoring scheme to 'replace' the Bergen op Zoom locality (Fig. 1). However the fish fauna caught in the new locality differs substantially from the fauna at Bergen op Zoom. The data for the Oesterdam were thus not included in the analysis. Repeating all the analyses without the time series data for the Bergen op Zoom locality did not affect the results in any substantial way. All species present in the fyke or the weir were noted. The presence of a fish species in each station was expressed as the percentage of catches in which the species occurred (frequency of occurrence) in a single year. The frequency of occurrence data for the five sampling localities as reported by Meijer (1989), were averaged to yield an annual frequency of occurrence for the 'Oosterschelde'. This annual frequency of occurrence was subjected to the correspondence analysis option in the package CANOCO (Ter Braak, 1987) in order to quantify total community variation. On the basis of the result of this analysis the same data were subjected to the principal component analysis (PCA) option in CANOCO. The frequency of occurrence data were arc sin transformed for normalisation (Sokal & Rohlf, 1981) prior to the PCA. The analysis was repeated after elimination of the species which occurred only in a single year and the species inadequately sampled or difficult to identify. Nilsson's pipefish *Syngnathus rostellatus* Nilsson 1855 and sand goby *Pomatoschistus minutus* Pallas 1769 are too small to be reliably retained and detected in the net. Sand gobies are also liable to predation by larger fish in the catch. Grey gurnard *Eutrigla gurnardus* Linnaeus 1758 was only identified by the fishermen operating the weir. Solenette *Buglossidium luteum* Risso 1810 is rather difficult to identify and may be overlooked in large catches. A Wilcoxon two sample test (Sokal & Rohlf, 1981) was performed on the annual frequency of occurrence data of the individual species to look for significant differences between the clusters of years distinguished along the first PCA axis. For ranking the frequency of occurrence data differences of less than 1% between years were discarded and indices for those years were considered to be ties.

The product-moment correlation coefficient and Kendall's rank correlation coefficient (Sokal & Rohlf, 1981) were calculated between the arc sin transformed frequencies of occurrence and yearclass strength indices for a number of species.

Results

A total of 67 species were recorded (Table 1). Ten species occurred on average in more than 50% of catches, 11 species occurred in 25–50% of catches, 12 species in 10–25% of catches, 7 species in 5–10 % of catches and 7 species in 1–5% of catches. Another 12 species occurred in less than 1% of catches but were found in at least two years, 9 species were only recorded in a single year.

The correspondence analysis showed that community variation is within a narrow range (less than 1.5 standard deviation units) along the first axis (eigenvalue = 0.03). Thus a linear method (PCA) is more appropriate in this case (Ter Braak & Prentice, 1988).

The results of the PCA on the same data are depicted in Fig. 2. In the sample plot (Fig. 2 bottom) two main clusters can be found in the plane formed by the first two axes (eigenvalues 0.33 and 0.19) with a group 1981–1983 clearly separated from the group 1985–1987 along the first (horizontal) axis. The years 1979 and 1980 are quite far apart within the left upper quadrant of the plane and are clearly separate from the 1981–1983 group along the second (vertical) axis. The year 1984 lies close to the origin. Towards the right of the plane the years 1985–1987 form a rather compact group. The year 1988 lies in the right upper quadrant.

The species plot (Fig. 2 top) is the output for

500

Table 1. List of species caught in Oosterschelde estuary, ranked according to frequency of occurrence.

Species occurring in > 50% of catches

Platichthys flesus Linnaeus 1758	0.96
Anguilla anguilla L. 1758	0.90
Zoarces viviparus L. 1758	0.82
Pleuronectes platessa L. 1758	0.80
Myoxocephalus scorpius L. 1758	0.76
Trisopterus luscus L. 1758	0.66
Clupea harengus L. 1758	0.63
Solea solea L. 1758	0.60
Limanda limanda L. 1758	0.52
Merlangius merlangus L. 1758	0.52

Species occurring in 25–50% of catches

Gadus morhua L. 1758	0.47
Mugilidae species	0.45
Atherina presbyter Cuvier 1829	0.44
Sprattus sprattus L. 1758	0.40
Dicentrarchus labrax L. 1758	0.38
Belone belone L. 1758	0.38
Trachurus trachurus L. 1758	0.32
Alosa fallax Lacépède 1758	0.31
Syngnatus acus L. 1758	0.28
Ciliata mustela L. 1758	0.27
Pholis gunnellus L. 1758	0.26

Species occurring in 10–25% of catches

Scophthalmus rhombus L. 1758	0.24
Scomber scombrus L. 1758	0.23
Pollachius pollachius L. 1758	0.21
Cyclopterus lumpus L. 1758	0.20
Engraulis encrasicolus L. 1758	0.18
Agonus cataphractus L. 1758	0.17
Ammodytes tobianus L. 1758	0.16
Trigla lucerna L. 1758	0.15
Gasterosteus aculeatus L. 1758	0.14
Oncorhynchus mykiss Walbaum 1792	0.13
Callionymus lyra L. 1758	0.13
Liparis liparis L. 1758	012

Species occurring in 5–10% of catches

Entelurus aequoreus L. 1758	0.09
Osmerus eperlanus L. 1758	0.08
Dasyatis pastinaca L. 1758	0.08
Sardina pilchardus Walbaum 1792	0.06
Scophthalmus maximus L. 1758	0.06
Salmo trutta L. 1758	0.05
Taurulus bubalis Euphrasen 1786	0.05

Species occurring in 1–5% of catches

Pollachius virens L. 1758	0.02
Pomatoschistus minutus Pallas 1769	0.02

Table 1. (Continued)

Mullus surmuletus L. 1758	0.02
Raniceps raninus L. 1758	0.02
Eutrigla gurnardus L. 1758	0.01
Gobius niger L. 1758	0.01
Petromyzon marinus L. 1758	0.01

Species occurring in < 1% of catches

Lampetra fluviatilis L. 1758
Conger conger L. 1758
Microstomus kitt Walbaum 1792
Syngnathus rostellatus Nilsson 1855
Spondyliosoma cantharus L. 1758
Melanogrammus aeglefinus L. 1758
Galeorhinus galeus L. 1758
Labrus bergylta Ascanius 1772
Buglossidium luteum Risso 1810
Scyliorhinus canicula L. 1758
Scomberesox saurus Walbaum 1792

Species recorded only in a single year

Pomatoschistus microps Krøyer 1838
Trachinus draco L. 1758
Squalus acanthias L. 1758
Crenilabris melops L. 1758
Arnoglossus laterna Walbaum 1792
Hyperoplus lanceolatus le Sauvage 1824
Trisopterus minutus L. 1758
Balistes carolinensis Gmelin 1789
Salmo salar L. 1758

the dataset without the species occurring in only a single year and without the inadequately sampled species. The plot with all species is similar except that a number of species around the origin can not be depicted because of lack of space. Species situated around the extremes of the figure along the first axis are the ones which have shown the biggest changes in frequency of occurrence between the years 1979–1984 and the years 1985–1988. Towards the left of the figure (species typical for the years before the barrier impact) are five bearded rockling *Ciliata mustela* Linnaeus 1758, brill *Scophthalmus rhombus* Linnaeus 1758, sardine *Sardina pilchardus* Walbaum 1792, allis shad *Alosa fallax* Lacépède 1758, Mugilidae species (presumably both thick-lipped grey mullet *Chelon labrosus* Risso 1826 and thinlipped grey mullet *Liza ramada* Risso 1826), sand-smelt *Atherina presbyter* Cuvier 1829, rainbow trout *On-*

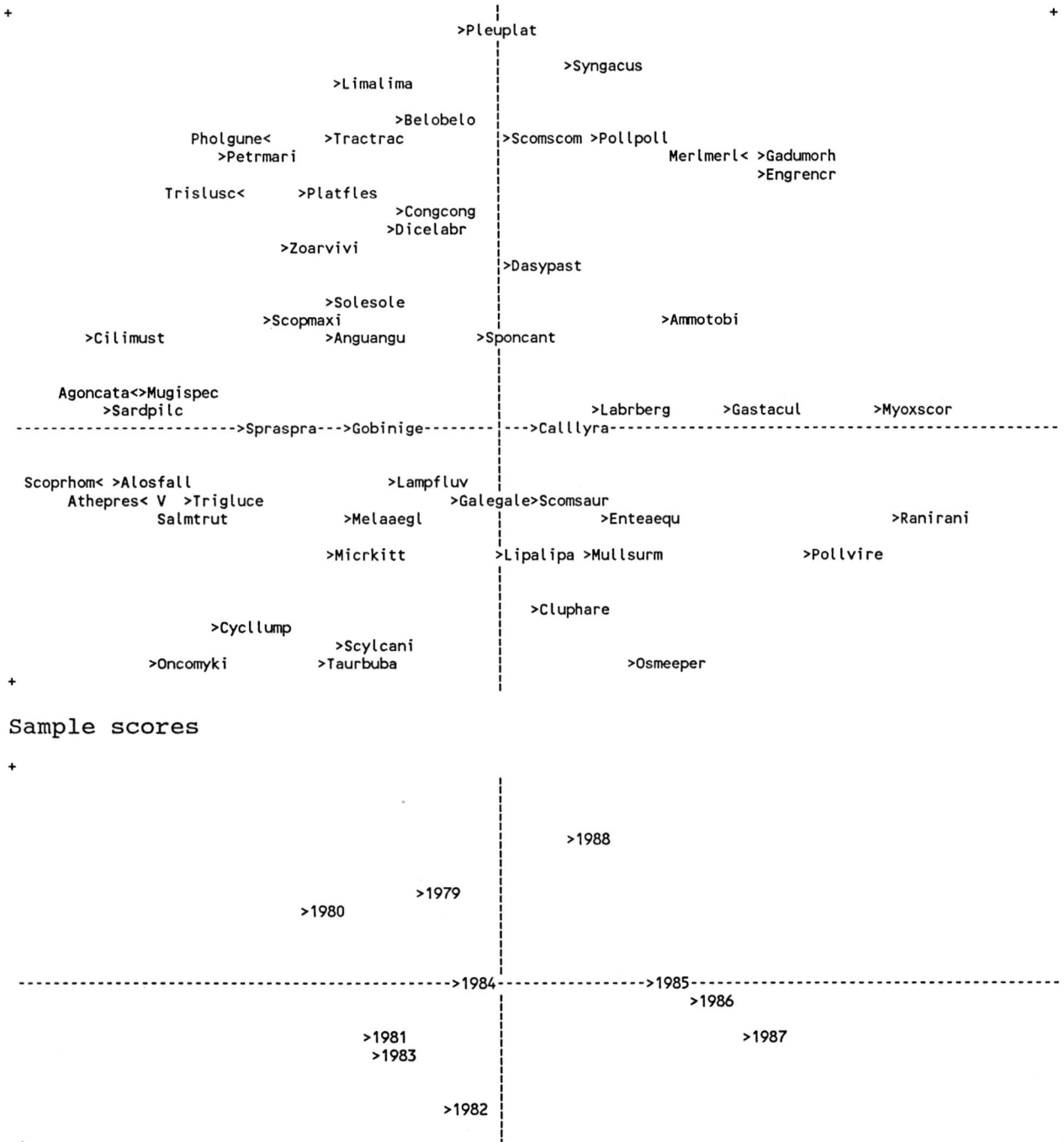

Fig. 2. Result of the Principal Component Analysis for the first two axes. Species scores and sample scores are depicted on the same scale. Species names have been shortened to the first four letters of the genus name plus the first four letters of the species name.

corhynchus mykiss Walbaum 1792, sea trout *Salmo trutta* Linnaeus 1758, tub gurnard *Trigla lucerna* Linnaeus 1758, lumpsucker *Cyclopterus lumpus* Linnaeus 1758, sea lamprey *Petromyzon marinus* Linnaeus 1758, sprat *Sprattus sprattus* Linnaeus 1758, etc. Towards the right of the figure we find the species typical for the years after the barrier impact: tadpole-fish *Raniceps raninus* Linnaeus 1758, bull-rout *Myoxocephalus scorpius* Linnaeus 1758, saithe *Pollachius virens* Linnaeus 1758, cod *Gadus morhua* Linnaeus 1758, anchovy *E. encrasicolus*, whiting *Merlangius merlangus* Linnaeus 1758, three-spined stickleback *Gasterosteus aculeatus* Linnaeus 1758, sandeel *Ammodytes tobianus* Linnaeus 1758, etc.

Significant differences for individual species (Wilcoxon two sample test) between the year blocks as distinguished by the PCA (1979–1984 versus 1985–1988) are shown in Table 2. Significant decreases are found for allis shad *A. fallax*, sea trout *S. trutta*, rainbow trout *O. mykiss*, five-bearded rockling *C. mustela*, sand-smelt *A. presbyter*, tub gurnard *T. lucerna*, lumpsucker *C. lumpus*, sand goby *P. minutus* and brill *S. rhombus*. Significant increases are found for cod *G. morhua*, tadpole-fish *R. raninus* and bull-rout *M. scorpius*.

Table 2. Results of the Wilcoxon two-sample test for species showing significant differences in frequency of occurrence (FO) between clusters of years as distinguished by PCA.

	Average FO 1979–1984	Average FO 1985–1988	
Species with decreased FO			
Alosa fallax	0.38	0.20 **	$p < 0.02$
Salmo trutta	0.07	0.02 **	$p < 0.02$
Oncorhynchus mykiss	0.18	0.05 *	$p < 0.05$
Ciliata mustela	0.33	0.18 *	$p < 0.05$
Atherina presbyter	0.51	0.32 *	$p < 0.05$
Trigla lucerna	0.19	0.09 ***	$p < 0.01$
Cyclopterus lumpus	0.24	0.14 *	$p < 0.05$
Pomatoschistus minutus	0.03	0.01 **	$p < 0.02$
Scophthalmus rhombus	0.29	0.16 *	$p < 0.05$
Species with increased FO			
Gadus morhua	0.38	0.63 **	$p < 0.02$
Raniceps raninus	0.01	0.03 *	$p < 0.05$
Myoxocephalus scorpius	0.69	0.88 ***	$p < 0.01$

Discussion

The use of multivariate techniques in ecology is a rather controversial point at present (review in James & McCulloch, 1990). We have tried to follow their advice as closely as possible and have only used the techniques as a descriptive tool in an exploratory analysis. Following Jongman et al., (1987) we first explored the data using a correspondence analysis (CA), this being a very general (unimodal) model. This gives us the level of variability in the data, which is low, *i.e.* all samples (years) are located within less than 1.5 standard deviations. Ter Braak & Prentice (1988) suggest that non-linear models (like the CA) are inappropriate in these circumstances and linear models (like the PCA) should be used. Following James & McCulloch (1990) we transformed the data for normalisation and used the variance-covariance matrix instead of the correlation matrix. Despite the fact that the Wilcoxon two sample test does not really provide information that is not already apparent from the multivariate analysis it is included in this study because many scientists still feel uncomfortable about the fact that no statistical significance levels are given in the multivariate techniques.

The qualitative dataset, as presently available, is of limited value to describe the changes that occurred in the fish fauna of the Oosterschelde during the construction period. The data can only be used for an exploratory analysis and are certainly not appropriate for any causal investigation. The incompleteness of the time series for the easterly stations may have affected the results.

First of all it is impossible to establish if the fauna caught in the fish traps is in any way representative for the fish fauna of the Oosterschelde. It most certainly cannot be considered to be a random sample: the 5 localities investigated were purposely chosen by the fishermen who, from their experience, selected the best spots to catch either eel or anchovy. Although the fishermen have recorded a crude measure of catch size (1–5, 6–20 and > 20) these data were not available for analysis. For some species we know only single individuals were caught, e.g. allis shad *A. fallax*. As the size of the catches is unknown for all other species the frequency of occurrence is certainly not a measure of fish abundance. Abundant species or species for which the gear is rather selective, *i.e.* eel *A. anguilla* and flounder *Platichthys flesus* Linnaeus 1758 will be almost always present in the catches and changes in abundance would have to be extremely drastic to become apparent. Eel catches are a case in point. Though the frequency of occurrence has decreased slightly the change is not significant. Both in Britain (Swaby & Potts, 1990) and the German Wadden Sea (Tiews, 1990) eel catches have been declining in recent years. The same phenomenon has probably occurred in the Oosterschelde as one of the commercial eel fishermen, involved in the sampling scheme, has had to give up because the fishery was no longer profitable. The other fishermen have had to diversify their fishing activities.

Similarly for schooling species like herring *Clupea harengus* Linnaeus 1758 and sprat *S. sprattus* frequency of occurrence seems not to be a sensitive measure for change. It is a well established fact that with the restoration of herring stocks in the North Sea during the 1980's sprat stocks have gone down again to their pre-1970 levels (Garrod, 1988, Daan *et al.*, 1990). Although a trend in the 'right' direction (increase in herring, decrease in sprat) is recorded in the frequency of occurrence of both species, these trends are not significant.

Catches of the allis shad *A. fallax* always refer to single adult individuals. For this species a change in the frequency of occurrence either reflects a change in phenology, *i.e.* the season in which the fish is present in the Oosterschelde has expanded or contracted or the change reflects a genuine change in abundance. The significant decrease in frequency of occurrence in *A. fallax* thus probably reflects a true change in abundance linked to the reduction of freshwater inflow in the Oosterschelde (Nienhuis & Smaal, 1994). Decreases in other anadromous species *i.e.* sea lamprey *P. marinus*, lampern *Lampetra fluviatilis* Linnaeus 1758 and sea trout *S. trutta* may have occurred for the same reason.

The changes in frequency of occurrence of both lamprey species are not significant mostly because these species are very rarely caught. This means there are a lot of zeroes in the time series that turn up as ties in the Wilcoxon two sample test. Still, frequency of occurrence is seven times lower for sea lamprey and four times lower for lampern in the years 1985–1988. All four species are rare or vulnerable in most of Northwest Europe because of pollution problems in their riverine habitat and engineering works on their migration routes (Aprahamian & Aprahamian, 1990; Swaby & Potts, 1990). For the structure and function of the fish community in the Oosterschelde ecosystem the decline in these species is probably unimportant. However, for the species themselves the marginal habitat offered by the Oosterschelde may have been one of their last refuges from which recolonisation of true estuaries, like the Westerschelde, could have originated after the restoration of water quality in these habitats.

A major problem in the interpretation of the observed changes is to distinguish changes possibly linked to the construction works in the Oosterschelde from natural fluctuations in the size of fish populations over wider areas. This can only be done for species for which we have data on year class strength or for species for which we have reliable data from other areas. In the German Wadden Sea the bull-rout *M. scorpius* increased in the bycatch of shrimp fisheries from 1970 until 1986. Since then it is on the decline again (Tiews, 1990). The increase in frequency of occurrence in the Oosterschelde therefore possibly bears no relationship to the construction works except if the population would remain at the present high level while it is declining elsewhere. From the beam trawl data it appears that *M. scorpius* is much less abundant in 1989 than in the years before (Hostens, unpubl. data). Similarly the decline in the tub gurnard *T. lucerna* is observed both in the Wadden Sea (Tiews, 1990) and the Oosterschelde and may therefore be part of a more general trend in this species.

The rainbow trout *O. mykiss* is not indigenous to the Delta but was introduced into the brackish lake Veerse Meer for 'sports' fisheries. Veerse Meer communicates with the Oosterschelde and the decrease probably reflects population changes in Veerse Meer.

The decrease in frequency of occurrence of the five-bearded rockling *C. mustela* is not matched by a similar decrease in the Wadden Sea. It is a species with a highly variable abundance and has not shown any consistent trend over the past 35 years (Tiews, 1990). A longer time series for the Oosterschelde and more information on the ecological requirements of the species are needed to judge if there may have been an impact of the construction works.

For the sand-smelt *A. presbyter* and the lumpsucker *C. lumpus* the decrease in frequency of occurrence does not seem to have a straightforward explanation. Although lumpsucker declined quite strongly in the Wadden Sea during the 1960's the population seems to have stabilised since then, with even a slight increase during the 1980's (Tiews, 1990). For the sand-smelt there are no data from the Wadden Sea. Both species and the garfish *Belone belone* Linnaeus 1758, which has also decreased (N.S.), spawn in the Oosterschelde and attach their eggs to algae (Wheeler, 1969). If a link exists between the decline in these three species and the construction works it is unclear in what way the impact operated. Needless to say there may be absolutely no link to the construction works, nor any link between the decline in the separate species. One hypothesis may be that the ice floes that moved around the Oosterschelde during the winters of 1984–1985, 1985–1986 and 1986–1987 temporarily damaged the spawning habitat.

For the brill *S. rhombus* no time series for other areas are available. It is therefore difficult to speculate about possible causes.

The increase in the tadpole-fish *R. raninus* seems to be a colonisation phenomenon. The species was first caught in 1982, before the start of the construction works. Initially it was confined to the most seaward stations. It reached the more easterly stations in 1986. It is a species which recently colonised the neighbouring saline Grevelingenmeer (Doornbos, 1985). The lake communicates with the sea through a sluice which may be the source for the increase in tadpole-fish in the Oosterschelde. Possibly the colonisation was helped by the decreased wave action and the reduced current velocities in parts of the Oosterschelde (Vroon, 1994).

For a number of species of commercial fish the Oosterschelde estuary is a nursery area and changes in frequency of occurrence may mainly reflect fluctuations in year class strength. For several species good estimates of yearclass strength for the North Sea covering the time period of this investigation have been published (Rijnsdorp *et al.*, 1991). A notable example is cod *Gadus morhua*. This species has clearly increased in the Oosterschelde and at first sight there seems to be no correlation between frequency of occurrence and year class strength of the same year (Product-moment correlation coefficient $r = 0.09$, N.S.), nor with year class strength of the year before (product-moment correlation coefficient $r = -0.38$, N.S.) (Fig. 3). Thus we might conclude that juvenile cod is using the Oosterschelde more intensively since the construction works. An *ad hoc* explanation could then be that the increased transparency of the water is advantageous for visual predators such as cod and whiting. The next step would be to link the decrease in the sand goby *P. minutus*, one of the preferred prey of juvenile cod and whiting (Vea Salvanes, 1986) to the increase of their predators. However, there is considerable spatial variability in the distribution of the juvenile cod stocks and the strong year class of 1985 was concentrated in the Southern Bight (Rijnsdorp *et al.*, 1991). Thus the high frequency of occurrence in 1985 was probably mainly caused by this phenomenon. Also, in the absence of length-frequency distributions of the fish caught, it is impossible to know which year class is being sampled. From the beam trawl data it appears that small 0-group cod were already present in the Oosterschelde in June 1985 and 1988. In other years, with smaller year classes it is mainly the 1-group cod that visits the Oosterschelde in winter (Hostens & Hamerlynck, 1994). A further complication is due to the fact that the data are at present only available on an annual basis. The 0-group and 1-group cod from a strong year class affect the frequency of occurrence data

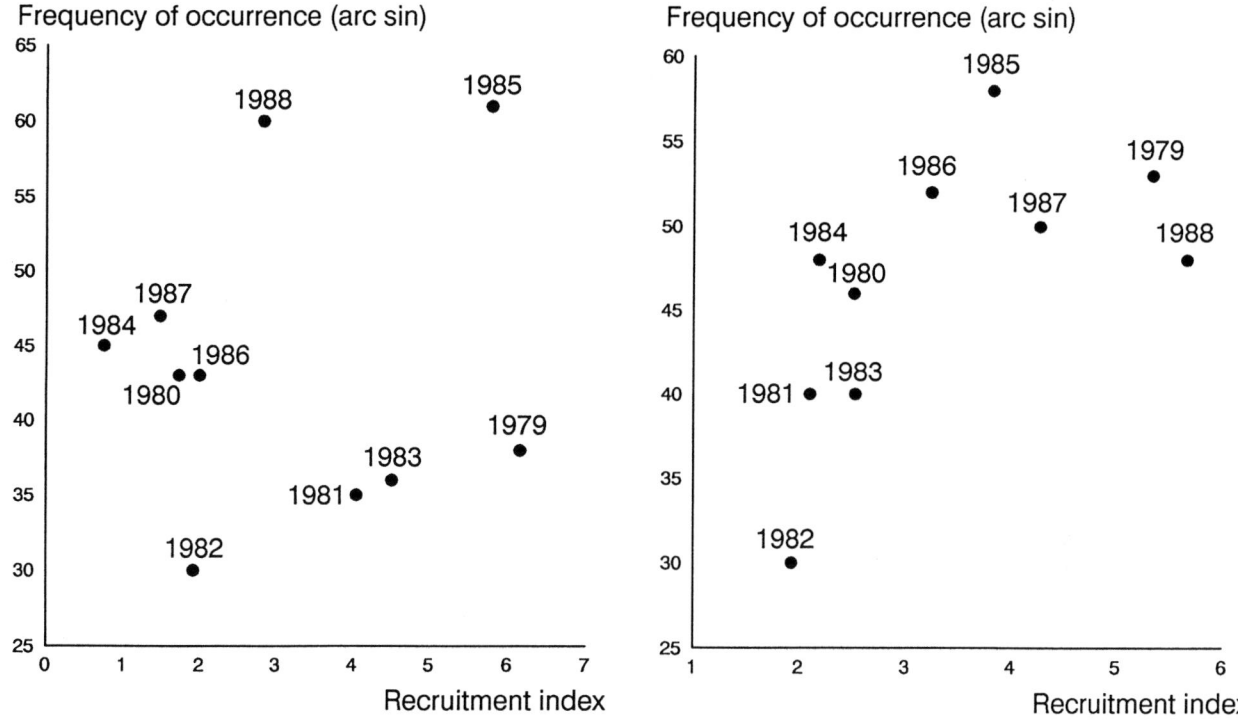

Fig. 3. Relationship between year-class strength expressed as a recruitment index (Anonymous, 1990) and frequency of occurrence in the Oosterschelde for cod (*G. morhua*) and whiting (*M. merlangus*).

of at least two years if they remain present in the Oosterschelde throughout the winter. We believe therefore that the increase in frequency of occurrence in cod is mainly due to the more southerly distribution of recent strong year classes (1985, 1988). For whiting *M. merlangus* an increase in recent years was suggested by the result of the PCA, though it was not significant in the Wilcoxon two sample test, mainly because of the high frequency of occurrence in 1979 (0.53), the second highest on record. For whiting there is a much better correlation between year class strength as estimated by the MSVPA (Anonymous, 1990) and frequency of occurrence in the Oosterschelde (product-moment correlation coefficient $r = 0.61$, $p < 0.1$, Kendall coefficient of rank correlation $\tau = 0.511$, $p < 0.05$) (Fig. 3). For whiting the increase in frequency of occurrence in recent years is thus probably mainly due to the

exceptionally poor year classes from 1981 through 1984.

Conclusion

The main part of the observed changes in 'community structure' seems to have no relationship to the construction works and the associated changes in the Oosterschelde estuary. Only the decrease in frequency of occurrence of the anadromous fish seems likely to be linked to the decrease of fresh water inflow in the Oosterschelde. Still, catches in the fykes and the weir seem to reflect changes in abundance of many species quite well. The decrease of *A. cataphractus* and *S. rhombus* and the increase of *M. merlangus* were also observed in the beam trawl surveys (Hostens & Hamerlynck, 1994). For other

506

species such as *A. tobianus* the two studies have conflicting results. The value of the data recorded by the fishermen would be greatly increased if data on size of the catches and length-frequency distributions could be included. Perhaps this would mean too much work for the commercial fishermen. Separate fish traps owned and operated by research institutes could provide relatively cheap and very valuable data on long-term trends in fish species. Care should be taken to assure that gear, location and operation remain extremely constant in time (see also Van der Veer *et al.*, 1992). Longer time-series, especially on non-commercial fish species, are an absolute requirement for the assessment of the ecological impact of civil-engineering works or other human activities on the fish community. Baseline studies from many different areas are needed to be able to distinguish natural fluctuations in abundance levels from those caused by man. For example the mere fact that the winters of 1984–1985, 1985–1986 and 1986–1987 were relatively severe may have been important in causing many of the changes observed within the relatively short time period considered here.

For most species it would make good biological sense to analyse the data using a 'year' starting on March 21th. Very few 0-group will occur before that date and few 1-group or older using the Oosterschelde in winter will still be present around that time.

For ecologists it would be a great help if fisheries research institutes could publish time series of abundance indices for all the species they catch in their annual surveys. These data have been collected in a standardised way over extensive areas for about twenty years and could be immensely valuable for the interpretation of local changes.

Analysing the frequency of occurrence data with descriptive multivariate statistical techniques was successful in summarising the structure in the data, thus allowing us to suspect a slight change in community structure after 1984 and pointing out the species that underwent the most important changes.

Acknowledgements

The survey was designed and directed by Bureau Waardenburg, Culemborg. Special thanks are due to the fishermen who cooperated in the project and provided the raw data reported in a concise form by Bureau Waardenburg: A. A. K. van Hoek, A. Bakker, J. van Westenbrugge, A. Verkamman, W. Verkamman, G. Verkamman, C. J. van Dort and C. van Dort. This study was supported by Rijkswaterstaat Dienst Getijdewateren, Middelburg. Contribution No. 707 of NIOO-CEMO, Yerseke, The Netherlands.

References

Anonymous, 1990. Report of the Roundfish Working Group. ICES CM 1990/Assess: 7 (mimeo).

Aprahamian, M. W. & C. D. Aprahamian, 1990. Status of the genus *Alosa* in the British Isles; past and present. J. Fish Biol. 37 (Suppl. A): 257–158.

Daan, N., P. J. Bromley, J. R. G. Hislop & N. A. Nielsen, 1990. Ecology of North Sea fish. Neth J. Sea Res. 26: 343–386.

Doornbos, G., 1985. The fish fauna of Lake Grevelingen (SW Netherlands: the role of fish in the food chain of a man-made saline lake some ten years after embankment of a former estuary. Ph. D. Thesis, University of Amsterdam, 169 pp.

Garrod, D. J., 1988. North Atlantic cod: fisheries and management to 1986. In J. A. Gulland (ed.), Fish population dynamics, second edition. John Wiley & Sons, Chichester: 185–218.

Gislason, H, 1989. The influence of variations in recruitment on multispecies yield predictions in the North Sea. ICES CM 1989/MSM: 15 (mimeo).

Hill, M. O., 1979. TWINSPAN a FORTRAN program for arranging multivariate data in an ordered two-way table by classification of individuals and attributes. Cornell University, Ithaca, 90 pp.

Hinz, V., 1989. Monitoring the fish fauna in the Wadden sea with special reference to different fishing methods and effects of wind and light on catches. Helgoländer Meeresunters. 43: 447–459.

Hostens, K. & O. Hamerlynck, 1994. The mobile epifauna of the soft bottoms in the subtidal Oosterschelde estuary: structure, function and impact of the storm-surge barrier; Hydrobiologia 282/283: 479–496.

James, F. C. & C. E. McCulloch, 1990. Multivariate analysis in ecology and systematics: panacea or Pandora's box? Annu. Rev. Ecol. Syst. 21: 129–166.

Jongman, R. H. G., C. J. F. ter Braak & O. F. R. van Tongeren, 1987. Data analysis in community and landscape ecology. Pudoc, Wageningen, 299 pp.

Meijer, A. J., 1989. Monitoring-onderzoek aan de visfauna van de Oosterschelde. Rapportage Resultaten 1988. Bureau Waardenburg bv, Culemborg, 30 pp.

Nédélec, C., 1982. Definition and classification of fishing gear categories. FAO Fish. Techn. Pap. 222: 51 pp.

Nienhuis, P. H. & A. C. Smaal, 1994. The Oosterschelde estuary, a case study of a changing ecosystem: an introduction. Hydrobiologia 282/283: 1–14.

Rijnsdorp, A. D., N. Daan, F. A. van Beek & H. J. L. Heesen, 1991. Reproductive variability in North Sea plaice, sole, and cod. J. Cons. int. Explor. Mer 47: 352–375.

Ruth, M. & R. Berghahn, 1989. Biological monitoring of fish and crustaceans in the Wadden Sea – potential and problems. Helgoländer Meeresunters. 43: 479–487.

Sokal, R. R. & F. J. Rohlf, 1981. Biometry (second edition). W. H. Freeman, San Francisco, 959 pp.

Swaby, S. E. & G. W. Potts, 1990. Rare British marine fishes – identification and conservation. J. Fish Biol. 37 (Suppl. A): 133–143.

Ter Braak, C. J. F., 1987. CANOCO – a FORTRAN program for canonical community ordination by (partial) (detrended) (canonical) correspondence analysis, principal components analysis and redundancy analysis (version 2.1). ITI-TNO, Wageningen, 95 pp.

Ter Braak, C. J. P. & I. C. Prentice, 1988. A theory of gradient analysis. Adv. ecol. Res. 18: 271–317.

Tiews, K., 1990. 35-Jahres-Trend (1954–1988) der Häufigkeit von 25 Fisch- und Krebstierbeständen an der deutschen Nordseeküste. Arch. Fish Wiss. 40: 39–48.

Van der Veer, H. W., J. IJ. Witte, M. A. Beumkes, R. Dapper, W. P. Jongeman & J. van der Meer, 1992. Intertidal fish traps as a tool to study long-term trends in juvenile flatfish. Neth. J. Sea Res. 29: 119–126.

Vea Salvanes, A. G., 1986. Preliminary report from a comparative study of four gadoid fishes in a fjord of western Sweden. ICES C.M. 1986/G:71 (mimeo).

Vroon, J., 1994. Hydrodynamic characteristics of the Oosterschelde in recent decades. Hydrobiologia 282/283: 17–27.

Wheeler, A., 1969. The fishes of the British Isles and north west Europe. MacMillan, London, 613 pp.

Witte, J. IJ., H. W. van der Veer & J. van der Meer, in press. Intertidal fish traps as a tool to study long-term trends in juvenile flatfish. Neth. J. Sea Res.

Hydrobiologia **282/283**: 509–524, 1994.
P. H. Nienhuis & A. C. Smaal (eds), The Oosterschelde Estuary.
© 1994 *Kluwer Academic Publishers. Printed in Belgium.*

Changes in the waterbird populations of the Oosterschelde (SW Netherlands) as a result of large-scale coastal engineering works

Hans Schekkerman [1,2], Peter L. Meininger [3] & Patrick M. Meire [4,5,1]
[1] *Netherlands Institute of Ecology, Centre for Estuarine and Coastal Ecology, Vierstraat 28, 4401 EA Yerseke, The Netherlands;* [2] *present address: IBN-DLO, P.O. Box 23, 6700 AA Wageningen, The Netherlands;* [3] *National Institute for Coastal and Marine Management/RIKZ, P.O. Box 8039, 4330 AE, Middelburg, The Netherlands;* [4] *University of Ghent, Ledeganckstraat 35, B-9000 Gent, Belgium;* [5] *Institute for Nature Conservation, Kiewitdreef 3, B-3500 Hasselt, Belgium*

Key words: carrying capacity, coastal engineering, estuary, habitat loss, waders, waterbirds

Abstract

Between 1982 and 1987, the construction of a storm-surge barrier and two secondary dams in the eastern and northern parts of the Oosterschelde/Krammer-Volkerak area resulted in the loss of 33% of the 170 km^2 of intertidal area in the estuary. Consequences for non-breeding waterbirds were evaluated on the basis of monthly high-tide counts during five seasons before and three seasons after the construction period.

In the entire Oosterschelde/Krammer-Volkerak area, numbers of wintering waders decreased but those of ducks increased. Peak numbers and total number of bird-days changed little, but the seasonal pattern shifted from a midwinter maximum to a peak in autumn.

In the Oosterschelde (excluding the Krammer-Volkerak), where 17% of the tidal flats disappeared, species feeding mainly on open water remained stable or increased. Species dependent on intertidal areas for foraging (mainly waders and dabbling ducks) generally decreased. Total density of intertidal foragers decreased slightly. In most intertidal species, the Oosterschelde wintering population showed a stronger decrease, or smaller increase, than was shown during the same period by numbers in Britain and Ireland which were taken as an index of the total W-European winter populations. Changes varied considerably between species, and were correlated with their distribution within the estuary. Species concentrated in the eastern sector, where most habitat loss occurred, declined more than species with a more westerly distribution.

Results indicate that intertidal foragers forced to move from the enclosed parts of the estuary were not generally able to settle into the remaining intertidal areas. Both dispersal to adjacent areas (mainly by dabbling ducks) and mortality during severe winter weather (in some wader species) may have contributed to the declines. Populations of intertidal foragers apparently were (and consequently still are) close to carrying capacity, and further changes in capacity, as foreseen from geomorphological changes still under way in the estuary, are likely to be reflected in bird populations.

Numbers of waders moulting in the Oosterschelde in late summer declined strongly compared to numbers in other seasons. Increased disturbance due to recreational activities may have played a role during this time of the year.

510

Introduction

Estuarine habitats throughout the world are affected by human activities like fisheries, disturbance, pollution, manipulations of the tidal regime and reclamation (Smit *et al.*, 1987). In many cases, such activities reduce the suitability of the habitat for birds, through a decrease in the feeding area or the food stocks available, or through a reduction in potential feeding time. Besides effects on bird populations on a local scale, such events may also affect total population size (Goss-Custard, 1977, 1985, 1990), although our present knowledge of the population dynamics of most species is not yet sufficient for a quantitative evaluation (Goss-Custard, 1980, 1981; Goss-Custard & Durell, 1990; Sutherland & Goss-Custard, 1991).

An impact of loss of habitat or feeding time on local waterbird populations can only be expected if bird numbers are close to carrying capacity in the area concerned (Goss-Custard, 1977, 1985). Here, carrying capacity is defined as the density at which the addition of any further birds would result in other birds dying or leaving the area because they fail to achieve adequate intake rates as a result of increased interference and/or depletion of prey stocks (Goss-Custard, 1985, Sutherland & Goss-Custard, 1991). Whether or not this critical density is reached or approached in estuarine areas used by waterbirds has important implications for nature conservation and management.

One possible approach to this issue is to study the impact of habitat loss on bird numbers in estuarine areas. If bird numbers were close to carrying capacity prior to the habitat changes, one would expect 1) that numbers decline after habitat loss, 2) that this decline is a local event unrelated to changes in total population size, and 3) that differing responses of individual species can be understood in view of their distribution and/or diet. So far, few case studies of estuarine habitat loss have been reported. Evans *et al.* (1979) and Evans (1981) described the impact of reclamation of part of the Tees estuary, Britain, on waders and Shelduck, showing some reduc-

tions in numbers. Laursen *et al.* (1983) documented the effects of the reclamation of 11 km^2 of salt marsh and tidal flats in the Danish and German Waddensea. A local redistribution of birds was found, but an impact on numbers on a larger scale was not clear.

This paper describes changes in numbers and species composition of non-breeding waders and other waterbirds coinciding with loss of one third of 170 km^2 of intertidal area in the Oosterschelde/Krammer-Volkerak, SW. Netherlands, in 1982–1987. This area is one of the major wintering haunts for estuarine birds in western Europe, in addition to being an important staging and moulting site in spring and autumn (Leewis *et al.*, 1984, Meininger & van Swelm, 1989). Lambeck *et al.* (1989) showed that numbers of some wader species on a tidal flat in the Oosterschelde increased abruptly after the closure of the nearby Grevelingen estuary in 1971, which suggested that carrying capacity had not been reached at that moment (but see discussion). Further increases in wader numbers on the estuary have occurred afterwards, and densities on the Oosterschelde in the period shortly before the engineering works were among the highest in western Europe. Were all waders using the intertidal areas lost in 1982–1987 to have moved to the remaining feeding areas in the Oosterschelde, this would have resulted in a 49% increase in overall densities. Meire & Kuyken (1984) predicted this to be unlikely.

Methods

Study area

The Oosterschelde/Krammer-Volkerak was the main estuarine area in the Delta region of the SW-Netherlands, with a total area of 452 km^2 and 170 km^2 of intertidal flats prior to the coastal engineering project. Between 1982 and 1986, a storm surge barrier was built in the mouth of the estuary (Fig. 1), causing a considerable reduction of the tidal volume. In order to retain a sufficient tidal amplitude, two compartmentation dams were completed in the eastern part in 1986–1987,

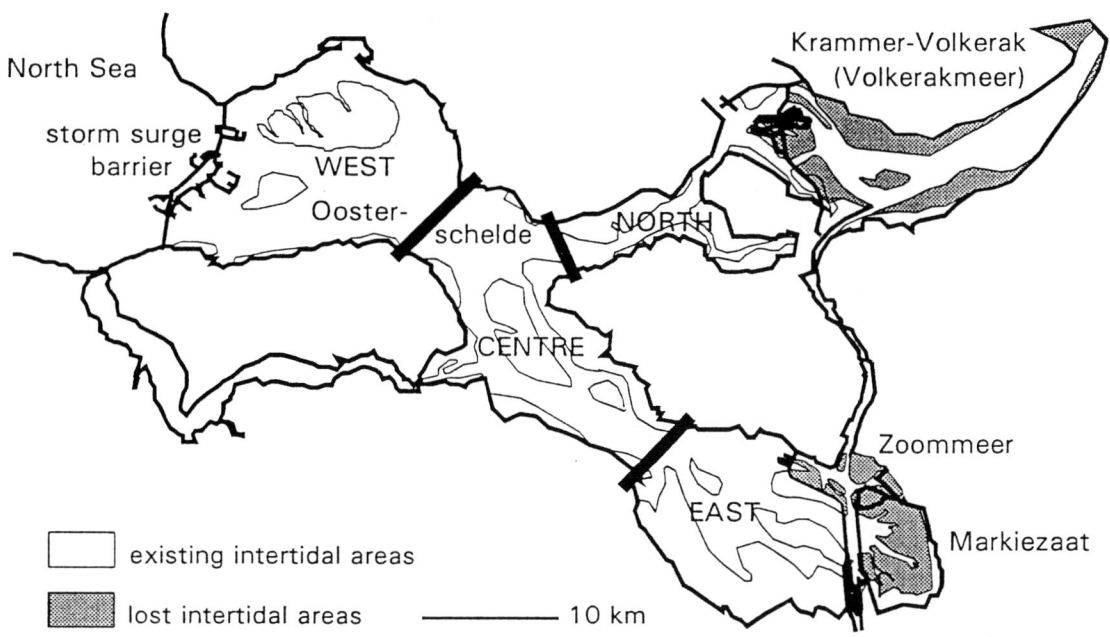

Fig. 1. Map of the Oosterschelde/Krammer-Volkerak estuary showing location of coastal engineering works, lost and remaining intertidal areas and boundaries of sectors defined within the area.

reducing the volume of water accommodated within the estuary. The areas behind the dams, including 50 km^2 of intertidal flats (about 30% of the former intertidal area), became freshwater lakes: the present Zoommeer/Markiezaat and Volkerakmeer. In addition, 6 km^2 (3%) of flats was lost in the remaining estuarine part of the Oosterschelde, through a 0.45 m reduction of the tidal range. The post-barrier intertidal area is 114 km^2. For a general description of the area's geomorphology and more details on the coastal engineering works see Mulder & Louters (1994), Smaal *et al.* (1991), and Nienhuis & Smaal (1994).

Changes in total waterbird numbers and species composition are described for the whole Oosterschelde/Krammer-Volkerak area, including the present Volkerakmeer (*i.e.* the former Krammer-Volkerak), Zoommeer and Markiezaat. Mainly because counts from the Krammer-Volkerak are less complete than from the Oosterschelde, more detailed analyses of changes in bird numbers, bird-days and densities are made for the Oosterschelde only. The latter area excludes

the Krammer-Volkerak, but includes for the pre-barrier period the (now closed) Zoommeer/Markiezaat. This apparent inconsistency was necessary because in the pre-barrier period birds feeding in the present Zoommeer/Markiezaat used the same high-tide roosts as birds feeding on adjacent tidal flats that still remain, and thus could not be separated. The coastal engineering works caused a decrease of the area of intertidal flats of 33% for the entire Oosterschelde/Krammer-Volkerak, and of 17% for the Oosterschelde proper.

Within the Oosterschelde, four sectors have been defined: west, centre, east and north (Fig. 1). The boundaries are chosen so that nearly all birds roost at sites in the same sector as their intertidal feeding areas. The area of tidal flats in each of the sectors before the engineering works was 24.5 (18% of the total), 34.8 (25%), 59.4 (44%) and 18.0 km^2 (13%), respectively. The effect of the engineering works was most pronounced in the eastern sector: 33% of the intertidal area was lost here, compared with 3%, 4%, and 9% in the western, central and northern sectors.

Count data

Monthly censuses of all waterbirds present in the Oosterschelde/Krammer-Volkerak were carried out by teams of professional and amateur ornithologists. Counts in tidal areas were performed during high tide, when birds concentrate on roosts. Count dates were set close to mid-month, but depended on the occurrence of high tide during daylight hours. Counts in different parts of the estuary were spread over a period of about a week around the requested counting dates, but neighbouring roosts between which bird movements were frequent were normally counted on the same day. Counts in non-tidal areas were made from boats and from the shore.

All waterbird species observed were included in the totals for the Oosterschelde/Krammer-Volkerak area. Gulls were counted only before the construction of the coastal engineering works; their numbers are assumed to have remained stable in intertidal areas in this paper. For a more detailed analysis of waterbird numbers in the Oosterschelde proper, 23 species were selected which are characteristic for estuarine habitats, and occur in the Oosterschelde in significant numbers (Table 1). These species comprise 74–80% of all waterbirds wintering on the Oosterschelde. Scientific names and acronyms for species mentioned in the text and figures are given in an appendix.

Counts during five seasons before (1978/79–1982/83; pre-barrier period) and during three seasons after the construction of the engineering works (1987/88–1989/90; post-barrier period) have been used (seasons running from July through June). Mean numbers per month were calculated for both periods. Bird-days per season were calculated by multiplying the sum of the monthly totals by 30. Three season-related functions of the estuary for waterbirds were distinguished: (1) wintering (mean maximum in December–February), (2) spring staging (mean maximum number during the period of spring migration; this period varies per species), and (3) autumn moulting/staging function (for species which moult flight-feathers in the area, mean

number in the month(s) in which the largest proportion of the birds present is in moult (usually August); for others the mean maximum number in the month(s) of autumn migration). Differences in means between periods were tested using Student's *t* statistic. In view of the large variability of bird numbers and the short study periods, a significance level of 10% has been used.

As an index of total population size to which changes in bird numbers in the Oosterschelde can be compared, population trends in Great Britain and Ireland have been used. Mean indices based on mid-winter counts (Kirby *et al.*, 1990) were calculated for the same periods as the Oosterschelde figures. Although for some species the breeding origin of the birds wintering in the British and Irish estuaries may be partly different from that of the Oosterschelde population, these estuaries are the only areas for which published data allow this approach. Together they hold 40–50% of the total number of waders wintering along the Atlantic coasts of Europe (Smit & Piersma, 1989). For some species (Great Crested Grebe, Cormorant, Kentish Plover), no British indexes were available. For the Brent Goose, the change in total population size is known from yearly counts (28% increase, Madsen *et al.*, 1991), and has been used instead of the British index.

Results

Bird numbers and species composition in the Oosterschelde/Krammer-Volkerak

During the pre-barrier period, a mean maximum of 292 000 waterbirds were present in the Oosterschelde/Krammer-Volkerak area in January (Fig. 2). These included 182 000 estuarine waders (65%), 57 000 ducks, geese and swans (19%), 40 000 gulls (14%) and 3000 other waterbirds (Fig. 3). Eight wader species wintered in numbers exceeding 1000: Oystercatcher, Grey Plover, Dunlin, Knot, Bar-tailed Godwit, Curlew, Redshank and Turnstone. Ducks and geese were dominated by Barnacle Goose, Brent Goose, Shelduck, Wigeon, Mallard, Pintail and Teal.

Table 1. Changes in numbers of non-breeding waterbirds in the Oosterschelde. Given are mean (x) and standard deviation (sd) of yearly bird-days, and mean peak numbers in winter, spring and autumn in the pre-barrier (1978/79–1982/83) and post-barrier situation (1987/88–1989/90) and changes expressed as % of pre-barrier mean (Diff). For spring and autumn, months on which figures are based (means of maximum counts within these months) are indicated (autumn: July, August, September, October, November; spring: March, April, May). Additional functions: m species moults in Oosterschelde; s important mass gain during spring migration. Significance levels (Student's *t*): * $p < 0.10$; ** $p < 0.05$; *** $p < 0.01$.

Species	Bird-days (×1000) 1978–82 x	sd	1987–89 x	sd	Diff.	Winter 1978–82 x	sd	1987–89 x	sd	Diff.	Autumn mo.	1978–82 x	sd	1987–89 x	sd	Diff.	Spring mo.	1978–82 x	sd	1987–89 x	sd	Diff.
Great Crested Grebe	26	27	51	10	+92	90	86	170	47	+89	SO	124	166	386	218	+211*	AY	19	17	56	30	+195*
Cormorant	20	5	146	30	+639***	41	21	260	64	+534**	ASO	224	67	1129	205	+404*	MAY	25	12	286	79	+1044**
Brent Goose	1493	497	2299	490	+54*	7999	2770	10950	2383	+37	ON	8143	2607	12561	4149	+54	Ys	3568	1513	5487	927	+54
Shelduck	1001	179	709	151	−29*	7605	1990	3625	803	−52	SO	905	315	1850	1177	+104	Y	1594	245	1155	358	−28*
Wigeon	2020	224	1860	291	−8	15387	4574	13517	5540	−12	SON	16931	3741	13873	2451	−18	M	1447	335	2922	1513	+102
Teal	213	26	120	46	−44***	1283	299	622	440	−52**	SON	1921	1048	930	321	−52**	M	340	92	260	139	−24
Pintail	461	164	322	74	−30	5526	1355	2897	183	−48**	SON	2735	1158	1669	710	−39	M	505	296	323	71	−36
Shoveler	213	58	102	43	−52*	1599	778	627	431	−61*	SON	2046	541	823	173	−60*	MA	393	175	243	243	−38
Goldeneye	29	11	60	30	+105	405	162	634	280	+57	ON	103	68	319	239	+210	M	131	125	139	90	+6
Red-breasted Merganser	40	13	38	8	−3	330	158	201	61	−36	ON	167	106	360	218	+116	M	229	55	167	55	−27
Oystercatcher	19810	726	19117	42	−3	98137	10007	75656	9891	−23**	Am	60161	14008	76228	5890	+27	AY	18832	5037	20028	1553	+8
Avocet	155	9	92	20	−41***	153	77	222	204	+45	Am	427	160	303	306	−29	MA	799	195	437	110	−45
Ringed Plover	112	20	133	40	+19	51	14	57	29	+12	ASm	1595	368	1474	753	−8	Ys	317	220	606	272	+91
Kentish Plover	64	13	37	13	−42**	0		0		–	JASm	828	193	598	168	−28	Y	116	68	124	67	+7
Grey Plover	1695	98	1432	190	−16*	5875	1287	5892	1157	+0	Am	4879	1197	2671	1152	−45*	Ys	6517	1340	6200	775	−5
Knot	2219	369	1895	303	−15	13426	5073	12523	1234	−7	Am	5235	2773	3452	1213	−34	Y	2381	1092	2119	1882	−11
Dunlin	9408	1349	6448	1392	−31**	53322	11652	37295	1126	−30*	Am	10615	2713	4260	2713	−60**	Ys	10315	4868	14431	13639	+40
Bar-tailed Godwit	1802	331	1788	206	−1	7307	1684	6647	500	−9	Am	7878	2718	4471	851	−38	Ys?	7818	792	6999	3323	−10
Curlew	2086	140	2240	280	+7	8568	1201	7262	278	−15	Am	9858	1423	9225	1178	−6	MAs	5320	966	5478	286	+3
Spotted Redshank	105	24	121	26	+15	5	4	23	15	–	ASm	1620	718	1341	352	−17	AYs	68	38	136	26	+100**
Redshank	721	102	542	72	−25**	2118	133	1492	73	−30**	JAm	3603	808	2393	304	−34*	As?	2252	1016	1203	157	−47
Greenshank	48	20	49	8	+2	1		2		–	Am	579	313	473	96	−18	AY	20	16	129	102	+545
Turnstone	328	43	284	11	−13	1347	282	1077	169	−20	Am	1354	359	1235	315	−9	Ys	793	274	788	169	−1

514

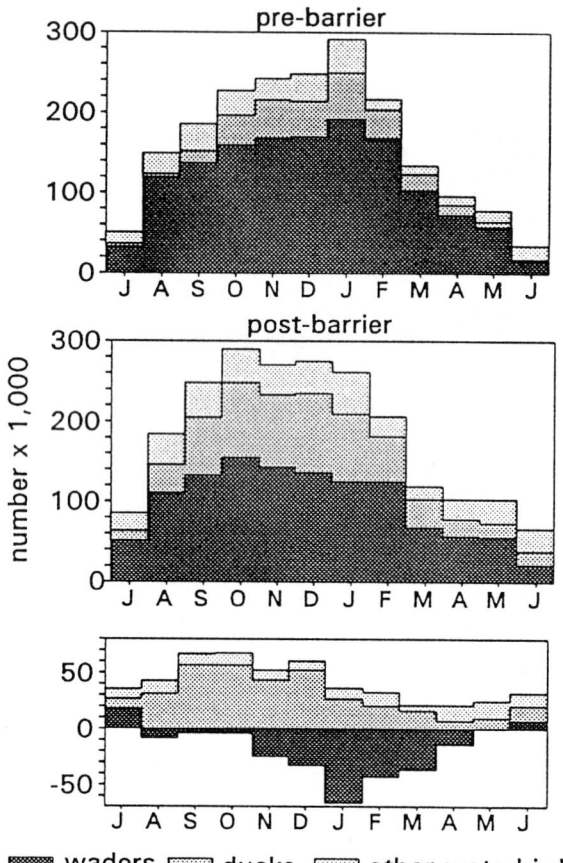

Fig. 2. Monthly mean numbers of waterbirds in the Ooster-
schelde/Krammer-Volkerak area during pre-barrier (upper
panel) and post-barrier (middle panel) periods, and changes
in numbers between these periods (lower panel).

In the post-barrier period, mean maximum
numbers of waterbirds have hardly changed
(Fig. 2). However, the seasonal peak shifted from
January to October. The area's significance as a
wintering site decreased, but the total number of
bird-days per season increased slightly. Changes
in the species composition occurred, but only in
the Volkerakmeer and Zoommeer/Markiezaat
(Fig. 3).

The average number of waders in the entire
Oosterschelde/Krammer-Volkerak area in January decreased to 125000 (31% less), and the
number of bird-days of waders decreased by 15%
compared to the pre-barrier situation. During the
pre-barrier period, only about 8000 waders wintered in the Krammer-Volkerak, 5% of the total

number in the Oosterschelde/Krammer-Volkerak
area. Therefore, although 90% of the waders wintering in the Krammer-Volkerak disappeared
after the closure of the Philipsdam, their contribution to the decrease in the total Oosterschelde/
Krammer-Volkerak area was only 15%, and the
majority of the waders which disappeared originated from the Oosterschelde. In spring and autumn, the importance of the Krammer-Volkerak
for waders was greater, with 20% and 10% of the
total numbers, respectively.

The maximum number of ducks in the Ooster-
schelde/Krammer-Volkerak area increased from
57500 to 98600, mainly in the new freshwater
habitats and during autumn; the number of bird-
days roughly doubled. Numbers in the Oosterschelde decreased by 21% in January, but increased during autumn. In many duck species
post-barrier numbers in the Volkerakmeer and
the Markiezaat/Zoommeer exceeded those in the
former intertidal areas, but numbers of wintering
Shelduck and Pintail were lower than before.

Bird numbers and densities in the Oosterschelde

Table 1 summarizes changes in the use of the
Oosterschelde proper (pre-barrier situation including present Zoommeer/Markiezaat) for
23 characteristic waterbird species. Significant
($p < 0.1$) changes in the total annual number of
bird-days have occurred in ten species: an increase in Cormorant and Brent Goose, and decreases in Shelduck, Teal, Shoveler, Avocet,
Kentish Plover, Grey Plover, Dunlin, and Redshank. Non-significant (although considerable, *i.e.*
25% or more) changes occurred in Great Crested
Grebe and Goldeneye (increase) and Pintail
(decrease).

The average winter maximum of Cormorant
showed a significant increase, those of Teal,
Pintail, Shoveler, Oystercatcher, Dunlin and
Redshank a significant decrease. Non-significant
changes exceeding 25% occurred in Great
Crested Grebe and Avocet (increase), and in
Shelduck and Red-breasted Merganser (de-
crease).

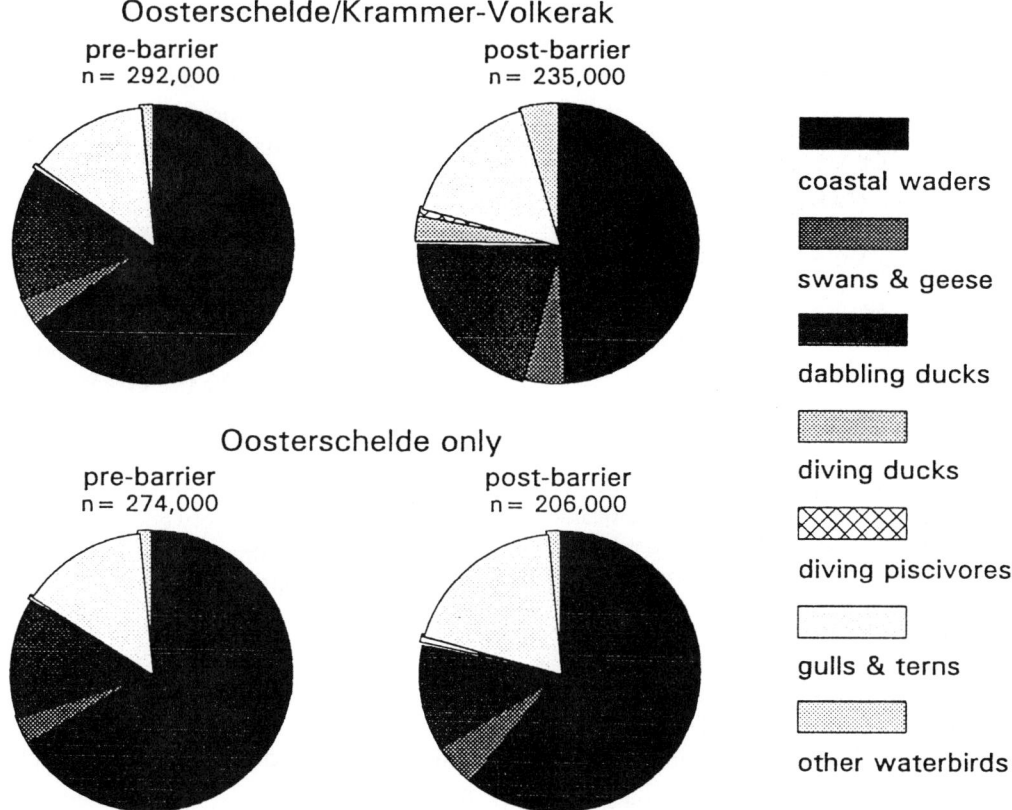

Fig. 3. Species composition of waterbirds wintering in the entire Oosterschelde/Krammer-Volkerak area and in the Oosterschelde proper during pre-barrier and post-barrier periods, based on mean January numbers. Note that gulls were not counted in the post-barrier period, but were assumed to have remained stable in numbers in the intertidal areas.

Significant increases in autumn were found in Great Crested Grebe and Cormorant, significant decreases in Shoveler, Grey Plover, Dunlin and Redshank. A 25% or stronger decrease in average number during the wing moult period occurred in Avocet, Kentish Plover, Grey Plover, Knot, Dunlin, Bar-tailed Godwit and Redshank, although most of these changes are statistically not significant.

In spring, few significant changes were found: increases in Great Crested Grebe, Cormorant and Spotted Redshank, and a decrease in Shelduck. Notable non-significant changes ($> 25\%$) were found in Brent Goose, Wigeon, Ringed Plover, Dunlin (increase) and in Avocet (decrease).

With the exception of the Avocet, which is very scarce in winter, and the Grey Plover, none of the species foraging in intertidal areas (see appendix)

showed the 20% or more increase in winter density that would be necessary to compensate for the area of intertidal flats lost (Fig. 4). In contrast, densities of several species showed a considerable decline. The total midwinter density of intertidal foragers declined by 8% from 16.2 to 14.9 birds ha^{-1}.

Changes in the Oosterschelde in relation to total population size

Figure 5 compares changes in winter numbers in the Oosterschelde with those in Britain and Ireland, taken as an indication of changes in the size of the total W-European winter population. Only four out of 17 species for which British indices were available are 'doing better' in the

516

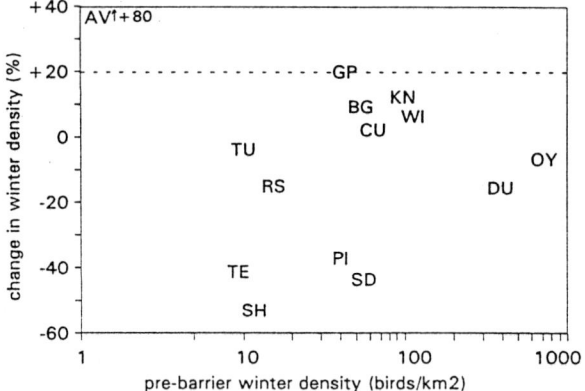

Fig. 4. Changes in mean winter density in the Oosterschelde of waterbirds foraging in the intertidal areas, expressed as % deviation from pre-barrier density. The broken line marks the density increase necessary to accommodate all birds displaced from Zoommeer/Markiezaat. Acronyms identify species (see appendix for clarification).

Oosterschelde, while 13 species are 'doing worse'. When the changes for each species in the Oosterschelde and in Britain and Ireland are used as paired observations in a Wilcoxon matched-pairs signed-ranks test, the Oosterschelde wader population appears to have fared significantly worse than the British and Irish one ($T = 22$, $N = 17$, $p < 0.01$). The difference is especially apparent in

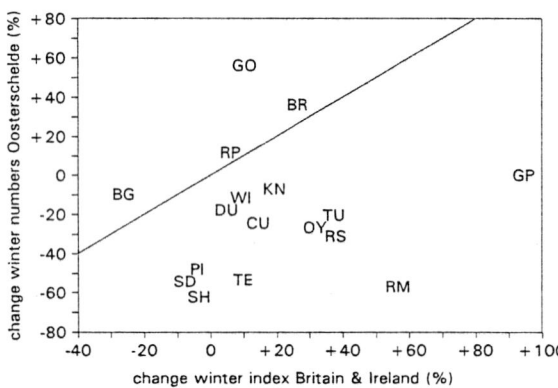

Fig. 5. Comparison of changes in mean midwinter numbers of waterbirds between pre-barrier and post-barrier periods in the Oosterschelde and simultaneous changes in midwinter index of abundance in Britain and Ireland. The line indicates equal change. Acronyms identify species (see appendix). For the Brent Goose, the change in the total dark-bellied population has been used instead of the British index.

intertidally foraging species: out of eleven strictly intertidal foragers wintering in the Oosterschelde in numbers exceeding 2000, only the Bar-tailed Godwit showed a more favourable trend in the Oosterschelde than in the British Isles ($T = 2$, $N = 11$, $p < 0.01$).

The increase in the Oosterschelde of Great Crested Grebe and Cormorant, for which no British indexes were available for comparison, coincided with an increase in the W-European winter populations. No recent population estimates are available for the Great Crested Grebe, though it is unlikely that the total population doubled as it did in the Oosterschelde. The European breeding population of Cormorant increased by ca 300% between 1980 and 1988 (Osieck, 1991), whereas Oosterschelde numbers increased by 600%.

Population changes in relation to distribution

The population development of species foraging in the intertidal zone showed considerable variation, ranging from a 52% decline to a 19% increase in birddays (Table 1). This variation appeared to be related to differences in within-estuary distribution. Sediment composition and macrobenthos vary locally and, modified by preferences for certain substrates or prey types, are reflected in the distribution of the bird species. Consequently, the loss of intertidal areas, which occurred mainly in the easternmost part, did not affect all species to the same extent. Species preferring soft mudflats bordering saltmarshes, which are found mainly in the northern and eastern sectors (e.g. Shelduck, Pintail, Shoveler, Redshank), suffered a considerably larger loss of feeding area than species feeding in the more sandy areas in the western and central parts of the estuary (e.g. Bar-tailed Godwit).

To evaluate this effect, 'distribution-corrected habitat loss' (DHL) was calculated and compared to changes in population size. For each species, the percentage of the pre-barrier total occurring in each sector of the Oosterschelde was multiplied by the percentage of intertidal area lost in that

Table 2. Distribution (% of total in each section) in the pre-barrier situation, and distribution-corrected habitat loss (DHL) of intertidal bird species in the Oosterschelde, based on yearly bird-days and midwinter maxima. For some species absent or very scarce in winter, distribution is based on bird-days only. W west, C centre, E east, N north.

Species	% of bird days				DHL (%)	% of winter maximum				DHL (%)
	W	C	E	N		W	C	E	N	
Shelduck	10	13	52	25	20	4	13	58	24	22
Wigeon	19	25	32	24	14	23	19	31	26	14
Teal	23	19	26	32	13	27	19	24	30	12
Pintail	2	2	81	15	28	2	3	79	16	28
Shoveler	13	5	68	14	24	13	5	73	9	25
Oystercatcher	25	30	26	18	12	25	31	28	16	13
Avocet	37	8	21	34	11					
Ringed plover	8	29	28	36	14					
Kentish plover	21	40	6	34	7					
Grey plover	18	28	37	16	15	11	31	43	14	17
Knot	32	24	26	18	12	21	22	38	19	16
Dunlin	12	26	47	15	18	11	31	47	10	18
Bar-tailed godwit	36	28	11	25	8	32	31	8	29	7
Curlew	20	19	44	16	17	18	22	44	16	17
Spotted redshank	23	33	14	30	9					
Redshank	14	22	46	17	18	12	31	41	16	17
Greenshank	6	25	44	25	18					
Turnstone	22	36	19	22	10	13	35	29	23	13

section. The sum of these terms for the four sections constitutes DHL (Table 2). Thus, a large DHL was calculated for a species mainly occurring in the eastern part of the Oosterschelde, and a small one for species occurring only in small numbers here.

In eleven species of which on average more than 2000 individuals fed in the intertidal areas of the Oosterschelde during pre-barrier winters, there is a significant relation between DHL and changes in wintering numbers (Fig. 6A). In these eleven species, DHL accounts for 45% of the total variance in the changes (linear regression, $F = 8.51$, $p = 0.02$). By combining DHL and population changes in the British Isles in a multiple regression analysis, the proportion of vari-

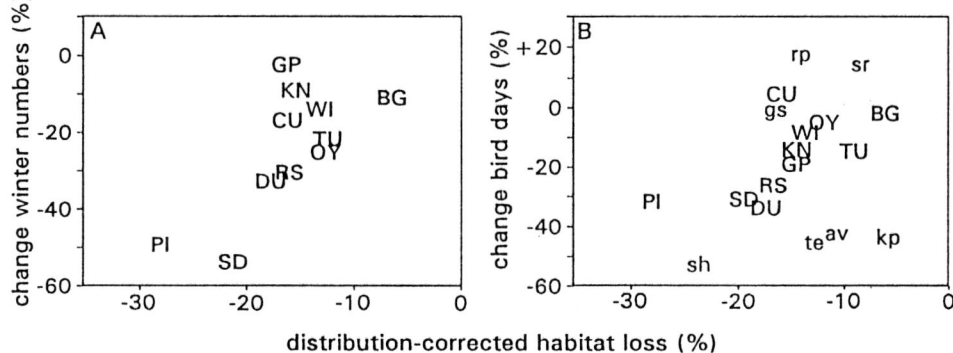

Fig. 6. Relationship between distribution-corrected habitat loss (see text) and changes in abundance of waterbirds feeding in the intertidal zone, based on (A) midwinter numbers and (B) total annual bird-days. In A, only species wintering with more than 2000 individuals are shown; in B these species only are indicated in capitals. Acronyms identify species (see appendix).

ance explained increased to 64% for these species ($F = 8.89$, $p = 0.01$; % change Oosterschelde = $1.67*(DHL) + 0.28 * (\%$ change British index) $- 2.55$). The relationship also exists when the calculation is based on bird-days. However, species which are common in the Oosterschelde only in autumn and spring (Avocet, Ringed Plover, Kentish Plover, Greenshank and Spotted Redshank), or winter in numbers under 2000 (Teal, Shoveler), tend to deviate from it (Fig. 6B).

Discussion

General considerations

Waterbird numbers are affected by many factors. There is a considerable variation in numbers between years, partly as a result of wide fluctuations in breeding productivity. In addition, some degree of inaccuracy is inherent to large scale waterbird counts (Rappoldt et al., 1984). Standard deviations are therefore in most cases considerable (Table 1). Differences between means, tested using Student's t, were significant in only relatively few cases, even when using a significance level of 10%. This is partly due to the rather short study periods: five (pre-barrier) and three years (post-barrier). However, environmental planners, management institutions and politicians generally cannot wait for an evaluation of projects such as the Oosterschelde works as long as would be desirable from a statistical point of view.

The pre-barrier period of five seasons included two cold (1978/79 and 1981/82) and three mild winters; the post-barrier period included three successive mild winters. The weather conditions may have affected the numbers of some species which are sensitive to severe winter weather (Meininger et al., 1991). The relatively high winter numbers of Avocet and Ringed Plover during the post-barrier period might be explained in this way. However, the weather effect is unlikely to have affected the general patterns observed in this study.

Notwithstanding these additional sources of variation, important changes in waterbird num-

bers have become apparent in the Oosterschelde. In general, species feeding in open water have increased, while the area became less important for intertidal foragers. Because the latter group is most important numerically, total bird numbers in the estuary declined.

Numbers of migratory birds are not only influenced by circumstances within the study area, but also by factors operating elsewhere. The main external factor to be considered is a change in overall size of the populations using W-European wetlands. A change in total population size may affect bird numbers in a particular area, even when local conditions remain unaltered (e.g. Moser, 1988). However, the comparison of changes in waterbird numbers in the Oosterschelde with those in Britain and Ireland indicated that changes in the Oosterschelde in most species did not merely reflect large scale population trends.

Pelagic foragers: population growth and water transparency

Out of the four species which do not feed in intertidal areas but in open water, Great Crested Grebe, Cormorant and Goldeneye increased in the Oosterschelde, while the Red-breasted Merganser showed a non-significant decrease in winter, but remained stable in bird-days due to a concomitant increase in autumn. Three of these species are piscivorous and locate their prey visually; only the Goldeneye mainly feeds on benthic and epibenthic invertebrates. For the increased species, the total W-European population is also growing, but the rate of increase in the Oosterschelde seems to be higher. It is possible that the increased water transparency within the estuary (Bakker & Vink, 1994; Wetsteyn & Kromkamp, 1994) is an additional explanation for the increase. Mean winter (minimum) water transparency in the eastern sector, measured using a Secchi disc, increased from approximately 0.8 m in 1980–85 to 2.5 m in 1987–90, causing a 30-fold increase in the volume of water in which prey can be detected by pursuit-diving predators (Eriksson, 1985). The water transparency in the eastern sector now approaches that in the nearby brackish-saline lakes Grevelingen and Veerse Meer, which are characterised by an abundance of piscivores (Doornbos, 1984; Meire *et al.*, 1989).

Intertidal foragers: carrying capacity of tidal flats

The majority of intertidally foraging species showed a decrease in winter numbers and/or bird-days, which was significant ($p < 0.1$) in Shelduck, Pintail, Teal, Shoveler, Oystercatcher, Avocet, Kentish Plover, Grey Plover, Dunlin, and Redshank. A significant increase was found only in the Brent Goose, which is not strictly intertidal but takes a considerable proportion of its food from agricultural areas. In waders, increasing trends were in no case significant and occurred only in species which occur mainly in spring and autumn and are very scarce in winter (Avocet, Ringed Plover, Spotted Redshank, and Greenshank). For the decreasing species, there are no indications that a similar decline occurred in the total western European populations, except for the Kentish Plover (Jönsson *et al.*, 1990). In addition, several species showed an increase in the British Isles, but not at all, or hardly so, in the Oosterschelde.

The fact that most strictly intertidal species decreased after the loss of 17% of the feeding area means that this loss was not compensated by higher bird densities in the remaining part of the Oosterschelde. In fact, overall densities of several species were lower in the post-barrier period than before. In the same way, birds displaced from feeding areas in the adjacent Krammer-Volkerak were unable to settle into the Oosterschelde. The observation that species mainly occurring in the eastern part of the estuary were more affected than others suggests that it was indeed the loss of feeding habitat which caused the decline.

Additional evidence that displaced birds were not generally able to settle into the remaining intertidal areas is presented by Lambeck (1991). Of Oyster catchers colour-marked in the Krammer-Volkerak in 1984–87, only 40% were still present in the Oosterschelde and the nearby

Westerschelde in the winter of 1989/90, while this figure was significantly higher (67%) for birds marked in other sectors of the Oosterschelde.

These observations suggest that numbers of the most abundant intertidal foragers were at least close to carrying capacity in the period before the coastal engineering works. Goss-Custard (1985) has pointed out that bird numbers can be said to be strictly at capacity only when every new individual entering the estuary (or every loss of habitat equivalent to one bird's means of living) causes the emigration or death of another bird. Demonstrating that this condition is met is practically impossible in case studies like that described here. It would require an assessment of the proportion of suitable habitat that is lost for each species much more precise than our estimate of DHL. However, reaching carrying capacity should not be considered as an abrupt process, but as the result of a gradually increasing pressure on individuals to look out for alternative settling options. Even before carrying capacity *sensu stricto* is reached, there is a trajectory of bird densities at which adding more birds, or removing part of the habitat, causes *some* birds to emigrate or die. In such cases habitat loss will reduce the number of birds, and the question as to whether or not bird numbers are strictly at capacity becomes academic from a conservation point of view.

Possible mechanisms of population decline

The level of pressure at which the local system starts to overflow when carrying capacity is approached and birds start leaving the area is likely to be influenced, in addition to factors affecting feeding efficiency within the estuary, by the presence or absence of alternative areas nearby where additional birds can still be accommodated. In the Oosterschelde, former intertidal feeding areas behind the compartmentation dams were replaced by freshwater lakes. Extensive shallow areas in these lakes and the development of a saltmarsh-like vegetation on the permanently exposed former tidal flats offered good feeding opportunities for several species of dabbling ducks. This development may help explain why these species showed such drastic declines compared with most waders, and with the area of tidal flats lost. Judging from numbers wintering in the Volkerakmeer and Zoommeer/Markiezaat in the postbarrier period, these lakes may have accommodated the majority of Wigeon, Teal and Shoveler displaced from the lost intertidal areas, but only about half of Pintail and less than 10% of Shelduck.

As an alternative to dispersal, the decrease in bird numbers may have been brought about by mortality. In normal winters, only few dead birds are found in the Oosterschelde area (Meininger, unpubl.). However, during three successive cold winters in 1985–87, a total of 12 900 starved waders and 2400 dead ducks were found during searches in the whole Delta area (Meininger et al., 1991). Although not all of these belonged to the Oosterschelde population, ringing recoveries suggest that the majority did. Although a large proportion of the frost victims were first-year birds which are likely to have had a lower survival than adults anyway, the numbers of Redshank, Turnstone and Oystercatcher found even within the boundaries of the Oosterschelde were substantial compared with the decrease in wintering

numbers (Table 3). For these species, mortality during these severe winters, possibly aggravated by manipulations of the tidal regime during the construction of the coastal engineering works (Lambeck, 1991, Meininger et al., 1991), may have been an important mechanism in (though not a cause of) the decrease. For the other species, this type of mortality was probably unimportant.

A historical comparison

An interesting comparison can be made between the events described here and those following the closure in 1971 of the Grevelingen estuary, 15 km north of the Oosterschelde, which held about 50 000 wintering waders (Wolff, 1967). A sudden increase in the numbers of Oystercatcher and Bar-tailed Godwit using the Roggenplaat, a tidal flat in the western sector of the Oosterschelde, was shown to coincide with this (Van Latesteijn & Lambeck, 1986, Lambeck et al., 1989). Wader numbers in the adjacent Krammer-Volkerak increased substantially, while total numbers in the Delta region as a whole did not clearly decrease (Leewis et al., 1984). Apparently, waders displaced from the Grevelingen feeding areas could

Table 3. Numbers and age composition of frost victims found in the total Delta area and in the Oosterschelde after cold spells in the winters of 1985–87, compared to the decrease in mean numbers wintering in the Oosterschelde between pre- and post-barrier periods. Data on frost victims from Meininger et al. (1991) and Meininger.

Species	No. of victims found		% adults among victims	Decrease of winter numbers Oosterschelde (B)	A/B (%)
	Total Delta (A)	Oosterschelde (A)			
Shelduck	1114	538	–	3980	14–28
Wigeon	53	28	–	1870	1–3
Teal	16	10	–	661	2
Pintail	190	97	–	2629	4–7
Oystercatcher	9811	6500	57%	22481	29–44
Grey Plover	643	424	55%	–	–
Knot	237	176	67%	903	19–26
Dunlin	604	280	59%	16027	2–4
Bar-tailed Godwit	39	26	57%	660	4–6
Curlew	451	252	71%	1306	19–34
Redshank	792	437	60%	626	70–100
Turnstone	230	169	55%	270	63–85

be accommodated elsewhere in the Delta. However, it would not be entirely safe to conclude that carrying capacity was not reached in the Oosterschelde/Krammer-Volkerak estuary before that time, since the closure of the Grevelingen brought about important changes. Benthic food stocks and feeding opportunities in the Krammer-Volkerak and part of the Oosterschelde are likely to have increased as a result of the higher salinity and larger tidal amplitude following the closure of the Volkerakdam in 1969. Wolff (1971) showed that diversity of macrobenthos in the Krammer-Volkerak increased substantially. At the same time, mussel cultures became established. A further increase was noted in several waterbird species in the Oosterschelde during the 1970s. However, already during this period some indications were found that bird numbers might be approaching an upper limit, as the seasonal pattern of several species on the Roggenplaat showed a shift from winter towards autumn (Lambeck *et al.*, 1989). This phenomenon was later reflected by numbers of Oystercatchers in the entire estuary (Lambeck, 1991).

Factors determining bird numbers in the estuary

Changes in carrying capacity for intertidal waterbirds in the Oosterschelde estuary could be caused by changes in three local factors, acting singly or in combination: the total surface of intertidal feeding area, the biomass per unit area of available macrobenthos, and the exposure time of tidal flats which limits potential feeding time for birds. Of these, the amount of feeding area has been affected most strongly by the coastal engineering works, and this has reduced both the total amount of food available to birds and the space available for their foraging behaviour.

The importance of benthic biomass and depletion of prey stocks in determining carrying capacity in the Oosterschelde is explored elsewhere in this volume (Meire *et al.*, 1994). The total intertidal macrobenthic biomass per unit area (excluding the strongly fluctuating cockle *Cerastoderma edule* and mussel *Mytilus edulis*) has not changed significantly since 1985 (Seys *et al.*, 1994). For some specialised foragers however, food availability may have changed. For instance, an important decrease in the mudsnail *Hydrobia ulvae* was noted in the central and eastern sectors of the Oosterschelde (Coosen *et al.*, 1994), and may have affected feeding conditions for species like Shelduck (Meininger & Snoek, 1992). Low biomasses of cockles and mussels during the past few years, partly caused by intensive cockle fisheries, are likely to have caused difficulties for Oystercatchers (Lambeck, 1991). The numerical relationship between Oystercatcher feeding densities and that of their main bivalve prey has not changed (Meire, 1991).

The coastal engineering works have affected potential feeding time of intertidal foragers within the remaining part of the Oosterschelde. Changed hydrodynamic conditions (Vroon, 1994) have caused a net transport of sediments from the tidal flats into the gullies (Mulder & Louters, 1994). As a result, an increasing proportion of the intertidal area is situated in the lower reaches of the tidal range, and thus exposed for shorter periods during low tides. This process is aggravated to some extent by a decrease in the tidal amplitude from 3.70 to 3.25 m, which causes intertidal areas above mid-tide level to be exposed for longer, but areas below mid-tide level for shorter periods than before. The result is a decrease of feeding potential (integration of area and time of exposure) of 15–30% on some flats in the western and central sectors between 1984 and 1989. Loss of feeding potential due to this process has been much smaller in the northern and eastern sectors.

It is not clear at present whether, and to what extent, reduced exposure time has affected bird numbers in the Oosterschelde. However, it may have contributed to the inability of the remaining area to accomodate birds displaced from lost feeding areas behind the compartimentation dams, and to the slightly lower feeding densities in the 'new' estuary. As the morphological changes in the estuary are expected to continue during the next few decades, and to eventually result in the loss of a further 15% of the intertidal area (Mulder & Louters, 1994), a further decline in intertidal waterbird numbers can be expected.

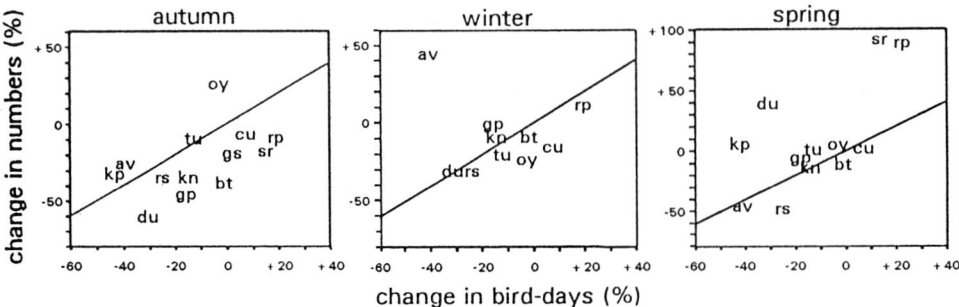

Fig. 7. Comparison of changes in season-related functions of the Oosterschelde for waterbirds and changes in the total use of the area (annual number of bird-days). Lines indicate equal change. Acronyms identify species (see appendix).

Decrease in numbers of moulting waders

When the changes in wader numbers are evaluated separately for the three seasons distinguished (autumn, winter, spring), the decrease in numbers wintering is in proportion to the decrease in bird-days (Fig. 7). Changes in numbers during spring are, for some species, much smaller. In contrast, a comparatively strong decline is seen in late summer. This suggests that the Oosterschelde lost part of its significance as a moulting area for waders.

It seems unlikely that a decrease in feeding habitat or food stocks would have a stronger impact on bird numbers in late summer than in winter. Total numbers and energy demand of the birds are relatively low, while the potential daily energy intake is likely to be higher than in winter as a result of higher macrobenthic biomass and production, longer daylight feeding time and higher surface activity of macrobenthic organisms (e.g. Evans, 1976). Since the regulation of bird densities in intertidal areas is likely to act through the possibility of maintaining a balanced energy budget over a period of days or months (Evans, 1976; Goss-Custard, 1985), these factors would be expected to result in a higher carrying capacity of the mudflats in spring and summer than in winter. This presumption is consistent with the observation that the number of Oystercatchers in the Oosterschelde increased during summer and early autumn, while wintering numbers have decreased markedly.

Another indication that factors other than hab-

itat loss have caused the disproportionate decline of moulting waders is the fact that Dunlin and Knot, among the more heavily affected species in late summer, at this time of the year hardly use the eastern sector where most habitat loss occurred (5 and 0% of the total in east). In addition, extensive sampling of macrobenthos in August 1985 and 1989 showed no important changes in benthic biomass in the main moulting area, apart from cockles and mussels (Seys et al., 1994).

The peak of recreational activities in the Oosterschelde (July–August) overlaps with the most important moulting period. Recreation in the Oosterschelde, especially boating and wind-surfing, increased considerably during the past years. For instance, the number of harbour sites for boats increased from 1200 to 2100 between 1982 and 1989. In addition, the coastal engineering works made several formerly isolated mud-flats easily accessible to people. It is not unlikely that increased disturbance has contributed to the decrease of moulting waders in the Oosterschelde.

Acknowledgements

The main basis for this paper is formed by the extensive series of bird counts, initiated and organised in the early 1970s by G. J. Slob (State Forestry Service) and H. J. M. Baptist (Rijkswaterstaat), and carried out by many professional and amateur ornithologists. Post-barrier counts of Krammer-Volkerak and Zoommeer were kindly made available by H. Smit and E. H. van

Nes (Rijkswaterstaat, RIZA). H.S.'s employment at the Centre for Estuarine and Coastal Ecology was funded by the Tidal Waters Division of Rijkswaterstaat.

We wish to thank A.-M. Blomert, E. Kuyken and E. C. L. Marteijn for stimulating discussions. R. H. D. Lambeck, B. J. Ens and two anonymous referees suggested improvements to the manuscript. C. M. Berrevoets and R. C. W. Strucker assisted in many ways during data processing and the preparation of the paper. This is communication No. 708 of NIOO-CEMO, Yerseke, The Netherlands.

Appendix

Acronyms and scientific names of bird species mentioned in text and figures. Species considered dependent on the intertidal flats for food are marked with an asterisk *.

–	Great Crested Grebe	*Podiceps cristatus*
–	Cormorant	*Phalacrocorax carbo*
–	Barnacle Goose	*Branta leucopsis*
br	Brent Goose	*Branta bernicla*
sd	*Shelduck	*Tadorna tadorna*
wi	*Wigeon	*Anas penelope*
–	Mallard	*Anas platyrhynchos*
te	*Teal	*Anas crecca*
pi	*Pintail	*Anas acuta*
sh	*Shoveler	*Anas clypeata*
go	Goldeneye	*Bucephala clangula*
rm	Red-breasted Merganser	*Mergus serrator*
oy	*Oystercatcher	*Haematopus ostralegus*
av	*Avocet	*Recurvirostra avosetta*
rp	*Ringed Plover	*Charadrius hiaticula*
kp	*Kentish Plover	*Charadrius alexandrinus*
gp	*Grey Plover	*Pluvialis squatarola*
kn	*Knot	*Calidris canutus*
du	*Dunlin	*Calidris alpina*
bg	*Bar-tailed Godwit	*Limosa lapponica*
cu	*Curlew	*Numenius arquata*
sr	*Spotted Redshank	*Tringa erythropus*
rs	*Redshank	*Tringa totanus*
gs	*Greenshank	*Tringa nebularia*
tu	*Turnstone	*Arenaria interpres*

References

Bakker, C. & M. Vink, 1994. Nutrient concentrations and planktonic diatom-flagellate relations in the Oosterschelde (SW Netherlands) during and after the construction of a storm-surge barrier. Hydrobiologia 282/283: 101–116.

Coosen, J., J. Seys, P. Meire & J. A. M. Craeymeersch, 1994. Effect of sedimentological and hydrodynamical changes in the intertidal areas of the Oosterschelde estuary (SW Netherlands) on distribution, density and biomass of five common macrobenthic species: Spio martinensis (Mesnil), Hydrobia ulvae (Pennant), Arenicola marina (L.), Scoloplos armiger (Muller) and Bathyporeia sp. Hydrobiologia 282/283: 235–249.

Doornbos, G., 1984. Piscivorous birds on the saline Lake Grevelingen, The Netherlands: abundance, prey selection and annual food consumption. Neth. J. Sea Res. 18: 457–479.

Evans, P. R., 1976. Energy balance and optimal foraging strategies in shorebirds: some implications for their distributions and movements in the non-breeding season. Ardea 64: 117–139.

Evans, P. R., 1981. Reclamation of tidal land: some effects on Shelduck and wader populations in the Tees estuary. Verh. orn. Ges. Bayern 23: 147–168.

Evans, P. R., D. M. Herdson, P. J. Knights & M. W. Pienkowski, 1979. Short-term effects of reclamation of part of Seal Sands, Teesmouth, on wintering waders and Shelduck. Oecologia 41: 183–206.

Goss-Custard, J. D., 1977. The ecology of the Wash. III. Density-related behaviour and the possible effects of a loss of feeding ground on wading birds (Charadrii). J. appl. Ecol. 14: 721–739.

Goss-Custard, J. D., 1980. Competition for food and interference amongst waders. Ardea 68: 31–52.

Goss-Custard, J. D., 1981. Role of winter food supplies in the population ecology of common British wading birds. Verh. orn. Ges. Bayern: 125–146.

Goss-Custard, J. D., 1985. Foraging behaviour of wading birds and the carrying capacity of estuaries. In R. M. Sibley & R. H. Smith (eds), Behavioural ecology ecological consequences of adaptive behaviour. Blackwell Scientific Publications, Oxford: 169–188.

Goss-Custard, J. D. & S. E. A. Durell, le V., 1990. Bird behaviour and environmental planning: approaches in the study of wader populations. Ibis 132: 273–289.

Eriksson, M. O. G., 1985. Prey detectability for fish-eating birds in relation to fish density and water transparency. Ornis Scand. 16: 1–7.

Jönsson, P. E., P. L. Meininger, R. Schultz & T. Székely, 1990. The WSG Kentish Plover Project. Wader Study Group Bull. 60: 1–3.

Kirby, J. S., R. J. Waters & R. P. Prys-Jones, 1990. Wildfowl and wader counts 1939–90. Wildfowl & Wetlands Trust, Slimbridge, 80 pp.

Lambeck, R. H. D., 1991. Changes in abundance, distribution and mortality of wintering Oystercatchers after habitat loss in the Delta area, SW Netherlands. Acta XX Congr. Int. Orn.: 2208–2218.

Lambeck, R. H. D., A. J. J. Sandee & L. de Wolf, 1989. Long-term patterns in the wader usage of an intertidal flat in the Oosterschelde (SW Netherlands) and the impact of the closure of an adjacent estuary. J. appl. Ecol. 26: 419–431.

Latesteijn, H. C. van & R. H. D. Lambeck, 1986. The analysis of monitoring data with the aid of time-series analysis. Envir. Monitoring & Assessment 7: 287–297.

Laursen, K., I. Gram & L. J. Alberto, 1983. Short-term effect of reclamation on numbers and distribution of waterfowl at Ho/jer, Danish Wadden Sea. Proc. 3rd Nordic Congr. Ornithol. 1981: 97–118.

Leewis, R. J., H. J. M. Baptist & P. L. Meininger, 1984. The Dutch Delta. In P. R. Evans, J. D. Goss-Custard & W. G. Hale (eds), Coastal waders and wildfowl in winter. Cambridge University Press, Cambridge: 253–260.

Madsen, J., A. St.Joseph & M. O'Brian, 1991. The status of Brent Goose Branta bernicla 1986–89. IWRB Goose Res. Group Bull. 1: 4–7.

Meininger, P. L. & N. D. van Swelm, 1989. Biometrisch en ringonderzoek aan steltlopers in de Oosterschelde in het voorjaar van 1984 en 1985. Rijkswaterstaat Dienst Getijdewateren nota GWAO-89.1009, Middelburg/Oostvoorne, 103 pp.

Meininger, P. L., A.-M. Blomert & E. C. L. Marteijn, 1991. Watervogelsterfte in het Deltagebied, ZW-Nederland, gedurende de drie koude winters van 1985, 1986 en 1987. Limosa 64: 89–102.

Meininger, P. L. & H. Snoek, 1992. Non-breeding Shelduck Tadorna tadorna in the SW-Netherlands: effects of habitat changes on distribution, numbers, moulting sites and food. Wildfowl 43: 139–151.

Meire, P., 1991. Effects of a substantial reduction in intertidal area on numbers and densities of waders. Acta XX Congr. Int. Orn.: 2219–2227.

Meire, P. & E. Kuyken, 1984. Barrage schemes- predicting the effect of changes in tidal amplitude. In P. R. Evans, H. Haffner & P. l'Hermite (eds), Shorebirds and large waterbirds conservation. Commission of the European Communities, Brussels: 79–89.

Meire, P. M., J. Seys, T. Ysebaert, P. L. Meininger & H. J. M. Baptist, 1989. A changing Delta: effects of large coastal engineering works on feeding ecological relationships as illustrated by waterbirds. In J. C. Hooghart & C. W. S. Posthumus (eds), Hydro-ecological relations in the Delta waters of the South-West Netherlands. TNO/CHO, The Hague: 109–145.

Meire, P. M., H. Schekkerman & P. L. Meininger, 1994. Consumption of benthic invertebrates by waterbirds in the Oosterschelde estuary, SW Netherlands. Hydrobiologia 282/283: 525–546.

Moser, M. E., 1988. Limits to the numbers of Grey Plovers Pluvialis squatarola wintering on British estuaries: an analysis of long-term population trends. J. appl. Ecol. 25: 473–485.

Mulder, J. P. M. & T. Louters, 1994. Changes in basin geomorphology after implementation of the Oosterschelde estuary project. Hydrobiologia 282/283: 29–39.

Nienhuis, P. H. & A. C. Smaal, 1994. The Oosterschelde estuary, a case study of a changing ecosystem: an introduction. Hydrobiologia 282/283: 1–14.

Osieck, E. R., 1991. Cormorant & man: a conservation view. In M. R. van Eerden & M. Zijlstra (eds), Proceedings workshop 1989 on Cormorants Phalacrocorax carbo. Rijkswaterstaat Directie Flevoland, Lelystad: 244–247.

Rappoldt, C., M. Kersten & C. Smit, 1984. Errors in large scale shorebird counts. Ardea 73: 13–24.

Seys, J. J., P. M. Meire, J. Coosen & J. Craeymeersch, 1994. Long-term changes (1979–89) in the intertidal macrozoobenthos of the Oosterschelde estuary: are patterns in total density, biomass and diversity induced by the construction of the storm-surge barrier? Hydrobiologia 282/283: 251–264.

Smaal, A. C., M. Knoester, P. H. Nienhuis & P. M. Meire, 1991. Changes in the Oosterschelde ecosystem induced by the Delta works. In M. Elliott & J.-P. Ducrotoy (eds), Estuaries and coasts: spatial and temporal intercomparisons. ECSA 19 Symposium, Olsen & Olsen, Fredensborg: 375–384.

Smit, C. J. & T. Piersma, 1989. Numbers, midwinter distribution and migration of wader populations using the East-Atlantic Flyway. In H. Boyd & J-Y. Pirot (eds), Flyways and reserve networks for water birds. IWRB Spec. Publ. No. 9, Slimbridge: 24–63.

Smit, C. J., R. H. D. Lambeck & W. J. Wolff, 1987. Threats to coastal wintering and staging areas of waders. In N. C. Davidson & M. W. Pienkowski (eds), The conservation of international flyway populations of waders. Wader Study Group Bull. 49, Suppl.: 105–113.

Vroon, J., 1994. Hydrodynamic characteristics of the Oosterschelde in recent decades. Hydrobiologia 282/283: 17–27.

Wetsteyn, L. P. M. J. & J. Kromkamp, 1994. Turbidity, nutrients and phytoplankton primary production in the Oosterschelde (The Netherlands) before, during and after a large-scale coastal engineering project (1980–1990). Hydrobiologia 282/283: 61–78.

Wolff, W. J., 1967. Watervogeltellingen in het gehele Nederlandse Deltagebied. Limosa 40: 216–225.

Wolff, W. J., 1971. Changes in intertidal benthos communities after an increase in salinity. Thalassia Jugoslavica 7: 429–434.

Hydrobiologia **282/283**: 525–546, 1994.
P. H. Nienhuis & A. C. Smaal (eds), The Oosterschelde Estuary.
© 1994 *Kluwer Academic Publishers. Printed in Belgium.*

Consumption of benthic invertebrates by waterbirds in the Oosterschelde estuary, SW Netherlands

Patrick M. Meire[1,2,3], Hans Schekkerman[2] & Peter L. Meininger[4]
[1]*Rijksuniversiteit Gent, Ledeganckstraat 35, B-9000 Gent, Belgium;* [2]*Netherlands Institute of Ecology-Centre for Estuarine and Coastal Ecology, Vierstraat 28, 4401 EA Yerseke, The Netherlands;* [3]*Ministry of the Flemish Community, Institute of Nature Conservation, Kiewitdreef 5, B-3500 Hasselt, Belgium;* [4]*National Institute for Coastal and Marine Management/RIKZ, P.O. Box 8039, 4330 EA Middelburg, The Netherlands*

Key words: waterbirds, predation pressure, carrying capacity, estuary, benthic consumption

Abstract

The number of waders in the Oosterschelde, S.W. Netherlands, declined after a reduction in intertidal area due to the construction of a storm surge barrier and secondary dams, suggesting that the carrying capacity had been reached (Schekkerman *et al.*, 1993). In this paper we present data on consumption and predation pressure by birds to explore whether the reduction in their numbers is due to prey depletion or to other factors.

The total annual consumption of benthic invertebrates by birds in the Oosterschelde amounted to 1573×10^3 g ADW y^{-1} in the period before the coastal engineering works (pre-barrier) and 1500×10^3 kg ADW y^{-1} in the post-barrier period. More than half of the total amount of biomass is eaten by the Oystercatcher, and only seven (pre-barrier) or even six (post-barrier) bird species together take 90% of the total.

Although the consumption by individual species may vary considerably among years, the total consumption was remarkably stable, with a CV of only 3–4% of the mean, especially compared to the variability of the prey populations. In the pre-barrier period, consumption was lowest in mid summer, increased sharply from August onwards until a peak was reached in January. A sharp decrease took place in March. In the post-barrier period, consumption peaked in October.

The total consumption per unit area per year does not differ much between different sectors of the Oosterschelde, apart from a distinctly lower value in the eastern part. Of the total amount of food taken by birds, only 0.1–0.4% is taken in the subtidal compartment. In several study plots on an individual tidal flat, there was a clear relation between consumption and benthic biomass.

The predation pressure was 13 and 23% of the standing stock, in the post- and pre-barrier period respectively. When cockles, mussels and their main predator, the Oystercatcher, are excluded from the calculations, the predation pressure of the other species was 30 and 37% of the biomass, respectively.

Predation pressure of Oystercatchers in individual study plots varied from less than 10% to more than 70% of the standing stock. On cockle beds the predation pressure was positively related to the average length of the cockles present.

Based on these results and a comparison with the literature we conclude that, at least for several species that feed intertidally, carrying capacity could be limited by the stocks of food. This does not mean that birds face food shortage each season. As the variability of the benthos populations is much higher than that of the bird densities it is likely that at some times food is not limiting, at other times it is. On the other hand, consumption is very low in the subtidal compartment and species feeding here could potentially increase substantially in numbers in the Oosterschelde.

Introduction

Within the Oosterschelde, a large estuary in the SW Netherlands, important coastal engineering works have taken place in the 1980s (Nienhuis & Smaal, 1994). As part of the ecological studies investigating the impact of these works, a simulation model of the Oosterschelde ecosystem (SMOES; Klepper *et al.*, 1994) has been made, which calculated the major carbon-flows between different components of the ecosystem. The higher trophic levels, especially fish and birds, however, were not included in this model, as their role in the overall C-balance of the estuary was considered to be negligible.

Despite the relatively unimportant trophic role of birds in the overall C-balance of the Oosterschelde ecosystem, the Oosterschelde is nevertheless of great significance for bird populations, especially waders, ducks and geese (Schekkerman *et al.*, 1994). This significance has played a prominent role in deciding whether the Oosterschelde should be closed or remain tidal (Nienhuis & Smaal, 1993). Eventually, a storm-surge barrier has been built, which has resulted in a considerable loss of the intertidal area. For more details of the coastal engineering works see Nienhuis & Smaal (1994).

Wader densities in the Oosterschelde used to be high compared to other Western European intertidal areas (Smaal & Boeije, 1991) and the major question has been posed as to whether a reduction in intertidal area would cause a drop in bird numbers or not. Habitat loss in estuarine areas is a widespread phenomenon all over the world, but its effects on waders have been studied in only a few occasions. The construction of the storm surge barrier in the Oosterschelde provided an opportunity to test whether or not the carrying capacity of the area had been reached. Carrying capacity is defined here as the density at which the addition of any further birds results in

other birds dying or leaving the area because they fail to achieve adequate intake rates due to increased interference and/or depletion of prey stocks (Sutherland & Goss-Custard, 1991). Schekkerman *et al.* (1994) showed that the number of waders in the Oosterschelde declined after the construction of the coastal engineering works as predicted by Meire & Kuijken (1987), suggesting that the carrying capacity had been reached. In this paper we present data on consumption and predation pressure by birds on benthic invertebrates in the Oosterschelde. By comparing the biomass consumed by birds with the total benthic biomass present in late summer, we explore the question whether carrying capacity could be directly limited by the size of the potential food stocks. Alternatively, carrying capacity could be determined by the spatial needs of the birds, resulting from specific foraging techniques or social factors such as interference and territoriality.

Material and methods

Consumption by birds

Total Oosterschelde: overall estimate
Consumption by birds in the whole Oosterschelde was determined on the basis of monthly high-tide counts of birds in the whole estuary (Schekkerman *et al.*, 1993). Data from two periods, one pre-barrier (five seasons, 1978/79–1982/83), and the other post-barrier (three seasons, 1987/88–1989/90) are used in the analysis. Gulls were only counted during the pre-barrier period; it was assumed that total numbers and seasonal patterns were the same in the post-barrier period. All counts refer to the Oosterschelde, excluding the brackish Krammer-Volkerak. For the pre-barrier period, the counts include the birds counted in the now fresh and stagnant Zoommeer and Mar-

kiezaat, which were dammed in 1983–1986. The total intertidal area was 13669 ha in the pre-barrier, and 11365 ha in the post-barrier period, so that 17% of the intertidal area was lost due to the engineering works. For some of the analyses the intertidal area of the Oosterschelde was divided into four different sectors (west, centre, east and north) (Fig. 1).

The total monthly consumption per species was calculated (Table 1) using the equation:

$$C = (30 \times N \times 3 \times BMR$$
$$\times (1/Q)/F) * 1000$$

in which C = total monthly consumption by species (kg Ash-free Dry Weight ADW); N = the number of birds present; BMR = Basal Metabolic Rate (kJ day^{-1}); Q = the assimilation efficiency of the food, and F its energy content in kJ g^{-1}.

BMR, the energy consumption of a resting bird at thermoneutrality, was estimated using the equations:

BMR = $5.06 \times LW^{0.729}$ for waders (Kersten & Piersma, 1987);
BMR = $3.56 \times LW^{0.734}$ for other species (Aschoff & Pohl, 1970),

in which LW = the lean (fat-free) weight of the species in g. Lean weights were used because fat stores are largely energetically inactive. Lean weights of waders were estimated from wing length, using the formulae given by Davidson (1983). For gulls, the lean weight was obtained by subtracting 10% from their weight in the breeding period, on the basis of the body composition of Herring Gulls (see Table 1 for scientific bird names) reported by Norstrom et al. (1986). The lean weights of grebes were based on Piersma

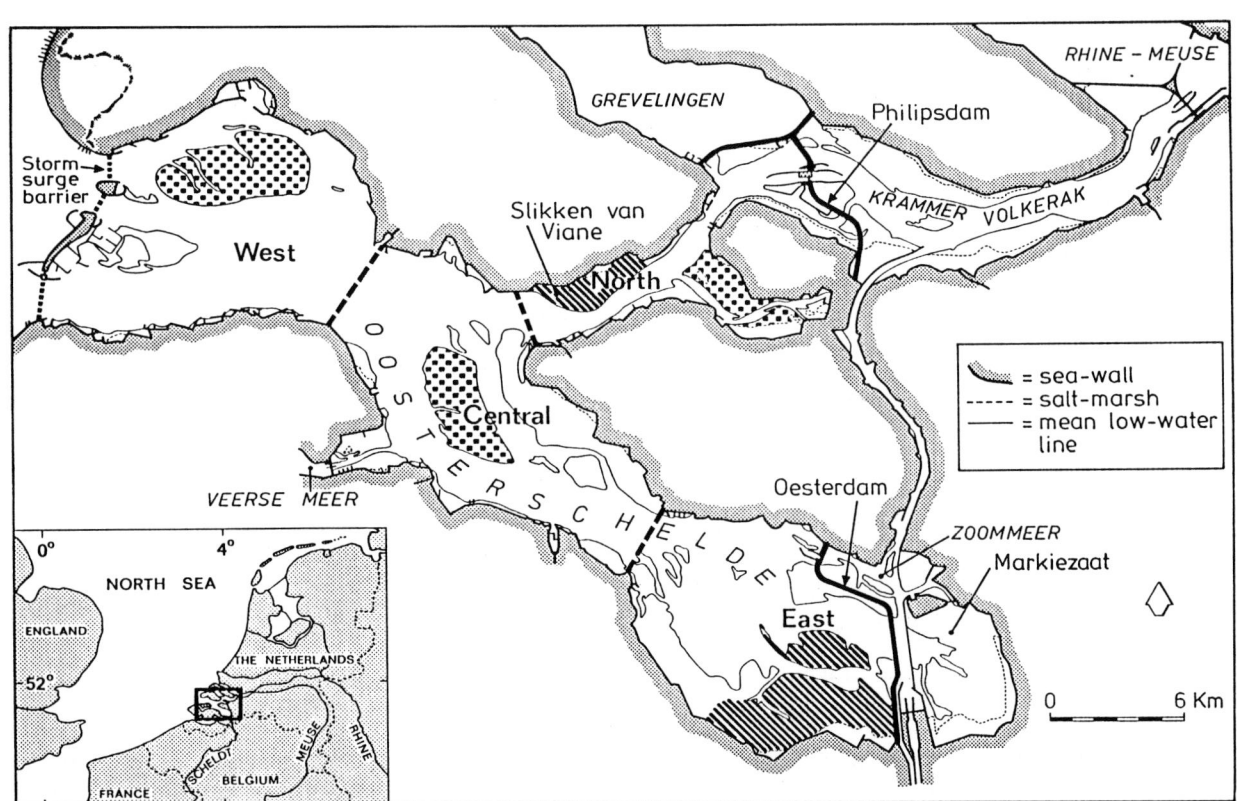

Fig. 1. Map of the Oosterschelde estuary with the location of the four compartments, Vianen mudflats, and the sampled areas in both 1985 and 1989 (stippled) and in 1985 (hatched) only.

Table 1. Basic assumptions and total consumption by benthivorous birds in the Oosterschelde.

Species	Lean weight (kg)	BMR (kJ d^{-1})	Daily intake (gAWD d^{-1})	Bird-days/year ($n \times 1000$) 78-82	Bird-days/year ($n \times 1000$) 87-89	Consumption gADW m^2 y^{-1} 78-82	Consumption gADW m^2 y^{-1} 87-89	% of total benthic cons. 78-82	% of total benthic cons. 87-89
Subtidal areas									
Eider (*Somateria mollissima*)	1.90	505	81	8	42	0.005	0.029	0.0	0.2
Goldeneye (*Bucephala clangula*)	0.75	255	41	28	52	0.008	0.018	0.1	0.1
Tidal flats									
Oystercatcher (*Haematopus ostralegus*)	0.53	275	44	20085	19901	6.381	7.605	55.4	57.6
Herring Gull (*Larus argentatus*)	0.95	303	49	4824	4824	1.690	2.032	14.7	15.4
Curlew(*Numenius arquata*)	0.70	337	54	2212	2277	0.861	1.066	7.5	8.1
Dunlin (*Calidris alpina*)	0.05	49	8	9556	6602	0.543	0.451	4.7	3.4
Bar-tailed Godwit (*Limosa lapponica*)	0.27	168	27	1909	1811	0.371	0.423	3.2	3.2
Shelduck (*Tadorna tadorna*)	1.08	333	54	1045	721	0.402	0.333	3.5	2.5
Black-headed Gull (*Larus ridibundus*)	0.23	107	17	1888	1888	0.233	0.281	2.0	2.1
Knot (*Calidris canutus*)	0.14	104	17	2367	1824	0.285	0.264	2.5	2.0
Grey Plover (*Pluvialis squatarola*)	0.19	130	21	1711	1439	0.257	0.260	2.2	2.0
Pintail (*Anas acuta*)	0.85	280	45	463	328	0.150	0.127	1.3	1.0
Redshank (*Tringa totanus*)	0.14	104	16	747	556	0.090	0.080	0.8	0.6
Common Gull (*Larus canus*)	0.35	146	23	381	382	0.064	0.077	0.6	0.6
Shoveler (*Anas clypeata*)	0.55	203	33	225	103	0.053	0.029	0.5	0.2
Turnstone (*Arenaria interpres*)	0.08	69	11	357	289	0.029	0.028	0.2	0.2
Great Black-backed Gull (*Larus marinus*)	1.44	412	66	46	46	0.022	0.026	0.2	0.2
Avocet (*Recurvirostra avosetta*)	0.24	154	25	161	94	0.029	0.020	0.2	0.2
Spotted Redshank (*Tringa erythropus*)	0.14	104	17	119	123	0.014	0.018	0.1	0.1
Ringed Plover (*Charadrius hiaticula*)	0.05	49	8	116	135	0.007	0.009	0.1	0.1
Greenshank (*Tringa nebularia*)	0.18	125	21	52	50	0.007	0.009	0.1	0.1
Lesser Black-backed Gull (*Larus fuscus*)	0.73	250	40	14	14	0.004	0.005	<0.1	<0.1
Sanderling (*Calidris alba*)	0.05	49	8	31	49	0.002	0.003	<0.1	<0.1
Kentish Plover (*Charadrius alexandrinus*)	0.05	49	8	64	37	0.004	0.003	<0.1	<0.1
Whimbril (*Numenius phaeopus*)	0.41	228	37	5	3	0.001	0.001	<0.1	<0.1

Totals

Species	Total consumption *10^{3}kg ADW 78-82	Total consumption *10^{3}kg ADW 87-89	Total consumption gADW m^{-2} y^{-1} 78-82	Total consumption gADW m^{-2} y^{-1} 87-89	% of total benthic cons. 78-82	% of total benthic cons. 87-89
Subtidal areas total	1.6	5.4	0.006	0.020	0.1	0.4
Tidal flats total	1571.7	1494.6	11.49	13.15	99.9	99.6
Ducks	82.8	55.7	0.60	0.49	5.3	3.7
Oystercatcher	872.2	846.3	6.38	7.61	55.5	56.4
Other waders	341.7	299.5	2.50	2.64	21.7	20.0
Gulls	275.2	275.2	2.01	2.42	17.4	18.3

(1984). For the remaining species, the lower values from the range of weights given by Cramp & Simmons (1977, 1983) were used to estimate lean weight. The obtained BMR value was converted in KJ· d^{-1}. Total daily energy expenditure (DEE) was assumed to amount to three times BMR (Drent *et al.*, 1978; Kersten & Piersma, 1987; Smith, 1975; Castro *et al.*, 1992). For fish and benthic invertebrates, a digestibility of $Q = 0.85$ was used (Kersten & Piersma, 1987; Zwarts & Blomert, 1990) and an energetic value of $F = 22kJ$ g^{-1} ADW (Zwarts & Blomert, 1990). To obtain yearly consumption monthly consumption was summed. For comparisons with other areas, consumption was expressed in gADW m^{-2} y^{-1}.

The method gives a rather crude estimation of total consumption for several reasons. Firstly, it was assumed that all birds were feeding exclusively in the benthic compartment of the Oosterschelde. Gulls however, may have taken part of their food from the pelagic compartment, or even outside the boundaries of the Oosterschelde (e.g., at rubbish tips). Secondly, species classified as benthivores were assumed to forage exclusively on this type of food. Pintail and Shoveler may, however, have included a significant proportion of vegetable matter in their diet. Furthermore, no adjustments were made for variations in energy expenditure within the annual cycle due to physiological processes like thermoregulation, deposition of energy reserves for wintering and migration, moult, gonad development or egg-formation. Finally, the amount of food taken from the estuary to feed chicks has not been taken into account. It is expected, however, that the latter assumption will have relatively little effect on the estimated total consumption, as the number of birds feeding young is small in comparison with the numbers present in the non-breeding season.

The consumption was calculated separately for the subtidal (below mean low water) and intertidal areas. It was assumed that Eider & Goldeneye were feeding in the subtidal part of the Oosterschelde.

Slikken van Vianen: detailed estimate
The foraging behaviour of waders was studied in detail on the Slikken van Vianen, a small intertidal area in the middle part of the Oosterschelde (Fig. 1) (Meire & Kuijken, 1984; Meire, 1987). Birds were counted at both low and high tide. At low tide, numbers were counted in permanent plots (0.5–1 ha) during an entire tidal cycle on 220 days between 1979 and 1990. For each day the average density of foraging birds and bird

feeding minutes were calculated. Until 1985 14 plots were studied, six of these also in the remaining years. Days with very short exposure time, caused by manipulation of the storm surge barrier or storms, were omitted from the analyses.

The intake rate (mg ADW ingested/minute of feeding) of Oystercatchers was estimated from visual observations in all study plots (see Meire & Ervynck, (1986) for details). For the other bird species an average intake rate was estimated from the known daily consumption of each species (see Table 1) and an estimate of the total feeding time per tide (Meire, unpublished data). This estimate is probably an overestimation of the real intake rate in plots with a low biomass and an underestimation in plots with a large biomass. In order to correct the intake rate in plots with a biomass less than 10 g ADW m^{-2} (excluding cockles and mussels) the intake rate was multiplied by 0.5, in plots with a biomass higher than 30 g ADW m^{-2} it was multiplied by 1.5. For Bar-tailed Godwits, the values obtained in this way did not differ significantly from the field data (Meire, unpublished data). Consumption per plot was then calculated by multiplying the number of feeding minutes and the intake rate. It was thereby assumed that both the number of feeding minutes and the intake rate were similar at night and during the day. The annual consumption and predation pressure of Oystercatchers was calculated for two seasons: 1984/85 and 1986/87. For all species the consumption was calculated for the months September/October 1984 and expressed as gADW m^{-2} d^{-1}.

Benthic biomass

The estimate of macrozoobenthic biomass in the entire Oosterschelde is derived from two large scale surveys carried out in August 1985 and 1989 (Meire *et al.*, 1994; Seys *et al.*, 1994). At this time of the year benthic biomass reaches its maximum values (Beukema, 1974). In winter growth and reproduction are small, so these values can be considered as the maximum potential food source available to the birds during the next winter season. The survey of 1985 covered most of the intertidal areas of the Oosterschelde: the Roggenplaat, Galgenplaat, Verdronken Land van Zuid-Beveland, Slikken van Vianen and Krabbekreek (Fig. 1). In 1989 only the Roggeplaat, Galgeplaat and Krabbekreek were sampled. These biomass data are used to compare with the consumption of birds as no information on benthic production is available.

In all permanent plots on the Slikken van Vianen, macrozoobenthos was sampled each year in September. Density and biomass of all species in the samples were determined and all molluscs were measured to the nearest mm (Meire & Dereu, 1990).

Results

Consumption by benthivorous birds in the Oosterschelde

Total consumption
The total annual consumption of benthic invertebrates by birds in the Oosterschelde amounted to 1573×10^3 kg ADW y^{-1} in the pre-barrier period (Table 1). Similar results were obtained by Meire *et al.* (1989) who estimated the consumption of benthivorous bird species in the Oosterschelde to be 1448×10^3 kg ADW y^{-1} for the period 1976–1984. Although methods of calculation and division of bird species into functional groups differed slightly from those used in this paper (Meire *et al.*, 1989), the results are very similar and can be used to compare consumption by benthivores to that of herbivorous and piscivorous birds. Total consumption by piscivores was estimated at 8.7×10^3 kg ADW y^{-1} and of herbivores at 520×10^3 kg ADW y^{-1} (Meire *et al.*, 1989). Compared to these figures, consumption by benthivores in the Oosterschelde ecosystem is very high.

In the post-barrier period, total benthic consumption was estimated at 1500×10^3 kg ADW y^{-1}, a reduction of about 4% compared to the pre-barrier period. The decrease is not evenly spread over the different species. For Oystercatchers the decrease is 3%, for 'other waders' it

is 12.3%. Consumption by ducks decreased by 32.8%.

Share of individual species
A striking feature in the breakdown of consumption of benthic invertebrates over the species is the dominance of only a few bird species (Table 1). More than half of the total amount of biomass is eaten by the Oystercatcher, and only seven (pre-barrier) or even six (post-barrier) species together take 90% of the total. The most important species are Oystercatcher, Herring Gull, Curlew, Dunlin, Bar-tailed Godwit, Shelduck and Black-headed Gull. There was very little difference between the two study periods in the order and relative contribution of individual species.

Temporal patterns of bird predation

Interannual variations in consumption and benthic biomass
Waterbird populations are known to show considerable year to year variation in numbers, due to factors such as variation in local food supply and breeding success. In order to establish the among-year variations in total consumption by waders and dabbling ducks in the Oosterschelde, consumption was calculated separately for each year and the coefficient of variation (CV) for both study periods determined. It should be noted that due to the method of calculation, between-year variation in the estimated consumption is due only to variations in bird numbers, not to other factors such as varying winter temperatures. Table 2 shows that, although consumption by particular species may vary considerably among years, the total consumption was remarkably stable, with a CV of only 3–4% of the mean. This stability was mainly caused by the Oystercatcher, which takes more than half of the total consumption. The stability of Oystercatcher consumption is remarkable in view of the highly variable biomass of mussels and cockles (Coosen *et al.*, 1994a; Van Stralen & Dijkema, 1994). Although the pattern of consumption of benthic invertebrates for all

Table 2. Yearly variations in benthic concumption by waders and dabbling ducks in the Oosterschelde in 1978/79–1982/83 and in 1987/88–1989/90. Given are minimum, maximum (in gADW m^{-2}y^{-1}) and coefficient of variation (cv = 100·sd/mean) for each period.

Species	1978/79-1982/83 (n = 5)			1987/88-1989/90 (n = 3)		
	Min	Max	cv (%)	Min	Max	cv (%)
Oystercatcher	6.110	6.720	3.3	5.992	6.274	2.1
Curlew	0.764	0.892	6.0	0.806	1.013	10.2
Dunlin	0.497	0.661	12.8	0.285	0.442	17.6
Bar-tailed Godwit	0.298	0.425	15.9	0.309	0.388	9.4
Shelduck	0.296	0.490	16.0	0.216	0.334	17.4
Knot	0.198	0.319	14.9	0.189	0.254	13.0
Grey Plover	0.244	0.276	5.0	0.188	0.245	10.8
Pintail	0.075	0.207	31.8	0.079	0.125	18.6
Redshank	0.075	0.106	12.7	0.056	0.073	10.8
Shoveler	0.032	0.068	24.4	0.012	0.031	34.7
Turnstone	0.025	0.027	2.9	0.021	0.027	12.2
Spotted Redshank	0.010	0.017	20.8	0.012	0.018	17.4
Greenshank	0.004	0.010	36.1	0.006	0.008	12.7
Avocet	0.003	0.003	5.4	0.001	0.002	17.5
Ringed Plover	0.001	0.001	16.2	0.001	0.01	24.1
Total	8.884	9.644	3.5	8.446	9.146	3.3
Excl. Oystercatcher	2.164	3.251	13.3	2.187	2.871	11.2

other bird species varied more, the overall variability is still rather small (Table 2; CV 11 to 13%) especially compared to the variability in the prey populations. This is exemplified in Table 3 which shows the CV of total density and biomass of different trophic groups, based on eight or nine late autumn samplings in the period 1979–1989 on six permanent plots on the Slikken van Vianen. It is clear that the variability of the benthos is much larger than that of the predation by birds, with the combined CV ranging between 20 and 76% for different trophic groups.

Seasonal pattern of predation

Within-year variation of consumption is shown in Fig. 2. Again, consumption as calculated reflects only variations in bird numbers, not effects of wheather conditions, moult and deposition of fat

Table 3. Yearly variations in benthic invertebrates in 6 study plots on the Slikken van Vianen. The coefficients of variation (%) for 8 or 9 autumn biomass values (years) are given for total density and biomass, and the biomass of deposit feeders, filter feeders, grazers and omnivores.

Plot	Total Density	Total Biomass	Biomass Deposit feeders	Biomass Filter feeders	Biomass Grazers	Biomass Omnivores
10 (*n* = 9)	41.8	42.6	47.1	45.7	109.5	62.7
13 (*n* = 9)	68.6	110.6	66.9	121.1	183.7	74.7
22 (*n* = 8)	36.8	34.7	64.7	40.3	69.2	106.4
32 (*n* = 8)	43.2	30.0	39.5	52.2	148.6	99.1
39 (*n* = 8)	51.9	36.2	44.9	50.7	130.1	75.6
60 (*n* = 9)	39.0	59.6	50.2	62.2	133.1	80.2
Total (*n* = 8)	28.9	36.1	20.2	42.3	75.5	42.7

532

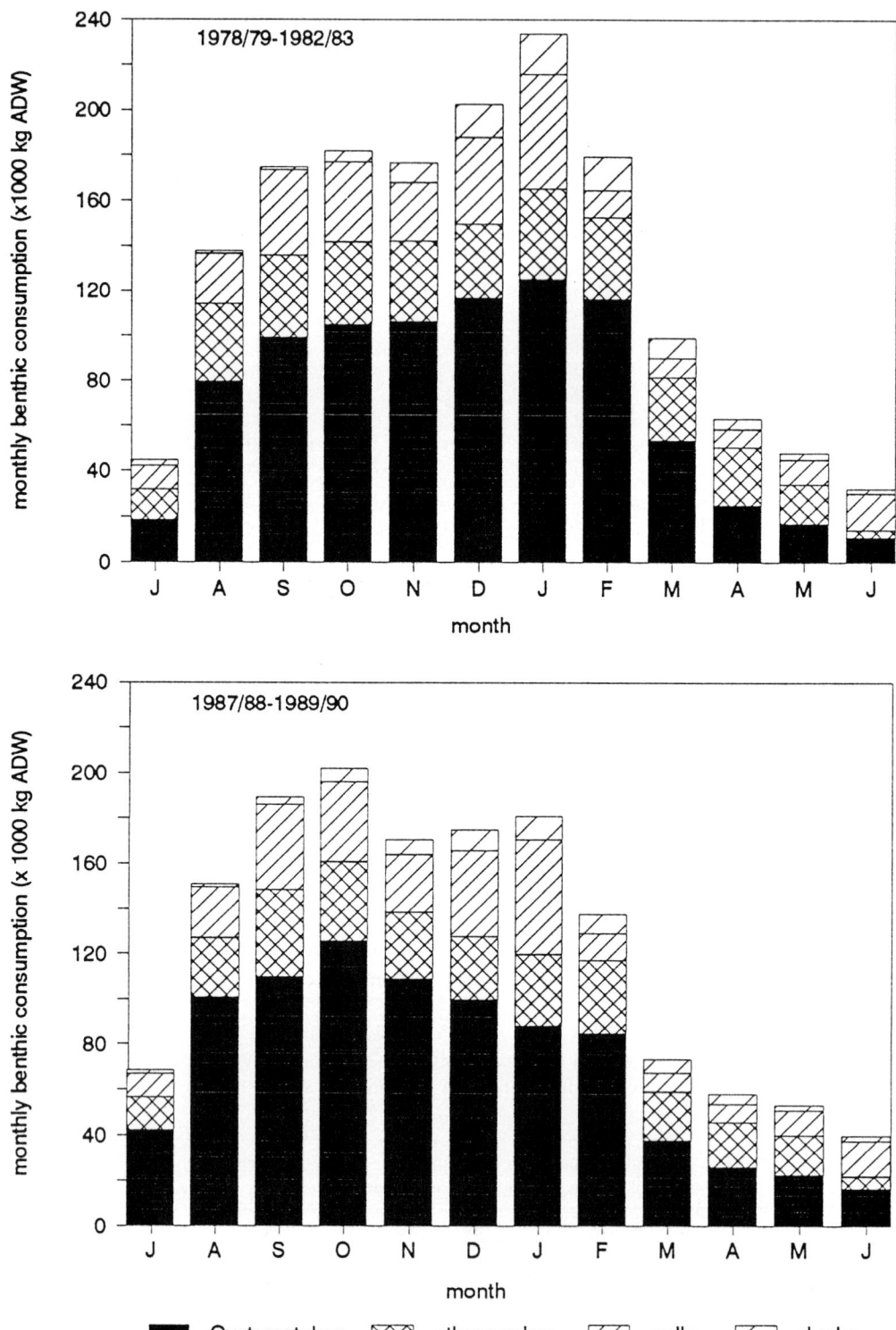

Fig. 2. Seasonal pattern of consumption by waterfowl. The consumption of different groups of benthivorous birds is given per month for the pre-barrier (upper panel) and the post-barrier period (lower panel).

stores. In general, energy requirements will be above the 3 BMR used in this paper during the winter months, while they will fall below this during summer. In the pre-barrier period, consumption was lowest in mid summer, but in August a sharp increase occurred with the arrival of large numbers of waders, especially Oystercatchers. A further increase occurred until a peak was reached in January. A sharp decrease took place in March, when most Oystercatchers left the area for the breeding grounds. In the post-barrier period, the consumption pattern was somewhat different, mainly caused by different seasonal occurrence of the Oystercatcher (Lambeck, 1991). Numbers, and consequently consumption, in late summer were higher as compared to the pre-barrier period. Instead of an increase until January, numbers and consumption decreased after a peak in October.

Spatial pattern of consumption by birds in the Oosterschelde and relation with macrozoobenthos

Consumption by birds in different sectors of the Oosterschelde

Total consumption was calculated for each of the four sectors of the intertidal area of the Oosterschelde estuary, using the total number of bird days per year per sector as a basis. These figures were not available for gulls; therefore gulls were assumed to be distributed homogeneously over the intertidal area. The results for the 12 most important bird species are presented in Table 4. Total consumption is given including and excluding gulls. The total consumption per unit area per year does not differ much among sectors, apart from a distinctly lower value in the eastern part. This lower value is almost totally caused by the lower densities of Oystercatchers in this area,

Table 4. Total benthic consumption (gADW m^{-2} y^{-1}) by birds in four sectors of the intertidal area of the Oosterschelde. W = west, C = centre, E = east, N = north, Pre, Post = pre and post-barrier.

Species	1978/79–1982/83				1987/88–1989/90				Total	
	W	C	E	N	W	C	E	N	Pre	Post
Oystercatcher	8.900	7.513	3.820	8.722	8.693	10.37	4.750	7.895	6.381	7.605
Curlew	0.960	0.642	0.872	1.046	0.914	1.090	1.210	0.885	0.861	1.066
Dunlin	0.364	0.554	0.588	0.619	0.365	0.508	0.474	0.406	0.543	0.451
Bar-tailed Godwit	0.745	0.408	0.094	0.704	0.766	0.404	0.096	0.762	0.371	0.423
Shelduck	0.224	0.205	0.482	0.764	0.270	0.125	0.568	0.323	0.402	0.333
Knot	0.509	0.268	0.171	0.390	0.314	0.414	0.157	0.183	0.285	0.264
Grey Plover	0.258	0.283	0.219	0.313	0.285	0.311	0.236	0.198	0.257	0.260
Pintail	0.017	0.012	0.279	0.170	0.024	0.013	0.285	0.123	0.150	0.127
Redshank	0.070	0.078	0.095	0.116	0.073	0.107	0.073	0.072	0.090	0.080
Shoveler	0.038	0.010	0.082	0.056	0.060	0.009	0.028	0.028	0.053	0.029
Turnstone	0.035	0.040	0.013	0.048	0.033	0.032	0.011	0.054	0.029	0.028
Avocet	0.059	0.009	0.014	0.074	0.068	0.005	0.005	0.017	0.029	0.020
Spotted Redshank	0.018	0.019	0.005	0.033	0.024	0.025	0.012	0.010	0.014	0.018
Ringed Plover	0.003	0.007	0.004	0.018	0.011	0.013	0.008	0.016	0.007	0.009
Greenshank	0.003	0.007	0.008	0.014	0.006	0.012	0.009	0.005	0.007	0.009
Kentish Plover	0.004	0.006	0.001	0.009	0.003	0.003	0.001	0.004	0.004	0.003
TOTAL	14.23	12.08	8.76	15.11	14.34	15.86	10.35	13.40	11.49	13.15
Ducks	0.28	0.23	0.85	0.99	0.36	0.15	0.88	0.48	0.60	0.49
Oystercatcher	8.90	7.51	3.82	8.72	8.69	10.37	4.75	7.90	6.38	7.61
Other waders	3.04	2.32	2.08	3.38	2.88	2.92	2.29	2.61	2.50	2.64
Gulls	2.01	2.01	2.01	2.01	2.41	2.41	2.41	2.41	2.01	2.42
Total excl gulls	12.22	10.08	6.75	13.10	13.93	13.45	7.94	10.99	9.48	10.74

534

which in turn can probably be explained by the small surface area of intertidal mussel beds and lower biomass of cockles, as illustrated by data from 1985 in Fig. 3a. The biomass value for the northern sector is probably an underestimate as some musselbeds were not covered in the survey. The lower consumption by other wader species in the central and eastern sector coincides with lower biomass (total biomass – biomass cockles and mussels) values here in 1985 (Fig. 3b).

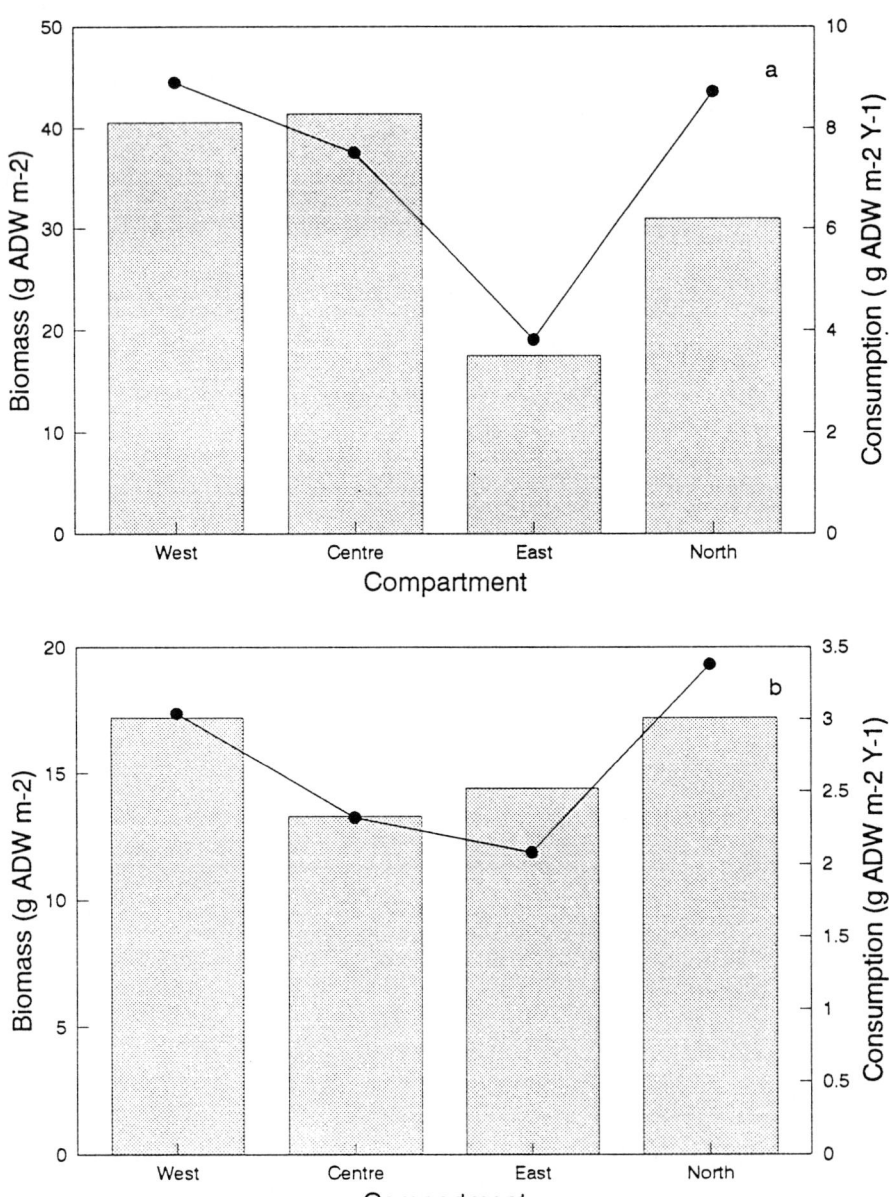

Fig. 3a. Consumption by Oystercatchers (dots) in the pre-barrier period and benthic biomass of suspension feeders (bars) in August 1985 in each compartment of the Oosterschelde.

Fig. 3b. Consumption by waders (minus Oystercatchers) (dots) in the pre-barrier period and benthic biomass (minus suspension feeders) (bars) in August 1985 in each compartment of the Oosterschelde.

Consumption in relation to the tidal and subtidal zone, and benthic biomass

Of the total amount of food taken by birds, only 0.1–0.4% is taken in the subtidal compartment (Table 1). In fact, it is even less than that since Goldeneye and Eider, the only diving benthivores that regularly occurred in the estuary, take part of their food from intertidal areas during high tide. Compared with the Wadden Sea where diving ducks, mainly Eider, take *ca* 30% of the total consumption (Smit, 1981), the absence of avian subtidal benthivores in the Oosterschelde is very noteworthy.

On the tidal flats, the birds are not distributed at random but aggregate on certain parts of the tidal flats. Consumption therefore varies considerably among sites. This is exemplified with the data from Vianen from the 1984/85 season. In Fig. 4 the total number of feeding minutes of all wader species in different study plots is given, showing variation between plots by a factor 20. There is a clear relation between consumption and benthic biomass when all species are considered (Fig. 5a) ($r^2 = 0.8$; $N = 15$; $p < 0.01$ after re-

moving plots 23 and 20). The two aberrant points, plots 20 and 23, are both situated on a very muddy mussel bed, low in the intertidal area and characterised by large pools at low tide. This relation also holds when leaving out consumption by Oystercatchers and the biomass of cockle and mussel, although a remarkably high consumption was seen in plots 6, 10 and 22, three plots situated on mussel beds (Fig. 5b) ($r^2 = 0.74$, $N = 14$, $p < 0.01$ after removing plots 6, 10 and 22).

Predation pressure

General. Assuming that all predation by gulls is confined to the intertidal areas, the yearly consumption of benthic invertebrates in the intertidal part of the Oosterschelde was estimated at 11.5 g ADW m^{-2} y^{-1} pre-barrier and 13.2 g ADW m^{-2} y^{-1} post-barrier, a 14.4% increase (Tables 1 & 5). In the subtidal zone there was a 233% increase from 0.006 to 0.02 g ADW m^{-2} y^{-1} (Table 1). Notwith-standing this increase, overall consumption here remains very low.

In Table 5 the consumption by birds in the intertidal area is compared to the benthic biomass

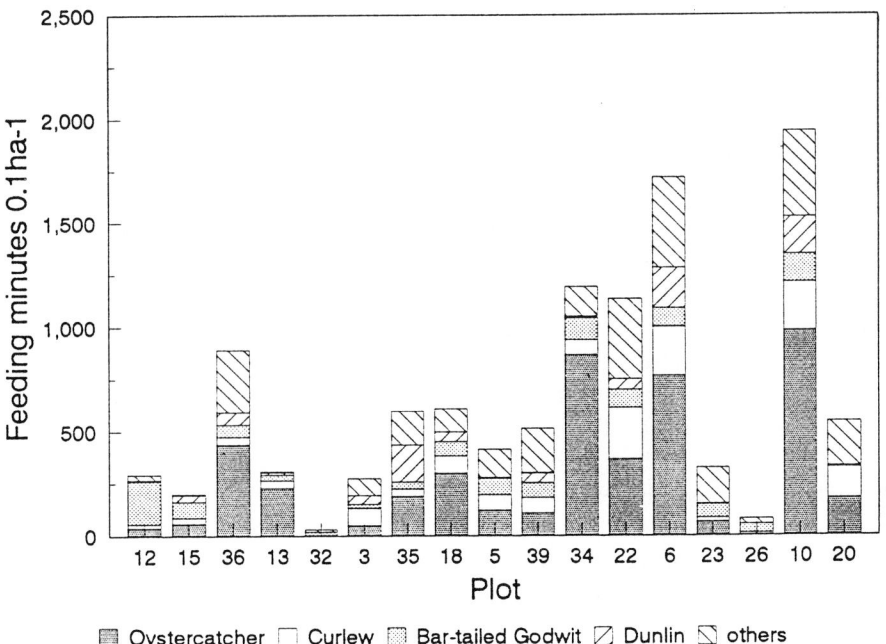

Fig. 4. Feeding minutes of waders per plot (average for the period August–September 1984) on the Vianen mudflat. Plots are ranked according to benthic biomass, being lowest in 12 and highest in 20.

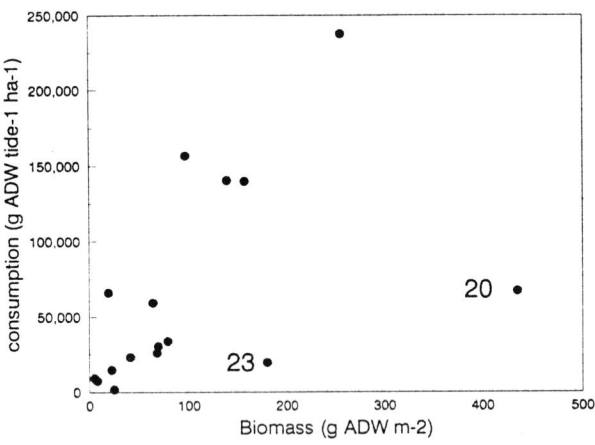

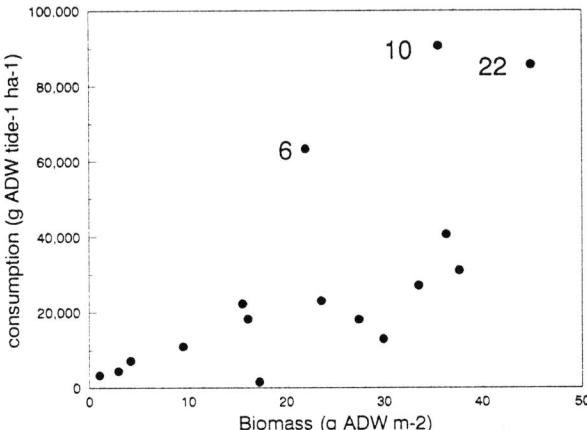

Fig. 5. Consumption of waders in relation to benthic biomass on different study plots of the Vianen mudflat. (a) for all species and (b) excluding Oystercatchers, cockles and mussels. The consumption was calculated based on observations from September/October 1984, benthic biomass was sampled in September 1984.

Table 5. Total macrobenthic biomass in the Oosterschelde and estimated predation pressure by birds in a pre-barrier (1985) and a post-barrier (1989) year. The percentage of benthic biomass removed by birds is given between brackets.

Biomass, Consumption and Predation Pressure		
Year	1985	1989
Total biomass (gADW m^{-2})	49.3	99.3
Total consumption (gADW m^{-2} y^{-1})	11.5 (23.3%)	13.2 (13.3%)
Biomass (excluding cockles & mussels) (gADW m^{-2})	17	14.9
Consumption (excluding Oystercatcher) (gADW m^{-2}y^{-1})	5.11 (30.1%)	5.54 (37.2%)
Biomass (cockles and mussels) (gADW m^{-2})	32.3	84.4
Consumption by Oystercatchers (gADW m^{-2} y^{-1})	6.38 (19.7%)	7.61 (9.0%)

biomass of cockles and mussels is given in Table 5 and amounted to 20% in 1985 and 9% in 1989.

Based on observations of Oystercatchers at low tide in permanent plots at the Slikken van Vianen during the seasons 1984/85 and 1986/1987, predation pressure per plot was estimated and plotted in Fig. 6. Predation pressure varied from less than 10% to more than 70% of the standing stock. The data suggest a large scatter at low biomass values and an average predation pressure of about 30% at higher biomass values (musselbeds), without a correlation between biomass and the percentage taken. The large scatter in the data from plots outside musselbeds is to a large extent dependent on the average length of the cockles present as shown in Fig. 7. In plots with larger cockles the predation pressure was significantly higher ($r = 0.74$, $n = 11$, $p < 0.01$).

The seasonal pattern of Oystercatcher numbers changed in the Oosterschelde in the post-barrier period (Lambeck, 1991; Schekkerman *et al.*, 1994). Numbers present in July–September increased, but from October onwards they de-

to estimate the predation pressure, which was found to be 23 and 13% of the standing stock, in the pre- and post-barrier period respectively. Total biomass showed large yearly variations. This is mainly due to the biomass of the filter-feeding cockles and mussels. When excluding cockles and mussels and Oystercatchers, their main predator, from the calculations, the predation pressure of the other species was 30 and 37% of the biomass (Table 5).

Predation pressure by Oystercatchers. The predation pressure by Oystercathers in relation to the

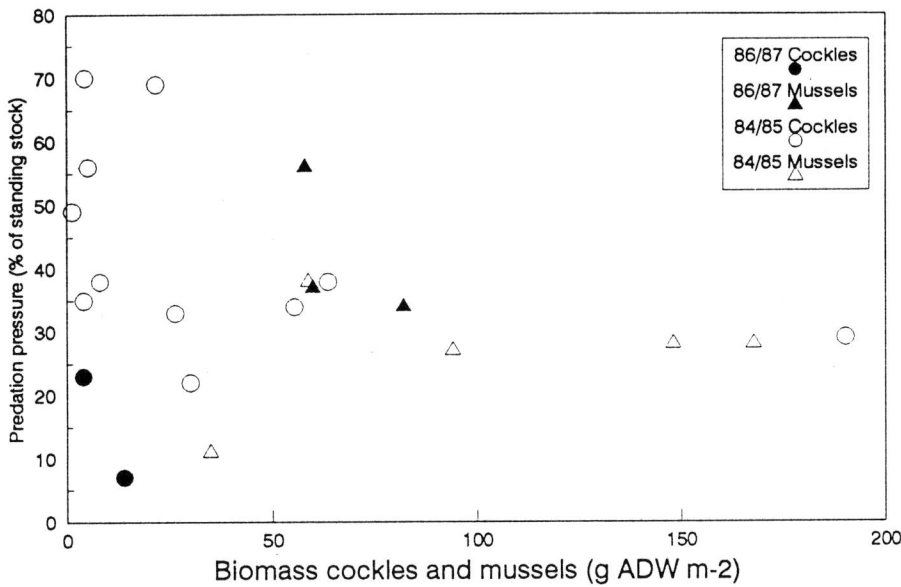

Fig. 6. Predation pressure (percentage of standing stock of cockles and mussels measured in August removed during one year) by Oystercatcher in relation to the biomass of cockles and mussels. Plotted are the data from several study plots on the Slikken van Vianen for both the season 1984/85 and 1986/87.

creased. Midwinter numbers were 23% lower in the post-barrier period. One possible explanation could be that in autumn, the food supply was depleted so rapidly by the higher numbers of birds, that birds were forced to leave. This seems unlikely. The difference in numbers is not large enough to explain a sudden depletion. Another explanation could be a change in the food supply.

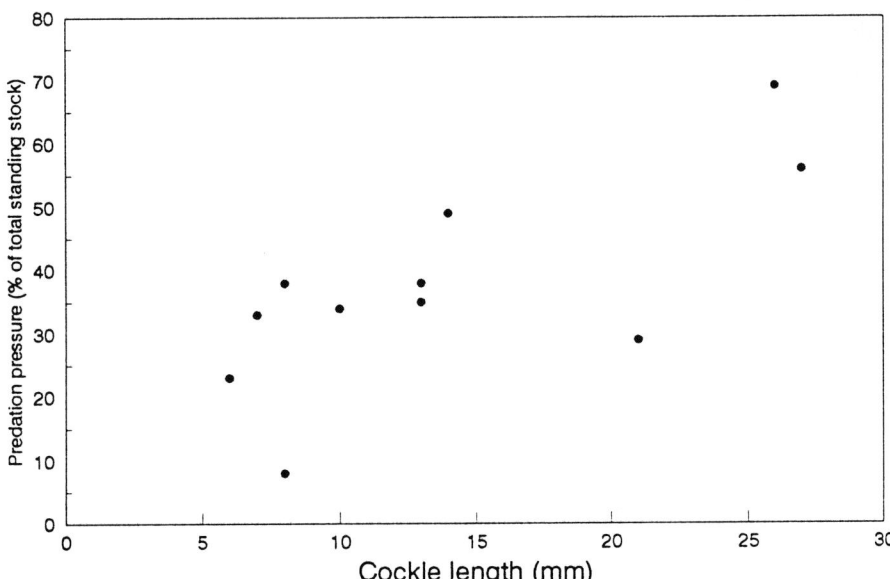

Fig. 7. Relation between predation pressure of Oystercatchers on cockles and the average length of cockles. Data from the season 1984/85 and 1986/87 from several study plots on the Slikken van Vianen are given.

In Fig. 8 the estimated total stock of cockles in August (based on Coosen *et al.*, 1994b) and the amount of cockles removed by cockle-fishers is plotted. Cockle fisheries (M. Van Stralen, pers. comm.) increased dramatically in the post-barrier period and coincided with low cockle stocks in recent years. As cockle fisheries removed most animals between October and December, the drop in Oystercatcher winter numbers could well be related to this.

Prey availability

One of the effects of the construction of the storm surge barrier in the Oosterschelde is the erosion of the intertidal flats (Mulder & Louters, 1994). This is shown in Fig. 9 for one of the major intertidal flats. The total food availability for birds can be expressed as the product of surface, benthos biomass and exposure time. Although benthos biomass did not change in relation to tidal elevation the overall food availability index (product of surface, benthos biomass and exposure time) decreased by 17% between 1984 and 1989, mainly due to the decreased tidal elevation

of the flat. If this trend continues, as expected, this will further reduce food availability.

Discussion

Based on the calculations presented in this paper, we can estimate the carbon flow from benthos to birds in the post-barrier period at 1.8 g C m^{-2} y^{-1}, assuming a conversion factor from g ADW to g C of 0.4348 and a total surface of 35 100 ha. The fluxes, calculated in SMOES, of carbon from phytoplankton and labile detritus to zooplankton is in the order of 100 g C m^{-2} y^{-1}, to suspensionfeeders about 120 g C m^{-2} y^{-1} (Van der Tol & Scholten, 1993). The role of birds in the overall C-balance of the estuary is indeed rather small. Their impact on benthic populations may however be important.

Consumption of benthos by birds

Schekkerman *et al.* (1994) have shown that in the Oosterschelde the numbers of waders declined after the reduction of the intertidal area and that

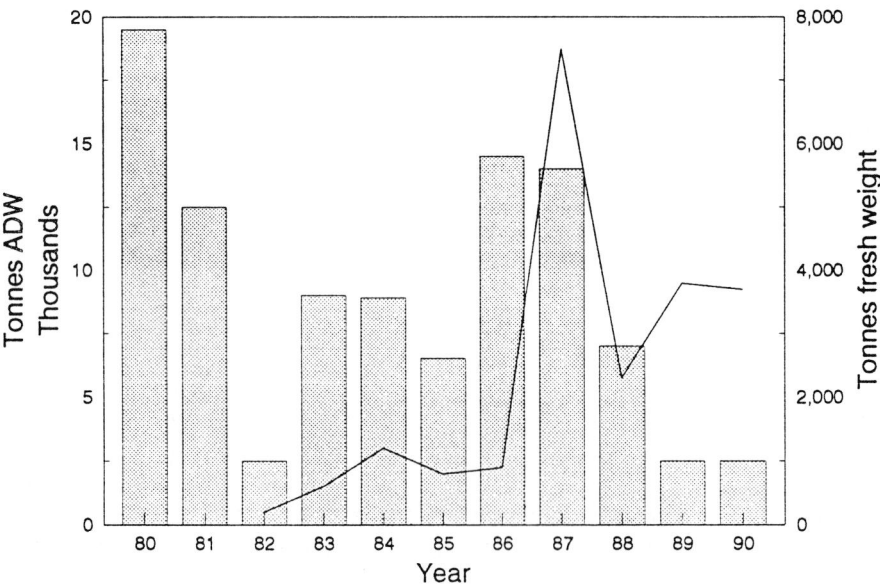

Fig. 8. Cockle standing stock in tonnes ADW (bars) and amount of cockles in tonnes fresh weight removed by fisheries (line) in the Oosterschelde

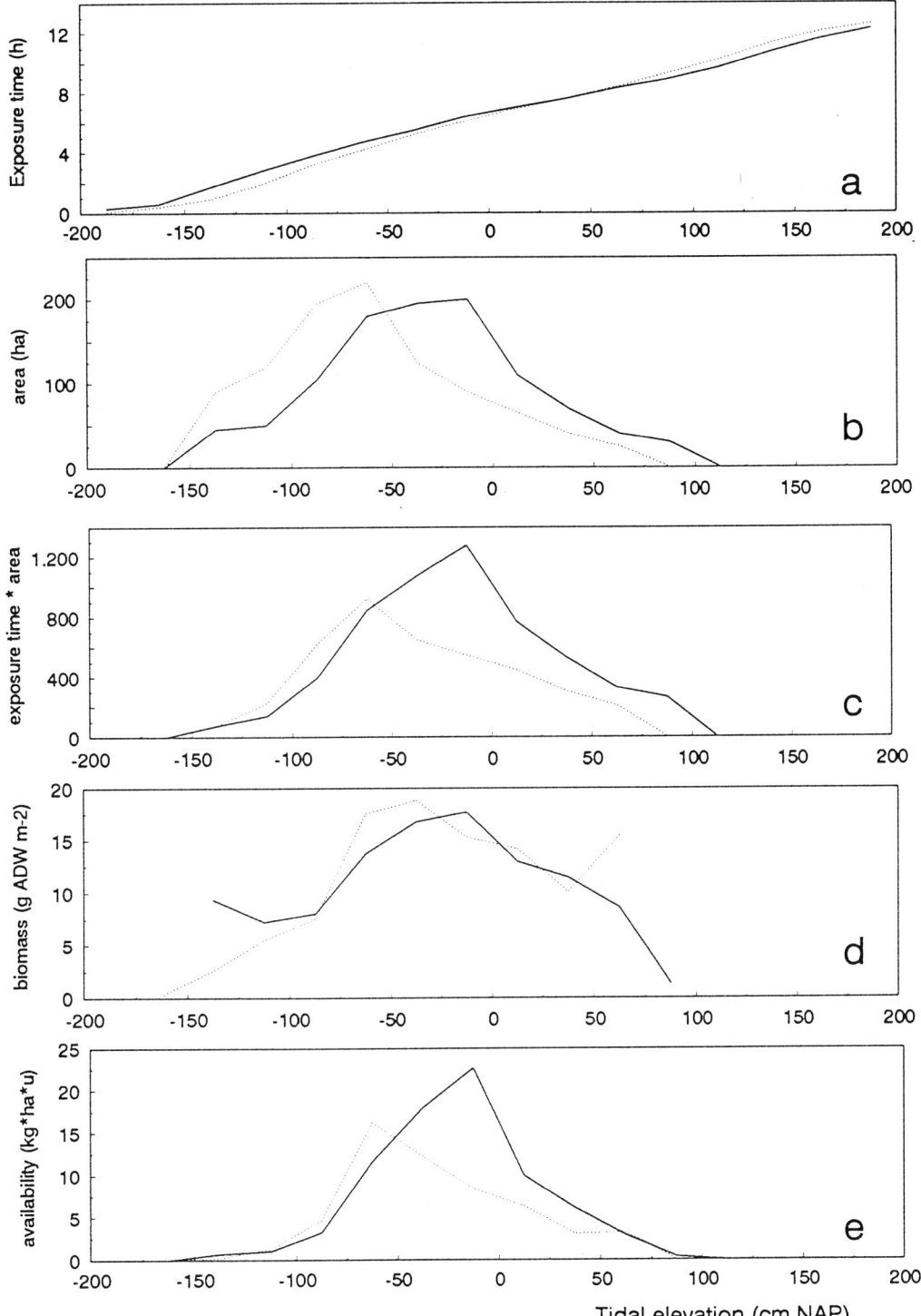

Fig. 9. Exposure time, surface, benthic biomass and food availability, expressed as the product of exposure time, surface and benthic biomass at different tidal elevations of the Roggenplaat in 1985 (solid line) and 1989 (dotted line).

the decrease of each species was related to the loss of its specific habitat, indicating that carrying capacity had been reached. As it is shown that the consumption of benthivorous birds is high in comparison to other trophic groups (Meire *et al.*, 1989), the question is whether or not the use of the area by waders was limited by the benthic food supply (Evans & Dugan, 1984; Goss-Custard, 1985). The benthic food supply will be limiting if birds consume the same amount or more than can be replaced by production and immigration. Although it is possible in some cases, measuring the production of harvestable biomass is very difficult and not feasible for all species (Piersma, 1987; Zwarts *et al.*, 1992). As, for the Oosterschelde, no data on benthic production and harvestability are available, the estimated consumption was compared with the standing stock in order to investigate whether or not the food supply could be an important factor limiting carrying capacity or not.

The results presented in this paper indicate that, depending on location, predator and prey species, between 5 and 70% of the benthic food supply present in August is taken annually by birds. This can be compared to other studies where predation pressure by birds on benthic invertebrates was measured. This is done either based on knowledge of bird numbers and metabolic requirements (as in this study) or by using exclosure experiments. In some exclosure studies no effects of birds on macrobenthos could be determined (Raffaelli & Milne, 1987; Wilson, 1991) but other experiments have found a moderate to high predation pressure by birds (Goss-Custard, 1977; Schneider, 1978; Schneider & Harrington, 1981; Boates & Smith, 1979; Reise, 1985).

The methods used in this paper to estimate the amount of food removed by waterbirds are similar to those used in other studies (Table 6). Although studies differed in assumptions or methods of calculations, with some caution the results can nevertheless be compared. For instance, Smit (1981) used Lasiewski & Dawson's (1967) or Aschoff & Pohl's (1970) formula for estimating BMR of waders, which results in a lower figure than the equation given by Kersten & Piersma (1987). On the other hand, he used mean

Table 6. Consumption, estimated from bird numbers and metabolic requirements, of waders and ducks in different estuaries. The consumption in gADW m^{-2} y^{-1}, the autumn or mean annual biomass in gADW m^{-2} and annual production in gADW m^{-2} y^{-1} and the predation pressure (between brackets) are given.

Site	Consumption	Autumn biomass (% consumed)	Mean biomass	Production	Reference
Oosterschelde	11.5–13.15*	49.3(23.3%)–99.3(13.3%)*			this paper
Westerschelde	4.1	15.1(27.2%)			Meire *et al.* 1989; Stuart *et al.*., 1989
Gevelingen	8.2		72 (11.4%)		Wolff *et al.* 1976; Wolff & De Wolf, 1977
Waddenzee	4.7		26.6 (17.6%)	28 (17%)	Smit, 1981; Beukema, 1976, 1981; De Wilde & Beukema, 1984
Ythan estuary	22		59.8 (36.7%)	111.4 (19.7%)	Baird & Milne, 1981
Tees estuary	17.2	44.41(38.3%)		38.6 (44%)	Evans *et al.*, 1979
Ventjager	5.5	11.63(47.3%)–16.18(34%)			Zwarts 1974
Langebaan	6.4			32 (20%)	Summers, 1977; Baird *et al.*, 1985
Berg River estuary	26.7			109.4 (26%)	Kaletja, 1992
Banc d'Arguin	11.5		14.5 (82.7%)	27 (42.6%)	Wolff & Smit, 1990

* Pre- and post-barrier data (see Table 1)

annual body mass including fat stores to estimate BMR instead of lean weights, and a factor 5 instead of $(1/0.85 \times 3) = 3.5$ to convert BMR into DEE; both factors resulting in a higher consumption. Calculating the consumption in the Wadden Sea using the number of birds given by Smit and the body masses and formulae applied for the Oosterschelde, the yearly consumption does not differ much from Smit's original figure: 4.3 gADW m^{-2} y^{-1} versus 4.7 gADW m^{-2} y^{-1}. We used 0.85 as a value of assimilation efficiency Q, while 0.80 (Wolff & Smit, 1990) or 0.75 (Castro et al., 1989) might be a more realistic value. Use of $Q = 0.85$ causes an underestimation of the consumption by 11% compared to $Q = 0.75$.

Another ground for differences in results lies in the conversion of predicted BMR to an estimate of DEE in the field. We used a conversion factor of 3, following Drent et al. (1978), Smith (1975), Kersten & Piersma (1987) and Castro et al. (1992). Recently, Wiersma & Piersma (1993), using climatic data, estimated the energy expenditure under field conditions for Knots in the Dutch Wadden Sea, taking into account effects of temperature, solar radiation and wind. Compared to a constant rate of 3 times BMR, the total annual energy expenditure per bird was estimated by Wiersma & Piersma was 19% higher. Because Knots, like most species, are most numerous in the Oosterschelde in winter when thermostatic costs are high, the resulting estimate of consumption would be as much as 28% higher than ours. The true difference is probably somewhat smaller as the winter climate of the Oosterschelde is more benign than that of the Wadden Sea. The difference as found in Knots cannot be assumed to apply to all species occurring in the Oosterschelde, since the thermostatic cost relative to BMR decreases with increasing body size (Wiersma et al., 1993). Thus the true energy expenditure of birds larger than Knots, which are most important in determining the total amount of food removed by birds in the Oosterschelde, is expected to be closer to the level estimated by our assumptions of 3 × BMR than that of Knots.

The data from Table 6 show that waders are able to remove a substantial part of benthic bio-

mass or production, as already found by Baird et al. (1985). The values for the Oosterschelde, including cockles, mussels and Oystercatchers are on the low side, but excluding them they are comparable to, or higher than those for other areas.

Within an estuary the consumption is not evenly spread over the flats. Data from the permanent plots at Vianen indicate a clear positive relationship between consumption and benthic biomass. Zwarts (1988), working in the Wadden Sea and Sidi Moussa (Morocco) found a similar relationship between consumption and benthic biomass, although the observed relationship differed significantly between study areas. In the Oosterschelde, predation pressure by Oystercatchers did not show any relation with the overall biomass. It is likely, however, that predation pressure is related to the amount of prey that can be economically harvested as suggested by the relationship between predation pressure and average cockle length. In plots with large cockles, which are both available and profitable for Oystercatchers, predation pressure is very high (up to 70%).

Besides consumption by birds, other epibenthic predators take their share of the macrobenthic food supply. Smit (1981) and Beukema (1981) estimated the consumption by crabs, shrimps and fish for the Wadden Sea at 10 g ADW m^{-2} y^{-1}, twice as large as the total consumption by birds. Sanchez-Salazar et al. (1987) found the consumption of adult crabs Carcinus maenas on cockles to be 25 times more important in numbers or twice as much in biomass than that of Oystercatchers. In addition to epibenthic predators, also several infaunal predators are present in the sediment. Although their role is less well understood they can have an important impact on the other benthic species (Ambrose, 1991).

Based on the evidence presented above it is clear that epibenthic predators and birds together must consume a substantial part of the benthic biomass. Furthermore, the percentage of the prey populations which are predated by birds were calculated based on the total amount of biomass present on the flats and not on the biomass harvestable by the birds. It is known that due to prey escape behaviour, burying depth, prey-size, coverage by barnacles etc. the biomass harvestable to birds at any one moment is much lower than the total biomass present (Durell & Goss-Custard, 1984; Esselink & Zwarts, 1989; Evans, 1976; Evans, 1987; Meire, 1991; Meire & Ervynck, 1986; Zwarts & Wanink, 1984, 1989). Zwarts *et al.* (1992) found on average over 10 years only 13.5% of the total biomass of bivalves to be harvestable by Knots. Meire & Ervynck (1986) estimated that, depending on the mussel bed, on average 30% of the mussels of the size classes taken by hammering Oystercatchers were available to the birds. Predation pressure on harvestable biomass must therefore be higher than the figures given above. To prove carrying capacity is reached due to shortage of food, it is crucial to know the amount of harvestable prey which is removed in relation to production of harvestable biomass (Piersma, 1987). These data are not available. The date discussed above do suggest that birds consume a substantial part of the total food supply, which moreover they have to share with other epibenthic and endobenthic predators consuming at least the same quantity of food.

Carrying capacity

Recent studies showed that bird numbers reach plateau values in some estuaries or feeding areas (Meire & Kuijken, 1984; Moser, 1988; Zwarts, 1974) and Schekkerman *et al.* (1993) showed that wader numbers decreased in the Oosterschelde in response to the reduction of intertidal area. We believe this provides circumstantial evidence that, at least for several species, the decline in numbers in the Oosterschelde (Schekkerman *et al.*, 1994) could be due to food shortage. This does not mean that birds face food shortage each season. As the variability of the benthos is much higher than that of the birds it is likely that food will not be limiting at some times, while at other times it does. During severe winters, food shortage in combination with cold stress can cause the death of many birds (Lambeck, 1991; Meininger *et al.*, 1991). As the variability of cockle and mussel

biomass is very high it is likely that at some times, when biomass is very high, much more is harvestable by Oystercatchers than is actually taken. At other times when biomass is low (caused by natural variation or by fisheries) it might limit Oystercatchernumbers. If this is true, any further loss of intertidal area, food supply, or availability will result in a further reduction of bird numbers. In recent years populations of both mussels and cockles have been very small in the Wadden Sea causing high mortality in Eider and decreasing numbers of Oystercatchers (Beukema & Swennen, pers. comm.). Whether or not this will have an effect on population level is another question (Goss-Custard & Durell, 1990).

The results from the intertidal area are in contrast with those from the subtidal compartment. Here consumption is very low compared to other areas as the Wadden Sea (Smit, 1981) or the saline lakes of the Delta area (Meire *et al.*, 1989). Although no data on benthic invertebrates of the subtidal compartment are available for the years analysed in this paper, present investigations by Craeymeersch (pers. com.) indicate that benthic biomass is comparable to that of the Wadden Sea (Dekker, 1989), resulting in a very low predation pressure in the Oosterschelde. If not limited by other factors, bird species like Eiders, feeding in the subtidal compartment, could probably increase substantially in numbers in the Oosterschelde, a trend which seems to have started already.

Historical perspective

The present results contrast with previous findings in the Delta area. There is no evidence that wader numbers in the whole Delta area of South-West Netherlands decreased after the closure of the estuaries Veerse Gat, Haringvliet and Grevelingen (Saeijs & Baptist, 1977; Leewis *et al.*, 1984; Meininger *et al.*, 1984). After the closure of the Grevelingen, an estuary adjacent to the Oosterschelde, wader numbers increased substantially in the Krammer-Volkerak, the northern branch of the Oosterschelde (Leewis *et al.*, 1984;

Meininger *et al.,* 1984) and numbers of Oyster-catchers and Bar-tailed Godwits increased abruptly on a large tidal flat in the mouth of the Oosterschelde (Lambeck *et al.,* 1989). In the Krammer-Volkerak, however, important changes occurred due to the coastal engineering works. Tidal amplitude and current velocity increased, chlorinity rose from 0.5–5 to 9–13%. In response to these abiotic changes, especially the chlorinity, the diversity of macrozoobenthos increased substantially (Wolff, 1971) and mussel cultures became established in the area. Although no data are available we can reasonably assume that in the Krammer-Volkerak the food availability increased substantially, increasing the carrying capacity for waders. If in the Oosterschelde (excluding Krammer-Volkerak) and Westerschelde, the two remaining estuaries, the benthic biomass did not change (increase) it seems that carrying capacity had not yet been reached in the 1960s. This might also hold in other Western European estuaries where numbers of some wader species increased in the last decades (Smit & Piersma, 1989) notwithstanding a reduction in intertidal area. In the past decades the benthic production might have increased due to eutrophication as is shown for the Wadden Sea (Beukema & Cadée, 1986) and hence carrying capacity as suggested by Van Impe (1985) to explain an increase in bird numbers in a part of the Westerschelde estuary. Tubbs *et al.* (1992) suggested that the increase in Dunlin numbers in the Solent since the 1950s reflects release from hunting pressure. A combination of both factors might indeed explain the increasing population sizes of most species in a period of decreasing feeding areas. For several species the balance now seems to have been reached and any further loss in intertidal habitat or deterioration in food supplies will ultimately result in a decrease of wader numbers.

Acknowledgements

The results presented in this paper are based to a large extend on bird counts collected by many professional and amateur ornithologists. H. J. M. Baptist and G. J. Slob initiated and coordinated the counts for many years. R. H. D. Lambeck, E. Kuijken, T. Ysebaert, G. Degelder, J. J. Beukema and D. McLusky suggested many important improvements to the manuscript.

Communication No. 709 of NIOO-CEMO, Yerseke, The Netherlands.

References

Ambrose, W. G., 1991. Are infaunal predators important in structuring marine soft-bottom communities? Am. Zool. 31: 849–860.

Aschoff, J. & H. Pohl, 1970. Der Ruheumsatz von Vögeln als Funktion der Tageszeit und der Körpergröße. J. Orn. 111: 38–47.

Baird, D. & H. Milne, 1981. Energy flow in the Ythan estuary, Aberdeenshire, Scotland. Estuar. coast. Shelf Sci. 13: 455–472.

Baird, D., P. R. Evans, H. Milne & N. W. Pienkowski, 1985. Utilization by shorebirds of benthic invertebrate production in intertidal areas. Oceanogr. Mar. Biol. annu. Rev. 23: 573–597.

Beukema, J. J., 1974. Seasonal changes in the biomass of the macrobenthos of a tidal flat area in the Dutch Wadden Sea. Neth. J. Sea Res. 8: 94–107.

Beukema, J. J., 1976. Biomass and species richness of the macrobentic animals living on a tidal flat area in the Dutch Wadden Sea. Neth. J. Sea Res. 10: 236–261.

Beukema, J. J., 1981. The role of the larger invertebrates in the Wadden Sea ecosystem. In W. J. Wolff (ed.), Ecology of the Wadden Sea. Balkema, Rotterdam: Vol 3: 211–221.

Beukema, J. J. & C. Cadée, 1986. Zoobenthos responses to eutrophication of the Dutsch Wadden Sea. Ophelia 26: 55–64.

Boates, J. S. & P. C. Smith, 1979. Length-weight relationships, energy content and the effects of predation on *Corophium volutator* (Pallas) (Crustacea: Amphipoda). Proc. N. S. Inst. Sci. 29: 489–499.

Castro, G., N. Stoyan & J. P. Myers, 1989. Assimilation efficiency in birds: a function of taxon or food type? Comp. Biochem. Physiol. 92A: 271–278.

Castro, G., J. P. Myers & R. E. Ricklefs, 1992. Ecology and energetics of Sanderlings migrating to four latitudes. Ecology 73: 833–844.

Coosen, J., J. Seys, P. Meire & J. A. M. Craeymeersch, 1994a. Effect of sedimentological and hydrodynamical changes in the intertidal areas of the Oosterschelde estuary (SW Netherlands) on distribution, density and biomass of five common macrobenthic species: *Spio martinensis* (Mesnil), *Hydrobia ulvae* (Pennant), *Arenicola marina* (L.), *Scoloplos armiger* (Muller) and *Bathyporeia* sp. Hydrobiologia 282/283: 235–249.

544

Coosen, J., F. Twisk, M. W. M. van der Tol, R. H. D. Lambeck, M. R. van Stralen & P. M. Meire, 1994b. Variability in stock assessment of cockles (Cerastoderma edule L.) in the Oosterschelde (in 1980–1990), in relation to environmental factors. Hydrobiologia 282/283: 381–395.

Coosen, J., J. Seys & P. M. Meire, 1993a. Effect of sedimentological and hydrodynamical changes in the intertidal areas of the Oosterschelde on distribution, density and biomass of some common macrobenthic species: Spio martinensis (Mesnii), Hydrobia ulvae (Pennant), Arenicola marina (L.), Scoloplos armiger (Muller) and Bathyporeia sp.. Hydrobiologia (in press).

Cramp, S. & K. E. L. Simmons, 1977. Handbook of the birds of Europe, the Middle East, and North Africa: the birds of the Western Palearctic. Vol 1: Ostrich-Ducks. Oxford University Press, Oxford.

Cramp, S. & K. E. L. Simmons, 1983. Handbook of the birds of Europe, the Middle East, and North Africa: the birds of the Western Palearctic. Vol 3: Waders to gulls. Oxford University Press, Oxford.

Davidson, N. C., 1983. Formulae for estimating the lean weight and fat reserves of live shorebirds. Ring. & Migr. 4: 159–166.

Dekker, R., 1989. The macrozoobenthos of the subtidal western Dutch Wadden Sea. I. Biomass and species richness. Neth. J. Sea Res. 23: 57–68.

De Wilde, P. & J. Beukema, 1984. The role of zoobenthos in the consumption of organic matter in the Dutch Wadden Sea. In R. W. P. J. Laane & W. J. Wolff (eds), The role of organic matter in the Wadden Sea. NIOZ Publ. Ser. 10: 145–158.

Drent, R., B. Ebbinge & B. Weijand, 1978. Balancing the energy budgets of arctic-breeding geese throughout the annual cycle: a progress report. Verh. Orn. Ges. Bayern 23: 239–263.

Durell le V. dit, S. E. A. & J. D. Goss-Custard, 1984. Prey selection within a size-class of mussels, Mytilus edulis, by Oystercatchers, Haematopus ostralegus. Anim. Behav. 32: 1197–1203.

Evans, A., 1987. Relative availability of the prey of wading birds by day and by night. Mar. Ecol. Prog. Ser. 37: 103–107.

Evans, P. R., 1976. Energy balance and optimal foraging strategies in shorebirds: some implications for their distribution and movements in the non-breeding season. Ardea 64: 117–139.

Evans, P. R. & P. J. Dugan, 1984. Coastal birds: numbers in relation to food resources. In P. R. Evans, J. D. Goss-Custard & W. G. Hale (eds), Coastal waders and wildfowl in winter. Cambridge University Press, Cambridge: 8–28.

Evans, P. R., D. M. Herdson, P. J. Knights & M. W. Pienkowski, 1979. Short-term effects on reclamation of part of Seal Sands, Teesmouth, on wintering waders and shelduck. Oecologia 41: 183–206.

Esselink, P. & L. Zwarts, 1989. Seasonal trend in burrow depth and tidal variation in feeding activity of Nereis diversicolor. Mar. Ecol. Prog. Ser. 56: 243–254.

Goss-Custard, J. D., 1977. The ecology of the Wash. III Density-related behaviour and the possible effects of a loss of feeding grounds on wading birds (Charadrii). Appl. Ecol. 14: 721–739.

Goss-Custard, J. D., 1985. Foraging behaviour of wading birds and the carrying capacity of estuaries. In R. M. Sibly & R. H. Smith (eds), Behavioural Ecology. Blackwell Scientific Publications, Oxford: 169–188.

Goss-Custard, J. D. & S. E. A. Le V. dit Durell, 1990. Bird behaviour and environmental planning: approaches in the study of wader populations. Ibis 132: 273–289.

Kalejta, B., 1992. Time budgets and predatory impact of waders at the Berg River Estuary, South Africa. Ardea, 80: 327–342.

Kersten, M. & T. Piersma, 1987. High levels of energy expenditure in shorebirds; metabolic adaptations to an expensive way of life. Ardea 75: 175–187.

Klepper, O., M. W. M. van der Tol, H. Scholten & P. M. J. Herman, 1994. SMOES: a simulation model for the Oosterschelde ecosystem: Hydrobiologia 282/283: 437–451.

Lambeck, R. H. D., A. J. J. Sandee & L. De Wolf, 1989. Longterm patterns in the wader usage of an intertidal flat in the Oosterschelde (SW Netherlands) and the impact of the closure of an adjacent estuary. J. appl. Ecol. 26: 419–431.

Lambeck, R. H. D., 1991. Changes in abundance, distribution and mortality of wintering Oystercatchers after habitat loss in the Delta area, SW Netherlands. Acta XX Congr. int. ornithol.: 2208–2218.

Lasiewski, R. C. & W. R. Dawson, 1967. A re-examination of the relation between standard metabolic rate and body weight in birds. Condor 69: 13–23.

Leewis, R. J., H. J. M. Baptist & P. L. Meininger, 1984. The Dutch delta area. In P. R. Evans, J. D. Goss-Custard & W. G. Hale (eds), Coastal waders and wildfowl in winter. Cambridge University Press, Cambridge: 253–260.

Meininger, P. L., H. J. M. Baptist & G. J. Slob, 1984. Vogeltellingen in het Deltagebied in 1975/1976–1979/1980. Rijkswaterstaat Deltadienst, Middelburg & Staatsbosbeheer Zeeland, Goes, Nota DDMI-84.23.

Meininger, P. L., A.-M. Blomert & E. C. L. Marteijn, 1991. Watervogelsterfte in het Deltagebied, ZW-Nederland, gedurende de drie koude winters van 1985, 1986 en 1987. Limosa 64: 89–102.

Meire, P. M., 1987. Foraging behaviour of some wintering waders: prey selection and habitat distribution. In A. C. Kamil, J. R. Krebs & H. R. Pulliam (eds), Foraging behavior. Plenum Press, New York.

Meire, P. M., 1991. Effects of a substantial reduction in intertidal area on numbers and densities of waders. Acta XX Congr. int. Ornithol.: 2219–2227.

Meire, P. M. & J. Dereu, 1990. Use of the abundance/biomass comparison method for detecting environmental stress:

some considerations based on intertidal macrozoobenthos and bird communities. J. appl. Ecol. 27: 210–223.

Meire, P. M. & A. Ervynck, 1986. Are Oystercatchers (*Haematopus ostralegus*) selecting the most profitable mussels (*Mytilus edulis*)? Anim. Behav. 34: 1427–1435.

Meire, P. M. & E. Kuijken, 1984. Relations between the distribution of waders and the intertidal benthic fauna of the Oosterschelde, Netherlands. In P. R. Evans, J. D. Goss-Custard & W. G. Hale (eds), Coastal waders and wildfowl in winter. Cambridge University Press, Cambridge: 57–68.

Meire, P. M. & E. Kuijken, 1987. A description of the habitat and wader populations of the Slikken van Vianen (Oosterschelde, The Netherlands) before major environmental changes and some predictions on expected changes. De Giervalk 77: 283–311.

Meire, P. M., J. Seys, J. Buijs & J. Coosen, 1994. Spatial and temporal patterns of intertidal macrobenthic populations in the Oosterschelde: are they influenced by the construction of the storm-surge barrier? Hydrobiologia 282/283: 157–182.

Meire, P. M., J. Seys, T. Ysebaert, P. I. Meininger & H. J. M. Baptist, 1989. A changing Delta: effects of large coastal engineering works on feeding ecological relationships as illustrated by waterbirds. In J. C. Hooghart & C. W. S. Posthumus (eds), Hydro-ecological relations in the Delta waters of the South-West Netherlands. Lakerveld B.V., The Hague: 109–143.

Moser, M. E., 1988. Limits to the numbers of Grey Plovers *Pluvialis squatarol a* wintering on British estuaries: an analysis of long term population trends. J. appl. Ecol. 25: 473–485.

Mulder, J. P. M. & T. Louters, 1994. Changes in basin geomorphology after implementation of the Oosterschelde project. Hydrobiologia 282/283: 29–39.

Nienhuis, P. H. & A. C. Smaal, 1994. The Oosterschelde estuary, a case study of a changing ecosystem: an introduction. Hydrobiologia 282/283: 1–14.

Norstrom, R. J., T. P. Clark, J. P. Kearney & A. P. Gilman, 1986. Herring Gull energy requirements and body constituents in the Great Lakes. Ardea 74: 1–23.

Piersma, T., 1984. Estimating energy reserves of Great Crested Grebes *Podiceps cristatus* on the basis of body dimensions. Ardea 72: 119–126.

Piersma, T., 1987. Production by intertidal benthic animals and limits to their predation by shorebirds: a heuristic model. Mar. Ecol. Prog. Ser. 38: 187–196.

Raffaelli, D. & H. Milne, 1987. An experimental investigation of the effects of shorebird and flatfish predation on estuarine invertebrates. Estuar. coast. shelf Sci. 24: 1–13.

Reise, K., 1985. Tidal flat ecology – An experimental approach to species interactions. Springer-Verlag, Berlin Heidelberg.

Saeijs, H. L. & H. J. M. Baptist, 1977. Wetland criteria and birds in a changing Delta. Biol. Cons. 11: 251–266.

Sanchez-Salazar, M. E., C. L. Griffiths & R. Seed, 1987. The interactive roles of predation and tidal elevation in structuring populations of the edible cockle, *Cerastoderma edule*. Estuar. coast. shelf Sci. 25: 245–260.

Schekkerman, H., P. L. Meininger & P. M. Meire, 1994. Changes in the waterbird populations of the Oosterschelde (SW Netherlands) as a result of large-scale coastal engineering works. Hydrobiologia 282/283: 509–524.

Schneider, D. C. & B. A. Harrington, 1981. Timing of shorebird migration in relation to prey depletion. The Auk 98: 801–811.

Seys, J. J., P. M. Meire, J. Coosen & J. A. Craeymeersch, 1994. Long-term changes (1979–89) in the intertidal macrozoobenthos of the Oosterschelde estuary: are patterns in total density, biomass and diversity induced by the construction of the storm-surge barrier? Hydrobiologia 282/283: 251–264.

Smaal, A. & R. Boeije, 1991. Veilig getij, de effecten van de waterbouwkundige werken op het getijdenmilieu van de Oosterschelde. Nota GWWS 91.088 DGW/directie Zeeland, Middelburg.

Smit, C. J., 1981. Production of biomass by invertebrates and consumption by birds in the Dutch Wadden Sea area. In W. J. Wolff (ed.), Ecology of the Wadden Sea. Balkema, Rotterdam: Vol. 3: 290–301.

Smit, C. J. & T. Piersma, 1989. Numbers, midwinter distribution, and migration of wader populations using the East Atlantic flyway. In H. Boyd & J.-Y. Pirot (eds), Flyways and reserve networks for water birds. IWRB special Publication No. 9, Slimbridge: 24–63.

Smith, P. C., 1975. A study of the winter feeding ecology of the Bar-tailed Godwit *Limosa lapponica*. Ph. D. Thesis, University of Durham.

Stuart, J. J., P. L. Meininger & P. M. Meire, 1989. Watervogels van de Westerschelde. University of Gent report WWE nr. 14; RWS nota AO 89.1010.

Summers, R. W., 1977. Distribution, abundance and energy relationships of waders (Aves: Charadrii) at Langebaan Lagoon. Trans. r. Soc. S. Afr. 42: 483–494.

Sutherland, W. J., 1982b. Spatial variation in the predation of cockles by Oystercatchers at Traeth Melynog, Anglesey II. The pattern of mortality. J. anim. Ecol. 51: 491–500.

Sutherland, W. J. & J. D. Goss-Custard, 1991. Predicting the consequence of habitat loss on shorebird populations. Acta XX congr. int. Ornithol.: 2199–2207.

Tubbs, C. R., J. M. Tubbs & J. S. Kirby, 1992. Dunlin *Calidris alpina alpina* in the Solent, southern England. Biol. Cons. 60: 15–24.

Van der Tol, W. M & H. Scholten, 1993. Response of the Eastern Scheldt ecosystem to a changing environment: functional or adaptive? Neth. J. Sea Res. 30: 175–190.

Van Impe, J., 1985. Estuarine pollution as a probable cause of increase of estuarine birds. Mar. Poll. Bull. 16: 271–276.

Van Stralen, M. R. & R. D. Dijkema, 1994. Mussel culture in a changing environment: the effects of a coastal engineering project on mussel culture (*Mytilus edulis* L.) in the Oosterschelde estuary (SW Netherlands). Hydrobiologia 282/283: 359–379.

546

Wiersma, P. & T. Piersma, (in press). Living exposed and in the cold: seasonal changes in the thermostatic costs of temperate-wintering Red knots. Condor.

Wiersma, P., L. Bruinzeel & T. Piersma, 1993. Energie besparing bij wadvogels: over de kieren van de Kanoet. Limosa. 66: 51–52.

Wilson, W. H. Jr., 1991. The foraging ecology of migratory shorebirds in marine soft-sediment communities: the effects of episodic predation on prey populations. Am. Zool. 31: 840–848.

Wolff, W. J., 1971. Changes in intertidal benthos communities after an increase in salinity. Thalassia Jugoslavica 7: 429–434.

Wolff, W. J. & L. De Wolf, 1977. Biomass and production of Zoobenthos in the Grevelingen estuary, The Netherlands. Estuar. coast. Mar. Sci. 5: 1–24.

Wolff, W. J., A. M. M. Van Haperen, A. J. J. Sandee, H. J. M. Baptist & H. L. F. Saeijs, 1976. The trophic role of birds in the Grevelingen estuary, The Netherlands, as compared to their role in the saline Lake Grevelingen. In G. Persoone & E. Jaspers (eds), Proceedings of the 10th European Symposium on Marine Biology. Universa Press, Wetteren Vol. 2: 673–689.

Wolff, W. J. & C. J. Smit, 1990. The Banc d'Arguin, Mauritania, as an environment for coastal birds. Ardea 78: 17–38.

Zwarts, L. 1974. Vogels van het brakke getij-gebied. Bondsuitgeverij van de jeugdbonden voor natuurstudie, Amsterdam.

Zwarts, L., 1988. Numbers and distribution of coastal waders in Guinea-Bissau. Ardea 76: 42–55.

Zwarts, L. & A.-M. Blomert, 1990. Selectivity of Whimbrels feeding on fiddler crabs explained by component specific digestibilities. Ardea 78: 193–208.

Zwarts, L. & J. Wanink, 1984. How Oystercatchers and Curlew successively deplete clams. In P. R. Evans, J. D. Goss-Custard & W. G. Hale (eds), Coastal waders and wildfowl in winter. Cambridge University Press, Cambridge: 69–83.

Zwarts, L. & J. Wanink, 1989. Siphon size and burying depth in deposit- and suspension feeding benthic bivalves. Mar. Biol. 100: 227–240.

Zwarts, L., A.-M. Blomert & J. H. Wanink, 1992. Annual and seasonal variation in the food supply harvestable for Knot *Calidris canutus* staging in the Wadden Sea in late summer. Mar. Ecol. Prog. Ser. 83: 128–139.

Hydrobiologia **282/283**: 547–555, 1994.
P. H. Nienhuis & A. C. Smaal (eds), The Oosterschelde Estuary.
© *1994 Kluwer Academic Publishers. Printed in Belgium.*

The harbour seal, *Phoca vitulina*, in the Oosterschelde: decline and possibilities for recovery

J. Mees[1] & P. J. H. Reijnders[2]
[1]*Marine Biology Section, Institute of Zoology, University of Ghent, Ledeganckstraat 35, B-9000 Gent, Belgium;* [2]*Aquatic Ecosystems, Institute for Forestry and Nature Research, P.O. Box 167, NL-1790 AD Den Burg, The Netherlands*

Key words: seals, *Phoca vitulina*, Oosterschelde

Abstract

Within a timespan of a few decades, the harbour seal almost completely disappeared from the estuaries in the south-west of the Netherlands. In 1960 a population of around 350 animals still lived in the Oosterschelde and Westerschelde area. About a quarter of this population lived in the Oosterschelde. At present less than 17 animals can be regularly observed in the whole area. Human influences are responsible for the rapid decline of the population. Initially a high hunting pressure and later environmental pollution are the main causes. Loss of habitat and disturbance at the resting places are additional important factors. The Oosterschelde still is a suitable habitat for seals. A short term natural development of a viable population in the area is not to be expected. Only with human help through active management, *i.e.* reintroduction of rehabilitated seals (preferably originating from that area) and strict conservation of the extant Oosterschelde seal population, accompanied by environmental sanitation of the neighbouring waters, can the current southern Dutch harbour seal population increase.

Introduction

As top predators seals play a significant indicative role for the quality of the marine ecosystem they live in. The integrity of marine ecosystems is of value to us and so, by implication, is the survival and well-being of marine mammals. This gives them a value for society which is hard to quantify but is nonetheless real (Holt, 1986). In most parts of their range European seal species are in need of active conservation to prevent their numbers diminishing (Summers *et al.*, 1978). This is certainly the case for the southern Dutch harbour seal population (Reijnders, 1985a, Reijnders *et al.*, 1990). In a management policy statement of the Dutch government for the Oosterschelde in

1982, it is clearly stated that the natural values of the system should be kept intact and, if necessary, improved. One of the elements important in this context is the restoration of a viable harbour seal population in the area.

The time period under consideration in this study is 1960 till present: from 1960 onwards, regular aerial surveys of the seal population in the Delta area have been carried out. This period coincides with the start of the major construction works aimed at safeguarding SW Netherlands from disastrous storm surges. A review of the changes in population size is presented. The causes for the decline and the possibilities for recovery are discussed.

General ecology of the harbour seal

The harbour seal occurring in the Oosterschelde belongs to the subspecies *P. vitulina vitulina*, which inhabits the northeastern part of the Atlantic Ocean from Iceland and Finland to the North Sea, the Irish Sea and the Baltic Sea. The most important populations are found around Iceland, in the Wash (UK), around the Orkney Islands, in the Wadden Sea and in the Kattegat-Skagerrak/ southern Norway area (Bonner, 1972, Reijnders & Lankester, 1990). Except some small colonies in France, the Dutch Delta area can be considered as the southern limit of its former geographical distribution. It is the only breeding marine mammal in these waters in recent times.

The animals are known to frequent estuaries and coasts where offshore sandbanks or rocks are regularly exposed at low tide. In areas with no tidal rhythm, a different regime more related to daylight can be observed. Common seals haul out on these spots to give birth and suckle their young, to rest and to moult. In most areas they are seen at the haul-out sites throughout the year, but their abundance varies in relation to factors such as season, time of day, tidal cycle and weather conditions (e.g. Godsell,1988; Boulva & McLaren, 1979; Thompson, 1989). Numbers are generally highest in summer.

It is difficult to obtain population estimates and other parameters necessary for accurate population dynamical studies (Thompson & Harwood, 1990). Even aerial counts seem to be too blunt an instrument as was shown by the recent seal plague when more animals died in the German Wadden Sea than the best estimates of the size of the local population size (Reijnders, 1988a, Harwood, 1989). Therefore all published figures have to be interpreted as minimum estimates of the total population size and can only indicate a trend in population size.

Seals are carnivorous mammals which subsist largely on fish though, at times, molluscs and crustaceans may form a significant part of their

Table 1. Prey species found in stomach and colon of 63 seals shot in the Delta area. Calculations from the data of Havinga (1933).

Species		Numbers	Weight (g)	Weight (%)
Platichthys flesus	< 14 cm	307	4605	15.6
	14–18 cm	62	2480	8.4
	> 18 cm	53	4505	15.2
Pleuronectes platessa	< 10 cm	145	722	2.4
	10–15 cm	24	480	1.6
Solea solea		13	569	1.9
Limanda limanda		28	1167	3.9
Myoxocephalus scorpius				
+ Agonus cataphractus	< 15 cm	103	2066	7.0
	> 15 cm	57	3990	13.5
Zoarces viviparus	< 13 cm	148	444	2.3
	> 13 cm	62	930	3.1
Clupea harengus	< 16 cm	38	190	> 1
	> 16 cm	7	87	< 1
Merlangius merlangus		17	1105	3.7
Gadus morhua		46	2658	9.0
Osmerus eperlanus		29	725	2.8
Pomatoschistus minutus		1567	1668	5.6
Belone belone		6	620	< 1
Engraulis encrasicolus		2	620	< 1
Ammodytes tobianus		53	620	< 1
Alosa fallax		4	620	< 1
Crangon crangon		512	540	1.8

diet. They are opportunistic feeders, eating a wide variety of fish species (Havinga, 1933, Sergeant, 1951; Rae, 1968; Harkonen, 1987; Sievers, 1989; Pierce *et al.*, 1990). No recent data on food preference of seals in the Oosterschelde are available. Havinga (1933) examined stomach contents of 63 seals shot in the Delta area throughout the year (Table 1). The main prey species of the harbour seal in the Delta area was flounder *Platichthys flesus* Linnaeus 1758 (39% on a weight basis). Other important species in their diet are bull rout *Myoxocephalus scorpius* Linnaeus 1758, and hooknose *Agonus cataphractus* Linnaeus 1758 (20%), cod *Gadus morhua* Linnaeus 1758 (9%), and gobies *Pomatoschistus species* (6%). Further, small quantities of whiting *Merlangius merlangus* Linnaeus 1758, plaice *Pleuronectes platessa* Linnaeus 1758, dab *Limanda limanda* Linnaeus 1758, sole *Solea solea* Linnaeus 1758, and a variety of other fish were found.

History of the common seal in the Oosterschelde

The estuaries in the south of the Netherlands are traditionally used by seals. Around 1900, approximately 11 500 seals lived in the Delta area, some 25% of this population in the Oosterschelde (Reijnders, 1992a). Van Bemmel (1956a) estimated the population at about 1000 animals in 1954, 25–30% of which lived in the Oosterschelde. He already noted that the numbers were decreasing (Van Bemmel, 1956b). Reijnders (1985a), reviewing Van Bemmels figures, agrees that during his research the population was already on the decline. However, in a later study it was concluded that Van Bemmels figures have to be upgraded to assess the actual numbers present (Reijnders, 1992a). From 1959 onwards, aerial surveys (van Haaften, 1974) showed a rapid decline in numbers of seals in the Delta area. Based on those actual surveys Reijnders (1985a) estimated the population size on 350 in 1959, down to 40 in 1969 and to less than 25 animals since 1974 (Fig. 1). In 1991 only 17 seals were counted in the Delta area, 7 of which in the Oosterschelde.

Possible causes of changes in the size of the population

Interactions with fisheries

Seals are regularly drowned after becoming entangled in fishing nets (drift nets, ghost nets). Nets made of synthetic fibre are more resistant than natural fibre nets (Bonner, 1972). Avoiding to discard old nets in the sea will help to reduce the incidence of these deaths.

Van Bemmel (1956b) already concluded that weir-fishery prevents the seals from reaching habitual resting places in the Oosterschelde. Especially fyke nets (also used in the Oosterschelde) seem to present a problem. In 1987 six seals drowned in the Ems estuary (De Boer, 1989). A simple solution like placing a large meshed 'keerwant' (allowing the fish to pass but preventing seals to enter) in the mouth of the fykes seems to reduce the risks for the seals significantly (Reijnders, 1985b).

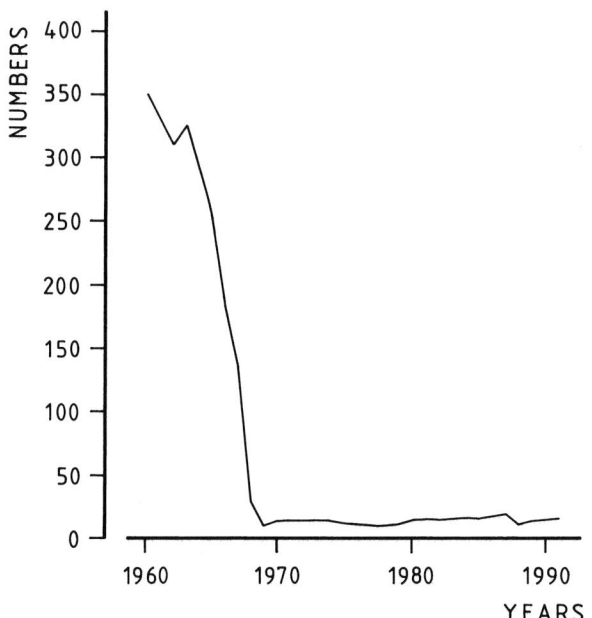

Fig. 1. The decline of the number of seals in the Delta area.

Hunting pressure

Seals have been persecuted by man for centuries, and also in the Delta area hunting was carried out on a large scale. It has been widely accepted that marine mammals, if they are exploited anyway, should be treated as renewable resources, and that only sustainable yields should be taken from their populations. Practice has usually been far from this ideal. The southern Dutch harbour seal population has clearly been over-exploited. This was stimulated by payment of bounties on the grounds of fisheries protection. In the early 1900s, when the number of common seals in the Delta was estimated at about 11 500 (Reijnders, 1994), the annual hunt was very high (*ca* 1000 animals). This slowly dropped to approximately 400 animals from 1920 till 1950. Then an increase took place to 700 animals in the mid 1950s, whereafter it declined again (Reijnders, 1994). After 1949 the bounty system was abolished and the demand for train-oil had ended. The result was that the interest in adults diminished and only pups were killed for the fur industry (Reijnders, 1976). This led to an increased population decline as was also noted by Van Bemmel who thought that if the population of about 1000 animals in the area was stable, the existing breeding stock could cope with an annual killing of 15% of the total population. In 1954 and 1955 the annual killing rate became much higher (up to 50%) and a sudden decrease in numbers became apparent (Van Bemmel, 1956b). In 1954 seals were included in the provisions of the hunting law and could only be taken under licence. The harbour seal became totally protected in 1962. The high hunting pressure caused a collapse of the population from 11 500 animals in 1900 to 350 in 1960 (Reijnders, 1994).

Environmental pollution

It is well established that environmental contaminants accumulate in marine mammals. For reviews of pollutants in seals we refer to Holden (1978) and Reijnders (1988b). Seals are particularly vulnerable to pollution because they are at the top of the marine food chain and therefore accumulate toxins. Furthermore, organochlorine compounds like the persistent polychlorinated biphenyls (PCBs) and DDT tend to accumulate extra in lipid tissue. Therefore concentrations of organochlorines are highest in organs or tissues with the highest lipid content. Muscle tissue and livers of the most important prey species of seals (fish) and the blubber and liver of seals itself are proven to act as storage sites of a wide variety of contaminants (Holden & Marsden, 1967; Ten Berge & Hillebrand, 1974; Reijnders 1980; Van Haren & Marquenie, 1988). Since aquatic mammals have a relatively high proportion of their body weight in the form of subcutaneous fat (blubber) they carry relatively high burdens of organochlorines. In the mid seventies high concentrations have been found in ringed seals *Phoca hispida* Schreber 1775, grey seals *Halichoerus grypus* Nilsson 1820, and harbour seals *Phoca vitulina* Linnaeus 1758 from the Baltic area and the Swedish west coast (Helle *et al.*, 1976a, b). In seals from the North Sea high concentrations have been detected in specimens from the Dutch coast (e.g. Koeman *et al.*, 1975), the east coast of Britain (e.g. Holden, 1978), and the German Wadden Sea (Drescher *et al.*, 1977). Organochlorine residues in male seals increase with the age of the animals (Addison & Brodie, 1977; Reijnders, 1980). This relationship is not established for female seals in low to medium polluted areas, suggesting that these can excrete the residues by some route(s) closed to the male (Addison & Brodie, 1977; Boon *et al.*, 1987). The most obvious mechanisms are parturition and lactation.

The contaminants are released when the animals mobilise their lipid reserves at times of stress, such as the pupping and lactation season or during the moult (Harwood & Reijnders, 1988). The recorded concentrations may be sufficient to result in some additional stress to the animals in case of fat mobilisation during illness or starvation (Law *et al.*, 1989).

High concentrations of certain organochlorines and metals have since long been associated with reproductive failure and increased juvenile mortality in marine mammals (De Long *et al.*, 1973; Helle *et al.*, 1976a, b). The most extensive study of these effects was carried out by Reijnders in the Dutch Wadden Sea. The harbour seal population in the Dutch part of the Wadden Sea has declined from about 17 000 in 1900 to about 900 animals in 1959 (Reijnders, 1976, 1992). Like in the Delta area, this decline was only caused by overhunting. However, after hunting was restricted and finally banned the population continued to decrease after an initial rise from 1959 to 1964. This decline was predominantly caused by a very low reproductive rate of seals from this part of the Wadden Sea (Reijnders, 1978, 1980). Reijnders (1978) showed that in the Dutch Wadden Sea population pup production was lower and juvenile mortality was higher than for the stable populations of Schleswig-Holstein and Denmark. The observed decrease in reproductive success correlated strongly with the high concentrations of PCBs in the tissues. Thus these environmental pollutants, mainly originating from the heavily polluted rivers Rhine and Meuse, have been implicated as possible causative agents of the drastic reduction in the harbour seal population in the Dutch Wadden Sea (Reijnders, 1980, 1981a; Reijnders *et al.*, 1981). Later Reijnders (1986) was able to demonstrate a causal relationship between naturally occurring levels of PCBs and a physiological response of the seals. He showed that harbour seals fed on PCB-contaminated fish from the polluted part of the Wadden Sea had significantly lower reproductive success than seals fed 'clean' fish. In a cooperative study an important role of vitamin A in PCB-induced reproductive failure in common seals was found. Vitamin A plays an important role in resistance to infections. PCB effects may thus be accompanied by a weakened condition and increased susceptibility to viral infections (Brouwer *et al.*, 1989).

Since the closing of some upstream tributaries there is no direct input of polluted fresh water in the Oosterschelde. Consequently water quality is relatively good. Yet, in the period 1980–1987 concentrations of PCB-153 in Oosterschelde mussels were so high that the reproduction of seals could be expected to be disturbed (Vonk, 1990). Since then the situation is improving. It should be noted here that the neighbouring Westerschelde estuary is still heavily polluted (Van Eck *et al.*, 1991). A clean Oosterschelde seems to be of limited value to the seals without an equally clean Westerschelde. Since both systems are very close together and since some traditional resting places of the Southern Dutch harbour seal population are (were) located in the Westerschelde, important migrations of animals from Ooster- to Westerschelde are to be expected once an Oosterschelde population is established.

Disturbance and loss of habitat

After the total protection of the harbour seal in 1962 no increase was found in the Delta area, similar to what happened in the Wadden Sea (Reijnders, 1976). Besides the effects of pollution, an additional factor could be the start of the engineering works within the frame of the Delta plan, which destroyed much of the habitat and caused great disturbance. However, the impact of these activities and the recreational disturbances are presumably only important for the last phase in the decline (1965 and onwards) because a major part of the population had already disappeared before the works started. A more important aspect of disturbance is probably the negative influence on recolonization possibilities. Many of the traditional resting places disappeared. In the mouth of the Grevelingen resting seals were often observed before the start of the construction works in 1965. Soon thereafter they disappeared. They might have moved to the heavily polluted Westerschelde that year, as numbers increased there (Benschop & Van Haperen, 1988).

The influence of disturbance from the construction works has decreased since 1986, but disturbance due to recreational activities has increased significantly during the last decade (Reijnders, 1985a). Besides this, there is also disturbance from professional traffic such as fisheries, shipping and research vessels. Marine mammals are also tourist attractions. There is some fear that increasing tourism will interfere with the vital activities of the animals. The critical period for the seals is mid June to the end of August. During the whelping and nursing period seals are highly dependent on the sandbanks during low tide. Seals are often forced to leave the tidal flats when disturbed by boats. This results in less time for nursing and resting and possibly in increased pup mortality. Disturbance also plays a significant role in the occurrence of skin lesions. Mechanical injury to the nearly healed umbilicus with consequent infection of the wound may also occur (Reijnders, 1981b).

Chances for recovery

Effects of the seal plague

In the summer of 1988 more than 17000 common seals died around the northwest coasts of Europe (Dietz et al., 1989). Some 60% of the Wadden Sea population is believed to have died. In 1991, approximately 6100 seals were counted in the Wadden Sea, 750 in the Dutch part (Reijnders, 1992). The primary cause of the seal plague has been identified as infection with a previously undiscribed virus from the morbilli genus, the phocine distemper virus (Cosby et al., 1988; Kennedy et al., 1988; Mahy et al., 1988; Osterhaus 1988; Osterhaus & Vedder, 1988; Osterhaus et al., 1988). These and several other authors have related the severity of the outbreak to pollution in the North Sea (Holt, 1989). As mentioned above, there is strong evidence that the high levels of organochlorines which have been recorded in the blubber of seals in the Dutch Wadden Sea and the Baltic Sea have contributed to the low fertility which has been observed in these areas. In the recent disease outbreak attention has again focused on PCBs. PCB effects may be accompanied by a weakened condition and increased susceptibility to viral infections (Brouwer et al., 1989). These compounds have been demonstrated to have an immunosuppressive effect and thus to increase vulnerability to infection (Wasserman et al., 1979). Whatever the proximate or ul-

timate causes of the epidemic were, any positive trends in the population developments of the common seal have been abruptly interrupted. The plague not only was a catastrophe for the entire Wadden Sea population itself, but especially for areas with lower birth rates, i.e. the Dutch Wadden Sea and the Delta area. The chances for a short term natural recolonization of the Oosterschelde are seriously reduced, since the establishment of a breeding stock of seals in the Oosterschelde is highly dependent on the immigration of young animals from the Dutch Wadden Sea. In the next few years there will also be fewer animals available from the nursery stations, limiting possibilities for large scale reintroduction programs (Reijnders et al., 1990).

Furthermore the Delta area itself was not isolated from the virus. The effects were not massive because of the very low population density, but still some animals were found in that region which died because of virus infection. The well-being of a possible future Oosterschelde population will largely depend on the evolution of the disease-cycle. If the plague is recurrent, the influx will be limited and only a marginal recovery will occur, if at all.

Management and conservation

In the period 1980 to 1988 there was a clear recovery of the seal population in the Dutch Wadden Sea. This was mediated by an increased immigration from the German and Danish parts where hunting ceased in the mid seventies (Reijnders, 1983, 1988a) and where pup production was higher than in the Dutch part. The protection of resting areas, the release of rehabilitated seals by nursery stations and the release of animals born in captivity contributed significantly to population growth by improving recruitment. The slight increase in the number of seals in the Delta area observed in these years is believed to be predominantly connected with increased numbers in the Wadden Sea. Because of the already mentioned dependency on dispersal of young animals from the Wadden Sea, the fate of the Oosterschelde population is closely related to the condition of Wadden Sea population.

The perspective of a natural recovery to a viable population is temporarily ruined by the epidemic in 1988. A short term natural development of this population is not to be expected because of the seal plague in the North sea in 1988 and consequent massive death (Reijnders et al., 1990). Long term perspectives for natural recovery are good providing the Wadden Sea population increases and water quality of the North sea, the Rhine, the Meuse and the Westerschelde improves. PCB levels in the Delta area are not dramatically high. The situation is not worse than in the Wadden Sea but still they are high enough to hamper reproduction.

The release of rehabilitated animals into the Oosterschelde seems to be a feasible way for a recovery on the short term. Eventually, a growing seal population in the Oosterschelde could be a centre for dispersion to other parts of the Delta area. A first experiment has been conducted (Reijnders et al., 1990). The introduction was successful but the authors conclude that disturbance by boats, airplanes and other professional and recreational activities in the Oosterschelde are of such a level that the quality of the habitat is too low to continue without accompanying additional conservational measures. Rest for the animals has to be guaranteed. Management of the seal population should include undisturbed rearing areas for the pups and reserves where the animals can rest at all times. Measures like increasing minimal flying height may also prove beneficial (De Vlas, 1988). Reijnders et al. (1990) also stress the importance of a 'keerwant' in all fykes to prevent drowning of seals.

Conclusions

The Oosterschelde still is a potentially suitable habitat for the harbour seal. Water quality is not poor. Although some of the traditional resting places disappeared due to the engineering works, numerous large sandbanks are still present and there is a rich food supply. The deep gullies on the seaward side of the storm surge barrier will not disappear. The barrier itself seems not to be a physical problem for the seals (Reijnders et al., 1990). In the Oosterschelde and the adjoining part of the Voordelta there is in principle room for 2000–2500 seals. Negative prospects for a successful recolonisation are the increasing recreation pressure and the ongoing heavy pollution in the Westerschelde. If the environmental conditions of neighboring waters improve and resting places in the Oosterschelde and parts of the Voordelta would be safeguarded, there is little doubt that a breeding population in the Delta area can develop. A short term natural colonization leading up to a viable seal population is, in view of the recent seal plague, not to be expected. A well planned reintroduction of rehabilitated seals in the area (preferably originating from there) can be a good strategy providing the conditions mentioned above are fulfilled and the introduction is accompanied by appropriate conservation measures.

Acknowledgements

We would like to thank Henk Zandstra for recent information on the number of seals and Hans Schoonheden for the drawing.

References

Addison, R. F. & P. F. Brodie, 1977. Organochlorine residues in maternal blubber, milk, and pup blubber from grey seals (Halichoerus grypsus) from Sable Island, Nova Scotia. J. Fish. Res. Bd Can. 34: 937–941.

Benschop, H. & A. Van Haperen, 1988. Zeehonden in de Zeeuwse wateren. Stichting Natuur- en Recreatieinformatie, Middelburg, 31 pp.

Bonner, W. N., 1972. The grey seal and harbour seal in European waters. Oceanogr. Mar. Biol. annu. Rev. 10: 461–507.

Boon, J. P., P. J. H. Reijnders, J. Dols, P. Wensvoort & M. Th. J. Hillebrand, 1987. The kinetics of individual polychlorinated biphenyl congeners in female harbour seals (Phoca vitulina), with evidence for structure-related metabolism. Aquat. Toxicol 10: 307–324.

Boulva, J. & I. A. McLaren, 1979. Biology of the harbour seal, Phoca vitulina, in eastern Canada. Bull. Fish. Res. Bd Can. 200: 24 pp.

Brouwer, A., P. J. H. Reijnders & J. H. Koeman, 1989. Polychlorinated biphenyl (PCB)-contaminated fish induces vitamin A and thyroid hormone deficiency in the common seal (*Phoca vitulina*). Aquatic Toxicology 15: 99–106.

Cosby, S. L., S. McQuaid, N. Duffy, C. Lyons, B. K. Rima, G. M. Allan, F. McNeilly & C. Craig, 1988. Characterization of a seal morbillivirus. Nature 336: 115–116.

De Boer, J., 1989. Verdrinking van zeehonden in fuiken. Waddenbulletin 24: 90–91.

De Long, R. L., W. G. Gilmartin & G. Simpson, 1973. Premature births in California sea lions. Association with high organochlorine pollutant residue levels. Science 181: 1168–1170.

De Vlas, J., 1988. Gevolgen van een virusepidemie: konsekwenties voor het beheer? Waddenbulletin 23: 204–207.

Dietz, R., M.-P. Heide-Jorgensen & T. Harkonen, 1989. Mass deaths of harbour seals (*Phoca vitulina*), in Europe. Ambio 18: 258–264.

Drescher, H. E., E. Huschenbeth & U. Harms, 1977. Organochlorines and heavy metals in the harbour seal (*Phoca vitulina*) from the German North Sea coast. Mar. Biol. 41: 99–106.

Godsell, J., 1988. Herd formation and haul-out behaviour in harbour seals (*Phoca vitulina*). J. Zool. 215: 83–98.

Harkonen, T. J., 1987. Seasonal and regional variations in the feeding habits of the harbour seal, *Phoca vitulina*, in the Skagerrak and Kattegat. J. Zool. 213: 535–543.

Harwood, J., 1989. Lessons from the seal epidemic. New Scientist 18 February 1989: 38–42.

Harwood, J. & P. Reijnders, 1988. Seals, sense and sensibility. New Scientist 15 October 1988: 28–29.

Havinga, B., 1933. Der Seehund (*Phoca vitulina* L.) in den Holländischen Gewässern. Tijdschr. ned. dierk. Ver (Leiden) 3: 79–111.

Helle, E., M. Olsson & S. Jensen, 1976a. DDT and PCB levels and reproduction in ringed seal from the Bothnian Bay. Ambio 5: 188–189.

Helle, E., M. Olsson & S. Jensen, 1976b. PCB levels correlated with pathological changes in seal uteri. Ambio 5: 261–263.

Holden, A. V., 1978. Pollutants and seals-a review. Mammal rev. 8: 53–66.

Holden, A. V. & K. Marsden, 1967. Organochlorine pesticides in seals and porpoises. Nature 216: 1274–1276.

Holt, S. J., 1986. Mammals in the sea. Ambio 15: 126–133.

Holt, S. J., 1989. Pollution and the seal epidemic. The Ecologist 19: 124.

Kennedy, S., J. A. Smyth, P. F. Cush, S. J. McCullough, G. M. Allan, 1988. Viral distemper now found in porpoises. Nature 336: 21.

Koeman, J. H., W. S. M. Van de Ven, J. J. M. De Goey, P. S. Tjioe & J. L. Van Haaften, 1975. Mercury and selenium in marine mammals and birds. Sci. Total Envir. 3: 279–287.

Law, R. J., C. R. Allchin & J. Harwood, 1989. Concentrations of organochlorine compounds in the blubber of seals from eastern and north-eastern England, 1988. Mar. Pollut. Bull. 20: 110–115.

Mahy, B. W. J., T. Barrett, S. Evans, E. C. Anderson & C. J. Bostock, 1988. Characterization of a seal morbillivirus. Nature 336: 115–116.

Osterhaus, A. D. M. E., 1988. Seal death. Nature 334: 301–302.

Osterhaus, A. D. M. E. & E. J. Vedder, 1988. Identification of a virus causing recent seal death. Nature 335: 20.

Osterhaus, A. D. M. E., J. Groen, P. De Vries, F. G. C. M. UytdeHaag, B. Klingeborn & R. Zarnke, 1988. Canine distemper virus in seals. Nature 335: 403–404.

Pierce, G. J., P. R. Boyle & P. M. Thompson, 1990. Diet selection by seals. In: Trophic Relationships in the Marine Environment, Proc. 24th EMBS, M. Barnes & R. N. Gibson (eds): 222–238.

Rae, B. B., 1968. The food of seals in Scottish waters. Mar. Res. 2: 1–23.

Reijnders, P. J. H., 1976. The harbour seal (*Phoca vitulina*) population in the Dutch Wadden Sea: size and composition. Neth. J. Sea Res. 10: 223–235.

Reijnders, P. J. H., 1978. Recruitment in the harbour seal (*Phoca vitulina*) population in the Dutch Wadden Sea. Neth. J. Sea Res. 12: 164–179.

Reijnders, P. J. H., 1980a. Organochlorine and heavy metal residues in the harbour seals in the Dutch Wadden Sea and their possible effects on reproduction. Neth. J. Sea Res. 14: 30–65.

Reijnders, P. J. H., 1980b. De zeehond als indikator voor het milieu. Waddenbulletin 15: 92–94.

Reijnders, P. J. H., 1981a. Threats to the harbour seal population in the Wadden Sea. In: Reijnders, P. J. H. & Wolff, W. J. (eds), Marine mammals in the Wadden Sea. Balkema Rotterdam: 38–47.

Reijnders, P. J. H., 1981b. Management and conservation of the harbour seal, *Phoca vitulina*, population in the international Wadden Sea area. Biol. Conserv. 19: 213–221.

Reijnders, P. J. H., 1983. The effect of seal hunting in Germany on the further existence of a harbour seal population in the Dutch Wadden Sea. Z. Säugtierkunde 48: 50–54.

Reijnders, P. J. H., 1985a. On the extinction of the Southern Dutch harbour seal population. Biol. Conserv. 31: 75–84.

Reijnders, P. J. H., 1985b. Verdrinking van zeehonden in fuiken. RIN-rapport 85/19. Rijksinstituut voor Natuurbehoud, Texel.

Reijnders, P. J. H., 1986. Reproductive failure in common seals feeding on fish from polluted coastal waters. Nature 324: 456–457.

Reijnders, P. J. H., 1988a. Gevolgen virusuitbraak voor zeehonden in het internationale waddengebied. Waddenbulletin 23: 201–203.

Reijnders, P. J. H., 1988b. Accumulation and body distribution of xenobiotics in marine mammals. In: W. Salomons, B. Bayne, E. Duursma & U. Foerstner (eds), Pollution of the North Sea: an assessment. Springer Verlag, Heidelberg: 596–603.

Reijnders, P. J. H., 1992. Retrospective analysis and related future management perspectives for the harbour seal (*Phoca vitulina*) in the Wadden Sea. In: N. Dankers, C. J. Smit & M. Scholl (eds), Proceedings of the 7th International Wadden Sea Symposium, Ameland, The Netherlands, 22–26 Oct 1990. Neth. Inst. Sea Res., Spec. Publ. Ser. nr. 20: 193–197.

Reijnders, P. J. H., 1994. Historical population size of the harbour seal, *Phoca vitulina*, in the Delta area, SW.Netherlands. Hydrobiologia 282/283: 557–560.

Reijnders, P. J. H., H. E. Drescher, J. L. Van Haaften, E. Bogebjerg Hansen & S. Tougaard, 1981. Population dynamics of the harbour seal in the Wadden Sea. In: P. J. H. Reijnders & W. J. Wolff (eds), Marine mammals of the Wadden Sea, Balkema, Rotterdam: 19–32.

Reijnders, P. J. H. & K. Lankester, 1990. Status of marine mammals in the North sea. Neth. J. Sea Res. 26: 427–435.

Reijnders, P. J. H., I. M. Traut & E. H. Ries, 1990. Verkennend onderzoek naar de mogelijkheden voor het terugzetten van gerevalideerde zeehonden, *Phoca vitulina*, in de Oosterschelde. RIN–rapport 90/10, Rijksinstituut voor Natuurbeheer, Texel: 33 pp.

Sergeant, D. E., 1951. The status of the common seal (*Phoca vitulina*) on the East Anglian coast. J. mar. biol. Ass. U.K. 29: 707–717.

Sievers, U., 1989. Nahrungsökologische Untersuchungen an Seehunden (*Phoca vitulina*, Linne 1758) aus dem schleswigholsteinischen Wattenmeer. Zool. Anz. 222: 249–260.

Summers, C. F., W. N. Bonner & J. L. Van Haaften, 1978. Changes in the seal populations of the North Sea. Rapp. P.-v. Réun. Cons. perm. int. Explor. Mer 172: 278–285.

Ten Berge, W. F & M. Hillebrand, 1974. Organochlorine compounds in several marine organisms from the North sea and the Dutch Wadden Sea. Neth. J. Sea Res. 8: 361–368.

Thompson, P. M., 1989. Seasonal changes in the distribution of common seal *Phoca vitulina* haul-out groups. J. Zool. 217: 281–294.

Thompson, P. M. & J. Harwood, 1990. Methods for estimating the population size of common seals, *Phoca vitulina*. J. Appl. Ecol. 27: 924–938.

Van Bemmel, A. C. V., 1956a. Planning a census of the harbour seal (*Phoca vitulina* L.) on the coasts of the Netherlands. Beaufortia 54: 121–132.

Van Bemmel, A. C. V., 1956b. Zeehonden in Nederland. De Levende Natuur 59: 1–12.

Van Eck, G. T. M, N. de Pauw, M. van den Langenbergh & G. Verreet, 1991. Emissies, gehalten, gedrag en effecten van (micro)verontreinigingen in het stroomgebied van de Schelde en het Schelde-estuarium. Water 60: 164–181.

Van Haaften, J. L., 1974. Zeehonden langs de Nederlandse kust. Wet. Med. K.N.N.V. 101: 36 pp.

Van Haren, R. J. F. & J. M. Marquenie, 1988. Pas op voor de zeehonden, een prognose voor 1995. Rijkswaterstaat, Dienst Getijdewateren, Nota GWAO-88.011.

Wasserman, M., D. Wasserman, S. Cusos & H. J. Miller, 1979. World PCBs map: storage and effect on man and his biologic environment in the 1970s. In: W. J. Nicholson & J. A. Moore (eds), Ann. N. Y. Acad Sci. 120: 69–124.

Hydrobiologia **282/283**: 557–560, 1994.
P. H. Nienhuis & A. C. Smaal (eds), The Oosterschelde Estuary.
© *1994 Kluwer Academic Publishers. Printed in Belgium.*

Historical population size of the harbour seal, *Phoca vitulina*, in the Delta area, SW Netherlands

P. J. H. Reijnders
Institute for Forestry and Nature Research, Department of Aquatic Ecology, P.O. Box 167, 1790 AD, Den Burg, The Netherlands

Key words: seals, *Phoca vitulina*, population trends, Dutch Delta area

Abstract

Seals in the Dutch Delta area have been subject to hunting pressure for centuries, promoted by a bounty system which generated a sort of hunting statistics. Hunting mortality is used to estimate historical population size. Based on ranges for most likely net recruitment rates, corresponding population trajectories are back calculated from an assessed population size of 350 seals in 1960. It is concluded that the size of the harbour seal population in the Delta area in 1900 will have been close to 11 500 animals. Significant loss of habitat has occurred due to closing off parts of the larger estuaries and the enlargement of the entrance to the harbour of Rotterdam in the 1960s and early 1970s. It is estimated that about 4000 harbour seals could inhabit the remaining Delta area under tidal influence. This outcome, based on retrospective population analysis, will be an important reference in defining management objectives for the recovery of the harbour seal population in the Delta area, which amounted to 18 seals in 1992.

Introduction

Only 18 harbour seals are known to inhabit the Dutch Delta area in 1992. However, for an adequate ecologically based management of the seals in this area, it is important to know how many animals would naturally occur there. Therefore it is necessary to assess the historical population size for a period when the pressure of recent threats such as pollution, landreclamation, damming up and disturbance was much less. Aerial surveys started in 1960 and from then onwards more accurate population assessment became possible. Therefore, changes in the size of the harbour seal population in the Delta area will be estimated from the beginning of this century until 1960.

Data on abundance and distribution provide a reference value for future management policy. As to how far that reference value will become the target value, is *i.a.* depending on possible changes in environmental conditions (see Dankers *et al.*, 1990). Furthermore, many areas have become more multifunctional in use and this implies that a decision about the final target value will be based on political policy.

Methods and data

Basic equations of demography require a detailed knowledge on age-specific survival and fecundity parameters (Lotka, 1907; Leslie, 1945). Information on age structure for the harbour seal population in the Delta area is lacking and therefore a simplified population model is applied analogous to calculations carried out for the seal population in the Wadden Sea (Reijnders, 1992). The equation reads:

$$N_{t+1} = N_t - K_T + R_T(N_t - FK_T).$$

The population size N at a certain time in year $t + 1$ equals the population size at the same time in year t, minus hunting mortality K_T plus the net recruitment of the population R in year T, times population size in year t minus FK_T. This last term is introduced to account for the number of females assumed to reproduce before being shot in that year. In the equation, N_t is a state variable, K_T is a forcing function and R_T is a parameter vector determined by the environmental conditions and population size. The equation is used in this study to carry out back-calculations, starting with an initial state of the population in 1960. To calculate former population trends, the net recruitment which is equal to the rate of increase, the initial size of the population, and the hunting mortality are needed. The factor F is calculated as follows: If catch i is the fraction of the annual catch in a given month i, and $pups_i$ is the fraction pups that is born in a month i, then the fraction pups that is born from month m onwards equals:

$$F = \sum_{m=1}^{12} (catch_m \sum_{i=m}^{12} pups_i).$$

The catch has to be averaged between a given month m and the preceeding month m-1, since an estimate of the catch at the beginning of each month is relevant. The factor F is then calculated according to the equation:

$$F = \sum_{m=1}^{12} \left[\frac{(catch_m + catch_{m-1})}{2} \sum_{i=m}^{12} pups_i \right].$$

The factor F used in this study is calculated to be 0.77, based on data from hunting records of Denmark (Søndergaard, 1976) and the Netherlands (Havinga, 1933).

The data used for calculations are derived from hunting statistics, which are based on a bounty system and available from about 1900 onwards (Havinga, 1933; Meyer, 1974). In cases where no data are available, the average cull from the surrounding periods is used. The population size in 1960 is estimated to amount to 350 animals (Reijnders, 1985).

Results and discussion

The recorded hunting data can not directly be used as hunting mortality since not all killed seals are retrieved and not all retrieved seals are reported. The loss rate due to sinking is assumed to be similar over the years, because the majority of seals was shot during the same period of the year. Based on information from former seal hunters in the Wadden Sea and Delta area and from experience elsewhere (Härkönen, 1987), a minimum loss rate of 25% due to sinking is calculated and the hunting data are accordingly adjusted. However, they still are an underestimation since the number of retrieved and not reported seals is unknown and no correction factor could be applied to account for that. The estimated numbers of seals killed in the Delta area between 1900 and 1960, are given in Fig. 1. Certain trends can be observed. High culls in the beginning of the century gradually decreased until about 1920, followed by stabilization until 1940. Thereafter some

Fig. 1. Estimated number of harbour seals hunted in the Delta area.

decrease occurred until the mid-1950s, when during some years an increase is observed followed by a sharp decline towards 1960. The sensitivity of the model for changes in the estimated initial population size in 1960 as well as the correction factor F, was tested. With an assumed underestimation of the initial population size (aerial survey in 1960) by 30% (based on Tougaards correction factor, see further on) and a constant F, the difference in calculated population size in 1900 was only 0.01%. With three different F values (0.7, 0.8 and 0.9) and a constant initial population size, the difference in calculated population size in 1900 was 1.7% at maximum. The calculations are obviously strongly governed by the hunting mortality and changes in both variables have a minor influence on the obtained population size in 1900.

The rate of increase for an exploited harbour seal population is estimated to vary between 0.05 and 0.12 (Härkönen, 1987). The rate of increase is probably density dependent and will have changed during the period 1900–1960. For the Delta population around 1900 it is assumed to be similar to the value (0.05) found for the Wadden Sea (Reijnders, 1992) and the Kattegat-Skagerrak (Härkönen, 1987). For the 1950s, the rate of increase in the Dutch seal population can not be used as a reference because pup production is too low due to the impact of pollution (Reijnders, 1986). Therefore the value (0.09) calculated for the seal population in the northern (Schleswig Holstein) Wadden Sea after cessation of hunting, is used instead. It is assumed that R_T will have been close to 0.05 at the beginning of the century, arriving via a sliding scale to 0.09 in the early 1950s. Therefore three population trajectories for R_T respectively 0.05, sliding from 0.05 to 0.09 and 0.12, are shown in Fig. 2. If the rate of increase had been 0.05, the estimated population size in 1900 would have been around 11 700 animals. With a rate of increase sliding from 0.05 to 0.09, the corresponding size is 11 500 animals. Should the rate of increase have been 0.12, the population would have numbered 6500 specimen. There is a population estimate (Havinga, 1933) that enables to check the likelihood of the different trajectories. In order to compare the figure provided

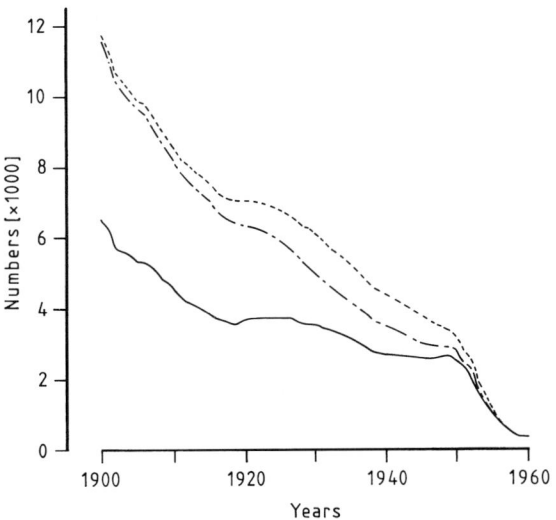

Fig. 2. Calculated historical population size of harbour seals in the Delta area, for different values of population rate of increase (R_T). – – –: $R_T = 0.05$; –––– –: $R_T = 0.05$ in 1900 sliding to 0.09 in 1960; –––––: $R_T = 0.12$.

by Havinga with the data from this study, Havinga's figure has to be corrected. Firstly by a factor to arrive at the correct age-class composition (Mohr, 1952) followed by an additional correction for the percentage of animals that are not present during surveys (Tougaard, in prep.). The revised estimate fits best (less than 10% difference) with those results from this study where a sliding scale for R_T is applied. Another indication can be obtained by relating the rate of decline in hunting records with the rate of decline in population size. If hunting interest and efficiency did not change over a certain period, the number of seals shot would be proportional to the numbers available. It is assumed that for the period 1900–1930, hunting interst did not change much (Havinga, 1933). However, after World War I, hunting efficiency has increased because of improved rifles as well as ammunition. This could explain the increase in number shot despite a declining population. Therefore the rate of decline in both the hunting record and the population size between 1900 and 1920 are compared and it was found that the population trajectory with the sliding R_T scale fits better in the mentioned proportional hunting concept. It is concluded that the trajectory with the sliding scale for R_T could reflect the changes in the harbour seal population in the Delta area between 1900 and 1960.

Large parts of the Delta area are no longer available for harbour seals. The actual habitat loss in square kilometers is about 50%. If the distribution and number of seals in all areas of the Delta area between 1930–1960 is taken into consideration and compared with the present available areas, a reduction by approximately 60% has occurred. Taking that habitat loss into account and the fact that tidal flats in the back of the Oosterschelde area seem to have become more muddy, a population size of around 4000 seals could inhabit the Delta area. Whether an increase from the present low numbers, 18 in 1992, to that figure will be realised is depending on many environmental conditions as explained in Mees & Reijnders (1994). The figure of 4000 seals should therefore not *per se* be interpreted as a management goal, but its order of magnitude is an important reference in defining management objectives for the recovery of the harbour seal population in the Delta area.

References

Dankers, N., K. S. Dijkema, P. J. H. Reijnders & C. J. Smit, 1990. De Waddenzee in de toekomst – waarom en hoe te bereiken? RIN-rapport 90/19. RIN-Texel.

Härkönen, T., 1987. Feeding ecology and population dynamics of the harbour seal (*Phoca vitulina*) in Kattegat-Skagerrak. Thesis Univ. Göteborg, Sweden.

Havinga, B., 1933. Der Seehund in den holländischen Gewässern. T. ned. dierk. Ver. 3: 79–111.

Leslie, P. H., 1945. On the use of matrices in population mathematics. Biometrika 33: 183–212.

Lotka, A. J., 1907. Relationship between birth rates and death rates. Science 26: 21–22.

Mees, J. & P. J. H. Reijnders, 1994. The harbour seal, *Phoca vitulina*, in the Oosterschelde: decline and possibilities for recovery. Hydrobiologia 282/283: 547–555.

Meyer, F. W., 1964. De zeehond aan onze kust. Ned. Jager 69: 626–627.

Mohr, E., 1952. Die Robben der europäischen Gewässer. P. Schöpps, Frankfurt/Main. 283 p.

Reijnders, P. J. H., 1985. On the extinction of the Southern Dutch harbour seal population. Biol. Conserv. 31: 75–84.

Reijnders, P. J. H., 1986. Reproductive failure in common seals feeding on fish from polluted coastal waters. Nature 324: 456–457.

Reijnders, P. J. H., 1992. Retrospective population analysis and related future management perspectives for the harbour seal *Phoca vitulina* in the Wadden Sea. In: N. Dankers, C. J. Smit & M. Scholl (eds), Proceedings of the 7th International Wadden Sea Symposium, Ameland, The Netherlands, 22–26 Oct., 1990. Neth. Inst. Sea Res., Publ. Ser. No 20: 193–197.

Søndergaard, N. O., A. H. Joensen & E. Bøgebjerg Hansen, 1976. Saelernes forekomst og saeljagten i Danmark. Danske Vildtunders. 26, 80 p.

Theme VIII:

The Oosterschelde estuary: an evaluation of changes

Hydrobiologia **282/283**: 563–574, 1994.
P. H. Nienhuis & A. C. Smaal (eds), The Oosterschelde Estuary.
© *1994 Kluwer Academic Publishers. Printed in Belgium.*

Policy planning in the Oosterschelde estuary

C. J. van Westen [1] & C. J. Colijn [2]
[1] *National Institute for Coastal and Marine Management/RIKZ, P.O. Box 8039, 4330 EA Middelburg, The Netherlands;* [2] *Till February 1992: Rijkswaterstaat, Directorate Zeeland; at present: Province of Zeeland, Department of Environment, Transport, Public Works and Water Management, P.O. Box 165, 4330 AD Middelburg, The Netherlands*

Key words: policy planning, decision making, environmental impact assessment, water management

Abstract

During the execution of the Delta project the awareness of the environmental values of estuaries was growing strongly. This awareness has led to a reconsideration of the original proposal to close off the Oosterschelde, and to an alternative view of the management of the estuarine areas and of the civil engineering structures. Now, the completed storm-surge barrier and two compartmentation dams, together with the reinforced dikes, guarantee a sufficient safety against flooding.

After the decision to build a storm-surge barrier it was soon recognized that a coherent plan would be needed for the management and development of the Oosterschelde. A policy plan was accordingly produced in which the broad outlines of the policy were set out regarding the various potential uses of the estuary and the possible means of effecting them. Because of the periodic improvement and adjustment of the policy plan an environmental research programme was set up. Also, environmental research played a significant role in the decision-making processes for the completion of the barrier and the compartmentation dams as well as for the management of the storm-surge barrier.

The coping-stone of the Delta Project marks the beginning of a new period in the water management of the Netherlands: the period of integrated water management.

Introduction

The decision to construct a storm-surge barrier in the mouth of the Oosterschelde has led to the drafting of a policy plan to safeguard the natural and fishery values of the area. The main objective of this chapter is to outline the characteristics of the policy plan and the role of environmental research in the improvement and adjustment of the plan. This environmental research programme was also conducted:

– to advise on the closure method of the compartmentation dams and on the completion of the storm-surge barrier, and

– to prepare guidelines for the management of the storm-surge barrier.

The role of the environmental research in these decision-making processes will be presented in this paper. The results of environmental studies and observations are discussed elsewhere (Nienhuis & Smaal, 1994). The use of these results by technicians and politicians in the execution of a civil engineering project will be treated in this paper.

The Oosterschelde project marks a new period in the water management of the Netherlands: integrated water management, the integrated care for the condition and the use of water systems.

The main features of this integrated water management and its consequences for estuarine management will be highlighted.

Delta Project

The Delta Region is the common name for the Southwestern part of The Netherlands, where the Rivers Rhine, Meuse and Scheldt flow into the North Sea. In the Fifties, the region was sparsely populated; urbanisation and industry only took place along the extreme Northern and Southern fringes, where the ports of Rotterdam and Antwerpen have to be mentioned. The islands were almost entirely used for agricultural purposes. In the dunes, forming the natural sea defence at the western ends of the islands, recreation took place only to a limited extent. The rest of the islands was protected against the sea by dikes having a total length of more than 1000 kilometres. During the depression of the Thirties and during the Second World War the maintenance of the dikes was more or less neglected.

On 1 February 1953 a long and violent storm, combined with a spring tide, burst the dikes in many places, especially in the delta region and caused widespread flooding – 1835 people died in the floods. About 72 000 people had to be evacuated, 200 000 hectares of land were flooded and 500 kilometres of dikes were completely or almost completely destroyed.

It was a national disaster. Three weeks later, the Government set up the Delta Committee to plan ways of preventing any repetition of the disaster as soon as possible. Its objective was not only to secure the land against flooding, but to consider all aspects of water control.

According to the advice of the Delta Committee the Government decided to close off the tidal inlets Haringvliet, Grevelingen and Oosterschelde at the mouth. The tidal basins would be changed into large fresh water lakes. More inland, so called secondary dams would be built to separate Lake Veere and Lake Grevelingen. The Westerschelde and the Nieuwe Waterweg, two important shipping ways to the ports of Antwerpen and Rotterdam, would not be closed off. Here, the required safety should be achieved by reinforcements of the dikes.

This plan had a rather one-sided technical approach. It was focused on the engineering measures in order to banish the unpredictable evil of the sea for ever. Other aspects were considered. However, they hardly played an explicit role during the preparation phase. The loss of important shell fisheries in the estuaries were considered as a negative side effect. This, however, should be compensated in a financial way. Ecological values were fully understood, but did not play any role.

All the other consequences of the implementation of the Delta Plan were positively valuated; to be mentioned are:

1. Improved water management. Salt intrusion in the Nieuwe Waterweg could be pushed back in favour of the horticulture.
2. Large freshwater lakes. The fresh water could be used for storage of drinking water and also for the improvement of the agriculture.
3. Improved infrastructure. The accessibility of the area would be improved.
4. Maintenance of sea defence. Instead of maintaining about 800 kilometres of dikes, only 30 kilometres of dams were to be observed.

Until the Grevelingen was dammed off, the Delta Project has been carried out with little or no public debate. But, during the late Sixties and in the early Seventies, the awareness of the environmental risks was growing strongly. More and more the exponential technical growth was debated. It hardly could be avoided that the Delta Project, still under construction, became subject of this debate. It was in particular during the early Seventies that a strong lobby started questioning the unchanged execution of the rest of the Delta Project. And because the Haringvliet and the Grevelingen estuaries were just closed off, this lobby was focused on the last phase of the Delta Plan, the closing off of the Oosterschelde estuary (Fig. 1).

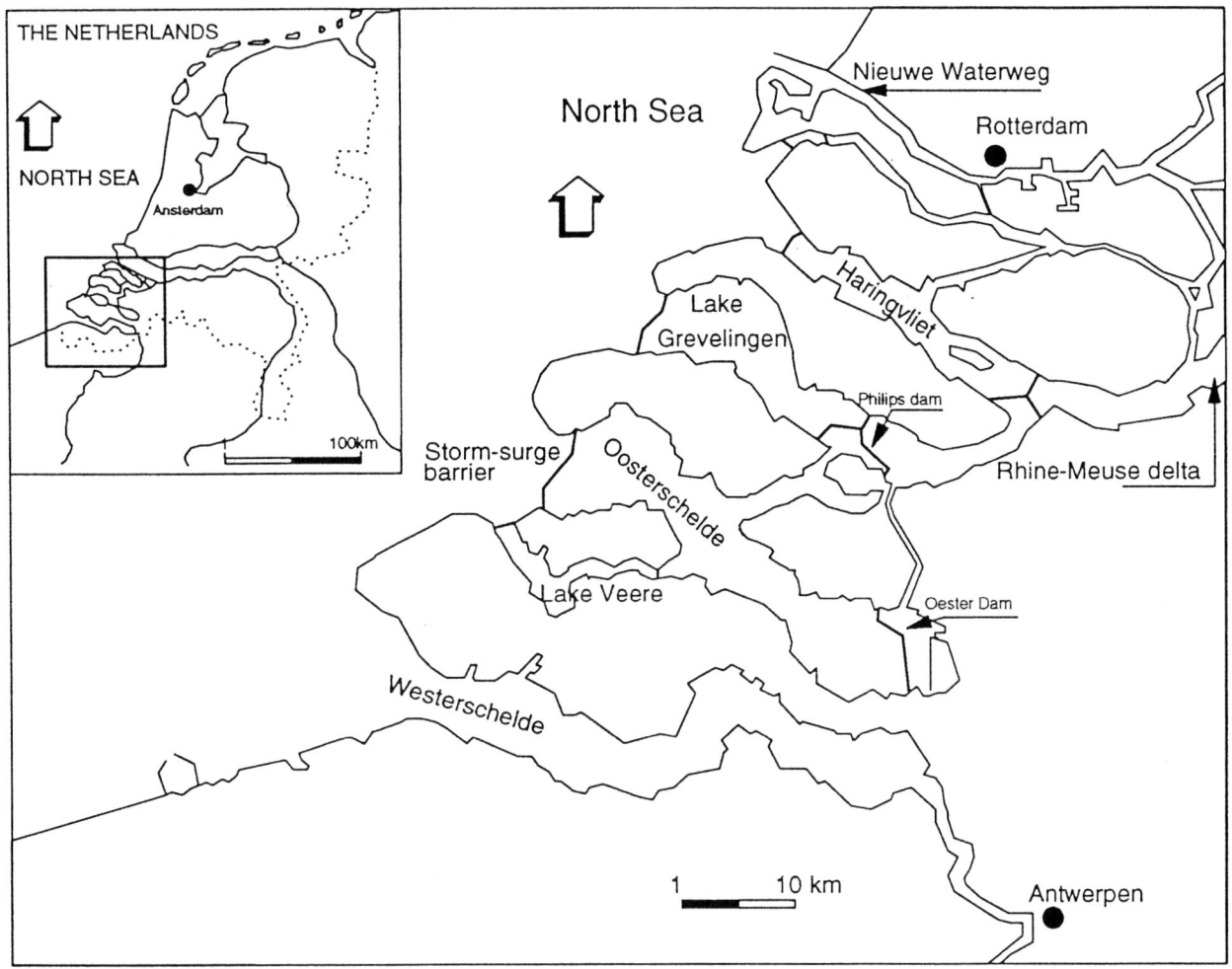

Fig. 1. The Delta Region.

Reconsideration of the Delta Project

For the saving of the tidal character of the Oosterschelde a campaign was conducted by the shellfish cultivators and the nature conservation organizations. The last phase of the Delta Project became subject of debate in the House of Parliament and the country was faced with serious political problems. It must be said that the aim of the campaign was not to discuss safety. In fact, the question was to look for a balance between safety against floods and ecology.

The debate concerning the closure of the Oosterschelde began in the mid Sixties with the publication of articles referring to the exception-

ally rich flora and fauna. The opponents of closure eventually came together in the Open Oosterschelde Committee, which urged that the dikes be raised to Delta level rather than the estuary be closed off completely. Because of increasing pressure the Minister of Transport and Public Works ordered to set up an independent Oosterschelde Committee in 1973. Its members were experts in the fields of hydraulic engineering, economics, environmental protection, biology, fishing and land-use planning. Its task was to examine the various possibilities and to select the one which best safeguarded the interests of both safety and environment.

After a difficult and complicated decision-making process, in 1976 the Government decided to build a storm-surge barrier in the mouth of the Oosterschelde with as main boundary conditions

- a minimum mean tidal range at Yerseke of 2.70 m;
- a salinity not below 15.5 Cl l^{-1} in the eastern part of the estuary;

and to build two compartmentation dams in the east for the following reasons:

- the Scheldt-Rhine Connection had to be made non-tidal;
- to create a fresh water lake for agricultural use;
- to limit the tidal reduction caused by the construction of the barrier in the mouth of the Oosterschelde.

The integrated policy plan for the Oosterschelde

In 1976 the Government decided to invest thousands of millions of guilders extra in the Oosterschelde area by building a storm-surge barrier instead of a closed dam as originally planned. It was recognised that developments after the completion of the storm-surge barrier should be managed in such a way that the unique environment and the fishing activities in the area will be preserved. In a country as industrialised and densely populated as the Netherlands, it is no simple matter to achieve objectives of this nature. All kinds of activities, of which recreation is one of the most important, will lay claims to the area. If no agreement can be reached on policy for the area among the various interest groups and the different relevant authorities, the assets that should be protected will be affected, in spite of the great increase in environmental awareness. A further complicating factor for the development of a policy for the Oosterschelde was the difficulty of predicting the effects of building a storm-surge barrier in the estuary. This resulted in the following policy requirements:

- inclusion of all existing functions in the area, as well as any that it might realistically acquire;
- flexibility in reaction to changing circumstances

and knowledge relating to the effects of the construction of the barrier;
- broad-based administrative and public support and a need to involve all levels of government control in the Netherlands – central, provincial, local – as well as the water control boards.

To enable a joint policy to be drawn up and implemented, a broadly based steering committee was set up in 1977. This group, the Oosterschelde Steering Committee, formulated the main objective for the development and management of the area as 'the conservation and if possible strengthening of the existing natural features and functions of the area, with due regard to the social interests – notably including fishing – concerned' (Stuurgroep Oosterschelde, 1982).

The plan drawn up on the basis of this aim resembles a process more than a static plan, providing sufficient flexibility to allow it to be adjusted if ecological evaluations or social developments make this necessary.

The plan contains the following elements (Knoester *et al.*, 1984):

A policy for the main functions of the area

Policy is aimed at securing the natural and fishing functions of the estuary. This means that the estuary's natural functions determine the nature, extent and locations of its utilization. Subject to the requirements of conservation, the fishing industry had a high priority. All areas suitable for shellfish beds are to be protected by precluding any development which might limit their future use as shellfish beds. New fishing activities or expansion of existing ones will be permitted only when there are no harmful consequences for the environment.

Development and management plan

The development plan indicates those parts of the estuary in which recreational or fishing activities can expand. All developments are governed by an integrated management system in which individual functions are subordinated to the functioning of the area as a whole. In general, given the character of the Oosterschelde, management measures are of greater importance than measures in the sphere of development.

Research plan

The construction of a storm-surge barrier in the mouth of the Oosterschelde was intended not only to guarantee safety but also to preserve the natural features and fishing interests of the area. Consequently, environmentalists should be involved in the project's implementation, especially since many decisions still had to be taken concerning, for example, the size of the wet cross-section of the storm-surge barrier, the location of the compartmentation dams, the season in which these dams should be completed and the way in which this should be done. Since the necessary environmental knowledge was almost entirely lacking, an extensive research programme was set up by the Rijkswaterstaat and a number of scientific institutes.

Hydraulic studies were carried out to support the project's implementation. In addition, a programme was drawn up for the periodic approval and adjustment of the policy plan, comprising the following elements:

– the new basic situation which will arise as a result of the construction of the storm-surge barrier and the compartmentation dams;
– the potential for the development of the fishing industry;
– the development of the recreational use of the area;
– the impact of human activities on the environment;
– possible and desirable management measures and their effects.

Research was also conducted to facilitate the management of the Oosterschelde barrier, concentrating on the effects of the barrier's closure on dikes and salt marshes in relation to the length of time it is closed and the water levels in the estuary.

The above-mentioned studies were more or less specific project studies and were based on general research into the various types of environment in the Oosterschelde and the changes which would occur in them after the completion of the storm-surge barrier. The principal research aspects included the morphology of channels and shallow areas, the impact of changes in the duration of flooding on the salt marshes, the basic food chain and organic communities, such as birds and other biota on the soft or hard substrata.

Amendment procedure

The policy plan for the Oosterschelde calls for a system of annual progress or evaluation reports. These reports include the following elements:

1. a survey of the developments which have taken place in the year to which the report relates;
2. the research results relevant to the policy plan;
3. a summary of the necessary changes or adjustments to policy which are needed because of points 1 and 2.

The integrated policy plan was agreed to by all of the authorities involved and was finalised in May 1982 after the general public and all interest groups concerned had been given ample opportunity to make their views known. Where possible, the plan was adjusted to accommodate the views put forward. In some cases, this even involved dropping certain of the initial aims, but if the plan was to be implemented successfully, it had to have wide public support. This could best be achieved if the plan was endorsed by as many interested parties as possible even though this might sometimes require that certain elements be scrapped.

Interim management policy for the Oosterschelde barrier

During the final construction stage of the project, from July 1985 to April 1987, there was a reduction of the tidal motion to an extent greater than the reduction at present (Vroon, 1994). The reasons for this temporary extra reduction were:

– the secondary dams were closed after completion of the storm-surge barrier;
– during the completion stage of the barrier and the final stage of construction of the secondary dams the tide was reduced by the completed parts of the storm-surge barrier; the tidal reduction during the final stages of the closure operations of the dams, allowed the remaining closure gaps to be completely filled in with sand (which yielded a saving of 80 million guilders).

During the planning stage for the construction work several different options were considered, although the degree of freedom was limited among other things by the following agreed principles:

– the lowering of a number of gates in the storm-surge barrier during the period from November 1985 to October 1986 would be permitted, to allow certain construction work to be completed;
– the storm-surge barrier was to be used to reduce the flow velocities during completion of the Oesterdam and Philipsdam;
– the Philipsdam could not be completed before the Oesterdam, for hydraulic engineering reasons, since this would have resulted in excessive flow velocities in the Scheldt-Rhine Canal;
– the completion of the Oesterdam and Philipsdam had to be carried out in different years to allow a better phasing of the capital expenditure on the project.

A number of boundary conditions were subsequently formulated based on a detailed understanding of the environment in the Oosterschelde, which were intended to minimise the potential damage the construction work might cause:

– the mean tidal range at Stavenisse was not allowed to drop below 2.30 m, (meaning that up to about 4200 m^2 of the open area of the barrier could be closed) except during the final phase of the construction work, when the entire barrier was closed for short periods;
– during the final phase of the dam's construc-tion, the barrier was not to be completely closed for longer than two days.

After extensive impact studies it was agreed that spring or autumn would be the most acceptable periods for the completion of the Oesterdam and the Philipsdam, both from an environmental point of view and in the interests of the fishing industry in the area.

Although extensive investigations have been carried out prior to the hydraulic engineering work in the Oosterschelde and assessments have been made of the environmental implications of such operations, it was not possible to give firm guarantees that permanent damage could be avoided (Van Westen & Leentvaar, 1988). The major uncertainties concerned the weather and the possibility of setbacks occurring in the engineering work. This meant that the situation had to be reconsidered at each stage of the project. A close watch was therefore kept on developments in the estuary to ensure a ready supply of up-to-the-minute information for making the necessary decisions.

During the transitional phase from the original to the new tidal situation, a large number of parameters were carefully monitored. This monitoring exercise was primarily intended to serve as an early warning system. More detailed investigations were instituted if it was discovered that readings from the field lay outside the natural variations in the original conditions.

Observations and measurements were taken on a more or less continuous basis during the final stages of the construction of the compartmentation dams. The information collected was processed and interpreted at once, so as to allow immediate corrective action to be taken if necessary. The way in which the storm-surge barrier was used, was modified several times as a result of these observations. For instance, during a later construction phase, a severe storm took place which coincided with the completion of the Oesterdam in October 1986. As a result delays were encountered at a time when the tide had already been greatly reduced. It was observed that birds were no longer foraging in the extremely inclement and cold weather conditions, while at the same time the size of the feeding grounds had been severely restricted. Under these circumstances, a significant increase in the mortality rate among the birds was anticipated. It was therefore decided to deviate from the original planning and temporarily reduce the water level in the Oosterschelde before starting work on the final stage of the dam's construction. As a consequence, the availability of foraging areas greatly increased, providing an adequate supply of food leading to a situation of birds making effective use of the opportunity presented.

The fact that the secondary dams were completed later than the storm-surge barrier caused a few adverse effects on the environment. Parts of the salt marshes in the Oosterschelde suffered some damage as a result of dehydration in the summer of 1986. Changes in the soil structure due to drying-out and settling were reported in some places. In the winter of 1986/1987 these changes resulted in more erosion of the salt marshes than was expected. Later on, the erosion stabilised on the original rate of 4 ha yr^{-1} (De Jong *et al.*, 1994).

Management of the Oosterschelde barrier

Under normal circumstances the storm-surge barrier remains open to allow water to flow in and out freely. The barrier is closed only if the expected water level exceeds a predetermined 'predicted critical level' (van Westen *et al.*, 1988). In such a situation the actual water level in the North Sea will continue to rise until the storm surge has reached its peak. Meanwhile the water in the Oosterschelde will become virtually semi-stagnant. In 1986 the Minister of Transport and Public Works decided to adopt a predicted critical level of Mean Sea Level (MSL) + 3.25 m for closure of the barrier. However, for the evaluation period (1987–1991) a predicted critical level of MSL + 3 m was used in order to have the opportunity to gain some experience in operating the barrier.

The decision to close the barrier on a predicted

rather than an actual water level allows the authorities to fix the desired water level in the Oosterschelde by choosing the appropriate moment to close the barrier. In the case of storm-surges predicted to peak more than once, the water level in the Oosterschelde will remain stagnant during the whole period of the storm if the barrier is kept closed. This is the so-called Closure at Inner Water Level strategy. There are two alternatives:

1. inner water level during the 3 high water peaks at MSL + 1 m, and
2. inner water level during the 3 high water peaks at MSL + 2 m.

On the other hand it is possible to open the storm-surge barrier after the first and second high water peaks and to close the barrier again before the next high water occurs, thereby fixing a new water level in the Oosterschelde (the so-called Alternating Strategy). Four alternatives are present:

3. 1st peak: inner water level MSL + 0 m; 2nd peak MSL + 1 m; 3rd peak: MSL + 2 m);
4. 1st peak: inner water level MSL + 1 m; 2nd peak MSL + 2 m; 3rd peak: MSL + 3 m);
5. 1st peak: inner water level MSL + 1 m; 2nd peak MSL + 2 m; 3rd peak: MSL + 1 m);
6. 1st peak: inner water level MSL + 2 m; 2nd peak MSL + 1 m; 3rd peak: MSL + 2 m.

Other possibilities (e.g. 0-2-0) were invalid for hydraulic reasons.

The various management strategies have been examined in detail with regard to safety. Some of the strategies did not comply with the required safety level specified for this region (Table 1).

A final choice between the remaining alternatives that satisfied the required safety criterion was made on the basis of environmental concerns and of the likely impact on fishing in the area (Table 2). After analyzing the various strategies, it was concluded that the alternating mode of operation offered significant advantages in terms of the environment and the preservation of fishing interests, as compared with other strategies based on the maintenance of fixed levels. This led to the 1-2-1 alternating strategy being adopted as the basis on which the system would be operated.

Table 1. Scorecard safety. For three possible predicted critical levels it was defined whether the use of the closure strategies leads to a probability of flooding equal or lower than the safety criterion (+), or higher (−).

Strategy	Predicted critical level		
	MSL + 3.25 m	MSL + 3.00 m	MSL + 2.75 m
1-1-1 Closure at inner water level	+	+	+
2-2-2 Closure at inner water level	−	−	−
0-1-2 Alternating strategy	+	+	+
1-2-3 Alternating strategy	−	−	−
1-2-1 Alternating strategy	+	+	+
2-1-2 Alternating strategy	+	+	+

MSL: Mean sea level
0-1-2: 1st peak: inner water level MSL + 0 m; 2nd peak: MSL + 1 m; 3rd peak: MSL + 2 m
Predicted critical level: the barrier will be closed when the predicted water level is higher than this value

It was anticipated that this mode of operation would prevent serious erosion occurring in the intertidal area, since such phenomena are mainly associated with water levels close to MSL. Furthermore, this strategy should limit the amount of disruption inflicted upon the environment by parts of the salt marshes being washed away.

It is also possible to close the Oosterschelde barrier for reasons other than safety, as was done during the completion of the Philipsdam and Oesterdam. Following a thorough analysis of these potential reasons for such closures, it was decided to pursue a very restrained policy in this regard, mainly in view of the provisions of the policy plan. The Oosterschelde barrier will be closed entirely or partially only to prevent disasters, e.g. following dike subsidence or serious storm damage.

In the period from 1 May 1987 to December 1991 the Oosterschelde Barrier was closed (Table 3) several times to prevent excessive water levels in the estuary. In most cases the water level in the Oosterschelde was successfully fixed at the desired level. This was even true for the cases when the storm-surge barrier had to be closed for storm-surges that peaked more than once. However, the duration of stagnation (4–9 hours) during several closures was longer than the expected 5–6 hours.

Evaluation of management of the storm-surge barrier and policy plan

During the period April 1987 to April 1991 an assessment was made of the effects of operating the storm-surge barrier and the presence of a new infrastructure (Van Westen, 1991). The aim of this evaluation was to consider the main aspects of water management and safety in relation to the original forecasts.

The safety evaluation concentrated on the strength of the barrier and the dikes, and the wave forces acting on them during the period of closure for flood protection. Calculations showed that a rise of the predicted critical level to MSL

Table 2. Scorecard environment and fisheries. A fixed water level can be most favourable (+), favourable (0) or unfavourable (−) for the different environmental aspects.

Aspects	Inner water level		
	MSL	MSL + 1 m	MSL + 2 m
Bottom sediments	+	+	+
Intertidal zone	−	−	0
Salt marshes	+	−	−
Hard substrata	−	−	+
Birds	+	+	+
Fisheries	+	+	0
Ohter functions	+	+	+

Table 3. Closures of the Oosterschelde Barrier during storm surges in the period May 1987–December 1991.

Date	Expected water level sea side	Highest water level sea side	Highest water level halfway estuary	Maximum duration of stagnancy
181286	MSL + 2.90 m*	MSL + 2.73 m	MSL + 1.03 m	6 hours
191286	MSL + 2.85 m*	MSL + 2.71 m	MSL + 0.97 m	5.2 hours
140289	MSL + 3 m	MSL + 3.17 m	MSL + 1.70 m**	5.5 hours
270290	MSL + 3.00 m	MSL + 3.17 m	MSL + 1.02 m	7.3 hours
270290	MSL + 3.29 m	MSL + 3.69 m	MSL + 2.06 m	4.5 hours
280290	MSL + 3.14 m	MSL 3.25 m	MSL + 1.06 m	7.5 hours
010390	MSL + 3.20 m	MSL + 3.25 m	MSL + 1.07 m	7.3 hours
210990	MSL + 3.05 m	MSL + 2.92 m	MSL + 1.02 m	6.5 hours
121290	MSL + 3.05 m	MSL + 2.80 m	MSL + 0.93 m	9.5 hours
131290	MSL + 3.06 m	MSL + 2.65 m	MSL + 1.98 m	4.5 hours

* Predicted critical level of MSL + 2.75 m.
** Prediction was not available in time.

+ 3.25 m, as originally planned, would in fact exceed the safety criterion. This finding results from two factors, the tidal range and the frequency of storm-surges that peak more than once. Both factors are higher than was expected. For this reason the predicted critical level of MSL + 3.00 m will be used for operating the barrier.

Following each period in which the storm-surge barrier was closed, civil engineers and environmentalists carried out observations to ascertain whether dikes, intertidal areas and fishing areas had sustained greater damage than they would have done if the barrier had not been closed. Some extra erosion of the higher parts of the intertidal areas and the salt marshes, and the washing away of eelgrass have been reported. Some evidence suggests that these impacts occurred after a stagnant water level of MSL + 2 m. Nevertheless the evidence is not very firm and there is no current need to select a different operating strategy. However it remains necessary to continue vigorous monitoring of the developments.

In 1993 the policy plan for the Oosterschelde was evaluated. The results of the evaluation studies (Coördinatiegroep Oosterschelde, 1992) showed that the main features of the policy plan are still valid. One of the most important changes in the policy plan will be the placing of a greater emphasis on the development of natural values by closing off some vulnerable intertidal areas and possibly by restoring the links between the nature areas behind the dikes and the Oosterschelde. The development of fisheries and recreational activities will remain limited.

Some national and international developments in management

Towards a multi-track approach for national water management

During the execution of the Delta Project opinion grew that the newly formed water systems provided new possibilities. A systems approach or an area approach appeared to be the right answer to the huge changes which were taking place. The discussion of the policy plan for the Oosterschelde supported this.

The present water management in the Netherlands is distinguished by integrated care for the condition and use of water systems, comprising the media water, beds and banks or shores, with their physical, chemical and biological components, in relation to their relevant surroundings: integrated water management. On a national level this was set in motion in the memorandum Living with water (Ministerie van Verkeer en Waterstaat, 1985). The third National Policy Document on Water management (Ministerie van Verkeer en Waterstaat, 1989) outlines target situations, the attainment of which guarantees a sustainable ecological development of water systems (preservation of production, diversity of species and self-regulation) as well as a sustained use by man. To reach the target situations a multi-track approach is formulated. The two main characteristics of this approach are an accelerated reduction of pollution (about 50 percent emission reduction of pollutants relative to 1985 and more than 50 percent reduction of organic micro-pollutants) and a guided use of the water systems.

International developments on estuarine management

In the last decades a changing attitude towards estuarine management has emerged. In general water management used to be directed towards the protection of land against water and towards keeping sufficient water available for the various user groups. In most cases water management only tried to cope with the most visible and troublesome pollution of surface waters. Attention for the natural environment itself is more recent, as is the objective of creating a healthy aquatic environment. An example of this is found in the Westerschelde estuary (Fig. 1), where management used to be restricted to for instance, dredging shipping lanes and harbours, maintaining dikes and sluices, beaconing the navigable passages etc. Nowadays the estuary is managed as a multi-functional system in which issues such as water quality performance and natural values are also taken into account.

As a result of the growing awareness of the deterioration of environmental quality, the ecological aspect has become a major item in water management during the last ten years. Attention

can no longer be restricted to water problems only. Water has to be seen in relation to its physical environment: surface water and its bottom, its banks and its related surrounding. Besides, water management can not be considered as an isolated activity. A great number of relations concerning other policy fields is taken more and more into consideration by managers involved (Glasbergen, 1991). However, a thorough search in the literature (Van Westen, unpublished) revealed that this conclusion is valid only for a few countries. In most cases the attention of scientists first of all goes out to the ecological processes while management strategies are dealt with superficially. For this reason the Organisation for Economic Cooperation and Development (1991) and NATO's Committee on the Challenges on Modern Society (Van Westen, unpublished) have made a comparison of management strategies in different countries. Based on the results the OECD developed a conceptual framework for integrated coastal zone management. This framework is almost similar to the conceptual framework for the policy planning for the Westerschelde estuary (Bestuurlijk Klankbordforum Westerschelde, 1986; see also the text about the policy plan for the Oosterschelde). It contains 7 phases:

1. Establishing administrative and political coordination and the creation of the institutional mechanism;
2. Generation of information (analysis and planning);
3. Reassessment of present policies, legislative requirements and legal/judicial action;
4. Preparation of alternative options and analysis of implications and risk /uncertainty;
5. Selection of the final plan, involving public participation;
6. Implementation;
7. Monitoring and evaluation to feed-back into planning.

The Pilot Group Estuarine Management of NATO's CCMS will go a step further by conceiving a model allowing authorities involved to compare local conditions with some basic guidelines for estuarine management. It is important that such planning guidelines go beyond the existing general description of abstract planning rules.

Conclusions

Integrated policy plans for estuarine areas provide an effective instrument in ensuring the sustainability of the systems. It is essential for all of the functions fulfilled by these areas to be listed, indicating their interrelationships especially with regard to the conflicts that exist between them. We have learned that plans that do not take all functions into account will have to be adapted at a later stage, which can have drastic consequences for the functions of the area and may create the need for expensive extra works.

It is also essential that authorities at all levels are involved and therefore share responsibility. Only this way the success of the plan can be guaranteed.

Discussion and compromise are part of the planning process, as is consultation with interested parties. Otherwise, no matter how good a plan may be technically, it will not exceed the mere status of letters printed on a page.

Integrated planning requires research. Evidently, knowing everything is not an absolute prerequisite before a policy can be formulated. Monitoring programmes concentrating on critical factors in an integrated policy plan can provide sufficient information to enable plans to be adapted. Experience has shown that it is perfectly possible to develop flexible plans of this kind.

Environmental research enables technicians to review more alternatives for the execution of a project and makes it sometimes possible to choose cheaper solutions.

According to Smith (1991) we are witnessing the beginnings of an approach to the development and management of the world oceans and seas based less on individual uses, state and industrial interests, and more on relationships among these, as these relationships become progressively more important. This changed approach fits in the concept of sustainable development, which is described in the Brundtland report Our Common Future (1987) as 'development that meets the needs of the present without compromising the ability of future generations to meet their own needs'.

Therefore, as far as rivers and estuaries are concerned, the entire basin area should be taken into account for conceiving plans, even though this may entail an international approach. If, however, this would delay the policy-making process for too long, there is no reason why an integrated plan should not be drawn up for part of the area and supplementary conditions be created for the remaining parts. These conditions can then serve as a basis for international consultations.

References

Bestuurlijk Klankbordforum Westerschelde, 1986. Aanpak en taakverdeling beleidsplan Westerschelde (Plan of work for the policy plan for the Western Scheldt). Notitie RFO-86.070, Rijkswaterstaat, directie Zeeland, Middelburg, 3 pp.

Coördinatiegroep Oosterschelde, 1992. Evaluatie Beleidsplan Oosterschelde (Evaluation of the policy plan for the Oosterschelde). Provincie Zeeland, Middelburg, 83 pp.

De Jong, D. J., Z. de Jong & J. P. M.Mulder, 1994. Changes in area, geomorphology and sediment nature of salt marshes in the Oosterschelde estuary (SW Netherlands) due to tidal changes. Hydrobiologia 282/283: 303–316.

Glasbergen, P., 1991. Drie voorwaarden voor integraal waterbeheer (Three conditions for integrated water management). Waterschapsbelangen, 7: 240–244.

Knoester, M., J. Visser, B. A. Bannink, C. J. Colijn & W. P. A. Broeders, 1984. The Eastern Scheldt Project. Wat. Sci. Tech. 16: 51–77.

Ministerie van Verkeer en Waterstaat, 1985. Omgaan met water (Living with water). Den Haag, 63 pp.

Ministerie van Verkeer en Waterstaat, 1989. Derde Nota waterhuishouding (third National Policy Document on Water management). SDU uitgeverij, Den Haag, 297 pp.

Nienhuis, P. H. & A. C. Smaal, 1994. The Oosterschelde estuary, a case-study of a changing ecosystem: an introduction. Hydrobiologia 282/283: 1–14.

Organisation for Economic Co-operation and Development, 1991. Report on coastal zone management: integrated policies and draft recommendation of the council on integrated zone management. Paris, 195 pp.

Smith, H. D., 1991. The regional bases of sea use management. Ocean and shore line management, 15: 273–282.

574

Stuurgroep Oosterschelde, 1982. Beleidsplan Oosterschelde (Policy plan for the Oosterschelde). Provincie Zeeland, Middelburg, 36 pp.

Van Westen, C. J., T. Pieters & L. Boom, 1988. The management of the Eastern Scheldt Barrier. Proceedings 16th Congress ICOLD, San Francisco: 391–402.

Van Westen, C. J., 1991. Evaluation of the management of the Eastern Scheldt Barrier. IABSE Colloquium, Nyborg: 63–70.

Van Westen, C. J. & J. Leentvaar, 1988. Ecological impacts during the completion of the Eastern Scheldt Project. In W. Salomons, B. L. Bayne & E. K. Duursma (eds), Pollution of the North Sea. An assessment. Springer-Verlag, Heidelberg, 687 pp.

Vroon, J., 1994. Hydrodynamic characteristics of the Oosterschelde in recent decades. Hydrobiologia 282/283: 17–27.

World Commission on Environment and Development, 1987. Our common future. Oxford University Press.

Hydrobiologia **282/283**: 575–592, 1994.
P. H. Nienhuis & A. C. Smaal (eds), The Oosterschelde Estuary.
© 1994 *Kluwer Academic Publishers. Printed in Belgium.*

The Oosterschelde estuary: an evaluation of changes at the ecosystem level induced by civil-engineering works

P. H. Nienhuis[1], A. C. Smaal[2] & M. Knoester[3]
[1] *Netherlands Institute of Ecology, Centre for Estuarine and Coastal Research, Vierstraat 28, 4401 EA Yerseke, The Netherlands*
[2] *National Institute for Coastal and Marine Management – RIKZ, P.O. Box 8039, 4330 EA Middelburg, The Netherlands*
[3] *National Institute for Coastal and Marine Management – RIKZ, P.O. Box 20907, 2500 EX Den Haag, The Netherlands*

Key words: Oosterschelde estuary, civil-engineering works, storm-surge barrier, hydrographical changes, geomorphological changes, long-term ecological changes, community and ecosystem responses, habitat availability, productivity, carrying capacity

Abstract

The interest in storm-flood protection has recently gained momentum, owing to the wide international discussion on the impact of sea-level rise on society. The Oosterschelde project is technically and scientifically unique. The storm-surge barrier represents an important breakthrough in marine civil engineering. The project also offered ample opportunities to perform integrated physical, chemical, geological and biological research. Integration of the knowledge gained, raised the entire project to the level of a case study of a changing estuarine ecosystem, and demonstrated the effects of human interference in a non-polluted estuary. Notwithstanding considerable changes in the environment, the Oosterschelde has retained most of its favourable abiotic factors, labeling the estuary as a high quality marine system. The water quality in the post-barrier period more closely resembles that of the North Sea than in the period before. Significant changes in erosion and sedimentation and the consequent redistribution of fine sediments, are continuing. The ecosystem has shown responses to various factors. Effects of severe winters and impact of mussel- and cockle fisheries could be distinguished from other factors. The physical response of the ecosystem to the civil-engineering project could be quantified in terms of changes in habitat availability, maintenance of biological productivity, and restricted maintenance of the carrying capacity as an internationally recognized wetland and fisheries area.

Introduction

After the devastating coastal flooding of southwest Netherlands in February 1953, the Delta Project was conceived in 1958 by an act of the Dutch parliament, as an answer to the continuous risk of flooding. Between 1960 and 1987 a large-scale coastal engineering project has been carried out in the SW Netherlands, resulting in the isolation of several estuaries from the North Sea by massive sea-walls. The original plan for Oosterschelde estuary called for a dam across the mouth of the estuary, a distance of 9 km, to be finished in 1978. The accumulated evidence

576

showed that the tidal basin would have been changed into a stagnant lake filled with polluted and eutrophicated water from the river Rhine and Meuse. However, through the 1960's and early 1970's conservationists provoked an awareness in many people of the need to protect the Oosterschelde's outstanding natural resources and its unique tidal habitat, including an extensive shellfish industry, the only one in the Netherlands (Smies & Huiskes, 1981).

The Dutch government decided to change the Delta Project design in 1974, and the final form of the present barrier differs drastically from the simple dam that has been envisaged originally. Between 1979 and 1986, a storm-surge barrier was constructed in the seaward mouth of the estuary (Fig. 1). The barrier contains a system of floodgates which will be closed only during severe storm-flood conditions on average once or twice a year. The building of the storm-surge barrier

meant a compromise solution: the barrier allows the tides to enter the estuary freely, thus safeguarding the tidal ecosystem, including the plant and animal communities. On the other hand the barrier guarantees safety for the human population and their properties when storm-floods threaten the area and the flood-gates are closed (Smaal *et al.*, 1991).

Although topographically separated from the main sea-wall, two auxiliary compartmentalisation dams, built between 1977 and 1987, form an indissoluble entity with the storm-surge barrier. The Oesterdam is an 11-km long dam in the rear end of Oosterschelde estuary, separating the saline sea-arm from the eastern fresh-water compartment, that functions as an important shipping route from Rotterdam to Antwerp. The Philipsdam is 6-km long and separates the northern branch of Oosterschelde estuary from the freshwater lake Krammer-Volkerak. This dam con-

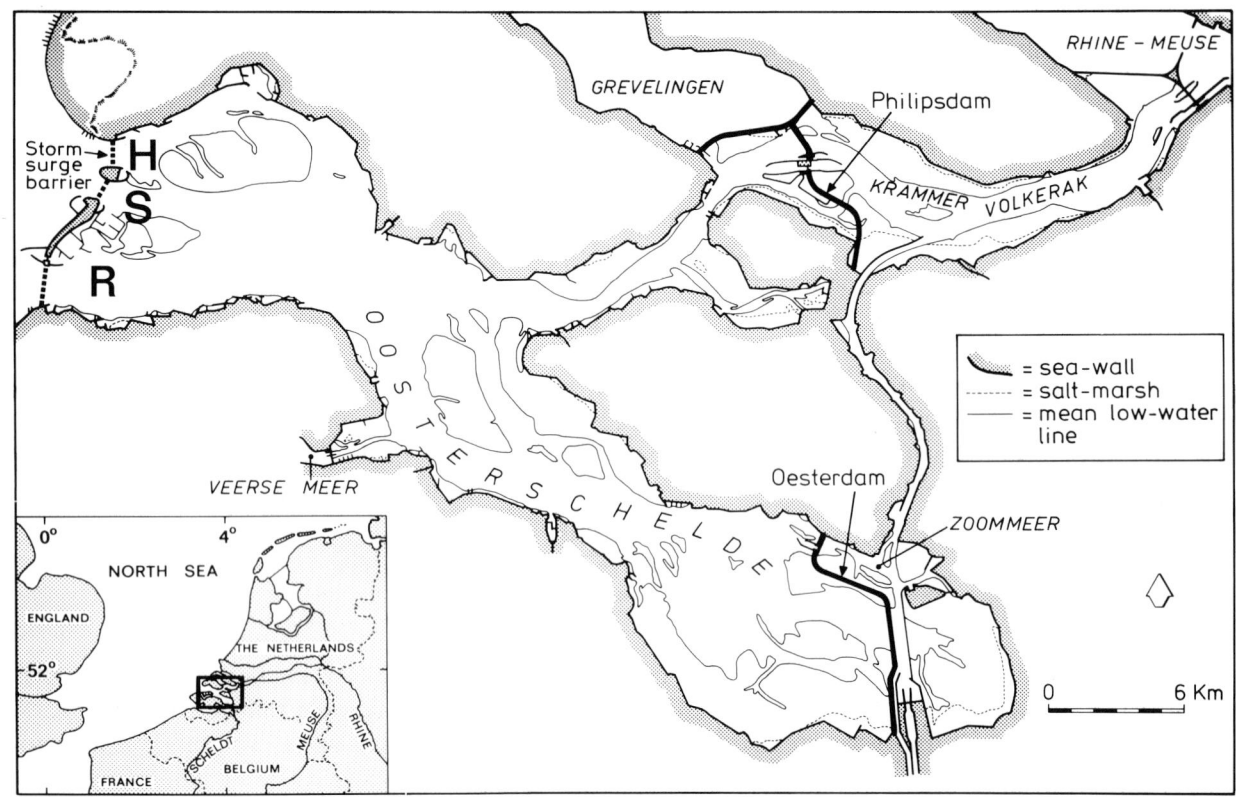

Fig. 1. Oosterschelde estuary in the SW Netherlands. R = Roompot; S = Schaar; H = Hammen. Krammer-Volkerak contains river water, Oosterschelde contains sea water. Locks in Philipsdam and Oesterdam indicated.

tains a number of shipping locks, allowing a very restricted volume of fresh water from the river Rhine ($10 \text{ m}^3 \text{ s}^{-1}$) to enter the estuary (Fig. 1).

Three periods are chosen to distinguish major shifts in tidal range: (1) the pre-barrier period (1980–1984); (2) the barrier construction period (1985–April 1987) in which prefabricated elements were positioned in the entrance of the estuary, and in which the two auxiliary dams were finally prepared; (3) the post-barrier period (1987–1990) in which the ecosystem is supposed to adjust itself to the changed hydrodynamics of the environment (Fig. 2)

During the same period, four cold to severe winters occurred, of which 3 consecutive severe winters (1984–1985; 1985–1986; 1986–1987) coincided with the construction of the barrier.

Several large research projects have been carried out in the past 10 years, to study the changes in the structure and functioning of Oosterschelde ecosystem before and during the construction and after the completion of the flood-control works.

Owing to many concerted research actions during the period 1980–1990, a unique, long-term data set came into existence, demonstrating the effects of interference of man in combination with the effects of natural stress on Oosterschelde ecosystem functioning.

The aim of this paper is to give an evaluation of the changes in the Oosterschelde estuary induced by civil-engineering works, and to discriminate between the effects of human-induced and natural stress.

Changes in the abiotic environment

An evaluation of changes of an ecosystem presupposes an interactive approach of physical, chemical and biological disciplines. In a highly dynamic environment like an estuary, however, the interaction is to a great extent in a one-way direction: the abiotic factors predominantly determine the conditions for the biotic communities,

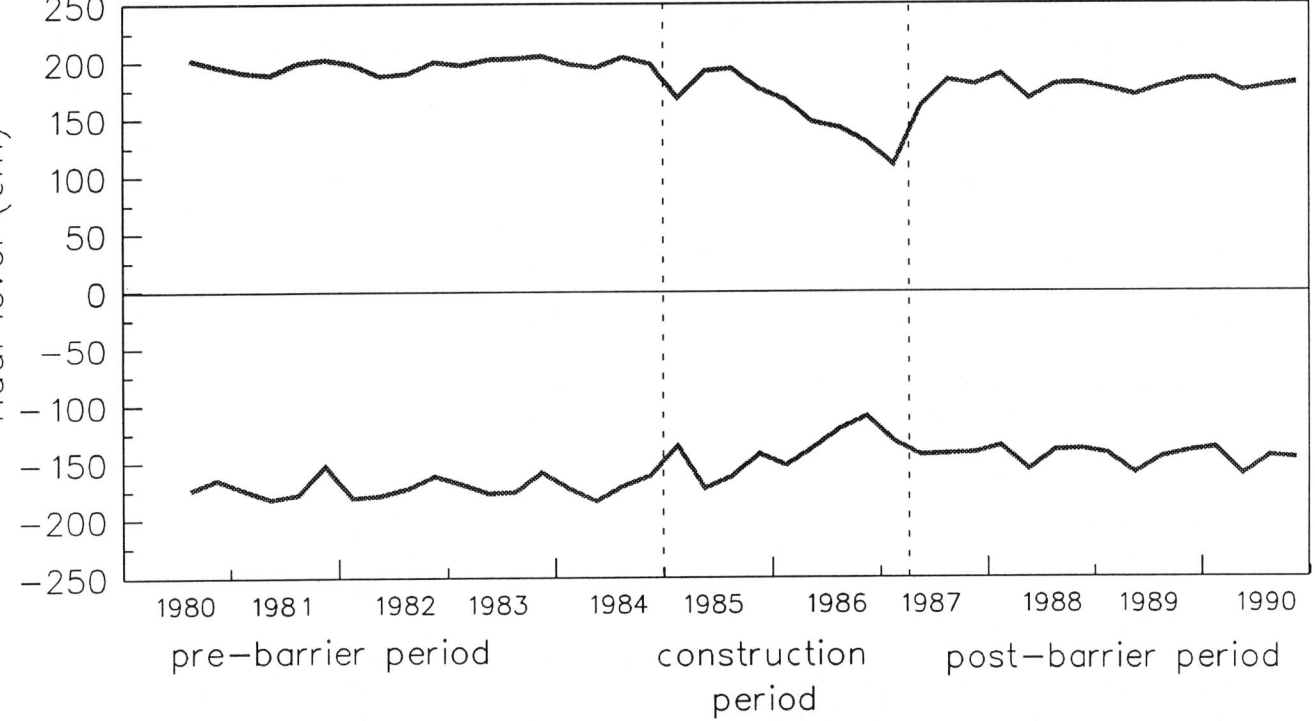

Fig. 2. Running average of mean high-water (MHW) and mean low-water (MLW) data, measured daily at Yerseke (central Oosterschelde). Data from Tidal Waters Division, Middelburg. The pre-barrier, the barrier construction and the post-barrier period are indicated.

whereas the abiotic environment is marginally affected by the biota. On the other hand the changes in physical and chemical properties in the Oosterschelde are directly related to the civil-engineering works. Hence, the evaluation of changes in the Oosterschelde ecosystem starts in considering the abiotic environment. We will distinguish between hydrodynamical, geomorphological and water quality aspects, after first having summarized the main changes in boundary conditions for the system.

Boundary conditions

The construction of the storm-surge barrier in the mouth of the Oosterschelde and the compartmentalisation dams in the rear of the estuary has had the most obvious consequences with respect to the tidal regime: the tidal volume being reduced by 30% and the tidal range by 13% (Vroon, 1994). During the construction period the reduction of the tidal range amounted up to about 25%, being a critical value for the upper eulittoral and supralittoral zones.

Under storm-surge conditions, with expected high water levels of > 3 m + Dutch Ordnance Level (DOL), the gates of the barrier are being closed; the average frequency of the operation is about 3 times a year. The periods of stagnancy in the Oosterschelde vary between 1 and 9 hours, mostly at a level of about 1 m + DOL, but other levels do occur due to circumstances of non-predicted or long-continued storm-surges (Table 1). No obvious effects of the stagnant periods have been observed till now (Smaal & Boeije, 1991), but there are indications that the erosion of cliffs on the salt marshes will increase, in particular in the cases of higher levels of stagnancy. Therefore the policy is to maintain the closure frequency and duration at a minimum (Van Westen & Colijn, 1994).

The compartmentalisation dams, built for reasons to restrain the reduction in tidal range in the estuary and to obtain a stagnant shipping canal between Rotterdam and Antwerp, have cut off the Oosterschelde from the input river water.

Table 1. Closure operations of the Oosterschelde storm-surge barrier from 1989 through 1993. The most remarkable event was the storm of February 1990 which made closure of the barrier during 3 consecutive high tides necessary; in between the level in the basin was changed in order to minimize negative environmental effects. DOL = Dutch Ordnance Level, approximately Mean Sea Level.

Date	Outside high water level (m + DOL)	Inside stagnant level (m + DOL)	Duration (h)
14–02–'89	3.17	1.70	6
27–02–'90	3.17	1.02	7
27–02–'90	3.69	2.06	4.5
28–02–'90	3.25	1.06	7.5
01–03–'90	3.35	1.07	7
21–09–'90	2.92	1.02	6.5
12–12–'90	2.80	0.93	9.5
12–12–'90	2.65	1.98	4.5
11–11–'92	3.15	2.39	2.5
24–01–'93	3.25	2.40	2
21–02–'93	3.13	1.35	8.5
16–11–'93	3.37	1.00	7.5
16–11–'93	3.37	2.02	4

Hence, an evolution of decreasing riverine influence which was started already more than a century ago was now completed, turning the estuary into a blind tidal bay. Another consequence of the compartmentalisation dams is that the tidal flat habitat was reduced from 18,300 to 11,800 ha, and the salt marsh area from 1725 to 643 ha.

Hydrodynamics

The above-mentioned reduction of the tidal volume implies a decrease of the tidal range as well as the tidal current velocity. The tidal range reduction means that the highest zone of the remaining salt marshes is no longer flooded.

The reduction in tidal current velocity is proportionate to the distance from the barrier, increasing to up to more than 70% close to the compartmentalisation dams. This reduction, combined with the absence of high floods in the post-barrier Oosterschelde, is in favour of sedimentation processes. An overall effect of the barrier and the compartmentalisation is that the residence time of the water mass in the Oosterschelde has increased by a factor 2 on average (Vroon, 1994). In spite of that the basin remained its character of a well-mixed, non-stratified estuarine system.

Geomorphology

The construction of the barrier and the compartmentalisation dams has considerable consequences for the geomorphological processes. The cross-sections of the present tidal channels are still more or less in accordance with the former tidal volume. In relative terms the former tidal gullies are too deep now, and the slopes of the intertidal flats are too steep. The decrease in tidal volume means that a new equilibrium has to develop towards channels with a smaller cross-section. Because of a shortage of imported sand this is a very slow process, which is predicted to take hundreds of years (Mulder & Louters, 1994).

Meanwhile, as a result of the tidal reduction, the wave energy dissipation is concentrated in a smaller vertical zone of the tidal flats, resulting in an increased erosion. This erosive process is no longer counterbalanced by sedimentation processes due to the diminished tidal energy. The result is a continuous net erosion of the intertidal area, and a sedimentation of sand and silt on the bottom and slopes of the channels and gullies. This process will lead to an estimated loss of 15% of the present surface area of the intertidal flats and the salt marshes, over a period of 30 years (Oenema, 1988). Net erosion of the salt marshes also occurred in the pre-barrier period, but the increased rate in the present situation is, apart from wave attack, ascribed to factors (De Jong *et al.*, 1994) as decreased sediment supply to the marshes, and reduced strength of the cliffs, the latter being due to the absence of inundation during a period in the construction phase of the compartmentalisation dams.

A consequence of the new hydrodynamic regime is that the Oosterschelde has changed from a sediment-exporting system into a sediment-importing system. This imported material, mainly fine sediments ($< 50 \ \mu m$), is deposited in the tidal channels, where the shear stress is not sufficient for resuspension (Ten Brinke *et al.*, 1994). Locally these fine deposits are negatively affecting the habitat conditions for benthic organisms.

Water quality

Due to the already formerly existing high salt content in the Oosterschelde, this parameter was assigned a crucial role in the assessment of alternatives for water management strategies during the construction period of the works. The average salinity – expressed as chloride ion concentration – increased only slightly: from 16.2 promille Cl^- to 16.9 promille Cl^- and the salt content at the rear side has become almost as high as in the mouth (Nienhuis & Smaal, 1994). Hence, a brackish zone is lacking in the post-barrier Oosterschelde, but due to water management measures in former decades, the brackish areas in the pre-barrier basin had already lost most of their importance.

Due to the isolation of the estuary from the river input the average nutrient concentration has decreased by 20–60%, but also there is a tendency for 'reversed' nutrient gradients in the basin where now depletion of the nutrients is observed from the mouth to the rear end of the basin. This tendency could become stronger if depletion of the intertidal and subtidal sediments will continue.

Contamination with inorganic and organic pollutants has never been a major problem in the Oosterschelde. The contents of the organic micropollutants in water and sediments were found to be mostly below the so-called effect level. As far as data were available, as was the case for the PCB-153 contents in mussels, the values turned out to be low compared with the reference situation, apart for the mouth of the basin due to the direct influence of the North Sea (Smaal & Boeije, 1991). When considering the concentration of heavy metals as an indication for water quality, improvements can be detected along the entire range towards values which are close to natural background values for marine waters. Only locally, higher values may occur due to human activities, e.g. in harbours. However, as shown by Absil (1993) bioavailability of trace metals may be relatively high in the Oosterschelde, due to low organic matter concentrations and hence low complexation of the trace metals.

Conclusion

It can be concluded that the Oosterschelde has retained most of its favourable abiotic factors with regard to a high quality estuarine system. But still considerable changes have occurred. To a certain extent the water quality in the post-barrier period more closely resembles that of the North Sea than in the period before. Significant changes with respect to the physical processes of erosion and sedimentation are to be taken into account. The relevance of these changes is to be considered in relation to the entire tidal ecosystem.

Changes in the ecosystem

The civil-engineering project has shown effects on many aspects of the Oosterschelde estuary. It therefore offers unique possibilities to evaluate the response to anthropogenic stress at the scale of an ecosystem. General variables to describe ecosystems are diversity, productivity and carrying capacity. These variables have been used in a modified way in this study: diversity of species is related to habitat availability, and therefore effects on habitats were used as additional variables. Productivity was defined as the net production of organic matter by primary producers, hence as the basis of the food chain. Carrying capacity was used to evaluate the effects on the functions of the Oosterschelde as wetland, for shellfish culture and fisheries and for recreational use.

Habitat and species diversity

All the main habitats in the 'old' Oosterschelde estuary: (i) gullies and channels, (ii) intertidal sand and mudflats, (iii) salt marshes and (iv) artificial littoral and sublittoral rocky shores, are still present. Yet, changes have occurred in the various habitats, and the population response will be evaluated in comparison to other factors involved, such as severe winters.

Gullies and channels

The civil-engineering project turned the estuary into a tidal bay, comprising restricted brackish habitats, no salinity fluctuations, increased residence times of the watermasses, and a 'reversed' nutrient gradient. Turbidity has decreased and water transparency shows less seasonality in the post-barrier period, compared to the pre-barrier period. As a response to the changed light-nutrient-salinity regime the phytoplankton community has shifted from the dominance of heavily silicified diatoms to smaller diatoms and flagellates. Summer species are now already observed in spring and spring species decreased in abundance; the original seasonal trend of the phytoplankton assemblage has disappeared (Bakker et al., 1990, 1994). Owing to the reduced nutrient concentrations an increased dominance of flagellates over diatoms has been observed, especially in the eastern compartment. This was related to the increased importance of 'regenerated' over 'new' production. The previously existing West-East gradient in phytoplankton biomass has disappeared (Bakker & Vink, 1994). Decreased abundance of *Skeletonema costatum* and increase of *Dytilum brightwelli* seemed to be a response to the reduced salinity fluctuations (Bakker et al., 1994). This phenomenon of 'lagoonization' of the estuary is corroborated by the increase of the rotifer *Synchaeta* spp and the copepod *Acartia clausi* and a decrease of *Acartia tonsa;* the latter being a typical euryhaline species (Bakker, 1994).

Zooplankton biomass increased in the eastern compartment: the original West–East gradient disappeared. This has been related to a decrease in silt concentration and hence an increased food quality, and prolonged residence time of the water mass in the eastern compartment (Bakker & Van Rijswijk, 1994; Tackx et al., 1994). Separated from the effects of the civil engineering project, the occurrence of severe winters might explain the dominance of specific phytoplankton species, e.g. the diatoms *Biddulphia aurita* and *Detonula confervacea* during the construction period of the barrier (Bakker *et al.*, 1994). No effects of severe winters on zooplankton were observed; the impact of low winter temperature on *Temora* development was probably counteracted by an increased food quality (Bakker & Van Rijswijk, 1994).

Frequency of fish occurrence in the Oosterschelde estuary has been monitored for ten years since 1979 in 4 fykes and a weir. In total 67 species were recorded. The community structure showed a slight difference when comparing annual average frequencies before and after 1985. Since 1985 a decrease in a number of anadromous fish species has been observed, as an effect of the reduction in fresh-water inflow. Other changes in species occurrence, such as the increase of visual predators, might be explained by changes in population density at a wider geographic scale (Hamerlynck & Hostens, 1994). In general, the engineering project has had a limited effect on the occurrence of fish in the Oosterschelde.

The subtidal soft bottoms are changing as a consequence of the geomorphological adaptations. In the pre-barrier period many locations were not a suitable habitat for benthic species due to high current velocities. The reduced current velocities now result in large scale deposition of silt, which does not offer a favourable habitat as well (Ten Brinke *et al.*, 1994). Yet, analysis of the mobile epifauna revealed no clear response to the engineering project. Only the increase of flatfish in the western compartment and the decrease of the sand eel may be the result of lower current velocities; the decrease of the brown shrimp is ascribed to reduced input of nutrient rich freshwater (Hostens & Hamerlynck, 1994).

Tidal flats

In addition to the reduction of tidal flat habitat, also the silt content of the intertidal sediments has decreased. The consequences for microphytobenthos, macrozoobenthos and birds have been studied in long-term monitoring programmes and in synoptic surveys.

According to a long-term monitoring programme the annual average microphytobenthos biomass has increased from 115 to 195 mg Chlorophyll-*a* m^{-2} in the post-barrier situation, owing to decreased current velocities, increased water transparency, and possibly increased availability of limiting nutrients such as inorganic carbon (De Jong *et al.*, 1994). The importance of microphytobenthos in the system's productivity has increased.

Extensive surveys of 300 stations in 1985 (pre-barrier period) and in 1989 (post-barrier period) have shown changes in the distribution, abundance and biomass of macrozoobenthos populations on the tidal flats. Statistical analysis of the survey data showed a clear relationship between sediment composition, water depth and benthos distribution. However, the range of the values of the independent variables (sediment, height) encountered in the sampling stations were similar in the pre- and post-barrier situation. It is suggested that the species have a broad tolerance relative to the change in the variables studied, and the changes in distribution and abundance cannot be attributed to the effects of the civil-engineering project (Meire *et al.*, 1994a).

Analysis of long-term changes in macrozoobenthos composition, based on a monitoring programme at 14 stations twice a year over 10 years, showed no trend related to the engineering project. Temporary changes in biomass and abundance have been observed at higher elevated stations after extended air exposure in the construction period (Seys *et al.*, 1994; Coosen *et al.*, 1994a). Dynamics of the benthos populations, however, were found to be associated with the occurrence of mild and severe winters. Low biomass, high densities and higher 'stress' values, expressed as abundance to biomass ratio, were observed after severe winters while the opposite was observed after mild winters (Seys *et al.*, 1994).

The impact of climate factors dominates the benthos dynamics and no significant short-term effect of the civil-engineering project could be established. Yet, long-term effects of the engineering project cannot be excluded due to possible changes in habitat, food availability, larval dispersal and settlement (Seys *et al.*, 1994; Meire *et al.*, 1994a).

Separate studies were conducted to describe the dynamics of the main macrobenthic species cockle, *Cerastoderma edule*, and mussel, *Mytilus edulis*. Biomass of cockles and mussels in the post-barrier period is in the same range as in the pre-barrier period (Smaal & Nienhuis, 1992). The standing stock and distribution of the cockle, *Cerastoderma edule,* was highly variable and related to effects of severe winters and fisheries. No effect of the civil-engineering project has been observed (Coosen *et al.*, 1994b). The mussel biomass was mainly determined by the activities of mussel farmers (Van Stralen & Dijkema, 1994).

Following a programme of monthly high-water bird-censuses from 1978–1990 in the Oosterschelde area, which included the new freshwater lakes behind the compartmentalisation dams, it was concluded that the overall bird numbers and numbers of bird-days have not changed significantly. However, a comparison of the pre- and post-barrier period showed that the total number of waders has decreased from 181,000 to 125,000, while ducks increased from 57,500 to 98,600; this attributed to the change of intertidal habitat into freshwater lakes and the increase of transparency in the Oosterschelde. The seasonal pattern shifted from a midwinter maximum to a peak in autumn. The average winter density of waders in the Oosterschelde proper is now 14.9 ha^{-1} intertidal area, which is comparable to the pre-barrier value for the same area (16.2 ha^{-1}). Apparently, populations of intertidal foragers were close to carrying capacity, and settlement of displaced waders has not occurred. Further reduction in habitat availability will probably be reflected in wader density. This is to be expected because the sediment transport from the higher to the lower parts of the tidal flats, as a consequence of geomorphological changes, results in reduced exposure time and consequently reduced foraging time for wading birds (Schekkerman *et al.*, 1994). In addition to the impact of the engineering project, disturbance by recreational activities may explain the decline

in numbers of moulting birds in late summer. As the pre-barrier observations coincided with severe winters and the post-barrier period with mild winter, no climate effect could be distinguished.

The harbour seal *Phoca vitulina* population in the total Delta area has decreased from 350 in 1960 to 18 individuals at present. Retrospective population analysis shows that 4000 individuals could potentially inhabit the remaining Delta area under tidal influence (Reijnders, 1994). The decrease in numbers is due to habitat loss by the various engineering projects in the Delta and to other disturbances by human activities.

Salt marshes
The marsh vegetation has adjusted to the changed inundation frequency. Most species moved down the marsh elevation gradient (De Leeuw *et al.*, 1994). The absence of inundation in the interim period (1985/1987) has resulted in changes in the soil structure and die-off of vegetation. This was enhanced by the occurrence of severe winters; already in 1984/85 a decline of *Halimione* and *Spartina* populations was observed. Extended periods of drought in combination with frost damage resulted in a large scale decline of marsh vegetation in that period. Yet, the vegetation structure has recovered, but still shows anomalies: in the basins 'levee species' are now dominant. Further changes in vegetation structure are to be expected (De Jong & van der Pluijm, 1994).

Rocky shores
The decrease in current velocity enabled species to colonize other, previously inaccessible regions: dominant species from adjacant stagnant lakes are now also observed in the Oosterschelde; the deposition of silt locally caused a loss of habitat and a dominance of resistent species; increase of the photic zone allowed algal communities to extend to deeper areas. Generally a shift in distribution of communities occurred, following the prevailing tidal currents for the aphotic communities and light attenuation for communities in the photic zone. In some areas species richness increased, in other areas it decreased (De Kluijver & Leewis, 1994). Plant and animal biomass

showed large variations; it decreased in all stations since 1988, but this change might only partly be related to the civil-engineering project. An extensive development of the brittlestar, *Ophiotrix fragilis*, was observed in years after mild winters, and this population outcompetes individuals of other species, leading to a decline in overall biomass (Leewis *et al.*, 1994). Hence climate factors obscure the impact of the civil-engineering project on the rocky shore communities and biomass. In general it can be said that the benthic biomass showed a delayed response to external stress; further changes are to be expected.

The littoral rocky shore communities are characterized by a typical zonation of macro-algae. The reduction in tidal range resulted in a downward shift of the upper zones consisting of *Entophysalis deusta* and *Blidingia* spp communities. Within a few years the intertidal rocky shore communities have adapted to the new situation. As an additional impact coinciding with the construction phase of the storm-surge barrier, dike-reinforcement works have shown a dramatic impact on a number of locations, where various types of rare communities, such as the *Pelvetia* community, almost disappeared (Meijer & Waardenburg, 1994).

Productivity

Primary production

The productivity of the Oosterschelde is estimated on the basis of the primary production of phytoplankton, which contributed 85% to the total primary production of the pre-barrier ecosystem (Scholten *et al.*, 1990). Other primary producers are macrophytes and microphytobenthos; on the basis of the increased biomass it was estimated that the contribution of microphytobenthos to the total primary production should have increased (De Jong *et al.*, 1994). The net import of organic matter from adjacent systems, predominantly the North Sea is insignificant: in terms of organic matter and food availability the Oosterschelde was, and still is, a self-sustaining ecosystem (Scholten *et al.*, 1990).

Due to the engineering project the main factors governing the primary production, e.g. light and nutrient concentrations, have changed in opposite directions: light conditions improved and nutrient concentrations decreased. In contrast to earlier reports (Wetsteyn & Bakker, 1991) further analyses have shown that the phytoplankton biomass, measured as chlorophyll-*a* concentration, has increased in the eastern and northern compartments; it did not change in the central and western compartment (Wetsteyn & Kromkamp, 1994). The annual integrated primary production has decreased in the western compartment, and has increased in the northern section. In the central and eastern compartments there was no change. Hence, the previously existing geographical West–East gradient vanished. It was concluded that the overall primary production was maintained in the post-barrier period (Wetsteyn & Kromkamp, 1994).

Carbon and nutrient fluxes

The ecosystem response to the engineering project has been integrated into a mathematical simulation model (SMOES), which describes the main carbon and nutrient fluxes in the pre- and post-barrier period, considering a spatial scale of 10–20 km and a time scale of 1 day. Besides the SMOES model, average annual carbon budgets have been presented (Klepper *et al.*, 1994; Scholten & Van der Tol, 1994). According to the annual budgets, main fluxes in the pre- and post-barrier period have not shown significant differences, except for zooplankton biomass and consumption, which have increased. Zooplankton consumption is now in the same range as consumption by benthic suspension feeders. The increased consumption by zooplankton is partly explained by the reduced silt content of the water and hence the improved food quality. The increased zooplankton consumption and biomass may as well be an effect of the lower biomass of benthic suspension feeders in the post-barrier period.

It can be seen from the carbon budgets that the productivity of the Oosterschelde shows only minor changes as a response to the civil-engineering works: the Oosterschelde ecosystem demonstrated a resilient response. This reaction might be explained by various adaptive responses in the food web. The phytoplankton composition has changed towards a dominance of species that previously occurred mainly in summer, and there is an extension of the growth season both to earlier and later in the year (Bakker *et al.*, 1990, 1994). These changes in the structure of the phytoplankton community are considered to be adaptations to the increased transparancy and the decreased nutrient concentrations (Van der Tol & Scholten, 1992). In addition, as an effect of the increased residence time of the water masses, the enhanced benthic-pelagic coupling and consequent benthic nutrient regeneration, may have played a role in maintaining the primary production. *In situ* measurements revealed the significant contribution of mussel buds to the regeneration of nutrients. The contribution of mussel buds to nitrogen mineralisation can equal 50% of the total nitrogen mineralisation (Prins & Smaal, 1994). The release of nutrients by the mussel beds may stimulate the primary production of phytoplankton, which partly compensates for the uptake of phytoplankton by the mussels (see Smaal & Prins, 1993, for review).

Carying capacity

In the policy plan the functions of the Oosterschelde have been given the following priority (1) protected wetland, (2) shellfish cultivation and fisheries and (3) recreational use (Van Westen & Colijn, 1994). Carrying capacity of the ecosystem is therefore defined as the capacity of the system to sustain optimal use for a number of specific functions without damage to characteristic features of the system (Smaal & Boeije, 1991).

Mussel cultivation

Bottom cultivation of mussels is executed in the Dutch Wadden Sea and in the Oosterschelde; the use of both areas by the farmers is interrelated and determined by the availability of spat, sheltered cultivation plots, and food. The average annual yield from the Oosterschelde was 1675 and 1350 10^3 kg ash-free dry weight (AFDW) in the pre- and post-barrier period respectively. The yield per unit of standing stock in the period June–November, at the beginning of the harvesting season of mussels from the Oosterschelde, was 35% and 32% in the pre- and post-barrier period respectively, which do not differ significantly (Smaal & Nienhuis, 1992). The standing stocks are lower now, due to the limited availability of juvenile mussels from the Wadden Sea: since 1988 there has been a very restricted spatfall.

At mean current velocities over 60 cm s^{-1} there is a risk that mussels are swept away by the currents. In the pre-barrier period only a limited area could therefore be used for mussel cultivation (Dijkema, 1988). Owing to the increased overall shelter of the Oosterschelde, areas suitable for mussel cultivation have extended, and more biomass can potentially be cultivated in the post-barrier period. The carrying capacity for the cultivation of mussels is now determined mainly by the availability of food, which depends on primary production, grazing pressure, e.g. biomass of suspension feeders and horizontal advection. Van Stralen & Dijkema (1994) have demonstrated a clear relation between mussel growth, stock sizes of mussels and cockles, and phytoplankton production and biomass. The quality of harvested mussels, expressed as meat yield, showed strong correlation with the phytoplankton production in the preceding summer. In years with dense bivalve stocks, phytoplankton production and meat yields were relatively low compared to years with lower standing stocks. It was concluded that increase of standing stocks to extend the mussel cultivation, may result in a decrease of primary production and hence a decrease of growth rates of suspension feeders in the estuary, including the cultivated mussels. Model simulations support this conclusion: a decrease of the bivalve standing stock results in an increase of primary production and mussel growth and a decrease of phytoplankton turnover; an increase of the bivalve standing stock had the opposite effect (Van Stralen & Dijkema, 1994).

The reduced mixing energy urged for redistribution of culture lots according to the new hydraulic conditions. On a local scale deposition of silt and limitation of food supply by reduced horizontal advection has been observed (Ten Brinke et al., 1994; Van Stralen & Dijkema, 1994). Cultivation experiments on selected locations showed good results.

It is concluded that the carrying capacity of the Oosterschelde ecosystem for the cultivation of mussels has been maintained. Owing to the civil-engineering project, adaptation of the cultivation practice to the new hydraulic conditions is required. There is no potential for extension of the mussel standing stock without negative effects on the suspension feeders.

Cockle fisheries

In contrast to mussels, cockles are harvested from not cultivated wild stocks. The annual yield of cockle fisheries has increased dramatically since 1986 owing to increased market demand and the development of more efficient gear and vessels. This is illustrated by the mean annual yield of the fisheries, which was 156 and 914×10^3 kg AFDW in 1982/85 and 1987/89, respectively. This corresponds with 3 and 20% of the total standing stock of cockles. Taking into account the required minimum density for efficient fisheries of 30 cockles m^{-2}, the yield has increased from 8% to 53% of the stock of available cockles. The increase of the harvested stock is related to both increased fisheries and decreased stocks, the latter being only partly an effect of fisheries, since from 1988 onwards recruitment failure plays a role as well (Smaal & Nienhuis, 1992).

The carrying capacity for cockle fisheries is not only determined by the availability of cockles. Meire et al. (1994b) and Schekkerman et al. (1994) have shown that intertidally foraging birds probably experienced food shortage in the post-barrier situation. The competition between cockle fisheries and waders has become obvious, and the function of the Oosterschelde as a protected wetland sets a limit to cockle fisheries.

The competition of waders and man for shellfish can be calculated for the oystercatcher (Haematopus ostralegus), which is the main predator of intertidal cockles and mussels. The annual consumption is ca. 1000×10^3 kg AFDW of cockles and mussels. In the pre-barrier period 50% of the oystercatchers were estimated as being mussel eaters and 50% as cockle eaters. In the post-barrier period this figure has changed. Due to the scarcity of juvenile mussels, the use of intertidal cultivation plots in the Oosterschelde was reduced. In the post-barrier period only 20% of the oystercatchers are eating mussels, and 80% depend on cockles corresponding with an annual consumption of 800×10^3 kg AFDW of cockles. The low standing stock of available cockles in the post-barrier situation, especially in 1989/91 has resulted in a strong competition between oystercatchers and cockle fisheries. In this period the numbers of oystercatchers have reduced most dramatically in winter, indicating food shortage. This effect is enhanced by the reduced foraging time in the intertidal habitat as an effect of geomorphological change (Meire et al., 1994b).

It has therefore been decided to protect the intertidal areas against disturbance, such as recreational activities; since 1992 these are restricted to certain areas and periods of the year. Measures to protect food stocks have been taken since 1993 by regulation of cockle fisheries: some areas are permanently closed for fisheries. Depending on the available cockle stocks, closed areas are extended. The civil-engineering project had no direct impact on the carrying capacity for cockle fisheries: low cockle stocks are a combined effect of recruitment failure and fisheries. Yet, reduced exposure time of intertidal foraging areas has increased the competition between waders and fisheries. As a consequence of the priority for the wetland function, carrying capacity for fisheries has been reduced in favour of the waders.

Man-induced changes and natural variability

The Oosterschelde study basically is a comparison of two stages in the development of an ecosystem, the pre-barrier and the post-barrier period, and fits in the tradition of comparative analysis of ecosystems (Cole et al., 1991). It matches with some of the problems mentioned by

Downing (1991), regarding (i) the choice and the number of variables, (ii) the comparison of eco-systems in a non-formalised way and (iii) the role of coincidences in making interpretations. In addition, we have faced (iv) the problem of the non-equilibrium state of the Oosterschelde as an estuarine ecosystem. In the Oosterschelde studies the choice of the variables has been subject to an extensive discussion, based on an explorative analysis of the ecosystem (Bigelow et al., 1977). It was decided to focus the pre- and post-barrier studies on the main carbon fluxes, and on the dominant populations and habitats. A formalised approach of the comparison of the ecosystems has been achieved by using uncertainty and sensitivity analysis of the pre-barrier model (Klepper et al., 1994), and the execution of a hindcast analysis of the post-barrier model in comparison with the pre-barrier model (Scholten & van der Tol, 1994). However, the responses of the ecosystem resulting from the engineering project, may coincide with the effects of other factors. Estuarine ecosystem comparisons in particular can be obscured by the large natural variability in these ecosystems. It is thereby noticed that the geomorphological changes will continue for many years and consequently the ecosystem will continue to adapt to these changes. Yet, after four years of study, the impact of the engineering project can be evaluated in a preliminary way, in view of other effects on the ecosystem.

Four different stress factors, operating during the study period can be distinguished: (i) the civil-engineering project proper, (ii) fisheries (iii) severe winters (iv) various external (unknown) factors (Nienhuis & Smaal, 1993). The effects on habitat and species diversity, productivity and carrying capacity can be summarised as follows: The engineering project has changed the estuary into a tidal bay with 'lagoonal' characteristics: previously existing gradients in salinity, nutrient concentrations, phytoplankton and zooplankton biomass, and growth of mussels disappeared. The growth season for phytoplankton has extended to earlier in spring and later in autumn, and the species adapted to 'summer' conditions of high water transparency and low nutrients are now dominant throughout the growing season. Species with a preference for sheltered, lagoonal conditions increased in abundance and extended their distribution (Nienhuis & Smaal, 1993).

The occurrence of severe winters corresponds with the blooming of the diatom *Biddulphia aurita* in 1985–87, the high numbers of the cockle *Cerastoderma edule* in 1979, 1982 and 1985, and low densities of the brittle star *Ophiothrix fragilis* in 1983, 1985 and 1986; after the mild winters in the post-barrier period, densities increased tenfold (Nienhuis & Smaal, 1993). These examples show the impact of severe winters for some species; other populations might as well have reacted to the low temperature. Combined effects of severe winters and the increased exposure during the completion phase of the engineering project may explain the limited spatfall of cockles in 1986 (Coosen et al., 1994b) and the die-off of salt marsh vegetation (De Jong & Van der Pluijm, 1994).

The impact of shellfisheries primarily comprises the standing stocks of the mussel and the cockle. The effects of man-induced high densities of mussels have been evaluated by model simulations of reduced mussel standing stock; as shown before this resulted in an increase in primary production, phytoplankton and zooplankton biomass and a decrease in the phytoplankton turnover. Herman & Scholten (190) demonstrated by model simulations a potentially stabilizing effect of benthic suspension feeding on phytoplankton dynamics. Prins & Smaal (1994) showed the potential of mussel beds in regenerating limiting nutrients which can promote phytoplankton production. It is therefore concluded that mussel cultivation affects ecosystem processes by competition for food with other suspension feeders; this may increase the turnover of phytoplankton and exert a stabilizing influence on the phytoplankton dynamics.

The cockle stocks are extremely variable from year to year owing to climatic factors (severe winters and storm events). Cockle fisheries in years with low standing stocks obviously can affect the food availability of waders.

Some external factors affecting Oosterschelde populations have been identified such as the introduction of *Crassostrea gigas* and *Sargassum muticum* by oyster breeders, and the disturbance of intertidal rocky shore communities by dike-reinforcement works (Nienhuis & Smaal, 1993).

Unexplained variations in densities regard the sudden appearance of the diatom *Leptocylindrus minimus* since 1988 in the Oosterschelde; from 1989 onwards the species also occurs abundantly in the Dutch coastal zone and in the Wadden Sea, bearing no relation with the civil-engineering project.

The mudsnail *Hydrobia ulvae* showed large annual biomass fluctuations and a significant decline in the post-barrier situation in some areas, but an increase in other areas. These fluctuations did not show any relation with any of the factors tested, such as climate, current velocity, sediment composition and tidal elevation; it was suggested that parasitic infestations might have played a role (Coosen *et al.*, 1994a).

In conclusion, the Oosterschelde ecosystem has shown responses to various factors. Effects of severe winters and mussel- and cockle-fisheries could be distinguished from other factors. The response of the ecosystem to the civil-engineering project could be quantified in terms of changes of habitat availability, maintenance of productivity and restricted maintenance of carrying capacity as a wetland and fishery area.

Discussion and conclusions

The interest in storm-flood protection has recently gained momentum, owing to the wide international discussion on the impact of sea-level rise on society. It is estimated on the basis of the observed changes since the beginning of this century that global warming of 1.5°C to 4.5°C would lead to a sea-level rise of 0.20 to 1.40 m at the end of the next century (Villach Conference, 1985; see Wind, 1987). Estuarine areas are extremely vulnerable to effects of sea-level rise. On the terrestrial side there is continuous land claim for a variety of economical purposes, which forces to the building of sea-defences and barrages. On the aquatic side there is a continuous process of channel dredging and sedi-

ment extraction. Consequently the coastal zone gets steeper and steeper and hence more difficult to protect. This 'estuarine squeeze' (Mitchell, 1993) forces estuarine managers to anticipate reliable shore protection for the generations to come. The concern for storm-flood protection is worldwide. Along the coast of Europe alone there are several designed or realized projects in the larger estuaries or coastal embayments. The storm-surge barrier in the Thames estuary, England, has been realized some 15 years ago. The barrier in the Nieuwe Waterweg to Rotterdam, The Netherlands, is under construction (Rigo, 1991). From 1979 onwards a project dealing with the flood protection of St. Petersburg is underway: a 25-km-wide storm-surge barrier in the Neva river is under design and partly under construction (Sevenard, 1991). The flood protection of Venice lagoon is already a matter of concern for more than 60 years (Passino & Todisco, 1984; Hedegaard *et al.*, 1991).

In some cases storm-flood protection is not the primary motive for barrier design. Desk studies on tidal power resources, as can potentially be distracted from the Severn estuary (Corlett, 1984) and the river Mersey schemes (Pinkney & Wilson, 1991), may illustrate this principle. In all designs and realized projects it is obvious that a number of conflicting interests should be satisfied, the most significant ones being safety for the human populations and their properties and environmental issues. Barriers in rivermouths are disrupting the estuarine gradient, thus blocking the continuous flow of riverwater into the sea and hence may accumulate pollutants and nutrients in undesired concentrations. The economic costs and risks of barrier constructions are high. The construction of a storm-surge barrier is expensive. The technical life-span of marine structures in general is only 50–100 years and in the case of storm-surge barriers, considering their high costs, 100 years or more is acceptable (200 years for the Oosterschelde storm-surge barrier in The Netherlands) (Rigo, 1991).

The Dutch have a long tradition in fighting against the sea. From the Middelages on land has been reclaimed on the sea, and consequently nowadays more than half of the country is situated below the level of the sea, and it is only by artificial means that the human population keep their feet dry. Concerning the Oosterschelde, early initiatives (Saeys & Bannink, 1978; Knoester *et al.*, 1984) already tried to integrate ecological aspects in coastal engineering projects. The Oosterschelde project, finished in 1987, became a well documented case study in which a balanced compromise between safety for the human population against storm floods and conservation of the natural resources, including fisheries, has been reached.

The Oosterschelde civil-engineering project is not the standard solution for problems of storm-flood protection, but it is one of the ways how to deal with safety risks in a socio-economically highly developed area containing large natural values. As such it has been presented as one of the case studies at the World Coast Conference 1993 (Hillen *et al.*, 1993). The Oosterschelde case-study fits into the recommendations made at the World Coast Conference, with regard to research on coastal zone management. The recommendations made, deal with the enhancement of the capabilities for systems modelling, with special reference to the distinction between human-induced and natural changes in coastal processes.

The Oosterschelde project is technically and scientifically unique. The storm-surge barrier represents an important breakthrough in marine civil engineering. The project also offered ample opportunities for integrated physical, chemical, geological and biological research. Integration of the knowledge gained, raised the entire project to the level of a case study of a changing estuarine system, and demonstrated the effects of human interference in a non-polluted estuary. There was a search for long-term responses of animal and plant communities to environmental stress. It this context discrimination between responses to human-induced stress, the construction of the seawalls, and responses to natural stress (e.g. temperature) have clearly been established in some cases, but in others it appeared difficult to quantify. The Oosterschelde project encompassed a considerable number of field descriptions. The majority of the cause–effect relations described during the project originate from correlative, descriptive research, which was not intended to offer causal explanations. A considerable number of documented fluctuations in population dynamics of the dominant plant and animal species cannot directly be attributed to the construction of the civil engineering works, and have to be related to other human-induced factors (e.g. fisheries) or natural factors (e.g. temperature). Considering the size of the project and employment of a considerable number of scientists, an impressive document came into existence. The papers contain useful information for both scientists and managers involved in comparable projects, dealing with major environmental changes in estuaries.

The main argument to build the storm-surge barrier was storm-flood protection, combined with conservation of the existing values of landscape and nature, fisheries included. The estuary is also recognized as an international wetland, including sand- and mudflats, salt marshes and inland brackish localities with specific brackish waterplants and animals. The Oosterschelde estuary was and still is an internationally recognized nature reserve for migratory waterfowl, waders, ducks and geese, feeding on the extensive intertidal sand- and mudflats, containing rich macrozoobenthos communities. The Oosterschelde is one of the few localities in the Netherlands where the hard stone substrates of man-made seawalls and dikes, the artificial rocky shore, harbours a diversified community of marine animals and plants.

Roughly 20% of the original surface area of Oosterschelde estuary is now situated beyond the eastern and northern compartment dams (Zoommeer-Krammer-Volkerak; Fig. 1). This stretch of water is an important shipping route between Antwerpen and Rotterdam. The water of Zoommeer-Krammer-Volkerak is fresh, derived from the rivers Rhine and Meuse. The area contains potential values of nature, especially the former salt marshes and now vegetated wetlands, which will increase when eutrophication can be adequately controlled.

A major characteristic of the Oosterschelde ecosystem is the robustness of the biological communities in reaction to the physical and chemical changes induced by the execution of the storm-surge barrier and the compartment dams. Oosterschelde estuary was isolated from the imput of river water and the rear ends of the estuary were separated from the cosystem by seawalls. The consequent decrease in tidal exchange with the North Sea induced sheltered circumstances: the ecosystem changed from a turbid estuary into a tidal bay. The physically accomodated biological communities showed robust and resilient responses to the environmental changes. The pelagic communities adjusted within 2 or 3 years to the new environmental regime. Annually integrated primary production of phytoplankton did not change significantly, notwithstanding the fact that the phytoplankton assemblages shifted from the heavily silicified diatoms to smaller diatoms and flagellates. The salt-marsh communities presumably take 5 to 10 years to readjust to the new physical circumstances. The annually integrated secondary production of zoobenthos did not change much, although large year to year variations blur the general picture. The biomass distribution of zoobenthos considerably changed, as a consequence of the changed geomorphology and hydrology of the estuary. Physically exposed and hence previously not accessible habitats, became accessible after 1986 for specific benthos communities. The opposite occurred also: originally sandy subtidal localities silted up after 1986 and became no longer suited for macro-infauna assemblages.

In general terms the qualitative structure of the pelagic and benthic foodweb did not change. The studied changes are of a quantitative nature: the ecosystem is resilient and remained within the potential limits of its existence. A restriction of our investigations is the dedication to biological processes instead of biodiversity. Consequently, the question whether some rare species have disappeared and some others have shown up, cannot be answered.

The adjustment of geomorphological processes in Oosterschelde estuary will take more than 100 years. A new equilibrium between the strongly decreased tidal volume of the estuary and the cross section of the channels of the estuary is still developing. The former tidal gullies are too deep and the slopes of the intertidal and subtidal sand flats are too steep now. A continuous erosion of the intertidal flats is occurring and excess sand and silt is being deposited in the tidal gullies. This gradually retarding process will continue for decades to come, and has considerable consequences for the distribution and species composition of the benthic communities and hence for the tertiary consumers (waders; fish) feeding on the benthos.

The extreme physical impact of northwesterly storms is no longer allowed in Oosterschelde estuary. When extremely high tides of more than 3 m above Mean Sea Level are predicted, the gates in the storm-surge barrier will be lowered into the water, to block off the estuary from the North Sea. According to the management strategy chosen, this will occur once or twice per year. As a consequence of this management strategy Oosterschelde ecosystem changed into a more sheltered system. This process of 'lagoonisation' can be witnessed by quantitative shifts in the abundance of specific plants and animals and the accumulation of silt at subtidal localities.

Our conclusions are based on only four years of post-barrier experience. The question remains whether the robustness of the Oosterschelde remains for the decades to come. It cannot be excluded that unexpected long-term changes might finally alter the functional relations in the Oosterschelde foodwebs; e.g. the exhaustion of inorganic nutrient resources; the introduction of new species; the gradual lowering of the intertidal flats.

Oosterschelde estuary can be classified as a man-manipulated coastal plain estuary, to be closed off by a storm-surge barrier during extreme storm floods. Salinity is high and the estuary carries a truly marine-estuarine flora and fauna. Marine migratory animals, including the harbour seal, may enter and leave the estuary undisturbed. Oosterschelde is a mesotidal estuary. The water is very clear: the input of riverine silt has almost completely ceased and marine silt, brought in by

590

tidal currents into the estuary will settle onto the subtidal slopes and bottoms of tidal gullies. This is in contrast with processes in macrotidal European estuaries, like the Gironde in France, the Severn in England and the Westerschelde in The Netherlands where a tremendous amount of silt is continuously kept in suspension by tidal currents and turbulent mixing, decreasing the transparency of the water column and, consequently, decreasing the primary production of phytoplankton, the dominant primary producer.

It can be concluded that, notwithstanding the considerable physical and hydrochemical changes, the carrying capacity of Oosterschelde estuary has been maintained, with regard to the main functions of the ecosystem, viz. a valuable, protected tidal wetland, high quality shellfish cultivation and local fisheries and extensive water-recreation.

Acknowledgement

We thank Dr Peter Herman for critically reading the manuscript. This paper is Communication No. 710 of NIOO-CEMO, Yerseke, The Netherlands.

References

Absil, C., 1993. Bioavailability of heavy metals for the deposit feeder *Macoma balthica* with special emphasis on copper. Ph.D. Thesis Wageningen University.

Bakker, C., P. M. J. Herman & M. Vink, 1990. Changes in seasonal succession of phytoplankton induced by the storm-surge barrier in the Oosterschelde (SW Netherlands). J. Plankton Res. 12: 947–972.

Bakker C., 1994. Zooplankton species composition in the Oosterschelde (SW Netherlands) before, during and after the construction of a storm-surge barrier. Hydrobiologia 282/283 (Dev. Hydrobiol. 97): 117–126.

Bakker, C., P. M. J. Herman & M. Vink, 1994. A new trend in the development of the phytoplankton in the Oosterschelde (SW Netherlands) during and after the construction of a storm-surge barrier. Hydrobiologia 282/283 (Dev. Hydrobiol. 97): 79–100.

Bakker, C. & P. van Rijswijk, 1994. Zooplankton biomass in the Oosterschelde (SW Netherlands) before, during and after the construction of a storm-surge barrier. Hydrobiologia 282/283 (Dev. Hydrobiol. 97): 127–143.

Bakker C. & M. Vink, 1994. Nutrient concentrations and planktonic diatom-flagellate relations in the Oosterschelde (SW Netherlands) during and after the construction of a storm-surge barrier. Hydrobiologia 282/283 (Dev. Hydrobiol. 97): 101–116.

Bigelow J. H., J. C. deHaven, C. Dzitzer, P. Eilers & J. C. H. Peeters, 1977. Protecting an estuary from flood: a policy analysis of the Oosterschelde; vol. IV; Assessment of long-run ecological effects, Rand, Santa Monica.

Cole, J. J., G. M. Lovett & S. E. G. Findlay, 1991. Comparative analysis of ecosystems: patterns, mechanisms and theories. Springer, New York.

Coosen, J., J. Seys, P. M. Meire & J. A. Craymeersch, 1994a. Effect of sedimentological and hydrodynamical changes in the intertidal areas of the Oosterschelde estuary (SW Netherlands) on distribution, density and biomass of five common macrobenthic species. Hydrobiologia 282/283 (Dev. Hydrobiol. 97): 235–249.

Coosen, J. C., F. Twisk, M. W. M. van der Tol, R. H. D. Lambeck, M. R. van Stralen & P. M. Meire, 1994b. Variability in stock assessment of cockles (*Cerastoderma edule* L.) in the Oosterschelde estuary (in 1980–1990), in relation to environmental factors. Hydrobiologia 282/283 (Dev. Hydrobiol. 97): 381–395.

Corlett, J., 1984. Tidal power from the Severn estuary. Wat. Sci. Tech. 16: 253–268.

De Jong, D. J., Z. de Jong & J. P. M. Mulder, 1994. Changes in area, geomorphology and sediment nature of salt marshes in the Oosterschelde estuary (SW Netherlands) due to tidal changes. Hydrobiologia 282/283 (Dev. Hydrobiol. 97): 303–316.

De Jong D. J., P. H. Nienhuis & B. J. Kater, 1994. Microphytobenthos in the Oosterschelde estuary (The Netherlands), 1981–1990; consequences of a changed tidal regime. Hydrobiologia 282/283 (Dev. Hydrobiol. 97): 183–195.

De Jong, D. J. & A. M. van der Pluijm, 1994. Consequences of a tidal reduction for the salt-marsh vegetation in the Oosterschelde estuary (The Netherlands). Hydrobiologia 282/283 (Dev. Hydrobiol. 97): 317–333.

De Kluijver, M. J. & R. J. Leewis, 1994. Changes in the sublittoral hard substrate communities in the Oosterschelde estuary (SW Netherlands), caused by changes in the environmental parameters. Hydrobiologia 282/283 (Dev. Hydrobiol. 97): 265–280.

De Leeuw, J., L. P. Apon, P. M. J. Herman, W. de Munck & W. G. Beeftink, 1994. The response of salt marsh vegetation to tidal reduction caused by the Oosterschelde storm-surge barrier. Hydrobiologia 282/283 (Dev. Hydrobiol. 97): 335–353.

Dijkema, R., 1988. Shellfish cultivation and fisheries before and after a major flood barrier construction project in the Southwestern Netherlands. J. Shellfish Res., 7(2): 241–252.

Downing, J. A., 1991. Comparing apples with oranges: methods of interecosystem comparison. In: Cole, Lovett & Findlay (eds), Comparative analysis of ecosystems: pattern, mechanisms and theories. Springer, New York: 24–45.

Hamerlynck O. & K. Hostens, 1994. Changes in the fish fauna of the Oosterschelde estuary – a ten-year time series of fyke catches. Hydrobiologia 282/283 (Dev. Hydrobiol. 97): 497–507.

Hedegard, I. B., K. Mangor & G. Cecconi, 1991. Safegarding of Venice. In IABSE Colloquium Nyborg, DK 1991: The interaction between major engineering constructions and the marine environment. IABSE Reports 63: 39–52.

Herman, P. & H. Scholten, 1990. Can suspension feeders stabilize estuarine ecosystems. In: M. Barnes & R. N. Gibson (eds), Trophic relationships in the marine environment. Aberdeen Univ. Press, Aberdeen: 104–116.

Hillen, R., A. C. Smaal, E. J. Van Huijssteeden & R. Misdorp, 1993. The Dutch Delta: aspects of coastal zone management. World Coast Conference – Delta's, pp. 1–7. Noordwijk, The Netherlands.

Hostens, K. & O. Hamerlynck, 1994. The mobile epifauna of the soft bottoms in the subtidal Oosterschelde estuary: structure, function and impact of the storm-surge barrier. Hydrobiologia 282/283 (Dev. Hydrobiol. 97): 479–496.

Klepper, O., M. W. M. van der Tol, H. Scholten & P. M. J. Herman, 1994. SMOES: A simulation model for the Oosterschelde ecosystem. Part I: Description and uncertainty analysis. Hydrobiologia 282/283 (Dev. Hydrobiol. 97): 437–451.

Knoester, M., J. Visser, B. A. Bannik, C. J. Colijn & W. P. A. Broeders, 1984. The Eastern Scheldt Project. Wat. Sci. Tech. 16: 51–77.

Leewis, R. J., H. W. Waardenburg & M. W. M. van der Tol, 1994. Biomass and standing stock on sublittoral hard substrates in the Oosterschelde estuary (SW Netherlands). Hydrobiologia 282/283 (Dev. Hydrobiol. 97): 397–412.

Meire, P. M., H. Schekkerman & P. L. M. Meininger, 1994b. Consumption of benthic invertebrates by waterbirds in the Oosterschelde estuary, SW Netherlands. Hydrobiologia 282/283 (Dev. Hydrobiol. 97): 525–546.

Meire, P. M., J. Seys, J. Buijs & J. Coosen, 1994a. Spatial and temporal patterns of intertidal macrobenthic populations in the Oosterschelde estuary: are they influenced by the construction of the storm-surge barrier? Hydrobiologia 282/283 (Dev. Hydrobiol. 97): 157–182.

Meijer, A. J. M. & H. W. Waardenburg, 1994. Tidal reduction and its effects on intertidal hard substrate communities in the Oosterschelde estuary. Hydrobiologia 282/283 (Dev. Hydrobiol. 97): 281–298.

Mitchell, R., 1993. Marine Scene. Joint nature Conservation Committee, No. 2, Peterborough, UK.

Mulder, J. P. M. & T. Louters, 1994. Changes in basin geomorphology after implementation of the Oosterschelde estuary project. Hydrobiologia 282/283 (Dev. Hydrobiol. 97): 29–39.

Nienhuis, P. H. & A. C. Smaal, 1993. The Oosterschelde (The Netherlands), an estuarine ecosystem under stress: discrimination between the effects of human-induced and natural stress. ECSA/ERF proceedings, in press.

Nienhuis, P. H. & A. C. Smaal, 1994. The Oosterschelde estuary, a case-study of a changing ecosystem: an introduction. Hydrobiologia 282/283 (Dev. Hydrobiol. 97): 1–14.

Oenema, O., 1988. Early diagenesis in recent fine-grained sediments in the Eastern Scheldt, Ph.D. Thesis, Utrecht University.

Passino, R. & A. Todisco, 1984. Environmental aspects connected with 'high water' protection of Venice. Wat. Sci. Tech. 16: 319–336.

Pinkney, M. & E. A. Wilson, 1991. Tidal power and the environment on the river Mersey. In IABSE Colloquium Nyborg, DK 1991: The interaction between major engineering constructions and the marine environment. IABSE Reports 63: 53–60.

Prins, T. C. & A. C. Smaal, 1994. The role of the blue mussel *Mytilus edulis* in the cycling of nutrients in the Oosterschelde estuary (The Netherlands). Hydrobiologia 282/283 (Dev. Hydrobiol. 97): 413–429.

Reijnders, P. J. H., 1994. Historical population size of the harbour seal, *Phoca vitulina*, in the Delta area (SW Netherlands). Hydrobiologia 282/283 (Dev. Hydrobiol. 97): 557–560.

Rigo, P., 1991. Estuary protection by storm surge barriers. In IABSE Colloguium Nyborg, DK 1991: The interaction between major engineering constructions and the marine environment. IABSE Reports 63: 31–38.

Saeijs, H. L. F. & B. A. Bannink, 1978. Environmental consideration in a coastal engineering project. The Delta project in the South-Western Netherlands. Hydrobiol Bull. 12: 180–202.

Schekkerman, H., P. L. M. Meininger & P. M. Meire, 1994. Changes in the waterbird populations of the Oosterschelde (SW Netherlands) as a result of large-scale coastal engineering works. Hydrobiologia 282/283 (Dev. Hydrobiol. 97): 509–524.

Scholten, H., O. Klepoper, P. H. Nienhuis & M. Knoester, 1990. Oosterschelde estuary (SW Netherlands): a self-sustaining ecosystem? Hydrobiolgia 195 (Dev. Hydrobiol. 55): 201–215.

Scholten, H. & M. W. M. van der Tol, 1994. SMOES: a simulation model for the Oosterschelde ecosystem. Part II: Calibration and validation. Hydrobiologia 282/283 (Dev. Hydrobiol. 97): 453–474.

Sevenard, Y., 1991. Leningrad Flood Protection. In IABSE Colloquium Nyborg, DK 1991: The interaction between major engineering structures and the marine environment. IABSE Reports 63: 71–78.

Seys, J. J., P. M. Meire, J. Coosen & J. A. Craeymeersch, 1994. Long-term changes (1979–1989) in intertidal macrozoobenthos of the Oosterschelde estuary: are patterns in total density, biomass and diversity induced by the construction of the storm-surge barrier? Hydrobiologia 282/283 (Dev. Hydrobiol. 97): 251–264.

Smaal, A. C. & R. C. Boeije, 1991. Veilig getij, de effecten van de waterbouwkundige werken op de getijdemilieu van de Oosterschelde. nota GWWS 91.088 DGW/directie Zeeland, Middelburg (in Dutch).

592

Smaal, A. C., M. Knoester, P. H. Nienhuis & P. M. Meire, 1991. Changes in the Oosterschelde ecosystem induced by the Delta Works. In: M. Elliot & J. P. Ducrotoy (eds), Estuaries and Coasts: spatial and temporal intercomparisons. ECSA 19 Symposium, Olsen & Olsen, Fredenborg: 375–284.

Smaal, A. C. & P. H. Nienhuis, 1992. The Eastern Scheldt (The Netherlands), from an estuary to a tidal bay: a review of responses at the ecosystem level. Neth. J. Sea Res. 30: 161–173.

Smaal, A. C. & T. C. Prins; 1993. The uptake of organic matter and the release of inorganic nutrients by bivalve suspension feeder beds. In: R. F. Dame (ed.), Bivalve filter feeders in estuarine and coastal ecosystem processes. Springer Verlag, Berlin-Heidelberg: 271–298.

Smies, M. & A. H. L. Huiskes, 1981. Holland's Eastern Scheldt estuary barrier scheme: some ecological consideration. Ambio 10: 158–165.

Tackx, M. L. M., P. M. J. Herman, P. van Rijswijk, M. Vink & C. Bakker, 1994. Plankton size distributions and trophic relations before and after the construction of the storm-surge barrier in the Oosterschelde estuary. Hydrobiologia 282/283 (Dev. Hydrobiol. 97): 145–152.

Ten Brinke, W. B. M., J. Dronkers & J. P. M. Mulder, 1994. Fine sediments in the Oosterschelde tidal basin before and after partial closure. Hydrobiologia 282/283 (Dev. Hydrobiol. 97): 41–56.

Van Stralen, M. R. & R. D. Dijkema, 1994. Mussel culture in a changing environment: the effects of a coastal engineering project on mussel culture (*Mytilus edulis* L.) in the Oosterschelde estuary (SW Netherlands). Hydrobiologia 282/283 (Dev. Hydrobiol. 97): 359–379.

Van Westen, C. J. & C. J. Colijn, 1994. Policy planning in the Oosterschelde estuary. Hydrobiologia 282/283 (Dev. Hydrobiol. 97): 563–574.

Van der Tol, M. W. m. & H. Scholten, 1992. Response of the Eastern Scheldt ecosystem to a changing environment: functional or adaptive? Neth. J. Sea Res. 30: 175–190.

Vroon, J., 1994. Hydrodynamic characteristics of the Oosterschelde in recent decades. Hydrobiologia 282/283 (Dev. Hydrobiol. 97): 17–27.

Wetsteyn, L. P. M. J. & C. Bakker, 1991. Changes of abiotic characteristics and consequences for the phytoplankton development during and after a large scale coastal engineering project in the Oosterschelde (The Netherlands): a preliminary evaluation. In: M. Elliot & J. P. Ducrotoy (eds), Estuaries and coasts: Spatial and temporal intercomparisons. Olsen & Olsen, Fredensborg, Denmark: 365–373.

Wetsteyn, L. P. J. M. & J. C. Kromkamp, 1994. Turbidity, nutrients and phytoplankton primary production in the Oosterschelde estuary (The Netherlands) before, during and after a large-scale coastal engineering project (1980–1990). Hydrobiologia 282/283 (Dev. Hydrobiol. 97): 61–78.

Wind, H. G. (ed.), 1987. Impact of sea level rise on society. Balkema, Rotterdam, 191 pp.

Hydrobiologia **282/283**: 593–597, 1994.
P. H. Nienhuis & A. C. Smaal (eds), The Oosterschelde Estuary.

General index